International Conference on Mechanical, Aeronautical and Automotive Engineering (ICMAA 2017)

MATEC Web of Conferences Volume 108 (2017)

Malacca, Malaysia
25 - 27 February 2017

Editors:

Haider F. Abdul Amir
Khiew Poi Sim

ISBN: 978-1-5108-4115-4

Printed from e-media with permission by:

Curran Associates, Inc.
57 Morehouse Lane
Red Hook, NY 12571

Some format issues inherent in the e-media version may also appear in this print version.

Printed by Curran Associates, Inc. (2017)

For additional information, please contact EDP Sciences – Web of Conferences
at the address below.

EDP Sciences – Web of Conferences
17, Avenue du Hoggar
Parc d'Activité de Courtabœuf
BP 112
F-91944 Les Ulis Cedex A
France

Phone: +33 (0) 1 69 18 75 75
Fax: +33 (0) 1 69 28 84 91

contact-edps@webofconferences.org

Additional copies of this publication are available from:

Curran Associates, Inc.
57 Morehouse Lane
Red Hook, NY 12571 USA
Phone: 845-758-0400
Fax: 845-758-2633
Email: curran@proceedings.com
Web: www.proceedings.com

TABLE OF CONTENTS

ISSN (electronic edition): 2261-236X

CONFERENCE COMMITTEES

Conference Chairs

Prof. Poi Sim Khiew, University of Nottingham, Malaysia

Prof. Ho Jee Hou, University of Nottingham, Malaysia

Ir. Prof. Dr. Sim Kok Swee, Multimedia University, Malaysia

Program Chair

Prof. Haider F. Abdul Amir, University Teknologi Mara in Shah Alam, Malaysia

Technical Committee

Dr. Mohd Nizam Bin Sudin, Universiti Teknikal Malaysia, Melaka, Malaysia

Dr. Eugene Larkin, Tula State University, Russia

Dr. Mohammad Khan, Abu Dhabi University, United Arab Emirates

Dr. Özgür Ekici, Hacettepe University, Turkey

Dr. Sivakumar A/L Dhar Malingam, Universiti Teknikal Malaysia Melaka, Malaysia

Dr. Loh Ser Lee, Universiti Teknikal Malaysia Melaka, Malaysia

Dr. Seung Hyun Kim, The Ohio State University, Columbus

Dr. Prabhudev M. S, Government Polytechnic, Kalgi, India

Dr. Kodakkal Kannan Viswanathan, Kuwait College of Science and Technology, Kuwait

Dr. N. ETHIRAJ, Dr. M. G. R Educational and Research Institute, India

Dr. Cheng See Yuan, Universiti Teknikal Malaysia Melaka, Malaysia

Dr. Adelle Waterworth, University of Huddersfield, UK

Preface

2017 International Conference on Mechanical, Aeronautical and Automotive Engineering took place in Malacca, Malaysia, from February 25-27, 2017. The conference was assisted by the University of Nottingham Malaysia and University Teknologi Mara in Shah Alam.

The objective of the conference was to address and deliberate on the latest technical status and recent trends in the research and applications of mechanical, aeronautical and automotive engineering. This conference has been designed with the view to provide an opportunity for the scientists, engineers, industrialists, students and other professionals from all over the world to interact and exchange their new ideas and research outcomes for future collaboration.

The proceedings of ICMAA 2017 consists of 122 selected papers from the conference which are presented orally at the conference or via poster sessions, and provide up-to-date, comprehensive and worldwide state-of-the art knowledge in this field. Each contributed paper was rigorously peer-reviewed by international reviewers who were drawn from the organizing and advisory committee members, and external reviewers in related fields from all over the world. The proceedings cover 17 different specific areas, including functional materials and performance testing, material analysis and processing, textile industry and technology, mechanical engineering, design and development of robot system, modern manufacturing, vehicle and ship engineering, power system and mechatronics, modern electronic information technology, control theory and technology, thermal theory and application, applied mechanics and dynamics, biomedical engineering, energy engineering and application, image processing and information system, civii and structural engineering, logistics and transportation

We'd like to express our heartful gratitude to all the reviewers for their great professionalism and efforts. And thank all the participants and sponsors for their valuable contributions and support.

Prof. Poi Sim Khiew, University of Nottingham Malaysia
Prof. Haider F. Abdul Amir, University Teknologi Mara in Shah Alam, Malaysia

February 2017

Statement of Peer review

In submitting conference proceedings to MATEC Web of Conferences, I certify to the Publisher that I adhere to the **Policy on Publishing Integrity** of the journal in order to safeguard good scientific practice in publishing.

1. All articles have been subjected to peer review administered by the proceedings editors.
2. Reviews have been conducted by expert referees, who have been requested to provide unbiased and constructive comments aimed, whenever possible, at improving the work.
3. Proceedings editors have taken all reasonable steps to ensure the quality of the materials they publish and their decision to accept or reject a paper for publication has been based only on the merits of the work and the relevance to the journal.

Title, date and place of the conference

2017 International Conference on Mechanical, Aeronautical and Automotive Engineering

Malacca, Malaysia

February 25-27, 2017

Proceedings editor(s):

editors:

Prof. Haider F. Abdul Amir, University Teknologi Mara in Shah Alam, Malaysia

Prof. Dr. Khiew Poi Sim, University of Nottingham Malaysia

Date and editor's signature 20/4/2017

17, avenue du Hoggar - PA de Courtabœuf – BP 112 - 91944 Les Ulis Cedex A (France)
Tél. : 33 (0)1 69 18 75 75 - Fax : 33(0)1 69 07 45 17 – www.edpsciences.org

Plane strain bending under tension as an ideal flow process in pressure – dependent plasticity

Sergei Alexandrov[1] and Prashant Date[2]

[1]*Russian Academy of Sciences, Institute for Problems in Mechanics, 199526 Moscow, Russia*
[2]*IIT Bombay, Department of Mechanical Engineering, 400076 Mumbai, India*

Abstract. Ideal plastic flows are those for which all material elements follow minimum work paths. Ideal flow solutions are widely used as the basis for inverse methods for the preliminary design of metalworking processes. The present paper provides the first ideal flow solution in pressure-dependent plasticity. In particular, the process of bending under tension is considered and it is shown that there are relations between the bending moment and tensile force that result in ideal flow paths.

1 Introduction

Ideal plastic flows are those for which all material elements follow minimum work paths. The ideal flow theory has long been associated with the Tresca yield criterion [1, 2]. Recently, it has been demonstrated that ideal flow solutions exist in anisotropic plasticity [3]. In particular, the theory of anisotropic plasticity proposed in [4] has been adopted. A comprehensive overview of the ideal flow theory and corresonding solutions has been provided in [5]. This theory has been widely used as the basis for inverse methods for the preliminary design of bulk metal forming processes driven by minimum plastic work (see, for example, [6]). In particular, optimal dies for extrusion (or drawing) have been found in this paper. Necessary conditions for the optimality of extrusion dies for rigid plastic materials obeying strictly convex yield criteria have been derived in [7]. The development of simplified methods for analysis and design of metal forming processes is of importance for applications since standard finite element simulations are too slow for many purposes [8]. It is worthy of note that the ideal flow theory results in rather a general method of analyzing and designing metal forming processes, whereas many other simplified methods deal with a specific process (see, for example, [9] for plate rolling and [10] for wire drawing). All the available ideal fow solutions are for pressure-independent models. On the other hand, pressure-dependency is an important property of some metals, for example [11 - 14]. The present paper provides the first ideal flow solution in pressure-dependent plasticity. In particular, the double shearing model proposed in [15] is adopted. The solution describes the process of bending under tension satisfying the ideal flow conditions. Non ideal flow solutions for this process have been given in [16 - 20] for various material models. The present solution is based on the mapping between Eulerian and Lagrangian coordinates proposed in [21].

2 Statement of the problem

The process of plane strain bending under tension transforms an initial rectangular $A_0B_0C_0D_0$ into a shape determined by two circular arcs, AB and CD, and two straight lines, AD and BC. The initial shape is shown in Fig.1 and the final shape in Fig. 2. The initial thickness of the sheet is H and the initial width is L. The sheet is loaded by bending moment M and force F per unit length. A consequence of the equilibrium equations is that some pressure p should be applied to surface CD. Assuming that this pressure is uniformly distributed over the surface it is possible to find that

$$F = r_{CD}p \tag{1}$$

where r_{CD} is the current radius of surface CD. Kinematics of this process has been described in [21]. For completeness, the basic equations are presented in this section. Choose an Eulerian Cartesian coordinate system (x, y) such that the initial rectangular is defined by the equations $x=0$, $x=-H$ and $y=\pm L$ and a Lagrangian coordinate system (ζ,η) such that $x=H\zeta$ and $y=H\eta$ at the initial instant. The mapping between these coordinate systems is

$$\frac{x}{H}=\sqrt{\frac{\zeta}{a}+\frac{s}{a^2}}\cos(2a\eta)-\frac{\sqrt{s}}{a},\qquad \frac{y}{H}=\sqrt{\frac{\zeta}{a}+\frac{s}{a^2}}\sin(2a\eta) \tag{2}$$

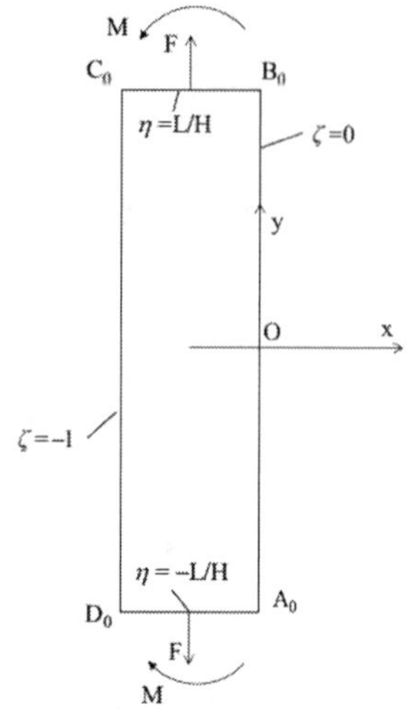

Figure 1. Initial configuration.

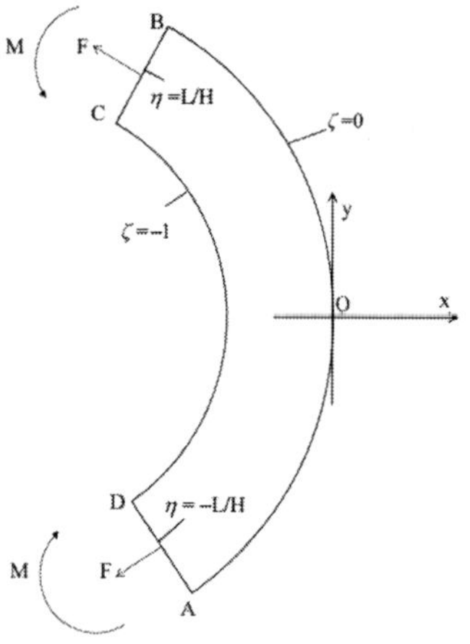

Figure 2. Final configuration.

where a is a function of the time, t, satisfying the condition $a=0$ at $t=0$ and s is a function of a satisfying the condition

$$s=\frac{1}{4} \tag{3}$$

at $a=0$. The function $s(a)$ should be found from the solution. It is possible to verify by inspection that Eq. (2) satisfies the equation of incompressibility. Moreover, the Lagrangian coordinate system is orthogonal and its coordinate curves coincide with trajectories of principal strain rates. It is also convenient to introduce a moving plane polar coordinate system (r,θ) with its origin at $x=-H\sqrt{s}/a$ and $y=0$. Then, it follows from Eq. (2) that

$$\frac{r}{H}=\frac{\sqrt{\zeta a+s}}{a},\quad \theta=2a\eta. \tag{4}$$

In this coordinate system, curves AB and CD are defined by the equations $r=r_{AB}$ and $r=r_{CD}$, respectively, and lines CB and AD by the equations $\theta=\pm\theta_0$. It follows from Eq. (4) that

$$\frac{r_{AB}}{H}=\frac{\sqrt{s}}{a},\quad \frac{r_{CD}}{H}=\frac{\sqrt{s-a}}{a},\quad \theta_0=\frac{2aL}{H}. \tag{5}$$

Using these equations the bending moment is expressed as

$$M=\int_{r_{CD}}^{r_{AB}}\left(\sigma_\theta-\frac{F}{h}\right)rdr. \tag{6}$$

Here σ_θ is the circumferential stress (σ_r will stand for the radial stress) and $h=r_{AB}-r_{CD}$. The principal strain rate components, ξ_ζ and ξ_η, can be found from Eq. (2) as

$$\xi_\zeta=\xi_{rr}=-\frac{(\zeta+\mathrm{d}s/\mathrm{d}a)}{2(\zeta a+s)}\frac{\mathrm{d}a}{\mathrm{d}t},\qquad \xi_\eta=\xi_{\theta\theta}=\frac{(\zeta+\mathrm{d}s/\mathrm{d}a)}{2(\zeta a+s)}\frac{\mathrm{d}a}{\mathrm{d}t}. \tag{7}$$

Here ξ_{rr} and $\xi_{\theta\theta}$ are the normal strain rates in the plane polar coordinate system. In the case under consideration the equivalent strain rate is

$$\xi_{eq}=\sqrt{\frac{2}{3}\left(\xi_\zeta^2+\xi_\eta^2\right)}=\frac{|\zeta+\mathrm{d}s/\mathrm{d}a|}{\sqrt{3}(\zeta a+s)}\frac{\mathrm{d}a}{\mathrm{d}t}. \tag{8}$$

Here Eq. (7) has been taken into account.

It is assumed that the material obeys the double shearing model [15]. In particular, the yield criterion in the plane polar coordinate system reads

$$|\sigma_r-\sigma_\theta|+(\sigma_r+\sigma_\theta)\sin\varphi=2k\cos\varphi. \tag{9}$$

In the case of the original double shearing model proposed for granular materials, k and φ are material constants. In the case of metals, it is natural to assume that k depends of the equivalent strain, ε_{eq} [13]. The equivalent strain is defined by the following equation

$$\frac{d\varepsilon_{eq}}{dt}=\xi_{eq}. \tag{10}$$

The equations that connect stresses and strain rates are [15]

$$\xi_{rr}+\xi_{\theta\theta}=0,\qquad \xi_{r\theta}\cos2\psi-\frac{1}{2}(\xi_{rr}-\xi_{\theta\theta})\sin2\psi+\sin\varphi\left(\omega_{r\theta}+\frac{d\psi}{dt}\right)=0. \tag{11}$$

Here $\xi_{r\theta}$ and $\omega_{r\theta}$ is the shear strain rate and material spin in the plane polar coordinate system, respectively, ψ is the angle which the axis corresponding to the major principal stress makes with the r – direction, and d/dt denotes the convected derivative. The constitutive equations should be supplemented by the equations of equilibrium. It will be seen later that there is one non-trivial equation in the form

$$\frac{\partial\sigma_r}{\partial r}+\frac{\sigma_r-\sigma_\theta}{r}=0. \qquad (12)$$

3 General solution

The first equation in Eq. (11) is the equation of incompressibility. This equation is satisfied by the mapping given by Eq. (2). Consider the second equation in Eq. (11). It follows from Eq. (2) that coordinate curves of the (r,θ) coordinate system are trajectories of the principal strain rates. Therefore, $\xi_{r\theta}=0$. Since these coordinate curves are also trajectories of the principal stresses, $\psi=0$ or $\psi=\pi/2$. In either case $\sin 2\psi=0$ and $d\psi/dt=0$. Finally, $\omega_{r\theta}=0$ since coordinate curves of the (r,θ) coordinate system coincide with coordinate curves of the Lagrangian coordinate system. Thus the second equation of Eq. (11) is satisfied. The equivalent strain rate in Eq. (10) can be eliminated by means of Eq. (8). Then, in domains where ξ_r (and ξ_θ) does not change its sense, Eq. (10) can be integrated using the initial condition $\varepsilon_{eq}=0$ at $a=0$ and Eq. (3) to give

$$\varepsilon_{eq}=\frac{1}{\sqrt{3}}\left|\ln\left[4(\zeta a+s)\right]\right|. \qquad (13)$$

Replacing in Eq. (12) differentiation with respect to r with differentiation with respect to ζ by means of Eq. (4) yields

$$\frac{\partial\sigma_r}{\partial\zeta}+\frac{a(\sigma_r-\sigma_\theta)}{2(\zeta a+s)}=0. \qquad (14)$$

In order to find the stress solution, it is necessary to consider the cases $\sigma_r-\sigma_\theta>0$ and $\sigma_r-\sigma_\theta<0$ separately. Assume that $\sigma_r-\sigma_\theta<0$. Then, solving Eq. (9) for σ_θ results in

$$\sigma_\theta=\frac{(1-\sin\varphi)}{(1+\sin\varphi)}\sigma_r+\frac{2k\cos\varphi}{(1+\sin\varphi)}. \qquad (15)$$

The inequality $\sigma_r-\sigma_\theta<0$ corresponds to $\ln\left[4(\zeta a+s)\right]>0$. Therefore, eliminating in Eq. (14) the stress σ_θ by means of Eq. (15) and replacing differentiation with respect to ζ with differentiation with respect to ε_{eq} by means of Eq. (13) leads to

$$\frac{\partial\sigma_r}{\partial\varepsilon_{eq}}=\frac{\sqrt{3}k\cos\varphi}{(1+\sin\varphi)}-\frac{\sqrt{3}\sin\varphi}{(1+\sin\varphi)}\sigma_r. \qquad (16)$$

Since k is a given function of ε_{eq} this equation can be immediately integrated.

The case $\sigma_r-\sigma_\theta>0$ can be treated in a similar manner. As a result,

$$\frac{\partial\sigma_r}{\partial\varepsilon_{eq}}=-\frac{\sqrt{3}k\cos\varphi}{(1-\sin\varphi)}+\frac{\sqrt{3}\sin\varphi}{(1-\sin\varphi)}\sigma_r. \qquad (17)$$

4 Ideal flow solution for bending under tension

The ideal flow conditions require that the neutral line coincides with a material line. It is seen from Eq. (8) that the neutral line is determined by the equation $ds/da=-\zeta$. Since ζ is a Lagrangian coordinate, the ideal flow requirement is $ds/da=-\zeta_0$ where ζ_0 is constant. Then, integrating with the use of Eq. (3) gives

$$s=\frac{1}{4}-a\zeta_0. \qquad (18)$$

It is evident that $\sigma_r-\sigma_\theta<0$ in the range $0\ge\zeta\ge\zeta_0$. Therefore, Eq. (16) is valid in this range. The boundary condition to this equation is $\sigma_r=0$ for $\zeta=0$. Then, solving Eq. (16) with the use of this boundary condition gives

$$\sigma_r=\frac{\sqrt{3}\cos\varphi}{(1+\sin\varphi)}\exp\left[-\frac{\sqrt{3}\sin\varphi}{(1+\sin\varphi)}\varepsilon_{eq}\right]\times \int_{\varepsilon_1}^{\varepsilon_{eq}}k(z)\exp\left[\frac{\sqrt{3}\sin\varphi}{(1+\sin\varphi)}z\right]dz. \qquad (19)$$

Here ε_1 is the value of ε_{eq} at $\zeta=0$ (or $r=r_{AB}$) and z is a dummy variable of integration. In the case under consideration, $\varepsilon_{eq}=0$ at the neutral line. Therefore, it follows from Eq. (19) that

$$\sigma_n=\frac{\sqrt{3}\cos\varphi}{(1+\sin\varphi)}\int_{\varepsilon_1}^{0}k(z)\exp\left[\frac{\sqrt{3}\sin\varphi}{(1+\sin\varphi)}z\right]dz \qquad (20)$$

where σ_n is the value of σ_r at $\zeta=\zeta_0$. Equation (17) is valid in the range $\zeta_0\ge\zeta\ge-1$. The boundary condition to this equation is $\sigma_r=\sigma_n$ for $\zeta=\zeta_0$ (or $\varepsilon_{eq}=0$). Then, solving Eq. (17) with the use of this boundary condition gives

$$\sigma_r = \exp\left[\frac{\sqrt{3}\sin\varphi}{(1-\sin\varphi)}\varepsilon_{eq}\right]\times \left\{\sigma_n - \frac{\sqrt{3}\cos\varphi}{(1-\sin\varphi)}\int_0^{\varepsilon_{eq}} k(z)\exp\left[-\frac{\sqrt{3}\sin\varphi}{(1-\sin\varphi)}z\right]dz\right\}. \quad (21)$$

Using this equation it is possible to find p as

$$p = -\exp\left[\frac{\sqrt{3}\sin\varphi}{(1-\sin\varphi)}\varepsilon_2\right]\times \left\{\sigma_n - \frac{\sqrt{3}\cos\varphi}{(1-\sin\varphi)}\int_0^{\varepsilon_2} k(z)\exp\left[-\frac{\sqrt{3}\sin\varphi}{(1-\sin\varphi)}z\right]dz\right\}. \quad (22)$$

Here ε_2 is the value of ε_{eq} at $\zeta = -1$ (or $r = r_{CD}$). In Eqs. (21) and (22) it is necessary to eliminate σ_n by means of Eq. (20). Using Eqs. (5) and (22) it is possible to find F from Eq. (1). The distribution of the circumferntial stress can be determined from Eqs. (9), (19) and (21). Then, the bending moment is calculated by means of Eq. (6).

The final shape is completely determined by the values of r_{CD} and θ_0 at the end of the process. Assume that these values are equal to r_f and θ_f, respectively, and a_f is the value of a at the end of the process. Then, it follows from Eqs. (5) and (18) that

$$a_f = \frac{H\theta_f}{2L}, \quad \zeta_0 = \frac{L}{2H\theta_f}\left(1 - \frac{r_f^2\theta_f^2}{4L^2}\right) - 1. \quad (23)$$

5 Summary

A new ideal flow solution has been found. Its distinguished feature is that the plasticity model adopted is pressure-dependent. The solution is for the process of bending under tension. The ideal flow paths in this process are obtained if the neutral line is a material line. It has been shown that the position of the neutral line can be found if the initial and final shapes are prescribed (see Eq. (23)). Then, the distribution of the radial stress is given by Eqs, (19) and (21). This distribution depends on two variables, a and ζ. In order to find the dependence of the radial stress on physical quantities, it is necessary to use Eqs. (4), (5) and (18). Once the distribution of the radial stress has been determined, the distribution of the circumferential stress follows from Eq. (9). Then, the bending moment and tensile forces are found from Eqs. (1), (5), (6) and (22).

Acknowledgment

The research described was supported by the grants RSCF-16-49-02026 (Russia) and INT/RUS/RFBR/P-214 (India).

References

1. R. Hill, J. Mech. Phys. Solids **15**, 223 (1967)
2. O. Richmond, S. Alexandrov, Acta Mech. **158**, 33 (2002)
3. S. Alexandrov, Y. Mustafa, E. Lyamina, Meccanica **51**, 2235 (2016)
4. I.F. Collins, S.A. Meguid, Trans. ASME J. Appl. Mech. **44**, 271 (1977)
5. K. Chung, S. Alexandrov, Appl. Mech. Rev. **60**, 316 (2007)
6. O. Richmond, H.L. Morrison, J. Mech. Phys. Solids **15**, 195 (1967)
7. H.F. Weinberber, Meccanica **38**, 547 (2003)
8. C.J. Cawthorn, E.G. Loukaides, J.M. Allwood, Prog. Eng. **81**, 2451 (2014)
9. E. Orowan, Proc. Inst. Mech. Eng. **150**, 140 (1943)
10. G.-L. Zhang, Z-W. Wang, S.-H. Zhang, M. Cheng, H.-W. Song, Proc. IMechE, Part B: J. Eng. Manufact. **227**, 1023 (2013)
11. W.A. Spitzig, R.J. Sober, O. Richmond, Metallurg. Trans. **7A**, 1703 (1976)
12. A.S. Kao, H.A. Kuhn, W.A. Spitzig, O. Richmond, Trans. ASME J. Eng. Mater. Technol. **112**, 26 (1990)
13. C.D. Wilson, Trans. ASME J. Appl. Mech. **69**, 63 (2002)
14. P.S. Liu, Mater. Sci. Eng. **422A**, 176 (2006)
15. A.J.M. Spencer, J. Mech. Phys. Solids **12**, 337 (1964)
16. M. Yoshida, F. Yoshida, H. Konishi, K. Fukumoto, Int. J. Mech. Sci. **47**, 1885 (2005)
17. M.H. Parsa, S.N.A. Ahkami, Int. J. Mater. Form. **Suppl. 1**, 173 (2008)
18. M.A. Guler, F. Ozer, M. Yenice, M. Kaya, Steel Res. Int. **81**, 801 (2010)
19. S. Alexandrov, K. Manabe, T. Furushima, Arch. Appl. Mech. **81**, 1935 (2013)
20. S.A. Kagzi, A.H. Gandhi, H.K. Dave, H.K. Raval, Mech. Adv. Mater. Struct. **23**, 80 (2015)
21. S. Alexandrov, J.-H. Kim, K. Chung, T.-J. Kang, J. Strain Anal. Eng. Des. **41**, 397 (2006)

DOI: 10.1051/matecconf/201710801002

Finite element analysis on the wire breaking rule of 1×7IWS steel wire rope

DU Wenzheng, MA Baozhu, XIE Zheng, CAO Dazhi and WU Peng

Xi'an High Technology Research Institutute, China

Abstract. Taking the wire rope of 1×7+IWS structure as the research object, the influences of the number of broken wires on the stress distribution under the same axial load were simulated and analysed, and it also explored the rule of wire breaking of steel wire ropes. Based on the SolidWorks software, the three-dimensional model of the wire rope was established. Importing the model into the ABAQUS, the finite element model of the steel wire rope was established. Firstly 5000 N axial tension was placed on the rope, the stress distribution was simulated and analysed, and the steel wire with the largest stress distribution was found out. Then one steel wire was truncated with the load unchanged, and the finite element simulation was carried out again, and repeated the steps several times. The results show that, with the increase of the number of broken wires, the Von-Mises stress of the wire rope increases sharply, and the stress distribution is concentrated on the rest of the unbroken wires, which brings great challenges to the safety of the wire rope.

1 Introduction

The wire rope is widely used in industrial production and daily life, its performance has important influence on the safety [1]. But the steel wire rope is of space spiral structure which is very complex, therefore the traditional stress analysis can not accurately solve the mechanical properties [2]. At the same time in the process of ascension, due to wear, corrosion, fatigue, impact and other reasons[3], the wire rope is easy to occur the phenomenon of wire breaking .In the fracture failure process of wire ropes, the number of broken wires increases gradually, until the whole wire rope breaks. Therefore, the study of the stress change and wire breaking rules in the process of wire breaking under the same load is very important to prevent the wire rope breaking accidents and improve the level of the use and maintenance of the wire rope. Li Lunyou analysed the reasons of failure of 6*36sw-25.0mm steel wire rope for field bridge through the electron microscope observation[4]. The fatigue damage of the steel wire rope bearing load under different pre tension was studied by fatigue test in Jia Xiaofan[5]. Zhang Lun used scanning electron microscope, energy spectrum instrument for the failure analysis of wire rope for the electric door and window elevator [6].Wei Huaichang elaborated the phenomenon and reason for abnormal wire breaking of the floor type friction wheel winch hoisting rope[7]. Hu Zhihui studied the failure mechanism of double broken line type multilayer winding steel wire rope[8].

In this paper, taking 1×7+IWS steel wire rope as the research object, using the finite element analysis software ABAQUS, simulation analysis was carried out on the wire breaking of the wire rope under axial load of 5000N[9].

2 Finite element model of 1×7+IWS steel wire rope

2.1 Geometry model of the steel wire rope

In this paper, the three-dimensional solid model of 1 * 7 IWS structure round strand wire rope was established by SolidWorks software. The twist moment is 42.3mm, the diameter of the rope core and the side strand wire are both 2.34mm. The steel wire rope structure is shown in Figure 1.

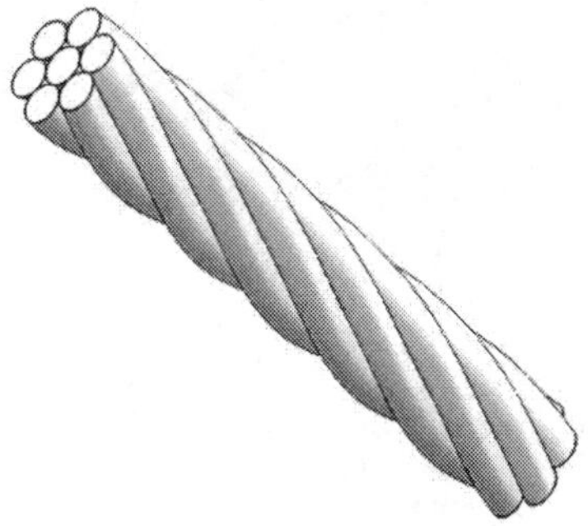

Figure 1. The geometric model of the 1×7+IWS steel wire rope

In order to reduce the amount of calculation, using wire rope cyclic symmetric structure characteristics, 1/6

 DOI: 10.1051/matecconf/201710801002

twist length (7.05mm)of the wire rope was intercepted as the research object, which was imported directly into ABAQUS software through the SolidWorks and ABAQUS plug-in interface. According to the result of each simulation, the steel wire which has the maximum stress is cut off in turn, to create a new assembly, and the simulation analysis is carried out again.

2.2 Material properties

The material of steel wire rope core and side wires is a high-quality carbon steel, for isotropic materials, and the specific parameters are shown in Table 1 [10].

Table 1. The material properties of the steel wire rope

Density ρ (kg/m^3)	Poisson's ratio γ	Elastic modulus$E$(Gpa)	Yield strength σ_s (Mpa)	Limitssing intensity σ_b (Mpa)
7800	0.3	200	1670	1803.5

2.3 Element mesh division

The rope core and side strands wires all used C3D8I unit (8-node linear hexahedral elements, incompatible mode). By means of the method of the neutral axis, the wire rope was meshed [11]. In the process of finite element simulation, in order to ensure the convergence results, the mesh need to be gradually refined to determine appropriate grid density according to the calculation results of different grid density. Through the calculation and test, the finite element model of steel wire rope was determined, as shown in Figure 2. The total cell number of the finite element model of the whole wire rope is 139264, and the total number of the nodes is 155382.

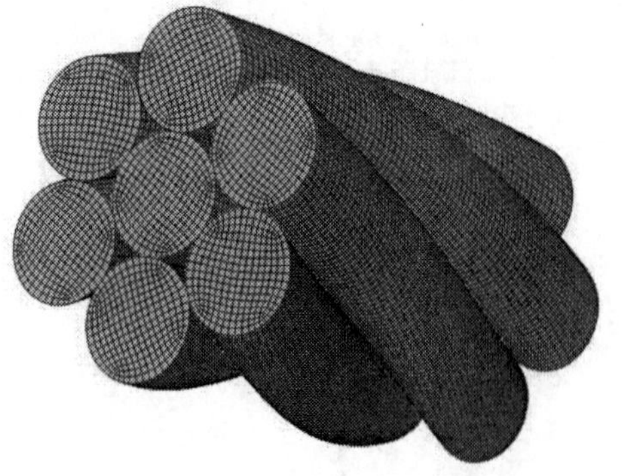

Figure 2. The finite element mesh of the 1×7+IWS wire rope

2.4 Contact attributes

As shown in Figure 3, the master-slave contact algorithm is applied to transfer load between the two contacts. According to the principle that slave surfaces can be penetrated, but main surfaces can not be penetrated [12], each wire can be either the master surface or slave surface to establish contact pairs between adjacent wires. Also finite sliding mode was selected between two contact surfaces. The friction coefficient between the steel wires was defined as 0.1 [13], and the friction type is penalty function.

(a) The contact pairs of the wire and straight strand rope core

(b) The contact pairs of the wires

Figure 3. The contact model of the wire rope

2.5 Boundary conditions and loading methods

Reference points are established respectively in the axis of the rope core and they are located at the center of the front surface and back surface of the wire rope to couple all the nodes of the corresponding cross-section. Each reference point has six degrees of freedom, namely, three translational degrees of freedom and three rotational degrees of freedom. In finite element analysis of wire rope, the model of coupling movement establishes a relationship of motion constraints between each node and reference point in the bound area. And it makes the reference point and the nodes of the bound area achieve the same displacement [14]. In the finite element analysis, all the degrees of freedom of one point were constrained and 5000N axial tensile load was placed on the other reference point.

DOI: 10.1051/matecconf/201710801002

3 Finite element simulation

3.1The first simulation (0 broken wires)

The 5000N axial tension was applied at the reference point 2, and the finite element simulation was carried out, and the analysis results are shown in Figure 4 and Figure 5.

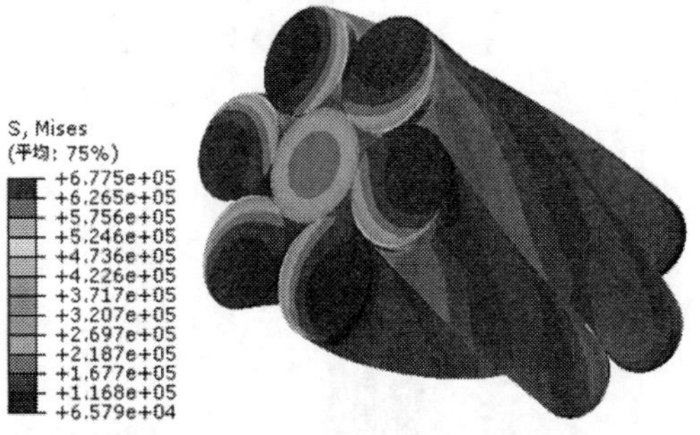

Figure 4. The Von-Mises stress distribution of the wire rope

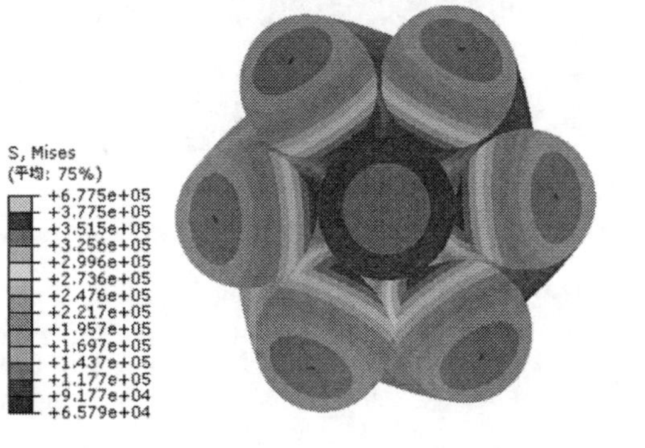

Figure 5. The stress distribution at cross section of the rope

On the whole wire rope, due to the constraints of the coupling movement on both ends of the wire rope, the stress on both ends is large while the stress in the central area is small. On the end face of the wire rope, the stress distributes radially around the rope core, the stress on the contact of wires is larger than other area. On the face of the rope core, the stress distributes like concentric circle. The central stress is the minimum while the outer stress is the maximum. Besides, the six side wires share the same stress distribution on the surface, which is consistent with the symmetry of the geometry model. On the side of the wire rope, the stress presents a helix staggered distribution along the surface of the spiral steel wire.

In order to reduce the influence of coupling constraint on the results, the intermediate section of the wire rope is taken as the research object, and the von-mises stress distribution is observed, as shown in Figure 5. For the whole cross section, the von-mises stress distribution around the center of the wire rope is symmetrical, and the stress of the rope core is the largest of the cross section. The stress on the rope core presents concentric circular distribution, and the stress on the contact area between the core and the side strand wire is 373.470MPa, which is the largest of the whole wire rope. For each side wire, the stress distribution is layered. From the center to the outside, the stress decreases gradually.

3.2 The second simulation (1 broken wires)

It is concluded from the first simulation that the stress on the rope core is the largest, and the the rope core is easy to produce fatigue fracture in the process of using. Assuming that the rope core is broken, the rope core is cut off in the three-dimensional model, and the finite element simulation results under the same 5000N axial load are shown in Figure 6 and 7.

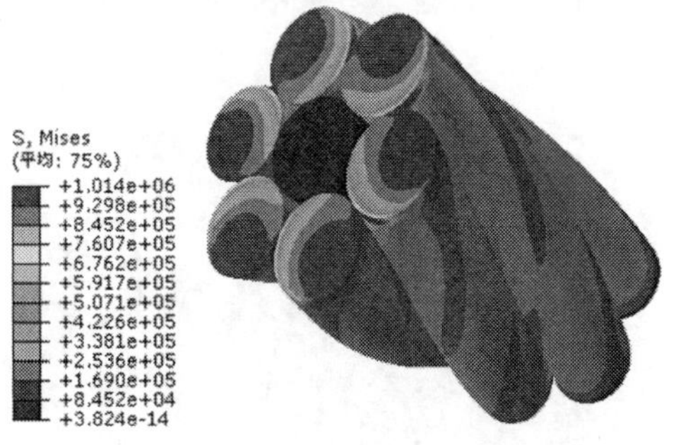

Figure 6. The Von-Mises stress distribution of the wire rope

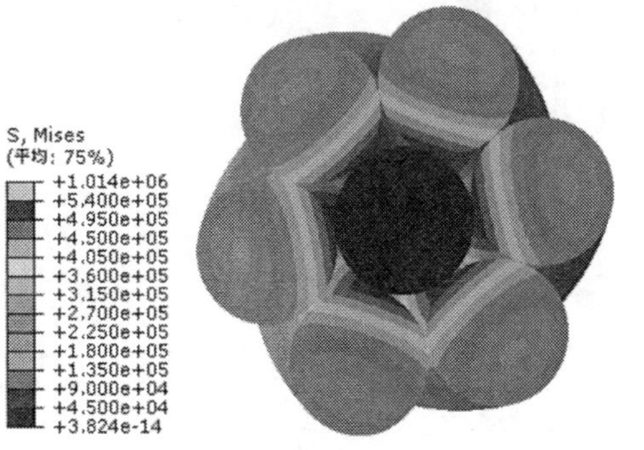

Figure 7. The stress distribution at cross section of the rope

Because the rope core is broken and it can not bear the load, the stress of the whole rope core is close to 0. The rest six side strand steel wires bear the axial tensile strength of 5000n, therefore the stress is increased significantly. On the middle cross section of the steel wire rope, the stress of the contact area between the side strand and the rope core is the largest which is 528.695MPa.

3.3 The third simulation (2 broken wiress)

It is concluded from the second simulation that the stress on the side strand is the largest of the rope, therefore it is easy to break during the working process. Assuming that a strand of wire rope is fractured, it is cut off in the three-dimensional model of the wire rope as well. Under the same 5000N axial load, the finite element simulation results are shown in Figure 8 and 9.

 DOI: 10.1051/matecconf/201710801002

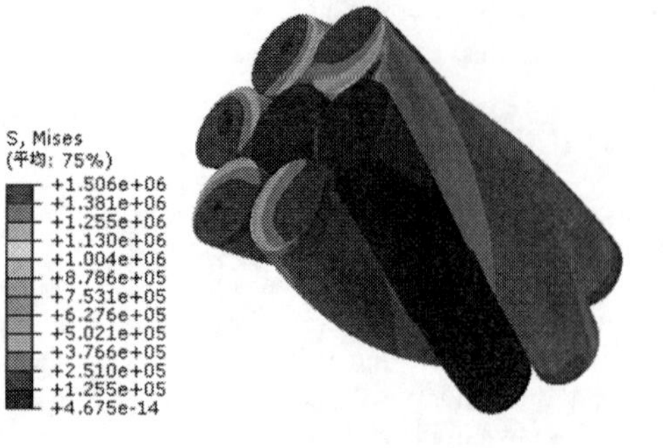

Figure.8 The Von-Mises stress distribution of the steel wire rope

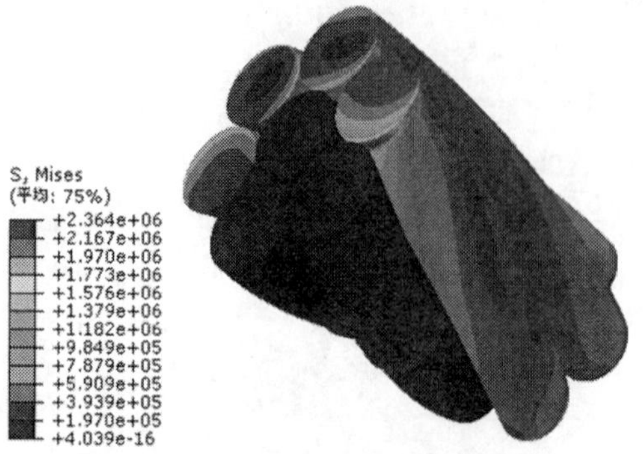

Figure 10. The Von-Mises stress distribution of the wire rope

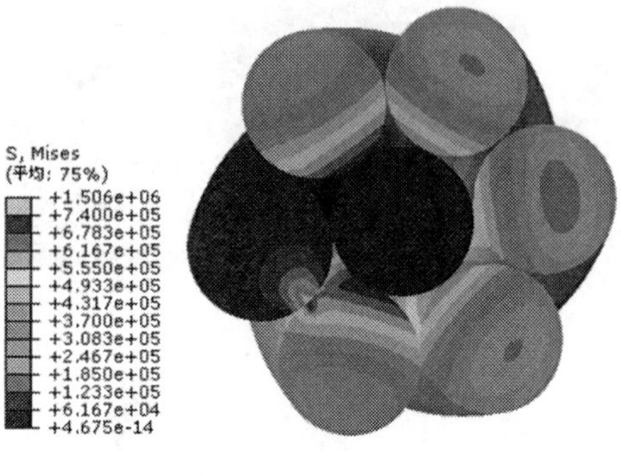

Figure.9 The stress distribution at cross section of the wire rope

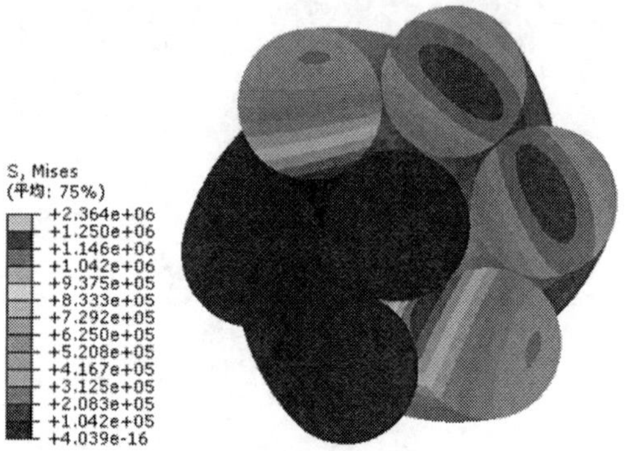

Figure 11. The stress distribution at cross section of the rope

The rope core and a side strand wire are broken, which can not bear the load, only play a supporting role, therefore its stress is close to 0.The axial tension of 5000N is fully suffered by 5 side strand steel wires and the stress distribution on each side strand is different, and the upper limit of the stress is increased significantly. From Figure 9, we can find that the maximum stress is distributed on the two side strand wires which are adjacent to the fractured side strand and the maximum stress is 719. 964MPa.

3.4 The fourth simulation (3 broken wires)

It is concluded that the maximum stress is distributed on the two side strand wires which are adjacent to the fracture side strand wire from the third simulation. Assuming that one side strand wire breaks in the course of work, it is cut off in the three-dimensional model of the wire rope, and the finite element simulation results under the same 5000N axial load are shown in figure 10 and 11.

The rope core and the two side strand wires are all broken, which can not bear the load, only play a supporting role, therefore its stress is close to 0. The axial tension of 5000N is fully suffered by 4 side strand steel wires, and the stress on the side strand wire increases significantly. The maximum stress is distributed on the two side strand wires which are adjacent to the fracture side strand wire, which is 1207.99 MPa. Compared with the third simulation results, the stress is greatly increased.

3.5 The relationship between stress,deformation and the number of broken wires

The four simulation results show that the Von-Mises stress is increasing with the increase of the number of broken wires, under the same load. The maximum stress σ_{max} and the maximum axial deformation U_{max} of the four simulations are distilled, as Table 2 shows, and the relation curves are shown in Figure 12 and 13.

Table 2. The corresponding value of σ_{max} $U_{max}$and μ

Number of broken wires n	0	1	2	3
Von-mises σ_{max}/MPa	373.470	528.695	719.964	1207.99
Axial deformation U_{max}/mm	0.0179	0.02654	0.04512	0.1078

DOI: 10.1051/matecconf/201710801002

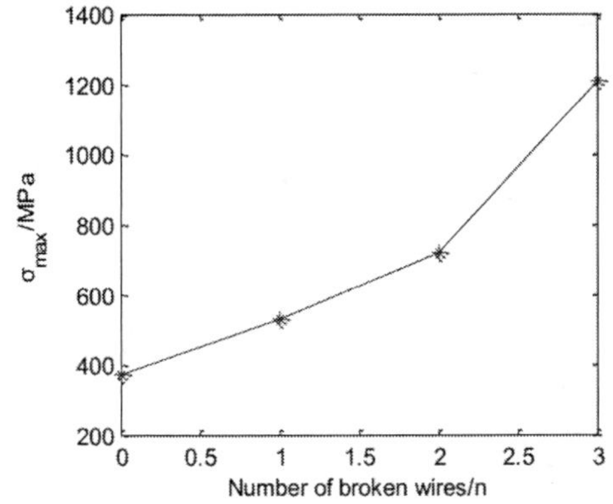

Figure 12.The relation curve of σ_{max} and n

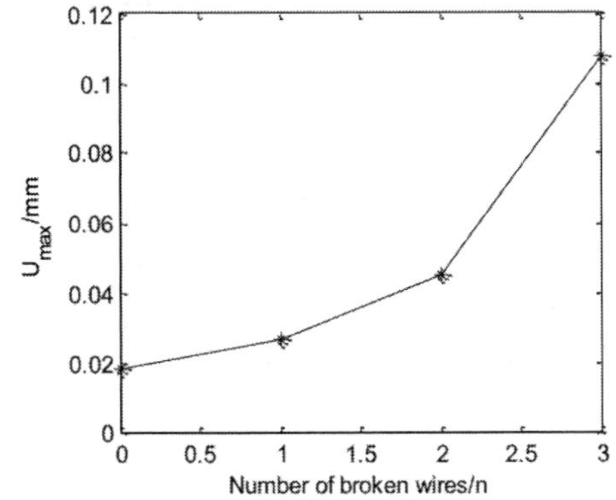

Figure 13.The relation curve of U_{max} and n

As Figure 12 and 13 show , under the same load, the maximum von Mises σ_{max} and axial deformation U_{max} are increased with the increase of the number of broken wires n, and the increasing speed is faster and faster. Therefore, with the increase of the number of broken wires n, anti load capacity of the wire rope decreases sharply, and the wire rope is more easy to be broken.

4 Conclusions

This paper analysed the influence of the number of broken wires on the stress distribution and the axial deformation of 1×7+IWS steel wire rope under the same axial load and explored the rule of the broken wires. The conclusions can be drawn as follows:

1). With the increase of the number of broken wires n, the von-mises stress of the steel wire rope increases rapidly under the same load. The greater the stress, the more serious the wear of the wire rope, and the shorter the fatigue life.;

2). With the increase of the number of broken wires n, the axial deformation is also increasing rapidly under the same load, which bring adverse effects on the steel wire rope;

3). The broken wires almost no longer bear the axial load, only play a supporting role, and the steel wire adjacent to the broken wire will bear larger load. It is easy to produce stress concentration, which accelerates the damage process.

References

[1] Li Zhanfang, Xiao Xingming, Liu Zhengquan, et al. Study on dynamics of mine hoisting rope [J]. Coal mine safety, 10:11-14, (2007)

[2] Hruska F.H.Radial forces in wire ropes[J].Wire Produets, 2(7): 459-463, (1952).

[3] Lv Xiaohui, Chen Zhiping, Wang Songgen, et al. Study on the failure mechanism of steel wire rope [J]. Metal products, 39 (5): 1-5, (2013).

[4] Li Lun you, Zhang Xiaoyan, Zhang Qi. Failure analysis of broken wire of 6*36sw-25.0 mm steel wire rope used by a bridge [J]. Guizhou industry university journal, 35 (1): 43-45, (2006).

[5] Jia Xiaofan, Zhang Dekun. Research on bending fatigue damage of bearing steel wire under different pre tension [J]. Journal of mechanical engineering, 47 (24): 31-37, (2011).

[6] Zhang Lun, Liu Hongfang, Sun Jie. Fatigue fracture analysis of steel wire rope for electric door and window elevator [J]. Metal products, 39 (1): 56-59, (2013).

[7] Wei Huaichang. Analysis and solution of the broken wire of winch lifting wire rope [J]. Coal mine machinery, 28 (11): 176-177, (2007).

[8] Hu Zhihui, Hu Yong, Hu Jiquan, et al. Study on the failure mechanism of the double broken line multilayer winding steel wire rope [J]. China Mechanical Engineering, 24 (23): 3195-3199, (2013).

[9] Zhang Min, Kou Ziming, Li Ting. Torque analysis of steel wire rope based onABAQUS [J].Coal mine safety, ,47 (2): 113-115, (2016).

[10] Zhang Jin, Chen Xiangyang, Tang Wenting, et al. Finite element analysis of.1 * 7 IWS structural steel wire rope [J]. Journal of Anhui University of Science and Technology, 34 (3):83-86, (2014).

[11] Zhang Jianwei. Abaqus 6.12 finite element analysis from entry to the master [M]. Beijing: Mechanical Industry Press, 87-132.

[12] Shi Yiping, Zhou Yurong. Abaqus finite element analysis of detailed examples [M]. Beijing: Mechanical Industry Press, 2015:136-140, (2015).

[13] Wang Dagang. Fretting and fretting fatigue life prediction of steel wire rope [D]. Xuzhou: China University of Mining and Technology, (2012)

[14] Zhu Wenmin, Zheng Gang, Liu Zhicai, et al. Analysis of stress and fatigue of steel wire rope based on Ansys Workbench [J]. Lifting and transportation machinery, (2): 34-38, (2016)

Preparation of a self-curable unsaturated polyester with vinyl double bonds and Investigation of its self-curing properties

Haiyan Li, Fangfang Zhu, and Chongwen Li

Institute of Materials Science and Engineering, Ocean University of China, Songling Road, Number238, Qingdao 266100, China

Abstract. In this paper, a kind of self-curable unsaturated polyester PGA-UP with vinyl groups on the end of molecule chains was synthesized through esterification reaction of glycidol with acrylic acid. The self-curable function of unsaturated polyester was put forward as a new solution to air pollution problems caused by the cross-linking monomer's volatilization in unsaturated polyesters (UP) industry. The synthetic method, reaction process, molecular structure and self-curing properties were studied. Investigation demonstrates that the unsaturated polyester resin PGA-UP could be synthesized through a two-step process and the liquid product was a mixed outcome of different extent esterification. The unsaturated polyester resin could be self-cured through the free radical reaction of vinyl groups on the end of the molecules when they were initiated by 1% benzoyl peroxide (BPO) at 80℃ for 1 hours and 120℃ for 2 hours with no cross-linking monomers. The intension strength of the self-cured PGA-UP is 13.5Mpa, close to the 15.6MPa of an usual unsaturated polyester S-UP made in our lab. The self-curable PGA-UP presents good potential application and the research will contribute to improve the environmental problems caused by volatile of cross-linking monomers in unsaturated polyester applications. 1H-NMR and FT-IR were used to characterize.

1 Introduction

The thermosetting unsaturated polyester (UP) is one of the most widely used resin in modern plastic industry. The double bonds on the main chain of UP is unable to be cured through free radical homo-polymerization and it is usually dissolved in some vinyl monomer and thermally crosslink through free-radical copolymerization with vinyl monomers initiated by peroxide [1]. However, styrene (St), the commonly used cross-linking monomer, produces toxic volatile during the curing process. In past year, a lot of measures have been taken, and most of them are focused on how to lower the amount of the volatile or to use lower toxic or non-toxic monomers as substitute and all demonstrate kinds of limits. Few of them are focused on the structure and function of unsaturated polyesters [1].

On different view from above, in our lab, it is believed that the unsaturated polyester with new molecule structures and self-curable properties can present a new solution to the air pollution caused by cross-linking monomers. Some new unsaturated polyesters with self-curable function have been designed and prepared, which could self crosslink without other monomers and of course the air pollution problem is avoid [2-3].

At present, self-curable unsaturated polyesters have attracted great attentions in the research field of medical polymer and tissue engineering [2-7]. Among them, the majority are the unsaturated polyesters with nucleophilic pendant groups on side chain, such as hydroxyl, thiol or amino group, which usually was prepared with malefic anhydride (MA) as one of the raw materials[2,3,6]. As the double bonds located on main chain present low free radical activity, it is difficult for such UPs to self-crosslink through the free radical homo-polymerization. Differently, they come into self-crosslinking through the nucleophilic addition reaction of nucleophilic pendant groups with the double bonds [2-3]. However, the self-curable UP resin is difficult to process because of its high viscosity and the high temperature for curing. Meanwhile, much fault existing in the product after cured and the low curing extent result in restricted applications. At present, kinds of self-curable unsaturated polyesters are designed and synthesized for biomaterials, but it is much insufficient for potential applications. The types and applications of self-curable unsaturated polyesters need great expansion.

In order to solve the air pollution problem and improve the curing process and properties after cured to expand the application of self-curable UPs, one kind of unsaturated polyester (PGA-UP) with vinyl bonds located at the end of molecular main chain was made in our lab, through esterifization of glycidol (GL) with acrylic acid (AC), with the existance of radical inhibitor. The PGA-UP obtained presented low viscosity at room temperature and could be thermally self-cured without the crosslinking monomer. Compared with the double bonds in the middle of molecular chain of common unsaturated polyesters, the vinyl bonds on the end of molecular chains presented higher molecular mobility, lower steric hindrance and, consequently, present higher self-crosslinking activity. In the synthesis of the PGA-UP, in order to avoid the radical homo-polymerization of the vinyl double bonds during the the esterifization process, temperature was controlled strictly under 118℃ which was much lower than the usual ersterifizaion temperature. As a result, the key point during the sythesis is how to promote the esterfization reaction rate under the low temperature. Here benzene and toluene mixture was added as water carrying agent and its mixing ratio was adjusted to control the reflux temperature of the reaction system. The synthetic technological process, reaction mechanism, molecular structure and self-curing properties were studied.

2 Experimental

2.1 Materials

Acrylic acid (A.R), toluene (A.R), p-hydroxyl phenol (A.R), anhydrous zinc chloride (A.R), xylene (A.R), benzoyl peroxide (BPO) (A.R) and N，N-dimthylben - zylamine (99%), were obtained from Tianjin Bodi Chemical Reagent Factory (Tianjin City, China) without further treatment. Glycidol (GL, 96%) was obtained from Sinopharm Chemical Reagent Co., Ltd. (Shanghai City, China) without further treatment.

2.2 Experimental

2.2.1 Synthesis of the self-curable resin PGA-UP

In this reaction, one molecule Glycidol (GL) is designed to be esterified by three Acrylic acid molecules. In order to make the esterifying reaction complete, a third excessive molar dose of AC was added. The experimental process was shown as follows:

GL22.44g (0.30mol), AC 3.603g (0.05mol), N, N-dimthylbenzylamine 0.02g(1% of GL amount), p-hydroxyl phenol 0.014g (HP,0.05% of GL amount) and anhydrous zinc chloride 0.108g as esterification catalyst were placed into a 250ml, four-necked, round-bottomed flask equipped with an agitator, a thermometer, a N_2 tube, a reflux condenser pipe and a constant pressure funnel. The system was run and slowly heated and kept at 74-80℃. AC 28.824g (0.40mol) in the constant pressure funnel was dropped into the reaction system at a constant speed during 5 hours. The reaction extent was monitored by the change of viscosity and the acid value of the system.

When the acid value in reactants solution could not decrease further, the reflux condenser pipe was displaced by water segregator with full of water carrying agent, the mixture of benzene and toluene (volume ratio being 3:1). The reaction system was heated slowly and the water carrying agent was also added into the system to keep the reflux temperature of the system band to 110-112℃.The residual AC 43.236g (0.60mol) was added into the constant pressure funnel and slowly dropped into the system for about 5 hours. During the reaction process, the system kept reflux to drive away the byproduct water into the water segregator for enhancing the esterification conversion further. With the reaction going further, the reflux temperature increased slowly. In order to control the reflux temperature varied between 110-118℃, the water carrying agent was added occasionally as dependent. The acid value was monitored every hour. When the acid value did not decrease further, the water segregator was displaced by a negative pressuredistillation system with water-ring vacuum pump. The excessive AC was distilled slowly under 68-72℃ for about 1.5 hours until no distillate out. The reaction product was extracted to remove the solid byproduct and distilled with oil-ring vacuum pump. A distillate of 58℃ was collected as our product, PGA-UP, a transparent light brown liquid at room temperature. ^{1}H-NMR and FT-IR were used to characterize the product. Both the acid value and hydroxyl value of the product were tested.

2.2.2 Curing experiments of PGA-UP resin

One PGA-UP sample mixed with the free radical initiator BPO being 1.0% of the PGA-UP amount, was named as PGA-UP-1. After the dissolution of BPO, The mixture was mounted into 7 molds respectively and heated respectively at different constant temperature as 60℃, 80℃, 90℃, 100℃, 120℃, 140℃, 150℃ for enough time to self crosslink. Another PGA-UP sample with no BPO addition was named as PGA-UP-2 and was carried on the same experiments as PGA-UP-1. Some of the Samples before and after cured were characterized by FT-IR spectra.

2.3 Analysis and Characterization

^{1}H-NMR spectra were carried out on a JEOL JNM-ECP 600 NMR spectrometer (Japan Electronics Co., Ltd.), with the resonance frequency of 600MHz. Infrared spectra were performed on a Bruker TENSOR27 Fourier transform spectrometer (Bruker Instruments Inc., Germany) at room temperature, measuring range 400~4 000cm^{-1}.

The intension strength were measured on an Instron 1122 testing instrument with 80mm× 10mm × 3mm specimens in accordance with GB/T2569-1995 at a drawing rate of 5mm/min. The samples for intension strength test were made as follows: The PGA-UP liquid mixed with BPO being 1.0% of the PGA-UP amount, was mounted into the molds, and was thermally cured at 80℃ for one hour and then 120℃ for 2 hours. After the curing process, the samples were polished and their dimensions were measured for the intension strength test next.

3 Results and discussions

3.1 Preparation of the PGA-UP resin

During the esterification of the GL and AC, If AC was added fast or the reaction temperature was high above 80℃, the reaction heat could not dissipate well, local overheating would lead to gelatin and the synthesis failed. In order to make the esterificaiton carry on smoothly without gelatin, free radical inhibitor was necessary, the reaction temperature must be strictly controlled below 80℃, AC must be dropped slowly into the reaction system and the system should be stirred strongly to dissipate the reaction heat. During the reaction step at 74-80℃, the viscosity of the system increased gradually and at last it increased no more. No water product was observed.

When the reaction was not able to go further below 80℃, the reaction temperature was heated slowly to 110℃.With the existence of the water carrying agent, the system reflux at 110-112℃ and water byproduct began to be distilled into the water segregator. The viscosity of the system began to increase gradually again. In the second step reaction, with the reaction's going on, the reflux temperature will slowly increase. In order to control the reflux temperature, appropriate dose water carrying reagent was added occasionally to control the temperature varied between 110-118℃. If AC was added faster or the reflux rate was fast or the reaction temperature was higher than 118℃, gelatin may occur and solid grain byproduct would come into being. Therefore, it is very important to control the reaction temperature and the reaction rate.

Based on the reaction behavior, we assume that the reaction carries on by two steps as shown in Figure 1. Firstly, GL reacted with AC through the ring opening reaction for forming ester group at 80°C; secondly, the intermediate products with two hydroxyl groups further reacted with two ACs to form another two ester groups at 110°C-118℃. As shown in Figure 1, one GL molecule needs to react with three AC molecules to produce the unsaturated polyester pre-polymer with three vinyl bonds on the end of molecule chain, named as PGA-UP. In order to make the hydroxyl produced in the reaction process being esterifed completely, the molar ratio of GL and AC is 1:4 with AC excessive.

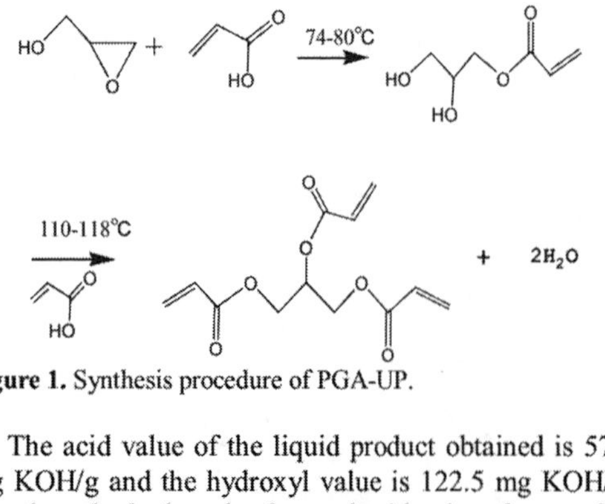

Figure 1. Synthesis procedure of PGA-UP.

The acid value of the liquid product obtained is 57.8 mg KOH/g and the hydroxyl value is 122.5 mg KOH/g. Based on the hydroxyl value and acid value, the reaction conversion extent of hydroxyl group is about 80.7%, the conversion rate of the carbonyl group is about 81.5%.

The above result indicates that the hydroxyl of GL has not been esterified completely. We assumed that the yellow liquid product was a mixture of compounds produced by GL with different esterifying reaction degree with AC, which is verified by the 1H-NMR and FT-IR spectra. The molecule structures are listed in Figure 2. The 1H-NMR with acetone-d6 as a solvent and FT-IR are shown in Figure 3 and Figure 4.

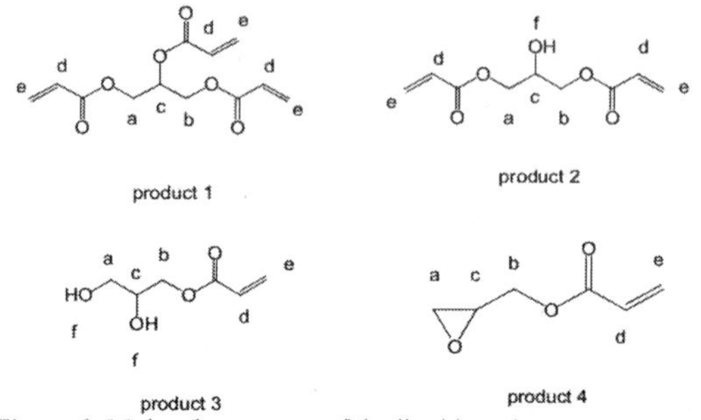

Figure 2. Molecule structure of the liquid product.

As shown in Figure 2, product1 is the target product PGA-UP as desired which is synthesized through the esterifying reaction of one molecule GL with three molecules of AC. As shown in the 1H-NMR spectrum, the peaks at 6.35ppm and 5.89ppm are designated for the protons e, from the vinyl double bonds on the end of the molecular chain, and the peak at 6.10ppm is designated for the proton of the vinyl double bonds named as d. The peaks at 4.39ppm is designated for the protons a and b, and the peaks at 5.21ppm is designated for the protons c.

Product 2 is produced through the esterifying reaction of one molecule GL with two molecules of AC, with a residual hydroxyl group unreacted on the molecule structure. The peaks at 4.24-4.21ppm are designated for the protons of *a* and *b* in Product 2. The peaks produced by the proton *c* of product 2 are the triplet peaks at 4.39ppm which are overlapped by the peaks of the protons *a* and *b* of product 1. The peaks produced by the protons d and *e* of product 2 are at the same location with those d and *e* of product 1.

Product 3 and 4 are produced through the esterifying reaction of one molecule GL with one molecule AC. Product 3 is produced by the ring-opening reaction of GL with AC. As shown in Figure 3, the peaks for the protons *a* are at 3.56-3.59ppm, the peaks for the protons *b* are at 4.16-4.12ppm, and the multiplet peaks for the protons *c* are at 3.9ppm. The peaks produced by the protons *d* and e of the vinyl double bonds of product 3 are at the same location with the protons *d* and *e* of product 1 and product 2. The peak at 4.72ppm are designated for the proton *f* of hydroxyl groups unreacted in product 2 and 3.Product 4 is produced by the esterifying reaction of the hydroxyl group on GL with the carboxyl group on AC. The protons *a* and *c* are located on the epoxy ring. The peaks for the two protons *a* are at 2.47-2.52ppm and 2.75ppm respectively, and the peaks for the protons *c* are at 3.30ppm. The peaks for *b* are at the same location with the proton *b* of product 3. The peaks produced by the protons *d* and *e* are at the same location with those of product 1. Based on the peak intensity, we can estimate

DOI: 10.1051/matecconf/201710801003

that the liquid product is mainly composed by product 1, 2 and 3, and the amount of product 4 is much less than the other three.

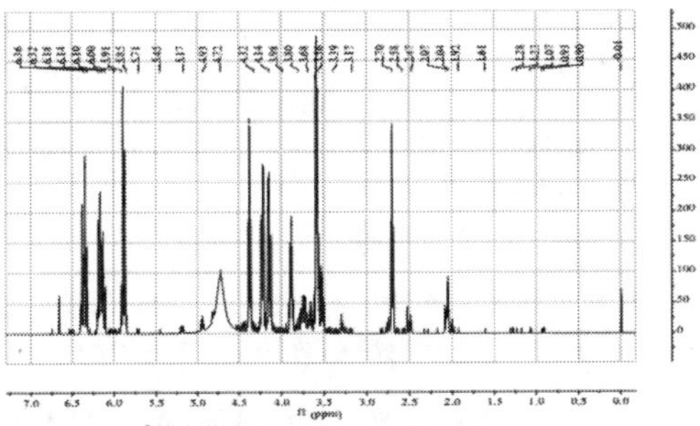

Figure 3. The ^{1}H-NMR spectra of PGA-UP.

As shown in the FT-IR spectrum in Figure 4, the peaks at 1635cm^{-1}and1618cm^{-1} are due to stretching vibration of C=C double bonds; the small peak at 3034cm^{-1}also shows the existence of C=C double bonds; the absorption at 1410cm^{-1} is caused by in-plane bending vibration of the C-H of vinyl double bonds; the absorption at 985cm^{-1} and 810cm^{-1} are caused by out-plane bending vibration of the vinyl double bonds on the chain end. All the above absorptions indicate the existence of the vinyl double bonds. The peak at 1722cm^{-1} is for the C=O stretching vibration; Peaks at 1165cm^{-1} and 1215cm^{-1} are caused by stretching vibration of C-O. The absorption at 3335cm^{-1} is caused by stretching vibration of residual hydroxyl group unreacted of product 2 and 3; the small absorption at 915cm^{-1} is attributed to epoxide group from product 4. The result is consistent with what shown in the ^{1}H-NMR spectrum.

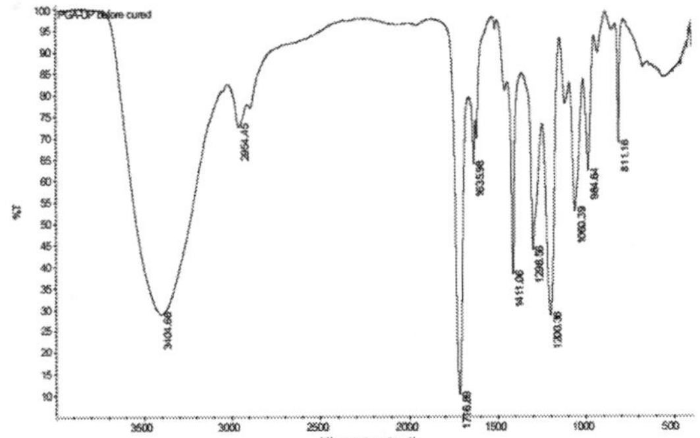

Figure 4. FT-IR spectra of PGA-UP before cured.

3.2 The self-curable properties of PGA-UP

After the curing experiments at different heating temperature, we found that the samples of PGA-UP-1, containing1%BPO initiator, could be thermally self-cured into a light yellow transparent solid after heated at 80°C for 1 hour and became more hard after heated at 120℃ for 2 hours .However, the samples PGA-UP-2 without BPO as initiator could be slowly cured until heated at 150℃ and completely cured until heated at 180℃ for 2 hours. Obviously, the existence of the radical initiator BPO greatly induced the self cross-linking reaction rate of PGA-UP.

As shown in the FT-IR spectrum of PGA-UP after cured in Figure 5,the peaks due to the vinyl double bonds at 1635cm^{-1},1618cm^{-1}, 3037 cm^{-1}, 985cm^{-1} and 810 cm^{-1} all disappeared and other peaks changed little compared with those before cured in Figure 4. Based on the above experiments, we assumed that the double bonds of PGA-UP can self crosslink through radical homo-polymerization due to the higher radical polymerization reactivity when thermally initiated by BPO at 80℃. PGA-UP can be self cured without other monomers. The self curing reaction schematic diagram is shown as Figure 6.

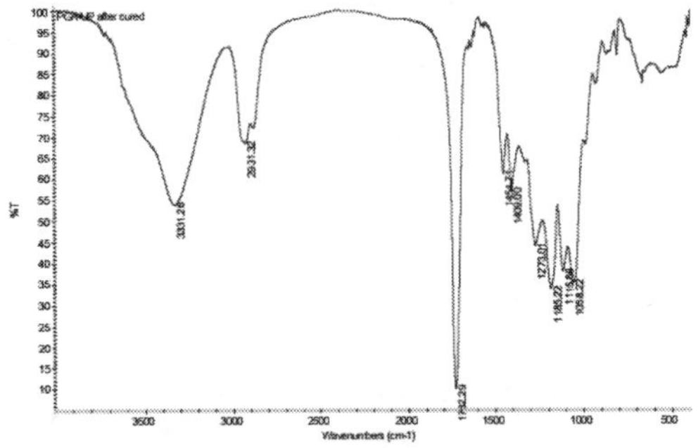

Figure 5. FT-IR spectra of PGA-UP after cured.

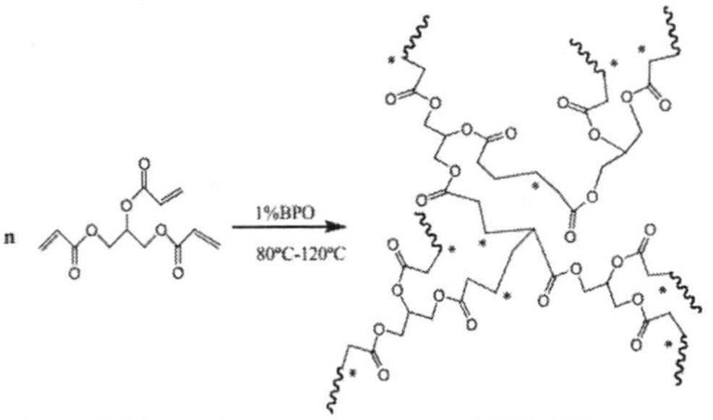

Figure 6. The curing process scheme of PGA-UP.
Note: * present the reaction point of vinyl double bonds.

The intension strength of PGA-UP after self-cured was 13.5Mpa, close to the 15.6MPa of the one of an usual unsaturated polyester S-UP, which was synthesized with propanediol, m-phthalic acid and maleic anhydride as the raw materials(molar ratio as 2.2:1:1) in our lab [7]. The potential application of PGA-UP is inspiring, especially in the research field of medical polymer and tissue engineering, because of its self-curable properties.

4 Conclusions

One kind of novel unsaturated polyester PGA-UP with vinyl double bonds on the end of molecule chain is synthesized with GL and AC as the raw materials under very low reaction temperature. The PGA-UP product is a liquid mixture produced by different esterifying reaction degree of GL with AC. It demonstrated that the unsaturated polyester resin PGA-UP with 1% BPO could be self-cured when heated above 80°C without additional cross-linking monomers. It was speculated that the self cross-linking process carried on through the radical homopolymerization of the vinyl double bonds. The resin after self-cured presents good intension strength close to the usual unsaturated polyesters and its potential application is inspiring. This research also contributes to improve air pollution caused by volatile of cross-linking monomers.

Acknowledgement

Of China for the Youth (Grant No.51303164) and Shandong Province Outstanding Young and Middle-Aged Scientists Research Awards Funds (Grant No.BS2011CL039).This work is sponsored by Natural Science Foundation

References

1. K. Q. Shen: *Unsaturated Polyester Resin and its Application* (Chemical Industry Press, Beijing, China, 2005)
2. N. Bicak, B. Karagoz, U. Tunca, J. Polym. Sci. Polym. Chem. Vol. 41 (2003) ,p.2549
3. N. Alemdar, B. Karagoz, A.T. Erciyes, N. Bicak*, Prog. Org. Coat. Vol. 60 (2007), p.69
4. S.Wang, M.J.Yaszemski, J.A.Gruetzmacher, L.Lu*,Polymer Vol. 49 (2008),p.5692
5. L.Cai, S. Wang, Polym. Vol. 51 (2010) ,p. 164.
6. N. Alemdar, B. Karagoz, A.T. Erciyes, N. Bicak*,J. Appl. Polym.Sci. Vol. 116 (2010),p.165
7. J.H. Meng, H.Y. Li *, Y.Y. Chen, Adv. Mat. Res. Vol. 699 (2013), p. 161.

Cracking behavior of thin gold strip deposited on polycarbonate plate under Cyclic and stepwisely-increased tension

Naoya Tada[1], Yafei Hu[2], Takeshi Uemori[1] and Toshiya Nakata[1]

[1]*Okayama University, Graduate School of Natural Science and Technology, Okayama, Japan*
[2]*Qingdao University of Science and Technology, College of Mechanical and Electronical Engineering, Qingdao, Shandong, P.R. China*

Abstract. Thin metal films deposited on polymers have been attracting much attention in flexible electronics. Various combinations of film and polymer substrate are expected to be developed. In this paper, deformation and fracture of thin metal film on polymer substrate was briefly analysed and the influencing factors were summarized. After that, focusing on the shape and thickness of thin metal film on polymer substrate, cracking behavior of small rounded-rectangular thin gold films called "strips" with a thickness less than one micrometer was observed under cyclic and stepwisely-increased trapezoidal tensions. The strips were deposited on a polycarbonate plate specimen, and wide and narrow strips with different thickness were prepared for the tests. The cracking behaviour was successfully observed and the increasing rate of crack number with the load or strain was examined. Since most cracks were initiated from the small defects, the crack initiation behavior was correlated with the distribution of defects.

1 Introduction

Thin metal films on polymer substrates are commonly used in flexible electronics and expected to be developed not only in the electronics but also in biological and medical fields [1, 2]. The mechanical performance of a thin metal film on polymer substrate is mainly determined by their individual properties such as elastic modulus, yield stress, ductility, thermal expansion coefficient. The geometrical factors of the film such as length, width, thickness, shape are also important factors to be taken into consideration. Various combinations of thin metal film and polymer substrate are possible and new materials are being developed actively. However, the fundamental investigation of thin metal film/polymer substrate structures is still important.

In this paper, after reviewing the deformation and fracture of thin metal film on polymer substrate briefly, cracking behavior on the surface of small rounded-rectangular thin gold (Au) films called "strips" deposited on plate specimens of polycarbonate (PC) was observed during cyclic and stepwisely-increased trapezoidal tensions. The effect of strip width and thickness on the cracking behavior and the crack initiation from the initial defect were discussed.

2 Deformation and fracture of thin metal film on polymer substrate

Deformation and fracture of thin metal film/polymer substrate structure seem to be very complex because various phenomena, such as cracking, necking, buckling, delamination and their combinations, occur. Firstly, these phenomena are briefly reviewed, and the influencing factors and their relations are clarified from the viewpoint of mechanical behavior of materials.

2.1 Deformation, cracking and delamination of thin metal film

Figure 1 shows the deformation of thin metal film under tension in two cases: (a) freestanding film and (b) film bonded on the surface of substrate. When the exact mechanical properties of thin film is required, mechanical tests using freestandning films are preferable [3, 4]. When a freestanding film is subjected to tension, deformation of the film is not constrained and the localized necking easily occurs. On the other hand, if the film is bonded on the substrate, the local deformation is constrained by the substrate and uniform deformation is continued [5]. But if the substrate is very soft, deformation of the film is possibly similar to that of freestanding one. It can be understood that the deformation of the film is strongly affected by the mechanical properties of substrate.

Typical fracture phenomena which occur in thin metal film on the substrate are shown in Figure. 2. Consider a case where the tensile stress acts on the thin metal film/polymer substrate structure shown in Figure 2(a), and the stress reaches to a certain level at which the film does not stand (*i.e.*, further deformation of the film is impossible). Cracking occurs in a perpendicular direction to the principal stress, as shown in Figure 2(b). With increasing the tension, lots of cracks are initiated in

 DOI: 10.1051/matecconf/201710801004

similar directions, and the crack sometimes propagates into the substrate and leads to the total fracture. The edge of film is also important part in film/substrate structure because a difference of the normal stress in the film and substrate brings the shearing stress, and it takes the maximum value at the edge. Then delamination occurs there when the shearing stress on the interface reaches its shearing strength, as shown in Figure 2(c). When the compressive stress acts on the film/substrate structure, it brings the interface delamination and the film is buckled, as shown in Figure 2(d). Even when the delamination does not occur due to relatively large interface strength, the film may show a wavelike buckling pattern on the surface [6].

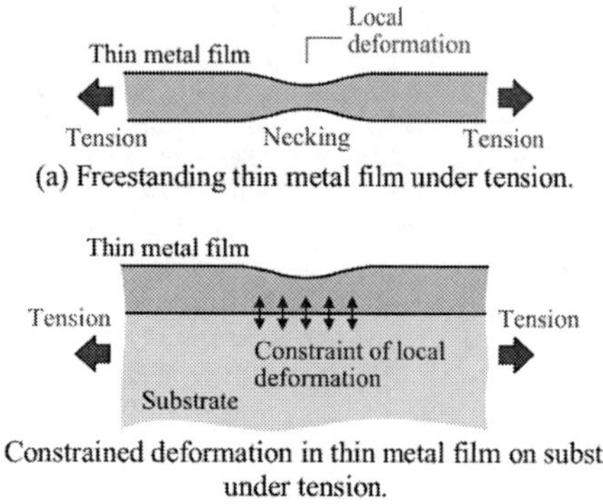

(a) Freestanding thin metal film under tension.

(b) Constrained deformation in thin metal film on substrate under tension.

Figure 1. Comparison of deformation of freestanding thin metal film and that bonded on substrate.

2.2 Relationship between influential factors in the fracture of thin metal film

Since various phenomena occur in the deformation and fracture of thin metal films on polymer substrate, their combination has to be taken into account when investigating the reliability of actual products. For example, when a crack is initiated in the thin metal film, as shown in Figure 2(b), stress concentrated at the tip promotes the local delamination of film/substrate interface. When large tension is given to the film/substrate structure, parallel cracks are initiated at almost equal distance in the film. Tensile stress in the film is released by these cracks. If the interface delamination occurs as shown in Figure 2(d), cracks are easily initiated on the tension (or outer) side of the bended thin film. The deformation, cracking, delamination are influenced each other and it makes the fracture of film/substrate structures complicated.

Figure 3 shows the relationship between influential factors in the fracture of thin metal film. Regarding the thin metal film, the thickness, shape, direction, and microstructure are typical influential factors. The microstructure influences on the mechanical properties, coefficient of thermal expansion (CTE), and strength of film. A combination of the mechanical properties and CTE determines the residual stress after deposition of the film. In order to investigate the fracture of thin metal film, each phenomenon has to be considered separately as the first step of fundamental research.

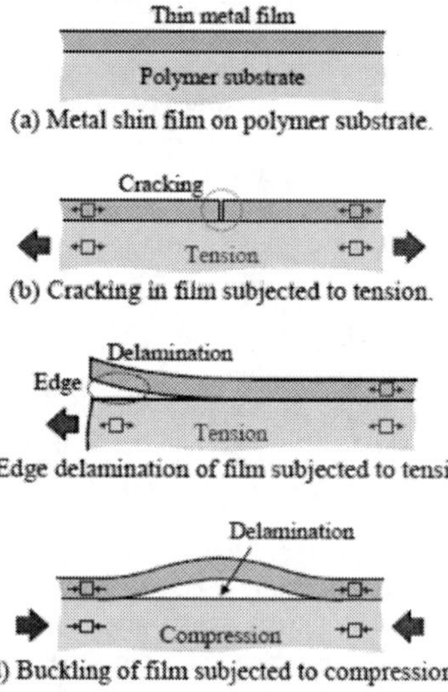

(a) Metal shin film on polymer substrate.

(b) Cracking in film subjected to tension.

(c) Edge delamination of film subjected to tension.

(d) Buckling of film subjected to compression.

Figure 2. Fractures occurred in thin metal film on polymer substrate.

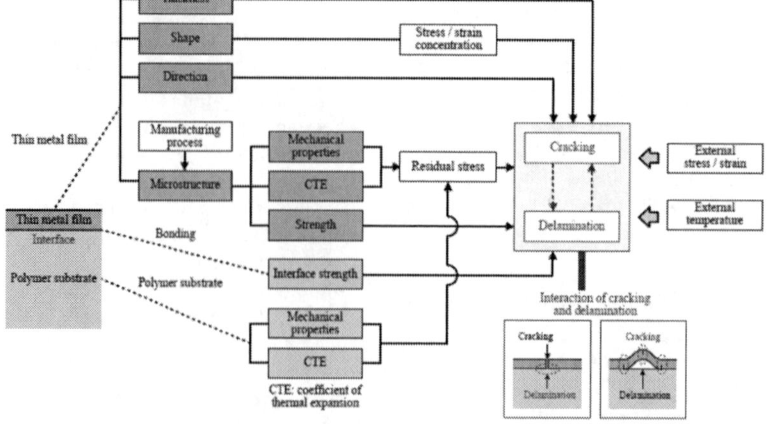

Figure 3. Relationship between influential factors in the fracture of thin metal film on polymer substrate.

DOI: 10.1051/matecconf/201710801004

3 Experimental procedure

3.1. Specimen and gold strips

Shapes and sizes of specimen (substrate) and strip are shown in Figure 4. Transparent colorless polycarbonate (PC) plates were used for substrate and two small rounded-rectangular strips (wide strip: 2 [mm] × 1 [mm] and narrow strip: 1.5 [mm] × 0.2 [mm]) were deposited at a parallel part of PC substrate by ion coater IB-5 (EIKO Engineering Co., Ltd.). The shape of strip was determined by putting slot grids as masking jigs. Deposition of 600 [s] with ionization current 5.5 [mA] and 300 [s] break were repeated until the required thickness was obtained. Specimens and strips used for the tests are shown in Table 1. Since it was found that most of cracks were initiated at small defects on the strip, the relation between defects and cracks was investigated using specimens S3 and S4. The average thickness of strip, h_f, was evaluated based on the height profile of each strip measured at three lines across the strip by surface-shape measurement microscope VF-7500 (KEYENCE Corporation). The defects and cracks were observed by materials science microscope BX41M LED and confocal laser scanning microscope OLS1100 (Olympus Corporation).

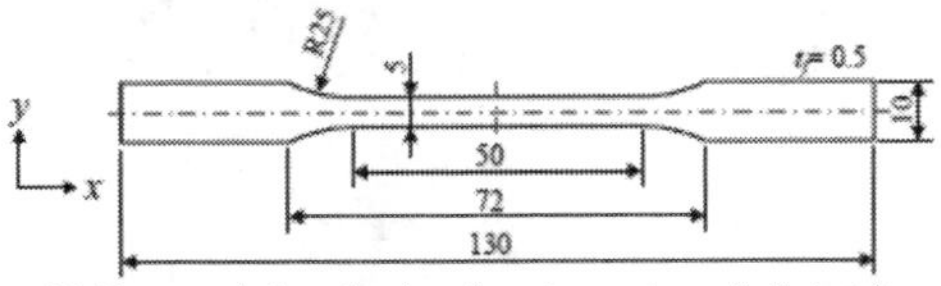

(a) Shape and size of polycarbonate specimen (substrate).

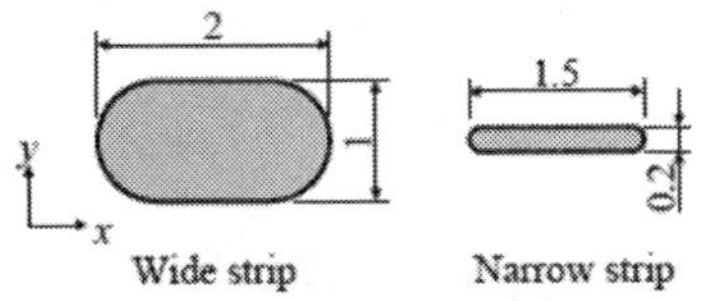

(b) Shape and size of wide and narrow strips of gold.

Figure 4. Polycarbonate (PC) specimen and thin gold (Au) strip on the surface.

Table 1. Strips on each specimen.

Specimen	Strip	Shape of strip	Strip thickness	Average strip thickness h_f [nm]
S1	W-L	Wide	Large	621
	N-L	Narrow	Large	632
S2	W-S	Wide	Small	324
	N-S	Narrow	Small	316
S3	W-L	Wide	Large	627
	N-L	Narrow	Large	645
S4	N-S-1	Narrow	Small	338
	N-S-2	Narrow	Small	314

3.2 Test conditions

Each specimen was subjected to stepwisely-increased trapezoidal tension. The loading and unloading rate was 1 [N/s], and the maximum load was gradually increased step by step as P_{max}=70, 90, 110 [N]. The increment of the maximum load was decreased to 5 [N] after these three steps. It was further decreased to 3 and 1 [N] when very large deformation was observed. The tests were carried out using a table-top type universal tester AGS-J (Shimadzu Corporation). The observation of strip surface was carried out after each unloading.

4 Experimental results and discussions

4.1 Cracking behaviour on gold strip

Examples of cracking on wide and narrow gold strips on PC substrate are shown in Figures.5 and 6, respectively. Cracks perpendicular to the loading direction were initiated with increasing the maximum load P_{max}. In the wide strip shown in Figure.5, small cracks shorter than the strip width appeared and grew to penetrate its total width. At the large maximum load, cracks were almost equally distant from each other. On the other hand, in the narrow strip shown in Figure 6, penetrated cracks appeared at the initiation and its number increased with the maximum load P_{max}.

(a) After tension of P_{max}=145 [N] (ε=4.0 [%]).

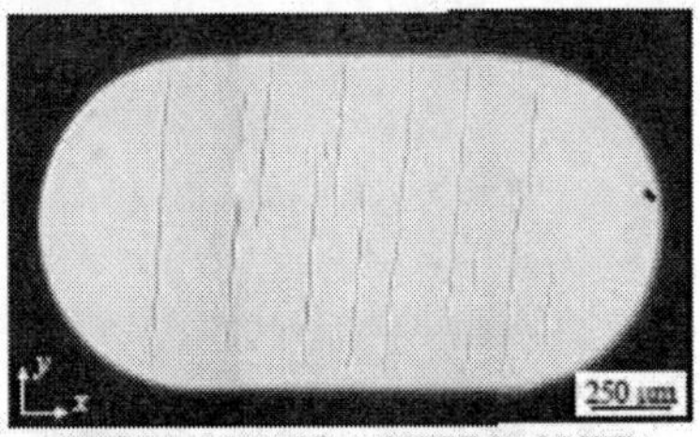

(b) After tension of P_{max}=150 [N] (ε=4.4 [%]).

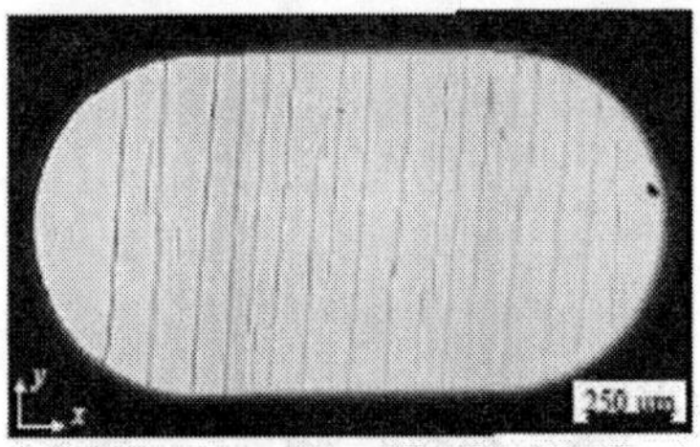

(c) After tension of P_{max}=160 [N] (ε=5.3 [%]).

Figure 5. Cracking behavior on a thick wide strip S1-W-L.

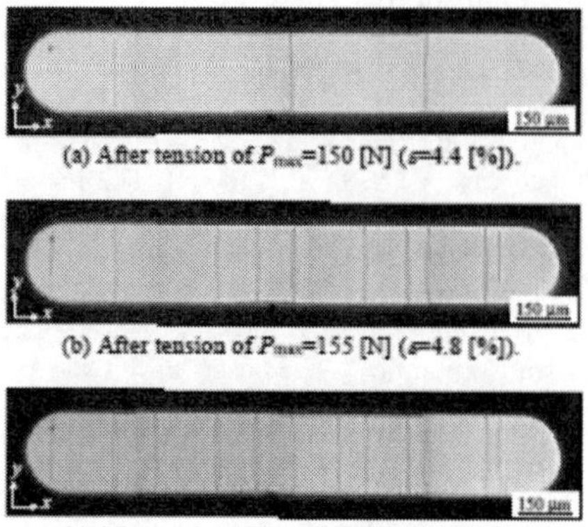

(a) After tension of P_{max}=150 [N] (ε=4.4 [%]).
(b) After tension of P_{max}=155 [N] (ε=4.8 [%]).
(c) After tension of P_{max}=160 [N] (ε=5.3 [%]).

Figure 6. Cracking behavior on a thick narrow strip S1-N-L.

Figure 7 shows the change in crack linear density ρ_c with the tensile strain of strip ε. ρ_c [μm^{-1}] was evaluated by the number of cracks N_c divided by the longitudinal length of the strip l_f (l_f =2 [mm] for wide strip and l_f =1.5 [mm] for narrow strip). Use of ρ_c minimizes the effect of the length of strip on the evaluated number of cracks. The strain ε was evaluated using the finite element analysis of strip/substrate structure which was carried out to adjust the load P-displacement δ curve with the experimental result. It was found from Figure.7 that cracks were initiated around ε=4 [%] in thick strips (S1-W-L and S1-N-L) and ε=6 [%] in thin strips (S2-W-S and S2-N-S). The effect of the strip thickness on the crack initiation strain is larger than that of strip width. On the other hand, the increasing rate of ρ_c against ε for thin strips is much larger than that for thick strips. It seems that the crack linear density curves for thin and thick strips converges on the same curve at large strain so that the strain determines the average distance between cracks.

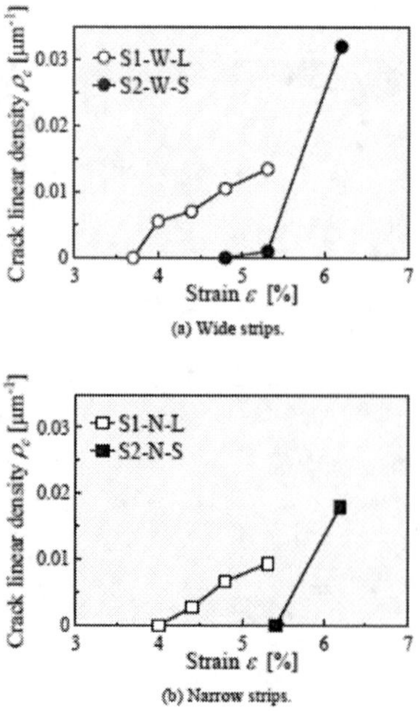

(a) Wide strips.
(b) Narrow strips.

Figure 7. Change in crack linear density on strips.

4.2 Defects and crack initiation

Magnified view of crack initiated on strip S3-N-L is shown in Figure.8. A crack was initiated from a circular defect perpendicularly to the loading direction (x-axis). Most of cracks were initiated from circular defects which were thought to be formed during the deposition process. Fine vertical stripe in y-axis direction, which was found before and after the tension, is very small undulations on the surface of PC substrate.

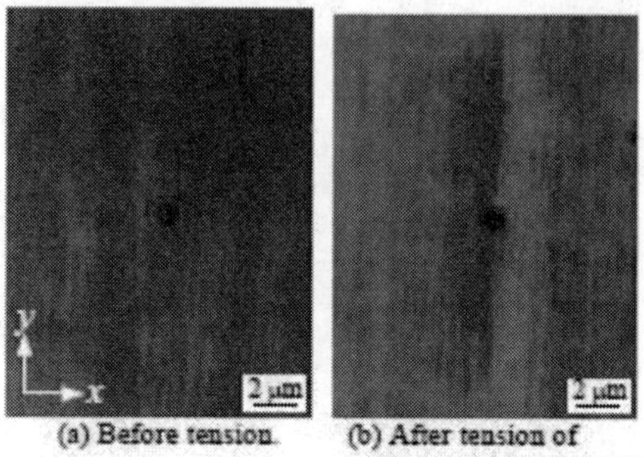

(a) Before tension (b) After tension of P_{max}=151 [N] (ε=4.9 [%]).

Figure 8. Crack initiated at a defect on strip S3-N-L.

The density of defects m [μm^{-2}], which is defined as the number of defects in a unit area on thin gold strip, is shown in Figure.9. An additional thick narrow strip S5-N-L was used to obtain the distribution of defects. The density of defects was about 800 [μm^{-2}] and constant independently of the thickness of strips.

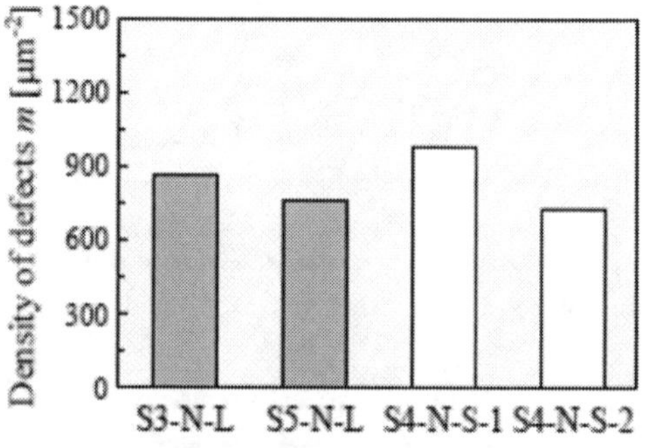

Figure 9. Density of initial defects on strips.

The distribution of equivalent diameter of defects, d_{eq} [μm], is shown in Figure.10. d_{eq} is defined as the diameter of circle whose area is equal to the defect. While the density of defects m was almost constant in thick and thin strips, the size distribution is different. Large defects were found only on thick strips. This is one of the reasons why cracking in thick strip occurred at lower strain than the thin strips. Considering the tendency that the crack linear density for thin and thick strips converged with each other at large strain, it is concluded that cracks are easily initiated at larger defects but its number is finally controlled by the strain.

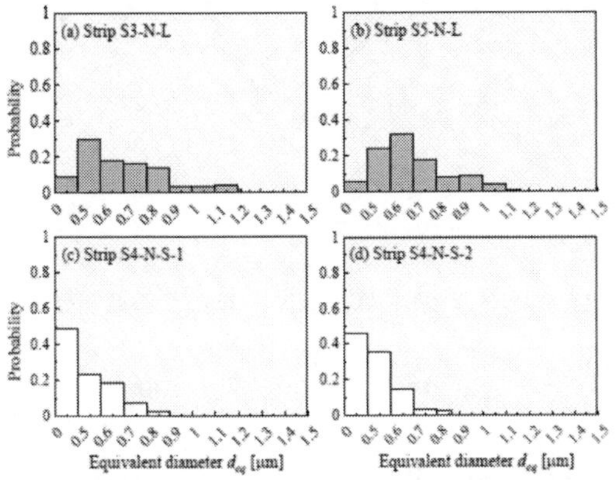

Figure 10. Distribution of equivalent diameter of initial defects on thick and thin strips.

5 Conclusions

Small rounded-rectangular thin gold films called "strips" were deposited on plate specimens of polycarbonate. Cracking behavior of the strips was observed during cyclic and stepwisely-increased trapezoidal tensions. Parallel cracks were observed in every strip and the initiation in thick strip was earlier (at smaller tensile strain) than in thin strip. The reason is considered to be the difference of defects' size on the strip which were formed during the deposition process.

References

1. V. J. Lumelsky, M. S. Shur, S. Wagner, IEEE Sensors Journal, **1** 41 (2001)
2. J. A. Rogers, T. Someya, Y. Huang, Science, **327** 1603 (2010)
3. S. J. Lee, J. M. Park, S. W. Han, S. M. Hyun, J. H. Kim, H. J. Lee, Exp. Mech. **50** 643 (2010)
4. B. Merle, M. Göken, J. Mater. Res. **29** 267 (2014)
5. T. Li, Z. Suo, Int. J. Solids Struct., **43** 2351 (2006)
6. N. Bowden, S. Brittain, A. G. Evans, J. W. Hutchinson, G. M. Whitesides, Nature, **393** 146 (1998)

DOI: 10.1051/matecconf/201710801005

Research on vibration noise of the composite laminated plate structure based on the FEM-BEM

Yang Xu, Dongyan Shi, Xiaoyan Teng and Hongxi Ren

Harbin Engineering University, Harbin, P. R. China

Abstract. Based on the FEM-BEM hybrid method, the effects of the angle of composite layer, number of layers and the opening position of a composite material and opening form are studied in order to get these three aspects' influence law on the vibration and acoustic radiation properties of the composite laminated plate structure, which can provide some reference for the engineers to design the composite laminated plate structure.

1 Introduction

The composite laminated plate structure is widely used in the fields of navigation, vehicle engineering, aerospace and other important engineering fields due to its excellent performance. Zhao Fei et al. theoretical calculated the natural frequencies and modes of the composite laminated plates based on the finite element method and the layered theory, and then through the experimental verification, the displacement mode of the laminated plates is solved by the layered finite element model, so as to solve the natural frequency[1]. Based on the semi-analytical method of the theory of state vector equation, a new numerical simulation method was proposed by Guanghui Qingden et al. to analyze the free vibration properties of the reinforced laminated plates [2]. U.N. Band, Y.M. Desai presents a unique finite element method for the analysis of the static and free vibration properties of thick plates and sheets. This method combines the 3D hybrid layered and the equivalent monolayer theory [3]. X.W. Yin et al. studied the acoustic radiation law of the concrete double-period parallel reinforced laminated plates under the point loading and the infinite fluid loading by the theoretical method [4]. Cao Xiongtao gives the five kinds of plate-shell structure' vibration and acoustic radiation prediction method, and the five structures includes the two groups of reinforced laminated plates, the ring rib reinforcement of constrained damping laminated cylindrical shells, the multi-periodic laminated cylindrical shells and so on[5].

Due to the characteristics of the composite structure, their mechanical properties not only depends on the form of the structure, but also on the mechanical properties of the each single-layer materials that form the macroscopic structure and the relationship between the layers. After a lot of paper-researching about the related published literature, it is not difficult to find out that most of the researchers has confined to focus on the research of form of the composite material structure, the theoretical modeling and the experimental study on the performance of related materials, etc. Therefore, the angle of composite layer, number of layers and the opening position of a composite material and opening form have great significance to the effect of the structure vibration noise. This paper is based on the FEM-BEM hybrid method, the effects of the angle of composite layer, number of layers and the opening position of a composite material and opening form are studied in order to get these three aspects' influence law on the vibration and acoustic radiation properties of the composite laminated plate structure, which can provide some reference for the engineers to design the composite laminated plate structure.

2 FEM-BEM analysis model

Figure 1 shows the flow chart of the structural analysis model based on the FEM-BEM hybrid method in this paper. It can be divided into the following three steps:

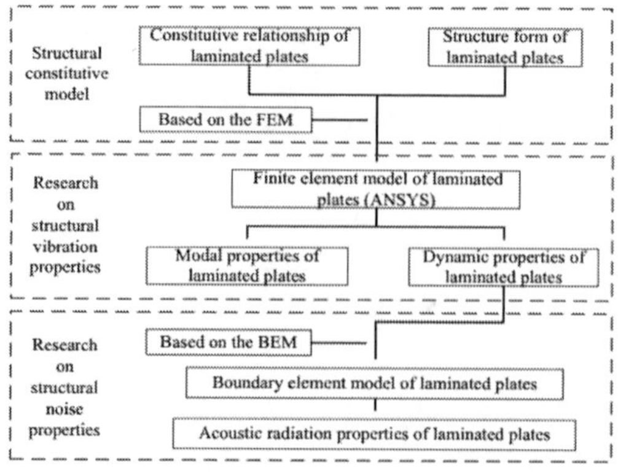

Figure 1. The flow chart of the structural analysis model.

DOI: 10.1051/matecconf/201710801005

2.1 Classical laminated plate theory

The difference between the composite laminated structure and the homogeneous material structure is that the laminated structure has macroscopic heterogeneity in the thickness, so the analysis of the laminated plates needs to be based on each single layer material. Only clearing the stiffness properties of each single layer material can obtain the overall stiffness properties of the laminated plate structure. Understanding the strength properties of each single layer can determine the overall strength properties of the laminated plates. Therefore, the macroscopic mechanical analysis of single layer materials is the basis of the laminated structure analysis. In classical theory, the Kichhoof hypothesis of the plate and the Kirchhoff-Love hypothesis of the shell are proposed [6]. The laminated structure of composite material is mostly orthotropic materials, the stress-strain relationship of the orthotropic materials will be described below on the basis of the Kichhoof hypothesis. The unidirectional composite material is not generally used alone, but is used as a base unit in a laminated plate, in combination with other materials. And the Structural width to thickness ratio is generally large in the practical engineering applications. Therefore, the mechanical properties of the single layer composite materials can be regarded as the plane stress problem, that is:

$$\sigma_3 = \tau_{23} = \tau_{31} = 0 \tag{1}$$

The relationship of the stress and strain in the main direction of the material are as follows:

$$\begin{bmatrix} \varepsilon_1 \\ \varepsilon_2 \\ \gamma_{12} \end{bmatrix} = \begin{bmatrix} S_{11} & S_{12} & 0 \\ S_{12} & S_{22} & 0 \\ 0 & 0 & S_{66} \end{bmatrix} \begin{bmatrix} \sigma_1 \\ \sigma_2 \\ \tau_{12} \end{bmatrix} \tag{2}$$

Where the S_{ij} is the compliance coefficient, their relationship with the engineering elastic constants are as follows:

$$S_{11} = \frac{1}{E_1}, S_{22} = \frac{1}{E_2}, S_{66} = \frac{1}{G_{12}}, S_{12} = -\frac{\mu_{12}}{E_1} = -\frac{\mu_{21}}{E_2} \tag{3}$$

Where the μ_{12} is the ratio of the strain $(-\varepsilon_2 / \varepsilon_1)$ caused by the individual strain σ_1, μ_{21} is the ratio of the strain $(-\varepsilon_1 / \varepsilon_2)$ caused by the individual strain σ_2.

From equation (3), the elastic constants of the orthotropic thin plate includes: E_1, E_2, G_{12}, μ_{12} and μ_{21}. From the reciprocal theorem,

$$\frac{\mu_{12}}{E_1} = \frac{\mu_{21}}{E_2} \tag{4}$$

So there are four independent elastic constants. Inverse the equation (2), the constitutive equation of the composite laminated plates can be obtained:

$$\begin{bmatrix} \sigma_1 \\ \sigma_2 \\ \tau_{12} \end{bmatrix} = \begin{bmatrix} Q_{11} & Q_{12} & 0 \\ Q_{12} & Q_{22} & 0 \\ 0 & 0 & Q_{66} \end{bmatrix} \begin{bmatrix} \varepsilon_1 \\ \varepsilon_2 \\ \gamma_{12} \end{bmatrix} \tag{5}$$

Where the Q_{ij} is the reduced stiffnesses, their relationship with the engineering elastic constants are as follows:

$$Q_{11} = \frac{E_1}{1-\mu_{12}\mu_{21}}, Q_{22} = \frac{E_2}{1-\mu_{12}\mu_{21}}, Q_{66} = G_{12}, Q_{12} = \mu_{12}Q_{11} \tag{6}$$

On the basis of the constitutive equation, the corresponding finite element analysis model will be established, and reasonable material and geometric parameters will be set according to the constitutive equation.

2.2 Establishment of the structure finite element analysis model based on the FEM

The mechanical foundation of finite element analysis is the elastic mechanics, and the solution principle of the equation is the weighted residual method or the functional extreme value principle. The realization of the method is the numerical discrete technology, and the technical carrier is the finite element analysis software. In this paper, ANSYS software will be used for analysis. First, a finite element model is needed. According to the geometry of the laminated plates, the thickness ratio of the laminated plates is more than 10, so the SHELL99 element is used in ANSYS software to establish the finite model [7]. In order to be closer to the actual engineering application, this paper selects the T300/BMP316 (the reinforcement is the carbon fiber T300, the Substrate is the BMP316), which is commonly used in engineering, to establish the model and the following analysis of the vibration acoustic radiation properties. The properties of T300 / BMP316 laminated plates and the geometrical parameters used in this paper are shown in Table 1 and Table 2.

Table 1. T300 / BMP316 material performance parameters.

$E_1 = 128.8\text{GPa}, E_2 = E_3 = 8.3\text{GPa}, G_{12} = G_{13} = G_{23} = 4.1\text{GPa}$
$\mu_{12} = \mu_{13} = \mu_{23} = 0.355$, ρ=1578kg/m^3

Table 2. Lamination size parameters.

Material	Length L × width B(mm)	Single layer thickness × number of layers (mm)
T300/BMP316	180 × 48	0.6 × 3
Layer angle (°)	Pore size	Opening position
[0/θ/0]	0.024	(90, 24)

In the case of using different unit to grid, not all of the performance singular need to be clear during the definition of material performance parameters In the process of modeling. For example: the SHELL99 element is used in this paper, as this unit is shell element, the thickness ratio is generally large. Therefore, the normal tensile modulus G_{23} of the material does not affect the natural frequency of the laminated plates, that is to say, the value of G_{23} can be arbitrary under the premise that the software can calculate. At the same time, the values of Poisson's ratio μ_{13} and μ_{23} can be omitted and the software will default to 0, which does not affect the modal analysis of the laminated plates. Finally, the finite element model of the structure is obtained by dividing the grid, as shown in Figure 2.

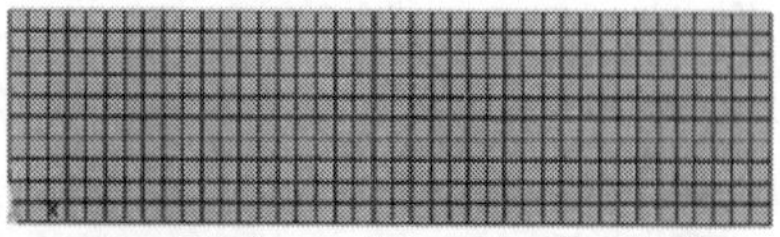

Finite element model of non - damage laminated plate.

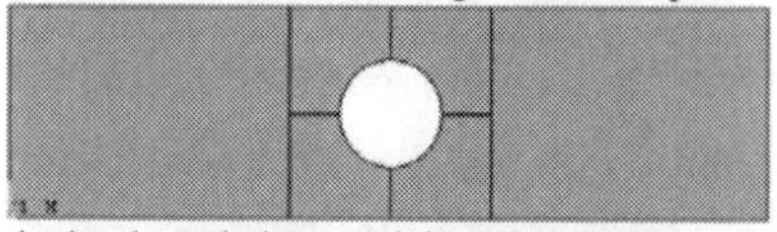

Split for the laminated plate model easy to mesh.

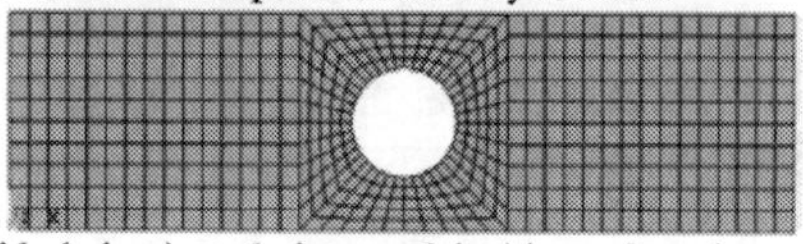

A round hole laminated plate model with good mesh.

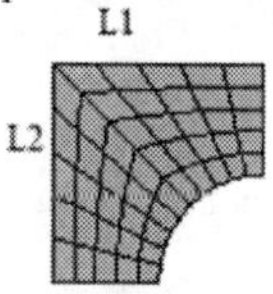

A 1/4 area mesh model around the round hole.

Figure 2. The finite element analysis model in this paper.

2.3 Establishment of the structure finite element analysis model based on the BEM

Boundary element method is a numerical calculation method for structural acoustic wave analysis, this method integrates the differential equation into boundary integral equation, and then discretizes the boundary into finite elements, and obtains the equations only with the unknown nodes on the boundary to solve the problem [8]. In this paper, the vibration response of the structure under the harmonic excitation is analyzed, and the acoustic simulation is also carried out. Under the action of the harmonic excitation, the acoustic pressure p generated by the structural vibration in the medium satisfies the Helmholtz equation:

$$\nabla^2 p + k^2 p = 0 \tag{7}$$

Where the p is the sound pressure, k is the number of the sound wave, $k = \omega / c$, ω is the excitation circular frequency, c is the sound velocity in the medium, in this paper, $c = 343\text{m/s}$. Structural vibration external field acoustic radiation problem belongs to the Neumman problem, that is to say, the surface normal velocity is given as $\partial p / \partial n = i\omega\rho v_n$, where the ρ is the medium density, v_n is the structural surface normal velocity. At the same time, the sound pressure must also satisfy the condition that the sound pressure is zero at infinity, using the Green's function can obtain the Helmholtz integral equation:

$$C(\mathbf{r}_a)p(\mathbf{r}_a) = \iiint\limits_{\Omega}\left[G(\mathbf{r}_b,\mathbf{r}_a)\frac{\partial p(\mathbf{r}_b)}{\partial n} - \frac{\partial G(\mathbf{r}_b,\mathbf{r}_a)}{\partial n}p(\mathbf{r}_b)\right]d\Omega(\mathbf{r}_b) \tag{8}$$

Where the $\mathbf{r}_b$ is the point on the surface of radiation, $\mathbf{r}_a$ is the point in space.

$C(\mathbf{r}_a)$ is depending on the location of $\mathbf{r}_a$ in the sound field. For the smooth surface Ω, when $\mathbf{r}_a$ is within the radiator, $C(\mathbf{r}_a) = 0$; when $\mathbf{r}_a$ is the nonsingular point on the radiation surface, $C(\mathbf{r}_a) = 2\pi$; when $\mathbf{r}_a$ is in the radiation field, $C(\mathbf{r}_a) = 4\pi$.

The boundary element method is used to solve the Helmholtz equation. Boundary element method is divided into the direct boundary element method and the indirect boundary element method. The direct boundary element method requires that the acoustic grid be closed, and the mesh of the indirect boundary element method can be closed or not. The direct boundary element method needs to discretize the boundary Ω_a into a number of elements (Ω_{ae}) and nodes. The sound pressure p and the normal velocity v_n at any point within each element can be determined by the acoustic pressure a_{pi}, the normal velocity a_{vi} on the node belonging to the element and the shape function N_i^e of the element. In this paper, the normal velocity a_{vi} is a known quantity obtained by the finite element method. Which is:

$$p(\mathbf{r}_a) = \sum_{i=1}^{n_e} N_i^e(\mathbf{r}_a)\cdot a_{pi},\ \mathbf{r}_a \in \Omega_{ae} \tag{9}$$

$$v_n(\mathbf{r}_a) = \sum_{i=1}^{n_e} N_i^e(\mathbf{r}_a)\cdot a_{vi},\ \mathbf{r}_a \in \Omega_{ae} \tag{10}$$

Where the n_e is the number of nodes on an element Ω_{ae}.

The element shape function N_i^e is 1 on the node i and 0 on the other nodes of the element. The element shape function N_i^e can be assembled into a global shape function N_i, which is 1 on the node i and 0 on the other nodes. So on the entire boundary element mesh Ω_a, there is:

$$p(\mathbf{r}_a) = \sum_{i=1}^{n_a} N_i \cdot p_i = \mathbf{N}_i\{p_i\},\ \mathbf{r}_a \in \Omega_a \tag{11}$$

$$v_n(\mathbf{r}_a) = \sum_{i=1}^{n_a} N_i \cdot v_{ni} = \mathbf{N}_i\{v_{ni}\},\ \mathbf{r}_a \in \Omega_a \tag{12}$$

Where the n_a is the number of nodes for all boundary element mesh.

At present a lot of commercial software already have mature acoustic boundary element analysis method, this

DOI: 10.1051/matecconf/201710801005

paper uses the Virtual.lab software to analyze the structure and obtain the acoustic radiation properties of the structure [9].

3 Numerical structure analysis

3.1 The verification of the modal

It's necessary to verify the correctness of the finite element analysis model established in this paper before the vibration noise analysis, and then gives a verification example, the material and structure parameters are shown in Table 3, where the L and B are the length and width of the plate, H is the thickness of the plate, θ is the layer angle. The mesh size of the example is 80×80×6, and respectively calculated with the SHELL99 and SOLID46 element, the results are shown in Table 4. In order to verify the correctness of the proposed method, the exact solution results [10] are also listed in Table 4. The Table 4 shows that the results from the FEM analysis model based on the finite element software ANSYS is in good agreement with the results of the literature. The calculation results of the SOLID46 element is more accurate, and the error of the SHELL99 element is relatively large, but the error is also controlled within 1%, which shows the correctness of this method.

Table 3. The material parameters and geometrical dimensions of the example.

$E_1=15E_2, E_2=E_3=10\text{GPa}, G_{12}/E_2=G_{13}/E_2=0.6, G_{23}/E_2=0.5$
$\mu_{12}=\mu_{13}=\mu_{23}=0.25, \rho=1450\text{kg/m}^3, L=B=1\text{m}, H/B=0.01, \theta=30^\circ$

Table 4. Comparison of the calculation results.

Ord-er	Literature results[10] (Hz)	SOLID46		SHELL99	
		Frequen-cy (Hz)	Error (%)	Frequen-cy (Hz)	Error (%)
1	9.45	9.4484	0.0169	9.3979	0.551
2	21.28	21.279	0.0047	21.167	0.531
3	46.06	46.064	0.00868	45.812	0.55
4	60.73	60.722	0.0132	60.289	0.726

3.2 The vibration acoustic radiation properties analysis

On the basis of validating the finite element analysis model, this paper will further study the vibration acoustic radiation properties of the composite laminated plates. In order to study the effects of the layer angle and the number of layers on the vibration acoustic radiation properties of the laminated plates, several models were established for comparison as shown in Table 5. The boundary condition of all the laminated plates are the one-end constraint, and the analysis process is shown in Figure 1. Firstly, the harmonic response of each plate is calculated and analyzed by the acoustic boundary element method to obtain the acoustic radiation curves of each plate, as shown in Figure 3. It can be seen from Figure 3 that the acoustic radiation properties of the structure will produce a maximum peak value of the sound pressure at the position corresponding to the natural frequency with the change of the excitation frequency, especially at the fundamental frequency position. After the fundamental frequency position, the increase of the sound pressure value is not particularly evident. In addition, it also can be seen from Figure 3 that the peak sound pressure of the structure is increased with the increase of the number of layers. When m exceeds 2, the trend gradually tend to be stable. In order to study the effects of the layer angle and the number of layers on the vibration acoustic radiation properties of the laminated plates comprehensively, the dB-Frequency curves of different layer angle and the number of layers are combined to draw a three-dimensional surface, as shown in Figure 4. For the angle of the layer, when the angle value is constant and observe the trend of the sound pressure dB with the change of frequency, we can see that all the curves obtain the peak sound pressure near the first and the third order natural frequency values, which shows that all the plates produce resonance near the first and the third order natural frequency values, and the noise generated thereby is maximized so that the sound pressure value dB is peaked. When the frequency is constant and observe the trend of the sound pressure dB with the change of the layer angle, we can see that when the frequency is not in the vicinity of the plate natural frequency, the sound pressure dB with the change of the layer angle is not obvious. For the number of layers, when the number of layers is 2, the excitation frequency corresponding to the peak sound pressure has obvious differences with the other laminated plates. When the number of layers increases, the excitation frequency corresponding to the peak sound pressure is consistent.

Table 5. Model parameters of the different number of layers.

Code	Size (mm)	Single layer thickness × number of layers	Layer angle
$A_n(n=0,1,2,\ldots,18)$	180×48	0.6×3	[0/5n/0]
$B_m(m=1,2,3,\ldots,15)$	180×48	(2/2m) ×2m	$[0/90]_m(m=1,2,3,\ldots,15)$

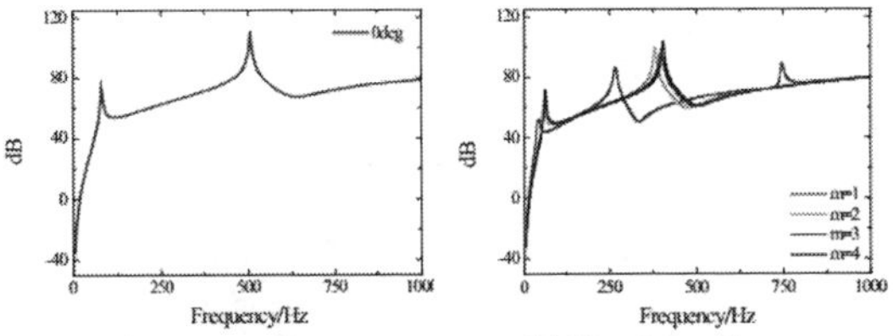

(a) Layer angle. (b) Number of layers.
Figure 3. The vibration acoustic radiation curve of the structure.

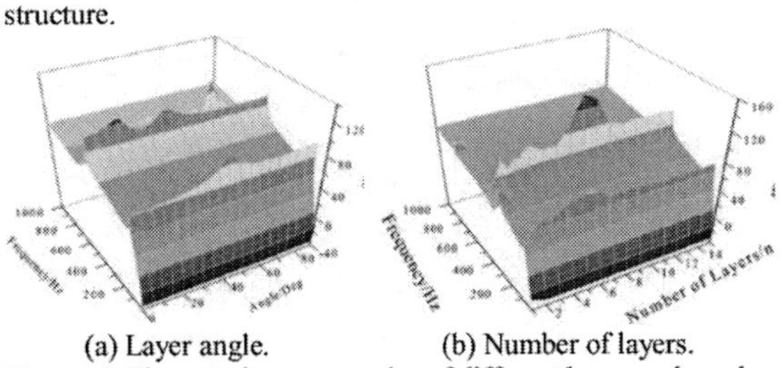

(a) Layer angle. (b) Number of layers.
Figure 4. The sound pressure value of different layer angle and

DOI: 10.1051/matecconf/201710801005

the number of layers curves with the change of the excitation frequency.

Furthermore, the effect of the opening structure on the vibration acoustic radiation of the structure is studied, the results are shown in Figure 5.

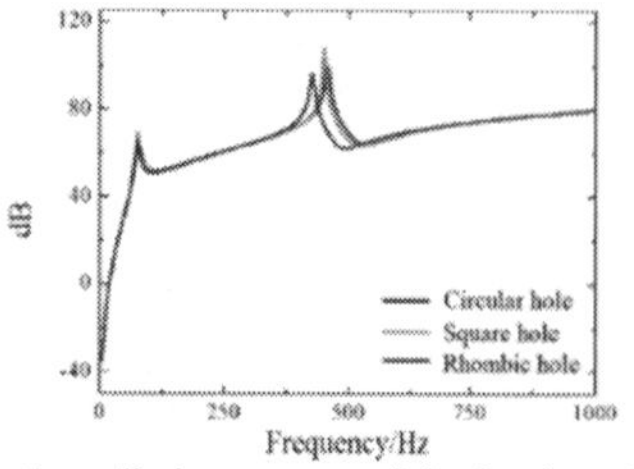

(a) The acoustic radiation curves of the laminated plates with different opening forms.

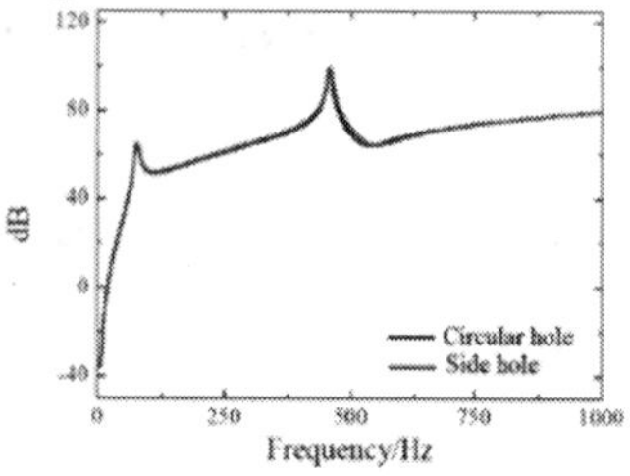

(b) The acoustic radiation curves of the laminated plates with different opening positions.

Figure 5. the acoustic radiation curves of the laminated plates with different opening.

The Figure 5 (a) shows that the third order natural frequency of the laminated plates with the rhombic hole is lower, so the excitation frequency corresponding to the peak sound pressure is smaller, and the second is the laminated plates with the square hole, the laminated plates with the circular hole is the largest. And the sound pressure of the laminated plates with the square hole at the natural frequency is the largest, the second is the laminated plates with the circular hole, and the laminated plates with the rhombic hole is the smallest. In Figure 5 (b), the excitation frequency corresponding to the peak sound pressure of the two curves are closed, which illustrate that the natural frequency of the two plates are closed. Among them, the sound pressure value of the curve of the laminated plates with the side hole is slightly larger than that of laminated plates with the circular hole between the first order natural frequency value and the third order natural frequency value, which indicates that the plate vibration can produce a larger noise caused by the side hole. In practical engineering, the influence cannot be neglected when there are many side holes.

4 Conclusions

Based on the FEM-BEM hybrid method, the effects of the angle of composite layer, number of layers and the opening position of a composite material and opening form are studied in this paper, and the influence law of these three aspects on the vibration acoustic radiation properties of the composite laminated plate structure are as follows:

(1) The vibration radiation noise of the composite laminated plate structure changes with the change of the angle, especially between the angle of 20 to 80 degree;

(2) For the number of layers, when the number of layers is 2, the excitation frequency corresponding to the peak vibration radiation noise has obvious differences with the other laminated plate structure. When the number of layers increases, the excitation frequency corresponding to the peak sound pressure is consistent.

(3) The opening structure has little effect on the peak value of the vibration radiation noise of the composite laminated plates, only affecting its peak position. The excitation frequency corresponding to the peak sound pressure of the laminated plates with the rhombic hole is smaller, and the second is the laminated plates with the square hole, the laminated plates with the circular hole is the largest.

The above conclusions are only qualitative analysis of the results obtained. In the actual engineering applications, it's necessary to comprehensively consider the various factors so as to choose the reasonable angle of composite layer, number of layers, the opening form and so on.

References

1. Zhao F, Wu J W, Zhao L S. Analysis for Natural Frequencies and Mode Shapes of Laminated Composite Plates Using Layerwise Theory. Noise and Vibration Control, **34**, 34-40 (2014)
2. Qing G, Qiu J, Liu Y. Free vibration analysis of stiffened laminated plates. International Journal of Solids and Structures, **43**, 1357-1371 (2006)
3. Band U N, Desai Y M. Coupled higher order and mixed layerwise finite element based static and free vibration analyses of laminated plates. Composite Structures, **128**, 406-414 (2015)
4. Yin X W, Gu X J, Cui H F, et al. Acoustic radiation from a laminated composite plate reinforced by doubly periodic parallel stiffeners. Journal of Sound and Vibration, **306**, 877-889 (2007)
5. Cao X T. *Vibrational and Acoustic Prediction Method of Laminated Plates and Shells* (2007)
6. Zhang J, Zhao W Z, Zhang W Y. Research on acoustic-structure sensitivity basing on FEM and BEM. Journal of Vibration Engineering, **18**, 366-370 (2005)
7. Wang X M, Li Y Q, Xu H W. *ANSYS Structural Analysis Unit and Its Application*, 200-300 (2011)
8. Tan J A. The Application of Boundary Element Methods in Environmental Acoustics. Noise and Vibration Control, 20-22 (2002)
9. Li Z G, Zhan F L. *Virtual.Lab Acoustic-Acoustic Simulation Computing Advanced Application Examples*. 105-160 (2010)
10. Qu Y, Yuan G, Wu S, et al. Three-dimensional elasticity solution for vibration analysis of composite rectangular parallelepipeds. European Journal of Mechanics - A/Solids, **42**, 376-394 (2013)

Oblique crushing performances of hybrid woven Kenaf fibre reinforced aluminium hollow cylinder

Al Emran Ismail, Mohd Norihan Ibrahim, Saifulnizan Jamian, Kamarul-Azhar Kamarudin, Muhammad Khairudin Awang, and Mohd Khir Mohd Nor

Crashworthiness and Collision Research Group (COLORED), Mechanical Failure Prevention and Reliability (MPROVE), Faculty of Mechanical and Manufacturing Engineering, Unviersiti Tun Hussein Onn Malaysia, Batu Pahat, 86400 Johor, Malaysia

Abstract. This paper presents crushing performances of hybrid woven kenaf fibre reinforced aluminium tubes under axial and oblique compression. According to open literature, lack number of works conducted to study the crushing responses when kenaf fibre in the form of woven is used to reinforce aluminium tubes. There are important parameters are used such as fibre orientations and oblique compression angles. Kenaf fibre in the form of woven is firstly wetted with polyester resin before it is wrapped around the tubes. After composite hardened, the hybrid tubes are quasi-statically compressed and force-displacement curves are recorded. Energy absorption performances are then determined and discussed in term of different fibre orientations and oblique angles. It is found that [±30°] fibre orientations capable to produce higher force-displacement responses and hybrid tubes are produced lower force ratio indicating the role of natural fibre in increasing the capability of energy absorptions.

1 Introduction

Crashwothiness is an engineering field considering safety aspects of vehicle integrity during collisions and impact. It is not only concentrated on the safety of passengers or occupants inside the vehicles but it is also considered the safety of pedestrians. In general metallic structures for example steel is widely used to fabricate the vehicle components and recently aluminium alloy is also used since it is offered high ratio of strength to weight. In front of vehicles, most of designers placed cashboxes to ensure during collusion some of kinetic energy will be absorbed and therefore reducing injuries among occupants.

During collision, plastic deformation is the only mechanism contributed to absorb impact energy and tremendous number of publications can be found elsewhere [1-4]. Eyvazian et al. [1] investigated axial crushing behaviour of corrugated tubes made of aluminium tubes. They used five different geometries of corrugations and simple tube for comparisons. It is found the crushing behaviour is affected by the corrugated geometries since the introduction of design defect capable to control the collapse behaviour of the tubes. A mathematical aspect of corrugated tube under axial compression can be found in [2].

Huang et al. [3] investigated dynamic crushing characteristics of high strength steel cylinder with elliptical geometric discontinuities. Finite element analysis is used to study the influence of cut-out locations, cut-out shapes and symmetry of cut-out on the crushing performances. They found that cut-out capable to significantly affect the energy absorption capability. If the cut-out is positioned in the middle, the tube tend to experience a global buckling whereas it is placed at an upper or bottom ends progressive collapse can be obtained.

Galib and Liman [4] experimentally and numerically investigated circular aluminium tubes under static and dynamic loading. They have shown that the comparison of the mean force from different solutions gave good concordance between them. They added that this type of analysis is expected to be a useful approach in the design before actual experimentations can be conducted.

Some researchers used synthetic fibre reinforced composites and it is wrapped around the energy absorbing device. The purpose is to enhance the performances of crushing behaviour. For examples Kalhor et al. [5] wrapped steel square tube with glass/epoxy composites before it is quasi-statically crushed to investigate the responses of force versus displacement curves. Numerical modelling is also developed in order to obtain optimum crashworthiness aspects.

El-Hage et al [6] numerically characterized square aluminium-composite hybrid tubes under axial compression. Fibre orientations are varied with respect to the axis of tubes. It is found that hybrid tubes produced higher energy absorption than aluminium tube alone. Shin et al. [7] found similar behaviour where hybridizing the tube with synthetic fibre capable to improve the crushing performances.

Recently, due to environmental awareness among

DOI: 10.1051/matecconf/201710801006

researchers and industrialists, natural fibres are given a bright opportunity to replace synthetic fibres. Some review papers on the potential of natural fibre can be found in [8-11]. Hybridizing metallic tubes with natural fibres are very difficult to find. There are several papers conducted to investigate the crushing responses can be found [12-19]. Most of these papers found that natural fibre capable to increase the capacity of energy absorption.

This paper present crushing performances of woven kenaf fibre mat wrapped around aluminium tubes. Different fibre orientations are used and it is quasi-statically compressed in order to obtain the force versus displacement curves. During crushing processes, the samples are not only crushed axially but obliquely. Energy absorption capability is calculated and discussed in correlating with fibre orientations and oblique compression.

2 Methodology

An industrial grade aluminium hollow tube is used with thickness of 1mm and 50mm external diameter. The length of tube is 100mm where it is long enough to prevent global buckling. As-received kenaf yarn (as in Figure 1(a)) is weaved into a plain woven mat before it is wrapped around the outer surface of aluminium tube. The weaving process is conducted using an in-house facility as in Figure 1(b) and 1(c). While Figure 1(d) revealed the completed woven kenaf mat.

Woven kenaf mat is firstly immersed into a resin bath where a proper care is given in ensuring resin penetrates into the fibres uniformly. The wet fibre mat is wrapped around the tube twice and each layer oriented according to the specific orientation as tabulated in Table 1.

Figure 2 shows the completed hybrid woven kenaf/aluminium hollow tube. This sample is positioned vertically between two flat rigid plates. The bottom plate is stationary while the upper one is moved downward. The sample is compressed quasi-statically with a speed of 1.5 mm/min of cross-head displacement. The force versus displacement curves are recorded automatically where the area under the curves represent the capability of energy absorption, E and it can be represented as:

$$E = \int_0^L P ds \quad (1)$$

where P is a force, ds is a small element on the curve and L is the crushed length. Instead of energy absorption performance, other important parameter such as force ratio, ψ is also determined as in Equation (2).

$$\psi = \frac{P_{mean}}{P_{peak}} \quad (1)$$

where P_{mean} is the average force during the crushing process and P_{peak} is the maximum or peak force just after the elastic deformation. Final crushed samples are also observed in order to study the effect of variations of fibre orientations on the crushing performances.

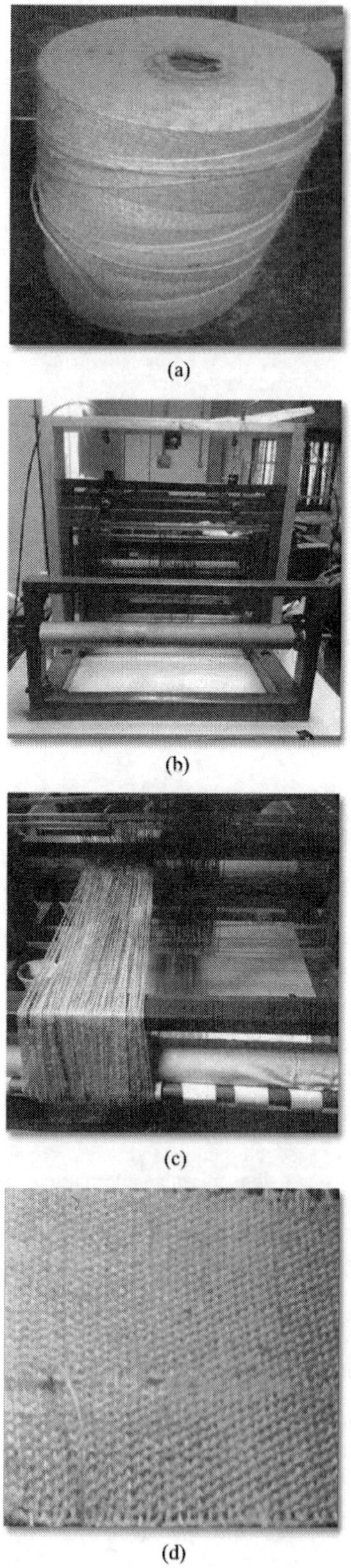

Figure 1. (a) As-received kenaf yarn, (b) Weaving machine, (c) Weaving process in progress and (d) Completed woven kenaf mat.

DOI: 10.1051/matecconf/201710801006

Table 1. Sample parameters and compression conditions.

No.	Orientations	Layers	Oblique Angles
1		-	0°
2	Empty tubes	-	5°
3		-	10°
4		-	15°
5		2	0°
6	[±0]	2	5°
7		2	10°
8		2	15°
9		2	0°
10	[±15]	2	5°
11		2	10°
12		2	15°
13		2	0°
14	[±30]	2	5°
15		2	10°
16		2	15°
17		2	0°
18	[±45]	2	5°
19		2	10°
20		2	15°

Figure 2. Hybrid woven kenaf/aluminium hollow tube.

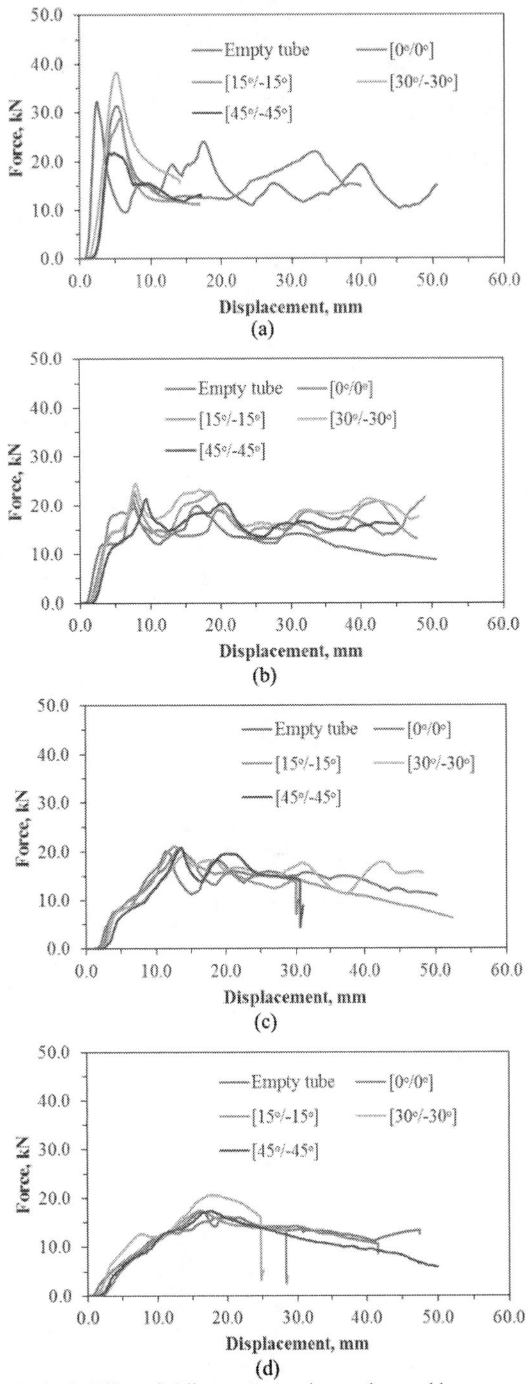

Figure 3. Effect of oblique compression on the crushing performances when fibre orientations are varied, (a) 0°, (b) 5°, (c) 10° and (d) 15°.

3 Results and discussion

Figure 3 shows the responses of force versus displacement curves of hybrid tubes under normal and oblique compression. Figure 3(a) shows the force-displacement curves under axial compression. It is observed that the curves followed a typical diagram where in the first stage force is linearly related with displacements until it is reached maximum forces. Once plastic deformation firstly occurred, the maximum forces dropped drastically before it is fluctuated when the tubes experienced progressive collapses. This region is also known as a plateau stage.

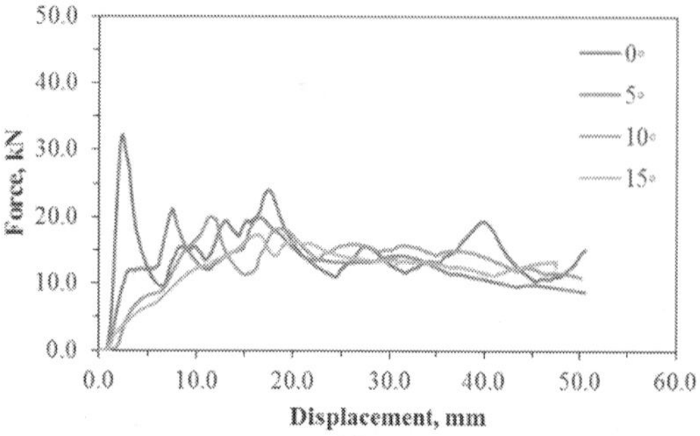

(a)

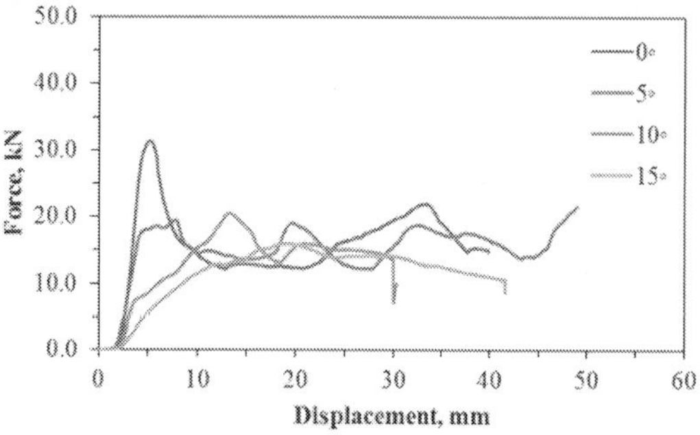

(b)

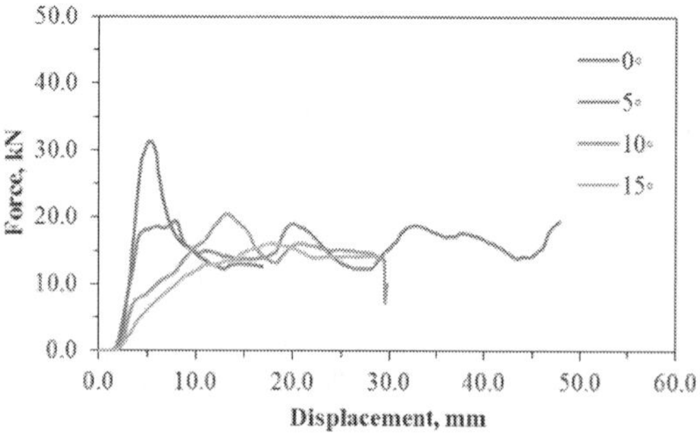

(c)

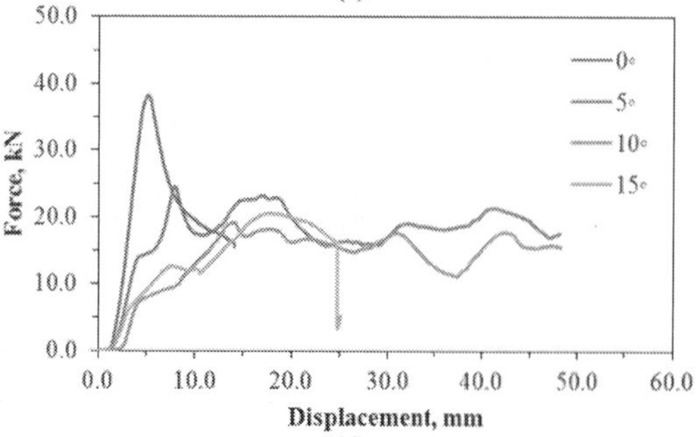

(d)

Figure 4. Effect of fiber orientations on the crushing performances when oblique compression are varies, (a) Empty tube, (b) [15°/-15°], (c) [30°/-30°] and (d) [45°/-45°].

For all cases of oblique compression as in Figures 3(b)-3(d), the present of oblique compression insignificantly contributed to the enhancement of the force-displacement responses. This is due to the fact that the introduction of oblique angles contributed to the creation of stress concentration resulted of bending moments. It is observed that the tubes are not fully experienced progressive collapses due to the eccentricity of compressions.

Figure 4 obviously reveals the roles of oblique angles on the responses of force versus displacement curves. Under axial compression, it is observed that increasing the fibre orientations to [±45°] capable to increase the peak forces when compared with empty tubes as in Figure 4(a). Once fibre reinforced composites are introduced around the outer surface of tubes, the force-displacement curves are observed to slightly increase as compared with empty tubes as in Figure 4(b)-4(d) especially in the first deformation stages. For the second stage, there is not significant deference between the curves indicating that the composites are not played an important role in increasing the mean forces. This is probably the plastic strain of tubes are higher than the composites where the composites do not have enough strength to resist the tube deformations.

(a)

(b)

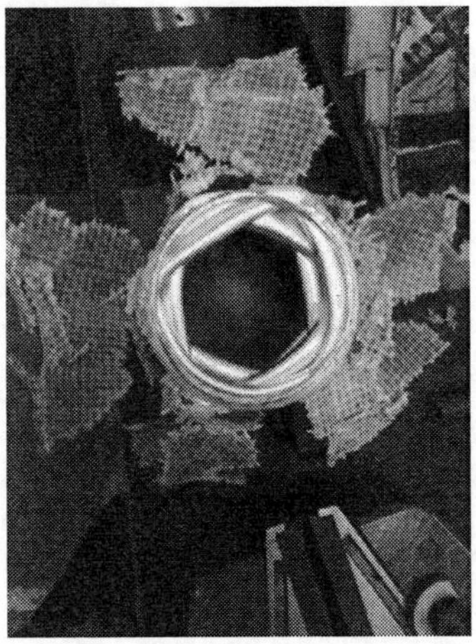

(c)

Figure 5. Crushing mechanisms of [±45°] fibre orientations wrapped around the aluminium tubes, (a) initial stage deformation, (b) final failure of hybrid tube and (c) area view of (b) under axial compression.

Figure 5 displays the crushing mechanisms of hybrid tubes under axial compression. The tube is wrapped with 45° fibre orientations. Figure 5(a) shows the deformation captured just after the peak force is achieved. It can be observed that composites experienced large fragmentations. This indicated that the composites unable to absorb crushing energy effectively. This behaviour is similar with the results with Figure 4 where there is not much different between hybridized and empty tubes. Large fragmentations are also observed for the case of final crushed condition as in Figures 5(b) and 5(c). This crushing mechanisms gave a strong evidence that natural fibre can be used to reinforce tubes however it is not absorbed crushing energy efficiently.

4 Conclusion

One of important conclusion can be obtained from these experimental works is by wrapping the tubes with woven kenaf fibre reinforced composites capable to enhance the responses of force versus displacement curves where both maximum and mean forces increased mostly for axial compressions. However when oblique compressions are introduced, there is no obvious peak force has occurred fortunately mean force is remain the same as the results obtained from axial compression. Based on the observations of mechanical deformation, there is large composite fragmentations occurred. This indicated that woven kenaf fibre composites are unable to absorb crushing energy efficiently.

References

1. A. Eyvazian, M.K Habibi, A.M. Hamouda, R, Hedayati. Materials and Design **54**, 1028-1038 (2014).
2. A. Eyvazian. MATEC Web of Conferences **40**, 02010 (2016).
3. M.Y. Huang, Y.S. Tai, H.T. Hu. Theoretical and Applied Fracture Mechanics 54, 44-53 (2010).
4. D.A. Galib, A. Limam. Thin-Walled Structures **42**, 1103-1137 (2004).
5. R. Kalhor, H. Akbarshahi, S.W. Case. Composite Structures **147**, 231-246 (2016).
6. H. El-Hage, P.K. Mallick, N. Zamani. Composite Structures **73**, 505-514 (2006).
7. K.C. Shin, J.J. Lee, K.H. Kim, M.C. Song, J.S. Huh. Composite Structures **57**, 279-287 (2002).
8. M.R. Sanjay, G.R. Arpitha, B. Yogesha. Materials Today **2**, 2959-2967 (2015).
9. V. Mittal, R. Saini, S. Sinha. Composite Part B: Engineering **99**, 425-435 (2016).
10. T. Gurunathan, S. Mohanty, S.K. Nayak. Composite Part A: Applied Science and Manufacturing **77**, 1-25 (2015).
11. N. Saba, M.T. Paridah, M. Jawaid. Construction and Building Materials **76**, 87-96 (2015).
12. A.E Ismail. International Journal of Engineering and Technology **6**, 2662-2668 (2015).
13. A. E. Ismail. *International Journal of Engineering and Technology* **7**, 1098-1104 (2015).
14. S. N. A. Khalid, A.E Ismail, M.H. Zainulabidin. Applied Mechanics and Materials **773**, 134-138 (2015).
15. M. A. Ismail, A.E. Ismail. International Journal of Mechanical and Mechatronics Engineering **15**, 100-104 (2015).
16. A. E. Ismail. Jurnal Teknologi **78**, 45-50 (2016).
17. A. E. Ismail. International Journal of Mechanical and Mechatronics Engineering **16**, 49-53 (2016).
18. A. E. Ismail. International Journal of Mechanical and Mechatronics Engineering **16**, 65-69 (2016).
19. A. E. Ismail, A. L. Mohd Tobi. ARPN Journal of Engineering and Applied Science **11**, 8668-8672 (2016).

The working zone in the interpolar area of the Faraday balance: an approach to testing the magnetic force factor stability criterion

Alexander V. Sandulyak, Anna A. Sandulyak , Maria N. Polismakova, Dmitry O. Kiselev, Darya A. Sandulyak , Vera A. Ershova

Moscow Technological University, 107996 Moscow, Stromynka, 20, RF

Abstract. Due to rapid development of the scientific and applied research in magnetic control and magnetophoresis of ferro- and ferromagnetic disperse fraction of various technological, natural and industry-specific media, the Faraday method is again in high demand as it is mainly aimed at defining magnetic susceptibility of solid and heterogeneous samples of small volumes. Based on the appearing (and then measured) ponderomotive force impacting the sample, the method allows accurate determining of single particles magnetic susceptibility by using the data of the sample represented by a conglomerate of the particles of such a fraction. In addition, it is mentioned that to date there is still a great gap in the methodology of the Faraday method as there are no exact recommendations on choosing both the form of the polar pieces of the Faraday balance and the positioning of the sample (the location of the working zone) in the interpolar area. Owing to these drawbacks, the well-known and long-time used Faraday method cannot be considered substantiated to a satisfactory degree. Thus, in our point of view, the treatment of the results obtained earlier with the help of the method should be cautious. In our work, we experimentally defined and substantiated an approach to identifying a working (local) zone, viz. the zone with stable values of the magnetic force factor – the product of the field intensity and induction by its gradient. The approach features the relative phenomenological analysis and is exemplified by polar pieces of non-traditional spherical form. It has been demonstrated that in order to state the fact of mere existence of this zone (and its location) in the interpolar area, the coordinate (usually nonlinear one) characteristic of intensity or induction, which is obligatorily obtained in an experiment, should have an inflexion, which guarantees a functionally extreme view of the following coordinate characteristics both of the gradient and that of the force factor. We also established the coordinates and the length of the working zone in a specific interpolar area.

1 Introduction

Faraday ponderomotive method is one of the classical techniques used to define magnetic susceptibility χ of solid and heterogeneous samples of low volume V. The method is based on measuring the magnetic (ponderomotive) force F affecting the studied sample that is placed in an inhomogeneous field between the polar pieces of the Faraday balance; the method presupposes employing the well-known dependence $F = \mu_0 \cdot \chi \cdot V \cdot H gradH = \chi \cdot V \cdot B gradB / \mu_0$ (where $\mu_0 = 4\pi \cdot 10\text{-}7$ H/m is a magnetic constant) to define values χ.

Despite available alternative yet not always liveable solutions, this method can be given a new perspective of its targeted practical application. We are talking about the advancing applied scientific research in magnetophoresis and magnetic control of highly dispersed ferro- or ferrimagnetic particles (ferrofractions of many technological, natural and industry-specific media) [1-6], where one of the key issues is to obtain the data on the particles magnetic susceptibility.

So, by the yielded (almost direct) results of the susceptibility of the sample represented by a conglomerate of the particles experimentally separated from some medium and necessary to realise the Faraday method, we can quite definitely estimate the susceptibility of single particles by dividing these data by the values of their packaging bulk density (under condition of their sufficient mutual separation).

However, despite the Faraday method having been used by many generations of researchers, it still remains underdeveloped and thus not accurate enough. Thus, in our opinion, the attitude and treatment of the data array on the Faraday method and its outcomes accumulated and elucidated in the special literature should be cautious.

The point is when implementing the Faraday method, as a rule, one of the key requirements is not observed, the one that calls for rigorous positioning of the studied sample in the zone of the Faraday balance interpolar area. Unlike somewhat simplified opinion according to which this zone should just feature a constant value of the intensity H gradient, i.e. *gradH* (or the value of induction B gradient $B=\mu_0 H$, i.e. *gradB*), a more fastidious attitude to the working zone would assume preserving the constant value of the united parameter – a product of *H·gradH*, or *B·gradB*. However, there is unfortunately very little consideration given either to the issue of substantiated choice of the polar pieces shape (that are capable of ensuring the existence of the working zone), or to the matter of experimental and/or theoretical defining of the location of this zone, e.g. a zone with a stable value of the force factor. The features prerequisite for defining these conditions are virtually absent; the available recommendations are quite vague and hardly suitable for specific practical application.

2 Conceptual approach

To give an unbiased estimation on the acceptability of the polar pieces choice for the Faraday balance (their capacity to create a working zone), we must have at our disposal an expanded coordinate of the force factor characteristic (in the chosen direction *x* permeating the interpolar area e.g. in the line of ponderomotive force). In this case, if we detect any stable (constant or close to constant) values of *HgradH* or *BgradB,* we can clearly state the presence of the working zone of our interest in the interpolar area. Hence, it would simultaneously prove the acceptability of the chosen polar pieces for the Faraday balance, if otherwise, the polar heads should be deemed inadequate for this application. The coordinates of location of the detected segment would be the ones which are to be observed when positioning the sample to perform the measurements by the Faraday balance.

Obtaining the needed coordinate characteristic of the force factor *HgradH* or *BgradB* is possible after getting the corresponding coordinate characteristic of parameter *H* or *B*. Indeed, such a characteristic, specifically, parameter *B*, defined e.g. by small stepwise movement of the Hall sensor in the chosen direction *x* and represented in graphic and/or analytical form, allows us to find the coordinate characteristic of the gradient, i.e. parameter *gradB* = *dB*/*dx,* and then the analyzed parameter *B·gradB.*

Alongside with that, basing on that requirement for the working zone as the one of stable values of the force factor (at the definite, even comparatively short segment of the coordinate characteristic of that factor), we cannot but acknowledge the following condition, which greatly hampers satisfying this requirement. So, for polar pieces of traditional and specific shapes, the coordinate characteristics of *B* or *H* in the expanded view, i.e. the ones obtained when *x* varies in the range of the interpolar area as a whole, are objectively nonlinear as a rule. This means that expanded characteristics of *gradB* and *BgradB* are highly unlikely to be linear and all the more so self-similar or close to self-similar ones.

This observation can create an impression that it is problematic or even fundamentally impossible to obtain such a coordinate characteristic *BgradB* at which we would explicitly see the desired limited segment (commensurable to the sample dimensions) featuring comparatively stable values of *BgradB* or *HgradH.*

Meanwhile, it is possible to provide a rational solution to this seemingly complex problem already at the stage of initial diagnostics of the tested polar pieces, i.e. at the stage of obtaining preliminary key coordinate characteristics *B* or *H*. Thus, judging by the layout of any of the yielded characteristics, it is easy to take a decision on, first and foremost, the first issue mentioned, that is the choice (and principal applicability) of the polar pieces. It makes sense to move on to the next question only if we get positive results on this one. Only then, we can start defining characteristics *gradB, BgradB,* and finally, employing the required characteristic *BgradB* we can determine the positioning of the working zone.

Unconditionally positive this solution can be only in the case when the obtained characteristic *B* or *H* is sinuous, i.e. it has an inflexion. Then its segment (short relatively to the entire expanded nonlinear characteristic) in the vicinity of the inflexion point can be considered close to a linear one, and the values *gradB (gradH)* will be practically stable. In this case, it is highly probable to expect that the force factor characteristic, i.e. the product *B·gradB*, will have an extremum (the abscissa of which is surely different from the abscissa of the *gradB* characteristic extremum), in the proximity of which the values *BgradB* (*HgradH*) are virtually stable as well, as has been said earlier.

The example is nonlinear characteristics of the field intensity *H* with a pronounced inflexion in the gap areas between opposing spherical surfaces in the direction *x*, normal towards the centre-to-centre line of balls chain (with its magnetization by the field with intensity $H_0 \ll H$). At the characteristic *gradH* = *dH*/*dx* in the point corresponding to the abscissa of the characteristic *H* inflexion, there is observed a pronounced extremum; moreover, characteristic *HgradH* really becomes also extreme (let us note that the extremum abscissa is shifted to the area of lesser values of *x*) [7-9].

From the perspectives of a guaranteed creation of a working zone fulfilling the requirement for preserving the values of the force factor *HgradH* (or *BgradB*) stable, the considered illustrative example evidently bespeaks the full applicability of the spherical polar pieces use for the Faraday balance (clearly, as mutually disconnected semi-spherical bodies, shown on Fig.1).

3 Research results and their discussion

The practicability of choosing spherical polar pieces for the Faraday balance is proven by the results of special experiments conducted when using 100mm semi-spherical polar pieces (Fig.1) mutually distanced at 10mm (along the axial line of the poles) in the magnetic loop of the Faraday balance, with the magnetising force of the electromagnet equalling 12000A.

Fig. 2*a* (dots) shows the data on measuring induction *B* obtained with the help of the Hall sensor at different distance *x* from the axial line. The inflexion of the obtained coordinate characteristic *B* (by *x*) is clearly observed. It means, according to the afore-stated conceptual approach, that in the vicinity of the inflexion

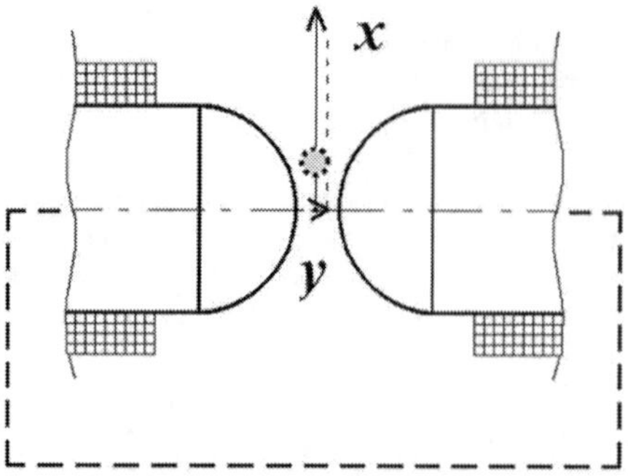

Figure 1. An illustration of one of the polar tips for Faraday balance. The spherical shape allows to create local (settling down in the neighborhood of an extremum of a power factor) a working zone.

point of dependence B on x, it is possible to linearly approximate the B data, as is shown in Fig. 2*a* by dashed lines (here for the illustration purposes they are a little bit longer than the segment of the linear approximation). It at the same time testifies the extreme view of characteristics *gradB* and quite probably, can be indicative of the extreme view of characteristics $BgradB = dB/dx$.

To prove this and to obtain the necessary sufficient case-based quantitative evidence, we first performed analytical (phenomenological) approximation of the nonlinear characteristic B obtained experimentally (Fig. 2*a*, dots) by the 4th degree polynom:

$$B = a_1x + a_2x^2 + a_3x^3 + a_4x^4 + a_0 \,. \qquad (1)$$

With the established (with the help of appropriate software, e.g. Excel, Advanced Grapher, MathCad etc.) individual values of coefficients a_1, a_2 a_3, a_4, and a_0, the initial experimental data (Fig. 2*a*, dots) can be described quite efficiently (Fig. 2*a*, line).

Then, using expression (1) and differentiating it, we can easily get an expression for describing coordinate characteristic $gradB = dB/dx$:

$$gradB = a_1 + 2a_2x + 3a_3x^2 + 4a_4x^3 \,. \qquad (2)$$

This characteristic *gradB* illustrated in Fig. 2*b*, being quite objective for a big range of x, is really extreme (except for the initial and final segments which are of no interest and the unusual flow of which is stipulated by the peculiarities of the chosen initial polynom, however, they do not influence the result of the formulated problem).

The revealed fact of *gradB* characteristic extreme view allows us to expect the presence of a similar phenomenon in a more interesting for us characteristic *BgradB*, for which we should use the following expression:

$$B\cdot gradB = B\cdot dB/dx = (a_1x + a_2x^2 + a_3x^3 + a_4x^4 + a_0)\times$$
$$\times\cdot(a_1 + 2a_2x + 3a_3x^2 + 4a_4x^3), \qquad (3)$$

which is the product of expressions (1) and (2).

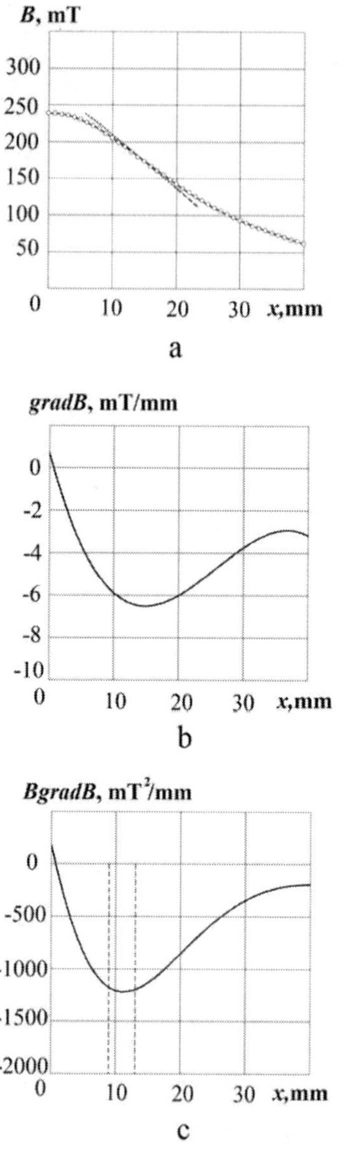

Figure 2. Coordinate characteristics of the field induction (a), its gradient (b), and the force factor (c) between the polar pieces, dots are the experimental data, lines are the calculations.

Fig. 2*c* illustrates the coordinate characteristic *BgradB* obtained with the help of (3). As has been expected, this characteristic also has an extreme view. Let us note at that, the value of the extremum abscissa (Fig. 2*c*) of characteristic *BgradB*, which is 11.2mm, is 1.33 times less than the values of the extremum abscissas *gradB*, which is 14.9mm

Therefore, the zone near the extremum of coordinate characteristic *BgradB*, as the area of relatively stable values *BgradB*, can be considered a working one, i.e. responsible for positioning of the sample studied. At that in the area examined (between the given spherical polar pieces) the size of the working zone per se, i.e. the zone of virtually stable values of parameter *BgradB* varies (by *x*) from 9mm to 13mm. At that, the proper size of the working zone, and consequently, the sample placed in it, should not exceed 4 – 4.5mm. It can be clearly seen in Fig. 2*c*, where the corresponding segment *x* is specially highlighted.

The established sizes of the working area testify only to its "longitudinal" size, namely along the axis *x* lying in the plane of symmetry of interpolar area (Fig.1). However such information isn't rather complete, it is necessary to have the relevant data on the found zone in its "cross", perpendicular an axis *x* direction (Fig.1). Receipt of such data requires accomplishment of the researches described above, but in the direction parallel "central" (the symmetry of interpolar area lying in the plane) to an axis *x*; on Fig.1 it is illustrated by the shaped line remote from the "central" axis *x* at *y* distance.

The measurements of induction *B* taken in the planes parallel to the plane of symmetry of interpolar area and remote from it at various distances of $y \leqslant 3$mm showed the following. The received characteristics of *B* were very close to the found characteristic shown on Fig.2*a*. The excess of each of them testified to a possibility of linear approximation of data in the neighborhood of an excess, showed an extremum of characteristics of *gradB* received by the polynomial described characteristics *B* (1) and their differentiation. Such excess show also the availability of extremum of characteristics of *BgradB*. At the same time, the proximity of values of abscissa and ordinate of extremum of characteristics *BgradB* was noted up to y = 2-2,2mm. Therefore, the working area in the "cross" direction can constitute at least the size $2y$ = 4-4,4mm, i.e. practically as well as in the "longitudinal" direction.

4 Conclusion

The paper substantiates and exemplifies by the appropriate choice of perspective polar spherical pieces the approach that has not been analysed yet. This approach allows identifying and diagnosing a local working zone with rigorous spatial constraints aimed at positioning the small-sized samples and studying their magnetic susceptibility in the interpolar area of the Faraday balance, the zone having a stable magnetic force factor at that. It comes down to choosing such a form of polar pieces (in particular, a spherical one) for the interpolar area, that would ensure the received coordinate characteristics of induction (intensity), which are usually nonlinear, are in this case sinuous, i.e. have an inflexion. Then for the zone close to the inflexion point and yielding to linearization, the corresponding derivative would feature an extremum, which in the final result ensures the extreme view of the force factor (and obtaining a working zone in the vicinity of the extremum). It has been established that in the area between spherical polar pieces with diameter of 100mm and mutually distanced at 10mm, the location of the working zone itself, i.e. the zone of practically stable values of parameter *BgradB,* varies (by *x*) from 9mm to 13mm. At that, the proper size of the working zone, and consequently, the sample placed in it, should not exceed 4 – 4.5mm.

Acknowledgement

The research is conducted with financial support from RFFI within the frameworks of research projects № 16-38-60034 mol_a_dk, № 16-58-10049 KO_a and from Russian Federation Ministry of Education and Science №9.9626.2017.

References

1. Patent 4492921 US. Method of Determining the Quantity of Solid Fraction of Ferromagnetic Matter in a Fluid. Sandulyak A.V., Garaschenko V.I., Korkhov O.J. Published 1985.
2. K.Nandy, S.Chaudhuri, R.Ganguly, I.K.Puri. Analytical model for the magnetophoretic capture of magnetic microspheres in microfluidic devices, Journal of Magnetism and Magnetic Materials, **320** (2008).
3. V.Murariu, J.Svoboda, The applicability of Davis tube tests to ore separation by drum magnetic separators, Physical Separation in Science and Engineering, **12** (2003).
4. S.Das, S.Chakraborty, Sushanta. K. Mitra. Magnetohydrodynamics in narrow fluidic channels in presence of spatially non-uniform magnetic fields: framework for combined magnetohydrodynamic and magnetophoretic particle transport, Microfluidics and Nanofluidics, **13**, 5 (2012).
5. M.Kawano, H.Watarai. Two-dimensional flow magnetophoresis of microparticles, Analytical and Bioanalytical Chemistry, **403**, 9 (2012).
6. A.A.Sandulyak, A.V.Sandulyak, B.M.Fethi Belgacem, D.O.Kiselev. Special solutions for magnetic separation problems using force and energy conditions for ferro-particles capture, Journal of Magnetism and Magnetic Materials, **401** (2016).
7. A. A. Sandulyak, V. A. Ershova. D.V.Ershov, A.V.Sandulyak. On the properties of "short" granular magnets with disordered chains of grains: A field between grains, Physics of the Solid State. **52**, 10 (2010).
8. A. V. Sandulyak, A. A. Sandulyak, V. A. Ershova. On the model of channel-by-channel magnetization of a granular medium (with a radial permeability profile of a quasi-continuous channel), Technical Physics, **54**,5 (2009).
9. A. V. Sandulyak, A. A. Sandulyak, V. A. Ershova. Magnetization curve of a granulated medium in terms of the channel-by-channel magnetization model (new approach), Doklady Physics. **52**, 4 (2007)

MATEC Web of Conferences **108**, 01008 (2017) DOI: 10.1051/matecconf/201710801008
ICMAA 2017

Enhancement of photocatalytic property on ZnS/MoS2 composite under visible light irradiation

Jiushan Cheng[1,2], Lu Han[1], Yen Wei[2] and Qiang Chen[1]

[1]*Laboratory of Plasma Physics and Materials, Beijing Institute of Graphic Communication, Beijing 102600, China*
[2]*Department of Chemistry and the Tisnghua Center for Frontier Polymer Research, Tsinghua University, Beijing, 10084, China*

Abstract: In this paper, the composite ZnS/MoS_2 was obtained via two steps including solvothermal methods. The as-synthesized sample was characterized by X-ray diffraction (XRD), scanning electron microscopy (SEM) and UV-Vis. diffuse reflectance spectra (DRS). The photocatalytic activity of the product was evaluated through photocatalytic degradation of Rhodamine B (Rh B) under UV-Vis. light irradiation; the electrical conductivity of ZnS/MoS_2 composites was significantly improved compared to ZnS, MoS_2, respectively. The results showed that the ZnS/MoS_2 composite photocatalyst possesses better photocatalytic activity in degrading Rh B than the single ZnS or the single MoS_2. The better photocatalytic properties may be due to the synergetic effect of two semiconductors, because of which electrons and holes were separated effectively. And its specific microstructure played an active role in evaluating photocatalytic performance.

1 Introduction

Environmental degradation is one of the challenging phenomenons due to the increasing rate of the industrialization in the world wide. Especially, organic dyes have been wildly used in the plastic, paper, textile, and other chemical industries in our lives [1-3]. And it can cause environmental pollution.

However, photocatalysis is one of the promising methods to remove the organic components from waste water. Semiconductor photocatlysts have attracted a subject of extensive interest due to their fundamental importance applications for solving energy and pollution problems. Recently, molybdenum disulfide (MoS_2), shows great potential for visible light irradiated photocatalysis applications. More and more researchers have been extensively investigated MoS_2 as a promising candidate for many materials, due to its excellent electrical carrier mobility, high chemical reactivity and optical properties [4-7]. For instance, Xiuquan Gu [8] et al. obtained MoS_2/Ag_3PO_4 hybrids by incorporating MoS_2 nanosheets during the preparation process of Ag_3PO_4 microparticles. The paper pointed out that the photocatalytic activity of MoS_2/Ag_3PO_4 hybrid got improved owning to compositing with trace amounts MoS_2. The group of Aiyong Zhang [9] prepared layer-controlled MoS_2 on three-dimensional Bi_2S_3 micro-flower through sacrificial templating strategy. The paper showed that 3D flower-like hybrid nanostructures make MoS_2/ Bi_2S_3 to display great photocatalysis under visible light irradiation.

Similarly, among the plenty of metal sulfide photocatalysts, especially, ZnS, have been paid more attention from the scientific research owing to their optical properties and unique catalytic functions. However, ZnS is responsive only in ultraviolet light region. To address this issue, many means have been used, such as nanomaterials morphology modification and semiconductor compositing. These methods are proved to be effective, and the best way is meeting the energy demands of ZnS with wide band-gap to maintain the catalytic activity [10-11]. Longxin Hu [12] et al. reported that SnO_2/ZnS nanocomposite was prepared by one step hydrothermal method. It was found that the compound showed the great photocatalytic activity. The photocatalytic degradation of RhB (Rhoadmine B) reached 95% at given time. In Baibiao Huang's group [13], they prepared ZnS microspheres with Mn-doping by a facile solvothermal route. And it was found that Mn-doping ZnS exhibited better photocatalytic performance than pure ZnS.

Owing to there have been few studies related to ZnS/MoS_2 composite. In this paper, we successfully prepared semiconductor of MoS_2 decroated with ZnS as catalyst for efficient degradation of dye. Therefore, we pay attention to the ZnS nanoparticles coated MoS_2 nanosheets system connecting two chemically stable and inexpensive multifunctional semiconductor materials together. In this system, MoS_2 nanosheets work as narrow band gap p-type semiconductor and ZnS serves as wide band-gap n-type semiconductor. The experimental reagents are simple and easy to get and the reaction conditions are simple and controllable. Experiment results indicate that ZnS composite MoS_2 exhibit superior photocatalyst performance towards RhB.

2 Experimental

In a typical procedure, 1.5 mmol Na_2MoO_4 and 15 mmol $C_3H_7NO_2S$ were dissolved in 80 ml deionized water. Then, the solution was transferred into a Teflon-lined stainless-steel autoclave, which was sealed and heated at 220℃ for 24 h, and then cool down to room temperature. The as-prepared precipitates were washed by ethyl alcohol and distilled water three times respectively. Then the black powder was collected and dried in vacuum at 80℃ for 12 h. The black precipate was denoted as MoS_2. 0.1g thioacetamide(CH_3CSNH_2), 0.6758g zinc acetate ($Zn(CH_3COO)_2$) and 0.245g Sodium Sulfide (Na_2S) were dissolved in 20 ml deionized water. The mixture was full stirred for 2 h at room temperature. Then, the solution was transferred into a Teflon-lined stainless-steel autoclave, which was sealed and heated at 170℃ for 24 h, and then cool down to room temperature. Then, the precipitates were collected and washed by ethyl alcohol and distilled water three times respectively. Then obtained powder was dried in vacuum at 60℃ for 12 h. The final powder was called ZnS.

0.1 as-prepared MoS_2, and 0.2g Sodium dodecyl sulfate(SDS), 1.3516g zinc acetate, 0.49g Sodium Sulfide were dissolved in 60 ml deionized water. Then, the solution was transferred into a Teflon-lined stainless-steel autoclave, which was sealed and heated at 170℃ for 2 h, and then cool down to room temperature. Then, the precipitates were collected and washed by ethyl alcohol and distilled water three times respectively. Then obtained powder was dried in vacuum at 60℃ for 12 h which denoted as ZnS/MoS_2.

The as-prepared products were characterized by powder X-ray diffraction (XRD) in Bruker D8 diffractometer using Cu Kα radiation. The surface morphology of the samples were observed by a HITACHI S3500 scanning electron microscopy (SEM). The light absorption properties were measured using UV–Vis. diffuse reflectance spectrophotometer U-4100 (DRS) with a wavelength range of 300~800 nm.

The photocatalytic activity of the samples was determined by degradation of Rhoadmine B (RhB) in an aqueous solution under visible light irradiation. The photo degradation experiments were carried out under UV-Vis. light irradiation whose source was a 300W Xe lamp. The distance between the xenon lamp and the sample was about 50 cm. In the each experiment, 20 mg photocatalyst was added into 20 mL RhB solution (C_0=10 mg/L). Prior to irradiation, the suspension was sonicated for a few minutes and then magnetically stirred in dark for hours to obtain desorption– adsorption equilibrium. At given time intervals, 5 mL of the sample solution was collected and separated through centrifugation. The HITACHI U-4100 UV–Vis. spectrometer was used to determine the concentration of RhB solution during the photocatalytic degradation process.

The electrochemical properties of all the samples were studied using electrochemical workstation (PGZ402, VoltaLab, France) with a three-electrodecon figuration. The samples were then used to modify screen-printed carbon electrodes (SPCEs) to immobilize antibodies and enhance the electrochemical signals.1 μ l of 1mg/ml ZnS、MoS_2 and ZnS/MoS_2 were spread on the working area of SPECs using a micropipette and the film was dried at room temperature, resepectively. All the cyclic voltammetry (CV) measurements were recorded in 0.1M phosphate buffer (PBS, pH7.4) containing 5mM $K_3[Fe(CN)_6]$ and $K_4[Fe(CN)_6]$ in the potential range of -1000 to +1000mV. A reproducible voltammogram was obtained under steady-state conditions after about five cycles. All the electrochemical impedance spectroscopy (EIS) measurements were carried out in 0.01M PBS containing 5mM $K_3[Fe(CN)_6]$ and $K_4[Fe(CN)_6]$ in a frequency range from 0.1 to 104 Hz at 220mV.

3 Results and discussion

The XRD patterns of the samples are presented in Figure. 1. There are no other phases and peaks of impurities, demonstrating that all the samples are pure. MoS_2 and ZnS possess the phases corresponding to JCPDS No. 37-1492(a = 3.161 Å and b = 3.161 Å, c = 12.299 Å) and 65-5476 (a=5.404Å, b=5.404 Å, and c=5.404 Å) respectively. The XRD pattern of the ZnS/MoS_2 nanohybrid shows MoS_2 major peaks at corresponding to (100), (103), (110) and there are peaks, like (111)、(220)、(311) observed related to ZnS. And, no other phase is detected in the ZnS/MoS_2 composite before and after the photocatalytic degradation, suggesting that the composite is stable and there is no chemical reaction between MoS_2 and ZnS.

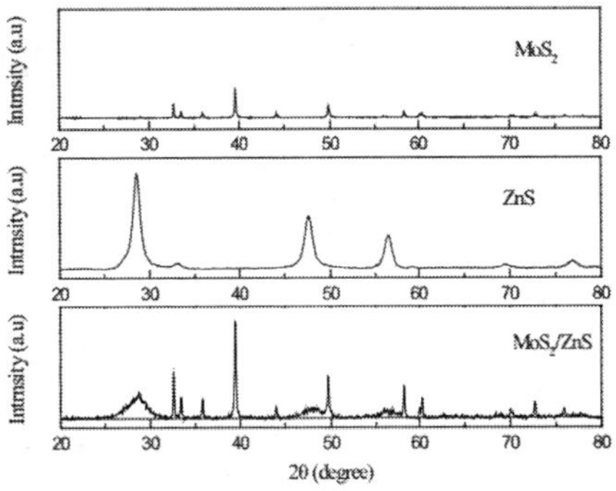

Figure 1. XRD pattern of the samples MoS_2, ZnS, ZnS/MoS_2.

The morphologies of the pure MoS_2, ZnS and ZnS/MoS_2 composite are observed by SEM. Figure.2a exhibits a representative SEM image of pure MoS_2. The highly monodispersed rough ball-like spheres are obtained which is composed of thin plates as shown in Figure.2b. The nanoplates align radically and tightly to assemble into the uniform spheres. The surfaces of nanoplates are smooth. The SEM micrograph of ZnS in Figure2c show a large number of plates arranged irregularly together. It can be seen the shapes of plate is the block shaped and the length is about 1 μm on average. Regarding to ZnS/MoS_2 composite, it could be seen that the microspheres are organized by hundreds of thin nanoplates and some ZnS plates are together with sphere with a diameter of 1μm (Figure.2d).

Figure 2. The SEM images of samples (a,b) MoS_2, (c) ZnS, (d) ZnS/MoS_2.

The absorption property is one of the most important properties to characterize the optical property of semiconductor. Figure.3 shows the UV-Vis. diffuse reflectance spectra of ZnS and ZnS/MoS_2 composite. It could be seen that the pure ZnS has an absorption edge occurred at about 375 nm. The absorption edge of the pure ZnS/MoS_2 is approximately at 390 nm. From the picture inset Figure.3, it can be seen UV-Vis. diffuse reflectance spectra of MoS_2. The absorption of ZnS/MoS_2 composite during visible light range is much improved may be owing to MoS_2 which plays an important role.

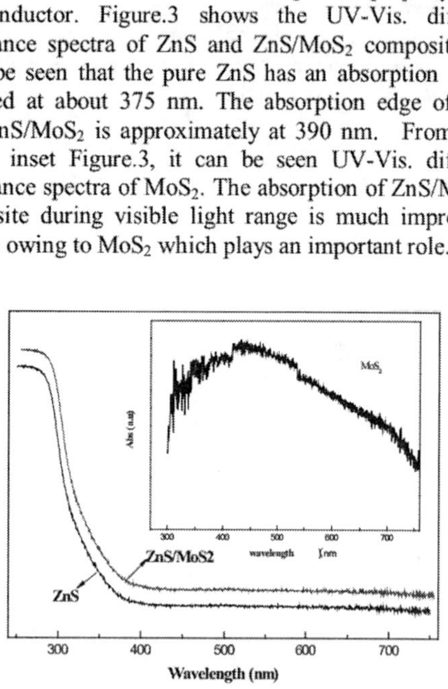

Figure 3. UV-Vis. DRS spectra of the samples ZnS, ZnS/MoS_2, the inset of Figure is UV-Vis. DRS of sample MoS_2.

The photocatalytic activities of as-prepared samples were evaluated by degrading Rhodamine B (Rh B) under UV-Vis. light irradiation. The photodegradation of the RhB as a function of the irradiation time over ZnS/MoS_2 composite is presented in Figure 4. We can see there is no significant change in the concentration curve with time when the photocatalysis is absent, which means Rh B is very stable without photocatalysis only under UV-Vis. light. From the curves of the concentration changes of RhB dye, it can be seen that the photocatalytic degradation rate of all as-prepared samples are increased with the increasing of the reaction time. In 80 minutes, nearly 70% RhB is degraded by ZnS/MoS_2. whereas in 80 minutes, the MoS_2 and ZnS degraded the RhB concentration by 64%, and 82%, respectively. It could be seen that the ZnS/MoS_2 composite exhibits the best photocatalytic activity among the samples.

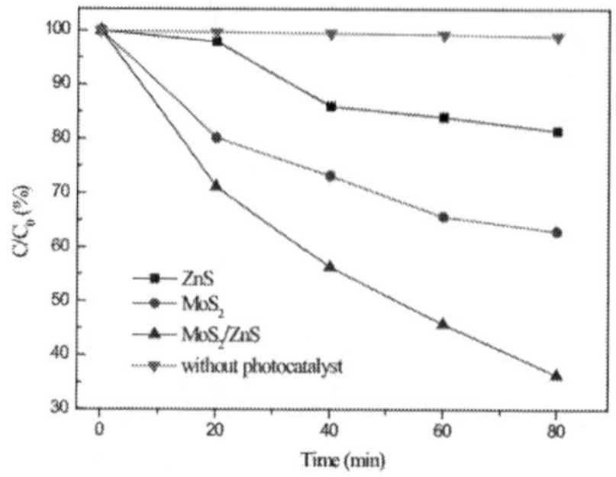

Figure 4. Photocatalytic degradation efficiency of the Rh.B over no photocatalyst, MoS_2, ZnS, ZnS/MoS_2.

In order to further confirm the enhanced conductivity of ZnS/MoS_2, CV curves of ZnS/MoS_2, ZnS, MoS_2 were measured in 0.1M phosphate buffer (pH 7.2) containing 5mM $K_3[Fe(CN)_6]$ and $K_4[Fe(CN)_6]$ as electrolyte. In Figure.5 a pair of symmertric and well-separated redox peaks was observed, indicating that the measured electroconductivity was based on the redox mechanism [14-17]. It was also noted that ZnS/MoS_2 had another cathodic peak at -700 mV may due to the redox of MoS_2. ZnS/MoS_2 exhibited higher current density than single ZnS and MoS_2 due to the accelerated electron transfer by MoS_2. This observation indicates an irreversible electrochemical reduction of MoS_2 when CV is conducted in the inorganic electrolyte at the potential range from 1000 to – 1000 mV. It means that the MoS_2 was successfully reduced.

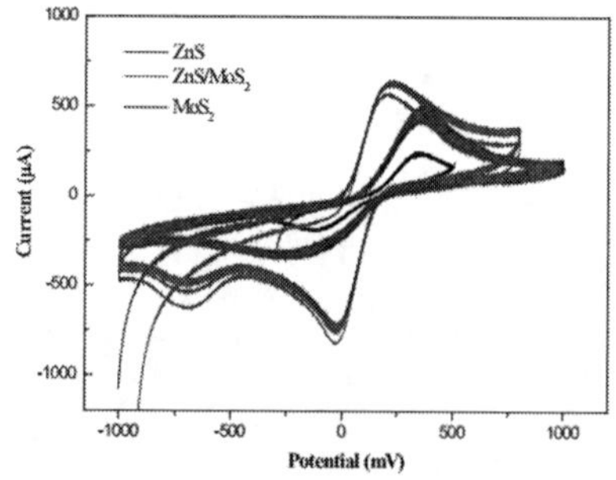

Figure 5. CV curves of MoS_2, ZnS, ZnS/MoS_2 at the scan rate of 100 mV s1 in 0.1 M PBS (pH 7.4) at RT.

It is obvious that ZnS/MoS_2 composite nanostructures possess more powerful photocatalytic capacity for the degradation of Rh B. As a consequence, a possible mechanism is proposed. According to some papers [18-21], we deduce that MoS_2 and ZnS have matching band potential.

Under irradiation, the photocatalyst could generate electrons and holes. The conduction band and valence band edge positions of MoS_2 are associated with those of ZnS, so both electrons and holes are transferred to MoS_2. Holes transfer to MoS_2 nanoflakes, which increases charge separation. Thus, MoS_2 acts as an oxidation site. In another words, the band potentials of both MoS_2 and ZnS semiconductors are conductive to prevent the recombination of electrons and holes in the ZnS/MoS_2 composite. In general, the ZnS/MoS_2 composite with special nanostructure and interface increases the visible light absorption and restrains the recombination of electron-hole pairs.

4 Conclusion

In summary, ZnS/MoS_2 composite photocatalyst was got by two steps which in order were solvothermal methods. The microsphere of the composite was composed of hundreds of MoS_2 thin nanoplates and some ZnS plates were together with the sphere about a diameter of 1μm from the view of morphology. From the photodegradation of the Rh B organic dye we found that the ZnS/MoS_2 composited photocatalyst displayed better photocatalytic activity than both single MoS_2 and single ZnS. As a proof of concept, the ZnS/MoS_2 with good conductivity, superior electron transfer rate and high electrochemical activity is successfully used to detect photocatalytic property.

Acknowledge

This work was financially supported by a project founded by Beijing Institute of Graphic Communication (No. 27170115005/020), Project supported by Beijing Postdoctoral Research Foundation (No. 10000200272), Beijing college students' scientific research and entrepreneurial action plan (No. 22150116005/046).

References

1. Ying Wang, Mingxuan Sun, Yalin Fang, Shanfu Sun, Jia He, J. Mater. SCI, **51**, 779 2016
2. Shaohui Gao, Xuanhua Li, Jinmeng Zhu, Tengteng Tong, ingqing Wei, Small, **12**, 5692 2016
3. Shijing Liang, Zhouming Zhou, Xiuqin Wu, Shuying Zhu, Jinhong Bi, Limin Zhou, Minghua Liu, Ling Wu, Molecules, **21**. 213 2016
4. M.Nath, A. Govindaraj, C.N.R. Rao, , Adv Mater, **13**, 283, 2001
5. Lei Huang, Xiuli Wang, Jinhui Yang,Gang Liu, Jingfeng Han, Can Li, J Phys Chem C, **117** 11584 2013
6. Guoping Chen, Dongmei Li, Fan Li, Yuzun Fan, Haofei Zhao, Yanhong Luo, Richeng Yu, Qingbo Meng, Appl Catal A-Gen, **443-444**, 138 2012
7. Xiaojie Xu, Linfeng Hu, Nan Gao, Shaoxiong Liu, Swelm Wageh, Ahmed A. Al-Ghamdi, Ahmed Alshahrie, Xiaosheng Fang, Adv Functi Mater, **25**, 445 2015
8. Shengping Li, Xiuquan Gu,Yulong Zhao, Yinghuai Qiang, Shuang Zhang, Mrirong Sui, J Mater Sci-Mater EL, **27**, 386 2016
9. Lulu Long, Jiejie Chen, Xing Zhang, Aiyong Zhang, Yuxi Huang, Qing Rong, Hanqing Yu, NPG Asia Materials, **8**, doi:10.1038/am/2016.46
10. S.V Prabhakar Vattikuti, Chan Byon, Sora Jeon, Physica B, **502**, 103 2016
11. Ch. Venkate Reddy, Jaesool Shim, Migyung Cho, J Phy Chem Solids, doi.org/10.1016/j.jpcs.2016.12.011
12. Longxing Hu, Feiyan Chen, Pengfei Hu, Lianpei Zou, Xing Hua, J Mol Catal A-Chem, **411** 203 2016
13. Lu Wang, Peng Wang, Baibiao Huang, Xiaojuan Ma, Gang Wang, Ying Dai, Xiaoyang Zhang, Xiaoyan, Qin, Appl Surf Sci, **391**, 557 2017
14. Lu Han, Chengmei Liu, Shilei Dong, Caixia Du, Xiaoyong Zhang, Luhai Li, Yen Wei, Biosens Bioelectron **87**, 466 2017
15. Zhebo Chen, Dustin Cummins, Benjamin N. Reinecke, Ezra Clark, Mahendra K. Sunkare, Thomas F. Jaramill, Nano Lett, **11** 4168 2011
16. Jun Chen, Nobuhiro Kuriyama, Huatang Yuan, Hiroyuki T. Takeshita, Tetsuo Sakai, J Am Chem Soc, **123** 11813 2001
17. Shixin Wu, Zhiyuan Zeng, Qiyuan He, Zhjuan Wang, Shijie Wang, Yaping Du, Zongyou Yin, Xuping Sun, Wei Chen, Hua Zhang, Small, **8**, 2264 2012
18. Nan Tian, Zhen Li, Danyun Xu, Yang Li, Wenchao Peng, Guoliang Zhang, Fengbao Zhang, Xiabin Fan, Industrial & Engineering Chemistry Reseach, **55**, 8726 2016
19. Herme G. Baldovi, Marcos Latorre-Sancez, Ivan Esteve-Adell, Anish Khan, Abdullah M. Asiri, Samia A. Kosa, Hermenegildo Garcia, J Nanopart Res, **18**, 240 2016
20. Quanjun Xiang, Jiaguo Yu, Mietek Jaroniec, J Am Chem Soc, **134**, 6575 2012
21. Xu Zong, Hongjian Yan, Guopeng Wu, Guijun Ma, Fuyu Wen, Lu Wang, Can Li, J Am Chem Soc, **130**, 7176 2008

Temperature Effects on Mechanical Properties of Woven Thermoplastic Composites for Secondary Aircraft Structure Applications

Yue Wang, Jipeng Zhang, Jiazhen Zhang and Zhengong Zhou

Center for Composite Materials and Structures, Harbin Institute of Technology, 15001 Harbin, China

Abstract. The effect of temperature on the mechanical behavior of 8-H satin woven glass fabric/polyethylene sulfide (GF/PPS) was investigated in this paper. Static-tensile tests were both conducted on notched and unnotched specimens at typical temperatures (ambient, 95°C and 125°C) based on the glass transition temperatures (T_g) of the neat resin and composite, their strength and moduli were obtained and compared. The damage patterns of failed specimens of notched and unnotched were examined with the aid of high-definition camera and stereomicroscope. The results of stress-strain relationships showed that the slight nonlinearity of the curves were observed for these two specimens, which was associated with the plastic deformation of localized resin. The damage patterns of notched and unnotched specimens at different temperatures proved that damage and plastic deformation were two simultaneous mechanisms and it was prominent in the notched. It was the overstress accommodation mechanism that led to a relative high strength rentention for the notched and a reduction of the hole sensitivity. The results obtained in this paper indicated that GF/PPS can be used as secondary aircraft structures at elevated temperatures higher than its T_g.

1 Introduction

"Light weight and high strength" is always the goal for the aircraft manufacturers and scientists. Composite materials have gained widely attentions due to the prominent advantages of weight reduction and improved durability compared to metals [1]. At present, the fraction of usage amount of composites to the total weight of aircrafts has been an important criterion for evaluating the progressiveness of civil aircrafts. For example, the composite materials make up 25%, 52% and 50% of the structural mass of A380, A350 and Boeing 787, respectively [2]. Despite thermosetting resin-based composite materials are regarded as the favored materials in designing structures, the use of high-performance thermoplastic composites is continuously increasing because of the advantages of unlimited shelf life, recyclability and improved processing technologies over conventional thermosetting composites [3]. In addition, excellent damage tolerance, corrosion and impact resistance they offered further expand their application fields. Until now, thermoplastic composites have been used as wing fixed leading edges, ribs, clips, angles and cleats [4], as Figure 1 shown. In oder to make full use of thermoplastic composites, they are considered as candidates for special aircraft structures, such as aircraft engine nacells that subjected to thermal effect in service, and it is necessary to clarify their mechanical responses and damage patterns at elevated temperatures, expecially when the environment temperature is higher than the glass transion temperature of the composite.

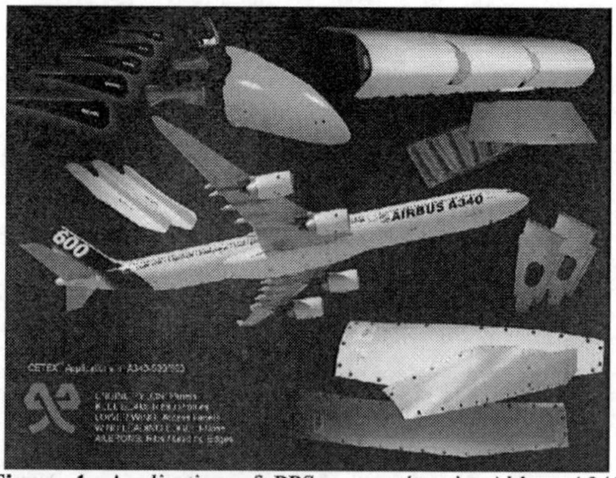

Figure 1. Application of PPS composites in Airbus A340 supplied by TenCate Company

The glass transition temperature is a critical temperature that amorphous phase of the polymer shifts from glass state to rubbery state. For the fiber reinforced polymer composites, the effect of T_g is always characterized by nonlinear behavior, especially for off-axial plies [4]. The nonlinear response is attributed to the shear deformation of the polymer matrix along the fibers, which will be enhanced at $T > T_g$. Some researchers investigated the mechanical properties of thermoplastic composites at elevated temperatures, comprsing of tensile, compressive, flexural and fatigue properties [5-8]. Most of them found that both of the strength and modulus present an obvious decline trend when the temperature was around T_g. But all of them focused the glass transition temperature of composite (T_g^c), and that of the neat resin (T_g^r) also plays a dominant role in mechanical response.

Furthermore, it is inevitable to drill holes when the thermoplastic composites are used in joint structures, such as angles and cleats. Then the failure behavior of the notched specimen will be dominated by the damage around the hole due to the stress concentration. Reference [9-11] investigated the notched tensile behavior and gave a deep insight into damage progression in the vicinity of the hole. They observed that the inital damage generated from the edge of the hole, leading to a redistribution of stress. Even at elevated temperatures, two simultaneous mechanisms of damage and plastic deformation will result in a more complex failure behavior [12].

In view of this, the tensile properties of notched and unnotched woven GF/PPS composite were investigated experimentally. The research is focused on the comparison of mechanical response and damage patterns between the two components at typical temperatures.

2 Materials and methods

2.1 Materials

The composite material in this work was a thermoplastic 8-harness satin weave composite laminate supplied by TenCate Advanced Composites, which has EC9 glassfibre fabric as reinforcement and Fortron 0214 PPS as a matrix. The volume fraction of fibers is 52.5% in GF/PPS laminate. The composite laminate was manufactured using hot pressing technique according to a quasi-isotropic lay-up: $[(\pm45)/(0,90)/(\pm45)/(0,90)]_{2s}$. Sixteen plies were stacked in the laminate and a thickness of 3.84mm was obtained.

2.2 Methods

Static tensile tests were performed on Instron 5569 universal mechanical testing machine, which was equipped with a 50kN capacity load cell and an environmental chamber. Tests on unnotched and notched specimens were carried out according to ASTM D3039 and ASTM D5766, respectively. The tests were conducted in displacement control mode at a same crosshead speed of 1 mm/min. In order to obtain the whole strains of the specimens, laser extensometer and strain gauge were used in notched and unnotched tensile tests, respectively. The geometries of specimens are shown in Figure 2.

Prior to the mechanical testing, a combination of thermal analyses was used to establish a relationship between the thermal property and mechanical property. A differential scanning calorimetry performed by Pyris Diamond DSC was conducted on the GF/PPS composite. Both the heating and cooling rate used were 10 °C /min. Three-point bending dynamic mechanical analysis was carried out by Perklin Elmer Pyris Diamond DMA. Temperature scans from room temperature to 250 °C at a heating rate of 5°C and the test was performed in Nitrogen atmosphere with a frequency of 1Hz.

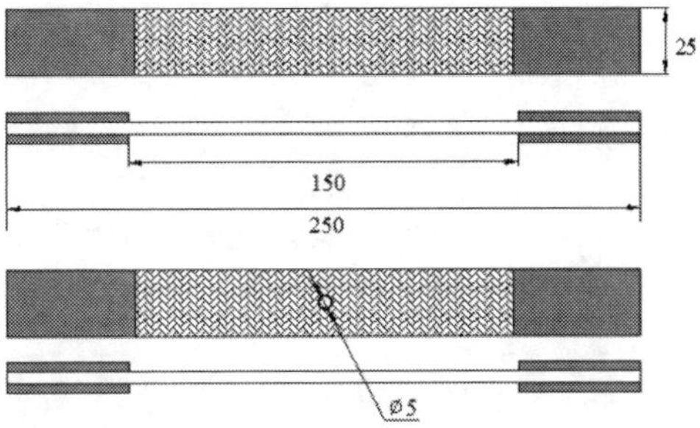

Figure 2. Geometries of the specimens. (Dimensions in mm)

3 Results and discussion

3.1 Thermal analysis of GF/PPS composite

The glass transition temperature ($T_g^c = 101^{\circ}\mathrm{C}$) is obtained by DMA, and the crystallinity (52%) is calculated by DSC using a value of heat of fusion of a 100% crystalline sample of PPS of 76.5 J/g, as shown in Figure 3. It is nearly the same with [1] and the tiny difference is probably due to the type of reinforced fibres. According to TenCate material parameter, $T_g^r = 90^{\circ}\mathrm{C}$ is obtained for neat resin. The deviation of the crystalline peak and the improvement of glass transition temperature are mainly attributed to the addition of fibres.

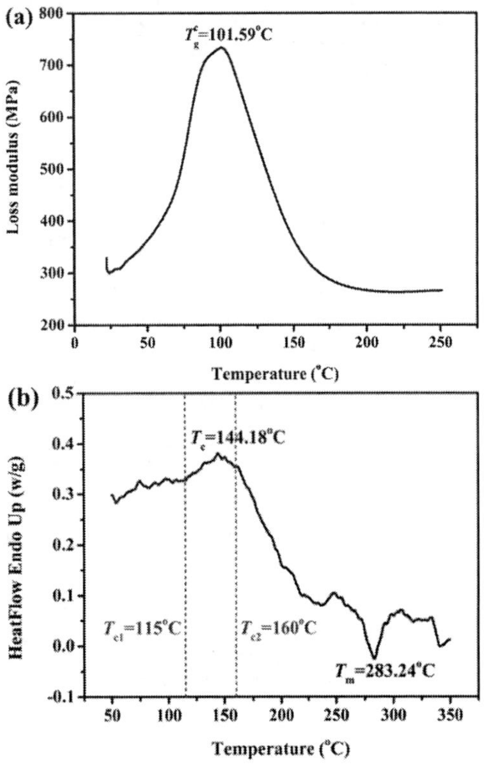

DOI: 10.1051/matecconf/201710801009

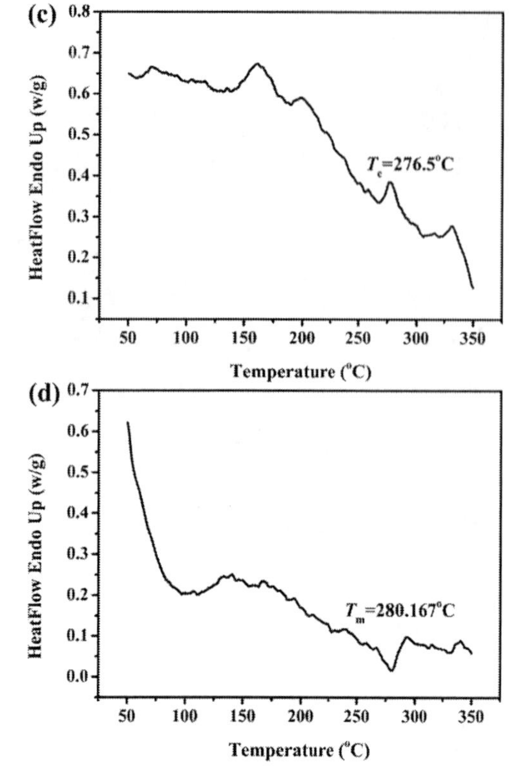

Figure 3. DMA and DSC curves of GF/PPS composite: (a) loss modulus curve obtianed by DMA; (b) first heating cycle, (c) cooling cycle and (d) second heating cycle. (b), (c) and (d) were obtained by DSC measurement. (T_g, T_c and T_m denote glass transition temperature, crystallization temperature and melting temperature. T_1 and T_2 denote the temperature of the initiation and end of the peak.)

3.2 Stress strain relationships

The whole strain of a notched specimen can not be obtained by strain gauge due to the stress concertration in the vicinity of the hole, thus a laser extensometer is in needed. In order to make the strain accurate, a calibrating experiment was conducted on the unnotched specimen at 95°C considering the effect of temperature, then the result is shown in Figure 4.

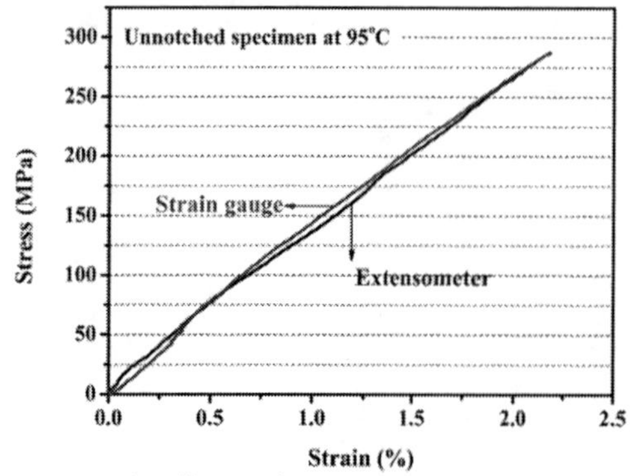

Figure 4. Strain calibration between extensometer and strain gauge.

It can be found that both of the two stress-strain curves obtained by extensometer and strain gauge are consistent and they also present some fluctuations while the extensometer one is more obvious. The reason for this may be associated with the slight plasitc deformation in (±45°) ply at the surface of the specimen.

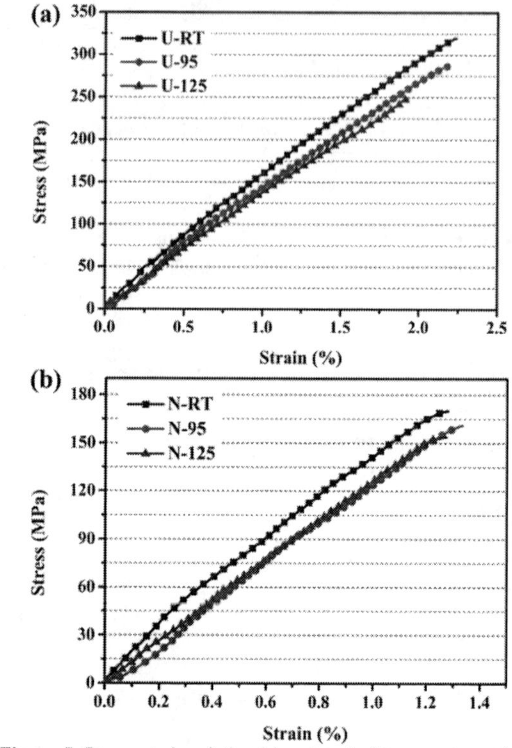

Figure 5. Stress-strain relationships at typical temperatures: (a) unnotched; (b) notched.

The axial tensile tests were conducted on nothed and unnotched specimens at RT, 95°C and 125°C, the results are shown in Figure 5. For the unnotched specimens, the stress-strain curve is almost linear at RT, which is similar with CF/PPS [13] and GF/epoxy [9], indicating that the polymer chains are frozen and the matrix is brittle at this temperature. When temperature reaches 95°C ($T_g^c > T > T_g^r$), the curve present a low degree of nonlinearity as the load increasing which is related with the characteristic of weave fabric and (±45°) plies. During the manufacturing of the fabric, crimps were formed at the crossovers of weft and warp fiber bundles while lots of resin-rich regions are also formed around the crimps during the hot pressing. Part of the resin in these regions are far away from fibers so that the glass transition temperature of the resin is still the same with T_g^r [14]. Thus, the slight nonlinearity of the curve is resulted from the plastic deformation of the localized resin in the resin-rich regions and ±45° oriented plies. At 125°C, the stress-strain curve is very closed to the one at 95°C, but the nonlinearity is more obvious due to the softening of matrix. In addition, a fluctuation is observed near the end of the curve, which is associated with the straightening of warp fiber in the loading direction [15]. In general, for

 DOI: 10.1051/matecconf/201710801009

the fiber reinforced polymers, the majority of tensile load are carried by 0° fibers, which results in a temperature-independent behavior [12].

For the notched specimens, the stress state is more complex than the unnotched due to the presence of a hole. As can be seen in Figure 5(b), the stress-strain curve of the notched specimen at RT present a similar linear trend as that of the unnotched one in spite of some fluctuations. While a low degree of nonlinearity is observed near the end, it is suggested that some damage generate before the final fracture. It is worth noting that the curves of notched specimens at 95°C and 125°C are nearly the same with a higher stress in former, and both of the two curves exhibit obvious nonlinearity near the end. As mentioned above, although the plastic deformation of matrix occures at $T > T_g$, it has a weak influence on the mechanical response. Accordingly, the enhancement of the nonlinearity of the curves are resulted from the damage progression in the vicinity of the hole.

3.3 Damage patterns

In order to investigate the effect of temperature on the damage patterns of the notched and unnotched specimens and make a comparison between the two components, the failed specimens were examined with the aid of high-definition camera and stereomicroscope, the results are shown in Figure 6. For the unnotched specimens, the specimens at RT failed in a brittle mode, which are characterized by fibre bundles fracture and little amount of delamination at the surface ply where a higher interlaminar stress usually presents [9]. Judging from the extent and degree of damage, the initial damage is generated from the free edge. This is consistent with [15], transverse weft cracking and fibre fracture that obtained by scanning electronic microscope led to the final fracture, and meanwhile the damage was observed at a relative high applied load close to the specimen fracture. This is the reason why slight nonlinearity is observed at the end of the stress-strain curve. When temperature increases to 95°C, the specimen fails in two parts and some filaments are present, as well as a bulge originated from delaminations and splitting. This phenomenon indicates that the increase of temperature enhances the ductility of matrix and weakens the fibre-matrix interface property. Whereas four specimens are tested at 125°C and they all failed near the tabs, the damage patterns present in Figure 5 are considered as reference which also exhibit a large amount of delaminations.

For the notched specimens, the damage patterns are substantially different from the unnotched ones, reflecting in the damage zone limited in the vicinity of the hole. The damage patterns of the specimen at RT are characterized by axial fibres fracture and a small amount of delaminations. As some publications about the damage progression of the notched specimen reported, transverse cracks as the initial damage generated from the edge of the hole and then propagated [10]. Thus, all of the notched specimen present transverse failure in Figure 6. As temperature increases, both of the length and the number of fibres pulled out increase, as well as splitting and delamination. Besides, part of fibres of notched specimen at 125°C rotate to the loading direction, which suggests the damage progression and plastic deformation are simultaneous at this condition.

3.4 Comparisons of mechanical response

Comparisons of strength and modulus have been made between notched and unnotched specimens as functions of temperature, the results are shown in Figure 7. It can be found that both of the ultimate tensile strength of the two components present decline trend as temperature increasing, especially at $T > T_g^c$, the strength decrease faster due to the softening of matrix. In addition, it should be noted that the tensile strength retention rate at 95°C and 125°C are 91% and 78% for the unnotched specimen. The relative lower retention of the latter may be originated from the effect of the tabs, although the tests at 95°C tell an error of 1% between the specimens that fractured at different locations. In spite of this, the strength retention rates of 93% and 90% for the notched are still higher than the unnotched. As mentioned in Section 3.3, the damage and plastic deformation in the vicinity of the hole plays an important role in the overstress accomdation.

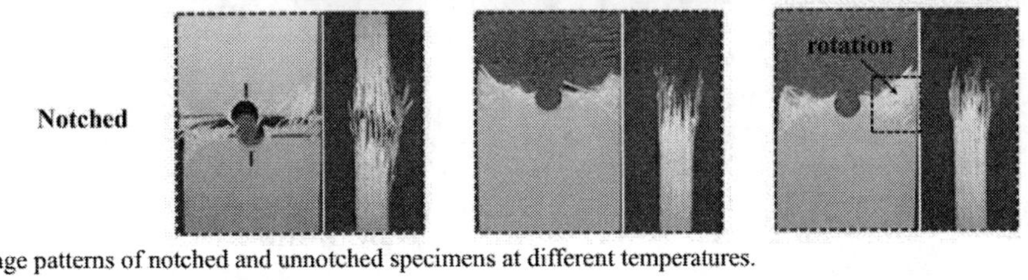

Figure 6. Damage patterns of notched and unnotched specimens at different temperatures.

The hole factor C_h reprensents the hole sensitivity of the laminate ultimate strength. Its definition is

$$C_h = \frac{\sigma_u^{notched}}{\sigma_u^{unnotched}} \quad (1)$$

where
$$\sigma_u^{notched} = \frac{F_u^{notched}}{t \cdot w} \quad (2)$$

$$\sigma_u^{unnotched} = \frac{F_u^{unnotched}}{t \cdot w} \quad (3)$$

Among them, $\sigma_u^{notched}$ and $\sigma_u^{unnotched}$ denote ultimate tensile stress for the notched and unnotched specimens, t and w are thickness and width. As can be seen in Figure 7(a), C_h increases obviously as temperature increasing, which indicates that the delamination and splitting at elevated tempertures make a contribution to stress relief. Moreover, the softening of matrix and damage progression in the vicinity of hole also make an effect on the reduction of hole sensitivity (being the reverse of the hole factor).

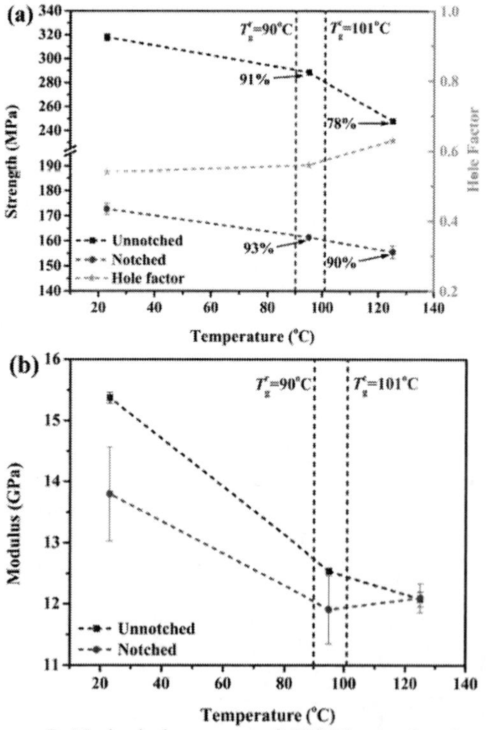

Figure 7. Mechanical response of CF/PPS as a function of temperature: (a) strength; (b) modulus.

It is expected that the modulus would present a similar trend with the strength, but it is found that the modulus of the notched specimen at 125°C is higher than that at 95°C, while the modulus of the unnotched specimen only present a slightly decrease at 125°C than 95°C. This is consistent with [15] and [7], the modulus start increasing in the 120°C- 130°C temperature range, probably due to the cold crystallization at 125°C. Actually, it is observed in Figure 3(b) that a crystallization peak exsit at 115°C-160°C and in this temperature range, a crystallization phenomenon of the amorphous phase of the polymer occurs, leading to an increase on crystallinity as well as the tensile modulus.

4 Conclusions

The temperature dependence of tensile properties for notched and unnotched 8-H satin woven GF/PPS were investigated at typical temperatures based on the glass transition temperatures of the neat resin and composite. The results obtained can be concluded as follows:

(1) The nonlinearity of the stress-strain curves for unnotched and notched specimens was enhanced as temperature increasing, which was mainly associated with plastic deformation of localized resin in the resin-rich regions and (±45°) plies. and For the notched specimens, the nonlinearity observed at the end of the curves was dominated by the initiation of damage in the vicinity of the hole.

(2) With increase in temperature, the damage patterns transformed from brittle mode to ductility mode, reflecting in the increase of delamination, splitting and fibres pulled out, which suggested the fibre-matrix interface bonding strength was weakened. For the notched specimens, part of fibres located in (±45°) plies rotated to the loading direction at 125°C, indicating that damage and plastic deformation were two simultaneous mechanisms which dominated specimens fractured.

(3) Obvious reduction of strength can be found for the two components due to the softening of matrix, especially $T > T_g^c$ the strength decreased fast. While it is relative gradual for the notched attributed to the overstress accomodation mechanism, leading to a reduction of hole sensitivity. Besides, the modulus decreased firstly and then slightly increased at 125°C, it was related with cold crystallization. Finally, the relative high strength retention suggested that GF/PPS can be used at elevated temperatures higher than its T_g under static loadings.

Acknowledgement

The present work was supported by the National Science Foundation of China under Grant Nos. 51271067, 11272105 and 11572101.

References

1. P. Carnevale, *Fiber-matrix Interfaces in Thermoplastic Composites* (PhD, Delft University of Technology, 2014)
2. S.Y. Du, Z.D. Guan, Acta Mater. Compos. Sin. **25**, 2 (2008)
3. D.R. Meyer, P. Carnevale, H.E.N. Bersee, Plast. Rubber Compos. Macromol. Eng. **39**, 122 (2010)
4. W. Albouy, B. Vieille, L. Taleb, Compos. Part A **49**, 165 (2013)
5. B. Vieille, J. Aucher, L. Taleb, Mater. Des. **35**, 707 (2012)
6. J.A. Grape, V. Gupta, Mech. Mater. **30**, 165 (1998)

DOI: 10.1051/matecconf/201710801009

7. D.R. Meyer, H.E.N. Bersee, A. Beukers, *49th Structures, Structural Dynamics, and Materials Conference* (Schaumburg, America, 2008)
8. L.A.L. Franco, M.L.A. Grac, F.S. Silva, Mater. Sci. Eng. A **488**, 505 (2008)
9. R.M. O'Higgins, M.A. McCarthy, C.T. McCarthy, Compos. Sci. Technol. **68**, 2770 (2008)
10. F. Pierron, B. Green, M.R. Wisnom, S.R. Hallett, Composites Part A **28**, 2321 (2007)
11. E.V. Iarve, D. Mollenhauer, R. Kim, Composites Part A **36**, 163 (2005)
12. B. Vieille, L. Taleb, Compos. Sci. Technol. **71**, 998 (2011)
13. S. Daggumati, I.D. Baere, W.V. Paepegem, J. Degrieck, J. Xu, S.V. Lomov, I. Verpoest, Compos. Sci. Technol. **74**, 20 (2013)
14. B.M. Walther, *An Investigation of the Tensile Strength and Stiffness of Unidirectional Polymer-Matrix, Carbon-Fiber Composites under the Influence of Elevated Temperatures* (Master, Virginia Polytechnic Institute and State University, 1998)
15. S. Rasool, *Fatigue of woven thermoplastic composites the effect of the fibre matrix interface* (PhD, Delft University of Technology, 2015)

 DOI: 10.1051/matecconf/201710801010

Tightening torque and washer size effects on bearing strength of fiber metal laminate bolted-joints

Shiming Zu, Zhengong Zhou, Lu Che, Yanfeng Zhang and Jiazhen Zhang

Center for Composite Materials and Structures, Harbin Institute of Technology, 150001, China

Abstract. The study investigated the effect of tightening torque and washer size on the mechanical properties of double-lap, single bolt fiber metal laminate (FML) joints. The tightening torque under the examination was 0.1Nm, 3 Nm, 5 Nm, 10 Nm and 15 Nm, and the washer size was changed by varying the value of outer diameter of washer (Dwo=11mm, Dwo=13mm, Dwo=15mm, Dwo=17mm and Dwo=19mm). A three-dimensional (3D) failure finite element model was used to predict the bearing behaviour of FML joints. The comparison between simulation and experimental results showed a good agreement in strength. The results showed that the effect of bolt tightening torque on the failure loads of bolt-loaded FML joints was obvious, and the effect of washer size on the load-bearing capacity was not sensitive. The present work provided supporting information for the design of FML joints.

1 Introduction

Composite materials have been widely used in aircraft industry due to the high strength and stiffness-to-weight ratios. Fiber metal laminate (FML) is a special kind of composite material which originated at Fokker/TU Delft in the Netherlands[1]. FML is a class of hybrid composites consisting of alternating layers of thin metal sheets and polymer matrix composite. Incorporating composite layers into a metal sheets system greatly enhances the overall performance of the FML which combines the advantages of both conventional composites and metals. In previous studies, FML has been proved to have many excellent mechanical properties such as outstanding fatigue resistance, high specific static properties, superior impact resistance, good residual and blunt notch strength, better bearing strength, fire resistance and corrosion properties [2]. Currently, glass-reinforced aluminium laminate (GLARE) has been selected for the upper fuselage skin structures of Airbus A380, as shown in Figure 1 [3].

Due to advantages in inspection, replacement and reliability, the mechanically fastened joints are of great need in aircraft structures. However, bolted and riveted joints represent potential weak points in a structure due to the highly localized stress, and their efficient design is therefore essential to the load-carrying capacity and the overall weight of the structure[4].

The bearing properties of FML joints have been experimentally researched by scholars [4-9].Wu and Slagter. [5]found that the bearing strength depends on failure mode as influenced by specimen geometry. They concluded that a minimum edge distance-to-diameter ratio (E/D) of 3 and width-to-diameter ratio (W/D) of 4 should be used to develop full bearing strength for pin-loaded FML joints. Yeh et al. [8] investigated the bolt type bearing tests on various fiber metal laminates. They found that the bearing strength and the failure mode of FML joints strongly depend on parameters such as the ratio of E/D, metal volume fraction, and fiber orientation. Frizzell et al. [9] examined the progression of damage from initiation up to ultimate failure in pin-loaded GLARE joints by using microscopy. They also found that GLARE 2 joints displayed significantly higher ultimate bearing strength and strain than GLARE 3 joints.

Figure 1. Application of GLARE laminate as the upper fuselage structures for A380 [3].

For complex composite material structure, the use of numerical techniques can be used to get the detailed results efficiently and reduce the number of tests required. Numerical simulation studies of FML joints have been conducted on pin and bolt configurations [7, 10-13]. Van

Roijen et al. [7] used a finite element model to describe the bearing behaviour of FML joints with varying edge distance radio. It was seen that the model returns a good fit with experimental data for several GLARE laminates. Frizzell et al. [11] investigated the effects of geometry on the mechanical behaviour of pin-loaded FML joints by a numerical study. A continuum damage mechanics model was used to examine the specimens designed to fail in bearing, net-tension and shear-out. The model gave a very good agreement with experimental results for the range of laminate geometries.

In spite of few works focused on the bearing properties of bolt-loaded FML joints, to the best of my knowledge, there is still no published study about the influence of washer size on the bearing properties of the FML joints. This paper aims to investigate the effect of tightening torque and washer size on bearing strength of bolt-loaded FML joints by a three-dimensional (3D) damage model.

2 Material and methods

The material used in this study is GLARE 3-3/2-0.4 fiber metal laminate, which includes three layers of 0.4mm thick 2024-T3 aluminium alloy and four layers of 0.15mm thick HS4 glass fiber/epoxy reinforced prepreg. The resin content in the prepreg was 25.3% by weight. Mechanical properties of unidirectional HS4 glass fiber/epoxy laminate used in this study are given in Table 1.

Table 1. Mechanical properties of unidirectional HS4 glass fiber/epoxy laminate.

Symol	Value	Property
E_{11}	54.6GPa	Longitudinal stiffness
E_{22}	10.5GPa	Transverse stiffness
E_{33}	10.5GPa	Out-of-plane stiffness
$\upsilon_{12}, \upsilon_{13}$	0.252	Poisson's ratio
υ_{23}	0.32	Poisson's ratio
G_{12}, G_{13}	3.5GPa	Shear modulus
G_{23}	3GPa	Shear modulus
X_t	1735MPa	Longitudinal tensile strength
X_c	1037MPa	Longitudinal compressive strength
Y_t	62.2MPa	Transverse tensile strength
Y_c	144MPa	Transverse compressive strength
S_{12}, S_{23}, S_{13}	76.1MPa	Shear strength
Z_t	50MPa	Out-of-plane tensile strength

The FML was cured according to the stacking layup by [Al/0/90/Al/90/0/Al] at 120°C under the pressure of 0.5MPa by using vacuum bag/autoclave molding technique. The measured thickness of the cured laminate is 1.8mm.

The geometry of pin-loaded bearing specimen is shown in Figure 2(a). As shown in this figure, a FML rectangular specimen of length L, width W, thickness t and edge distance E with a pin hole of diameter D is considered. Machining techniques should be used in the specimen fabrication. Testing was carried out on an INSTRON-5569 machine in tensile mode. A double shear experimental setup, as shown in Figure 2(b), was used for the pin-loaded bearing test. A 6mm diameter steel bolt was inserted into the holes of specimen and pin supports and then loaded under displacement control. The loading speed was 0.5mm/min.

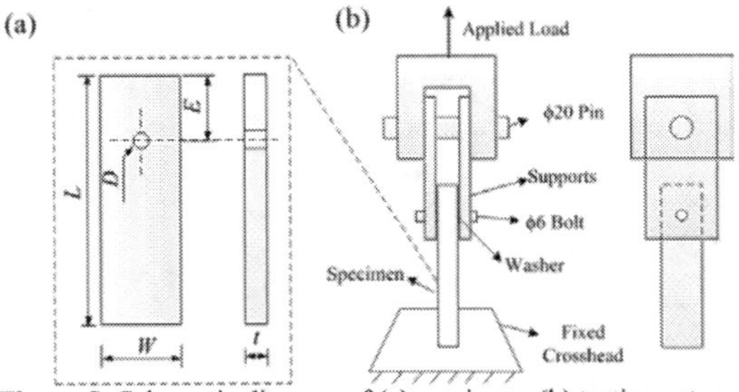

Figure 2. Schematic diagram of (a) specimen, (b) testing setup.

The joint under investigation is a double-lap, single bolt configuration, subjected to a quasi-static tensile load. The specimen geometry is based on the radios of W/D=4 and E/D=4. The outer diameter of washer is D_{wo}=13mm. The tightening torque under the examination is 0.1Nm, 3 Nm, 5 Nm, 10 Nm and 15 Nm. A 3D damage finite element model was used to investigate the effect of washer size (D_{wo}=11mm, D_{wo}=13mm, D_{wo}=15mm, D_{wo}=17mm and D_{wo}=19mm) on bearing behaviour of FML joints.

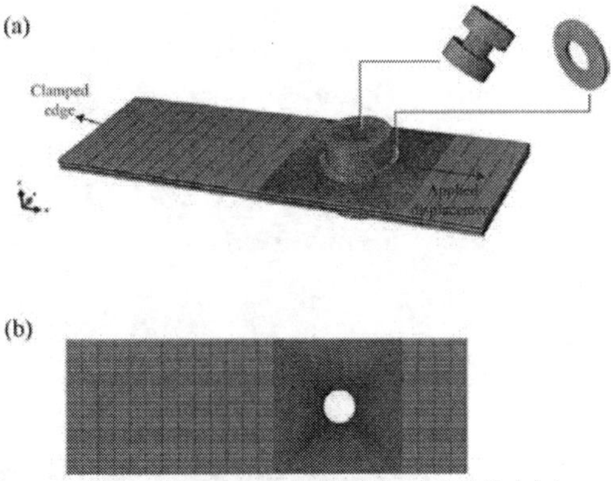

Figure 3. Finite element model of bolt-loaded FML joint.

3 Finite element model

In order to reveal the progressive damage process of fiber reinforced polymer layer in FML, a 3D damage model based on the Hashin failure criteria has been conducted by using the explicit finite element calculation software ABAQUS/Explicit.

The boundary and loading conditions of the one-quarter model are shown in Figure 3. All degrees of freedom were set to zero at the fully clamped laminate end. In order to simulate the realistic loading conditions, an incremental displacement load was applied to the pin in the positive direction. In order to reduce solution runtime, the elements were refined in the vicinity of the pin-hole and reduced further from the hole, as shown in Figure 3(b).

The bolt and washer was meshed using 8-node linear brick, reduced-integration element (C3D8R). The properties of bolt and washer were $E =$ 210 GPa and $\nu = 0.3$. The pre-tightening force of bolt was achieved by changing the temperature of washer. The expansion of washer was $1.35 \times 10^{-5}/°C$. A friction coefficient of 0.1 between bolt and FML was chosen as recommended by Frizzell et al. [11]. The aluminum was meshed using 8-node linear brick, reduced-integration elements (C3D8R). The properties of bolt were $E =$ 70.42 GPa and $\nu = 0.33$. The fiber-reinforced epoxy layer, of 0.15mm thickness, was meshed using 8-node linear brick, reduced-integration element (C3D8R).

3 Results and discussions

The comparison between simulation and experimental load-displacement curves are presented in Figure 4. The specimen geometry was based on the ratios of W/D=4 and E/D=4. The outer diameter of washer is D_{wo}=13mm and the bolt tightening torque was 5 Nm. A good agreement between numerical prediction and experimental data could be found. The failure load of joint simulated by the finite element model is 11061N. Compared with the average failure load of experimental results, 11450N, the error of simulation result is only about -3.4%.

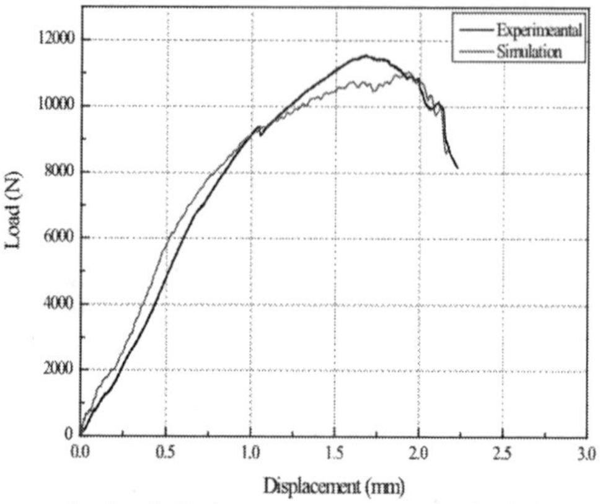

Figure 4. Load-displacement curves of simulation and experimental result (T=5 Nm).

The internal fiber failure prediction for bolt-loaded FML joints under the 5 Nm torque is shown in Figure 5. It can be seen that the internal fiber failure was localized the bearing plane, which mainly included the 0° fiber failure, 90° fiber failure, 0° matrix failure and 90° matrix failure. The deformation of the FML joint was localized very near the hole edge, and bearing failure occurred.

3.1 Effect of tightening torque

The effect of tightening torque on the bearing behaviour of double-lap, single bolt FML joints was examined by varying the tightening torque force (T=0.1 Nm, T=3 Nm, T=5 Nm, T=10 Nm and T=15 Nm).The comparison between simulation and experimental ultimate failure loads are presented in Figure 6.

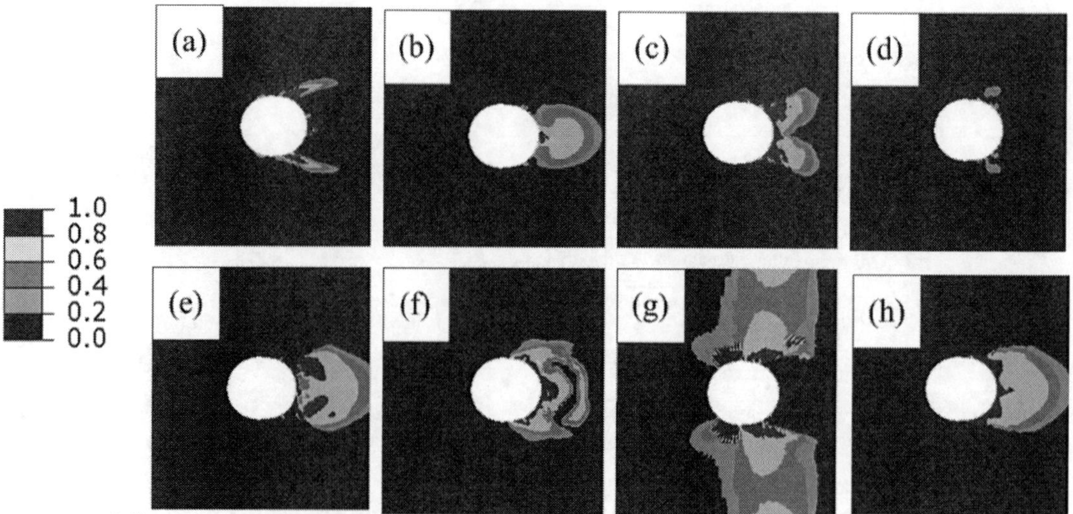

Figure 5. Internal fiber failure prediction for bolt-loaded FML joints (T=5 Nm), (a) 0° fiber tension failure,(b) 0° fiber compressive failure, (c) 90° fiber tension failure, (d) 90° fiber compressive failure, (e) 0° matrix tension failure, (f) 0° matrix tension failure, (g) 0° matrix compressive failure, (h) 90° matrix compressive failure.

DOI: 10.1051/matecconf/201710801010

It can be seen that the 3D damage finite element model give a very good agreement with experimental results for the range of bolt tightening torque. The ultimate failure loads of the FML joints increase observably with the increasing values of bolt tightening torque from 0.1Nm to 10Nm. The increase of ultimate failure loads is not directly when the values of bolt tightening torque increase from 10Nm to 15Nm. The maximum increase in failure loads occurs on T=15Nm with the rate 20%, in comparison with the results of T=0.1Nm. The effect of bolt tightening torque on the failure loads of bolt-loaded FML joints was obvious. Appropriate adding pre tightening force is the effective measures to increase the load-bearing capacity of double-lap, single bolt FML joints.

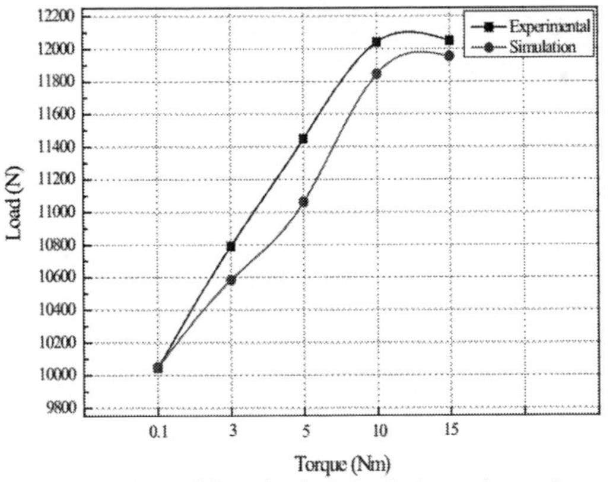

Figure 6. Ultimate failure load of simulation and experimental results.

3.2 Effect of washer size

Due to the 3D damage finite element model showed a good agreement with experimental results, the effect of washer size on the bearing behaviour of FML joints was studied by numerical technique. The change of washer size was achieved by changing the outer diameter of washer (D_{wo}=11mm, D_{wo}=13mm, D_{wo}=15mm, D_{wo}=17mm and D_{wo}=19mm). Figure 7 shows the comparison between simulation and experimental ultimate failure loads of FML joints under the T=0Nm and T=10Nm.

It can be seen that the ultimate failure load of FML joints increase to the maximum and then decrease with the increasing the outer diameter of washer. The Figure 7(a) shows that the maximum increase in failure loads of FML joints (T=0Nm) occurs on D_{wo}=15mm with the rate of 2.5% in comparison with the results of D_{wo}=11mm. When the bolt tightening torque force is T=10Nm, the maximum increase in failure loads occurs on D_{wo}=17mm with the rate of 2.6% in comparison with the results of D_{wo}=11mm. Due to the ultimate failure loads changed little, it can be concluded that the washer size is not sensitive to the bearing strength of double-lap, single bolt FML joints.

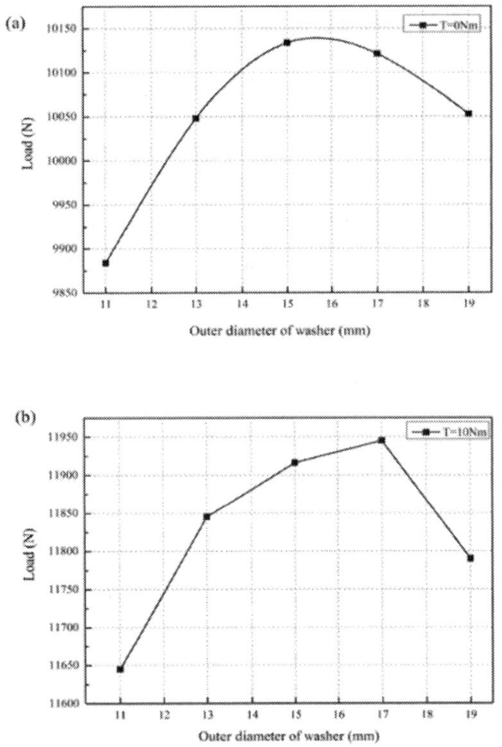

Figure 7. Ultimate failure load of simulation and experimental results (a) T=0 Nm, (b)T=10Nm.

4 Conclusions

The present work has investigated the effect of temperature and geometry parameters on the mechanical properties of pin-loaded FML joints. A 3D failure finite element model was used to predict the mechanical behaviour of double-lap, single bolt FML joints with varying the bolt tightening torque force and outer diameter of washer. From the results of this study, the conclusions can be drawn as follows:

(1) The 3D damage finite element model had a good agreement with the experimental results.

(2) The effect of bolt tightening torque on the failure loads of bolt-loaded FML joints was obvious. The ultimate failure loads of the FML joints increase observably with the increasing values of bolt tightening torque from 0.1Nm to 10Nm. The increase of ultimate failure loads was not directly when the values of bolt tightening torque increasing from 10Nm to 15Nm. The maximum increase in failure loads occurs on T=15Nm with the rate 20%, in comparison with the results of T=0.1Nm.

(3) The effect of washer size on the failure loads of bolt-loaded FML joints was not sensitive. The ultimate failure load of FML joints increased to the maximum and then decreased with the increasing the outer diameter of washer. The maximum increase in failure loads of FML joints with the bolt tightening torque force (T=0Nm and T=10Nm) was only 2.5% and 2.6%.

Acknowledgements

The present research is supported by the National Science Foundation of China under grand Nos. 11272105, 11572101 and 51271067.

References

1. A. Vlot, J. W. Gunnink, *Fibre metal laminates: an introduction* (Springer, 2001).
2. G. Wu, J. Yang, JOM-US, **57**, 72 (2005).
3. A. Vlot, *History of the development of a new aircraft material* (Springer, 2001).
4. R. M. Frizzell C. T. McCarthy, M. A. McCarthy, Composites Part B: Engineering **39**, 907 (2008).
5. H. F. Wu, W. J. Slagter, J. Aircraft. **31**, 936 (1994).
6. G. Caprino, A. Squillace, G. Giorleo, L. Nele, L. Rossi, Composites Part A: Applied Science and Manufacturing **36**, 1307 (2005).
7. R. Van Rooijen, J. Sinke, T. J. De Vries, S. Van Der Zwaag, J. Compos. Mater. **40**, 5 (2006).
8. P. Yeh, P. Chang, J. Wang, J. Yang, P. H. Wu, M. C. Liu, Compos. Struct. **94**, 3160 (2012).
9. R. M. Frizzell, C. T. McCarthy, M. A. McCarthy, Compos. Sci. Technol. **68**, 3314 (2008).
10. P. P. Krimbalis, C. Poon, Z. Fawaz, K. Behdinan, J. Compos. Mater. **41**, 1137 (2007).
11. R. M. Frizzell, C. T. McCarthy, M. A. McCarthy, Compos. Struct. **93**, 1877 (2011).
12. R. M. Frizzell, C. T. McCarthy, M. A. McCarthy, Compos. Sci. Technol. **71**, 1225 (2011).
13. G. Kolks, K. I. Tserpes, Composites Part A: Applied Science and Manufacturing. **56**, 51 (2014).

Experimental Analyses for The Mechanical Behavior of Pressed All-Ceramic Molar Crowns with Anatomical Design

Liliana Porojan[1], Sorin Porojan[2], Lucian Rusu[3], Adrian Boloș[2] and Cristina Savencu[1]

[1]*Department of Dental Prostheses Technology, School of Dentistry, V. Babeş University of Medicine and Pharmacy, 9 Revolutiei 1989 Blv., 300041 Timişoara, Romania*
[2]*Department of Oral Rehabilitation, School of Dentistry, V. Babeş University of Medicine and Pharmacy, 9 Revolutiei 1989 Blv., 300041 Timişoara, Romania*
[3]*Department Mechanics and Vibrations, Faculty of Mechanical Engineering, Politehnica University, 1 Mihai Viteazu Street, 300222 Timisoara, Romania*

Abstract. Ceramic restorations show considerable variation in strength and structural reliability regarding to the type of material, and design characteristics. The fracture of ceramics occurs with little or no plastic deformation, with cracks propagated in an unstable manner under applied tensile stresses. The aim of the study was to assess experimental analyses of pressed monolithic ceramic crowns with anatomical design used in the posterior areas in order to understand their mechanical behavior before following their clinical use. Experiments were conducted on a complete molar crown preparation. Experiments show different modes of fracture for the tested samples. Digital images from the fractured pieces of the crowns were used to verify the fragments in all cases final fracture occurred by splitting into two and often more parts. The graphically representation of the displacement depending on the load highlights a series of peaks that can be correlated with cracks occurred in crowns. The development of well-designed mechanical experiments could be useful to help to predict clinical survival of these new all-ceramic restorative techniques and materials. Because failure is often accompanied by complete cracking of the crowns, preliminary research should represents a compulsory goal.

1 Introduction

Despite progress of ceramic materials and their technologies made in the last decade, all-ceramic restorations still have lower clinical longevity compared to metal-ceramic restorations, due to their brittle nature [1]. All-ceramic restorations are amongst the most aesthetically prosthetic restorations because they exhibit life-like appearance due to the lack of metallic substructure.

The fracture of ceramics occurs with little or no plastic deformation, with cracks propagated in an unstable manner under applied tensile stresses. Ceramic restorations show considerable variation in strength and structural reliability regarding to the type of material, and design characteristics [2,3].

Different microstructures, depending on the presence and volume of the crystalline phase, mean particle size and distribution throughout the glassy matrix affect crack propagation and also the mechanical behavior [4]. It is also important to mention that other factors such the thickness and shape of restoration, flaws inserted during the manufacturing and the presence of residual stresses can also influence clinical longevity, making a direct correlation between the parameters of crack growth and an estimate of lifetime for a ceramic restoration more difficult [5].

Due to the complex geometry of the tooth crown and contact during mastication, fracture evaluation of ceramic all-ceramic crowns have often been simplified in different studies as a flat layered structure. Other studies considered the influence of surface curvature [6,7].

Recently monolithic all-ceramic crowns are increasingly favored over traditional bilayered crowns, which are associated with complex fabrication, presence of residual stresses, and chipping of the veneers [8-11].

Sintered feldspathic and leucite reinforced ceramic are used for the fabrication of metal-free restorations using traditional laboratory based fabrication techniques. With the development of hot-pressed and CAD/CAM technologies, different ceramic materials can now be used for aesthetic applications.

Ceramics are brittle and susceptible to fatigue fracture in repetitive function. Although occlusal loading is compressive, some tensile stresses in the restorations are inevitable. Cracks tend to follow paths where tensile stresses are greatest. Fracture during static load differ from those resulted after an extended period of lower loads [12, 13].

Even their many advantages, glass ceramics are brittle materials. Crack initiation and propagation can result in compromise of the restorations. The compromised structural integrity indicated by a significant reduction in fracture strength will be reflected in a poor clinical performance. Therefore the clinical significance of these results is to give an indication of the strength of hot-pressed ceramic crowns to be used in the molar areas as monolithic restorations.

The aim of the study was to assess experimental analyses of pressed monolithic ceramic crowns with anatomical design used in the posterior areas in order to understand their mechanical behavior before their clinical use.

2 Materials and method

Experiments were conducted on a complete crown preparation made on a first maxillary molar from a typodont model. The axial undercuts were eliminated by a reduction of 1.5 mm, while the occlusal surface was reduced by 1.5-2 mm. A 1 mm deep chamfer finish line and a convergence angle of 6 degree were chosen. Finally all sharp angles were rounded.

This preparation was replicated and alloy dies were obtained by casting, using the lost-wax technique. The dies were scanned using Cercon Eye system and a dedicated CAD (computer aided design) software Cercon Art (Degudent, Hanau, Germany) was used for the design of the full anatomically crowns.

In order to obtain identical samples, eight wax-ups were milled in Cercon base cast material. These were invested, and specific steps were covered according to the manufacturer's instructions in order to obtain heat pressed ceramic crowns using Cergo pressed ceramics (Degudent, Hanau, Germany). The restorations thickness was verified by periodic measurements at several locations. The crowns were glazed and conventionally cemented on the metallic dies.

A digital compressive load was applied to the occlusal surface of each crown during setting to ensure complete seating and maintained for 24 hours before loading. The specimens were mounted into an Instron 8874 (Instron, Norwood, MA, USA) testing device and subjected to failure testing by compressive loading.

Each crown was positioned under a stainless steel ball with a diameter of 6 mm fixed to the upper crosshead of a universal testing machine. Due to the radius of the contact sphere, the actual contact positions were located adjacent to the central fossa. A thin rubber foil of 0.2 mm was inserted between the tested crowns and the antagonists in order to reduce peak stresses at the contact points.

A compressive load was applied at a crosshead speed of 2 mm/min, and failure was recorded. The vertical load was aligned with the tooth's axis.

The initial crack observed in the monolithic materials occurred at the same time as the ultimate failure of the crown, so the crack origin could not be exactly determined visually, but in all cases it started from the contact area with the stainless sphere (Fig. 1).

Digital images from the fractured pieces of the crowns were used to verify the fragments in all cases final fracture occurred by splitting into two, three and often more parts (Fig. 2).

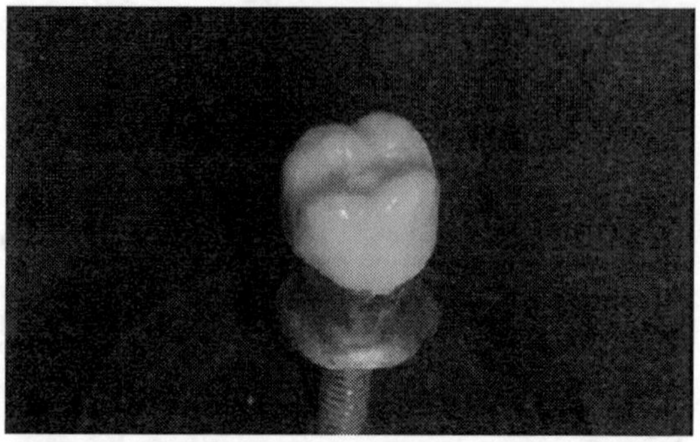

Figure 1. The propagation of cracks in the crown under occlusal loading.

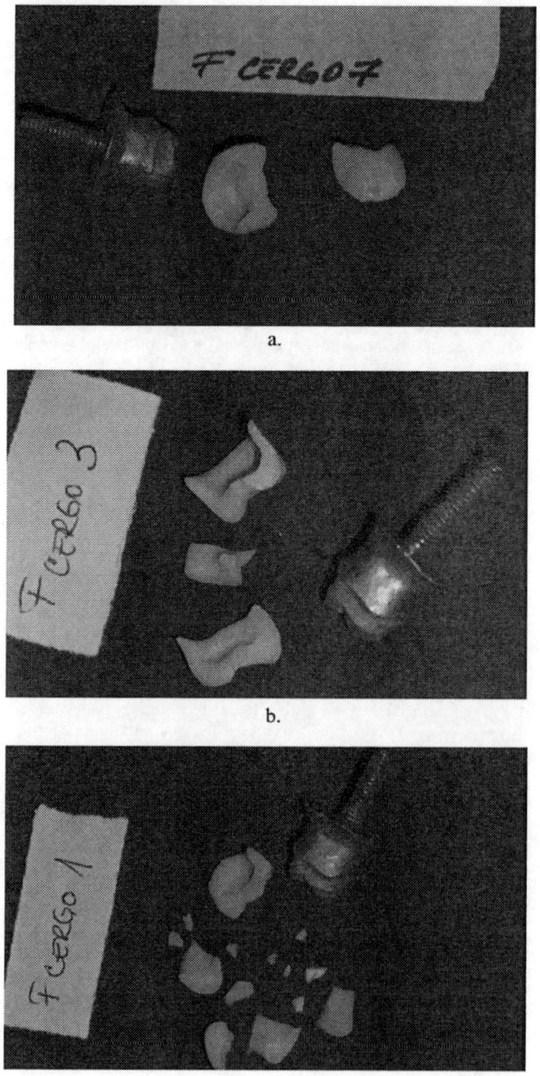

a.

b.

c.

Figure 2. Fracture components of the tested samples: a. sample 7, sample 3, c. sample 1.

DOI: 10.1051/matecconf/201710802001

3 Results and discussions

Experiments show different modes of fracture for the tested samples. The graphically representation of the displacement depending on the load highlights a series of peaks that can be correlated with cracks occurred in crowns. An increased number of peaks are associated with multiple fragments of the fractured samples (Fig. 3).

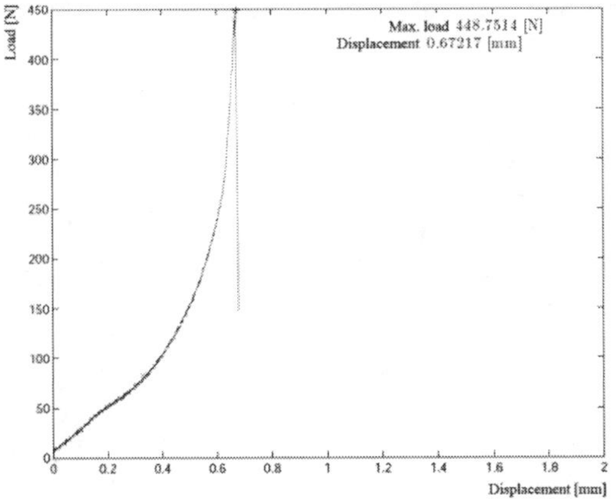

a.

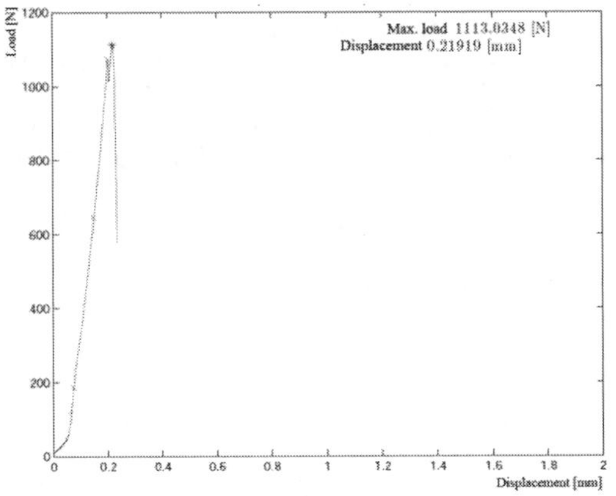

b.

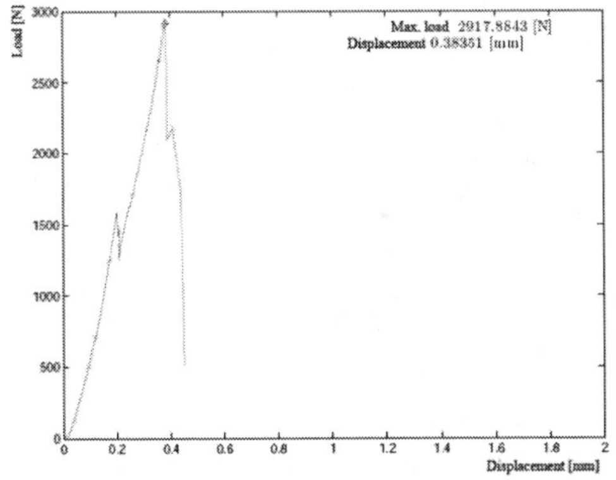

c.

Figure 3. Graphical representation of peaks recorded during mechanical compression tests: a. sample 7, sample 3, sample 1.

Critical radial cracks initiate from the occlusal contact area with the testing indenter. These cracks ramify and multiple fragments result.

Studies have shown that fracture initiation begins under the indenter tip when a hard spherical antagonist is used for fracture strength testing [14,15].

The mean value of the compression load was 1529.54 N, and of the displacements 0.36 mm necessary to cause full splits in each monolithic crown system (Fig. 4, 5). Previous mean values of fracture strength in similar studies have varied: 3147 N for lithium disilicate (1.5 mm thickness) [16], 743 N for 2-mm-thick lithium disilicate [17], 1912 N for monolithic lithium disilicate (1.5 mm thickness) [18], above 1400 N of 1.5-mm-thick monolithic lithium disilicate crowns [19].

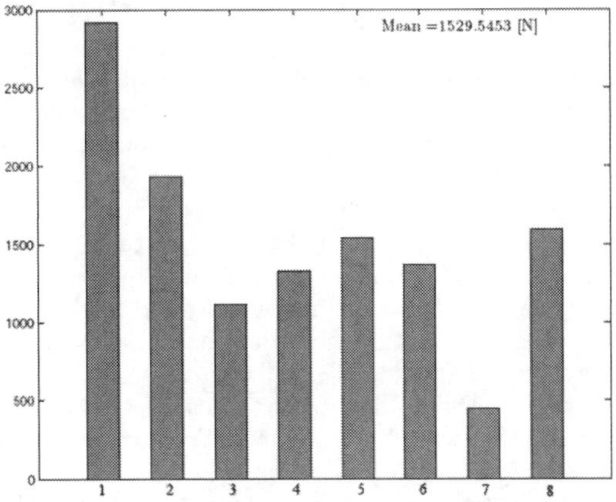

Figure 4. Graphical representation of the maximal compression loads.

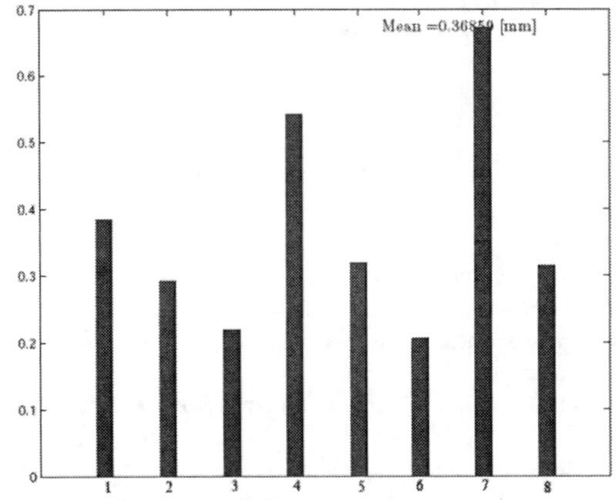

Figure 5. Graphical representation of the maximal displacements.

In order to minimize clinical failure, restorations should be fabricated with consideration of their constituent material properties. The flexural strength of lithium disilicate is 365 MPa [20]. Based on these values, manufacturers recommend tooth reduction for posterior monolithic restorations be 1.5 mm for lithium disilicate [19]. Different studies evaluated failure of glass ceramic crowns of thicknesses between 1 and 2 mm. Crowns with 2-mm occlusal thickness sustained 1.7 times more cycles and 1.5-mm occlusal thickness, 1.5 times more cycles than 1-mm thick crowns [21].

Ceramic materials are known for their relatively high fracture resistance and improved aesthetics, but brittleness remains a concern [22]. Failures are often associated with complete cracking of the crown. Regarding the forces wich lead to failure, it is known that the maximum bite force for adults may reach 700–800N [23]. The common occlusal load in oral conditions ranges below 400 N, ceramic crowns rarely undergo rupture due to fracture strength as a result of function in the initial service time. However, rupture could happen due to the accumulation of fatigue stress [24]. Therefore further mechanical analyses simulating the fatigue behavior of this kind of restorations are requires for a more natural reproduction of oral conditions. The stressed area of the material is a key factor for the failure analysis, and numerical simulations may play an important role in the understanding of the behavior of all-ceramic restorations.

Even their many advantages, glass ceramics are brittle materials, and crack initiation and propagation can result in compromise of the restorations. The structural integrity indicated by a significant reduction in fracture strength will be reflected in a poor clinical performance. Therefore the clinical significance of these results is to give an indication of the strength of hot-pressed ceramic crowns to be used in the molar areas as monolithic restorations. Further studies are required to assess the durability of such restorations by different experimental methods, before clinical use [25].

New materials are often introduced into the market and especially in the current practice, without a basic understanding of their clinical performance because long term controlled clinical trials are required, which are both time-consuming and expensive [26].

4 Conclusion

Glass ceramics are brittle materials, and crack initiation and propagation can result in compromise of the restorations during functions.

The development of well-designed mechanical experiments could be useful to help to predict clinical survival of these new all-ceramic restorative techniques and materials.

The experiments have to be conducted and interpreted reported to the brittle behavior of ceramic systems.

Because failure is often accompanied by complete cracking of the crowns, preliminary research should represents a compulsory goal.

Acknowledgment

This work was supported by a grant of the Romanian National Authority for Scientific Research and Innovation CNCS-UEFISCDI, project number PN-II-RU-TE-2014-4-0476.

References

1. F.J. Burke, P.S. Lucarotti, J Dent. **37**,12–24, (2009).
2. D.J.Green, Cambridge: University Press; (1998).
3. J.E.Ritter, Dent Mater. **11**, 142–6, (1995).
4. H.N. Yoshimura, P.F. Cesar, W.G. Miranda, C.C. Gonzaga, C.Y. Okada and H. Goldenstein, J Am Ceram Soc. **88**, 1680–3, (2005).
5. C.C. Gonzaga, P.F. Cesar, W.G. Miranda Jr and H.N. Yoshimura, Dent Mater. **27**(4), 394-406, (2011).
6. T. Qasim, M.B. Bush, X. Hu and B.R. Lawn, J Biomed Mater Res B. **73**, 179–185, (2005).
7. T. Qasim, C. Ford, M.B. Bush, X. Hu, and B.R. Lawn, J Biomed Mater Res B. **76**, 334–339, (2006).
8. G.J.Christensen, J Am Dent Assoc. **130**, 409–11, (1999).
9. Y. Zhang, I. Sailer and B.R. Lawn, J Dentistry. **41**,1135–47, (2013).
10. Y. Zhang, JJ-W. Lee, R. Srikanth and B R. Lawn, Dent Mater. **29**, 1201–8, (2013).
11. G.J.Christens, JAm Dent Assoc. **145**, 192–4, (2014).
12. Y. Zhang, I. Sailer, B.R..Lawn, J Dent. **41**(12):1135-47 (2013).
13. J.R..Kelly, Dental Clinics of North America. **48**:513–530 (2004).
14. E.D. Rekow, G. Zhang, V. Thompson, J.W. Kim, P. Coehlo and Y. Zhang, J Biomed Mater Res B Appl Biomater. **88**, 436-46, (2009).
15. J.O. Clausen, M. Abou Tara, M. Kern, Dent Mater **26**:533-8 (2010).
16. K. Nakamura, A. Harada, R. Inagaki, T. Kanno, Y. Niwano, P. Milleding, U.Örtengren, Acta Odontol Scand. **1**, 1-7, (2015).
17. M. Dhima, D.A. Assad, J.E. Volz, K.N. An, L.J. Berglund, A.B. Carr and T.J.Salinas, J Prosthodont. **22**, 256-60, (2013).
18. T. Sun, S. Zhou, R. Lai, R. Liu, S. Ma, Z. Zhou, S. Longquan, J Mech Behav Biomed Mater. **35**, 93-101, (2014).
19. B. Baladhandayutham, N.C. Lawson, J.O. Burgess, J Prosthet Dent. **114**(2), 266-71, (2015).
20. W.S. Lin, C. Ercoli, C. Feng, D.Morton, **21**, 353-62, (2012).
21. M. Dhima, A.B. Carr, T.J. Salinas, C. Lohse, L. Berglund, K.A. Nan, J Prosthodont. **23**, 353-7, (2014).
22. Y. Zhang, Z. Mai, A. Barani, M. Bush, B.Lawn, Dent Mater. **32**(3), 442-9, (2016).
23. K. Nishigawa, E. Bando, M. Nakano, J Oral Rehabil **28**, 485–491, (2001).
24. O. El-Mowafy, J.F. Brochu, "Longevity and clinical performance of IPSEmpress ceramic restorations—a literature review," J Canad Dent Assoc B, 233–237, (2002).
25. L. Porojan, S. Porojan, A.Cernescu, C. Savencu, Int J Syst Appl, Eng & Dev, **10**, 266-269, (2016).
26. S. Nasrin, N. Katsube, R.R. Seghi and S.I. Rokhlin, J Biomech. **10**, pii:S0021-9290(16)30550-4, (2016).

DOI: 10.1051/matecconf/201710802002

Utilization of slaughter plant waste proteins in gypsum retarder: A case study

Chen Xiong-mu [1], Zhang Meng-meng[1], Li Ling-xiao[2] and Zhao Feng-qing[1,2]

[1]*Department of Chemical Engineering, Hebei University of Science & Technology, Shijiazhuang, 050018, China*
[2]*Hebei Engineering Research Center of Solid Waste Utilization, Shijiazhuang, 050000, China*

Abstract. Slaughter plant waste proteins contain animal hair, bones, nails etc. Improper treatment will pollute the environment and result in waste of resources. A process is proposed to prepare gypsum retarder from such waste, firstly through hydrolysis at acidic condition, and then neutralized with alkalis. The optimal conditions for acid hydrolysis are obtained: concentration of HCL 4.8 mol/L, reaction temperature 98℃, reaction time 7 h. After neutralization (to pH=7), the product can be used as gypsum retarder (retarder A). Compared with commonly used retarder, citric acid, the mechanical strength loss of gypsum paste containing retarder A is much lower. It is a cost-effective approach for safe utilization of slaughter plant waste proteins.

1 Introduction

Slaughter plant waste proteins include animal hair, bone and nail, which contain a lot of proteins. Most of these by-products of meat industry are not directly edible and have less appeal in fresh and frozen market. At present, it is mainly used for producing animal feed or feed additives, pesticides, fertilizers and pharmaceuticals. But there is a drawback for these: the toxic substances could be transmitted along the food chain, and would be a potential threat to human health.

Considering that these wastes contain plenty of proteins, they can be used to produce gypsum retarder, which not only reduce the risk to the health of human and animals, but also have economic benefit [1,2]. Much work has been done in this field these years and the hydrolysis of proteins is often the key to prepare gypsum retarder [3-5]. The final hydrolysis products are mostly amino acids. The current methods of preparing amino acids by hydrolysis of protein include acid hydrolysis [6], alkali hydrolysis [7], enzymatic hydrolysis, high temperature hydrolysis and high pressure hydrolysis. They all have their own advantages and disadvantages. However, using slaughter plant waste proteins as material to prepare gypsum retarder has not yet been reported.

The present study was undertaken to treat slaughter plant waste proteins firstly by acid hydrolysis, then alkali neutralized to produce a gypsum retarder, investigating the effect on setting time.

2 Experimental

2.1 Materials

Slaughter plant waste proteins were bought on market; concentrated hydrochloric acid, formic acid and caustic soda are analytical reagents. The FGD gypsum used in this work was provided by Shijiazhuang Xibaipo Power Plant with the composition: SO_3-49.27%, CaO-33.42%, SiO_2-0.42%, MgO-1.08%, Fe_2O_3-0.32%, Al_2O_3-0.82%. The properties of the calcined hemihydrate gypsum (Prepared by heating the FGD gypsum at 160-180 ℃ for 1.5 hr.) are shown in Table 1.

Table 1 Physical and mechanical properties of gypsum

Setting tine [min]		Bending strength [MPa]	Compressive strength [MPa]	Standard consistency
Initial setting time	Final setting time			
9	12	6.3	20.6	0.57

2.2 Methods

Slaughter plant waste protein is first treated to remove impurities, then mixed with acid solution according to the desired proportion, through hydrolysis, to obtain compound amino acid solution. Add alkali and adjust pH=7, to obtain final product, gypsum retarder A. See Figure 1..

Six prismatic 40×40×160 mm specimens were made from each formulation. The normal consistency and setting time of hemihydrate gypsum were determined as per GB/T17669.3-1999. The specimens were cast at normal consistency, cured for 3 days, then dried at 60 °C

 DOI: 10.1051/matecconf/201710802002

to constant mass, and tested for compressive and flexural strength values.

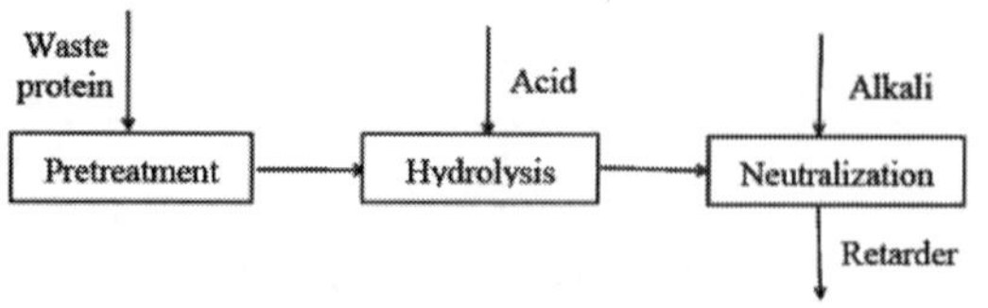

Figure 1. Process of gypsum retarder from waste protein

3 Results and discussion

3.1 Selection of acid for hydration

Based investigative experiment, hydrochloric acid, mixture of hydrochloric acid and folic acid were selected for hydration of Slaughter plant waste proteins, with various dilution ratio, time and temperature. See Figure 2.

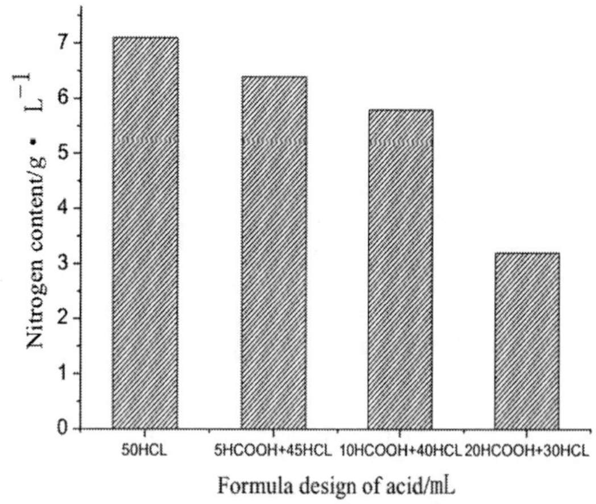

Figure 2. Amino acid nitrogen content of different cases

The results show that the content of formic acid in mixed acid has negative effect on amino acid nitrogen content in hydration product. Pure hydrochloric acid behaves best with the amino acid nitrogen content of 7.1 g/L.

3.2 Concentration of HCL

Initial experiments proved that the concentration of acid has obvious effect on the dissolving rate of waste proteins. It is necessary to find the optimal concentration of hydrochloric acid. Thus the effects of the concentration of hydrochloric acid on the maximum dissolved raw materials and the corresponding amino acid nitrogen content were investigated under the condition of reaction temperature 95 ℃ and reaction time 4 h. See Figure 3.

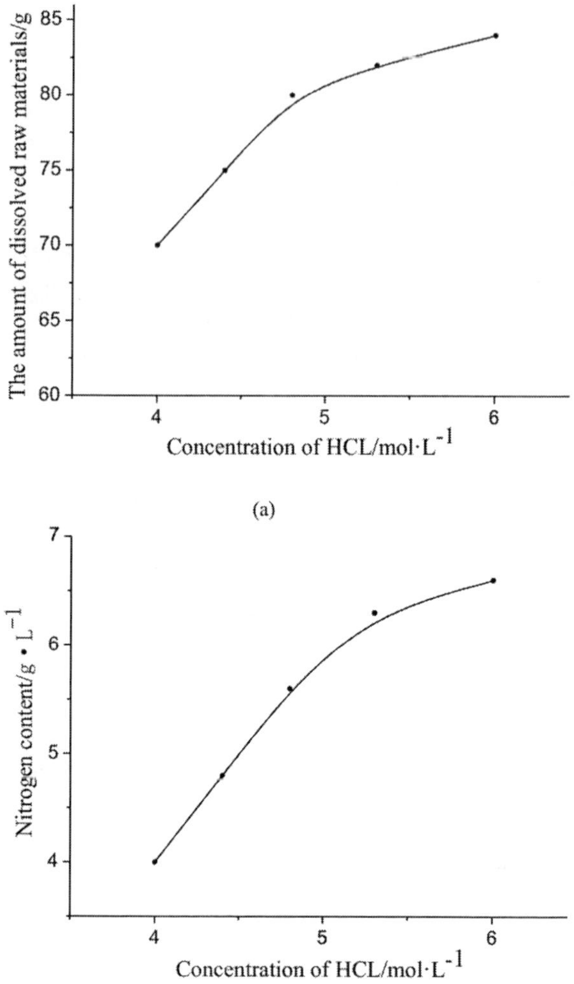

(b)

Figure 3. Effect of concentration of HCL on dissolved amount of raw material l(a) and nitrogen content (b)

It can be known from Figure 3(a) that the amount of dissolved raw material rapidly increases with the increase of the concentration of HCL. When the concentration of hydrochloric acid increases, the amount of the dissolved raw materials increases steadily. Figure 3(b) shows that with the increase of HCL concentration, the content of amino nitrogen also increases. Higher concentration of HCL will consume much more alkali in the later neutralization process. Therefore, the proper concentration of HCL is 4.8 mol/L.

3.3 Reaction temperature

With the concentration of HCL 4.8 mol/L and reaction time of 4 h, reaction temperature as control factor, the amount of dissolved raw material and the amino acid nitrogen content were measured.

It can be known from Figure 4(a) that the amount of dissolved raw material is increased with the increase of temperature. With the increase of the reaction

DOI: 10.1051/matecconf/201710802002

temperature, the molecular activity was improved, which promote the catalytic activity of HCL. When reaction temperature is above 98 ℃, the amount of dissolved raw material tends to be gentle. Figure 4(b) shows that with the increase of temperature, the content of amino nitrogen increases. When the temperature is above 98 ℃, part of the amino acid decompose, accompanied by unpleasant odours. Therefore, the optimal acid hydrolysis temperature is 98 ℃.

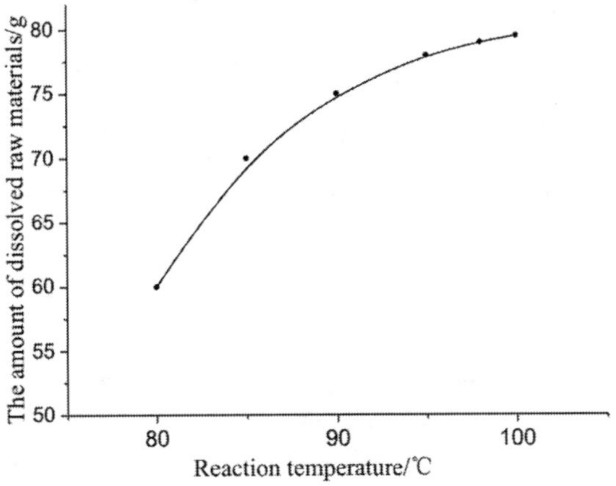

(a)

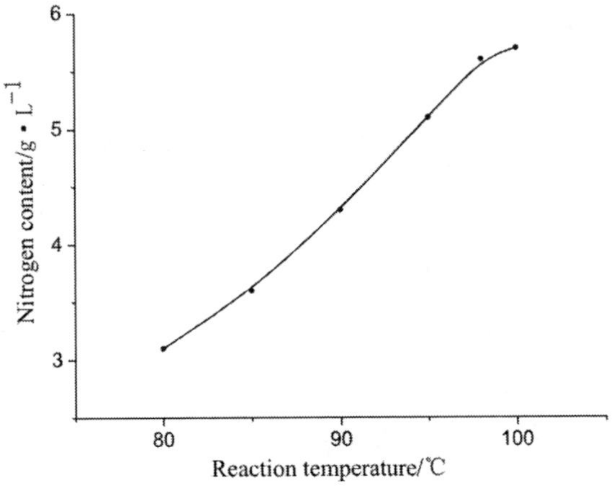

(b)

Figure 4. Effect of temperature on the dissolved amount of raw material (a) and nitrogen content (b)

3.4 Hydrolysis time

With mass ratio of slaughter plant waste protein and HCL solution 1:2, concentration of HCL 4.8 mol/L, reaction temperature 98℃, the amino acid type nitrogen content was measured. The results are shown in Figure 5.

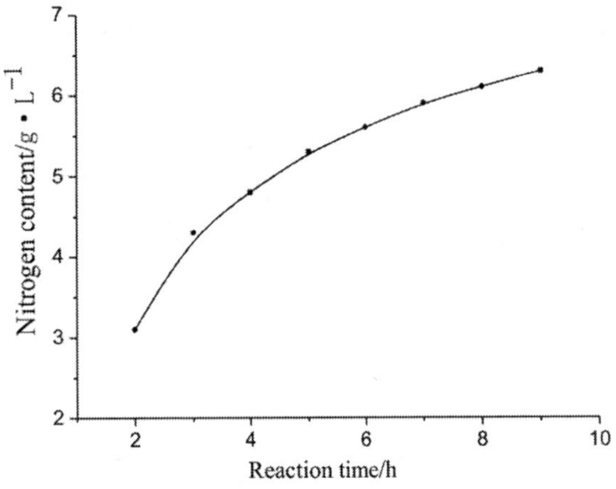

Figure 5. Content of amino nitrogen vs. reacton time

The result show that the content of amino acid type nitrogen gradually increases with the increase in reaction time. Considering that longer hydrolysis time will result in loss in efficiency and increase in cost. The suitable reaction time is 7 h.

In summary, the optimal conditions for acid hydrolysis of slaughter plant waste protein were obtained: HCL concentration 4.8 mol/L, reaction time 7 h, and the reaction temperature 98℃.

3.5 Application in gypsum paste

The compound amino acid solution was mixed with caustic soda, adjusting pH to 7, to obtain gypsum retarder A. The set-retarding performance of retarder A was investigated. The results are shown in Figure 6. We can see that retarder A has significant retarding effect for gypsum paste. When the dosage is 1.25%, the initial and final setting time was 78 min. and 121 min. respectively, which can meet the ordinary uses as gypsum retarder. If needed, the final retarding time of gypsum paste can be longer than 180 min. with the dosage above 1.5%.

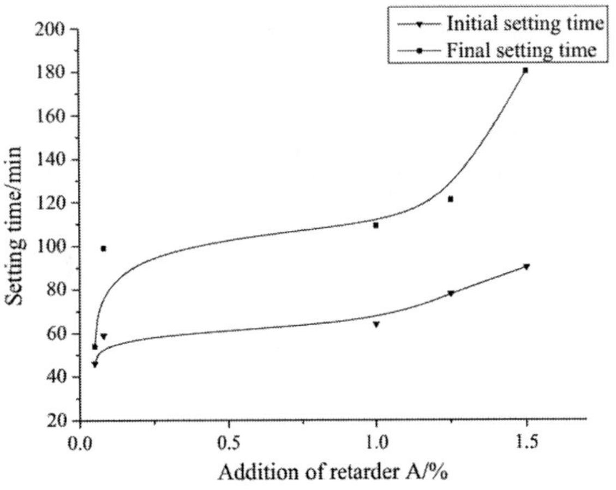

Figure 6. Effect of retarder A on setting time of FGD gypsum

With the roughly same controlled initial setting time of 80 min., the mechanical strength loss values of gypsum paste using retarder A and citric acid were measured. The results are shown in Table 2.

Table 2 Effect of different retarder on mechanical strength

Retarder	Mechanical strength of gypsum paste gypsum paste（MPa）	
	Bending strength	Compressive strength
0	6.3	20.6
Retarder A	3.3	9.7
Citric acid	2.7	7.5

The results show that all the set retarders have negative effect on the mechanical strength of gypsum paste, while prolonging the setting time. However, different retarder behaves different. It is clear that the mechanical strength loss of the gypsum paste containing citric acid is much higher than that of gypsum paste containing retarder A.

4 Retarding mechanism

The hydration process of hemi-gypsum in paste is mainly composed of three steps:

- Solving of gypsum in water, producing Ca^{2+}, SO_4^{2-}
- Formation of dihydrate gypsum crystal nucleus
- Growth of gypsum crystal and formation of binding.

As for the retarding mechanism, it may attribute to the complexation of active groups in retarder A with calcium in the paste. Compound amino acid is adsorbed on the surface of the gypsum particle by this complexation, forming protective colloid layer on the gypsum crystal nucleus, wrapped up the particle. Thus the growth of gypsum crystal was inhibited to some extent, delaying the setting time of gypsum paste. See Figure 7 [1].

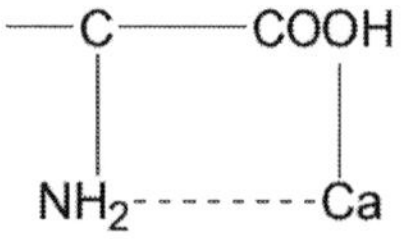

Figure 7. Complexation of retarder A with Ca^{2+}

5 Summary

Slaughter plant waste proteins (animal hair, bone and nail etc.) can be used to prepare gypsum retarder by acid hydrolysis and then treated with caustic soda. It is a safe approach for utilization of slaughter plant waste protein.

The proper conditions for acid hydrolysis were: concentration of HCL 4.8mol/L, reaction temperature of 98 ℃, reaction time 7 h. The compound amino acid was then neutralized (to pH=7) with caustic soda, to obtain gypsum retarder A. When it is used as gypsum retarder, the proper dosage is 1.5% in the mixture. It also shows that, when needed, the final retarding time of gypsum paste can be longer than 180 min. with the dosage above 1.5%.

Compared with commonly used retarder citric acid, the mechanical strength loss of gypsum paste containing retarder A is much lower. The mechanical loss of different cases is probably because of different effects on the gypsum crystal growth.

References

1. F.Q. Zhao, X.Q Ren, Q.Q. Hu, Study on Preparation of desulfurization gypsum retarder by penicillin waste mycelium, Environmental science and technology. **2**, 134-137 (2014)
2. H.J. Yang, Y. Mu, X.Q. Ren, et al, Study on Preparation of building gypsum retarder by waste mycelium, Journal of Hebei University of Science and Technology. **34**, 10-15 (2013)
3. J. Liang, Z Li, X.Yan, et al, Study on Gypsum Plasters with Modified Waste Mycelium as Retarder, Proceedings of International Symposium on Computers and Informatics. **13**, 2251-2258 (2015)
4. M.M. Chen, M.M. Zhang, F.Q. Zhao, Preparation of Gypsum Retarding Material and Activated Carbon from Waste Mycelium, Key Engineering Materials. **703**, 297-300 (2016)
5. M.M. Zhang, M.M. Chen, et al, Improving Water proof Property of Gypsum Block with Organic-Inorganic Compound Materials, Proceedings of 2nd International Conference on Material Engineering and Application, 707-711 (2015)
6. G.F. Ren, Progress in production and application of amino acid, J. Journal of amino acids. **3** (1987) 19-28.
7. Y.C. Wang, Study on hydrolysis of keratin, J. Chemical World. **9**, 422-423 (1988)

Direct Precipitation Method of Nano-CuO

Mingyu Li, Kefu Ding, Zhihao Li, Zeng Qingxuan

[1]*State Key Laboratory of Explosion Science and Technology, Beijing Institute of Technology, Beijing 100081, P. R. China*
**Corresponding author: Li Mingyu. E-mail: mingyuli@163.com*

Abstract: Different reaction conditions were studied through orthogonal experiment and single factor experiment. Using Cu(NO3)2 as the reactant, water-alcohol solution (H2O:CH2CH3OH was 3:1) as solvent at 20℃ and reaction for 45min could get nano-CuO with good dispersion and uniform size. Different kinds of characterization method were used to characterize the resultant. The average diameter of nano-CuO was 20.38nm after calculated using some relevant formula, the surface area was about 21.988 m2/g, the pore diameter was about 3.079 nm and the pore volume was 1.349×10-1 cc/g . So the method is a feasible, low cost way to industry..

1. Introduction

With small structure units, high specific surface area and unique mechanical, optical, magnetic properties, nanocrystalline materials have become a focus of common attention in the 21st century [1-4]. Among nano materials, nano-CuO has played an important role in catalytic technology, bio-medicine technique, fine chemical industry and other fields and it is likely to have a good future [5, 6]. However, bulk particle is easy to form during the preparation and application losing characteristics of nano materials [7, 8], so the preparation process is very essential. Commonly, solid state reaction method, hydrothermal method, alcohol-thermal method, precipitation method are the methods used in industry and laboratory, recently other methods [9-12] are found such as micro-emulsion method and laser heating gas-evaporation method. Compared with other methods, direct precipitation method has the advantages of simple preparation and low cost, so it can be widely used in industrial production [11].

In the work, different kinds of reactants, temperature, reaction time and ratio of water's volume (V_{water}) and alcohols' volume ($V_{R\text{-}OH}$) were studied through orthogonal experiment and single factor experiment, then the nano-CuO generated was characterized using X-ray diffraction (XRD), fourier transform infrared (FTIR), thermogravimetric analysis (TG), scanning electron microscopy (SEM) and particle surface area measurements to find the best preparation condition. What's more, the diameter, surface area, pore diameter and the pore volume of the CuO generated under best condition were measured and calculated.

2. Experiment

2.1 Materials

Materials included $CuSO_4{\cdot}5H_2O$, C_2H_5OH produced by Beijing chemical plants; $CuCl_2{\cdot}2H_2O$, $Cu(NO_3)_2{\cdot}3H_2O$, Na_2CO_3 produced by Beijing Yili Fine Chemical Co; $C_4H_6CuO_4{\cdot}H_2O$ by Sinopharm Chemical Reagent Beijing Co. All of the chemicals and reagents used were analytical grade and were used without further purification.

2.2 Equipments

Equipments included an AR2140 electronic balance provided by Ohaus Co. (Shanghai, People's Republic of China), a 81-2 isothermal magnetic agitator provided by Cile Instruments (Shanghai, People's Republic of China), a 80-2 centrifugal precipitator provided by Xinkang Yiliaoqiye Co.(Kangyan, Jiangsu Province, People's Republic of China), a DZF-6020 vacuum drying oven provided by Shanghai Bluepard instrument Co.(Shanghai, People's Republic of China), a stopwatch provided by Jun SD (Shenzhen, Guangdong Province People's Republic of China).

2.3. Experimentalscheme

2.3.1. Precipitation process

Some different kinds of copper salts ($CuCl_2$, $CuSO_4$, $(CH3COO)_2Cu$ and $Cu(NO_3)_2$) were put in the solvent and heated to a certain temperature, then some $NaCO_3$

solution was added under magnetic stirring condition for a period of time.

2.3.2. Filtration, washing, dying and roasting

Put mixture into centrifuge tubes and centrifuged **for** a period of time, then washed the precipitate using ethanol twice and put it into vacuum drying oven for 3h at 50 ℃ to get $Cu_2(OH)_2CO_3$ precursor. Heated $Cu_2(OH)_2CO_3$ to 400℃ for 2h with heating rate 2 ℃·min^{-1}.

2.4 Characterization

X-ray diffraction (XRD), fourier transform infrared (FTIR), thermogravimetric analysis (TG), scanning electron microscopy (SEM) and particle surface area measurements were used to analyse chemical composition, structure and properties of the nano-CuO generated.

3. Results and discussions

3.1. Factors influence the experimental process

The volume of CH_3CH_2OH was studied in single factor experiment and then the orthogonal test method was adopted to explore the influences of 4 factors on the products in order to study the factors that influenced nano-CuO. Influence of single factor was given in table 1and orthogonal factors were given in table 2.

Table 1. Experimental design with single factor

sequence	reactants	temperature	solvent (CH_3CH_2OH : H_2O)	time
I	$CuCl_2$	20℃	pure water	15min
II	$CuCl_2$	20℃	1：1	15min
III	$CuCl_2$	20℃	1：2	15min
IV	$CuCl_2$	20℃	1：3	15min

Table 2. Experimental design of orthogonal experiment

sequence	reactants	solvent (CH_3CH_2OH : H_2O)	temperature	time
1	①$(CH_3COO)_2Cu$	①pure water	①20℃	①15min
2	①$(CH_3COO)_2Cu$	②1:1	②40℃	②30min
3	①$(CH_3COO)_2Cu$	③1:2	③60℃	③45min
4	①$(CH_3COO)_2Cu$	④1:3	④80℃	④60min
5	②$CuCl_2$	①pure water	②40℃	③45min
6	②$CuCl_2$	②1:1	①20℃	④60min
7	②$CuCl_2$	③1:2	④80℃	①15min
8	②$CuCl_2$	④1:3	③60℃	②30min
9	③$CuSO_4$	①pure water	③60℃	④60min
10	③$CuSO_4$	②1:1	④80℃	③45min
11	③$CuSO_4$	③1:2	①20℃	②30min
12	③$CuSO_4$	④1:3	②40℃	①15min
13	④$Cu(NO_3)_2$	①pure water	④80℃	②30min
14	④$Cu(NO_3)_2$	②1:1	③60℃	①15min
15	④$Cu(NO_3)_2$	③1:2	②40℃	④60min
16	④$Cu(NO_3)_2$	④1:3	①20℃	③45min

In experiment above, the concentration of Cu^{2+} is 0.1mol/L, the volume of dispersant that contained CH_3CH_2OH and H_2O is 40 mL. The concentration of precipitant (Na_2CO_3) is 0.1mol/L and the solvent is 40 mL.

3.2. SEM analysis of Nano-CuO

Tested nano-CuO above using Hitachi S-4700 under operation voltage of 30 kV. Figure 1 was SEM micrographs of nano-CuO under different conditions with one factor changing. (a, b were the micrographs under condition I; c, d were the micrographs under condition II; e, f were the micrographs under condition III; g, h were the micrographs under condition IV). Figure 2 and 3 were SEM micrographs of nano-CuO of orthogonal experiment (micrograph A to P was sequence 1 to 16 respectively).

DOI: 10.1051/matecconf/201710802003

Figure 1. SEM micrographs of Nano-CuO under different conditions (I , II, III and IV)

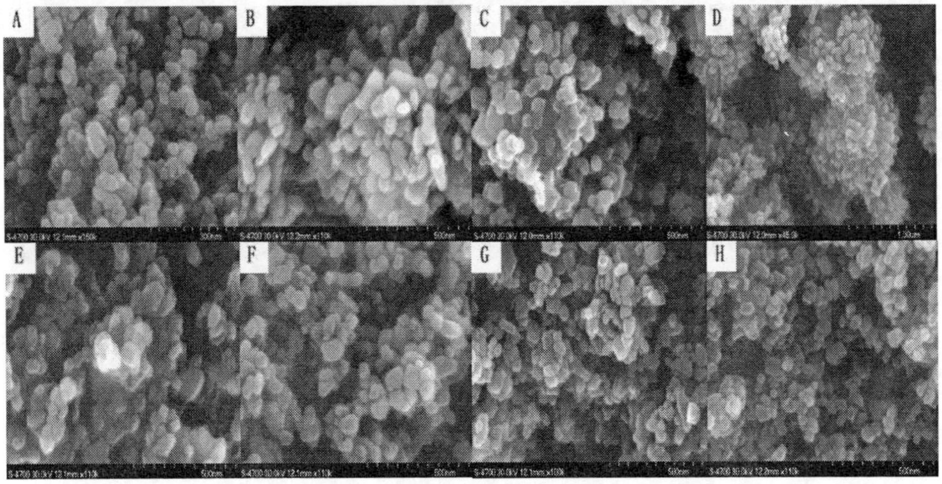

Figure 2. SEM micrographs of Nano-CuO under different conditions (1 to 8)

Figure 3. SEM micrographs of Nano-CuO under different conditions (9 to 16)

From the micrographs above, figure A and P that correspond to condition 1 and 16 were best among the figures because the nano-CuO had good dispersion and uniform size. The dispersion of nano-CuO became better when proportion of CH_3CH_2OH increased. Reaction temperature and reaction time had effect on the reactants' morphology and size, and the nano-CuO generated by $CuCl_2$ was best. Figure 4 was the micrographs under condition 1 and 16 with different magnifications.

Figure 4. SEM micrographs of nano-CuO with different magnifications under condition 1 and 16

From figure 4, the nano-CuO generated under condition 16 was better than that under condition 1.

3.3. FTIR analysis of nano-CuO

Tested the nano-CuO prepared under condition 16 using FTIR using TENSOR 27. Figure 5 gave the results of final resultant and the precursor.

From figure5, peaks around 536cm^{-1} for strong bands of the Cu-O stretching vibrations and 582 cm^{-1} for strong bands of the lattice vibration were observed, and peaks of the O-C-O stretching vibrations were found at 1379cm^{-1} and 1481cm^{-1}. Peaks at 839cm^{-1} and 754cm^{-1} for the O-H stretching vibrations were observed as well, and in addition, peak of Cu-OH around 487 cm^{-1} was noticed. This observation provides evidence that the final resultant and the precursor were CuO and $Cu_2(OH)_2CO_3$ respectively.

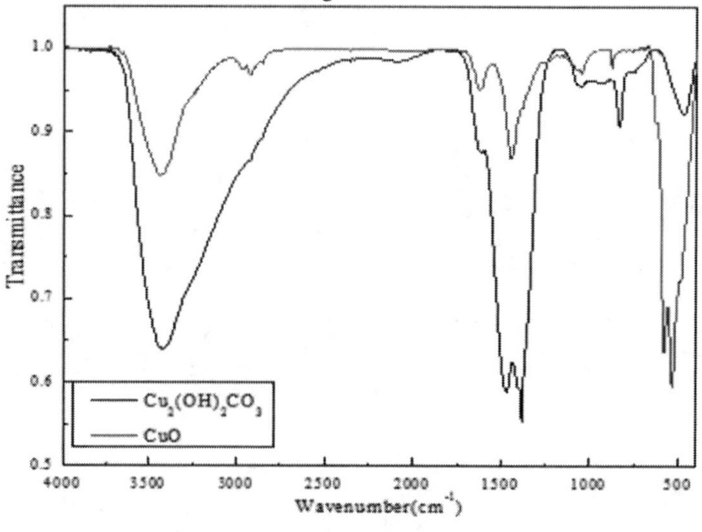

Figure 5. FTIR spectra of $Cu_2(OH)_2CO_3$ and CuO under condition 16

3.4. TG and DTG of nano-CuO

Tested nano-CuO above using Seiko TG/DTA 6300 with temperature varying from 30 ℃ to 480 ℃ and the heating rate was 5 ℃/min. Figure 6 gave TG and DTG of $Cu_2(OH)_2CO_3$ under condition 16.

From figure 6, $Cu_2(OH)_2CO_3$ began to decompose

rapidly at about 145℃ and the strongest peak of DTG appeared at 327 ℃, which meant the fastest decomposition temperature was 327 ℃. The mass of $Cu_2(OH)_2CO_3$ decreased from 30 ℃ to 250 ℃ for free water's losing and $Cu(OH)_2$'s decomposing, and the decomposing of $CuCO_3$ resulted in the rapid decreasing from 250 ℃ to 340 ℃.

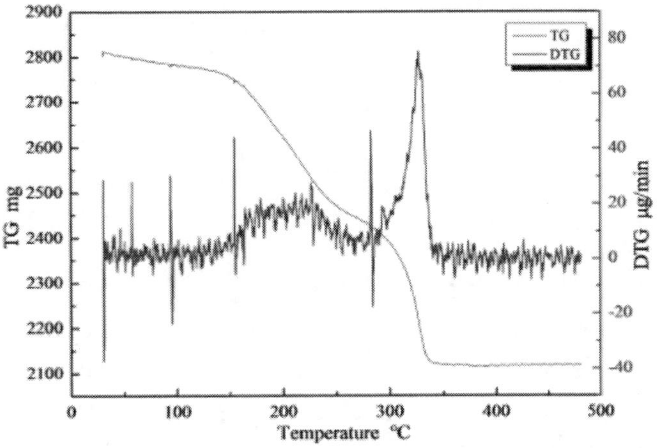

Figure 6. TG and DTG of $Cu_2(OH)_2CO_3$ under condition 16.

3.5. XRD analysis

CuO generated under condition 1 and 16 and its precursor were analysed using XRD (laser power 40 kV×40 mA). From figure 7 and 8, characteristic diffraction peaks appeared where 2θ is 14.701°,17.507°, 24.084°, 31.266° and 35.557°, and compared with JCPDS card, the material was $Cu_2(OH)_2CO_3$. The strongest peak appeared at 2θ=31.266°, d=2.85849nm and the material corresponded to $Cu_2(OH)_2CO_3$ crystal plane (020), (021), (022), (-102), (042), (125), (123) and (135), which meant the $Cu_2(OH)_2CO_3$ has crystallized completely and its component was single.

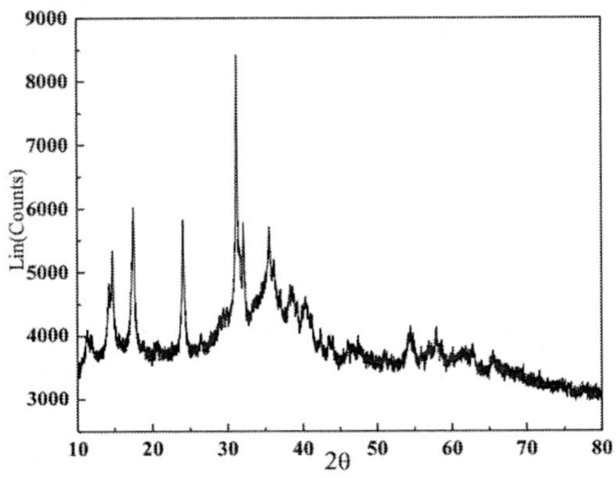

Figure 7. XRD spectrum of precursor under condition 1

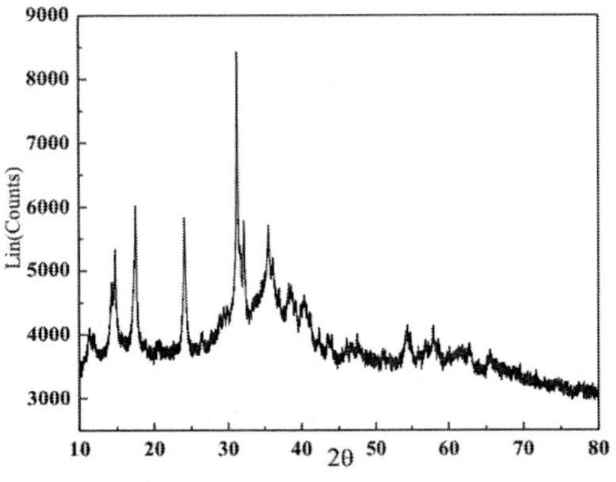

Figure 8. XRD spectrum of nano-CuO under condition 1

From figure 9 and 10, characteristic diffraction peaks appeared where 2θ=32.65 °, 35.47 °, 38.86 °, 48.99 °, 53.57 °, 58.37 °, 61.70 °, 66.43 °, 68.13 °, and compared with JCPDS card, the material was CuO, and the strongest peak appears at 2θ=35.471°, d=2.52872 nm and the material corresponded to CuO crystal plane (110), (-111), (111), (-202), (020), (202), (113), (-311) and (220).

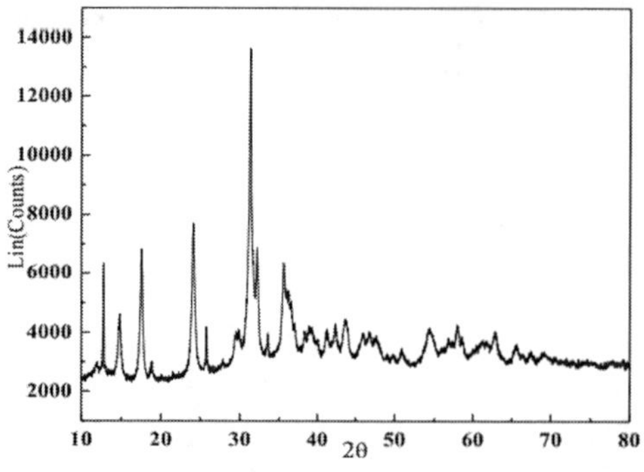

Figure 9. XRD spectrum of precursor under conditioı

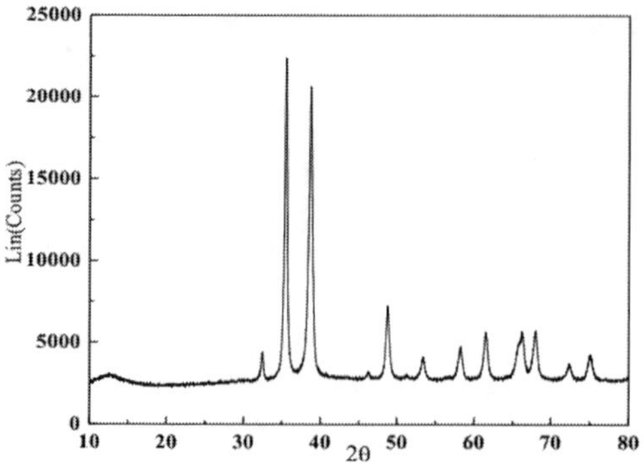

Figure 10. XRD spectrum of nano-CuO under condition 16

Calculated particle size of CuO and $Cu_2(OH)_2CO_3$ using Deby-Xeror formula as follows:

$$D_C=K\lambda/(\beta\cos\theta) \quad (1)$$

Where D_C is the size of crystal (nm), K is the Xeror constant (to cube particles K=0.94, to spherical particles K=0.89), λ is the wavelength of X-ray (using CuKa radiation, λ=1.5406A≈0.154nm), β is full width at half maximum, θ is the diffraction angle of the peak. Particle size of CuO and $Cu_2(OH)_2CO_3$ under condition 1 and 16 were given in table 3.

Table 3. Particle size of CuO and $Cu_2(OH)_2CO_3$ under condition 1 and 16

$Cu_2(OH)_2CO_3$	Size (nm)	CuO	Size (nm)
1	27.58	1	16.20
16	85.40	16	20.38

From table 3, materials' diameter became smaller after calcinations and the size of CuO generated using $Cu(NO_3)_2$ changes more significantly than that of other materials after calcinations.

3.6. Particle surface area measurements

CuO generated under condition 16 was analysed using NOVA 4200 to get its surface area, average pore diameter and total pore volume. The nitrogen adsorption-desorption curve of nano-CuO under condition 16 was given in figure 11.

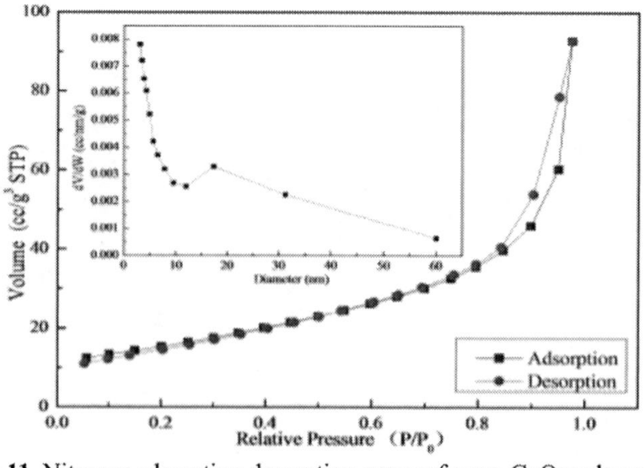

Figure 11. Nitrogen adsorption-desorption curve of nano-CuO under condition 16

After dealing with some procedures, the pore diameter was got, which was about 3.079 nm using BJH method. The surface area is 21.988 m^2/g using multipoint BET method and the pore volume was 1.349×10^{-1} cc/g using BJH method.

4. Conclusions

1 Direct precipitation method of nano-CuO using $Cu(NO_3)_2$ and Na_2CO_3 is feasible way to industry that is low cost and easy to achieve.

2 After orthogonal test, single factor experiment and characterization, best process conditions are got: reactant is $Cu(NO_3)_2$, temperature is 20℃, precipitant's (Na_2CO_3) concentration is 0.1mol/L, H_2O-CH_3CH_2OH ratio is 3:1, reaction time is 45min, roasting time is 2h and temperature is 400℃.

3 Nano-CuO generated using best condition above is spherical, uniform and with good dispersivity, the average diameter of nano-CuO is 20.38 nm, the surface area is about 21.988 m/g ,the pore diameter was about 3.079 nm and the pore volume was 1.349×10^{-1} cc/g .

Acknowledgements

The financial support of the project supported by a foundation item from the General Armament Department of PLA (9140A05071l1BQ0107), the Project of State Key Laboratory of Explosion Science and Technology, P. R. China (YBKT12-03) and the Specialized Research Fund for the Doctoral Program of Higher Education (20131101110009).

References

1. Liu Jiping, *Inorganic Nano-Materials* (Beijing: Science Press, 2003)
2. Li Xiaojie, Qu Yandong, Yan Honghao, Rare Metal Mat. Eng., **36(12)**, 2069(2007)
3. Chu Huaqin, Lu Yunfeng, Chinese J. Anal. Chem., **38(3)**, 442(2010)
4. R. V. Kumar, Y. Diamant, A. Gedanken, Chem. Mater., **8**, 2301(2000)
5. Hong Weiliang, Liu Jianhong, Chen Pei, J. Propul. Techno., **22(3)**, 254(2001)
6. A. Bhaumik, A. M. Shearin, R. Patel, K. Ghosh, Phys. Chem. Chem. Phys., **22(16)**, 11054(2014)
7. Zhu Chuanfeng, Wang Jun, Micronanoelectronic Techno., **44(6)**, 273(2007)
8. Ye Zhao, Zhang Hanhui, Spectrosc. Spect. Anal., **23(3)**, 487(2003)
9. Wu Xiuwen, Jiang Xin, Chem. Equipment Techno., **25(4)**, 46(2004)
10. Yan Bo, Wang Xin, Shao Chunhong, Chinese J. Inorg. Chem., **23(11)**, 1869(2007)
11. Luo Yuanxiang, Li Dan, Yang Juan, Chinese J. Explos. Propell, **25(3)**, 53(2002)
12. Liu Kehui,Yan Liushui, Luo Guoan, Chinese J. Anal. Chem., **33(6)**, 847(2005)

DOI: 10.1051/matecconf/201710802004

Contact stress analysis and optimization of single crystal turbine blade tenon/disk mortise structure considering thermal-solid coupling

Lei Li[1] , Zhonghao Tang[1], Dingyuan Deng[2], Mengchuang Zhang[1], Jiawei Yu[1]

[1]*Department of Engineering Mechanics, Northwestern Polytechnical University (Chang'an Campus), Xi'an 710129, P.R. China*
[2]*Institute of Shanghai Satellite Engineering, Shanghai 201100, P.R. China*

Abstract. Contact stress analysis and optimization design of single crystal turbine blade tenon/disk mortise structure considering thermal-solid coupling is proposed in this paper. Contact thermal conductivity analysis of turbine blade/disk mortise structure is carried out to obtain temperature distribution. Contact stress of mortise structure considering temperature influence is analyzed by FEM method. On basis of contact stress analysis, the optimization design method considering thermal-solid coupling is proposed. Broaching angle, wedge angle and gap distance are chosen as optimization design variables. The minimum Mises stress, average tensile stress and average compressive stress are chosen as optimization objectives. A three fir tree tenon/mortise structure is optimized to decrease the maximum Mises stress 14% by proposed method.

1 Introduction

The single crystal (SC) alloy has been widely applied for aircraft engine high pressure turbine blades, due to its fatigue and creep characteristics in high temperature. SC turbine blade always connect powder metallurgy (PM) disk with fir tree tenon/mortise structure. The blade tenon/disk mortise structure works at about 700 ℃ environment. High temperature working environment would reduce material properties of tenon/mortise structure. Meanwhile, blade tenon/disk mortise structure thermal deformations can cause contact load assign among blade tenon. So, analysis and design of SC turbine blade/PM disc fir tree tenon/mortise structure should consider thermal-solid coupling. Because contact stress is sensitive to size of tenon/ mortise structure , so it is important to optimize tenon/mortise structure to obtain minimum contact stress.

Cui[1] used finite element method to analyze contact stress of a coattail-type tenon/mortise structure at room temperature. Wang[2] studied the effect of processing tolerance on contact by finite element analysis. Wang[3] investigated effects of contact gap on contact properties of turbine blade tenon /mortise. Liao[4] introduced a collaborative optimization strategy to optimize a typical fir-tree mortise structure. Shen[5] optimized a three pairs of teeth fir-tree tenon/mortise structure and decreased the maximum equivalent stress by 12%. Yang[6] employed multi-island genetic algorithm and sequential quadratic programming to optimize turbine tenon/mortise. It can be seen from current literatures that most analysis and design of tenon/mortise structure do not considering effects of temperature.

This paper aims to propose a SC blade tenon/PM disk mortise structure optimization method considering thermal-solid coupling. The main works include two sections. Firstly, the contact stress analysis considering thermal-solid coupling is carried out. Secondly, optimization of blade tenon/ disc mortise structure is studied. Finally, a SC blade tenon/PM disk mortise structure is optimized by method proposed

2 Contact analysis of SC blade tenon/PM disk mortise structure considering thermal-solid coupling

The tenon/mortise structure which connected single crystal turbine blade and powder metallurgy disk is shown in Fig.1.The centrifugal load blade suffered is transferred to disk through contact surface of tenon/mortise structure. Because SC turbine blade tenon/PM disk mortise structure works at a high temperature about 700 ℃ , so the contact analysis considering temperature is necessary. In this work, the contact thermal analysis and contact stress analysis are carried out sequentially.

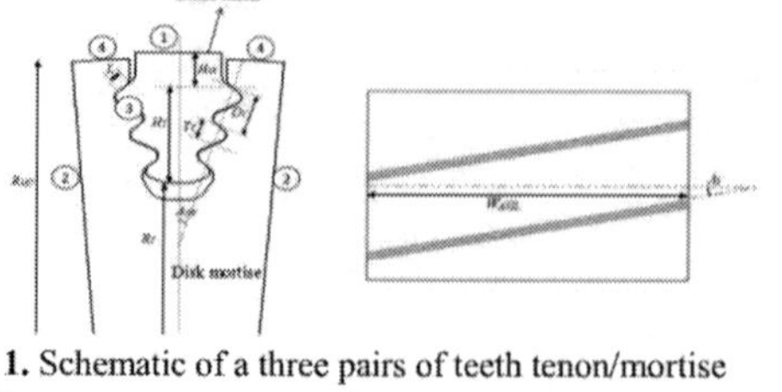

Figure 1. Schematic of a three pairs of teeth tenon/mortise structure

2.1 Contact thermal analysis

Turbine blade work around the high temperature combustion gas. Heat will transfer from blade to disk through contact surface. Relative low temperature air is used to reduce working temperature which flows through gaps between tenon and mortise.In this paper, the contact thermal analysis is carried out by FEM method. Thermal conductivity and specific heat capacity of blade material CMSX-2 and disk material Rene95 are shown in Table1-4.

Table 1. Thermal conductivity at different temperature of CMSX-2

Temperature/℃	300	500	700	900	1000	1100
Thermal conductivity $(W/(m\cdot°C))$	13.9	20.2	24.6	30.0	33.2	10.2

Table 2. Specific heat capacity at different temperature of CMSX-2

Temperature/℃	300	500	700	900	1000	1100
Specific heat capacity $(J/(kg\cdot°C))$	490	507	557	662	733	821

Table 3. Thermal conductivity at different temperature of Rene95

Temperature/℃	200	300	500	700	900	1000
Thermal conductivity $(W/(m\cdot°C))$	10.5	11.7	13.9	15.7	18.4	20.0

Table 4. Specific heat capacity at different temperature of Rene95

Temperature/℃	200	300	400	500	600
Specific heat capacity $(J/(kg\cdot°C))$	443.4	447.4	453.2	458.9	516.7

The SC blade tenon/ PM disk mortise structure is divided to mesh grids for contact thermal analysis. Thermal contact coefficient $TCC=3000$ is applied on contact surface. Convective heat transfer coefficient $60W/m^2\cdot K$ is applied for heat exchange analysis when cooling air flows through gaps between tenon and mortise. The first thermal boundary condition $800°C$ is applied on the top surface of blade tenon. In this paper, The FEM software Ansys is employed to contact thermal analysis. Fig.2-3 give the temperature distribution of blade and disk after contact thermal analysis. It can be shown that the top surface of blade tenon has the highest temperature. The maximum temperature of disk is about 750℃, which locatedat region of mortise. The structure temperature decreases as radius decreases.

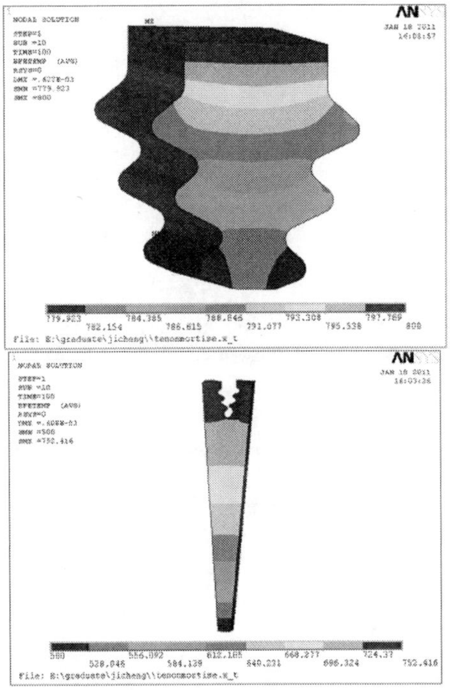

Figure 2. Temperature distribution of blade and disk after contact thermal analysis

2.2 Contact stress analysis of SC blade tenon/PM disk mortise

Nickel-base single crystal alloy having face centered cubic is a classical anisotropy material. The elastic stress strain relationship of single crystal alloy can be expressed by following Eq.

$$\sigma' = C\varepsilon'$$

where, σ is stress vector, and $\sigma' = \{\sigma_{11}', \sigma_{22}', \sigma_{33}', \tau_{12}', \tau_{13}', \tau_{23}'\}$. ε is strain vector and $\varepsilon' = \{\varepsilon_{11}', \varepsilon_{22}', \varepsilon_{33}', \gamma_{12}', \gamma_{13}', \gamma_{23}'\}$.

The matrix C can be expressed as

$$[C] = \begin{bmatrix} C_{11} & C_{12} & C_{12} & 0 & 0 & 0 \\ C_{12} & C_{11} & C_{12} & 0 & 0 & 0 \\ C_{12} & C_{12} & C_{11} & 0 & 0 & 0 \\ 0 & 0 & 0 & C_{44} & 0 & 0 \\ 0 & 0 & 0 & 0 & C_{44} & 0 \\ 0 & 0 & 0 & 0 & 0 & C_{44} \end{bmatrix}$$

where, C_{11}, C_{12}, C_{44} is a function of elastic modulus E, Poisson ration υ and shear modulus G.

$$C_{11} = \frac{2(1-\upsilon)E}{(1+\upsilon)(1-2\upsilon)}$$

$$C_{12} = \frac{2\upsilon E}{(1+\upsilon)(1-2\upsilon)}$$

$$C_{44} = G$$

The material constants of CMSX2 alone <001> crystallographic orientation is shown in Table 5. The material constants of disk material Rene95 is shown in Table 6.

Table 5. The material constants of CMSX2 along <001> crystallographic orientation

Temperature(°C)	E (GPa)	υ	G (GPa)
20	115	0.313	137
700	89	0.322	113
800	85	0.326	107
850	78	0.330	104

Table 6. The material constants of disk material Rene95

Temperature(°C)	E (GPa)	υ
20	211	0.288
350	188.5	0.192
400	187.5	0.189
450	184	0.183
500	179	0.175
550	174.5	0.168
650	169	0.167

FEM software Ansys is employed to analyze contact stress of single crystal turbine blade tenon/power disk mortise structure. Periodic boundary conditions are applied on the surface besides disk mortise in order to simulate the entire turbine disk. A equivalent pressure load $p = 255.2MPa$ conversd by blade centrifugal load is applied on the top surface of blade tenon. Because contact stress has large gradient and sensitive to mesh grids on contact surface, so grids on contact surface is meshed very small. And the meshed grid is checked by grid-independent test in order to analyze accurately.

Fig.3 gives the Mises stress distribution of SC blade tenon. It can be seen that the maximum contact stresses locate at the boundary regions of contact surface. In the three pairs of turbine blade tooth, the third pairs of tooth have the maximum Mises stress 809MPa. The contact stresses of first and second pairs of tooth are lower than the third pairs of tooth. The first pairs of tooth maximum stress is 493MPa, the second pairs of tooth maximum stress is 624MPa.

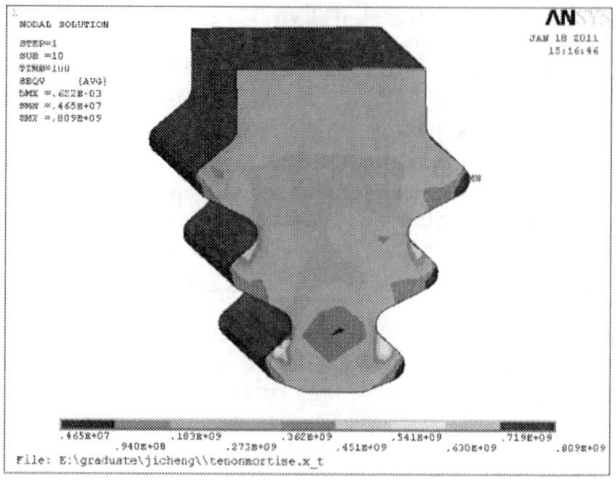

Figure 3. Mises stress distribution of single crystal blade tenon

Fig.4 gives the Mises stress distribution of PM disk mortise. It can be seen that the maximum contact stresses also locate at the boundary regions of contact surface. The first, second and third pairs of tooth maximum Mises stress are 622MPa, 711MPa and 711MPa.

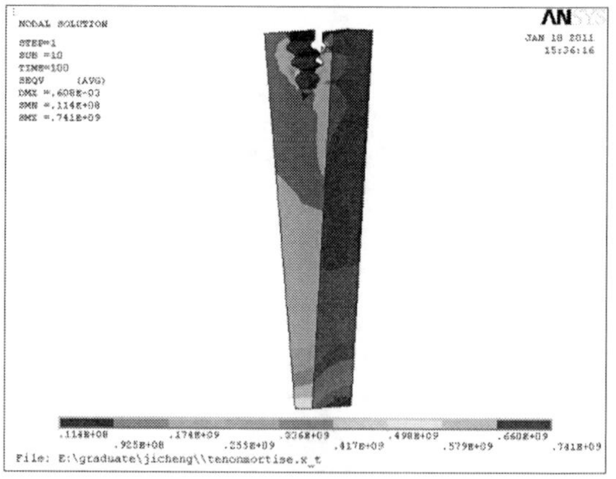

Figure 4. Mises stress distribution of disk mortise

3 Optimization

Optimization of SC blade tenon/PM disk mortise structure to reduce structural stress is carried out on basis of contact thermal and stress analysis. Fig.5 gives the optimization flowchart of single crystal blade tenon/disk mortise structure optimization. Firstly, the parameterized blade tenon/disk mortise structure is created. Then, the automated contact thermal analysis and contact stress analysis are carried out orderly. Optimization algorithm is used to optimize turbine tenon/mortise.

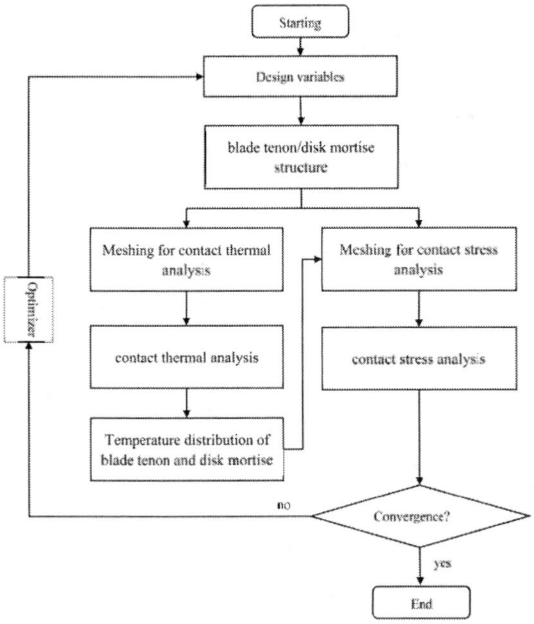

Figure 5. Flowchart of SC blade tenon/PM disk mortise structure optimization

The parametric tenon/mortise structure is shown in Fig.1. There are several design variables for tenon/mortise structure, such as contact line L, tooth distance D_T, broaching angles A_S, wedge angles A_{SW}, gap distance G_T and so on. The broaching angles A_S, wedge angles A_{SW}, gap distance G_T and tooth distance D_T are chosen as optimization design variables which impact contact stress significantly. Blade tenon/disk mortise structure could occur extrusion damage and tensile failure. So the minimum Mises stress, average tensile stress of tenon and mortise, average compressive stress of teeth are chosen as optimal objectives. The optimization of single crystal blade tenon/disk mortise structure can be formulated as following.

$$\begin{cases} find & A_S, A_{SW}, G_{Ti}, D_T \\ \min & f = F(\sigma_T, \sigma_M, \sigma_{TA}, \sigma_{MA}, \sigma_{TC}) \\ & A_S^{\,l} \le A_S \le A_S^{\,u} \\ s.t. & A_{SW}^{\,l} \le A_{SW} \le A_{SW}^{\,u} \\ & G_{Ti}^{\,l} \le G_{Ti} \le G_{Ti}^{\,u} \\ & D_T^{\,l} \le D_T \le D_T^{\,u} \\ & \sigma \le [\sigma] \end{cases}$$

Where, $F(\sigma_T, \sigma_M, \sigma_{TA}, \sigma_{MA}, \sigma_{TC})$ is multi-objective function of stress σ. σ_T is the maximum Mises stress of single crystal blade tenon. σ_M is the maximum Mises stress of disk. σ_{TA} and σ_{MA} are average tensile stress of tenon and mortise. σ_{TC} is the average compressive stress of blade tenon tooth. [ ] means allowable values. $i = 1,...,5$ mean gap between tenon and mortise of each teeth.

In this paper, Multi-island genetic algorithm is used to optimize turbine tenon/mortise. After optimization, the maximum Mises stress of SC blade tenon is decrease 14% relatively. The maximum Mises stress of disk mortise is decrease 3.2% relatively.

Table 7. Comparison of objectives between original and optimized

	constraint	objectives	original	optimized
σ_T /MPa	860	↓	808.5	692.1
σ_M /MPa	1095	↓	741.3	717.2
σ_{TA} /MPa	344	↓	243.5	242.4
σ_{MA} /MPa	438	↓	174.5	182.4
σ_{TC} /MPa	473	↓	500.8	456.3

Table 8. Comparison of design variables between original and optimized

Design variable	Low limit	Up limit	original	optimized
A_S /°	0	14	9	3.81
A_{SW} /°	12.5	20	19.76	13.42
G_{T1} /°	0.007	0.02	0.007	0.015
G_{T2} /°	0.01	0.05	0.04	0.033
G_{T3} /°	0.01	0.03	0.01	0.025
G_{T4} /°	0.01	0.05	0.04	0.022
G_{T5} /°	0.01	0.03	0.02	0.012
D_T /°	2.6	2.7	2.67	2.7

4 Conclusion

The blade tenon/disk mortise structure suffers centrifugal and thermal load at the same time. So, the contact analysis considering thermal-solid coupling is necessary. An optimization method which analyzing contact thermal and contact stress sequentially is introduced in this paper. A three fir tree SC blade tenon/PM mortise structure is optimized to decrease the maximum Mises stress 14% by proposed method

Acknowledgements

National Natural Science Foundation of China (Grant No. 51575444), China Postdoctoral Science Foundation (Grant No. 2014M562281), Aerospace Technology Support Foundation (2014-HT-XGD) support this work.

References

1. Cui, H.-t., H.-q. Ma, and W.-d. Wen, *Reliability-Based Shape Optimization of Tenon/Mortise in Aero-engines with Contact.* Chinese Journal of Aeronautics, 2003. **16**(4): p. 193-197.
2. Wang, P., et al., *Effects of One-side-deviation on the Stress Distribution for the Aero-Engine Turbine Joint Structure.* Gas Turbine Technology, 2014.
3. Wang, S.F., et al., *Effects of contact gap on contact properties of turbine blade tenon /mortise.* Tuijin Jishu/journal of Propulsion Technology, 2015. **36**(1): p. 112-118.
4. Liao, H., et al. *Multidisciplinary Optimization Technology Research on Typical Turbine Assembly Structure.* in *ASME 2014 International Mechanical Engineering Congress and Exposition*. 2014.
5. Shen, X.L., et al., *SHAPE OPTIMIZATION OF AERO ENGINE FIR-TREE TENON/MORTISE STRUCTURE.* Engineering Mechanics, 2011. **28**(12): p. 231-237.
6. Yang, M.C. and S.Y. Sun, *Structural optimization of turbine tenon/mortise.* Journal of Aerospace Power, 2010. **25**(8): p. 1876-1882.

A study on the application prospect of functional shirt fabrics in the uniform industry

Xing Xue[1], Gu Wen[2], Li Li[1], Liu Xiaogang[2]

[1] *Donghua University, NO.1882 West Yanan Road, Shanghai, 20051, China*
[2] *Donghua University, Shanghai Fashion Design & Value Creation Collaborative Innovation Center, NO.1882 West Yanan Road, Shanghai, 20051, China.*

Abstract. This paper analyses the value of functional shirt fabrics in the application of uniform industry. Through the functional research of the professional wear shirt fabric, we analyzed the industry needs and demands strength, and on the basis of the analysis above, we combed the industry forefront research, and then we put forward two development directions of functional shirt fabrics to provide some suggestions for the development of uniform shirt fabric.

1 Introduction

Uniform has the characteristic of strong wearing. Since shirt is a common category in uniform industry, the material is an important program which needs to research. In recent years, functional fabric is too numerous to mention, but there are still some blind spots in how to use them in uniform industry. This paper analysed the value of functional fabric applications, through the investigation of shirt fabric functional requirements to get a direction of research.

2 The value of functional shirt fabrics in uniform industry

2.1 Social values

The main values of companies in 21th century consider that the value of a company is not to pursuit of the interests maximize, but to realize value maximization. That means companies should create wealth effectively for community. [1] Dr. Manuela Weber of the University of Lüneburg described the CSV as a component of CSR benefits (CSR value appreciation), stakeholder relations (KPI), and sustainable development strategy elements for corporate social responsibility. Based on this theory, domestic scholars constructed the social value system of the enterprise, they pointed out that the content of the CSV should be the responsibility performance of the enterprise to the stakeholders, which can be divided into market contribution value, environmental contribution value and social contribution value. [3] Therefore, social value and social responsibility are inseparable to enterprises. To fabric enterprises, new functional shirt fabrics are not only to meet the needs of market consumers, and some environmental fabrics also can save environmental resources, pollution reduction and emission reduction ,which can realize the value of the enterprise's environmental contribution at the same time. In recent years, with the concept of environmental protection clothing fabrics heated up, there are some important issues such as how to deal with the large numbers waste of uniforms, and how to use environmentally friendly recycled fiber, which can ensure the healthy development of uniform industry. These also can reflect enterprises' social responsibilities.

2.2 Economic values

Economic values of functional shirt fabric can reflect on both manufacturers and users. For the shirt fabric manufacturers, through technical means, uniform functional shirt fabrics give new blood for the market, and that improve the entire shirt fabric product system at the same time, which can provide more opportunities for the market. On the other hand, Scientific and technological innovation can be free from the vicious competition in price competition, making enterprises in a benign operation and construction. For the users, High-quality functional shirt fabric can increase the shirt life cycle, which can reduce unnecessary wear in the uniform cost. In addition, Functional fabrics help employees to improve work efficiency, promoting business operations thereby.

2.3 Management values

Under the conditions of socialist market economy, as the main body of economic activities, enterprises should not only pursue economic benefits, but also pursue the "moral character" of enterprise development and realize the moral management of enterprises. As Zhang Jun, a scholar of economics said that "An enterprise produces 'private goods', but its productivity is a 'public good'. But self-interest motive cannot guarantee the efficiency of enterprises; enterprises also need moral values, such as team spirit and corporate culture. Therefore, the moral plays a very important role." [4] The use of functional shirt fabric is a humanistic concern for workers actually, and it is a way to manage from the moral level, which can approach to help the workers to build loyalty to enterprise and promote the spiritual and cultural construction at the same time.

2.4 Cultural values

The core of enterprise culture is the enterprise values, and it is the terns made of values of things. Value is a relationship category, and human existence is the prerequisites, so it is meaningless analyzing enterprise value except human beings. Enterprise culture is comprehensively expressed with "corporate citizen"; itself is a "people-oriented" concept. [5] It is the enterprise which has "people-oriented" concept in itself that the functional shirt fabrics have adopted.

Edgar H Schein, a professor at MIT, built a corporate culture model, which is made of three levels. The surface is visible "behavior, rules, and some of the organization's typical products (artifacts)". The middle is the values advocated by the enterprise. The bottom is the basic assumptions of employees. Surface culture is the result of the underlying culture; and the reason why there will be visible surface of the cultural phenomenon is the underlying culture plays. [6] In terms of the functional fabric manufacturers, the persistence of new materials and environment-friendly technology can affect the employees' values, morality, behavior and way of thinking, affecting the institutional arrangement and evolution mechanism of enterprises. New functional materials' application is also an important way for enterprises to set up characteristic business model, characteristic products and promote good cultural values.

3 Functional requirements of shirt fabrics for uniform industry

Uniform shirt has a wide range of functional fabrics; the functional requirements of different industries must be different. With the high-tech in fabric research and development, Environmental protection, health care and other new functional fabrics have begun to emerge. But these new types of functional fabrics are still lack of understanding in uniform industry. Nowadays, the market demand is still based on basic functionality. Therefore, new functional shirt fabrics still have a large space for development in uniform industry.

3.1 Research design

In this research, Industry division is based on uniform shirt basic needs and style features divided into bank financial, government agencies, education and training, health care, hotel catering, entertainment, logistics, aviation, IT, advisory bodies, cultural media, road building Maintenance, real estate, department stores and other industries. This research is aimed at the working population. A total of 22 questions were designed to around with six directions. These are the basic information of the working population, the opinions on the existing shirt fabric, and the seasonal basic functional requirements of the shirt fabric, the demand for the new functional shirt fabric, acceptable price and uniform shirt dressing cycle. Through the questionnaire issued, the aim is to get the needs of market and understanding the needs of different sectors, key functional requirements, the new fabric prospects and R & D direction.

3.2 Questionnaire credibility

A total of 223 valid questionnaires were collected from the survey. The Cronbach's Alpha coefficient is 0.944, and the Cronbach's Alpha coefficient is 0.945. Because the Alpha coefficient is greater than 0.8, so the credibility of the questionnaire is satisfactory. The detail coefficient is shown as Table 1 and Table 2.

Table 1. Case handling summary

		N	%
cases	effective	220	100.0
	Excluded	0	.0
	total	220	100.0

Table 2. Reliability statistics

Cronbach's Alpha	Based on standardized items Cronbachs Alpha	Number of items
.944	.945	38

3.3 Result and discussion

3.3.1 Analysis on functional requirements of fabrics for uniform shirt

(1) Analysis on basic functional requirements of uniform shirt

Basis functional requirements mean to meet the basic needs of wearing comfortable. According to the research result, most respondents dealt with the existing uniform shirt fabric satisfaction as the "general satisfaction" rating areas. There is an urgent need to improve the fabric permeability and wrinkle resistance, which demand more than 70%, followed by wash ability, stain resistance, wear resistance; The key demand for functional fabrics in summer is permeability, cool feeling, followed by anti-permeability; winter functional requirements focus on skin-friendly comfort, followed by anti-staic function. The detail propotion of basic functional requirements is shown as Figure 1 and Table 3.

(2) Analysis on new functional requirements of uniform shirt

New functions as a non-essential function can improve the wearing feelings and working efficiency, which comes from care of employees. This article combined with the forefront of the industry, and listed functions such as the body skin care, plant health care, antibacterial immunity, safety reflection, intelligent temperature control, special luster, induction appearance, environmental regeneration, radiation protection, high-strength stretch, moisture-proof, cool touch, warm function 13 uniform shirt new fabric functions. Among them, the antibacterial immune, radiation protection moisture-proof, cool touch and warm function are higher than others, reaching 44.39%, 41.26%, 42.15%, 37.67% and 36.32% respectively. The detail propotion of new functional requirements is shown as Table 4.

3.3.2 Analysis on the difference of Industry demands

According to the popularity of uniform shirts, the paper analyzes the eight industries of banking, government, education, entertainment, hotel catering, media, consultancy and shopping malls. The results of industry needs are as the table below (Table 5), which can help to develop the functional products for uniforms.

The Table 5 describes the core needs of shirt fabric functionality in the eight industries. There are commonalities and differences in the functional requirements for fabrics among industries. There are the common core needs of the industry such as anti-wrinkle, washability, permeability, skin-friendly, moisture-proof and cotton component. But according to the working environment and other social factors, there are differences in functional requirements of the fabric. The main difference among industries lies in the new function of preferences.

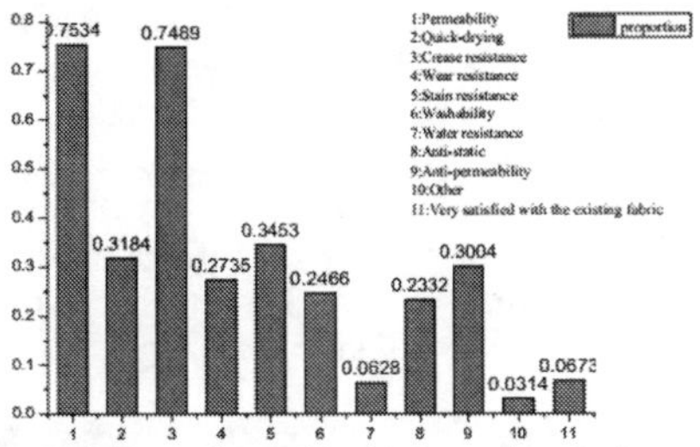

Figure 1. Existing uniform shirt fabric needs to be improved.

Table 3. Basic functional requirements of uniform shirt fabric

functions	Complete-ly unnecessary	Unnecessa-ry	Necessa-ry	Very necessa-ry
Anti-wrinkle	6.73%	5.83%	25.56%	61.88%
Washable	6.73%	7.62%	34.53%	51.12%
Stain resistance	7.62%	10.31%	38.57%	43.5%
Wear resistance	7.63%	11.21%	38.58%	42.6%
Waterpro-of	18.39%	28.25%	32.74%	20.63%

Table 4. New functional requirements of uniform shirt fabric

New functions	Completely unnecessary	Unnecessary	No opinion	Necessary	Very necessary	Average score(total 5)
Body skin care	14.8%	9.42%	21.97%	20.63%	33.18%	3.48%
Plant Health Care	13%	12.11%	23.32%	21.97%	29.6%	3.43%
Antibacterial immunity	9.42%	6.28%	19.28%	20.63%	44.39%	3.84%
Safety reflective	17.94%	13.45%	29.15%	17.94%	21.52%	3.12%
Intelligent temperature control	14.8%	10.76%	29.6%	15.7%	29.15%	3.34%
Special luster	16.14%	13.45%	38.57%	17.49%	14.35%	3%
Inductive appearance	17.94%	16.14%	27.8%	18.83%	19.28%	3.05%
Environmental regeneration	8.97%	10.76%	27.8%	26.46%	26.01%	3.5%
Radiation protection	8.52%	3.59%	23.77%	22.87%	41.26%	3.85%
High-strength stretch	10.76%	10.31%	40.36%	18.83%	19.73%	3.26%
moisture-proof	7.17%	4.93%	20.18%	25.56%	42.15%	3.91%
Cool touch	7.17%	4.04%	23.32%	27.8%	37.67%	3.85%
Warm function	7.17%	5.38%	26.01%	25.11%	36.32%	3.78%

3.3.3 Analysis on the requirement of environment-friendly fabric

Environment-friendly fabrics are loved by the majority of employees. The points are mostly between "general" and "very necessary", among which "very necessary" and "necessary" occupied 38.12% and 24.66%.people who opposed only occupied 10.76%. However, the attitude of enterprises stood in neutral attitude, and this occupied 40.81%. Opponents and supporters accounted for 23.31% and 35.87 %. The adoption of environmentally friendly fabrics mostly depends on the cost of fabrics.The attitude towards environment-friendly fabrics of employees and enterprises are shown as Figure 2 and Figure 3.

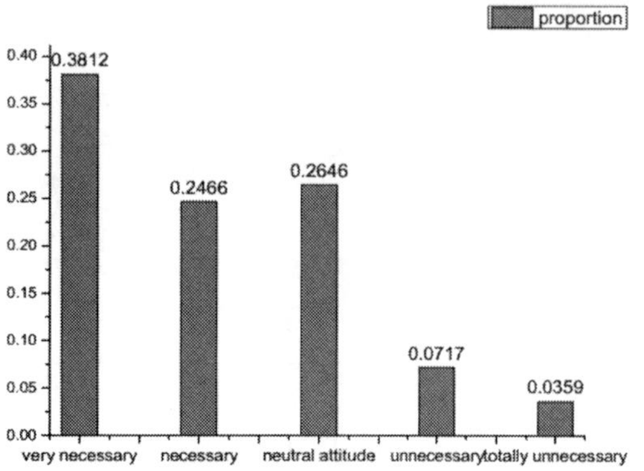

Figure 2. Employees' attitude towards eco-friendly fabrics.

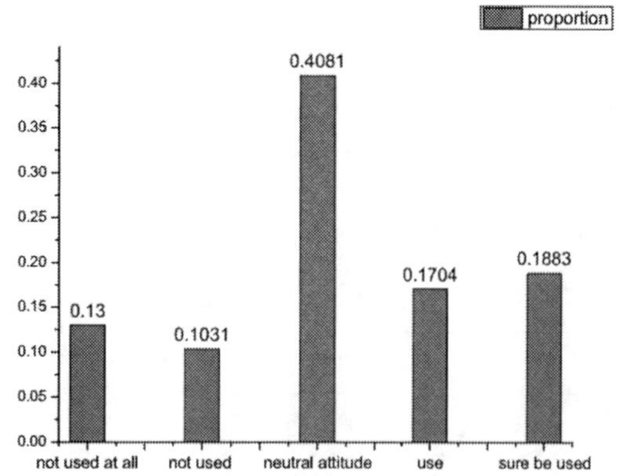

Figure 3. Enterprises' attitude towards eco-friendly fabrics.

4 Frontier research of shirt functional fabric

Based on the result of market research, this paper combined the frontier research of some core demands like permeability, anti-wrinkle, anti-penetration, stain resistance, wear resistance, skin-friendly, moisture-proof, radiation protection, antibacterial immunity, cool touch and warm function.

4.1 Frontier research of shirt basic functional fabric

The research on the frontier of functional fabrics is expounded from the aspects of Permeability, Anti-wrinkle, anti-penetration, Stain resistance; wear resistance and skin comfort. In the frontier research, natural fibers have been studied most extensively. Natural fibers have been used for thousands of years in clothing. However, since the last century, natural fibers have the characteristics of skin-friendly and comfortable, but they are easy to be wrinkled and difficult to take care. In recent years, with the development of fabric technology, the natural fiber began to become thin, soft. Through the sizing, calendaring and other processes to improve the shortcomings of its easy to fold [7]. To be more consistent with the modern fast-paced life, the "fog polymerization" method can improve the wear resistance of cotton fabrics [8]. SLIPS-type fabric is the most prominent research result for Stain resistance, through the SLIPS self-healing coating which has a super smooth feature, to resist the cotton fabric and polyester fabric from any water, ice, oil and bacteria and other substances [9]. Breathable determines the body's contact with the outside world, permeability can help the skin breathe, in addition to the traditional cotton fabric and blended fabrics; the new fabric called "porel" has been loved by its good breathable and quick-drying characteristics [10]. Aiming at the problems of how can the fabric be both thin and opaque, Scientists used the theory of "mie" scattering in the latest research. They innovated a large number of high refractive index powder thin-walled diaphragm, the macro-reflective interface and numerous microscopic reflection interface of the complex fiber, so that the light in any radial incidence will be intercepted, which can improve the anti-penetration effect [11].

Table 5. The Difference of Functional Requirements of Shirt Fabrics

Content	Banking	Government	Education	Entertain-ment	Hotel catering	Media	Consultancy	Shopping malls
Core Require-ment	Anti-wrinkle; Washable	Anti-wrinkle; Washable; Wear resistance	Anti-wrinkle; Washable; Stain resistant	Anti-wrinkle; Washable; Stain resistant	Stain resistant	Anti-wrinkle; Washable; Stain resistant	Anti-wrinkle; Washable	Anti-wrinkle; Washable
Existing problems	Per-meability	Per-meability	Permeability	Anti-wrinkle	Per-meability	Anti-wrinkle	Anti-wrinkle	Per-meability
Core demands in summer	Per-meability;	Per-meability; Stain resistance	Per-meability; Anti-penetration; Cool feeling	Per-meability; Anti-perspective; Cool feeling;	Per-meability	Per-meability; Cool feeling	Per-meability; Anti-perspective	Cool feeling; Per-meability

DOI: 10.1051/matecconf/201710802005

Core demands in winter	Skin-friendly and comfortable	Skin-friendly comfortable	Skin-friendly and comfortable	Skin-friendly and comfortable	Skin-friendly and comfortable	Anti-static; Stain resistant; Skin-Friendly and comfortable	Skin-Friendly and comfortable	Anti-static; Skin-friendly and comfortable
Pre-ferences of the new func-tional shirt fabric	Warm function; moisture-proof; Radiation protection; Environ-mental regeneration	Anti-bacterial immunity; Moisture-proof; Warm function; Cool touch	Radiation protection; Moisture-proof; Anti-bacterial immunity; Anti-perspective	Moisture-proof	Moisture-proof ; Intelligent temperature control	Moisture-proof; Radiation protection; Cool touch	Moisture-proof; Radiation protection	Moisture-proof; Cool touch
Fabric com-position pre-ferences	Cotton	Cotton	Cotton; Cotton and chemical fiber blended	Cotton	Cotton	Cotton; Cotton and linen blend	Cotton	Cotton

4.2 Frontier research of shirt new functional fabric

According to the results of research, the new functional moisture resistance, radiation resistance, antibacterial immunity, cool feeling and warm feeling have a relatively high demand. Through the moisture-proof treatment, the fabric can better prevent the oxidation reaction and hydrolysis reaction, so that the bacterial growth will be reduced. As the frontier research, antimicrobial textile technology can be divided into two categories: the first is to build antibacterial fiber system, and then make various types of antibacterial fabrics. Bamboo fiber has the function of antibacterial. Silver nanoparticles coating on bamboo pulp fabric modified with dopamine is not only antibacterial but also environmentally friendly and easy to carry out [12]. The second is to use antibacterial agents to reach the antibacterial function. For example, using Tragacanth gum/nano silver hydrogel on cotton fabric to In-situ synthesize antibacterial properties [13]. In modern society, electronic products used more frequently, so that electromagnetic radiation and light radiation damage human body seriously; therefore, radiation protection fabrics are being more and more attention. At present, the functional fiber material containing ceramic powder [14] and the silver-silver fiber fabric [15], which fuse nano-silver ions are the main research results in light-proof and electromagnetic radiation. In the study of cool functional fabrics, the new fabric called "Topcool" and "coolmax" is due to its better water absorption and wetting function makes the more comfortable and cooler feeling, the market was well, [16] which have resolved the problem of cotton fiber permeability decreased caused by sweating. The low temperature in winter makes the thin and warm fabric also stand on to opposite side, but the "softwarm" fever fiber, "thermogear" fiber and "eks" fiber can get warm function by absorbing moisture into heat. [17]

5 Expanding direction of uniform shirt fabric

5.1 Environmentally friendly and economical functional shirt fabric

Nowadays, "sustainable" is a major trends in textile and garment industry. Textile and garment manufacturers are increasingly concerned about sustainable products, which can meet environmental and social requirements at the same time. [18] Since the growing rate of polyester production and lacking of sound recovery system, Recycling of recycled fiber is a small fraction of the total fiber production. The China government made a promise that carbon dioxide emission would decrease by 40% to 45% in per unit of GDP in 2020 compared to 2005 [19]. Therefore, recyclable fabrics will become a trend from the national policy. Uniform will be phased out after the bulk procurement, so it will produce a large number of waste fabrics. The recycling problem in uniform industry should be attended and the adoption of recycled fabrics will become a new trend.

In addition to the use of environmentally friendly recycled fiber, some new types of fiber which can reduce carbon dioxide emissions also will become trends. For example, the permanent anti-static environmental fabrics which both have environmental protection and anti-static function, the breathable fabric which both have environmental protection and breathable function [20]. Some fabrics made by plants fibers also can protect the environment and improve the wearing comfort.

According to the survey of this paper, enterprises stand in neutral attitude most in the use of environmentally friendly fabrics, and the main factor is fabric costs. Uniform shirts cost can be accepted in the range of 200-400 yuan to the most proportion. Therefore, how to create economic functional fabric, which can both satisfy the basic needs and environmental protective function is the develop direction for uniform shirt in the future.

5.2 Exclusive functional fabrics based on the differences among industries

Due to the differences of working environment, there are some differences on industries' needs. The main difference is about the new function requirements. In terms of the basic function needs, most industries seemed to be similarity. Fabric functional design can be based on industry characteristics to construct industry exclusive shirt fabric. For example, for the long-term using of electronic equipment crowd such as banking and financial industry, fabrics can be added radiation protection and health care functions like blood circulation promotion; outdoor working industries such as logistics and transportation can be added UV protection in the fabric; for some students especially the teenagers, shirt fabric can be added anti-bacterial immune function. Function design of the uniform shirt fabric should according to the characteristics of the industry, and base on the economic, environmental protection, security and other multi-angle sides.

Reference on the Maslow's hierarchy of needs theory, functional uniform shirt fabric design can rely on a four-level mining process (Figure 4). The first layer of fabric functional demands on the basic functional needs, such as permeability, wrinkle resistance, skin-friendly. The second layer pays attention to human safety, such as radiation protection, anti-ultraviolet function, and security alerts function; The third layer focuses on personal care, such as plant health care and warm function; the forth layer can stand for the social value of enterprises, such as environmental protective function. Exclusive functional fabrics can be based on enterprise needs and economic strength or its social value. Functional requirements are gradually optimized from the grass-roots level to the top level.

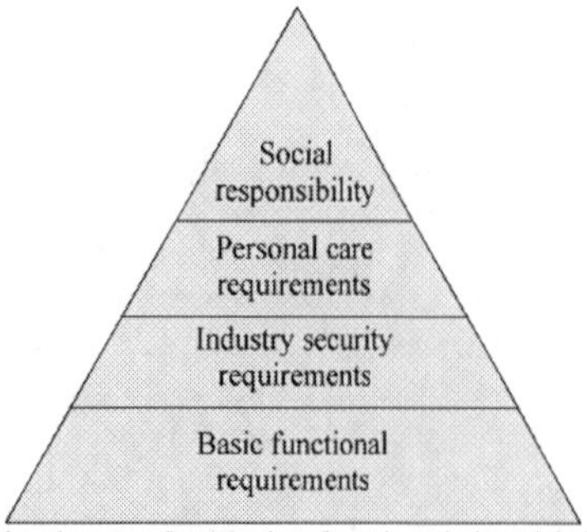

Figure 4. Design levels of exclusive functional shirt fabrics.

6 Conclusions

The value of functional fabrics used in uniform industry can be expressed in social value, management value, economic value and cultural value, and its significance cannot be underestimated. Anti-penetration, anti-wrinkle and other basic functions are the most prominent functions. New functional shirt fabrics have differences between industries. Environmentally friendly fabrics have a certain degree of recognition, but because of the cost, enterprises take a neutral attitude. Therefore, the uniform shirt fabric function design can be combined in economic side and environmental friendly side. Environmentally friendly and economical functional shirt fabrics need to be developing in the future. Material designer can abbey four levels to design according to the needs of enterprise, those are: basic functional requirements, industry security requirements, personal care requirements, social responsibility.

Acknowledgments

The authors would like to appreciate the fund projects below:

1. Shanghai Fashion Design & Value Creation Collaborative Innovation Center. (ZX201311000031)
2. Chinese Central Communist special funds for basic research projects in Donghua University (NO. 16D110715), which focusing on the research and application of Fashion trend from uniform perspective.

References

1. See Blair, M. M. (1998). For whom should corporations be run? An economic rationale for stakeholder management. Long Range Planning, 32(2), 195-200.
2. Thorpe, J. (2003).Developing value: The business case for sustainability in emerging markets. Greener Management International, 44:17- 32.
3. MAI Sheng, WANG Keyi, KUANG Haibo. (2011).A Study on the evaluation of enterprise social value. Science Research Management, 6, 11.
4. WANG Xiaoxi. (2011). Economic Value of Morality].Social Sciences in China, 4, 55-66.
5. BAO Lifeng.(2011). The value of people oriented corporate culture environment and construction. Northeast Normal University.
6. FENG Qiaosi, FENG Yuan. (2013). Construction of corporate culture and environmental management value system. Accounting research, 8, 24-31.
7. ZHU Yuansheng. (2002). New development of natural fiber fabrics. Journal of Ningbo Vocational and Technical College, 1, 10-12.
8. FAN Wanchao. (2016). Study on "Fog Polymerization" Method for Strengthening Wear Resistance of Textile Fabrics. Zhejiang University of Science and Technology.
9. MENG Xianguo. (2014, May 3). SLIPS-type super anti-fouling fabric research has been successful developed. China Textile News.
10. YI De. (2013). Porel,a breathable quick-drying fabric. Textile Decoration Technology, 4, 23.
11. WEN Weijian. (2013, April 9) you will never worry about "emptied"when you wear light-colored clothes.
12. PENG Linghui; Guo Ronghui; Lan Jianwu. (2016). Microwave-assisted deposition of silver nanoparticles on bamboo pulp fabric through

dopamine functionalization. Applied surface Science, 386,151-159.
13. Montazer, M.; Keshvari, A.; Kahali, P. (2016).Tragacanth gum/nano silver hydrogel on cotton fabric: In-situ synthesis and antibacterial properties. Carbohydrate polymers, 154, 257-26.
14. Qi Lu. (1999). Research on ceramic powder of PP fiber material, 4,152-154.
15. SONG Kexin. LIU Lianjun. SUN Guodong. (2011). Production of bamboo silver antimicrobial radiation fabrics. Cotton Textile Technology, 8, 53-56.
16. LIU Peng, LIU Yanjun, FENG Lanqing. (2012). Application of wicking fiber in knitting, 2, 59-60.
17. LI Ling, MENG Jiaguang. (2010). Application of wicking fiber in knitting. The Sixteenth National Fancy Yarn and its textile technology progress seminar Proceedings, CA.
18. CHE Yinghong. (2015). New environmentally friendly textiles. Printing and dyeing, 22, 56-57.
19. SHAO Wei. (2015). High-performance fibers and green fibers to play the important task of the industry, Textile and Apparel Weekly, 42, 34-35.
20. LIANG Liping. (2011 October 6) Focus on low - carbon environmental protection - Nanshan new functional fabric. China Textile News, 1.

DOI: 10.1051/matecconf/201710802006

Research on the Test Method of Machining Stress Based on the Instrumented Indentation Test

Linlin Zhang , Xiaoyang Li and Rong Chen

College of Mechanical Engineering and Applied Electronics Technology, Beijing University of Technology, Beijing, 100124, China

Abstract. The main objective of this study is to measure the distribution of machining stress by the instrumented indentation test (IIT). Finite element method (FEM) was performed to investigate the influences of tensile stress and compression stress on the indentation response (Load-displacement curves, average contact pressure and contact area). The thicknesses of machining stress layer under turning and grinding processing ways were obtained by layer stripping method based on instrumented indentation technique. Results indicate that the machining stress brings complex stress state. Under the same load, the corresponding indentation displacement increases while the indentation plastic work, contact area and average contact pressure decrease with the increases of machining stress (from compressive stress to tensile stress). Tensile stress does just the opposite. In addition, the layer thicknesses of the machining stress for turning and grinding members were measured by instrumented indentation method, results agree with X-ray method and demonstrate the effectiveness of the proposed approach.

1 Introduction

Metal cutting process is very complicated, the residual stress will produce inside workpiece during the turning, grinding and other different cutting processes, which affects the mechanical properties of materials. The interior or exterior residual stress of metal components has importance effect on mechanic performance [1]. A quantity of friction heat generates instantaneously in the cutting process, few of the heat passes workpiece interior and almost all heat gathers at the surface of workpiece. The difference of temperature distribution makes the metal surface produce tensile stress or compressive stress. That is the reason of residual stress caused by mechanical processing [2, 3]. A series of domestic and overseas research have been conducted to investigate the effects of machining parameters on working stress [4]. Lots of residual stress measurement methods have been developed at present, familiar ones are hole drilling method and X-ray diffraction method [5]. But these methods have certain damages to the specimen and they are relatively difficult to implement.

Some scholars have developed a new method for measuring machining stress by using micro hardness test [6-8]. Tsui and Oliver conducted a series of nano indentation experiments on AL 8009 alloy under uniaxial and biaxial stress conditions and calculated the indentation hardness by using Oliver-Pharr method [9]. Bolshakov investigated the influence of maching stress on the indentation surrounding material's pile-up using FEM analysis [10]. Cheng introduced the dimensional method into the indentation analysis, which provieded a new idea to establish the relation between indentation response and material properties [11]. Xu et al studied the effect of residual stress on the unloading behaviors for indentation samples [12]. 2012, Jin researched the relation between residual stress and indentation hardness based on energy method [6]. The instrumented indentation test (IIT) can characterize materials' micro mechanical properties and it is used more and more widely due to its high precision, easy operation and non-damage detection.

This study proposes a new method to measure the machining stress distribution based on FEM and IIT. For turning and grinding specimens, the machining stress states (tensile stress or compressive stress) and machining stress distribution along the strip layer depth are measured respectively. This study provides the reference and guides for the engineering application of the machining stress's measurment.

2 Materials and methods

2.1 FEM model

Figure 1 shows the finite element model of ABAQUS/Standard used in the present study. The spherical indenter was modeled as an analytic rigid surface with radius R=1mm in cylindrical coordinates, while the metal material was a half-space axisymmetric model with radius r=5mm and height h=15mm. According to the principle of Saint Venant, the effect of boundary on the stress distribution of the indentation area can be eliminated. Contact status between the rigid sphere and the deformable medium was assumed to be frictionless. Element type is the explicit linear model CAX4R quadrilateral element. And the local refined mesh size is 0.003mm.

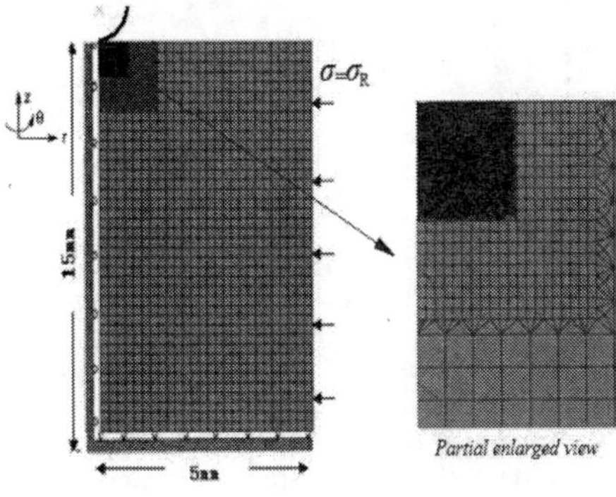

Figure 1. FEM model

The constitutive material model was assumed as elastic-perfectly plastic, its property parameters were expressed as E=200GPa, υ=0.3, yield strength σy=350MPa. Yielding is determined by the von Mises yield criterion [13], expressed as:

$$\sigma_{eq} = \sqrt{\frac{3}{2} S_{ij} S_{ij}} = \sigma_Y \tag{1}$$

Where σeq is the von Mises equivalent stress and Sij represents components of the deviatoric stress tensor. The evolution of plasticity in the indented half-space is tracked by the equivalent plastic strain $\bar{\varepsilon}_p$, defined as:

$$\bar{\varepsilon}_p = \int_{\Omega} \sqrt{\frac{2}{3} d\varepsilon_{ij}^p d\varepsilon_{ij}^p} \tag{2}$$

Where Ω is the strain path used to track the accumulation of plasticity and $d\varepsilon_{ij}^p$ denotes increments of plastic strain.

The load and displacement boundary conditions are as shown in Figure 1, σR is defined as the uniaxial machining stress, its negative value presents compressive stress and positive value presents tensile stress. According to the values of σR, the calculations were divided into five groups, which represented compressive stress state, unstressed state and tensile stress state (Table 1). Besides, the rigid indenter load F=-20N.

Table 1. The machining stress state σ_R (MPa).

No.	1	2	3	4	5
Values	-300	-150	0	150	300
Stress state	compressive stress		unstressed	tensile stress	

2.2 Experiment

To measure the thickness of surface machining stress layer, the IIT was conducted based on layer stripping method.

Figure 2 shows the instrumented indentation testing machine Germany's Zwick ZHU 2.5 system. It works by measuring the load-displacement curves continuously in loading and unloading processes to calculate material's mechanical parameters. The resolution of the displacement measurement system is 0.02μm. A spherical indenter with diameter *D*=1mm was used, and the level of the loading force was HBW2.5: *F*=24.52N [14].

The experiment was divided into two groups: turning group and grinding group, each group included three specimens. The specimens were cylinders with diameter *d*=10mm and height *h*=25mm. Their material was 45 steel with *E*=200GPa, *υ*=0.3 and yield strength σ_y=350MPa. The distribution law of the max. Indentation depth h_m with the layer depth *d* was measured based on layer stripping. Each layer included 3-5 testing data, the 2000cc sandpaper was used to strip layer and the digital dial gauge with resolution 0.001mm was used to control layer depth.

(a) Zwick ZHU 2.5 machine

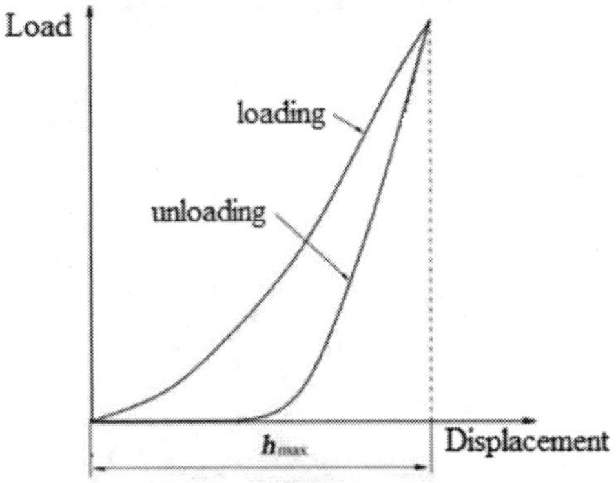

(b) Principle of the instrumented indentation test

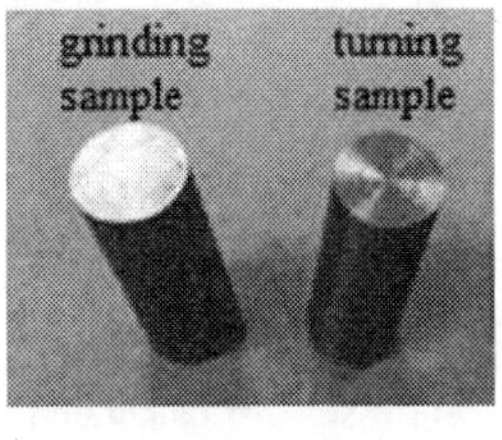

(c) samples

Figure 2. Experimental facility

 DOI: 10.1051/matecconf/201710802006

3 Results and discussions

3.1. Equivalent stress distribution

For spherical indenter presses into the bulking materials with elastic-plastic hardening characteristics in loading process, the load and mechanical parameters have the following relation [9]:

$$F = f_L(E, v, E_i, v_i, \sigma_y, \sigma_R, n, h, R) \tag{3}$$

Where E, v, E_i, v_i are the elastic modulus and possion's ratio of tested material and indenter; σ_y, σ_R are the yield strength and machining stress; n, h, R are the material's hardening exponent, indenter displacement and indenter radius.

Figure 3 shows the equivalent stress distribution in tensile stress state with σ_R=300MPa and unstress state respectively. It is obvious that the equivalent stress distribution is more complicated compared with unstress state when the machining stress exists. Besides, the plastic zone increases with the increase of indentation depth.

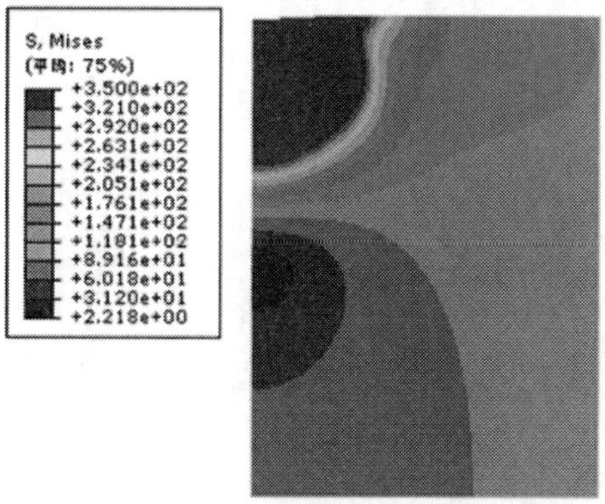

(a) Tensile stress state

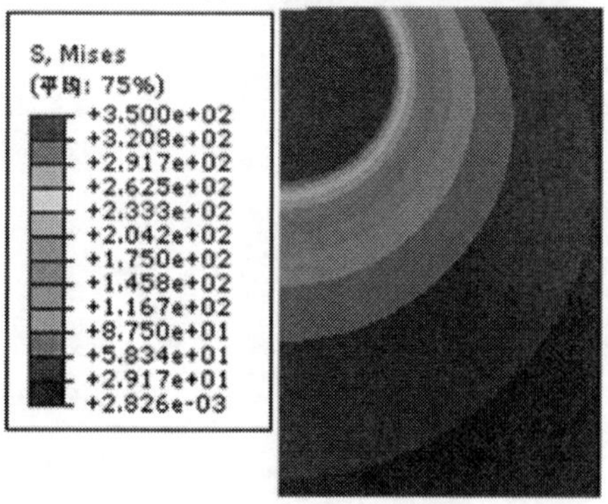

(b) Unstressed state

Figure 3. Equivalent stress distribution

3.2 Load-displacement curves

According to Table 1, the load-displacement curves (F-h curves) in different cases are calculated, their results are shown in Figure 4.

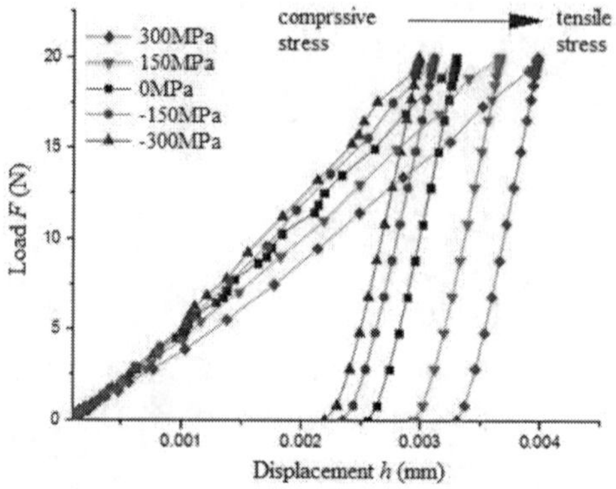

Figure 4. Load-displacement curves with different machining stresses

It is observed that the indentation depth h_m increases when machining stress turns from compressive stress to tensile stress. Under a constant load, compressive stress will make the F-h curve move towards a lower h_m while tensile stress will make the F-h curve go towards a higher h_m. Besides, h_m decreases with the increase of compressive stress and increases with the increase of tensile stress. In addition, by comparing the slopes of loading and unloading, it is discovered that the loading process is more sensitive to machining stress. Also, tensile stress effect on the F-h curve is larger than compressive stress.

3.3 Average contact pressure and contact area

By assuming that the pressure distribution on contact area is parabolic shape, then the expressions of contact pressure and contact radius are [13]:

$$p = p_0\sqrt{1-(r/a)^2}, \quad a = \left(\frac{3FR}{4E^*}\right)^{\frac{1}{3}} \tag{4}$$

Where p_0 is the max. the contact pressure in the center of contact area and r is the radial distance, E^* is reduced elastic modulus that defined as $1/E^* = (1-v^2)/E + (1-v_i^2)/E_i$, where E, E_i, v and v_i are elastic modulus and possion's ratio of two contact bodies.

In indentation process, the loading curve integral area is the *total* work W_t [14] while the unloading curve integral area is the elastic work W_e, then the plastic work W_p is:

$$W_p = W_t - W_e \tag{5}$$

Table 2. Relationship between the plastic work and the machining stress

Machining stress (MPa)	-300	-150	0	150	300
Plastic work (10^{-5}J)	2.02	2.16	2.34	2.94	3.01

According to Eq. (5), the plastic works with different machining stresses are calculated (Table 2). It shows that W_p increases gradually when machining stress turns from compressive stress to tensile stress, which because of the integral path $h_c > h_t$.

Figure 5 shows the effect of machining stress on average contact pressure and contact area. The average contact pressure and contact area decrease with the increase of machining stress. That is because the plastic deformation increases with σ_R increasing (Table 2). Under the same load, material is prone to sink-in with displacement increases [6], therefore the contact area decreases. Besides, as the slope $k_c > k_t$, so the effect of compressive stress on the contact area is larger than tensile stress.

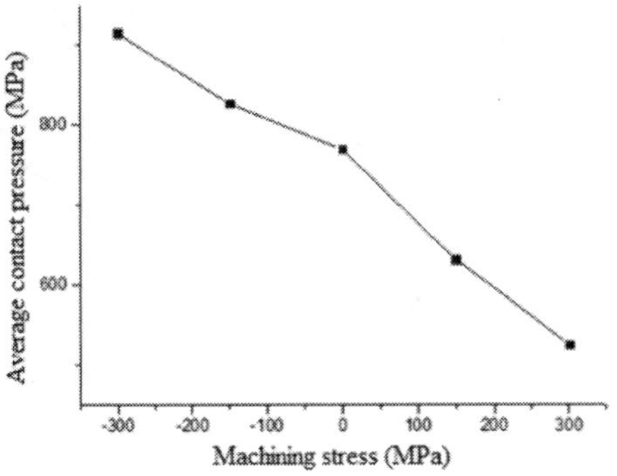

(a) Average contact pressure

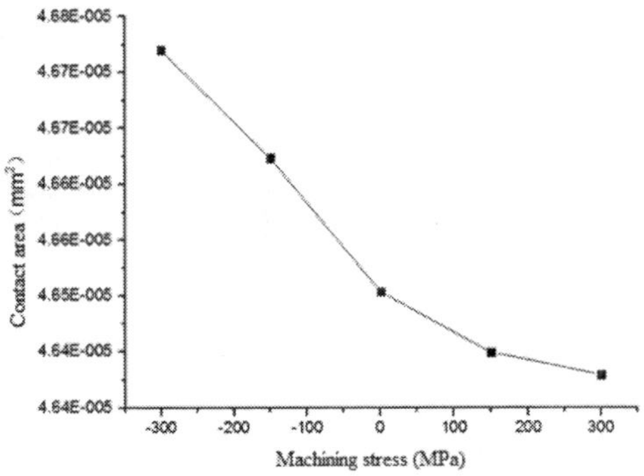

(b) Contact area

Figure 5. Effect of machining stress on average contact pressure and contact area.

3.4 Machining stress thickness

Figure 6 indicates the relationship between layer depth d and the indentation depth h_m, Fig 6a shows the turing specimen with real surface roughness Ra=3.27 while Figure 6b shows the grinding one with Ra=0.31.

The max. indentation depth of parent metal h_0=6.42μm. Figure 6a shows that h_m decreases at first, increases subsequently and tends to be stable finally. $h_m > h_0$ when stripping depth d<20μm, $h_m < h_0$ when 20μm<d<50μm and $h_m = h_0$ when d>50μm.The change rule of h_m reflects the specimen's machining stress thickness. The initial range with $h_m > h_0$ is tensile stress zone; h_m decreases with depth increasing and when $h_m < h_0$, material gets into compressive zone; and when there's no significant difference between h_m and h_0 means that the machining stress disappears and material becomes parent metal.

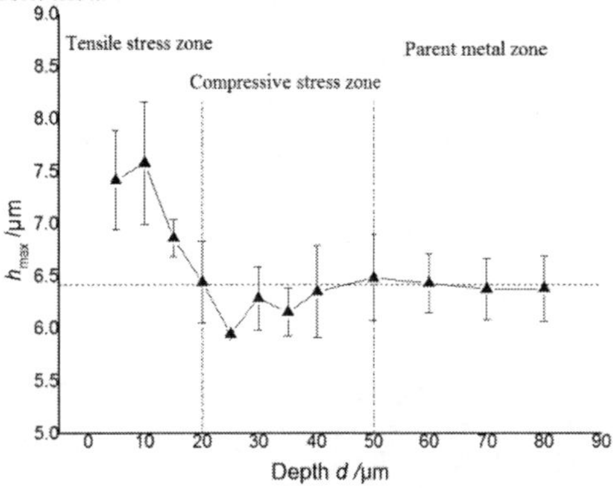

(a) Turning specimen

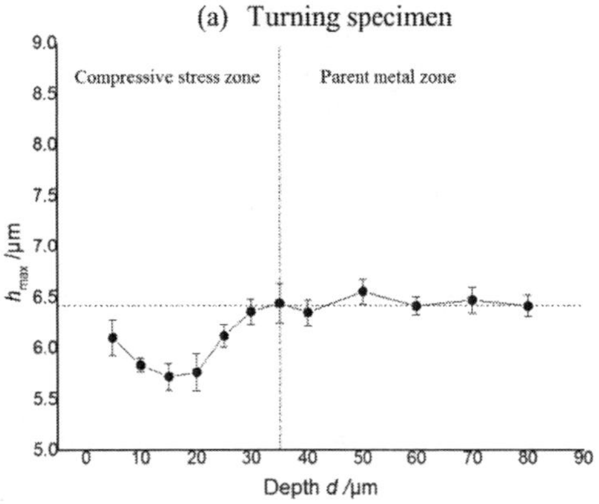

(b) Grinding specimen

Figure 6. Machining stress layer thickness based on layer stripping method

These results presented here are identical with the values obtained by X-ray method in reference [4]. Similarly, in Figure 6b, the ranges of d<35μm and d>35μm represent cmpressive stress zone and parent metal respectively. In conclusion, the measured truning machining stress layer based on IIT consists of tensile stress and compressive stress and its stress layer thickness is about 50μm. The measured grinding machining stress layer is characterized by compressive stress and its thickness is about 35μm.

4 Conclusions

In this work, the test method of machining stress based on the instrumented indentation test is investigated. The influences of tensile stress and compression stress on the indentation response are studied by FEM. Under the same indentation load, the indentation displacement increases with the increasing of machining stress (from compressive stress to tensile stress). The stress state with machining stress is more complex. Compressive stress reduces the contact area, average contact pressure and plastic work, tensile stress will do just the opposite. Additionally, the machining stress layer thickness under turning and grinding processing ways are obtained by layer stripping method based on instrumented indentation technique, these results demonstrate the effectiveness of the proposed approach.

References

1. G.M. Mi, The production and counter measure of residual stress (Mechanical Industry Press, Beijing, 1983)
2. Q.H. Kong, Y.H. Yu, J. Tongji Univ. **27**, 549-552 (1999)
3. Y.B. Guo, D.W. Yen, Wear. **256**, 393-399 (2004)
4. Y.L. Zhang, H.R. Huang, X. Li, J. B. Univ. Technol. **32**, 582-586 (2006)
5. C.Z. Zhang, D.M. Li, S.P. Pang, S.J. Guan, J. Jilin. Inst. Chem. Techno. **18**, 73-75 (2001)
6. H.P. Jin, W.Y. Yang, L. Yan, Philos. Mag. **92**, 480-499 (2012)
7. X. Chen, J. Yan, A.M. Karlsson, Mat. Sci. Eng. A-Struct. **416**, 139-149 (2006)
8. W.C. Oliver, G.M. Pharr, J. Mater. Res. **7**, 1564-1580 (1992)
9. T.Y. Tsui, W.C. Oliver, G.M. Phar, J. Mater. Res. **11**, 752-759 (1996)
10. A. Bolshakov, W.C. Oliver, G.M. Pharr, J. Mater. Res. **11**, 760-768 (1996)
11. Y.T. Cheng, C.M. Cheng, J. Appl. Physics. **84**, 1284-1291 (1998)
12. Z.H. Xu, X. Li, Philos. Mag. **86**, 2835-2846 (2006)
13. K.L. Johnson, *Contact mechanics* (Cambridge university press, Cambridge, 1985)
14. GB/T 21838.1-2008. *Metallic materials-Instrumented indentation test for hardness and materials parameters-Part 1, Test method.*

Effect of air Recirculation on Moisture Removal Efficiency in Combined Hydrolytic–Aerobic Rotary Bio-Drying Process

Khajon Somsai[1], Tusanee Tondee[1] and Somrat Kerdsuwan [2]

[1]*Rattanakosin College for Sustainable Energy and Environment (RCSEE) Rajamangala University of Technology Rattanakosin, Nakhon Pathom 73170, Thailand*
[2]*Department of Mechanical and Aerospace Engineering King Mongkut's University of technology North Bangkok Bangkok 10800, Thailand*

Abstract. Rotary bio-drying is the process to reduce moisture from Municipal Solid Waste (MSW) by using the heat from aerobic bio-degradation. This study aimed to investigate the effects of air recirculation on moisture removal efficiency in combined hydrolytic–aerobic rotary bio-drying process for drying high initial moisture content MSW, allowing satisfied moisture content to convert a waste into biofuel energy. The results from the laboratory scale experiments were indicated that increasing the exhaust air recirculation ratio, more moisture increased in the end product. However, while the air recirculation ratio was equals or greater than 50 %, the moisture content of the end product was not significant different.

1 Introduction

Most of MSW for small municipality in Thailand was disposed in non-regulated open dump. Only 9.4% of the MSW generated was recycled while about 10% of the MSW in Thailand was properly treated in sanitary landfills and other proper technology facilities [18] that may be cause of the outbreak of the gastrointestinal disease and the environmental problem. MSW was generally characterized as high water content (up to 75%) due to the relatively high proportion of food waste over 60 %. High water content will reduce the efficiency of energy recovery from their direct combustion in incinerator and reduce the ability for mechanical sorting utilization.

Bio-drying technology goal is to improve the quality of MSW by relying on the heat generated from the decomposition of organic matter, microbes in the reactor to reduce the moisture content of the waste [9, 10], while preserving calorific value [1]. After a bio-drying process, the MSW can be used as a source of solid fuel [1, 13]. Moisture content was reducing form MSW without the use of fossil fuels and offers minimal power consumption that suitable for safe and economical combustion in a biomass boiler by bio-drying process [7]. It was a significant alternative way for treating MSW [11], which a good solution to reduce the amount of water and converted a waste into energy (Refuse Derived Fuel; RDF) for solved the energy recovery problem from high water content of MSW.

The moisture content in the MSW was reduced through two main steps: (1) water molecules evaporate from the surface of MSW into the surrounding air by change of phase from liquid phase to gas phase ; and (2) Then removed the evaporated water from the MSW pile by the air for the exhaust gases [13]. Limited amounts of free water may seep through the MSW pile and be collected at the bottom of the bio-drying reactor as leachate [11].

The air flow was influence the temperature biomasses drying process, while the level of decomposition of organic matter degradation affects the calorific value and stability of the final product [1]. The daily inversion of airflow could achieve MSW drying in very short times (8–9 days) and leads to a homogeneous final product [13]. Zhang [16,17] improved the moisture and accelerated the decomposition of waste with high moisture content by increasing the circulating leachate during hydrolytic degradation prior to aerobic decomposition. There were a few studies in aerobic rotary bio-drying reactor of MSW [2, 12, 14, 15]. Khajon [5] reported that combined hydrolytic-aerobic in rotary bio-drying process could produce homogeneous RDF from MSW. It was also found that the increased airflow rate was not linearly proportional to the weight loss. With the first 2 days of hydrolytic stage before supplied high air flow in aerobic stage could more increase moisture removal efficiency for rotary bio-drying process than increase double air flow rate only [5].

The main objective of this study was to investigate the effect of direct air recirculation on moisture removal in combined hydrolytic–aerobic rotary bio-drying for high moisture content MSW, allowing to be obtain biofuel of satisfactory energy content.

2 Methods

2.1 Characteristics of MSW feedstock

The raw MSW was collected from dumping area the Administrative Organization of Tumbon (AOT) Kaerai, kratumban samutsakorn province, Thailand. The raw MSW was grinded by hammer mill then sorted with screening size 50 mm. The remains on the screen were collected for preparing feedstock and the grinded MSW which smaller than 50 mm were rejected. The bulky waste such as carpet, yard waste whose size that bigger than 250 mm was rejected by hand. Only MSW size between 50 – 250 mm was used as feedstock for laboratory process.

2.2 Experimental equipment setup

The MSW feedstock was filled until the half of reactor which the total weight approximately 250 kg. The retention time of bio-drying process was 8 days [2]. The reactor turning frequency was fixed at 1.5 rpm. Air was supplied into the reactor at the centre core shaft of horizontal tank reactor by air blower.

The scheme of the laboratory scale rotary bio-drying process used for developing the experiment was presented in Figure 1.

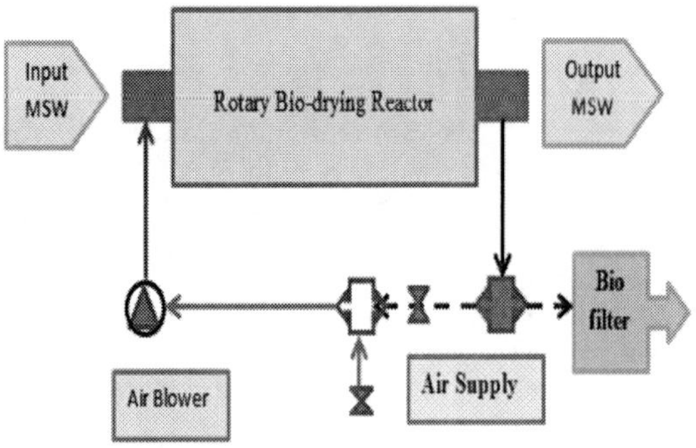

Figure 1. Scheme of the Rotary bio-drying Process

The temperature and humidity of inlet and outlet air were measured twice a day by humidity/temperature meter model DIGICON HT-776.

The temperature of composting biomass and inner reactor was continuously monitored in every 15 minutes by using digital thermometer which installed at three different locations inside the reactor; at its centre and at its two ends. The experimental data was recorded by data logger.

The experiment operation was consists of two stages; the hydrolytic and the aerobic stages. Which automatically controlled by a timer program. For hydrolytic stage, the reactor was only rotate without aeration and the operating period was 2 day. Following by aerobic stage with ventilation frequency at 48 times/ day (run 20 minute, stop 10 minute) and the operating period was 6 day. The air recirculation ratio 0%, 25%, 50%, 75% and 100% respectively, was investigated.

2.3 Sampling, Parameter Analysis and Statistic Analysis

A 100 g of samples were collected from three different locations, mainly from the mid span and two end terminals of the reactor, using a composite sampling method to minimize disturbance of the adjacent materials.

The samples were dried in an oven at 105 °C for 24 hr. in order to determine their moisture content following the ISO5068-1 standard.

Statistical analysis was performed using ANOVA. Test for significant difference in each condition were defined by $P<0.05$.

3 Results and discussions

3.1. Moisture content and moisture removal efficiency

The results were showed in the form of moisture content and the moisture removal efficiency of the trial. The results of the moisture content for five different constructional solutions were presented.

Table 1. The rotary bio-drying experiment results

Trial	Air recirculation ratio %	Hydrolytic Stage moisture content %		Aerobic Stage moisture content %	
		Initial	Final	Initial	Final
A	0%	64.7	62.2	52.6	18.6
B	25%	74.6	72.1	61.5	27.1
C	50%	73.2	70.6	65.4	29.5
D	75%	71.9	68.4	63.3	32.7
E	100%	70.6	66.2	61.2	35.8

All trials moisture content were decreasing, trial A, B, C, D and E, from 64.7% to 19.6%, 74.6% to 27.1%, 73.2% to 29.5%, 71.9% to 32.7% and 70.6 % to 35.8 %, respectively. From the results, all trials show the percentage of moisture content reduced to lower than 50%. However, only the final product for trial A showed the acceptance percentage for RDF or energy purposes which is less than 25 %. The end products for reactor B to E did not achieve RDF requirements [8] as showed in Table 1.

DOI: 10.1051/matecconf/201710802007

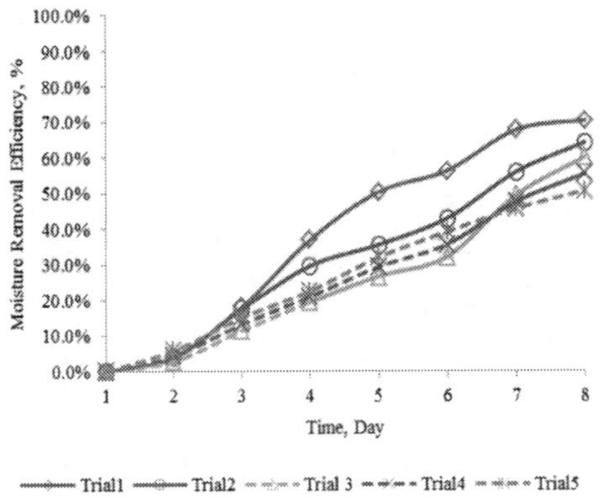

Figure 2. The moisture removal efficiency

Fig. 2, showed the moisture removal efficiency of trial A to E. The moisture removal efficiency for rotary bio-drying with combined hydrolytic and aerobic process was higher more than 50.0% on 8th day trial. Only the operation condition without air recirculation can achieved the moisture content below 25%. This condition proved that air ventilation or aeration was playing the important role for bio-drying assistance. The more increased direct air recirculation ratio was more decreased the moisture removal efficiency and more moisture content in the end product inversely.

3.2 Temperature and relative humidity

The temperature was the key parameter for water evaporation and organics degradation in bio-drying process. During the hydrolytic stage, the temperatures inside the reactor were higher than the ambient air temperature about 1-5 °C. The temperatures in the aerobic stage were inversely proportional to the temporal span time of the hydrolytic stage.

Trial A, a combined hydrolytic-aerobic bio-drying without air recirculation for the first 2 days of operation, the exhaust temperature and relative humidity were higher than ambient temperature and relative humidity, respectively. The exhaust relative humidity was declined when the air was supplied to the reactor (Fig. 3(a)).

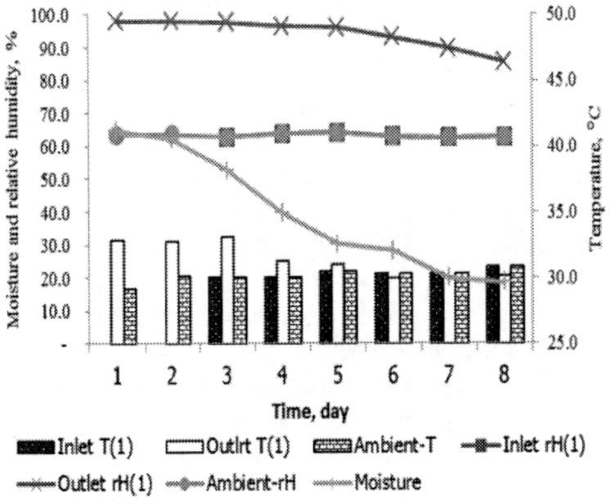

Figure 3(a) Temperature, Moisture and relative humidity profile of Trial A

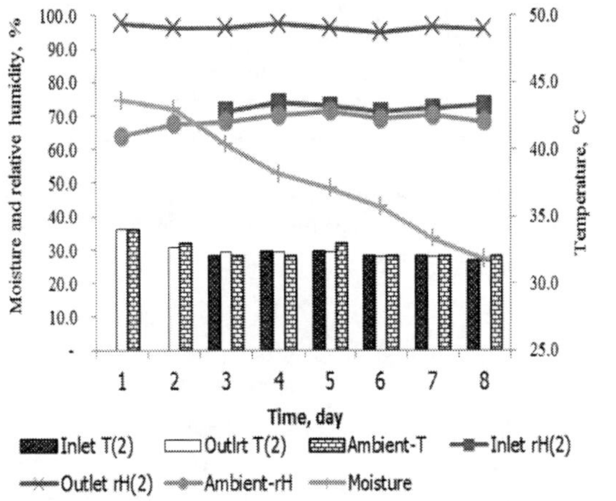

Figure 3(b) Temperature, Moisture and relative humidity profile of Trial B

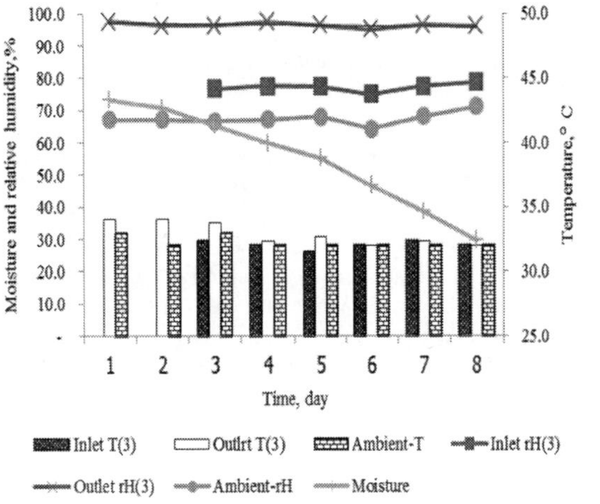

Figure 3(c) Temperature, Moisture and relative humidity profile of Trial C

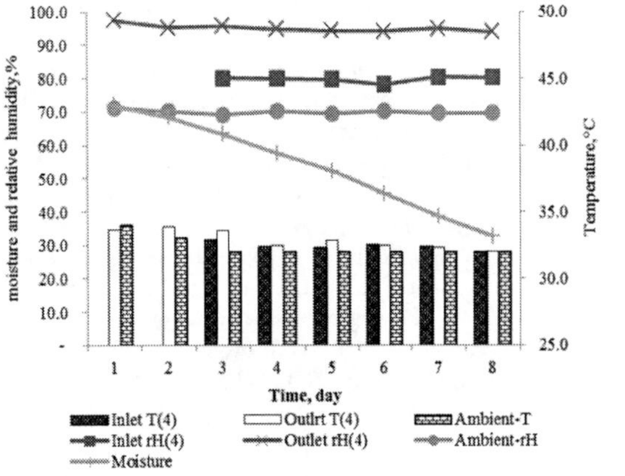

Figure 3(d) Temperature, Moisture and relative humidity profile of Trial D

The proper moisture content for decomposing was between 40% - 60 % and the highest decomposition rate was achieved at 50% of moisture content [4, 6]. At Trial A, the moisture content was higher than 40% within 4 days of operation and decreasing based on the daily rotation the rotary drum and supply air without recirculation (Figure 3a). For Trial B to E that operated with air recirculation, the moisture content was higher than 40% for a long time that proper for organic decomposition process.

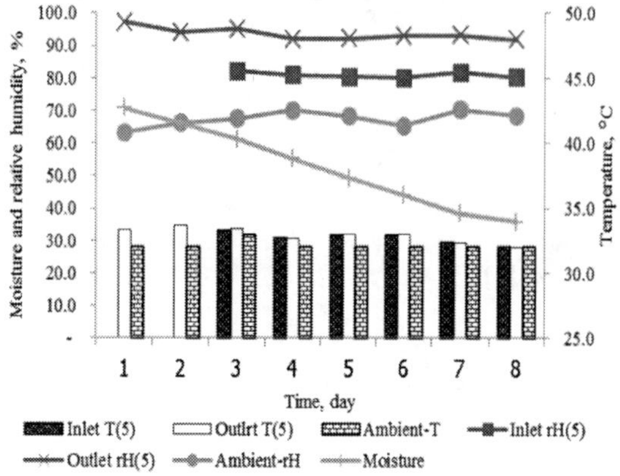

Figure 3(e) Temperature, Moisture and relative humidity profile of Trial E

For Figure 3b to 3e, an inlet relative humidity was higher than ambient relative humidity. An inlet relative humidity was decreased when increased the recirculation ratio. The exhaust relative humidity was stable when supplied the air recirculation into the reactor. The moisture content was decreased inversely of recirculation ratio increase.

Exhaust gas leave from rotary bio-drying process will be near saturated and almost the same temperature of reactor. If the difference between inlet and outlet air was greater than 25°C, the relative humidity of inlet air will a minor effect in determining overall moisture removal. The relative humidity will be increased base on recirculation ratio due to the high relative humidity of exhaust air. According to increasing of inlet relative humidity, the difference between inlet and exhaust air will be decreased while the moisture removal efficiency was decreased. Although the different of an inlet and exhaust air temperature and relative humidity were little. However, the drying process in rotary bio-drying could be occurred even in high inlet relative humidity. That was consistent with studied of Huag [3] which the drying could be occurring even in climate with high ambient humidity in air. Drying time increase slightly for 8-12% with recirculating 50% when exhaust air higher than ambient air at 15 °F (8.33 °C) [21].

Inversely, if there was added some the energy into the reuse air in mixing proportions superior to 10 % increased the thermal efficiency [19]. The evaporation rate for the airflow rate control trials was in average 15% higher compared to the conventional process, which indicates that the airflow rate control could be used to enhance also the performance of the dryer [20].

4 Conclusions

The effects of air recirculation on moisture removal efficiency in combined 2 days hydrolytic– 6 days aerobic rotary bio-drying process for drying high initial moisture content MSW was observed in this work. The air recirculation ratio was 0%, 25%, 50%, 75% and 100%. The moisture content of the end product was 18.6%, 27.1%, 29.5%, 32.7% and 35.8%, respectively. While increasing the exhaust air recirculation ratio, more moisture increases in the end product. However, while the air recirculation ratio was greater than or equals to 50 %, the moisture content in the end product did not has significant different. Only the operation condition without air recirculation can achieved the moisture content below 25% for RDF standard. When recirculate the exhaust air with a high relative humidity, moisture content of the end product did not meet the requirement of RDF standard.

References

1. Adani, F., Baido, D., Calcaterra, E., Genevini, P., Bioresource Technology. **83** (2002)
2. Bartha, B.K., *PhD Thesis*, Technische Universität Dresden, Dresden, (2008)
3. Haug RT, *Practical Handbook of Compost Engineering,* Lewis Publisher, 1993
4. Jolanun B1, Tripetchkul S, Chiemchaisri C, Chaiprasert P, Towprayoon S., Environ Technol, **26**(2005)
5. Khajon S, Tusanee T, Somrat K, Journal of Teknologi, **78** (2016)
6. Nabam Rich, Ajay Bharti, International Research Journal of Engineering and Technology (IRJET), **02**(2015)
7. Navaee-Ardeh S, Bertrand F, Stuar P R, Bioresource Technology, **101**(2010)
8. NA Ab Jalil, H Brasri, NE Ahmad Basri, Mohammed FM. Abushammala, Environmental Engineering Resource, **21**(2016)
9. Rada E.C., Ragazzi M., Panaitescu V., UPB Scientific Bulletin, Series D: Mech.Eng**.**, **71**(2009)
10. Sadaka S, Van Devender K, Costello T, Sharara M, University of Arkansas Division of Agriculture. 2010
11. Shao L M, Ma Z H, Zhang H, Zhang D Q, He P H, Waste Management, **30**(2010)
12. Skourides, I., Theophilou, C., Loizides, M., Hood, P., Smith, S.R., Waste 2006-41 Sustainable waste and resource management, Stratford-upon-Avon, UK 19-21 September 2006
13. Sugni, M., Calcaterra, E., Adani, F., Bioresource Technology **96**(2005)
14. Velis C A, Longhurst P J, Drew G H, Smith R, Pollard S T J, Bioresource Technology, **100**(2009)
15. Zawadzka A, Krzystek L, Ledakowicz S, Chemical Papers, **64**(2010)

16. Zhang D Q, He P J, Shao L M, Jin T F, Han JY, Journal of Environmental Sciences, **20**(2008)
17. Zhang D Q, He P J, Yu L Z, Shao L M, Bioresource Technology, **100**(2009)
18. Cherdsatirkul C, Earth Engineering Center Columbia University, (2012)
19. Juliana T. C. L. Toneli, Lucas B. Monteiro, Marco Aurelio J. Briso, Deovaldo Moraes Jr, Chemical Engineering Transactions, 32(2013)
20. T. Jokiniemi, T. Oksanen and J. Ahokas, Agronomy Research 13(1), 89–94, 2015
21. Jim Young, Jim Thompson and Tom Rumsey, Sun-Diamond Grower, (1981)

 DOI: 10.1051/matecconf/201710802008

Modelling of Thermal and Gas-Dynamic Processes in the Porous Structures

M.N. Gaynullin, V.A. Sotova, P.A. Chertykovtsev , E.O. Barbonov, S.V. Zaika

Samara University, Heat engineering and heat engines Department, Samara, Russia

Abstract. In this article is considered the use of porous material in heat exchangers. The programming model with a porous regenerator heat exchanger was compiled in the program ANSYS Fluent complex. The paper does comparative analysis of influence of different material porosity on flow parameters.

1 Introduction

In recent years, there is an important issue about creating a highly efficient compact heat exchangers for different purposes in many branches of industry including aviation, space, missile and laser.[1,2] At the same time, emerging tasks can be successfully solved only in case of the intensification of heat transfer processes [3].

One of the most promising and effective ways of heat intensification exchange processes is the use of heat exchangers porous metals. The physical basis of this method is in the high intensity of heat transfer between the metal frame and flowing coolant through it due to the high surface area of contact and effective mixing in the pores [4,5,6].

A wide range of porous materials properties, the ease of fabrication of these structural elements, the high intensity of heat – all these aspects enable to use the porous materials in diversified conditions.

Currently using of various technologies gives reasons for the creation of porous materials with different structures. For the above-mentioned areas the highly heat-carrier structures are in the great interest of technology.

Thermal and hydraulic calculation of heat exchangers on the basis of porous structures is carried out using mass conservation equations, pulse strength and energy with appropriate initial and boundary conditions. The mathematical model in general formulation is rather complicated and has no exact analytical solutions. That is the reason various kinds of assumptions and empirical relations for transfer coefficients in the equations and boundary conditions are introduced.

One of the most important parameters is the porosity of the material which is determined by the ratio of void volume to total volume of the porous body. The ideal environment is called a continuous porous plate which is pierced in the direction of the filtering system of cylindrical channel of the same diameter; its length is equal to the plate thickness. Real porous structure has significant features. The pores usually constitute a curved channel variable cross-section, both in form and in size.

For grabbing information about the hydraulic resistance mechanism and intensity of heat transfer at the moment of moving cooler in porous structures for the closure of a mathematical model, extensive experimental studies are needed. And if the characteristics of the structures of small and medium porosity and heat and mass transfer processes in these studied enough, these questions are not studied deeply enough for the high porosity structures.

2 The dynamic processes in porous structures calculation methodology

At the present time the CAE-systems are used for modeling of thermal and gasdynamic processes and for calculation of machines made of porous materials. [7]

Table 1. Calculated model initial parameters

Element	Radius, m	Length, m	Wall thickness, m	Material
The cold heat exchanger	$1 \cdot 10^{-2}$	$2{,}5 \cdot 10^{-2}$	$1{,}07 \cdot 10^{-3}$	Copper
Regenerator	$8 \cdot 10^{-3}$	$14 \cdot 10^{-2}$	$1{,}07 \cdot 10^{-3}$	The glass tube
The hot heat exchanger	$2{,}4 \cdot 10^{-2}$	$2 \cdot 10^{-2}$	$1{,}65 \cdot 10^{-3}$	Copper
Receiver	$2{,}6 \cdot 10^{-2}$	$1{,}49 \cdot 10^{-1}$	$5 \cdot 10^{-3}$	Stainless steel

DOI: 10.1051/matecconf/201710802008

2.1 The programme modelling in the program ANSYS Fluent complex

The programming model with a porous regenerator heat exchanger in the program ANSYS Fluent complex was compiled on the basis of a study of dependencies operating parameters of the design and other work. The model is a two-dimensional model of a longitudinal section, built in the "Compass 3D» (Figure 1). The model geometric parameters are presented in Table 1. It is needed to be mentioned that the system is closed. Pressure within the system is 0.1 MPa. Air in the system under the influence of the alternator Q-drive enters the cold heat exchanger, whereby the temperature drops.

For model work it is necessary to set the input parameters values of the calculated installation, such as the average pressure, frequency, volume rate, etc. (Table 2). The crank radius is set to 4 mm and the piston stroke - 8 mm.

Table 2. The model initial parameters at the entrance.

Parameters	Designation ANSYS Fluent	Value
Frequency, Hz	Freq	50
The initial gas temperature, K	K	300
The initial pressure value, Pa	Pa	$0.101325 \cdot 10^6$

Then air is drawn into a glass tube, wherein the porous insert is mounted. After insertion of the porous air enters the hot heat exchanger, where the temperature rises. Then, the air enters the receiver. The pressure in the receiver increases, while the pressure in the cavity to cool heat exchanger is relatively small.

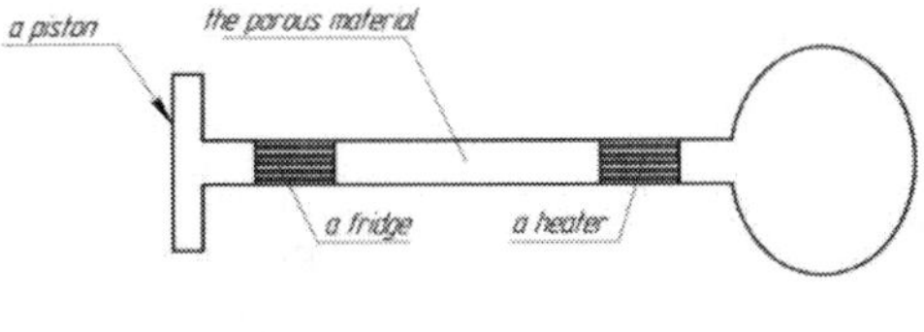

Figure 1. Estimated model regenerator.

Ansys Fluent software was used to modeling the processes in the cooler. There was built two-dimensional model, was made the appropriate settings of the boundary conditions. The model with a dynamic mesh was used to simulate the compressor piston vibrations. To set the piston motion in Ansys Fluent was used option «In-Cylinder», allowing to set the required piston motion law. Using this option the calculation step has been set according to the crank angle and was 1 degree. Using PC with a processor Intel Core I7-4770K 3.5 GHz, 32 GB, SSD 240GB calculation time of each calculation model was about 60 hours. On figure 1 and figure 2 shows the fields of total pressure in top dead centre and in bottom dead centre tor stainless steel material with porosity 0,6.

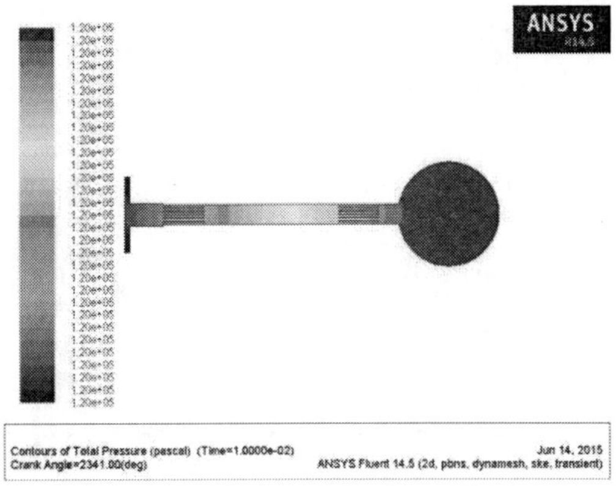

Figure 2. The field of total pressure in top dead centre (material: stainless steel, Porosity: 0,6)

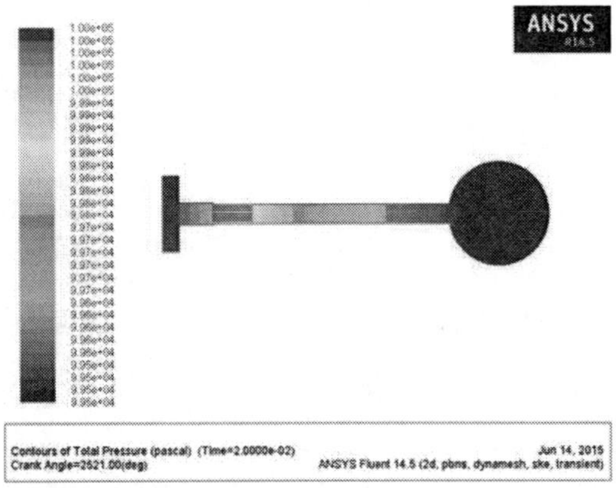

Figure 3. The field of total pressure in bottom dead centre (material: stainless steel, Porosity: 0,6)

2.2 Calculations and air flow parameters' dependence of the porosity

The following formula was used for thermal conductivity in a porous medium ratio calculation:

$$\lambda = \varepsilon\lambda_f + (1-\varepsilon)\lambda_s \tag{1}$$

According to the results of calculation, there are the graphs of pressure dependencies of the Reynolds number and velocity of the porosity of the material (Figures 4-6).

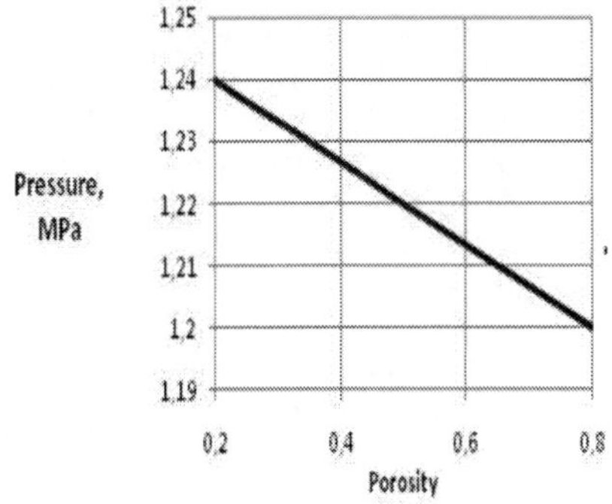

Figure 4. Pressure dependence of the porosity of the material

 DOI: 10.1051/matecconf/201710802008

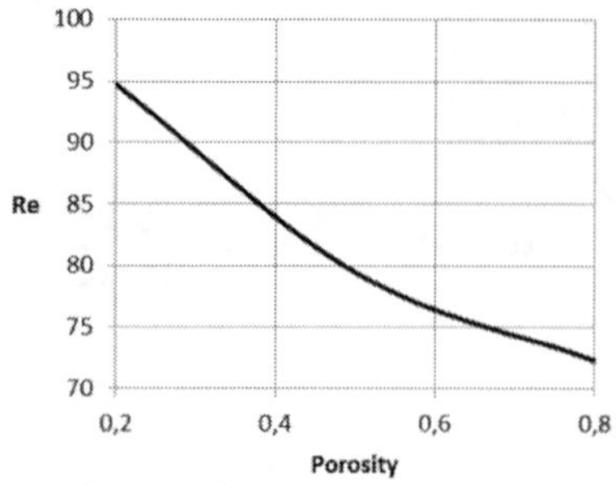

Figure 5. Dependence of Reynolds number on the porosity of the material.

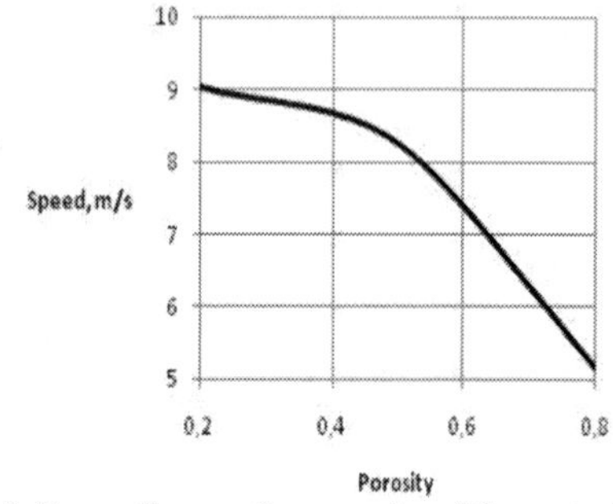

Figure 6. Depending on the porosity of the material speed.

For the most detailed research of the porosity effectiveness of the gas-dynamic processes, additional calculations were carried out using such materials as bronze, brass, stainless steel. During the calculation the coefficient of local loss, the coefficient of friction losses, the differential pressure on the porosity of the materials were obtained (figures 7-11).

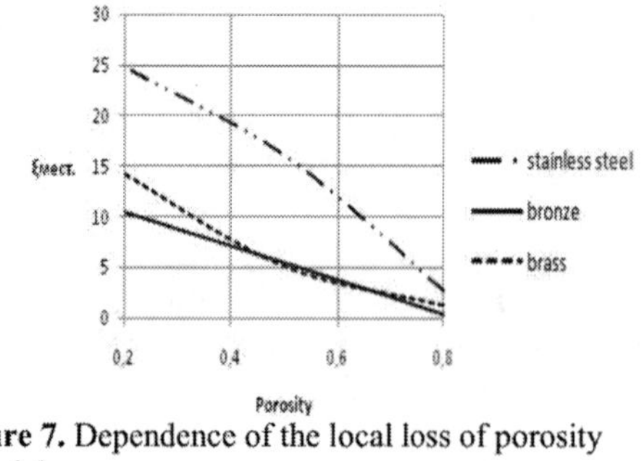

Figure 7. Dependence of the local loss of porosity materials.

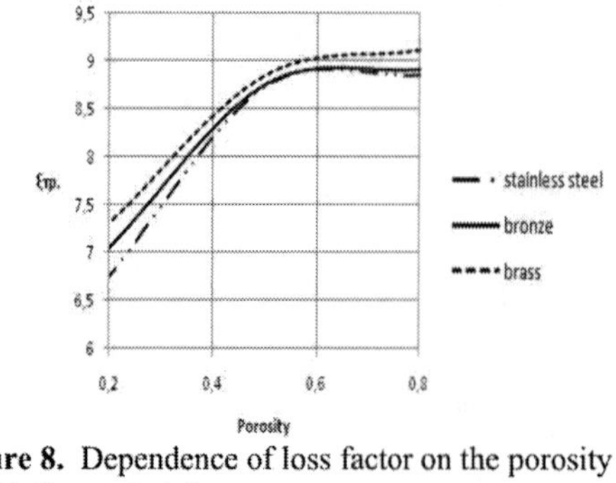

Figure 8. Dependence of loss factor on the porosity of the friction material.

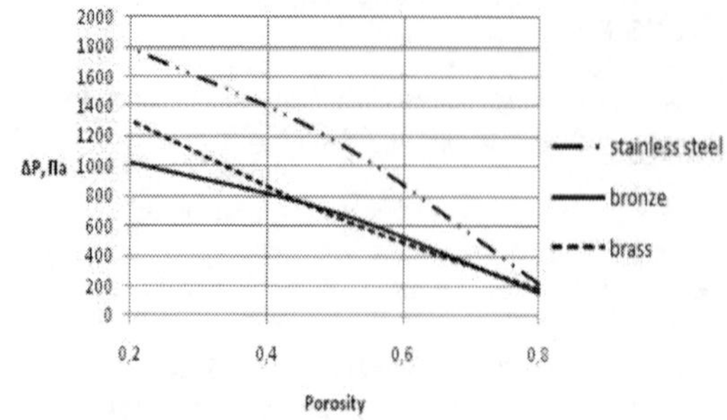

Figure 9. Differential Pressure dependence of the porosity of the material.

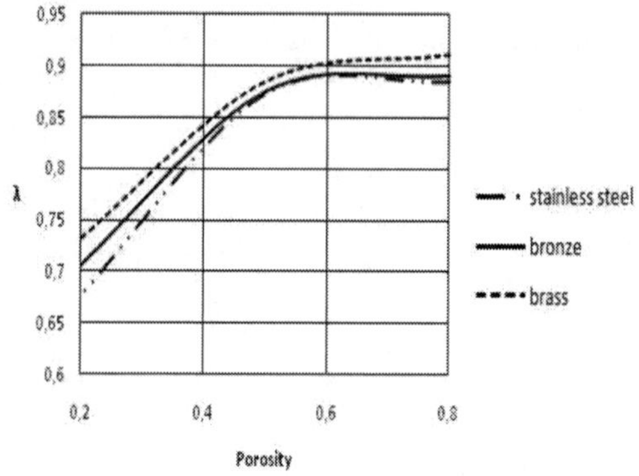

Figure 10. Dependence of the coefficient of hydraulic friction losses on the porosity of materials.

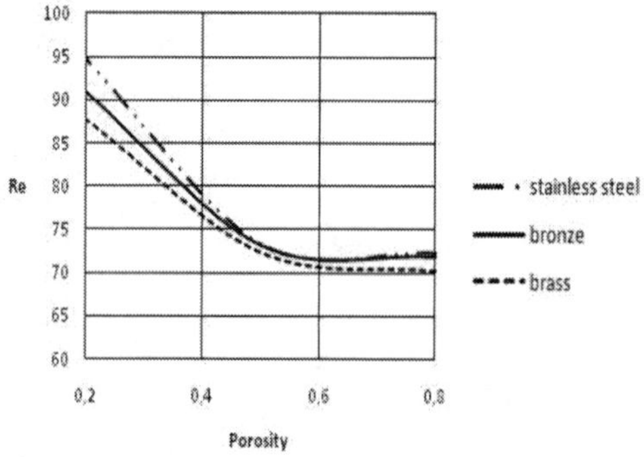

Figure 11. The dependence of the Reynolds number on the porosity of materials

3 Conclusions

The following conclusions are based on the results of the calculations:

1. With increased porosity materials differential pressure drops 7-8 times.

2. The loss factor decreases by 97%, taking into consideration the increasing porosity materials, whereas the coefficient of friction loss increases by 25% with increasing porosity from 0.2 to 0.5, and changes slightly with increasing porosity from 0.5 to 0.8.

3. The hydraulic loss factor increases with porosity from 0.2 to 0.5 by 30%, while the Reynolds number falls quarter. Further changes are minor.

References

1. T. Trollier, Air Liquid space cryocooler system, Cryogenics **48**, Pp. 267-273, (2008).
2. T. Trollier, T. Ravex, I. Charles, L. Duband, J. Mullié, P. Bruins, T. Benschop, M. Linder, Miniature 50 to 80 K Pulse Tube Cooler for Space Applications, International Cryocooler Conference, (2002).
3. A. M. Arkharov, Cryogenic systems: basic theory and calculation, Engineering, 464, (1988).
4. I. A. Popov, Gidrodinamika and heat transfer in porous heat exchange elements and devices, 240, (2007).
5. C. S. Kirkconnell, R. C. Hon, T. Roberts Raytheon Stirling/Pulse Tube Cryocooler Maturation Programs Cryocoolers **15**, 31-37, (2009).
6. W. Scheirer, T. Roberts, E. Pettyjohn, Multistage Pulse Tube Refrigeration Characterization of the Lockheed-Martin RAMOS cryocooler, Cryocoolers **15**, 39-47, (2009).
7. T. R. Ashwin, G. S. V. L. Narasimham and Subhash Jacob CFD analysis of high frequency miniature pulse tube refrigerators for space applications with thermal non-equilibrium model Applied Thermal Engineering, **30**, 152-166, (2010).

Utilization of abrasive water jet for cutting parts of intricate shapes

Safwan M. A. Al-Qawabah[1], Adnan I. O. Zaid[2] and Nasser Abdellatif[3]
[1]*Mechanical Engineering Department, Tafila Technical University, Amman, Jordan*
[2,3]*Mechanical and Industrial Engineering Department, Applied Science University, Amman, Jordan*

Abstract. As early as 1974 the British Hydrodynamic Research Association, BHRA, held the First International Conference on Cutting by Water Jets. The subject was at its early stages. Since then a large amount of research work has been carried out and the process has been greatly developed. In this paper, utilization of water jets for flexible cutting parts of intricate shapes in steel plates and granite is presented and discussed.

1 Introduction

The abrasive water jet (AWJ) cutting technique is one the most rapidly improving technological methods of cutting materials. In this cutting technique, a thin, high velocity water jet accelerates abrasive particles that are directed through an abrasive water jet nozzle at the material to be cut. Advantages of abrasive water jet cutting include the ability to cut almost all materials, no thermal distortion, and high flexibility, small cutting forces and being environmentally friendly. Because of these capabilities, this cutting technique is more cost-effective than traditional and some non-traditional machining processes [1]. The surface roughness is mainly a result of various controllable or uncontrollable process parameters and it is harder to attain and track than physical dimensions are. A considerable number of studies have researched the effects of the feed rate, standoff distance, water pressure, abrasive grain size, and other factors on the surface roughness [2, 3, 4]. Experimental measurements of vibrations and acoustic emissions accompanying AWJ cutting were conducted and their results investigated in-depth by Hloch et al. [5, 6] to provide a practical method of monitoring the process and predicting surface roughness. It is a well-established and recognized fact that the developments which underlie our present civilization e.g. the modern cars, locomotives, electrical trains, space crafts and modern house ware which we rely for our pleasure and livelihood have been made possible largely by the continued improvement in the surface quality of the produced parts by various casting, forming and machining processes. Accurate manufacturing and the production of near net shape parts of closely mating parts or duplicate parts to small dimensional limits with small tolerances is inseparably bound with high qualitymachined surfaces. Any effort to advance in the field of surface quality, however, leads inevitably to study the nature of the produced surface, its surface roughness and metallurgical structure. Hence, we have experienced in recent years outstanding developments in producing and measuring means for economic attainment of high quality produced surfaces. In this paper, six parts of intricate shapes, (five are made of mild steel and one made of granite) were produced by cutting with water jet. The surface roughness at the edges of the produced parts aremeasured and the obtained results are presented and discussed. Furthermore, the Vickers micro-hardness at their edges and at different distances were determined and discussed. Finally, the advantages of this method as compared with flame cutting and high explosive shaped charges are discussed.

2 Materials and experimental procedures

2.1 Materials

Two materials were used one ductile, ASTM A36 mild steel of the chemical composition shown in Table 1; from which parts A, B, C, D and E, were produced, Fig. 2 and a brittle material, granite, part F in Fig. 2.

Table 1. Chemical composition of ASTM A36 mild steel

Element	C	Cu	Fe	Mn	P	Si	S
Wt.%	0.25-0.29	0.2	98.0	1.03	0.04	0.28	0.05

2.2 Equipment

An abrasive water jet machine type (CMS EASYLINE 7785), shown in Fig. 1 was used for cutting six parts of intricate shapes four of them are made of ductile material, mild steel, A, B, C, D and E of the mechanical behavior shown in Table 1; and one is made of brittle material, granite, F. They are all shown in Fig. 2.

The microhardness test was performed using Microhardness machine tester model HWDM-3, where the average surface roughness (Ra) measurement was based on cut off distance = 0.8 and ISO 13565 (Rk), after each cutting test using surface tester type (kosaka surfcorder SE3500).

Figure 1. Abrasive water jet machine type (CMS EASYLINE 7785).

Figure 2. Flexible shapes, (A, B, C, D, and E) are made from mild steel and (F) is made from granite which were machined using the abrasive CNC water jet machine.

2.3 Experimental procedures

The experimental procedure started by adjusting the abrasive water jet machine to the working conditions defined in Table 2, followed by placing the work piece, the plate, and fixing it in position, preparing the abrasive water jet, adjust the CNC program, CREW software, and finally start the machine at the working conditions specified in Table 2.

3 Results and discussion

In this section, The surface roughness of the six produced parts was measured at five different locations on each part, Table 3, from which the average RMS was determined, the obtained results are presented, discussed and compared with values reported in the literature for different machining processes including finishing and super finishing processes, Table 4.

3.1 Comparison between surface quality of parts produced by Abrasive water jet and other Methods

Table 4 shows the values of the average applied values of surface roughness, Ra, and the less frequent produced by different machining processes, included for comparison purposes. It can be seen from Tables 3 and 4 that the surface quality of the intricate parts produced by abrasive water jets is comparable to the average applied surface quality produced obtained from chemical machining, CM, and electric discharge machining, EDM and better than other machining process illustrated in Table 4. Furthermore, it can also be observed in Table 3 that the parts of intricate shapes which are geometrically similar around the horizontal and vertical center lines produced better surface quality than the other parts as Ra(av.) falls between 1.92 μm and 2.24 μm for the first and between 4.08 μm and 4.65 μm for the latter respectively.

Table 3. Average surface roughness of the different produced parts by water jets.

Shapes	R_a	R_a	R_a	R_a	R_a	Ra (μmm)	Ra (μin)
A	1.79	1.97	2.0	2.03	1.81	1.92	75.59
B	1.86	2.16	1.8	2.1	1.95	1.98	77.95
C	3.90	3.3	4.6	4.4	4.20	4.08	160.63
D	3.82	4.80	4.36	4.91	4.13	4.40	173.23
E	4.22	4.65	4.28	4.97	5.30	4.65	183.07
F(granite)	2.65	2.06	2.38	2.22	2.41	2.24	88.19

Table 2. Constant parameters and their values

Constant parameters	Orifice diameter (mm)	Focusing tube diameter (mm)	Water jet pressure (MPa)	Abrasive type	Abrasive size (grit no)	SOD (mm)
Value	0.25	0.762	200	GMT (TiO_2)	80 mesh (200gr/min) Size: 5 μm, ρ =5 mg/m^3	2

Table 4. The average applied frequent methods of different machining processes [7]

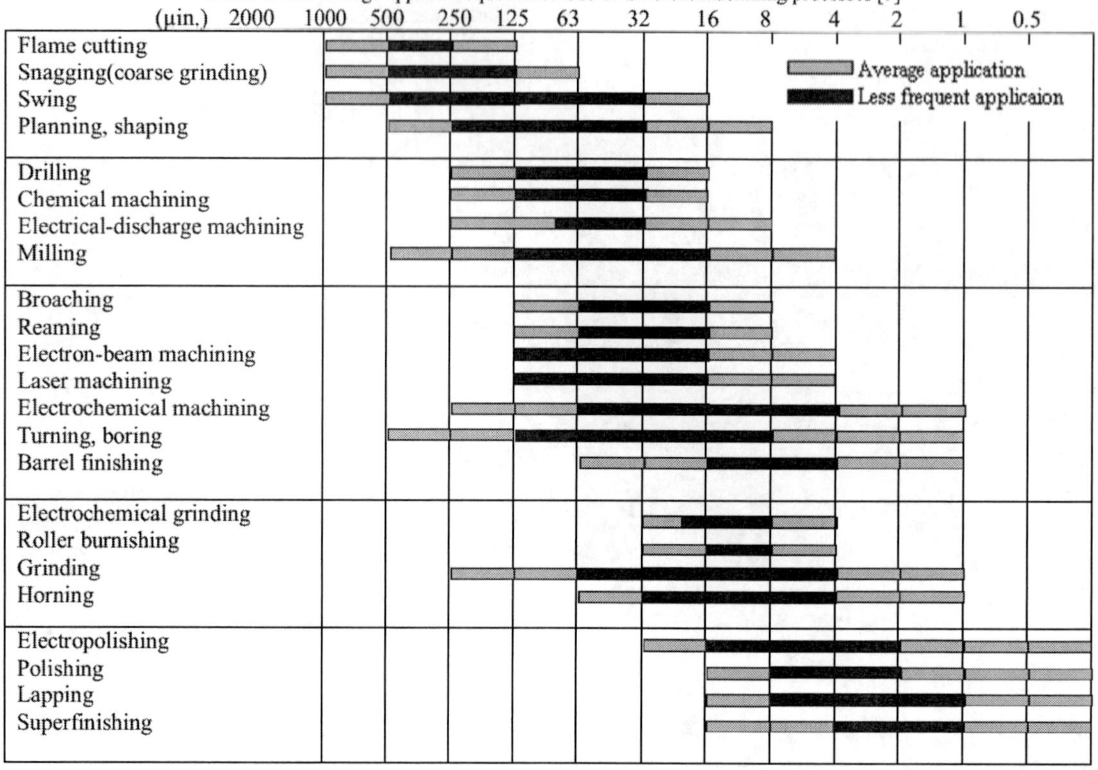

3.2 Advantages of cutting by abrasive water jets

The following are its advantages over other cutting processes:

i. Very good surface quality can be obtained of the produced parts; better than some other cutting (material removal processes) e.g. flame cutting, oxyacetylene cutting and cutting by high explosive charges.
ii. Very little or no material loss as chips.
iii. It produces parts effectively from both ductile and brittle materials.
iv. It is capable of producing parts of intricate shapes with very small tolerances which is almost like the original part.
v. It does not require highly qualified technicians like the other machining processes e.g. turning, milling, shaping, grinding.
vi. Although its capital cost is higher than some other machining processes, it is outweighed by the the operational low cost, hence it can be considered economical process.
vii. Little or no temperature rise in the produced parts or the remaining parent part due to the low pressure of the jet; hence no metallurgical changes are expected to take place in any of them.
viii. Wide range of sizes, weights and shapes of different materials. Thickness up to 222 mm
ix. Less precautions and no preparation of the machined parts are required as compared with other cutting and machining processes.

3.3 Effect of cutting by abrasive water on vickers micro-hardness

To investigate the effect of cutting by the abrasive water jet on the hardness, the produced parts were sectioned across their geometrical center lines and the Vickers microhardness was measured at seven locations at the edges of the cut and at equal distances of I mm away from it, from which the average microhardness was determined and the results are presented in Fig. 3a) and b) for mild steel and granite respectively. It can be seen from these two figures that very slight variation in their Vickers microhardness has taken place with a maximum increase of 1.23 % and of 1.8 % decrease in the mild steel work pieces and retained their original hardness after 5 mm from the cutting edge. Regarding the granite work piece no drop in its microhardness was observed and the increase in it varied between maximum increase of 2.59 % and 2.81 %. These values are much smaller than the increase in cutting by other cutting means like oxyacetylene, plasma, laser or high explosive shaped charges where more than 25 % has taken place [8].

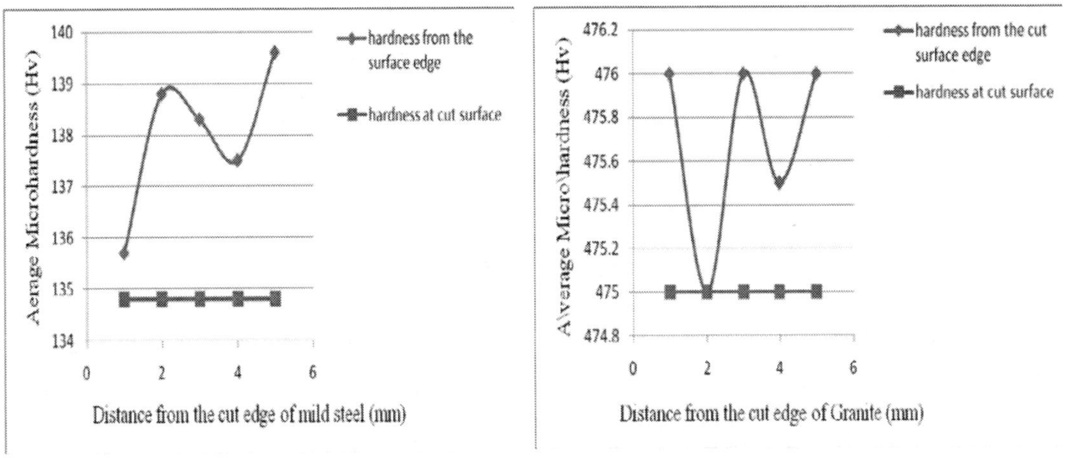

a) Mild steel b) Granite

Figure 3. Variation of the Vickers micro-hardness at the edge of the cut and a distance from it.

4 Conclusions

Based on this research work and within its limitations; the following points are concluded:

i) The abrasive CNC water jet can produce net intricate shapes with small tolerances and very high surface quality comparable to the super finishing processes. The maximum average RMS of the produced parts is 4.65 µm and the minimum average value is 1.98 µm.

ii) Slight increase in the Vickers microhardness, HV, at the edges of the produced parts but it retains its original hardness at 5 mm from the cutting edge.

iii) Little or no temperature rise in the produced parts or the remaining parent part due to the low pressure of the jet; hence no metallurgical changes are expected to take place in any of them.

iv) Cutting with abrasive CNC water jet may be considered economical process although the capital investment is relatively high compared with other machining means. However, it is outweighed by the low operating costs.

Acknowledgement

The first author would like to express his thanks to technician Amer Amayerah and to Tafila Technical University support.

References

1. D. Krajcarz: *Comparison metal water jet cutting with laser and plasma cutting*, (Proceeding of 24th DAAAM International Symposium on Intelligent Manufacturing and Automation, Procedia Engineering 69, 2014) p. 838-843.
2. R. Kovacevic: Surface texture in abrasive water jet cutting Journal of Manufacturing System Vol. 10 (1991), p, 32-40.
3. A. Akkurt, M.K. Kulekci, U. Seker and F. Ercan: Effect of feed rate on surface roughness in abrasive water jet cutting applications, Journal of Materials Processing Technology Vol. 147 (2004), p. 389-396.
4. D.A. Axinte, D.S. Srinivasu, M.C. Kong and P.W. Butler-Smith: Abrasive water jet cutting of polycrystalline diamond; A preliminary investigation, International Journal of Machine Tools & Manufacture Vol. 49 (2009), p.797-803.
5. D. Kinik, B. Ganovska, S. Hloch, P. Monka, K. Monkova, and Z. Hutyrova: On-line monitoring of technological process of material abrasive water jet cutting. Tehnički vjesnik Vol. 22 (2015), p. 351-357.
6. P. Hreha, S. Hloch and V. Perzel: Analysis of acoustic emission recorded during monitoring of abrasive water jet cutting of stainless steel AISI 309. Tehnički vjesnik Vol.19 (2012), p.355-359.
7. S. Kalpakjian, S.R.Schmid: *Manufacturing Processes for Engineering Materials*, 4th Edition, Prentice Hall Publishers, (2003).
8. A.I.Zaid, Cutting plates by high explosive shaped charges, M.Sc. Dissertation, UMIST, (1971).

Paste Backfilling Used for Coal Fire Control in Thailand

Manoon Masniyom and Dang Vu Hai

Department of Mining and Materials Engineering, Faculty of Engineering, Prince of Songkla University, Hat Yai, Songkhla 90112, Thailand

Abstract. The environmental effects of spontaneous combustion coal seam fires in Mae Moh Lignite Mine in Thailand are subject to a wide array to land surfaces disturbance and air pollution. Air pollution in coal fire is mainly due to the fugitive emission of particulate matter and gases including sulphur dioxide (SO_2) and oxides of nitrogen (NOx). The traditional methods, executed by fire fighting teams by covering the coal fire areas with soil, excavated burning coal outcrops and injecting water in the coal seam fires are continuously improved. Initiatives to introduce modern techniques, such as backfill placement at fracture and coal seam to cool down the burning coal and cut off air supply. This study is being conducted to investigate the properties of cement and cement kiln dust as backfill materials. As part of this investigation, mixture of materials were prepared by combining cement and cement kiln dust in various ratios of 0.35, 0.4, 0.45, 0.5, 0.55 and 0.6, to determine their optimal mix proportions for compressive strength development and permeability. The compressive strength of the mixtures of cement and cement kiln dust specimens continued to increase with increasing curing time. It was found the maximum compressive strength of the mixture of 20 % cement and 80 % cement kiln dust was 6.74 MPa mixing a water to solid ratio 0.46. The permeability of mixtures were decreases when cement kiln dust increases.

1 Introduction

Mae Moh Lignite Mine is located at Lampang, the northern city of Thailand. It's the largest open-pit mine in Thailand with mining area about 28 km^2. The mine geological reserve is about 1,140 million tons of coal (814 million tons mineable). Since 1969, 411 million tons of coal has been extracted. The mine and its mine-mount power plant are operated by Electricity Generating Authority of Thailand. About 16 million tons of coal is harvested annually and supplied to the power plant. This amount of coal can produce electricity accounting for about 10 % of Nation's electric power generation. [1]

The environmental effects of spontaneous coal seam fires in Mae Moh Mine are subject to a wide array to land surfaces disturbance and air pollution. Air pollution in coal fire is mainly due to the fugitive emission of particulate matter and gases including sulphur dioxide (SO_2) and oxides of nitrogen (NO_x) are shown on Figure 1. Locally, coal fires pose serious health and safety risks from toxic fumes and land surface collapses that destroy infrastructure [2]. On a global basis, they contribute large quantities of greenhouse gases to the atmosphere without providing any of the benefits from energy consumption. The traditional methods, executed by fire fighting teams by covering the coal fire areas with soil, excavated burning coal outcrops and injecting water in the coal seam fires are continuously improved. Initiatives to introduce modern techniques, such as paste backfill placement at fracture and coal seam to cool down the burning coal and cut off air supply. Paste backfill is a uniform, low permeability, generally consisting of high solids density mixture. Due to compaction, paste backfill constituent particles do not settle out of suspension at zero flow rates with the fine particles of the matrix forming an annulus around a plug of coarser particles to act as a form of lubrication thereby greatly reducing pipeline frictional resistance.

Figure 1. Coal fire in Mae Moh Lignite Mine

Paste backfill uses very small particles of backfill to lubricate the pipeline and therefore requires less water for transportation. Since there is considerably more solid materials in paste backfill, water drainage from the backfill in the stope is generally negligible. A paste backfill may contain between 75 % and 88 % solids (by weight). Cement must always be added to paste backfill

DOI: 10.1051/matecconf/201710802010

to prevent the risk of backfill liquefaction [3]. First used during the 1980's, paste backfill is typically a low permeability, high density solids mixture of aggregate solids and water containing at least 15 % by weight fines material at sizes ranging between minus 20 µm (625 mesh) to 45 µm (325 mesh) content. Because of its higher density state, less water is required to manufacture paste backfill than slurry backfill. When pumped, paste backfill's constituent particles do not settle out of suspension at zero flow rate. This is due to the fact that, once backfill volumetric concentration increases above 50 % solids, particles are no longer free to settle and flow within a pipe becomes that of a plug. Plug flow is the mechanism used to transport highly concentrated paste backfill mixtures. In this state, the bulk of material flows as a core of interlocked and water saturated solid particles surrounded by a thin lubricating layer consisting of a homogeneous mixture made up of water and very fine particles. This high density backfill is able to form a plug as a result of a careful mix design governed by the content of minus 325 mesh fines and the grading of the coarser aggregates.

Cement kiln dust (CKD) is a by-product of cement manufacturing therefore, sometimes considered an industrial waste. A medium size cement plant may produce up to 30,000 tons of CKD annually. Most of the material is disposed off on-site without any further reuse or reclamation [4]. El-Sayed et al. have investigated the effect of CKD on the compressive strength of cement paste and on the corrosion behaviour of embedded reinforcement. The study reported that up to 5 % substitution of CKD by weight of cement had no adverse effect on cement paste strength and on the reinforcement passivity. A similar conclusion was reached in an investigation carried by Batis et al. [5]

2 Materials and methods

The materials used in this investigation consisted of Portland cement type I and cement kiln dust from cement plant. Cement kiln dust is a fine, highly alkaline powder. This laboratory test program was implemented to study the relation between the water and solid ratio, compressive strength development and permeability. Particle size analyses were conducted on the backfill sample using a laser particle size analyser and the results are shown on Figure 2. Particle size distribution is one of the most important characteristics of any backfill material is a well-graded particle size distribution [6]. The distribution has the strongest effect on backfill porosity and delivery. In the case of paste backfill, it is the portion of the materials finer than 20 µm, which seems to have the most pronounced effect on pump ability.

Portland cement type I was replaced by cement kiln dust at percentages of 0, 5, 10, 15 and 20 by weight of cementitious materials, and all mixture proportions of paste backfill are shown in Table 1.

The backfill materials, cement, cement kiln dust and water were first mixed and homogenized in a double spiral mixer to produce the paste backfill mixtures. After that, slump tests, according to ASTM C 143-90, were done to evaluate the transportability of the backfill. The pulp density and the water/solid ratios (w/s) were also determined to evaluate the water demand of the paste backfill. The paste backfill mixtures produced were then poured into curing cylinders, 10 cm in diameter and 20 cm high. These samples were then sealed and cured in a humidity chamber at approximately 80 % humidity and 23 oC. After curing times of 28 days, the uniaxial compressible strength (UCS) according to ASTM C 39, the specific gravity, particle size, bulk density, water content and permeability of each paste backfill sample were determined.

Table 1. Mixture proportions of paste backfill

Materials	Mix 1	Mix 2	Mix 3	Mix 4	Mix 5
Cement (%)	0	5	10	15	20
CKD (%)	100	95	90	85	80

3 Results and discussions

3.1. Paste backfill materials characteristics

Table 2 and Figure 2 show the physical characterize of materials; specific gravity determination, particle size measurement, bulk density, water content and permeability. During the investigation water and solid ratios (w/s) 0.6, 0.55, 0.5, 0.45, 0.4 and 0.35 (by weight) were chosen for various mixtures by vicat test. Water and solid ratio below 0.35 and above 0.55 could not be mixed in a satisfactory way.

Table 2. Physical characteristics of backfill materials

Samples	Bulk density (g/cm^3)	Specific gravity (g/cm^3)	Permeability K_0 (m^2)	Water content (%)	d_{50} (µm)
Cement	1.53	3.03	4.36 E-17	1.52	12.5
CKD	1.14	2.38	3.8 E-16	1.71	80.5

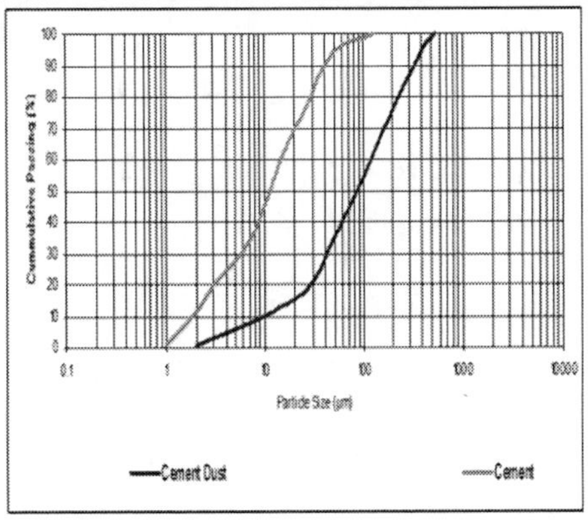

Figure 2. Particle size distribution of backfill materials

3.2 Chemical properties

Chemical composition was determined by using X-ray fluorescence spectrometer (XRF). The results are shown in Table 3. Total lime (CaO) constitutes more than 50 % of CKD composition. It is relatively high in alkaline content (K_2O and Na_2O).

Table 3. Chemical composition of backfill

Chemical composition (%)	Cement dust	Cement
SiO_2	13.5	21.3
Al_2O_3	4.34	4.64
Fe_2O_3	2.72	3.57
CaO	52.33	62.45
MgO	2.71	2.43
Na_2O	0.15	0.2
K_2O	2.98	0.5
MnO_2	-	0.04
P_2O_5	-	0.04
TiO_2	-	0.02
SO_3	1	2.7

3.3 Uniaxial compression tests

For the study, A total of 90 specimens were tested by uniaxial compression machine to investigate compressive strength and subjected to uniaxial compression tests using a mechanical press having a normal loading capacity maximum of 1,000 kN. The results are shown in Figure 3. The compressive strength of the cement and mixtures of cement and cement kiln dust specimens continued to increase with increasing curing time. It was found the maximum compressive strength of the mixture of 20 % cement and 80 % dust filter (Mix 5) was 6.74 MPa mixing a water to solid ratio 0.46.

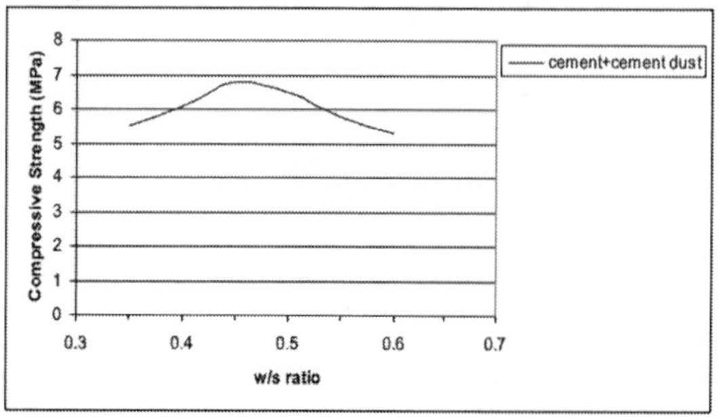

Figure 3. Compressive strength of paste backfill mixtures at 28 day

3.4 Application of Backfill in coal fires

The mixtures of backfill are mixed at the mixing area to be used to fight the coal fire. The paste backfill was placed to coal seam fracture and distributed around the source of the coal fire are shown in Figure 4. The basic behaviour of coal fires depends on the transport and supply of oxygen in the subsurface and therefore mainly on the permeability, thermal conductivity and specific heat capacity of the material. Because of this, systematic experiments, carried out in the laboratory on backfill samples, yield the functional dependency of these parameters of the material. The experiments have shown that if backfill mixture of cement, cement kiln dust and water are placed into high temperature zones of burning coal or even used for rewetting an oxygen isolating sand cover, water will evaporate and the cement will solidification. Besides the cooling effect, the pores of the permeable rock will be partially closed by binder. The resulting lower permeability will prevent further oxygen transport.

Figure 4. Paste backfill placed to coal seam fracture

In general, higher backfill strengths are achieved by adding more binder, so costs are higher. However, any required strength can usually be achieved using a number of different mixes, so testing is necessary to identify the optimum mix from the available materials. The relative prices of portland cement and cement kiln dust, the cost-effectiveness of chemical additives and the potential savings from disposal of tailings or industrial waste as backfill must all be incorporated into the determination of operating costs.

4 Conclusions

The paste backfill mixture is a very low permeability and high density that can placed to the coal seam and distributed around the source of the coal fire. The resulting lower permeability will prevent further oxygen transport and control coal fire. The permeability of mixtures were decreases when cement kiln dust increases. The compressive strength of the mixtures of cement and cement kiln dust specimens continued to increase with increasing curing time. Furthermore backfilling voids enhances the stability and improving safety of the mine.

References

1. A. Laksanayothina and W. Ariyawong, Groundwater treatment of Thailand's Mae Moh lignite mine, Proceeding of International Conference on Environmental Issues in Mining and Natural Resources Development (EMNR):November 2016, pp.2-6 (2016)
2. L. Jing, S. Voight, C. Kunzer, Y. Bol, Z. Jianzhoung, Z. Yianmin, K. Bing, and Z. Songmeil, The Progress in Detecting of Coal Fire on Remote Sensing, The First Result of The Joint SINO-German Research Project on Innovative Technologies for Exploration, Extinction and Monitoring of Coal Fires in North China, Proc. 2005 Dragon Symposium "Mid-Term Results", Santorini, Greece 27 June – 1 July, 7 p. (2005)
3. Y. Potvin, E. Thomas and A. Fourie, Handbook on Mine Fill, Australian Centre for Geomechanics, 179 pp. (2005)
4. H.A. El Sayed, N.A. Gabr, S. Hanafi and M.A. Mohran, Reutilization of by-pass kiln dust in cement manufacture, International Conference on Blended Cement in Construction, Sheffield, UK, September 1991.
5. G. Batis, E. Rakanta, E. Sideri, E. Chaniotakis and A. Papageorgiou, Advantages of simultaneous use of cement kiln dust and blast furnace slag, International Conference on Challenges of Concrete Construction, University of Dundee, Dundee, UK, September. (2002)
6. A.G. Grice, Recent Minefill Developments in Australia, Proceedings of the 7th international symposium on mining with backfill, Seattle, Washington, pp. 351-358. (2001)

Material Research of Rotary Sealing Device for Combined Cutting System

Zeng Rui[1,2], Zhang Yong[2], Lin Zhenrong[2] and Wang Lulu[1]

[1] *College of Field Engineering, PLA Univ. of Sci. & Tech., Nanjing 210007, China*
[2] *Department of Airfield Engineering, Air Force Logistics College, Xuzhou 221000, China*

Abstract. In order to solve the rotary sealing problem of rotary shaft in drum shearer combined cutting system, the material and structure of combined cutting system rotary sealing device needs to be selected and designed. In the paper, the rotary sealing structure of four grades in series was designed first, and then the material of NBR-40 and PTFE 4FT-4 under the separate static and dynamic sealing tests were carried out on the combined tooth-slip-ring sealing test-bed. The tests show that the NBR-40 O-Ring with PTFE 4FT-4 tooth-slip-ring has no low-pressure leakage problem and low leakage in the sealing progress, the sealing effect of which is the best.

1 Introduction

The high pressure water jet which is supplied to the nozzle of combined cutting drum needs to flow through the hollow rotary shaft of the drum. The leakage of high pressure water can lead to a significant reduction in water pressure, which can not reach the requirements of cutting hard coal rock, and the leakage of water flow into the rocker gear box will greatly affect the gear transmission[1]. The sealing form of drum shearer & water jet combined cutting system's cutting part is a typical high pressure dynamic rotary sealing, and to solve the problem of high pressure water sealing is very important. On the premise of meeting the requirements of the high sealing pressure, the starting performance and the lubrication effect of the drum rotary shaft must be considered.

2 Structure of rotary sealing device

The rotary sealing device structure design requirements of combined cutting system are as follows:

(1) The structure is simple and the size is small, and should be matched with rotary shaft parameters.

(2) Can withstand more than 40MPa sealing pressure.

(3) Can withstand 50r/min rotational wear.

(4) The sealing device is reliable, long service life and easy to replace.

The overall structure of the rotary seal designed in this paper is shown in Figure 1. First of all, the rotary sealing device did not change the basic structure of spray system of shearer, and only the end of the rotary shaft was lengthened 30mm, which could be modified in the coal mining machine directly. The high pressure water flowing through the hollow rotary shaft could take away a lot of friction heat, which effectively improved the seal working environment and increased the sealing life. Secondly, the four grades in series sealing structure model was adopted. The four grades in series sealing was separated by sealing sleeves and the internal contact pressure balance was adjusted, which avoided the rotary shaft of uneven wear. If the first grade sealing failed, part of the high pressure water could drop from the first leakage point to give a warning, and then the second grade sealing began to work, and so on. four grades in series sealing structure can extend the service life of the rotary sealing effectively, and reduce the times of sealing replacement. The wear of the sealing ring can be detected at any time, and the damaged sealing also has the role of resisting the pressure for the next grade sealing.

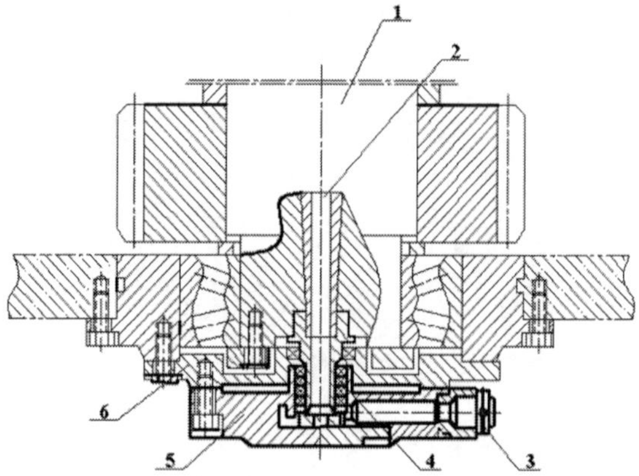

1-drum shaft; 2- stainless steel pipe; 3- pipe connection; 4- four grades in series sealing; 5- end cap; 6- inner end cap

Figure 1. Structure of rotary sealing device

After comprehensive comparison and selection of various types of rotary sealing[2~4], the KNOR361 series tooth-sliding-ring of Xuzhou Kono seal Co. Ltd was chosen to use, shown in Figure.2.

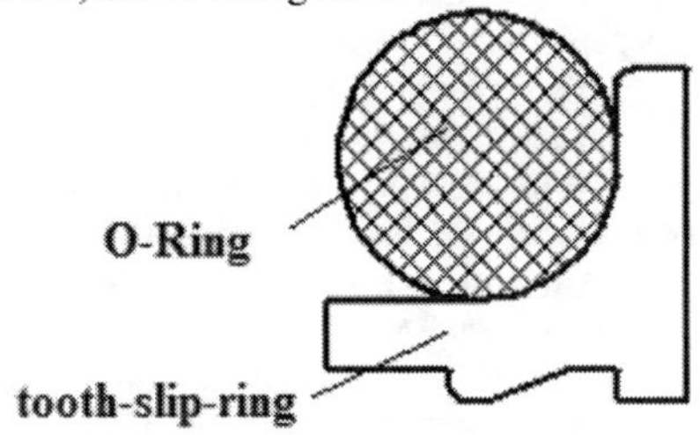

Figure 2. Combined tooth-slip-ring sealing

The KNOR361 series tooth ring lip is thin and easy to be deformed. It can produce enough deformation to compensate for wear under the elastic force of O type ring. The tooth ring side can block the O ring to prevent from being squeezed out, which improves the overall sealing capacity of tolerance and pressure.

3 Material test and result analysis

3.1 Building test-bed of combined sealing

3.1.1 Structure of test-bed of combined sealing

In the combined cutting system, rotary sealing department working environment is high pressure, medium rotation rate and room temperature, so the composite materials of high pressure and strong wear resistance tend to be chosen. According to the criterion of HG/T 2579-2008 and the practical experience[2,5,6] , the material of NBR-40 and PTFE 4FT-4 are consistent with the requirements of the part of rotary sealing. The test-bed of combined tooth-slip-ring sealing was built to test the practical results of the NBR-40 and PTFE 4FT-4, shown in Figure.3.

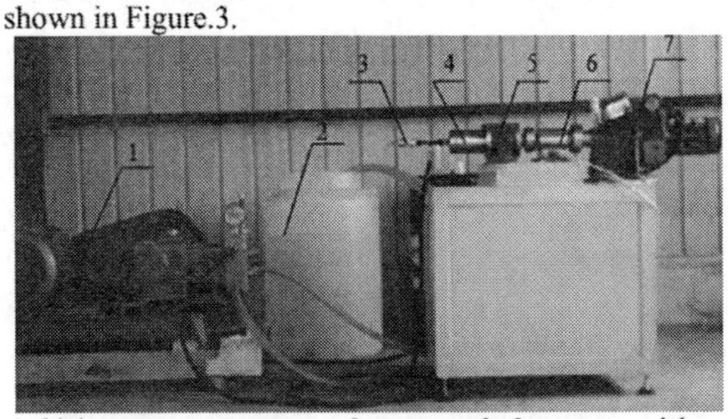

1- high pressure water pump; 2- water tank; 3- pressure pickup; 4- rotary sealing part;5- Coupling part; 6- torque sensor ; 7- electric motor

Figure 3. Test-bed of combined tooth and slip-ring sealing

3.1.2The components and functions of the test-bed

(1) The 3SP40-A series high-precision high pressure pump provides a stable 0~80MPa water jet pressure, and the pressure of the pump's outlet can be adjusted.

(2) The water tank is used to provide experimental water.

(3) The measurement range of JNBP-2-100MPa pressure pickup is 0~100MPa. Its output signal is the voltage signal to measure the pressure of the water outlet

(4) The rotary sealing part adopts the four grades in series sealing structure model, and single sealing is tooth-sliding-ring combined sealing, shown in Figure.4. The main purpose of the test is to test the sealing performance of the rotary sealing.

(5) The coupling part is composed of transmission shaft and transmission cylinder, which is connected with the rotary sealing part to transfer power to the rotary sealing system and plays the role of supporting and protecting.

(6) The measurement range of JN-DNJ-100 torque sensor is 0~100N·m. Signal output is a kind of low voltage standard signal, which is used to measure the torque of the output shaft.

(7) The Y2-B590-4 electric motor is the power device of the system, which is used to simulate the rotation of the drum shaft.

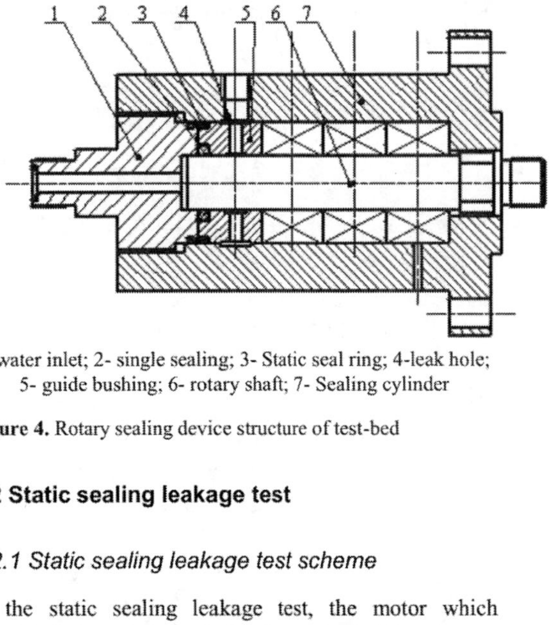

1-water inlet; 2- single sealing; 3- Static seal ring; 4-leak hole; 5- guide bushing; 6- rotary shaft; 7- Sealing cylinder

Figure 4. Rotary sealing device structure of test-bed

3.2 Static sealing leakage test

3.2.1 Static sealing leakage test scheme

In the static sealing leakage test, the motor which simulated the rotation of the drum shaft did not start, while the high pressure water pump started. The material of NBR-40 (group A) and PTFE 4FT-4 (group B) rotary shaft combined tooth-slip-ring sealing were test to check the sealing effect. In order to facilitate the assessment of the sealing effect, the sealing parts were sealed continuously after 4 hours and the seal leakage per hour under different water pressure as the standard. It was found that when the leakage of water was more than 5500 mL/h, high pressure pump would have obvious pressure and power fluctuations, which could seem to be sealing failure.

3.2.2 Static sealing leakage test result analysis

Under static sealing, when the pressure was in the middle and low pressure stage of 10MPa and 20MP, the group A and group B were all reliable sealing in 4 hours of continuous static sealing, but the group B had a small amount of leakage and the total leakage is about 65 mL/h. This is because the hardness of the PTFE 4FT-4 is relatively large, and it is difficult to produce enough deformation to form the tight sealing tape in low pressure. But the leakage too small to effect on the sealing performance.

When the pressure was in the high pressure stage of 30MPa and 40MPa, group B formed a tight sealing tape gradually and its leakage phenomenon disappeared. After 4 hours' continuous sealing, there was no leakage basically. In the high pressure stage, The first grade sealing ring of group Acould only maintain about 20 minutes of stable seal, and about 60minutes later the overall sealing failed.

In order to solve the problem of group A and group B, the NBR-40 O-Ring with PTFE 4FT-4 tooth-slip-ring was adopted as group C. In the static seal leakage test, the group C demonstrated excellent sealing performance. In the middle and low pressure stage, Total leakage of group C was only 8 mL/h, and in the high pressure stage, there was no leakage as group B.

3.3 Dynamic sealing leakage test

3.3.1 Dynamic *sealing leakage test scheme*

In dynamic sealing leakage test, the high pressure water pump was started and the pressure was adjusted to 40Mpa. The motor is started to simulate the rotation of the drum shaft, and 7 different speed segments are set in turn. Because the group A was not suitable for the high pressure seal of the rotating sealing part, only group B and group C were test in the dynamic sealing leakage test. The group B and the group C were continuously work for 4 hours at 7 different rotational speeds, and the relation of rotate speed and leakage was shown in Figure.5

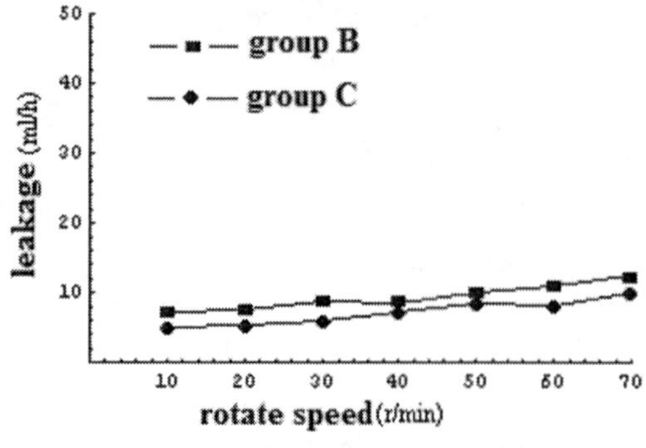

Figure 5. Relation of rotate speed and leakage under dynamic sealing

3.3.2 Dynamic *sealing leakage test result analysis*

Both group B and group C could meet the requirements of dynamic sealing leakage test, and Leakage was only about 10mL/h, which was belong to a minor leak, and the slight leakage could be beneficial to the seal itself.The sealing leakage rate of group B was slightly higher than group C. This showed that the NBR-40 seal ring produced more pressing force to tooth ring under the pressure of 40MPa, which caused the lower rotary seal leakage and better sealing effect of group C.

The signals collected by the torque sensor were sent to the dynamic signal acquisition instrument which was DASP-V10, shown in Figure.6. Tte statistical of the two groups of signals was processed by DASP-V10 with its own signal analysis software, and the average torque of group B was 34.732 N·m and group C was 39.401 N·m. The resistance of the group B to the rotation of the drum shaft was slightly less than that of the group C, because the friction factor of PTFE 4FT-4 was less than the NBR-40.

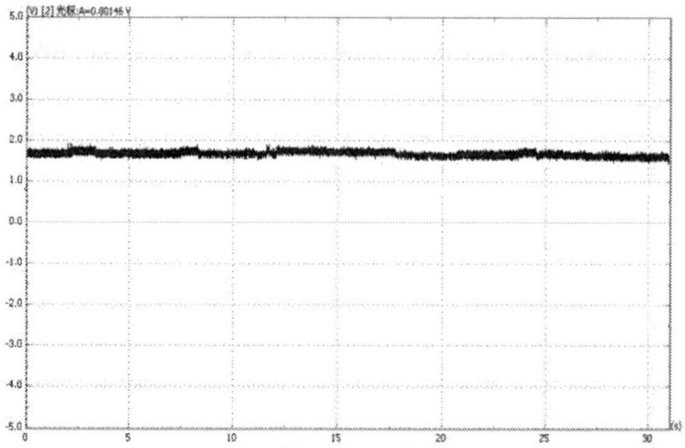

(a) Group B

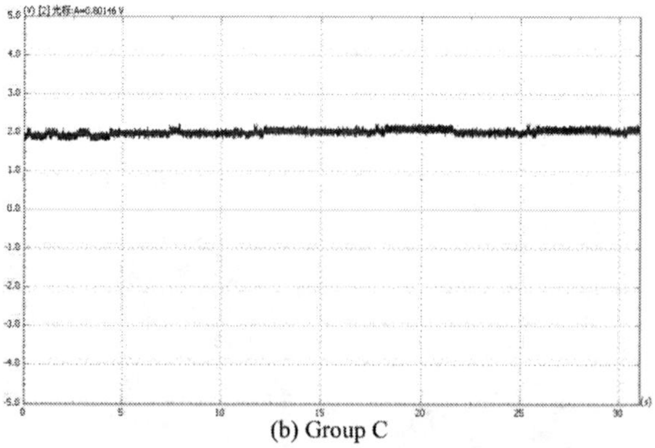

(b) Group C

Figure 6. DASP-V10 output curve of dynamic sealing

3.4 FEM Analysis of combined tooth-slip-ring sealing

In the process of finite element analysis, in view of the complexity of the boundary conditions, the rotating shaft and seal guide sleeve of the sealing structure are analyzed as a whole, which is shown in Figure.7.

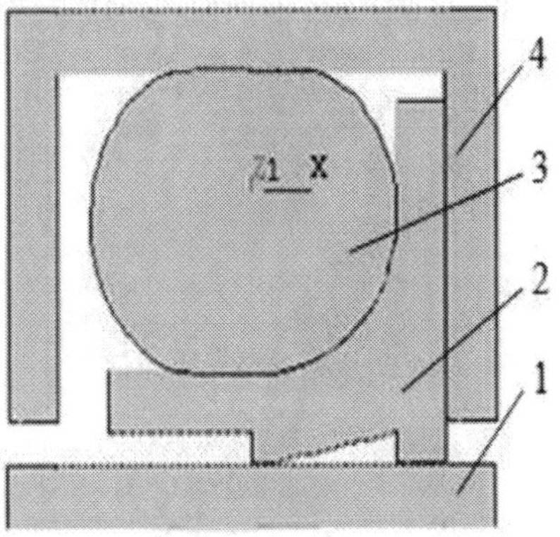

1- Rotation axis; 2- Toothed ring; 3-O ring; 4- Seal guide sleeve
Figure 7. Geometric model of combined seal

According to the contact of the definition of standards, the O ring is the contact body, and the slip ring is the target. Rotary shaft seal guide sleeve and the slip ring use the linear solid element PLANE42 simulation, and rubber unit adopts super elastic element PLANE182 simulation, which is shown in Figure 8.

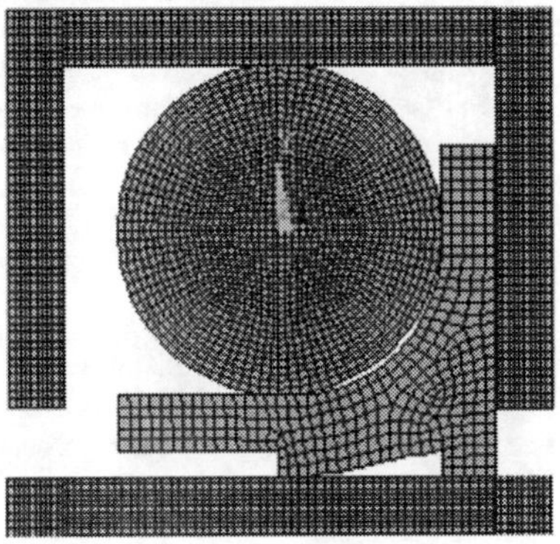

Figure 8. FEM model of combined seal

Under the loading condition, the equivalent stress distribution of tooth ring is shown in Figure 9 and the tooth contact pressure nephogram of slip ring combined seal is shown in Figure 10.

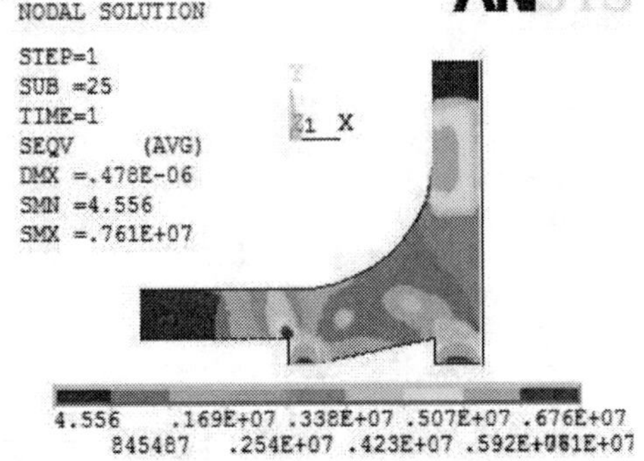

Figure 9. Von-mises stress distribution of sliding ring

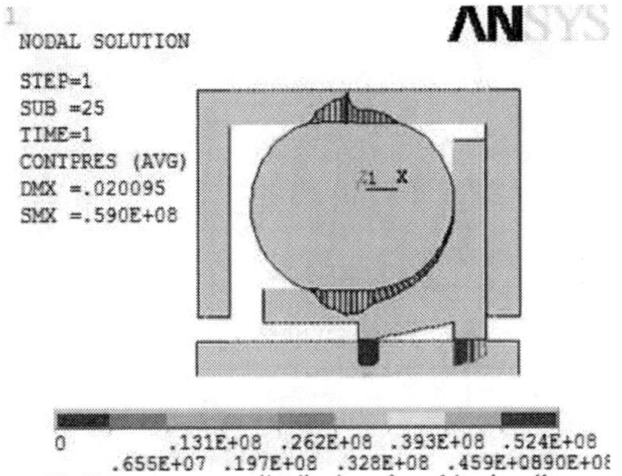

Figure 10. Contact pressure distribution of combined sealing

From the Figure 9 and 10, we can see that the main sealing lip and a secondary sealing lip at the tooth ring exist some stress concentration. The stress distribution of other positions is relatively uniform, and the stress value is small. From the sealing point of view, due to the stress concentration of the sealing lip, the contact surface produced a high contact pressure to achieve good sealing effect.

4 Conclusions

(1) The rotary sealing device did not change the basic structure of spray system of shearer, and only the end of the rotary shaft was lengthened 30mm, which could be modified in the coal mining machine directly.

(2) The rotary shaft tooth-slip-ring made by NBR-40 could not meet the sealing requirements of combined cutting work pressure, while the PTFE 4FT-4 could, but a small amount of leakage occured at low pressure.

(3) The NBR-40 O-Ring with PTFE 4FT-4 tooth-slip-ring had no low-pressure leakage problem and low leakage in the sealing progress. Although the resistance to rotation of the drum shaft is larger than full PTFE 4FT-4 material, comprehensive consideration of various factors, the group C should be selected.

References

1. M.S. Cui, J.J. Sun. Coal Industry, **2**,21(1993).
2. Q. Dong. Journal of China University of Mining and Technology, **4**, 54(2013).
3. J.C. Zhang, Q.M. Wang, H.F. Li. Lubrication Engineering, **36**, 59(2011).
4. S.H. Chen. Mechanical Engineer, **8**, 19(2000).
5. Markus, Henzler. Lubrication Engineering. **35**, 93(2010).
6. Z.F. Li, C.H. Gao, Y.X. Lin. Journal of plastics industry, **33**,13(2005).

Local Reinforcement in Notched Fiber Metal Laminates by Bonding Titanium Patch

Jipeng Zhang, Yue Wang, Jiazhen Zhang and Zhengong Zhou

Center for Composite Materials and Structures, Harbin Institute of Technology, Harbin 150001, China

Abstract. In order to reduce the stress concentration in notched fiber metal laminate (FML) and improve its damage tolerance, the local reinforcement concept was employed in present paper. Two different reinforced structures were proposed by bonding titanium patch on single side or double sides of the standard glass fiber reinforced aluminum laminate (GLARE). Tensile tests were conducted on these reinforced GLARE, and their mechanical properties were compared for evaluating the reinforced effect of titanium patch, as well as the damage behavior which dominated the mechanical response was analysed later to capture the failure mechanism of the reinforced GLARE. The reinforced effect of titanium patch on mechanical response of GLARE laminate was usually characterized by the improvements of modulus and yield strength, as well as the significant increase of ultimate strain and strength. Cohesive failure and adhesive failure were detected in the adhesive between titanium patch and GLARE laminate, where the adhesive failure corresponded to the weak reinforced case. In addition, the variation of delamination in GLARE laminate was also closely related to the reinforced effect on mechanical response. It is expected that the present investigation can provide technical support for damage tolerance design of fiber metal laminates.

1 Introduction

With the increasing requirements of damage tolerance in aeronautics, FMLs which consist of thin metal sheets and fiber reinforced polymer composites have gained more attention, especially after the successful application of GLARE in the A380 aircraft [1]. The FML concept was originally proposed for fatigue critical applications. While with its development, some potential advantages over the conventional metal alloys and fiber reinforced composites in structural components applications have also been recognized, which usually in terms of the high strength and excellent resistance to impact, corrosion and moisture [2-5].

In structural design, the blunt notches are usually inevitable, which usually come in the form of rivet holes or bolt holes. The mechanical behavior of structure components with holes should be paid particular attentions, since the undesired stress concentration and its accompanying properties degradation or even premature failure possibilities will constitute a challenge to the service life. A few investigations have been carried out to evaluate the notched behavior of FMLs. The effects of notch geometry and size on notched strength and notch sensitivity were investigated experimentally, from which it was concluded that the sizing effect was more significant [6, 7]. Moreover some notched FMLs have also been tested under off-axis loading, in which the off-axis dependence of notched behavior was evaluated [8, 9]. Except the experimental investigations, numerical simulations were also performed on the notched FMLs, where the damage patterns of the constituents in FMLs and their corresponding damage evolution processes were analysed [10-12].

The damage tolerance design requires the damaged structural components to sustain the limit load in service life, and then the reinforcement concept should be paid more attention, for the purpose of preventing the premature failure. In past decades, reinforcements on conventional metallic structures and composite structures have been proposed, such as the bonded patch and tear strap method, as well as the interlaminar doubler solution. The bonded tear strap method was also adopted in FML [13], and on basis of it an internal tear strap concept was proposed benefiting from the hybrid nature of FML [14]. Moreover, a structure termed as CentrAl has also been developed [15]. It should be noted that the present investigations were only focused on the improvement of fatigue resistance of FML, while almost no studies about the reinforcement of notched FML have been carried out.

Based on the previous investigations, the bonded patch reinforced concept was employed in notched GLARE in present paper. Mechanical response and damage behavior of both the single patch reinforced GLARE (SPR-GLARE) and double patches reinforced GLARE (DPR-GLARE) were investigated. For reference, those of the standard GLARE were also presented. Accordingly, the reinforced effect of titanium patch on notched behavior of GLARE laminate was clearly illustrated.

DOI: 10.1051/matecconf/201710802012

2 Experimental procedures

2.1 Materials and specimens

The FML under investigation in present paper is the standard GLARE 3-3/2 laminate, in which three 2024-T3 aluminium layers and two S4C9-glass/SY-24-epoxy glass fiber reinforced epoxy composite (GFRP) layers were alternatively bonded up, and each GFRP layer consisted two cross-ply laminas. Thus a symmetric configuration was obtained for GLARE 3-3/2 laminate, which was given as [Al/0/90/Al/90/0/Al], as shown in Figure 1. In order to achieve a well bonding performance, the aluminium sheets were treated by means of the phosphoric acid anodizing method following the ASTM D3933, and after drying an epoxy primer was sprayed on them. The stacked laminates were cured in an autoclave at 120 °C for two hours, where a heating rate of 2 °C/min was adopted. At approximately 70 °C, the autoclave was pressured to 0.5 MPa, and it was maintained till the end of the curing process. It should be noted that, the GLARE laminate under curing was kept in a vacuum bag with a vacuum degree of 0.1 MPa.

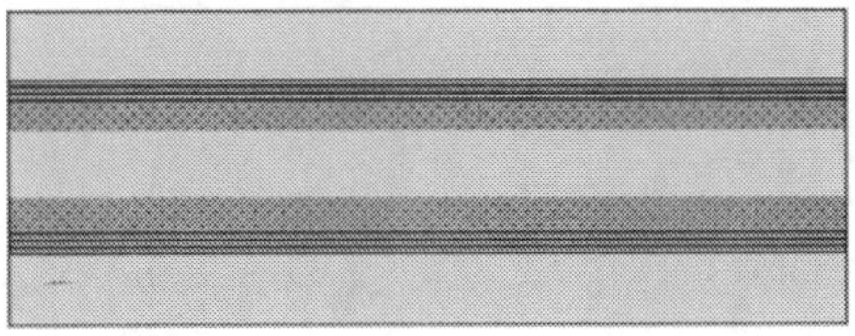

Figure 1. Configuration of the GLARE 3-3/2 laminate.

All test specimens were cut from the GLARE panels with dimensions of 370 mm × 370 mm by using the water jet cutting method. The length (*L*) and width (*W*) of the specimen was 250 mm and 25 mm, respectively, as well as the thickness (*t*) was 1.8 mm, which was calculated based on the thickness of the constituents (0.4 mm for aluminium sheet and 0.15 mm for GFRP lamina). Circular open hole with a diameter (*d*) of 5 mm in the center of the specimen was also made by the water jet cutting method. Titanium patches with diameters (d_r) of 10 mm, 12.5 mm, 15 mm, 17.5 mm and 20 mm were cut from a 0.57 mm titanium sheet with using the wire electrical discharge machining method, and center circular holes with the same diameter as that in GLARE specimen were also made. The titanium patches were bonded on single or double sides of the GLARE specimens, thus two different reinforced structures were obtained, referred to as the single patch reinforced GLARE and double patch reinforced GLARE, respectively. The adhesive used for bonding the patch was an epoxy film adhesive (J272), which should be cured at 120°C. Prior to bonding, the patches and GLARE specimens were well roughed with the sandpaper. The geometries and dimensions of the titanium reinforced GLARE specimens are shown in Figure 2, in which the *x* direction denotes the longitudinal or 0° fiber direction, while the *y* direction denotes the lateral or 90° fiber direction.

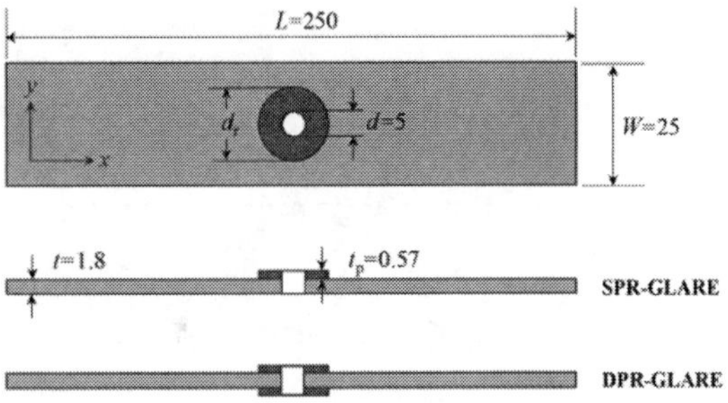

Figure 2. Geometries and dimensions of titanium reinforced GLARE specimens (dimensions in mm).

2.2 Test methods

Quasi-static tensile tests were conducted on the titanium patch reinforced GLARE specimens at room temperature with employing a 50 kN INSTRON 5569 testing machine. For reference, the standard or non-reinforced GLARE (NR-GLARE) specimen was also tested. Tensile load was applied with displacement control in *x* direction at a cross-head speed of 1.0 mm/min, following the standard ASTM D3039. The displacement in the loading direction was measured with the LE-05 laser extensometer, in which the gauge length was specified to stride over the local reinforced region.

After test, the damage behaviour of the adhesive films between patches and GLARE laminates were examined optically with employing a Canon 700D camera. In addition, the outer aluminium layers of the post-failure specimens were chemically removed by means of the etching method, which was carried out in a sodium hydroxide solution. After this etching process, the damage patterns of the GFRP layers were examined, for the purpose of illustrating the effects of bonded patches and their dimensions on failure behaviour of the GLARE laminates.

3 Results and discussion

3.1 Stress-strain relationships

In order to evaluate the reinforced effect of titanium patch on mechanical response of GLARE laminate, the stress-strain curves of SRP-GLARE and DPR-GLARE were firstly investigated, as shown in Figure 3. For reference, that of the standard GLARE or NR-GLARE is also present. It should be noted that the stress mentioned here is the gross stress, which is calculated by dividing the applied load by the overall width of the specimen. Similarly, the notched strength discussed in later sections is the gross notched strength.

As can be seen that, the stress-strain curves of the reinforced GLARE laminates present a typical bilinear behaviour as that of the NR-GLARE showed, in spite of the slight descending in the second parts which are marked in the dashed boxes in Figure 3. Unlike the stress-strain curve of the unnotched GLARE laminate, slow transition zones between the first and second parts are detected in stress-strain curves of these notched specimens, which correspond to the local plastic deformations of aluminium in the net sections that resulted from the stress concentration in the vicinities of the holes.

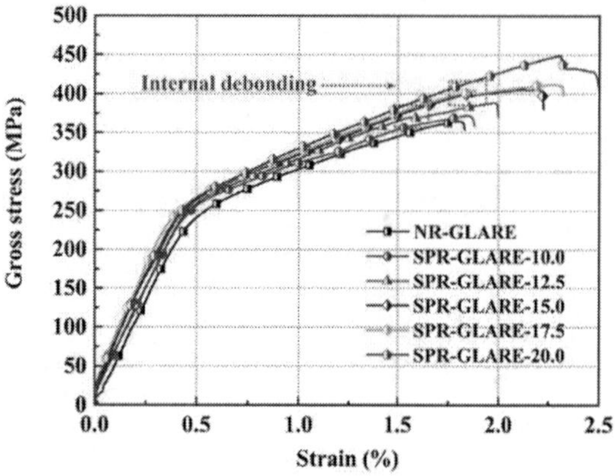

(a)

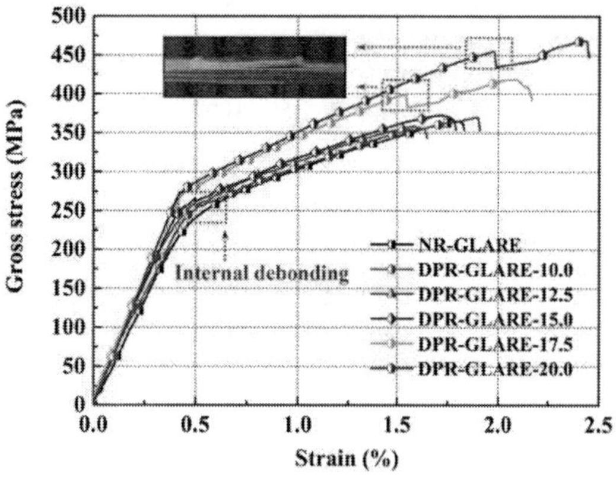

(b)

Figure 3. Typical stress-strain curves of (a) SPR-GLARE and (b) DPR-GLARE.

In addition, it can be seen that remarkable increase of the elastic modulus and yield strength are obtained by the reinforced GLARE laminates, especially for the DPR-GLARE with large titanium patches. Moreover, the tangent modulus in second parts of the stress-strain curves increase as well, meanwhile significant increase of the ultimate strain is also detected. The increase of modulus and ultimate strain is attributed to that the load applied on the GLARE specimen is partly shared by the titanium patch, benefiting from its relative higher elastic modulus as compared with those of the constituents in GLARE laminates.

However, as observed from Figure 3(b), almost no increase of the tangent modulus (in second parts) and ultimate strain are achieved by DPR-GLARE with small patches (10 mm, 12.5 mm and 15 mm). It indicates that less work has been done by small patches in the later loading stage, namely that debonding between these small patches and GLARE laminates are preliminarily encountered. As expected, slight descending of the stress-strain curves are detected just after the elastic parts in Figure 3(b), which are also audible in experiment. Note that this debonding is an internal case, for which cannot be detected visually in experiment. Similarly, the slight descending of stress-strain curves of SPR-GLARE with large patches are also caused by this internal debonding. For DPR-GLARE with larger patches, obvious descending of the stress-strain curves are detected just before fracture. They are associated with the edge debonding, as observed from the small image in Figure 3(b), which was captured during the experiment. It is interesting to note that significant rise of the stress-strain curves are detected after these descending, while those of the SPR-GLARE are negligible, which suggests that the double patch reinforced method is more effective.

3.2 Notched strength

For further evaluating the reinforced effect of titanium patch on mechanical response of notched GLARE laminates, the notched strength of SPR-GLARE and DPR-GLARE were compared and their corresponding strength increasing percentages over the NR-GLARE were also present, as shown in Figure 4.

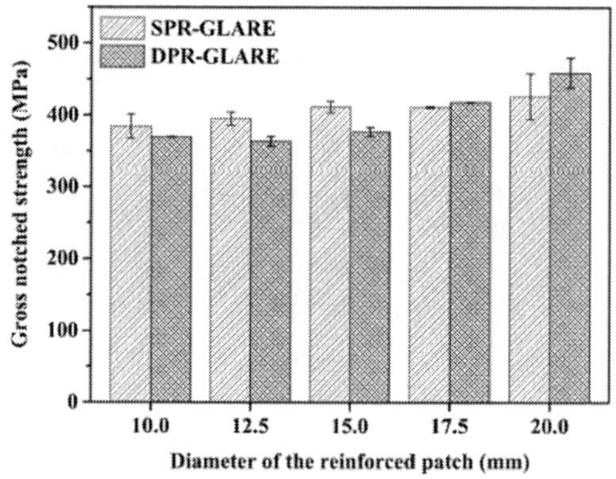

(a)

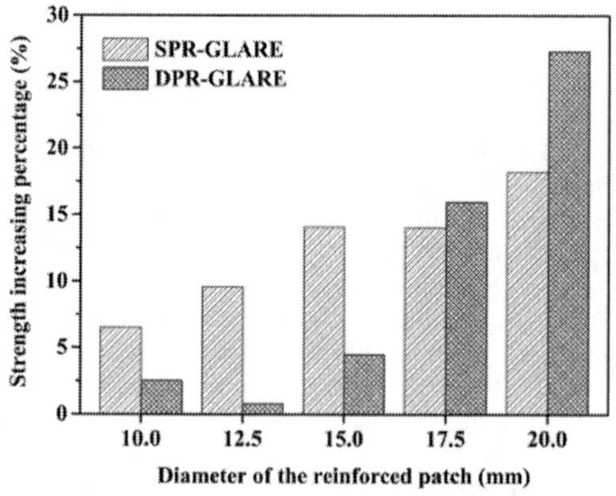

(b)

Figure 4. (a) Notched strength of reinforced GLARE and (b) the corresponding strength increasing percentages.

It can be seen from Figure 4, general upward trends of notched strength with increasing the patch diameter are obtained by both the SPR-GLARE and DPR-GLARE.

When small patches (10 mm, 12.5 mm and 15 mm) are adopted, the reinforced effects of titanium patches on notched strength of GLARE laminates are more significant for the single patch reinforced cases, while tiny strength increasing percentages are obtained by the DPR-GLARE, which are less than 5%. This phenomenon is in contradiction with the general mind, which may be attributed to the introduction of thermal residual stress, for which the bonding process of titanium patch on GLARE laminate was operated at an elevated temperature of 120°C, as well as their thermal expansion coefficients are mismatch. For double patch reinforced case, the thermal residual stress effect is more remarkable, thus the reinforced effect is seriously weakened.

As the patch diameter exceeds 17.5 mm, the reinforced effect of DPR-GLARE becomes more prominent, and a strength increasing percentage of 27.31% is obtained by the DPR-GLARE with two 20 mm titanium patches. While less increase of the notched strength of SPR-GLARE is achieved as further increasing the patch diameter from 15 mm to 20 mm. With the increasing of patch diameter, the thermal residual stress effect becomes negligible as compared with the significant increase of local stiffness in the reinforced region, accordingly a higher load bearing capacity will be achieved and which is more remarkable for DPR-GLARE. In addition, due to the asymmetric configuration of SPR-GLARE, a bending effect will be introduced at the application of external load and it becomes more serious as the patch diameter increasing. The bending effect will constitute a risk to the loading bearing capacity of the SPR-GLARE, since it can result in an increase of the peeling stress in the adhesive film, as well as a higher tensile stress can be formed at the un-reinforced side of the laminate. Based on the above discussions, it should be noted that attention must be paid on both the thermal residual stress and bending effect when the secondary cocure reinforcement is adopted.

3.3 Damage behavior

For the purpose of illustrating the reinforced mechanism of titanium patch on GLARE laminate. The damage behavior of reinforced GLARE which played a dominant role in the mechanical response was emphatically investigated in present paper. Firstly, the failure behavior of the adhesive film between the patch and GLARE laminate was examined. It is of utmost importance, since the adhesive function as a medium for load transferring between the patch and GLARE laminate, accordingly it will be the most direct reflection of the reinforced effect of titanium patch. Secondly, the damage behavior of the GLARE laminate was characterized, it mainly focused on that of the inner GFRP layer, for which was the determinant of the final fracture of the reinforced GLARE.

3.3.1 Damage behavior of adhesive

Figure 5 shows the damage behavior of adhesive films between titanium patches and GLARE laminates in SPR-GLARE and DPR-GLARE.

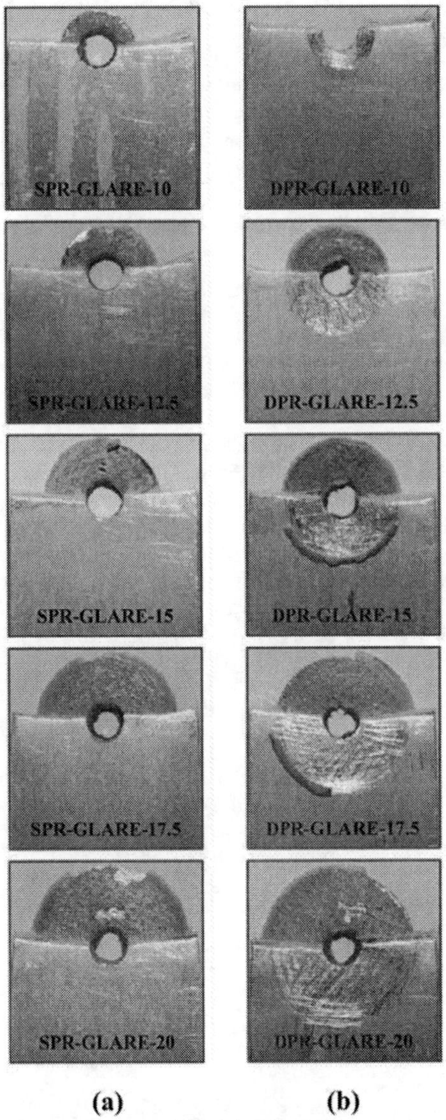

Figure 5. Damage behavior of the adhesive between patch and GLARE laminate: (a) SPR-GLARE; (b) DPR-GLARE.

As observed from Figure 5 (a), the damage patterns of adhesive films between patches and GLARE laminates in SPR-GLARE are characterized by the combination of adhesive failure and cohesive failure, which denote respectively the weak and strong load transfer capacity. For small patch case, the cohesive failure occurs at the free edge of the patch, it is also detected in the vicinity of the hole when the patch is large enough, as well as some punctate cohesive failure can also be detected. Positions where cohesive failure take place suggest that higher shear stress and peel stress present at the free edge of the patch, as well as in the vicinity of the hole, which is similar to that in previous investigation [16].

For DPR-GLARE with small patches, typical adhesive failure behavior are detected, while for large patch cases the same cohesive failure as that in SPR-GLARE are obtained. The adhesive failure present in small patch case suggests that less applied load has been born by the titanium patch, and it is the reason that results in the weak reinforced effect, as above described in Figure 4. This phenomenon is also related with the thermal residual stress, as well as with the smaller bonding area. Consequently, it can be concluded that the bonding performance is another critical factor that associates with the reinforced effect.

3.3.2 Damage behavior of the GLARE laminate

Figure 6 shows the damage behavior of GLARE laminates in SPR-GLARE and DPR-GLARE, in which the damage patterns are examined after removing the outer aluminium sheets by means of the chemical etching method. As can be seen from Figure 6, damage patterns of the reinforced GLARE are similar to that of the standard one, which are usually characterized by the fiber breakage, splitting and delamination, and the delamination size is consistent with the splitting level. However, the degree of splitting or delamination is different between SPR-GLAER and DPR-GLARE, as well as between different patch diameter conditions.

For DPR-GLARE, a triangular delamination shape is detected, regardless of the patch diameter. The delamination sizes of DPR-GLARE with patches less than 15 mm are the same with that of the standard GLARE laminate, but significant increase of the delamination size can be obtained with increasing the patch diameter to 17.5 mm and 20 mm. As previous investigation showed, the delamination was benefit for stress redistribution, namely that a larger delamination will suggest a higher load bearing capacity [17]. It is in accordance with the variation of the notch strength present in Figure 4, where negligible strength increasing ratio is obtained for DPR-GLARE with small patch while it is more remarkable when the patch is large enough. Similar upward trend is obtained by the delamination size of SPR-GLARE when increasing the patch diameter, however an analogous trapezoid delamination shape is detected, which is more serious than that in DPR-GLARE, especially for the large patch cases. It is mainly attributed to the bending effect that resulted from the asymmetric configuration, which has introduced a higher out-of-plane stress. It should be noted that the delamination caused by the bending effect plays a less role in stress redistribution.

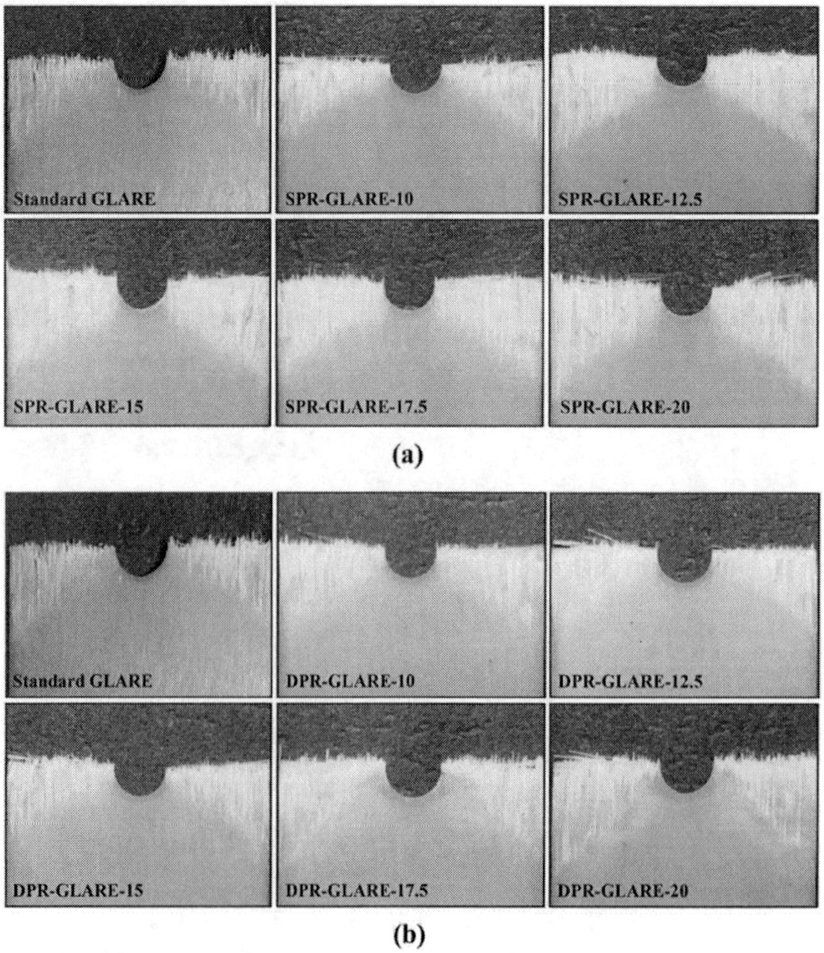

Figure 6. Damage behavior of the GLARE laminate: (a) SPR-GLARE; (b) DPR-GLARE.

4 Conclusions

The purpose of present paper is to investigate the reinforced effect of titanium patch on notched behavior of GLARE laminate. The mechanical response analyses revealed that the modulus and yield strength of GLARE laminate were increased by bonding the titanium patches. As well as significant increase of notched strength was obtained, especially for the DPR-GLARE with 20 mm patches, which has achieved a strength increasing percentage of 27.31%. However, negligible reinforced effect could be obtained by DPR-GLARE with small patches due to the thermal residual stress effect. Moreover the notched strength of SPR-GLARE increased less when the patch diameter reached to some certain values, which was attributed to the bending effect. The comparison suggests that the thermal residual stress and bending effects are of utmost importance in the case of secondary cocure reinforcement, and they will be further investigated experimentally or numerically in our future work.

In addition, the damage behavior of the adhesive between patch and GLARE laminate was examined, which was usually characterized by adhesive failure for SPR-GLARE and DPR-GLARE with large patches, while an adhesive failure was obtained by DPR-GLARE with small patch, which was associated with the negligible reinforced effect of titanium on mechanical response. Furthermore, the damage behavior of GLARE laminate was examined later, where the delamination sizes of SPR-GLARE and DPR-GLARE both increased with increasing the patch diameters, which was benefit for the strength improving. Meanwhile undesired trapezoid delamination was also detected in SPR-GLARE with large patche due to the bending effect.

It is expected that present investigations can provide some useful information on further structural design and application of the FMLs in aeronautics.

Acknowledgement

The present work was supported by the National Science Foundation of China under Grant Nos. 51271067, 11272105 and 11572101.

References

1. R.C. Alderliesten, Mater. Des. **66**, 421 (2015)
2. T. Sinmazçelik, E. Avcu, M.Ö. Bora, O. Çoban, Mater. Des. **32**, 3671 (2011)
3. H. Nakatani, T. Kosaka, K. Osaka, Y. Sawada, Compos. Part A **42**, 772 (2011)
4. P. Manikandan, G.B. Chai, Compos. Struct. **117**, 17 (2014)
5. Y.C. Zhong, C.S. Joshi, Mater. Des. **87**, 138 (2015)
6. E. Sadeghpour, M. Sadighi, S. Dariushi, J. Reinf. Plast. Compos. **32**, 1143 (2013)
7. G.C. Wu, Y. Tan, J.M. Yang, Mater. Sci. Eng. A **457**, 338 (2007)
8. M. Hagenbeek, C. Van Hengel, O.J. Bosker, C.A.J.R. Vermeeren, Appl. Compos. Mater. **10**, 207 (2003)
9. M. Kawai, Y. Arai, Compos. Part A **12**, 1990 (2009)
10. I. Lapczyk, J.A. Hurtado, Compos. Part A **11**, 2333 (2007)
11. J.F. Chen, E. Morozov, K. Shankar, Compos. Struct. **114**, 64 (2014)
12. D.D. Du, Y.B. Hu, H.G. Li, C. Liu, J. Tao, Compos. Part B **91**, 65 (2016)
13. R. Rodi, R.C. Alderliesten, R. Benedictus, *The 24th ICAF Symposium* (Naples, Italy, 2007)
14. C. Rans, F. Moriniere, R. Rodi, R.C. Alderliesten, R. Benedictus, J. Aircraft **48**, 2122 (2011)
15. G.H.J.J. Roebroeks, P.A. Hooijmeijer, E.J. Kroon, M.B. Heinimann, *First International Conference on Damage Tolerance of Aircraft Structures* (Delft, Netherland, 2007)
16. R.D.S.G. Campilho, M.F.S.F. de Moura, J.J.M.S. Domingues, Compos. Sci. Technol. **65**, 1948 (2005)
17. G.D. Lawcock, L. Ye, Y.W. Mai, C.T. Sun, Compos. Sci. Technol. **57**, 1609 (1998)

MATEC Web of Conferences **108**, 03001 (2017)
ICMAA 2017
DOI: 10.1051/matecconf/201710803001

Comparison of sodium dithionite and glucose as a reducing agent for natural indigo dyeing on cotton fabrics

Laksanawadee Saikhao[1], Jantip Setthayanond[2], Thitinun Karpkird[1] and Potjanart Suwanruji[1]

[1]*Department of Chemistry, Faculty of Science, Kasetsart University, Bangkok, 10900 Thailand*
[2]*Department of Textile Science, Faculty of Agro-Industry, Kasetsart University, Bangkok, 10900 Thailand*

Abstract. A traditional reducing agent in an indigo dyeing process with cotton fabrics is sodium dithionite ($Na_2S_2O_4$) which is environmentally unfavorable because the resulting by-products cause various problems to the disposal wastewaters. In this research, glucose was used as a possible replacement of $Na_2S_2O_4$ in indigo dyeing. The comparison of reduction power of $Na_2S_2O_4$ and glucose for natural indigo dyeing on cotton fabrics based on reduction potential was analyzed. The optimum reduction temperature for natural indigo dye of both reducing agents was at 70°C. The reduction time did not have a significant effect on the reduction potential under the condition studied. $Na_2S_2O_4$ could give higher color strength than glucose. However, wash fastness of the fabric samples from a glucose reduction was slightly better than $Na_2S_2O_4$ ones. Hence, glucose virtually has a potential to be used as a green reducing agent in natural indigo dyeing.

1 Introduction

Natural indigo is an ancient vat blue dye that can be extracted from native indigo-bearing plants, especially from the leaves. *Indigofera*, *Isatis tinctoria* and *Persicaria tinctoria* are examples of indigo-bearing plants. Indican is hydrolyzed by enzyme to indoxyl and glucose. When indoxyl is exposed to oxygen by vigorously whisked, the indoxyl is converted to indigo blue [1] as shown in Figure 1.

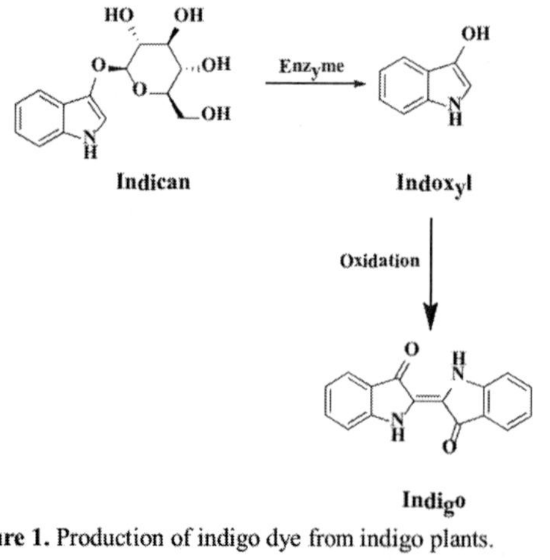

Figure 1. Production of indigo dye from indigo plants.

In Thailand, indigo is called "Kramm" and indigo plants, *Indigofera tinctoria*, are grown mainly in the north-east areas such as Surin, Udon Thani, Kalasin and Sakon nakhon [2]. The water insoluble indigo pigments consist of a conjugated dicarbonyl system [3] that can be reduced to a leuco form by a reducing agent in an alkaline condition (pH 11-14) before applying on cellulosic fibers [4]. Only air oxidation is enough to convert a leuco dye to the original indigo pigment inside the fibers [5]. Figure 2 shows the oxidation and reduction reactions of indigo dyes.

The traditional reducing agent used is sodium dithionite ($Na_2S_2O_4$) because of its effective reduction for indigo dyes [6]. It is also called sodium hydrosulphite or hydros [7]. Typically, the amount of $Na_2S_2O_4$ that is industrially used in the reduction reaction of indigo dyeing is more than the stoichiometric amount, resulting in a large amount of by-products from the process e.g. sulfite (SO_3^{2-}) and sulfate (SO_4^{2-}) ions [8] as shown in Equation (1).

$$S_2O_4^{2-} + O_2 + 2OH^- \longrightarrow SO_3^{2-} + SO_4^{2-} + H_2O \quad (1)$$

These ions can cause ecological problems, for example, SO_3^{2-} can be easily oxidized to SO_4^{2-} and the high concentration of SO_4^{2-} can cause corrosion in concrete pipes [9]. Thus, our approach in this work is to use glucose as a green reducing agent instead of $Na_2S_2O_4$ in natural indigo dyeing process. Glucose is a reducing sugar that has an aldehyde functional group which can be oxidized to carboxylic acid and simultaneously reduce indigo pigments to leuco dyes. The comparison of glucose and $Na_2S_2O_4$ was studied in terms of their reduction potential and optimal dyeing conditions. The color strength (K/S) and wash fastness of dyed cotton fabrics were also examined.

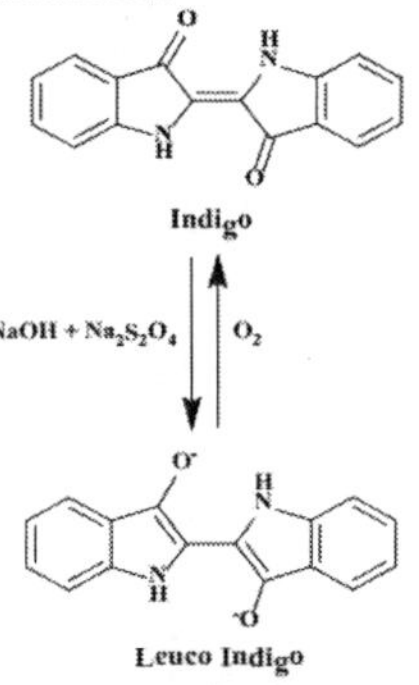

Figure 2. Oxidation-reduction of indigo.

2 Experimental

2.1 Materials

Natural indigo paste was purchased from Ban Tam Tao, Sakon Nakhon, Thailand. Cotton fabrics was supplied by P.S.Jevasy. Co. Ltd, Thailand. $Na_2S_2O_4$, glucose and NaOH used were AR grade.

2.2 Methods

2.2.1 Preparation of indigo dye and cotton fabrics

Natural indigo paste was vacuum filtered and oven-dried at 60°C for 24 hrs. The dried indigo powder was milled and stored at room temperature. Cotton fabrics were scoured at 90-95°C for 45 min in a bath containing 4 g/L NaOH and 1 g/L wetting agent using 30:1 liquor ratio. The scoured cotton was rinsed with hot and cold water, respectively, and then air-dried.

2.2.2 Reduction and dyeing conditions

Cotton fabrics (2 g) were dyed with natural indigo (4 g) using a liquor ratio of 50:1, 12 g/L NaOH, 10 g/L $Na_2S_2O_4$ or 20 g/L glucose. The reduction temperature (T_R) was set from 30 to 90°C with the reduction time (t_R) varied from 0 to 60 min. The cotton fabrics were put into dyebath and the dyeing temperature (T_D) was set at 30°C and dyeing time (t_D) was carried out at 10, 20 and 30 min as shown in Figure 3. The dyed fabrics were washed in distilled water until the water was neutral and then air-dried.

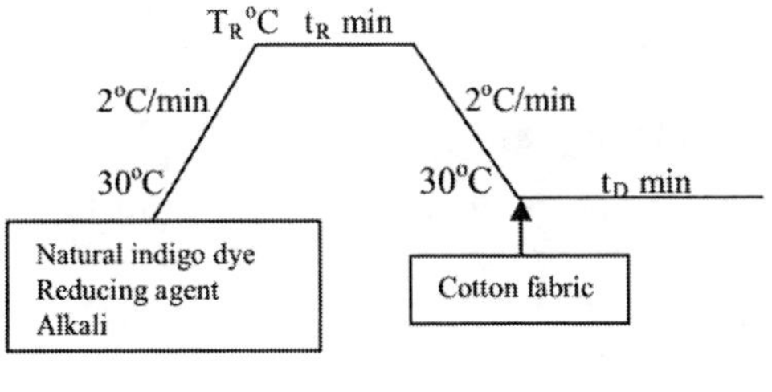

Figure 3. Dyeing process of natural indigo dye on cotton fabrics.

2.2.3 Reduction potential measurement

Reduction potential of all reducing dye solutions was measured by an oxidation-reduction potential (ORP) platinum electrode and the Ag/AgCl reference electrode with KCl electrolyte, connected to a pH meter (Metter Toledo, Inlab) and recorded in (mV) [10]. The experiment set up is shown in Figure 4. The reducing power of reducing agents was compared.

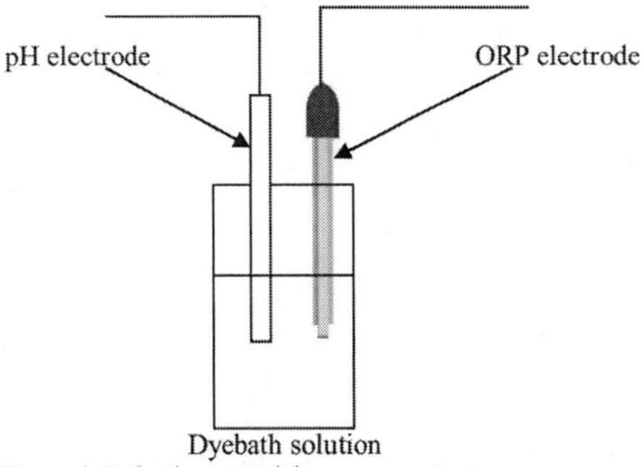

Figure 4. Reduction potential measurement set up.

2.2.4 Color measurement

The determination of color strength (K/S) values of all dyed cotton fabric samples was carried out on a Macbeth ColorEye 7000 Spectrophotometer under Illuminant D_{65} at 10° observer position. The K/S values were calculated from Kubelka-Munk Equation (2).

$$K/S = \frac{(1-R)^2}{2R} \tag{2}$$

where R is the reflection of all dyed cotton fabric samples [11].

2.2.5 Wash fastness

The dyed samples were evaluated for the wash fastness according to the ISO 105-C06: 1994. The fastness properties were compared among the fabric samples using different reducing agents [12]. The color change of the dyed fabric samples and staining on the multifiber fabric were assessed by a grey scale for color change and staining, respectively. The level of grey scale is rated from 1 to 5. Level 1 refers to the highest degree of shade change or staining, while level 5 refers to no shade change or no staining.

3 Results and discussion

3.1 Reduction temperature (T_R)

Figure 5 shows the plot between measured reduction potential in mV and T_R in °C. The optimum reduction temperature of indigo was at 70°C for both $Na_2S_2O_4$ and glucose. The redox potential of the indigo dye was reported about -700 mV [13]. From the experiment, the redox potentials at 70°C of $Na_2S_2O_4$ and glucose were -924 mV and -722 mV, respectively which were more negative than that of indigo. Both of them were able to use in the reduction of indigo dyes. Hence, glucose can be an alternative reducing agent in the indigo dyeing process. In the case of $Na_2S_2O_4$, as the temperature increased the reduction potential slightly increased in the negative values which is in agreement with Meksi, 2012. On the other hand, the reduction potentials of glucose above 70°C tended to decrease.

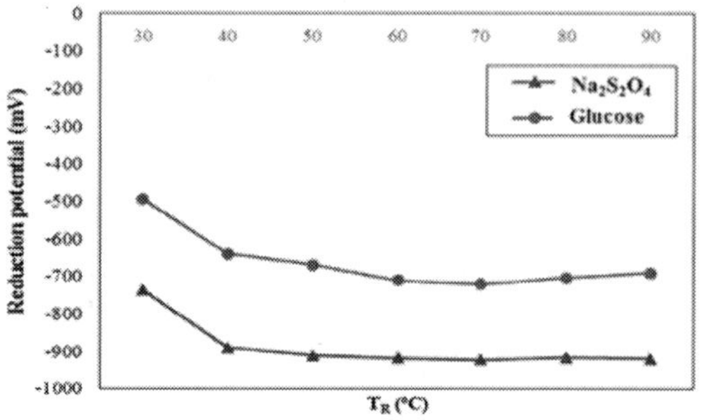

Figure 5. Reduction potentials of indigo dyeing using $Na_2S_2O_4$ and glucose as reducing agents at the designated reduction temperature.

3.2 Reduction time (t_R)

Figure 6 shows the effect of reduction time (t_R) on the reduction potential of $Na_2S_2O_4$ and glucose. The reduction potentials were recorded every 10 min from 0 - 60 min. The results demonstrated that reduction time did not have a significant effect on the reduction potential under the conditions studied. The results indicated that it was not necessary to reduce indigo dyes longer than 10 min because the reduction potentials were about constant. The reduction potentials of $Na_2S_2O_4$ and glucose at 10 min were -906 and -736 mV, respectively. Even though $Na_2S_2O_4$ gave higher negative reduction potential, glucose provided a negative reduction potential that was effective enough to reduce the indigo pigment to be in a soluble leuco form.

3.3 Dyeing time (t_D)

Leuco dyes from $Na_2S_2O_4$ and glucose reduction were applied on cotton at room temperature for 10, 20 and 30 min. After dyeing for 10 min, the reduction potentials of the dyebath showed the least negative values as shown in Figure 7. The reduction potentials of $Na_2S_2O_4$ and glucose as shown in Table 1 were -890 and -629 mV, respectively. These reduction potentials also corresponded to the K/S values of the dyed cotton fabrics as shown in Table 2. Dyeing cotton fabrics for 10 min gave the highest K/S values. $Na_2S_2O_4$ yielded K/S that was higher than glucose about 1 unit. However, glucose could effectively reduce the indigo dye and gave a comparable color strength on cotton fabrics. The remaining dyebath would have less environmental effect than that of $Na_2S_2O_4$.

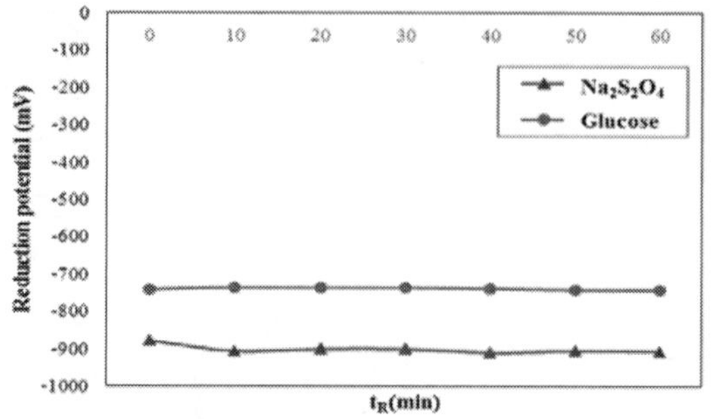

Figure 6. Reduction potentials of $Na_2S_2O_4$ and glucose in dyebath at various reduction times.

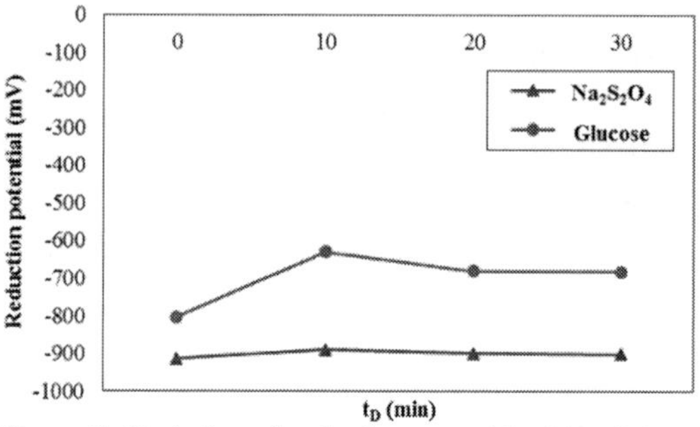

Figure 7. Evolution of reduction potential of $Na_2S_2O_4$ and glucose as reducing agents in indigo dyeing process at various times.

Table 1. Reduction potential of $Na_2S_2O_4$ and glucose as reducing agent in indigo dyeing process at various times.

Reducing agent	t_D (mV)	Reduction potential (mV)	
		Before dyeing	After dyeing
$Na_2S_2O_4$	10	-914	-890
	20	-914	-899
	30	-914	-901
Glucose	10	-804	-629
	20	-804	-680
	30	-804	-681

3.4 Wash fastness

Table 3 shows the wash fastness results of dyed cotton using $Na_2S_2O_4$ and glucose as reducing agents. In general, the color change and staining on multifiber were similar comparing between the fabric samples from

$Na_2S_2O_4$ and glucose. Staining on nylon and acetate was less in the case of glucose.

Table 2. Color strength of cotton fabrics at dyeing time of 10, 20 and 30 min.

Reducing agent	t_D (min)	Dyed samples	K/S
$Na_2S_2O_4$	10		7.41
	20		6.39
	30		7.29
Glucose	10		7.41
	20		6.08
	30		6.33

Table 3. Wash fastness of indigo dyed cotton fabrics.

Reducing agents	Wash fastness						
	SW	SAcr	SP	SN	SC	SAce	Alt
$Na_2S_2O_4$							
t_D =10 min	5	5	5	3-4	5	3-4	4
t_D =20 min	5	5	5	3-4	5	3-4	4
t_D =30 min	5	5	5	3-4	5	3	4
Glucose							
t_D =10 min	5	5	5	4	5	4	4
t_D =20 min	5	5	5	4	5	3-4	4
t_D =30 min	5	5	5	4	5	3-4	4

*SW: staining on wool, SAcr: staining on acrylic, SP: staining on polyester, SN: staining on nylon, SC: staining on cotton, SAce: staining on acetate and Alt: change of color

4 Conclusions

The reduction power of two reducing agents, $Na_2S_2O_4$ and glucose for natural indigo dyeing process was compared. The optimum reduction temperature was found to be at 70°C, giving reduction potential of $Na_2S_2O_4$ and glucose at -924 mV and -722 mV, respectively. The optimum reduction duration was 10 min. Reduction potentials of $Na_2S_2O_4$ and glucose corresponded to K/S values of the dyed cotton fabrics. The K/S values of $Na_2S_2O_4$ samples were higher than those of glucose samples about 1 unit while the wash fastness properties of the latter one were slightly better than the former one. The results indicated that glucose could effectively reduce indigo dyes to a leuco form and gave a comparable dyeing properties as compared to $Na_2S_2O_4$. Moreover, glucose was considered as a eco-friendly reducing agent in natural indigo dyeing process.

Acknowledgement

The authors are indebted to Kasetsart University Research and Development Institute, Bangkhen Campus, Faculty of Science, The Graduate School and International Affairs Division, Kasetsart University, Thailand for financial support of this research. A great acknowledgement is also given to Department of Chemistry, Faculty of Science and Department of Textile Science, Faculty of Agro-Industry, Kasetsart University.

References

1. J. Balfour-Paul, Indigo, British Press, London (1998).
2. M. Moeyes, Natural dyeing in thailand, Cheney White Lotus, Bangkok. (1993).
3. N. Meksi, M.B. Ticha, M. Kechida and M.F. Mhenni. Journal of Cleaner Production, 24 (2012): 149-158.
4. S. Komboonchoo, A. Turcanu and T. Bechtold. Dyes and Pigments, 83(1) (2009): 21-30.
5. B. Mojca and V. Kokol. Dyes and Pigments, 46 (2008) : 299-309.
6. M.B. Ticha, N. Meksi, M. Kechida and M.F. Mhenni. International Journal of Environmental Research, 7(3) (2013) : 697-708.
7. V. Anne, Reduction and analysis methods of indigo. Doctoral dissertation, Turku University (2008).
8. M.B. Ticha, N. Meksi, M. Kechida and M.F.Mhenni. Chemical Industry and Chemical Engineering Quarterly, 20(4) (2014): 463-470.
9. R. M. Khan, Z. Uddin and D. Hossain. Middle-East Journal of Scientific Research, 22(7) (2014):1090-1095.
10. B.S. Richard and A. Harvey. Environmental Science and Technology, 38(14) (2004): 4034-4039.
11. S. Younsook, M. Choi and D.I. Yoo. Fashion and Textile, 1(1) (2014).
12. J. L. Jung, W.S. Shim, I. S. Kim and J. P. Kim. Fibers and Polymers, 6(3) (2005): 244-249.
13. J. N. Chakraborty: Indian Journal of Fibre and Textile Research, 39 (2004): 100-109.

Eco-friendly coloration of linen to ameliorate its practical approach

Md. Nahid Pervez[1,2], Umarsharif Y. Inamdar[1], Md. Eman Talukder[1], Sakil Mahmud[1], Md. Ahsan Habib[1], Md. Kamruzzaman[1] and Yingjie Cai[3,4]

[1]*School of Chemistry and Chemical Engineering, Wuhan Textile University, Wuhan 430073, China*
[2]*Research institute of flexible materials, School of Textiles & Design, Heriot-Watt University, Galashiels TD1 3HF, UK*
[3]*Hubei Provincial Engineering Laboratory for Clean Production and High Value Utilization of Bio-based Textile Materials, Wuhan Textile University, Wuhan 430073, China*
[4]*Engineering Research Centre for Clean Production of Textile Dyeing and Printing, Ministry of Education, Wuhan Textile University, Wuhan, 430073, China*

Abstract. Reactive dyes are the perfect choice for dyeing of cellulosic fibers due to produce a wide gamut of bright colors with excellent color fastness, however environmental pollution is increasing in recent days seeing that of making use of inorganic salt and alkali. This paper offers a new avenue for eco-friendly dyeing of linen fiber by incorporating biodegradable alkaline organic salts in presence of three commercial reactive dyes and amongst them organic salt i.e. disodium maleate showed superior result in terms of color strength (K/S), dye exhaustion (%), color fastness, tensile strength and crease recovery angle parameters that were evaluated by traditional technique.

1 Introduction

The eco-friendly fabric dyeing has received more and more interest due to the elevated recognition of environmental pollution likes discharge of nonbiodegradable inorganic salts, alkalis, different processing aids and organic subject similar to dyes to the effluent for the duration of the arena [1]. The foremost assignment that now faces the cloth industry is to change creation at a competitive price by means of utilizing secure dyes and chemical substances as good as via decreasing treatment fee [2]. Reactive dyes are one of the lots of accepted classes of dye stuffs. They can fix to fabric by the accumulation of a covalent bond and accordingly accept boundless photolytic adherence and affable wash fastness on textile fibers, which is an advantage over added classes of dyes, such as direct dyes that are immobilized via physical adsorption or mechanical entrapment (1) Reactive dyes with the dichlorotriazine reactive group accumulation are frequently acclimated in the dyeing of affection fabrics (2) Dye fixation involves the formation of a covalent bond between a hydroxyl group, beneath alkaline altitude in the fabric, and the dye molecule, by the nucleophilic displacement of chloride ion [3, 4]. The inorganic salt, similar to sodium chloride or sodium sulphate, for dye transfer to and penetration into the fibre, and inorganic alkali, reminiscent of sodium bicarbonate, sodium carbonate or sodium hydroxide, for dye-fibre reaction, are required in significant portions to achieve the dyeing system. Without reference to the dyeing approach utilizing reactive dyes, nearly the entire salt and alkali is drained to effluent. Such effluents are characterised through high phases of dissolved solids which is environmentally undesirable [5, 6]. In up to date years, alternative of nonbiodegradable inorganic salt with biodegradable natural and organic salt is attracting the concentration of many researchers in this area. It is pronounced that inorganic salts will also be efficaciously replaced by using the organic salt in the textile process with larger dye uptake than sodium chloride in the reactive dyeing system [7]. Linen may be the oldest fabric known, is an important textile material with different characteristics. With the trend of appearance appear natural, comfortable yet elegant fabrics, linen and linen-blended fabrics accept acquired authority and added in acceptability attributable to the fabric qualities of comfort, hygiene, and elegance recently. Linen reduces γ-radiation nearly by bisected and protects the human organism towards sun radiation. It is an array of nice actual for advantageous applications due to its amore transfer, permeability to air. Heat conductivity of linen is five times as high as that of wool and 19 times as that of silk, additionally, the tensile strength of linen thread is twice as high as that of cotton and three times that of wool. Nevertheless, dyeability is the main barrier that unable its wide applications in the real life [8]. Therefore, development of fiber modification and finishing treatments are required to improve the performance of linen fabric to promote its production as well as its usage.

Present research aims to establish a platform for eco-friendly dyeing of linen in presence of biodegradable salts as exhausting agents in order to reduce the discharge pollution of wastewater. To the best knowledge of authors up to date no directly related research has been reported.

DOI: 10.1051/matecconf/201710803002

2 Experimental

2.1 Materials

Plain, woven, bleached, and scoured linen fabric (210 g/m^2) was used throughout this work. The main dyes used for this study were C.I. Reactive Yellow 176 (Dye 1), C.I. Reactive Yellow 3 (Dye 2), and C.I. Reactive Blue 19 (Dye 3). Sodium chloride, sodium carbonate, sodium sulfate, sodium citrate, sodium edate, sodium oxalate and disodium maleate were analytical of commercial grade and used without further purification.

2.2 Dyeing procedures

All dyeing processes were carried out using a liquor ratio of 30:1. The dyebath was prepared by adding the three dyes (1% o.w.f.) to distilled water at 60°C and wet fabric and left for 15 min, then The temperature was raised subsequently to 80°C, and fixation was conducted using sodium carbonate and left for 30 min. The dyed fabrics were rinsed thoroughly in hot water and soaped in a soap solution 5 min at 90°C, and then rinsed thoroughly in hot tap water and air-dried.

2.3 Measurements

The color strength (*K/S*) and fastness values were determined according to our previous work [9]. The dye exhaustion percentage of samples (E%) was calculated according to the following equation:

$$E\%=100\%\times (C_1-C_2)/C_1 \quad (1)$$

Where C_1 and C_2 are the concentrations of dye in the dyebath before and after dyeing, respectively. Tensile strength of fabric was measured by following ASTM-D-5035(1995) method using Instron tensile tester (Model No. 5565) [10]. The standard ASTM D 1269 test method was used to determine the crease recovery angle (CRA) using the Shirley Crease Recovery tester.

3 Results and discussion

3.1 Color strength

Different electrolytes and three reactive dyes with the same mass concentration (30 g/L) were used for pad-steam dyeing of linen fabric with 1% shade, in order to get the suitable biodegradable salt for eco-friendly dyeing of linen fibers is shown in Figure 1.

Salt is used for increasing dye levelness in the course of the fibre and the consequent colour-yield, in pad-steam dyeing. The response between the reactive dye molecules and the hydroxyl organizations of cellulose occur at alkaline pH conditions. The dye can also react with the hydroxide ions in an aqueous dye resolution [11, 12]. Figure 1 shows the comparative color strength of the three dyed fabrics amongst these bought using sodium chloride, sodium sulfate, sodium citrate, sodium edate, sodium oxalate and disodium maleate. It was once clear that sodium citrate, sodium edate, sodium oxalate and disodium maleate considered as suitable electrolytes to interchange inorganic salts. Among the three dyed samples utilizing organic salts, sodium maleate showed the absolute best K/S value on dye 3 sample. The major influence of the electrolytes on the dyes, moreover to its extinguishing outcomes on the negative charge on cellulose fibers, is to increase the aggregation of the dye molecules in solution via the common ion effectiveness. The electrolyte reduces the solubility of dye in the dyebath and modifies the equilibrium in desire of action of dye molecules from the approach to the fiber via suppressing the ionization of the dye molecule in solution. This is highly massive broaden comparing to the feasible shrink color-yield within the typical reactive dyeing of cellulose by pad-steam system for the reason that of immoderate dye hydrolysis and cut back ultimate dye-fixation levels [13]. This may be attributed to the higher ionic-force and better stability (at elevated pH) of disodium maleate for reaction between vinylsulphone derivatives (dye three) and cellulose. It indicated that as far as the K/S value of the dyed samples had been concerned; disodium maleate might be a compatible substitution for inorganic salts worth one more is taught.

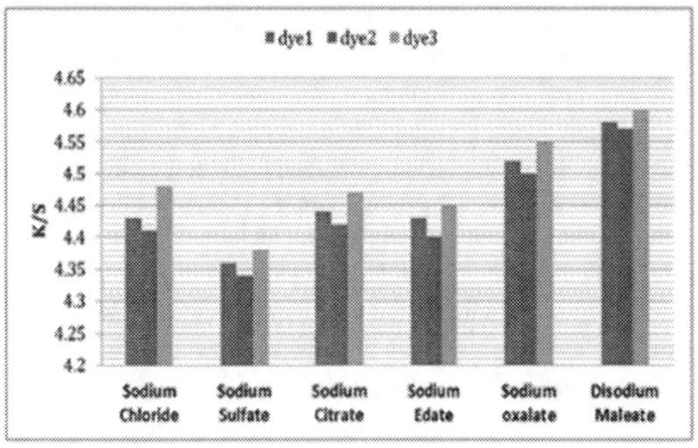

Figure 1. Effect of Different Electrolytes on Color Strength.

3.2 Color Fastness

As proven in Table 1, color fastness results that indicate that all the organic and inorganic salts dealt with fabrics have applicable color fastness results. The organic salts materials exhibit reasonably better color fastness results than the inorganic salts handled fabrics because of dye-mordant complex may be firmly connected to the substrate ensuing into a greater color yield and low awareness of alkali, the extent of formation of nucleophilic cellulose O^- moiety of the fabric is low, this ends up with low measure of nucleophilic addition reaction with the vinyl sulphone reactive dye. Furthermore, the fastness properties improved due to presence of benzene ring and conjugated system in the colorant [14]. Overall results confirmed that disodium maleate salt has exhibit satisfactory result after dyeing with vinyl sulphone based (dye 3).

3.3 Dye exhaustion

As shown in Figure 2, the exhaustion percent (E %) effectivity of dyed materials is plotted in y-axis. For the same color percent of dye, three dyes deviate in dye absorb with fiber. The effect suggests that dye1 and dye3 at all times had a greater and just about identical percent of exhaustion. But dye 2 had a negative exhaustion percentage in evaluation with the others. It is visible that outcome of organic salts is superior on inorganic salts efficiency and disodium maleate as organic salt has been showed absolute best exhaustion rate amongst all dealt with samples in this study. As we know cellulose consists of a negative charge in pure water and dye is also anion consequently negatively charged, there's electrostatic anion-anion repulsion between dye and cellulose.

As a consequence, by including electrolyte to the dyebath, electrostatic barrier as appeared and often called the Donnan potential, can be mostly suppressed, facilitating dye/fabric contact and allowing better interaction of the Yoshida and van der Waal, forces and as a consequence making improvements to substantivity. The diffusion coefficient of the dye is as a consequence a function of each dye and electrolyte concentration [15]. The extent to which reactive dyes are littered with the addition of electrolytes to the dyebath is often called salt sensitivity. It is viewed that the initial rapid rise is due to the response of dye to the reducing of electrical potential barrier to diffusion as the concentration of electrolyte raises. Their hydrogen bonding ability can also be aided through their rectangular and flat molecular structure which allows for them to lie alongside a cellulose chain in register with hydroxyl corporations. Some studies proven salt free dyeing of cellulose material using cationization of the fabric itself in order that the dye will likely be migrated to the fabric and constant with none addition of salt to the dyeing tub [16-18]. However, majority of the chemical substances used for introducing cationic sites in cellulose material are themselves now not secure for the environment [19]. As a consequence, this record tried to compensate the alternate-off via utilizing decrease quantity of salt with higher dye fiber fixation and slash environmental have an effect on with the aid of utilizing biodegradable salt.

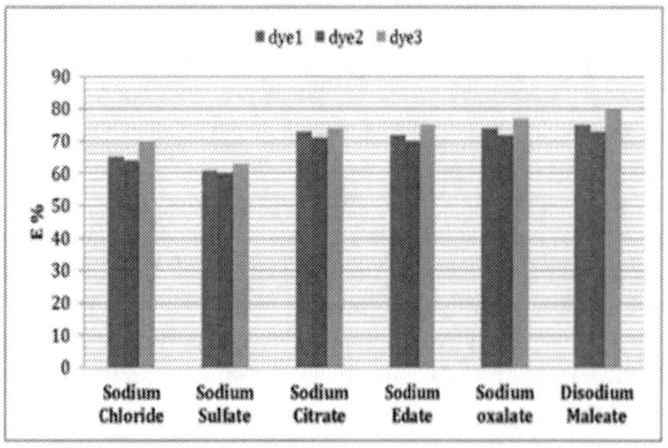

Figure 2. Effect of Different Electrolytes on Dye Exhaustion.

3.4 Tensile strength

Figure 3 shows the tensile strength values of three reactive situated dyed fabrics with inorganic and organic salts. It can be visible that the tensile strength of organic salts dealt with material suggests comparatively better than inorganic salts dealt with. On this case also disodium maleate salt carried out good those others with dye 3. The feasible rationale for increase in tensile strength is the alkaline pH of disodium maleate, which can provide force to fabric and the network structure and excellent adhesion on the interface was liable for high tensile strength [20].

Table 1. Color Fastness Properties of Dyed Samples.

Dyes	Different salts	Rubbing fastness		Washing fastness		Light fastness
		Dry	Wet	Color change	Staining	
Dye 1	Sodium chloride	3-4	3	3-4	2	5
	Sodium sulfate	3-4	3	3-4	2	4
	Sodium citrate	4	3-4	4	3	6
	Sodium edate	4-5	3-4	4	3	6
	Sodium oxalate	4-5	3-4	4	3-4	6
	Disodium maleate	5	4	5	4	7
Dye 2	Sodium chloride	3	3	3	3	5
	Sodium sulfate	3	3	3	2	4
	Sodium citrate	4	3-4	4	3-4	5
	Sodium edate	4-5	4	4	3-4	6
	Sodium oxalate	4	4	4-5	3-4	6
	Disodium maleate	5	4	4-5	4	6
Dye 3	Sodium chloride	3-4	3-4	3	3	6
	Sodium sulfate	3	3	3-4	3	5
	Sodium citrate	4	4	4	4	6
	Sodium edate	4-5	4	4-5	4	6
	Sodium oxalate	5	4	4-5	4-5	6
	Disodium maleate	5	4-5	5	4-5	7

3.5 Crease recovery angle

Crease or creasing recovery is defined as the property of a fabric that enables it to balance from folding deformations. The most accepted adjustment of measuring crease recovery is to bend a strip of fabric by a folding force, advance it in the angled agreement for a accustomed period of time, and then measure the angle of recovery after the bending load is removed [21]. The higher the wrinkle recovery angle of a fabric, the better is its crease resistance. For automated applications, it has to be ensured that the biodegradable salts are not alone assuredly able in dyeing but as well they are accordant with the finishing agent. Consequently, the effects of adding the organic salts to the formulation of the crease resist finishes as well as the finish durability were evaluated. There are two theories for the apparatus of anti-wrinkle finishing of fabrics. The deposition theory proposes that small particles of finishing agents are dispersed in amorphous region of the fibres where they might be deposited after resinification and form H-bonds with the fibre molecules. This may bind the movement of the atomic chains and abatement anamorphosis of the fabrics. The cross-linking approach proposes that finishing agents form covalent bonds with fibre molecules, and affix the adjoining atomic chains within the fibres. This gives the recoverability accessory of the deformed fibres and increases the wrinkle resistance. Also, as can be observed from Figure 4, disodium maleate treated fabric showed extensive wrinkle resistant value with dye 3. This advance was mainly attributed to the actuality that the reactive dye could acknowledge with cellulose fibres through their H-bonds, which led to the cross-linking of cellulosic fibres and bound the about movement of the atomic chains [22]. It was found that both aforementioned theories are adaptable for the wrinkle resistant properties of dyed treated linen fabrics.

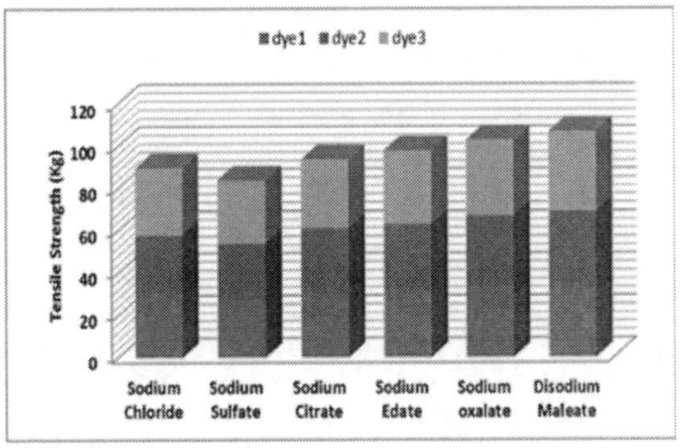

Figure 3. Effect of Different Electrolytes on Tensile Strength

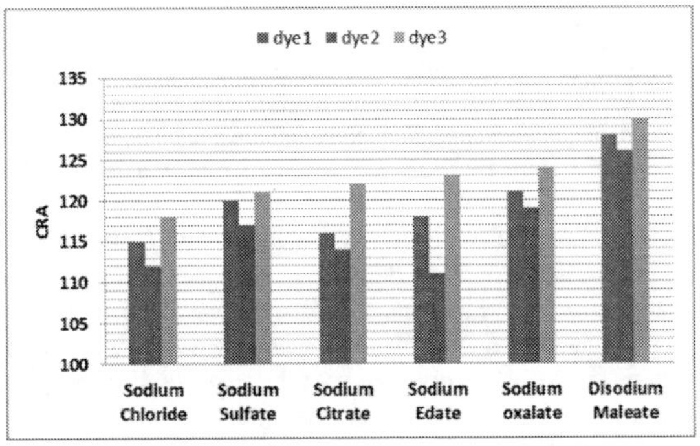

Figure 4. Effect of Different Electrolytes on crease recovery angle

4 Conclusion

This is the first report about disodium maleate used as biodegradable organic salt on linen by pad steam dyeing with reactive dyes. On the basis of the results given above, it can be concluded that in phrases of color strength, exhaustion (%), color fastness, tensile strength and crease recovery angle, organic salts are inferior to inorganic salts and it explores the viability of exchanging the traditional inorganic salts through incorporating biodegradable organic salts.

Acknowledgement

This work was financially supported by the China National Textile & Apparel Council (2013 "Textile Vision" Applied Basic Research, 2013-153); Hubei Province Science and Technology Support Program (Grant No. 2013BAA043) and the Collaborative Innovation Plan of Hubei Province for Key Technology of Eco-Ramie Industry (2014-8).

References

1. R. Chavan, IJFTR. **26**, 93 (2001)
2. N. Ul-Haq, H. Nasir, J. Text. I. **103**, 304 (2012)
3. N. Meksi, M.B. Ticha, M. Kechida, M.F. Mhenni, J. Clean. Prod. **24**, 149 (2012)
4. T. K. Yip, E. R. Smith, J. Booth, R. G. Compton, C. M. Brennan, J. H. Atherton. *J. of coll. and int. Sci.* **186**, 387 (1997)
5. P.J. Dolby, Textile Chemist & Colorist. **9**, 264 (1977)
6. J. Aspland, Textile Chemist & Colorist. **24**, 35 (1992)
7. H.G. Prabu, Color. Technol. **118**, 131 (2002)
8. A. Hadj Taieb, S. Msahli, F. Sakli, J. Nat. Fibers. **7**, 251 (2010)
9. M.N. Pervez, M.A. Rahman, IJSER. 6, 660 (2015)
10. A. Khan, M.N. Pervez, I.A. Khan, S. Ahmad, R. Masood, T. Hussain, F. Telegin, Asian J. Chem. **28**, 1191 (2016)
11. D. Phillips, Advances in Colour Science and Technology. **1**, 1 (1998)
12. J. Shore, *Cellulosics dyeing*. (1995).
13. A.D. Broadbent, *Basic principles of textile coloration*. (2001)
14. T. Gulzar, S. Adeel, I. Hanif, F. Rehman, R. Hanif, M. Zuber, N. Akhtar, J. Nat. Fibers. **12**, 494 (2015)
15. A. Johnson, *The theory of coloration of textiles*. (1989)
16. K. Srikulkit, P. Santifuengkul, Color. Technol. **116**, 398 (2000)
17. D. Chattopadhyay, IJFTR. **26**, 108 (2001)
18. Y. Zhang, B. Zhang, Dyeing and Finishing. **8**, 47 (2001)
19. S. Giolando, R. Rapaport, R. Larson, T. Federle, M. Stalmans, P. Masscheleyn, Chemosphere. **30**, 1067 (1995)
20. C. Thongpin, J. Srimuk, P. Wachirapong. *IOP Conference Series: Materials Science and Engineering*. **87**, (2015)
21. A. Sevda, E. Yılmaz. *Fibers and Polymers* **17**, 644 (2016)
22. H. Emadaldin, S. Shahidi, P. Zolgharnein. *A. Res. J.* **15** 104 (2015)

DOI: 10.1051/matecconf/201710803003

Electrolyte influence on sorption behaviours of Direct Blue 71 dye on ramie fibre

Chi Yuan[1], Kankan Lou[1,2], Liping Yu[3], Md. Nahid Pervez[4], Siwei Su[1], Jun Huang[2], Xiongyi Peng[2] and Yingjie Cai[1,2,5]

[1]*School of Chemistry and Chemical Engineering, Wuhan Textile University, Wuhan 430073, China*
[2]*Hubei Provincial Engineering Laboratory for Clean Production and High Value Utilization of Bio-based Textile Materials, Wuhan Textile University, Wuhan 430073, China*
[3]*Zhejiang Institute of Modern Textile Industry, Shaoxing 312000, China*
[4]*Research institute of flexible materials, School of Textiles & Design, Heriot-Watt University, Galashiels TD1 3HF, UK*
[5]*Engineering Research Centre for Clean Production of Textile Dyeing and Printing, Ministry of Education, Wuhan Textile University, Wuhan, 430073, China*

Abstract. Ramie loose fibre was dyed using Direct Blue 71 dye at 70, 80, 90 and 100°C without and with NaCl electrolyte in order to investigate the distinction of dye sorption behaviours. The results show that the dye exhaustion increases with addition of NaCl and shortens the equilibrium dyeing time. The dye adsorption process of dyeing without and with NaCl followed pseudo second-order kinetics, but the rate constant of sorption is larger for the latter compared to the former.

1 Introduction

Cellulose is the most abundant biopolymer and of specific interest in accoutrement renewable, sustainable, biodegradable biopolymers for automated applications. Cellulose is a linear 1, 4-β-glucan polymer where the units are able to form rather ordered systems, this occurs as a result of sizable interplay through intra- and intermolecular hydrogen bonding of the three hydroxyl corporations in each cellulose assemblage [1]. Ramie ranks as an archetypal fibre in that it has been acclimated in abounding countries since time immemorial, aswell accepted as "China grass,". It is the longest, most powerful and maximum notably orientated cellulosic fibre; appropriately degummed and bleached fibre is nearly natural cellulose, like cotton, and includes approximately high percentage of cellulose (65-75 %) and low proportion of lignin (1-2 %) in comparison with wood (40-50 % of cellulose and 15-35 % of lignin) [2-5]. Earlier advisers [6] have in comparison the size and shape of ramie, flax, and cotton fibers and accept approved that the adjustment of cross-sectional size among the three fibers is ramie > flax > cotton. Ramie fabrics address warmth, take in moisture and have better resistance to mould than other cellulosic fibers. Ramie is an abiding plant. The stems grow from rhizomes and are harvested five years afterwards planting. The plants can yield fibre for 30 years and can be harvested three times a year. There has been an incredible deal of interest inside the studies and utilisation of ramie in contempo years [7-9].

Direct dyes are about inexpensive, simple to apply, and accessible in a full ambit of shades. They are aswell tinctorially stronger than different cellulose dyes, and very dark shades can be acquired economically with low dye concentrations. Furthermore, a few direct dyes have excellent light fastness [10]. Direct dyes are accepted in dyeing of cellulosic fibres due to its anionic and water soluble (sodium salts of sulfonic acids) attributes [11]. Differences from the covalent bonds of reactive dyeing on cellulose, direct dyes are loosely associated with cellulosic substance by substantivity including the Van der Waals forces and hydrogen bonds. It is aswell believed that cellulosic fibers access a negative charge when immersed in aqueous media due to their decreased dielectric constant connected with that of water and that this has the aftereffect of against the analogously charged ion of the substantive dye. But it is claimed that electrolytes have a tendency to abate or extinguish the allegation on the fiber, appropriately facilitating an alteration of dye from. Earlier workers had explained the aftereffect of electrolytes on the base of an abatement in the abrogating apparent abeyant of the fiber [12-14].

Water pollution has apparently influenced on the sustainable development of textile industries, owing to an increase of the cost for treating the waste water before discharging. In order to obtain high dye exhaustion, normally, increase of dyeing temperature or addition of electrolyte is widely applied, and both operations alter the dye sorption behaviours during dyeing [15].

Therefore, this paper aims to discover the distinction of dye sorption behaviours under varying dyeing temperatures with or without electrolyte addition, and this achievement potentially assists the dyeing plants to seek a suitable dyeing profile for direct dyeing of cellulosic fibre.

2 Experimental

2.1 Materials

Ramie loose fibre was obtained from Hunan Huasheng Zhuzhou Cedar Company, China. Direct Blue 71 dye (commercial grade), was purchased from the Shanghai Jiaying Chemical Company, China and utilized directly without purification, which molecular structure is shown in Figure 1. Sodium chloride (electrolyte, 99.5%) was supplied from the Sinopharm Chemical Reagent Company, China.

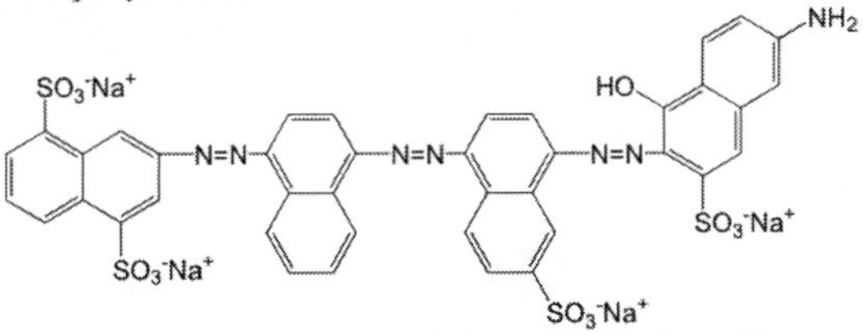

Figure 1. Molecular structure of Direct Blue 71 dyestaff

2.2 Procedures

The ramie loosely fibre was dyed with 3% dye o.m.f at a liquor ratio of 50:1 at 70, 80, 90, and 100°C in a shaker dyeing machine (HB-24, Huibao Ranzheng Jixie Company, China). The dye bath was heated to the desired temperature (70, 80, 90, and 100°C), subsequently ramie loosely fibre was immersed in the dye bath. After 15 min, NaCl of 30 g L^{-1} was added in, and dyeing was conducted until dye absorption equilibrium.

2.3 Measurement

The dye exhaustion percentage (%E) is the ratio of the amount of dye adsorbed by a fiber after dyeing with the initial amount of dye. The light absorbance of dye bath for different dyeing time was measured using an UV/visible spectrophoto-meter, (UV1100, Beijing Rayleigh analytical instrument Corp., China) at the wavelength of maximum absorbance (570 nm for Direct Blue 71). The dye exhaustion rates for varying dyeing time were calculated by Equation 1.

$$E\% = \frac{A_0 - A_t}{A_0} \times 100\% \quad (1)$$

Where E% is the dye exhaustion rate, A_0 and A_t are the light absorbency values of the dye bath before and after dyeing respectively.

The amount of adsorbed dye per unit mass of substrate, $[D]_f$ (g g^{-1}), was calculated using Equation 2.

$$[D]_f = \frac{E\% \times M_d}{M_f} = E\% \times o.m.f\% \quad (2)$$

Where M_d (g) is the mass of dye used in dyeing, M_f (g) is the mass of ramie loosely fibre, o.m.f% is the value that dye quantity on the mass of fibre, which is 3% in this research work.

3 Results and discussion

The dyeing exhaustion rates of Direct Blue 71 on ramie loose fibre without and with addition of NaCl are displayed in Figure 2. For dyeing temperature at 70 to 100°C, the dye exhaustion rates for without addition of NaCl, the values are similar, which are within 60 to 65%; while for addition of salt, that are in the range of 80 to 94%. It is interesting that both the exhaustion rates for without and with addition of NaCl at 100°C are the lowest one among the dyeing temperatures. It is possible that during dyeing, the dye exhaustion is a reversible process due its physical sorption, thus at high temperature, the desorption rate is higher than at low temperature thereof the better colour uniformity for dyed substance.

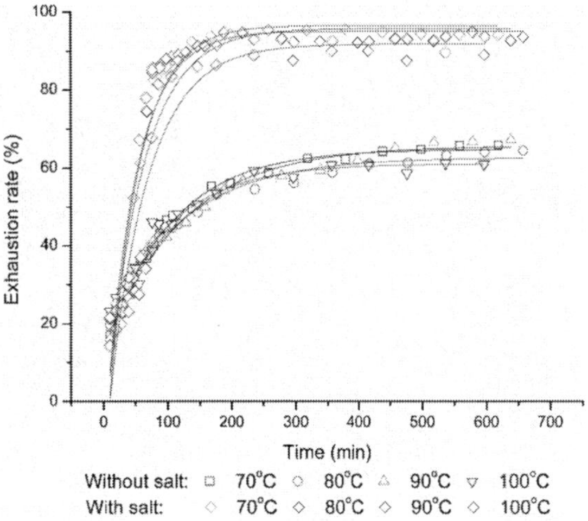

Figure 2. Exhaustion rates of Direct Blue 71 without and with addition of NaCl

As mention above, direct dyes were adsorbed on substance by substantivity, which can be calculated by Equation 3 [16, 17]. It represents the relationship between substantivity, liquor ratio, and the percentage of exhaustion at equilibrium.

$$K = \frac{\%E_e \times L}{100 - \%E_e} \quad (3)$$

Where K is the substantivity of dye; $\%E_e$ is the percentage of exhaustion at equilibrium; and L is the liquor ratio.

The K value indicates the efficiency of dye transferred from the dye bath to the substrate. A greater value of K means that the dye is favored to stay in the fiber. Table 1 shows the K values of dyeing with addition of NaCl are greater than those of without addition. Therefore, it could be concluded that addition of NaCl is an effective method to promote dye exhaustion in direct dyeing. It is apparently shown that dyeing at 90°C in with or without addition of salt, it has the highest exhaustion thereof the highest substantivity.

Table 1. Substantivity values of different dyeing conditions

Temperature (°C)	L	Without NaCl		With NaCl	
		%E_e*	K	%E_e*	K
70	50	64.7	91.6	95.8	1140.8
80	50	62.5	83.3	91.8	562.2
90	50	65.4	94.4	96.6	1419.0
100	50	61.0	78.2	95.1	970.6

* %E_e is calculated by the OriginLab software

In direct dyeing process, kinetic model analyse is assistant to interpret the dynamic interaction between ramie fibre and the direct dyes, which normally uses pseudo first-order model and pseudo second-order model [18, 19] to predict the sorption mechanism. The pseudo first- and second-order model can be expressed by Equation 4 and Equation 5.

$$\log([D]_{f,e} - [D]_{f,t}) = \log[D]_{f,e} - \frac{K_1}{2.3}t \quad (4)$$

Where k_1 is the pseudo first-order rate constant (s^{-1}) of adsorption; $[D]_{f,e}$ and $[D]_{f,t}$ (g g^{-1}) are the amounts of dye adsorbed per unit mass of ramie yarn at equilibrium and time t (s), respectively. The value of $\log([D]_{f,e}-[D]_{f,t})$ can be calculated from the experimental results and plotted against t (s).

$$\frac{t}{[D]_{f,t}} = \frac{1}{K_2[D]_{f,e}^{\,2}} + \frac{t}{[D]_{f,e}} \quad (5)$$

Where K_2 is the pseudo second-order rate constant of adsorption (g g^{-1}). The slope and intercept of the linear plot $t/[D]_{f,t}$ vs. t yield the values of $[D]_{f,e}$ and k_2.

The plots of the pseudo second-order with both dyeing conditions are displayed in Figure 3, which apparently shows that the value of $t/[D]_{f,t}$ are sharply dropped after addition of NaCl. However, the dye sorption of kinetic model does not change, which still fits the pseudo second-order, and the K_2 values are larger with addition of NaCl than that without addition of NaCl. The experimental and calculated $[D]_{f,e}$ values, pseudo second-order rate constants, and regression coefficient (R^2) values are listed in Table 2, and the data for addition of salt are after addition of NaCl.

Table 2 and Table 3 show that for adsorption of both dyeing conditions, the correlation coefficients of the pseudo second-order model are better than that of the pseudo first-order model. The better correlation between the equilibrium adsorption capacities of calculation ($[D]_{f,e}$, cal.) and the experimental values ($[D]_{f,e}$, exp.) implies that the pseudo second-order model is more likely to represent the adsorption kinetics of both dyeing conditions.

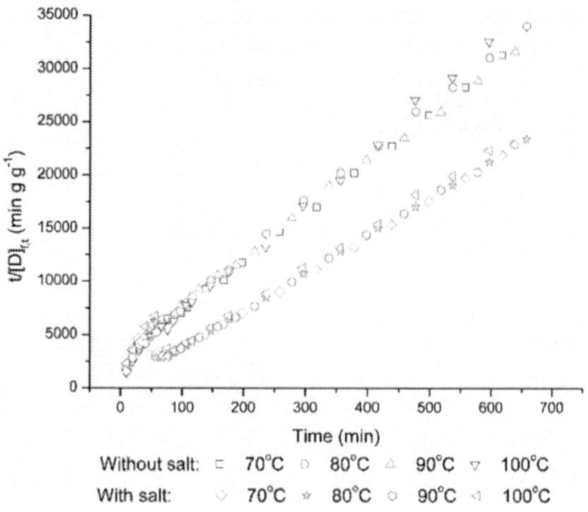

Figure 3. Plots of pseudo second-order without and with addition of NaCl

The direct dyeing process was a solid/liquid phase process, which proceeded by the diffusion of the dye molecule through the dye bath (liquid phase) to the ramie fibre (solid surface) as a result of the affinity of the two, followed by the dye diffusion into the fibre. As a result, the first solution diffusion process would be swift and the second diffusion process into the solid fibre would be slow.

Table 2. Kinetic parameters of pseudo first-order for adsorption of dyes on ramie yarns

Conditions	Temp. (°C)	$[D]_{f,e}$ (exp) (g g^{-1})	K_1 (s^{-1})	$[D]_{f,e}$ (cal) (g g^{-1})	R^2_1
Without NaCl	70	0.0194	0.0097	0.0157	0.9702
	80	0.0188	0.0074	0.0125	0.9830
	90	0.0196	0.0085	0.0165	0.9035
	100	0.0183	0.0108	0.0132	0.8723
With NaCl	70	0.0287	0.0076	0.0084	0.6063
	80	0.0276	0.0359	0.0486	0.9666
	90	0.0290	0.0051	0.0090	0.5388
	100	0.0285	0.0044	0.0119	0.5597

Half-dyeing time ($t_{1/2}$) is the time that 50% of the dye amount adsorbed at equilibrium by fiber. It can indirectly express the diffusion rate of dye in a fiber and is used to compare the behavior of dyes applied under the same dyeing conditions. The half-dyeing time of the different dyeing conditions, as listed in Table 4, were calculated using Equation 6 based on Equation 5 ($t=t_{1/2}$, $[D]_{f,t}=[D]_{f,e}/2$).

$$t_{1/2} = \frac{1}{K_2 \times [D]_{f,e}} \quad (6)$$

Table 3. Kinetic parameters of pseudo second-order for adsorption of dyes on ramie yarns

Conditions	Temp. (°C)	$[D]_{f,e}$ (exp) (g g^{-1})	K_2 (g g^{-1} s^{-1})	$[D]_{f,e}$ (cal) (g g^{-1})	R^2_2
Without NaCl	70	0.0194	0.8486	0.0215	0.9987
	80	0.0188	0.9409	0.0206	0.9978
	90	0.0196	0.6971	0.0221	0.9973
	100	0.0183	1.2424	0.0197	0.9964
With NaCl	70	0.0287	1.9123	0.0295	0.9973
	80	0.0276	3.2050	0.0286	0.9996
	90	0.0290	4.5646	0.0284	0.9993
	100	0.0285	3.7138	0.0273	0.9988

Table 4. Half-dyeing times and diffusion coefficients

Conditions	Temp. (°C)	K_2 (g g^{-1} s^{-1})	$[D]_{f,e}$ (cal) (g g^{-1})	$t_{1/2}$ (s)	D (×10^{-10} m^2 min^{-1})
Without NaCl	70	0.6475	0.0194	79.59	1.48
	80	1.6859	0.0187	31.64	3.72
	90	2.8307	0.0196	18.01	6.53
	100	7.1619	0.0183	7.63	15.42
With NaCl	70	1.9123	0.0287	18.19	6.47
	80	3.2050	0.0275	11.33	10.389
	90	4.5646	0.0290	7.56	15.56
	100	3.7138	0.0285	9.44	12.47

The kinetics of the adsorption of the dyes by cellulosic fibers is usually described by the Hill equation for diffusion [20]. The diffusion coefficient at half-dyeing time, as given in Table 4, was calculated using Equation 7 which was deduced from the Hill equation [19].

$$D = \left(\frac{[D]_{f,t}}{[D]_{f,e}}\right) \times \frac{100 \times d^2}{t} \quad (7)$$

Where D (m^2 min^{-1}) is the diffusion coefficient at half-dyeing time; t (min) is the half-dyeing time, $t_{1/2}$; d is the diameter of ramie fiber, 40.2 μm.

In Table 4, the half-dyeing time and diffusion coefficient at half-dyeing time for dyeing without NaCl decrease and increases with increasing of dyeing temperature respectively; and it is almost similar with the dyeing with NaCl excepting for the dyeing at 100°C. At 100°C dyeing with NaCl, the half-dyeing time increases from 7.56 min at 90°C to 9.44 min and the diffusion coefficient at half-dyeing time reduces from 15.56 ×10^{-10} m^2 min^{-1} at 90°C to 12.47×10^{-10} m^2 min^{-1}.

4 Conclusion

Additon of NaCl electrolyte promotes the sorption of Direct Blue 71 on ramie loosely fibre, and shortens its equilibrium time. The dye adsorption process of dyeing without and with NaCl followed pseudo second-order rate kinetics, but the rate constant of sorption is larger for the latter compared to the former. Besides, in contrast with the dyeing without addition of NaCl, the half-dyeing time and the diffusion coefficient for $t_{1/2}$ of dyeing with NaCl is shorter and higher respectively.

Acknowledgement

This work was financially supported by the China National Textile & Apparel Council (2013 "Textile Vision" Applied Basic Research, 2013-153); Hubei Province Science and Technology Support Program (Grant No. 2013BAA043) and the Collaborative Innovation Plan of Hubei Province for Key Technology of Eco-Ramie Industry (2014-8).

References

1. D. Klemm, B. Heublein, H.P. Fink, A. Bohn, Angew. Chem. Int. Ed., **44**, 3358 (2005)
2. S. N. Pandey, Textile Progress, **39**, 1 (2007)
3. Z. Li, Z. Li, R. Ding, C. Yu, Textile Research Journal, **86**, 451 (2016)
4. D. Fengel and X. Shao, Wood Sci. and Technol., **18,** 103 (1984)
5. S. Nam, A. N. Netravali, Fibers and Polymers, **7,** 372 (2006)
6. R. Thakur, C.R. Sarkar, R. Sarmah, Indian Journal of Fiber & Textile Research, **24,** 276 (1999)
7. L. Cheek, L. Roussel, Textile Research Journal, **59,** 541 (1989)
8. D. Das, A.K. Samanta, P.C. Dasgupta, Indian Journal of Fiber & Textile Research, **22,** 53 (1997)
9. L. Cheek, Clothing and Textiles Research Journal, **8** 38 (1990)
10. M. Kanik, P. J. Hauser. Textile research journal, **74,** 43 (2004)
11. A.O. Noah, C. M. Martins, J. A. Braimah, Journal of applied polymer science, 32, 5841 (1986)
12. R. B. Chevan, J. V. Rao, Text& Res. J., **39,** 292 (1983)
13. H. F. Willis, J. O. Warwicker, H.A. Standing, A.R. Urquhart, Transactions of the Faraday Society, 45, 506 (1945)
14. G.M. Simu, S. Funar-Timofei, L. Kurunczi, Chemical Bulletin of Politehnica University of Timisoara. **53,** 153 (2008)
15. J.R. Aspland, Textile Chemist & Colorist. **23,** 13 (1991)
16. C.-w. Kan, AUTEX Research Journal. **8,** 255 (2007)
17. M. Razvigorova, T. Budinova, N. Petrov, V. Minkova, Water Res. **32,** 2135 (1998)
18. Y.S. Ho, G. McKay, Process Biochem. **34,** 451 (1999)
19. G.E. Krichevsk, I.M. Movshovich, Text Res J. **40,** 793 (1970)
20. M.N. Michael, S.H. Samaha, S.F. Lbrahim, Colourage. **44,** 23 (1997)

A novel study on UV protection and antibacterial properties of washed denim garment

Md. Nahid Pervez[1, 2], Md. Anisur Rahman[3], Liping Yu[4] and Yingjie Cai [1, 5]

[1]*School of Chemistry and Chemical Engineering, Wuhan Textile University, Wuhan 430073, China*
[2]*Research Institute of Flexible Materials, School of Textiles & Design, Heriot-Watt University, Galashiels TD1 3HF, UK*
[3]*School of Textile Science and Engineering, Wuhan Textile University, Wuhan, 430073, China*
[4]*Zhejiang Institute of Modern Textile Industry, Shaoxing 312000, China*
[5]*Hubei Provincial Engineering Laboratory for Clean Production and High Value Utilization of Bio-based Textile Materials, Wuhan Textile University, Wuhan 430073, China*

Abstract. On this planet, many investigations are applied to switch conventional chemical cloth techniques via eco-pleasant and economically attractive bioprocesses using enzymes. The present study offers an enzymatic washing system using enzyme (Cellzyme SPL H/C) for boosting the ultraviolet and antimicrobial undertaking of denim garments. Experimental results showed that the 4.0% o.w.f enzyme awareness furnished a greater UPF than the other concentrations and before washed. Results divulge that enzyme (Cellzyme SPL H/C) not handiest preserve the fabric surface from UV degradation but also performed extended degree of antibacterial endeavour in opposition to some species of bacteria that leading to act as a nice antibacterial agent on the denim materials. The enzyme washing healing diminished the skin hairiness and accelerated the skin evenness of the denim fibres as shown by means of SEM measurements.

1 Introduction

As it is known, denim (jeans) is among the trendiest, dynamic, progressive fabric products and it is the handiest material that can be worn throughout any season [1]. Brands are now watching for more technological, sustainable, and alleviation fabrics and jeans to preserve and give a boost to their progressive line. Now-a-days specific types of systems are applied and washing is known as one of the crucial greatly used finishing healing that has colossal utilization in cloth sectors on account that of creating specific appearance and making trendy and wear at ease garments of the state-of-the-art world [2, 3]. Essentially the most frequently used denim washing methods are enzyme wash, bleach wash, acid wash, usual wash, stone wash, and so forth. Among the many washing methods, enzymatic methods are essentially the most broadly used procedure in industry as a result of its low damaging nature on clothes floor and eco-friendly nature compared to others [4-8]. In enzyme washing, cellulases attack notably surface of the clothes, leaving the interior of the clothes as it is by doing away with the dye gift in the floor layer of clothes [9]. Nonetheless, extraordinary types of materials and finishing remedies are anticipated to have a large impact on sensible characteristics. Aside from the innovations in fabrication, application of useful finishes to clothes, corresponding to UV safety, antimicrobial finish, fragrance conclude and antifungal, add up more value to the top product in the present market situation. Among them UV-security and antimicrobial conclude appeal to much attention in up to date years. As a consequence, purchasers have turn out to be more and more aware of the necessity for sun safety, which is regarding the incidence of solar brought on dermis harm and its relationship with an expanded exposure to UV light. In case of antimicrobial conclude presents the more than a few advantages of controlling the infestation by means of microbes safeguard textiles from staining, discoloration, and fine deterioration and prevents the odour formation [10, 11]. Mainly, finishing is applied for already sewn garments, hence, it is vitally major to know the influence made by way of particular washing on the stitching garment material in an effort to keep the exceptional of a sewing garment.is applied for already sewn garments, therefore, it is very important to know the impact made by particular washing on the sewing garment fabric in order to preserve the quality of a sewing garment.

In this study, the idea of imparting antibacterial property into cotton denim material along with UV protection was involved, aiming to boost aged-regarded denim by means of using cellzyme in washing process which based authors abilities have not ever done before.

2 Experimental

2.1 Materials

DOI: 10.1051/matecconf/201710803004

Indigo dyed denim garment used for this experiment is commercially available 100% cotton, twill weave (3/1), construction: 72 (Ends/inch) X 41 (Picks/inch) / 8 Ne (Warp yarn count) X 12 Ne (Weft yarn count). Amylase, Cellzyme SPL H/C was supplied by Jeans Corporation, Australia.

2.2 Method

Samples were desized for 15 min at 70°C, pH =7, Liquor to Goods Ratio 1:50, required amount of amylases, soda ash, detergent, anticreasing and desizing agent had been adjusted and then, washed thoroughly. The bathing system was once applied liquor containing acetic acid (1.0 g/l) at pH (4-5.5) and material to liquor ratio of 1:30 at concentration of Cellzyme SPL H/C of (2-6) % o.w.f at (40-60°C) temperature for 60 min. The process was once terminated through growing the pH to 8 and the temperature to 80oC for 10 min then washed thoroughly. Finally, garments were treated with cationic softener at 40°C for 10 min. Subsequently, clothes were treated with cationic softener at 40°C for 10 min. For hydro-extraction procedure, treated denim samples had been squeezed in a laboratory scale hydro-extractor machine at 200 rpm for 4 min and dried in a steam drier at 75°C for 40 min.

2.3 Characterization and functional properties

Scanning electron microscope (SEM) measurements of the samples were carried out on Hitachi-S 3400 N, (Japan) [12]. The ability of a fabric to block UV light is given by the ultraviolet protection factor (UPF) values. The UPF values are calculated according to AATCC test method 183-2004. The antibacterial activity of enzyme washed denim fabric samples (washed and unwashed) was evaluated qualitatively by using the AATCC-147 method, using Staphylococcus aureus (gram-positive) and Escherichia coli (gram-negative) bacteria.

3 Results and discussion

Since the primary venture of the present work is to browsing for the correct cure formulations and stipulations for upgrading and embellishing the UV protective and antibacterial functional houses of washed cotton denim fabrics for producing fabric articles with better defense potential, a large variety of causes were examined. Outcome bought together with their right dialogue follow.

3.1 UV-Protection

Researchers showed that ultraviolet radiation (UVR) from the solar would be a principal cause of skin melanoma [13, 14]. The bigger the UPF worth, the longer a person can stay within the sun until the area of epidermis underneath the material turns into red. However, many of the reports have eager about the material parameters with woven materials only, whereas there were few stories regarding denim fabrics. In order to analyse the have an effect on of various solution parameters such as enzyme concentration, temperature and pH on UPF values is reported in Table 1. Results of UPF measurements of pristine sample just is not good which shows that without any treatment do not have UV protecting potential as their UPF values are round 11.5 (<15) which can not be labeled as just right safety. Nevertheless, enzymatic desized fabric promotes a very good level of protection 20.2 (15-24). However, there is a brilliant develop in UPF values once the materials are washed with enzyme. The washed materials verified superb UV security rating (25-39). In phrases of solution parameters, UPF values of materials have been progressively stronger with enzyme attention increased however after 4% found low UPF rating. If we take into consideration about effect of temperature, generally speaking, enzyme reaction raises with temperature however it is only activated within a temperature range in which the enzyme constitution keeps steady and unchanged. Beyond this optimum range, the enzyme activity will decrease sharply as the protein structure of enzyme is tangled through thermal agitation. The enzyme activity will slash sharply as the protein structure of enzyme is tangled by means of thermal agitation. Enzyme performs its maximal endeavour at a specific range of pH values. Deviation of the foremost pH variety will alter the interior constitution of enzyme as a result of the electrostatic interactions within the enzyme. As a result, the enzyme will not participate in its natural perform [15]. The alterations discovered in UPF values are attributed to material shrinkage which, in flip, leads to reduced porosity of the materials and excessive absorptive potential of the cotton fibre which allows the buildup of enzyme on cotton materials and due to as anticipated, a larger quantity of shrinkage occurred in the cotton materials. Also, the materials thickness was accelerated after enzyme washing treatment, lowering fabrics porosity, producing a closer weave; preventing UVR transmission via the compact material surfaces for this reason multiplied UPF. As we all know, smoother fibre surface reflects extra UV rays and the hydration or moisture content could be one more element accountable for increasing UPF values after enzyme washed [16]. The cellzyme utilized in enzyme washing may just effect in swelling of fibers as a result of accelerated moisture absorption, hence lowering the interstices and increases the imply UPF rating [10], which means treated sample 3 furnished excellent UV safeguard and chosen as the optimized washing conditions.

Table 1. UPF mean values of enzyme washed denim

Samples	Enzyme concentration (%)	Temperature (°C)	pH	Mean UPF
Pristine	-	-	-	11.5
Desized	-	-	-	20.2
Treated 1	2	40	4.2	31.2
Treated 2	3	45	4.8	33.3
Treated 3	4	50	5.1	35.6
Treated 4	5	55	5.3	34.7
Treated 5	6	60	5.5	34.5

DOI: 10.1051/matecconf/201710803004

3.2 Antibacterial activity

The antibacterial activity of washed fabric used to be evaluated quantitatively with the aid of measuring the number of colonies of gram-positive (S. aureus) and gram-negative (E. coli) bacteria as per the experiment approach mentioned in the experimental part. Results are presented in Table 2. As expected, there used to be no noticeable development of both bacteria on pristine fabric. In addition, there was a development in micro-organism recovered from the pristine material indicating that both strains can use the material as a suitable substrate. It was additionally discovered that enzymatic desized fabric confirmed bacterial growth of around 20 %. In terms of solution parameters, the antibacterial endeavour of the handled samples had been evaluated and it was located that with an increase within the awareness of enzyme (up to 4 %), the bacterial development improved as much as 40 % however after that material showed much less antibacterial activity in opposition to both types of bacteria. This may be due to the improved weight add-on % of the enzyme lively ingredients on the material with growing enzyme awareness. The enhancement in antimicrobial recreation of treated substrate is a direct outcome of utilising an electrochemical mode of motion to penetrate mobile wall of microorganism, thereby inflicting leakage of essential metabolites and bodily disrupting other key phone features on contact to kill it and thoroughly leaching antimicrobial moieties to enter or react chemically with the microorganisms performing as a poison or inhibitor [17].

Table 2. Antibacterial activity of enzyme washed denim

Enzyme concentration (%)	Temperature (°C)	pH	*S. aureus*	*E.coli*
-	-	-	-	-
-	-	-	20	21
2	40	4.2	31	32
3	45	4.8	33	34
4	50	5.1	38	40
5	55	5.3	36	37
6	60	5.5	35	38

3.3 Fibre morphology

SEM measurements have been applied with a view to be trained the outcome of the enzyme washed on the skin morphology of the fibres. Pristine fabric is smooth and there is no coating and deposition (Figure 1 d) and in case of desized fabric (Figure 1 e) is embedded with tiny particles of sizing agent and the surface of fabric is irregular due to emphasize of sizing agent and waxes have been gently eliminated from the material floor, leaving it intact with applicable whiteness. It may be observed from (Figure 1 f) that entanglements of fibre surfaces of handled pattern with combination of enzymatic desized and cellzyme are broken not up to handled samples with amylases. This can also be due to lowering of cellzyme activity when mixed with amylase and hydrolytic removal of protruding fibrils and important partitions of the fibres during washing system [18]. Also, which means the cellzyme have not any hazard to enter to the yarn building during the enzymatic processing time to hydrolyse the internal fibers and total regarded that, it may lessen the fiber surface damage.

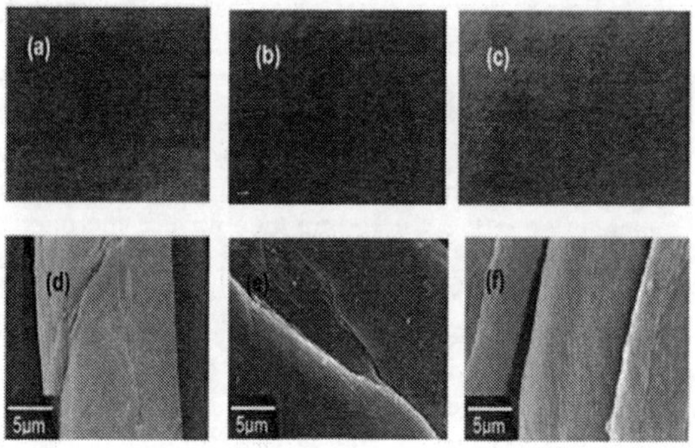

Figure 1. Digital photos of (a) pristine, (b) enzymatic desized and (c) enzyme washed and SEM images of (d) pristine, (e) enzymatic desized and (f) enzyme washed (optimum condition: enzyme concentration 4%, pH 5.1, temperature 50°C)

3.4 FT-IR analysis

FT-IR, an apparent acute address was acclimated to study the chemical modification on the surface of fabrics are illustrated in Figure 2. For all samples, a broad band between 3100 and 3700 cm^{-1}, centred on 3360 cm^{-1}, illustrated characteristics of OH functional groups in cellulose. A strong adsorption band with a maximum at 1030 cm^{-1} is an aftereffect of the overlapping bands attributed to the functional groups of cellulose, namely the C–C, C–O and C–O–C stretching vibrations [19]. Some apparent differences observed in the assimilation characteristics in the 900-1100 cm^{-1} arena of the treated sample with enzyme after the washing. This region is found to be very sensitive to the crystalline as compared to the amorphous structure of the cellulose of denim fabrics since the hydrolysis of 1, 4 β-glucosique of the cellulose molecule link breaks the molecule into several pieces (i.e. broadening of the band reflects a higher degree of the disordered structure). Due to the disorder of the cellulose structure is caused by the angle changes around the ß-glucosidic linkage and rearrangement of hydrogen bonds [20], a marked increase is shown in intensity.

DOI: 10.1051/matecconf/201710803004

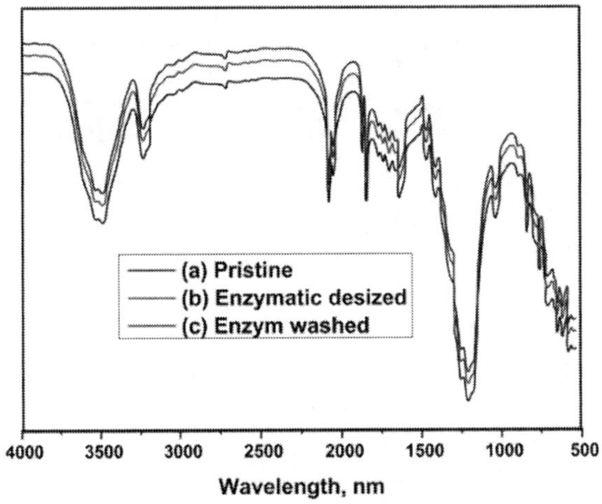

Figure 2. FT-IR spectra of (d) pristine, (e) enzymatic desized and (f) enzyme washed (optimum condition: enzyme concentration 4%, pH 5.1, temperature 50°C)

3.5 Crease recovery angle

Creases are authentic as the fabric deformations based on its viscoelastic properties. This affectionate of anamorphosis not alone has abominable furnishings on the fabrics' actualization but aswell would advance to abatement in the in the fabric wear quality [21]. Since the consumers usually prefer to wear such clothing which are resist to be creased during wear, fabrics performance related to the crease recovery property is of the main importance. Figure 3 shows the crease recovery angle values of the denim fabrics. It should be acclaimed that the washing followed by the enzymatic treatment has the highest values compared with the other treatments. This is due to the swelling of cellulose fibers which decrease the cellulose binding and improve this fiber to return to its original form. That adumbrated that the fabrics become less rigid and stiff, easier to amplitude and looser in structure. This may be regarded to cellulase enzyme treatment optimize the surface properties [22].

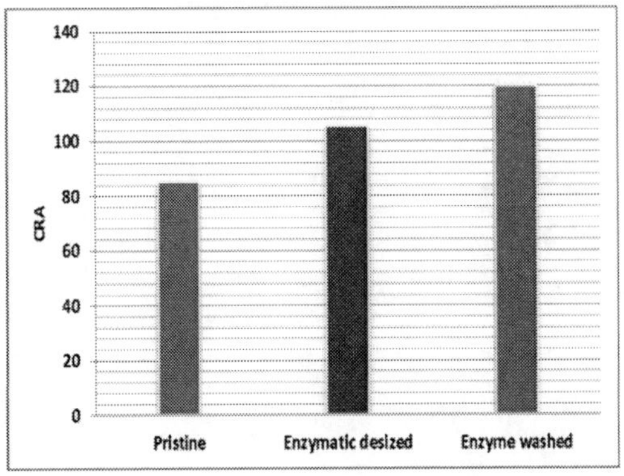

Figure 3. Crease recovery angle of (d) pristine, (e) enzymatic desized and (f) enzyme washed (optimum condition: enzyme concentration 4%, pH 5.1, temperature 50°C)

4 Conclusion

Biotechnology in textiles is among the progressive methods to promote fabric discipline. Consequently, our outcome awarded a potential new biotechnology with a fascinating antibacterial and appropriate UV protection residences via denim washing, situated on the use of Cellzyme. As a conclusion of this experimental study, it is determined that the healing of cellzyme on denim clothes washing may just create useful functional property in special phases by using obtaining optimum parameters like enzyme concentration-4.0% owf (on weight of fabric), pH 5.1 and temperature 50°C. All these factors make them incredibly exciting from the factor of view of an industry that strives to scale down its environmental impact and to appeal to the environmentally conscious purchaser.

Acknowledgement

This work was financially supported by the China National Textile & Apparel Council (2013 "Textile Vision" Applied Basic Research, 2013-153); Hubei Province Science and Technology Support Program (Grant No. 2013BAA043) and the Collaborative Innovation Plan of Hubei Province for Key Technology of Eco-Ramie Industry (2014-8).

References

1. D. Miller, S. Woodward, Social Anthropology. **15,** 335 (2007)
2. R.M. Tyndall, Textile Chemist & Colorist. **24,** 23 (1992)
3. W. Zhao, Dyeing Finish. **34,** 34 (2008)
4. N.K. Pazarlioğlu, M. Sariişik, A. Telefoncu, Process Biochemistry. **40,** 767 (2005)
5. M. Sariisik, AATCC review. **4,** 24 (2004)
6. Y. Yu, J. Yuan, Q. Wang, X. Fan, X. Ni, P. Wang, L. Cui, Carbohydrate polymers. **95,** 675 (2013)
7. T. Bechtold, A. Turcanu, R. Campese, P. Maier, W. Schrott, Journal of Applied Electrochemistry. **36,** 287 (2006)
8. M. Jucienė, V. Dobilaitė, G. Kazlauskaitė, Materials Science. **12,** 355 (2006)
9. C. Carr. *Chemistry of the textiles industry* (Springer Science & Business Media, 1995)
10. D. Saravanan, AUTEX Research Journal. **7,** 53 (2007)
11. Y. Gao, R. Cranston, Textile Research Journal. **78,** 60 (2008)
12. A. Khan, M.N. Pervez, I.A. Khan, S. Ahmad, R. Masood, T. Hussain, F. Telegin, Asian Journal of Chemistry. **28,** 1191 (2016)
13. D.L. Narayanan, R.N. Saladi, J.L. Fox, International journal of dermatology. **49,** 978 (2010)
14. R.N. Saladi, A.N. Persaud, Drugs of Today. **41,** 37 (2005)
15. C. Kan, C. Yuen, W. Wong, Journal of applied polymer science. **120,** 3596 (2011)

16. Y. Dong, Z. Bai, L. Zhang, R. Liu, T. Zhu, Journal of applied polymer science. **99,** 286 (2006)
17. R. Goyal, Colourage. **52,** 81 (2005)
18. M. Montazer, A.S. Maryan, Journal of applied polymer science. **110,** 3121 (2008)
19. M. N. Pervez, F. Shafiq, M. M. Jilani, Z. Sarwar, Y. Cai, Material Science Forum, **893**, 71, (2017).
20. J. Blakwell. *ACS Symposium Series,* **48,** 206, (1977).
21. Y. Hassan, M. E., EL-Salmawy, A. Almetwally, Indian Textile Journal, **120**, 22 (2010).
22. M.M. Eladwi, R.M. Kotb 2015. International Journal of Textile and Fashion Technology, **5,** 1, (2015)

A novel route to process rationalisation on cellulose dyeing

Md. Nahid Pervez[1,2,3], Mohammed Akramul Hoque[3], Liping Yu[4], Md. Moshfique Hossain Rinkon[3], Md. Salman Amin[3], Md. Nazmul Ehsan[3] and Yingjie Cai[1, 5]

[1]*School of Chemistry and Chemical Engineering, Wuhan Textile University, Wuhan 430073, China*
[2]*Research Institute of Flexible Materials, School of Textiles & Design, Heriot-Watt University, Galashiels TD1 3HF, UK*
[3]*Department of Textile Engineering, Southeast University, Dhaka 1213, Bangladesh*
[4]*Zhejiang Institute of Modern Textile Industry, Shaoxing 312000, China*
[5]*Hubei Provincial Engineering Laboratory for Clean Production and High Value Utilization of Bio-based Textile Materials, Wuhan Textile University, Wuhan 430073, China*

Abstract. This research work presents an altercation on dyeing machine controlling parameters, particularly in terms of the relation between the effect of different factors i.e. reel speed (m min^{-1}), cycle time (min), pump pressure (%) and nozzle position (%) on color strength (K/S) property of different type's cotton fabric, so only the process controlling parameters were varied; the material used for the study remained the aforementioned throughout. The optimum reel speed, cycle time, pump pressure and nozzle position were found to be 240 m min^{-1}, 2.5 min, 60% and 70% respectively for winch dyeing machine with respect to cellulose materials. This work is useful in selecting optimum values of various controlling parameters that would not alone approve the process rationalisation but as well reduce the fabric consumption and advance the superior of the dyed fabric on applied imminence.

1 Introduction

Dyeing is the third important but a lot of circuitous action and it is a part of the key causes in the boastful trading of textile products than weaving and spinning, which is added an artwork than a science [1, 2] and it encompass three steps together with the diffusion of dye through the aqueous dye bath on to the fiber, the adsorption of dye into the outer layer of the fiber and the diffusion of dye from the adsorbed surface into the fiber interior. It goes after adage that, use of affection knitted fabric has been accretion in apple wide. People around each corner of the world, each custom, casts feel comfortable wearing cotton knitted fabric due to able applicable properties, softer handle backdrop and high extension at low tension compared to woven manufactured fabric [3]. By means of these excessive traits, knitted surfaces were apparently acclimated to woven materials. Reactive dyes are soluble anionic dyes which, in solution, are repelled by the negatively charged surface of the cotton fibre; this dyeclass has become more accepted [4]. That is mainly due to high levels of washing fastness, a wide gamut of bright colours, and versatility for altered appliance methods. The high fastness to washing of reactive dyes is due to their unique reactive groups, which forms covalent bonds with the hydroxyl groups of the cotton cellulose under alkaline pH conditions. It aswell supports dye assimilation into the fibre autogenously (diffusion) which leads to bigger dye fixation A salt such as sodium chloride or sodium sulphate is added as an electrolyte to advance the dye alteration (exhaustion) to the fibre. There are greater than 50% of cellulosic fibers are dyed with reactive dyes. Market share of reactive dyes among all textile dyes captured about 29%. As a result of their strong bonding with many surfaces of synthetic and usual materials, reactive dyes are also used for dyeing wool, nylon, silk, and modified acrylics [5-7]. In an effort to achieve the specified dyeing first-rate, all causes that may impact the dyeing system need to be exactly controlled. As a consequence, it is wanted to improve productiveness and profitability of the field via making a choice on the various process controlling parameters. The principles of system manipulate measures are becoming preferred now days due to two main factors: (I) growing competitions and increasing cost of creation. (II) Fabric enterprise now a day is dealing with awfully stiff competitors. This competition is at macro and micro degree. Underneath these situations, cost and first-rate turns into fundamental standards, as a consequence, approach manipulate becomes more predominant, for reaching overall success. In literature, only a limited number of studies have been published about dyeing parameters and their effect on fabric properties [8-10]. However, no report yet published about influence of dyeing machine controlling parameters on dyeability of cotton fabric.

The current paper is intended to fill the missing link in the above papers and to discuss the application of control equipment; in particular parameters i.e. reel speed (m min^{-1}), cycle time (min), pump pressure (%) and nozzle position (%) effect on cellulose dyeing on lab scale.

2 Experimental

2.1 Materials

Commercially available three different knit structures based cotton fabrics were used throughout the study. Remazol Red RR (MCT-VS dye), Na_2SO_4, Na_2CO_3, NaOH, soaping agent and acetic acid were supplied by supplied by local chemicals company and are laboratory grade.

2.2 Dyeing procedure

The optimized dyeing process of fabrics with Remazol Red RR dye was performed at 5% shade and 1: 50 liquor ratio in exhaust process via winch dyeing machine (Figure 1). The fabrics were bathed inside the dyeing solution at room temperature followed by addition of 30 g L^{-1} of salt. In the meantime, the temperature was kept at 25°C and the dyeing process was continued for 30 minutes. Then 20 g L^{-1} of NaOH was added to the dyeing bath and stayed for additional 30 minutes. The final dyed fabrics were neutralized with 2.5 g L^{-1} of acetic acid followed by rinsing with excess water until the final rinse become colorless.

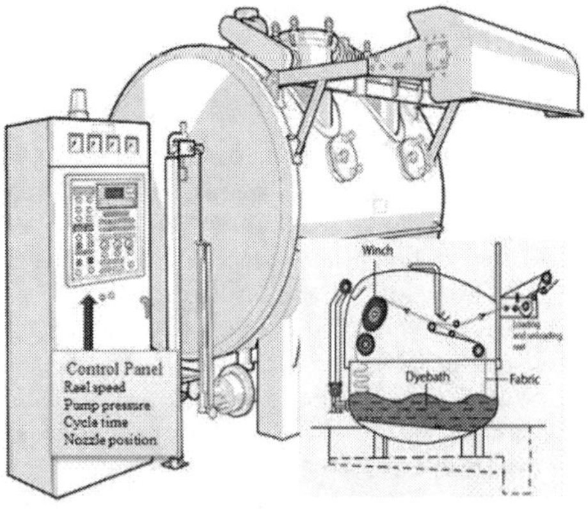

Figure 1. Schematic diagram of Winch Dyeing M/c (Inside view)

During optimized dyeing process, for the optimization of dyeing machine controlling parameters i.e. reel speed (m min^{-1}), cycle time (min), pump pressure (%) and nozzle position (%) were carried out at different variables. At first, experiments were carried out to optimize dyeing reel speed varies at 220-260 m min^{-1}. To understand the effect of cycle time, another set of experiments were carried out in optimized cycle time for different time periods that are 2.20-2.60 minutes. Another set of experiments were also carried out at different pump pressure such as 40-60 %. In case of nozzle position, the experiments were also accelerated in different variables at 40-80 %. After dyeing, the dyed samples were rinsed with cold water and washed in a bath with a liquor ratio of 1: 50 using 1 g L^{-1} of the soaping agent at 60°C for 10 minutes and then they were again rinsed and finally dried in a dryer. Based on the color strength (K/S) values of the dyed samples, optimum dyeing machine parameters such as reel speed, cycle time, pump pressure and nozzle position were selected.

2.3 Colour strength evaluation

The colour strength and the reflectance spectra of all the dyed fabrics were measured using a UV vis spectrophotometer. The colour strength (K/S) value was calculated using Equation 1 given below

$$\frac{K}{S}=\frac{(1-R)^2}{2R} \tag{1}$$

Where K is the coefficient of absorption, S is the coefficient of scattering; and R is the reflectance value of the fabric at max.

3 Results and discussion

For purpose to describe the process expediently and to determine the optimum parameters we assess the effect of each parameter at variable values. The K/S values which are considered as representations of color strength were measured and the results are discussed in order hereinafter.

3.1 Effect of reel speed on color strength (K/S)

In winch dyeing machine, winch reel not only controls the rate of movement of the fabric rope, but also the configuration of the rope in the dye bath. It does not grip the fabric positively, but by the weight of the wet fabric and the friction between the reel and fabric. The speed of reel can be calculated according to Equation 2.

$$\text{Reel speed}=\frac{\text{Fabric weight}}{\text{GSM}\times\text{Fabric Width}\times\text{Cycle Time}\times\text{Nozzle}} \tag{2}$$

As shown in Figure 2, it is clear that the color strength values had been accelerated with expand of reel speed and reaches a highest value at 240 m min^{-1} then it decreases slowly because excessive speed explanations a gigantic lengthwise stretching stress and pull on the fabric being dyed which is unsafe to best [11]. The effect of reel speed of dye bath can also be attributed as a result of special resonating varieties of dye molecules i.e. heterogeneous distribution of dyes and the interaction between coloring component and pore size of cotton fibers used and the better kinetic vigour of the dye molecule, the fiber swelling characteristic and availability of the significant number of energetic binding sites onto the outer surface of cotton molecules. Additionally, a big amount of water exists within the interfibres and interyarns or even on the fabric surface at this time, which may cause severe reactive dye hydrolysis [12] and further reduce the K/S value. The dyestuff is present in the solution as single molecules as well as aggregates. Increasing the speed leads to the breaking down of aggregates. So, the numbers of single molecules existing within the solution emerge as raises. When single molecule absorbed through the fiber, additional hydroxyl groups consider from the aggregates, which can be, taken up via material, leading to a complete even dyeing [13].

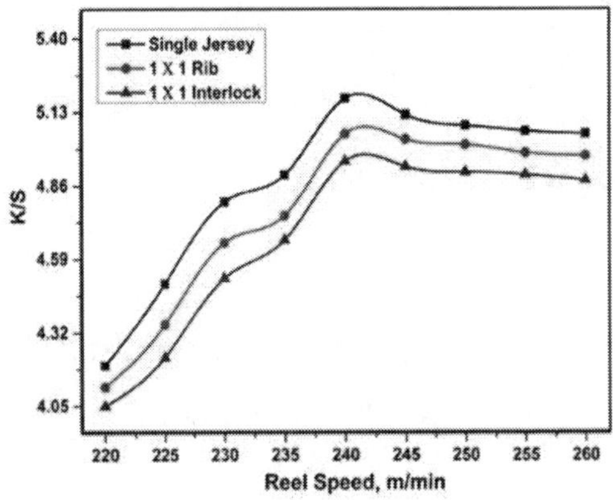

Figure 2. Effect of reel speed on color strength (K/S)

3.2 Effect of cycle time on color strength (K/S)

In exhaust dyeing method cycle time is important factor. The time required to move the whole loop is termed as cycle time and it can be measured by Equation 3.

$$\text{Cycle time} = \frac{\text{Meters per Chamber}}{\text{Reel speed}} \quad (3)$$

The relationship between dye adsorption and cycle time in cellulose dyeing is demonstrated in Figure 3. It is seen that dye uptake expressed as color strength (K/S) increased with increasing cycle time and there is decrease in the colour strength after surpasses time 2.5 min. The decline in colour strength is attributed to the shift in equilibrium of colouring component from fabric into dye bath during longer cycle time. In case of the advance in K/S values reflected the absolute accept an effect on of increasing cycle time which concluded in an acceptable and compatible adsorption of the dye via the fibers, as acceptable as compatible assimilation and circulation of the dye into the material, which resulted in the accessory of the uptake of the dye into the material [14]. However, this result was not consistent with all fabrics, since the K/S value of the single jersey plain knit fabric is higher than that of the other two type's fabric as shown in Figure 2. This may be attributed to the decomposition of the dye with prolonged time leading to a lower K/S value and high stretching strain damage the material and tend to impart permanency to the wrinkles, and this often necessitates full width dyeing [11].

3.3 Effect of pump pressure on color strength (K/S)

The pump generates static stress, which may also be infinitely adjusted and remains constant throughout treatment, irrespective of the temperature. Machine topping up may also be completed at any time making use of the stress pump. The effect of pump pressure (%) on the colour strength of the dyed cotton fabrics is illustrated in Figure 4. It can be seen clearly that both the K/S value increased markedly with increasing pump pressure when it was shorter than 60%. Thus, the access in the K/S ethics may be activated with the access in the apparent acerbity and wettability of a fabric, which may access its water swelling capability and its affection for acknowledging dyes. However, if the pump burden exceeded 60%, the accretion trend of the K/S amount began to decline. Results can be explained from thermodynamic attempt acquiescent the clapeyron effect. A pressure was activated assimilate the fabric during dyeing to ensure activity manual and bonding amid the fabric surfaces. This pressure creates a hot compaction on the fibers that affects the bonding amid the fabric layers. It is sometimes abominably believed that high pressure is essential to accomplish a high liquor flow through the material. The rate of flow of liquor is determined by the accommodation of the pump and by the aberration in pressure immediately before and beyond the wound package, known as the differential pressure. In general, the lower the attrition of the actual to an assertive pump, the lower the differential pressure and the higher the flow [15]. At low pressure poor transmission of energy takes place, and consequently, fabric color strength does not ability to the appropriate akin and the fiber acclimatization is as well not maintained due to abridgement of constraint. Under acceptable pressure, fibers are aseptic from shrinking and loosing acclimatization consequently, application the fiber strength and announcement able bonding [16].

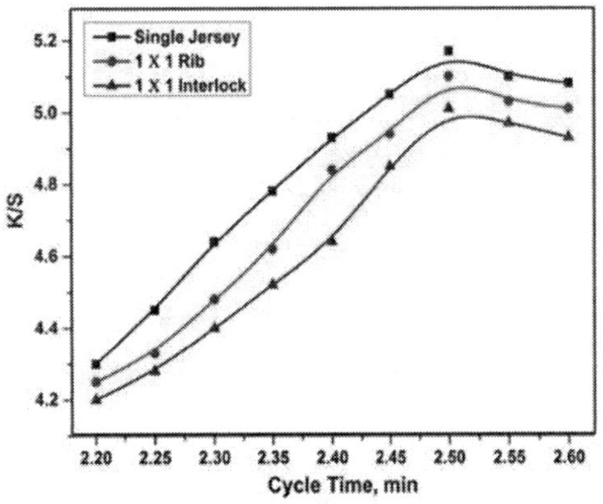

Figure 3. Effect of cycle time on color strength (K/S)

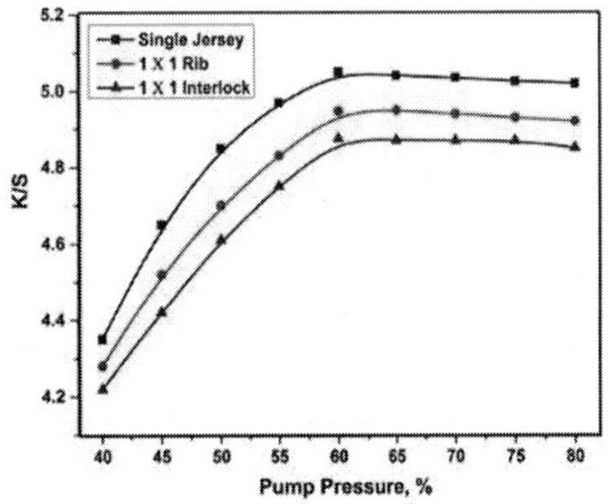

Figure 4. Effect of pump pressure on color strength (K/S)

DOI: 10.1051/matecconf/201710803005

3.4 Effect of nozzle position on color strength (K/S)

A nozzle accumulation for a fabric dyeing apparatus comprises a tubular nozzle affiliate through which fabric travels, the nozzle affiliates authentic by four ancillary walls and accepting an ellipsoidal cross-section, and a nozzle gap in an ancillary bank of the nozzle affiliate through which aqueous can be activated to fabric in the nozzle affiliate to actuate the fabric along a travel path [17]. The effect of nozzle position (%) on the colour strength of the dyed cotton fabrics are shown in Figure 5. It can be seen clearly that both the K/S value increased markedly with increasing nozzle position when it was shorter than 60% but abruptly decreased values after 60% and claim maximum k/s values at 70%. The design of such a nozzle can be challenging. On one hand, the nozzle should not absorb too abundant space, as this would yield up too abundant aqueous aggregate within the machine and affect the liquor ratio (ratio of the volume of dye liquor to the weight/amount of fabric that is required to produce a desired dyeing result). To accomplish an even dyeing effect, it is adorable that the fabric braiding opens beyond the amplitude of the fabric during dyeing, and changes position each time it passes through the nozzle. This provides an able and even appliance of dye liquor assimilate and into the fabric. However, a circular nozzle tends to clasp and abbreviate the fabric casual through it rather than acceptance the fabric to extend width-wise. The assay helps in compassionate the beginning results. The nozzle position of the ejector influences the ejecting accessory a lot. The optimum nozzle position of the subsonic ejector exists if the flow and added geometry ambit are fixed. The ejecting accessories will ability its best when setting the nozzle at the optimum position, and abatement if deviating. The optimum nozzle position of an ejector is not constant; but is a variable, afflicted abundantly by the flow ambit [18].

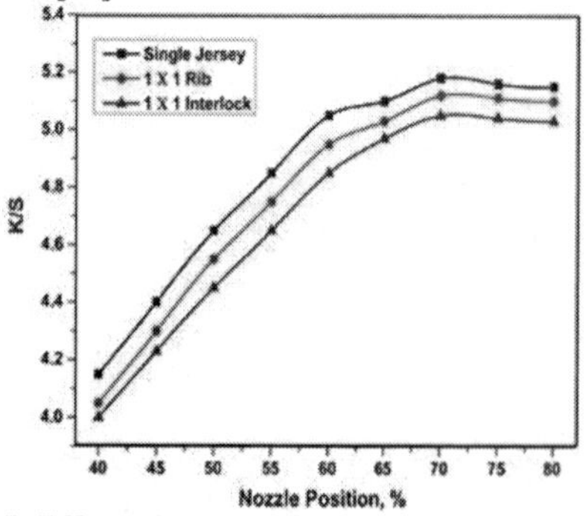

Figure 5. Effect of nozzle position on color strength (K/S)

4 Conclusion

The study reveals a cogent access of process variables, namely reel speed, cycle time, pump pressure and nozzle position on knit structures based cotton fabric color strength property. The after-effects acquired in this plan adumbrated that the optimization of machine controlling parameters can be an important footfall in acceptable textile dyeing with respect to the energy savings that implemented by an aggregate of action process parameters using dyeing principles. This embodiment provides meaningful production data for the effective development of added assorted dyeing industry accompanying products along with the localization of accomplishment for absolute garments.

Acknowledgement

This work was financially supported by the China National Textile & Apparel Council (2013 "Textile Vision" Applied Basic Research, 2013-153); Hubei Province Science and Technology Support Program (Grant No. 2013BAA043) and the Collaborative Innovation Plan of Hubei Province for Key Technology of Eco-Ramie Industry (2014-8).

References

1. T. Hussain, R. Wardman, R. Shamey, Coloration Technology. **121,** 53 (2005)
2. M.N. Pervez, M.A. Rahman, International Journal of Scientific & Engineering Research. **6,** 660 (2015)
3. N. Anbumani, *Knitting Fundamentals, Machines, Structures and Developments*, New Age International, (2007)
4. Holme, I., Dyeing cellulosic fibres. Text. Mag. **31** 8 (2004).
5. M.O. Bulut, K. Akçali, Oktav Bulut ve Akçalı, Erciyes Üniversitesi Fen Bilimleri Enstitüsü Dergisi. **30,** 39 (2016)
6. L. Lin, M.Z. Rahman, S. Wen, M.N. Pervez, J. Li, Y. Cai. Improvement of color fastness for deep black shade of cotton fabric. in 4th International Conference on Machinery, Materials and Computing Technology. China: Atlantis press. (2016)
7. M.M.R. Khan, A. Islam, M.A. Habib, M.N. Pervez, International Journal of Scientific & Engineering Research. **6,** 19 (2015)
8. I. Hossain, A. Hossain, I.A. Choudhury, The Journal of the Textile Institute. **107,** 1 (2016)
9. K. Ojha, R. Sharma, S. Pathak, Journal of Mechanical Design and Vibration. **2,** 74 (2014)
10. M.A. Jahid, M.N. Pervez, American Journal of Materials Science. **5,** 57 (2015)
11. V.C. Bohm, Dyeing apparatus. 1955, Google Patents: USA.
12. Z. Mao, H. Yu, Y. Wang, L. Zhang, Y. Zhong, H. Xu, Industrial & Engineering Chemistry Research. **53,** 8927 (2014)
13. V. Shenhnai, *Technology of textile processing* (Sevak publications, 1976)
14. P.J. Hauser, *Textile technology* (Intech open access publisher: Croatia, 2011)
15. P. Horn, Journal of the Society of Dyers and Colourists. **81,** 262 (1965)
16. J. Ibar, Polymer Engineering & Science. **38,** 1 (1998)
17. A.T. Jr, Apparatus for wet-treating fabrics. 1977, U.S. Patents 4,020,658.
18. T.W.T. Ming, Nozzle for jet fabric dyeing machine. 2007, European Patents 1988205 A1.

DOI: 10.1051/matecconf/201710803006

Making of Natural Dyeing Scarves by Tie-Dyeing Technique

Jin Soun Jung

Major of Fashion Culture Design, School of Convergence Design, Semyung University, Jecheon, South Korea

Abstract. This research aims to make a natural and eco-friendly scarves of Korean and modern style. For the purpose, natural dyeing with safflower, sappan wood, gromwell, amur cork, and areca nut was carried out. As a result, traditional multi-colored scarves were created. And unique scarves of patterns were made by various tie-dyeing methods like tying, folding, wrapping, pleating, and twisting.

1 Introduction

Natural dyes obtained from plants, insects/animals and minerals, are renewable and sustainable bio resource products with minimum environmental impact and known since antiquity for their use, not only as food ingredients [1] and cosmetics [2] but also in textile colouration [3]. Natural dyes have also been used for printing and dye-sensititized solar cells [4]. It is reported that some natural dyes not only dye with unique and elegant colors, but also provide antibacterial, deodorizing and UV protective functions to fabrics [5]. With the level of interest rising about accessory today, scarf get out of ancilary role of clothes and its proportion is gradually expending to a leading position where it leads the overall atmosphere of fashion. But as Korean scarf design is not getting away from the mimic stage to foreign brand, Koreans should develop their own creative scarf design now. It needs to develop scarf design of Korean and modern style through natural dyeing using tie-dyeing technique. Therefore the purpose of this study is to make a natural and eco-friendly scarves by various tie-dyeing methods using natural materials such as safflower, sappan wood, gromwell, amur cork, and areca nut.

2 Theoretical background

2.1 Safflower

Safflower (*Carthamus tinctorius* L.), which is an annual plant of chrysanthemum, contains yellow and red pigments (carthamin) in its petal. The safflower pigments have been traditionally used in cloth dyeing and natural cosmetic material as early as 4500 BC. The yellow dyestuff (water-soluble) was used for dyeing mordanted wool and silk while the red dyestuff (water-unsoluble) was used as a direct dye, i.e. without mordanting, or as a red organic pigment for painting purposes [6]. Carthamin is the major red pigment in safflower and it has antioxidant activity [7]. The basic structure of those pigments is a C-glucosyl quinochalcone [8]. Figure 1 showed the chemical structure of carthamin contained in safflower.

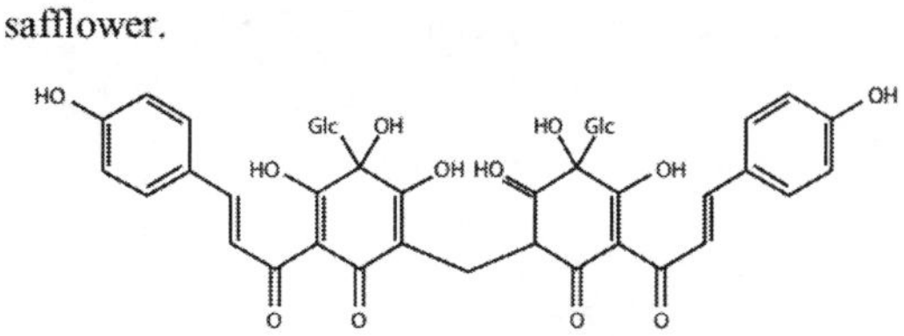

Figure 1. Chemical structure of carthamin contained in safflower

2.2 Sappan wood

Caesalpinia sappan L. is a plant of Leguminosae family, commonly known as Brazil or Sappan wood. *Caesalpinia sappan* L. (Leguminosae) is distributed in Southeast Asia. Brazilin is the major compound naturally occurring in the *Caesalpinia sappan* L. heartwood and is used as a red dye for histological staining [9]. In traditional Chinese medicine, brazilin is used for treatment of increased blood circulation, promotes menstruation and exhibit analgesic and anti-inflammatory potentials [10]. Figure 2 showed the chemical structure of brazilin contained in sappan wood.

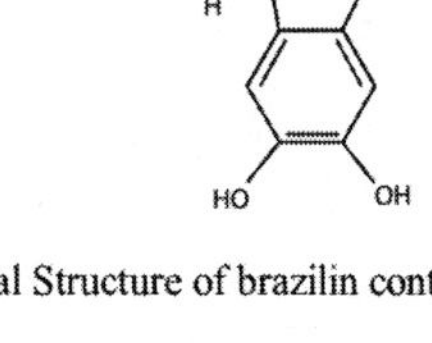

Figure 2. Chemical Structure of brazilin contained in sappan wood

DOI: 10.1051/matecconf/201710803006

2.3 Gromwell

Lithospermum erythrorhizon Sieb. et Zucc. has been used traditionally as medicine in China and as a dye for staining fabrics and food colorants [11]. The roots of *Lithospermum erythrorhizon* Sieb. et Zucc. accumulate shikonin derivatives, which are red naphthoquinone pigments. Shikonin, a naphthoquinone pigment, has been used in traditional Chinese medicines for thousands of years. Figure 3 showed the chemical structure of shikonin contained in gromwell.

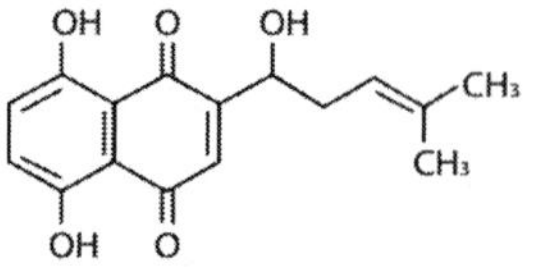

Figure 3. Chemical structure of shikonin contained in gromwell

2.4 Amur cork

Phellodendron amurense Rupr is an important associated species in the forest, in Northeast China, and also a pharmaceutical resource plant for its phloem, a Chinese traditional medicine [12]. *Phellodendron amurense Rupr* contains a number of alkaloids such as berberine, palmatine, phellodendrine etc [13], and used as anti-inflammatory agent, promoting blood circulation, dysentery, diuretic, hemorrhage and blood glucose regulating etc [14]. Figure 4 showed the chemical structure of berberine contained in amur cork.

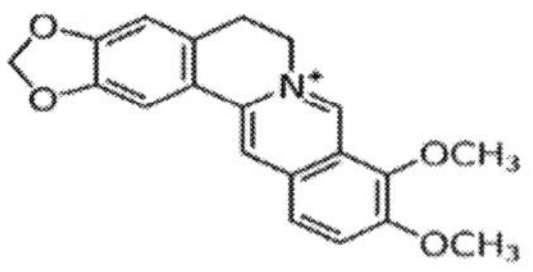

Figure 4. Chemical structure of berberine contained in amur cork

2.5 Areca nut

A. catechu Linn. commonly known as betel nut, a slender palm from Arecaceae family, which is widely distributed in south-east Asia and India in coastal regions of Maharashtra to Kerala and Tamil Nadu. The nuts of *A. catechu* exhibits multiple therapeutic properties. Areca nut largely contains sugars, lipids and polyphenols [15]. Polyphenols in areca nut comprise of condensed tannins, hydrolysable tannins, and flavonols. Polyphenols are the most bio-active components of areca nut. Figure 5 showed the chemical structure of catechuins contained in areca nuts.

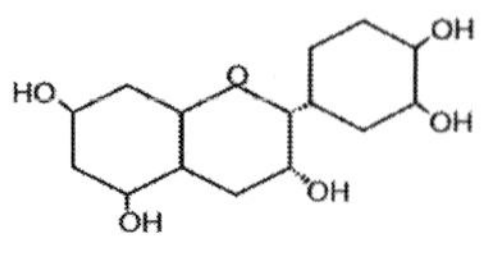

Figure 5. Chemical structure of catechin contained in areca nuts

3 Materials and methods of tie and dyeing

3.1 Materials

3.1.1 Natural materials

The plant materials such as safflower, sappan wood, gromwell, amur cork and areca nut were purchased from traditional medical herb market at Jecheon.

3.1.2 Fabrics for scarves

Satins and chiffons of 100% silk were used for scarves.

3.1.3 Mordants and reagents

Ferrous sulfate, copper sulphate, aluminium acetate, ethanol, potassium carbonate, and acetate acid etc was used first-class goods on the market.

3.2 Methods of tie dyeing

3.2.1 Tie methods

The tie-methods [16, 17] used in this study were shown in Table 1.

Table 1. The tie methods

The tie methods		**The images tied**
1	Grab the center of the fabric and tie it. By varying the number of ties or intervals, it is possible to make a variety of tie expression.	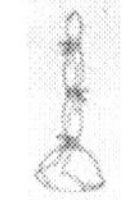
2	After broad stitching the fabric with a round circle, pull carefully the threads and wrap the part three times with the pulled thread. Then wrap the thread up at 3~5㎜ intervals and rewind it to make the knot.	
3	Tie a small portion of the fabric with thread or the like.	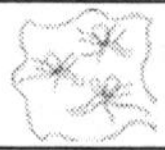
4	Put objects such as coins and beads in fabric and tie them with thread.	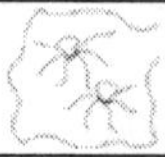
5	First fold the fabric into folding screen. Then tie it with thread.	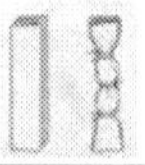
6	Grasp both ends of the fabric and braid, then tie the twisted fabric with thread or rubber band.	
7	Grab both ends of the fabric and braid, then tie with thread in the middle.	
8	Fold the fabric into folding screen and then knot. At this time, put wooden bar between the knots, to easily untie the knot after dyeing.	

DOI: 10.1051/matecconf/201710803006

9	After the fabric is crumpled, sew the crumpled part, pull tightly and then tie up.	
10	Wrap the fabric and tie it with thread so that it does not loosen.	

3.2.2 Dyeing methods

Safflower dyeing

Extraction: In order to extract the yellow color of the safflowers, the flowers were soaked in 20 times of water for 24 hours. Rinsed with water while changing water several times. This process was repeated until removing of yellow water. Poured the 1% solution of potassium carbonate (pH 10.8) kept at about 30 ° C into the vivid red safflowers that yellow colors were removed and rub for 20 minutes. Put them in the bag and squeezed them. Repeated this four times. The first extract was discarded. The 2nd to 4th extracts were collected and used as the red coloring extract.
Dyeing Method: The red pigment extracts obtained by extracting with 1% potassium carbonate solution were added with warm water of 10 times the amount of the extract to reach pH 7.5 with 1% acetic acid and it was used as a solution for dyeing. The tied fabric was dyed for 20 minutes while maintaining the temperature of dye solution at 40 to 45 ° C.

Sappan wood dyeing

Extraction of natural dyes: Sappan woods were soaked in the 20 times of water, heated and extracted for 60 minutes, and filtered. This process was repeated two times. The first and second extracts were combined and used as a solution for dyeing.
Dyeing Procedure: Fabrics were dyed at 90℃ for 30min. while maintaining a bath ratio 1:50. The dyed fabrics were then washed and air dried. In the case of mordanting, postmordanting was carried out using 0.1% copper sulphate(w/w with respect to the fabric) at 40℃ for 10min. The postmordanted fabrics were squeezed, washed and air dried.

Gromwell dyeing

Extraction of natural dyes: Gromwells were soaked in the 20 times of ethanol, heated and extracted at room temperature for 24 hours, and filtered. the filtered solution was diluted with the 5 times of water and it was used as a solution of dyeing.
Dyeing Procedure: Fabrics were dyed at room temperature for 30min. while keeping a bath ratio 1:50. The dyed fabrics were then washed and air dried. In the case of mordanting, premordanting was taken with 0.1% aluminium acetate at room temperature for 10min. The fabrics premordanted were squeezed, washed and air - dried.

Amur cork dyeing

Extraction of natural dyes: Amur corks were soaked in the 20 times of water, heated and extracted for 60 minutes, and filtered. This process was repeated two times. The first and second extracts were combined and used as a solution for dyeing.
Dyeing Procedure: Fabrics were dyed at 90℃ for 30min. while remaining a bath ratio 1:50. The dyed fabrics were then washed and air dried. In the case of mordanting, postmordanting was carried out with 0.1% ferrous sulfate(Fe) at 40℃ for 10min. The postmordanted fabrics were squeezed, washed and air dried.

Areca nut dyeing

Extraction of natural dyes: Areca nuts were soaked in the 20 times of water, heated and extracted for 60 minutes, and filtered. This process was repeated two times. The first and second extracts were combined and used as a solution for dyeing.
Dyeing Procedure: Fabrics were dyed at 90℃ for 30min. while maintaining a bath ratio 1:50. The dyed fabrics were then washed and air dried. In the case of mordanting, postmordanting was carried out using 0.1% ferrous sulfate(Fe) at 40℃ for 10min. The postmordanted fabrics were squeezed, washed and air dried.

4 The natural dyeing scarves dyed by tie dyeing technique

4.1 Making of the work 1

a. Takes a 90×90cm square of white silk fabric of satin and mark some circles on the fabric using water soluble pen.
b. Take and pull the center of each circle, then tie tightly each part indicated with rubber band.
c. Yellow was dyed with amur cork, red was dyed with sappan wood and, deep red was post-mordanted with copper sulphate after dyeing with sappan wood(work 1).

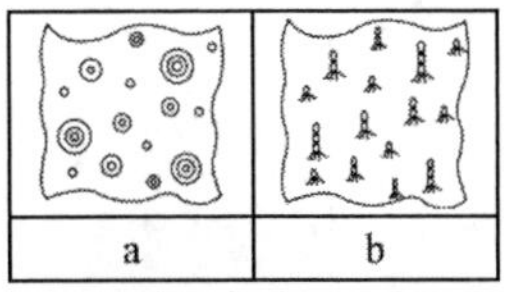

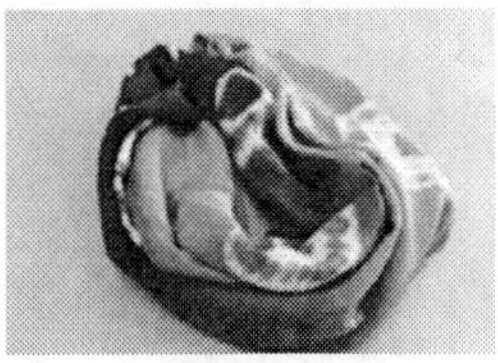

Work 1

4.2 Making of the work 2

DOI: 10.1051/matecconf/201710803006

a. Takes a 90×130cm square of white silk fabric of chiffon.
b. Fold the fabric into folding screen.
c. The center of the folded fabric is knotted. At this time, put wooden bar between the knots, to easily untie the knot after dyeing. And bind tightly both the end of fabric with rubber bands.
d. Wrap firmly between the knots and the bound end. Secure with a knot.
e. Yellow was dyed with amur cork, brown was dyed with areca nut and, khaki was post-mordanted with ferrous sulfate after dyeing with amur cork(work 2).

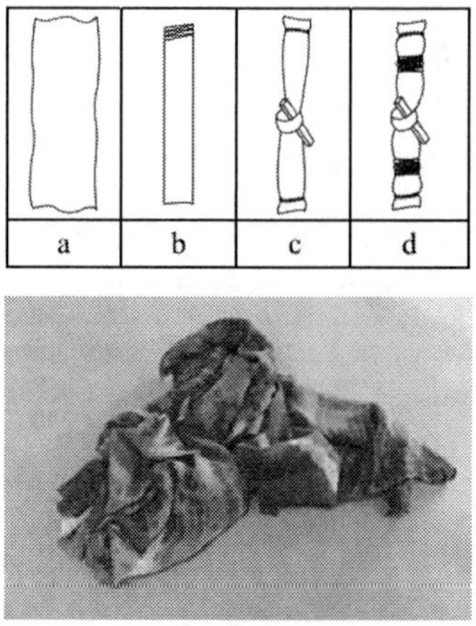

Work 2

4.3 Making of the work 3

a. Takes a 90×130cm square of white silk fabric of chiffon and, mark dots on the fabric with water soluble pen.
b. Put coins on the back of the marked spot.
c. Tie them with rubber bands.
d. Khaki was post-mordanted with ferrous sulfate after dyeing with amur cork and gray was post-mordanted with ferrous sulfate after dyeing with areca nut(work 3).

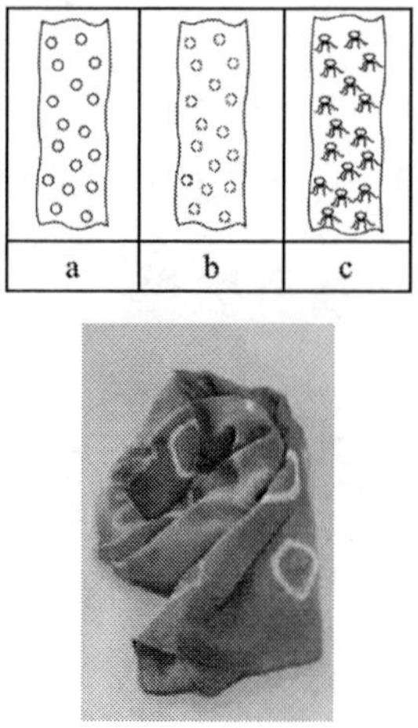

Work 3

4.4 Making of the work 4

a. Takes a 90×130cm square of white silk fabric of chiffon.
b. Fold the fabric into folding screen.
c. Braid it, then tie with rubber bands in the middle.
d. Brown was dyed with areca nut and gray was post-mordanted with ferrous sulfate after dyeing with areca nut(work 4).

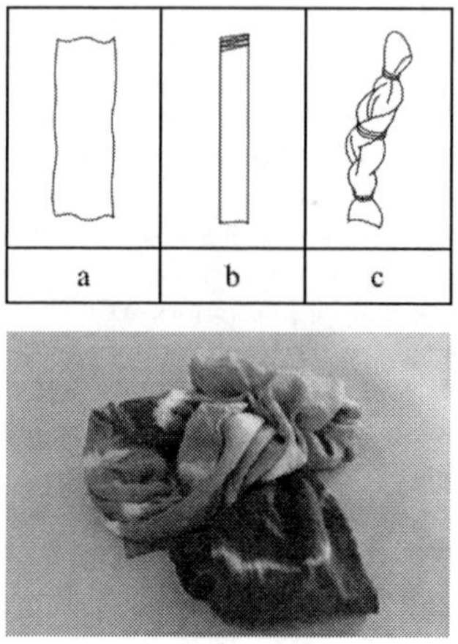

Work 4

4.5 Making of the work 5

a. Takes a 90×90cm square of white silk fabric of satin and mark some circles on the fabric using water soluble pen.
b. Broad stitch the marked place using a double thread in needle. After broad stitching, cut leaving at least 30~50cm of thread.
c. After broad stitching the fabric with a round circle, pull carefully the threads and wrap the part three times with the pulled thread. Then wrap the thread up at 3~5mm intervals and rewind it to make the knot.
d. Brown was dyed with areca nut and gray was post-mordanted with ferrous sulfate after dyeing with areca nut(work 5).

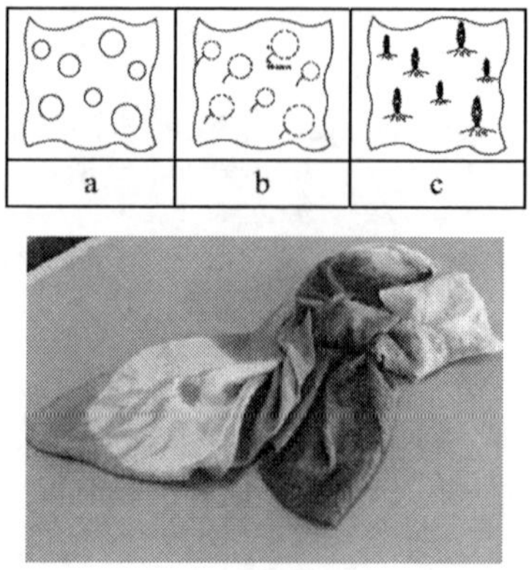

Work 5

4.6 Making of the work 6

a. Takes a 90×90㎝ square of white silk fabric of satin.
b. Wrap the fabric.
c-d. Sew the crumpled part, pull tightly and then tie up.
e. Pink was dyed with safflower and purple was dyed with gromwell after pre-mordanting with aluminium acetate(work 6).

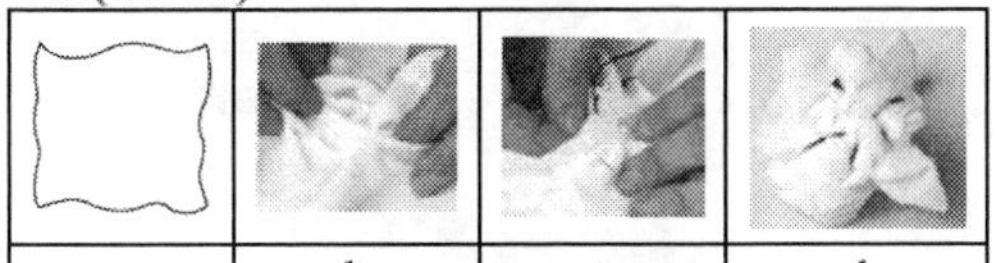

a	b	c	d

Work 6

4.7 Making of the work 7

a. Takes a 90×90㎝ square of white silk fabric of satin.
b. Crumple the fabric and bind the crumpled fabric with thread.
c. Purple was dyed with gromwell after pre-- mordanting with aluminium acetate(work 7).

a	b

Work 7

5 Conclusion

In this study, seven eco - friendly scarves were made with natural materials such as safflower, sappan wood, gromwell, amur cork, and areca nut using various tie-dyeing methods. Safflower was used to dye pink. Sappan wood was provided for red. Gromwell was supplied for purple. And Areca nut was used to dye brown. Also colorful scarves were able to make using aluminum, copper and iron mordant. The soft and natural scarves could be made through natural dyeing.

Acknowledgement: This paper is the product of research that was carried out the support of campus academic research funds of Semyung university in 2015.

References

1. A.C. Dweck, Int. J. Cosmet. Sci., **24**, 287-302 (2002).
2. D. Frick, The coloration of food. Rev. Prog. Color, **33**, 15-32 (2003).
3. S.J. Kadolph, Delta Kappa Gamma Bull., **75**, 14-17 (2008).
4. K. Wongcharee, V. Meeyoo, S. Chavadej, Sol. Energy Mater. Sol. Cells, **91**, 566-571 (2007).
5. D. Grifoni, L. Bacci, G. Zipoli, L. Albanese, F. Sabatini, Dyes Pigm., **91**, 279-285 (2011).
6. Hofenk de Graaf JH., *The colorful past: origins, chemistry and identification of natural dyestuff*, London: Archetype Publications, (2004).
7. Y. Takahashi, N. Miyasaka, S. Tasaka, I. Miura, S. Urano, M. Ikura, et al., Tetrahedron Lett., **23**, 5163-5166 (1982).
8. Y. Jin, X.L. Zhang, H. Shi, Y.S. Xiao, Y.X. Ke, X.Y. Xue, et al., Rapid Commun. Mass Spectrom., **22**, 1275-1287 (2008).
9. I.K. Bae, H.Y. Min, A.R. Han, E.K. Seo, S.K. Lee, Eur. J. Pharmacol., **513**, 3, 237-242 (2005).
10. China Pharmacopoeia Commission, *The Chinese pharmacopoeia*, Beijing: China Medical Science Press, (2010).
11. M. H. Cho, Y. S. Paik & T. R. Hahn, J. Agric. Food Chem., **47**, 4117-4120 (1999).
12. X. Li, Y. Wang, X. Yan, Acta Ecol. Sin., **27**, 1, 58-64 (2007).
13. Y.M. Lee, H. Kim, E.K. Hong, B.H. Kang, S.J. Kim, J. Ethnopharmacol., **73**, 429-436 (2000).
14. H. Yan, X.D. Sun, S.G. Sun, S.H. Wang, J. Zhang, R. Wang, P. An, F. Yang, W. Kang, Int. J. Biol. Macromol, **48**, 720-725 (2011).
15. P. Wetwitayaklung, T. Phaechamud, C. Limmatvapirat, S. Keokitichai, Naresuan Univ. J., **14**, 1-14 (2006).
16. Kyoto University Art and Design, *Learn how to dye*, Kadokawa Shoten, (1998).
17. Library of Congress Cataloging-in-Publication Data, *Exploring Textile Arts*, Creative Publishing international, (2002).

MATEC Web of Conferences 108, 04001 (2017) DOI: 10.1051/matecconf/201710804001
ICMAA 2017

Study of Different Factors Influence on Thermocompressor Performance

Evgeny Blagin , Alexandr Dovgyallo and Dmitry Uglanov

[1]Heat engineering department, Samara National Research University, 443086 Samara, Russian Federation

Abstract. This article deals with investigation of different factors influence on thermocompressor. These factors include regeneration efficiency, dead volume, working frequency, heat transfer coefficient, thermocompressor diameter and fins application. Calculation shows that most influential factors are regeneration efficiency, dead volume and diameter. It was also found that fins application is not effective for thermocompressor performance enhancement.

1 Introduction

Thermocompressor (TC) is a device which transform heat directly to a potential energy of pressure. There are two principal types of thermocompressors. The first one is steam control device that uses high-pressure steam, referred to as motive steam, to induce flow from a lower pressure steam source and discharge the mixture at an intermediate pressure [1]. The second one is mechanical device which represents cylinder where piston moves and pushes the gas through the regenerator. Regenerator can be external [2] or integrated into displacer. Such regenerators can be wired [3] or annular [4].

This article deals with mechanical thermocompressor with wired regenerator integrated with displacer.

First concept of thermocompressor was introduced by V. Bush in 1939 [5]. Recent studies of thermocompressor are pointed on research of compressor with annular regenerator or thermocompressor operation within different devices. For example, thermocompressor operation as a drive for pulse tube was investigated in [6] and application of thermocompressor for pneumatically actuated ankle-foot orthosis was investigated in [7]. Thermodynamic analysis of thermocompressor with non-ideal component characteristics was made in [3]. However it does not take into account temperature change inside the thermocompressor change. Investigation of the thermocompressor performance with non-isothermal chambers was made in [8]. This research shows significant influence of temperature change on thermocompressor performance.

The results achieved in [8] raise a question of other factors influence on thermocompressor performance. These factors include:

1. Regeneration efficiency.
2. Geometric characteristics (heat exchange area, length of the cylinder which will affect the piston velocity and consequently gas flow velocity).
3. Time factor which will influence by means of working cycle frequency.
4. Dead volume.
5. Application of heat transfer intensifications such as cylinder finning.
6. Non-isothermal conditions of chambers walls.

The influence of first 5 factors is investigated in this article.

Thermocompressor can be used in LNG cryogenic complexes for LNG cold energy utilization during its gasification to obtain an increased gas pressure and decrease costs for its compressing. Cold energy utilization efficiency are described in [9, 10].

2 Physical model of thermocompressor and its operation principle

Physically, considered thermocompressor consists of cylinder with piston-displacer inside. Displacer separates cold chamber from hot chamber. When piston moves in direction of chamber. it pushes the working fluid of this chamber through regenerator into another chamber. Chambers have a different temperatures which causes pressure oscillations due to Amontons's Law. Principal scheme of considered thermocompressor is presented on Figure 1.

Working cycle of thermocompressor starts from high dead centre when both valves are closed and volume of the cold chamber is maximal (point 1). Piston starts to move downside pushing working fluid through regenerator to a hot chamber. Due to Amontons's Law of Pressure-Temperature pressure is increasing because mean temperature of the working fluid inside the thermocompressor is increasing. When the pressure is high enough, output valve opens and working fluid with given pressure starts to be charged to a customer. When piston reaches low dead centre, output valve closes (point 3) and working fluid stops to be charged to a customer.

Piston starts to move upside pushing working fluid back to a cold chamber. Pressure is decreasing due to decrease of the mean temperature inside the thermocompressor. When pressure drops low enough, input valve opens, charging new portion of working fluid inside the thermocompressor. Working cycle of thermocompressor is presented on Figure 2.

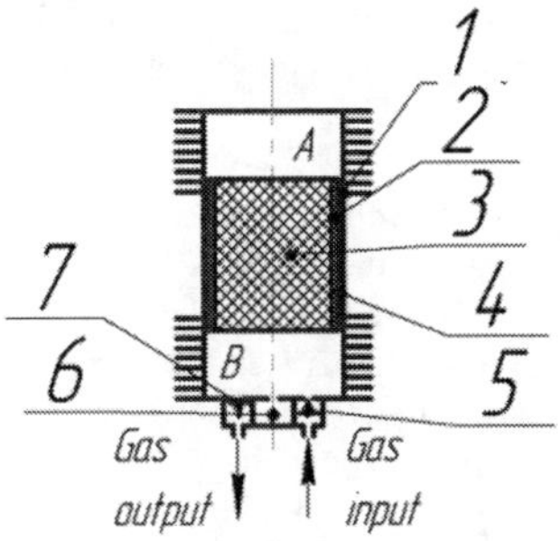

Figure 1. Principal scheme of the mechanical thermocompressor:
1 – thermocompressor shell; 2 – displacer-regenerator shell; 3 – regenerator; 4 – gap; 5 – input valve; 6 – rod; 7 – output valve; A – hot chamber; B – cold chamber.

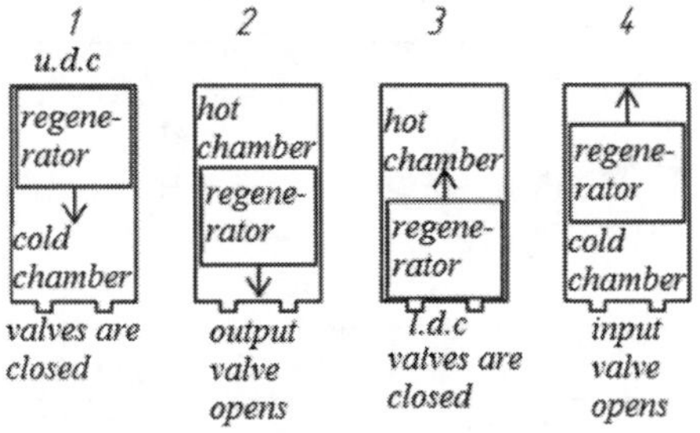

Figure 2. Working cycle of the thermocompressor.

Thermocompressor, investigated in [8] was taken as a prototype for given research. Its characteristics are presented in Table 1.

Table 1. Characteristics of considered thermocompressor.

Parameter	Value
Input pressure, P_{in}	101325
Pressure ratio, π_{tc}	1.5
Cold chamber wall temperature T_{wc}, K	100
Hot chamber wall temperature T_{wh}, K	300
Dead volume of chambers, m^3	$2{,}5\cdot10^{-4}$
Displaceable volume, m^3	$5\cdot10^{-3}$
Thermocompressor cylinder diameter, m	0.15
Regenerator volume, m^3	$3{,}55\cdot10^{-3}$
Frequency, Hz	1
Working fluid	Nitrogen

3 Research of different factors influence

Mathematical model for thermocompressor calculation was presented in [8]. Thermocompressor delivery and arithmetic mean temperature deviations was used as comparison criteria.

3.1 Drive frequency influence on thermocompressor performance

Change of the thermocompressor drive frequency will affect the operation of thermocompressor in different ways. On the one side increase of the working frequency decreases the heat exchange duration and vice versa. One the other side increase of the working frequency increases the velocity of the gas flow and consequently intensify the heat exchange. Also, increase of the frequency causes increases of the friction in regenerator and heating of the gas in both chambers due to this which can cause the thermocompressor overheating, when pressure inside the thermocompressor does not drop below the pressure of input valve opening and input valve does not open. This effect is called choking of thermocompressor.

The results of modelling for frequency range from 0.1 Hz to 5 Hz is presented in table 2.

Table 2. Thermocompressor drive frequency influence on thermocompressor performance.

Frequency f, Hz	TC delivery, W	TC delivery per cycle, J	Temperature deviations	
			Cold chamber ΔT_c, K	Hot chamber ΔT_h, K
0.1	8.74	87.43	3.21	-3.58
0.3	25.79	85.98	3.95	-4.28
0.5	42.58	85.16	4.44	-4.69
1	83.89	83.89	5.1	-5.63
2	164.8	82.39	5.9	-6.68
3	244.2	81.41	6.41	-7.4
5	400.2	80.05	7.51	-7.44

The results of the modelling shows that increase of the working frequency negatively influence thermocompressor delivery in one cycle and temperature deviations. However, this influence is rather small (TC delivery per cycle is decreased on 8.43% but number of cycles is increased in a factor of 50). This causes almost linear dependency of thermocompressor delivery per unit of time of working frequency. It allows to make a conclusion that influence of the working frequency on processes in thermocompressor is negligible and working frequency may be high.

3.2 Geometric characteristics influence of thermocompressor performance

Geometric configuration of thermocompressor will also affect the performance of thermocompressor. Let's consider a thermocompressor with constant displaceable volume. Its geometric configuration then can be determined by one parameters – its diameter.

Geometric characteristics have a two principle factors which will influence thermocompressor performance. The first one is a heat exchange area and increase of the this area causes better heat exchange and better thermocompressor delivery. With constant volume heat exchange area will be determined by sum of side and end surfaces:

$$F_h = \frac{\pi D^2}{4} + \frac{4\pi V_h}{D};$$
$$F_c = \frac{\pi D^2}{4} + \frac{4\pi V_c}{D}. \quad (1)$$

Functions $F_h(D)$ and $F_c(D)$ have a certain minimum which correspond to a minimal heat exchange area. Configuration of thermocompressor must be selected so that diameter of the compressor is differ than diameter which corresponds to a minimal area.

The second factor is connected with increase of the piston velocity with increase of thermocompressor cylinder length. Piston must overcome the longer distance during the same time so piston velocity as well as gas flow velocity increases. Effect of the velocity increase is described above. It should be noted that increase of the cylinder length causes increase of the displacer length because it must separate hot chamber from the cold one.

Together, these factors make a complex effect which is represented in Table 3.

Table 3. Thermocompressor diameter influence on thermocompressor performance.

Diameter D, m	TC delivery, J	Temperature deviations	
		Cold chamber ΔT_c, K	Hot chamber ΔT_h, K
0.05	84.69	4.98	-4.56
0.1	77.44	9.59	-9.38
0.15	73.62	10.5	-12.19
0.18	72.99	10.72	-12.73
0.2	73.1	10.65	-12.73
0.25	74.45	10.03	-11.97
0.3	76.39	10.01	-12.72

A minimal value of compressor delivery can be seen and it is caused by proximity of diameter value to a diameter corresponding to a minimal heat exchange area. This means that first geometric factor makes a more significant influence than a second one. Difference between maximal and minimal values of compressor delivery here is equal to 13.8%. This means that geometric factor must be taken into account during designing of the thermocompressor.

3.3 Regeneration efficiency influence on thermocompressor performance

Regenerator is a key element of thermocompressor. Without it thermocompression is theoretically possible but it would take much longer time. Regenerator efficiency shows the degree of chambers temperature approach to temperatures of the chambers walls. Ideal regeneration means that gas discharging from regenerator will have a temperature equal to a temperature of the corresponding wall. So there will be no heat exchange between gas and walls because there is no temperature difference between them. Ideal regeneration is a primary condition for isothermal model of thermocompressor. Non-ideal regeneration means that gas discharging from regenerator will have a temperature higher (in case of cold chamber) or lower (in case of hot chamber) than temperature of the wall. This will cause temperature change in chambers. The lower regenerator efficiency – the higher temperature difference will be.

Effect of the regeneration efficiency η_{reg} influence on thermocompressor delivery is presented in table 4.

Table 4. Thermocompressor diameter influence on thermocompressor performance.

Diameter D, m	Thermocompressor delivery with regeneration efficiency equal to, J			
	0.8	0.85	0.9	0.95
0.05	47.84	55.43	63.34	71.45
0.1	24.96	36.63	49.23	62.37
0.15	13.65	27.61	43.1	59.38
0.2	11.95	26.27	42.25	59
0.25	15.65	29.34	44.4	60.09
0.3	21.19	33.85	47.53	61.65

Regeneration efficiency is one of the main factors which influence thermocompressor performance. Decrease of the η_{reg} value on 15% (from 95% to 80%) causes losses of delivery equal to 77% in case with diameter equal to 0.15 m. Thermocompressor stops operating properly with low η_{reg} value lower than 70-75%. Designing of the regenerator with high efficiency is a key moment during thermocompressor designing.

3.4 Heat transfer coefficient influence on thermocompressor performance

Heat transfer coefficient change influence the heat exchange intensity between working fluid inside the chambers and its walls (heat exchange in regenerator is determined by regeneration efficiency). Unsurprisingly, the higher value of heat transfer coefficient the more thermocompressor delivery and vice versa. For given research deviation of heat transfer coefficient equal to -

30%. -20%, -10%, 10%, 20% and 30% were investigated. The results of the investigation are presented in Table 5.

Table 5. Thermocompressor diameter influence on thermocompressor performance.

Heat transfer coefficient deviation	TC delivery, J	Temperature deviations	
		Cold chamber ΔT_c, K	Hot chamber ΔT_h, K
0.7α	78.9	7.64	-9.36
0.8α	80.26	6.98	-8.33
0.9α	81.33	6.43	-7.5
1α	82.27	5.96	-6.82
1.1α	83.04	5.56	-6.25
1.2α	83.7	5.2	-5.77
1.3α	84.28	5.15	-5.35

Calculation results are corresponds with prediction made above. However, difference between thermocompressor deliveries on different regime is rather small and during designing more attention must be focused on other issues that heat transfer intensity.

3.5 Dead volume influence on thermocompressor performance

Table 6. Dead volume influence on thermocompressor performance.

Dead volume percentage, %	TC delivery, J	Temperature deviations	
		Cold chamber ΔT_c, K	Hot chamber ΔT_h, K
1	83.47	5.96	-6.82
5	82.27	4.94	-5.66
10	76.91	4.49	-5.01
15	70.75	4.25	-4.54
20	64.29	4.21	-4.39
25	57.69	4.15	-3.77

Dead volume is a volume of the thermocompressor which is not included in displaceable volume. Presence of any dead volume negatively influence the performance characteristics of thermocompressor because dead volume of cold chamber decreases maximal pressure and dead volume of hot chamber increases minimal pressure. Unfortunately, presence of some dead volume is inevitable because it is caused by thermocompressor design elements. For example, volume of gas pockets inside regenerators are dead volume but they are needed for gas passing through regenerator.

Influence of the dead volume (not including regenerator volume) on thermocompressor performance is presented in Table 6. As it can be seen from the table, dead volume almost does not influence temperature change in cold chamber but it significantly influence thermocompressor delivery. The value of dead volume must be kept as low as possible. Together with regeneration efficiency it is one of the most influential factors which affects thermocompressor performance.

3.6 Application of finning for heat exchange intensification

Finning are used in heat engineering for heat exchange intensification. There are two principal types of fins location: lateral (Figure 3) and transverse (Figure 4). Both types of fins serves for increase of the heat exchange area. Transverse fins also forms a flow vortexes between them thus also intensifying heat exchange. However both types of fins add additional dead volume inside cylinder. Calculation of the thermocompressor with both types of fins modelling is presented in Table 7.

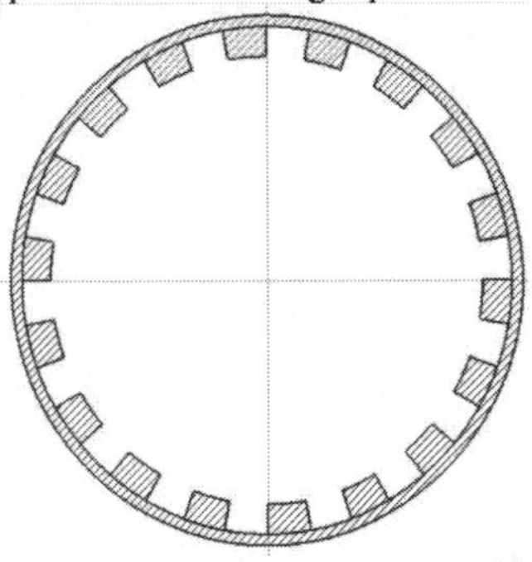

Figure 3. Cross section of thermocompressor with lateral fins

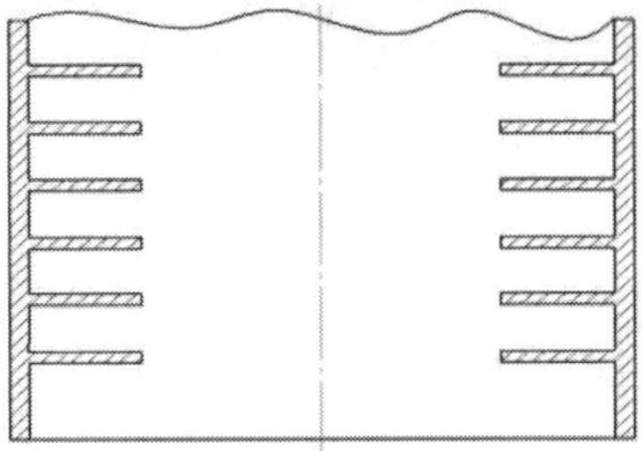

Figure 4. Axial section of thermocompressor with transverse fins

Table 7. Fins application influence on thermocompressor performance.

	Dead volume, m^3	TC delivery, J	ΔT_c, K	ΔT_h, K
No fins	$2.5\cdot10^{-4}$	73.6	10.1	-11.1
Lateral	$13.5\cdot10^{-4}$	53.1	9.4	-9.1
Transverse	$13.3\cdot10^{-4}$	48.7	6.3	-5.8

Calculation shows that fins application positively influence on temperature differences but increase of the dead volume compensates potential raise of compressor delivery. Conclusion can be made that fins application is not suitable for thermocompressor performance enhancement. Lateral fins can be used in case when piston is made with special form which will «streamline» the circuit of fins but manufacturing of such piston is very complicated and thermal expansion must be taken into account.

4 Conclusion

This article deals with estimation of different factors influence on thermocompressor performance. These factors include:

- Regeneration efficiency.
- Thermocompressor diameter.
- Working frequency.
- Dead volume.
- Heat transfer coefficient deviation.
- Fins application.

The two most influential factors are regeneration efficiency and dead volume. Decrease of the regeneration efficiency on 15% can cause decrease of the compressor delivery up to 77% (in case with thermocompressor diameter equal to 0.15 m). Increase of the dead volume percentage from 1% to 20% causes decrease of the compressor delivery equal to 22.98%.

Heat exchange area of the thermocompressor depends on its diameter and this dependency has minimal value. During designing, diameter of the thermocompressor must be chosen in such manner to avoid this minimal value.

Application of fins is not effective way to increase thermocompressor performance because increase of the dead volume due to fin application is more significant factor than intensification of heat exchange. In case of vertical fin, special displacer can be designed which will fill the space between fins, but it leads to a significant complications in displacer design.

The results of this work can be used during thermocompressor design optimization for receiving a better delivery. Optimization will results in operation costs decrease which is important part of energy saving.

Acknowledgement

This work was supported by the Ministry of Education and Science of the Russian Federation in the framework of the implementation of the Program «Research and development on priority directions of scientific-technological complex of Russia for 2014– 2020».

References

1. M. Soucy, G.L. Timm. Thermocompressor Design and Operation for High Efficiency, Pulp & Paper Canada, **111(5)**: (Sept/Oct 2010).
2. A. Kornhauser. Analysis of an idealized Stirling thermocompressor. In Proceedings of the 31st IECEC (1996).
3. M.J. Edwards, R.B. Peterson. Reciprocating thermocompressor performance with non-ideal component characteristics. Proc. of the Inst. of Mech. Eng., Part A: Journal of Power and Energy **221** (2007).
4. W.-Y. Lin, X.-H. Wu, J.-L. Yang, L.-W. Yang. Experimental study and numerical analysis of thermocompressors with annular regenerators. International Journal of Refrigeration. **36** (2013).
5. Bush, V. Apparatus for compressing gases. US Patent 2,157,229, 1939.
6. W. Ji, X. Xue, J. Wang, Y. Zhou, L. Chen, W. Zhu. Coupling study of a novel thermocompressor driven pulse tube refrigerator. App. Therm. Eng. **51** (2013).
7. N.S. Kumar, M.E. Hofacker, E.J. Barth. Design of a stirling thermocompressor for a pneumatically actuated ankle-foot orthosis. ASME 2013 Dynamic Systems and Control Conference, DSCC 2013. **1** (2013).
8. E.V. Blagin, A.I. Dovgyallo, D.A. Uglanov. Numerical modeling of non-stationary processes in cryogenic mechanical thermocompressor. MATEC Web of Conferences (to be published).
9. E.V. Blagin, A.I. Dovgyallo, D.A. Uglanov. About LNG Energy Utilization Efficiency Estimation. Proc. Eng. **152** (2016).
10. E.V. Blagin, A.I. Dovgyallo, D.A. Uglanov. Efficiency Criteria and Comparative Analysis of Combined Energy Plants Utilizing LNG Cold Energy. Proc. Eng. **152** (2016).

DOI: 10.1051/matecconf/201710804002

Fully automatic CNC machining production system

Jeng-Dao Lee , Cheng-Wei Tsai-Lin, Yi-Cheng Lee, Min-Che Liu, and Li-Yin Chen

National Formosa University, Automation Engineering Department, 632 Yunlin County, Taiwan

Abstract. Customized manufacturing is increasing years by years. The consumption habits change has been cause the shorter of product life cycle. Therefore, many countries view industry 4.0 as a target to achieve more efficient and more flexible automated production. To develop an automatic loading and unloading CNC machining system via vision inspection is the first step in industrial upgrading. CNC controller is adopted as the main controller to command to the robot, conveyor, and other equipment in this study. Moreover, machine vision systems are used to detect position of material on the conveyor and the edge of the machining material. In addition, Open CNC and SCADA software will be utilized to make real-time monitor, remote system of control, alarm email notification, and parameters collection. Furthermore, RFID has been added to employee classification and management. The machine handshaking has been successfully proposed to achieve automatic vision detect, edge tracing measurement, machining and system parameters collection for data analysis to accomplish industrial automation system integration with real-time monitor.

1 Introduction

The profit of automated production comes from a large machining of single species. By considerable production can be brought to the benefit of high effect, high yield, and low cost. The production pattern of low-volume customized is conferenced to uneconomic, slow payback period and difficult of production scheduling. The change of consumer's habit results in the product life cycle become faster. The importance of CTO is displayed at this time [1]. However, the pattern of the low-volume customized can't avoid complex pre-operation. Besides, the sudden situations are like the customer temporary increasing quantities and the customer return the unqualified product, etc. From the above, the product output the low yield if the pre-operation is more and more frequently. The problem deeply troubles the manufactures with the necessary waste. Therefore, Germany proposes the plan of industry 4.0.

The purpose of Industry 4.0 was built a highly flexible production and service model of customization and quantization, with real-time interactions between people, products, and devices during the production process. For example, a factory that accepts consumer orders and directly produces, and ships the required product will dispense of separate sales and circulation channels, which will have a strong impact on the traditional e-commerce sales model. Industry 4.0 will affect not only industry of Germany, or even international industrial development but will become a driving force [2].

Take the countries of relevant policy of Industry 4.0. For example, Germany has developed a strategic plan to implement Industry 4.0. The main points of this plan can be summarized as: building a network, researching two major themes, realization of three integrations, and achievement of eight planning objectives [3]. Japan has proposed a new strategy for the robot policy. It focuses on the coexistence of man-machine for the future factory. The technical strategy predominantly is based on the systematic integration that developing sensors system, controller system, driver system, cloud computing, and artificial Intelligence of robot. The method will use in the requirement of aged society and revitalized the economy of manufacturing. China proposed the policy of made in China 2025. It based on the intelligent manufacturing developing the new generation of mobile communications, internet of things, cloud computing and other industry. The policy will achieve mechanical equipment of high level and informatization of the manufacturing [4].

In view of above, industry 4.0 demonstrates the fundamentals of future manufacturing technology. The device must have the foundation of digital information and matched the hardware and software in order to optimize the flow of production. The production line achieved the optimization of the product By instantly transmit the communication between production and facility.The method will help manufacturers to product life management if industry 4.0 be achieved it. The manufacturer can advance planning of product design, production, and maintenance routine. Firms shortened the time to market of the product by the planning of intelligent manufacturing. Manufacturers can be a more accurately grasp the marketing of the product [5].

DOI: 10.1051/matecconf/201710804002

In conclusion, Industry 4.0 must rely on smart manufacturing, smart production, and smart product in series when the production is achieved the performance. In the industrial value chain, the CNC controller will undoubtedly be the hub because it can control machines and communicate between factory and the internet. The method not only thinks outside the box of CNC industry, also it brings the chance to upgrade for the machine-tool industry in Taiwan.

The research is to explore actual demand based on the machine-tool industry. The approach uses the CNC controller, servo motor, variable-frequency Drive, intelligent sensors, PLC, and manipulators. CNC controller accomplished the connection between devices through the serial transmission (RS485). At the same time, the overall system of the important data instantly transmits to the computer by Ethernet. Using the SCADA software from DELTA ELECTRONICS, INC. developed the man-machine interface to monitoring. It will alert and inform the relevant departments when the machine-specific parameter is abnormal.

2 System of Description

System block diagram shows as Figure 1. The CNC controller is the main controller in the system. Servo motors in the moving machining platform are connected the CNC controller via DMCNET. The parameters of other equipment are collected to SCADA software by RS485 and RS232 protocol. User can monitor machine on the computer by Ethernet. The PLC (programmable logic controller) sends command to the servo motor which is on the conveyor via CANopen protocol, and sends ASCII string、to the manipulator (Six-axis articulated robot arm) using RS232 protocol. User can immediately get machining data via the computer. The real-time monitor was accomplished the integration of the multi- automated system by SCADA software.

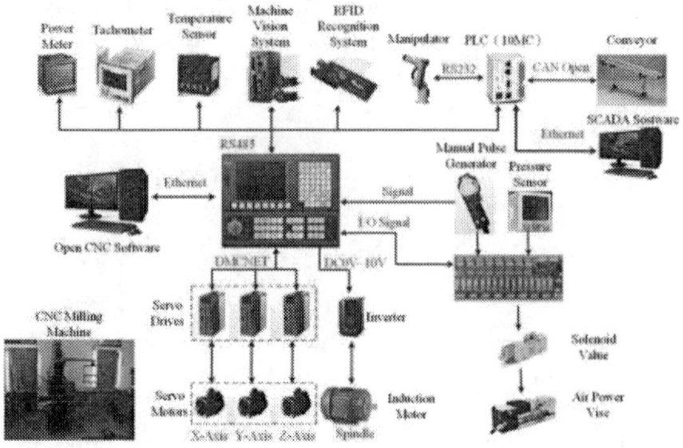

Figure 1.System structure

The manipulator is used to loading and unloading the material to accomplish automatic machining. The operational flow chart can be separated into three parts: vision inspect, manipulator and CNC machining.

The vision inspection conveyor operational flow chart is shown in Figure 2. The vision inspects the material whether on the conveyor or not when the conveyor is moving.

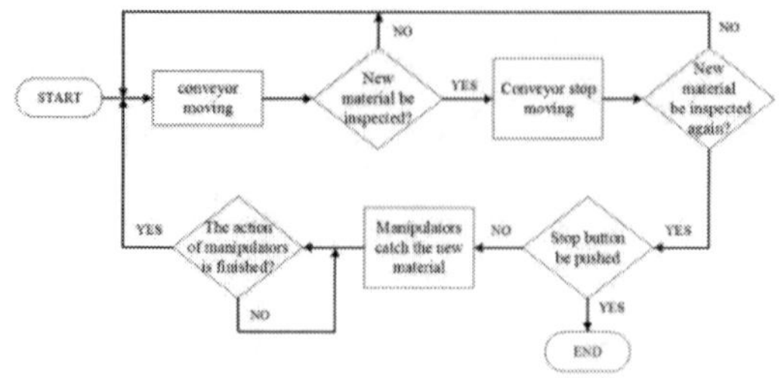

Figure 2. Vision inspection conveyor operational flow chart

The manipulator operational flow chart as Figure 3, the manipulator will catch the new material when the new material vision has been detected.

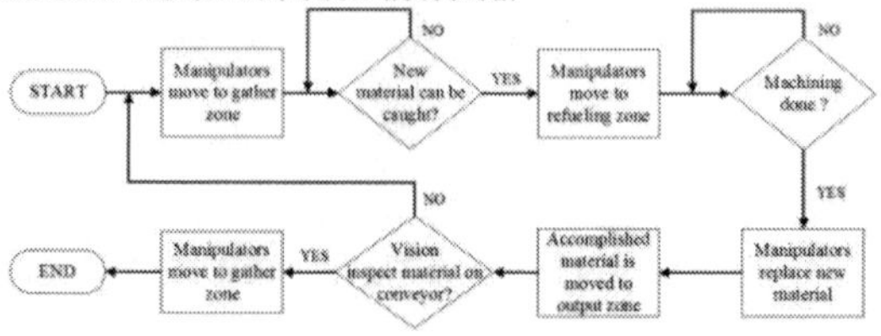

Figure 3.Manipulator operational flow chart

CNC miller is adopted in the fully automatic CNC machining production system and the operational flow chart is shown in Figure 4. CNC miller will move to the zone of edge finder and finding the edge of X coordinate. After done the inspection of edge, CNC miller began to the machining the material when it back to the origin. After machining is finished, the CNC will be waiting for manipulator to change the material.

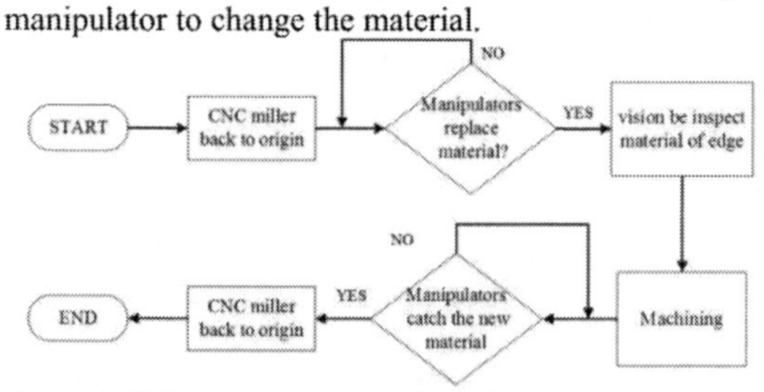

Figure 4.CNC miller operational flow chart

3. Vision Inspection

In the current industry of the automatic factory, the method is often replaced man-resource what using the manipulators to loading and unloading. In order to achieve the effect of automatic loading and unloading go well with the manipulators and vision inspection.

The following sections will point out the introduction for respective to edge finder and object finder.

3.1 Material Finder System

The research will go with the manipulator to loading and unloading so as to complete the automated production line. The position of the object is different when the conveyor was delivered the material every time. In order to avoid this problem, the vision inspection will go to determine the current position of materials.

Figure 5 is shown in the interface of the material finder. The analytical data of X-axis position, Y -axis position, and angle. The relevant parameters will be transferred to CNC controller when new material was been inspected via the material finder. In view of above, the transmission is important among the devices for the automatic production line. Next, following will discuss the delivery method with regard to the vision inspection system and the manipulator.

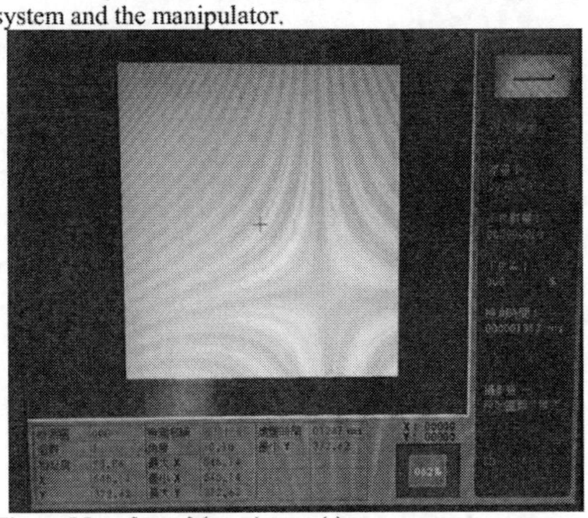

Figure 5. Interface of detection position

As Figure 6 shown, the slaves of devices delivered command by CNC controller, and slaves will act according to the special command by the CNC controller.

The vision inspection system will store numerical value to the internal memory when shooting was completed. The CNC controller will catch the numerical value from the internal memory of vision inspection system. At the same time, CNC controller sends the relevant parameter to the multi-axis controller of PLC. The numerical value changes to the ASCII of character string when the multi-axis controller received it. Finally, the changed character string will command manipulator to act. With the method, it achieves to the communication between the devices.

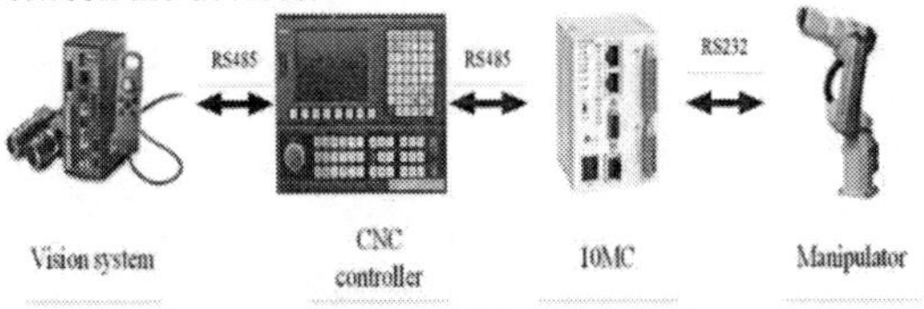

Figure 6. The communication architecture of vision system

3.2 Edge Finder System

The act will lead material to deviation when the pressure vise has gone on the work of clamping. So the machining has the different origin every time, and then makes the precision to reduce.

Therefore, the research was look for the edge of material by vision inspection in order to improve the precision of machining. As Figure 7 shown, the parameter from zone of edge finder has transmitted to CNC controller for compensating. The vision inspection system will compensate the error coordinate origin when automatic loading results in the different position.

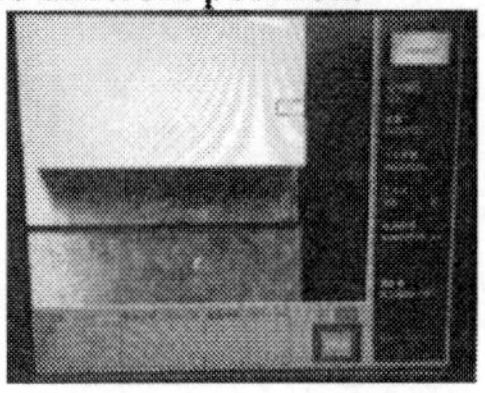

Figure 7(a).position of camera **Figure7(b).** Picture of edge finder

4. Monitor and Control

In order to achieving the function of the system to supervise, the following will explore the man-machine interface of Open CNC and SCADA.

In this part of Open CNC, it added to the function of the send email of alarm and real-time surveillance system. In the function of the send email, the Open CNC will notify the relevant department by the email when the machine have happened alarm. In the function of real-time surveillance system, it monitored the current image via webcam. The foregoing method will increase work efficiency.

In this part of SCADA software, the method of transmission connects the programmable logic controller (PLC) by the Ethernet. PLC will deliver CNC parameters like temperatures, currents, voltages…etc. to SCADA software. By the above method, achieving the purpose of prevented to failure in advance and improved efficiency. The users can beforehand prevent abnormal happened such as the tool wear and the machine is out of order...etc. by analysis parametric.

Figure 8 shown the status of the production process, the picture of SCADA will display where the current position of the mechanical production line.

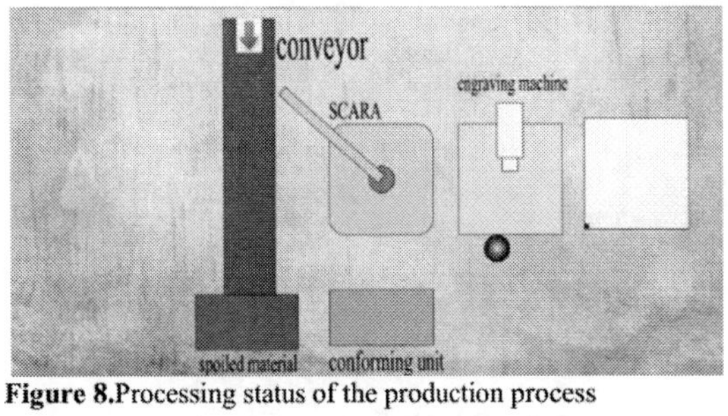

Figure 8. Processing status of the production process

Figure 9 shows the supervisory system and displayed the most important parameter on the human-machine interaction from PLC (10MC) deliver to CNC controller. At the same time, the system also planned the real-time display for the material detection region. The Figure 9 demonstrated the current position of material in the range of the camera. The parameter delivered to the real-time

DOI: 10.1051/matecconf/201710804002

graph from PLC (10MC), and it shows the current position on the conveyor.

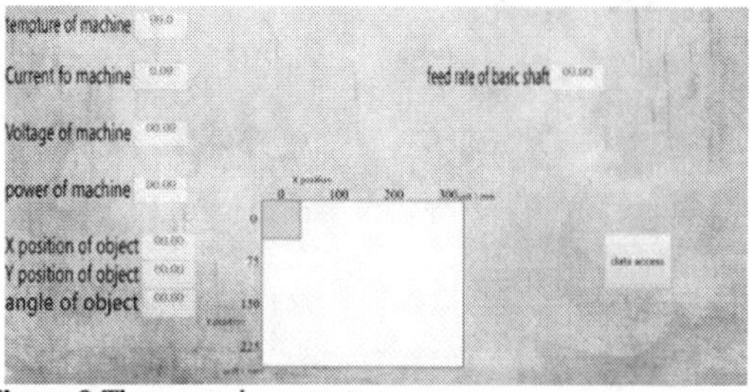

Figure 9. The supervisory system

As Figure 10 shown the interface of alarm records, the function will record the error and the form can export to excel. Users can focus on the detection into the machine according to the special alarm of history record. The method will promote the effect of work.

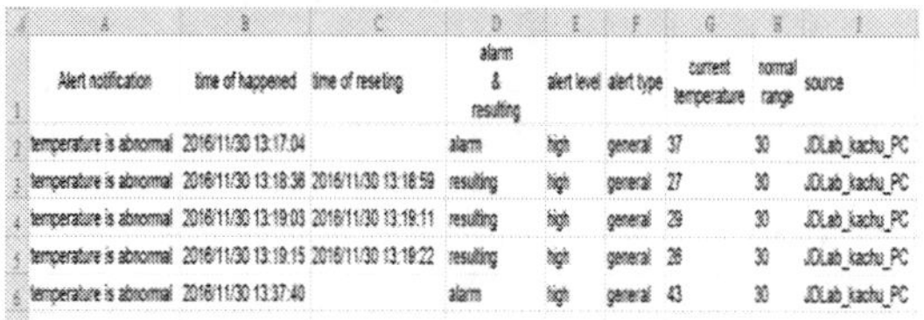

Alert notification	time of happened	time of reseting	alarm & resulting	alert level	alert type	current temperature	normal range	source
temperature is abnormal	2016/11/30 13:17:04		alarm	high	general	37	30	JDLab_kachu_PC
temperature is abnormal	2016/11/30 13:18:38	2016/11/30 13:18:59	resulting	high	general	27	30	JDLab_kachu_PC
temperature is abnormal	2016/11/30 13:19:03	2016/11/30 13:19:11	resulting	high	general	29	30	JDLab_kachu_PC
temperature is abnormal	2016/11/30 13:19:15	2016/11/30 13:19:22	resulting	high	general	28	30	JDLab_kachu_PC
temperature is abnormal	2016/11/30 13:37:40		alarm	high	general	43	30	JDLab_kachu_PC

Figure 10. Function of alarm records

Figure 11 shows the control software interface (Open CNC). The software is specially developed for CNC controller for Delta Electronics, Inc. The method provided the more choice for operation.

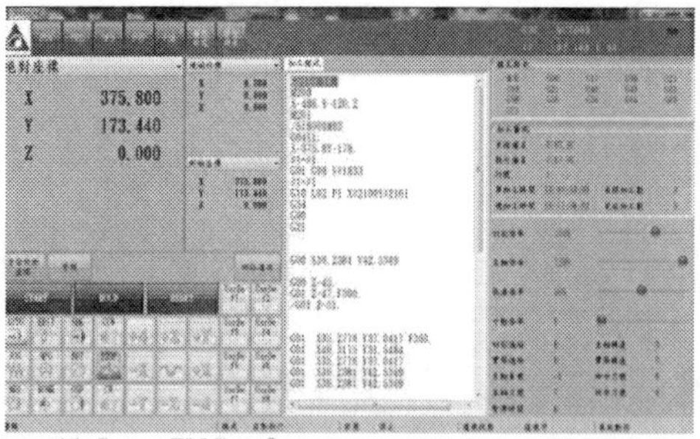

Figure 11. Open CNC software

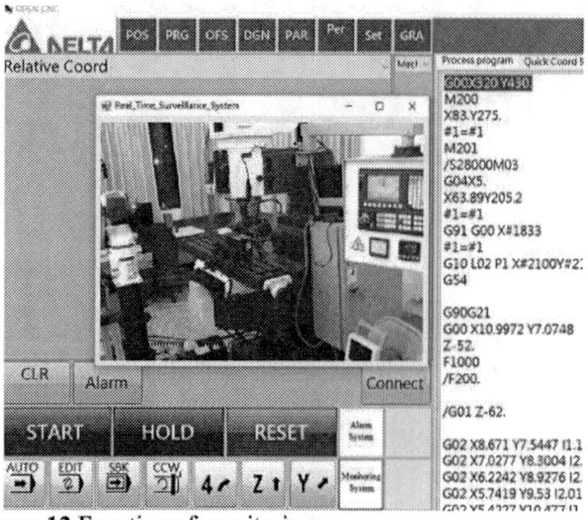

Figure 12. Function of monitoring

Figure 12 shows the real-time surveillance system. The function surveyed the situation of the current scene. Users can know whether have happened the situation of error, and they can immediately aid the field employees.

Figure 13 shows the function of sending a mail. The mail will be sent to the designated department when there is an error happened on the machine. The designated department resolve the error of machine in best time. The function can reduce the time of resolve the alarm and increase the work efficiency.

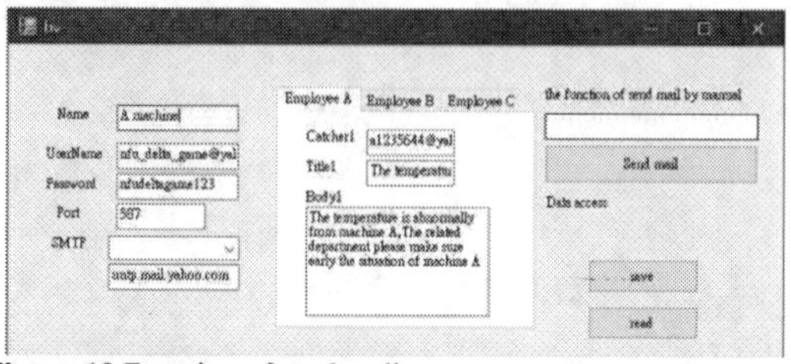

Figure 13. Function of send mail

The mail also set up the three patterns of Yahoo, Hotmail, and Gmail. Users can set up the pattern of transmission according to the habit of the company. The method achieves to the transmission of Extranet. It can deliver the message to the designated company when the machine happened the particular error.

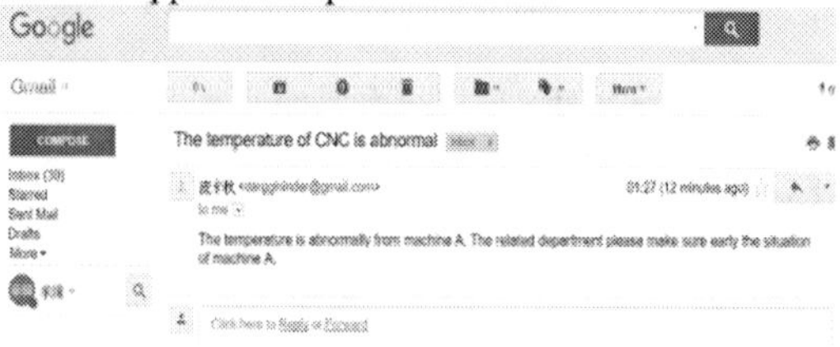

Figure 14. Warning email Gmail

Figure 15. Warning email Yahoo

5. Experiment Result

Figure 16. Material after machining

The using material is 70mm*70mm*30mm of ABS. With developed the automatic loading and CNC machining system via vision inspection. Users can process lots of same products, and they can alter the project when the requirement is different. As Figure 16, the material is completed machining by the automatic production line.

As Figure 17, the air pressure-clamp on the manipulator to loading and unloading automatically when the vision inspection system inspects material position and angle from the machining platform and the conveyor. CNC miller can machine the graph accurate by the edge find from vision inspection system.

Figure 17. Equipment Instruction

As Figure 18, compare the costing time between human and vision system in edge tracing. It displayed the edge tracing by human cost around 60 to 100 seconds. If the production line use edge tracing by vision, after filming, coordinate calculate and CNC bed moving just only cost 10 seconds to finish, and the time of machining is standardization. According to above of result, the project chosen approach of the vision inspection is more effective.

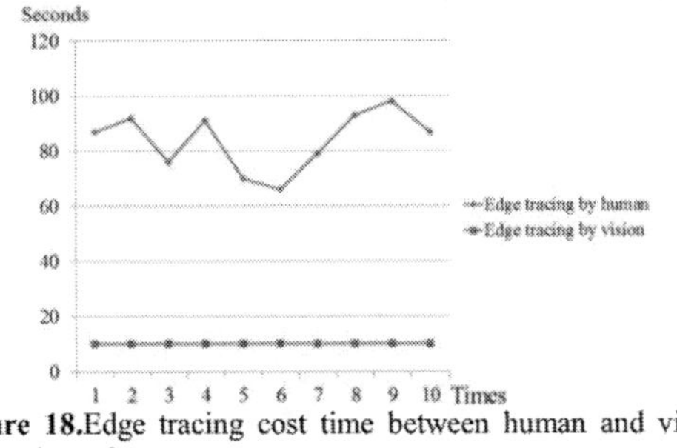

Figure 18. Edge tracing cost time between human and vision comparison chart

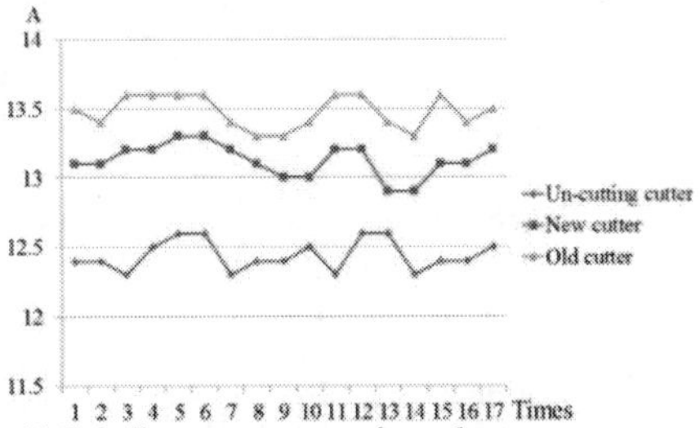

Figure 19. Spindle current comparison chart

In this research, SCADA software is used to get spindle current value to analysis —un-cutting, new cutter and old cutter in the same machining situation. As Figure 19, we can know current value difference among new cutter, old cutter and un-cutting cutter. If the cutter was worn, the current value is bigger than a new cutter. So users can know the cutter status and change cutter at the suitable timing by the parameter of current.

6 Conclusions

In this research, an automatic loading and unloading CNC machining system via vision inspection had been developed. CNC controller is the core of automatic production line, and it completed the communication among all devices. In the system of the object finder, the position and angle of incoming materials will be inspected and saved in the registers when the material photograph is passing vision inspection on the conveyor. The CNC controller sends those data to the programmable logic controller (10MC) by RS485 to send the command to control the manipulator, conveyor. The project completed an automatic production line with high-speed edge finding, the material which is on CNC miller moved to edge finding zone and it will be inspect after manipulators placing materials.

At the same time, the SCADA software collected the real-time data from the related sensors when the automatic production line is processing. The function of real-time monitor and alert email notification were built by Open CNC. The related worker will receive alert email or notification when alarm happened so that they can troubleshoot immediately. With SCADA and Open CNC, the alarms are prevented and reduce the repair times so that the production efficiency will be increased.

Acknowledgment

The authors would like to acknowledge the financial support of the Ministry of Science and Technology of Taiwan, R.O.C. through grant number MOST 105-2221-E-150-038.

References

1. "The synopsis of FMS be proposed it, "2011(in Taiwan)
2. J. Wan, H. Cai, and K. Zhou, "Industrie 4.0: enabling technologies." In Intelligent Computing and Internet of Things (ICIT), 2014 International Conference on. IEEE, 2015, pp. 135-140.
3. W. Luo, "Germany industrial 4.0 strategic enlightenment to China's industrial transformation," Fiber Reinforced Plastics/Composites, pp. 125-128, 2014 (in Chinese).
4. "Executive Yuan provide the developing scheme of industry 4.0, "2015(in Taiwan)
5. Y. Wang, "It from industry 4.0 is observed the challenge of productivity 4.0", 2015 (in Taiwan).
6. LI-JING industrial Co., Ltd-conveyor . 2016(in Taiwan).
7. DELTA ELECTRONICS, INC. 2016 (in Taiwan).
8. J. Chen, "the sever motor adjusts it from CNC, "May. 2015 (in Taiwan).

MATEC Web of Conferences **108**, 04003 (2017) DOI: 10.1051/matecconf/201710804003
ICMAA 2017

Formation and Control Technology of Top Burr in Micro-Milling

Zhiyang Zhang, Zhenjian Yuan and Guicheng Wang

School of Mechanical Engineering, Jiangsu University, Zhenjiang, China

Abstract. Burr formed during micro-milling slot has the most direct effect on surface quality and dimension accuracy of work-piece. Compared with other positional burrs, top burrs having the biggest dimension make the subsequent de-burring operation more difficult. To analyze the effect of spindle speed, cutting depth , feeding per teeth and blade radius on top burr size, Al6061-T6, used as research object, was conducted on a 5-axis machining center with the single factor method. Effect regularities of the process parameters on burr dimension were preliminarily revealed, active control of burr dimension during micro milling operation could be possible on the basis of the experiment.

1 Introduction

Burr is a special phenomenon during metal cutting process, it always occurs at corners and edges of the machined parts. Most of traditional and non-traditional manufacturing process generate burrs and the de-burring process is considered to be time consuming and non-productive. Because the burrs may reduce the fit of components and can be the source of small particles during information storage hardwire assembly, thus it is inevitable to add de-burring process [1]. Gillespie[2] pointed that de-burring and edge finishing on precision components may constitute as much as 30 percent of the outlay of the finished work-pieces. For the reason of critical cutting depth and size effect, burr dimension and processing feature size are relatively close to each other, and it makes subsequent de-burring operation more difficult. Mechanical engineering experts and scholars at home and abroad have studied burr formation mechanism and control technology, made a number of research results and promoted the development of the less burr processing technology. Gillespie[3] is one of the first experts to depict and classify burr types as follows: passion burr, rollover burr, tear burr and cut-off burr. He also provided valuable references for further study. Narayanaswami[4,5]proposed a new algorithm to minimize burr formation during face milling of arbitrarily shaped polygons and their simulation records showed that cutter position ,cutter radius ,and work-piece orientation could be changed for primary burr. Biermann[6] conducted micro-milling experiments on austenitic stainless steel X5CrNi18-10, studied the action law of cutting parameters, tool geometry , lubrication mode, etc.. It was revealed that these factors had significant affection on top burr size. Toropov[7] researched the mechanism of consecutive and intermittent burr in micro-milling and feeding direction burr was the focus of attention. A useful forecasting model was set up for burr dimension and its forming process. Compared with the turning experiment of Al6061-T6, the maximum prediction error was up to 30%. Le[8] explored burr phenomena in micro-cutting and the cutting direction –side burr-of prism pattern, cutting direction –exit burr –of pyramid patter were introduced. They proposed the predicting model for burr size in each case. Lee[9] carried out a study on micro burr formation and proposed a method to minimize burr size through process control in micro milling process with end mills. The results revealed that cutting speed, cutting edge radius and feed per teeth had impact on top burrs formation to some extend. Ravi[10]presented their experimental records aiming to investigate the influence of main cutting parameters on the formation of various burrs types. They also proposed an analytical model for exit burr height, which was based on burr formation geometry and the principle of continuity of work at the critical position nearing to the end of work-piece. Liu[11] established a mathematical model for calculating uncut thickness and describing the dynamic behaviors of micro milling forces components ,with influences of tool wear, edge plowing and critical uncut thickness included. Their experimental records showed that a small feeding speed and tool material not prone to stick to work-piece were desirable. Thepsonthi[12]obtained optimum cutting parameters with the aim of minimizing the burr dimension and surface roughness. Their experimental model was developed with statistically methods utilized in multi objective particle swarm optimization. Meanwhile, it was noticeable that factors ignored on macro scale showed significance in micro milling operation. e. g. vibration, micro structure and deflection.

The above literatures carried out burr researches from various aspects. While, as for the problem about how cutting parameters influence top burr size, there is yet no uniform and clear conclusions. It is necessary to further work on the topic. Analyzing the terms of influence, revealing the rules, making reasonable choices of parameters and planning edge path, the purpose of active control and reduction of burr dimension could be successfully achieved, maximization of economic be benefitted.

2 Experimental procedure

Micro-slot milling experiment of Al6061-T6 was conducted on a high speed 5-axis machining centre MIKRON CUP800. Maximum spindle speed up to 20000r/min, repeating position accuracy 3μm for X and Y respectively. Choosing fine grained cemented carbide as tool material, standard type GM-2ES-D0.5, spiral angle 30°, rake angle 10°, lag angle 5°, blade dimension 0.5mm in diameter with various blade arc radius. In order to avoid introducing disturbance to experimental results by surface oxide layer, finishing milling of top surface was done in advance. Measuring instrument OLYMPUS DSX500 optical microscope. Taking vibration into consideration, causing top burr size to be metabolic, selecting the mean values in stable machining condition as measured records, Figure. 1 and Figure. 2 are micro-milling tools with various blade radius and measuring apparatus and equipment used in the experiment.

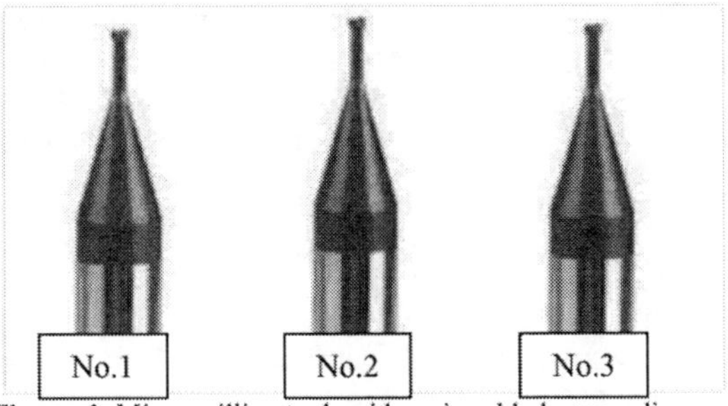

Figure. 1. Micro-milling tools with various blade arc radius

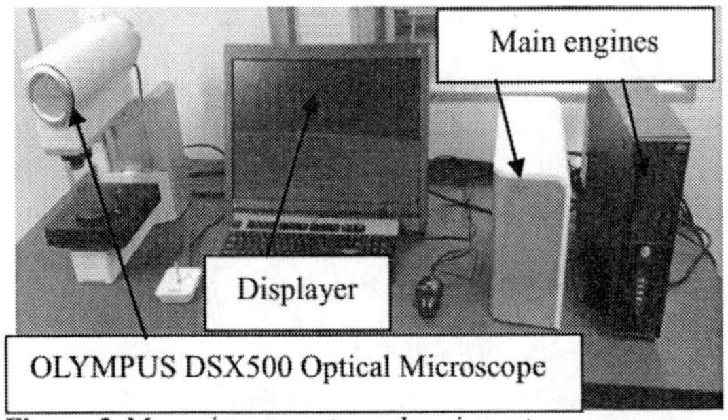

Figure. 2. Measuring apparatus and equipment

Firstly, setting up a group of benchmark parameters, using single factor method to analyze the changing attributes of top burr height along with different factors combination. Parameters are as follows in Table 1.

Table 1. Setting of cutting parameters.

Spindle Speed(r/min)	7500, 8500, 9500, 10500, 11500, 12500
Feeding speed(μm/z)	0.5, 1, 1.5 ,2.0 ,3.0 ,4.0
Cutting Depth（μm）	30, 40, 50, 60, 70, 80
Benchmark Value	10500r/min, 4.0μm/z, 40μm

3 Results and discussion

Work-piece was presented in Figure. 3 and micrographs of top burrs were shown in Figure. 4, respectively.

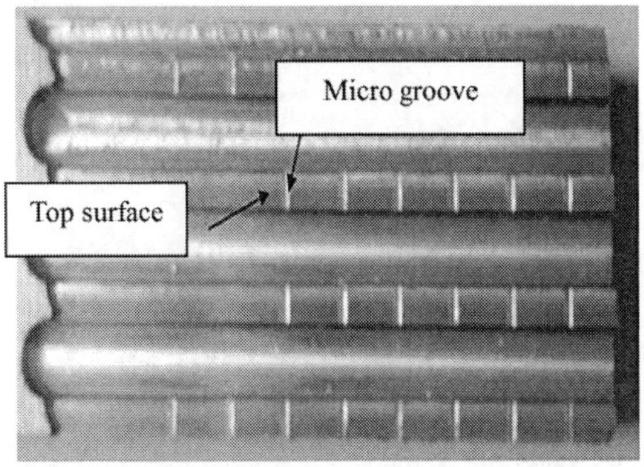

Figure. 3. Work-piece after micro-milling

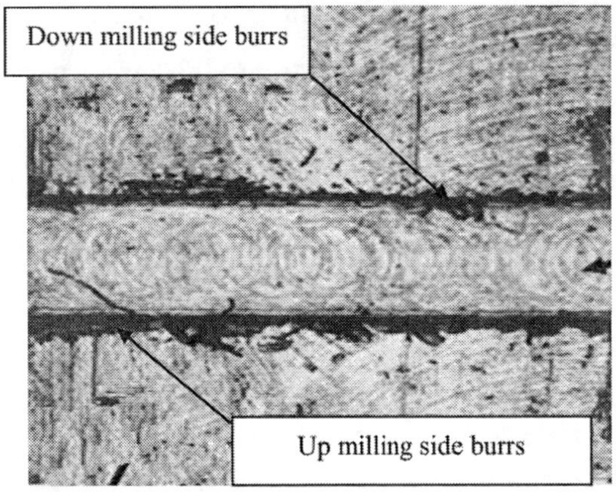

Figure. 4. Micrograph of top burrs

3.1 Effect of spindle speed on top burr size

Variation tendency of top burr dimension with mutative spindle speeds, fixed cutting depth 40μm and feed per teeth 4μm, is shown in Figure. 5.

From Figure. 5, it can be seen that the whole dimensions present rising trend, along with gradually incremental spindle speeds. From 7500r/min to 10500r/min, increasing trend-line is not obvious. The main reasons may be that increasing cutting speed solely leads to more material removal rate in the same time. Because crumb slot is very limited owing to its physical dimension. If the removed material can not be timely discharged, it will be pushed to top surface along up and down milling sides automatically and top burrs emerge on the work-piece. In the meantime, increased speed will quicken the deformation of the removed materials, accordingly contacting time between tool and work-piece be decreased, bring milling force to lower degree and eventually lead to smaller burrs, so the trend-line will be explained by the synthetic action. Further, with speed continuing to augment, material removal rate increases rapidly and gradually takes a dominant position during the machining process, leading the top burrs to increase in dimension significantly.

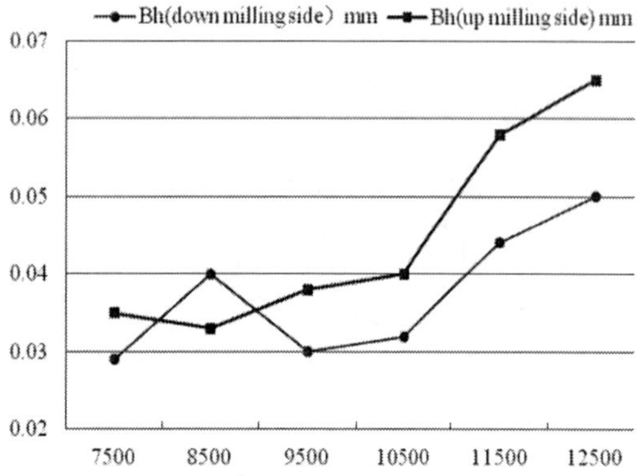

Figure. 5. Effect of spindle speed on top burrs

3.2 Effect of feeding speed on top burr size

On condition that spindle speed 10500r/min, milling depth 40μm, the trend-chart of top burr height along with feeding speed is described in Figure. 6.

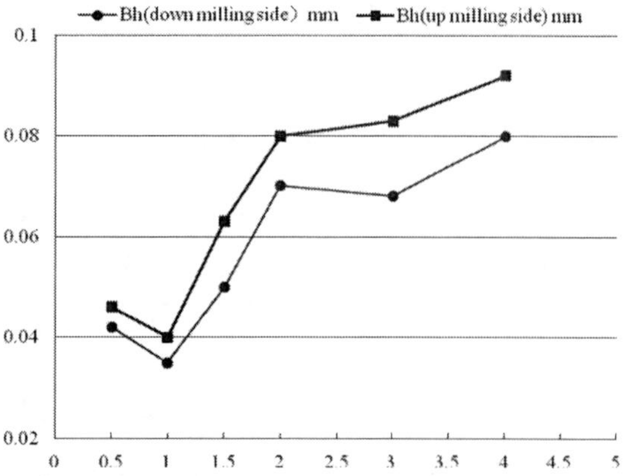

Figure. 6. Effect of feeding speed on top burrs

According to Figure. 6, the conclusion drawn is that when taking minimum value , milling thickness is near to critical layer, cutting and ploughing can act on material simultaneously in this case. Cutting layers flow onwards by forward extrusion of cutter blade and accumulate on rake face, eventually larger burr will not be avoided. Exceeding critical thickness, plow effect will no longer play a leading role and smaller dimension could be obtained. Meanwhile, with feeding speed constantly rising, thickness of burr is bound to have large augmenter as shown in above Figure. 6, resulting in jumbo size. What is more, edge wear could not be ignored, burr dimension presents increasing tendency accompanying with plough action.

3.3 Effect of depth of cut on top burr size

As shown in Figure.7, the trend-line of burr size continues to rise with changing cutting depth, on condition that feed per teeth 4μm, rotating speed 10500r/min, respectively.

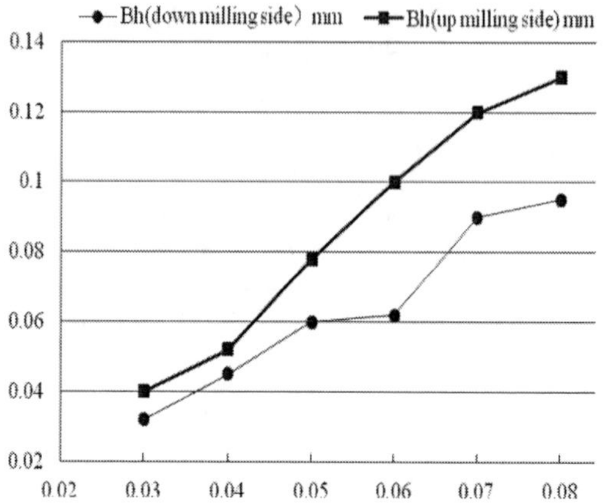

Figure. 7. Effect of depth of cut on top burrs

It is obvious that top burr size is more sensitive to axial milling depth. Increased depth contributes to much volume of removed material at the same time, contacting area between blade and work-piece rises to certain degree, and thus it makes friction force become larger, exacerbate tooth wear accordingly causing size effect to take a position that can not be ignored during micro-milling process. However, owing to its physical dimension, crumb slot can not hold so many scraps and remained chips may be crushed by subsequent ones not having absolute deformation. In the case of integrated effect of multiple factors, curly burrs can be formed. At the same time, along with the machining process, tool blade is bound to be obtuse, and it becomes easy to understand the phenomenon that down milling burr size is much smaller than up milling. It also can promulgate the fact that top burr dimension shows sensitive to degree of tool sharpness.

3.4 Effect of blade arc radius on top burr size

Compared with conventional milling process, edge wear quantity, feed per teeth and axial depth are roughly in the same order of magnitude. Edge radius of circular arc not only has significant impaction on surface quality and dimensional accuracy, but also on burr size.

Optical microscopic of three milling tools with different abrasion loss and corresponding micro-groove are presented in Figure. 8. Parameters employed in this case are as follows: spindle speed 10500r/min, feed per teeth 4μm and axial depth 40μm, respectively.

Figure.8 (a), (c), (e) represent tools No.1, No.2 and No.3, respectively. The abrasion loss increase in turn. According to Figure. 8 (b), (d), (f), it is obvious that

DOI: 10.1051/matecconf/201710804003

significant differentiation emerges among top burr sizes and types. The first tool has the most sharp edge and corresponding burrs have the smallest dimension and uniform distribution along two edges machined. No.2, owning a mild degree of abrasion loss of blade, makes the homologous burr dimension being large than the former distinctly. It is noted that the last has maximum dimensions and deteriorative shapes on top surface, thus resulting in much obstacle to subsequent de-burring operation.

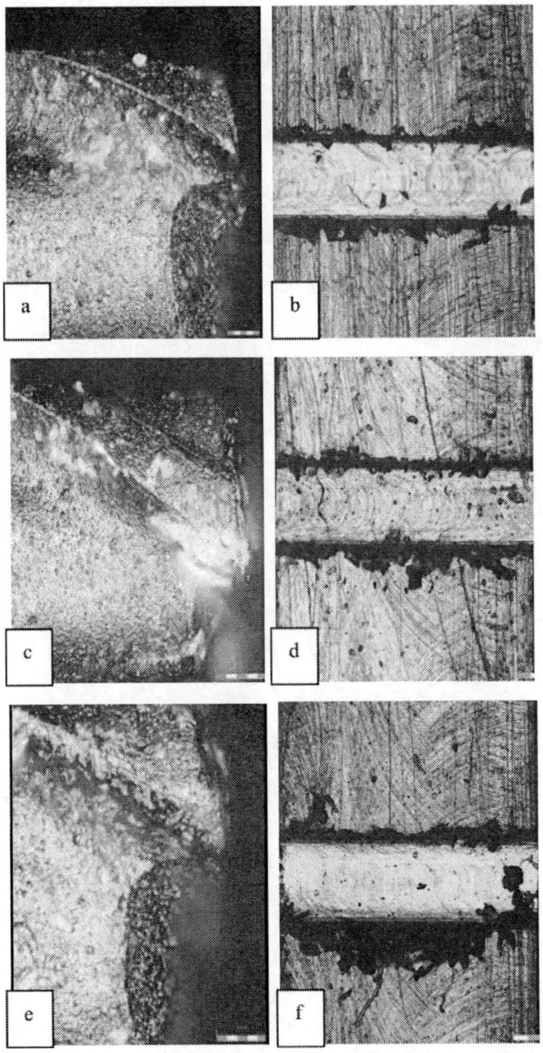

Figure. 8. Micro graphs of wearing edges and corresponding top burrs

4 Conclusion

Impactions of spindle speed, feeding per teeth, cutting depth and circular arc radius on micro-slot milling top burr size are investigated using single factor method. According to experimental results, the following conclusions are drawn:

(1) Among range of 7500r/min to 10500r/min, burr dimension lightly increased. When beyond 10500r/min, increasing trend-line is very obvious. It is worth noting that there is critical speed value leading to the smallest burr dimension.

(2) Generally speaking, with the augment of feeding speed, burr height shows an accessorial trend. There is a critical feeding value 1μm/z, which results in minimum dimension. Augmenter is apparent and significant.

(3) Influence of milling depth on burr size is more obvious. The dimension continues to rise linearly. On the satisfaction of production efficiency condition, in order to achieve the purpose of active control of burr size, choosing small milling depth is sensible.

(4) Top burr is susceptive on abrasion loss. With the increase of abrasion, burr size becomes larger and shape more scrambled.

In view of the limitations of the experimental conditions, high speed micro milling of aluminum alloy Al6061 - T6 was conducted diminutively. In order to guarantee the rationality of the results, it is of great significance to further expand the research scopes.

In subsequent research and analysis, it is consumingly recommended that more influential factors, such as types of tested materials , selection ranges of cutting process parameters, different kinds of tool materials and coating types, be taken into elaborative consideration. Moreover, machine tools vibration, work-piece excision angle and cutting tools geometry during micro milling, may also be influential factors on the response values, that is to say burr dimension and types.

At the same time, it is strongly suggested to explore other schemes to carry out micro milling experiments, e.g. orthogonal test which is widely used in scientific research. It is noticeable that other interesting methods, for instance, response surface, analysis of variance and multiple targets optimization, can also be utilized to further study the experimental records. Finally, the influence laws of cutting process parameters on different position burrs, micro milling force components and surface roughness , are also train of thought.

In conclusion, in order to achieve the goal of higher productivity, dimensional precision and surface quality simultaneously, micro milling cutting process parameters with optimum combinations and reasonable scientific analysis methods, should be taken into consideration scrupulously.

Acknowledgement

This research is based upon work by the NNSF of China (Grant No.51075192) and National key projects of P. R. China（Grant 2013ZX04009031）.

References

1. S.L. Ko, D.A. Dornfel, A Study on Burr Formation Mechanism, J. ENG. MATER, **113**, pp.75-87, 1991.
2. 2. L.K. Gillespie, *De-burring and Edge Finishing Handbook* (SME, New York, 1999).
3. L.K. Gillespie, The formation and properties of machining burrs, J. Manuf. Sci. E., **98**,

pp.66-74,1976.

4. R. Narayanaswami, D. Dornfeld, Burr minimization in face milling: a geometric approach , Trans ASME, J. Eng. Ind., **119**, pp. 170-177.
5. R. Narayanaswami, D. Dornfeld, Design and process planning strategies for burr minimization and deburring. Trans. NAMRi/SEM, **22**, pp.313, 1994.
6. D. Biermann, M. Steiner, Analysis of micro burr formation in austenitic stainless steel X5CrNi18-10, Procedia CIRP. **3**, pp.97-102, 2012.
7. A. Toropov, L.K. Sung, A model of burr formation in the feed direction in turning, Int. J. Mach. Tool. Manu., **46**, pp.1913-1920, 2006.
8. D. Le, J.M. Lee, S.J. Kim, et al., Burr analysis in micro-grooving, Int. J. Adv. Manuf. Tech. **50**, pp.569-577, 2010.
9. K. Lee, D.A. Dornfeld, An experimental study on burr formation in micro milling aluminium and copper, Trans NAMRI/SME, **30**, pp.255-262, 2002.
10. L. Ravi, B. Vivek, K.S. Ramesh and S.S. Joshi, Characterization and modeling of burr formation in micro-end milling, PRECIS. ENG., **35**, pp.625-637, 2011.
11. H. Liu, Y. Sun,Y. Geng and D. Shan, Experimental research of milling force and surface quality for TC4 titanium alloy of micro-milling, Int. J. Adv. Manuf. Technol., **79**, pp.705-716, 2015.
12. T. Thepsonthi, T. Özel, Multi-objective process optimization for micro-end milling of Ti-6Al-4V titanium alloy, Int. J. Adv. Manuf. Technol., **63**, pp.903-914, 2012.

Experimental Study on Fracture Failure of BRW 250 Pump Liquid Valve Mechanical Spring Surface

Zeng Rui[1,2], Zhang Yong[2], Lin Zhenrong[2] and Li Jie[2]

[1] *College of Field Engineering, PLA Univ. of Sci. & Tech., Nanjing 210007, China*
[2] *Department of Airfield Engineering, Air Force Logistics College, Xuzhou 221000, China*

Abstract. In this paper, the singularity analysis method based on the continuous wave for the vibration signal of the plunger pump liquid valve under different conditions was studied, and the LMS based weighted least square method with good robustness and validity was proposed to calculate the LPZ index, which was the judgment criterion for fault of liquid valve mechanical spring. Fault diagnostic test results showed that the method could overcome the singularity of the binary discrete wave in the detection and quantitative accuracy problem, realize the accurate positioning of the singular point in the signal, identify the liquid valve disc in the spring break state to the liquid valve seat or lift limit the impact of the moment, and determine the fault of the liquid valve mechanical spring effectively.

1 Introduction

Liquid valve mechanical spring is an easily damaged parts type of BRW250 high pressure plunger pump liquid valve components. When the plunger pump liquid valve mechanical spring fracture, the volume efficiency will drop sharply[1]. If the fault cannot be ruled out, it will cause the plunger pump running in low efficiency in the work area. Not only a waste of energy, but also cause serious production accidents. The test system to realize the automatic identification of the fault of the liquid valve mechanical spring is very important. Through measuring the vibration signal of liquid valve, the fault features of the vibration signal are picked up. The spring fracture state of liquid valve disc on the impact of the seat or the lift limiter time is identified and accurate positioning of the spring fracture fault of the liquid valve is located, so as to achieve the diagnosis of vibration signal of plunger pump liquid valve mechanical spring fracture fault.

2 Experiment of plunger pump liquid valve mechanical spring fracture fault

2.1 Experiment method

In order to avoid signal interference, the distortion and weakening of the transport was blocked, by which the measured signal can reflect the maximum liquid valve working condition. The feasibility according to the structural features and design of emulsion plunger pump head, processing special steel bolt screw plug through the piezoelectric acceleration sensor was firmly arranged at both ends the plunger pump head. The suction liquid valve end and the vibration signal of liquid discharge liquid valve end were measured respectively, shown in Figure.1.

Figure 1. Sensors layout

The main part of the spring breaking was in the middle, so the fault respectively from the middle plunger pump liquid valve mechanical spring to simulate liquid valve liquid valve mechanical spring fracture, shown in Figure.2.

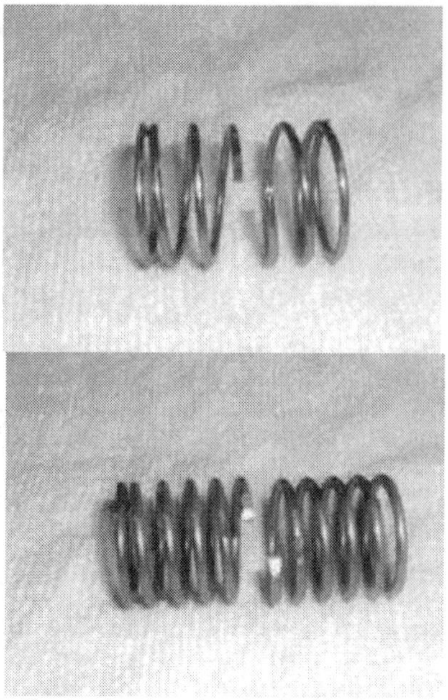

Figure 2. Fractured spring of suction liquid valve and discharge liquid valve

2.2 Original experiment dates

Vibration signal of original experiment dates were shown in Figure.3, Figure.4.

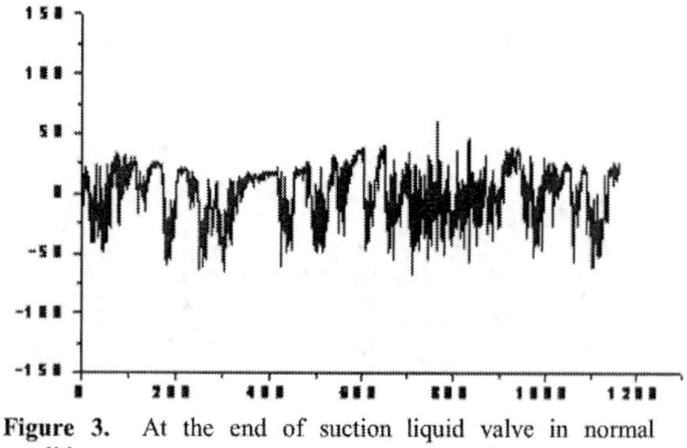

Figure 3. At the end of suction liquid valve in normal condition

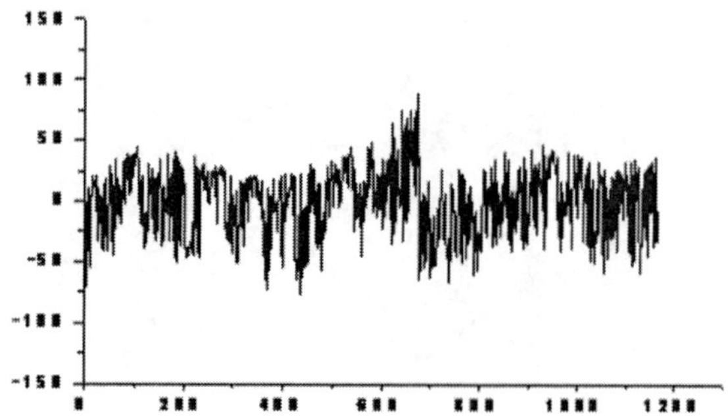

Figure 4. Suction liquid valve mechanical spring fracture

3 Test result analysis

Because of the vibration of the reciprocating motion of crank connecting rod mechanism of emulsion plunger pump and plunger caused by the coupling of emulsion in the cavity pressure pulsation and vibration between the components, the measured vibration signal of the complex get obvious non-stationary characteristics. The domain curve was difficult to directly extract fault features, which could not make accurate judgment the rupture failure of liquid valve mechanical spring.

Fracture liquid valve mechanical spring to open or close the liquid valve disc lag angle get significant changes. Due to the liquid valve disc to open or close the speed was far greater than the maximum allowable falling speed, and the impact on the liquid valve seat or liquid valve plate and lift limiter. Liquid valve mechanical spring fracture gets greatly reduced the volume efficiency of emulsion plunger pump. Plunger pump head assembled quality as a whole, but the impact would affect the plunger pump signal acquisition, and the signal generated local mutation. Mutation in the signal reflects the mechanical change of state at the corresponding time, and the change mechanism of signal through the sudden change of the detecting part could study the mechanical running condition and reveal the fault reason of machinery. This would combine the wavelet singularity analysis and fault diagnosis of plunger pump liquid valve mechanical spring fracture.

If a signal was differentiable at a certain point or in a certain interval, the signal was regular in that point or interval. On the other hand, it was singular. In mathematics, the Lipschitz exponent was used to describe the singularity of the function quantitatively, and it could also be used to describe the singularity of the signal. In the actual calculation, the continuous wavelet transform were discrete parameters, but it could choose a discrete level of detail mesh according to the actual analysis. The modulus maxima of the singular points of a signal in the time scale plane could be expressed as

$$\log_2 |WT_x(a,t)| \le \log_2 K + (\alpha + 0.5)\log_2 a \quad (1)$$

Assumed $c_1 = \log_2 K$, $c_2 = \alpha + 0.5$, $y = \log_2 |WT_x(a,t)|$, $x = \log_2 a$, the formula (1) can be changed as

$$y \le c_1 + c_2 x \quad (2)$$

Based on the least square estimation theory, the sum of squares of residuals between the estimated and observed values is as follows:

$$f(c_1,c_2) = \min \sum_{i=1}^{n} \varepsilon_i^2 = \min\left(\sum_{i=1}^{n} (y_i - c_1 - c_2 x)^2 \right) \quad (3)$$

However, when the residual error of the model does not obey the normal distribution or the outliers in the data, the value obtained by the least square method will have obvious deviation. In order to solve the problem, the objective function should be changed from the sum of squares of residuals to the median of the squared residuals, as follows:

$$f(c_1,c_2)=\min\left(\underset{i=1,2,\cdots,n}{\text{median}}\,\varepsilon_i^{\,2}\right)=\min\left(\underset{i=1,2,\cdots,n}{\text{median}}\left(y_i-c_1-c_2x\right)^2\right) \quad (4)$$

The formula (4) is called least median of squares(LMS). The method has good robustness, and the fitting results are almost not affected by outliers in the data.

According to the different working process of suction and discharge of pump and valve, the mathematical model of the differential equation describing the motion of the valve is derived as:

(1) $\alpha=\beta_0\sim\pi$

$$\left.\begin{aligned}
&\frac{V}{E}\frac{\mathrm{d}p}{\mathrm{d}t}=-|Av|+\left(Q_{ax}+\frac{\pi d_{a2}^2}{4}\frac{\mathrm{d}h}{\mathrm{d}t}\right)\\
&m_a\frac{\mathrm{d}^2h}{\mathrm{d}t^2}=\frac{\pi d_{a2}^2}{4}(p_a-p)-\left(1-\frac{\gamma}{\gamma_0}\right)m_ag-F_{a0}-k_ah\\
&Q_{ax}=\mu\pi d_{am}h\sin\theta\sqrt{\frac{2}{\rho}|p_a-p|}
\end{aligned}\right\} \quad (5)$$

(2) $\alpha=\pi+\beta_2\sim2\pi$

$$\left.\begin{aligned}
&\frac{V}{E}\frac{\mathrm{d}p}{\mathrm{d}t}=|Av|-\left(Q_{bx}+\frac{\pi d_{b2}^2}{4}\frac{\mathrm{d}h}{\mathrm{d}t}\right)\\
&m_b\frac{\mathrm{d}^2h}{\mathrm{d}t^2}=\frac{\pi d_{b2}^2}{4}(p-p_b)-\left(1-\frac{\gamma}{\gamma_0}\right)m_bg-F_{b0}-k_bh\\
&Q_{bx}=\mu\pi d_{bm}h\sin\theta\sqrt{\frac{2}{\rho}|p-p_b|}
\end{aligned}\right\} \quad (6)$$

where m_a is disc quality ; F_{a0} is spring preload ; k_a is spring stiffness ; d_{am} is disc diameter ; p_a is pump suction pressure ; V is total volume of liquid in hydraulic cylinder ; V_c is clearance volume ; A is Plunger cross-sectional area.

The various available formula is as follows :

$$\left.\begin{aligned}
&\frac{V}{E}\frac{\mathrm{d}p}{\mathrm{d}t}=K_1\left\{K_3\left[|Av|-K_4\left(Q_{ax}+\frac{\pi d_{a2}^2}{4}\frac{\mathrm{d}h}{\mathrm{d}t}\right)\right]\right\}\\
&\qquad+K_2\left\{K_5\left[|Av|-K_6\left(Q_{bx}+\frac{\pi d_{b2}^2}{4}\frac{\mathrm{d}h}{\mathrm{d}t}\right)\right]\right\}\\
&\frac{\mathrm{d}^2h}{\mathrm{d}t^2}=K_1\left\{\frac{K_4}{m_a}\left[\frac{\pi d_{a2}^2}{4}(p_a-p)-\left(1-\frac{\gamma}{\gamma_0}\right)m_ag-F_{a0}-k_ah\right]\right\}\\
&\qquad+K_2\left\{\frac{K_6}{m_b}\left[\frac{\pi d_{b2}^2}{4}(p-p_b)-\left(1-\frac{\gamma}{\gamma_0}\right)m_bg-F_{b0}-k_bh\right]\right\}\\
&Q_{ax}=\mu\pi d_{am}h\sin\theta\sqrt{\frac{2}{\rho}|p_a-p|}\\
&Q_{bx}=\mu\pi d_{bm}h\sin\theta\sqrt{\frac{2}{\rho}|p-p_b|}
\end{aligned}\right\} \quad (7)$$

where $K_1, K_2, K_3, K_4, K_5, K_6$ are comprehensive factors.

The valve disc parameters was put into the original state variable model, and the four order Runge Kutta method was used for programming calculation. The result were shown in Figure.5.

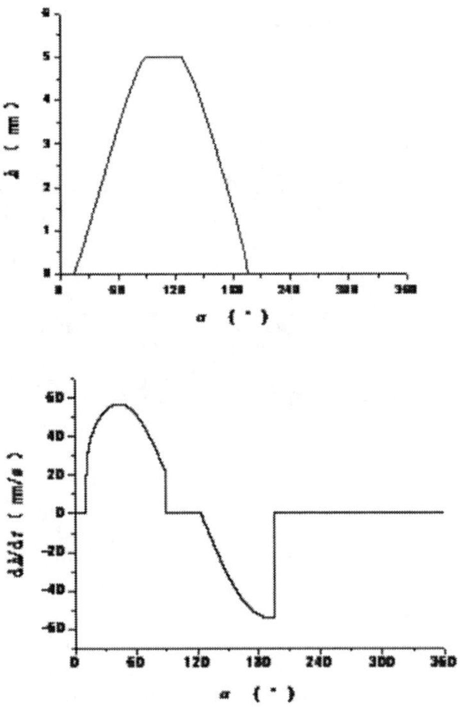

Figure 5. Curve of valve disc

3.1 Singularity analysis of plunger pump liquid valve signal in normal condition

The signal in the normal state is carried out, and the coefficients of the scale time plane are obtained, shown in Figure.6.

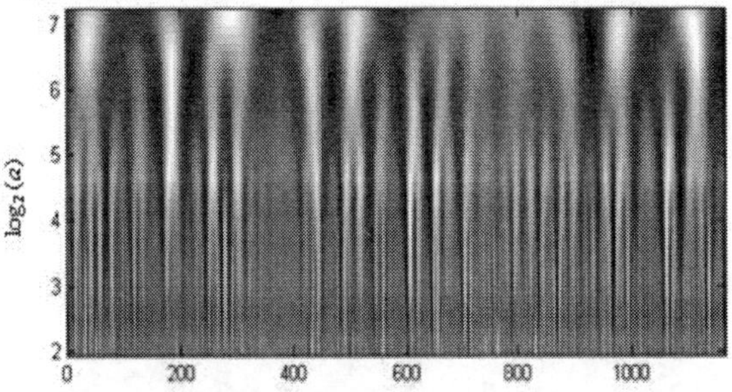

Figure 6. Normal condition

After the transform of the normal condition of the vibration signal, the coefficient chart showed a regular pattern of black and white. Each white line or black line exists only in some scale space, and not throughout the scale space, which was the performance of plunger pump head vibration signal with non-stationary[2-4]. The vibration signal of plunger pump head was not only affected by the movement of liquid valve plate and the emulsion pulse in high pressure chamber, but also influenced by the coupling action between the rotary motions of the crank connecting rod mechanism. The reciprocating motion of the piston and the vibration were caused by various motions. So, the vibration signal was complex and obvious non stability.

3.2 Singularity analysis of plunger pump liquid valve mechanical spring fracture signal

When the liquid valve mechanical spring was broken, the transform was shown in Figure.7.

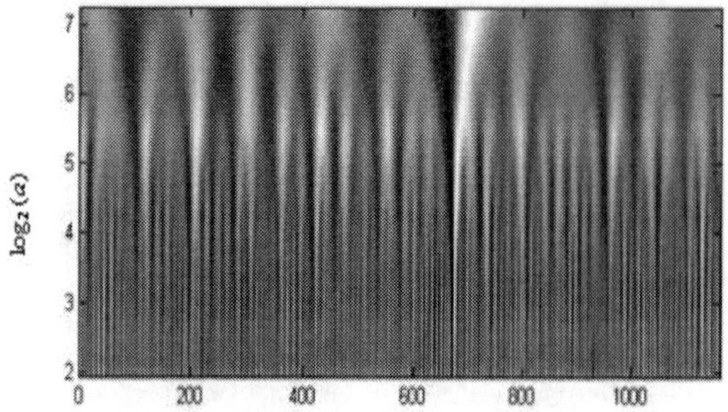

Figure 7. Fracture condition

From Figure.6, we could find that the transform corresponded to the black and white mutation, and was reflected in every scale continuously. The most dark represented negative maximum value, and the brightest represented the maximum value [5,6]. The wave coefficients were mapped by the brightest and most dark place corresponding along the scale together to constitute the modulus maxima line. In the process of large and small scale, the modulus maxima curve converged to the corresponding change time. At a certain scale, a maximum value point was found, and in the direction of the displacement at the direction of decreasing, the maximum value of the search mode was continued in the cone, the time to search for the modulus maxima at the scale close to zero should be the singular point or the peak value of the signal. However, the corresponding point of the modulus maxima curve was a singular point or a non singular point, which was determined by the attenuation of the modulus maxima curve. The slope of the modulus maxima logarithmic curve was corresponding to the LPZ index α. The logarithmic curve of modulus maxima was shown in Figure.8.

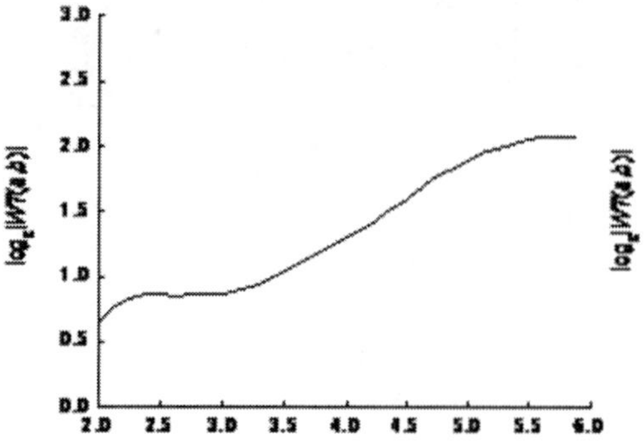

(a) negative curve

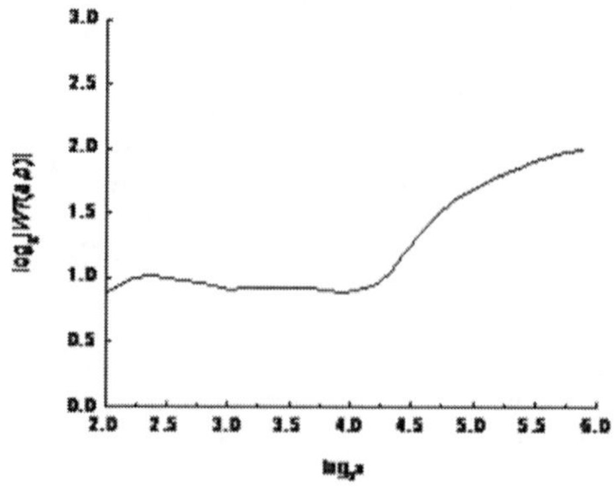

(b) positive curve

Figure 8. Logarithmic curve of modulus maxima

Because the signal at this point had a mutation, and this point the corresponding left and right sides of the signal was not the same degree of mutation, the calculated negative modulus maxima lines were not the same as the corresponding LPZ exponents of the positive modulus maxima lines. The corresponding point of the minimum value was singular point, and the moment that the vibration signal corresponded to this point had been changed.

After the observation of the original signal, it was found that the signal exist a small pulse. This showed that the disc impacted on the seat, and the vibration signal was caused by the suction liquid valve end of the measured mutation. So, the singularity analysis method using continuous wave with improved weighted least squares method based on LMS calculation of the LPZ index could detect the liquid valve seat of the impact of the moment and determined the failure of the liquid liquid valve mechanical spring.

4 Conclusions

(1) The detailed degree of the discrete grid according to the actual analysis could be selected by the continuous wave in the analysis of signal singularity to improve the accuracy of the calculation.

(2) The LPZ index could be obtained from the abnormal values of the data, which solved the singularity of the binary discrete wave in the detection and quantitative accuracy problem, realized the accurate positioning of the singular point in the signal, identified the liquid valve disc in the spring break state to the liquid valve seat or lift limit the impact of the moment, and determined the fault of the liquid valve mechanical spring effectively.

(3) The criterion of fault liquid valve mechanical spring fracture not only can judge the fault of plunger pump liquid valve mechanical spring fracture on the exact location of the fault location, but also can be used to determine the impact of the liquid valve seat or lift limiter to accurately identify the opening or closing of the liquid valve disc.

References

1. LIU Songyong, Ball A. Proceeding of ImechE, **220**79 (2006)
2. Figurearella T, Jansen M H, Mechanical Systems and Signal Processing, **21**,1212(2007)
3. Peng Z K, Chu F L, Mechanical Systems and Signal Processing, **21**, 780(2007)
4. Chen Y G, Tian Z P, Chemical Engineering Science, **59**, 3569(2004)
5. Mallat S, Huang W L, IEEE Transactions on Information Theory, **38**, 617(1992)
6. Gold C, Neurocomputing, **55**, 221(2003)

Model and Simulation of Globoidal Cam Mechanisms with Oscillating and Dual Stopping Followers

Rongyu Ge , Jin Wang and Chengliang Zhang

University of Jinan, School of Mechanical Engineering, 250021 Jinan, China

Abstract. Globoidal cam mechanisms with oscillating and dual stopping followers are necessary to accomplish the rise-dwell-fall-dwell periodic motion.The trajectory surface of roller axis of oscillating globoidal cam mechanism is the offset surface of working surface of the cam, and its mathematical model is relatively simple. According to the spatial geometric relationship, the coordinate equation of trajectory surface of roller axis is derived and the original angle positions for dual stopping followers are analyzed. As a result, the three-dimension (3D) model of globoidal cam mechanism with oscillating and dual stopping followers is generated based on the offset surface method based on the UG software. Finally, the dynamic model is realized using Adams software, and through the dynamic simulation, the oscillating angle displacement, angular velocity and angular acceleration of the double-stop swinging cam mechanism are studied. According to the simulation results, it can be concluded that the input speed of globoidal cam mechanism is an important factor that affects the output performance of the mechanism. Through the above research, it can provide some reference for the design of spatial cam mechanism.

1 Introduction

The globoidal cam mechanism, including driving cam mounted through input axis and following turret fixed through output shaft, which use physical parts to convert constant rotary motion of input shaft into intermittent motion of output shaft. Due to the advantages of compact structure, high load capacity, high reliability, low noise and low vibration, globoidal cam mechanisms, are widely used in many types of automatic machinery such as machine tool changer, automatic assembly lines, medicine machines, paper processing machines and packing machines [1]. So globoidal cam mechanisms have already become the most popular and precise intermittent device at present. In term of the shape, globoidal cam mechanisms is one of the most complicate, which can be classified cam mechanisms with indexing followers or with oscillating followers according to the output motion. Cam mechanisms with oscillating followers can arrive the reciprocating and oscillating output motion that is used to convey materials between two stations in production line. As is well known, the motion should be stopped for a portion of a cycle while an operation is performed on the production at every station, so globoidal cam mechanisms with oscillating and dual stopping followers are necessary to accomplish the rise-dwell-return-dwell periodic motion.

The most important task when designing and machining the globoidal cam mechanisms is to represent their working surfaces. The working surface of globoidal cams are the surfaces that contact with the roller surfaces. These surfaces are very complex and it is very difficult to directly create the. Up to now, a number of works dealing with finding the way to describe accurately these surfaces have been proposed. Mostly, researchers derived mathematically expressions for the surface geometry of the globoidal cam based on coordinate transformation, differential geometry, and theory of conjugate surfaces [2-6]. In that way, it is so complex to infer the equations requiring designers to have solid mathematical knowledge. While cam rotates, there is a swept surface of roller axis, which is called the trajectory surface of roller axis in this paper and it is the offset surface of working surface of the globoidal cam. In fact, from designing and machining point of view, the globoidal cam mechanism can be determined by the trajectory surface of roller axis. The objective of this paper is to introduce an effective modelling and analysing method for globoidal cam mechanism with oscillating and dual stopping followers, which can be implemented by using commercial CAD/CAM/CAE softwares.

2 Mathematical model of trajectory surface of roller axis

2.1 Calculation of the surface equation

The structure diagram of globoidal cam mechanism with oscillating and dual stopping followers is shown in Fig. 1. Here, the center distance between the globoidal cam and the oscillating turret is *C*, the pitch radius of oscillating turret is R_e, the roller radius is R, the roller length is m, the clearance between the bottom end of the roller and the cam body is e, the angle between the two roller axes is β, and P represents any trajectory point on the trajectory surface of roller axis, and the distance from P to the oscillating turret axis is l. The coordinate system $oxyz$ is attached on the cam, where the axis of globoidal cam is defined as the oz coordinate axis, the coordinate plane oxy passes through the axis of oscillating turret, and the coordinate axis oy represents the projected line on the coordinate plane oxy of the initial line during the oscillation period. In one moment, the angular displacement of globoidal cam is θ_i, when the oscillating angular displacement of lower roller is ϕ^1 and that of upper roller is ϕ^2. As is shown in Fig. 1, the angular displacement of rollers can be obtained as follows:

$$\phi^j = \phi_0^j + p \cdot \phi_i \quad j = 1,2 \tag{1}$$

$$\phi_i = \begin{cases} S \cdot \phi_f & \text{oscillating period} \\ \phi_f & \text{stopping period} \end{cases} \tag{2}$$

where, ϕ_0^j is the initial angular displacement of rollers, ϕ_i denotes the oscillating angular displacement of rollers relative to the cam rotation angle θ_i, ϕ_f is the oscillating angle range of the oscillating turret, S is the dimensionless angular displacement of oscillating turret, which is decided by various predetermined cam motion curves. p is a symbol coefficient, and when the oscillating turret makes a clockwise rotation, p is equal to 1, otherwise that is equal to -1.

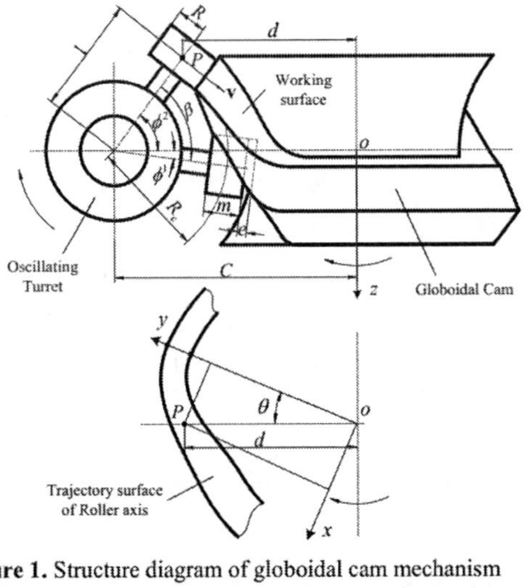

Figure 1. Structure diagram of globoidal cam mechanism

From Fig. 1, the distance d between the trajectory point P of roller axis and the coordinate axis oz can be expressed as:

$$d = C - l\cos\phi^j \tag{3}$$

Based on the geometry relations as shown in Fig. 1, the coordinate equation of trajectory surface of roller axis in the coordinate system $oxyz$ can be maintained as follows:

$$\mathbf{r}(l,\theta) = \begin{bmatrix} (C - l\cos\phi^j)\sin\theta \\ (C - l\cos\phi^j)\cos\theta \\ l\sin\phi^j \end{bmatrix} \quad j = 1,2 \tag{4}$$

Eq. (4) can also be described as

$$\mathbf{r}(l,\theta) = \begin{bmatrix} C\sin\theta \\ C\cos\theta \\ 0 \end{bmatrix} + l \cdot \begin{bmatrix} -\cos\phi^j \sin\theta \\ -\cos\phi^j \cos\theta \\ \sin\phi^j \end{bmatrix} \quad j = 1,2 \tag{5}$$

From Eq. (5), the equation of trajectory surface of roller axis accords with mathematical definition of the ruled surface:

$$r(l,\theta) = p(\theta) + lq(\theta) \tag{6}$$

So, the trajectory surface of roller axis of oscillating globoidal cam is a ruled surface. And the working surface is the offset surface of the trajectory surface of roller axis, but its derivation process of mathematical model is especially complex.

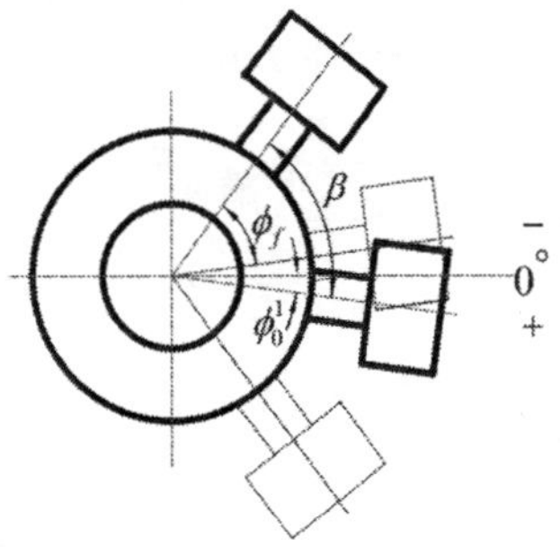

Figure 2. Schematic diagram of initial angle analysis

2.2 Analysis of initial angular displacement

For the globoidal cam mechanisms with oscillating and dual stopping followers, the convex body of two intermittent sections is commonly symmetrical structure, and it means that the oscillating bottom dead point of upper roller and the oscillating top dead point of lower roller are symmetric along the coordinate plane of oxy. As is shown in Fig. 2, the central plane oxy denotes the angular displacement with zero degree, and the clockwise is the positive direction. Then when both the two rollers are located in the upper dead point, they are considered with the initial angular displacement. According to the geometry relations, the initial angular displacement of two followers can be given by

$$\phi_0^j = \frac{(-1)^{j-1}\beta - t \cdot \phi_f}{2} \quad j = 1,2 \tag{7}$$

where t represents the symbol coefficient. For the oscillating section and intermittent section shown on the left side of Fig.2 when two rollers are located in the top dead point, t is equal to 1. For the other oscillating section and intermittent section when two rollers are located in the bottom dead point, t is equal to -1.

3 3D solid modeling based on UG

Working surface equation of the globoidal cam with oscillating followers is especially complex, so that it is difficult to directly establish its mathematical model in 3D software. At present, many kinds of advanced 3d softwares are integrated with the modeling method based on the offset surface, and in this way precise modeling on the complex regular surface can be indirectly realized. Here, UG software is used as a tool for the model of the globoidal cam mechanism with oscillating followers. The design parameters about the globoidal cam mechanism with oscillating follower is listed as follows: C=45 , R=4, R_c=22 , m=5 , e=1, β=45°, ϕ_f=20°. The motion cycle of oscillating followers incules four phrase, clockwise oscillation, dwell, anticlockwise oscillation, dwell and the corresponding cam angles are respectively 60°, 60°, 135° and 105°. That is, the rotating angles of globoidal cam during two oscillating period are θ_{f1}=60° and θ_{f2}=135°. And the widely used modified sine curve is chosen as the cam motion curve.

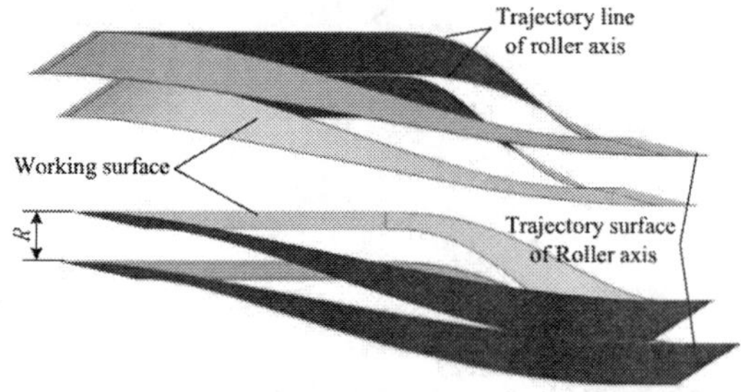

Figure 3. Model of trajectory surface of roller axis and working surface of globoidal cam mechanism

According to Eq. (4), if the parameter l is defined as a fixed value, then $\mathbf{r}(\theta)$ denotes a piece of spatial curve which is called trajectory curve of roller axis in this paper. Actually the trajectory surface of roller axis is an ruled surface that is generated by the sweeping of roller axis along two trajectory curve of roller axis. The working surface of the cam is the offset surface of ruled surface. Based on this idea, parameter l is respectively defined as two fixed value which is expressed as follows:

$$l_1 = R_e + {}^{m}\!/_{2} + e \quad (8)$$

$$l_2 = R_e - {}^{m}\!/_{2} \quad (9)$$

First of all, a UG expression table should be built including all the constants, variables and equations to accomplish the parametric modeling. Then by using the command 'By law curve' in the curve menu of UG software, two trajectory curves of roller axis are created according to Eq. (4), Eq. (8) and Eq. (9). Besides, using the command 'Offset surface' in the insert menu, two working are generated by offsetting two trajectory surfaces of roller axis with the offset distance roller radius R as is shown in Fig.3. In the same way, the inside surface and outside surface that is placed between the two working surfaces. Subsequently, the two working surfaces, the inside and outside surface are united together to the convex body of globoidal cam by using the command 'Sew'. Cam basis body is obtained by revolution. Eventually, the 3D model of globoidal cam mechanisms with oscillating and dual stopping followers is completed after the boolean operation to unite the basis body and convex body shown in Fig. 4.

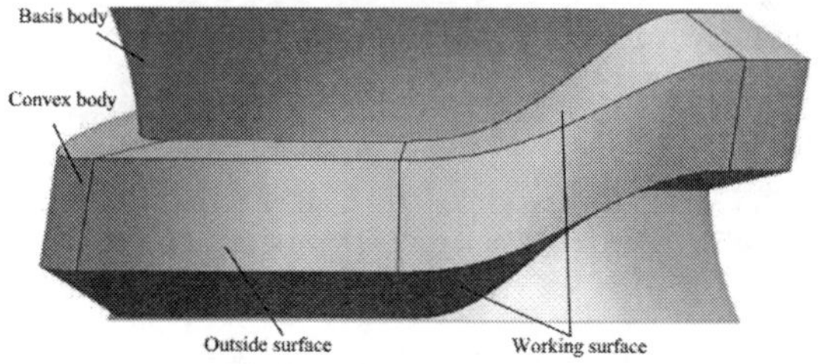

Figure 4. 3D model of oscillating globoidal cam

4 Dynamic simulation based on ADAMS

4.1 Construction of dynamic model

With MSC Adams software users can produce virtual prototypes to simulate complex mechanical systems and quickly analyze multiple design variations until an optimal design is achieved. MSC Adams is one of many computer programs for modeling and simulating multi body systems. To get the virtual prototype of cam mechanism, the 3D model of the oscillating turret with two rollers is established in the UG software and then assembled with the globoidal cam. The assembling file is saved as *.x_t parasolid file format and imported into ADAMS software.

Using ADAMS, firstly revolute joint (Two bodies-One Location, Normal To Grid) between the cam and the ground is built, and then revolute joint (Two bodies-One Location, Normal To Grid) between the oscillating turret and output axis is built, adding the rotary pair to the oscillating globoidal cam and the oscillating turret, so that it can rotate freely around the axis of rotation. And then adding the contact force between the cam and the oscillating turret. It is important and complex to create the contact force between the cam and the rollers.

For the software of ADAMS, the contact problem between two rigid bodies is looked as a nonlinear system with a damper and a spring. The contact force F_n can be calculated as follows [7]:

$$F_n = K\delta^e + C\dot{\delta} \quad (10)$$

where, δ and $\dot{\delta}$ are respectively the contact deformation and the deformation velocity, K is contact stiffness, C is damping coefficient, e is rigidity index. For the metal collision, e is equal to 1.5 generally. Contact stiffness K can be expressed by

$$K = \frac{4}{3} R^{\frac{1}{2}} E^{*} \tag{11}$$

$$\frac{1}{R} = \frac{1}{R_1} + \frac{1}{R_2} \tag{12}$$

where, R_1 and R_2 are the equivalent radius of contact point with the cam and roller. E^* can be calculated according to the following equation:

$$\frac{1}{E^{*}} = \frac{1-\mu_1^2}{E_1} + \frac{1-\mu_2^2}{E_2} \tag{13}$$

μ_1 and μ_2 are the poisson ratio with the material of the cam and the roller in this formula. E_1 and E_2 are the elastic modulus with the material of the cam and the roller.

ADAMS software uses the step function to define the damping coefficient. The formula is as follows [8]:

$$C = \text{step}[\delta, 0, 0, d, C_{max}] \tag{14}$$

where d is penetrating depth. When the contact stiffness, stiffness coefficient, damping coefficient, infiltration depth are inputted, the dynamic simulation model of globoidal cam mechanism with oscillating is completed, as is shown Fig. 5.

Figure 5. The dynamic simulation model of oscillating globoidal cam mechanism

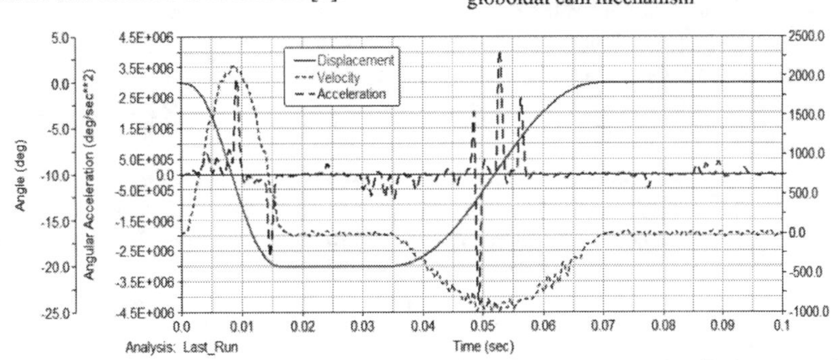

Figure 6. Output curve of angular displacement, angular velocity and angular acceleration of oscillating turret with cam rotation speed 600r/min

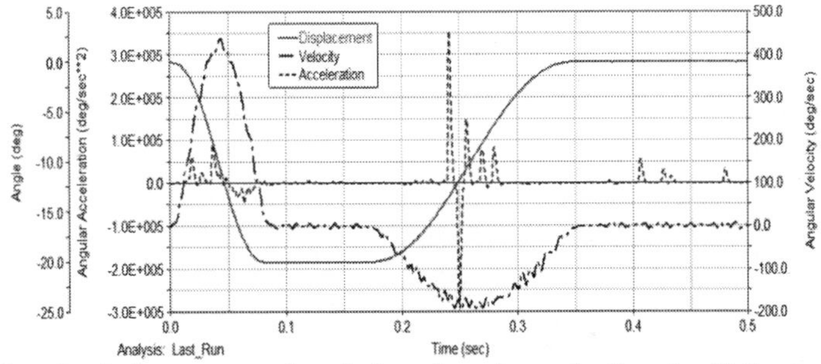

Figure 7. Output curve of angular displacement, angular velocity and angular acceleration of oscillating turret with cam rotation speed 120r/min

4.2 Dynamic simulation

The cam is made a rotary motion, the rotational speeds are 600r/min and 120r/min respectively, and a load with 500N·m is performed on the oscillating turret. And then the cam mechanism starts to carry on the dynamics simulation, and the simulation end time is 0.1s, the steps are 1000. During the process of the cam simulation, we can observe the real-time visualization, which can display high quality realistic picture was executed under the simulation and the result can be received as the same as real motioning. When the simulation is finished, the angular displacement, angular velocity and angular acceleration of oscillating turret with two kinds of cam rotational speeds are measured by using the post processing module of ADAMS. The results are shown in

Fig. 6 and Fig. 7. Fig. 6 shows the simulation results with cam rotation speed 600r/min, and Fig. 7 shows the simulation results with cam rotation speed 120r/min. From simulation results, the actual angular displacement curve of oscillating turret is smooth, and it is consistent with the theoretical output curve of modified sine cam cure, which illustrates that the dynamic model of globoidal cam mechanism is correctly built. The actual angular velocity curve of the oscillating turret has a slight oscillation and the actual angular acceleration curve has a big fluctuation. There are many reasons for the results. On the one hand, it is due to impact of mechanism stiffness and external load. On the other hand, the error caused by the transformation from UG model to ADAMS model. At the same time, if the cam speed is reduced the actual maximum acceleration will be fallen down by comparing the actual output curve of the two kinds of speed. Therefore, the input speed of globoidal cam mechanism is an important factor that affects the output performance of the mechanism.

5 Conclusion

The globoidal cam mechanism with oscillating and dual stopping followers is widely used in many types of automatic production line, but the introduction of comprehensive analysis is few for working surface. In this paper, the equation of the working surface is derived by using the method of offset surface, which greatly simplifies the process of solving and analyzing. The virtual prototype system framework of the cam mechanism is established. The three-dimension digital model of the cam is built with UG. Dynamic simulation analysis of the cam mechanism is studied by means of software ADAMS. By dynamic simulation, the data of cam follower's displacement, velocity and acceleration are obtained. It is can be conclude that cam rotation speed makes great effects on the performance to the cam mechanisms.

Acknowledgement

This research was financially supported by the Key Research Projects of Shandong Province in China (No. 2015GGX103005 and No. 2016GNC110025) and the National Natural Science Foundation of China (No. 5130 5164). In addition, valuable suggestions from the reviewers in improving the quality of the paper are highly appreciated.

References

1. R. L. Norton. *Cam design and manufacturing handbook,2nd Ed* (Industrial Press, New York, 2009)
2. V. Thinh Nguyen, D Joong Kim. Flexible cam profile synthesis method using smoothing spline curves, Mech Mach Theory, **42**, 825-838 (2007)
3. S Acharyya, T Kanti Naskar. Fractional polynomial mod traps for optimization of jerk and hertzian contact stress in cam surface, Computers & Structures, **86**, 322-329 (2008)
4. H Qiu, C Jun Lin, Z Ye Li, et al. A universal optimal approach to cam curve design and its applications, Mech Mach Theory, **40**, 669–692 (2005)
5. H S Yan, M C Tsai, M H Hsu. A variable speed method for improving motion characteristics of cam-follower systems, ASME J Mech Des, **118**, 250-257 (1996)
6. Y A Yao, H S Yan, C Zhang. Integration design of cam mechanism and servo control system, China Sci, **30**, 524-530 (2000)
7. H.M. Lankarani and P.E. Nikravesh. Continuous Contact Force Models for Impact Analysis in Multibody Systems, Nonlinear Dynamics, **5**, 193-207 (1994)
8. James B. McConville. *Introduction to Mechanical System Simulation Using Adams* (SDC Publications, 2015)

MATEC Web of Conferences **108**, 04006 (2017) DOI: 10.1051/matecconf/201710804006
ICMAA 2017

Initial Tool Orientation Set-up for 5-Axis Flank Milling Based on Faceted Models

Gandjar Kiswanto[1] , Ario Sunar Baskoro [1]and Eko Arif Syaefudin [2]

[1]*Dept. of Mechanical Engineering, University of Indonesia, Depok, Indonesia*
[2]*Dept. of Mechanical Engineering, Jakarta State University, East-Jakarta, Indonesia*

Abstract. One of the factors affecting the effectiveness of machining time of 5-axis miling is the method being used. By using flank milling method, as one of the optimized processes to make a workpiece, the time required for the process becomes shorter.This research is aimed at developing the method for determining the initial orientation of the tool for a sculptured surface on the basis of faceted model. By determining cc-point as the basis for positioning the tool on the surface of the workpiece, the cutting direction is formed from the nearest cc-point in the XY flat plane direction of the faceted model at the spatial coordinate. The positioning of the tool is initially based on the Local Coordinate System developed by the cross product between the normal vector *n*at each cc-point and cutting direction vector *F*from one cc-point to the other. The cross product resulted is a tangent vector *T*of the plane formed from the normal vector and cutting direction. The orientation of the tool is formed and defined by an inclination angle (α) and a screw angle (β). Maximizing the cutting volume and avoiding gouging at each cc-point during the flank milling are carried out through optimal adjustment of these two rotational angles. Furthermore, when the adjustment of rotational angles cannot resolve the gouging, appropriate tool lifting along the normal vector is conductedhis method is very much applicable for flank milling having the basis of data in the form of faceted models.

1 Introduction

Based on the current development of technology, the manufacturing of a product having a complex surface such as aircraft components, automotive parts, molds and dies, etc. an effective and efficient method is required. In order to respond such challenge, the use of 5-axis milling machining becomes one of the solutions mainly selected by manufacturers [1-3].

For processing of planar surfaces in 5-axis milling machine, normally the flank milling or so-called peripheral milling method is used. Although the material removal rate (MRR) for the machining process using this method is very huge compared to the other methods, until now this method still has many disadvantages. One of the disadvantages is its being uncapable of processing a workpiece having a *sculptured* surface [4-6].

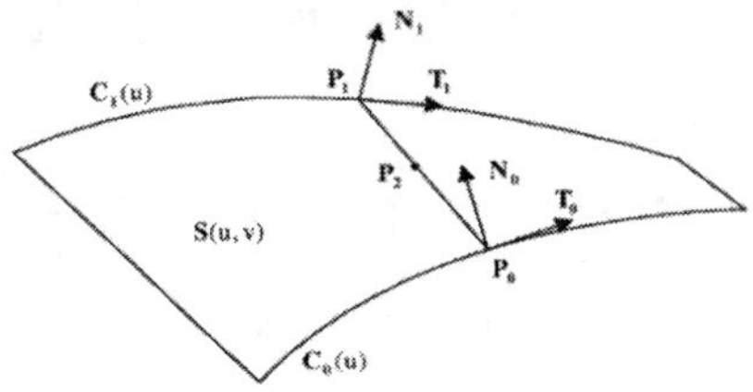

(a)

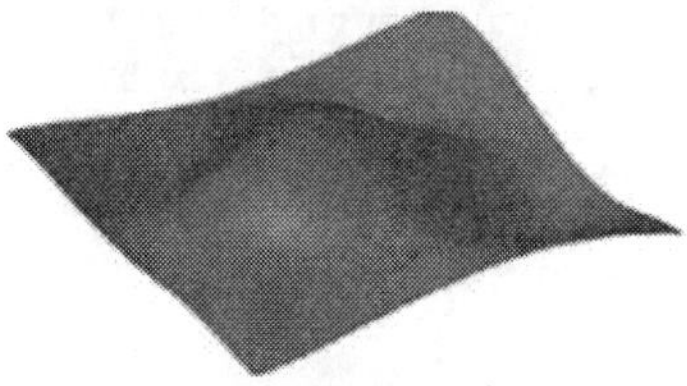

(b)

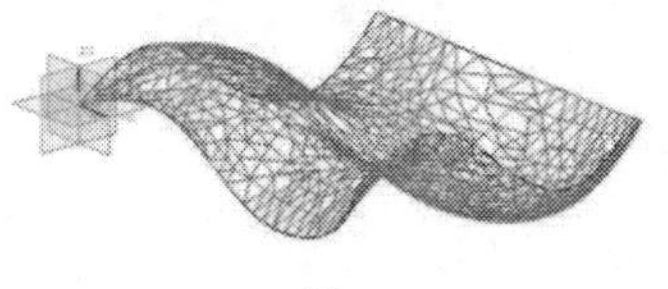

(c)

Figure 1. a. The development of tool track for flank milling [7], b. Sculptured surface, with many curve [7], c. STL of Sculptured Surface [8]

S Bedi and C Menzel [7] made a tool track for flank milling by controlling 2 constructing curves as shown in Figure 1. As for the other methods, that is SPO and DPO also control 2 reference pints to determine the position of the tool on the workpiece surface as described in Figure 2 below.

For a planar surface, *parametric* equation is commonly used for making the flank milling method by among others Chih-Hsing Chu and Li [3, 6].

In this study, the basis of data used is faceted model directly obtained from a CAD system with STL file format.The use of faceted model provides several advantages, which among others are easy to determine the normal vector at each triangle that willl later be used as the basis for calculating *initial tool orientation*at CC ccpoint for flank milling, where to calculation cc-point has be done on research before [8, 9]. To determine the normal vector at eact cc-point, calculation is made using the following equation and is illustrated in Figure 2 below

$$n = \frac{\sum_l s_l n_l}{\sum_l s_l} \tag{1}$$

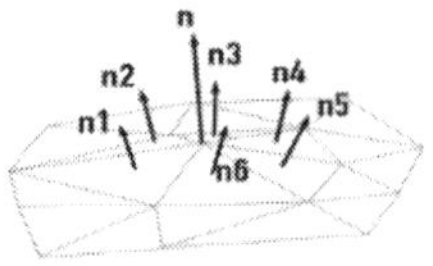

Figure 2. Calculating the cc-point normal vector (n), there is between the triangles

2 Methodology

In this study, determination of tool orientation for flank milling starts from calculation of the normal vector value at each cc-point. This can be described by the flow chart below.

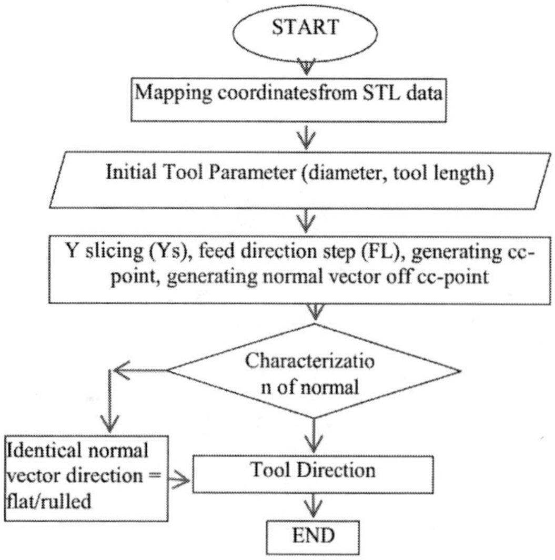

Figure 3. Flowchart of initial tool orientation set-up for flank milling.

To form a set of cc-points on a workpiece surface, the slicing plane method is applied in this study. The slice of the slicing plane (Ys) against each triable on the workpiece surface becomes the basis for the formation of this cc-point. Meanwhile, the position of cc-point at each triangle passed by the slicing plane will also be calculated based on the normal vector formed by each of the said triangles. This slicing planemethod will form am imaginary line, so that later the cutting direction in the machining process will follow this slicing plane. As described in Figure 4 below, the position of each cc-pointon the faceted model is formed from the intersection between the slicing plane and the faceted model and thus result the cc-point coordinates x, y and z using the following equation:

$$\frac{x-x1}{x2-x1} = \frac{y-y1}{y2-y1} = \frac{z-z1}{z2-z1} \tag{2}$$

$$x = \frac{y-y1}{y2-y1} x(x2 - x1) + x1 \tag{3}$$

$$z = \frac{y-y1}{y2-y1} x(z2 - z1) + z1 \tag{4}$$

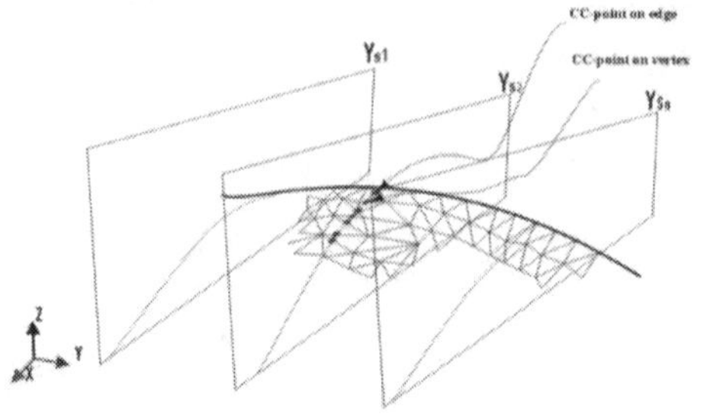

Figure 4. The development of cc-point using slicing plane method

This process can be computed by the *psudocode* below:

```
if_is_in_between(y, point_1, point_2)
    x = (y - point_1(2)) / (point_2(2) - point_1(2)) *
    (point_2(1) - point_1(1)) + point_1(1);
    z = (y - point_1(2)) / (point_2(2) - point_1(2)) *
    (point_2(3) - point_1(3)) + point_1(3);
    output = [x y z];
else
```

Meanwhile, to form a normal vector at each cc-point, based on the result of slicing plane method in the previous steps the *psudocode* used is as follows:

```
function normal = build_normal_on_edge(vertex_idx_1,
vertex_idx_2, triangles)
    %% find all triangles having vertex_idx_1 and
    vertex_idx_2 as its two of its vertices
      [row1, col] = find(triangles(:,1:3) == vertex_idx_1);
      [row2, col] = find(triangles(row1,1:3) ==
    vertex_idx_2);
    row = row1(row2);
    neighbor_triangle_normal_vectors = triangles(row, 4:6);
      %% vector sum
    if size(neighbor_triangle_normal_vectors,1) > 1
    normal = sum(neighbor_triangle_normal_vectors);
    else
    normal = neighbor_triangle_normal_vectors;
    end
    end
```

3 The determination of initial tool orientation for flank milling

Based on the previous explanation that the cc-point was obtained from the result of slicing at the triangles on the faceted model by the cutting plane (Ys) with the pre-determined step-over (in Figure 4), the cc-point coordinates can be obtained. Then the normal vector at the cc-point can be calculated based on the average value of the vectors surrounding it and are indexed like what has been performed by Lauwers, Kiswanto, et al and illustrated in Figure 2. Each of the cc-points produces a normal vector and also feed direction vector as described in Figure 5, where the feed direction vector is a vector pointing from the current cc-point to the next cc-point.Then, from each cc-point a Local Coordinate System can be formed with the axes being normal vectors $\vec{N}$, feed direction vector $\vec{F}$ and cross product between the normal vector and feed direction vector $\vec{T} = \vec{N}\ x\ \vec{F}$) [10, 11], as depicted in Figure 6.

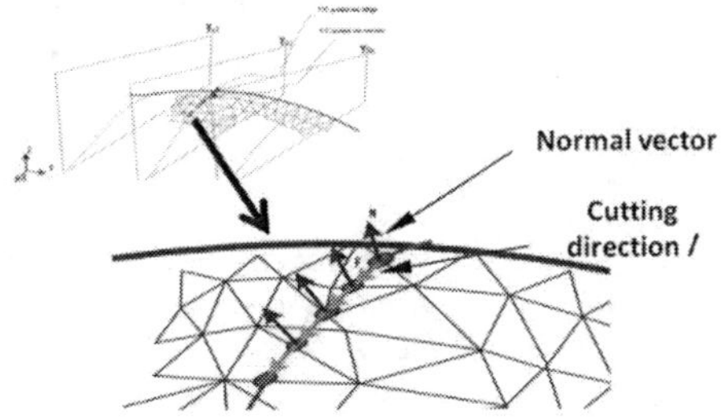

Figure 5. Determining *normal vector* (*N*) and *feed n direction vector* (*F*)

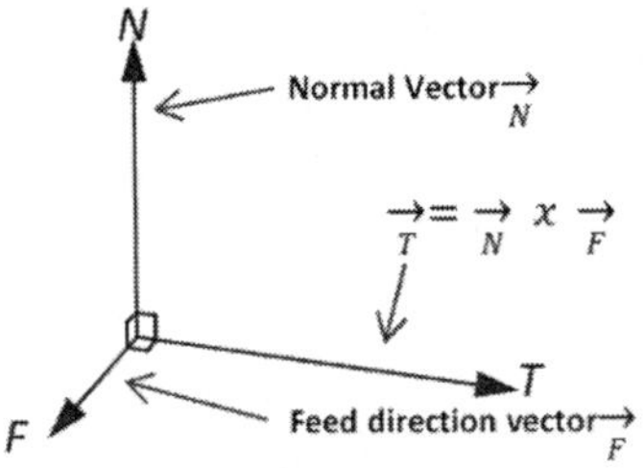

Figure 6. Local tool coordinate references

The tool orientation at this flank milling based on faceted model is formed by 2 angles, namely inclination angle (α) and screw angle (β). Inclination angle (α) is an angle between tool axis vector and the normal vector ($\vec{N}$), whilethe screw angle (β) is the angle tool axis vector and plane formed by Feed direction vector ($\vec{F}$) and normal vector ($\vec{N}$). Such angles (α and β) are the tool orientation angles during the flank machining process. In order to avoid gouging, *adjustments* conducted to both angles, so that the gouging can be eliminated. Description about the rotation angles can be seen in Figure 7. In the condition of initial position, the tool is placed in parallel with the Axis T, in this case α has a value of 90°.

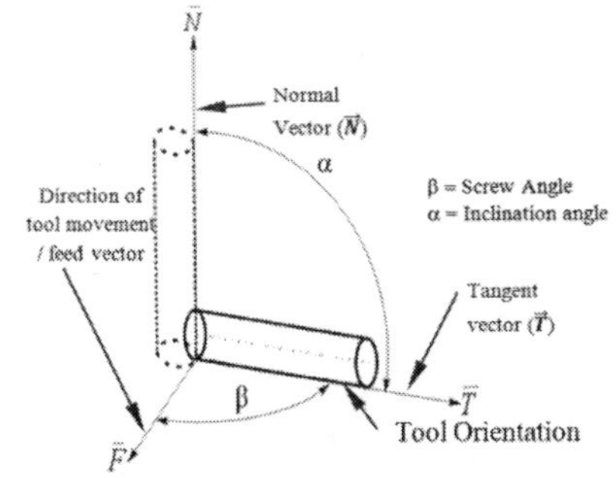

Figure 7. Tool orientation at the local coordinate system

Furthermore, in order to reduce the complexity in avoiding gouging during the flank machining operation, (β) angle is adjusted to be always 90°, so that the tool orientation is mostly performed by the inclination angle (α) as illustrated in Figure 8.

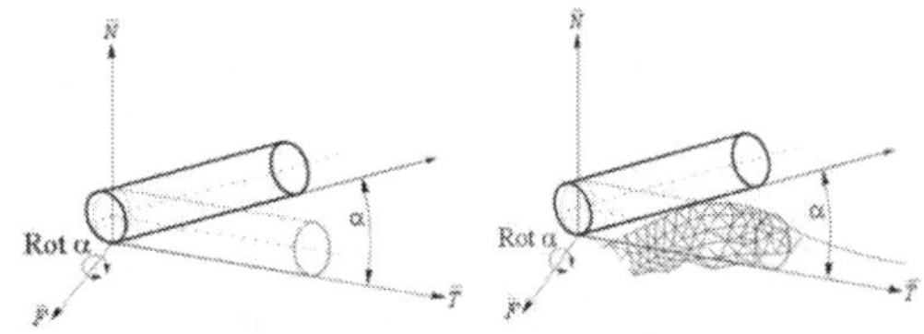

Figure 8. Tool inclination (α) to avoid *gouging*.

However, when the adjustment of tool orientation by inclination angle cannot resolve the gouging, an appropriate tool lifting along the normal vector must be conducted, as shown in Figure 9.

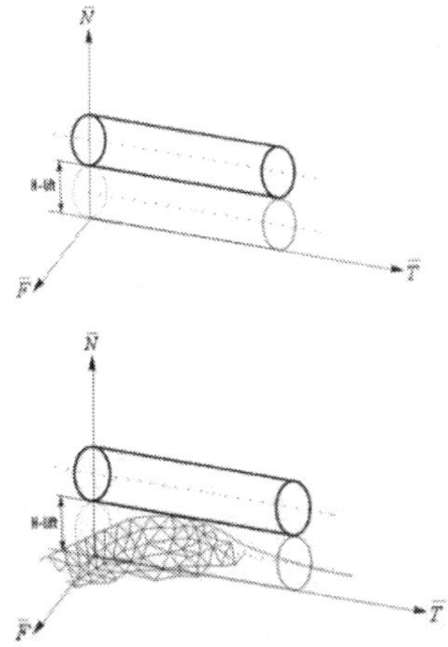

Figure 9. *Tool Lifting* to avoid *gouging*.

Figure 10 illustrates *tool* when avoiding *gouging* by inclining and lifting simultaneously.

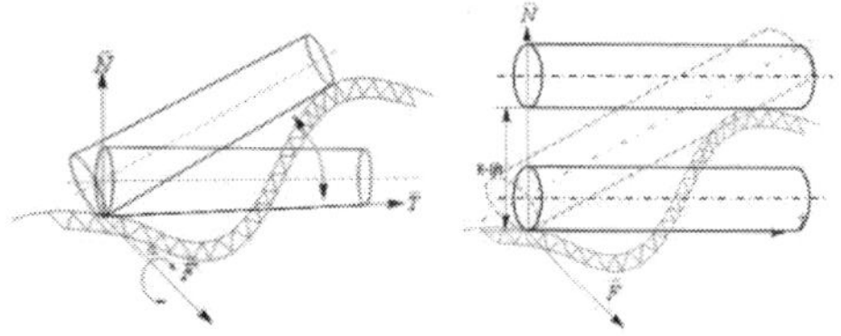

Figure 10. Tool conductinginclinationand lifting

Illustration of the result of the *initialtool orientaion methode flank milling* before *gouging avoidance* in this paper has been successfully and well applied in programming algorithm as described in Figure 11.

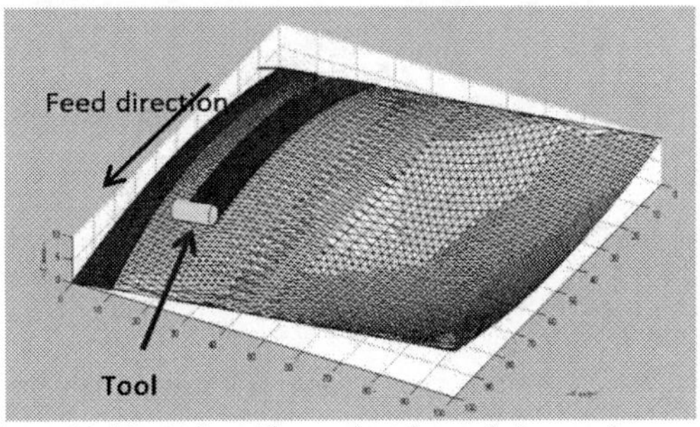

Figure 11. The tool position against the workpiece surface using flank mlling method.

4 Conclusions

By the use of the faceted model as the basis for the method for determining the tool direction for this flank milling, the determination of the normal vector and the tool position can become very easy to use and very much applicable.The tool orientation direction is a result of cross product or tangential value of 2 vectors, that is normal vector and vector from the cutting direction. Meanwhile, the tool position against Cc-point is the outer side of the contact between the tool and workpiece surface. Thus, the center position of the tool is equal to the Cc-point plus a half of the tool diamter against Z axis.

References

1. Li, L.L.B.C.F.L.C., *Complexity analysis and calculation for sculptured surface in multi-axis CNC machining based on surface subdivision.* Int J AdvManufTechnol, (2014). **71:** 1433-1444 DOI 10.1007(s00170-013-5544-2).
2. Chaves-Jacob, J., G. Poulachon, and E. Duc, *New approach to 5-axis flank milling of free-form surfaces: Computation of adapted tool shape.*Computer-Aided Design, **41**(12): p. 918-929 (2009)
3. Zhu, L., H. Ding, and Y. Xiong, *Simultaneous optimization of tool path and shape for five-axis flank milling.* Computer-Aided Design, **44**(12): p. 1229-1234 (2012)
4. Bedi S, M.S., Menzel C. *Flank milling with flat end milling cutters.* (2003). Computer-Aided Design.
5. Kruth, G.K.B.L.J.-P., *Gouging elimination through tool lifting in tool path generation for five-axis milling based on faceted models.* Int J Adv Manuf Technol (2007). DOI 10.1007/s00170-005-0338-9: p. 16.
6. Chih-Hsing Chu, H.-T.H., *Generation of reciprocating tool motion in 5-axis flank milling based on particle swarm optimization.* International Journal of Machine Tools and Manufacture © Springer Science+Business Media, LLC 2010, **23**: p. 1501 (2012)
7. Sanjeev Bedi, S.M., Cornelia Menzel, *Flank milling with flat end milling cutters.* Computer-Aided Design, Vol. **35**, (2005): p. 293-300
8. Kruth, G.K.B.L.J.-P., *Gouging elimination through tool lifting in tool path generation for five-axis milling based on faceted models.* Int J Adv Manuf Technol - DOI 10.1007/s00170-005-0338-9, (2007): p. 293–309
9. Lauwers, B.K., G. Kruth, J. P., *Development of a Five-axis Milling Tool Path Generation Algorithm based on Faceted Models.* CIRP Annals - Manufacturing Technology, Vol. **1**, (2003): p. 85-88
10. Zill, D.G. and M.R. Cullen, *"Definition 7.4: Cross product of two vectors".* Advanced engineering mathematics (3rd ed.), Jones & Bartlett Learning. p. 324. ISBN 0-7637-4591-X., 2006.
11. Zich, C.A.G.a.R.E., *Cross product in N Dimensions – the double wedge product.* Politecnico di Milano, Energy Department, via La Masa 34, 20156 Milan, MI, Italy, (August 26, 2014)

DOI: 10.1051/matecconf/201710804007

The spray characteristic of gas-liquid coaxial swirl injector by experiment

Chen Chen[1,2], Yan Zhihui[2], Yang Yang[2], Gao Hongli[1], Yang Shunhua[2] and Zhang Lei[2]

[1]*School of Mechanical Engineering, Southwest Jiaotong University, 610031Chengdu, China*
[2]*Air-breathing Hypersonic Technology Research Center, China Aerodynamics Research and Development Center, 621000 Mianyang, China*

Abstract. Using the laser phase Doppler particle analyzer (PDPA), the spray characteristics of gas-liquid coaxial swirl injector were studied. The Sauter mean diameter（SMD）, axial velocity and size data rate were measured under different gas injecting pressure drop and liquid injecting pressure drop. Comparing to a single liquid injection, SMD with gas presence is obviously improved. So the gas presence has a significant effect on the atomization of the swirl injector. What's more, the atomization effect of gas-liquid is enhanced with the increasing of the gas pressure drop. Under the constant gas pressure drop, the injector has an optimal liquid pressure drop under which the atomization performance is best.

1 Introduction

As the combustion performance and the uniformity of the flow field are closely related to the structure and working parameters of the injector, injector is a key part of the combustion chamber. By studying the breakup length and particle size distribution, it could provide basis for the ignition and combustion optimization, high or low frequency instability analysis and combustion performance prediction [1]. So it is necessary to conduct a comprehensive and systematic study.

At present, the gas-liquid coaxial swirl injector has been widely used in liquid rocket engine due to its excellent atomization and mixing performance. Wu et al. [2] found that the injection pressure drop change of will lead to different flow and mixing ratio distribution. There is an optimal value, at which the atomization performance is best [3-8]. The injection pressure drop of propellants has a big effect on the droplet diameter and size distribution [9-12]. Li et al. [13] contrasted atomization characteristics under cold and hot conditions. Some results have obtained as mentioned above, which systematically investigated the influence of structure parameters and working parameters on atomization characteristics. But most of them tried to describe the distribution characteristics by the average value of each measurement plane or points in a plane. So they lack of a detailed study of the spatial distribution of the whole flow field.

In order to determine whether the atomization performance of the injector can meet the requirement of application, it is necessary to carry out some experiments. Air and water were chosen as the propellant. Under different gas injection pressure drop and liquid injection pressure drop, the spray flow field was measured by PDPA. It analyzed the spatial distribution of gas-liquid coaxial injector in detail. And typical measuring plane were picked to investigate the effect of gas injection pressure drop and liquid injection pressure drop on the atomization characteristics, which is helpful to the gas-liquid coaxial injector fine design, ignition location selection, spray flow field calculation and analysis.

2 Experimental methods

The test gas-liquid coaxial swirl injector model is shown in Fig.1. Water and air were chosen as the simulation mediums. Water flows into the injector through the tangential hole. And the gas flows from the loop into the outer injector. It exists a recessing value L between the inner and outer injector.

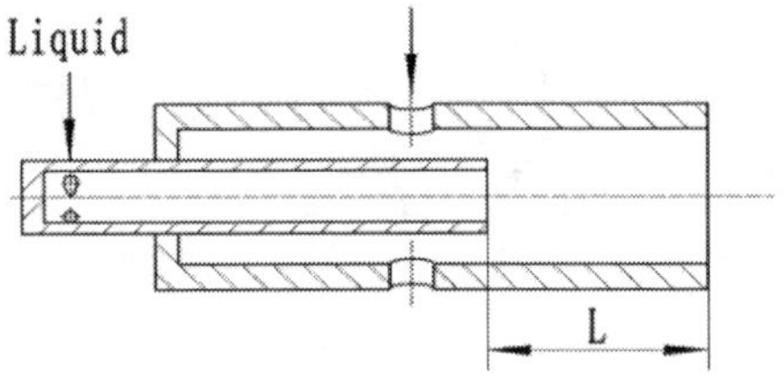

Figure 1. Schematic of gas-liquid coaxial swirl injector.

Figure 2 is the schematic diagram of the test system. The laser phase Doppler particle analyzer (PDPA) was used to measure the droplet diameter and axial velocity. The measurement range of speed is -150~1000 m/s, whose accuracy is 0.1%. And the measurement range of droplet diameter is 0.5-4000μm, whose accuracy is 0.5%.

DOI: 10.1051/matecconf/201710804007

The exit of outer injector is treated as the starting point. The measurement plane is XY plane while the Z direction is along the injection direction. Optical measurement conducted at Z = 15, 50, 100, 150, 200 and 300 mm in turn as shown in Fig. 3. In order to guarantee the accuracy of the test results, the number of collected droplets at each measured point is more than 5000 [14-15].

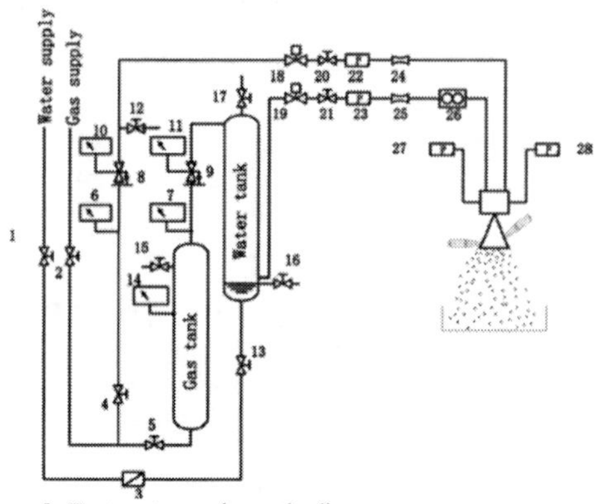

Figure 2. Test system schematic diagram.

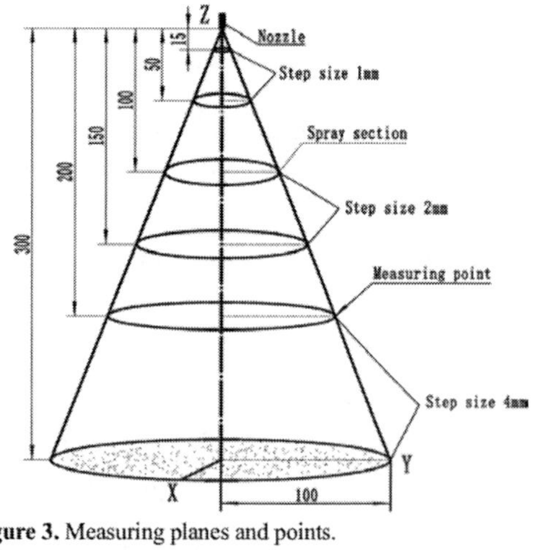

Figure 3. Measuring planes and points.

3 Results and discussion

3.1 The spatial distribution of a swirl injector

In order to better understand the distribution characteristics of gas-liquid coaxial swirl injector, it is essential to research the spatial distribution of inner injector. Figure 4 is the spatial distribution of diameter, axial velocity and data rate (particles number collected at the measurement points per second) under 0.8MPa. In literature, the use of liquid water content is usually adopted to reflect the density of the flow field. However, the liquid water content is calculated by particle number and particle size. And the particle number at the measurement point is related to the breakup and coalescences. So data rate is used to characterize the density at the measurement point in this paper.

After a series of test, we found that: (1) On the whole, the droplet size was larger while the axial velocity and the data rate are very low. So the atomization effect is poor. (2) With the increase of the axial distance, the amplitude of diameter variation in the curve plane increase. And axial velocity and data rate curve tend to smooth. Under the continuous interaction of ambient gas, the droplet keeps breaking. So the droplet diameter and velocity decreased. As the spray spread out, the data rate reduced. (3) The droplet size shows a rising trend along the radial direction. The formation of this phenomenon can be explained as follows:

The inertia force of large droplets is greater. So they were thrown to the spray periphery. It increased the coalescence probability, which leads to the increase of droplet diameter along the radial direction.

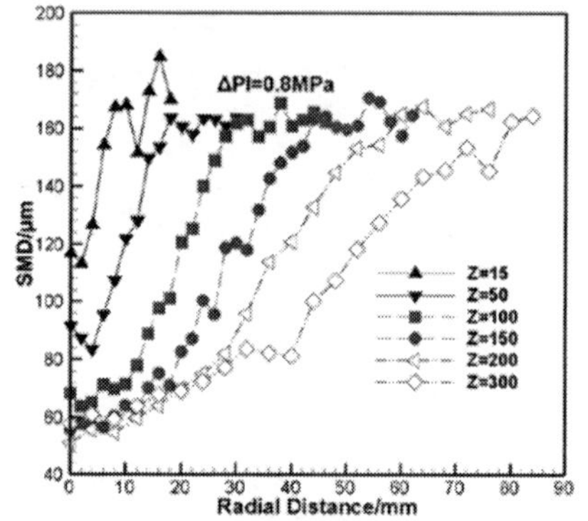

(a) SMD

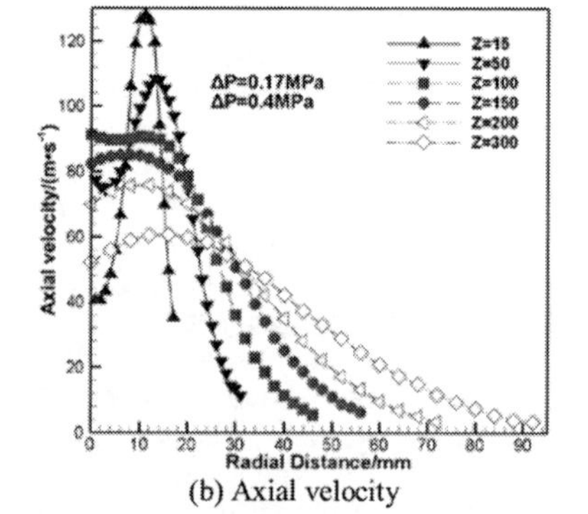

(b) Axial velocity

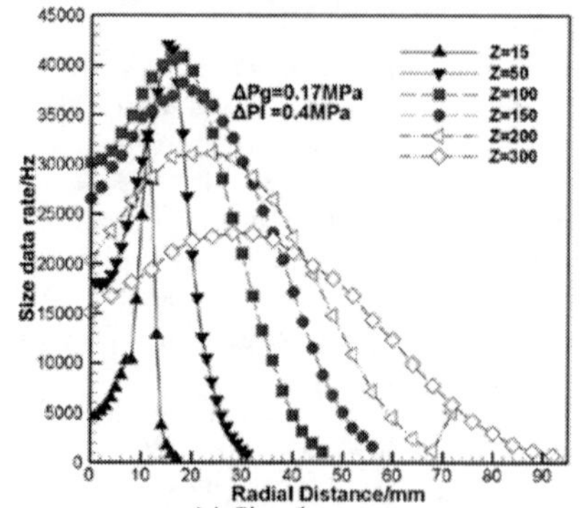

(c) Size data rate

Figure 4. Spatial distribution of the swirl injector.

DOI: 10.1051/matecconf/201710804007

3.2 The spatial distribution of a gas-liquid coaxial swirl injector

The spatial distribution of gas-liquid coaxial swirl injector is shown in Fig.5 when the gas and liquid injection pressure drop were 0.17MPa and 0.4 MPa respectively. With the increase of axial distance, the characteristic curves becomes more and more gently. At Z = 15 mm plane, the variation amplitude of SMD and axial velocity are 33.4 and 92.7 respectively. And at Z = 300 mm plane, the variation amplitude of SMD and axial velocity are 15.5 and 57.4 respectively. So the amplitude of SMD and axial velocity reduced by half. So the distribution is more uniform.

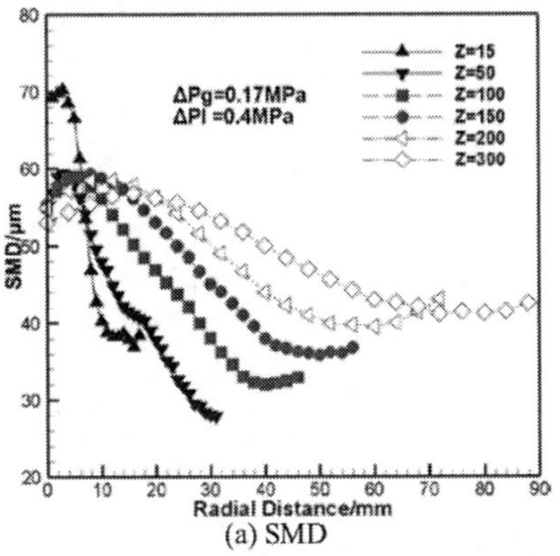

(a) SMD

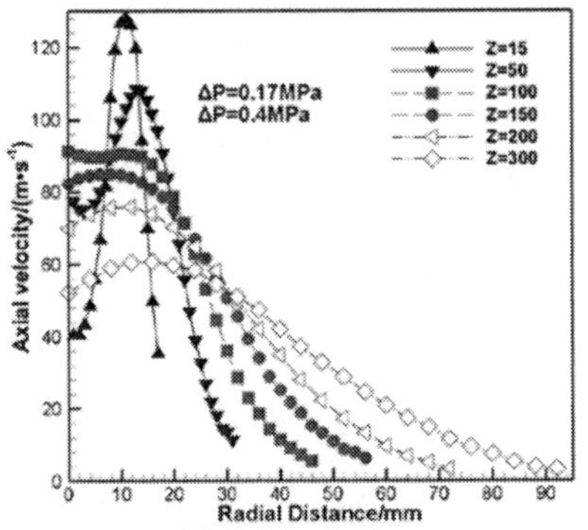

(b) Axial velocity

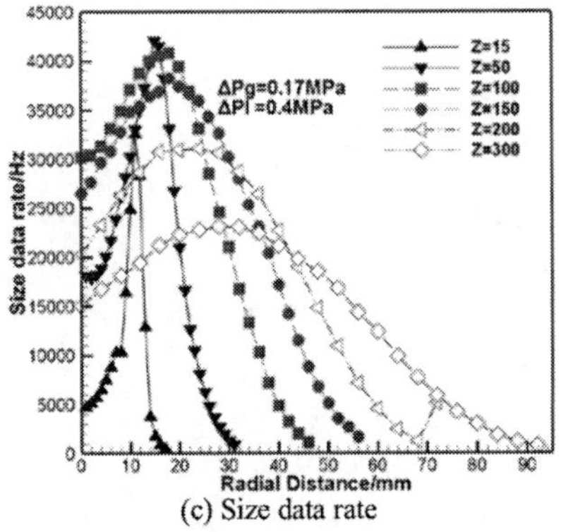

(c) Size data rate

Figure 5. Spatial distribution of gas-liquid coaxial swirl injector.

3.3 The effect of gas injection pressure drop

In order to better understand the influence of gas pressure drop, it is better to study the effect of gas existence to the atomization quality firstly. In order to avoid the randomness of single working condition, two groups of tests were conducted. And SMD is adopted to evaluate the atomization quality. As shown in Fig.6a and 6b, atomization quality was very poor when only the liquid injected. Most of the droplets are more than 100μm, and the peak is up to 170μm. The diameter variation amplitude in a plane is more than 100μm. And the maximum speed is less than 10m/s. When the gas injected, all the droplet diameter is less than 100μm with maximum amplitude of 50μm. And the minimum velocity is 20 m/s. The formation of this phenomenon can be explained as follows: The pneumatic nebulization and pressure atomization worked at the same time. In Fig. 6a and 6b, the radial measurement range obviously decreased. So it could be assumed that the spray was suppressed by the gas, which is consistent with the experiment observations.

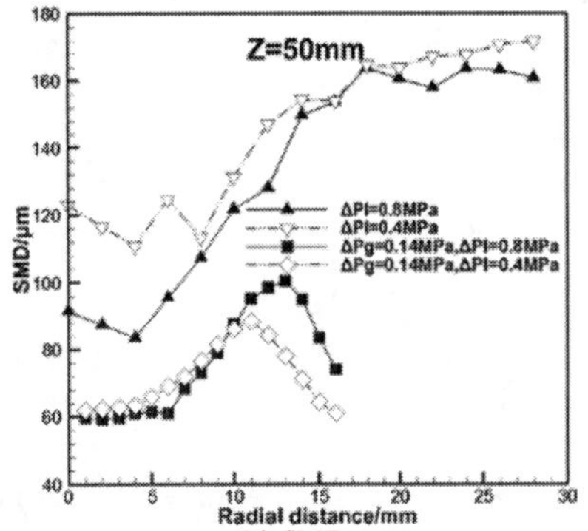

(a) SMD

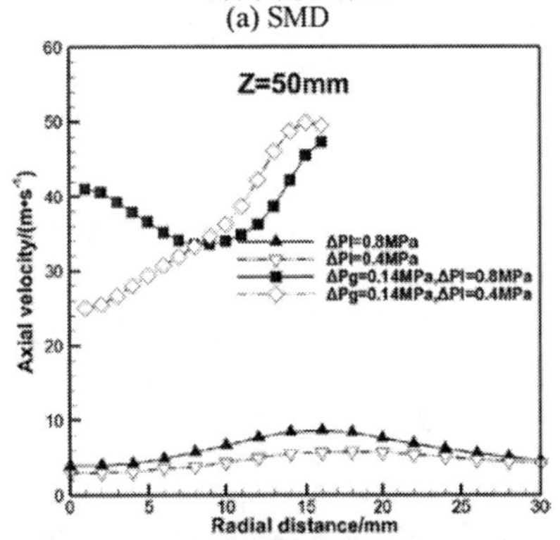

(b) Axial velocity

Figure 6. Gas presence on SMD and axial velocity.

As shown in Fig7a and 7b, the SMD decreased while the axial velocity increased at the same position with the increase of gas injection pressure drop. And the shape of the SMD and axial velocity curve are basically identical. When the gas injection pressure drop rises, the gas-liquid ratio increases. What's more, the relative velocity between gas and liquid increased. So it would increase the aerodynamic force. As a result, the droplet is more easily to break. So with the increase of injection pressure drop, the atomization effect strengthened.

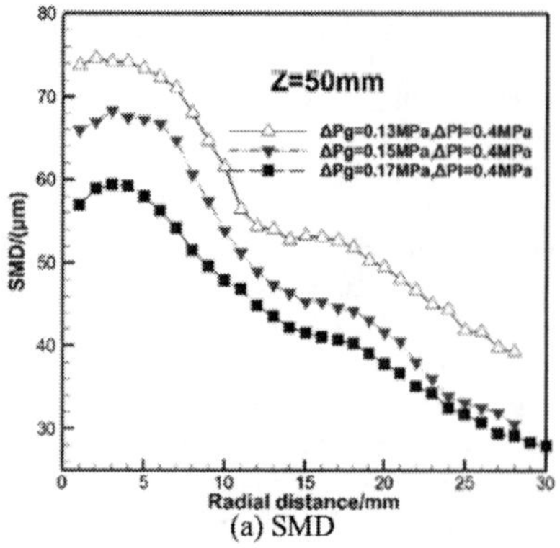

(a) SMD

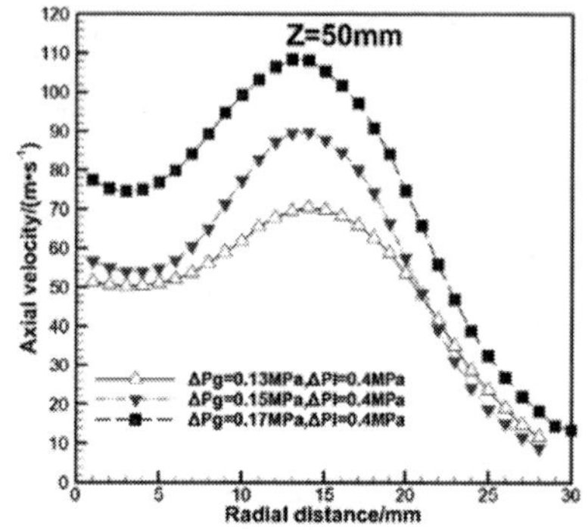

(b) Axial velocity

Figure 7. Spatial profiles of SMD and axial velocity under different gas pressure drop.

3.4 The effect of liquid injection pressure drop

As the liquid injection pressure drop rises, it would produce some effects. First of all, it would enhance the atomization effect of internal swirl injector. However, the relative velocity between air and water in the recess region decrease, which increase the initial droplet diameter. What's more, it lowers gas-liquid ratio. As a result, pneumatic atomization becomes poor as the liquid injection pressure drop increases. As shown in Fig. 8a and 8b, SMD showed a decreased trend with the increase of radial distance. On the whole, the droplet diameter and velocity value under 0.37 MPa are less than those under 0.22 MPa. The formation of this phenomenon can be explained as follows: Though the pressure atomization and pneumatic nebulization effect at the same time, the pressure atomization dominated. The higher is the liquid injection pressure drop, the smaller is the droplet diameter.

The curve shape of droplet diameter under 0.37 and 0.6 MPa is similar in Fig. 8a. But the droplet diameter under 0.6 MPa is bigger. The formation of this phenomenon can be explained as follows: On the one hand, excessive liquid injection pressure drop would lead to worse pneumatic atomization. On the other hand, with the increase of liquid injection pressure drop, the pressure atomization turns better. However, the improvement of atomization effect is not significant when the liquid injection pressure drop increases to a certain extent. It could be assumed that the pressure atomization under 0.6 MPa and 0.37 MPa is similar while the pneumatic atomization of 0.6MPa is far worse than 0.37MPa. So the droplet diameters under 0.6MPa are bigger.

Fig. 8c shows the droplet distribution of z=150 mm, at which the spray is fully developed. It is obviously that the atomization effect under the injection pressure drop of 0.37MPa is better than those under 0.22 MPa and 0.6 MPa. So it exist an optimal liquid injection pressure drop within the scope of 0.37 MPa and 0.6 MPa, at which the atomization effect is best with a minimum SMD. Further study would be conducted to investigate the effect of liquid injection pressure drop on the atomization model.

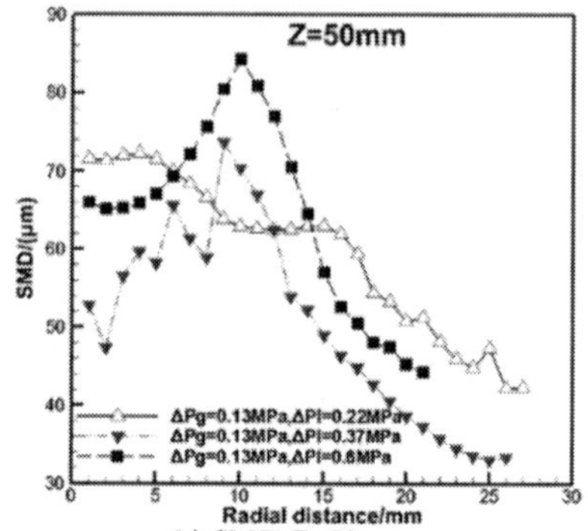

(a) SMD-Z=50mm

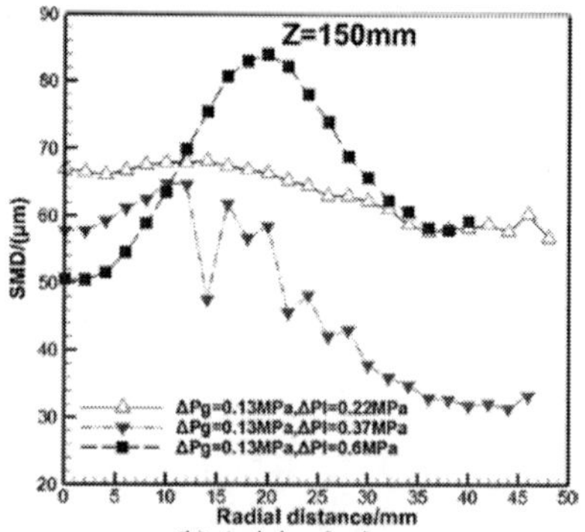

(b) Axial velocity

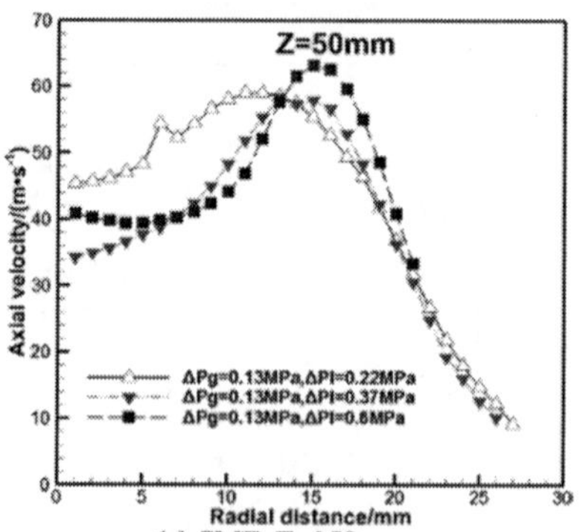

(c) SMD-Z=150mm

Figure 8. Spatial profiles of SMD and axial velocity under different liquid pressure drop.

4 Conclusion

The unique spatial distribution of a gas-liquid coaxial swirl injector was investigated. On the study of centrifugal injector has Further research of the external flow field would be carried out with the aid of PIV.

(1) Comparing to the single liquid injection, the droplet diameter rapidly decreased while the axial velocity greatly increased after the gas was injected. The existence of the gas has obvious optimization effect to spray atomization. What's more, it suppressed the expansion of the spray.

(2) At a fixed liquid injection pressure drop, injector, atomization quality was optimized as the gas injection pressure drop increase.

(3) An a fixed gas injection pressure drop, it exist a optimum liquid injection pressure drop, in which the atomization is best.

References

1. Zhang Mengzheng. The research and development of the atomizer in Liquid Rocket Engine. *Journal of Rocket Propulsion*, 1997, 6（3）: 17-32
2. Zhou Jin, Hu Xiaoping, Huang Yuhui, et al. A study of the Mixing Characteristics of Gas-Liquid Coaxial Swirling Injector of Liquid Rocket Engine. *Journal of National University of Defense Technology*, 1997, 19(4):9-13
3. J.H Seol, P.G Han, W.H Jeong, et al. Recess Effects on Spray Characteristics of Swirl Coaxial Injectors, *International Journal of Aeronautical and Space Sciences*, 2003,4(1):26-33
4. Sasaki M, Sakamoto H, Takahashi M, et al. Comparative Study of Recessed and Non-Recessed Swirl Coaxial Injector. *AIAA*. 97-2907
5. Tharakan T J, Gupta N K, Gupta N K. Effect of recess on the spray characteristics of gas-liquid coaxial swirl injectors. *International Conference on Aerospace Science and Technology*, Bangalore, India, 2008
6. Tian Zhangfu, Wu Jiping, Tao Yujing, et al. Experimental study on spray characteristic of gas-liquid coaxial injectors. *Journal of National University of Defense Technology*, 2006, 28(4):10-13
7. Yang Lijun, Ge Minghe, Zhang Xiangyang, et al. Effects of spout length on spray characteristics of swirl injector. *Journal of Propulsion Technology*, 2005, 26(3): 209-214
8. Yang Lijun, Ge Minghe, Zhang Xiang-yang, et al. Spray characteristics of recessed gas-liquid coaxial swirl injector. *Journal of Propulsion Technology*, 2006, 27(4):289-293
9. Zhang Mengzheng, Li Ao, Li Jinxian, et al. Experimental study on flow rate and spray characteristic of coaxial centrifulgal injector. *Journal of Propulsion Technology*, 2004, 25(1):19-22
10. Wang Zhenguo, Zhou Jin, Yang Xiaoqing, et al. Experimental investation on spray characteristics of coaxial swirling injector. *Journal of Propulsion Technology*, 1996, 18(1):7-11
11. Wang Zhenguo, Wu Jinxiang, Yang Xiaoqing, et al. Numerical simulation of spray flow process in coaxial swirling injector. *Journal of Propulsion Technology*, 1996, 17(3):43-49
12. Hu Xiaoping, Zhou Jin, Wang Zhenguo, et al. Experimental studies on atomization and flux distributions of gas-liquid coaxial injectors. *32nd Joint Propulsion Conference and Exhibit*, Florida: AIAA, 1996:96-3023
13. Li Jinxian, He Haobo, Yue Chunguo, et al. Experimental study on gas-liquid coaxial swirling injector of atomization and combustion under normal pressure. *Journal of astronautic*, 2008, 29(5): 1563-1569
14. Yuan Jiangtao, Yang Li, Zhang Jian, et al. Experimental study on the effect of sample size on measurement accuracy of water droplet size. *Chinese Journal of Scientific Instrument*, 2009, 30(10): 2112-2118
15. Azzopardi B J. Drops in annular two phase flow. *International Journal of Multiphase Flow*, 1997, 23(23):1-53.

Effect of contact angle and helix angle on slide-roll ratio under the uniform motion state of BSM

Kong Deshun, Wang Min and Gao Xiangsheng

Key Laboratory of Advanced Manufacturing Technology, Beijing University of Technology, Beijing 100124, China

Abstract. To study the effect of contact and helix angles on slide-roll ratio at the ball contact point under the uniform motion state of ball screw mechanism (BSM), a slide- roll ratio model was derived in the literature [9] were used in this paper. It also conducts a simulation analysis of the slide-roll ratio relationship between the uniform motion at the ball center and ball contact point under different contact angle and helix angle. As shown by the analysis, with the increase in the BSM's contact angle, the slide/roll ratio at the contact point decreased, and the contact angle had a relatively significant effect on the slide/roll ratio; however, with the decrease in the BSM's helix angle, the slide/roll ratio at the contact point decreased, and helix angle had a relatively insignificant effect on the slide/roll ratio.

1 Introduction

As a key component of a numerical control machine tool, the performance of BSM has great impacts on the processing quality of the workplaces. During the numerous parameters of the BSM, it has been found that the running status between the balls and raceway has more significance influence on the performance of the BSM. The running status is restricted to the friction torque between the nut and the screw, contact angle, helix angle between the ball and raceway, and the preload, etc. Therefore, it is very important to affect the wear and precision retaining ability of the BSM. Wei [1-3] introduced the theory of differential geometry of curve to analyze the motion rules of a ball considering the effect of the elastic deformation of contact angle and found that the self-spin and rotational angular velocities of the ball have an approximately linear relationship with the rotational velocity of the screw. Lin et al. [4-5] confirmed the existence of both the rolling and sliding motions relative to each other between the ball and raceway during the operation of BSM and proposed a calculation model to estimate the transmission efficiency. Gnanamoorthy and Govindarajan [6-7] identified the slide/roll ratio of the contact point as the key factor influencing the wear velocity of the contact area. Ma [8] affirmed the influencing rules of the contact and helix angles of the BSM on the slide-roll ratio and found that both contact angle and helix angles had an opposite trend of the slide-roll ratio and an increase in helix angle significantly decreased the slide/roll ratio of the nut contact point. Li [9] established the model of the dynamic contact angle of BSM combined with the geometric parameters of the ball and the raceway between nut and screw. He verified that the inner contact angles in screw raceway are always greater than the external angles in nut raceway under high velocity working conditions. Hu [10] found five different motion states between the ball, the screw of raceways, and the nut of raceways by analysis and also evaluated the effects of the motion velocity of the ball on the slide-roll ratio. In the above cases, the motion rules of the ball were explored when the BSM is under a uniform motion and quasistatic state. However, there is no quantitative analysis about contact angle and helix angle of BSM. The motion characteristics of the ball during the uniform motion of the BSM decreased the wear of the BSM during this process and increased its precision and service life.

In this study, the influence of contact angle and helix angle on the slide-roll ratio is analyzed on the basis of Hu studied. It was found that an increase in contact angle can decrease the slide/roll ratio of the contact point; in contrast, a decrease in helix angle can decrease the slide/roll ratio of the contact point.

2 Contact point position of ball

The contact angle of the ball screw determines the relative positional relationship between the ball and the screw as well as the nut can be showed in Fig. 1. The position vectors of contact point A between the ball and the screw of raceway and contact point B between the ball and the nut of raceway in the Frenet–Serret coordinate system $O_{\mathrm{H}}tnb$ is as follows:

$$\boldsymbol{P}_{\mathrm{A}}^{\mathrm{H}} = \begin{bmatrix} 0 \\ r_{\mathrm{b}}\cos\beta_{\mathrm{A}} \\ r_{\mathrm{b}}\sin\beta_{\mathrm{A}} \\ 1 \end{bmatrix} \tag{1}$$

$$\boldsymbol{P}_{\mathrm{B}}^{\mathrm{H}}=\begin{bmatrix}0\\-r_{\mathrm{b}}\cos\beta_{\mathrm{B}}\\-r_{\mathrm{b}}\sin\beta_{\mathrm{B}}\\1\end{bmatrix} \tag{2}$$

where β_{A} is contact angle between the ball and the screw of raceway, β_{B} is contact angle between the ball and the nut of raceway, and r_{b} is the radius of the ball.

Taking the right-hand screw as an example for the convenience of studying the BSM, the nut moves along the negative direction of Z_{W} axis when the screw rotates around the positive direction of Z_{W} axis, and contact angles β_{A} and β_{B} are positive.

3 Analysis of slide-roll ratio of BSM

There were probably have slide-roll ratio at the contact point A between the ball and the screw of raceway and contact point B between the ball and the nut of raceway, it can be obtained as [10]. The slide-roll ratio at contact point A and B can be written as:

$$\begin{aligned}S_A=abs\left(g\dot{\Omega}\left(\frac{r}{\cos\alpha}-r_b\cos\beta_0\cos\alpha\right)\right)/\left(abs\left(g\dot{\Omega}\right.\right.\\ \left(\frac{r}{\cos\alpha}-r_b\cos\beta_0\cos\alpha\right)+\dot{\Omega}\cos\alpha\left(r-r_b\cos\beta_0\right)\Big)\\ \left.+abs\left(\dot{\Omega}\cos\alpha\left(r-r_b\cos\beta_0\right)\right)\right)\end{aligned} \tag{3}$$

$$\begin{aligned}S_B=abs\left(\left(2v_{bt}-g\dot{\Omega}\left(\frac{r}{\cos\alpha}-r_b\cos\beta_0\cos\alpha\right)\right.\right.\\ \left.\left.+\dot{\Omega}\left(\frac{r}{\cos\alpha}+r_b\cos\beta_B\cos\alpha\right)\right)\right)/\Big(\mathrm{abs}\Big(\\ 2v_{bt}-g\dot{\Omega}\left(\frac{r}{\cos\alpha}-r_b\cos\beta_0\cos\alpha\right)\Big)+\\ abs\left(-\dot{\Omega}\left(\frac{r}{\cos\alpha}+r_b\cos\beta_B\cos\alpha\right)\right)\Big)\end{aligned} \tag{4}$$

In the BSM, the screw drives the ball which further drives the nut at the same time, therefore, the velocity of the ball at contact point A in the tangential direction t is less than that of the screw. Based on the relative velocity relationship expression $\left[\boldsymbol{v}_{AS}^{H}\right]=\begin{bmatrix}v_{bt}+\omega_{bn}r_b\sin\beta_A-\omega_{bb}r_b\cos\beta_A\\-\omega_{bt}r_b\sin\beta_A\\\omega_{bt}r_b\cos\beta_A\end{bmatrix}$ of contact point A between the ball and the screw of raceways on the relative velocity relationship expression $\boldsymbol{v}_{BS}^{H}=\begin{bmatrix}v_{bt}-\omega_{bn}r_b\sin\beta_B+\omega_{bb}r_b\cos\beta_B\\+\dot{\Omega}\left(\frac{r}{\cos\alpha}+r_b\cos\beta_B\cos\alpha\right)\\\omega_{bt}r_b\sin\beta_B+\dot{\Omega}r_b\sin\alpha\sin\beta_B\\-\omega_{bt}r_b\cos\beta_B-\dot{\Omega}r_b\sin\alpha\cos\beta_B\end{bmatrix}$ of contact point B between the ball and the nut of raceways,

$$\begin{aligned}-\dot{\Omega}\left(\frac{r}{\cos\alpha}+r_b\cos\beta_B\cos\alpha\right)\leq\\ v_{bt}-\omega_{bn}r_b\sin\beta_B+\omega_{bb}r_b\cos\beta_B\leq 0\end{aligned} \tag{5}$$

Considering the two ultimate uniform motion states of the ball, i.e., it is motionless relative to the nut $v_{bt}=-\dot{\Omega}\frac{r}{\cos\alpha}$ and screw $\begin{cases}v_{bt}=0\\\omega_{bt}=\omega_{bn}=\omega_{bb}=0\end{cases}$, the velocity range of the ball along helix of the screw is:

$$-\dot{\Omega}\frac{r}{\cos\alpha}\leq v_{bt}\leq 0 \tag{6}$$

Similarly, the linear velocity range of the ball in the tangential direction of helix path can be obtained using $\boldsymbol{v}_{AS}^{H}=\begin{bmatrix}v_{bt}+\omega_{bn}r_b\sin\beta_A-\omega_{bb}r_b\cos\beta_A\\-\omega_{bt}r_b\sin\beta_A\\\omega_{bt}r_b\cos\beta_A\end{bmatrix}=\begin{bmatrix}0\\0\\0\end{bmatrix}$ and $\begin{cases}v_{bt}=\omega_{bb}r_b\cos\beta_A-\omega_{bn}r_b\sin\beta_A\\\omega_{bt}=0\end{cases}$ as follows:

$$-\dot{\Omega}\sin\alpha\leq\omega_{bt}\leq 0 \tag{7}$$

Assuming that

$$\begin{aligned}\omega_{bn}r_b\sin\beta_0=-v_{bt}+\omega_{bb}r_b\cos\beta_0\\ +g\dot{\Omega}\left(\frac{r}{\cos\alpha}-r_b\cos\beta_0\cos\alpha\right)\end{aligned} \tag{8}$$

where $-1\leq g\leq 0$,

The slide-roll ratio of the ball contact point A and B becomes:

$$\begin{aligned}S_A=g\dot{\Omega}\left(\frac{r}{\cos\alpha}-r_b\cos\beta_0\cos\alpha\right)/\left(abs\left(g\dot{\Omega}\right.\right.\\ \left(\frac{r}{\cos\alpha}-r_b\cos\beta_0\cos\alpha\right)+\dot{\Omega}\cos\alpha\left(r-r_b\cos\beta_0\right)\Big)\\ \left.+abs\left(\dot{\Omega}\cos\alpha\left(r-r_b\cos\beta_0\right)\right)\right)\end{aligned} \tag{9}$$

$$\begin{aligned}S_B=\left(\left(2v_{bt}-g\dot{\Omega}\left(\frac{r}{\cos\alpha}-r_b\cos\beta_0\cos\alpha\right)\right.\right.\\ \left.\left.+\dot{\Omega}\left(\frac{r}{\cos\alpha}+r_b\cos\beta_B\cos\alpha\right)\right)\right)/\Big(\mathrm{abs}\Big(\\ 2v_{bt}-g\dot{\Omega}\left(\frac{r}{\cos\alpha}-r_b\cos\beta_0\cos\alpha\right)\Big)+\\ abs\left(-\dot{\Omega}\left(\frac{r}{\cos\alpha}+r_b\cos\beta_B\cos\alpha\right)\right)\Big)\end{aligned} \tag{10}$$

Based on a comprehensive consideration of Eq. (6), g should also satisfy the following condition:

DOI: 10.1051/matecconf/201710804008

$$g \geq \frac{v_{\mathrm{bt}}^{*} - \omega_{\mathrm{bb}} r_{\mathrm{b}} \cos\beta}{\ddot{\Omega}\left(\dfrac{r}{\cos\alpha} - r_{\mathrm{b}} \cos\beta \cos\alpha\right)} \tag{11}$$

By substituting the parameters shown in Table 1 into Eqs. (9) and (10) to arrive at the values satisfying Eqs.(7) and (8) and $-1 \leq g \leq 0$, the relationship between v_{bt} and ω_{bn} (Fig. 2) as well as those between slide-roll ratio and v_{bt} (ω_{bn}) (Fig. 3/Fig. 4) can be worked out.

Table 1. Parameters of the BSM.

Parameters	Value	Unit
Pitch radius r	16	mm
Ball radius r_b	2.9765	mm
Helix angle α	5.6833	degree
Contact angle β_0	44.5	degree
Pitch$L$	5	mm
Screw angular velocity $\dot{\Omega}$	0-196	rad/s

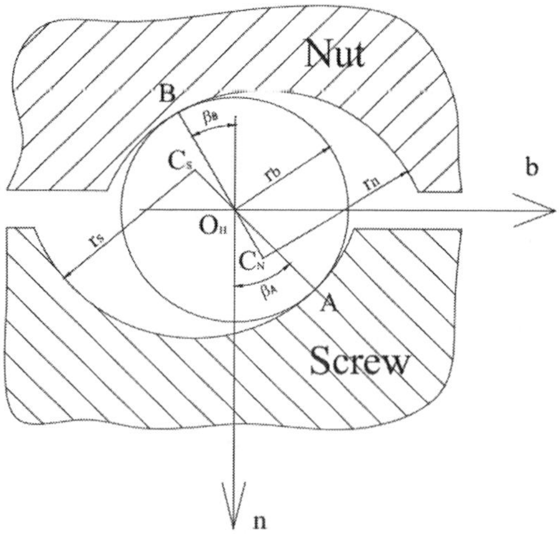

Figure 1. Contact angle of BSM.

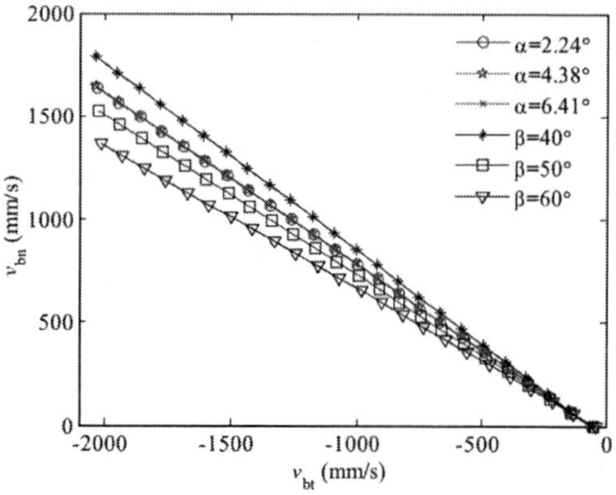

Figure 2. Relationship between v_{bt} and ω_{bn} .

4 Results and discussion

(1) When the BSM is under the same velocity, the greater the contact angles, the smaller value of $\omega_{\mathrm{bn}}/v_{\mathrm{bt}}$ in Fig.2; under the same velocity, the greater helix angles, the bigger value of $\omega_{\mathrm{bn}}/v_{\mathrm{bt}}$, but their values are affected by helix angle to a lesser degree. At the same time, with the increase in the value of v_{bt}, $\omega_{\mathrm{bn}}/v_{\mathrm{bt}}$ changes with the permanent amplitude under the constant contact angle and helix angle.

(2) Under the same v_{bt}, S_{A} and S_{B} decrease with the increase in contact angle as shown in Fig. 3. Under the same v_{bt}, S_{A} and S_{B} increase with the increase in helix angle, but their values are affected by helix angle to a lesser degree. Increasing the value of v_{bt}, S_{B} changes with the smaller amplitude.

(3) As Fig. 4 shown, under the same ω_{bn}, the greater contact angles, the bigger values of S_{A} first and the smaller later, but the decrease range is bigger. The greater contact angles, the smaller values of S_{B} . The greater helix angles, the smaller values of S_{A} and the bigger values of S_{B} , and with the variety of helix angles, the range of S_{A} and S_{B} are the smaller. At the same time, with the increase in ω_{bn}, S_{A} and S_{B} each present the same change trend as under the influence of v_{bt}.

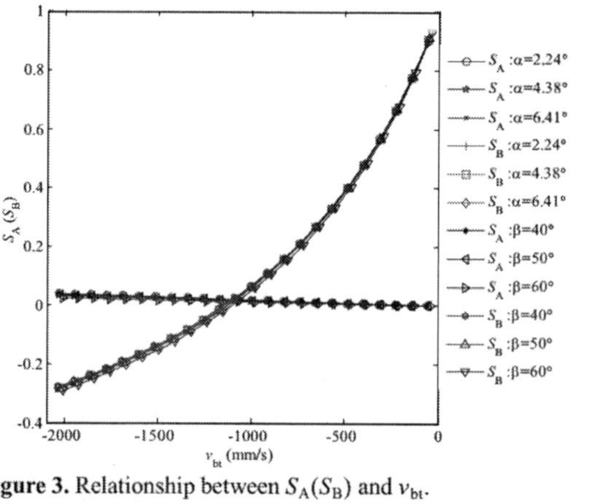

Figure 3. Relationship between $S_{\mathrm{A}}(S_{\mathrm{B}})$ and v_{bt}.

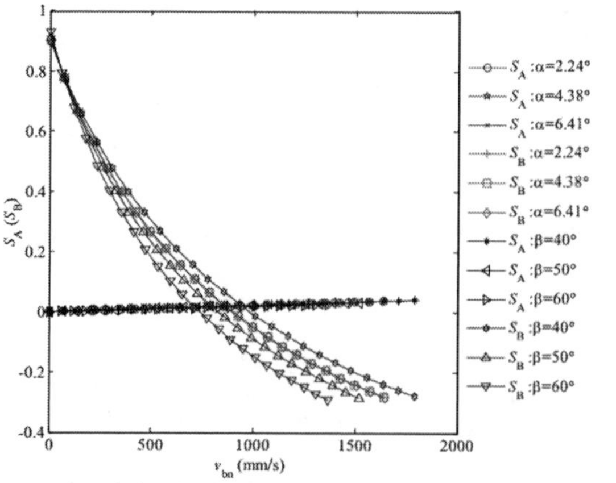

Figure 4. Relationship between $S_{\mathrm{A}}(S_{\mathrm{B}})$ and ω_{bn}

5 Conclusions

(1) Relationship between ω_{bn} and v_{bt} of the contact points of BSM are analyzed, it also explores the effects of contact angle and helix angle on ω_{bn} and v_{bt} .

(2) Relationship between S_A and S_B , ω_{bn} and v_{bt} in the course of motion of the BSM are analyzed, at the same time, the effects of contact angle and helix angle on S_A and S_B are also explored.

Acknowledgements

This work is supported by National Natural Science Foundation of China, Grant No. 51575014 and 51505012, and Natural Science Foundation of Beijing, Grant No. KZ201410005010. Beijing Postdoctoral Research Foundation, Grant No. 2015XX-13.The authors are grateful to other participants of the projects for their cooperation.

References

1. Wei C C, Lin J F. Kinematic analysis of the ball screw mechanism considering variable contact angles and elastic deformations[J]. ASME Journal of Mechanical Design, 2003, 125: 717-733
2. Wei C C, Lin J F, Horng J H. Analysis of a ball screw with a preload and lubrication[J]. Tribology International, 2009, 42: 1816-1831
3. Wei C C, Lin J F, Horng J H. Kinematical analyses and transmission efficiency of a preloaded ball screw operating at high rotational velocitys[J]. Mechanism and Machine Theory, 2011, 46: 880-898.
4. Lin M C, Ravani B, Velinsky S A. Kinematics of the ball screw mechanism[J]. ASME Journal of Mechanical Design, 1994, 116(3): 849-855
5. Lin M C, Ravani B, Velinsky S A. Design of the ball screw mechanism for optimal efficiency[J]. ASME Journal of Mechanical Design, 1994, 116(3): 856-861
6. Gnanamoorthy R, Govindarajan N. Effect of slide-roll ratio on the contact fatigue behavior of sintered and hardened steels[J]. Journal of Failure Analysis and Prevention, 2004, 4(2): 78-83
7. Govindarajan N, Gnanamoorthy R. Rolling/sliding contact fatigue life prediction of sintered and hardened steels[J]. Wear, 2007, 262: 70-78
8. Ma S J,Liu Geng,Fu Xxiaojun,et al. Rolling-sliding characteristics of planetary roller screw considering elastic deformation. Journal of Southeast University(Natural Sciences Edition),2015,45(3):461-468.
9. Li M Q,Yu H,Li D Y,et al. Modeling for dynamic contact angle of ball screw mechanism aimed to structural parameters. Transactions of the Chinese Society of Agricultural Engineering, 2016, 32(4):98-104.
10. Hu J Z, Wang M, Zan T, The kinematics of ball-screw mechanisms via the slide-roll ratio, Mech. Mach. Theory 79 (2014) 158–172.

Investigation on the design and the stiffness demonstration for a space square-truss deployable mast (STDM)

Hai-jun Luo[1,2], Qi-feng Cui[1,2], Ming Li[1] and Zhi-long Peng[1]

[1]*Aerospace System Engineering Shanghai, Shanghai, 3888 Yuanjiang Road, China*
[2]*The State Key Laboratory of Mechanical Transmissions, Chongqing University, Chongqing, 174 Shazheng Road, China*

Abstract. Space deployable mast is an important device or system, which can deploy and support spacecraft fittings on orbit, such as payload, solar array, antenna, and etc. In order to achieve the pre-function of spacecraft fittings and attitude control of spacecraft, space deployable mast of high stiffness is required. Therefore, it is necessary to analyze the stiffness of the mast to ensure the capability. STDM consisting of longerons, rigid frames, flex-battens, hinge joints, and etc has higher stiffness than coil or articulated deployable mast. The design of STDM is presented, under the demand of future large-scale spacecraft. The theory analysis of stiffness is achieved and the impact factor are found, which establish the theoretical baseline for the optimization design and application of the mechanism product.

1 Introduction

In recent decades, with the rapid development of space science and technology such as satellite communication, space exploration and earth observation, demands for large spacecraft such as the space station and large satellite is increasing[1-2]. As the important component which can change spatial position of spacecraft fittings relative to the mainbody or the overall topology configuration, one-dimensional space deployable mast need to have large size, high stiffness, high load capacity and can deploy and fold repeatedly. The length of STDM is typically greater than 10m, and the truss should have high stiffness, high positioning accuracy and high reliability. Currently, the mast has been successfully applied in the International Space Station to expand and support large-area flexible solar array[3-6].

Fundamental frequency and stiffness property of payload on orbit is directly determined by the stiffness characteristic of expanding space square truss mast[7]. Flex-battens, longerons and diagonal cables are key components of the truss, whose performance directly affects the stiffness property of mast. Therefore, investigating the stiffness characteristic of the truss in the initial stages of design, the axial stiffness, bending stiffness and torsional stiffness should be quantified, so that the key factors affecting the truss stiffness property can be identified in order to provide an effective baseline for optimization design of the following products.

2 Structural design and working principle

2.1 Structural design

STDM is mainly composed of deployable truss, collection system and drive-transmission system. The deployable state of mast is illustrated in Figure 1.

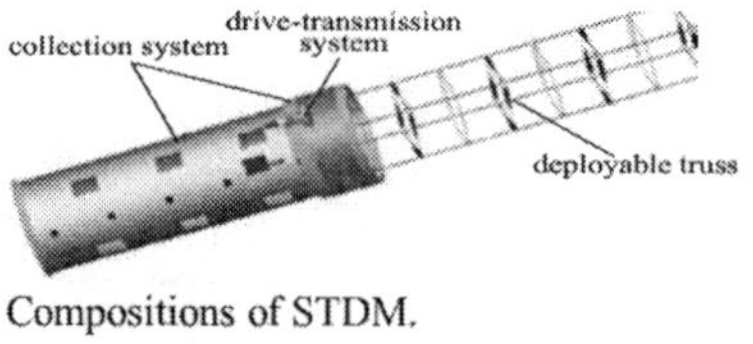

Figure 1. Compositions of STDM.

The truss is a square variable topology which can extend and retract repeatedly, mainly assembled by longerons, rigid frames, elbows, flex-battens, hinge joints, diagonal cables and etc, as illustrated in Figure.2. It's the main load-bearing mechanism of mast. Four flex-battens being post-buckling consist of a flexible frame, whose configuration transform between bend and stretch process lead to rectilinear motion of rigid frame, that's why the truss can deploy and retract. Relative to trilateral-shaped structure, quadrilateral structure has greater moment of inertia and bending stiffness, better mechanical property, which leads to a stable configuration.

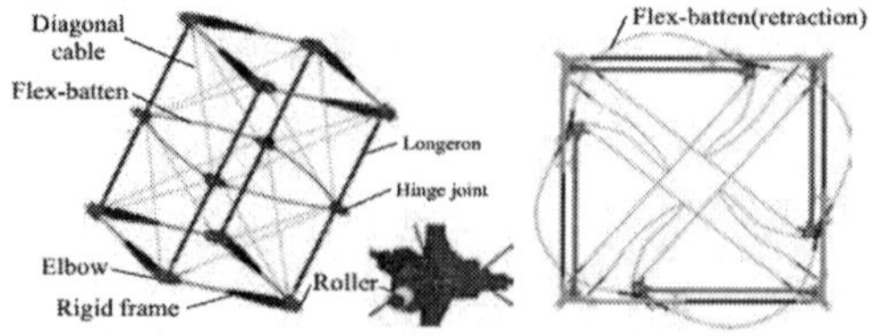

Figure 2. The truss being extended and retractable of STDM.

DOI: 10.1051/matecconf/201710804009

As illustrated in Figure.3, rigid frame can be seen as a rigid plane whose square shape remains invariant during the extension/retraction operations. In addition, the upper and lower longerons hinge with rigid frame and flexible frame, and then four nodes of flexible frame form a quadrangle whose side length increases or decreases during the extension/retraction operations of mast.

It should be noted that the outer envelope radius of flex-batten after deformation should not be greater than the radius of the truss R during the extension/retraction operations of mast. Thereby, kinematical trajectory of rigid frame $A_2B_2C_2D_2$ and flexible frame $A_3B_3C_3D_3$ is limited within a cylinder whose bottom is the circumscribed circle of four nodes of rigid frame $A_1B_1C_1D_1$, and the track of each node of $A_2B_2C_2D_2$ is a straight line in the axial direction of the cylindrical surface, while the track of each node of $A_3B_3C_3D_3$ is a spatial rotation curve whose trajectory radius r varies.

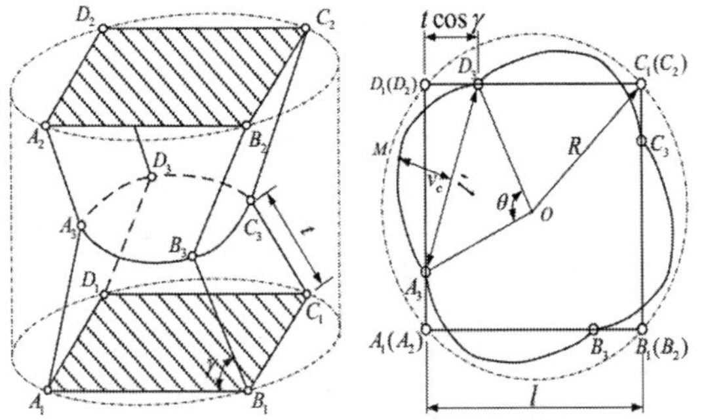

(a) (b)

Figure 3. Analysis of deployment-retract kinematic principle of flexible frame.

As shown in Figure.3(b), the relation between D_1D_3, A_3D_1, A_3D_3 can be expressed as,

$$\overline{D_1D_3}^2 + \overline{A_3D_1}^2 = \overline{A_3D_3}^2 \tag{1}$$

Where D_1D_3 and A_3D_1 is respectively the distance of corresponding or adjacent hinge joints between flexible frame and rigid frame within the flexible frame plane, and then it can be deduced that $\overline{D_1D_3} = t\cos\gamma$ and $\overline{A_3D_1} = l - \overline{D_1D_3} = l - t\cos\gamma$, A_3D_3 stands for the distance of adjacent hinge joints or the span across the post-buckling flex-batten after bended,

$$\begin{aligned} l' &= \sqrt{(t\cos\gamma)^2 + (l - t\cos\gamma)^2} \\ &= t\sqrt{(\sqrt{2}\cos\gamma - R/t)^2 + (R/t)^2} \end{aligned} \tag{2}$$

Where t is the span between upper and lower frames or the length of longerons, R is the outer envelope radius of the truss, $l = \sqrt{2}R$ is the side length of rigid frame or the span across the flex-battens when the truss bay is expanded state, γ stands for the included angle of longerons and hinged rigid frames during the extension/retraction operations of bay.

2.2 Working principle

With elastic bending deformation of flex-batten and the pin joint between upper and lower longerons, the truss can only completely stow; with rotation of drive nut and release of elastic strain energy that flex-batten stored, the truss can only achieve the overall extension.

Figure.4 shows how the upper and lower longerons can be treated as the legs of a card table that fold in perpendicular directions to each other. Figure.5 shows flex-battens movement during the extension/retraction operations of the truss.

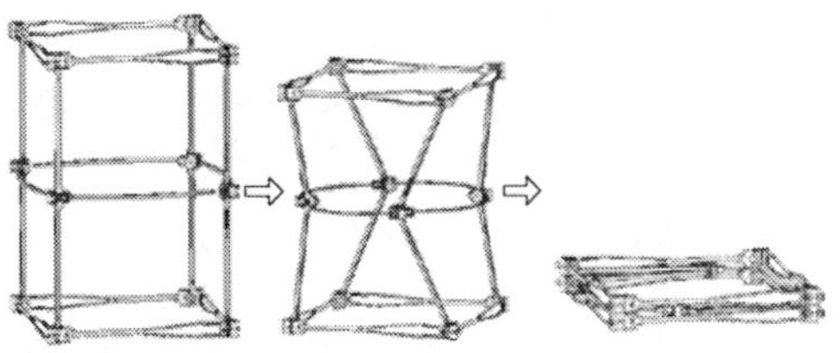

Figure 4. Longerons movement.

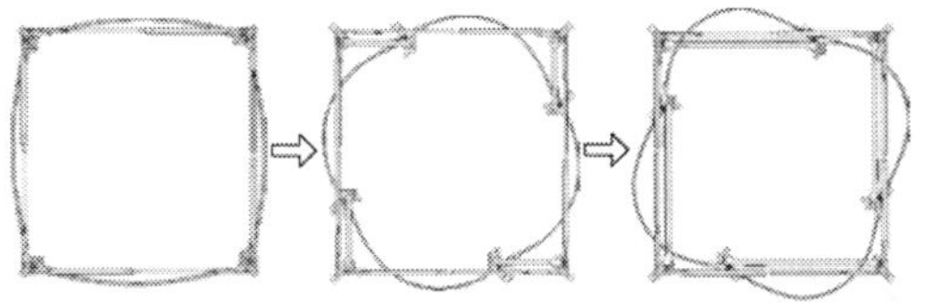

Figure 5. Flex-battens movement.

The truss is a mechanism consisting of numbers of bays, the component diagram which shows how the mast deploys and retracts is illustrated in Figure.6 according to the structure and principle.

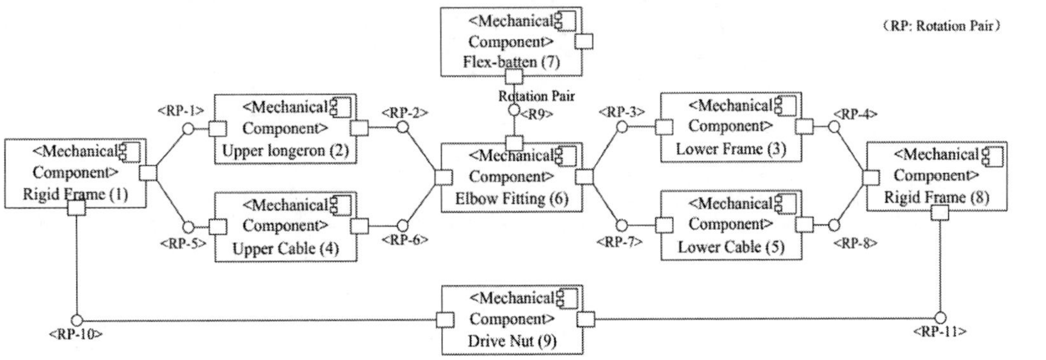

Figure 6. Component diagram of STDM.

DOI: 10.1051/matecconf/201710804009

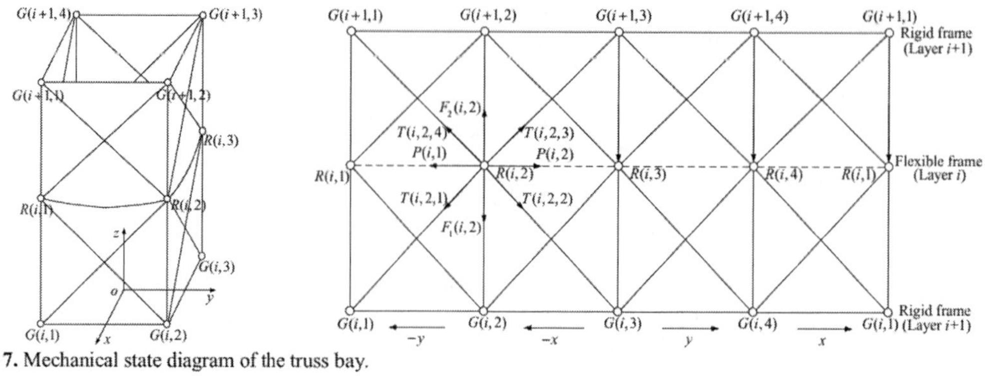

Figure 7. Mechanical state diagram of the truss bay.

3 Stiffness analysis of the truss

The truss is the main load-bearing parts of mast, by investigating the stiffness characteristic of the truss, the axial stiffness, the bending stiffness and the torsional stiffness can be quantified, furthermore, the key factors affecting the truss stiffness property can be identified. It is worth noting that the preload of the flex-batten is provided by the diagonal cable in truss bay. There are some assumptions following in order to simplify the stiffness analysis appropriately:

(1) Diagonal cables are regarded as the linear material, the initial preload of each cable is equal and the preload would not be reduced to zero under slightly deformation state for the truss. Then the truss can be considered with linear mechanical characteristics;

(2) Flex-batten will not further post-buckle when the truss is being deployed;

(3) The connecting gap of elbow or hinge joints are ignored.

Figure.7 illustrates the mechanical state of one of the truss bays, O stands for origin of fixed coordinate system and it's geometric centre of the rigid frame named layer i, there is three translational degrees of freedom at each node. Among others, $i=1,2,\cdots N$ is the layer number of rigid/flexible frame, $j=1,\cdots 4$ is the node number of four quadrants corresponding layer i, $k=1,\cdots 4$ is four diagonal cables number corresponding layer i and node j. $G(i,j)$、$R(i,j)$ are nodes of rigid/flexible frame respectively, $F_1(i,j)$、$F_2(i,j)$ are elastic internal force of upper and lower longerons respectively, $P(i,j)$ is the internal tension of flex-batten, $T(i,j,k)$ is the tension of diagonal cables being in rigid/flexible frame nodes.

3.1 Analysis of the axial stiffness

The cross section of the truss is quadrangle and it is intuitional that when an axial force is applied on the truss, the force of each node can be treated as $F/4$.

3.1.1 Analysis of the tension stiffness

Suppose that F is a tensile force, Figure.8 illustrates the axial mechanical analysis of longeron at node $R(i,2)$, $R'(i,2)$ stands for the position of $R(i,2)$ after the longeron is deformed.

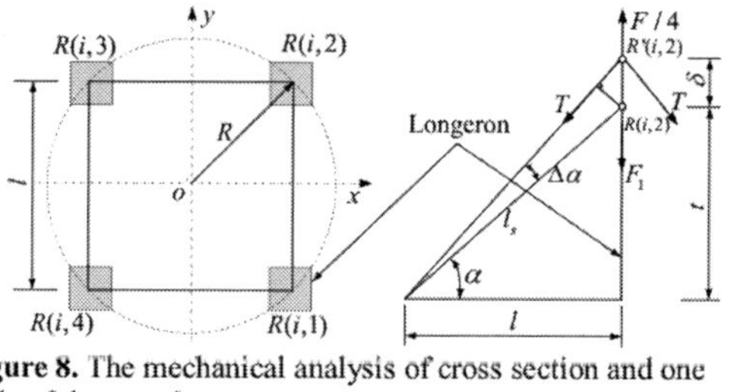

Figure 8. The mechanical analysis of cross section and one node of the truss bay.

As the above illustration shows that,

$$\begin{aligned}\frac{F}{4}&=2T\sin(\alpha+\Delta\alpha)+F_1\\&=\frac{2E_sA_s}{l_s}\delta\sin^2(\alpha+\Delta\alpha)+\frac{EA}{t}\delta\end{aligned}\tag{3}$$

Herein, $l_s=\sqrt{l^2+t^2}=\sqrt{2R^2+t^2}$,

$$F=\frac{8E_sA_s}{\sqrt{2R^2+t^2}}\delta\sin^2(\alpha+\Delta\alpha)+\frac{4EA}{t}\tag{4}$$

In the above equation, E_s、A_s stands for elasticity modulus and cross-sectional area of the cables respectively, E、A and t stands for elasticity modulus, cross-sectional area and length of the longeron respectively, R is the outer envelope radius of the four longerons, $l=\sqrt{2}R$ is the span between longerons or side length of frames and there is $l=\sqrt{2}R$, δ is the tensile deformation of longerons in axial direction. It can be deduced $\sin\alpha=t/l_s$, therefore, the tension stiffness can be obtained,

$$\begin{aligned}[EA]_2&=\lim_{\Delta\alpha\to 0}\frac{Ft}{\delta}=\lim_{\Delta\alpha\to 0}\left(\frac{8E_sA_st}{\sqrt{2R^2+t^2}}\sin^2(\alpha+\Delta\alpha)+4EA\right)\\&=8E_sA_s\left[2(R/t)^2+1\right]^{-\frac{3}{2}}+4EA\end{aligned}\tag{5}$$

DOI: 10.1051/matecconf/201710804009

The side length of the cross section of longeron is p and its modulus $E = 70\text{Gpa}$, the diameter of cables is d and its modulus $E_s = 60\text{Gpa}$. The curve varying from R/t of the tension stiffness of the truss is illustrated in Figure 9. It can be concluded that, with the increase of R/t the stiffness shows a non-linear decreasing trend; the change of the cable diameter hardly affects the stiffness; while increasing the cross-sectional dimension of longeron can enhance the stiffness significantly.

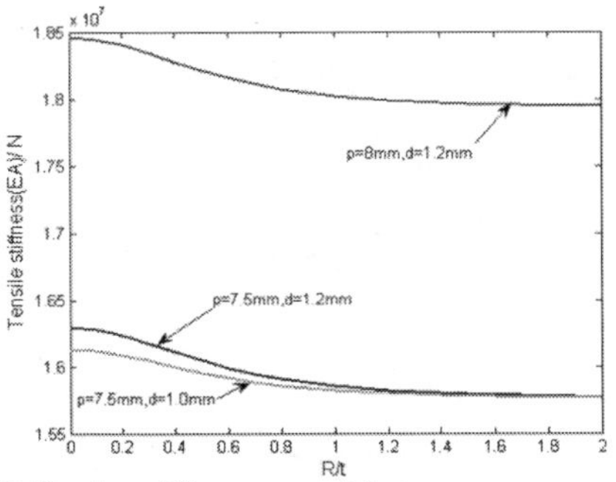

Figure 9. Tension stiffness curve of the truss.

The conclusion that the cable stiffness contribution to tension stiffness is negligible through the analysis and the calculation. In summary, the tension stiffness of the truss can be obtained,

$$[EA] = 4EA \tag{6}$$

3.1.2 Analysis of the compression stiffness

Suppose that F is a compressive force, the cable has no contribution to the compression stiffness of the truss for the cable cannot withstand the pressure. Then the compression stiffness is,

$$[EA]'=4EA \tag{7}$$

It shows that increasing the cross-sectional dimension of longeron can enhance the stiffness significantly.

3.2 Analysis of the bending stiffness

The stiffness contribution of cables to the bending stiffness of the truss is negligible when ignoring the rotation gap of elbow or hinge joints. Figure.10 shows the state that the truss bay is placed at any angle.

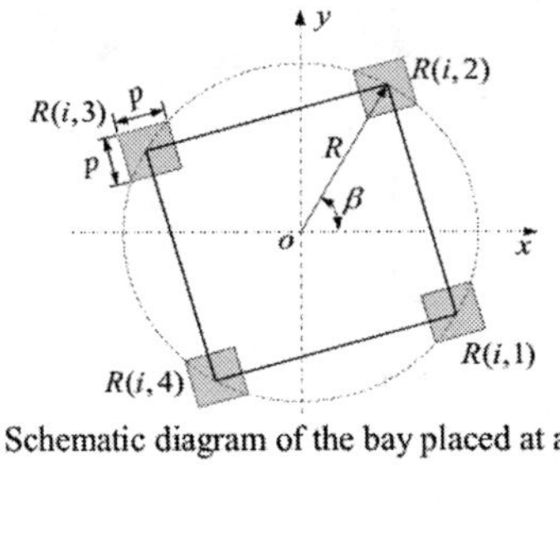

Figure 10. Schematic diagram of the bay placed at any angle.

According to the translation theorem and rotation-axis theorem, the moment of inertia of x axis is,

$$\begin{aligned} I_x &= 4\times\frac{p^4}{12} + p^2(R\sin\beta)^2 + p^2\left[R\sin(\beta+\frac{\pi}{2})\right]^2 + \\ &\quad p^2\left[R\sin(\beta+\pi)\right]^2 + p^2\left[R\sin(\beta-\frac{\pi}{2})\right]^2 \\ &= \frac{p^4}{3} + 2p^2R^2 \end{aligned} \tag{8}$$

Herein, β is the angle of the cross-section of the truss and axis x. It can be concluded from the equation that the moment of inertial is independent with β. In addition, the moment of inertia of longeron itself is negligible when $p \ll R$. Then is the bending stiffness of the truss,

$$EI = 2EAR^2 \tag{9}$$

Figure.11 illustrates the quadratic curve changed with the outer envelope radius R. From the curve, it can be summarized that the change of the cross-sectional dimension of longeron has little effect on the bending stiffness of the truss when R is small enough. Obviously, when R increases to a certain size, such as $R \geq 200$mm, the bending stiffness becomes more sensitive to changes in cross-sectional dimension. Also, the material of longeron is an important factor affecting bending stiffness of the truss, which means longerons with high elastic modulus can significantly improve the bending stiffness. For example, the bending stiffness of the truss will increase about 57% if the aluminium alloy is replaced by titanium alloy for longeron when $R = 300\text{mm}$.

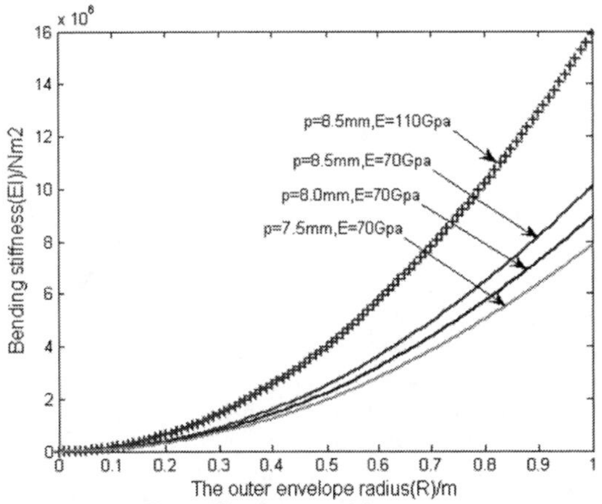

Figure 11. Bending stiffness curve of the truss.

3.3 Analysis of the torsional stiffness

The upper and lower frames will result in torsional deformation when the truss bears a torque T', as illustrated in Figure 12.

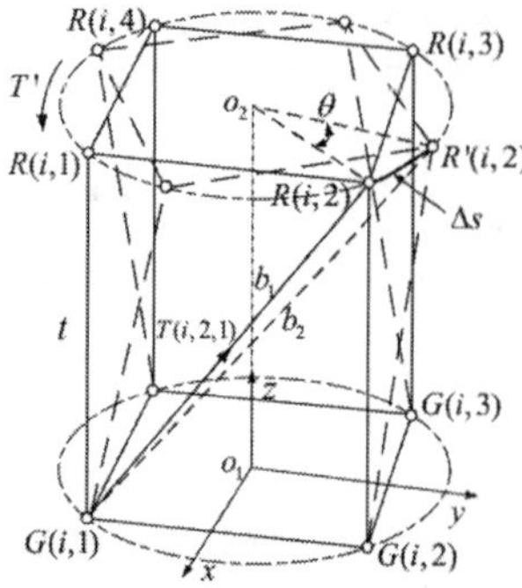

Figure 12. Schematic diagram of torsional deformation adjacent frames of the truss bay.

The truss bay results in small-angle θ torsion because of T', internal force of left diagonal cable becomes smaller and the right diagonal cable tightens at the same time the length extend to b_2 from initial length b_1, then the horizontal displacement in the yz plane of point $R(i,2)$. Small angle θ approximately is,

$$\Delta l \approx \Delta s = R\theta \tag{10}$$

The calculation model of the tensile deformation of cable is expressed in Figure.13.

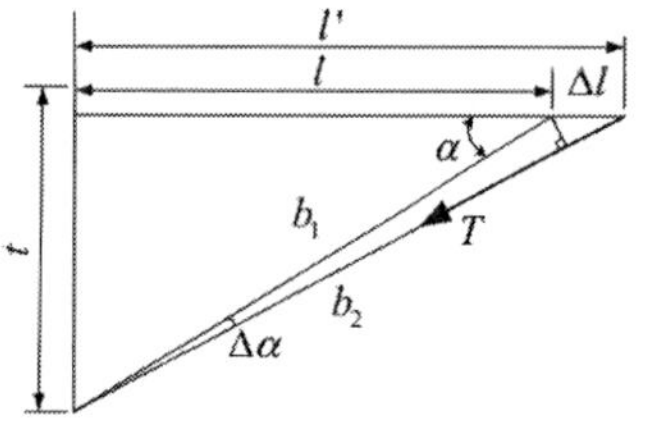

Figure 13. Calculation model of tensile deformation of cable.

The tensile force can be calculated from the torsion,

$$\begin{aligned} T &= \frac{E_s A_s}{b_1}(b_2 - b_1) = \frac{E_s A_s}{t}\sin\alpha \Delta l \cos(\alpha - \Delta\alpha) \\ &= \frac{E_s A_s R\theta}{t}\sin\alpha\cos(\alpha - \Delta\alpha) \end{aligned} \tag{11}$$

Considering that when the truss bears respectively torque load whose direction is positive or negative, the torsional stiffness contribution of longeron to the truss is zero because the longeron has rotational degrees of freedom in the direction of the truss stowing. Based on the analysis, the truss suffers a torque $T' = 4TR\cos\alpha$, then the torsional stiffness of the truss,

$$\begin{aligned} GI_p &= \lim_{\Delta\alpha \to 0} \frac{T't}{\theta} \\ &= \lim_{\Delta\alpha \to 0} 4E_s A_s R^2 \sin\alpha\cos\alpha\cos(\alpha - \Delta\alpha) \\ &= 4E_s A_s R^2 \left(1 + \frac{1}{2(R/t)^2}\right)^{-1} \left(1 + 2(R/t)^2\right)^{-\frac{1}{2}} \end{aligned} \tag{12}$$

Figure.14 illustrates the torsional stiffness curve of the truss changed with R/t. The curve shows that R has little effect on the torsional stiffness while diameter of the cable is the a key factor. Obviously, increasing the diameter of the cable is an effective way to improve the torsional stiffness, and with R/t increases, the torsional stiffness increases when $0 < R/t < 1$ and decreases when $R/t > 1$.

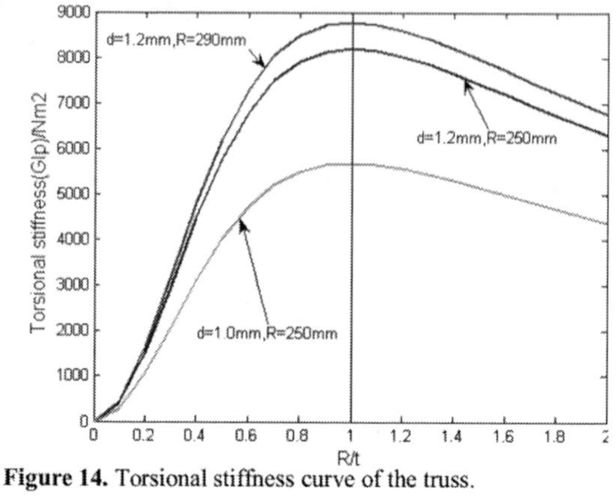

Figure 14. Torsional stiffness curve of the truss.

When $R/t = 1$,

$$[GI_p]_{\max} = \frac{8E_s A_s R^2}{3\sqrt{3}} \tag{13}$$

3.4 Discussion of the stiffness analysis

The aforementioned analyses of stiffness characteristic of mast are carried out under some assumptions, while the analysis can help identifying several key factors affecting stiffness. Under the conditions that the materials are certain and the weights meet the requirements of mast, the diameter of cable, the cross-sectional area of longeron and the ratio R/t all are important factors. In the engineering design, it needs to consider the overall mutual influence relations between the axial stiffness, the bending stiffness and the torsional stiffness so that the stiffness characteristics of mast can be optimized.

Comprehensive consideration of the conditions on orbit, when there is $R/t = 1$ that the stiffness of the truss is more appropriate, and then the whole structure of mast will be in a good coordination. Figure.15 shows structures of the quadrilateral and trilateral configuration and the stiffness comparison between them shows in Table 1.

Table 1. Stiffness comparison between the square truss and trilateral configuration structure[a].

Stiffness of truss	Quadrilateral configuration	Trilateral configuration	Ratio
Axial stiffness	$4EA$	$3EA$	133.33%
Bending stiffness	$2EAR^2$	$\frac{3}{2}EAR^2$	133.33%
Torsional stiffness	$\frac{8E_s A_s R^2}{3\sqrt{3}}$	$\frac{2E_s A_s R^2}{\sqrt{3}}$	133.33%

[a].Design parameters of the truss: $R = t$

DOI: 10.1051/matecconf/201710804009

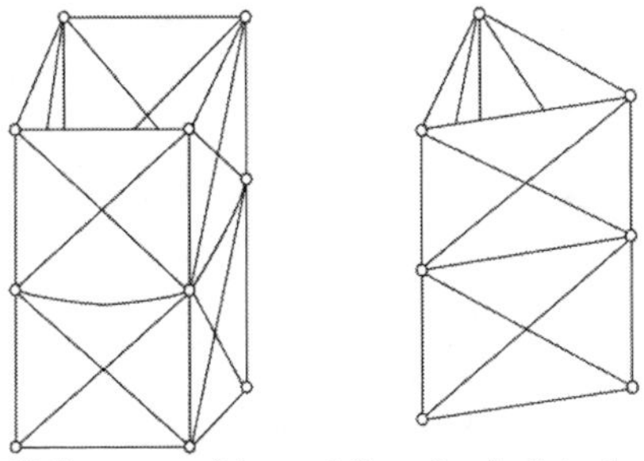

Figure 15. Structures of the quadrilateral and trilateral configuration.

Comparing with the trilateral configuration structure, the quadrilateral configuration has obvious advantages about stiffness under the same basic dimensions of the truss. In general, increasing the number of longerons can effectively improve the deployable stiffness of truss but weakening influence of weight of mechanism.

4 Conclusion

Large one-dimensional space deployable mechanism is widely used in space science presently, commonly used for long-distance extension and high-precision positioning of space payload. The working principle of the STDM is based on the extension and retraction of the truss so that the mast can deploy or fold. Obviously, the extension stiffness of the truss decides the fundamental frequency and work performance on orbit of spacecraft fittings. Firstly this paper proposed a design scheme of a quadrilateral mechanism, and analyzed the relationship of components during the extension/retraction operations of the truss. Then the paper analyzed the axial stiffness, the bending stiffness and the torsional stiffness with analytical method by simplifying certain models, and pointed out several effective measures enhancing the stiffness.

Parametric analysis of the stiffness can provide a valid basis in favor of structure optimization, performance improvement, and risk reduction of product development. Stiffness testing of principle prototype will be carried out in the following work to verify the analyses above.

Acknowledgement

The authors are grateful for the supports received from the open program from the state key laboratory of mechanical transmissions, Chongqing University (SKLMT-KFKT-201404).

References

1. Cui Qifeng, Peng Zhilong. Study on the design and stiffness for a space articulated deployable mast[C]. 3RD National Manned Space Congress, 1-2 (2014) (in Chinese)
2. Shen Bin. Design and Characteristic Analysis of the Coilable Lattice Mast[D]. Shanghai: Shanghai Academy of Spaceflight Technology, 1-3 (2007) (in Chinese)
3. Elliott, Kenny B. Model Calibration Efforts for the International Space Station's Solar Array Mast[R]. AIAA-NF1676L-13312, 1-3 (2012).
4. Knight, Norman F, Jr. FAST Mast Structural Response to Axial Loading: Modeling and Verification[R]. AIAA-NF1676L-13313, 1-2 (2012).
5. Karen M.B. Taminger. Solid Freeform Fabrication: an Enabling Technology for Future Space Missions. Keynote Lecture for 2002 International Conference, San Antonio, 4 (2002)
6. Bowden M, Benton M. Design of deployable-truss masts for space station [C]// AIAA/AHS/ASEE Aerospace Design Conference, Goleta ca, 1-10 (1993)
7. Deng Zongquan. Design of Space Deployable and Foldable Mechanisms[M], 10-12 (2013) (in Chinese).

Design of a ferrochrome – slag mechanical separator case study

Loice Keresia Gudukeya , Tawanda Mushiri, Charles Mbohwa

University of Johannesburg, Faculty of Engineering and the Built Environment, Auckland Park 2006, P.O Box 524, Johannesburg, South Africa

Abstract. This research focused on the design of a ferrochrome – slag mechanical separator for a company in Harare, Zimbabwe. The company's current production is 420 tonnes of alloy per month. The aim of project is to raise the production from 420 to 500 tonnes of ferrochrome production per month by introducing a mechanical separator at metal recovery plant that produces 80 tonnes. The difference in specific gravities of ferrochrome and slag is the key principle applied in achieving separation of the two products. The densities of ferrochrome and slag are 6800 kg/ $m^3$and 2800kg/ m^3 respectively hence gravity separation method in water medium can be done.

1 Introduction

The production target of the metal recovery plant at a Harare Company is 500 tonnes per month and the current output is 420 tonnes. The crushing, screening and all material handling processes in the plant are effective because they produces at least 10 tonnes of slag ferrochrome material to go for jigging per day and hence the problem leading to the failure of meeting the target is the final process which is jigging for ferrochrome and Slag separation as it separates 4 tonnes out of 10 tonnes per day. There is a single pneumatic air pulsed jig used for the separation process but it is failing to meet production demands on its own hence it needs the support of another mechanical separator in order to maximise production to the required target. Slag is discarded in large quantities as waste during the process of ferrochrome production. Reclaiming ferrochromium metal entrapped in slag seems to be economic. In order to recover metal, slag has been crushed and subjected to wet magnetic and gravimetric methods [1].

2 Literature review

2.1 Ferrochrome and slag production

Ferrochrome production is done by a direct carbothermic reduction of chromite on three phase submerged arc furnaces. Ferrochrome and slag are the main two products tapped off at regular intervals from tap holes at the bottom of the furnace. Cascade tapping is used to tap slag and metal through the same tap hole into the same metal vessel. Slag will float on the top due to its low density and eventually overflow via a secondary vessel through the ladle spout to the slag pot. Ferrochrome is sent to casting of specific products [3]. The ferrochrome slag products are granulated slag as well as classified slag products made by crushing and screening. The main part of slag is granulated. The crushed and screened slag products are produced from air cooled lumpy slag which is collected from various process phases, e.g. from ladles and launders. The slag is directly granulated during tapping where ferrochrome is tapped into ladles. The overflow from the ladles flows along the slag launder to the granulation pond, where high-pressure water breaks slag into small fractions and efficiently it cools down. The ferrochrome separation from the lumpy slag happens in a special process, which is based on a gravity and magnetic separation [4].

2.2 Methods of separation

2.2.1 Gravity separation

The gravity separation is a method of separating two materials of different specific gravity that is carried out by their relative motion in response to the combination of force due to gravity and one or more other forces. The other factors besides specific gravity which are shape, size, and weight of the particles affect the relative motion and hence the separation. The ease or difficult of separation is determined by the differences in these factors. The separation effectiveness can be measured by concentration criteria (cc)

$$CC=\frac{dh-df}{dl-df}$$

d_h = Specific gravity of heavy material
d_f =Specific gravity of the fluid
d_l = Specific gravity of lighter material

For relatively ease separation CC should be greater than 2.5 (whether positive or negative) and it becomes

DOI: 10.1051/matecconf/201710804010

difficulty as CC decreases and when CC < 1.25 the gravity concentration is not feasible. Besides specific gravity, particle motion is also affected by particle size. The gravity concentration efficiency increases with increase in particle size. For the small particle the motion is mostly affected by friction and has poor response to high capacity gravity separators. The types of gravity separations are spirals, floater density separator, shaking tables; water only cyclone, rotating cones and jigs [2].

2.2.2 Jigging separation process

The jigs follow a 3-stage process to separate the particles of different densities from each other. The processes are differential acceleration of particles, consolidating trickling and hindered settling. Water is used as the separating medium and this allows efficient and continuous recovery of ferroalloy from slag in a particle bed. Types of jigs include: Harz Jig, Denver Jig, Inline pressure Jig.

3 Generation of possible solutions

In the study carried out the jigging machines pulsation mechanism can be pneumatic air pulsed mechanism, high pressure water valve controlled pulsating mechanism but this research objective is to design a mechanical separator with a pulsating mechanism which can apply the use of a diaphragm, plunger or piston driven by an electric motor and a cam and follower actuating mechanism.

3.1 Possible concept 1: Diaphragm pulsed separator

System parts and functions , shown in Fig 1:

- Hutch separating vessel – that's where water pulsates and stratification occurs
- Screen- the screen is there to maintain the required feed size
- Cam – the cam is driven by a motor shaft through belt drives changing the rotation motion to the linear motion for the follower arm linked to the diaphragm
- Diaphragm- it is moved up and down giving pressure to the water thus producing pulsating currents

Advantages
- Cheap diaphragm material
- Reduced vibrations problems
- Low maintenance costs
- Efficiency jigging
- They are no water leakages
- Easy recovery of course material hence less requirements of much grinding process
- Open visual and physical inspection possible with many adjustments

Figure 1. Concept 1 Diaphragm pulsed separation.

Disadvantages
- Slow stratification
- Not suitable for recovery of fines

3.2 Possible concept 2: Plunger pulsed separator

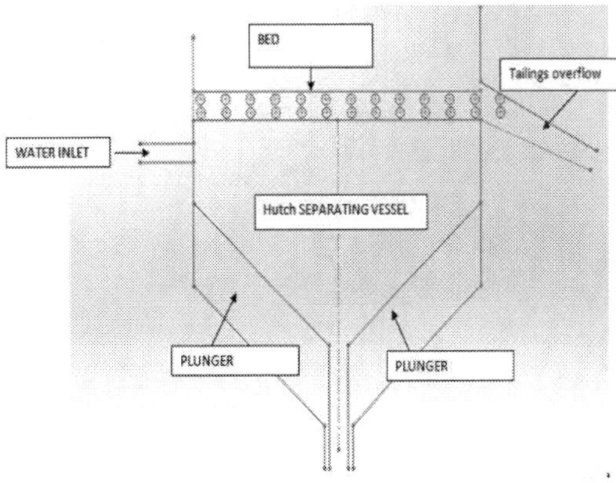

Figure 2. Plunger pulsed separator

System parts and functions

Some of the parts in concept 2 are the same as the parts in concept 1 such as hutch, and screen but concept 2 differs from 1 as it contains plungers instead of a diaphragm and the plungers are placed at the bottom of the separating vessel. Plungers are used to provide change in pressure to water in the hutch so that it pulsates to cause stratification.

Advantages
- High pressure hence high pulsation rate
- Quick stratification

Disadvantages
- There is problems of water leakages therefore requires proper sealing system.
- They are risks of vibrations
- Not suitable for recovery of fines

DOI: 10.1051/matecconf/201710804010

3.3 Possible concept 3: Diaphragm pulsed hutch rotated separator

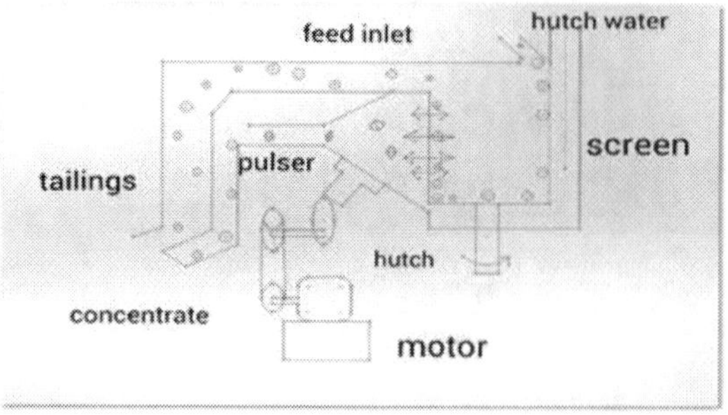

Figure 3. Diaphragm pulsed with centrifugal effect separator

System parts and functions

- Hutch -As in previous concepts, in the hutch that's where pulsation, dilation of the bed and stratification occurs but in this case the hutch is also rotated about a central fed pipe in a spinning bowel so that there will be centrifugal action of the bowel rotation causes feed to distribute on the ragging.
- Screen – used to maintain required feed size and in this concept the screen and the ragging are vertical
- Spinning bowl- Contains the hutch and the screen; it is driven by a motor and makes the hutch to rotate about a central feed pipe.
- Diaphragm- it is connected to the arm linked to the cam driven by the motor so that it induces pulses to the water in the hutch.

Advantages

- Efficiency jigging
- Can easily separate fine particles

Disadvantages

- The rotor may become unbalanced and cause severe vibrations that may damage the equipment
- It is very expensive high capital and operating costs
- High maintenance costs
- Requires more amount of water to be used

4 Concept selection

The following criteria will be used for comparisons in determining the best solution

A- Function, B- Reliability, C- Ease of maintenance, D- Easy of manufacture, E- Simplicity of layout, F- Efficiency, G- Ergonomics, H- Weight, I- Lifespan, J- Quality, K- Cost

In order to determine the best solution the different concepts was compared according to how much they meet the above criteria. The design criteria was ranked in order of their relative significance using the Binary Dominance Matrix method (BDM). Concept 1 of **diaphragm pulsed jig** was found to be the best solution.

5 Development of chosen solution

The design of diaphragm pulsed mechanical separator for ferrochrome and slag consists of sizing of the jug hutch compartment volume, stress analysis for hutch material, particle dynamics calculations, sizing of electric motor, design of power transmission systems (belt drives and pulleys), design of cam shaft and follower links, bearings, screen and diaphragm membrane pulsar.

5.1 Calculating vessel volume and dimensions

Hutch volume = volume of water required + volume of feed material +volume for air space

Calculating volume of the feed:

Feed amount per hour = 112 kg/ hr.

Densities in the feed are 2800 for slag and 6800 for ferrochrome

Maximum vessel volume is obtained by considering slag density $=\frac{112000}{2.8}=16339.8693\ cm^3$

Volume of feed $=16339869.3 mm^3$

For good concentration the feed amount should be take a range from 7-10% with water amount of 80 -90% of the jig volume

Taking the feed amount to be 7%

Total jig volume by proportion $= \frac{100}{7}\times 16339869.3 = 243000000 mm^3$

The hutch is square pyramidal in shape hence volume v $=\frac{1}{3}\times b^2\times h$

Taking b =h

$\frac{1}{3}b^3 = 243000000$

b=882 mm take design value 900 mm

Volume of water 90% of jig volume and the other 3% will be for water displacement allowance.

5.2 Calculating vessel pressure and fluidising velocity

Pressure drop due to layout weight of particles $\Delta P = (1\text{-}e) \times (þ_s\text{-}þ)\ l\ g$ (culson 2002) where:

- e is bed voidance or porosity
- *l is* depth ,g is acceleration due to gravity
- $þ_s$ and þ are densities of solids and liquid respectively

Assuming *l* = separating height vessel

According to (culson 2002) bed voidance for particles between 75μm to 20mm =0.41

ΔP due to ferrochrome = (1-0.416) × (6800-1000) ×900×9.81= **29906 KN/m²**

ΔP due to slag = (1-0.416) × (2800-1000) ×900×9.81= **9281 KN/m²**

Therefore pressure drop ranges from **9281 to 29906 KN/m²**

ΔP=29906 KN/m² is the maximum vessel pressure

The minimum **fluidising velocity** $U_{mf} = 0.005\ (\frac{emf^3}{1-emf})\ \frac{d^2(þs-þ)g}{\mu}$ (Carman-Kozeny equation) where:

- e_{mf} is the voidance of fluidised bed takes 0.4 according to (C. a. Richardson, Particle technology and separation processes 2002)
- μ is drag coefficient= 0.44 for water

$$U_{mf} = 0.0055\ (\frac{0.4^3}{1-0.4})\ \frac{0.02^2(6800-1000)\times 9.81}{0.44} = \mathbf{0.03}\boldsymbol{m/s}$$

After this the following were also designed: Power calculations and Electric motor sizing, Power Transmission V Belts, Cam Shaft, Cam Profile, Follower Push Rod, Follower Helical Spring, Selection of Bearings and The Jig Hutch. Assembly and Working drawings were done.

6 Recommendations and conclusion

6.1 Recommendations

After assessing the setup of the plant and its current operating condition the following recommendations were deduced:

Safety

- The operators should not overload the machine by putting a feed beyond its capacity
- There should be a good drainage system so that the working environment will not be muddy
- The operators should wear safety clothes at working environment

Quality control

- The operators should properly monitor the water levels for efficient separation
- Correct feed rate should be done
- The operators should monitor the tailings overflow and remove the tailings manually when necessary

6.2 Conclusion

The need for sustainable metal recovery from slag has been the drive to this project and I have appreciated and grasped a number of concepts in this field. A jigging machine or mechanical separator might be the foundation for recovery of valuable metals or mineral processing thereby reducing environmental impacts and ensuring good waste management of the company.

References

1. M Erdem. Journal of hazardas material b126, 2005: 176-182.
2. C. J. Evans, and E. J Mitchel. A review of mineral size and recovery methods. 1997.
3. P. Niemela, Helge and Pavi. NFM Breef Revised. 2014.
4. P Niemelä and M Kauppi. Production, Characteristics and Use of Ferrochromium Slags. INFACON XI 2007.

Decoupling Analysis on Pressure Fluctuation and Needle Valve Response for High Pressure Common Rail Injector

Wang Hao , Zhao Dongchang, Lv Wang

Automotive Data Center, China Automotive Technology and Research Centre, Tianjin 300300, China

Abstract. In the process of multiple injections, the influence of different injections makes the controlling of cycle fuel injection quantity more difficult. The high pressure common rail (HPCR) simulation model is established in AMESim environment. Through the method of combining numerical simulation and experiment test, it is found that the strong coupling of pressure fluctuation and needle valve response is the fundamental reason, which leads to the fluctuation of main injection fuel quantity (MIFQ) with dwell time (DT). The result shows that the largest fluctuation quantity is 3.6mm3 when the reference value of main injection is 60.0mm3. Non-damping LC hydraulic system model is also established. Through the analysis of the model, reducing the length-diameter ratio of internal oil duct and the delivery chamber volume are decoupling methods to the strong coupling.

1 Introduction

High pressure common rail is a complex system that coupling with electronic, magnetic, machine and liquid. Different parts will interact when the fuel injector is working [1-2]. Needle valve response is depending on delivery chamber pressure (DCP). The opening of needle valve can lead to pressure fluctuation in the delivery chamber, and the pressure fluctuation can impact the operating of needle valve as well. Especially in the process of multiple injections, dwell time is short between different injections, and the operating frequency is fast, which leads to the effect of coupling of pressure fluctuation and needle valve response can be particularly evident so that the actual fuel injection value cannot correspond with the theoretical injection value. As mentioned above, it can deteriorate the working condition of diesel engine, and the economic and emission performance can be influenced as well [3-4].

Simulation model of high pressure common rail was established in AMESim environment based to study the interactions between different injections in the process of multiple injections. In addition, the impact of the pilot injection to MIFQ and the fundamental reason of MIFQ fluctuating with DT are also studied and revealed. It also obtains the decoupling methods to the coupling of pressure fluctuation and needle valve response. Consequently, it provides a theoretical basis to the stability controlling of fuel quantity in the process of multiple injections.

2 Simulation model of HPCR system

As shown in Fig. 1, the simulation model of the HPCR system with electronic control is established in AMESim environment. It mainly consists of feed pump, common rail and electronic control injector. Main parameters of the model are shown in Table 1.

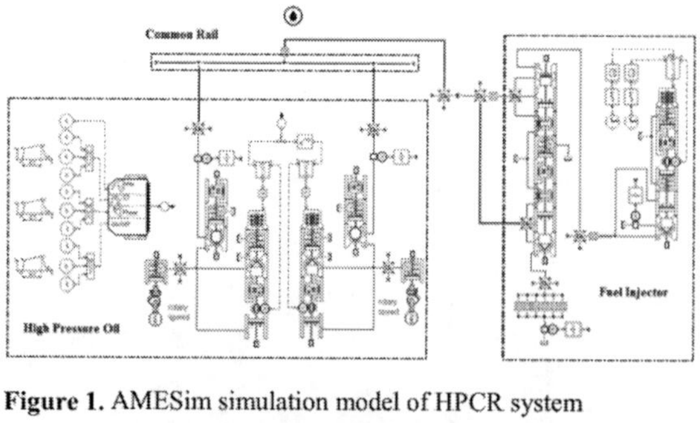

Figure 1. AMESim simulation model of HPCR system

Figure 2. HPCR system test bench

Table 1. The main parameters of the simulation model

Parameter	Value	Parameter	Value
Pump cam maximum lift /mm	9	Control chamber outlet /mm	0.24
Common rail pipe inner diameter/mm	10	Needle valve maximum lift/mm	0.35
Common rail pipe length /m	0.3	Needle valve spring pre-tightening force /N	30
Control valve maximum lift/mm	0.08	Spray hole diameter/mm	0.15
Control chamber inlet /mm	0.27	Spray hole flow coefficient	0.8

HPCR system test bench is showed in Fig.2. To ensure the accuracy of the simulation model, 1000r/min cam speed and 120Mpa common rail pressure shall be adopted to compare the result of simulation and experimental measurement. As shown in Fig.3, common rail pressure fluctuation keeps stable after several cycles. As shown in Fig.4, the fuel injection rate curve of simulation result highly corresponds with the experimental measurement. Thus, the simulation model can simulate the diesel HPCR system accurately, and it can cater the requirements of calculation and analysis of fuel injection system.

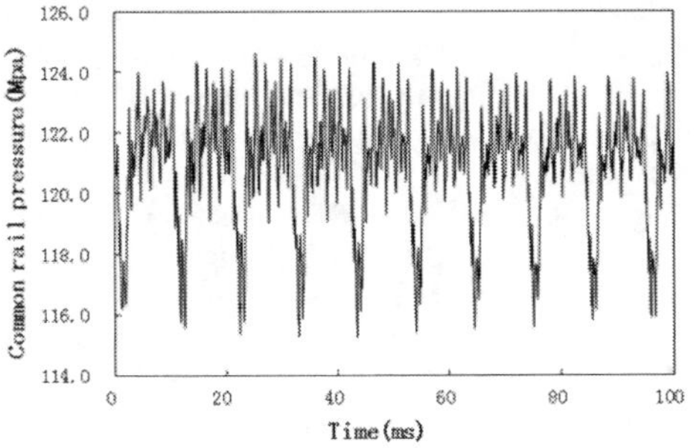

Figure 3. Curve of common rail pressure

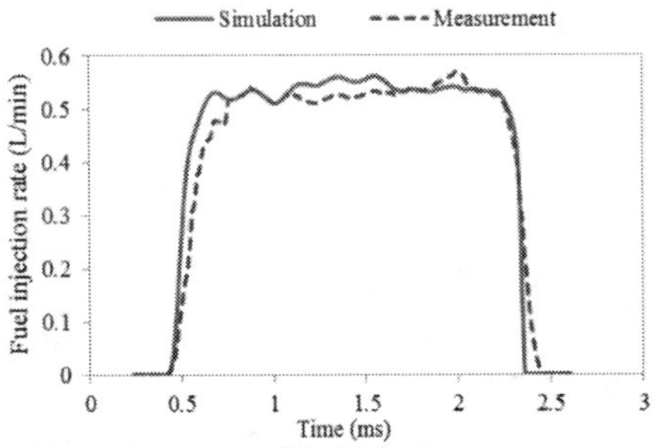

Figure 4. Comparison curve of fuel injection rate

3 Influence of pilot injection to main injection fuel quantity

Pilot and main injection fuel quantity were calibrated respectively as 6.2 mm^3 and 60.0mm^3. MIFQ fluctuation with DT from 2 °CA to 32 °CA in the process of pilot-main injection simulation experiments is shown in Fig.5. It shows that influences of the same pilot injection are different from the main injection with different dwell times. MIFQ varies with DT like cosine curve, and the minimum fuel quantity is 58.3mm^3 while the maximum is 61. 9 mm^3. The fluctuation amplitude is 3.6mm^3.

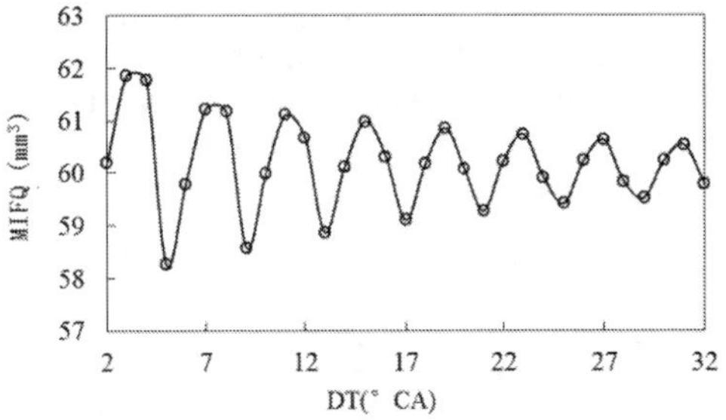

Figure 5. Fluctuation of MIFQ with DT

Simulation experiments of pilot-main injection with different dwell times conducted in this paper. Delivery chamber pressure and needle valve lift (NVL) curves are showed in Fig. 6 when dwell times are 4 °CA, 18 °CA and 32 °CA. It shows that pilot injection causes drastic pressure fluctuation in delivery chamber. Though the pressure fluctuations with the same pilot injection are identical, main injection fuel quantities are different because of needle valve's diverse opening and falling moments with different dwell times. So it is the coupling of needle valve response and pressure fluctuation caused by needle valve action that impacts the MIFQ.

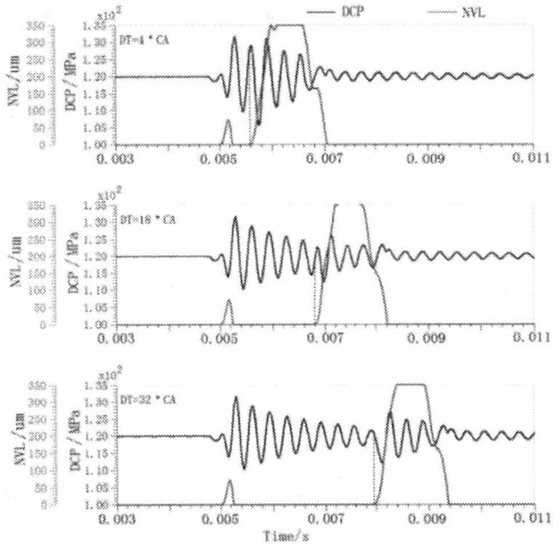

Figure 6. Delivery chamber pressure and needle valve response curve with different DTs

4 Analysis to the coupling of needle valve response and pressure fluctuation

Fuel injection quantity can be calculated by liquid flow formula (1) [5]:

$$Q = \int_0^t C_A A \sqrt{\frac{2}{\rho}(P1 - P2)} \qquad (1)$$

Where: Q = liquid flow(m^3); t = flowing time(s); A = Cross-sectional area(m^2); ρ = fluid density(kg/m^3); CA= flow coefficient; P1= inlet pressure(pa), P2= outlet pressure(pa).

By the nature of the integral calculation, when the ρ and CA are fixed, cycle injection fuel quantity is determined by the product of pressure difference between P1 and P2 with A in flowing time. In fuel injection system, P1 is delivery chamber pressure, P2 is cylinder pressure. Because P2<<P1, so P1-P2≈P1, cycle fuel injection quantity is related to the integral value of P1 in injection pulse width [6].

4.1 Opening stage of needle valve

Delivery chamber pressure and needle valve response are shown in figure 7 when dwell times are 9 °CA and 11 °CA. According to the simulation experiment, if the crankshaft speed set as 2000r/min, so the difference of needle valve action time should be about 0.167ms when DT differs 2 °CA. But the needle valve opening stage of two curves in Figure 7 has only 0.115ms time difference, in terms of this tiny time difference, it results from when DT=9° CA, needle valve opening time is at the wave trough of delivery chamber pressure, so the absolute value of pressure is lower(about 113MPa). To lift the needle valve, the control chamber needs longer oil drainage time to reach lower pressure to make the fuel in delivery chamber can get over the control of chamber fuel pressure and the initial tension of spring. The needle valve opening time is delayed, circulation time is shortened.

Meanwhile, when DT=9° CA, the main injection needle valve elevates at the wave trough of delivery chamber pressure, then the rising speed of delivery chamber pressure is obviously decreased, with a long time past, it will approach the new wave crest as shown in the area of the oval dotted line.

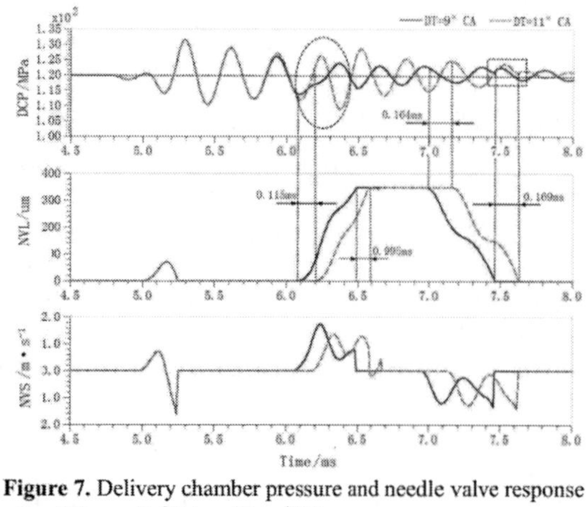

Figure 7. Delivery chamber pressure and needle valve response when DTs are 9 °CA and 11 °CA

4.2 Falling stage of the needle valve

As shown in Figure 7, when DT=9° CA, needle valve dropping time is under the wave crest of delivery chamber pressure(about 123MPa), while DT=11° CA, needle valve dropping time is at the wave crest of delivery chamber pressure(about 125MPa). The difference of dropping time between two curves is 0.164ms, it differs with the equivalent time of 2 °CA phase only 0.003ms. Here is the reasons: in the process of fuel inlet in control chamber, the oil inlet opens only, and the oil outlet is closed, the pressure building process is faster than the pressure leak process when the needle valve is lifting. Thus, it is able to form high pressure at control chamber in a short time which makes the influence of delivery chamber pressure difference shorter.

To sum up, the pressure fluctuation which caused by pilot injection can influence the cycle fuel injection quantity mainly through influencing the needle vale opening stage of main injection. Therefore, main injection fuel quantity depends on its location of delivery chamber pressure fluctuation, in other words, it depends on DT.

5 Decoupling analysis of the coupling between needle valve response and pressure fluctuation

Because the capacity of the hydraulic container is similar to the capacitor, the hydraulic pipeline has the same electrical inductance on the delay, choked flow and oscillation. Therefore, when one-dimensional unsteady flow dynamic of high pressure oil pipe is analyzed regardless of damping and throttling effect, the system can be regarded as a circuit system of capacitance and inductance [7-8].

Figure 8 shows the model of the LC common rail system without damping hydraulic system, where P_1 and C1 are common rail pressure and liquid capacity respectively, P_2 is injector inlet pressure, C_2 is the liquid capacity of high pressure fuel pipe and injector inlet chamber, P_3 is delivery chamber pressure, C_3 is the liquid capacity of the injector's internal oil duct and delivery chamber, L_{12} and F_{12} are fluid inductance and mass flow rate of high pressure oil pipe respectively, L_{23} and F_{23} are fluid inductance and mass flow rate of internal oil duct respectively.

From the analysis of Section 3, if the delivery chamber pressure fluctuation P_3 can be mitigated, the amplitude of the main fuel injection quantity can be reduced accordingly to decouple the coupling between needle valve response and pressure fluctuation. As shown in Figure 8, L_{23} and C_3 are directly related to P_3, by changing the structure of the injector itself, values of C3 and L_{23} can be changed, therefore the fluctuation of delivery chamber pressure (P_3) can be influenced. According to the formula (2) and (3), the aspect ratio of internal oil duct and the volume of delivery chamber can influence the physical parameters of L_{23} and C_3.

$$L=l/A \tag{2}$$

$$C=V/a^2 \tag{3}$$

where *l* and *A* respectively stands for the length (m) and cross-sectional area of the injector's internal oil duct (m^2),

V is the volume of the injector's internal oil duct (m^3)and the delivery chamber, a is the local velocity of sound (m/s).

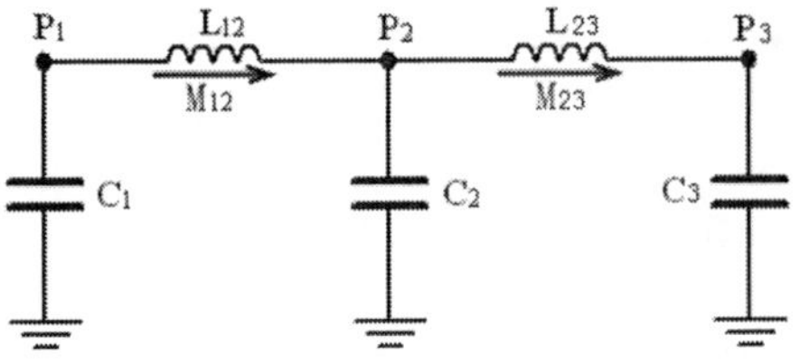

Figure 8. Non-damping LC hydraulic system model of HPCR

5.1 Aspect ratio of injector's internal oil duct

As shown in Figure 9, when injector's internal oil duct length contracts while the diameter expands, the amplitude of main injection fuel quantity with DT decreases obviously, and the speed of decay accelerates. When the length of oil duct extends from 2cm to 10cm, the diameter varies from 3mm to 2mm, the fluctuation amplitude of the main injection fuel quantity will decrease from 3.6mm^3 to 1.8mm^3, the descend range is as much as 50%.

Therefore, it can reduce the degree of coupling between needle valve response and pressure fluctuation by decreasing the aspect ratio of internal oil duct. This method can be realized by removing the control piston which is above the injector needle valve, making the control chamber pressure directly acting on the upside of the needle valve, reducing the distance between the injector inlet and delivery chamber.

5.2 Volume of the delivery chamber

On the basis of reducing the aspect ratio of internal oil duct in the 5.1section, the volume of delivery chamber decreases to 0.02cm^3. As shown in Figure 10, the amplitude of main injection fuel quantity with DT has reduced to 1.7mm^3, the coupling degree between needle valve response and pressure fluctuation is further reduced. Therefore, by reducing delivery chamber volume can decouple the coupling between the needle valve response and pressure fluctuation as well, but the effectiveness is relatively limited.

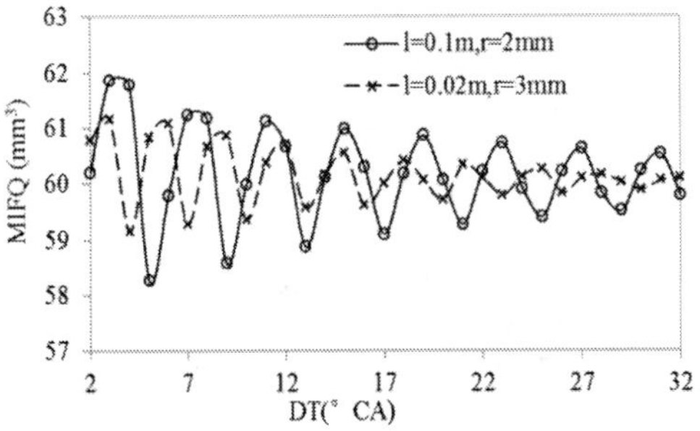

Figure 9. The influence of aspect ratio of internal oil to the main fuel injection quantity

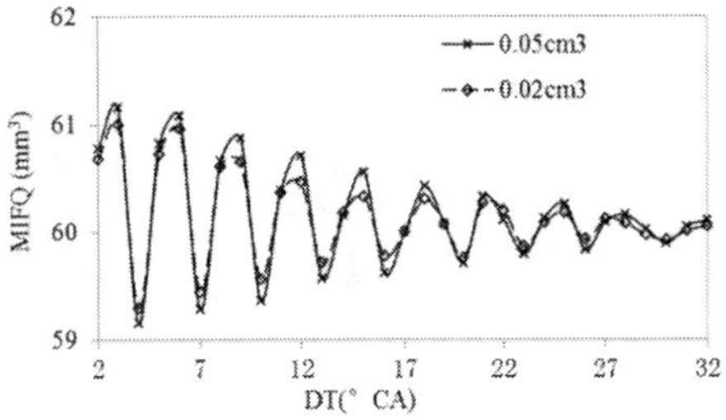

Figure 10. The influence of delivery chamber volume duct to the main fuel injection quantity

6 Summary

1. The main fuel injection quantity fluctuates with DT which is similar to the cosine curve at pilot-main injection, when the main injection base quantity is 60.0 mm^3, the fluctuation amplitude is 3.6 mm^3.

2. The pressure fluctuation in delivery chamber which caused by pilot injection is the fundamental reason of influencing the main fuel injection quantity. The pressure fluctuation can couple with needle valve response, and then change the action rules of the needle valve at opening stage, thereby it affects the main injection fuel quantity.

3. By reducing the aspect ratio of internal oil duct and delivery chamber volume can decouple the coupling between needle valve response and pressure fluctuation, and the effect of reducing aspect ratio of internal oil duct is very remarkable.

References

1. Ma Xiuzhen, Tian Bingqi, Fan Liyun, etal. Quantitative Analysis on the Effects of Parameters of Electronic Injector on the Cycle Fuel Injection Quantity Fluctuation of High Pressure Common Rail System[J]. Automotive Engineering 2015, Vol.37(1):55-61.
2. Saud Aldajah, Mohamed Y E Selim. Tribological Behavior of Dual Fuel Diesel Engine[J]. International Journal of Mechanical Engineering and Robotics Research, Vol. 2, No. 2:41-50, April 2013
3. Liu Zhen-ming, Shao Li-ming, OuYang Guang-Yao, etal. Optimization Study On Injection System Parameters Of Common Rail Diesel Engine Based On Approximate Model[J] .Chinese Internal Combustion Engine Engineering, 2011, 32(6):63-67,73.
4. Mikio Kumano, Kenzo Yano, Kanehito Nakamura, etal. Advanced Electronics for a Clean Diesel Engine Management System[C]. SAE Paper 2006-21-0059, 2006.
5. Zheng Jinbao, Liao Xuelong, Hong Jianmei, etal. Measurement and Analysis of Pressure of Nozzle Volume in Electric Solenoid Common-Rail System[J]. Transactions of CSICE, 2012, Vol.30(1):86-90.
6. Andrea E. Catania, Alessandro Ferrari, etal. Experimental Investigation of Dynamics Effects on

Multiple-Injection Common Rail System Performance[J]. Engineering for Gas Turbines and Power, MAY 2008, Vol. 130 / 032806.
7. Su Haifeng, Zhang Youtong, Luo Xu, etal. Experimental Investigation on Water Hammer Pressure Wave in High Pressure Common Rail System[J]. Transactions of CSICE, 2011, 29(2), 163-168.
8. Andrea E. Catania, Alessandro Ferrari, Ezio Spessa. Numerical-Experimental Study and Solutions to Reduce the Dwell-Time Threshold for Fusion-Free Consecutive Injections in a Multijet Solenoid-Type CR System[J]. Engineering for Gas Turbines and Power, MARCH 2009, Vol. 131 / 022804.

MATEC Web of Conferences **108**, 04012 (2017) DOI: 10.1051/matecconf/201710804012
ICMAA 2017

Analysis of windlass sprocket wheel intensity

Yang-yang Shi, Zhi-ming Wang , Jian Zheng and Ning Zhang

Institute of Mechanical Engineering, Nanjing University of Science and Technology, 210094 Nanjing, China

Abstract. The multi-body dynamics model of windlass was established by RecurDyn and Pro/E in this paper, and the multi-body dynamics simulation was also conducted. According to the results of the multi-body dynamics simulation, structural static analysis on sprocket wheel in dangerous conditions were executed by ANSYS Workbench. The results showed that there are three bearing situations in the meshing transmission process of sprocket wheel and chain: Single tooth bearing, two tooth bearing and carrying equal contact force and two tooth bearing but one tooth contact force reaches the maximum value. The equivalent stress value at the contact position of sprocket wheel and chain exceeds the yield stress of the chain wheel material, but it does not cause damage to chain wheel. In addition to the contact location, the equivalent stress value of the remaining parts of sprocket wheel are in the allowable stress range of sprocket wheel material, comply with the requirement of strength design.

1 Introduction

Windlass is an equipment to overcome the external force in the work phase of anchor and keep the stablility of the ship. It ensures the safety of the ship when it leave away from the dock and in case of emergency braking. Safe and reliable operation of the windlass directly affect normal operation and operational safety of the ship. The anchor of windlass is implemented by meshing transmission process of sprocket wheel and chain. Therefore, the normal work of sprocket wheel is essential for windlass.

Design and Verification on windlass components are based on empirical formulas and simple mechanical models currently [1], such as Zhou Zhong-wang, Fang Lian-xing [2-3]. Their calculation of the load in the sprocket shaft and the base by finite element analysis are different from the actual load when the windlass working. In this paper, the sprocket wheel was studied, and the multi-body dynamics model of windlass was established by RecurDyn and multi-body dynamics simulation was also conducted. The mechanical characteristics of sprocket wheel during operation were analyzed and the dangerous working conditions of sprocket wheel were determined. The strength of sprocket wheel were analyzed by ANSYS Workbench.

2 Brief introduction of sprocket wheel

Since the shape of sprocket wheel is so complex, it is generally made by casting fabrication [4], and chain are also blank parts, they both have large manufacturing errors. Therefore, there is slip phenomenon in the sprocket wheel nest in the meshing transmission process of sprocket wheel and chain. It caused abrasion in the tooth surface of sprocket wheel. The meshing transmission between sprocket wheel and chain belongs to circular chain meshing transmission. There are frequent shocks between sprocket wheel and chain in the meshing transmission process [5]. So the sufficient strength of sprocket wheel is required.

In this paper, the diameter of the hydraulic windlass anchor is 78mm. Sprocket wheel is a five gears of A type. Its material is GS-52, and the yield strength is 260MPa. Its tooth structure is shown in Fig. 1. According to People's Republic of China shipbuilding industry CB/T 3179-1996 <sprocket wheel> Standard [6], the main structural parameters are as follows: the pitch circle's diameter is 1062mm, namely when the sprocket wheel meshing with a chain, circumscribed circle diameter of the regular polygon formed by the center line of chain, can be used as the calculation diameter when sprocket wheel transfering torque. The speed calculation diameter is 993mm, used as the calculation diameter which determines the anchor's mobile line speed when anchor handling.

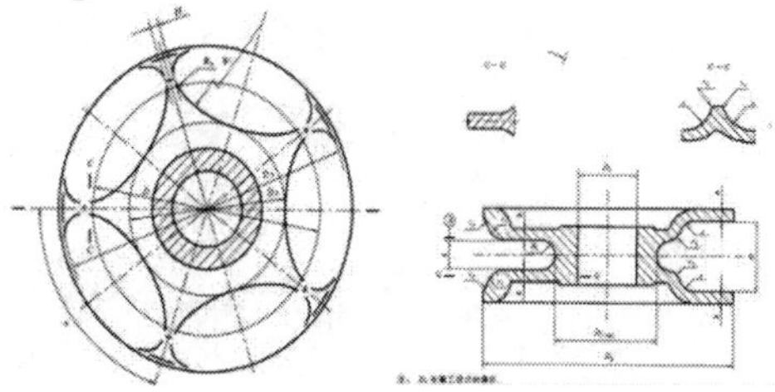

Figure 1. Tooth-shaped structure of five gears sprocket wheel.

3 Windlass multi-body dynamics model building and analysis of results

Since the shape of sprocket wheel is so complex, it is generally made by casting fabrication [4], and chain are also blank parts, they both have large manufacturing errors. Therefore, there is slip phenomenon in the sprocket wheel nest in the meshing transmission process of sprocket wheel and chain. It caused abrasion in the tooth surface of sprocket wheel. The meshing transmission between sprocket wheel and chain belongs to circular chain meshing transmission. There are frequent shocks between sprocket wheel and chain in the meshing transmission process [5]. So the sufficient strength of sprocket wheel is required.

3.1. Multi-body dynamics model building

3.1.1 Import and process of solid models of windlass

The windlass, chain and chain stopper are interference-free assembled in Pro/E [7] and save as a Parasolid file, then import it to RecurDyn. Set units to "MMKS", adjust the direction of gravity for the Z-axis positive direction, the remaining parameters keep the default settings. Use Merge function to merge every parts of windlass, the assembly model of windlass is shown in Fig. 2 after the merger. The name of the components are shown in table 1.

Table 1. Name of the components

Number	Name	Number	Name
1	Chain stopper	7	Spindle support base
2	Chain	8	Sprocket wheel base
3	Sprocket wheel	9	Sprocket wheel gear module
4	Pre-axis module	10	Roll module 1
5	Spindle module	11	Roll module 2
6	Spindle module	12	----------

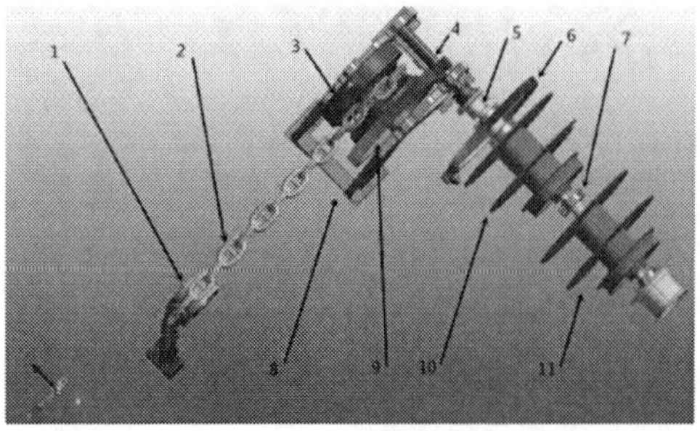

Figure 2. Assembly model of windlass.

3.1.2 Windlass model constraints exerts

According to the working principle of the various components of the windlass, constraints for various components of the windlass are added, they are shown in table 2.

Table 2. Constraints for various components

Constraint objects	Constraint name
Chain stopper and earth	Revolute Joint
Sprocket wheel and sprocket wheel base	Revolute Joint
Sprocket wheel and sprocket wheel gear	Perpendicular Joint
Pre-axis module and pre-axis pinion	Fixed Joint
Spindle and spindle bearing	Revolute Joint
Gearbox pinion and earth	Revolute Joint
Sprocket wheel base and earth	Fixed Joint
Sprocket wheel gear and sprocket wheel base	Revolute Joint
Pre-axis module and sprocket wheel base	Revolute Joint
Spindle support base and earth	Fixed Joint
Gearbox gear and spindle	Fixed Joint
Roll module 1,2 and spindle bearing	Fixed Joint

3.1.3 Windlass model exposure modeling

The contact between various components of the windlass use the Extended Surface To Surface in RecurDyn, such as the contact between chain and chain stopper, chain and chain, chain wheel and chain.

3.1.4 Windlass model-driven load

The speed of the anchor must be greater than or equal to 0.15m/sec in the working phase of anchor. The speed of gearbox pinion is 24 rad/s. To reduce the impact of sudden changes in velocity, load is applied by STEP function[8-9]. Gearbox pinion angular velocity-time curve graph is shown in Fig. 3.

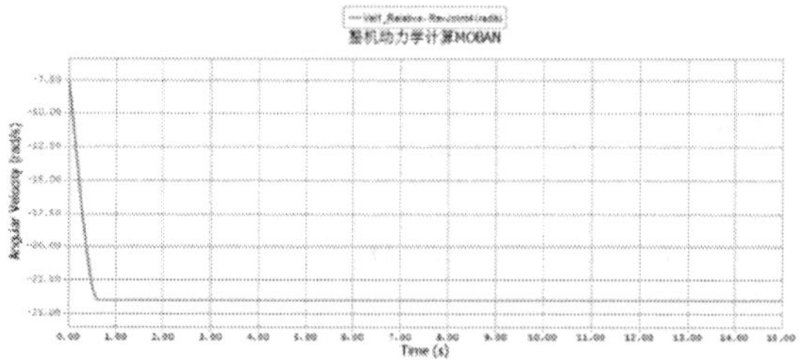

Figure 3. Gearbox pinion angular velocity-time curve graph.

DOI: 10.1051/matecconf/201710804012

3.1.5 Virtual prototype test

Virtual prototype needs to be tested after created. Pre-analysis results of windlass virtual prototype model are shown in Fig. 4. It can be drawn that windlass model established in this paper contains 67 parts, 111 surface contact (extended surface contact), 304 degrees of kinematics freedomand and no redundant constraints. Model validation is correct.

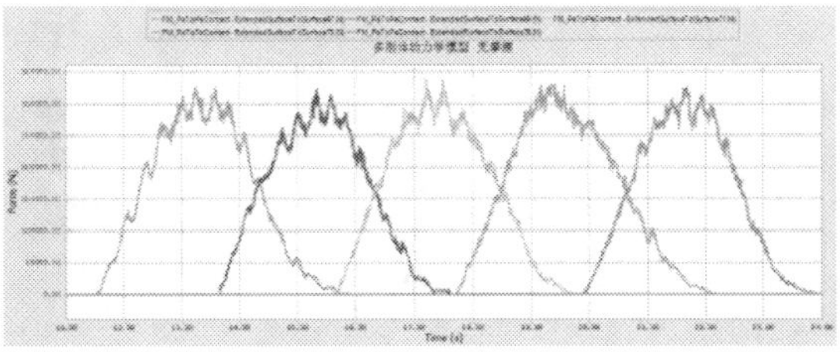

Figure 4. Pre-analysis results.

3.2. Analysis of Kinetic simulation results

Composition of contact force graph between sprocket wheel and chain during the working phase of anchor is shown in Fig. 5.

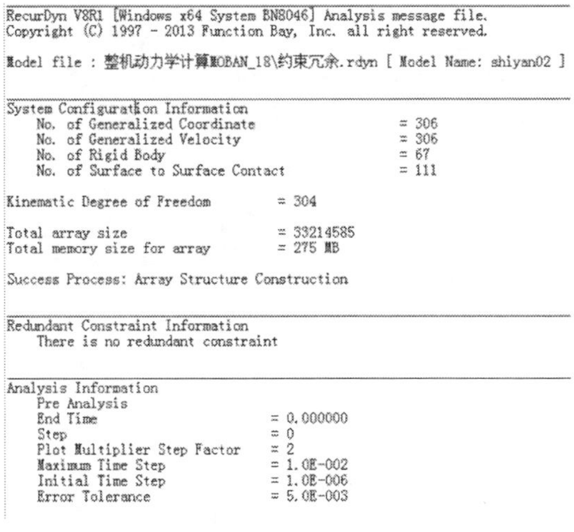

```
RecurDyn V8R1 [Windows x64 System BN8046] Analysis message file.
Copyright (C) 1997 - 2013 Function Bay, Inc. all right reserved.

Model file : 整机动力学计算MOBAN_18\约束冗余.rdyn [ Model Name: shiyan02 ]

System Configuration Information
    No. of Generalized Coordinate            = 306
    No. of Generalized Velocity              = 306
    No. of Rigid Body                        = 67
    No. of Surface to Surface Contact        = 111

Kinematic Degree of Freedom        = 304

Total array size                   = 33214585
Total memory size for array        = 275 MB

Success Process: Array Structure Construction

Redundant Constraint Information
    There is no redundant constraint

Analysis Information
    Pre Analysis
    End Time                       = 0.000000
    Step                           = 0
    Plot Multiplier Step Factor    = 2
    Maximum Time Step              = 1.0E-002
    Initial Time Step              = 1.0E-006
    Error Tolerance                = 5.0E-003
```

Figure 5. Composition of contact force.

The analysis of results from figure 5 are listed as follows:

1. The number of chain are no more than two during the meshing transmission process of sprocket wheel and chain;

2. Points of intersection of adjacent curves represent two teeth bearing and carrying equal contact force;

3. With the rotation of sprocket wheel, sprocket wheel tooth contact force increasing and the anchor chain nest slippage occurs, impact between sprocket wheel and chain appears, the contact force mutates and reaches the maximum value, namely the time when contact force of various sprocket wheel and chain tooth reaches the maximum in the diagram. In this case, two teeth bearing, but one tooth contact force reaches the maximum value.

4 Strength analysis of sprocket wheel

According to the results of contact force graph between sprocket wheel and chain, there are three bearing situations in the meshing transmission process of sprocket wheel and chain: Single tooth bearing, two tooth bearing and carrying equal contact force and two tooth bearing but one tooth contact force reaches the maximum value. The third case is a dangerous condition from the perspective of sprocket wheel strength. Teeth load values are listed as follows: FX1=880.92N, FY1=108268.8N, FZ1=331395.91N, FX2=466.63N, FY2=8110.92N, FZ2=6391.01N.

Sprocket wheel and Sprocket wheel gear are connected by five bolts and five cylindrical pins. The effective area of the bolt and cylindrical pin are 1152mm^2 and 1440mm^2. The torque of sprocket wheel in the working phase of anchor is T=F*D/2=290KN*1062mm/2=153990Nm. The results showed that the pressure value of cylinder pin hole and bolt hole are 12.98Mpa and 16.23Mpa.

In the ANSYS Workbench software, material properties on chain wheel are defined as follows: elastic modulus of 202Gpa, Poisson's ratio of 0.3, density of 7800kg/m^3. Mesh the procket wheel, add pressure load to bolts and cylindrical pins, then add tooth load, define boundary conditions and take the static analysis. Mesh model of sprocket wheel is shown in Fig. 6. Finite element analysis model of sprocket wheel is shown in Fig. 7.

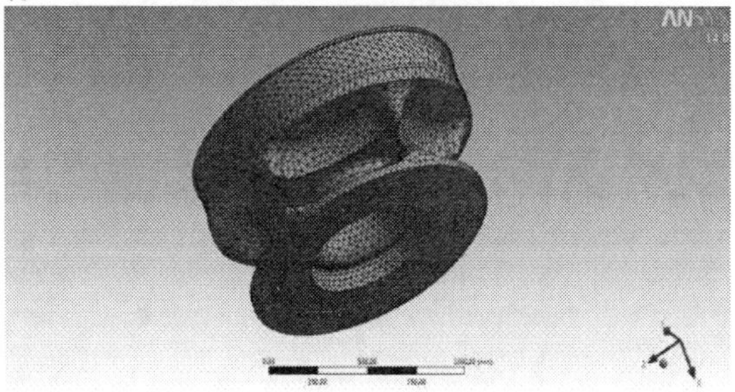

Figure 6. Mesh model of sprocket wheel.

Figure 7. Finite element analysis model of sprocket wheel.

According to the static structural analysis results of sprocket wheel, the equivalent stress figure of sprocket wheel in dangerous conditions is shown in Fig. 8.

Figure 8. The equivalent stress figure of sprocket wheel in dangerous conditions.

Strength analysis of sprocket wheel are listed as follows: the contact between sprocket wheel and chain belongs to point contact in theory, therefore, stress value at the contact position of sprocket wheel and chain is very large, the maximum equivalent stress reaches 1567.6Mpa. When the stress exceeds the yield limit of the material of sprocket wheel, plastic deformation occurs at the contact position of the sprocket wheel and chain. When the plastic deformation occurs, the contact area of the sprocket wheel and chain increases, the stress value declines sharply. Therefore, sprocket wheel will not be undermined in the actual work process. In addition to the contact position of the sprocket wheel and chain, the equivalent stress values of the remaining parts of the sprocket wheel do not exceed 111.97Mpa, less than the allowable stress of the sprocket wheel material 0.9 –234Mpa, meet the strength requirements of chain wheel.

5 Conclusion

The mechanical characteristics were analyzed in the meshing transmission process of sprocket wheel and chain by RecurDyn. The results showed that: there are three bearing situations in the meshing transmission process of sprocket wheel and chain: Single tooth bearing, two tooth bearing and carrying equal contact force and two tooth bearing but one tooth contact force reaches the maximum value. The third case is a dangerous condition from the perspective of sprocket wheel strength.

As can be seen from the results of static analysis, the equivalent stress value at the contact position of sprocket wheel and chain exceeds the yield stress of the chain wheel material, but it does not cause damage to chain wheel. In addition to this location, the equivalent stress values of the remaining parts of the sprocket wheel are in the allowable stress range of sprocket wheel material, comply with the requirement of strength design.

The equivalent stress cloud graph and the related conclusions of the sprocket wheel through the strength analysis provide a reference for structure optimized design and lightweight design of the windlass.

References

1. F. C. Hu. Wuhan University of Science and Technology, **1**, 8-10 (2007)
2. Z. W. Zhou, R. S. Wang, Manufacturing Automation, **33**, 151-153 (2011)
3. Z. X. Fang, L. D. Zhang, Guangdong Shipbuilding, **1**, 40-42 (2008)
4. D. L. Liu, Jiangsu Shipping, **19**, 37-38 (2002)
5. F. L. Chen, Journal of Shanghai Jiaotong University, **3**, 133-150 (1979)
6. Standardization Administration of China, Sprocket wheel, **3179**, 4-6 (1996)
7. Y. Lu, X. G. Ma, Q. L. Shu, Transactions of Shenyang Ligong University, **27**, 76-79 (2008)
8. X. G. Ma, W. Yang, X. M. You, Chinese Journal of Construction Machinery, **7**, 146-152 (2009)
9. X. G. Ma, Y. Lu X. M. You, China Mechanical Engineering, **20**, 1956-1959 (2009)

MATEC Web of Conferences **108**, 04013 (2017) DOI: 10.1051/matecconf/201710804013
ICMAA 2017

Research and Application of Device for Simulation Sports Training based on Cycle Condition

Jian Zhao
Jilin Engineering Vocational College, Siping, 136001, China

Abstract. the slope simulation method of the spiral mechanism, the rubbing transmission, the resistance simulation method of magnetic particle brake loading on the rear wheel is proposed by anglicizing the road sports conditions and resistance of the actual cycling. the structure design, loading design and control design of the road sports condition simulator is designed. the structure of putting the motor forward and forcing the big sprocket rotation through the chain transmission, and the method of the frequency converter regulates the rotational speed of the motor within the range of certain frequency, PLC programming the rotation rules is adopted. The complete device is applied to athletes sports training of the national bicycle team at ordinary times after debugging successfully.

1 Introduction

Scientific training methods should be combined with modern training equipment. In training, coaches should be able to master the basic parameters of athletes riding and physiological conditions in order to address the weak link, scientific and effective guidance training. Cycling training in the outdoor sports venues, the training of athletes by the weather and climate impact. Rain, wind, snow and other bad weather athletes are not normal training, long winter and rainy season greatly affected the coaches of the training program, the quality of the circle of training.

Sprint stage, is the most tired athletes, due to the lack of training equipment related to athletes usually can not effectively sprint special training, resulting in China's elite athletes in the race sprint stage slower than the other, losing the game. Therefore, the National Cycling Team urgently needs a high-quality cycling simulation trainer, improve training conditions, improve the quality of training, improve the game results.

The indoor cycling simulator can simulate different outdoor road conditions, and has the same body posture and force feeling, pedalling frequency and the exertion of the athlete as the outdoor training and competition. 1. Bike simulation trainer can be free to set the requirements of different coaches and their changes in traffic conditions, and changes in the simulation of different driving force of the athletes feel and body posture to be consistent with the actual outdoor training. 2. The simulator can also be free to set the athlete's pedalling force size and frequency. Athletes should be able to easily set the parameters of the above changes in the riding, but also by the trainer with a radio remote control of the above parameters. 3. The training device also need to have security protection device, when the pedal rotation speed exceeds the athlete's ultimate capacity, the safety clutch can be separated from their own, so athletes will not be pedal injury.

2 Introduction of Bike Simulator

Bicycle movement is a semi-mechanized project, bicycle simulation training equipment need to take into account the bike's own movement and human movement changes, so the problem is more complex. The domestic bicycle simulation training device is less developed, most of the training products are developed abroad.

Europe and the United States on the bicycle simulation training device research earlier, as early as the seventies and eighties there have been many bicycle simulation training device patented product. At present, Nanyang University in Singapore and Korea University of Science and Technology have developed an advanced interactive simulation training device.

The bicycle simulator developed by the Korea University of Science and Technology is shown in Figure 1, which mainly includes four sub-systems.

1. Motion generation simulation system

The system uses a six-degree-of-freedom Stewart platform consisting of six freely-extendable struts and an up-and-down platform. The platform is driven by a motor and has two processors (PC and DSP) for digital servo control of the motion platform. The two processors communicate with each other through the dual-channel memory RAM installed in the DSP chip and connect with the PC bus . For high-speed motion control, the amount of computation is very large.

2. Vehicle steering force simulation

There are two forces on the bicycle handlebar: active and passive. This system consists of an MR clutch and a brushless DC motor to produce these two forces. The connection of the bicycle to the clutch and the electric motor.

When an active force is generated, the motor is running and the MRF (Magneto-Rheological Fluid) clutch transmits the torque to the handlebar. When the resistance is generated, the motor is stopped and the MRF clutch acts as a damper. In both cases, the magnitude of the force is controlled by the current input to the MRF clutch. MRF clutch maximum output torque is 5Nm, the control frequency can reach 2Hz.

3. Pedal resistance system control

To feel real, a pedal resistance system is used to simulate the actual riding resistance, which includes ground friction, air resistance, and resistance to human inertia. The drag system also generates the motive force to simulate the accelerating force at the downhill. In the pedal drag system, the torque applied by the rider to the pedal is estimated, and the rotational speed of the speed control sprocket, which is calculated by the bicycle power module, is then routed. The pedal resistance system consists of an AC servo motor, a flywheel, and an MR brake. The motor simulates the accelerating inertia of the downhill and the brakes produce a drag torque. This resistance system is placed between the Stewart platform and the rear wheel.

4. Visual simulation system

In order to better produce simulation results, a virtual campus of the real scene. In this campus model, the geographic features and buildings are modelled separately. First, a 1: 5000 campus scene map was scanned as a basic image, and then the road and geographic features were made using the related software.

5. System integration

The integration of the system focuses on information exchange and coordination between images, platform movements, handlebar forces and foot-resistance systems to produce real visual effects.

3 Structural Design of Road Condition Simulation Trainer

The road condition simulator uses the racing frame as the basic frame, and uses the crank-slider mechanism to simulate the slope movement. In this mechanism, the frame for the connecting rod, the shaft support for the rocker, screw vice for the mobile pair. Motor by worm gear reducer drive screw rotation, the nut will move up and down along the rail, the body also with the swing, so that cyclists sense of slope movement. The flywheel is mounted on the rear frame to simulate the kinetic energy of the bicycle. A friction wheel is fixed above the flywheel. The friction wheel is coaxial with the magnetic powder brake and transmits the braking torque by the friction between the flywheel and the friction wheel. Structural design, including simulation of structural design and loading load simulation structure design, the overall layout shown in Figure 1.

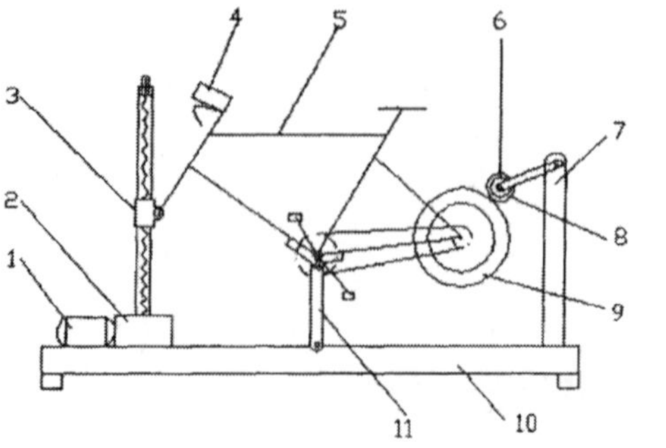

1-motor; 2-worm reducer; 3-front axle lifting device; 4-control display device; 5-frame; 6-magnetic powder brake; 7-loading support; 8-friction wheel; 9 – flywheel; 10-A bottom bracket; 11-a central shaft supporting device

Figure 1. The scheme of the road condition simulator.

4 Determination of Simulated Equations for Road Condition Simulator

For the sake of research, it is assumed that neither the road surface nor the wheels are deformed. When a person stepping on the car, a force couple moment is applied to the wheel by the transmission of the chain, trying to rotate the rear wheel. Due to the wheel and the ground between the role of adhesion, in the non-sliding rolling, the road will inevitably exert a force on the wheel, the direction of the same direction with the bike, which is driving the bicycle driving force, that is, the driving force. The main force is transmitted from the lower edge of the wheel to the axle of the wheel, trying to push the wheel forward. When the driving force is increased enough to overcome the resistance of the bicycle at rest, the rear axle begins to move forward, at which time the rear wheel rolls along the road while the front wheels roll and the bicycle begins to move forward. After the start of the bicycle, its driving condition depends on the driving force and the driving process of the relationship between the various resistance. When the bicycle is running, the value of the driving force must be equal to the total resistance of the bicycle. The main resistance to air resistance, rolling resistance, acceleration resistance and slope resistance.

In the test system, with the power meter measuring the torque applied to the crank, with the sensor to measure the crank speed, the product of the two people is the power applied to the bicycle. When a torque is applied to the crank, the strain gauge in the crank integrated device will feel a slight deformation. Sprocket integrated disc internal contact switch, crank every rotation, the contact switch is fixed in the frame of the magnetic sensor excitation time.

The measured torque and speed values are digitized in the crank integrated device and converted to a high frequency pulse width modulation electronic signal which is fed to the metal coil inside the sprocket assembly disc, the secondary coil fixed to the frame This signal is sensed and passed to the microprocessor in the power controller via a cable where the processor calculates the speed of rotation and the average torque of the crank in one rotation, multiplying the two to obtain the power value, On the LCD display shows the reading. The total mechanical energy calculation is obtained by integrating the power over the entire training and testing process.

When the wheel rotates one turn, the contact switch fixed to the front fork is excited once by the magnet mounted on the bar, thus obtaining the wheel speed. Through the power controller input wheel circumference, the microprocessor will calculate the speed of bicycle travel.

5 BT-ATS power car special ability training

We use the BT-ATS power car for cycling athletes special ability training, to overcome the previous traditional practice of some special means and models. Special ability is good or bad, to a certain extent, the results of cycling athletes have a direct impact. In the actual training and competition process, the most important for cyclists is the ability to special good or bad. Special ability is the special level reflected by the athletes in the riding process, often expressed as the athletes in the higher frequency of riding a long time to maintain a certain power demonstrated by the special ability. It is found that the speed of the bicycle can be increased more effectively by improving the pedalling technique of the athletes and improving the pedalling frequency to increase the transmission ratio and the lower limb strength while keeping the rhythm unchanged. At the same time doing work also increased the power output, improved cardiovascular function. The ability to improve the specific sports and training competition results are inseparable.

Pedalling action is the bicycle movement in the key technical action, is also the most complex and most difficult to master the action. In the use of BT-ATS power car training or testing when the athletes do not have a good pedal technology will make it strong in the means of content likely to cause physical harm, especially the cyclist foot joints, knees and skeletal joints, if Long pedalling technology is not standardized or even cause premature end of the sports career. So we must pay attention to standard pedalling technology. In the actual riding or training in the BT-ATS power pedal technology better athletes can achieve the minimum energy consumption of the output more power to achieve higher speed. So cycling athletes must improve pedal technology.

6 Method of Pedalling for Cyclists

Bicycle pedal there are three ways: freestyle, front toe down and rear heel press. Most of the excellent athletes in our country have adopted the freestyle pedalling method. This freestyle pedalling method, foot rotation in the course of a week, is based on different parts, step on the joint angle also changes with. Feet in the highest point A, the heel slightly drooping 8 degrees, and then forced down the force gradually increased to point B, the soles of the feet into parallel relationship with the ground, pedalling force the most. And then down, the force gradually reduced into the relaxation area down, the muscles began to relax, slightly raised his heels, to the C point, the heel gradually mentioned about 15 degrees. When the foot back to D point and parallel to the ground. Up, heel and then quickly brought up, loop into the A point. Freestyle pedal, in full compliance with the principles of biomechanics, forced the direction of rotation and pedal rotation formed when the tangent line, reducing the knee and thigh force amplitude, effectively improve the pedal frequency instantly through the relaxation area, to avoid death The presence of points. Thigh muscles and small muscle groups can also be an instant relaxation. However, this pedal method is not easy to grasp. Before the tiptoe pedal approach: In particular, relatively high speed athletes to use more than the front tiptoe pedal method of the overwhelming majority. This pedal mode is always downward during the entire pedal rotation. Foot range of activities is very small, particularly conducive to improve the pedal frequency, the athletes as long as the heart learned to learn it is easier, but the leg muscles will always be in a relatively tight state is not conducive to the relaxation of the leg muscle group recovery.

After the heel down pedal method is a little toe up, heel down 7-14 degrees. This method is rarely used in general athletic training and competition, only a small part of the players in the training or race riding excessive adjustment force will be used when the heel down pedal method. It is characterized by changes in muscle strength in a short time, get a short break, to restore the effect of muscle fatigue.

7 BT-ATS power car in the bicycle speed training on the edge

The change of bicycle speed training means has always been a major topic of concern to trainers. We use BT-ATS power car training to change the intensity, frequency and periodicity of the stimulus so as to make the nerve center of cycling athletes form good dynamic shaping, The speed of cycling athletes has greatly improved.

If too short in this distance will not play the role of speed, if it is too long with endurance characteristics, the loss of the meaning of bicycle speed training. Incubate the acceleration capability with a distance of at least the same distance as the acceleration phase. In the absolute speed training is generally used in the distance of 500m-1000m better, of course, in the BT-ATS power car can also use other speed means, such as the ability to train 3km sprint, speed training may also add a few 3km Sprint riding, but the athletes riding at this distance, the excitement of the nervous system is very high, relatively abundant physical strength, no significant reduction in riding speed, although with endurance characteristics, but the speed of the Development and improvement still has a forward role, but this speed training in BT-ATS power car can not repeat too many times, if repeated too much will be away from speed training this track.

Speed training, the sense of rhythm training in speed training, we realize the good action rhythm is also an important factor in achieving excellent results, when the rider pedaling required rhythm, will make him produce a sense of rhythm, and this A sense of rhythm that can make him ride in the ride to play a smooth speed, but also the relative savings in physical strength, improve excitability, to achieve the purpose of technical action to relax and rationalize. Through a sense of rhythm training, athletes acceleration, relaxation and cycling movements have a certain significant increase. This kind of exercise if arranged in the excitement of the athletes in general time, can play a role in mobilizing nervous system excitement. The optimal timing of speed training is to be arranged when the cyclist is in good athletic state because the flexibility of the nerve, the stretching of the muscle, the elasticity and the biochemical mechanism are important factors constituting the speed, so if the speed training is arranged in the athlete's fatigue, Not only the above factors are not fully displayed, and because the nervous system in a dark state of the deeper, lower muscle elasticity, loss of movement of the original frequency and amplitude, in this state for speed training, not only for speed Will not achieve the promotion of the role of the passage of time will also form a speed barrier to the entire training adverse effects.

Therefore, we in the entire training cycle arrangements, the general speed training courses scheduled for Monday or after the appropriate adjustments to resume the first training session. In short, as far as possible, the speed training courses arranged in the period of over-recovery, so that athletes can show a high level of competition, showing high energy, high excitability, thereby enhancing the cycling speed training effectiveness.

8 BT-ATS power car in the training means the advantages

BT-ATS power car 8 seconds training means This means the main purpose of training in the energy system, along with high resistance to improve lactic acid resistance. If you decide to use 36x26 gear ratio, then the muscle nervous system efficiency than muscle strength to give priority. The number of exercises can be started in situ 12 times at this stage, practice at this time for 6 seconds ++2 seconds (best effort) (can be within 7 seconds shift) using the smallest bicycle and BT-ATS Power car 26 teeth on the rear flywheel, recovery time in the practice interval of 2 minutes 45 seconds, each practice interval of 3 minutes. It is important to maintain the same posture while practicing a horse and maintaining a posture throughout the process. It is easy to do in the 3 minutes of the practice and should use the same gear ratio as the game. A series of low-frequency training in this training phase is the use of the largest bike tooth plate and BT-ATS power car 12-15 on the fly after the device. The training season can use high frequency training, using the smallest bike tooth plate and BT-ATS power car on the 26 teeth of the fly after each exercise should be in the upper training intensity effect will be very obvious.

BT-ATS power car 30 seconds training means This means the main purpose of training is lactic acid-resistant ability of the energy system and muscle nervous system strength.This is a high-intensity training and a variety of training forms of bicycle training. Training can be started in the practice of the number of times in this stage of 6 - 10 times, practice at this time for 30 seconds +3 seconds (best) (can be within 7 seconds shift) using the smallest bicycle and BT -ATS power 26-tooth car after the flywheel, recovery time in the practice interval of 5 minutes -20 minutes. Up to 2 times a week training. Ask for a practice mount and maintain a pose throughout the exercise. This training session begins with a 10-second period of low intensity followed by a 5-second increase in intensity to the last 5 seconds of the highest intensity. The same gear ratio as in the race should be used. This series of low-frequency exercises, Of the bicycle tooth plate and BT-ATS power car 12-15 teeth on the fly after the device. The training season can use high frequency training, using the smallest bicycle disk, and the BT-ATS power car on the 26-tooth fly after the device, using the test power data. Each exercise should be in the upper exercise intensity. BT-ATS power car 5 points Training means The main purpose of this training is to improve the anaerobic capacity of the energy system and the efficiency of the muscular nervous system, which is to improve the training intensity and variety of training forms of bicycle training.

The number of practice sessions is three to ten times during this period. The duration of the exercise is 5 minutes and the recovery time is 5-10 minutes in the interval. The interval between exercises is 10 to 15 minutes. Such training may take place once a week. It is important to maintain the same posture while practicing the mount and maintaining a posture throughout the exercise. The same gear ratio as the race should be used. The training phase can use high frequency training, the use of the smallest bicycle tooth plate, and BT-ATS power car on the 26-tooth fly after the device. The target should have a cadence of 120-140 beats per minute, and each exercise should be of medium to high intensity.

9 Conclusion

(1) On the basis of consulting a large amount of literature and data, the overall design of working simulator and sprint simulation trainer is put forward according to the requirement of development. The design of slope simulation simulator and simulation load structure And the design of the control principle. The frame structure design, the cadence control design and the security protection design of the sprint simulation trainer are completed. After the assembly of the prototype production, the prototype debugging is carried out, and the shortcomings are improved.

(2) Establishing the kinematics model of the working condition simulation structure, and obtaining the relation equation between the vehicle frame angle and the nut lift. On the basis of analyzing the actual running resistance of the bicycle, the experimental results show that the rotational speed of the crankshaft is between the air resistance and the rolling resistance torque. The dynamic model of the bicycle and the dynamic model of the

DOI: 10.1051/matecconf/201710804013

simulator were established, and the loading equation of the magnetic brake was deduced.

(3) In the design of road simulation trainer, a kind of original slope and resistance simulation method is adopted, and the corresponding matching load is realized by PLC programming and control. According to the domestic and international research, the simulation training of this project has not appeared The structure of the device, indicating that the device is an innovative product.

(4) sprint simulation trainer is designed according to the special needs of our athletes, which uses a small sprocket and clutch torque limiter, simplifies the transmission, the athletes to achieve effective security protection, the structural design can That is an innovation.

(5) The successful development of the simulation trainer can provide a set of advanced training equipment for the Chinese cycling team to solve the practical difficulties of the coaches and athletes in daily training. At the same time, the coaches have mastered the training of the athletes. To help coaches effective implementation of the training program, improve the athletes competition results, enhance the strength of the Olympic Games won the title played a role.

References

1. Xu zyong. Mechanical Design Manual Volume III. First Edition. Mechanical Industry Press, 1992: 22-115-120.
2. Chen Zuomo. "Mechanical Principles" (Sixth Edition). Higher Education Press, 2001,5: 1882.
3. Zheng Ping.Mechanical Control Technology of Electric Machines.First Edition.Chongqing University Press, 2003, 8: 145-147.
4. GAO Wen-da, MU Shan-xiang. Axial-driven Bicycle Booster. Internal Combustion Engine .2014, (3): 710.
5. Sun Ning, Huang Qiuhe. Bicycle Exercise and Mechanics. Modern Physics Knowledge, 2008, (63): 37-38.
6. DING Siyuan. Modal analysis of bicycle frame test. Henan Science .2013. 12 (4): 327-330.
7. Emst Albin Hansen, L, ars Vincents Jorgensen, Kurt Jensen, Benjamin, Jon Fregly Gisela Sjogaard. Crank inertia (load effects freely chosen pedal rate during cycling. Journal of Biomechanics. 2002 (35): 277-285.
8. D. PHeil.Body mass scaling of frontal ~ in competitive cyclists not using aero-handlebars. Eur J Appl Physiol. 2002 (87): 520-528.

DOI: 10.1051/matecconf/201710804014

Design of Pump as Turbine Experimental Test Facility

D. L. Zariatin, Damora Rhakasywi, Fahzul Ade and Abdi Setyo

Mechanical Engineering Department, Universitas Pancasila, Jakarta, Indonesia

Abstract. This paper presents the design process of experimental test facility for pump as turbine hydropower system. Three design possibilities that related to the PAT condition of operation was developed and analyzed by using CFD Software. It is found that the First Variant with a straight flow to the PAT will produce higher velocity, which is needed to generate more rotation of the shaft generator, in order to generate more electric power. The strength of PAT construction was analyzed by using FEM software. It was found that the maximum stress is 6 MPa and can be concluded that the construction is appropriate to the design requirement.

1 Introduction

Hydropower is the cleanest energy provided by nature. Utilization of hydropower system becomes worldwide, especially in the area that has the supporting resources. There are many types of turbine used in hydropower system, such as Pelton, Kaplan turbine, Francis reaction turbine, Pump as Turbine (PAT), etc. PAT technology that first introduced in 1931 [1] is one of a micro hydropower system that used an available commercial pump as a turbine in order to generate rotational movement. PAT has several advantages such as 53% cost equipment reduction compare to another hydropower system [2], the power source availability, simple design and easy to install [3], components are available in the market [3], and long life time up to 25 years.

In order to perform researches on PAT characteristic and technology, a laboratory scale PAT installation was required. Figure 1 shows an open PAT experimental test facility used other researchers. There are two types of test facility that used by other researchers. The first type is a test facility that has a direct flow from the pump to PAT, used by Jain et al. [4], Yang et al [5], etc. The second type is a test facility that has a tank which provide a water drop from a certain height, used by Giosio et al. [6], Singh and Nestmann. [7], Suarda et al. [8].

The second type of test facility which has a reservoir tank, is considered closer to the actual condition of PAT operation. However, there might be three actual conditions that happen. First is, the condition where the area has a waterfall and the PAT could be placed right below the waterfall. The second condition might happen when the PAT could not be placed below the waterfall. The third condition might happen when there is only a water stream with certain inclination. This paper describes the design and analysis of the three conditions (Variant) mentioned by using Computational Fluid Dynamic (CFD) Software. One of the variant, which has an optimum output flow parameter was then selected and developed.

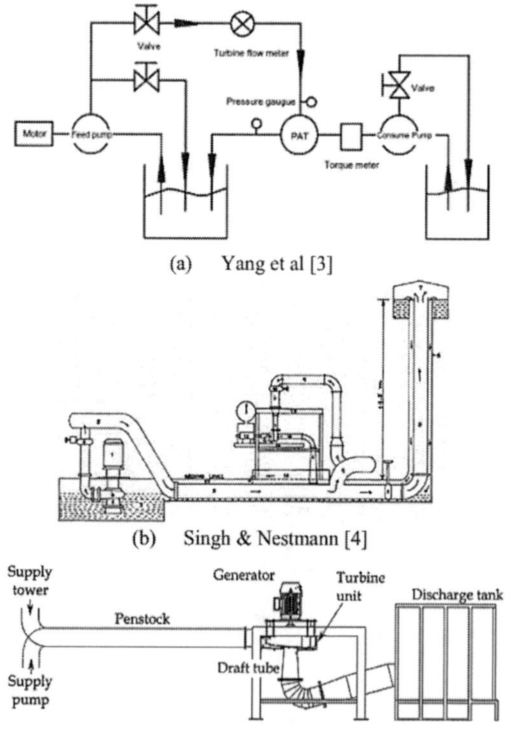

(a) Yang et al [3]

(b) Singh & Nestmann [4]

(c) Giosio et al [5]

Figure 1. Experimental test facilities used by other researchers

2 The design

Verien Deutscher Ingenieure 2221 (VDI 2221) design method was used in designing the PAT test facility. Three variants were developed in order to find the optimum installation design. The fluid flow of the design was analyzed by using CFD Software. Meanwhile the strength of PAT construction was analyzed by using a Finite Element Analysis Software.

2.1 First variant

Three variants were developed in order to get the best design for the experiment test facility. The first Variant has a straight flow from the tank to the pump, which used as a turbine, as shown in Figure 2. The first design condition might occur in a remote area which has a waterfall resource and the pump could place below the water reservoir tank. Clean water (assumed as H_2O Constant) flows from the PVC tank into the PVC pipe. Due to the space limitation in the laboratory, the tank was placed 3.7 meters' height from the ground level. The velocity of water drop from the tank was calculated by using Equation (1). The Top Tank was designed with a volume of 640 liters, and the water density is 1000 kg/m^3. The inlet of the pump has a diameter size of 3 inches. A 4 inches' diameter pipe with height of 1.2 meters was used to connect the top tank and a 3 inches diameter pipe before the water flow into the pump. V is the flow velocity; g is gravity of 9.81 m/s^2. The cross sectional area A of the pipes was calculated by using Equation (2) where d is the diameter of pipe. Q is flow-rate (m^3/s). The pressure and Reynold number were calculated by using Bernoulli Equation (4) and Equation (5).

$$\text{v} = \sqrt{2.g.Z_1} \tag{1}$$

$$A = \frac{\pi}{4} d^2 \tag{2}$$

$$Q = A\,\text{v} \tag{3}$$

$$P_1 + \frac{1}{2}\rho\ V_1^2 + g\,Z_1 = P_2 + \frac{1}{2}\rho\ V_2^2 + g\,Z_2 + \Delta p$$

$$Re = \frac{V.D}{\nu} \tag{5}$$

The calculation result found that the inlet parameters before entering the 4 inches pipe, which is velocity, pressure, and flow-rate was 8.52 m/s, 137,622 Pa, and 0.08094 m^3/s, respectively. These values were entered as the input parameter for flow model simulation in CFD.

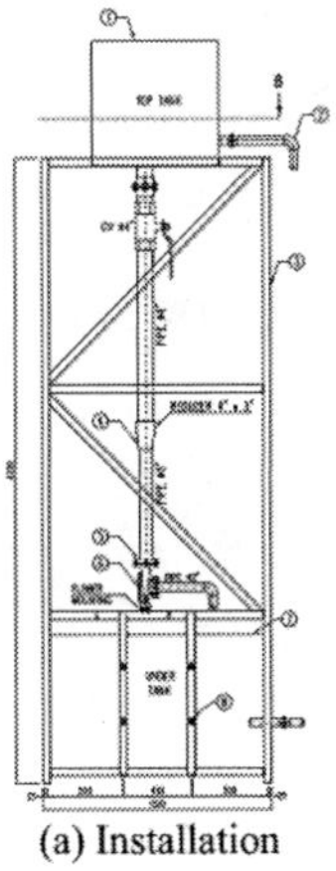

(a) Installation

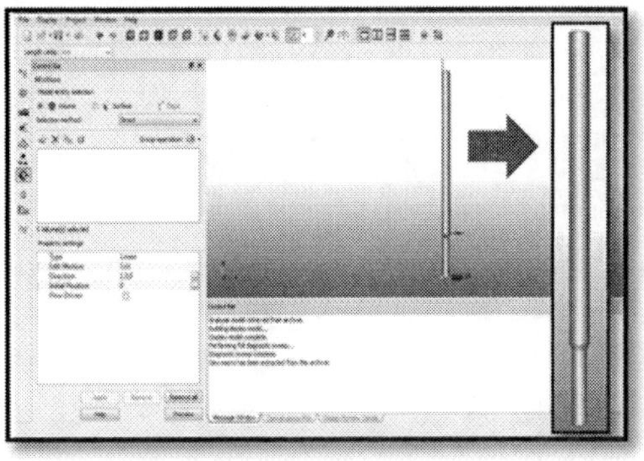

(b) Model for CFD analysis

Figure 2. First variant (a) installation, (b) model for cfd analysis

Figure 3 (a) shows the result of CFD analysis for the velocity of the water. It shows that the velocity of the water was increasing, especially at the 3 inches' diameter section (point b). Meanwhile Fig. 3 (b) shows the CFD analysis of pressure. The pressure drops in the 3 inches diameter pipe. The value of velocity, pressure and flow-rate at the end of the pipe (pipe outlet = pump inlet) is 13,2671 m/s, 2796.3 Pa and 0.0824295 m^3/s, respectively. The Reynold number was 1084750, which is greater than 4000, it is indicated the flow in the pipe was turbulent.

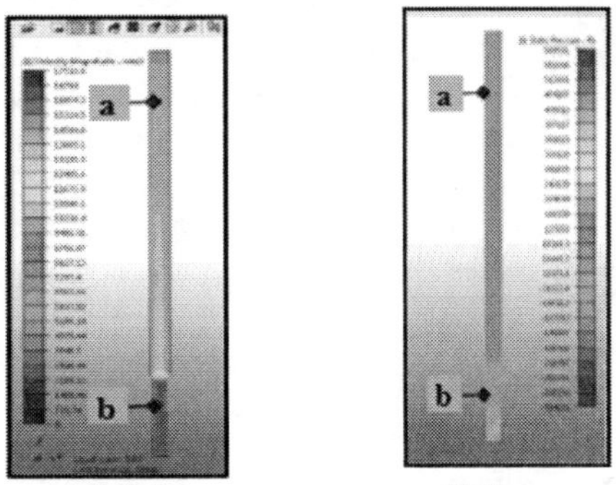

(a) Velocity (b) Pressure

Figure 3. Result of CFD analysis of first variant; (a) velocity, (b) pressure

2.2 Second variant

The Second Variant has an L shape pipe joint (elbow) between the tank and a pipe that flow to the pump, as shown in Figure 4. This design condition might be happening on a remote area which has a waterfall resource, but unable to place the pump below the water reservoir tank. The input parameter was calculated by using Equation (1) – Equation (5), the same as the input parameter for the first Variant, which is 8.52 m/s, 137,622 Pa, and 0.08094 m^3/s for velocity, pressure, and flow-rate was, respectively. The CFD analysis is shown in Figure 5.

Figure 5 (a) shows that at point a, the velocity was stable and it dropped when entering the elbow (point b). The velocity raised again after leaving the elbow and keep on rising until it reached point c and enter the pump, which is indicated by the red colour in the flow analysis.

Figure 5 (b) shows that the pressure was decreasing continuously. The CFD calculation result were 9.19954 m/s, 3939.7 Pa, 0.0489639 m^3/s for the velocity of fluid, pressure and flow-rate, respectively. The Reynold number was 700936, which indicate that the flow is turbulent.

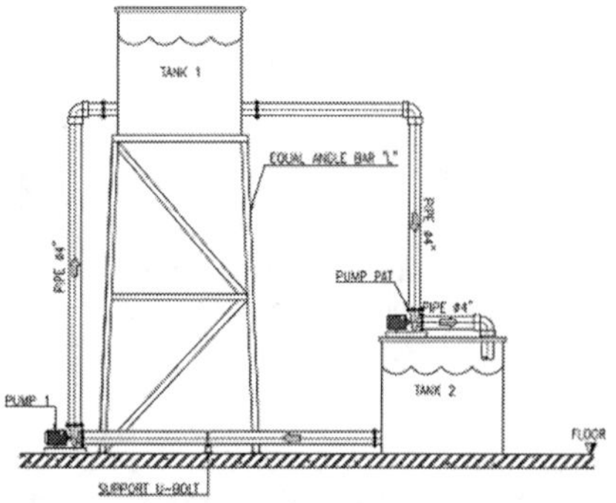

(a) Installation

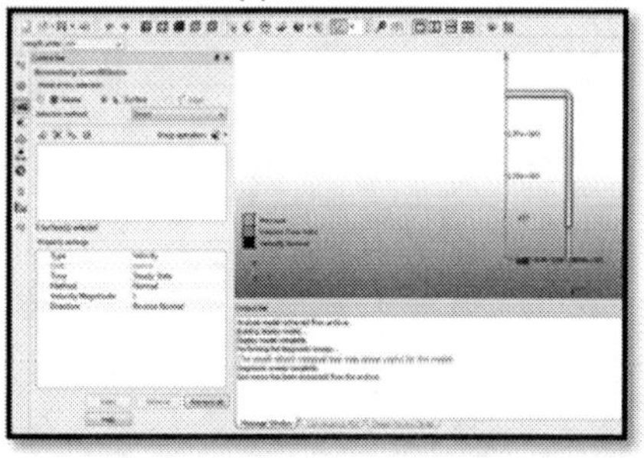

(b) Model for CFD Analysis

Figure 4. Second Variant (a) Installation (b) Model for CFD Analysis

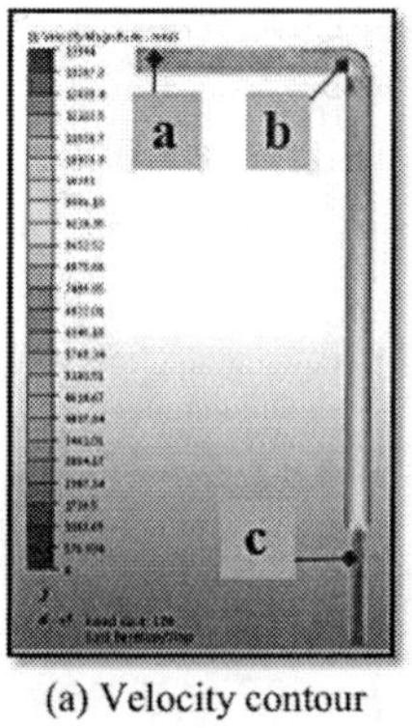

(a) Velocity contour

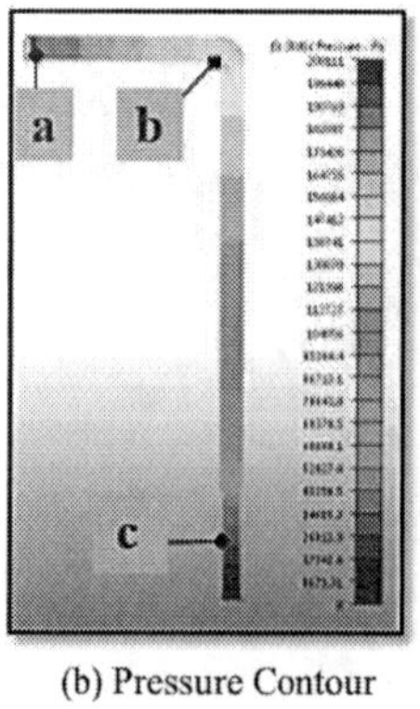

(b) Pressure Contour

Figure 5. Result of CFD analysis of second variant; (a) velocity, (b) pressure

2.3 Third variant

The Third Variant has a 45° inclination as shown in Figure 6. This design condition might be happening in a hilly area with a certain inclination on the landscape and has no waterfall resource.

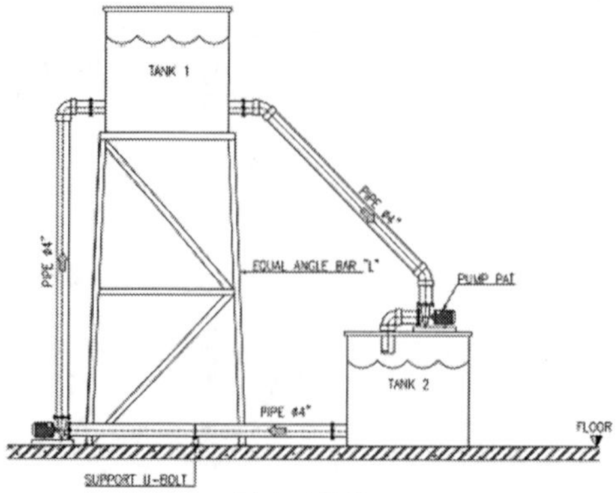

(a) Installation

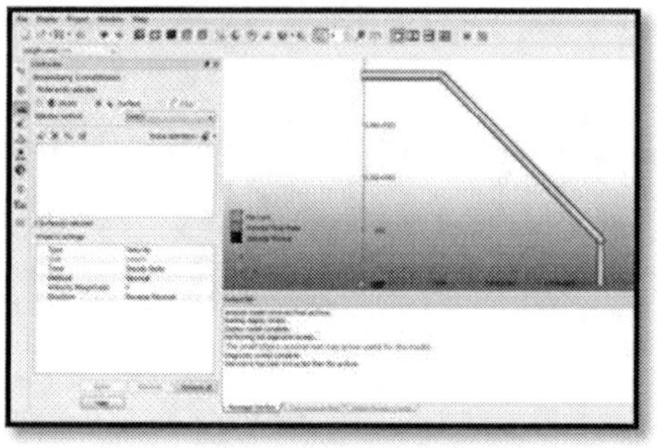

(b) Model for CFD Analysis

Figure 6. Third variant (a) installation (b) model for CFD analysis

Figure 7 (a) shows that the velocity increased as the fluid entering the elbow (point a). The greatest velocity occurred on point c as the cross sectional area reduced from 4 inches to 3 inches. Meanwhile Figure 7 (b) shows the pressure drop along the 45° inclined pipe. The velocity, pressure and flow-rate is 6,46684m/s, 9564.2 Pa, 0.0394708 m^3/s, respectively. The flow was turbulence which is indicated by the Reynold Number by 406768.

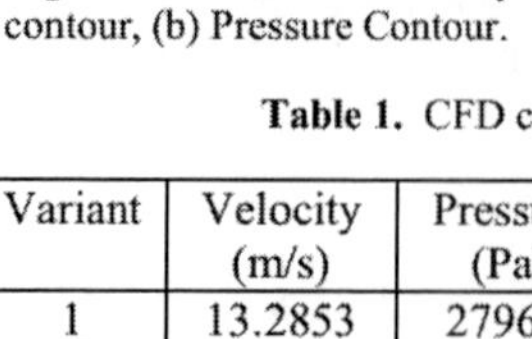

(a) Velocity

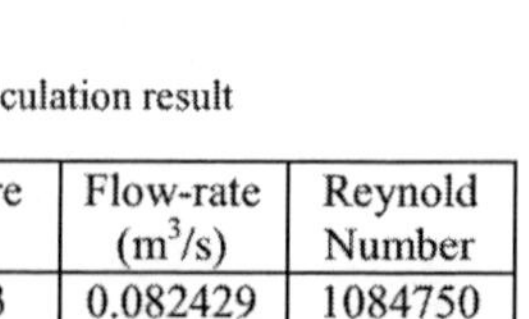

(b) Pressure

Figure 7. Result of CFD analysis of Third Variant; (a) Velocity contour, (b) Pressure Contour.

Table 1. CFD calculation result

Variant	Velocity (m/s)	Pressure (Pa)	Flow-rate (m^3/s)	Reynold Number
1	13.2853	2796.3	0.082429	1084750
2	9.19954	3939.7	0.048963	700936
3	6,46684	9564.2	0.039470	406768

3 Design implementation

Figure 8 shows the implementation of experimental test facility that designed and analysed by CFD software. Various condition was performed to acknowledge the power that generated by the PAT generator. The construction of the experimental test facility was made of St-37, L shape with dimension of 50x50 mm and thickness of 4.62 mm. In order to estimate the strength of the installation, FEM was performed as shown in Figure 8(b). It is found that the maximum stress on the construction is 6.62 MPa and considered safe for the experimental facility.

(a) Experimental test facility

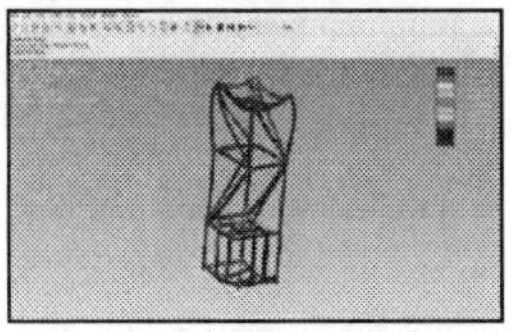

(b) Construction's strength analysis

(c) Measurement equipment

Figure 8. The PAT experimental test facility

4 Conclusion

In developing the experimental test facility, three variants that represent the possible conditions of the PAT working area were designed and analysed by using a CFD software and calculation. It is found that the first Variant, which has a straight flow is the best design among two others. First Variant has the highest velocity of 13.2853 m/s and flow-rate of 0.0824295 m^3/s. The second has a velocity of 9.19954 m/s and flow-rate of 0.0489639 m^3/s. Meanwhile the third Variant has velocity of 6,46684 m/s, and flow-rate of 0.0394708m3/s. The First Variant has the lowest pressure compare to second and third Variant, which are 2796.3 Pa, 3939.7 Pa, and 9564.2 Pa, respectively. All of three Variant has a turbulent flow, according to their Reynold number, which is 1084750, 700936, and 406768. By comparing these parameter, it concludes that the first Variant is the best choice.

The construction was made by using L shape scaffold from St 37 with 4.62 mm thickness and 50 mm width. The strength of the construction was analysed by using FEM software. It was found that the maximum stress is 6 MPa. It can be concluded that the construction is appropriate to the design requirement.

References

1. A. Elbratran, O. Yaakob, Y. M. Ahmed and H. Shabara, "Operation, performance and economic analysis of low head micro-hydropower turbines for rural and remote areas: A review," Renewable and Sustainable Energy Reviews, vol. **43**, no. 1, pp. 40-50, (2015)
2. M. Arriaga, "Pump as turbine - A pico-hydro alternative in Lao People's Democratic Republic," Renewable Energy , vol. **35**, pp. 1109-1115, (2010)
3. S. V. Jain and R. N. Patel, "Investigations on pump running in turbine mode : A review of the state-of-the-art," Renewable and Sustainable Energy Reviews, vol. **30**, pp. 841-868, (2014)
4. S. V. Jain, A. Swarnkar, K. H. Motwani and R. N. Patel, "Effect of impeller diameter and rotational speed on performance of pump running in turbine mode.," Energy Conversion and Management, vol. **89**, no. 1, pp. 808-824, (2015)
5. "Numerical Research on Effect of Splitter Blades to the Influence of Pump as Turbine," International Journal of Rotating Machinery, vol. **2012**, no. 1, p. 9, (2012)
6. D. Giosio, A. Henderson, P. Walker, J. Sargison and P. Gautam, "Design and perfomance evaluation of a pump-as-turbine micryo-hydro test facility with incorporated inlet flow control," vol. **78**, (2015)
7. P. Singh and F. Nestmann, "Internal hydraulic analysis of impeller rounding in centrifugal pumps as turbines," Experimental Thermal and Fluid Science, vol. **35**, no. 1, pp. 121-134, (Januari 2011)
8. M. Suarda, N. Suarnadwipa and W. B. Adnyana, "Experimental Work on Modification of Impeller Tips of a Centrifugal Pump as Turbine," in The 2nd Joint International Conference on "Sustainable Energy and Environment (SEE 2006)", Bangkok, Thailand, (2006)
9. Y. Sun-Sheng, K. Fan-Yu, J. Wan-Ming and Q. Xiao-Yun, "Effects of Impeller Trimming Influencing Pump as Turbine," Computer & Fluids, vol. **67**, pp. 72-78, (Agustus 2012)

Non Machinable Volume Calculation Method for 5-Axis Roughing Based on Faceted Models through Closed Bounded Area Evaluation

Gandjar Kiswanto[1], Himawan Hadi Sutrisno[2] and Jos Istiyanto[1]

[1] *Dept. of Mechanical Engineering, Universitas Indonesia, Depok, Indonesia*
[2] *Mechanical Engineering, Universitas Negeri Jakarta (State University of Jakarta), Indonesia*

Abtract. The increase in the volume of rough machining on the CBV area is one of the indicators of increased efficiencyof machining process. Normally, this area is not subject to the rough machining process, so that the volume of the rest of the material is still big. With the addition of CC point and tool orientation to CBV area on a complex surface, the finishing will be faster because the volume of the excess material on this process will be reduced. This paper presents a method for volume calculation of the parts which do not allow further occurrence of the machining process, particulary for rough machining on a complex object. By comparing the total volume of raw materials and machining area volume, the volume of residual material,on which machining process cannot be done,can be determined. The volume of the total machining area has been taken into account for machiningof the CBV and non CBV areas. By using delaunay triangulation for the triangle which includes the machining and CBV areas. The volume will be calculated using Divergence(Gaussian) theorem by focusing on the direction of the normal vector on each triangle. This method can be used as an alternative to selecting tothe rough machining methods which select minimum value of nonmachinable volume so that effectiveness can be achieved in the machining process.

1 Introduction

In the manufacturing industry, especially for manufacturers of dies and molds, efficiency in the process of machining becomesan issue for the machining cost reduction. The use of the 5-axis milling machine provides an opportunity for the single step processto form a product. In general, machining processare divided into 3 groups such as roughing, semi finishing and finishing. The rough machining process has biggerexcess materials if compared to the following two machining methods, although the results achievedhas a high level of roughness.

Until now, the development of the rough machining methods continues to be developed by several researchers. Some of them use faceted model as the database [1, 2], but some use parametric model as the basis of calculation [3, 4].

In fact, forms of the models that will be created using the process of machining extremely vary.This allows the process to continue even in certain forms containing non-machinable sections; in other words, there is an obstacle for the tool to enter the area [5, 6]. This part is commonly referred toas the close bounded volume (CBV area), where the methodfor detecting and determining the position of the CBV area on a workpiece model is as introduced by G. Kiswanto et al [6]. CBV area, generally a signal that in such area there is a hitch so that it requiresa special strategy such as the addition of CC point to the area [7]. However, the existence of evasion of damage due to over cuttingin the rough machining process is also important [8].

For the purpose of calculating the machiningvolume, the extract volume method for each position of the tool has been introduced by C. Park [9]. On the basis of the final part as the base of comparison, a part that is supposed to be disposed of is calculated using 3D booleans.Meanwhile ,the calculation of the effective area for rough machining process has also ever been introducedby Zhiwei Lin et al [10]. This method uses the slicing plane to get the particular Z-axis value as a layer in the rough machining process.

As the basis for calculation, the facet model used to obtain the total volume of material will be described in the procedure below.

2 Development of methods

From the literature study, the feature of a workpiece model and machining area of a final form of a product can be determined using several methods. [11]. Slicing plane method is used to identify primitive features in one model. In the development of the methods for calculating the volume of the remnant of the materials in this study, the slicing method is also usedto identify the final part, CBV and non CBV areas as described in Figure 1 below.

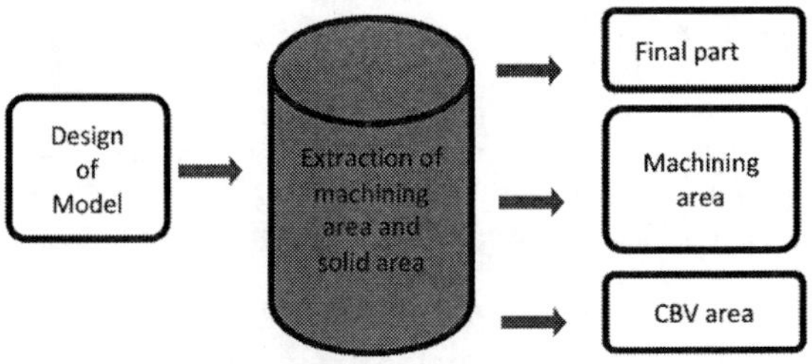

Figure 1. Extract of machining area and solid area

Facet model is used as the basis for the data obtained from the CAD system with STL files format. The volume of the area in which machining process cannot take place is the difference between the raw material and total machining area, as described in Figure 2 below.

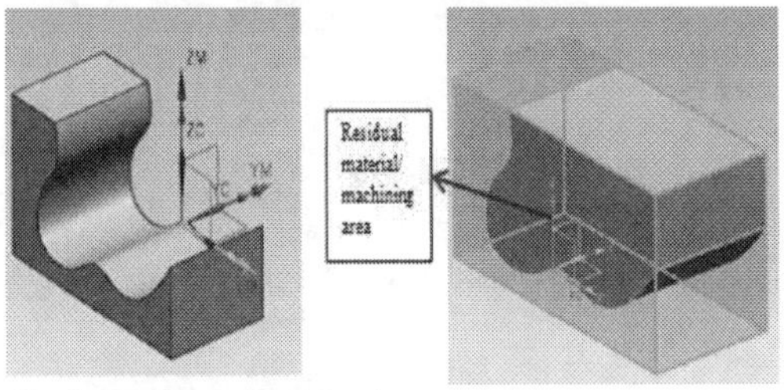

2a. Final part 2b. residual stock.

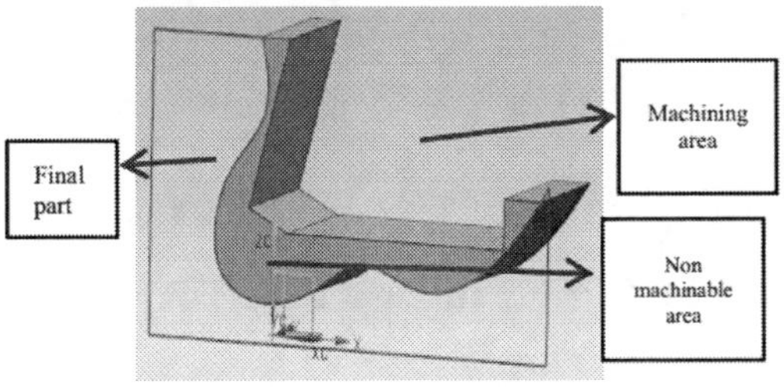

Figure 2. Separation of machining residual area.

2.1 Determining total volume of raw material

To begin with identification of the dimension of the object, in a 3 dimensional place, the location of the furthest point is identified as maximum valuesof X (X_max), Y (Y_max) and Z (Z_max) while the nearest to the axis is the minimuum values of X (X_min), Y (Y_min) and Z (Z_min). These points are determined from the triangle that forms models of STL files. The total volume of raw materialis as follows:

*raw material volume = length of part in X-axisdirection * length ofpart in Y-axisdirection * length of part in Z-axisdirection*

$$= (max\_x - min\_x) * (max\_y - min\_y) * (max\_z - min\_z) \quad (1)$$

As for the final part volume of a model with the basis of data in the form of faceted model using Gauss/Divergence theorem as described below:

$$Final\ part\ volume = \iint_G \vec{\rightarrow}_n ds = \iiint_G div \vec{\rightarrow} dV \quad (2)$$

The divergence theorem states that the outward flux of a vector field through a closed surface is equal to the volume integral of the divergence over the region inside the surface.

2.2 Determining machining volume in CBV area

To determine Close Bounded Volume (CBV area), grouping the point cloud in the machining area is doneby CBV evaluation concept[6] with the end result of this concept is CBV area and non CBV area or Open BoundedVolume (OBV area). This concept based on point cloud evaluation with slicing line method. Point cloud formation on all parts of the raw material, based on the point cloud formation in the XY plane by the distance between the point cloud is determined by the value of densityseen in figure 3a.Where previously the point cloud formed on XY plane base was also formed in each XY place with a certain Z value. In figure 3b, the entire point cloud that is formed (at each level of Z-axis) will later be used as a CC point for machining process.

By adopting the slicing line with fast, minimum storage ray/triangle intersection method [12] which is applied for every point cloud along XY plane on Z = 0 (psudocode below), intersection of lines on the workpiece surface tiangle is used as the basis for determining the CC point for machining process seen in Figure 3c. The point cloud defined as CC point is the one outside two intersections from the slicing line in the vertical slicing line that covers workpiece,

```
for i = 1:size(points_cloud, 1)
for j = 1:size(points_cloud, 2)
    x = points_cloud(i,j,1);
    y = points_cloud(i,j,2);
    z = points_cloud(i,j,3);
    z_max = max_min(1,3);
    line([x; x], [y; y], [z; z_max],
'Color','b','LineWidth',1,'LineStyle','-');
end
end
```

CBV area is area between two solid models in one slicing line as Figure 3d below. Data of the slicing line on the previous step is used to classify CC point on each surface machining layer (based on intersections of slicing line with triangle surface).

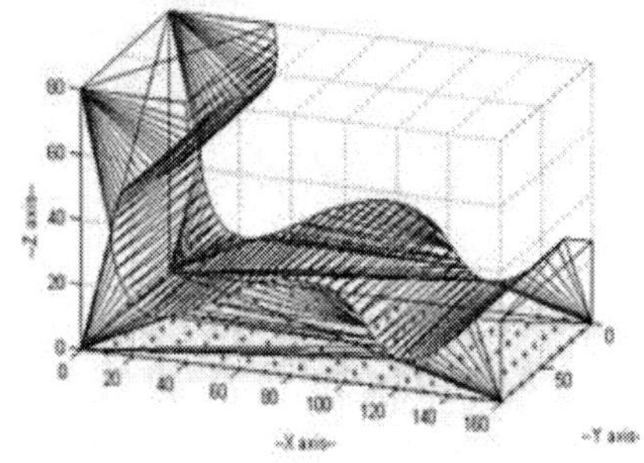

3a. The formation of point cloud in the XY plane

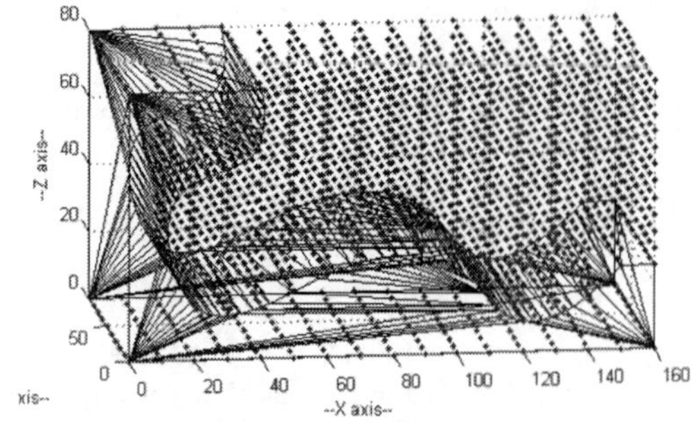

3b. The formation point cloud in the machining area

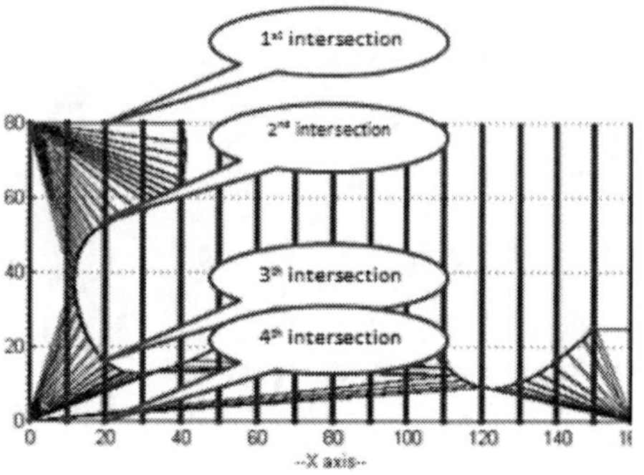

3c. licing line of every point cloud

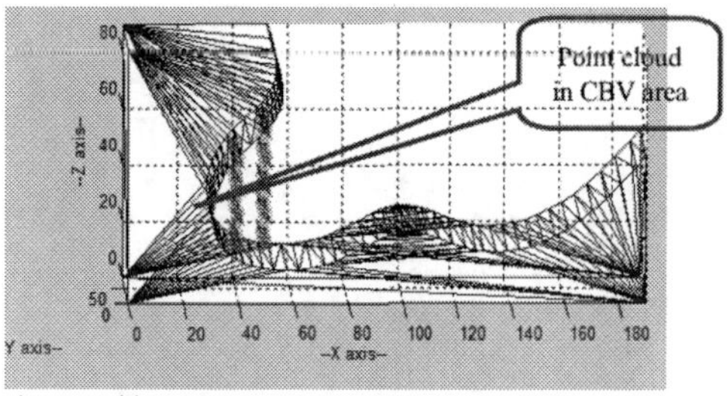

3d. Point cloud in CBV area

Figure 3. Formation of point cloud and Slicing line of every point cloud

Then, the position of the point cloud inside the final part, CBV area andoutside the CBV area can be identified[7].Then,the machining volume in CBV area depends on a dimension of tools as used and vector direction of CC point in the machining of CBV area. In every direction of machining (u and v direction) the volumeof each CC point will be summed up as shown in Figure 4 below.Whereas, the volume for CBV machining area equalsto feeding tools track from the nearest CC point outside CBV until CC point inside CBV with the inclination angle of Θ and the track is in the form of a cicular section. Therefore, the result of the calculation will be equal to thevolume of the circular section plus the semi-cylindrical volume (tool diameter) seen in Figure 5.

Algorithm: cilindrical volume calculation
Input: angle degree norm vektor, tool diameter
Output: CBV part volume
Define u or v formed the rotationangle between u and v =
　atan2 (norm(cross(u,v)),dotn(u,v))
cbv_ part_volume = angle/2π * area * heigth
intersected_cbv_volume = angle / (2 * pi) * (pi * tool_length - vertical_stepover) ^ 2) * (max_min(1,2) - max_min(2,2))
cbv_part_volume= cbv_part_volume - intersected_cbv_volume
total_cutting_volume = total_cutting_cbv_volume + obv_part_volume;

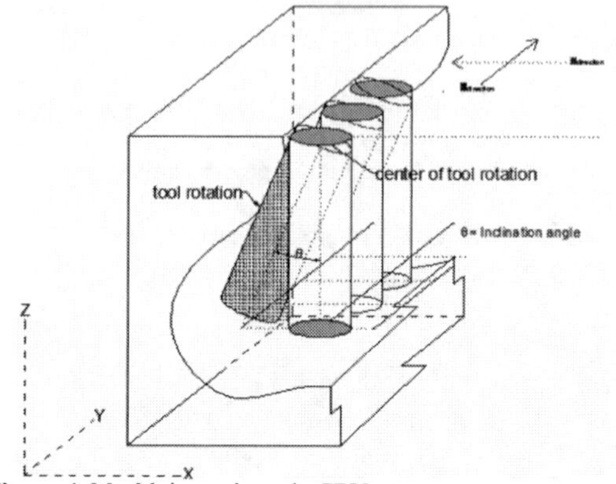

Figure 4. Machining volume in CBV area.

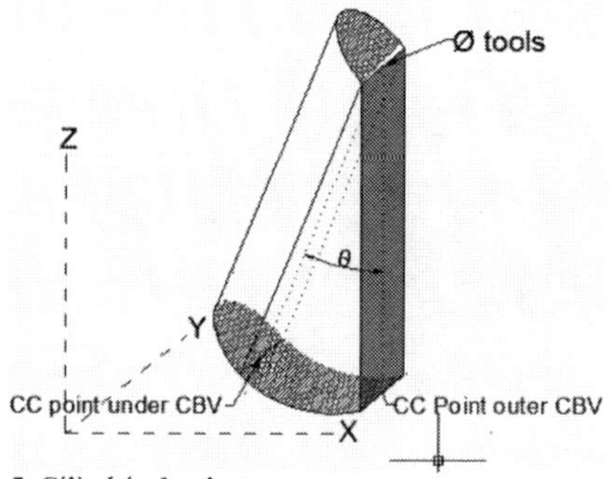

Figure 5. Cilindrical volume

2.3 Determining machining volume in non CBV area

In determining the volume of machining on a non CBV area(commonly called as OBV area), the volume of consumption will depend on number of CC point at each layer (Z-level) and diameter of the toolsis used. Due to the direction of orientation on OBV this area with z-axis normal then, volume for eachfour CC point will equal to the volume of a boxwith a length and width equal with density also a height equal with dept of cut.

*Totalmachiningvolume in OBV area = Σ cube(every four CCpoint *dept of cut or layering)* (3)

2.4 Volume of non machinable area

Based on the whole calculation of raw material, the final part and total volume of the machining area (CBV area plus OBV area), the calculation of non machinable volume can be calculated as the total volume of raw material less the volume of the final part along with the machining area. This can be expressed mathematically as follows:

Non machinable volume = raw material volume – (part volume + volume of machining area) (4)

Table 1. Implementation method

NO	Model with CBV Detection	Volume of Cutting Area Under CBV	Volume of Machining Area (mm^3)	Volume of non Machinable Area (mm^3)
1	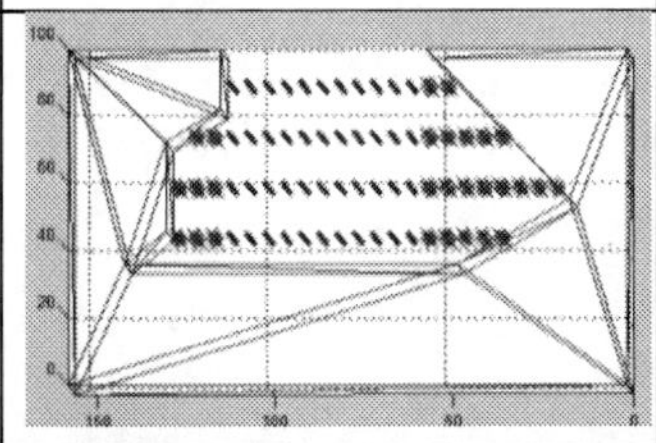	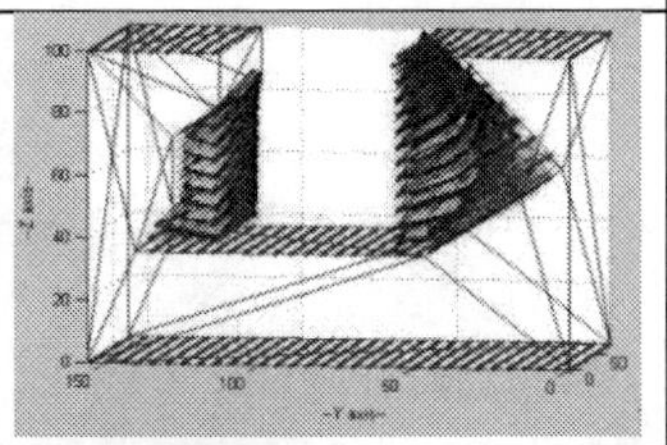	4.6206e+005	1.1722e+005
2	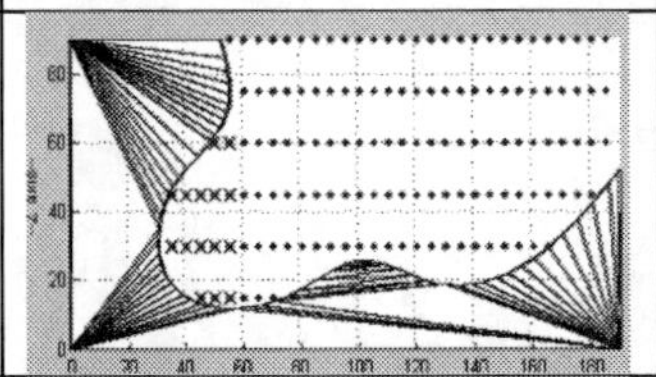	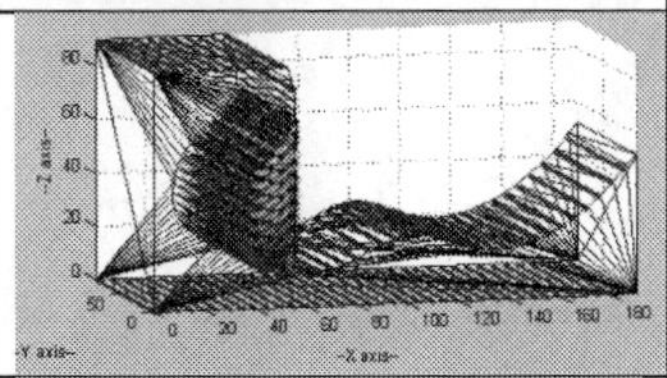	8.8888e+005	3.4595e+004
3	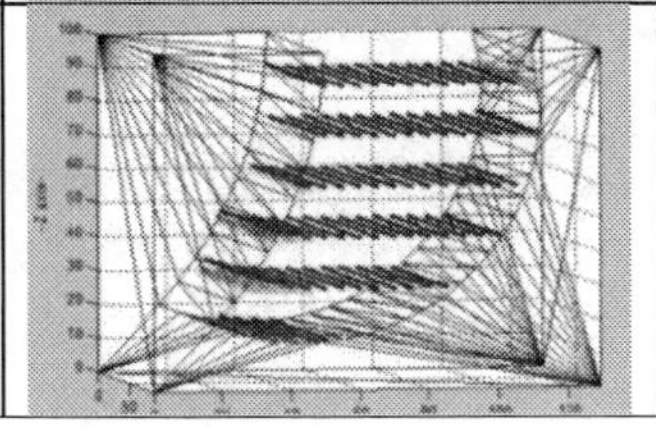	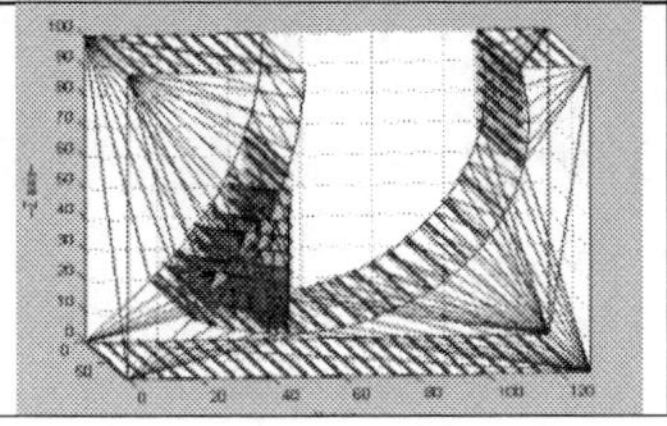	7.2045e+005	1.1699e+004

3 Implementation

In order to justify whether or not this method can be used for each model of the workpiece, particularly forcomplex forms containing CBV area, this method is tested on the forms of models below.

4 Conclusion

The results of simulation of some of workpieces and models show that the volume of any non machinable area can be exactly known; therefore, by utilizing this method can be used in general to select 5-axis rough machining methods that produce the smallest non-machinable. The less residual material left for 5-axis roughing, the finishing process will be faster so the machining time becomes more efficient.

References

1. M. Balasubramaniam, P. Laxmiprasad, S. Sarma, and Z. Shaikh, "Generating 5-axis NC roughing path directly from a tesselated representation," *Computer-Aided Design,* vol. **32**, p. 17, (2000)
2. B. L. G. K. J. P. Kruth, "Development of a Five axis Milling Tool Path Generation Algorithm based on faceted model," (2003)
3. H.-Z. Fan, W. Wang, and G. Xi, "A novel five-axis rough machining method for efficient manufacturing of centrifugal impeller with free-form blades," *The International Journal of Advanced Manufacturing Technology,* vol. **68**, pp. 1219-1229, (2013)
4. R. Qi, W. Liu, H. Bian, and L. Li, "Five-axis rough machining for impellers," *Frontiers of Mechanical Engineering in China,* vol. **4**, pp. 71-76, (2009)
5. K. Suthunyatanakit, E. L. J. Bohez, and K. Annanon, "A new global accessibility algorithm for a polyhedral model with convex polygonal facets," *Computer-Aided Design,* vol. **41**, p. 14, (2009)
6. G. Kiswanto. and A. Y. T. Panuju, "Development of closed bounded volume (CBV) grouping method of complek faceted model through CBV Boundaries identification.," *IEEE,* vol. **3**, p. 5, (September 2010)
7. G. Kiswanto, H. H. Sutrisno, and J. Istiyanto, "Development of Initial Tool Orientation Method At Close Bounded Area for 5-Axis Roughing Based On Faceted Models," *ICMM,* vol. **3**, (2017)
8. G. Kiswanto, B. Lauwers, and J.-P. Kruth, "Gouging elimination through tool lifting in tool path generation for five-axis milling based on faceted models," *Int. J Adv Manuf Technol,* vol. **32**, p. 21, (2007)
9. M. Chang and S. C. Park, "An algorithm to extract machining volumes," *Int. J Adv Manuf Technol,* vol. **36**, p. 8, 5 (January 2007)
10. Z. Lin, J. Fu, H. Shen, and W. Gan, "Efficient cutting area detection in roughing process for meshed surfaces," *The International Journal of Advanced Manufacturing Technology,* vol. **69**, pp. 525-530, (2013)

11. G. Kiswanto and M. Azka, "Automatic Part Primitive Feature Identification Based on Faceted Models," *international journal of computer science issues,* vol. **9**, p. 7, (2012)
12. T. Moller and B. Trumbore, "Fast MinimumStorage RayTriangle Intersection," (1997)

DOI: 10.1051/matecconf/201710804016

A Method to Determine System Parameters of Sdof Structural Systems by Applying a Step Load

Wenlung Li , and Kuo-Ching Liu

Dept. of Mechanical Engineering, National Taipei University of Technology, Taipei 10608, TAIWAN

Abstract. A novel method for identifying the physical parameters of structural systems has been proposed, unlike most existing methods are in terms of modal properties. Furthermore, the present method can be applied when structures are in services. The method is first theoretically derived by using single degree-of-freedom (sdof) systems, and will be generalized to multi-dof of lumped mass ones later. The present method requires a step external excitation, but the final expressions are independent of its input. Thus, it can be classified as an "output-only" method. In fact, it is the amplitudes of decaying spectrum, which is obtained by the so-called moving window Fourier transform (mwFT) from system signals is concerned, in addition to the steady-state deflection. Using the amplitude of this steady-state deflection, which can be obtained from the spectrum of the mwFT, as the quantity of calibration, the system parameters can be simultaneously estimated. The method is numerically verified by a model of bridge cranes.

1 Introduction

The so-called "inverse problems" are to identify the system parameters from the measured response signals or the outputs. This is quite important in engineering prac-tices since the results can be used to monitor the health of a system. Basically, to accomplish this goal is to retrieve the dynamic properties from the available vibration signals which may include outputs along or both of inputs and outputs. Thus, one calls this group of engineering problem as a parameter identification problem as well.

The health of a system may directly in terms of to its physical properties, or indirectly to its modal parameters. The latter includes like the eigen-frequencies [1], mode shapes and damping ratios [2, 3], the former like energy, mass, damping and stiffness of the system. However, the two are well correlated, though one may more sensitive than the other in some cases. For example, low sensitivity in changes of natural frequencies and mode shapes has been observed [4] when relating degradation of system stiff-ness. Therefore, one intention of the present report is to derive an identification method that is directly applicable of the physical properties, instead of the modal ones.

In fact, the authors have provided an output-only algorithm to identify the damping coefficients of sdof and mdof systems. The method was later successfully applied to estimate the modal damping ratios of machine tools. However, it was found that the method is still limited in the modal properties. It was further generalized together with a known mass to a new one [5]. The advantages of the new method are notable. Readers are referred to that report. Yet, it is inconvenient in some applications like bridge cranes. Motivated by such application practices need, the authors thus prepare this report. In addition, all notations are kept unchanged as the former [5].

2 Theoretical backgrounds

2.1 Basic concept

Theoretically, the system response of an sdof structural system with mass m, damping c and stiffness k under a unit-step force is governed by

$$m\ddot{x}+c\dot{x}+kx=f(t) \tag{1}$$

where $f(t)$ is a step function with a rise time at $t_0 = 0$ and the final value F_0, or

$$f(t)=F_0\cdot u(t-t_0)\,, \tag{2}$$

where $u(..)$ is the unit-step function. The responses of eqn (1) can be represented as [e.g., 6]

$$x(t)=x_0+e^{-\zeta\omega_n t}[C_1'\cos(\omega_d t)+C_2'\sin(\omega_d t)]\,, \tag{3}$$

where $x_0=F_0/k$, $\omega_d=\omega_n\sqrt{1-\zeta^2}$ is the damped natural circular frequency of the system, and

$$\omega_n^2=k/m \quad \text{and} \quad \zeta=c/(2m\omega_n)\,, \tag{4}$$

are the undamped natural circular frequency and the damping ratio of the system, respectively. Two constants in eqn (3), $C'_{..}$, can be determined from the initial conditions. For example, if both of initial conditions are zero, or $\dot{x}(0)=x(0)=0$, then the two constants are [e.g., 6]

$$C_1'=-F_0/k \quad \text{and} \quad C_2'=\frac{-F_0\zeta}{k\sqrt{1-\zeta^2}}, \tag{5}$$

respectively. For the reason of derivation completeness, however, the present report keeps them for now.

Depending on the type of sensors used, one is normally to measure either responses (or deflections) or accelerations if experimental measurements are carried out. Assuming that it is the structural deflections, $x(t)$, are of the main interest, one has the exponential form as

$$x(t)=x_0+e^{-\zeta\omega_n t}[C_1e^{j\omega_d t}+C_2e^{-j\omega_d t}], \tag{6}$$

and $C_{..}$ are the two complex constants, and $j=\sqrt{-1}$. Examining eqn (6) closely, there is no difficult to find that the system response can be treated as the linear combinations of the transient and steady-state responses. Nevertheless, applying the so-called moving window Fouier Transform (mwFT) to eqn (6), a windowed system response of eqn (6) is transformed by computing

$$X(\Omega,\tau)=\int_{-\infty}^{\infty}[x(t)\cdot w(t-\tau)]\exp(-j\Omega t)dt, \tag{7}$$

where $w(..)$ is the moving window function that located at τ in the time domain. That is, to virtually divide the transient signal $x(t)$, after discarded L_0, into several pieces by the window function, and individually take the FT for those windowed signals. Refer to the Parseval's power theorem [e.g., 6], the powers of the right-hand-side (RHS) and LHS of eqn (7) are equal within the individual windows. For example, if one considers a rectangular window centered at $t=\tau$, then the windowed $x(t)$ may be written as

$$x(t)\cdot w(t-\tau)=x(t)\cdot [u(t-(\tau-L))-u(t-(\tau+L))], \tag{8}$$

where L is the half-length of the window, cf. Figure 1. Note also that the length of the moving window L can be a critical parameter. If L is chosen too short, it may be not able to capture all essential temporal features. On the other hand, however, if L is too long, it may include too much noise at the right tail into transformation in the real engineering practices. Nevertheless, one computes the power spectrum from eqn (7) by using Mathematica, and obtains

$$X(\Omega,\tau)=a_0+\frac{j}{\sqrt{2\pi}}\left[C_1\frac{e^{-j(\tau-L)A_1}-e^{j(\tau+L)A_1}}{A_1}+C_2\frac{e^{-j(\tau-L)A_2}-e^{j(\tau+L)A_2}}{A_2}\right], \tag{9}$$

where A_1 and A_2 are both complex functions of ω_d and defined by

$$A_{1,2}(\Omega)=(\Omega\pm\omega_d)+j(\zeta\omega_n). \tag{10}$$

Notice also that the first term in RHS of eqn (9), a_0, is the mean of the oscillating windowed signal, cf. Figure 2. It is also real, time-invariant and the Fourier coefficient if one further compares eqn (9) with (3). Or, it can be read from any one of the power spectra of the windowed signals. Thus, it is no difficult to write

$$a_0=x_0=\frac{F_0}{k}, \tag{11}$$

from the average of $x(t)$ and the corresponding spectrum. That is, the stiffness is related to the mean (a_0 or dc) of a power spectrum. Note that the accuracy of x_0 is vital to the whole identification since it has been chosen as the calibration quantity. Care should be taken when to identifying a_0. Figure 1 graphically shows the concept to estimate a_0 of the Fourier coefficient.

On the other hand, if one restricts the damping ratio is reasonably small ($\zeta \ll 1.0$, says 10% for example) by examining eqn (9), then only the first term in the brackets governs the result when imposing $\Omega\rightarrow\omega_d$. The reason stems form that the denominator of this term is much smaller than the other, while the numerators are in the same order. That is, one can drop the second term in the brackets. Or, the power spectrum of the oscillating part in eqn (9) can be approximated by

$$X(\Omega=\omega_d,\tau)\approx\frac{C_1}{\sqrt{2\pi}}\cdot\frac{e^{\zeta\omega_n(\tau-L)}-e^{-\zeta\omega_n(\tau+L)}}{\zeta\,\omega_n}. \tag{12}$$

This can be also observed from Figure 2 that the power is maximal when $\Omega\rightarrow\omega_d$. In addition, referring Figure 1, the window length is set 2×L. How-ever, the total length obviously needs at least one complete cycle to get the right transformed ω_d. For this reason, $L\ge 2\pi/\omega_d$ is used as the minimum length of window functions.

Taking the ratio of the two consecutive powers between τ and (τ+Δ), which are evaluated right at ω_d, together with the minimum length for the window, one obtains

$$\left.\frac{X(\omega_d,\tau)}{X(\omega_d,\tau+\Delta)}\right|_{L=2\pi/\omega_d}\approx e^{\zeta\omega_n\Delta}=\exp(\frac{c\Delta}{2m}). \tag{13}$$

The upper neat form is readily derived and independent of constants C_1 and C_2, which indicate actually the two input energies, cf. Figure 2 for detail. In addition, the notations of the right most term in eqn (13) have been substituted by the system physical parameters. Notice that the center term of eqn (13) is the exact form of the inverse of the decaying exponent that appears in eqn (3). Therefore, eqn (13) provides additional information to identify the physical parameters c and m, if the system damping is small such that ω_n can be reasonably replaced by ω_d. And, it also provides a way to estimate the dam-ping ratio ζ of the system:

$$\zeta\approx\frac{1}{\omega_d\Delta}\ln\left[\frac{X(\omega_d,\tau)}{X(\omega_d,\tau+\Delta)}\right]. \tag{14}$$

DOI: 10.1051/matecconf/201710804016

It is also worthy to note that eqn (14) is neither dependent of the initial conditions, nor of the input loads. However, the general form of the approximate power spectrum, eqn (12), is function of C_1, or the initial conditions.

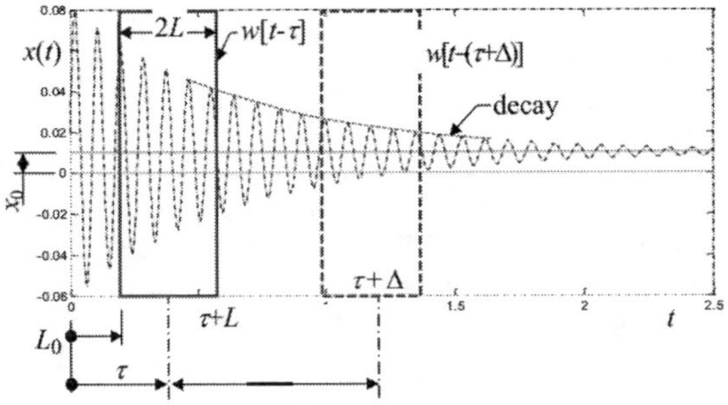

Figure 1. Concept of a moving window Fourier Transform.

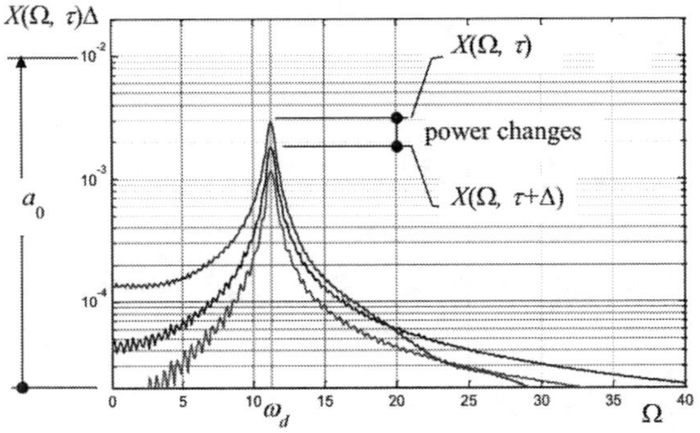

Figure 2. Response spectra of Figure 1 after transforming.

2.2 System parameters

Hence, taking the natural logarithms for the both sides of eqn (13), one can get the following relation equations

$$\begin{cases} \dfrac{c\Delta}{2m} = \ln\left[\dfrac{X(\omega_d,\tau)}{X(\omega_d,\tau+\Delta)}\right] = a_1 \\ \dfrac{k}{m} \approx \omega_d^2 \end{cases}, \tag{15}$$

where a short-hand notation a_1 has been adopted for the non-dimensional logarithmic value of this power spectra ratio. Using eqns (11) and (15), simultaneously solve the three system parameters, mass m, damping c and stiffness k, the following expressions can be obtained

$$\begin{cases} k = \dfrac{F_0}{a_0}, \\ m = \dfrac{F_0}{a_0 \cdot \omega_d^2}, \\ c = \dfrac{1}{\Delta}\dfrac{2a_1 \cdot F_0}{a_0 \cdot \omega_d^2}, \end{cases} \tag{16}$$

Therefore, in accordance with eqns (16), it is possible to estimate the parameters of sdof systems if the system is struck by a known external step excitation. Note that all parameters in the RHS of eqn (16) can be estimated from the signal power spectra. In addition, eqn (15a) can be further recast as

$$a_1 = \frac{1}{r}\ln\left[\frac{X(\omega_d,\tau)}{X(\omega_d,\tau+r\cdot\Delta)}\right], \tag{17}$$

where r is an integer. Or, it simply implies that Δ can be used as a parameter for tuning the sensitivity of identification.

2.3. Summary for complete procedures

Though the previous derivation has followed this procedure, complete procedures for applying the present method to determine the parameters of an sdof system is summarized as follows:

1) Apply a known external step excitation to the target structural system. Notice the system damping has to be small, for example $\zeta \le 10\%$.
2) To measure responses (displacement) at the locations of interest, says $x_i(t)$, $i = 1, 2, \ldots, n$. The total length of the signals should be long enough so that the next analyses can be accurately completed.
3) To identify the damped natural frequencies, ω_i', from the measured signals in the frequency domain. The averaged of these n ω_i' will be taken as ω_d.
4) To apply mwFT to these $x_i(t)$ signals and get their corresponding spectra $X_i(\Omega,\tau+r\cdot\Delta)$ for all $r = 1, 2, \ldots, \ell$ for individual measurements. It is recommended to discard the first few oscillations for all $x_i(t)$.
5) To complete identifying the system parameters using eqns (16), starting with system stiffness k. Appropriate statistic measures are recommended.

3 Numerical verifications

Following the procedure of the last section, numerical studies are performed to verify the present algorithm. In order to compare the identified parameters with the true input ones which are known *a priori*, all signals are generated by applying the Simulink Toolbox of MATLAB. Through the whole simulations, the numerical signals are acquired at the sampling frequency 500 Hz. Meanwhile, the shift and length of the moving windows, Δ and L, as well as the window function are all kept unchanged for all examples. The type of a hanning window is chosen since that it has been found [4] more robust and accurate. In addition, all these physical parameters are in consistent units and neglected hereafter.

3.1 Model of a bridge crane

The specification of the bridge crane which is taken as the example is listed in Table 1. In accordance with CAE results, the corresponding undamped natural frequency is 1.96 Hz. If the crane is to suddenly lift a weight, F_0, when the hoister is at the center of the girder, then this loading can be regarded as a step load. Or, neglecting the reinforcement inside the two girders, the equivalent sdof system has the equation of motion in its deflection $y(t)$:

$$m\ddot{y}+c\dot{y}+ky=F_0\cdot u(t)\,, \tag{18}$$

with m = 35,000 kg, k = 6,200 N/mm, and c is set to 46 N-s/mm or the approx. damping ratio is ca. 5%, which is believed practically reasonable. Figure 3 shows the 3D schematic diagram of the model. Notice also from Figure 3(c) where shows the corresponding mode shape.

Table 1. Specification of the bridge crane.

Rating Load	Span S	Speed			Lifting H
		Hoisting	Transv.	Travel	
m. Ton	m	(m/min)	(m/min)	(m/min)	(m)
51.6	33	40/80	150	100	17.9

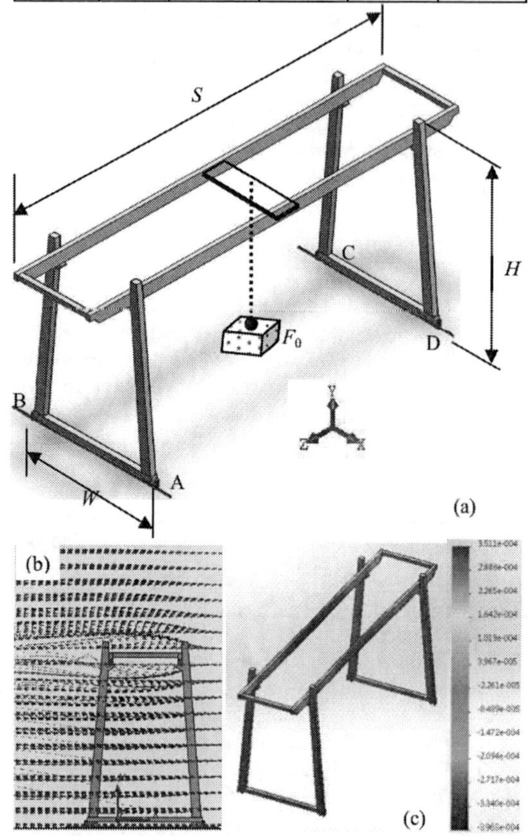

Figure 3. (a) A bridge crane and its sdof model; (b) the side view; and (c) 1st mode shape in y-direction.

3.2 Single-dof systems: parametric studies

In addition to verify the validity of the present method, the authors intend to show how those given parameters affect to the final results. These parameters include: The window length (L), the size of the step-load (F_0), the discarded length of signal in the time domain (L_0), and the shift (Δ) of window functions. All these parameters are in terms of non-dimensional quantities. That is, the lengths in the time domain are normalized by the period of the damped oscillation, and F_0 is divided by its weight of the system mass. Moreover, all errors are defined by

$$\text{relative error} = \frac{(\text{estimated value - true value})}{\text{true value}}\,. \tag{19}$$

3.2.1 The effects of the window length L

Firstly, one is interested in the choice of L to the estimation of the system natural frequency, f_d. Refer to Figure 4, it can be clearly seen if a reasonable accuracy is desired, one has to select the window length somewhat greater than 12 to 15 times of the period of the system oscillation. The plot also reveals that it always under-estimates f_d, or errors are in the negative side.

In addition to f_d, the steady-state response is also a key factor since it is used as a quantity of calibrations. It is found that the accuracy of a_0 (estimated from the spectra) is quadratically deceasing with L increases. However, to select $L\geq(12\sim15)/f_d$ can basically meet the engineering needs.

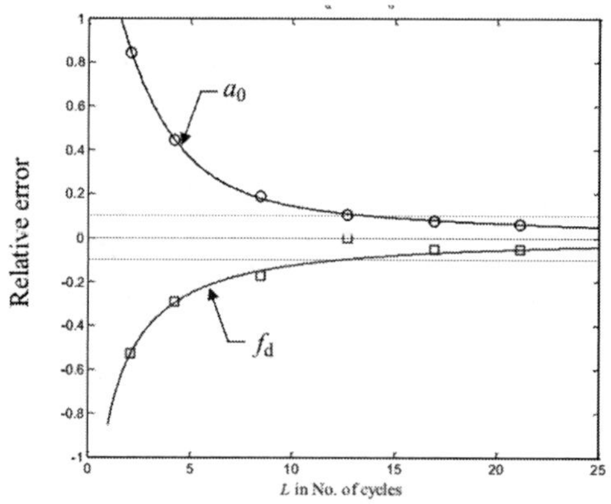

Figure 4. Effects of L to the accuracy of f_d and a_0.

3.2.2 The effects of the step-loads F_0

The applying step-load, F_0, is to generate the signals of oscillations. It may cause lower SNR signals, if F_0 is too small. Figure 5 shows that it requires at least ca. 4% ~ 5% of its system mass to generate a good calibration quantity, a_0. Note also that a larger of F_0-to-mg ratio tends to have better effects. On the other hand, however, it may not be practical in engineering practices. Refer to Figure 5 for detail.

3.2.3 The effects of the discarded length, L_0

It has been found that L_0 can be a factor which affecting the accuracy of identifying a_0. Figure 6 shows that the choice of L_0 to the accuracy of a_0. According this plot, at least ca. 3 first oscillating cycles of the signal after a step-load cannot be included to the present method. That is, depending on the initial conditions, users are not to include signals at least first $3/f_d$ to $4/f_d$ when signals are strongly influenced by those of initial conditions.

DOI: 10.1051/matecconf/201710804016

3.2.4 The effects of the window shift, Δ

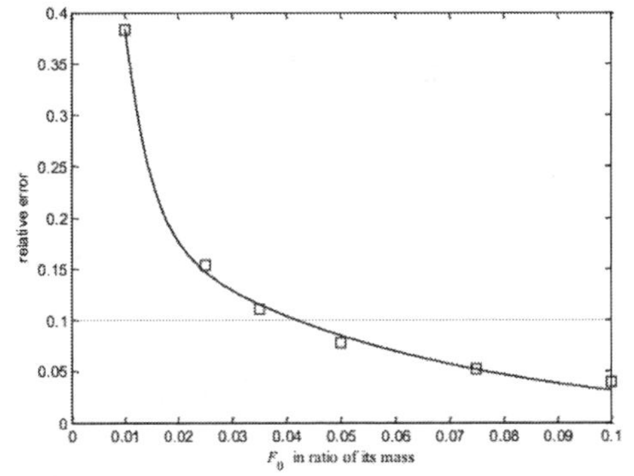

Figure 5. Effects of F_0 to the estimation of a_0.

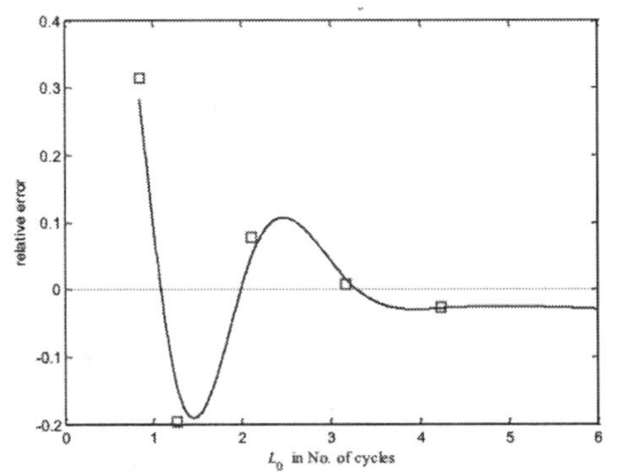

Figure 6. Effects of L_0 to the estimation of a_0.

Parameter Δ (the window shift) can be used as a tuning factor to adjust the sensitivity, according to eqn (13). Figure 7 displays the three system parameters vs. Δ. The diagram implies that one is to choose $\Delta \geq (0.9\sim1.1)/f_d$ so that the relative errors can be assured to minimum. Meanwhile, the trends of the three errors are quite similar.

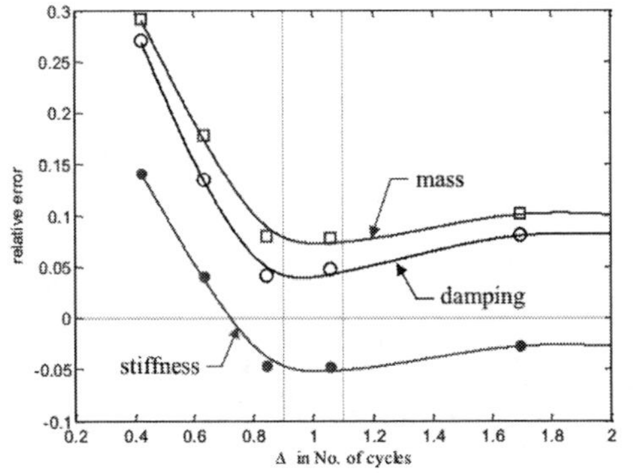

Figure 7. Effects of Δ to accuracies of the parameters.

3.2.5 The effects of Δ and L

The combined effects of window shift and its length L to the estimation of ζ are shown in Fig. 8. Similar to those of in Fig. 7, one is able to have quite good estimates as long as to select L is larger ca. 2 oscillations while Δ is somewhat larger than at least one oscillations.

4 Concluding remarks

A novel method that can be regarded as an "output-only" identification method has been proposed in the present report, though it requires a step-load as an external excitation force. The method is to measure both oscillating and the steady-state deflections after such step-loads. The next step is to apply the moving window Fourier transform to the former, while the latter is used as a calibration reference. The complete theoretical derivation as well as procedures has been well documented.

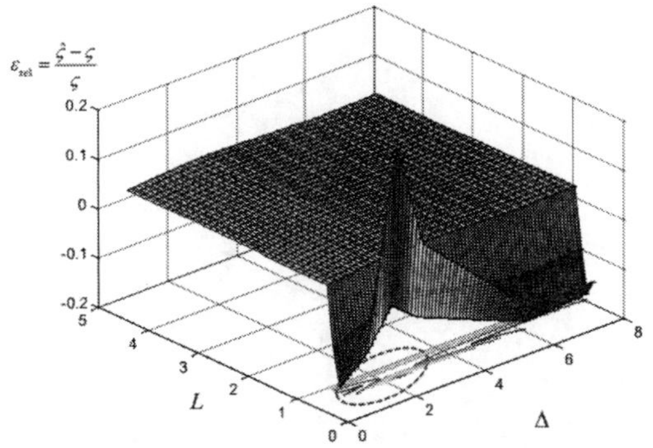

Figure 8. Effects of Δ and L to accuracies of the parameters.

In addition, numerical verifications have been carried out to substantiate the validity of the new method. A bridge crane of single-dof model has been used as the example of such parametric studies. The four parameters, including the window length (L)of the mwFT, the shift (Δ) of this window function, the size of the step-load (F_0), and the discarded length of signal in the time domain (L_0) are of the main concern. The simulation results show that the optimal parameter combinations for this new method are

$$\begin{cases} L \geq (12\sim15)/f_d \\ F_0 \geq (4\%\sim5\%)\times mg \\ \Delta \geq (1.0\sim1.1)/f_d \end{cases}$$

where f_d and m are the system damped natural frequency in Hz and the mass, respectively. Furthermore, it is also recommended that to discard at least first 3 to 4 oscillations, which are violently oscillating, after the external step-load is applied. By doing so, the influence of impulsive part of the step-load to the steady state response can be alleviated. Nevertheless, the method can be very convenient to monitor structural system like cranes.

References

1. Z. Feng, M. Liang, F. Chu, Mech. Syst. Signal Pr. **38**, 165 (2013)
2. W. Li, J. Sound Vib. **279**, 1181 (2005)
3. Feeny BF., J Vib Acoust ASME **131**, 041005-1 (2009)
4. R.Y. Hong, Master Thesis (in Chinese), Nat'l Taipei Univ. Tech (2012)
5. W. Li, K.-C. Liu, Y.-R. Lin, Int'l Conf. of Adv. Manuf., ICAM (2016)

 DOI: 10.1051/matecconf/201710804016

6. D.J. Inman, *Engineering Vibration*, 3rd Ed., Pearson Edu., (2007)
7. A. Palamides and A. Veloni, *Signals and Systems Laboratory with* MATLAB, CRC Press, (2011)

Minimum Position Based on Robot's Energy Consumption: A New Algorithm for Coordinated Multi-Robot Exploration

Abdenour Benkrid and Noura Achour

LRPE Laboratory, USTHB University, B.P 32 El Alia 16111 Bab Ezzouar Algiers, Algeria

Abstract. This paper addresses the multi-robot exploration problem in unknown environments. The proposed method aims to minimize the exploration time considering the robot's limited energy. The target area is defined as a frontier which represents the limit between an unknown and explored part of the environment. Thus, each robot from the team is assigned to the frontier for which it has the lowest rank among the other robots in term of energy consumption. An implementation and tests in computerized simulation as well as a comparison with some existing approaches has been performed. The results obtained demonstrate the validity and the efficiency of our method.

1 Introduction

The exploration of an unknown environment is one of the most important tasks in autonomous mobile robotics. The objective is to build an accurate representation of the world based on the set of observations performed by the robots within a minimal time. Thus, the produced map can be used by the human for achieving some dangerous missions like oceanic and planetary exploration [1, 2]. In this context, the use of multiple robots can be advantageous in many real world situations where the collaboration of several simple and not expensive robots proves to be faster and more efficient than using a large and complex robot [3]. However, these advantages are conditioned by the choice of an appropriate coordination strategy that assigns each robot to a specific target.

One of the famous solution to explore an unknown environment using a team of mobile robots is given by Yamauchi [4] who introduced the concept of frontier [5] defined as a limit between unexplored and explored accessible areas in an environment represented by occupancy grid map [6]. During an exploration process, each robot moves to the closest frontier, the cooperation between the team members is limited to sharing their positions and local maps. Consequently, the decision about where the robots should move is taken independently. This approach is known to be completely asynchronous (robots are not waiting for other robots), distributed and remain robust to the robot loss. As a consequence, it presents the problem of redundancy in term of frontiers assignment to the robots.

Simmons *et al* [7], performed an online approach that explicitly coordinates the robots by considering the estimated path cost to reach a frontier while evaluating its utility corresponding to the expected gain in information from it. Their work has been extended to deal with the problem of limited communication [8] and the problem of obstruction where a path of one robot is blocked by another one [9]. Recently, Bautin *et al* [10], proposed a new algorithm so-called MinPos (for Minimum Positions) based on the frontier concept. This approach takes into account the position or rank of the robot towards its target, which is defined as the number of robots closer to the frontier than the considered one. The position/or rank is evaluated based on the path distance needed to the robot for reaching the frontier. This distance is calculated using the wavefront propagation algorithm (WPA) [11]. The advantage of the MinPos algorithm is its ability to separate the robots in different directions which allows to effectively cover the environment and so, reduce the overall exploration time. Furthermore, its complexity is lower compared to previously cited algorithms. However, in the very distinct situations where several robots have the same rank toward a same target, the algorithm introduces redundancy in the affectation of robots.

The main contribution of this paper is a new distributed approach that introduces the energy consumed by the individual robots as the main criterion to assign them the exploration targets defined as frontiers between the unknown and the explored areas. Our assignment algorithm takes into account the position of a robot toward a frontier, which is defined as the number of robots that should consume less energy to reach it. The simulation results show the efficiency of our approach to distribute the robots on the environment which significantly reduces the overall exploration time.

The paper is outlined as follow: The next section presents some preliminaries and introduces the problem posed by exploring an unknown environment using several robots. Section 3, describes our coordination strategy based on the energy consumption for a team of mobile robots. The simulation and experimental results are presented in section 4. We conclude this work in section 5.

2 Preliminaries and problem statement

The environment to explore is a finite space in which several robots are deployed. The fleet of robots is assumed to be homogeneous, each of them is equipped with sensors that allow it to build a local map and localise itself in it using a simultaneous localisation and mapping algorithm [12]. In order to simplify our study, we consider that all robots can communicate between them wherever they are, the produced map is represented by an occupancy grid in which the frontiers cells to explore are identified. Thus, we focus in this paper on the coordination of robots by giving each of them the appropriate frontier to reach in order to explore the environment within a minimal time.

The distribution of robots over the frontiers can be considered as an optimisation problem where the number of possible assignments is equal to the number of permutations without repetition. While solving this problem optimally is intractable for large teams of robots, we propose a new distributed approach that deals with the following criteria:

• During exploration process, new frontiers can potentially be discovered. Therefore, no robot should be left without assignation,
• For deploying the robots in the environment we affect to each robot a frontier to explore,
• Regardless the number of robots used to explore the environment they must be distributed in a balanced manner on the frontiers,
• The distribution of robots over the frontiers involves the consideration of the energy ability of each robot to reach the frontier assigned to it for the exploration.
• Since no information about what behind the frontiers is available, the impact of a given assignment on the global system performance cannot be determined. Therefore, the assignment process must be performed in each iteration or extension of the map representing the environment.

3 Proposed assignment approach

3.1 Robot's energy consumption

A mobile robot's energy consumption can be divided into two parts: mechanical and electrical parts [13], the first one can be related to the energy consumed by the actuators (DC motors), the second part can be referred to the consumption of different electrical components embedded in the robot such as the microcontroller and the sensors. In addition to these components we can also consider the computer embedded in each robot. However, due to the significant variation of its energy consumption across different programs, we assume that it has its own battery and thus independent of the robot.

In general, the energy consumption of a robot Ec_r is the sum of the energy consumptions of its N components

$$Ec_r = \sum_{i=1}^{N} Ec_i, \tag{1}$$

where Ec_i is the component's energy consumption.

Let's consider the components cited previously, the equation (1) becomes:

$$Ec_r = Ec_m + Ec_{sr} + Ec_{mr}, \tag{2}$$

where Ec_m is the DC motors energy consumption, Ec_{sr} is the sensors energy consumption and Ec_{mr} is the microcontroller's energy consumption.

In order to get the model of the energy consumption of a given robot, we need to model the energy consumption of each one of its components. However assuming that all the team robots have the same microcontroller and the same sensors, we can fix their tasks over the time and thus, consider them as constant. Thereby, the energy consumption of a robot is a linear function of the DC motors energy consumption

$$Ec_r = Ec_m + cst. \tag{3}$$

The DC motors allow to convert input electrical power to output mechanical power while generating power loss caused by the armature resistance and the internal mechanical friction [14]. The power consumption P_{el} of the motor is calculated by dividing the mechanical power P_{mc} by the energy conversion efficiency ρ

$$P_{el} = \frac{P_{mc}}{\rho}. \tag{4}$$

In this paper we take into account the two stages of the robot motion, the first one is the acceleration step where the robot starts moving until reaching a desired velocity v with a constant acceleration a. In the second step, the robot keeps moving with the same constant velocity v until reaching the frontier.

The traction force, necessary for moving a robot's mass m at a velocity v and an acceleration a on the ground with a viscous friction constant μ is given by

$$F_{T1} = m(a + g\mu), \tag{5}$$

where g is the gravity constant.

Accordingly, the mechanical power delivered by the motor can be given according to the velocity, acceleration and the robot's mass:

$$P_{mc1} = m(a + g\mu)v. \tag{6}$$

The traction force, necessary to maintain a robot's mass m at a velocity v with a viscous friction constant μ is given by

$$F_{T2} = mg\mu. \tag{7}$$

Thus, the mechanical power delivered by the motor is given by:

$$P_{mc2} = mg\mu v. \tag{8}$$

Referring to the equation 4 we obtain the power consumption of the motor at each step of the robot's motion

$$P_{el1} = \frac{m(a + \mu g)v}{\rho}. \tag{9}$$

$$P_{el2} = \frac{m\mu g v}{\rho}. \quad (10)$$

By integrating the power consumption of the motor along the path for reaching the frontier position between a time t_0 when the robot starts to move toward a frontier and the time t_f when the robot reaches it or becomes close enough to it, we obtain:

$$Ec_r = \int_{t_0}^{t_f} P_{el} dt. \quad (11)$$

As the robot's motion has two steps, the robot's energy consumption becomes

$$Ec_r = \int_{t_0}^{t_v} P_{el1} dt + \int_{t_v}^{t_f} P_{el2} dt, \quad (12)$$

where t_v is the time needed for the robot to reach the velocity v.

Thus, the final expression of the robot's energy consumption is:

$$Ec_r = \frac{1}{\rho}\left[\frac{1}{2}mv^2 + m\mu g \Delta X\right] + cst, \quad (13)$$

where ΔX is the path length from the frontier to the robot determined by the wavefront propagation algorithm (WPA) [11] previously detailed in [10].

To take into account the ability of a robot to reach a frontier we compare the value of the energy consumption Ec_r of each robot of the team with the energy remaining in its battery Eb_r, if this value is bigger, we use a modified energy consumption defined as:

$$Ec_r^{'} = \begin{cases} Ec_r & if\, Ec_r \leq Eb_r \\ \infty & otherwise \end{cases} \quad (14)$$

Thus, considering the case where the energy consumption of a robot is bigger than the energy remaining in its battery, this robot should not be took into consideration at the frontier assignment to the robots using our algorithm.

3.2 Frontier assignment strategy

The proposed method is based on an assignment algorithm that uses the notion of position or rank of a robot toward a frontier based on its energy consumption. This position represents a number of robots that can consume less energy to reach a frontier than the considered one. Thus, a robot is affected to the frontier for which it has the lowest rank, this frontier should have the smallest number of robots that can consume less energy to reach it than the evaluated robot. Considering P_{rf} the position of a robot r towards a Frontier f such as:

$$P_{rf} = \sum_{\forall r' \in R, r' \neq r, Ec_f^{r'} < Ec_f^{r}} 1. \quad (15)$$

Algorithm 1 allows to assign an appropriate frontier $f^{'}$ to a specific robot r using the notion of position or rank and based on its energy consumption.

Algorithm 1. Minimum Position Based on Robot's Energy Consumption

Given a robot $r \in R$ and the energy consumption Ec_F which holds for each robot from the set R the energy consumption for reaching each frontier from the set F.

1: **For** each $f \in F$ do

2: $P_{rf} = \sum_{\forall r' \in R, r' \neq r, Ec_f^{r'} < Ec_f^{r}} 1.$ **end**

3: Frontier $f^{'}$ is assigned to the robot r with $f^{'} \in F \wedge f^{'} = \arg\min(P_{rf})_{\forall f \in F}$

4: If there is equality, choose the minimum energy consumption Ec_f^{r} among $\min P_{rf}$

The main specification of this algorithm is to separate the robots in different directions toward the frontiers, thus improving their spatial distribution in the environment. In fact, reasoning on the robot's position based on its energy consumption instead of only the distance allows to distribute two close robots toward frontiers where they are in the first position in term of their energy consumption. In addition, the low complexity of this algorithm ($O(NM)$ with N : number of robots and M number of frontiers) [10] allows its use in the case of robots with limited computational capabilities.

During an exploration process each robot of the team performs repeatedly and independently the following steps. First, it shares its position, energy remaining in its battery and its local map built from its perceptions and merged with the broadcasted maps of the other robots. Secondly, the generated partial map of the environment is used to identify the set of frontiers. Thus, the energy consumption matrix is calculated using positions of all robots and their energy autonomy. This matrix holds for each robot of the team, the expected energy consumption to reach the frontiers. The energy consumption matrix is then used to assign the appropriate frontier to the specific robot using the minimum position algorithm based on robot's energy consumption (Algorithm 1).

4 Simulation experiments and analysis

In this section, we performed several simulations with MATLAB software for evaluating the proposed coordination algorithm. The environments used in our experiments are depicted in two maps showed in Figure 1 which represent offices (Figure.1.a) and randomly generated mazes (Figure.1.b). The robots are assumed with identical size and have the same computational capability embedding identical sensors that allow them to scan their neighbourhood at 270 degrees around them with a parameterised range. During exploration process, the global position of each robot is assumed to be known and shared with its local occupancy grid map and the energy remaining in its batteries to the other team members. The communication between the robots is guaranteed by a simulated wireless network which is supposed to be perfect without limitations. Nevertheless, even if the information gathered from the other robots is taken into account, the decision about the assignment target is completely independent.

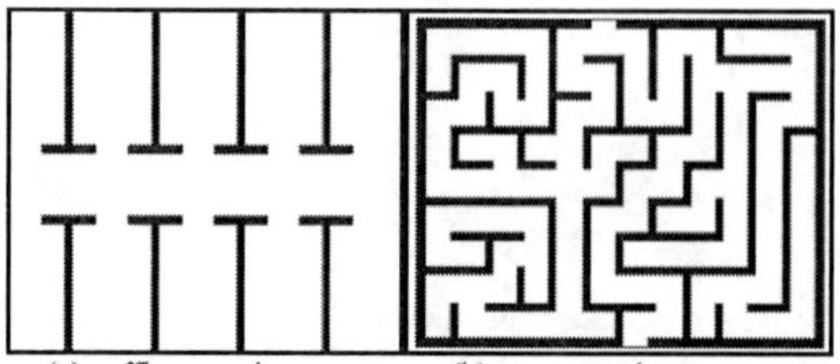

(a): offices environment (b): maze environment

Figure 1. Environments used for the simulation experiments.

In order to evaluate the ability of our assignment algorithm to reduce the overall exploration time, we compared it with two existing methods: the nearest frontier introduced by Yamauchi [4] and the cost-utility based approach performed by Burgard *et al* [8]. In each analyse of the three strategies in the both environments, several simulated exploration runs were performed while varying the number of robots from two to ten with the same random starting position. The objective of the experiments was to completely explore the environment. The exploration time is collected when there are no more frontiers to reach and thus, no unknown accessible areas to explore for all robots of the team. Figure 2 compares the exploration times provided in simulation steps according to the environment used in the experiments.

The results represent an average of 50 runs of each assignment algorithm given the number of robots used in the exploration process. The first observation that can be made in both environments is the reduction of the total exploration time (number of simulation steps) due to the increase in the number of robots. Moreover, it can be seen that, the nearest frontier based approach is less efficient compared to the two other methods that reduce approximately by 30% on average the number of simulation steps needed for completely explore the office environment, and by 29% for the maze environment. On the other hand, our energy consumption based algorithm outperforms the cost-utility algorithm proposed by Burgard *et al*, especially within the office building that can be considered as a structured and large environment where our approach takes advantage by separating the robots in different directions to unknown areas of the environment through the frontiers. Consequently, our approach is 12% more efficient on the average than the cost-utility based approach regardless to the number of robots. This performance decreases in the maze environment that can be regarded as being crowded and, where the energy consumption algorithm is 6% more efficient on average when the number of robots is small (from two to six robots). However, for the large team of robots (more than six robots) the cost-utility algorithm has close but slightly better performance.

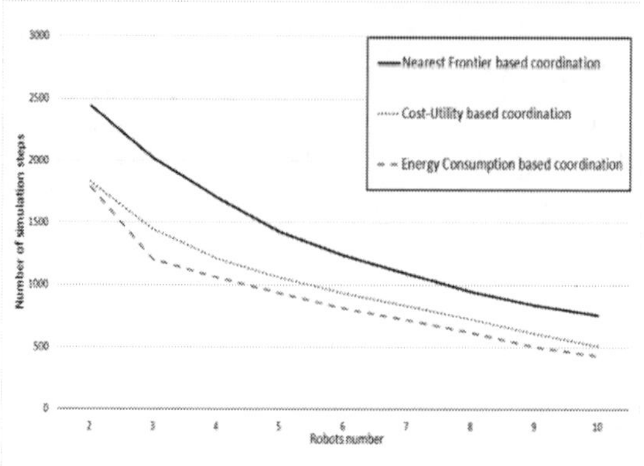

(a): offices environment

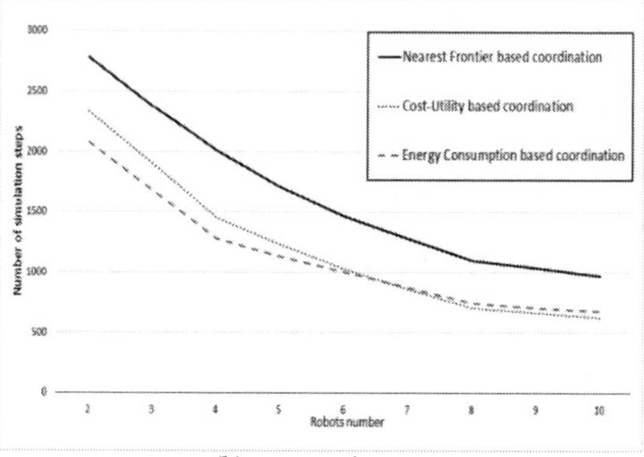

(b): maze environment

Figure 2. Exploration results for simulations experiments.

The explanation for this last result can be found in Figure 3. Based on the proposed algorithm (see Fig.3a), the robot 2 leaves the target T_4 (where it is in the 3rd position in term of its energy consumption) unsigned and moves toward the target T_1 where it is in the 2nd position. In this case, a cost-utility algorithm (see Fig.3b) performs better by assigning the robot 2 to the target T_4, thus avoiding the redundancy in the assignment of robots.

5 Conclusion

In this paper, the problem of coordinating a team of mobile robots in order to explore an unknown environment has been dressed. We introduced a novel algorithm based on the robot's energy consumption which allows to efficiently distribute the robots on the different unknown parts of the environment and thus, to reduce the overall time needed to explore it. The main difference with the existing methods is the ability consideration of each robot to reach a frontier according to the energy remaining in its battery, which is more realistic as a parameter for the robot's coordination. The performed simulation with various indoor environments and the obtained results demonstrate the ability of the proposed algorithm to reduce more efficiently the total exploration time compared to the nearest frontier and the cost-utility based approach.

DOI: 10.1051/matecconf/201710805001

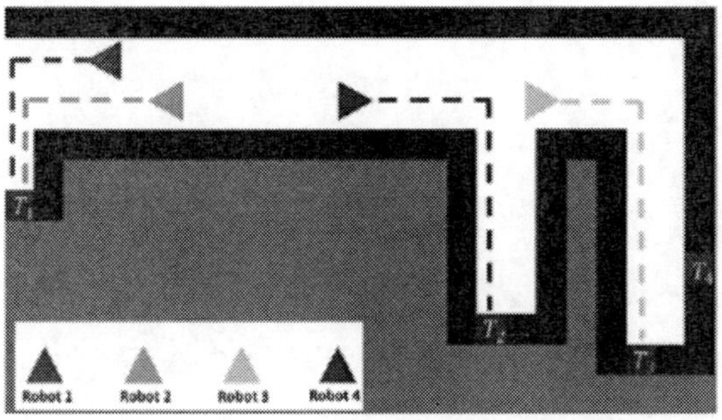

a): Minimum position algorithm based on robot's energy consumption

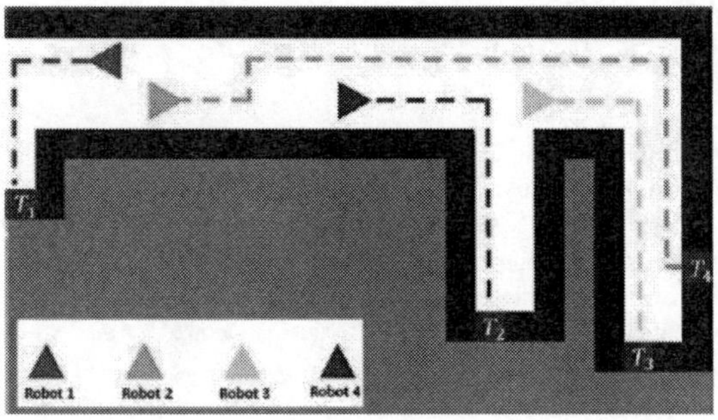

(b): Cost-utility algorithm

Figure 3. Resulting target assignments.

References

1. A. Pascoal, P. Oliveira, C. Silvestre, C. Sebastião, M. L. Rufino, V. Barroso, and P. Dando, *Oceans 2000 MTS/IEEE Conference and Exhibition*, **1**, 409-415 (2000)
2. D. S. Apostolopoulos, L. Pedersen, B. N. Shamah, K. Shillcutt, M. D. Wagner, and W. L. Whittaker, *IEEE International Conference on Robotics and Automation*, **4**, 4174-4179 (2001)
3. O. Simonin, O. Grunder, Mechatronics, **19**, 463-470 (2009)
4. B. Yamauchi, *Proceedings of the second international conference on Autonomous agents*, 47-53 (1998)
5. B. Yamauchi, *IEEE International symposium on computational intelligence in robotic and automation,* 146-151 (1997)
6. A. Elfes, Computer, **22**, 46-57 (1989)
7. W. Burgard, M. Moors, D. Fox, R. Simmons and S. Thrun, *Proceedings of the IEEE international conference on robotic and automation,* **1**, 476-481 (2000)
8. W. Burgard, M. Moors, and F. Schneider, IEEE Trans. Robot, **21**, 376-386 (2005)
9. A. Howard, M. J. Matarię, and G. S. Sukhatme, *IEEE/RSJ International Conference on Intelligent Robots and Systems*, **3**,2849-2854 (2002)
10. A. Bautin, O. Simonin, and F. Charpillet, *International conference on intelligent robotics and applications* ,**7507**, 496-508 (2012)
11. J. Barraquand, B. Langlois, and J. C. Latombe, IEEE Trans. Syst. Man Cybern, **22**, 224-241 (1992)
12. H. Durrant-Whyte and T. Bailey, IEEE Robotics Automation Magazine, **13**, 99-110 (2006)
13. Y. Mei, Y.H. Lu, Y. C. Hu, and C. S. G. Lee, *Proceedings of the IEEE International Conference on Robotics and Automation*, **5**, 4344-4349 (2004)
14. Y. Mei, Y.H. Lu, Y. C. Hu, and C. S. G. Lee, *the 12th International Conference on advanced robotics,* 492-497 (2005)

MATEC Web of Conferences **108**, 05002 (2017) DOI: 10.1051/matecconf/201710805002
ICMAA 2017

Blood Velocity Estimation Based on an Optimal Observer for Magnetic Nanorobots Navigation

Ton Duc Do

Department of Robotics and Mechatronics, School of Science and Technology (SST), Nazarbayev University, 53 KabanbayBatyr Ave, Astana Z05H0P9, Republic of Kazakhstan

Abstract. Magnetic nanorobots are recently used in a drug delivery system in order to perform minimally invasive medical procedures. One main difficulty for controlling nanorobots is the complicated behavior of hydrodynamic drag force and uncertain factors in their dynamics. The drag force depends on the blood velocity but unfortunately, it is apparently unknown. In previous feedback control system designs, the blood velocity was assumed to be known or set constant. This assumption did not reflect the reality and reduce the feasibility of the control systems. Considering this fact, this paper suggests an optimal observer for estimating blood velocity. The only information required for the proposed observer and overall control system is the position of nanorobots. The characteristic, stability, as well as gain tuning rules of the proposed optimal observer is discussed in detail. Simulation results are carried out in MATLAB/Simulink to demonstrate the effectiveness of the proposed optimal observer as well as the overall feedback control systems for nanorobot navigation in blood vessels.

1 Introduction

Recently, minimally invasive medical treatments have incredible growth since related techniques can approach the remote sites without surgical operation and improve targeting. To achieve these objectives, an interesting technique called magnetic drug targeting (MDT) that adds drugs to magnetic nanorobots and then applies external magnetic fields to steer and concentrate them at disease locations is used. Typically, the circulation of the blood will spread the drug throughout the body that may cause systemic toxicity in organs to which the drug is not targeted. Magnetic drug targeting is the desired technique to maximize the therapeutic efficiency and safety while avoiding medical side effects in healthy tissues [1]-[4]

There are two mains approaches in MDT that is opened-loop control [5], [6] and closed-loop control [7]-[14]. Each technique has its own advantages and drawbacks. The opened-loop control approach does not require any feedback information such as position of nanorobots, however, its efficiency is affected by a lot of factors such as vessel geometry, blood velocity, and magnetic field gradient as defined in [5]. Complicated mutual relations of these parameters need to be carefully considered or it will result in low efficiency [6]. One main disadvantage of closed-loop control approach is the requirement of real-time feedback information of nanorobots position. Fortunately, nowadays, this information can be achieved with monitoring devices such as magnetic resonance imaging (MRI) or magnetic particle imaging (MPI). Another drawback of most published closed-loop control methods [7]-[13] is that they only deal with aggregated particles in which the effective radii are about a few hundred micrometers. However, with the ability of navigating nanorobots to the correct sites precisely and effectively, closed-loop control techniques are a promising approach.

In all feedback control system designs, nanorobots depend on different forces whose modelling is vital to predict their behavior in blood vessels. However, the blood velocity, which is the key parameter of the drag force, is unknown. The measurement of blood velocity is a challenged task since it requires ultrasonic sensors. An alternative is using computational fluid dynamics software to achieve the numerical resolution of the blood velocity [15], [16]. But, this technique requires a lot of computation effort since it is based on Navier-Stokes equations. Therefore, it is not feasible for real-time applications such as in feedback control systems. The authors in [17] introduced analytical expressions of the blood velocity profiles near the bifurcation. However, the vessel geometry information is essential. In conclusion, online estimation of the blood velocity seems to be the most appropriate approach for avoiding the aforementioned difficulties. In [14], two observers are suggested for estimating the blood velocity. The simulation results show that these observers can precisely estimate the blood velocity and significantly enhanced the control performance.

This paper introduces an optimal observer for estimating blood velocity as an input to the adaptive sliding-mode control (ASMC) system. The stability and gain tuning rules of the proposed observer are discussed in detail. The numerous simulation results show that the proposed optimal observer can precisely estimate the blood. Simulations also prove that the proposed observer-based control system can attain high-performance under different scenarios.

2 Problem formulations

The state-space dynamic modelling of aggregate nanorobots in arterials [11]-[14] is as follows,

$$\begin{cases} \dot{x}_1 = x_2 \\ \dot{x}_2 = f(x_2, v_1) + \frac{\tau_m M}{\rho} u \\ y = x_1 \end{cases} \quad (1)$$

where x_1 and x_2 are the aggregate position and velocity, respectively, u is the magnetic field gradient considered as control input, v_1 is the blood velocity, τ_m is the ferromagnetic ratio, ρ is the aggregate density, M is the magnetization, and y is the output which is the aggregate position measured by MRI or MPI, respectively. The function f is expressed as,

$$f(x_2, v_1) = \frac{1}{m}\left[-sgn(x_2 - v_1)\left((a|x_2 - v_1|) + b(x_2 - v_1)^2 + c\frac{(x_2 - v_1)^2}{1 + d\sqrt{|x_2 - v_1|}} + V(\rho_f - \rho)g\right)\right] \quad (2)$$

where

$$a = \frac{6\pi\eta r}{\beta}, b = \frac{0.2\rho f\pi r^2}{\beta^2}, c = \frac{3\rho f\pi r^2}{\beta^2}, d = \sqrt{\frac{2r\rho_f}{\beta\eta}},$$

$$sgn(z) = \begin{cases} -1 \; if \; z < 0 \\ 0 \; if \; z = 0 \\ 1 \; if \; z > 0 \end{cases} \quad (3)$$

In (3), r is the aggregate radius, β is a dimensionless ratio related to the wall effect caused by the vessel occlusion by a spherical aggregate of radius r, η and ρ_f are the blood viscosity and density, respectively.

Note that the function f in (1) and (2) includes the expression of drag force and weight (the last term in the bracket is weight and the others are drag force) whereas the control input u is proportional the magnetic force. Also, in (1), only the nanorobot position x_1 is measurable, the nanorobot velocity and blood velocity are needed to be calculated and estimated.

In order to estimate the blood velocity, its dynamic equation should be clarified. According to [13], [15], the pulsatile blood velocity v_1 can be defined as an nth-order truncated Fourier series,

$$\begin{cases} \dot{v}_1 = v_2 \\ \dot{v}_2 = -\omega^2(v_1 - v_3) \\ \vdots \\ \dot{v}_{2k-1} = v_{2k} \\ \dot{v}_{2k} = -\omega^2(k^2 v_{2k-1} - v_{2k+1}) \\ \vdots \\ \dot{v}_{2n+1} = 0 \end{cases} \quad (4)$$

where ω is the pulsation of the blood flow assumed to be known and v_{2n+1} is the average value of the blood velocity.

3 Feedback control design

3.1 Adaptive sliding-mode control design

Let us consider the following ASMC law for the system (1),

$$\begin{cases} \sigma = e_1 + \alpha e_2 \\ u = -\frac{1}{\alpha}\frac{\rho}{\tau_m M}\left(\alpha f - \dot{x}_{1r} - \alpha\ddot{x}_{1r} + \theta sgn(\sigma)\right) \\ \dot{\theta} = |\sigma|/\gamma \end{cases} \quad (5)$$

where $e_1 = x_1 - x_{1r}$ and $e_2 = x_2 - \dot{x}_{1r}$ are the tracking errors, x_{1r} is the position reference command of the nanorobots, α and γ are the control gains. In (5), the adaptive control gain θ is online estimate by an adaptive law to avoid high control gains as well as with high control efforts. The SMC law in (5) may include chattering due to using a sign function. There are several ways to reduce the chattering, however, a simple method is the approximated sign function by the following approximated function:

$$sgn(\sigma) = \frac{\sigma}{|\sigma| + \varepsilon} \quad (6)$$

where ε is an very small positive number.

3.2 Nanorobot velocity calculation

As the nanorobot velocity is the time-derivative of the nanorobot position, it can be directly calculated from nanorobot position by the following discrete-time formula [16]-[18]:

$$x_2 = \frac{\varphi}{T+\varphi}x_2(k-1) + \frac{1}{T+\varphi}[x_1(k) - x_1(k-1)] \quad (7)$$

where k and k–1 denote two consecutive sampling instants, T is the sampling period, and φ is a sufficiently small time constant.

4 Blood velocity observer design

Combining (1) and (4), we achieve the following dynamic equations for estimating blood velocity,

$$\begin{cases} \dot{x}_1 = x_2 \\ \dot{x}_2 = -ax_2 + ax_3 + f'(x_2, x_3, u) \\ \dot{x}_3 = x_4 \\ \dot{x}_4 = -\omega^2(x_3 - x_5) \\ \vdots \\ \dot{x}_{2k+1} = x_{2k+2} \\ \dot{x}_{2k+2} = -\omega^2(k^2 x_{2k+1} - v_{2k+3}) \\ \vdots \\ \dot{x}_{2n+3} = 0 \\ y = x_1 \end{cases} \quad (8)$$

where $x_{i+2} = v_i \forall i \in \{1,2,\cdots n\}$ and

$$f'(x_2, v_1, u) = \frac{1}{m}\left[-sgn(x_2 - v_1)\left(b(x_2 - v_1)^2 + c\frac{(x_2 - v_1)^2}{1 + d\sqrt{|x_2 - v_1|}} + V(\rho_f - \rho)g\right)\right] + \frac{\tau_m M}{\rho} u$$

Equation (8) can be rewritten in the following state-space form,

$$\begin{cases} \dot{x} = Ax + Bf' \\ y = x_1 \end{cases} \quad (9)$$

where $x = [x_1 \ \cdots \ x_{2n+3}]^T, B = [0 \ 1 \ 0 \ \ldots \ 0]^T$,

$$A = \begin{bmatrix} 0 & 1 & 0 & 0 & & 0 & 0 & 0 \\ 0 & -a & a & 0 & & 0 & 0 & 0 \\ 0 & 0 & 0 & 1 & & 0 & 0 & 0 \\ 0 & 0 & -\omega^2 & 0 & & 0 & 0 & 0 \\ & & & & \ddots & & & \\ 0 & 0 & 0 & 0 & & 0 & 1 & 0 \\ 0 & 0 & 0 & 0 & & -n^2\omega^2 & 0 & \omega^2 \\ 0 & 0 & 0 & 0 & & 0 & 0 & 0 \end{bmatrix}$$

Then, the optimal observer model for estimating blood velocity can be represented as

$$\begin{cases} \dot{\hat{x}} = A\hat{x} + Ly - LC\hat{x} + Bf' \\ y = Cx \\ \hat{x}_3 = C_T\hat{x} \end{cases} \quad (10)$$

where $C = [1 \ 0 \ \ldots \ 0]^T, C_T = [0 \ 0 \ 1 \ 0 \ \ldots \ 0]^T$, $\hat{x}$ is the estimated vector of x, and L is the observer gain vector.

Fig. 1 illustrates the block diagram of the proposed blood velocity observer.

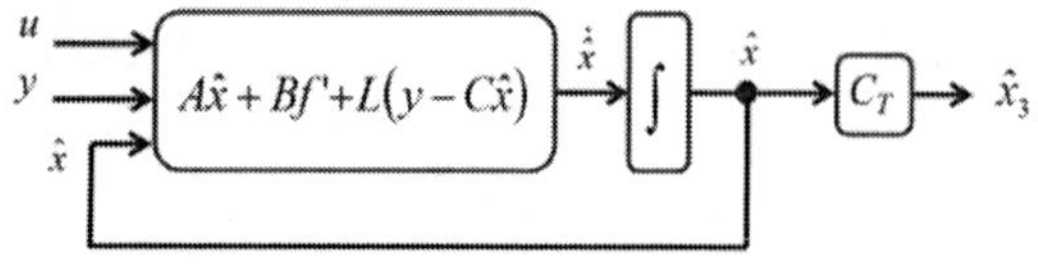

Figure 1. Block diagram of the proposed optimal blood velocity observer.

The error dynamics of the load current observer can be obtained as follows:

$$\dot{\tilde{x}} = (A - LC)\tilde{x} \quad (11)$$

where $\tilde{x} = x - \hat{x}$.

Theorem 1: Consider the following algebraic Riccati equation (ARE)

$$AP + PA^T - PC^TR^{-1}CP + Q = 0 \quad (12)$$

where $Q \in R^{(2n+3)\times(2n+3)}$ is a symmetric positive semidefinite matrix, $R$is a positive number, and $P \in R^{(2n+3)\times 1}$ is a solution vector. And the load current observer gain matrix $L$is given by

$$L = PC^TR^{-1} \quad (13)$$

Then, the estimation error converges exponentially to zero.

Proof: Consider the following Lyapunov function: $V(\tilde{x}) = \tilde{x}^T H \tilde{x}$, where $H = P^{-1}$. Its time derivative along the error dynamics (10) is given by

$$\dot{V} = \frac{d}{dt}\tilde{x}^T H \tilde{x} = 2\tilde{x}^T(HA - PC^TR^{-1}C)\tilde{x} \\ = \tilde{x}^T H(AP + PA^T - 2PC^TR^{-1}CP)H\tilde{x} \\ \leq -\tilde{x}^T HQH\tilde{x} \quad (14)$$

This implies that $\tilde{x}$ is exponentially stable.

Remark 1: The proposed blood velocity observer is the Kalman-Bucy optimal observer which minimizes the performance index $E(\tilde{x}^T\tilde{x})$ representing the expectation value of $\tilde{x}^T\tilde{x}$ for the following perturbed model

$$\begin{cases} \dot{x} = Ax + Bf' + d \\ y = Cx + v \end{cases} \quad (15)$$

where $d \in R^{2n+3}$, v is a scalarare independent white Gaussian noise signals with $E(d) = 0$, $E(v) = 0$, $E(dd^T) = Q$, and $E(v^2) = R$.

Remark 2: Typically, the observer performance is mostly influenced by the system model if the measurements are noisy (R large) and the input noise intensity is small (Q small). Therefore, L is small. This leads to a slow observer as measured by the location of its eigenvalues. However, if the measurements are good and the input noise intensity is large, the observer relies on the measurement. In this case, L is large, resulting in a fast observer with high bandwidth. Consequently, by assuming that the measurement is good, the fast observer is desirable. Lastly, the subsequent procedure summarizes the tuning process of the observer gain matrix L [19]:

1) Q and R are set as identity matrices;
2) Gradually, increase Q and decrease R, then calculate L as in (12) and (13);

If the observer performance is not satisfied, return to step 2 above. Otherwise, quit.

5 Results and discussion

Referring to [11]-[14], the modelling parameters are summarized in Table I. Without loss of generality, we consider the blood velocity as a second-order truncated Fourier series. In this paper, the angular frequency of the blood velocity is chosen as $\omega = 6.28\,rad/s$. With the guidelines in Remark 2 and via extensive simulation trials, the gains of controller and weighting matrices of observer are selected as follows:

$$\alpha = 100, \gamma = 0.2, \varepsilon = 0.001,$$
$$Q = diag(1, 1000, 1, 1, 1),$$
$$R = 1$$

Then the observer gain matrix are achieve as follows,

$$L = [8.016 \quad 31.622 \quad 0.7807 \quad -8.838 \quad 1.00]^T$$

Table I. Modelling parameters.

Parameter	Notation	Value	Unit
Blood density	ρ_f	1060	kg/m^3
Blood viscosity	η	0.0016	Pa.s
Aggregate radius	r	2.5×10^{-4}	m
Payload density	ρ_p	1500	kg/m^3
Ferromagnetic ratio	τ_m	0.75	-
nanorobot density	ρ_m	7500	kg/m^3
Vessel diameter	D	3	mm
Magnetization	M	1230	kA/m

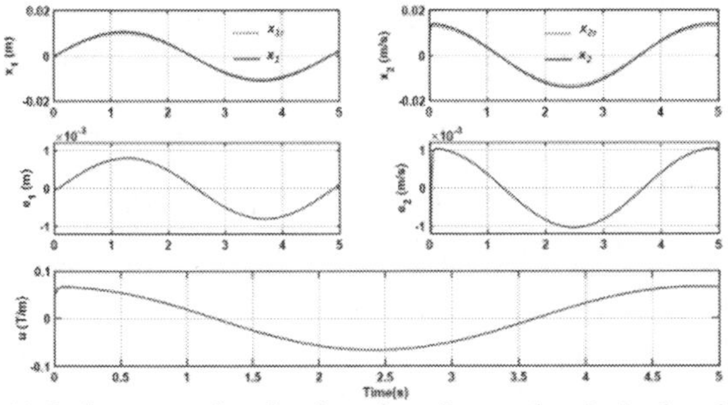

(a) References and real trajectories of nanorobot (x_{1r}in dotted blue and x_1 in red), real and calculated velocity of nanorobot (x_{2r}in dotted blue and $x_2$in red), position tracking error (e_1), nanorobot velocity calculation error (e_2), and control input – magnetic field gradient (u).

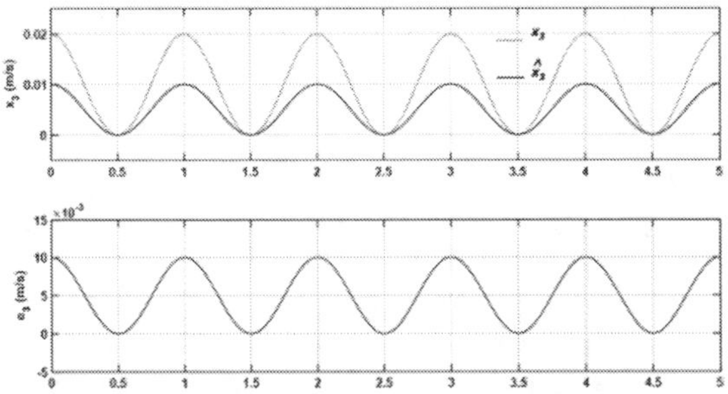

(b) Real and estimated blood velocities ($x_3$in dotted blue and $\hat{x}_3$ in red) and estimation error (e_3)

Figure 2. Simulation results without a blood velocity observer.

Fig. 2 shows the simulation results of the ASMC scheme without a blood velocity observer with a 50% of estimation error whereas Figs. 3 and 4 illustrate the simulation results with the proposed blood velocity observer. Also Figs. 2 and 3 are attained with the aggregate nanorobots with the radius size equal to 2.5×10^{-4} m whereas Fig. 4 is achieved with single nanorobot with radius of 400×10^{-9} m. Figs. 2-4 (a) demonstrates the references and real trajectories (x_{1r} and x_1), real and calculated velocity of nanorobot (x_{2r} and x_2), tracking error (e_1), nanorobot velocity calculation error (e_2), and control input – magnetic field gradient (u). And Figs. 2-4 (b) shows real and estimated velocities ($x_3$and $\hat{x}_3$) and estimation error (e_3). Note that only the real velocity of nanorobotx_{2r}, calculated nanorobot velocity x_2 and estimated blood velocity $\hat{x}_3$ are available, the blood velocity x_3 are not available; however, in simulations, we show x_3 in order to verify the effectiveness of the proposed optimal observer.

In Fig. 2, it should be noticed that the position tracking error without an observer is quite large (0.85 mm) and limit the access to the desired site of human body. Also it may not be feasible to control with such a big tracking error since the blood vessel diameter is just 3 mm. Whereas in Fig. 3, the estimated blood velocity is converge very fast (1.1 ms) with satisfied estimation error (0.19 mm/s)

By comparing the results from Figs. 2 and 3, we can see that with the proposed observer, the blood velocity is accurately estimated (the estimation error is 0.19 mm/s) and consequently the better position tracking error and nanorobot velocity calculation error are attained (proposed observer/without observer: position tracking error 0.85/0.079 [mm], nanorobot velocity calculation error: 1.03/0.95 [mm/s]).

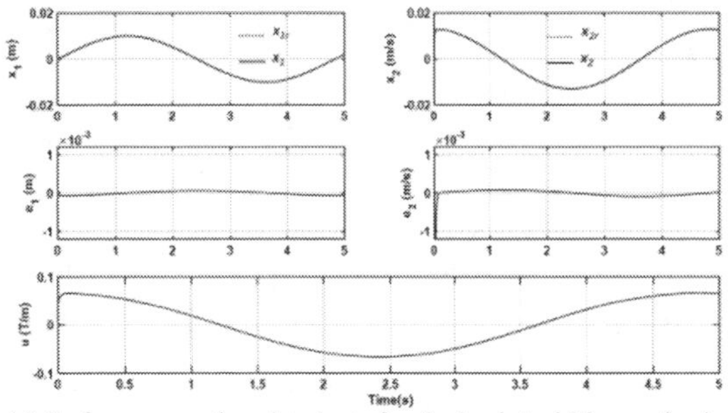

(a) References and real trajectories (x_{1r}in dotted blue and $x_1$in red), real and calculated velocity of nanorobot (x_{2r}in dotted blue and $x_2$in red), position tracking error (e_1), nanorobot velocity calculation error (e_2), and control input – magnetic field gradient (u).

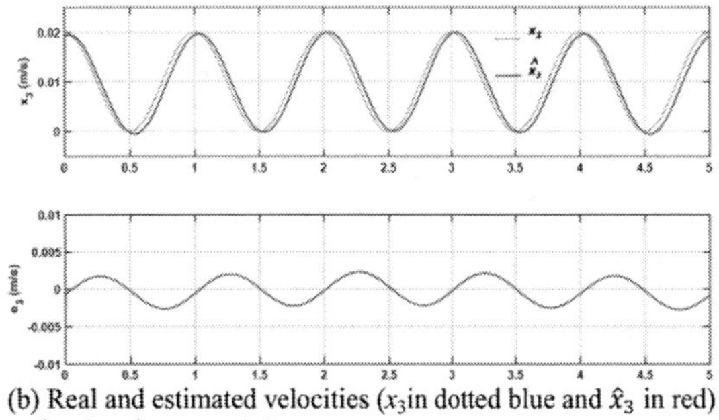

(b) Real and estimated velocities ($x_3$in dotted blue and $\hat{x}_3$ in red) and estimation error (e_3)

Figure 3. Simulation results with proposed blood velocity observer.

In order to further verify the feasibility of the proposed blood velocity observer and overall control scheme, in Fig. 4, we show the simulation results with reduced size of aggregation, i.e. we reduce the size of the object from hundreds micrometer to hundreds nanometer. The aggregate radius now is assumed to be 400 nm. Under the same gains as the previous case, in this case, the position tracking error is still in acceptable range whereas the estimation error (0.21 mm/s) is almost the same as previous case. Note that the control effort (2.8 T/m) is higher compared to the previous case. As presented in [14], an electromagnetic is able to comfortably generate this magnetic field gradient. Also with this result, we prove the proposed blood velocity observer as well as overall control scheme is effective in a wide range of the size of the aggregate nanorobots. In other words, the limitation indicated in [13] is solved.

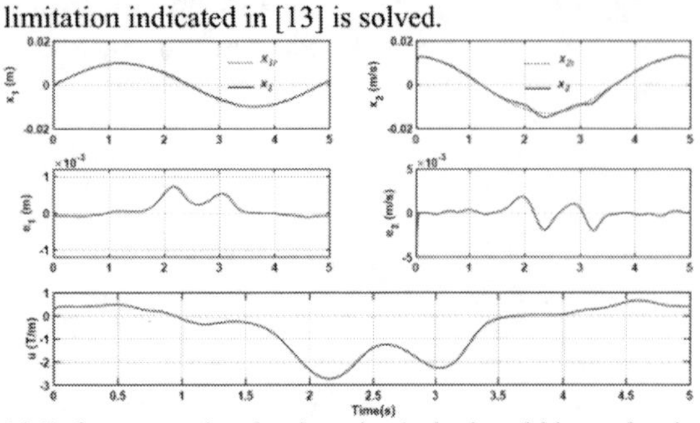

(a) References and real trajectories (x_{1r}in dotted blue and x_1 in red), real and calculated velocity of nanorobot (x_{2r}in dotted blue and x_2 in red), position tracking error (e_1), nanorobot velocity calculation error (e_2), and control input – magnetic field gradient (u).

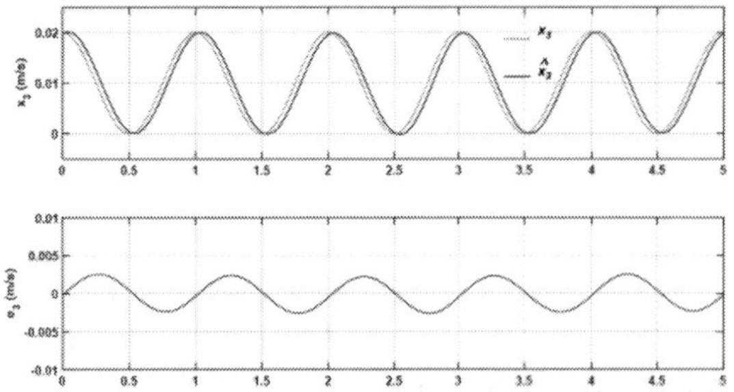

(b) Real and estimated blood velocities ($x_3$in dotted blue and $\hat{x}_3$ in red) and estimation error (e_3)

Figure 4. Simulation results with proposed blood velocity observer with single nanorobot.

6 Conclusions

In this work, we proposed an optimal observer for blood velocity estimation in blood vessels. Based on this observer, the real-time information of blood velocity is estimated precisely. Then this estimated signal was input to the controller to improve the performance of the overall control system. The only measured signal was position of aggregated nanorobot. We analyzed that the proposed optimal observer is asymptotically stable. The gain tuning rules for proposed observer was also detailed in this paper. Finally, simulation results have shown to prove the effectiveness of the proposed observer as well as overall feedback control system.

References

1. J. Bar, R. S. Herbst, A. Onn, ExpertOpin. Drug Deliv., **6**, 10, (2009)
2. V. P. Torchilin, Handb. Exp. Pharmacol., 197, (2010)
3. S. Martel, J. Nanopart. **17**, 2, (2015)
4. M. Hamdi, A. Ferreira, IEEE Trans. Robot., **30**, 1, (2014)
5. M. D. Tehrani, J. H. Yoon, M. O. Kim, J. Yoon, IEEE Trans. Biomed. Eng., **62**, 1, (2015)
6. T. D. Do, Y. Noh, M. O. Kim, J. Yoon, IROS 2015, Hamburg, Germany, Sep. 28-Oct. 03, (2015)
7. K. B. Yesin, K. Vollmers, B. J. Nelson, Int. J. Robot. Res., **25**, 527-536, (2006)
8. J. B. Mathieu, S. Martel, Biomed. Microdevices, **9**, (2007)
9. S. Martel, O. Felfoul, J. B. Mathieu, A. Chanu, S. Tamaz, M. Mohammadi, M. Mankiewicz, N. Tabatabaei, Int. J. Robot. Res., **28**, 9, (2009)
10. P. Pouponneau, J. C. Leroux, G. Soulez, L. Gaboury, S. Martel, Biomat., **32**, 13, (2011)
11. K. Belharet, D. Folio, A. Ferreira, IEEE Trans. Biomed. Eng., **60**, 4, (2013)
12. L. Arcese, M. Fruchard, A. Ferreira, IEEE Trans. Biomed. Eng., **59**, 4, (2012)
13. M. Fruchard, L. Arcese, E. Courtial, IEEE Trans. Robot., **30**, 1, (2014)
14. T. D. Do, Y. Noh, M. O. Kim, and J. Yoon, J. Nanosci. Nanotechnol., **16**, 6, (2016)
15. J. Womersley, J. Physiol., **127**, 3, (1955)
16. J. Astrom, B. Witternmark, *Computer-Controlled Systems – Theory and Design* (Englewood Cliffs, New Jersey: Prentice-Hall, 1990)
17. O. Vainio, IEEE Trans. Ind. Electron., **46**, 5, (1999)
18. T. D. Do, H. H. Choi, J. W. Jung, IEEE/ASME Trans. Mechatronics, **20**, 6, (2015)
19. B. Shahian and M. Hassul, *Control System Design Using Matlab* (Englewood Cliffs, New Jersey: Prentice-Hall Inc., 1993)

«Concurrency» in M-L-Parallel Semi-Markov Process

Eugene Larkin[1] and Alexey Ivutin [2]

[1]*Tula State University, Department of Robotics and Industry Mechanization, 300012, Tula, Russia*
[2]*Tula State University, Department of Computer Technology, 30012, Tula, Russia*

Abstract. This article investigates the functioning of a swarm of robots, each of which receives instructions from the external human operator and autonomously executes them. An abstract model of functioning of a robot, a group of robots and multiple groups of robots was obtained using the notion of semi-Markov process. The concepts of aggregated initial and aggregated absorbing states were introduced. Correspondences for calculation of time parameters of concurrency were obtained.

1 Introduction

Mobile robots are widely applied for environmental monitoring in counterterrorism, ecology and intelligence systems and in systems for rectification of consequences of technological disasters, etc. Despite the differences among the multiple environments, in which the robot can operate (air, ground, above-water, underwater ones) and despite the differences in the performed functions, the robot as a controlled object, has typical structure comprised of a sensor subsystem, on-board control computer, actuating devices and mechanisms, as well as of means of communication with a remote control station. Computer-based control over the robotic units and mechanisms comes down to the performance of the sequence of actions that unfold in time. Each action can be characterized by a random time period, measured from the beginning of the action to its end, as well as by random transition to another action of the sequence, if there is an alternative of continuation.

Thus, semi-Markov process is an adequate model of control over the single robot, which is characterized by multiple states, random time of being in the states and the stochastic nature of transition into the conjugated state. In practice, the use of separate mobile robots is ineffective, so the group use of robots using the so-called swarm technologies is an actively developing field. In this regard, the development of new principles and approaches to the organization of control over the swarm, performing a common task, for example, the development of model control process, which is carried out with the temporary concurrence of the functioning and organization of several modeling circuits. A special case of such control is when the commands are given to the subgroup of robots, included into the group, and within the execution of external commands, each unit of the swarm retains its autonomy, which means that the commands are reduced to the realization of one of the cyclograms, that were programmed in the onboard computer during the design stage [2-5]. Actions and status of each of the robots in the group, unfold in time, and switching from the current state to one of the possible following states is random for the external observer. Therefore, the same mathematical apparatus of semi-Markov processes can be used to model the functioning of organized swarms of robots [6-10].

2 ML-parallel semi-Markov

To summarize the above said, it could be noted that in this section we will focus on the modeling of the so-called M-L-parallel semi-Markov processes, in which there are M-parallel processes divided into L groups.

Let us define the elementary semi-Markov process as

$$\mu'_m = \{A'_m, \mathbf{r}'_m, \mathbf{h}'_m(t)\} \tag{1}$$

where A'_m is a number of states, among which there is at least one initial and one absorbing state; $\mathbf{r}'_m$ - adjacency matrix; $\mathbf{h}'_m(t)$ - semi-Markov matrix; $1 \leq m \leq M$.

Using the methods described in [4, 6, 9, 10], the process (1) can be transformed into an elementary process including one initial and one absorbing state,

$$\mu_m = \{A_m, \mathbf{r}_m, \mathbf{h}_m(t)\} \tag{2}$$

where $A_m = \{a_{1(m)}, a_{2(m)}\}$; $a_{1(m)}$ - the initial state; $a_{2(m)}$ - absorbing state; $\mathbf{r}_m = \begin{bmatrix} 0 & 1 \\ 0 & 0 \end{bmatrix}$, $\mathbf{h}_m(t) = \begin{bmatrix} 0 & f_m(t) \\ 0 & 0 \end{bmatrix}$ - the adjacency matrix and semi-Markov matrix respectively; $f_m(t)$ - distribution density of time of the process being in the state $a_{1(m)}$ until it switches to the state $a_{2(m)}$.

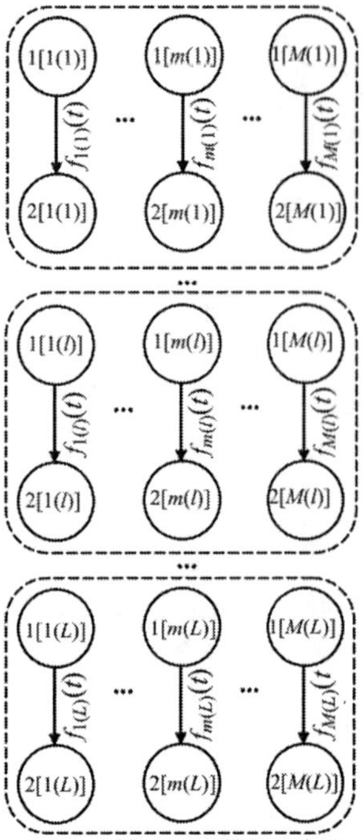

Figure 1. The structure of M-L-parallel semi-Markov process.

Processes (2) combine into L groups with M (l) elementary processes of type (2) in a group (Fig. 1). In the meantime $\sum_{l=1}^{L} M(l) = M$, $1(l) \le m(l) \le M(l)$.

l group of the semi-Markov processes can be represented as

$$^{l}\mu = \left\{ {}^{l}A,\ {}^{l}\mathbf{r},\ {}^{l}\mathbf{h}(t) \right\} \tag{3}$$

where

$$^{l}A = \left\{ \begin{array}{l} \{a_{1[1(l)]}, a_{2[1(l)]}\}, \ldots, \{a_{1[m(l)]}, a_{2[m(l)]}\}, \ldots, \\ \{a_{1[M(l)]}, a_{2[M(l)]}\} \end{array} \right\} \tag{4}$$

$$^{l}\mathbf{r} = \begin{bmatrix} 0 & 1 & & & & & \\ 0 & 0 & & & & 0 & \\ & & \ldots & & & & \\ & & & 0 & 1 & & \\ & & & 0 & 0 & & \\ & & & & & \ldots & \\ & 0 & & & & 0 & 1 \\ & & & & & 0 & 0 \end{bmatrix} \tag{5}$$

$$^{l}\mathbf{h}(t) = \begin{bmatrix} 0 & f_{1(l)}(t) & & & & & \\ 0 & 0 & & & & 0 & \\ & & \ldots & & & & \\ & & & 0 & f_{m(l)}(t) & & \\ & & & 0 & 0 & & \\ & & & & & \ldots & \\ & 0 & & & & 0 & f_{M(l)}(t) \\ & & & & & 0 & 0 \end{bmatrix} \tag{6}$$

M-L-parallel semi-Markov process has the following form:

$$^{M,L}\mu = \left\{ {}^{M,L}A,\ {}^{M,L}\mathbf{r},\ {}^{M,L}\mathbf{h}(t) \right\} \tag{7}$$

where

$$^{M,L}A = \bigcup_{l=1}^{L} {}^{l}A \tag{8}$$

$$^{M,L}\mathbf{r} = \begin{pmatrix} {}^{1}\mathbf{r} & & & & 0 \\ & \ldots & & & \\ & & {}^{l}\mathbf{r} & & \\ & & & \ldots & \\ 0 & & & & {}^{L}\mathbf{r} \end{pmatrix} \tag{9}$$

$$^{M,L}\mathbf{h}(t) = \begin{pmatrix} {}^{1}\mathbf{h}(t) & & & & 0 \\ & \ldots & & & \\ & & {}^{l}\mathbf{h}(t) & & \\ & & & \ldots & \\ 0 & & & & {}^{L}\mathbf{h}(t) \end{pmatrix} \tag{10}$$

It is believed that l group of semi-Markov processes reaches the aggregated absorbing state, when the last process in the group, for example, the $m(l)$, switches to the state $a_{2[m(l)]}$, and there are no other semi-Markov processes in the l group, which have not reached one of M(l) absorbing states. In this case, the group has a single absorbing state, which is the aggregated absorbing state of group.

We can assume without loss of generality that l group consists of elementary semi-Markov processes (2) with the numbers $\sum_{n=1}^{l-1} M(n) + 1 \le m(l) \le \sum_{n=1}^{l} M(n)$, and thus the indices m are recalculated according to the indices $m(l)$ using the correspondence:

$$m = m(l) + \sum_{n=1}^{l-1} M(n) \tag{11}$$

The distribution density of time taken by the l group of semi-Markov processes to reach the aggregated absorbing state is calculated with the correspondence [11-13]

$$^{l}f(t) = \frac{d}{dt}\left[\prod_{m=\sum_{n=1}^{l-1} M(n)+1}^{\sum_{n=1}^{l} M(n)} F_m(t) \right] \tag{12}$$

3 The total semi-Markov process

Correspondence (12) allows representing the l group of semi-Markov processes as an aggregated semi-Markov process, which includes the aggregated initial state $a_{1(l)} = \bigcup_{m(l)=1(l)}^{M(l)} a_{1[m(l)]}$ and the aggregated absorbing state $a_{w(l)} = \bigcup_{m(l)=1(l)}^{M(l)} a_{2[m(l)]}$. Consequently, *M*-*L*-parallel semi-Markov process can be represented as *L*-parallel semi-Markov process, which comprises $2L$ states

$\{a_{1(1)},\dots,a_{1(l)},\dots,a_{1(l)},\ a_{2(1)},\dots,a_{2(l)},\dots,a_{2(l)}\}$ and is described by semi-Markov matrix

$$ {}^{L}\tilde{\mathbf{h}}(t)=\begin{bmatrix} 0 & {}^{1}f(t) & & & & & \\ 0 & 0 & & & 0 & & \\ & & \dots & & & & \\ & & & 0 & {}^{l}f(t) & & \\ & & & 0 & 0 & & \\ & & & & & \dots & \\ & 0 & & & & 0 & {}^{L}f(t) \\ & & & & & 0 & 0 \end{bmatrix} \quad (13)$$

The distribution density of time taken by the l group of semi-Markov processes to be the first to reach the aggregated absorbing state of the semi-Markov processes (before the other groups have reached their absorbing states) is calculated according to the correspondence

$$ {}^{l}f_{w}(t)=\frac{{}^{l}f(t)\cdot\prod_{\substack{n=1\\n\neq l}}^{L}\left[1-{}^{n}F(t)\right]}{{}^{l}p_{w}} \quad (14)$$

where ${}^{l}p_{w}$ - the probability that the l group of semi-Markov processes the first one out of L-parallel processes to reach the aggregated absorbing state:

$$ {}^{l}p_{w}=\int_{0}^{\infty}{}^{l}f(t)\cdot\prod_{\substack{n=1\\n\neq l}}^{L}\left[1-{}^{n}F(t)\right]dt \quad (15)$$

Numerical characteristics of the distribution density (14) are calculated, according to the correspondences

$$ {}^{l}T_{w}=\int_{0}^{\infty}t\frac{{}^{l}f(t)\cdot\prod_{\substack{n=1\\n\neq l}}^{L}\left[1-{}^{n}F(t)\right]}{{}^{l}p_{w}}dt \quad (16)$$

$$ {}^{l}D_{w}=\int_{0}^{\infty}\left(t-{}^{l}T_{w}\right)^{2}\frac{{}^{l}f(t)\cdot\prod_{\substack{n=1\\n\neq l}}^{L}\left[1-{}^{n}F(t)\right]}{{}^{l}p_{w}}dt,\ 1\leq l\leq L.$$

The distribution density of time the aggregated absorption state is reached by the last l group of semi-Markov processes (after the other groups have reached their absorbing states) is calculated by the correspondence

$$ {}^{l}f_{\overline{w}}(t)=\frac{{}^{l}f(t)\cdot\prod_{\substack{n=1\\n\neq l}}^{L}{}^{n}F(t)}{{}^{l}p_{\overline{w}}} \quad (17)$$

where ${}^{l}p_{\overline{w}}$ - the probability that the l group of semi-Markov processes will be the last one out of L-parallel processes to reach the aggregated absorbing state:

$$ {}^{l}p_{\overline{w}}=\int_{0}^{\infty}{}^{l}f(t)\cdot\prod_{\substack{n=1\\n\neq l}}^{L}{}^{n}F(t)dt \quad (18)$$

Numerical characteristics of the distribution density (17) are calculated by correspondences:

$$ {}^{l}T_{\overline{w}}=\int_{0}^{\infty}t\frac{{}^{l}f(t)\cdot\prod_{\substack{n=1\\n\neq l}}^{L}{}^{n}F(t)}{{}^{l}p_{\overline{w}}}dt \quad (19)$$

$$ {}^{l}D_{\overline{w}}=\int_{0}^{\infty}\left(t-{}^{l}T_{\overline{w}}\right)^{2}\frac{{}^{l}f(t)\cdot\prod_{\substack{n=1\\n\neq l}}^{L}{}^{n}F(t)}{{}^{l}p_{\overline{w}}}dt,\ 1\leq l\leq L.$$

The latency period for the l group of semi-Markov processes, which has already reached the aggregated absorbing state, before k group reaches its aggregated absorbing state, is defined as:

$$ f_{l\to k}(t)=\frac{\eta(t)\int_{0}^{\infty}\frac{d}{dt}\left[\prod_{m=\sum_{n=1}^{k-1}L(n)+1}^{\sum_{n=1}^{k}L(n)}{}^{m}F(t+\tau)\right]\cdot d\prod_{m=\sum_{n=1}^{l-1}L(n)+1}^{\sum_{n=1}^{l}L(n)}{}^{m}F(\tau)}{\int_{0}^{\infty}\prod_{m=\sum_{n=1}^{l-1}L(n)+1}^{\sum_{n=1}^{l}M(n)}{}^{m}F(\tau)d\prod_{m=\sum_{n=1}^{k-1}L(n)+1}^{\sum_{n=1}^{k}L(n)}{}^{m}F(t+\tau)} \quad (20)$$

$$ T_{l\to k}=\int_{0}^{\infty}t\frac{\eta(t)\int_{0}^{\infty}\frac{d}{dt}\left[\prod_{m=\sum_{n=1}^{k-1}M(n)+1}^{\sum_{n=1}^{k}L(n)}{}^{m}F(t+\tau)\right]\cdot d\prod_{m=\sum_{n=1}^{l-1}M(n)+1}^{\sum_{n=1}^{l}L(n)}{}^{m}F(\tau)}{\int_{0}^{\infty}\prod_{m=\sum_{n=1}^{l-1}L(n)+1}^{\sum_{n=1}^{l}L(n)}{}^{m}F(\tau)d\prod_{m=\sum_{n=1}^{k-1}L(n)+1}^{\sum_{n=1}^{k}L(n)}{}^{m}F(t+\tau)}dt \quad (21)$$

$$ D_{l\to k}=\int_{0}^{\infty}(t-T_{l\to k})^{2}\frac{\eta(t)\int_{0}^{\infty}\frac{d}{dt}\left[\prod_{m=\sum_{n=1}^{k-1}M(n)+1}^{\sum_{n=1}^{k}M(n)}{}^{m}F(t+\tau)\right]\cdot d\prod_{m=\sum_{n=1}^{l-1}M(n)+1}^{\sum_{n=1}^{l}M(n)}{}^{m}F(\tau)}{\int_{0}^{\infty}\prod_{m=\sum_{n=1}^{l-1}M(n)+1}^{\sum_{n=1}^{l}M(n)}{}^{m}F(\tau)d\prod_{m=\sum_{n=1}^{k-1}M(n)+1}^{\sum_{n=1}^{k}M(n)}{}^{m}F(t+\tau)}dt \quad (22)$$

4 Reaching of the aggregated absorbing states by any K out of L groups of parallel processes

In case when the fact of reaching of aggregated absorbing states by any K of L, $K<L$, groups of parallel processes, is of primary importance, the density of time distribution of the specified event is calculated by the correspondence

$$ f_{ML}^{K}(t)=\frac{d\sum_{c_{L}^{K}=1}^{C_{L}^{K}}\prod_{l=1}^{L}\Phi\left({}^{l}f,\sigma_{l}^{c_{L}^{K}}\right)}{dt} \quad (23)$$

where ${}^{l}f(t)$ - the distribution density of time spent by the l group of semi-Markov processes to reach the aggregated absorbing state, calculated by the correspondence (12); c_L^K - the number of the figure in the range $N_L^K = \{n_1, \ldots, n_{c_L^K}, \ldots, n_{C_L^K}\}$; C_L^K - K binomial coefficient raised to the power of L; $n_{c_L^K} = \left\langle \sigma_1^{c_L^K}, \ldots, \sigma_l^{c_L^K}, \ldots, \sigma_L^{c_L^K} \right\rangle$ - the figure from the range N_L^K; $\sigma_l^{c_L^K}$ - l digit place of $n_{c_L^K}$, which can take two values in accordance with (3.29); $C_L^K = \frac{L!}{K \cdot (L-K)!}$; $\Phi\left({}^{l}f, \sigma_l^{c_L^K}\right)$ - function that can get values

$$\Phi\left({}^{l}f, \sigma_l^{c_L^K}\right) = \begin{cases} {}^{l}F(t), \text{ if } \sigma_l^{c_L^K} = 1; \\ [1 - F_l(t)], \text{ if } \sigma_l^{c_L^K} = 0 \end{cases} \quad (24)$$

For the distribution density (23) numeric characteristics can be found:

$$T_{ML}^K = \int_0^\infty t \cdot d \sum_{c_L^K = 1}^{C_L^K} \prod_{l=1}^{L} \Phi\left({}^{l}f, \sigma_l^{c_L^K}\right) \quad (25)$$

$$D_{ML}^K = \int_0^\infty \left(t - T_L^K\right)^2 d \sum_{c_L^K = 1}^{C_L^K} \prod_{l=1}^{M} \Phi\left({}^{l}f, \sigma_l^{c_L^K}\right) \quad (26)$$

To determine the cost of a losing (winning) the "concurrency" it is necessary to use the procedures set forth in [11-13]. In this case, the cost of waiting should be calculated for the case when one group of semi-Markov processes is waiting for all the semi-Markov processes of another group to reach the aggregated absorbing state.

5 Conclusion

Thus, the model of functioning of a swarm of robots, in the form of *ML*-parallel semi-Markov process has been developed. The correspondences, determining the time intervals of switching from the current aggregated state to the conjugated aggregated states have been obtained, which makes it possible to estimate the time of the performance of a command given to a swarm of robots, with calculation accuracy of distribution density. Further research is associated with the development of simple engineering techniques for synthesis of optimal algorithms to control a swarm, basing on the theoretical assumptions and using win/loss cost in case of uncoordinated actions of separate robots of the swarm as the major criterion for optimization.

Benefit of proposed technology of modeling of swarm is that the operability and an effectiveness of a soft under development may be evaluated on the stage of design without costly experiments on real swarm.

The reported study was partially supported by RFBR and Tula Region Government, research project No. 16-41-710160 r_a.

References

1. Ivutin A.N., Larkin E.V., Kotov V.V. Established Routine of Swarm Monitoring Systems Functioning // Advances in Swarm and Computational Intelligence. Lecture Notes in Computer Science. – Springer International Publishing, Switzerland, (2015) – Vol. **9141** – pp. 415-422. DOI: 10.1007/978-3-319-20472-7_45.
2. Ivutin A.N., Larkin E.V. Generalized Semi-Markov Model of the Algorithm for Digital Devices' Control // Izvestiya of Tula State University. Ser. Technical science. Vol. 1. Tula: Tula State University Publishing House, (2013). pp. 221-228.
3. Larkin E.V., Ivutin A.N. Estimation of Latency in Embedded Real-Time Systems // 3-rd Mediterranean Conference on Embedded Computing (MECO-2014). 2014. June 15-19. Budva, Montenegro, (2014). Pp. 236-239.
4. Larkin E.V., Ivutin A.N. Calculation of Time in the Control Algorithms // Izvestiya of Tomsk Polytechnic University. No. 5. Vol. **324**. (2014). pp. 6-12.
5. Klintsov G.N., Larkin E.V. Control of Tunnel-boring Systems Using SCADA Software // Izvestiya of Tula State University. Ser. Technical science. Vol. **9**. Part 1. Tula: Tula State University Publishing House, (2014). pp. 156-162.
6. Ivutin A.N., Larkin E.V. Time and Probability Characteristics of Transactions in Digital Control Systems // Izvestiya of Tula State University. Ser. Technical science. Vol. **1**. - Tula: Tula State University Publishing House, (2013). pp. 252-258.
7. Ivutin A.N., Larkin E.V. Predicting the Runtime of the Algorithm // Izvestiya of Tula State University. Technical science. Vol. **3**. - Tula: Tula State University Publishing House, (2013). pp. 301-315.
8. Larkin E.V., Kotov V.V., Kotova N.A. Evaluating the Efficiency of Robot's Software Using Petri-Markov net // Izvestiya of Tula State University. Technical science. Vol. **9**. Part 2. - Tula: Tula State University Publishing House, (2013). pp. 156-163.
9. Klintsov G.N., Larkin E.V. Semi-Markov Modeling of Application Perfomance Algorhythms for Computer // Izvestiya of Tula State University. Ser. Technical science. Vol. **9**. Part 1. (2015). pp. 143-149.
10. Larkin E.V., Privalov A.N., Klepikov A.K. Algorithm of Sequential Simplifications in Evaluation of Time Complexity of Algorithms with Semi-Markov Models // Izvestiya of Tula State University. Ser. Technical science. Vol. **9**. Part 2. Tula: Tula State University Publishing House, (2014). pp. 83-98.
11. Larkin E.V., Lutskov Yu.I., Ivutin A.N., Novikov A.S. Simulation of concurrent process with Petri-Markov nets // Life Science Journal. (2014). N. 11. pp. 506-511. (ISSN: 1097-8135). http.lifesciencesite.com.
12. Ivutin A.N, Larkin E.V. Simulation of Concurrent Games // Bulletin of the South Ural State University. Series: Mathematical Modeling, Programming and

Computer Software. Chelyabinsk. 2015. Vol. **8**. No. 2. pp. 43-54. DOI: 10.14529/mmp150204

13. Ivutin A.N, Larkin E.V. Simulation of Concurrent Games // Bulletin of the South Ural State University. Series: Mathematical Modeling, Programming and Computer Software. – Chelyabinsk, 2015. – Vol. **8**, no. 2. pp. 43-54. DOI: 10.14529/mmp150204

DOI: 10.1051/matecconf/201710805004

Modbus RTU protocol and arduino IO package: A real time implementation of a 3 finger adaptive robot gripper

Amirul Syafiq Sadun, Jamaludin Jalani and Jumadi Abdul Sukor

Faculty of Engineering Technology, Universiti Tun Hussein Onn Malaysia (UTHM), Batu Pahat, Johor, Malaysia

Abstract. Recently, the Modbus RTU protocol has been widely accepted in the application of robotics, communications and industrial control systems due to its simplicity and reliability. With the help of the MATLAB Instrument Control Toolbox, a serial communication between Simulink and a 3 Finger Adaptive Robot Gripper can be realized to demonstrate a grasping functionality. The toolbox includes a "to instrument" and "query instrument" programming blocks that enable the users to create a serial communication with the targeted hardware/robot. Similarly, the Simulink Arduino IO package also offers a real-time feature that enabled it to act as a DAQ device. This paper establishes a real-time robot control by using Modbus RTU and Arduino IO Package for a 3 Finger Adaptive Robot Gripper. The robot communication and grasping performance were successfully implemented and demonstrated. In particular, three (3) different grasping mode via normal, wide and pinch were tested. Moreover, the robot gripper's feedback data, such as encoder position, motor current and the grasping force were easily measured and acquired in real-time. This certainly essential for future grasping analysis of a 3 Finger Adaptive Robot Gripper.

1 Introduction

Nowadays, many robotic hardware has a built in function of the Modbus controller unit. An example of this function can be found in a 3 Finger Adaptive Robot Gripper developed by *Robotiq.* The Modbus protocol was developed in 1979 by Modicon where fundamentally the protocol is mostly applied in manufacturing/automation industries. Many recent industrial products such as PLC, I/O devices and instrument are equipped with the protocol due to its user friendly and easy to be used for hardware integrations [1]. The physical layer for the serial communications is normally, USB, RS-232, RS-422 and RS-485. The fundamental block diagram for Modbus RTU protocol is shown in Figure 1 [2].

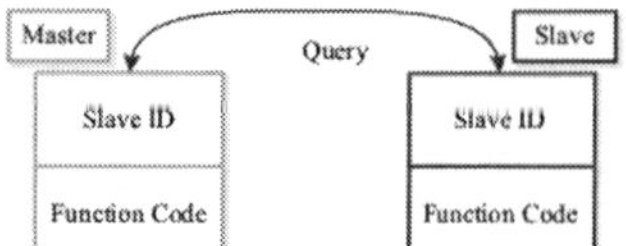

Figure 1. Modbus RTU Fundamental.

Modbus RTU (Remote Terminal Unit) is a type of Modbus protocol that uses serial communication to create a Master - Slave or Server- Client communication between intelligent devices [3]. Based on Figure 1, the basic concept of the Modbus RTU protocol is "query" and "response" between master and slave in which this process continuously occurs throughout the operation of the robot. The master devices send a query to slave devices while the slave devices send a response to the query with a specified code. Table 1 generally summarized the communication structure which consists of Slave ID, Function Code, Data, and Error Code (CRC).

Table 1. General explanation of Modbus RTU communication structure.

Structure	Purposes
Slave ID	Used to assign the address of Slave devices
Function Code	The function code tells the slave device what kind of action to be executed at particular "query".
Data	The data of the slave devices which are united into a network (i.e gripper status, & finger position request).
Error Code (CRC)	Cyclic redundancy check is used to check the error of the data transmission over the physical layer.

Therefore, the setup for Modbus RTU protocol is rather simple and easy to be used via serial communication (USB connection). Figure 2 shows the setup for a robot hand with a built in Modbus RTU function:

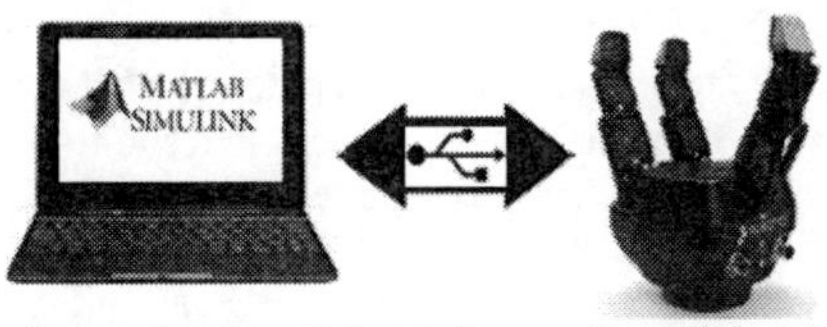

Figure 2. Connection for a Robot Gripper with Built in Modbus RTU Function.

The 3 Finger Adaptive Robot Gripper as shown in Figure 2 supports the Modbus RTU and TCPIP communication protocol. However, the robot comes with a default GUI that only suitable for demonstrations. To further uses the robot in the research environment (i.e. closed loop control system), a communication platform (i.e. by using MATLAB Simulink) is needed for future experimental and analysis. Moreover, to add the element of external sensory feedback to the robot system (i.e. force sensor), a DAQ device is compulsory. Referring to [4], the results for use of Arduino IO package in MATLAB Simulink showed a great potential whereas the Arduino hardware is capable of acting similarly to the DAQ device in real-time. This paper introduces a real-time robot control by using Modbus RTU platform with the integration of Arduino IO Package for a 3 Finger Adaptive Robot Gripper. The robot control is programmed mainly by using the MATLAB Instrument Control Toolbox. Additionally, the integration of robot control with the Arduino IO Package is possible by installing the supported library package to MATLAB Simulink. The study focused on developing a Simulink program to control and acquire the robot available feedback data in real-time such as finger position, actuator motor current and grasping force. It is important to establish the programming platform for the future study related to the 3 Finger Adaptive Robot Gripper (i.e grasping analysis or compliance control). The available input and output data of the robot gripper was recorded individually for each finger utilizing all the available robot modes (i.e. normal, wide and pinch). Furthermore, a graphical data analysis was done to observe the robot operation in the "open loop system".

2 Preliminary setup

2.1 Hardware setup

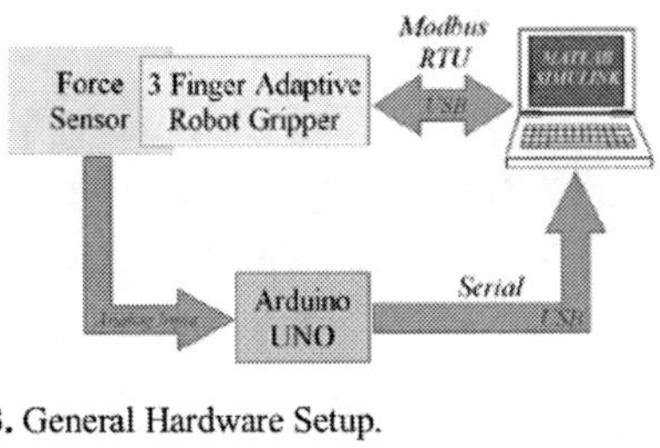

Figure 3. General Hardware Setup.

Figure 3 shows the general hardware setup for the experiment. The Modbus RTU communication protocol was done by using MATLAB Simulink where the robot is connected to a computer (laptop) via USB cable. The setup also consists of Arduino UNO which acts as a DAQ device. The analog input from sensors are connected to Arduino ADC (Analog to Digital) pin and the data were then acquired by using Simulink Arduino IO Package. The robot control and data acquisition were run in the same Simulink program by using 2 different COM Ports.

2.2 Force Sensor Setup

Previous studies for FSR sensors showed that that the FSR sensors are sufficient and reliable to measure and record the data once the force is detected even though a minor drawback has been identified regarding the fluctuating readings (sensor accuracy) [5]. In general, the FSR sensor is measured in voltage (V) which can be converted to Force (N) by using the linear model equation as presented in [6]. The force range detection is between 1N to 100N. However, a proper method of sensor placement has to be carefully designed and positioned so that the contact surface is adequately measured [5]. The better force reading can be obtained by spreading the pressure over a greater area. For this experimental purposes, the researcher has developed a 3D printed plastic cover for the FSR sensor as shown in Figure 4.

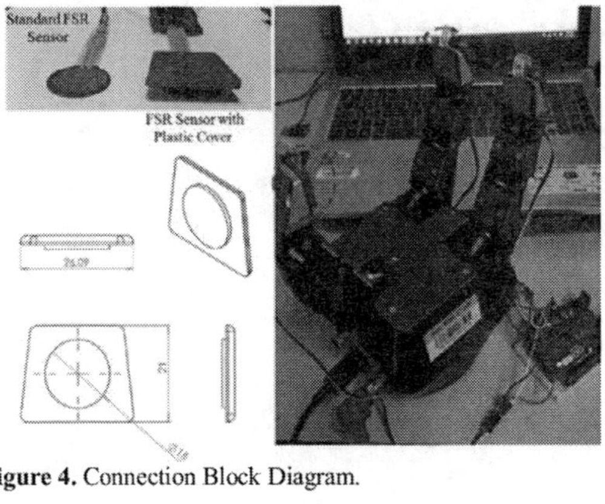

Figure 4. Connection Block Diagram.

3 The control and communication scheme

3.1 Cyclic redundancy check (CRC)

In Modbus RTU communication, is it compulsory to perform a Cyclic Redundancy Check (CRC) to ensure the data integrity in term of errors and abnormalities. During the robot gripper's communications, the data is sent and received as a string of 8 binary characters framed with a start bit, and a stop bit, making each byte 10 bits [7]. Thus, every byte in the data instructions/message is used to calculate the CRC either during the "request" or "query" of the robot gripper [8]. Figure 5 shows the flow chart of the CRC.

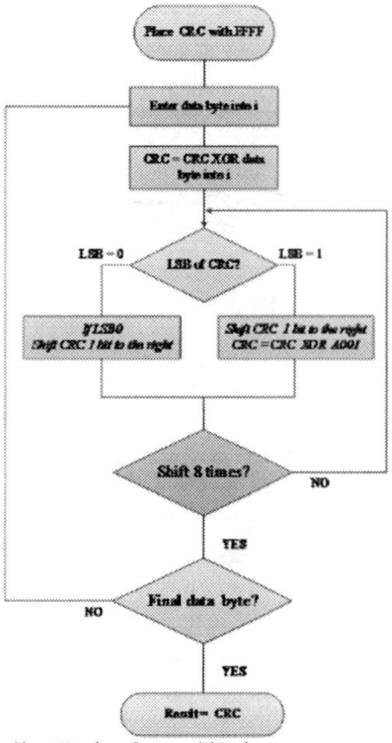

Figure 5. Cycling Redundancy Check.

Referring to Figure 5, the process is repeated for every new data transmission (send/receive) throughout the Modbus RTU protocol. If the results of the CRC are incorrect, communication errors occur between the robot gripper and MATLAB Simulink. Thus, the robot will stop responding to the requested instruction.

3.2 Programming design structure

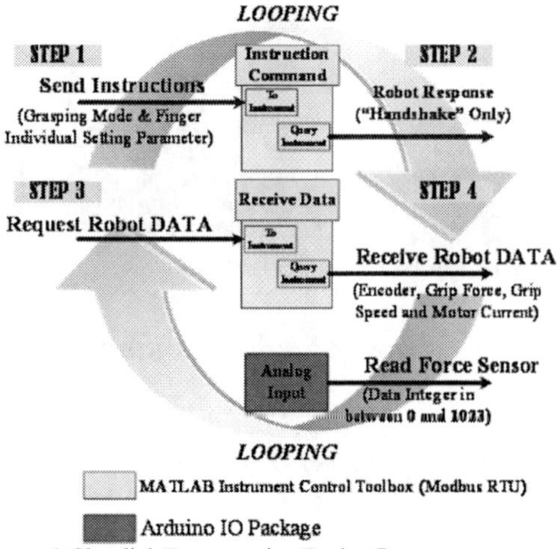

Figure 6. Simulink Programming Design Structure.

Figure 6 illustrates the programming design structure for the Modbus RTU communication protocol and the Simulink Arduino IO package. The programming techniques for Modbus RTU communication can be divided into 4 stages. The first step is to send the instructions to set the grasping mode and the finger's individual setting parameter. At this step, it is important to identify the correct parameter setting based on the robot gripper's user manual. Step 2 is to receive the robot gripper response. This "handshake" stage works as an initial indication to identify any abnormalities and to indicate the previous instruction data is well received and correct. Step number 3 was carried out to request the robot grippers data in real-time (i.e. the encoder position changes). At this stage, the data of the robot gripper is being requested for every 1 millisecond. Next, the robot gripper completed the request in step 3 by returning the data of the requested parameters (in step 4). These 4 steps are repeated during the program runs. Simultaneously, the analog data of the force sensor can be acquired by using the Arduino IO Package. Figure 7 shows the Simulink programming blocks.

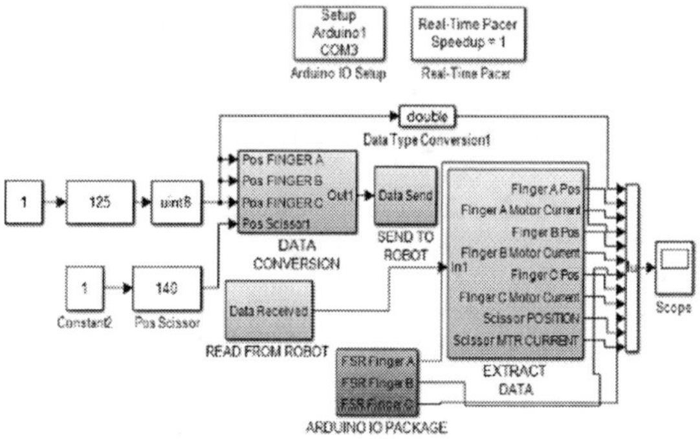

Figure 7. Programming Simulink Blocks.

4 Experimental results

4.1 Robot grasping

This section demonstrates the robot gripper's grasping by using the developed MATLAB Simulink program. The purpose of the test is to control the robot in "open loop system" while capturing the desired and actual finger position of the robot gripper. The test was implemented by using the 3 available modes of the robot gripper namely a normal, wide and pinch modes. Figure 8 (a), (b) and (c) shows the acquired data for desired versus the actual for all 3 modes.

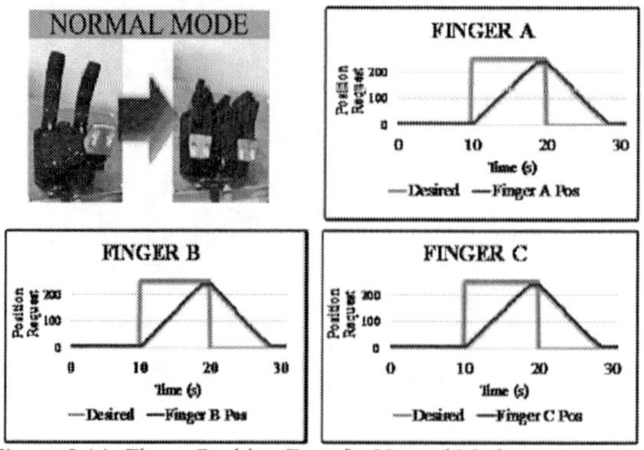

Figure 8 (a). Finger Position Data for Normal Mode.

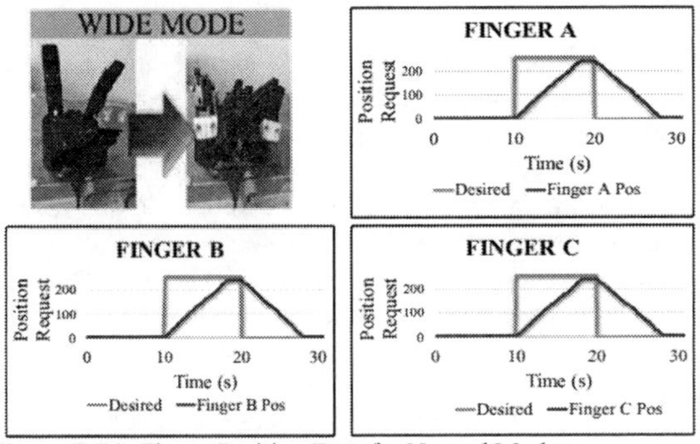

Figure 8 (b). Finger Position Data for Normal Mode.

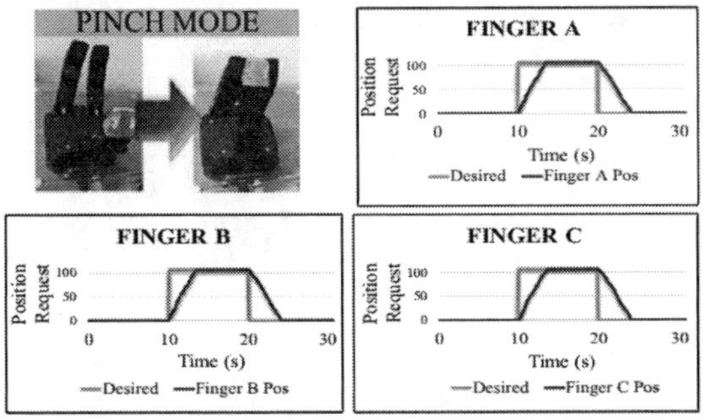

Figure 8 (c). Finger Position Data for Pinch Mode.

Figure 8 (a), (b), and (c) shows the desired (requested) range for the normal and wide mode in between 0 to 255 degrees while in pinch mode the movement is in between 0 to 100 degrees. During the operation, the encoder position was converted into a decimal to make it responding in the similar value range. Based on the results, it can be observed that the acquired data for normal, wide and pinch mode follows satisfactorily the desired position. On the other hand, the program also successfully acquired the robot gripper's actuator motor current during the operation. The data for motor current versus the actual robot finger position are shown in Figure 9.

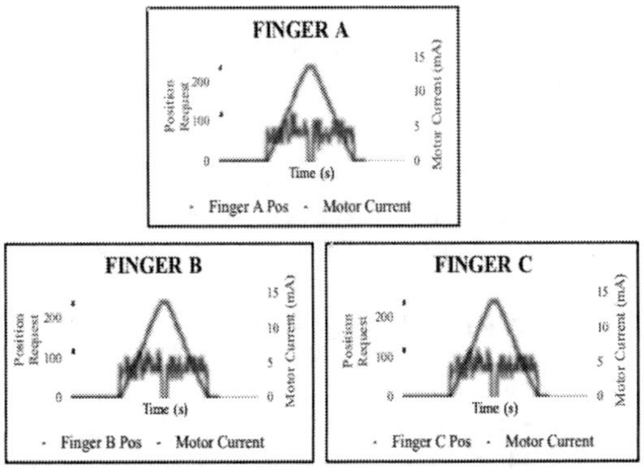

Figure 9. Actuator Motor Current vs Actual Robot Finger Position.

Based on Figure 9, the data for actuator motor current during grasping and ungrasping indicates a constant reading (no load/no object condition). For all 3 fingers, the motor current reading has an average of 4.5 mA during the motor movement, and once the desired position is reached, the current value drops to 0 mA (motor stops). The higher resistance during the grasping (i.e grasping a high stiffness object), the higher value of motor current can be expected.

4.2 Grasping with force sensing

This section demonstrates the grasping test with force sensing element by utilizing the Arduino IO package. The purpose of this test is to observe the capability of the program and the hardware setup to capture the data during the grasping operation (i.e finger position and force sensor value). The desired position was set from 65 to 80 (decimal) for all 3 fingers and grasping test was carried out with the robot grasping a low stiffness ball in wide mode (see Figure 10).

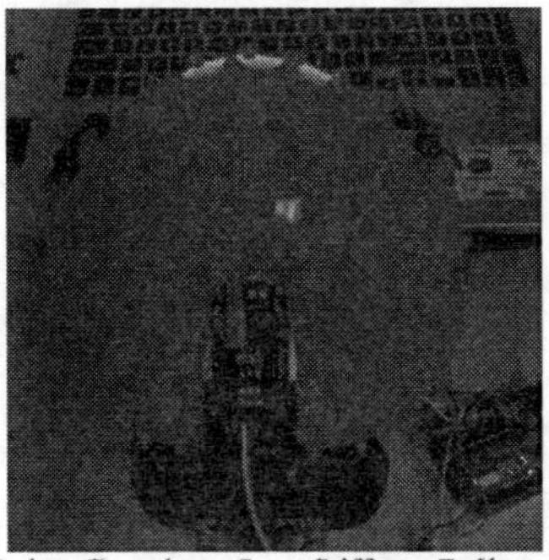

Figure 10. Robot Grasping a Low Stiffness Ball.

It is known that the robot gripper has the capability to grasp with the force ranging 15N to 60N. Thus, for this experiment, the force was set to maximum (60N) to observe data recording for finger position and force sensor. Figure 11 (a), (b) and (c) shows the data for the grasping test.

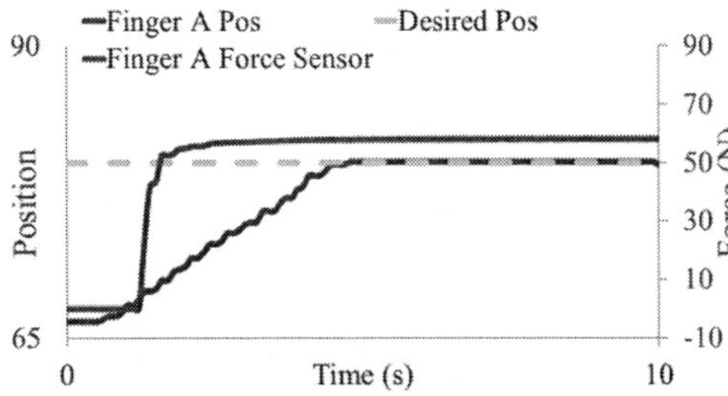

Figure 11 (a). Finger A Grasping Force Sensing.

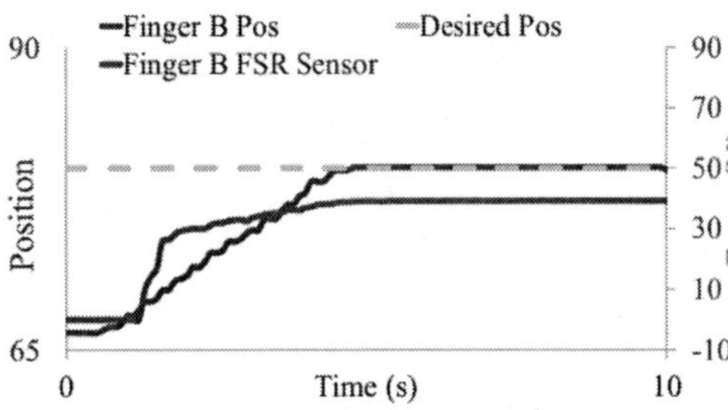

Figure 11 (b). Finger B Grasping Force Sensing.

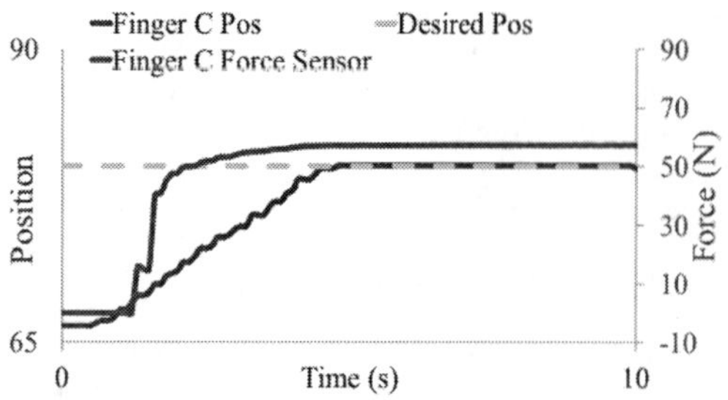

Figure 11 (c). Finger C Grasping Force Sensing.

Figure 11 clearly shows that the data for finger position and force sensor has successfully been recorded during the grasping. The data plot for force sensor indicates that the force gradually increases as the finger moving towards the desired position (grasping the object). The force values vary for each finger ranging from 40N to 58N. There are many factors contributing to the variation of the forces, the main factor is due to the underactuated mechanism of the robot gripper. Furthermore, the study also able to capture the data for actuator motor current versus force sensor as shown in Figure 12. The simple test is executed to see the fluctuation of motor current during the grasping contact with the plastic bottle.

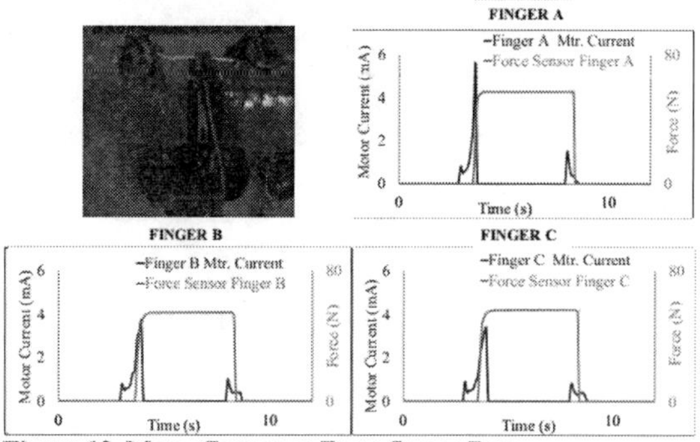

Figure 12. Motor Current vs Force Sensor Data.

Based on the results, the motor current for each finger increases as the resistance become higher (ranging from 3.8 mA to 5.8 mA) and the motor current drops to 0 mA when the motor stops. Additionally, the ungrasping operation indicates a low motor current values. Further discussion is needed on the topic of grasping analysis, however, it is not included in this study.

5 Conclusion

This paper established a real-time robot control by using Modbus RTU via MATLAB Instrument Control Toolbox and Arduino IO Package for a 3 Finger Adaptive Robot Gripper. The robot communication was successfully implemented and demonstrated for an individual finger by using all the available robot grasping modes. Based on the results of this study, Modbus RTU protocol was proven to be a reliable and stable platform for the robot control that integrates with the Arduino IO package. Moreover, the hardware setup for the FSR force sensor was able to detect the increment of force value during the object grasping. With the accuracy of 1 millisecond, the acquired force data are reliable for the analysis of an active compliance control for the future study. On the other hand, the study also able to record the actuator motor current during the grasping process. Based on the results, the motor current has and an average of 4.5 mA during "no load" condition and fluctuated up to 5.8 mA during the object grasping. Nevertheless, for future study, it is important to know the accurate angular proportional of the robot finger joint. An actual joint angular position should be used instead of the decimal input and output (0 to 255). Finally, the platform that has been created in this study can be used for the future study involving the 3 Finger Adaptive Robot Gripper especially for the application in the "closed loop system" environment.

Acknowledgement

This topic is part of a project funded by the Fundamental Research Grant Scheme (FRGS), Vot 1480. The authors also wish to thank the Faculty of Engineering Technology, Universiti Tun Hussein Onn Malaysia for providing a platform to carry out the research activities.

References

1. I. Modbus Organization, "Modbus Protocol." .
2. S. Tamboli, M. Rawale, R. Thoraiet, and S. Agashe, "Implementation of Modbus RTU and Modbus TCP communication using Siemens S7-1200 PLC for batch process," *2015 Int. Conf. Smart Technol. Manag. Comput. Commun. Control. Energy Mater. ICSTM 2015 - Proc.*, no. May, pp. 258–263, 2015.
3. G. B. M. Guarese, F. G. Sieben, T. Webber, M. R. Dillenburg, and C. Marcon, "Exploiting Modbus protocol in wired and wireless multilevel communication architecture," *Brazilian Symp. Comput. Syst. Eng. SBESC*, pp. 13–18, 2012.
4. A. S. Sadun, J. Jalani, and J. A. Sukor, "A Comparative Study on the Position Control Method of DC Servo Motor with Position Feedback by using Arduino," no. August, pp. 10–11, 2015.
5. A. S. Sadun and J. Jalani, "Force Sensing Resistor (FSR): Preliminary Results for Active Compliance Control of the Multifingered Robot Hand Faculty of Engineering Technology , Universiti Tun Hussein Onn Malaysia , 86400 Parit Raja , Batu," vol. 6, no. 0, pp. 2–5.
6. C. Lebosse, B. Bayle, M. De Mathelin, and P. Renaud, "Nonlinear modeling of low cost force sensors," *2008 IEEE Int. Conf. Robot. Autom.*, pp. 3437–3442, 2008.
7. simplymodbus, "Modbus ASCII vs Modbus RTU," 2015. [Online]. Available: http://www.simplymodbus.ca/.
8. H. Sideng, Z. Zhengming, Z. Yingchao, and W. Shuping, "A novel modbus RTU-based communication system for adjustable speed drives," *2008 IEEE Veh. Power Propuls. Conf. VPPC 2008*, 2008.

DOI: 10.1051/matecconf/201710805005

Research on Miniature Calibre Rail-Guns for the Mechanical Arm

Cao Ronggang , Zhang Guangwei and Su Ming

Beijing Institute of Technology5 South Zhongguancun Street, Haidian District, Beijing, 100081+86-1069818450, China

Abstract. Rail-gun should not only be used to military applications, but also can be developed as applications in the civilian aspects of the market. With the development of the electromagnetic launch technology, based on the similarity theory, using the existing rail-gun model to guide the construction of more economical miniature calibre rail-guns, and apply it in some machinery and equipment, this idea will open up a wider rail-gun application space. This article will focus on the feasibility of application of miniature calibre rail-guns in the mechanical arm. This paper designs the schematic diagram, then theoretical analyzes force conditions of the armature in the mechanical arm, calculates the possible range of the current amplitude and so on. The existing rail-gun model can be used to guides design the circuit diagram of the miniature calibre rail-gun. Based on the similarity theory and many simulation experiments, designed the experimental parameters of a miniature rail-gun and analyzed the current, Lorentz force, velocity, and location of the existing rail-gun and miniature rail-gun. The results show that the rail-gun launching technology applied to robot arms is feasibility. The application of miniature calibre rail-guns in the mechanical arm will benefit to the further development of rail-guns.

Nomenclature

Uc(t) = capacitors voltage,

I(t) = circuit current,

R_0 = equivalent initial resistor of circuit,

R(t) = rail resistor,

L0 = equivalent initial inductance of circuit,

L(t) = rail inductance,

E(t) = back electromotive force of armature,

F(t) = electromagnetic force,

L' = rail inductance gradient,

m = armature mass,

a(t) = armature acceleration,

Fz = armature drag force,

v(t) = armature velocity,

x(t) = armature displacement,

v_0 = initial velocity of the armature,

x_0 = resistance gradient of rail,

U_0 = charging voltage of the capacitors,

C = capacitors capacitance,

w = thick of the sub-scale rail-gun,

h = tall of the sub-scale rail-gun,

s = long of the sub-scale rail-gun.

1 Introduction

Electromagnetic launching technology as a new launching technology is a general technology following the traditional chemical energy firing technology. According to the principle it can be divided into electromagnetic rail-guns and coil-guns. Electromagnetic rail-gun has been extensively studied for many years. Its principles and concepts were put forward in the late 19th century. French military has corresponding theoretical exploratory research at the beginning of 20th century. German military got financial aid to do the experimental verification of principles during World War II. Be restricted by the industry at that time and the processing means of basic industry such as the material, American experienced a long-term repeated verification after the Second War in world, and temporarily shelved the research on military application of electromagnetic launching technology in 1960s. Until the late of 1970s, the results of successful experiments of Marshall are published, this new launching technology re-entered the military research program. And in 1980s started a new wave of study of rail-gun launching technology [1]. After entering the 1990s, with the development of rail-gun launching technology to the

DOI: 10.1051/matecconf/201710805005

depth, military and many researchers found some serious problems that it is difficult to solve in short time on basic theories and applications, such as the problems of power supply and launcher life. From this period to 2007 or so, the main study of electromagnetic launching technology is solving key techniques. Electromagnetic gun oriented to military requirement is mainly concentrated on rail-guns. In the recent 10 years [2], with American navy stimulated rail-guns on a large scale electromagnetic launching technology has obtained considerable development, especially electromagnetic launcher. By the issued literatures we can see that the launcher life of rail-guns has been basically solved. Recently electromagnetic launching technology is expected to rapidly progress. Figure 1 shows the schematic diagram of rail-gun.

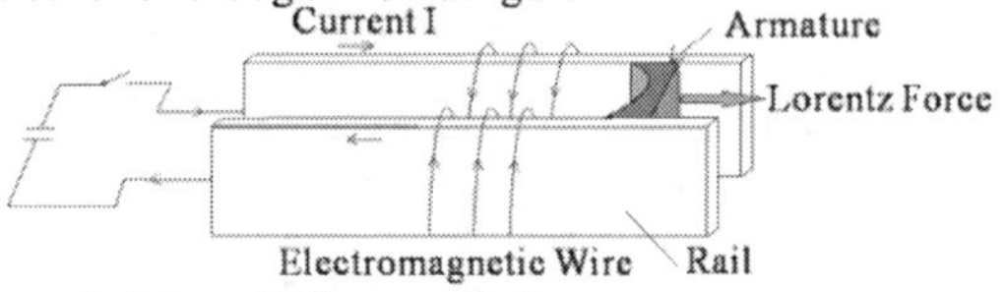

Figure 1. Schematic diagram of rail-gun

There are many kinds of the transmission mechanism of miniature manipulate at present. They can be roughly divided into electric drive, pressure drive, and internal combustion engine drive. Motor drive is the common way of electric drive, such as asynchronous and synchronous motor drive and stepper motor drive. The usual form of pressure drive is hydraulic or pneumatic actuators. And internal combustion engine drive generally used micro internal combustion engines. Rail-guns as the power mechanism of mechanical arm is belong to direct linear drives. The structural of the traditional linear motor or electron pusher mainly are coil-guns or are finished by the mechanical structure of straight line mechanism converted by rotating mechanism. The schematic diagram of linear motor is shown in figure 2. Comparing with rotating mechanism and coil-type linear motor, the rail-gun has the advantages of simple structure and easy control while is applied to drive manipulators.

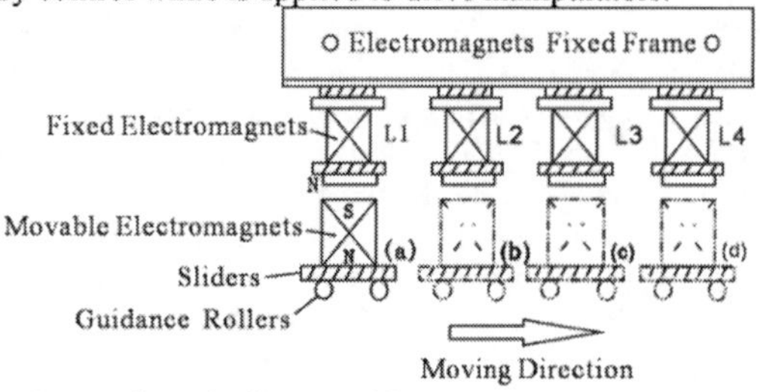

Figure 2. Schematic diagram of linear motor

Electromagnetic launching technology is mainly driven by the military demand. The research focus and the hotspot in current study are both placed on military applications. The study is developed from medium-small caliber to large caliber. This paper mainly studies the design and computation of the launching technology of micro caliber rail-gun launcher and launch system in the micro movable and manipulator. Based on design and analysis of the manipulator system of micro caliber rail-gun, this paper gained the appropriate electric parameters by researching and analyzing the characteristics, and theory analyzing the force condition of armature.

2 Analysis the circuit parameter of miniature calibre rail-gun

The storage energy of pulsed power supply system of rail-gun could reach mega-joules, the length of the launcher could be up to several meters, and the calibre of rail-gun was usually several tens of millimeters. With one experimental device as example, the parameters of the rail-gun is line in Table 1. For the electromagnetic force endured by armature during the acceleration is great, the friction force can be assumed to 0. Figure 3 shows the rail-gun equivalent circuit model of the system [3] [4]. And figure 4 shows the simplified equivalent circuit of rail-gun.

Table 1. Railgun parameters

Parameters	Values
Capacitance configuration	2mF
charge voltage	8KV
Rail length	5000 mm
Armature mass	0.1Kg
Pulse-shaping inductor	50μH
Capacitor resistance	20mΩ
Inductor resistance	10mΩ
Cable resistance	2mΩ/m
Inductance gradient	0.5μH/m
Resistance gradient	0.1mΩ/m

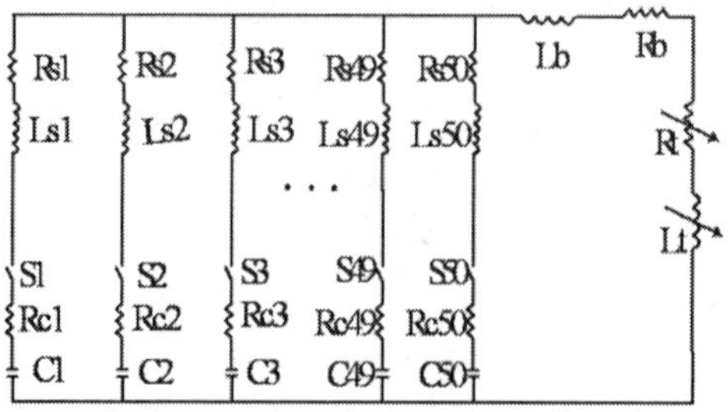

Figure 3. Rail-gun equivalent circuit model configuration

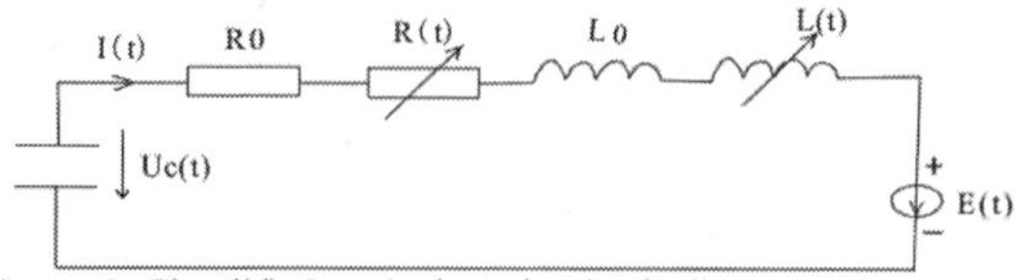

Figure 4. Simplified equivalent circuit of rail-gun

The equivalent circuit equation may be written:

DOI: 10.1051/matecconf/201710805005

$$U_c(t)-E(t)=(\mathrm{R}(t)+\mathrm{R}_0)\mathrm{I}(t)+\frac{d}{dt}[(\mathrm{L}(t)+\mathrm{L}_0)\mathrm{I}(t)] \quad (1)$$

The electromagnetic force on armature can be expressed as

$$F(t)=\frac{L'}{2}\mathrm{I}^2(t) \quad (2)$$

The acceleration equation can be written as

$$F(t)-F_z=ma(t) \quad (3)$$

The armature velocity v(t), the armature displacement x(t), the rail resistance R(t), and the rail inductance L(t) can be derived based on the equation (2)~(3)

$$\mathrm{v}(t)=v_0-\frac{F_z}{m}+\frac{L'}{2m}\int_0^t \mathrm{I}^2(t)dt \quad (4)$$

$$x(t)=x_0+v_0t-\frac{F_z}{m}t+\frac{L'}{2m}\int_0^t\left(\int_0^t \mathrm{I}^2(t)dt\right)dt \quad (5)$$

$$\mathrm{R}(t)=2\mathrm{R}'x_0+2\mathrm{R}'\left[v_0t-\frac{F_z}{m}t+\frac{L'}{2m}\int_0^t\left(\int_0^t \mathrm{I}^2(t)dt\right)dt\right] \quad (6)$$

$$L(t)=L'x_0+L'\left[v_0t-\frac{F_z}{m}t+\frac{L'}{2m}\int_0^t\left(\int_0^t \mathrm{I}^2(t)dt\right)dt\right] \quad (7)$$

For the back electromotive force of armature is proportional to the product of magnetic induction intensity and velocity, and the magnetic induction intensity is proportional to circuit current, so the back electromotive force of armature can be written as[5] [6]

$$E(t)=\frac{L'}{2}\mathrm{I}(t)\left(v_0-\frac{F_z}{m}+\frac{L'}{2m}\int_0^t \mathrm{I}^2(t)dt\right) \quad (8)$$

Based on the relation of the charge of capacitors, the voltage of capacitors can be written as

$$U_c(t)=U_0-\frac{1}{C}\int_0^t \mathrm{I}(t)dt \quad (9)$$

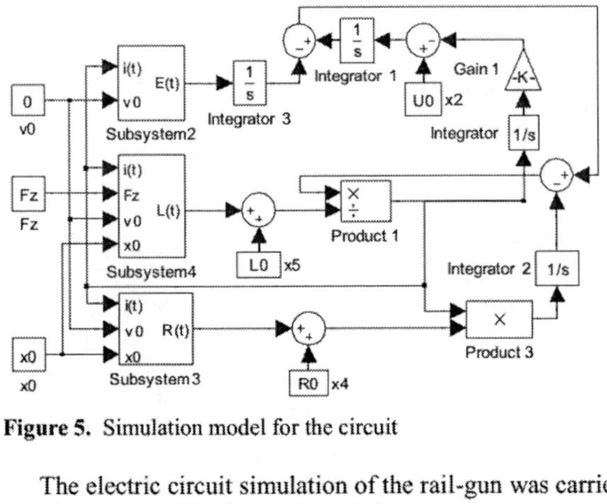

Figure 5. Simulation model for the circuit

The electric circuit simulation of the rail-gun was carried out by using Simulink. Simulink is a component of MATLAB, which provides an integrated environment for dynamic system modeling, simulation and comprehensive analysis [7]. Based on the modeling and analysis of electric circuit parameters, this paper deduced the integrate form of equation (1), and then putted up a simulation model of the circuit. Figure 5 shows the simulation model.

In Figure 5, Ls was the pulse-shaping inductance of circuit, Lb was the initial inductance in the rail, Rs was the total resistance in the capacitor branch, Rb was the initial resistance in the rail. Figure 6 shown the simulation results, included the current, velocity, and the location of armature. In figure 6, the first figure shows that the discharge time of the capacitor was ~1.5 ms, the peak current was ~0.8 MA. The second figure shows that the maximum Lorentz force of armature was about 1.65*105N. The third figure shown that the velocity of armature reached the maximum after the capacitor discharged. The fourth figure shown that there was a acceleration process of armature during the capacitor discharging, and then the armature was moving with uniform velocities in straight lines. The barrel residence time of armature was 4.8ms. As shown in figure 6, the simulation results did describe the experimental data quite well.

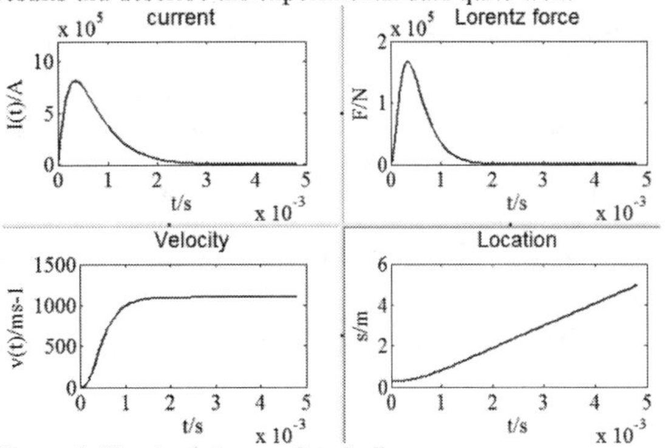

Figure 6. The simulation results of railgun

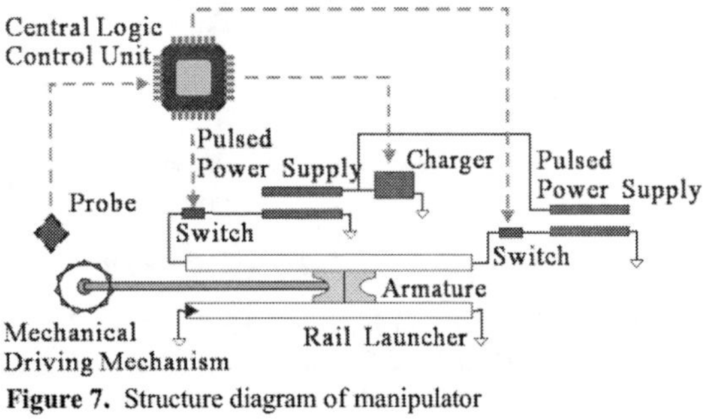

Figure 7. Structure diagram of manipulator

3 Design of manipulator system based on the electromagnetic launching technology

Rail-guns need extremely high pulse currents, and make strict demands on the power supply. Considering the miniature caliber rail-gun doesn't need high muzzle velocity and acceleration of armature compared with the rail-gun for

DOI: 10.1051/matecconf/201710805005

military requirements, this paper using the enhanced rail-gun and the enhanced armature of high field magnets. Be restricted by the future application requirement of manipulator, the method of connecting pulsed power supply with the muzzle and the breech respectively is applied to realize the reciprocate motion of armature. Figure 7 shows the structure diagram. And the system diagram of manipulator is shown in Figure 8.

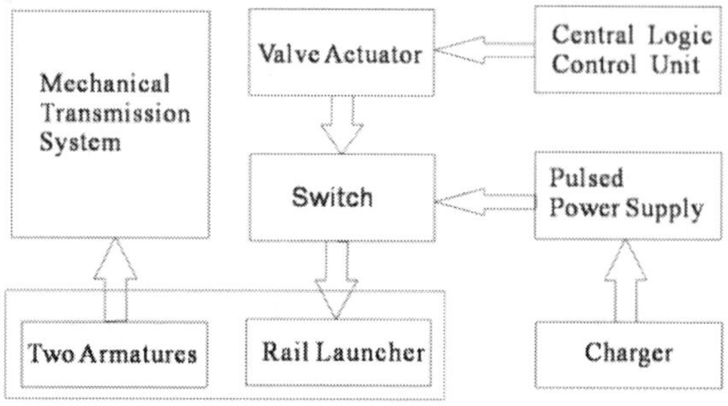

Figure 8. System diagram of manipulator

3.1 Parameter Calculation

The drive system based on rail-gun has the simple control mechanism and control Strategy for it only need to control the firing time, and don't need to keep accurate timing control just like linear motor and coil-gun. The manipulator structure which is suitable for the future engineering application used the method, which parallels many manipulators with redundancy control, to completed different extension angle and displacement of manipulators with different measurement range. By the statements as above, we can find out that the manipulator system with micro rail-gun has advantages of simple structure and simple control.

In order to study the feasibility of the rail-gun launching technology applied to robot arms, this paper derived the sub-scale rail-gun parameters based on the scaling method [8]. We obtained for subscale a 200-mm-long, 8-mm square-bore gun with rail dimensions of 8mm*2mm* 200mm. The scaled launch package mass was 0.8g, the armature resistance was 30mΩ. The resistance of conductor was 0.4mΩ/m, and the value of the pulse-shaping inductor was 20μH. Based on the equation $R' = R/L = \rho/S$ the resistance gradient was derived as 2.5mΩ/m.

In the case of high frequency magnetic field distribution the calculation formulas of the inductance gradient of the electromagnetic launcher with a rectangular aperture can be written as

$$L' = ((0.4406 - 0.0777\ln(F_1))\ln(F_2)) \qquad (10)$$

$$F_1 = 1 + 3.397(w/h) - 0.06603(w/h)(s/h) \qquad (11)$$

$$F_2 = 1.0077 + 2.7437(s/h) + 0.02209(w/h) + 0.02637(w/h)(s/h) \qquad (12)$$

The inductance gradient was derived as 0.5μH/m. It can be known by many simulations that increasing voltage and capacitive value can improve the bearing tension force of the miniature rail-gun when the shape and the materials of the rail-gun were determined. However, the oversize voltage would cause oversize current, and the oversize capacitance would cause over lowing energy utilization ratio. In this paper the sub-scale system consisted of 50 capacitors, the charging voltage of capacitor was set to 18.09V, and the capacitance of capacitor was set to 15mf, the resistance of capacitor was 11.6mΩ. The resistance force of armature was assumed to be 0.003N in order to reach the purpose that the armature could move to muzzle in 0.1second.

3.2 Simulation analysis

The simulation results of resistance force existing in the sub-scale railgun was shown in Figure 9, including the current, the resultant force, the velocity, and the location of armature.

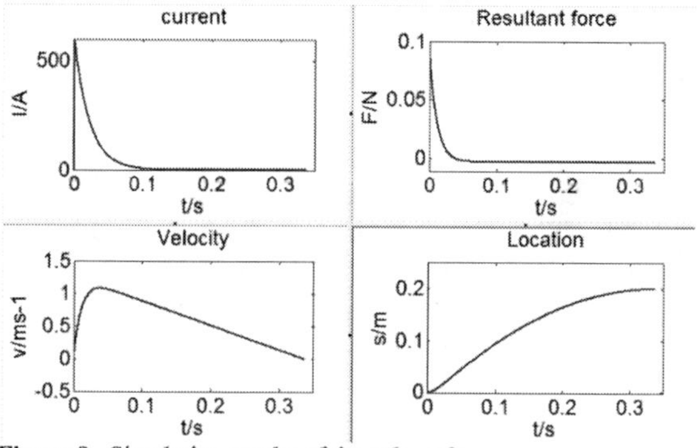

Figure 9. Simulation results of the sub-scale

As shown in the figure 9, (the first figure shows the discharge time of the capacitor, the second figure shows the resultant force of armature, the third figure shows the velocity of armature, the fourth figure shows the location of armature).the discharge time of the sub-scale rail-gun was ~0.1s, the peak current was ~550A, the maximum resultant force of armature was ~0.08N, when the resultant force reduced to 0N, the velocity of armature reached the maximum, which was 1.1m/s-1, the muzzle time of armature was 0.338s, it's also the time for the velocity reduced to 0m/s-1.

In this simulation, the peak current was about 550A, it was still great and may cause unsafe problem. While the tension force of the miniature rail-gun was too small. Increasing inductance gradient and current can improve the bearing Lorentz force of armature was known by the equation (2), in this way the tension force of the miniature rail-gun can increased. The inductance gradient can be increased by a change of the rail geometric parameters.

4 Conclusion

The theoretical analysis and the result of simulation show that the rail-gun launching technology applied to robot arms is feasibility. But the shape and the materials of the miniature rail-gun should be change. For the armature can easily get larger Lorentz force from coil-guns, change the rail-guns to coil-guns may can also suitable.

Acknowledgment

This work is supported by the National Natural Science Foundation of China under Grant 51407010.

References

1. H. D. Fair. Progress in Electromagnetic Launch Science and Technology. *IEEE Trans Magn*. **43**,**1**,93-98, (2007)
2. H. D. Fair. Advances in Electromagnetic Launch Science and Technology and its Applications. *IEEE Trans Magn.* **44**, **1**, 225-230, (2009)
3. N. Pan, X. J. Han. Electromagnetic Railgun Circuit Simulation Modeling. *Ship Science and Technology*. **35**, **10**, 101-105, (2013)
4. P. Z. Liu, X. J. Yu, J. Li and S. Z. Li. Energy Conversion Efficiency of Electromagnetic Launcher with Capacitor-based Pulsed Power System. *IEEE T Plasma SCI.* **41**, **5**, 1295-1299, (2013)
5. G. Q. Chen, H. Y. Zhang, Y. H. Wang, X. L. Wei and C. Z. Fang. The Circuit Theory Analysis and Dynamic Aimulation for the Capacitor Drived Railgun. *Electric Machines and Control.* **10**, **1**, 23-26, (2006)
6. G. Q. Chen, Y. H. Wang, X. L. Wei and H. Y. Zhang. Computer Simulation of Electromagnetic Process in the Capacitor Drive Rail gun. *Transactions of China Electrotechnical Society.* **21**, **4**, 68-71, (2007)
7. J. Yao and S. H. Ma. Simulink Simulation and Modeling. Xi'an: Xidian University press. (2007)
8. L. W. Jin, J. Li and B. Lei. Approximate Field Scaling of Railgun Launcher Under the Condition of Matching Projectile Dynamic Parameters. *IEEE T Plasma SCI.* **43**, **9**, 3286-3292, (2015)

DOI: 10.1051/matecconf/201710805006

An Experimental Study on Operability of Master-Slave Manipulator System using Human–in–the-Loop Type Simulator

Yosuke Tashiro[1], Hirofumi Ohtsuka[1], and Tomohiro Kawahara[2]

[1]*National Institute of Technology, Kumamoto College, 2659-2 Suya, Koshi, Kumamoto*
[2]*Kyushu Institute of Technology, 2-4 Hibikino, Wakamatsu-ku, Kitakyushu-shi, Fukuoka, Japan*

Abstract. Double arm Master-Slave (M-S) manipulator has been attracted the attention of the robotics researchers today. The micro surgery is one of the big targets which M-S manipulator is expected to effectively perform. This paper describes about the human in the loop (HIL) simulator consisted of the prototype master manipulator and the virtual simulator of slave manipulator. Today, we have a challenge injecting drugs to a chick embryo's blood vessels that are enough cultured in an artificial eggshell. The M-S manipulator is expected to apply for such a work like a micro surgery, ex. catching a blood vessel and sticking a cylinder. The embryo is extremely sensitive so that M-S manipulator should have both high operability and high accuracy movement. To evaluate the M-S manipulator quantitatively, we develop the human in the loop (HIL) simulator. The simulator is consisted with the prototype master manipulator and the virtual slave manipulator.

1 Introduction

A surgery requires a lot of precise operation such as sewing a diseased part. In those operations, there necessarily exists some possibility that a human error makes heavy damage to patient. Therefore, it should be accepted for licensed doctors so that they are expected to grow their skills and institutional knowledge as sufficient high level. For the purpose of burden applied to both the surgeon and the patient, a surgery robot, such as "Da Vinci" has been developed and used. An operator (surgeon) can operate the slave manipulator through the master manipulator. The slave manipulator moves with a reduction of the master manipulator's motion. In addition, "Da Vinci" gives feedback force information from the slave to the master manipulator. However, its high cost has become an obstacle to widespread use.

Furthermore, when expanding the field of view, there is demand for surgical manipulation of the fine structure in the area of research related to the in-vivo technology. Many researchers try new drugs to chick eggs because it is quite cheap and ethically acceptable [1]. A chick egg is covered by hard eggshell, which pass oxygen into the inside of eggshell. Moreover it can't observe from outside [2]. Hence, Kawahara et al. have cultivated in artificial eggshell that has transparent body (Figure 1). The membrane thickness is 0.3 mm, the length is 40 mm, and the total volume of the cubic eggshell is approximately 50 ml [3].

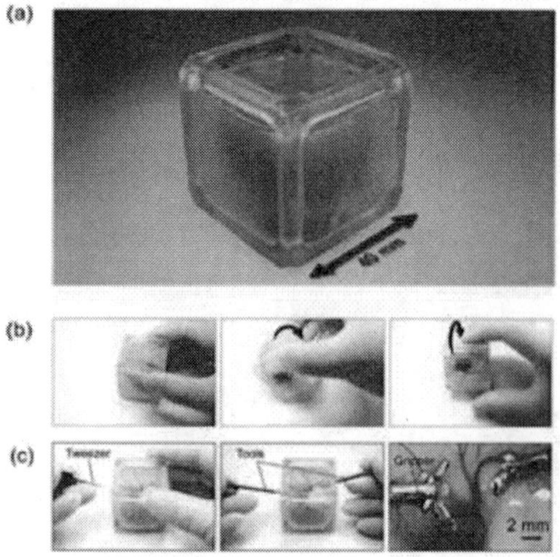

Figure 1. Developed cubic eggshell and the observation and manipulation of chick embryo.

It is clear that the chick embryo is delicate and the demand of robot is increasing. In this field, there are only few manipulators and these do not have enough operability. So, we have undertaken the M-S manipulator device which is capable of gripping the targeted blood vessel cultured in an artificial eggshell and on injecting a drug to it. For this target, intricate structure such as "Da Vinci" is not required. It is desirable that can be deferred on the experiment table, so that we suggest proto-type 3 links manipulator shown in Figure 2. Master manipulator is smaller than slave one but both of manipulators have similarity shape 3 links. The slave manipulator successfully controlled to track the motion of master manipulator. However, operator's assessment is quite low for operability of master manipulator. So, we have undertaken the M-S manipulator device which is capable of gripping the targeted blood vessel cultured in an artificial eggshell and on injecting a drug to it. For this target, intricate structure such as "Da Vinci" is not required. It is desirable that can be deferred on the experiment table, so that we suggest proto-type 3 links manipulator shown in Figure 2. Master manipulator is smaller than slave one but both of manipulators have similarity shape 3 links. The slave manipulator successfully controlled to track the motion of master manipulator. However, operator's assessment is quite low for operability of master manipulator.

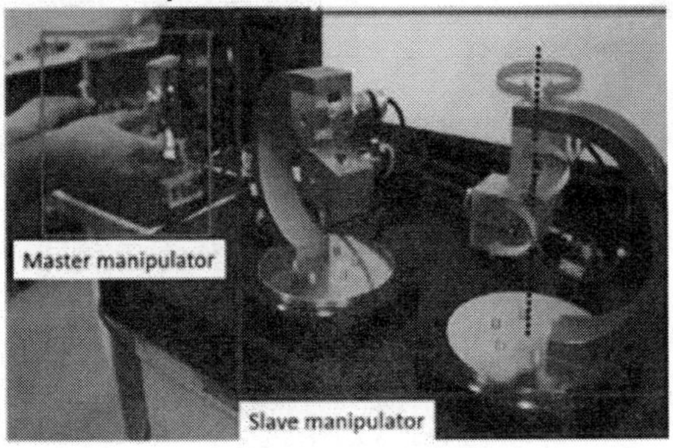

Figure 2. Proto-type of master slave manipulator

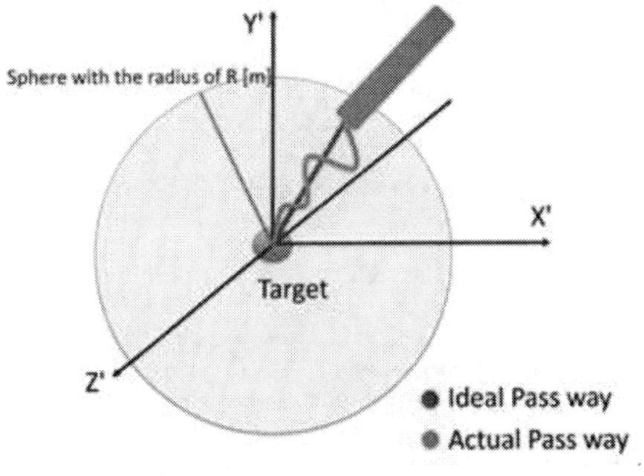

Figure 3. Pass way of slave manipulator tip position

2 Human-in-the-Loop type Master-Slave Manipulation Simulator

Main objective of this study is improving the operability of M-S manipulator shown in Figure 2. To design the high operability M-S manipulator system, the quantitative index for operability is required. Since the performance of M-S manipulator system depends on the movement of the slave manipulator's end effect joint, let us focus on the tip position of end effector of slave manipulator.

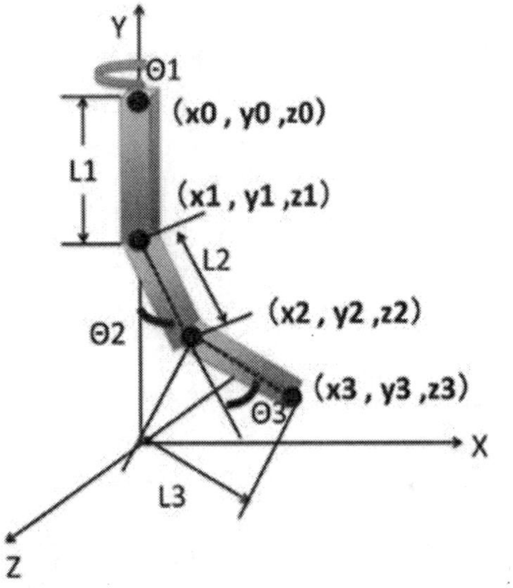

Figure 4. Physical model of the slave manipulator

When the 3 dimensional coordinate denoted in Figure 3, we consider the evaluation space as a sphere of the radius R[m] with the centre point as target position. And an orthogonal coordinate space, with origin as target position, is defined as shown in Figure 3. After the tip of the slave manipulator has entered the sphere space, its trajectory becomes an ideal pass way shown as red line in Figure 3 for the best operability. For less operability case, its trajectory shows the fluctuation around the ideal pass way such as "Actual Pass way" described by blue line.

2.1 Forward kinematics of manipulator

The slave manipulator has 3 links shown in Figure 2, so its schematic diagram can be described by Figure 4. The origin of orthogonal coordinate is different from Figure 3. Here, we define the arms from the top, arm 1, arm 2 and arm 3. And the fasten position is given as follows.

$$(x0, y0, z0) = (0, Y0, 0) \quad (1)$$

$$(x1, y1, z1) = (0, Y0 - L1, 0) \quad (2)$$

Now calculation is below. (Each arm's rotation is θ1, θ2 and θ3.)

$$x2 = L2 \sin\theta2 * \cos\theta1 + x1 \quad (3)$$

$$y2 = y1 - L2\cos\theta2 = Y0 - L1 - L2\cos\theta2 \quad (4)$$

$$z2 = L2 \sin\theta2 * \sin\theta1 + z1 \quad (5)$$

In the end, the coordinates of the tip position is calculated as below.

$$x3 = x2 + L3 * sin(\theta 2 + \theta 3) * cos\theta 1$$
$$= \{L2sin\theta 2 + L3\ sin(\theta 2+\theta 3)\} * cos\theta 1 \quad (6)$$

$$y3 = y2 - L3 * cos(\theta 2 + \theta 3)$$
$$= Y0 - L1 - L2cos\theta 2 - L3cos(\theta 2+\theta 3) \quad (7)$$

$$z3 = z2 + L3 * sin(\theta 2 + \theta 3) * sin\theta 1$$
$$= \{L2\ sin\theta 2 + L3 * sin(\theta 2+\theta 3)\} * sin\theta 1 \quad (8)$$

Now, let us denote a displacement between the tip position and the target position as:

$$d = \sqrt{p_x^{\,2} + p_y^{\,2} + p_z^{\,2}} \quad (9)$$

where (p_x, p_y, p_z) is a coordinate of tip position. The coordinate of tip position (p_x, p_y, p_z) can be obtained from eq.(6)-(8) by use of the coordinate transformation matrix among those orthogonal coordinate space.

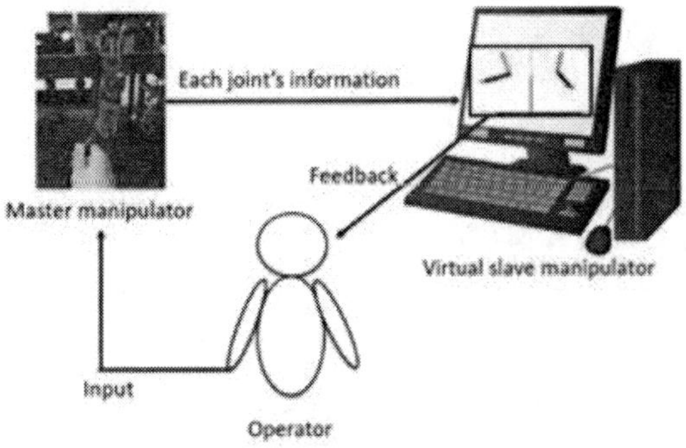

Figure 5. Human in the loop simulator

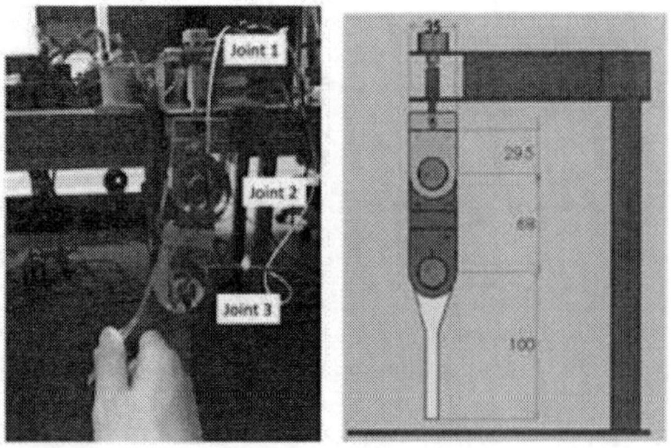

Figure 6. Overview of the master manipulator

2.2 Simulator

To evaluate the operability of M-S manipulator system, we built an experimental context which is shown as Figure 5. The mechanical design of master manipulator (Figure 6) has been done with a use of 3D-CAD software "Solid works". Rotational angles for all joints of master manipulator can be detected by potentiometer attached with the each axis. The voltage signals from them are input to the computer through the AD/DA converter interface board. Specs of the equipment are shown in Table 1, Table 2 and Table 3.

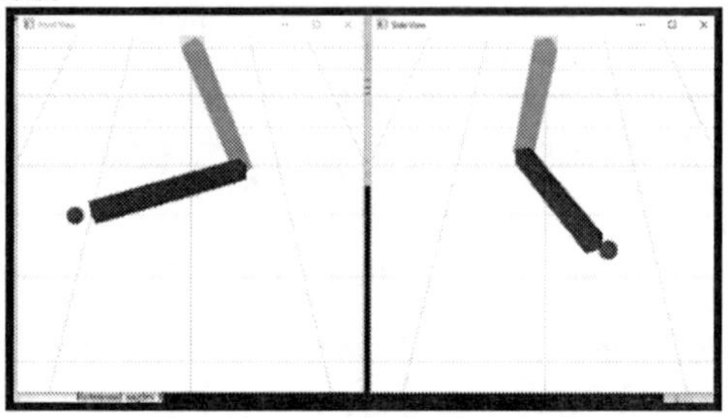

Figure 7. User interface of virtual slave manipulator

Table 1. PC Specs

Processor	Intel Core i5 - 4460 CPU 3.20GHz
RAM	16.0GB
Operating System	Windows 10 Pro 64bit

Table 2 D/A BOARD Specs

Model number	PEX-340416
Channels number	4 channels
Output voltage	? 0V
Resolution	16bit

Table 3 A/D board specs

Model number	PEX-361216
Channels number	Single ended:16 channels
	Differencial: 8 channels
Maximum sampling time	1 MSPS
Input voltage	? 0V

The operator can see the animation image of the virtual slave manipulator on the monitor screen (Figure 7). At the same time, the operator can handle and operate the master manipulator to achieve a task such as positioning the tip position to the target position. The motion of the slave manipulator is calculated in the computer by using the forward kinematics equations (1) - (9) as mentioned above. The tip positions of each link can be calculated, then wire-flame model are repeatedly drawn on the monitor screen. There will be two windows from the front and side viewpoint because it is generally hard to get the position of the target with only one viewpoint even it is a 3D image.

3 Quantitative evaluation for operability

3.1 The time for ending a work

Now, we propose the following 3 type of the quantitative index. The time to end an assigned work is one of the important factors showing the operability. It is clear that the time for ending tasks becomes longer the case of less operability than the case of high operability.

3.2 Average Absolute Error (AAE)

In the experiment for assessment of the operability, a subject is ordered to control the virtual S-manipulator which is shown on the screen through the M-manipulator. The system shows the target as moving red ball on the screen in real-time. The target moves linearly from the previous target position to next target position.

Then Ni second later (i = 0, 1, 2: i is the task number), the target will arrive the target position. After the target stops moving, the target can take next process when the tip position is near the goal. Now, we define the time series target position's x, y and z coordinate of axis as Xt(t), Yt(t) and Zt(t). And the time series tip position's x, y and z coordinate of axis as X(t), Y(t) and Z(t). Furthermore we suggest a parameter AAE to measure how a user could follow the target position as the equation (10).

$$AAEi = \sum_{t=0}^{Ni} \frac{\sqrt{(Xt(t)-X(t))^2+(Yt(t)-Y(t))^2+(Zt(t)-Z(t))^2}}{Ni} \tag{10}$$

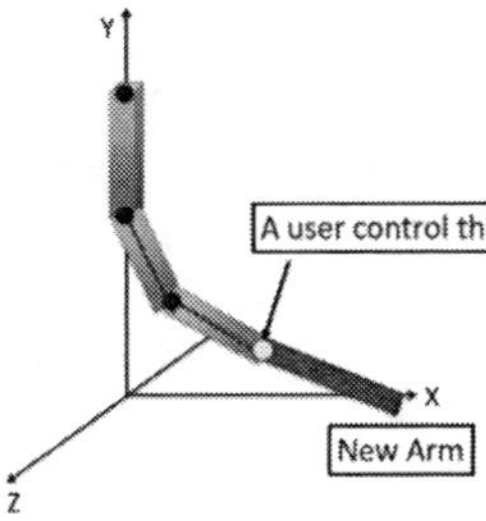

Figure 8. An arm downing operability

3.3 Vibration rate

Taking into the account that how the manipulator is accuracy near the target, we adopt an assessment value vibration rate ζ, which is defined as follow.

$$\zeta = P_l / R \tag{11}$$

When the tip position of the slave manipulator is in R [m] to the target, by recording the tip position, we can measure the approximate length of actual pass way Pl [m] as

$$P_l = \sum \sqrt{\Delta P_x{}^2 + \Delta P_y{}^2 + \Delta P_z{}^2} \tag{12}$$

where $\Delta P_* = p_*(t + \Delta_s) - p_*(t)$, $*$ is x, y or z. and Δ_s denotes the sampling period of simulator.

4 Experimental results

To certify the effectiveness of the parameters, we consider two types of master manipulator. We call these two manipulators the original type and less operability type. The second type is extended length of 3rd arm as shown in Figure 8.

4.1 Binomial test

A subject is ordered adjusting the tip position to a red point shown in the window. The target position moves linearly from the previous target position to the next target position in constant velocity. There are 3 target position and we call the order from the target position to position as work I. (I = 1, 2, and 3: number of target position.) Mainly we assess the 3 parameters that already defined section 3 however comparing the single value is not effective. Because the result basically has a dispersion caused by human controlling. For that reason, we need to measure how often a user get a value over a standard value. In this time, we use the each median value that is calculated from the designed M-manipulator at each of subject. Figure 9 is a binomial distribution with parameters 80 (number of experiment) and 0.5 (probability of occurrence).

4.2 Result

We have done experiment 20 times to 4 subjects and calculated the parameters of each work with new arm and without new arm. Then we collect the numbers how many time the data over the median value as Table 4. The number how many time the parameter over and below the median value are shown as good and bad. The median value is calculated individually from the experiments with the original M-manipulator.

Table 4. Result of experiment

	time1		time2		time3	
	good	bad	good	bad	good	bad
Subject A	13	7	12	8	7	13
Subject B	2	18	8	12	6	14
Subject C	10	10	6	14	11	9
Subject D	3	17	7	13	10	10
sum	28	52	33	47	34	46
Percentage	0.002407348		0.026361786		0.036441293	
	Vibration rate 1		Vibration rate 2		Vibration rate 3	
	good	bad	good	bad	good	bad
Subject A	7	13	7	13	1	19
Subject B	0	20	5	15	1	19
Subject C	7	13	5	15	6	14
Subject D	15	5	13	7	17	3
sum	29	51	30	50	25	55
Percentage	0.004316624		0.00733826		0.000300609	
	AAE1		AAE2		AAE3	
	good	bad	good	bad	good	bad
Subject A	7	13	16	4	6	14
Subject B	5	15	9	11	8	12
Subject C	2	18	1	19	10	10
Subject D	8	12	11	9	6	14
sum	22	58	37	43	30	50
Percentage	2.24073E-05		0.071194185		0.00733826	

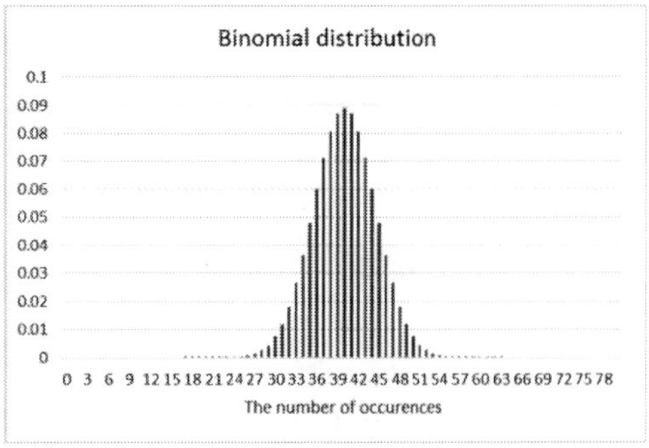

Figure 9. Binominal distribution of 80 time operation

The percentage, how many time occur the result over standard value below binominal distribution, is the percentage in the Table 4 [5].

5 Conclusion

We set the significance level as 0.05 for one side binominal test and the Figure 9 shows binominal distribution [6]. The parameters, time1, time2, time3, vibration rate1, vibration rate2, vibration rate3, AAE1 and AAE3, meet one side binominal test. AAE2 doesn't meet the one side binominal test however AAE2 meets the both side binominal test. Taking into account these result, we conclude the parameters are enough effective to assess the operability.

Consequently, we will quantitatively improve the operability of master manipulator based on these parameters.

References

1. Huang W, Arai F, Kawahara T (2015) Egg-inCube: Design and Fabrication of a Novel Artificial Eggshell with Functionalized Surface. PLoS IONE 10 (3);e0118624. Doi:10. 1371 / journal.pone.0118624
2. Tsai T, Yang JM, Lai CW, Cheng YH, Lin CC, Yeh CW (2006) Characterization and adsorption prop-erties of eggshells and eggshell membrane. Bioresour Technol 97: 488–493. PMID: 15896954
3. Wenjing Huang, Fumihito Arai, Tomohiro Kawahara(2015); Inducing blood vessel formation using cubic eggshell with patterned surface. IEEE 15872410 DOI: 10.1109/MHS.2015.7438255
4. ISO 9241-11:1998(en), Ergonomic requirements for office work with visual display terminals (VDTs) — Part 11: Guidance on usability; https://www.iso.org/obp/ui/#iso:std:iso:9241:-11:ed-1:v1:en(Access 05/10/2016)
5. Kei Takeuchi, Kazutake Fuhino (1981); Binomial Distribution and Poisson distribution. Japan University of Tokyo Press
6. Jakob Nielsen, Toshikazu Shinohara(translator): Usability Engineering(2002), Tokyo Denki University Luigi T.De Luca, *Propulsion physics* (EDP Sciences, Les Ulis, 2009) . F. De Lillo, F. Cecconi, G. Lacorata, A. Vulpiani, EPL, **84** (2008)

Preliminary Investigation of Micro-V-Bending

Gandjar Kiswanto [1], Aida Mahmudah [1] and Dedi Priadi [2]

[1]*Mechanical Engineering Department, Universitas Indonesia, Indonesia*
[2]*Metallurgy and Material Engineering Department, Universitas Indonesia, Indonesia*

Abstract. Bending in sheet metal forming is defined as uniformly straining process around a linear axis. A V-bending process is normally used in sheet metal working due to the tool and process simplicity. The main problem in bending process is a springback occurrence which causes an error in part dimension. In a micro level, the springback angle is difficult to be assessed because of the influence of size effects in all aspects of the system. The springback behavior has been investigated in many previous kinds of research. However, the springback correction is still needed to be studied extensively so that it can be implemented reliably in micro-bending. The purpose of this research is to investigate the influence of punch velocity and holding time to the springback angle. The process is performed to the 0.1 mm thickness of copper foil. The punch velocities are 0.1 mm/s, 1 mm/s, and 10 mm/s. While the variation of holding time, which is set during the bottoming stage, are 7 s, 9 s, and 11 s. The result shows that the punch velocity clearly influences the springback angle. In addition, the holding time during the bottoming stage is a potential factor in springback angle correction.

1 Introduction

With the increasing of micro-parts demand due to the trend towards miniaturization, many kinds of research have been encouraged to conduct a development of a micro-manufacturing system and investigate the process at the micro level. One of the favorite methods for producing a micro-part is micro sheet metal forming. Bending is a basic configuration forming process in producing 3D profiles of micro-parts such as micro-electric contacts/switches, mechanical sensors and 3D sheet metal profile for optical devices housing.

Bending in sheet metal forming is defined as a straining process of metal around a straight axis [1]. During bending, the top section of material may be stretched while the bottom section is compressed. The most widely used in bending process are U-bending and V-bending.

The main issue in bending process is springback phenomenon that occurs when the foil is unloaded. Springback is an elastic recovery in plastic deformation that causes deviation from the desired dimension so that it reduces the quality of bent-part and generate assembly difficulties.

The springback amount in micro bending is influenced by the *t/d* ratio. When the thickness of the material is less than 350μm, the springback angle increase with a decrease of t/d ratio [2]. Diehl et al. [3] found that springback behavior of SE-CU 58 was different for the thickness less than 100μm and larger than 100μm. This result caused by the different dominant factor that influences the amount of springback angle.

The grain properties also affect the springback behavior in micro level. Liu et al. [4] found that a large springback occurs with thickness decreasing or grain size increasing. Besides, an increasing of elastic anisotropy of surface grains with the reduction of grains along the thickness direction will cause the springback angle scattering [4, 5]. Chen et al. [6] discovered that yield strength and maximum punch force decrease with a decrease of *t/d* ratio. It was also found that when the t/d ratio is smaller than 2, spring-forward occurs.

Other researchers explained that deformation behavior in micro bending is mostly affected by plastic strain gradient hardening that requires the characteristic length, *l*, which is depended on the shear modulus, the yield strength and the Burgers vector of the material [7, 8].

The plastic deformation causes the increasing of material hardness which the highest value was indicated near the neutral axis [9, 10]. The grain properties distribution in the forming area induces a significant influence on the springback behavior. Thus, Fang et al. [11] considered the grain heterogeneity in On tessellation application to predict the springback during micro V-bending in finite element model.

It is necessary to control springback angle in obtaining a good part quality and facilitating the next micro-part assembly process. Zhen-Yuan [12] decreased the bending moment in bending area with a coining process so that and springback angle would be decreased. Besides, Ma et.al [5] stated that in a roll bending test, the springback angle, and its scattering can be reduced by controlling the sizing repetitions, sizing force, holding time and punch speed.

DOI: 10.1051/matecconf/201710806001

Furthermore, Zheng et al. [13] investigated an elevated temperature bending process with thin pure titanium foils. Since the flow stress of surface became lower than in room temperature, the springback angle decrease with decreasing foil thickness.

The previous controlling springback studies need an additional implementation to reduce the springback angle. Consequently, the properties of the micro-bent part may be changed. In this study, a preliminary investigation is held to observe the springback controlling by applying a higher punch velocity and holding time in a micro-V-bending process. A copper foil with 1 mm thickness is preferred as a material specimen.

2 Micro-V-bending

As well as in macro range, several basic techniques in a micro-bending process can be used to make a bent-shape. Because of the simplicity and the wider application in bending angle, V-bending is a popular method. Two common ways in V-bending are air-bending and coin-bending [14], as shown in Figure 1 and 2, respectively. In air-bending, the punch should always be in V-shaped, while the die may in U or V shaped (figure 1a and 1b). In the other hand, the punch and die should always be in V-shaped for coin-bending (Figure 2). With an air-bending application, various bending angles can be obtained with a single V-bending tool by controlling the length of punch trajectory.

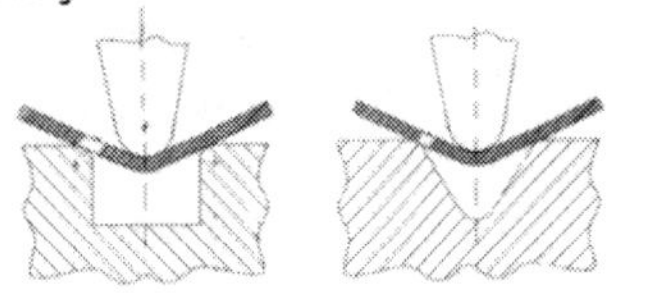

a) b)

Figure 1. Air-bending a) U-shaped die and; b) V-shaped die [14]

The coin-process in V-bending has four distinct stages to complete a bending process (Figure 2). The first stage is free-bending (Figure 2a). In the second stage, the end of the workpiece is touching the side walls of the die (Figure 2b). While, the end of the workpiece is touching the side walls of the punch in the third stage (Figure 2c). Finally, the fourth stage is a bottoming stage which all surfaces of the workpiece are touching the punch and the die (Figure 2d). In this stage, the bending process is completed.

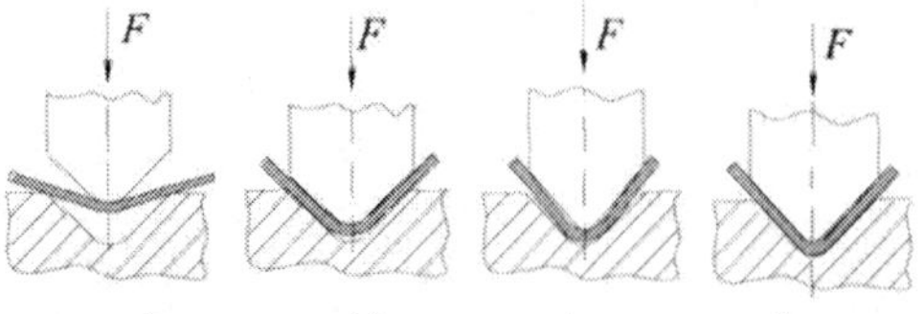

a) b) c) d)

Figure 2. Coin-bending in V-profile a) first stage; b) second stage; c) third stage; d) fourth stage [14]

3 Experimental setup

3.1 Equipment

The experiment is conducted using a micro-V-bending tool which is mounted on a micro-forming machine 5kN. The dimension of the specimen to be bent is 0.1x1x7.4 mm which has obtained from the previous blanking process. Bending angle measurement was carried on with Dyno-Lite digital microscope PRO2 AD-4113ZT. Figure 3 shows the experimental setup of the micro-V-bending.

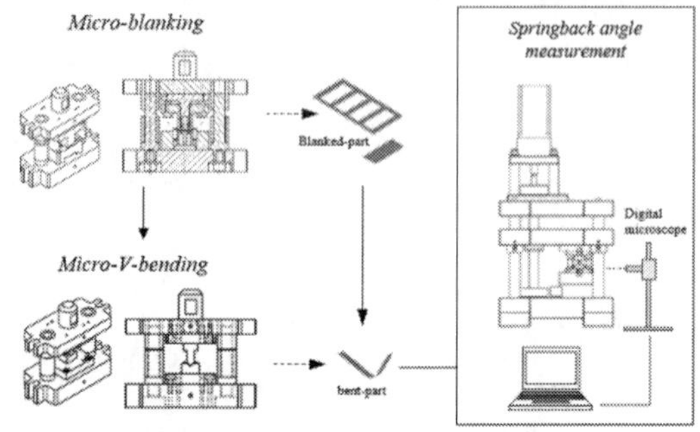

Figure 3. Experimental setup of micro-V-bending process

A die-set of the micro-V-bending tool is prepared to conduct the forming process. The punch and die are V-shaped with certain dimensions. Figure 4 shows a micro-V-bending tool with detailed V-shaped in the die. A partially caging system is used to locate the specimen to be bent.

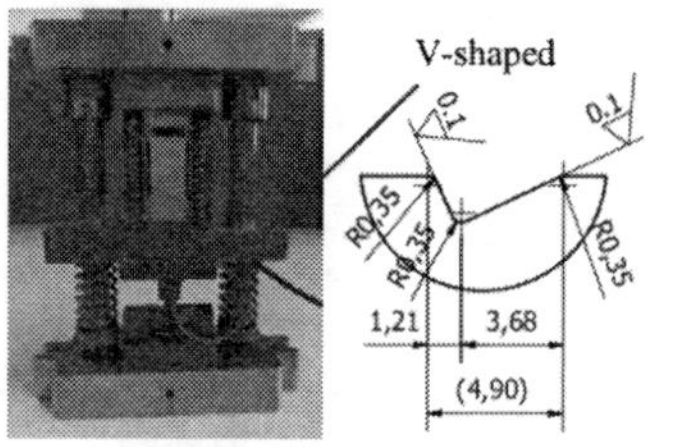

Figure 4. Micro V-bending tool

For the main component, i.e. punch, die and locator, the minimum requirements of material properties can be fulfilled by tool steel material, while the others can use machinery steel. The mechanical properties of the used-material in micro-V-bending tool components can be seen in Table 2 below. For groups of tool steel, it used AISI D2, while AISI 1045 for the machinery steel.

Table 1. Mechanical properties of AISI D2 and AISI 1045

		Cold Work ToolSteel (D2/SKD11)	Machinery Steel (1045/S45C)
Ultimate tensile strength	[MPa]		569
Yield Strength	[MPa]	1650	343
Elongation	[%]		20
Modulus of Elasticity	[GPa]	209.9	205
Poissons Ratio			0.29
Machinability	[%]		55
Shear Modulus	[GPa]		80

The μ-machine with 5kN capacity was driven by an Oriental servo motor NX940MS-PS10-3. The mechanical force is transferred to the ram using ball screw and change rotational motion into translational one [15]. The assembled micro forming machine is shown in Figure 5, and Table 2 illustrates the specification of it.

Figure 5. Micro V-bending tool

Table 2. μ-Forming machine 5kN specification

Maximum Load Capacity	[kN]	5
Servo motor: NX940MS-PS10-3		
Max holding torque	[Nm]	34.3
Moment of rotor inertia	[kg·m^2]	0,314·10^{-4}
Resolution	[P/R]	100 – 100000$^{*)}$
Gear ratio		10
Allowable speed range	[rpm]	0 to 300
Backlash	[°]	0.25
Resolution	[μm]	0.5 to 500
Working area	[mm]	L x w x h$^{**)}$ 100 x 200 x 120
Average deviation linear motion	[μm]	0.5 ÷ 4

$^{*)}$ factory setting 1000)
$^{**)}$ recommended at h = 70 to 90)

Copper foil with 0.1 mm thickness is chosen as a specimen to investigate the micro-V-bending process. Table 3 and 4 show the chemical composition and mechanical properties of the copper foil, respectively. The chemical composition was obtained from Optical Emission Spectrometer, and the tensile test was held on Servopulser Shimadzu 20T. Because the experiment observes with the given properties, no pre-heated process conducted to the material foil.

Table 3. Chemical composition of the copper foil

Cu (%)	Zn (%)	Pb (%)	Sn (%)	P (%)	Mn (%)	Fe (%)	Ni (%)
98.5	0.191	0.024	0.046	0.195	0.003	<0.005	0.111
Si (%)	**Mg (%)**	**Cr (%)**	**Al (%)**	**Ag (%)**	**Co (%)**	**Bi (%)**	**Sb (%)**
0.042	0.001	0.003	0.009	0.006	0.044	0.213	0.027

Table 4. Mechanical properties of copper foil

Tensile Strength	[MPa]	647
Yield Strength	[MPa]	-
Elongation	[%]	1.9

3.2 Experimental process parameters

The blank part is then bent in the micro V-bending tool to obtain a specified bending angle. One piece loading was also implemented in the bending process. The coin-bending method was held, due to the holding time variation was observed in decreasing springback angle. The experiment was conducted with five specimens for each variation of punch speed. i.e. 0.5mm/s 1mm/s, and 5mm/s.

Since the micro-tool construction used the wire spring, the punch need time to be unloaded, i.e. ht_0 = 7 seconds. Thus, in this research, the holding time is set as shown in Table 5.

Table 5. Mechanical properties of copper foil

	Holding time set		
	1	2	3
Time [s]	$ht_0 + 0$	$ht_0 + 2$	$ht_0 + 4$
	7	9	11

Bending angle measurement was carried on with digital microscope Dyno-Lite PRO2 AD-4113ZT, as soon as the bending part was unloaded. Figure 6 shows (a) bottoming position and (b) the measurement of bending angle.

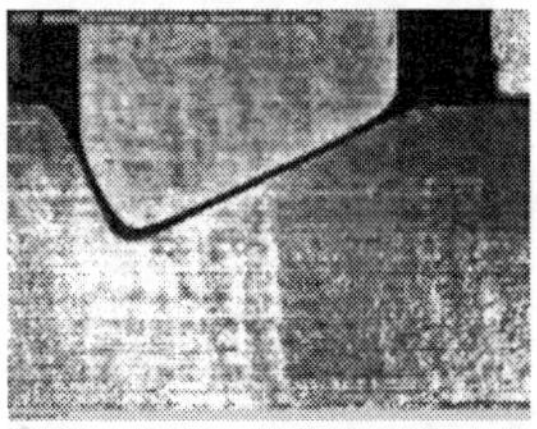

(a)

(b)

Figure 6. (a). Bottoming position (b) bending angle measurement

4 Result and discussion

In the micro V-bending experiment, the measured springback showed difference relating to the punch velocity and holding time variation. Figure 7 to 9 describe the occurring springback for each parameter variation. Good tendencies are showed with the 0.5mm/s and 1mm/s punch velocity (Figure 7 and 8). The inconsistent result is produced with 10mm/s punch velocity. The springback angle increased in 2nd holding time set and then decreased slightly with a longer holding time (Figure 9).

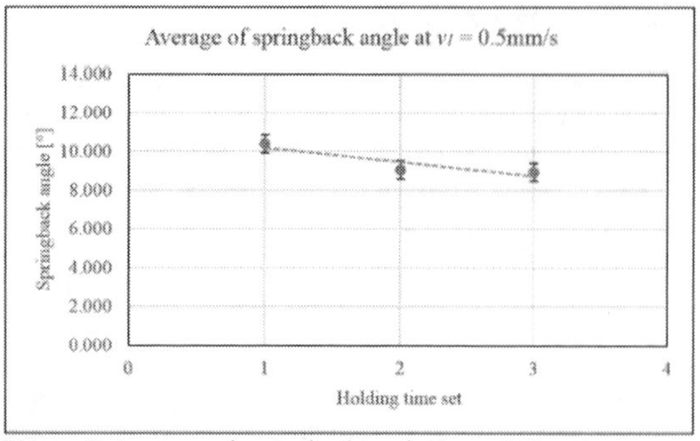

Figure 7. Average of springback angle at v_1 = 0.5mm/s

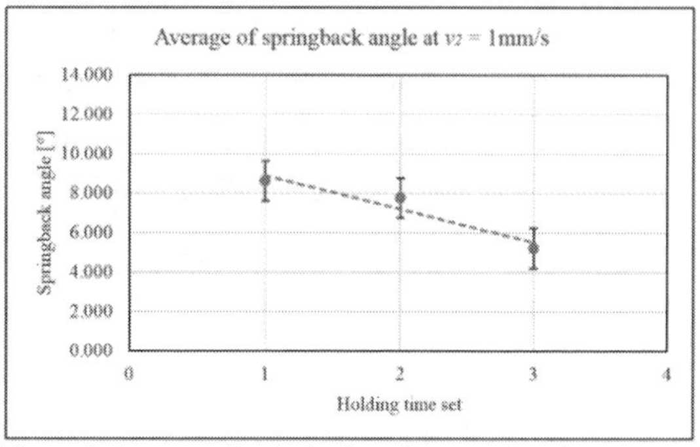

Figure 8. Average of springback angle at v_2 = 1mm/s

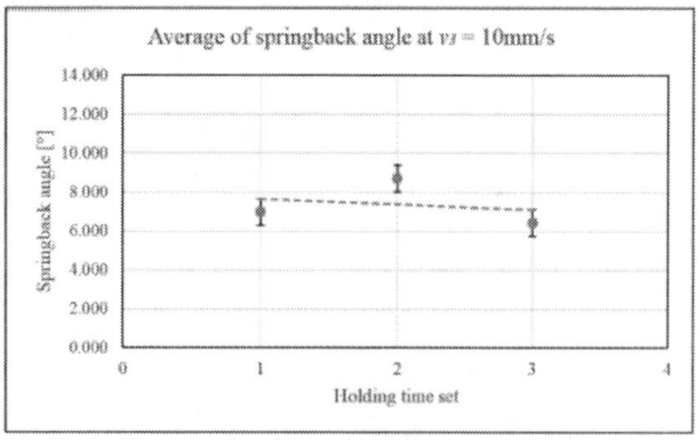

Figure 9. Average of springback angle at v_3 = 10mm/s

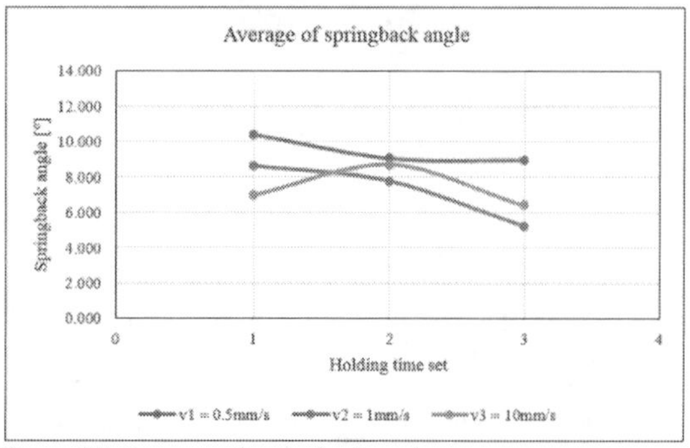

Figure 10. Average of springback angle

Figure 10 shows the average of springback angle for punch speed and holding time set variation. It can be seen obviously that the springback angle can be reduced by applying a higher punch velocity. With an appropriate holding time, the springback angle will be more reducible.

5 Conclusions

The results show that the determined process parameters can improve the product quality in the micro process. The springback angle indicates the differences when punch velocity variation was observed. The springback angle decreases with the higher punch velocity. Also, the variation of holding time during bottoming stage has a potential to reduce springback angle.

For future works, it needs more investigation of the holding time regarding decrease the springback angle. Another material should be observed to find out the characteristic of springback reduction with a variation of determined process parameters.

References

1. Qin, Y., *Micro-forming and miniature manufacturing systems — development needs and perspectives.* Journal of Materials Processing Technology, **177**(1–3): p. 8-18 (2006)
2. Gau, J.-T., C. Principe, and M. Yu, *Springback behavior of brass in micro sheet forming.* Journal of Materials Processing Technology, **191**(1–3): p. 7-10, (2007)
3. Diehl, A., U. Engel, and M. Geiger, *Influence of microstructure on the mechanical properties and the forming behaviour of very thin metal foils.* The International Journal of Advanced Manufacturing Technology, **47**(1-4): p. 53-61 (2010)
4. Liu, J., et al., *Influence of size effect on the springback of sheet metal foils in micro-bending.* Computational Materials Science, **50**(9): p. 2604-2614 (2011)
5. Ma, Z., et al., *Grain size effect on springback behavior in bending of Ti-2.5Al-1.5Mn foils.* Journal of Materials Processing Technology, **224**: p. 11-17. (2015)
6. Chen, C.-C., and C.-P. Jiang, *Grain Size Effect in the Micro-V-Bending Process of Thin Metal Sheets.* Materials and Manufacturing Processes, **26**(1): p. 78-83 (2011)
7. Zhu, H.X. and B.L. Karihaloo, *Size-dependent bending of thin metallic films.* International Journal of Plasticity, **24**(6): p. 991-1007 (2008)
8. Hezong, L., et al., *Determination of material intrinsic length and strain gradient hardening in microbending process.* International Journal of Solids and Structures, **48**(1): p. 163-174 (2011)
9. AHMET PARASIZ, S., R. VanBenthysen, and B.L. Kinsey, *Deformation size effects due to specimen and grain size in microbending.* Journal of manufacturing science and engineering, **132**(1) (2010)
10. Li, H.-z., et al., *Analysis of microbending of CuZn37 brass foils based on strain gradient hardening models.* Journal of Materials Processing Technology, **212**(3): p. 653-661 (2012)
11. Fang, Z., et al., *Study on springback in micro V-bending with consideration of grain heterogeneity.* The International Journal of Advanced

Manufacturing Technology, **78**(5-8): p. 1075-1085 (2015)

12. Huang, Z.-y., H. Zhang, and F. Ruan, *Control Method of the Spring Back in Bending Micro and Precise Stamping Parts*, in *Proceedings of the 2009 International Conference on Measuring Technology and Mechatronics Automation - Volume 01*2009, IEEE Computer Society. p. 782-784.
13. Zheng, Q., T. Shimizu, and M. Yang, *Scale effect on springback behavior of pure titanium foils in microbending at elevated temperature*. Journal of Materials Processing Technology, **230**: p. 233-243, (2016)
14. Boljanovic, V., *Sheet Metal Forming Process and Die Design*. New York: Industrial Press Inc. 184. (2004)
15. Kiswanto, G., A. Mahmudah, and S. Supriadi. *Development and Testing of 5 kN Micro Forming Machine for Micro Part Manufacturing*. in *13th Quality in Research*. Yogyakarta, Indonesia, (2013)

Rapid Time Response: A solution for Manufacturing Issue

N.Norazlin[1], A.Y. Bani Hashim[1], M.H.F.M.Fauadi[1], Teruaki Ito [2] and Zinvi Fu[1]

[1]*Faculty of Manufacturing Engineering,Universiti Teknikal Malaysia Melaka,Hang Tuah Jaya, 76100 Durian Tunggal, Melaka, Malaysia*
[2]*Institute of Technology and Science Tokushima University, Minami-Josanjima 2-1,Tokushima-shi, Tokushima, 77-8506 Japan.*

Abstract. Respond time in manufacturing give the major impact that able to contribute too many manufacturing issues. Based on two worst case scenario occurred where Toyota in 2009 made a massive vehicles call due to car complexity of 11 major models and over 9 million vehicles. The recalls cost at least $2 billion in cost of repair, lost deals and result in lost 5% of its market share in United State of America, while A380 was reported on missing target in new production and leads to delayed market entry due to their weak product life cycle management (PLM). These cases give a sign to all industries to possess and optimize the facilities for better traceability in shortest time period. In Industry 4.0, the traceability and time respond become the factors for high performance manufacturing and rapid time respond able to expedite the traceability process and strengthen the communication level between man, machine and management. The round trip time (RTT) experiment gives variant time respond between two difference operating system for intra and inter-platform signal. If this rapid time respond is adopted in any manufacturing process, the delay in traceability on every issue that lead to losses can be successfully avoided.

1 Manufacturing Issues

Based on two worst case scenario occurred in 2009 where Toyota made a massive vehicles call due to car complexity of 11 major models and over 9 million vehicles. The recalls cost at least $2 billion in cost of repair and lost deals. The recall result in lost 5% of its market share in United State of America and further drops foreseen [1]. While A380 was reported on missing target in new production and leads to delayed market entry due to their weak product life cycle management (PLM) [2]. These two cases give a sign to all industries to possess and optimize the facilities for better traceability in shortest time period.

The manufacturing issues are summarized in Figure 1 where it's divided into three main factors which are man, machine and management. To handle the complexity in manufacturing network, the future focused leadership and mind set is required. Furthermore the more intelligent equipment or machines occupied, the higher skill worker required to operate that. The challenge and manufacturing issues in Industry 4.0 is divided into three main factors which are man, machine and management. To handle the complexity in manufacturing network, the future focused leadership and mind set is required. Furthermore the more intelligent equipment or machines occupied, the higher skill worker required to operate that. Traceability become a main focused in this study where the usage of IoT (Internet of Things) is manipulating to track the product information and material used during manufacturing process. The traceability makes the whole manufacturing processes become visible and easy to manage.

Emerging manufacturing concerned on how fast the processes and information taken in order to meet the customer demand. To achieve that, the industry should deploy the new and innovative medium to be competitive and ready to face the market changing in anytime. Traceability and real time respond are indicators for effective communication in order to avoid any breakdown and ensure the raw material is continuously supply into manufacturing process. The more efficient of traceability system and rapid real time respond able to bring the manufacturing process into high performance of manufacturing. In this study, real time respond is shared where the time measured in millisecond (ms) through JAVA programming.

1.1 Issues in PLM

PLM emphasized the combination of people, process and data to be successfully implemented. However the implementation of PLM also contributes to several losses based on real case scenarios occurred around the world when it's neglected several factors. Figure 2 shows the six issues that been identified from the current study and the factors affected in sustainable PLM.

 DOI: 10.1051/matecconf/201710806002

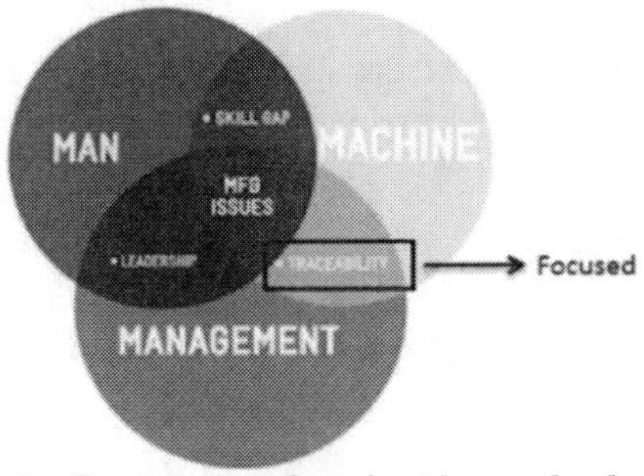

Figure 1. Current manufacturing issues faced. The challenge and manufacturing issues in Industry 4.0 is divided into three main factors which are man, machine and management. To handle the complexity in manufacturing network, the future focused leadership and mind set is required.

Green focused should provide an important competitive advantage instead of minimizing the environmental harm only. In to integrate environmental issues into new product development (NPD), the environmental factors must be considered in all stages of the manufacturing process [3]. The emerging of green technology involve two sides in manufacturing perspective, customers demand and supply from manufacturers which pressuring and responding to it. This point of view enforced the pre-production stage to consider the environmental issues in the design process [4]. The successful of green NPD involve a wide set of stakeholders and the suppliers plays a major role for NPD where it's begin from the design concept stage to the prototype development stage [3, 5]. Collaboration and communication are two main factors for green NPD. Collaboration is defined as coordination and alignment with project teams since the green NPD having a broad demand and various inputs and multifunctional product development, to meet market and environmental regulatory requirements become a main reason why the team needs to be coordinated. Effective communication between stakeholders is needed in order to provide information to produce green NPD. The information become extremely valuable in preproduction stage where its involve design and testing in order to ensure the NPD is meet the environmental regulations.

Complexity in NPD required a stable system to manage the development process. To manage the entire PLM is not an easy activity in order to meet the target such as customer demand, early market entry, new invention product and etc. The transforming of virtual production (designing, testing and simulation) into physical production is difficult to control during phase of life. The managerial complexity of PLM becomes cross-enterprise issues and even more challenging. Late market entry and exceed the targeted cost are the serious consequences faced if the company loses control in PLM [6]. It is proven by real cases reported when weak PLM was implemented.

Case 1: Toyota Company: Until year 2009, Toyota made a massive vehicles call due to car complexity of 11 major models and over 9 million vehicles. The recalls cost at least $2 billion in cost of repair and lost deals. The recall result in lost 5% of its market share in United State of America and further drops foreseen [1].

Case 2: Airbus Company: A380 reported on missing target in new production and leads to delayed market entry due to their weak product life cycle management (PLM) [2].

The fluctuation demand occurred when the awareness campaign on green product and keep the environment safe become effective. This point enforced company to create new product that comply with environment regulation. The interest in sustainable development growth rapidly when company start to consider mitigating the material used and waste product and any future weakness as well as inefficiencies can be avoided [7]. Organizing and managing the sustainable development and NPD become more complex and it's dependent on organized process and technology as a critical success factors [8, 9] PLM required technology as an integrative approach in order to manage the data and process for NPD towards sustainable and efficiently possible in new product process but not in development/design phase only [8].

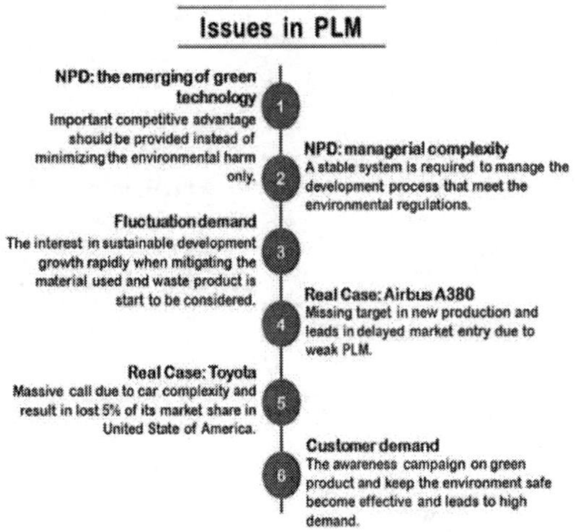

Figure 2. Issues in PLM. In PLM, there are three important factors that make it complete and works efficiently. People, process and data required good collaboration and by intervening the technology into PLM make it become more successful to merge the business globally.

Current practise of PLM is shows in Table 1 where the most application is neglected the technology invention in order to manage the data and process. The successful collaboration process can be achieved when PLM able to interact with coordination, information exchange, negotiation and solving conflicts [10]. In PLM, there are three important factors that make it complete and works efficiently. People, process and data required good collaboration and by intervening the technology into PLM make it become more successful to merge the business globally. To reduce the communication barrier cause by geographically factor and the used of web-based management or IoT seem the only way to make it successfully manage. Furthermore the manufacturing web-based only focus in early PLM stage which are pre-production and production only by neglecting the information trading between supplier, customer and retailer.

Gaps analysis as shown in Figure 3 has been concluded from the reviews on current PLM implementation and agent web-based application. Too focused on NPD is noticed as the first gap where current implementation or research put a lot of focused in pre-production stage in order to make sure the product development comply with the environment rules and regulations. Second gap reveal the poor information interaction and lack of data exploitation for the entire PLM and noticed that the sustainable NPD is only feasible to be done in pre-production only and hard to be implemented in the entire of PLM [8]. It is because the company lack of communication by ignoring the information exchange between supplier, customer and retailer become the third gap in current PLM practise. This point of view proved the idea proposed by [3] where the sustainable NPD should be involved with wide stakeholders. In order to achieve the sustainable PLM, the information exchange and trading is required in entire PLM. By emphasizing the IoT in PLM, it's able to make the idea of sustainable PLM happen with its ability to solve the complexity and expedite the process and secure communication network in management and production process.

Table 1. Studies reviewed on PLM. In PLM, there are three important factors that make it complete and works efficiently. People, process and data required good collaboration and by intervening the technology into PLM make it become more successful to merge the business globally.

Author	Issues Focused	Industrial Focused	Idea Proposed
[11]	Product Life Cycle Energy Management (PLEM): Energy consumption	Manufacturing Firms Area: design, production and serve process	IoT in PLEM
[10]	The interactions between SLM (Service Life Cycle Management) and PLM (Product Life Cycle Management)	Manufacturing Firms	Combining the PLM with SLM by using the IT technology. .
[8]	New Product Development (NPD)	Automotive companies	Sustainability in NPD
[12]	Communication in Sustainable PLM (Green Information System)	Manufacturing Firms	Intelligent Product Model for Sustainable PLM Applications
[13]	Closed-loop PLM	Production Management	Physical device i.e: sensor, timer etc. is required to achieve the level of intelligent system/process
[14]	Non-effective application of lifecycle management concepts	Manufacturing and Business	Process oriented framework to support effective PLM implementation

2 Industry 4.0

Online customization and purchasing is a new disruptive purchasing model that affected the manufacturing system and chain. This model required an evolution management while the operational levels become a huge challenge [15]. Terms of Big Data cannot be denied in Industry 4.0 where the only effective solution to manage and control the complexity and disturbances is by adapting the manufacturing networks [16]. Behind the manufacturing networks, IoT, data exchange, product life cycle management (PLM), business web, social web, computer hardware and software become the pillars. It is view by [16] in Figure 4 that incorporates the recent trends in internet technologies that able to give better support to the Industry 4.0.

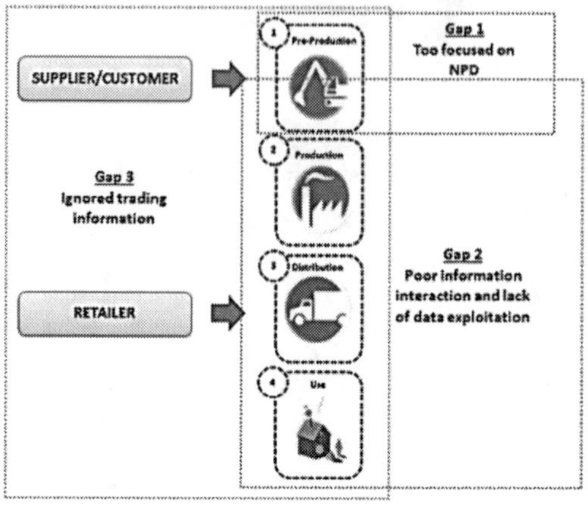

Figure 3. Gaps in current PLM practise. Current practise of PLM neglected the technology invention in order to manage the data and process. The successful collaboration process can be achieved when PLM able to interact with coordination, information exchange, negotiation and solving conflicts [10].

3 Round Trip Time

The purpose of the experiments in this section is to verify the optimal cost through round trip time (RTT). In manufacturing view, time plays the major in production and very significant with cost. Most of the manufacturing target emphasized the time period for every activities including communication. The RTT is measure by using socket communication between two or more computer.

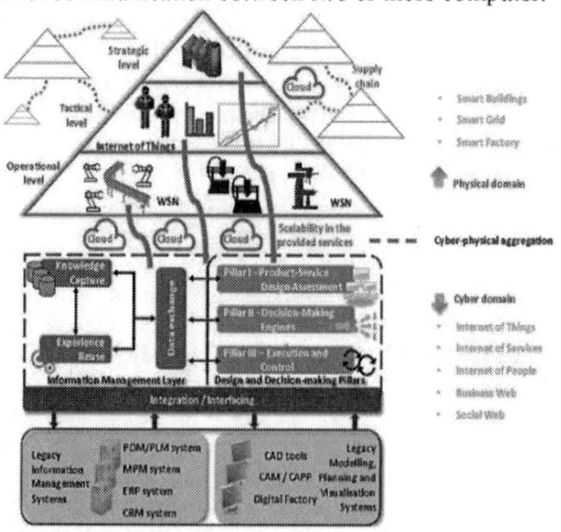

Figure 4. Manufacturing view in the future [16]. As an emerging manufacturing, the automation is not only automate the physical processes but data also include. [17] agreed that the automated of physical processes and information processing able to achieve a long term sustainable production. The automation processes become a

goal in deterministic manufacturing and one of the criteria for Industry 4.0.

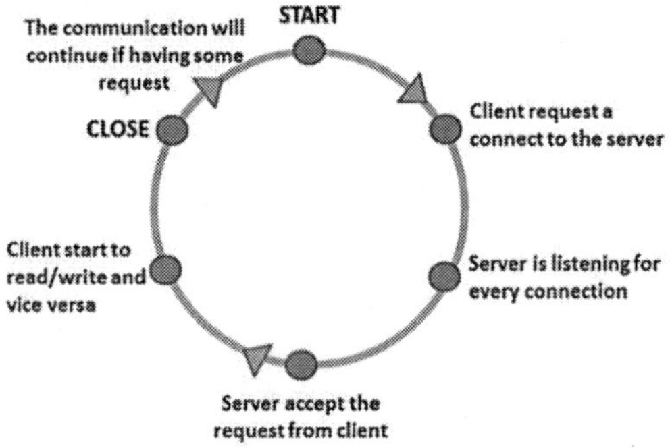

Figure 5. The general process cycle for one way communication through sockets and server. The activities are kept on happening until the server is disconnected.

There are two type of computer used which are local and remote computers that running over the network. The two-way communication between two programs running through socket that constitutes a client-server application. The connection process started with client send a request to the server on specific port. The server is on ready mode for listening and accepts the request from the client. Once the connection is accepted, the client able to use the socket to communicate with the server and the communication begin with read/write from their sockets. The process cycle is shown in Figure 5 below where the activities is keep on happening until the server is disconnected. The communication occurred in port 7999 on server network. The port can be changed and in this study, the port is randomly picked. Figure 6 show the general illustration for one way communication where the read and write is triggered by command "*skt.getInputStream()*" and "*skt.getOutputStream()*". While Figure 7 and Figure 8 shown the programming flow for Client and Server.

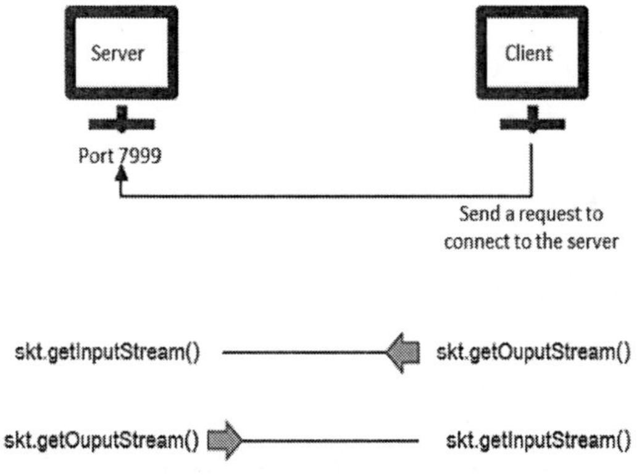

Figure 6. General illustration for one way communication on socket 7999. The communication occurred in port 7999 on server network. The port can be changed and the port can be picked randomly. Once the connection is accepted, the client able to use the socket to communicate with the server and the communication begin with read/write from their sockets.

3.1 Machines Detail

Most of the manufacturing target emphasized the time period for every activities including communication. The purpose of the experiments in this section is to verify the optimal cost through RTT. The RTT is measure by using socket communication between two or more computer involve client and server environment. The PC or workstation is referring to client which provides with friendly interface such as Windows. While a group of users is provide by server to client for sharing the server program [18].

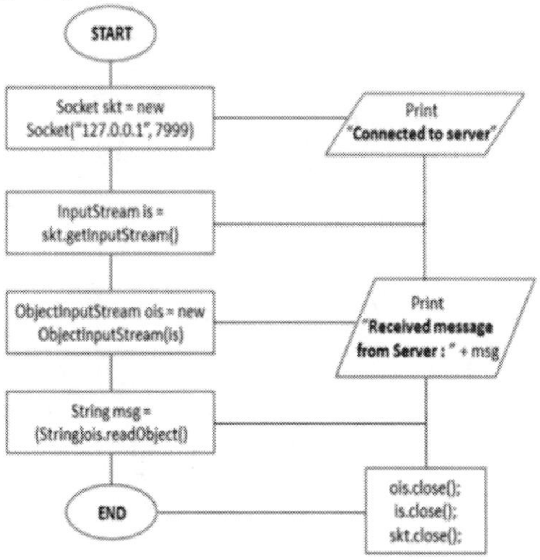

Figure 7. Programming flow for Client in RTT experiments. The communication where the read and write is triggered by command "skt.getInputStream()".

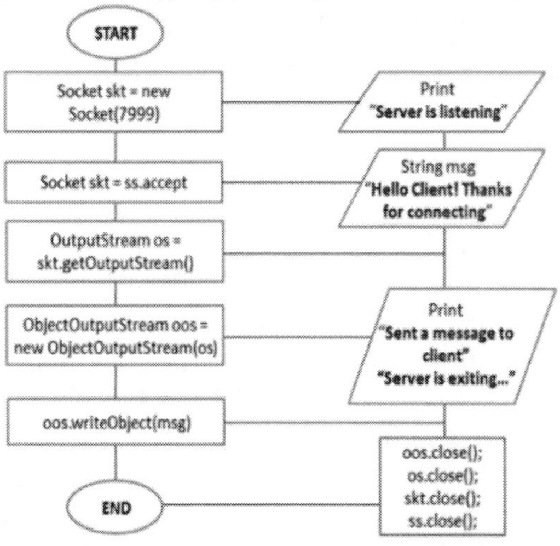

Figure 8. Programming flow for Server in RTT experiments. The communication where the read and write is triggered by command "skt.getOutputStream()".

There are two type of operating system used that running over the network where the details of machines (computers) used is shown in Table 2. The communication between two programs running through socket that constitutes a client-server application. The reason on choosing Windows operating system is because this Windows is widely used in industry and low minimum cost of investment. While Linux operating system is choose because of the level of security system and its performance for programming development and operation is better compared to Windows. The reasons why the industries do not want to adopt Linux into their environment because of complexity in term of usage and required a lot of training need to be provided for their employees.

Table 2. Machine details for RTT experiments. Windows and Linux as an operating system provide the communication link between users and the devices. The communication between two programs running through socket that constitutes a client-server application.

Machine name	Operating System	RAM	IP Address
V1	Windows 7	4.00 GB	120.17.42
V2	Windows 10	4.00 GB	120.17.92
V3	Windows 8	4.00 GB	120.17.45
V4	Linux Ubuntu	972.6 MB	120.17.194

4 Result and Discussion

4.1 Intra-Platform signal

Intra-platform communication occurred as internal signal respond in one machine as shown in Figure 9. Huge difference of respond obtained by V3 compare to V1, V2 and V4. The different probably cause by the machine itself where V3 is frequently used for programming development compare to others. Furthermore the other machine is rarely used. However, the speed of network also contributes to the long period taken for every machine to respond.

4.2 Inter-Platform Signal

For the inter-platform signal responds involved difference server for every testing. Overall performance showed that the server V3 in Figure 12 gives the fastest signal respond in 14ms compared to others. However, the V4 in Figure 13 shown the overall signal respond in 588ms where the Linux is an operating system for that machine. Linux give a lot of benefit for computer and network development but it less to be used for manufacturing purpose. For server V1 in Figure 10 and server V2 in Figure 11 shown the instability occurred with overall signal respond is 644ms and 645ms respectively.

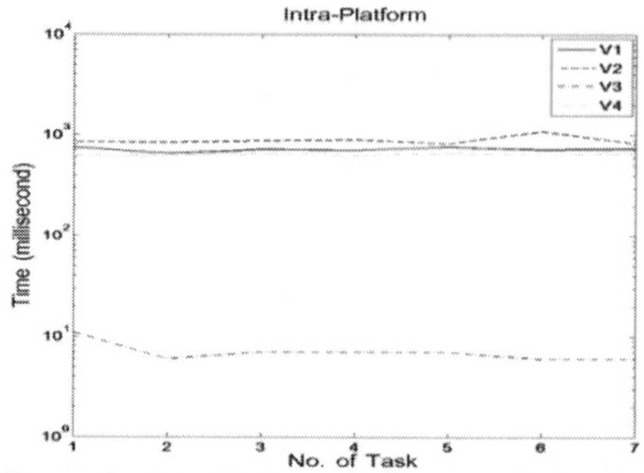

Figure 9. Intra-Platform signal respond in time (millisecond). Machine V3 obtained rapid respond compared to other machines. This frequency of used for programming may become a contribution factor.

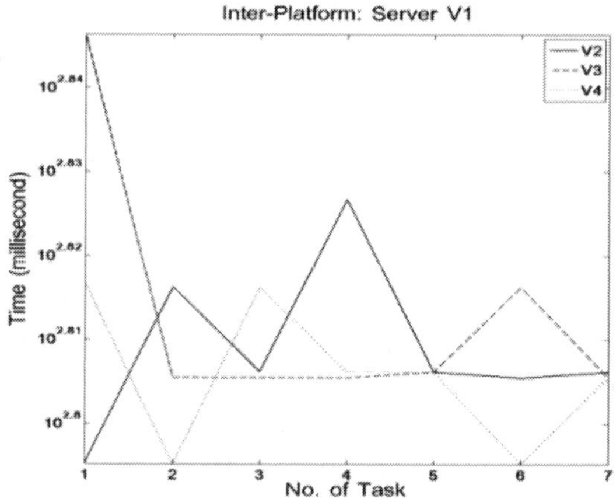

Figure 10. Inter-Platform signal respond in time (millisecond) for Server V1. Instability of connection show in server V1, however machine V3 shows stability in one way communication.

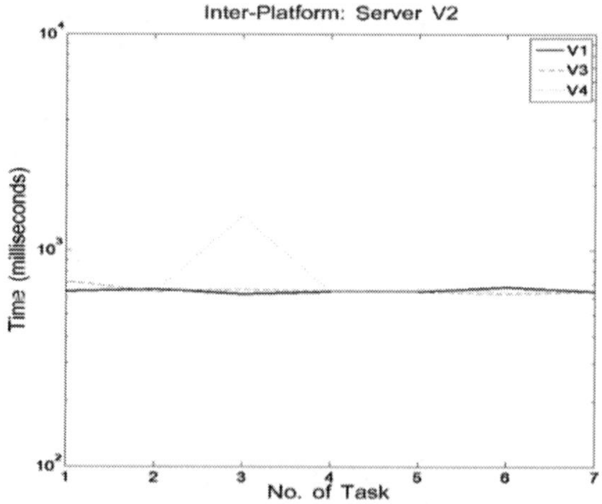

Figure 11. Inter-Platform signal respond in time (millisecond) for Server V2. Stable communication shown by most of the machine but the respond not so rapid compare to server V3 and V4.

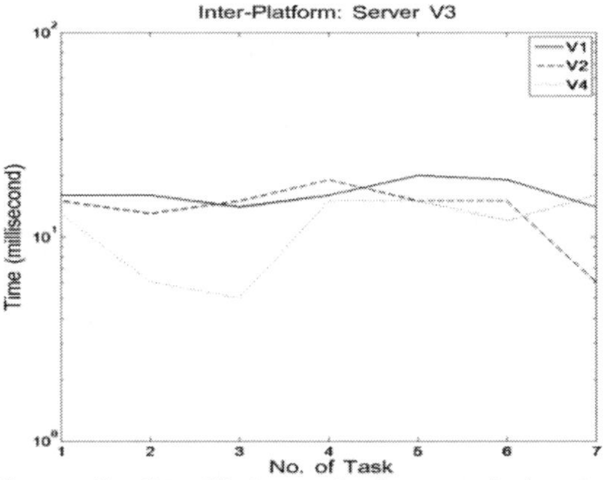

Figure 12. Inter-Platform signal respond in time (millisecond) for Server V3. The fastest respond is obtained by V3 compared to V1 and V2 that used same operating system which is Window. If all manufacturing process could have this time respond it may give benefit to the manufacturing management in delay traceability and expedite the contingency plan in order to meet the market demand.

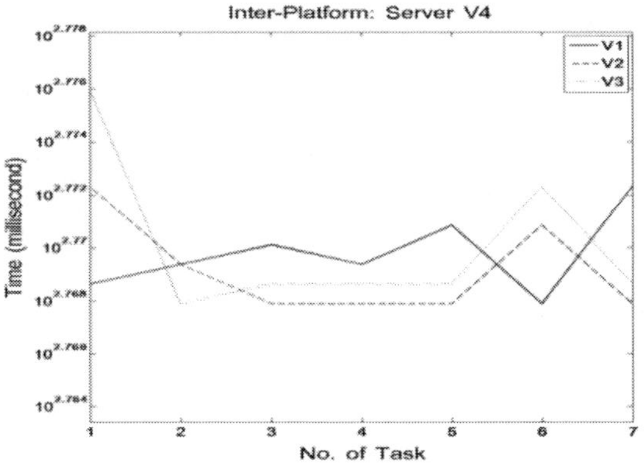

Figure 13. Inter-Platform signal respond in time (millisecond) for Server V4. Linux as an operating system for V4 also give better respond even the machine does not frequently used for programming development. Linux as an operating system offered the stability in system security compared to others operating system.

5 Conclusion

Stable communication showed by most of the machine but the respond not so rapid compare to server V3 and V4 however, instability of connections show in server V1 in one way communication. The fastest respond is obtained by V3 compared to V1 and V2 that used same operating system which is Window. Linux as an operating system for V4 also give better respond even the machine does not frequently used for programming development. Linux as an operating system offered the stability in system security compared to others operating system but because of the complexity of system to be used as well as operated become an occupied barrier.

Industry 4.0 required rapid respond for every issue to be measured by help from IoT. The fastest respond gained, the more efficient of manufacturing processes is achieve. However, in current practise of PLM and manufacturing management, it's difficult to obtain. The fastest time respond may give benefit to the manufacturing management in delay traceability and expedite the contingency plan to be implemented in order to meet the market demand. The IoT also give benefit to cost reduction in paperless management and sustainable development when market demand changed rapidly nowadays.

Acknowledgments

The researchers would like to thank the Malaysian Government, Universiti Teknikal Malaysia Melaka (UTeM) for their financial support and provision of facilities to carry out this study. Million thanks to Dr. Ahmad Yusairi Bani Hashim for the great thought and helpful comments. Also thank to The University of Tokushima Japan, for the supports. Their supports and comments are greatly acknowledged

References

1. Gu, Xiaoyuan. 2010. *Toyota Recalls: Revealing The Value Of Secure Supply Chain*. Diss. Massachusetts Institute of Technology.
2. Tech Drummer. 2008. *Case Study: IBM & Airbus: Tech Drummer*
3. Jay Polonsky, M., & Ottman, J. 1998. *Stakeholders' contribution to the green new product development process*. Journal of Marketing Management, 14(6), 533-557.
4. Baumann, H., Boons, F., & Bragd, A. 2002. *Mapping the green product development field: engineering, policy and business perspectives*. Journal of Cleaner Production, 10(5), 409-425.
5. Lee, K. H., & Kim, J. W. 2011. *Integrating suppliers into green product innovation development: an empirical case study in the semiconductor industry*. Business Strategy and the Environment, 20(8), 527-538.
6. Stark, J. 2015. *Product Lifecycle Management*. (pp. 1-29). Springer International Publishing. (2015)
7. Bevilacqua, M., Ciarapica, F. E., & Giacchetta, G. 2007. *Development of a Sustainable Product Lifecycle In Manufacturing Firms: A Case Study*. International Journal of Production Research, 45(18-19), 4073-4098.
8. Gmelin, H., & Seuring, S. 2014. *Achieving sustainable new product development by integrating product life-cycle management capabilities*. International Journal of Production Economics, 154, 166-177.
9. Johnson, M.E., Cochran, J.J., Cox, L.A., Keskinocak, P., Kharoufeh, J.P., Smith, J.C. 2010. *Product/Service Design Collaboration: Managing the Product Life*. Wiley Encyclopedia of Operations Research and Management Science. John Wiley & Sons, Inc.
10. Wiesner, S., Freitag, M., Westphal, I., & Thoben, K. D. 2015. *Interactions between service and product lifecycle management*. Procedia CIRP, 30, 36-41.

11. Tao, F., Wang, Y., Zuo, Y., Yang, H., & Zhang, M. 2016. *Internet of Things in product life-cycle energy management*. Journal of Industrial Information Integration, 1, 26-39.
12. Främling, K., Holmström, J., Loukkola, J., Nyman, J., & Kaustell, A. 2013. *Sustainable PLM through intelligent products*. Engineering Applications of Artificial Intelligence, 26(2), 789-799.
13. Kiritsis, D. 2011. *Closed-loop PLM for intelligent products in the era of the Internet of things*. Computer-Aided Design, 43(5), 479-501.
14. Schuh, G., Rozenfeld, H., Assmus, D., & Zancul, E. 2008. *Process oriented framework to support PLM implementation*. Computers in industry, 59(2), 210-218.
15. Mourtzis, D. 2016. *Challenges And Future Perspectives For The Life Cycle Of Manufacturing Networks In The Mass Customisation Era*. Logistics Research, 9(1), 1-20.
16. Mourtzis, D., Doukas, M., & Psarommatis, F. 2015. *A Toolbox For The Design, Planning And Operation Of Manufacturing Networks In A Mass Customisation Environment*. Journal of Manufacturing Systems, 36, 274-286.
17. Ståhl, J. E., Gabrielson, P., Stål, C., & Andersson, C. 2012. ACMFDAOALIDBM. In 45th CIRP Conference on Manufacturing Systems 2012.
18. Xue, M., & Zhu, C. 2009. *The Socket Programming and Software Design For Communication Based On Client/Server*. In Circuits, Communications and Systems, 2009. PACCS'09. Pacific-Asia Conference on 2009, (pp. 775-777). IEEE.

A Study of Resin as Master Jewellery Material, Surface Quality and Machining Time Improvement by Implementing Appropriate Cutting Strategy

Paryana Puspaputra

Department of Mechanical EngineeringIndonesia Islamic University. Jl. Kaliurang Km 14, Jogjakarta 55581 Indonesia

Abstract. This paper deals with a research about art and jewellery product machining that focused in the selection of appropriate material for jewellery master which is machined by CNC. CNC is used for better surface finish for no undercut design and more complex ornament. The need of production speed requires minimum process without reducing the quality of detail ornament significantly. Problems occur when high surface quality is required. In that condition high speed spindle is used with low feeding speed, as a result is high temperature in cutter-material area will melt the resin and build the build-up edge (BUE). Due to the existence of BUE, the cutting tool will no longer cut the resin, as a result the resin will then melt due to friction and the melt resin will then stuck on the relief and surface finish become worst and rework should be done. When required surface is achieved problem also occur in next going process, that is silicon mould making. Due to galvanization process for silicon at about 170°C, resin material may be broken or cracked. Research is then conducted to select appropriate resin type suitable for all production steps.
Keywords: master product, CNC, resin, micro-machining, jewellery, machining strategy.

1 Introduction

In a jewellery production which use investment casting process that is shown in Figure.1, a master is a most important part. The master product should have high quality either on surface or geometry. Traditionally a master is made by a high skilled craft man, but recently it is made by using a CNC or rapid prototyping machine.

In a product which have small sized as it is shown in Figure.2, complex and detail relief, high quality surface finished is very difficult to achieved due to material properties, brittleness and melting point and grain size is sometimes become obstacles.

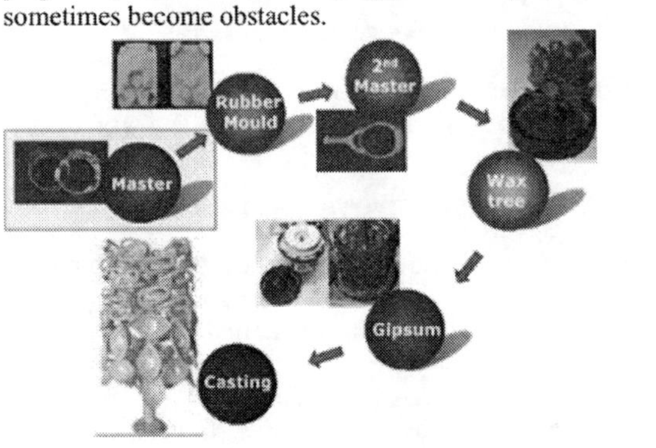

Figure 1. Jewellery production process using investment casting [1].

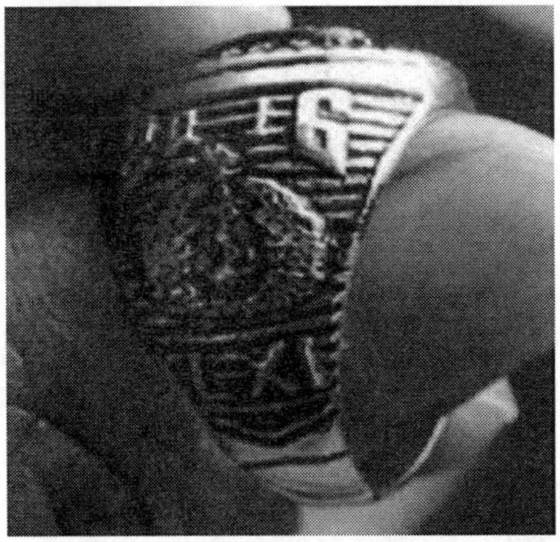

Figure 2. Sample of complex and detailed relief ring.

A simple method that usually used in jewellery production is by using rapid prototyping machine or 3D printer.

It is shown in Figure.3 that 3D printer can realized complex and detail relief, but if we see more detail, due to its process, the surface finished is restricted by the step or layer thickness. The thinner the layer, the better the surface finish but the longer process is needed, but however this thickness is limited.

Another alternative is by using CNC. Complexity of product that could be realized using the CNC are depend on the axis number of CNC, supporting jig and also machining strategy that defined by operator. The advantage is that the surface finish if better than one that produced by 3D printer, but complex surface with undercut is natural restriction. Figure 4 shows machining result using CNC engraver.

Figure 3. Jewellery master using rapid prototyping machine [2].

Figure 4. Jewellery master using CNC machine.

2 Master Model

Master model is one of the most important parts of the jewellery industry. A good model maker must know all the processes of jewellery manufacturing, including design, cast, polish, and stone setting [3].

All of those parameters should be taken into account, design, stone setting, and cast must be analysed in the design process either using engineering intuition or application software such as computer aided engineering (CAE). Since jewellery is made of precious metal, it is required to minimize polishing process to achieve its surface quality [1,4].

From the illustration above, high quality surface is very important for a jewellery master model. It is not only reducing the polishing chip, but also improving the relief quality

2.1 Types of Master Model Manufacturing

Master model of jewellery can be made by using the process shown in Figure. 5. Depends on the relief complexity and production schedule, a wax (or resin) master can be prepared by manual, CNC or rapid prototyping machine.

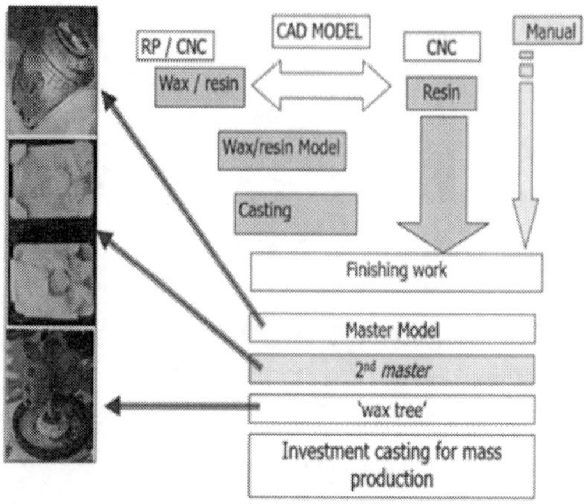

Figure 5. Jewellery master model making alternatives [1,5].

In CAD/CAM based application, jewellery making is started using modelling wax or resin as material after its model is developed using digital modelling. Most are using rapid prototyping (RP) machine since it is very simple in implementation. The disadvantage is that the price is high and surface quality depends on its layer thickness. Thinner layer is better in surface quality, but thinner layer is longer time processing which will affect production time.

Another way is using the CNC machine. Compared with RP machine, this machine price is not expensive but it need high capability operator to handle. It can also produce product with better surface quality than those with RP machine.

In process which surface quality is not good enough, master jewellery preparation need further process to achieve its surface criteria as shown in Figure. 5. Things that should be noted here is that there are possibilities in dimensional changes that may affect relief geometry as well as stone setting due to thermal deformation and finishing process.

In the utilization of CNC machine. Jewellery master can be produced directly after machining, so that deformation and changes in relief and geometry can be minimized [1,4,5,6].

2.2 Jewellery Master Model Material

In the utilization of CNC machine. Jewellery master can be produced from resin or mostly modelling wax.

2.2.1 Polyester and Polyurethane Resins

Polyester and polyurethane resins are the most commonly used casting material as shown in Figure.6. The polyester resin used in the moulding application is a viscous liquid requiring the addition of catalysts and accelerators to complete the curing process. Polyester resins are contact products which require no pressure to cure and can be cured from a fluid or solid state [7].

The advantage of polyester resin is that it is easy to use and is the lowest cost resin among resin products. However, there are a number of critical disadvantages in lack on mechanical properties and are less water resistant than polyurethane resins. The flash off of styrene during the curing process causes a fairly large degree of volumetric shrinkage, the full extent of which is often difficult to predict. Styrene emissions are also harmful and may necessitate the institution of specific, often expensive, environmental safety measures balance off ease of use, low cost, and positive physical characteristics. In many cases, polyurethane resin is produced in a liquid form and can be poured into moulds [7].

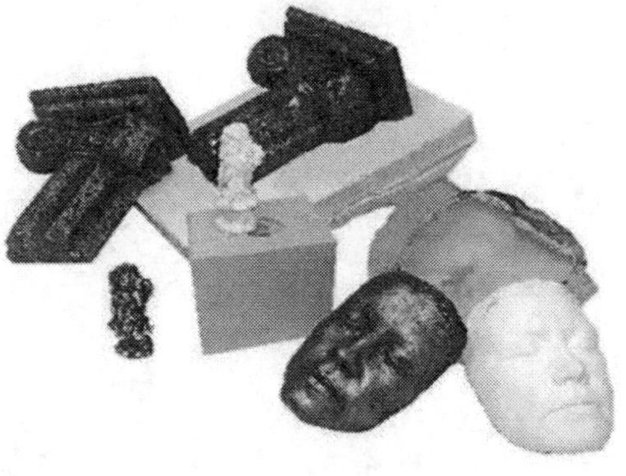

Figure 6. Sample of resin products [6].

There are many advantages to polyurethane resins. It can be formulated to create pliable rubber moulds and then be made into a hard plastic casting material or a water clear casting material [7].

Though polyurethane costs a bit more to manufacture, they have significantly better mechanical properties (stronger and more tear resistant), more importantly the emissions are extremely low when curing, and thus far safer to be around [7].

2.2.1 Jewellery Modelling Wax

Figure 7. Types of Jewellery modelling wax [8].

Most of commonly used material for Jewellery master is wax. Most CNC cannot produce product as it is designed especially undercut profiles, hence further process is needed to meet the design with real product. Using wax material, improvement can be easily done.

Disadvantages of jewellery wax is that there is electrostatic that is produced during machining process, so that the chip stuck in the product, and demagnetization is necessary to remove the stuck chips from product (see Figure.8). The other important is that wax is fragile, so that high clamping force will make it cracked or broken (see Figure.8).

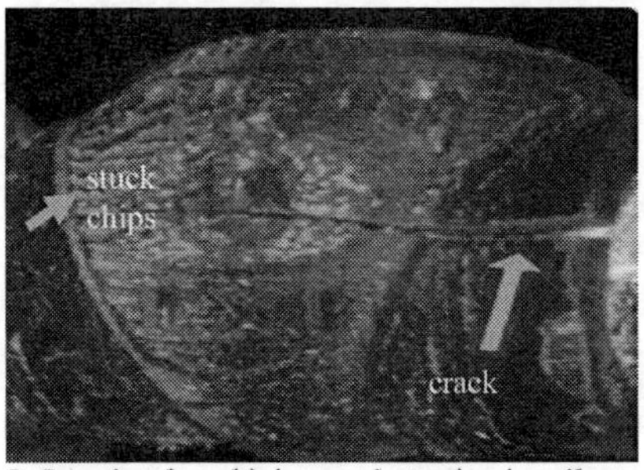

Figure 8. Sample of machining product using jewellery wax.

2.2.2 Micromachining

Micromachining is defined as fabrication of tiny devices: the techniques used in fabricating the miniaturized devices and moving parts into which microelectronic circuitry is integrated (Encarta, Encyclopedia).

Another definition as machining strategy where extremely small tool diameters are used. The tool characteristics are: diameter is in the range of 0.1 to 2 mm, small cutting length, high accuracy, coated [8].

Machine requirement are high spindle accuracy, high rpm, thermal stability against spindle growth. The micromachining area is production of cavity, slot, pocket, holes or engraving in many type of material [9].

According to references, micromachining is usually done in semiconductor manufacturing, bio mechanics, jewelry, etc. Micromachining usually deals with laser cutting, layer manufacturing, chemical machining, etc.

In this paper, micromachining is used in term of machining jewelry master (master) using CNC milling machine with tiny single lip tool as shown in 0

Figure 9. Conical tool [6].

2.3 Research and Experiments

JWX10 is a dry machine that not allow user to use liquid coolant or even lubrication for all slide ways. It has swing area of 66mm in diameter, it means that work-piece and fixture volume is limited. Work-piece is attached and clamped in rotary axis in overhung position (see Figure.10). Due to that situation, deflection in at the edge will affect machining result.

Figure 10. Machining setup.

Experiment is done by using Roland JWX10 CNC (Figure.10) and resin as work-piece (Figure.11). A detail model is illustrated in Figure.11. as a sample with 60mm in diameter and total 3mm thickness, double sided.

Machining is done under the condition shown in table 1., and use circular offset strategy as shown in Figure. 12. The simulation result is also seen in Figure.12.

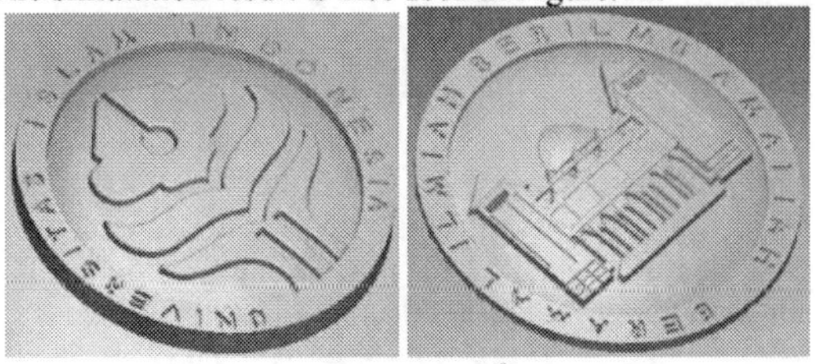

a. b.

Figure 11. Work-piece model, (a) Front-side, (b) Back-side.

Table 1. Machining condition.

	Roughing	Finishing
Tool	**End mill**	**Conical 40°**
Tip radius	3mm	0.5
Feed	7 mm/s	9 mm/s
Spindle	15000 rpm	19000 rpm
Allowance	0,35 mm	0 mm
Time	20 min	150 min

Figure 12. Machining strategy and simulation result.

That machining condition resulting surface as it is seen in Figure.13. It is shown that chips are collected and the adjacent area at the contact between tool and work-piece cannot moved away. That condition resulting temperature raised approaching resin's melting point (250°C) especially between tool and work-piece contact.

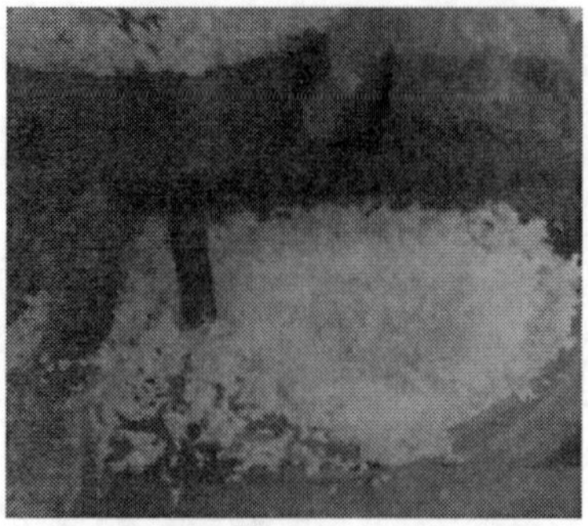

Figure 13. Machining situation, resin is collected, melt and stuck.

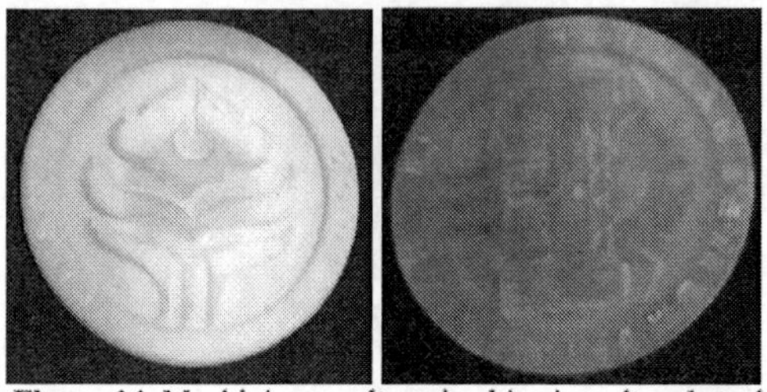

Figure 14. Machining result, resin chips is melt and stuck.

It is shown in Figure.14 that many part of machining result are in more white colour, showing that there are locations at which resin chip are melt and stuck on the surface and cannot be cleaned using tumbling or polishing process. Such abrasion process will affect surface finish quality or relief.

Resin material is a result of chemical reaction between resin solution and hardener. From this situation, cutting thin allowance will result very small particle of chip (see Figure.15). The tiny chips will then float and move with the air movement [11].

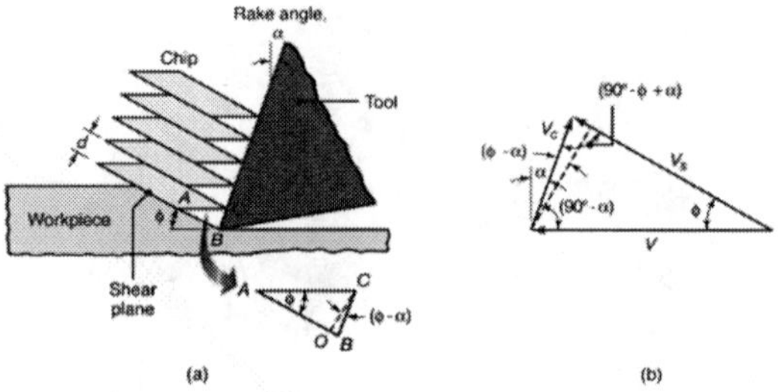

Figure 15. Shear phenomena in orthogonal cutting process [10].

With the very light weight the chip will move with the air from higher pressure (ambient pressure) into the spindle area. Spindle and tool rotate at very high speed, that make the air surrounding the area will have very low pressure, so that air will move into it.

During machining, most cutting energy will transferred into heat that will increase temperature in tool and work-piece area. When tiny chip is trapped in high temperature, it will be stuck in melted material and make poor surface finish quality.

It is clear from Figure.16 that the smaller step over the better surface finish, but the longer machining time. Bigger step over will also make the tool do not cut but hit the work-piece [5].

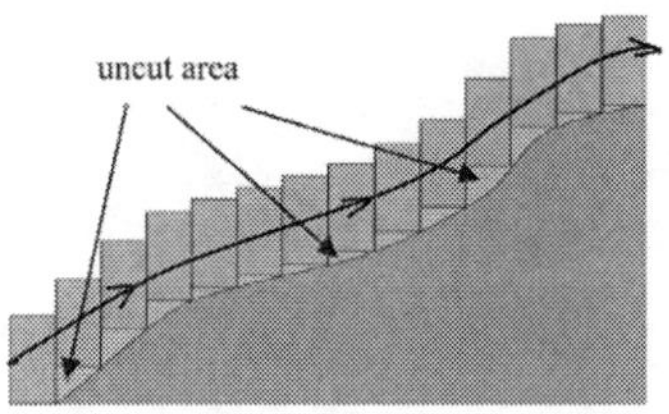

Figure 16. The effect of step over in tool movement to the surface quality [1,5,12].

From the situation illustrated in Figure.16, Improvement is then made by increasing chip size to bigger volume without increasing step over (increasing depth of cut). Bigger volume is selected to absorb heat so that temperature will not raise high and also allow chip moved away by centrifugal forces of the rotating tool. With the bigger mass, chip is easier to move away from tool-work-piece contact area and does not follow the air flow (see Figure.17).

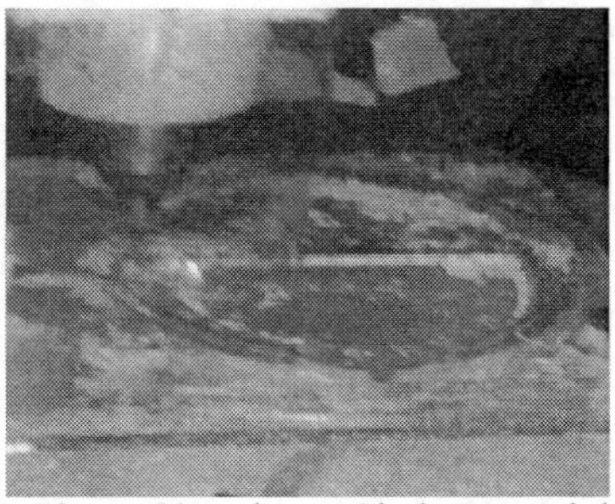

Figure 17. Bigger chip volume, chip is not stuck but moved away.

To make constant cutting forces, to avoid tool from broken, chip volume can be control by controlling the allowance in roughing step in machining. Using the allowance more than 1 mm resin will be cut with bigger chip sized form. This bigger sized chip will then throw centrifugally away from the tool and leave the surface area when it cut.

Since the chip are moved away, temperature is decreased, and nothing is stuck and better surface quality can be achieved (see Figure.18).

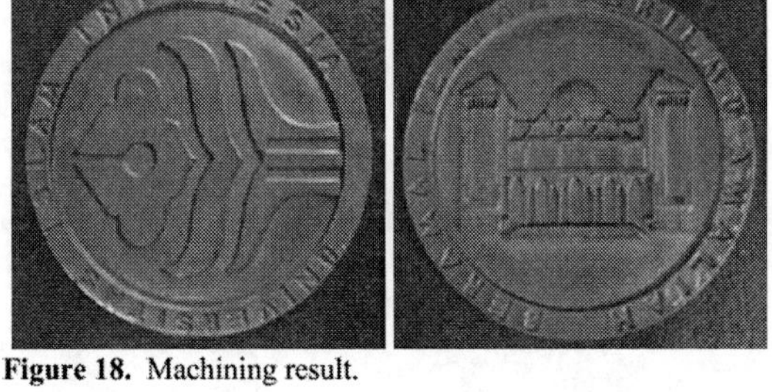

Figure 18. Machining result.

3 Conclusions

From the experiment, it is found that resin material that is suitable for jewellery model material has unique characteristic when it is machined, they are:

1. 200°C melting point make it critical to be machined since machining process will produce much.
2. Tiny sized chip from resin will float and flow with the air and may trapped into high temperature area and trapped and affect surface quality of machined resin.
3. Increasing depth of cut will increase chip volume and improve machine surface quality.
4. Increasing depth of cut will reduce machining step and improve total machining time significantly.

4 Future Works

After machining process, master model will be used to make rubber-mould. To do that, master will be placed in a silicon and vulcanized at about 170°C. Simple model found no serious problem, but complex and thin model will. It is necessary to make another researches to find:

1. better material than resin
2. better type of silicon rubber

References

1. P. Puspaputra. *Development of Machining Methode of Bitmap-based complex Surface.* DT.JTM-ITB. (2012).
2. Pinterest. https://www.pinterest.com/laleipsig/3d-printing-2d-lasercutting/ (2016)
3. Goulds Jewellery, (http://www.gouldsjewellery.com /services/master-model-maker). (2016).
4. P. Puspaputra *Micro-machining Problems in Undetermined Complex Surface (Case Study: Machining of Jewelry's Master Model)*, APIEMS C (2010).
5. P. Puspaputra. *The Application of CAD/CAM Technology for Small Industries in Developing Creative Cultural Design of Art and Jewelry.* APIEMS C. (2008).
6. P. Puspaputra. *Single-Lip Tool Analysis for Jewellery Product Machining.* RAPI-UMS.(2009).
7. Art Mould, https://www.artmolds.com/polyurethane. (2016).
8. Freeman Manufacturing, http://www.freemanwax.com/products.html. (2016)
9. Seco Tool, Jabro Catalog. (2006).
10. S.Kalpakjian, & S.R.Schmid. *Manufacturing, Engineering & Technology (5th edition).* P.E.Inc. NJ. (2006).
11. P. Puspaputra. *A Study of Resin as Master Jewelry Material, a New Alternative Material to Perform Higher Complexity and Surface Quality of Jewelry Master using CNC.* APIEMS C. (2016).
12. P. Puspaputra. *Introduction to Bitmap-based 3D Model and Its Machining Problems. Case Study: Brooch Machining.* (IJENS - IJET), Volume: **12**,(2012).

DOI: 10.1051/matecconf/201710806004

Cytotoxicity of three commercial orthodontic elastomeric ligature brands

Peerapong Tua-ngam[1], Nittaya Jira-anankul[2], Niwat Anuwongnukroh[3] , Surachai Dechkunakorn[4] and Pasaree Laokijcharoen[5]

[1]*Research Office, Faculty of Dentistry, Mahidol University, 6 Yothee Road, Rajthevee, Bangkok Thailand*
[2]*Postgraduate student of specialty training program in Orthodontics, Faculty of Dentistry, Mahidol University, 6 Yothee Road, Rajthevee, Bangkok Thailand*
[3]*Department of Orthodontics, Faculty of Dentistry, Mahidol University, 6 Yothee Road, Rajthevee, Bangkok Thailand*
[4]*Department of Orthodontics, Faculty of Dentistry, Mahidol University, 6 Yothee Road, Rajthevee, Bangkok Thailand*
[5]*National Metal and Materials Technology Center, National Science and Technology Development Agency, Ministry of Science and Technology*

Abstract. Objective: This study aimed to evaluate cytotoxicity of three commercial brands of orthodontic elastomeric ligatures. Materials and Methods: The three commercial orthodontic elastomeric ligature brands were C-brand, T-brand and U-brand from different manufacturing countries. The cytotoxic assay was performed using cell cultures (L929) which were subjected to the cell viability test with methyl-tetrazolium (MTT). The L929 cell line was grown in 96-well tissue culture plates ($1x10^5$ cells/cm^3). Elastomeric ligatures (0.2 g) from each brand was placed in 1 ml of Minimum Essential Medium (MEM) at 37°C with 5% CO_2 at 100% humidity in an incubator for 24 hours. After incubation, the cell mitochondrial activity was evaluated by the MTT test for detecting viable cells from optical density according to ISO 10993-5: 2009 (E) test procedure. Cytotoxic effects were determined by the percentage of cell viability. Results: Microscopic investigation showed that all commercial elastomeric ligatures caused alterations in the fibroblast cell morphology from an elongated to a spherical form. The cell viability of C-brand, T-brand and U-brand were 51.10, 73.04 and 93.04%, respectively. Conclusion: T-brand and U-brand orthodontic elastomeric ligatures were non cytotoxic potential. U-brand had the greatest cell viability and C-brand had the lowest cell viability at 24 hours.

1. Introduction

Elastomeric ligatures are commonly used nowadays in orthodontic treatment. They are used to engage the archwires to orthodontic brackets. The force exerted by an elastomeric ligature appears dependent upon the magnitude of the initial force, the duration that the force may be applied and the decay rate of the ligature. In addition to the initial extension force and residual forces, the size and dimension of elastomeric ligature also affect the force exerted on the archwire. Currently, orthodontists prefer to use elastomeric ligatures to stainless steel ligatures because the former can be easily used, patient-freindly in nature, esthetic appearance and less time is spent to engage archwires in the orthodontic brackets. Elastomeric ligatures are placed in direct contact with patient's soft tissues, often for months and years during the course of the orthodontic treatment. Moreover, elastomeric ligatures produced from polyurethane are cytotoxic to oral mucosal cells due to the presence of oligomers (low molecular weight polymers), unreacted monomers or other additives [1]. When a new product of elastomeric ligature is introduced in the orthodontic market, the manufacturer tends to promote the clinical performance but they are usually reluctant to inform about the composition, the manufacturing process, or the results of biocompatibility test. Therefore, it is necessary to determine the safety and biocompatibility of these products to prevent potential allergic or toxic reactions to the patients.

The toxicity test of dental products by using cell culture medium is a valid method to determine a material's biological behavior [2]. The suggested testing methods (ISO 10993; ISO 7405) use cell counting, dye-binding, metabolic impairment or membrane integrity as endpoints of the cytotoxicity test or assay. The ISO standard 10993-5 requires the biological evaluation of medical devices. The manufacturer is responsible for ensuring that their products are safe to be used and the required tests should be performed. The direct use of cell and colony counting as an assay endpoint is commonly used but it is probably the least reliable method [3].

Dechkunakorn et al. (2012) and Anuwongnukroh et al. (2013) [3,4] compared three commercially available elastomeric ligatures of different colors in relation to their initial and residual forces and also compared the dimensions and residual forces of various orthodontic elastomeric ligatures. However, there has been lack of study in terms of the cytotoxicity of these elastomeric ligatures. Nowadays, new elastomeric ligatures continue to be marketed, rapid and sensitive tests for examining their toxic effect are needed for patient safety. The purpose of this in vitro study was to compare the cytotoxicity of three different commercial elastomeric ligatures: C-brand, T-brand and U-brand.

2. Materials and methods

The three different commercial brands of orthodontic elastomeric ligatures (Table1) were tested according to ISO 10993-5: 2009(E) test procedure.

Table 1. The three commercial orthodontic elastomeric ligatures

Product	Manufacturer	Color
C-brand	China	Blue
T-brand	Thailand	Blue
U-brand	USA	Blue

All three commercial elastomeric ligatures were compared to the positive control, polyurethane film containing 0.1% zinc diethyldithiocarbamate (ZDEC). Plastic coverslip Lot. No.600562, was used as a negative control.

2.1 Cytotoxicity tests

2.1.1 Cell preparation (1st day)

The L929 cell line (Lot Number: 2869501) was the target cell used in this experiment. The cells were fed with a Minimum Essential Medium (MEM) 100 µL, completed medium, in 96-well tissue culture plates at the concentration of $1x10^5$ cells per well. Each well was maintained at 37°C with 5% CO_2 at 100% humidity in an incubator for 24 hours to obtain a monolayer cell growth.
1.2 Orthodontic elastomeric ligature preparation for specimen (2nd day)

Orthodontic elastomeric ligatures (0.2 g) from each manufacturer were sterilized by ultraviolet radiation for 40 minutes. Thereafter, 1 ml of 10% MEM was used to cover these sterile orthodontic elastomeric ligatures. This was maintained at 37°C with 5% CO_2 at 100% humidity in an incubator for 24 hours to obtain a solution of orthodontic elastomeric ligature. Positive control with a coverage of 6 cm^2/ml was prepared and negative control with a size of 6 mm was immersed in 1 ml of 10% MEM. This was maintained at 37°C with 5% CO_2 at 100% humidity in an incubator for 24 hours.

1.3 MTT assay (3rd day)

All samples of the orthodontic elastomeric ligatures along with the positive and negative and reagent controls were removed from the well tissue culture plates. Each well was washed with phosphate-buffered saline (PBS). Freshly prepared 50 µL of methyl tetrazolium (MTT) (0.1% without supplements and without phenol red) solution was dispensed into each well and incubated for two hours. Then MTT solution was removed and washed with PBS. One hundred microliters of isopropanol was dispensed into each well. The plates were agitated on a shaker to enhance the dissolution of formazan for 30 min and the spectrophotometric absorbance was measured at 570 nm using isopropanol as the blank.

The data generated was recorded in the raw data file. The results were presented in tabular form including experimental groups with the test samples and positive and negative and reagent controls.

A decrease in the number of living cells results in a decrease in the metabolic activity in the sample. This decrease directly correlates to the amount of blue-violet formazan formed, as monitored by the optical density (OD) at 570 nm. To calculate the reduction of cell viability compared to the blank, the following equation was used

$$\text{Cell viability } \% = \frac{100 \times OD_{570e}}{OD_{570b}}$$

OD570e represented the mean value of the measured OD of the 100% extracts of the test samples.

OD570b represented the mean value of the measured OD of the blanks.

2.2 Evaluation of cell viability

The lower the cell viability % value, the higher was the cytotoxic potential of the test samples. If cell viability was reduced to <70% of the blank, it was considered to have cytotoxic effects. If the 50% extract of the test samples had at least the same or higher cell viability than the 100% extract, the test was repeated.

3. Results

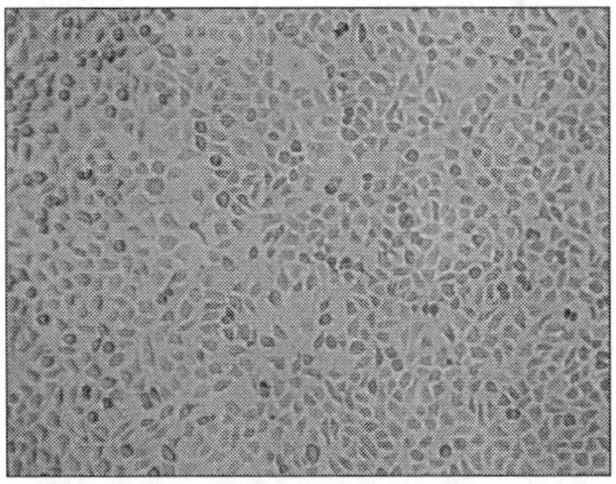

(a)

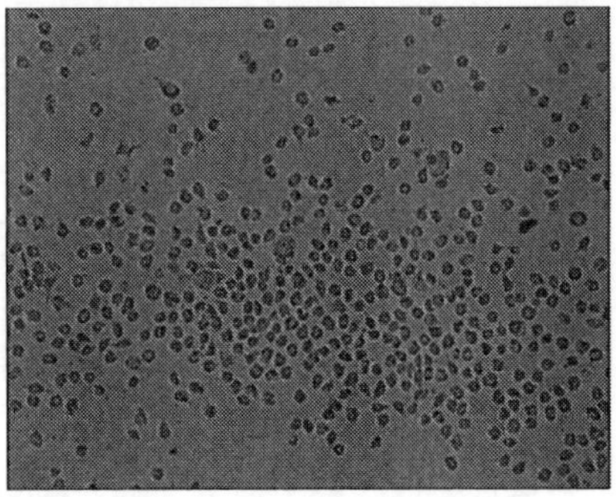

(b)

 DOI: 10.1051/matecconf/201710806004

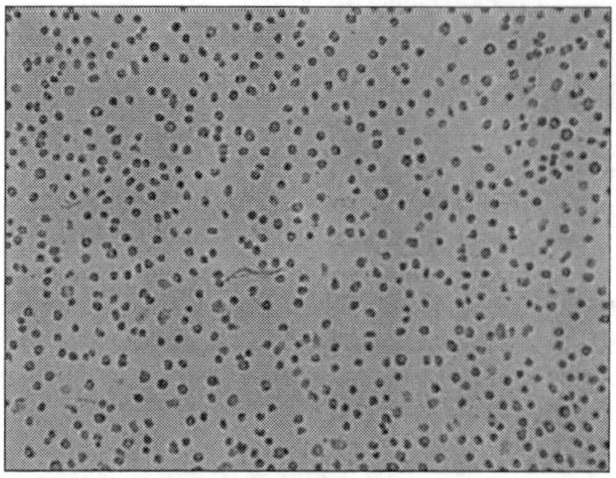

(c)

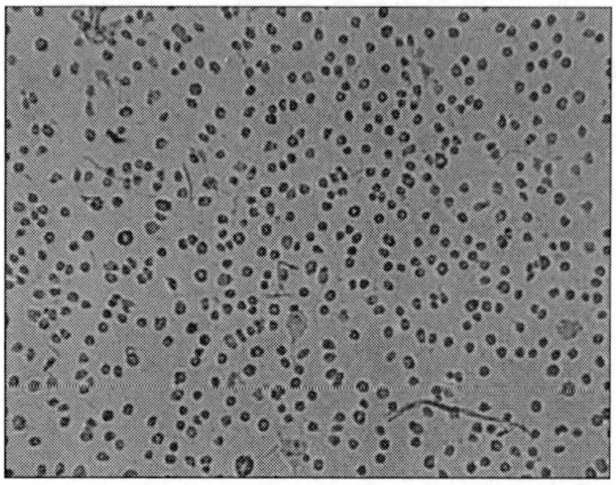

(d)

Figure 1. Microscopic investigation of morphological changes in L929 cells at day 3. (a) Control, (b) C-brand, (c) T-brand and (d) U-brand.

The cytopathic effect of elastomeric ligatures on the L929 cells at day 3 is shown in Figure 1. In the control group, the cells had typical multipolar fibroblast morphology. Cells had elongated shapes and were attached to the surface of microplate. However, C-brand, T-brand and U-brand elastomeric ligatures had altered cell morphology. A number of cells had lost their elongated shapes and had a spherical morphology.

Table 2. Descriptive result for cell viability of orthodontic elastomeric ligatures and control groups

Type of orthodontic elastomeric ligature	Positive control (%)	Negative control (%)	Cell Viability (%)
C-brand	9.74	93.82	51.10
T-brand	8.47	100.99	73.04
U-brand	8.10	104.20	93.04

The results showed the cell viability of C-brand, T-brand and U-brand were 51.10, 73.04 and 93.04%, respectively. C-brand had the lowest cell viability at 24 hours. The cell viability in the U-brand solution was 93.04% which signified that it had the lowest toxicity among all the other samples.

4. Discussion

Elastomeric ligatures are mostly made of polyurethane. They are synthesized by extending a polyester or a polyether glycol or polyhydrocarbon diol with a di-isocyanate [5]. The two main methods of processing the elastomeric ligatures are injection molding and die stamping. Injection molded ligatures are made by injection of liquefied elastomeric material into a mold and curing it. The die-stamped elastomers have been found to be more consistent in mechanical properties. In addition, the incorporation of the additives in the polymer can improve their mechanical properties; for instance, some company may incorporate fillers in the polymer matrix, in a similar manner to that for composite restorative resins in order to improve the strength of the elastomers [6].

Under clinical conditions, it is important that the elastomeric ligature provide sufficient force to maintain full engagement of the archwire within the bracket slot. Many researchers revealed a large force decay during the first 24 hours after placement of the elastomeric ligatures. Factors precipitating the rapid force decay are the oral environment, pH, temperature variations in the oral environment, along with the accumulation of plaque and formation of microbial colonies on the surfaces of the elastomeric liagatures. These factors may affect the structure, surface properties, and the conformation of the polyurethanes. Moreover, water also acts as a plasticizer by weakening intermolecular forces in the polyurethanes, leading to chemical degradation [7]. These polymerization chemicals can pose some risk to patients.

This study chose L-929 cell mouse fibroblasts to represent primary human gingival fibroblasts suggested by ISO 10993-5: 2009(E) [8].

In this study, we used the monolayer [9] of cell culture model and used dye-uptake technique for determining cytotoxicity with spectrophotometry.

Spectrophotometric essay allows rapid and reliable evidence of cell viability to be obtained based on the use of vital stain incorporated by viable cells [10,11]. Dead or damaged cells cannot incorporate vital satin and are thus not recognized on optical reading.

As sterilization is a prerequisite for cytotoxicity essays, ultraviolet radiation [12,13] was performed on each elastic surface used in this study for 40 minutes. It was observed that all elastics exhibited the same color aspect and malleability after sterilization with the UV light.

Cell viability in the tested group (Table 2) showed a range of 51.10-93.40 % which indicated that U-brand and T-brand were non-cytotoxic (cytotoxicity scale < 70%). Plastic coverslip, which was used as the negative control, did not produce any cytotoxic response when tested in accordance to ISO10993-5. On the other hand, the extracts from polyurethane film containing 0.1% zinc diethyldithiocarbamate (ZDEC), which was used as the positive control, showed cell viability of 8.10-9.74% signifying a cytotoxic response.

From the results of the study, it was shown that U-brand had the greatest cell viability (93.04%), whereas and C-brand had the least cell viability (51.10%). It was interesting to observe that all of the tested samples had non-cytotoxicity effects except C-brand. Furthermore, cell morphology of the L929 cells exposed with the solution of C-brand, T-brand and U-brand was also investigated and it was found that structure of the cells changed to a spherical form. This effect may have been related to the inhibition of cell growth after exposure of the cells to the elastomeric ligatures.

The composition and additives of the elastomeric ligatures as well as their manufacturing process may have a profound effect on the cytotoxicity [1]. Polyurethanes, which are the materials used for production of orthodontic elastomeric ligatures, can be dangerous due to the presence of oligomers (polymers of low molecular weight and bearing a short chain) or un-polymerized chains. These are formed during the manufacturing process and can later readily dissolve in the oral cavity [14]. The cell viability of C-brand, T-brand and U-brand were 51.10, 73.04 and 93.04%, respectively. The differences in cell viability may be due to the differences in manufacturing process. Furthermore, the composition of the elastomeric ligatures along with the additives could also play a major role in its cytotoxicity when compared to their manufacturing process. However, further investigations are necessary to validate this effect.

Another concern for cytotoxicity is the effects of polymer biodegradation [15]. Through exposure to light, air, warm temperatures and moisture, polymers suffer mechanical and chemical transformations. The effect of enzymes on polymers are variable and may differ among the products [16]. The accumulation of plaque [17] comprising both aerobic and anaerobic microbes may also degrade the polymers.

As these orthodontic ligatures are commonly used in orthodontics, emphasis should be given regarding their safety and biocompatibility, as they come in close contact with oral tissues. In patients with gingival hyperplasia and potential periodontal problems, the elastomeric ligatures with the lowest cytotoxicity or preferably metal ligatures should be used [18].

5. Conclusion

1) All orthodontic elastomeric ligatures in this study were non-cytotoxic potential except C-brand.
2) Among the tested elastomeric ligatures, U-brand showed the greatest cell viability in the time interval of 24 hours.
3) The cell viability of C-brand was lower than the other samples.

References

1. Thomas M Graber, Theodore Eliades, Athanasiou AE. Risk management in orthodontics. 1 ed. United States: Quintessence Publishing Co, Inc; (2004)
2. Schmalz G. J Dent. 22:S6-11. (1994)
3. Dechkunakorn S, Viriyakosol N, Anuwongnukroh N, Suddhasthira T, Laokijcharoen P, Churnjitapirom P, et al. Advanced Materials Research. 674-80. (2012)
4. Anuwongnukroh N, Jiraanankul N, Dechkunakorn S, Churnjitapirom P, Tua-ngam P. Advanced Materials Research. 315-20. (2013)
5. Wong AK. Angle Orthod. 46 :196-205.(1976)
6. Brantley WA, Eliades T. Orthodontic materials :scientific and clinical aspects 2001, Thieme Stuttgart, New York
7. Huget FH, Reynolds WG. Am J Orthod. 67 :554-62. (1975)
8. Schedle A, Samorapoompichit P, Rausch-Fan XH, Franz A, Fureder W, Sperr WR, et al. J Dent Res. 74(8):1513-20. (1995)
9. Tomakidi P, Koke U, Kern R, Erdinger L, Kruger H, Kohl A, et al. J Orofac Orthop. 61(1):2-19. (2000)
10. Pithon MM, dos Santos RL, Martins FO, Romanos MT, Araujo MT. Aust Orthod J. 26(1):16-20. (2010)
11. Rogério Lacerda Dos Santos, Matheus Melo Pithon, Márlio Vinícius de Oliveira, Gabriella da Silva Mendes, Maria Teresa Villela Romanos, Ruellas ACdO. Journal Oral Sciences. 7(24):1520-6. (2008)
12. Rogério Lacerda dos SANTOS, Matheus Melo PITHON, Gabriella da Silva MENDES, Maria Teresa Villela ROMANOS, RUELLAS4 ACdO. J Appl Oral Sci. 17(4):326-9. (2009)
13. Dos Santos RL, Pithon MM, Martins FO, Romanos MT, de Oliveira Ruellas AC. Orthod Craniofac Res. 13(1):28-33. (2010)
14. Toxicity and safe handling of diisocyanates and ancillary chemicals. A code of practice for polyurethane flexible foam and elastomer manufacture. (2000)
15. Huget EF, Patrick KS, Nunez LJ. J Dent Res. 69(2):496-501. (1990)
16. Santerre JP, Shajii L, Tsang H. J Dent Res. 78(8):1459-68. (1999)
17. Sukontapatipark W, el-Agroudi MA, Selliseth NJ, Thunold K, Selvig KA. Eur J Orthod. 23(5):475-84. (2001)
18. Lacerda Dos Santos R, Pithon MM, Romanos MT. Angle Orthod. 82(4):709-14. (2012)

Detecting Attacks in CyberManufacturing Systems: Additive Manufacturing Example

Mingtao Wu[1], Heguang Zhou[1], Longwang Lucas Lin[1], Bruno Silva[2], Zhengyi Song[1], Jackie Cheung[1] and Young Moon[1]

[1]*263 Link Hall, Department of Mechanical and Aerospace Engineering, Syracuse University, Syracuse NY 13244, USA*
[2]*15 Francisco Alexandre Almeida - Vila Westin, UNIFAE, São João da Boa Vista SP 13870-377, Brasil*

Abstract. CyberManufacturing System is a vision for future manufacturing where physical components are fully integrated with computational processes in a connected environment. However, realizing the vision requires that its security be adequately ensured. This paper presents a vision-based system to detect intentional attacks on additive manufacturing processes, utilizing machine learning techniques. Particularly, additive manufacturing systems have unique vulnerabilities to malicious attacks, which can result in defective infills but without affecting the exterior. In order to detect such infill defects, the research uses simulated 3D printing process images as well as actual 3D printing process images to compare accuracies of machine learning algorithms in classifying, clustering and detecting anomalies on different types of infills. Three algorithms - (i) random forest, (ii) k nearest neighbor, and (iii) anomaly detection - have been adopted in the research and shown to be effective in detecting such defects.

1 Introduction

CyberManufacturing System is a vision for future manufacturing where physical components are fully integrated with computational processes in a connected environment. CyberManufacturing system utilizes recent developments in Internet of Things, Cloud Computing, Fog Computing, Service-Oriented Technologies, Modeling and Simulation, Virtual Reality, Embedded Systems, Sensor Networks, Wireless Communications, Machine Learning, Data Analytics, and Advanced Manufacturing Processes. Manufacturing resources and abilities can be sensed and connected to each other directly or through the Internet, thus enabling intelligent behaviors of a manufacturing system or networked manufacturing systems such as self-awareness, self-prediction, self-organization, and self-configuration [1]. The concept can also extend to a network of shared CyberManufacturing resources over the Internet.

However, realizing the vision requires that its security be adequately ensured. A recent study by IBM reports that manufacturing is the second most frequently targeted industry in 2015 in terms of the number of cyber-attacks [2]. The consequence of cyber-attack on manufacturing systems could be defected parts, the cost of poor quality, damaged equipment, and loss of customer loyalty or even create customer safety issues [3].

Particularly, additive manufacturing systems have unique vulnerabilities to malicious attacks, which can result in defective infills but without affecting the exterior. This can lead to the production of malicious defective parts without any warning. Those defects can also result in a reduction in yield load, reduction in strain at failure, change of natural frequency, etc. that eventually affect product's safety and functionality. Image classification using machine learning techniques is adopted to detect such defects caused by deliberate attacks, from normal parts as well as common defects such as weak infill, gaps in thin walls, inconsistent extrusion, layer separation and splitting, and bed drop.

The previous research by Wu et al. [3] defined five different infill defect patterns: Seam, Irregular Polygon, Circle, Rectangle, and Triangle. The research used Naive Bayes Classifier and J48 Decision Trees classification algorithms to differentiate the defective and non-defective parts. The images were simulated from MakerBot software with the setting of Honeycomb infills. The result showed Naive Bayes Classifier has an accuracy of 85.26% and J48 Decision Trees has an accuracy of 95.51% for correct image classification.

This work builds on the previous research [3], but extends further by:

1) using real images collected by an Arducam OV2640 2 Megapixels Camera. Relevant changes are made to address the differences between the simulated images and real images.
2) using unsupervised machine learning methods in addition to supervised marching learning methods, thus covering not only classification but also clustering and anomaly detection mechanisms;
3) testing four additional types of infills (Diamond, Linear, Star and Catfill) in addition to a single

DOI: 10.1051/matecconf/201710806005

type of infill (Honeycomb), to test the infill type factor on accuracy of machine learning methods.

2 Image collection

Two methods have been used in collecting images: (i) by generating software simulated images, and (ii) by capturing real images using a real-time camera system that was designed and installed in a 3D printer to collect images under real environment.

2.1 Images from simulation

Simulated images have been generated from 3D printing software, MakerBot Desktop 3.9.1 preview function. For the feature extraction stage, a fixed size of images at 512 x 512 pixels was used. In total 3887 images have been generated for training classifier and testing. 532 images of non-defect parts have been generated and labeled as group A. Within group A, 34 images are from parts with honeycomb infill, 60 images are from parts with linear infill, 151 images are from parts with star infill, 140 images are from parts with catfill infill, and 147 images are from parts with diamond infill. Fig 1. shows five types of non-defective part infill. 3355 images of defective parts are captured and labeled as group B. Within group B, 102 images are from parts with honeycomb infill, 349 images are from parts with linear infill, 956 images are from parts with star infill, 946 images are from parts with catfill infill, 1002 images are from parts with diamond infill. Both the non-defective group A images as well as the defective group B images have been captured every 3-5 layers during the printing process, with infill density varying from 8%-12% to increase the diversity of training image.

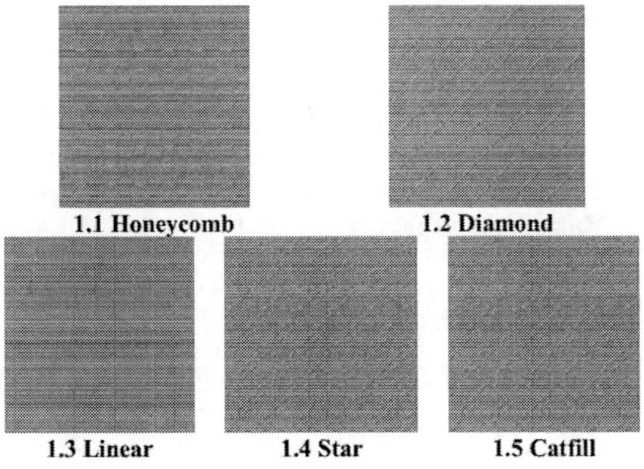

Figure 1. Five types of part infill.

2.2 Images from camera

To test and verify the image classification method in real environment, a camera-based vision detection system has been designed and installed on two different brands of 3D printers: MakerBot Replicator^{TM}2 and CubePro Duo. The camera used in the experiment is an Arducam Mini Module Camera Shield with OV2640 2 Megapixels Lens, compatible with Arduino UNO Mega2560 Board. The camera unit's dimensions are 3 x 2 x 1 inches, connected to the Arduino UNO via extended jumper wires. With programming in Arducam software, it can produce images any size scaling down from SXGA to 40×30 in jpeg format. As a result, the feature extraction code for previous 512x512 size images needed to be altered. Moreover, because the camera provides a color image instead of grayscale image from simulation, additional image preprocessing was necessary.

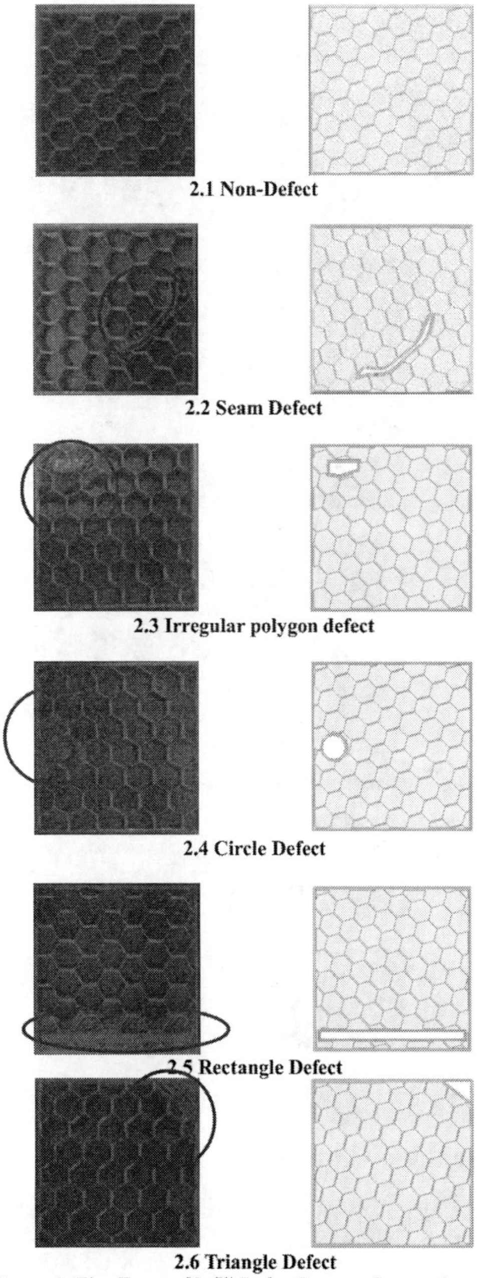

Figure 2. Five Types of Infill Defect Patterns Camera & Simulation View.

MakerBot Replicator^{TM}2 has the build envelope of 11.2 x 6.0 x 6.1" and can print at 100 μm per layer.

DOI: 10.1051/matecconf/201710806005

Installation of the camera on a MakerBot Replicator 2 is shown in Fig 3. There are two ways to install on MakerBot Replicator 2. One is mounting the camera right next to the intruder and move along with it, called 'moving camera.' The other is mounting the camera on the frame of the 3D printer, called 'static camera.' The CubePro Duo is capable of printing models up to 9 x 10.75 x 9.5" at the resolution of as fine as 70 μm. The camera is mounted on a cooling fan frame of CubePro Duo located next to the first intruder.

3.1 Moving camera mount

3.2 Static camera mount

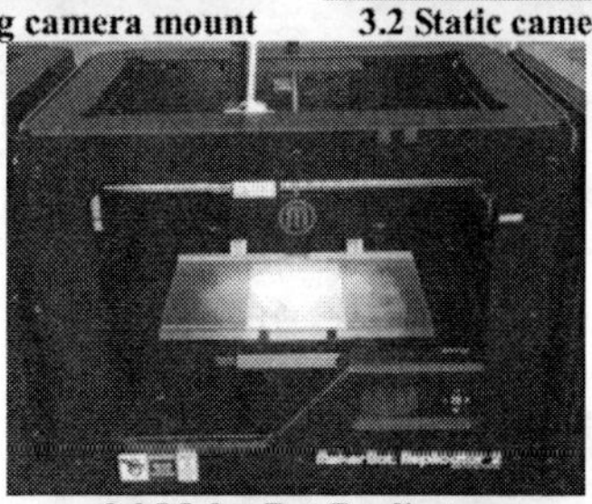
3.3 MakerBot Replicator

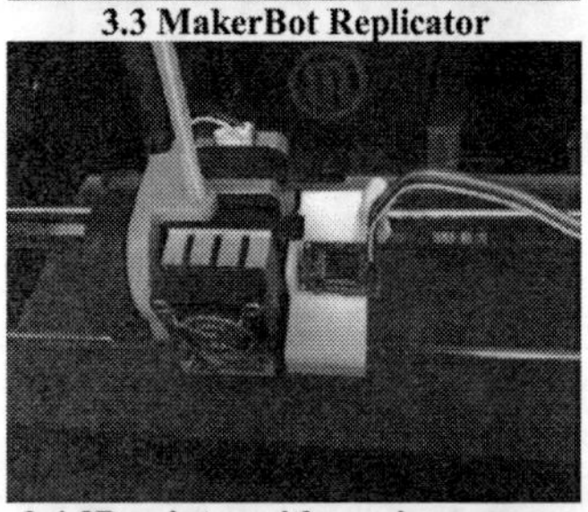
3.4 3D printer with moving camera

3.5 3D printer with static camera

Figure 3. MakerBot Replicator with moving and static camera

3 Machine learning methods

To improve the accuracy for malicious defect detection, this work extends from supervised machine learning methods to unsupervised machine learning methods. The reason is that even if a large number of sample experiments are used as training set, it is impossible to predict unseen malicious attack patterns. By implementing random forest, k-nearest neighbors (kNN) and anomaly detection machine learning algorithms, the research attempts to explore the optimal methods for 3D printing vision detection system.

Random Forest: A random forest classifier consists of a number of trees. The leaf nodes of each tree are labeled by estimating the posterior distribution over the image classes. Each internal node contains a test that best classifies the data. An image is classified by sending it down each tree and aggregating the reached leaf distributions [4]. In this research, three decision trees are used and each of them has five leaf nodes to classify. Compared to J48 decision tree algorithm, the random forest classifier with three decision trees was able to achieve a high accuracy with relatively shorter time to run the classifier.

k-Nearest Neighbors (kNN): kNN classifier is used to perform discriminant analysis when reliable parametric estimates of probability densities are unknown or difficult to determine [5]. K value is a positive integer, typically small, but normally larger than one. In this research, a heuristic rule that is k is calculated as the square root of "n" number of features, which is very likely to achieve the highest accuracy.

Anomaly detection: Anomaly detection techniques have been used to detect any changes in an image over time (motion detection) or in regions which appear abnormal on the static image [6]. Images from moving camera are often not so clear. Therefore, in this research, the anomaly detection method is combined with the random forest method to increase the accuracy. By embedding the anomaly detection method in the random forest method, detection accuracy was increased almost by 5% for the moving camera cases as shown in Table 2.

4 Result analysis

The analysis for the research focuses on three main issues: i) accuracy of machine learning methods on both simulated and real images, ii) whether a certain infill type has particular influence on accuracy of detection, and iii) whether the vision detection system can maintain acceptable accuracy levels in real environment and can be adopted for real-time detection applications. Accuracy is defined by the equation (1). Where *TruePositive* means images in class A that are predicted as class A, and *TrueNegative* stands for images in class B predicted as class B.

$$Accuracy = \frac{TurePositive + TureNegative}{Total} \quad (1)$$

According to (1), the accuracy for the Simulated Linear Infill Random Forest classification example in Table 1 is calculated as:

$$Accuracy = (57+330)/(57+3+19+330) = 94.6\%$$

DOI: 10.1051/matecconf/201710806005

Table 1. Confusion Matrix for Simulated Linear Infill Random Forest Classification Results.

Class		Predicted Class	
		A	B
Actual Class	A	57	3
	B	19	330

Table 2. Accuracy results (HC-Honeycomb, DM-Diamond, LN-Linear, ST-Star, CA-Catfill)

Accuracy		Machine Learning Method		
		Random Forest (%)	kNN (%)	Anomaly Detection (%)
Image from Simulation	HC	88.4	81.3	100.0
	DM	100.0	85.0	100.0
	LN	94.6	92.5	100.0
	ST	97.8	100	99.8
	CA	91.5	100	100.0
Image from Moving Camera	HC	68.4	68.75	72.5
Image from Static Camera		95.5	87.5	96.1

As shown in Table 2, images from simulation show that different type of infill have some influences on accuracy of kNN, but still within an acceptable range. The reason behind 100% accuracies by the anomaly detection method could be the simulated images are captured as clean image without any distortion. For the simulated data sets, the anomaly detection method still achieved the best accuracy. However, the images from 'moving camera' result in less accuracy compared to images from 'static camera.' This is because the machine learning system was trained based on simulated images that were captured from a full view of the object, as shown in **Fig. 4.1**, while the moving camera images contain only a part of the object as shown in **Fig. 4.2**. Also, the static view images are clearer because the moving camera creates blur images as illustrated in **Fig. 4.3** that decreases accuracy. The images captured by static camera as shown in **Fig. 4.4** has resulted an accuracy of 96.1%.

5 Conclusion

This work aims to detect malicious infill structure during 3D printing processes using machine learning techniques. This work is based on Wu et al. [3] but extends it with i) additional machine learning methods, ii) more types of infills and iii) use of a camera in collecting and processing real images from 3D printing machines, and iv) application of machine learning to detect malicious defect in real images. The steps of installation of a camera, image processing and detection results shows the potential for implementing this 3D printing vision-based detection system in a real CyberManufacturing Systems. The following can be achieved in future: i) research on the influence on the detection accuracy by different colors and types of materials, ii) implementation of the system in differentiating natural defects in 3D printing process rather than malicious defects, and iii) additional systems and methodologies to adopt for ensuring security in CyberManufacturing systems.

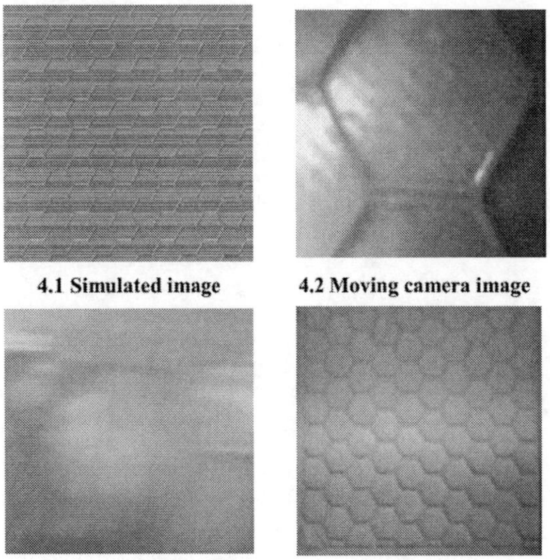

4.1 Simulated image **4.2 Moving camera image**

4.3 Blur Moving camera image **4.4 Static camera image**

Figure 4. Sample images.

References

1. Z. Song and Y.B. Moon, *Proceedings of the IFIP 13th International Conference on Product Lifecycle Management*, (in press).
2. IBM, *IBM X-Force Research Report*, 2016.
3. M. Wu, V. V. Phoha, Y. B. Moon, and A. K. Belman, *Proceedings of the ASME 2016 International Mechanical Engineering Congress and Exposition*, (in press).
4. A. Bosch, A. Zisserman, and X. Munoz, *2007 IEEE 11th International Conference on Computer Vision*, 1–8, (2007)
5. L. Peterson, *K-nearest neighbor*, Scholarpedia, 2009. [Online]. Available:http://www.scholarpedia.org/article/K-nearest_neighbor. [Accessed: 23-Jun-2016].
6. V. Chandola, A. Banerjee, and V. Kumar, ACM Comput. Surv., **41-3**, 1–58, (2009)

Empirical Study on the Measurement of the Technical Efficiency of Jilin Province's Five High-tech Industries

Jie Yao , Huan Ma

School of Economy and Management, Northeast Dianli University, 132012 Jilin, China

Abstract. This paper first analyzed the industrial distribution of Jilin province's high-tech enterprises, and then selected 13 input and output indicators based on analyzing the influence factors of high-tech industry. Through the correlation analysis between each indicator's relevant data, the paper constructed the technical efficiency evaluation model of Jilin Province's high-tech industries including 12 indicators. After that it measured the technical efficiency of Jilin Province's five high-tech industries from 2004 to 2013 by using DEA analysis method. Finally, this paper drew conclusions and gave suggestions.

1 Introduction

Since the 21st century, China's high-tech industry has made great achievements and become the pillar industry in China's national economy. According to the classification method in "China statistics yearbook on high technology" , high-tech industry can be divided into five major industries, including manufacture of medicines, manufacture of aircrafts and spacecrafts, manufacture of electronic equipment and communication equipment, manufacture of computers and office equipments, manufacture of medical equipments and measuring instrument.

At present, many scholars have carried out research work on the related issues of high-tech industry.

Some scholars have done some research work on the technical efficiency of high-tech industry, for example, Liu Junling (2016) has done empirical study on the efficiency of China's high-tech industry based on PCA and three-stage DEA model [1]. Based on factor analysis and data envelopment analysis method, Qi Yuan etc. (2014) have made empirical study on the technical efficiency of Beijing's high-tech industry from 1996 to 2010. Firstly, the index systems are established through methods of literature research and factor analysis, and then the technical efficiency is measured based on data envelopment analysis on Malmquist productivity index and C^2R model respectively. The results show that the combination method of factor analysis and data envelopment analysis is feasible, and the technical efficiency of Beijing has reached efficiency since 1999 [2]. Feng Peng ect. (2012) examined the relationship between technology imports, R&D investment and technical efficiency of Chinese high-tech industry using provincial panel data for the period between 2000 and 2009[3].

Some scholars have done some research work on the innovation efficiency and R&D efficiency of high-tech industry. For example, Zhao Shukuan etc. (2013) have constructed the evaluation index system and model of the innovation efficiency of high-tech enterprises, and evaluated the innovation activities of 151 high-tech enterprises in Jilin Province from four aspects, including efficiency, effectiveness, returns to scale and projection analysis [4]. Based on value chain perspective, Yu Yongze and his partners (2010) have divided the research and development process of high-tech industry into two stages, technology development and transformation of technological achievements, by using DEA model with slack variables, and have done empirical study on its efficiency and influence factors [5]. Han Jing (2010) analyzed the innovation efficiency of China's high-tech industry by using SFA method [6]. Huang Shun and Guan Yan (2010) have divided the whole process of technological innovation into two stages, including technology output and technology transformation, and have measured the technological innovation efficiency of China's high-tech industry from 1999 to 2007 by using two-stage DEA model [7]. Guan Jiancheng and Chen Kaihua (2009) have measured the technical efficiency, pure technical efficiency, scale efficiency and scale state of the technological innovation activities of China's high-tech industry by using data envelopment analysis model with slack variables and critical efficiency measurement model [8].

And some other scholars have done some research work on the competitiveness of high-tech industries in various regions. For example, Liu Yannan ect. (2011) have done empirical analysis on the competitiveness of Liaoning Province's High-tech Industry based on factor analysis method [9]. Liu Zhaode etc. (2009) have comprehensively evaluated the competitiveness of

Guangdong Province's and other 11 provinces' High-tech industry [10].

Due to the relatively few studies on the five major high-tech industries, this paper took it as a breakthrough point, and made empirical study on the technical efficiency of Jilin Province's five high-tech industries by using data envelopment analysis method.

2 Industrial distributions of High-tech enterprises in Jilin Province

After years of development, high-tech enterprises in Jilin province have taken shape, many enterprises have developed rapidly, and renowned both at home and abroad.

A number of pharmaceutical companies, such as Jilin Aodong Pharmaceutical Ltd, Xiuzheng Pharmaceutical Ltd, Jilin Changbai Mountain Medicine Company, and so on, have made Jilin become a large pharmaceutical manufacture province ahead of other provinces.

The enterprises, such as First Automotive Works, Huawei Electronics Ltd, Jilin Chemical Industry Group Corporation, Shenhua New Energy Corporation, and so on, have maintained strong growth momentum so that Jilin province has reached the leading domestic level in automobile manufacturing industry, microelectronics, chemical industry and new energy development.

The total number of high-tech enterprises in China has been 26894 by the end of 2013. And 394 enterprises of them are located in Jilin Province, accounting for only 1.47% of the total. Therefore, there is still a gap from the national average level.

The industrial distribution of high-tech enterprises in Jilin Province from 2007 to 2013 has been shown in Fig. 1.

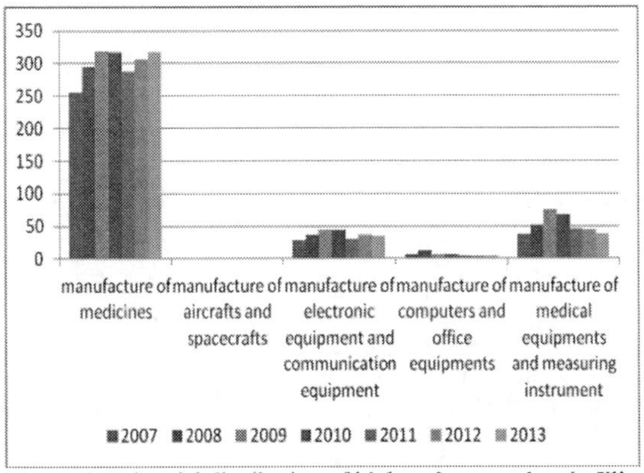

Figure 1. Industrial distribution of high-tech enterprises in Jilin Province.

It can be seen from Fig. 1 that the high-tech enterprises of Jilin Province mainly concentrate in the following industries, including manufacture of medicines, manufacture of electronic equipment and communication equipment, manufacture of medical equipments and measuring instrument. There are few enterprises in the next two industries, manufacture of aircrafts and spacecrafts, and manufacture of computers and office equipments. Therefore, the development of these two industries is slow.

3 Construction of technical efficiency evaluation model

3.1 Model construction method

In this research, data envelopment analysis (DEA) method was used. Although DEA methods are diverse, but generally there are two main types.

One is to set the output value in advance, and under this premise the purpose of the research work is to compute the input proportion which can be reduced to the utmost. This method is also known as input-oriented DEA method.

The second is just opposite to the first one. Considering input as fixed value, it wants to see how much of the output can be increased proportionally, so it is known as output-oriented DEA method.

Considering the actual situation, this article selects the output-oriented BCC model to measure and evaluate the technical efficiency of Jilin province's high-tech industry. BCC model is a very practical and effective efficiency measurement model. It can complete the measurement of the technical efficiency with multiple input and output indexes, so it is suitable for measuring high-tech industry' technical efficiency.

3.2 Construction of evaluation model

Considering the characteristics of high-tech industry, this paper argues that the development of high-tech industry is mainly influenced by the following four aspects, including its economic capacity, technological innovation capability, the support environment and its development scale.

During the process of selecting indexes, we mainly considered about the following two aspects. On one hand, the indexes can measure the actual capacity of high-tech industry comprehensively so that the government can predict the prospects of high-tech enterprises. On the other hand, they can be used in the dynamic study of technical efficiency and then be optimized.

According to the principle of index selection, among the factors affecting the development of high-tech industries, the input indexes selected include R&D Full-time Equivalent (man-year), Expenditure on New Product Development (100 million RMB yuan), Number of S&T Institutions (unit), Expenditure for Acquisition of Foreign Technology (100 million RMB yuan), Expenditure on Technical Renovation (100 million RMB yuan), Number of Enterprises (unit), Annual Average Number of Employed Personnel (person), Newly Increased Fixed Assets (100 million RMB yuan), which are represented by X_1, X_2, X_3, X_4, X_5, X_6, X_7, X_8 respectively.

And the output indexes selected include Revenue from Principal Business (100 million RMB yuan), Taxes and Profits (100 million RMB yuan), Exports (100 million RMB yuan), Sales Revenue of New Products

(100 million RMB yuan), Owning Inventive Patent (unit), which are represented by Y_1, Y_2, Y_3, Y_4,Y_5 respectively.

The input and output indexes above were made correlation analysis first by using SPSS22.0 software to ensure that those indexes final selected are logically correlated but not numerically correlated. The analysis result shows that the input index X_4 is not suitable, so we eliminate this factor and establish the evaluation model finally, as shown in Fig. 2.

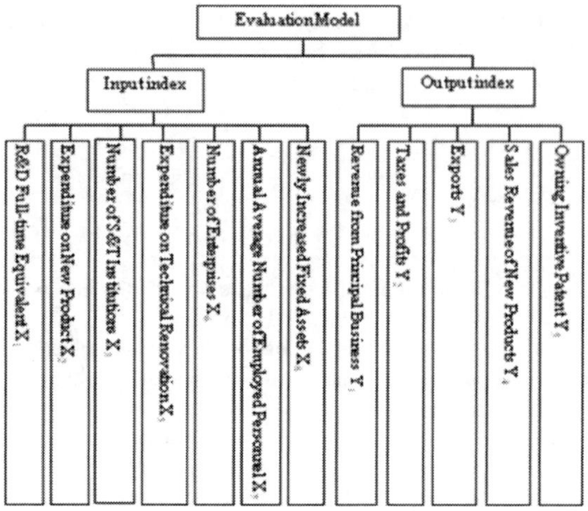

Figure 2. Evaluation model of the technical efficiency of Jilin Province's high-tech industry.

4 Empirical analysis of the technical efficiency of Jilin Province's five major High-tech industries

According to the evaluation model established, the paper analyzed each index's corresponding data of the five high-tech industries in Jilin Province from 2004 to 2013.

By using deap2.1 software, this paper finally got the measure results of technical efficiency, pure technical efficiency and scale efficiency of Jilin Province's five high-tech industries. (Note: The data are sourcing from "Statistical Yearbook of China's High-tech Industry from 2005 to 2014".)

4.1 Measure result of technical efficiency

The measure result of technical efficiency of Jilin Province's five major high-tech industries from 2004 to 2013 is shown in Fig. 3.

It can be seen from Fig. 3 that, under the condition of constant scale, the technical efficiency of manufacture of computer and office equipment has maintained to be 1, showing that this industry has been continuously technical effective among all the five major high-tech industries of Jilin Province.

The technical efficiencies of manufacture of aircrafts and spacecrafts, and manufacture of electronic equipment and communication equipment are also very high. They have not realized effective technical efficiency only once in ten years, with the average value to be 0.99 and 0.96 respectively.

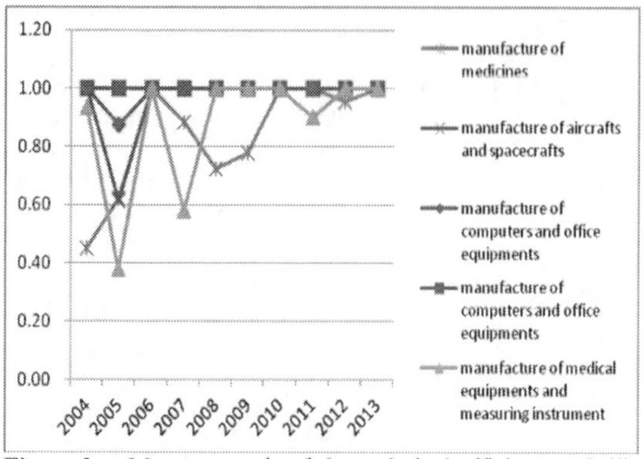

Figure 3. Measure result of the technical efficiency of Jilin Province's five major high-tech industries

Only the technical efficiency of manufacture of medicines and manufacture of medical equipments and measuring instrument are not stable. Although these two industries have realized effective technical efficiency for four years and six years respectively during the period of ten years, but they have shown a trend of fluctuations in general, indicating that these two industries still need to be further developed and enhanced.

4.2 Measure result of pure technical efficiency

The measure result of pure technical efficiency of Jilin Province's five major high-tech industries from 2004 to 2013 is shown in Fig. 4.

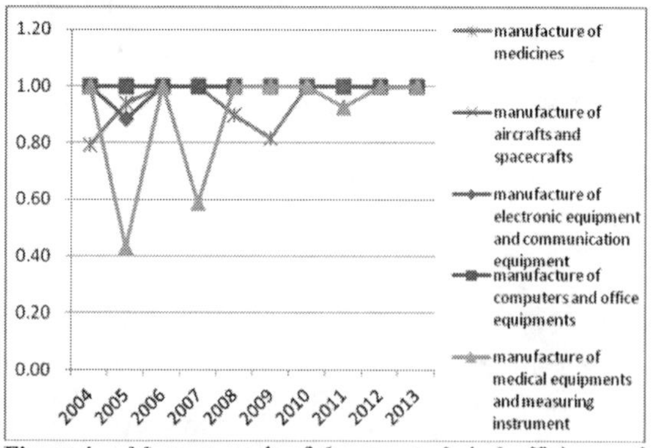

Figure 4. Measure result of the pure technical efficiency of Jilin Province's five major high-tech industries

As shown in Fig. 4, under the condition of variable scale, all the others of Jilin Province's five major high-tech industries have basically maintained to be effective in technical efficiency with their average efficiency value to be close to or equal to 1, except the following two industries, manufacture of medicines and manufacture of medical equipments and measuring instrument, with their average efficiency value to be 0.94 and 0.89 respectively.

It shows that, under the condition of variable scale, the technical efficiency of these three industries has been configured reasonably, the resources have been utilized effectively, and they have basically realized the sustainable developing mode.

4.3 Measure result of scale efficiency

The measure result of scale efficiency of Jilin Province's five major high-tech industries from 2004 to 2013 is shown in Figure 5.

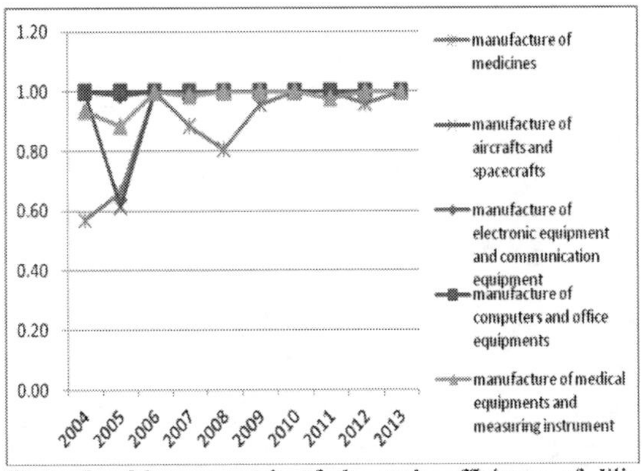

Figure 5. Measure result of the scale efficiency of Jilin Province's five major high-tech industries

It can be seen from Fig. 5 that, during the ten years, the scale efficiency value of the following two industries, manufacture of electronic equipment and communication equipment, and manufacture of computer and office equipment, has maintained to be 1, showing that these two industries have been continuously effective in scale.

The scale efficiency value of the next two industries, manufacture of medical equipments and measuring instrument and manufacture of aircrafts and spacecrafts, is also very high, with the average value to be 0.98 and 0.96 respectively.

Only the average value of manufacture of medicines is relatively low, it is only 0.88. And the overall volatility is relatively clear, indicating that the development of this industry is not stable enough.

5 Conclusion

Based on the empirical research above, this paper draws the following conclusions.

First, despite the good developing momentum of Jilin Province's high-tech industry, there still exists a phenomenon of wasting resources. Therefore, the government needs to formulate corresponding policies that are helpful for the development of high-tech industry, and each enterprise also needs to adjust its management system and resource inputs, striving to maintain a fast, scientific and sustainable developing trend.

Second, both technical efficiency and scale efficiency have been realized in the industries of manufacture of electronic equipment and communication equipment, and manufacture of computer and office equipment. So these two industries should continue to maintain their current developing level and further consolidate their pillar status in the development of Jilin Province's high-tech industries.

Third, manufacture of medicines and manufacture of medical equipments and measuring instrument are the weak links of Jilin Province's high-tech industries. With larger scale but insufficient output, these two industries should be paid close attention to and put under effective control and management by the government.

Finally, although manufacture of aircrafts and spacecrafts has maintained to be technically efficient since 2006, but its overall scale is still small. Therefore, the government should increase investment in this industry in order to support its development.

Acknowledgment

This research was financially supported by the Social Science Foundation of Jilin Province "Empirical study on the influence factors and efficiency measurement of wind energy development and utilization in Jilin Province" (Grant NO.2016JD58) and the 12th Five-Year planning project of Jilin Province's Education Department "Empirical Study on the Measurement and Evaluation of the Technical Efficiency of Jilin Province's High-tech Industry" (Grant NO.2015-171).

References

1. L. Junling, Liaoning Tech. Univ. J. (Soc. Sci. E.) **01**, 25-32 (2016)
2. Q. Yuan, Z. Yongan, Sci. and Tech. Mgmt. Res. J. **04**, 43-48 (2014)
3. F. Peng, L. Li, Proc. of 2012 IEEE 19th Int. Con. on Ind. Eng. and Eng. Mgmt. (IE&EM 2012) **10**, 2623-2625 (2012)
4. Z. Shu Kuan, Y. Haiqing, G. Shunlong, Sci. Res. Mgmt. J. **02**, 36-43 (2013)
5. Y. Yongze, W. Peng, L. Jianbing, Sci. Learn and Mgmt. of Sci. and Tech. J. **05**, 60-65 (2010)
6. H. Jing, Sci. Res. J. **03**, 467-472 (2010)
7. H. Shun, G. Yan, Indus. Tech. Econ. J. **01**, 92-97 (2010)
8. G. Jiancheng, C. Kaihua, QTY Econ. and Tech. Res. J. **10**, 19-33 (2009)
9. C. Hongchuan, Soft Sci. J. **08**, 21-23 (2010)
10. L. Zhaode, Z. Weiguo, H. Mingliang, Tech. and Econ. J. **01**, 46-49 (2009)

Analytical Investigation on the Damping Performance of Atorque Converter in an Automotive Driveline Model

Pınar Sayın and Jürgen Schöftner

Johannes Kepler University, Technical Mechanics Department, Linz, Austria

Abstract. Torque converters are avital part of automatic transmission vehicles. In this paper the influence of a torque converter on the lowest mode of driveline vibrations (shuffle mode) is investigated. The automotive driveline is modelled by lumped parameter models, in which a steady-state torque converter model is inserted between the engine and the gearbox. A detailed mathematical model of a two-degree of freedom and a three-degree of freedom dynamic systemis given. The resulting nonlinear differential equations are linearized to obtain system transfer functions and to study damping characteristics of the torque converter.Additionally, a dynamic system model with nonlinear equations is developed in *MATLAB/Simulink* to verify the linearized system. The simulations show that the high frequency vibrations from the engine side are damped due to the torque converter and the linearized model predicts the system response accurately.

1 Introduction

Torque converters play an important role in the driving comfort, fuel economy and performance of automatic transmission vehicles. Therefore, they have been investigated analytically, numerically and experimentally by many researchers in the last decades.

Hrovat et al. [1] derived four first order nonlinear differential equations by using moment of momentum equations for pump, turbine and stator inertias, including also the energy balance equation for the fluid inertia along with the power losses. In the study the forward flow mode and the reverse flow mode equations were derived. An analysis with the bond graph theory was conducted and verified by experiments. The given equations are suitable for modelling the transient characteristics of the torque converter and investigating the parameter sensitivity. Barglazan [2] presented a theoretical model by considering the torque converter as a machine with lumped parameters. In the study, harmonic sinusoidal perturbations on the parameters of torque converter were applied and consequently a linearized dynamic model was obtained. Inthe study by Walker [3] dual clutch transmission powertrain models with 4DOF (degree of freedom) and 15 DOF were compared to study the influence of engine torque harmonics on the transient response. Free vibration analysis and shift transient simulations were conducted. By using a lower order model, only the lowest natural frequencies have been seen in the powertrain response. Rosbi et al. [4] studied the effect of a dynamic absorber to reduce shudder in automatic transmission powertrain. A linear multi-degree offreedom system has been used to conduct a vibration analysis and the results were compared with the experiments. The shudder occurred in fifth gear of a six-speed transmission in the experiments; therefore, a theoretical model of the fifth gear for the powertrain with 13 DOF was built. By using an optimized dynamic absorber, the shudder was entirely suppressed.

The numerical modelling approach was utilized in the following works. Asl et al. [5] presented the damping characteristics of a dynamic torque converter for the torque multiplication (forward flow), coupling and reverse flow modes by *MapleSim*. The results showed that the high frequency disturbances are effectively damped from the engine/pump side to the transmission/turbine side in all three operation modes. In the work of Asl et al. [6] a math-based torque converter in the automatic driveline with a mean value engine model was modelled. By using the input of the throttle angle, the fuel consumption and mechanical power delivered to the pump of the torque converter was obtained. The gearbox is modelled by simple input-output torque with variable gear ratios. The vehicle's longitudinal dynamics were modelled considering the vehicle mass, final drive ratio, aerodynamic drag and rolling resistance. The study investigated the effect of a lock-up clutch on powertrain efficiency and ride comfort. Moreover, the engine braking and coasting phenomena were simulated.

Some researchers did experimental work on this subject such as:Castiglione et al. [7] investigated the transient torque converter behaviour and compared with the steady-state response measured in experiments. The main purpose of the study was the development of an empirical model which can describe the transient behaviour. The simulations were based on the nonlinear equations. Six different torque converters were tested with a test bench and reverse engineering was performed for the nonlinear equation parameters like blade angles and design path radiuses. The experiments showed similar transient performance characteristics with the simulations. A linearized frequency response study was also conducted. The transient fluid moment inside the torque converter has only been seen when the speed ratio near or above coupling point and torsional frequency was between 1 to 10 Hz.

2 Objectives

The present work uses analytical and numerical methods to study the damping characteristics of a torque converter in a linear multi-degree of freedom driveline model. In the analysis, a reduced driveline model with a steady-state torque converter model is considered to evaluate the system response. The torque converter is modelled by characteristic performance curves. Only the low frequency domain, which affects the shift comfort, has been investigated. This is achieved by linearizing the nonlinear differential equations around an equilibrium point. The performance of the linearized model is compared with dynamic simulations of the reduced nonlinear driveline model.

3 Driveline model

Asimple driveline model consisting of a steady state engine model, a torque converter, a two-gear transmissionand anoutput shaft is shown in Figure 1.

For the further analytical investigations; we will take advantage of a simpler, i.e. reduced order model with two- and three-degree of freedom (DOF). These models are explained in detail in sections 3.1 and 3.2.

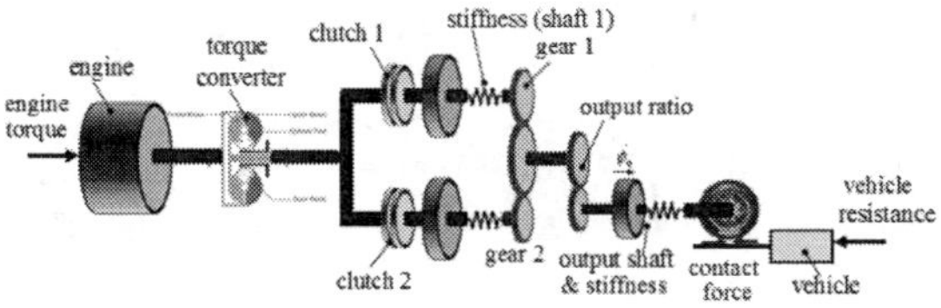

Figure 1. Driveline model.

3.1 Two-degree of freedom model

A reduced model of the driveline and the torque converter with two-degree of freedom is shown in Figure 2. The moment of inertias of the engine and of the pump side torque converter (primary side) are rigidly connected. The turbine side torque converter (secondary side) is connected to the mass of the vehicle, the wheels, the output shaft and the gearbox mass. These elements are combined and reduced to the shaft of the torque converter.

The 2DOF model is the most simple driveline model. Since no flexibility is taken into account (unlike the 3 DOF described in section 3.3), the fundamental torsional mode of the driveline does not occur anddriveline oscillations cannot be simulated with this model.

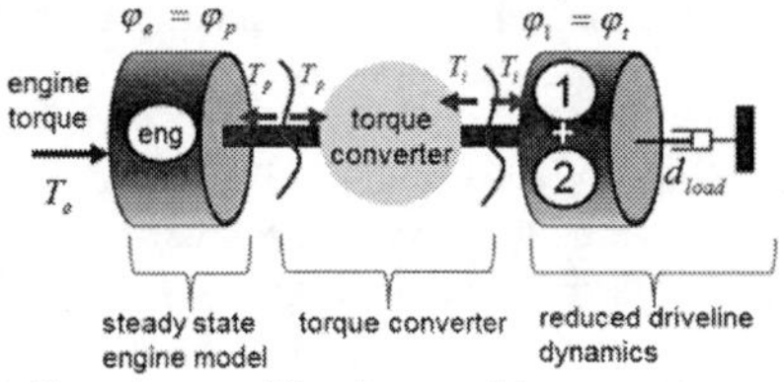

Figure 2. Two-degree of freedom model.

In general, this system can be described in the form of the differential equation system given in (1): M is the inertia matrix, D is the damping matrix, F is the external load matrix and φ is the angular displacement vector. In the equations, a single underscore denotes a vector and a double underscore denotes a matrix.

$$\underline{\underline{M}}\,\underline{\ddot{\varphi}} + \underline{\underline{D}}\,\underline{\dot{\varphi}} - \underline{F} = 0 \quad (1)$$

As given in Figure 2, the engine and the torque converter pump are rigidly connected and the angular displacement of the engine (φ_e) and the torque converter pump (φ_p) are equal. Similarly, the angular displacement of the reduced mass (φ_1) is equal to the angular displacement of turbine (φ_t). The governing equations defining the system in Figure 2 are presented in (2): I_e is the engine inertia, I_p is the torque converter pump inertia, I_t is the torque converter turbine inertia, I_1 and I_2 are the reduced driveline inertias, T_e is the engine torque, T_p is the torque converter pump torque, T_t is the torque converter turbine torque and d_{load} is the load damping constant.

$$\begin{aligned}
&\left(I_e + I_p\right)\ddot{\varphi}_e - T_e + T_p = 0 \\
&(I_t + I_1 + I_2)\ddot{\varphi}_1 + d_{load}\dot{\varphi}_1 - T_t = 0 \\
&\varphi_e = \varphi_p \\
&\varphi_1 = \varphi_t
\end{aligned} \quad (2)$$

3.2 Three-degree of freedom model

The three-degree of freedom model is shown in Figure 3. Thedriveline has flexibility, i.e. k_{12}. It is an extension of the 2 DOF model by separating the secondary side of the torque converter and the reduced vehicle mass by a spring.Therefore, this model is able to simulate driveline vibrations.

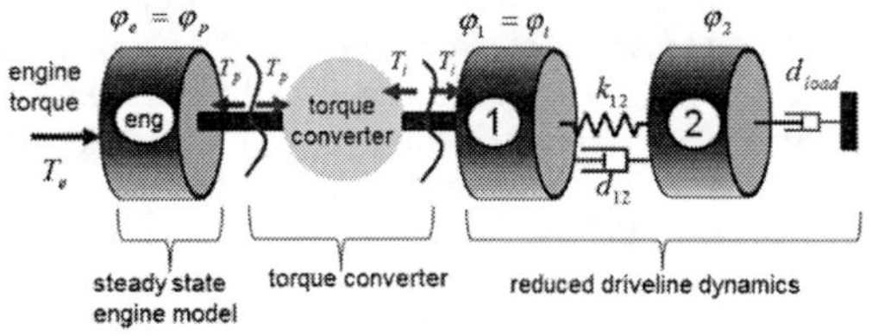

Figure 3. Three-degree of freedom model.

The system matrix is given in (3): In addition tothe inertia, M, the damping, D, and the external load, F, matrices, a stiffness matrix, K, is added to the system.

$$\underline{\underline{M}}\ddot{\underline{\varphi}}+\underline{\underline{D}}\dot{\underline{\varphi}}+\underline{\underline{K}}\underline{\varphi}-\underline{F}=0 \quad (3)$$

The governing equations defining the system in Figure 3 are presented in (4): φ_2 is the angular displacement of the reduced vehicle mass, k_{12} is the reduced stiffness constant and d_{12} is the reduced damping constant between the secondary side of the torque converter and the reduced vehicle mass.

$$\begin{aligned}
&\left(I_e+I_p\right)\ddot{\varphi}_e-T_e+T_p=0\\
&I_1\ddot{\varphi}_1+d_{12}\left(\dot{\varphi}_1-\dot{\varphi}_2\right)+k_{12}\left(\varphi_1-\varphi_2\right)-T_t=0\\
&I_2\ddot{\varphi}_2+d_{12}\left(\dot{\varphi}_2-\dot{\varphi}_1\right)+k_{12}\left(\varphi_2-\varphi_1\right)+d_{load}\dot{\varphi}_2=0 \quad (4)\\
&\varphi_e=\varphi_p\\
&\varphi_1=\varphi_t
\end{aligned}$$

3.3 Torque converter model

The easiest mathematical model of a torque converter is via characteristic maps, i.e. the torque ratio and the capacity factor curves.The characteristic curves for the torque ratio, $r(w)=T_t/T_p$, and the capacity factor, $\kappa(w)=\dot{\varphi}_p/\sqrt{T_p}$, are functions of the speedratio, $w=\dot{\varphi}_t/\dot{\varphi}_p$, which is the ratio of the turbine speed to the pump speed of the torque converter. A typical characteristic map of an automotive torque converter is shown in Figure 4.

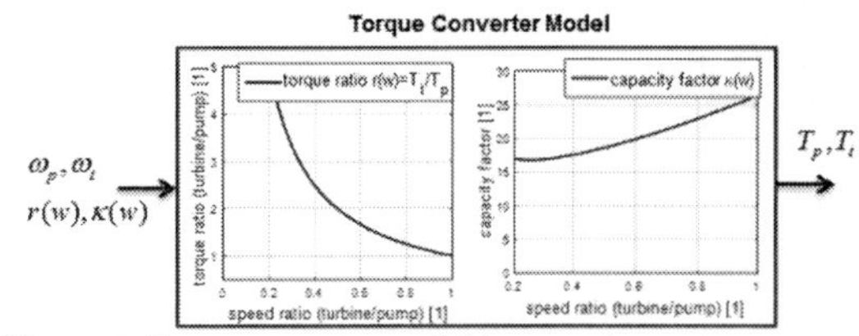

Figure 4. Torque converter model inputs and outputs.

The inputs of the torque converter model consist of the angular speeds of the pump, $\dot{\varphi}_p=\omega_p$ and the turbine, $\dot{\varphi}_t=\omega_t$.

From the characteristic curves, the torque acting on the engine T_p (pump torque) and the torque acting on driveline T_t (turbine torque) can be calculated as follows:

$$\begin{aligned}
T_p&=\dot{\varphi}_p^2/\kappa^2(w)\\
T_t&=T_p r(w)
\end{aligned} \quad (5)$$

As previously mentioned, this model only describes the behaviour of a torque converter in the low frequency domain. More refined models, which also take into account physical and geometrical parameters, can be found in [1] and [5-6].

4 Linearization

It can be seen that the torque converter equations are non-linear, and hence the damping performance of the 2 DOF and the 3 DOF model depend on the operation point. Therefore, the linearized equations are valid only at the chosen equilibrium point.

The linearization is conducted by setting up the equations for the reduced model of the driveline, which are then used to find equilibrium points for a given set of torque loads. By linearizing the differential equations and inserting the equilibrium point, a set of linear differential equations can be found. In the following, the linearization equations for the 3 DOF model is presented. They can be derived for the 2 DOF model analogous to the 3 DOF model.

First, the differential equations are transformed into the structure in (6): The functions f_1, f_2 and f_3 dependon the variables given in brackets.

$$\begin{aligned}
&f_1(\ddot{\varphi}_e,\dot{\varphi}_e,\dot{\varphi}_1,T_e,T_p)\approx 0\\
&f_2(\ddot{\varphi}_1,\dot{\varphi}_1,\varphi_1,\dot{\varphi}_2,\varphi_2,T_t)\approx 0 \quad (6)\\
&f_3(\dot{\varphi}_1,\varphi_1,\ddot{\varphi}_2,\dot{\varphi}_2,\varphi_2)\approx 0
\end{aligned}$$

Assuming continuous functions for f_1, f_2 and f_3, the Taylor series representationsat the equilibrium state, $(\dot{\varphi}_{e0},\dot{\varphi}_{10})$, can be obtained as in (7):

$$\begin{aligned}
&f_1\approx f_1(\varphi_{1_0},\varphi_{e_0})+\frac{\partial f_1}{\partial\ddot{\varphi}_e}\Delta\ddot{\varphi}_e+\frac{\partial f_1}{\partial\dot{\varphi}_e}\Delta\dot{\varphi}_e+\ldots\\
&\ldots+\frac{\partial f_1}{\partial\dot{\varphi}_1}\Delta\dot{\varphi}_1+\frac{\partial f_1}{\partial T_e}\Delta T_e+\frac{\partial f_1}{\partial T_p}\Delta T_p\approx 0\\
&f_2\approx f_2(\varphi_{1_0},\varphi_{e_0})+\frac{\partial f_2}{\partial\ddot{\varphi}_1}\Delta\ddot{\varphi}_1+\frac{\partial f_2}{\partial\dot{\varphi}_1}\Delta\dot{\varphi}_1+\ldots\\
&\ldots\frac{\partial f_2}{\partial\varphi_1}\Delta\varphi_1+\frac{\partial f_2}{\partial\dot{\varphi}_2}\Delta\dot{\varphi}_2+\frac{\partial f_2}{\partial\varphi_2}\Delta\varphi_2+\frac{\partial f_2}{\partial T_t}\Delta T_t\approx 0 \quad (7)\\
&f_3\approx f_3(\varphi_{1_0},\varphi_{e_0})+\frac{\partial f_3}{\partial\dot{\varphi}_1}\Delta\dot{\varphi}_1+\frac{\partial f_3}{\partial\varphi_1}\Delta\varphi_1+\ldots\\
&\ldots\frac{\partial f_3}{\partial\ddot{\varphi}_2}\Delta\ddot{\varphi}_2+\frac{\partial f_3}{\partial\dot{\varphi}_2}\Delta\dot{\varphi}_2+\frac{\partial f_3}{\partial\varphi_2}\Delta\varphi_2\approx 0
\end{aligned}$$

It should be noted that the deviation variables of the turbine torque, ΔT_t, and the pump torque, ΔT_p, are also a function of the angular speeds, $\dot{\varphi}_e$ and $\dot{\varphi}_1$. By using the chain rule in partial differentiation, these terms should be replaced in (7) by the expressions given in (8):

$$\begin{aligned}
\Delta T_t&=\frac{\partial T_t}{\partial\dot{\varphi}_e}\Delta\dot{\varphi}_e+\frac{\partial T_t}{\partial\dot{\varphi}_1}\Delta\dot{\varphi}_1\\
\Delta T_p&=\frac{\partial T_p}{\partial\dot{\varphi}_e}\Delta\dot{\varphi}_e+\frac{\partial T_p}{\partial\dot{\varphi}_1}\Delta\dot{\varphi}_1
\end{aligned} \quad (8)$$

4.1 Linearization – 2 DOF model

For the 2 DOF model, the functions,f_1 and $f_2$are shown in (9):

$$\begin{aligned}
f_1&=\left(I_e+I_p\right)\ddot{\varphi}_e-T_e+T_p=0\\
f_2&=(I_t+I_1+I_2)\ddot{\varphi}_1+d_{load}\dot{\varphi}_1-T_t=0
\end{aligned} \quad (9)$$

By applying (7) the linearized differential equations for the 2 DOF model at the equilibrium point are determined by (10):

$$\begin{bmatrix} I_p + I_e & 0 \\ 0 & I_t + I_1 + I_2 \end{bmatrix} \begin{bmatrix} \Delta\ddot{\varphi}_e \\ \Delta\ddot{\varphi}_1 \end{bmatrix} + \dots$$
$$\dots \begin{bmatrix} \frac{\partial T_p}{\partial \dot{\varphi}_e} & \frac{\partial T_p}{\partial \dot{\varphi}_1} \\ -\frac{\partial T_t}{\partial \dot{\varphi}_e} & -\frac{\partial T_t}{\partial \dot{\varphi}_1} + d_{load} \end{bmatrix} \begin{bmatrix} \Delta\dot{\varphi}_e \\ \Delta\dot{\varphi}_1 \end{bmatrix} = \begin{bmatrix} \Delta T_e \\ 0 \end{bmatrix} \tag{10}$$

4.2 Linearization – 3 DOF model

For the 3 DOF model, the functions, f_1, f_2 and f_3 are presented in (11):

$$\begin{aligned} f_1 &= (I_e + I_p)\ddot{\varphi}_p + T_p - T_e \\ f_2 &= (I_t + I_1)\ddot{\varphi}_t + d_{12}(\dot{\varphi}_t - \dot{\varphi}_2) + k_{12}(\varphi_t - \varphi_2) - T_t \\ f_3 &= I_2\ddot{\varphi}_2 + d_{12}(\dot{\varphi}_2 - \dot{\varphi}_t) + k_{12}(\varphi_2 - \varphi_t) + d_{load}\dot{\varphi}_2 \end{aligned} \tag{11}$$

Consequently, one obtains the linear differential equations by applying (7) as stated in (12):

$$\begin{bmatrix} I_p + I_e & 0 & 0 \\ 0 & I_1 + I_t & 0 \\ 0 & 0 & I_2 \end{bmatrix} \begin{bmatrix} \Delta\ddot{\varphi}_e \\ \Delta\ddot{\varphi}_1 \\ \Delta\ddot{\varphi}_2 \end{bmatrix} + \dots$$
$$\dots \begin{bmatrix} \frac{\partial T_p}{\partial \dot{\varphi}_e} & \frac{\partial T_p}{\partial \dot{\varphi}_1} & 0 \\ -\frac{\partial T_t}{\partial \dot{\varphi}_e} & -\frac{\partial T_t}{\partial \dot{\varphi}_1} + d_{12} & -d_{12} \\ 0 & -d_{12} & d_{12} + d_{load} \end{bmatrix} \begin{bmatrix} \Delta\dot{\varphi}_e \\ \Delta\dot{\varphi}_1 \\ \Delta\dot{\varphi}_2 \end{bmatrix} + \dots \tag{12}$$
$$\dots \begin{bmatrix} 0 & 0 & 0 \\ 0 & k_{12} & -k_{12} \\ 0 & -k_{12} & k_{12} \end{bmatrix} \begin{bmatrix} \Delta\varphi_e \\ \Delta\varphi_1 \\ \Delta\varphi_2 \end{bmatrix} = \begin{bmatrix} \Delta T_e \\ 0 \\ 0 \end{bmatrix}$$

4.3 Thestate space representation

The inertia and damping matrices for the 2DOF systemare given in (13): For the sake of simplicity,the Δ-symbol will be dropped in the equationsfrom now on.

$$\begin{aligned} &\underline{\underline{M}} \cdot \ddot{\underline{\varphi}} + \underline{\underline{D}} \cdot \dot{\underline{\varphi}} = \underline{F} \\ &\underline{\underline{M}} = \begin{bmatrix} I_p + I_e & 0 \\ 0 & I_t + I_1 + I_2 \end{bmatrix} \\ &\underline{\underline{D}} = \begin{bmatrix} T_{p,\dot{\varphi}_e} & T_{p,\dot{\varphi}_1} \\ -T_{t,\dot{\varphi}_e} & -T_{t,\dot{\varphi}_1} + d_{load} \end{bmatrix} \\ &\underline{F} = \begin{bmatrix} T_e \\ 0 \end{bmatrix} \end{aligned} \tag{13}$$

The inertia and damping matrices are given in (14) for the 3 DOF system:

$$\begin{aligned} &\underline{\underline{M}} \cdot \ddot{\underline{\varphi}} + \underline{\underline{D}} \cdot \dot{\underline{\varphi}} + \underline{\underline{K}} \cdot \underline{\varphi} = \underline{F} \\ &\underline{\underline{M}} = \begin{bmatrix} I_p + I_e & 0 & 0 \\ 0 & I_1 + I_t & 0 \\ 0 & 0 & I_2 \end{bmatrix} \\ &\underline{\underline{D}} = \begin{bmatrix} T_{p,\dot{\varphi}_e} & T_{p,\dot{\varphi}_1} & 0 \\ -T_{t,\dot{\varphi}_e} & -T_{t,\dot{\varphi}_1} + d_{12} & -d_{12} \\ 0 & -d_{12} & d_{12} + d_{load} \end{bmatrix} \\ &\underline{\underline{K}} = \begin{bmatrix} 0 & 0 & 0 \\ 0 & k_{12} & -k_{12} \\ 0 & -k_{12} & k_{12} \end{bmatrix}, \underline{F} = \begin{bmatrix} T_e \\ 0 \\ 0 \end{bmatrix} \end{aligned} \tag{14}$$

The torsional mode frequency is assumed to be the square root of the ratio of the output shaft stiffness and the inertia, $\sqrt{k_{out}/I_{out}}$, which is assumed to be at 5 Hz.

The resulting system of equations containing mass, damping and stiffness matrices are then transferred to a standard state space representation for continuous linear time invariant systems, which is shown in (15). The vector of state variables is indicated by x, u is the input vector, y is the output vector, A is the system matrix, B is the input matrix and C is the output matrix.From the state space representation, the matrices A, B and C can be found and the transfer functions can be calculated for each of the input and output parameters.

$$\begin{aligned} \dot{\underline{x}} &= \underline{\underline{A}}\underline{x} + \underline{\underline{B}}\underline{u} \\ \underline{y} &= \underline{\underline{C}}\underline{x} \end{aligned} \tag{15}$$

The sizes of the matrices depend on the number of the degrees of freedom.

The input vector u and the state vector $x$for the 2 DOF system are given in (16).

$$\begin{aligned} &\underline{x} = \begin{bmatrix} \underline{\varphi} \\ \dot{\underline{\varphi}} \end{bmatrix} = \begin{bmatrix} \varphi_p \\ \varphi_t \\ \dot{\varphi}_p \\ \dot{\varphi}_t \end{bmatrix}, \underline{u} = \begin{bmatrix} T_e \\ 0 \end{bmatrix} \\ &\underline{\underline{A}} = -\underline{\underline{M}}^{-1}\underline{\underline{D}}, \underline{\underline{B}} = \underline{\underline{M}}^{-1}, \underline{\underline{C}} = \begin{bmatrix} 0 & 0 & 1 & 0 \\ 0 & 0 & 0 & 1 \end{bmatrix} \end{aligned} \tag{16}$$

The input and output vectors for the 3DOF system are presented in (17): The identity matrix is represented by I.

$$\begin{aligned} &\underline{x} = \begin{bmatrix} \underline{\varphi} \\ \dot{\underline{\varphi}} \end{bmatrix} = \begin{bmatrix} \varphi_p \\ \varphi_t \\ \varphi_2 \\ \dot{\varphi}_p \\ \dot{\varphi}_t \\ \dot{\varphi}_2 \end{bmatrix}, \underline{u} = \begin{bmatrix} T_e \\ 0 \\ 0 \end{bmatrix} \\ &\underline{\underline{A}} = \begin{bmatrix} \underline{\underline{0}} & \underline{\underline{I}} \\ -\underline{\underline{M}}^{-1}\underline{\underline{K}} & -\underline{\underline{M}}^{-1}\underline{\underline{D}} \end{bmatrix}, \underline{\underline{B}} = \begin{bmatrix} \underline{\underline{0}} \\ \underline{\underline{M}}^{-1} \end{bmatrix} \\ &\underline{\underline{C}} = \begin{bmatrix} 0 & 0 & 0 & 1 & 0 & 0 \\ 0 & 0 & 0 & 0 & 1 & 0 \\ 0 & 0 & 0 & 0 & 0 & 1 \end{bmatrix} \end{aligned} \tag{17}$$

DOI: 10.1051/matecconf/201710807001

The transfer function can be calculated in the form of the state space representation by the Laplace transformation as in (18). A detailed explanation on the derivation of (18) can be found in [8].

$$G(s) = \frac{Y(s)}{U(s)} = \underline{\underline{C}}(s\underline{\underline{I}} - \underline{\underline{A}})^{-1}\underline{\underline{B}} \quad (18)$$

5 Simulations

In this study, the systemresponses due to variations of the engine torque are investigated. In particular, the response of the angular speeds ω_p and ω_t due to an increase of the engine torque by 1Nm (step function) or due to a Dirac impulse are compared. The simulations are performed with *MATLAB/Simulink*.

In the following Figures 5-10, the results denoted by "exact" are obtained from the dynamic *Simulink* model and the results denoted by "linearized" are obtained from the solution of linearized equations (see Eqs. (10), (12), (13), (14), (16), (17) and (18)).

In order to find the equilibrium point, a time-constant engine torque (or pump torque) of about $T_{e0} = T_{p0} = 30\ Nm$ and a constant load (or turbine torque)of $T_{t0} = 55.85\ Nm$are assumed. This yields the equilibrium points for the speeds $\omega_{p0} = 104.3\ \mathrm{rad/s}$ and $\omega_{t0} = 55.85\ \mathrm{rad/s}$.

Model parameters used in the analytical calculations and the simulations are given in Table 1.

Table 1. Model parameters.

I_e, engine inertia	0.5 kgm^2
I_p, pump inertia	0.092 kgm^2
I_t, turbine inertia	0.026 kgm^2
r_{whl}, wheel radius	0.35 m
m_{veh}, vehicle mass	2000 kg
$I_{veh} = m_{veh} r_{whl}^2$, vehicle inertia	245 kgm^2
$f_{shuffle}$, torsional mode frequency	5 Hz
I_{out}, output shaft inertia	24.5 kgm^2
k_{out}, stiffness of output shaft	24181 Nm/rad
d_{out}, damping of output shaft	6.28 Nms/rad
d_{load}, reduced load damping constant	1 Nms/rad
g_1, gear ratio of 1st gear	10
$I_1 = I_{out} / g_1^2$, reduced mass inertia	0.245 kgm^2
$I_2 = I_{veh} / g_1^2$, reduced mass inertia	2.45 kgm^2
$k_{12} = k_{out} / g_1^2$, reduced stiffness constant	241.8 Nm/rad
$d_{12} = d_{out} / g_1^2$, reduced damping constant	0.0628 Nms/rad

5.1 Unit-step response

In order to find the unit step response of the system, the input given in (19) is applied to the system. The function *H(t)* denotes the Heaviside step function, which is defined as in (20):

$$T_e(t) = T_{e0} + H(t) \quad (19)$$

$$H(t) = \begin{cases} 0, & t < 0 \\ 1, & t \geq 0 \end{cases} \quad (20)$$

The step responses of the angular velocities of the pump, the turbine and the additional mass for the 2DOF and the 3DOF models are shown in Figure 5 and Figure 6. In both figures zoomed views up to the first second of the simulation time are given under the full time range plots.

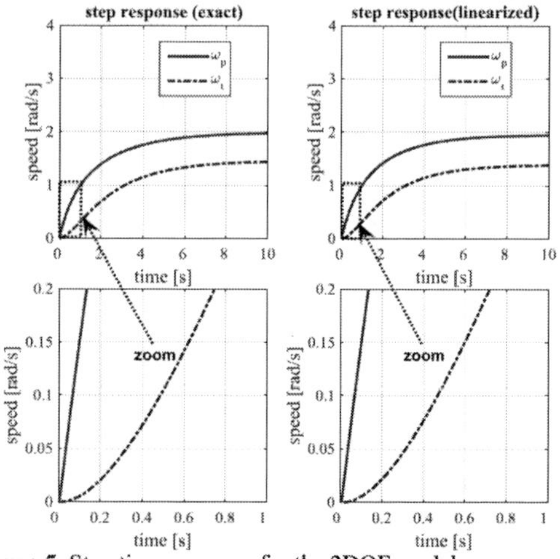

Figure 5. Step time response for the 2DOF model.

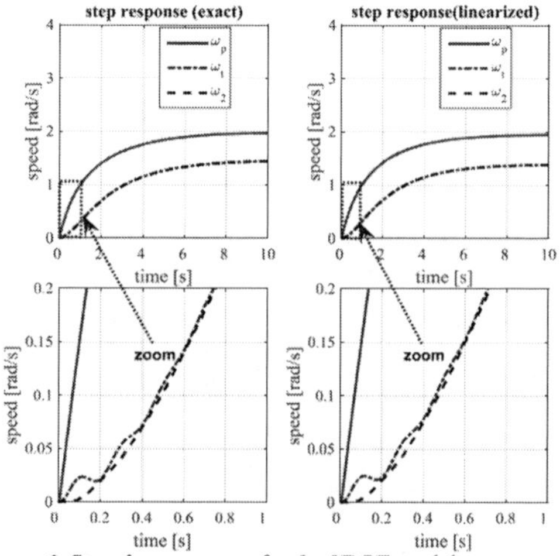

Figure 6. Step time response for the 3DOF model.

The linearized model and the exact model yield the same results in both 2 DOF and 3 DOF systems.

The steady state responses of the 2 DOF and 3 DOF models are also the same. The pump speed has a steady state value of 2 rad/s and the turbine speed has a steady state value of 1.4 rad/s after 10 seconds. However, it can be seen that the turbine side has speed oscillations because of the additional mass about 5 Hz in the 3 DOF model.Since the 2 DOF model has no flexibility, this effect cannot be observed in the 2 DOF model.

The vibrations on the turbine side of the 3 DOF model are eliminated after 2 to 3 time periods by the torque converter.

5.2 Impulse response

The impulse input is given in (21): The function $\delta(t)$ denotes the Dirac delta function.

$$T_e(t) = T_{e0} + \delta(t) \qquad (21)$$

Again, the dynamic simulation from the exact and the linearized models have the same impulse response (see Figure 7 for the 2 DOF and Figure 8 for the 3 DOF). Furthermore, the 2 DOF and the 3 DOF models display similar steady state behaviour.

The oscillations of the angular speed of the turbine can be observed only in the 3 DOF model as in the previous section 5.1 (see zoomed plots in Figure 8). The angular speed of the additional mass (ω_2) vanishes in the 2 DOF model.

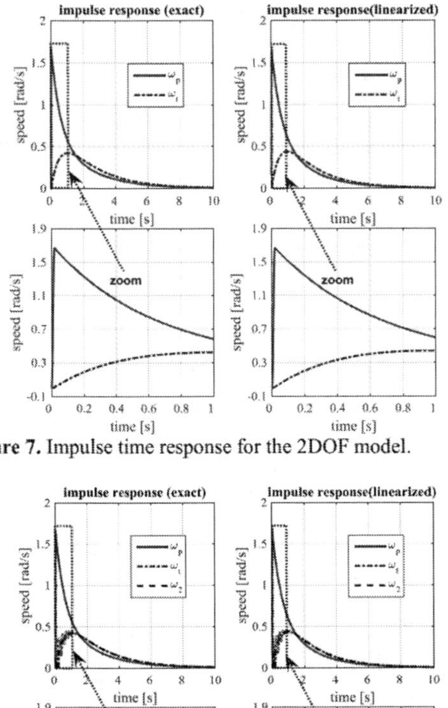

Figure 7. Impulse time response for the 2DOF model.

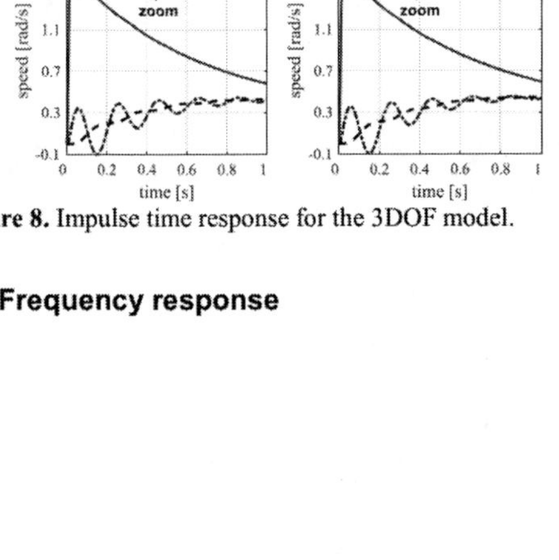

Figure 8. Impulse time response for the 3DOF model.

5.3 Frequency response

In this section the transient results due to time-harmonic excitation are presented. The harmonic engine torque input given in (22) is applied to the system.

$$T_e(t) = T_{e0} + \sin(2\pi f t) \qquad (22)$$

In order to compare the results for several frequencies, the frequency is varied in the simulations every 10 seconds, e.g. between 20 and 30 s, the excitation frequency is 10Hz (see Table 2).

Table 2. Frequency input time intervals.

Time range	Frequency
0 s < t < 10 s	f = 50 Hz
10 s < t < 20 s	f = 25 Hz
20 s < t < 30 s	f = 10 Hz
30 s < t < 40 s	f = 1 Hz

Figure 9 and Figure 10 show the results for the 2 DOF and 3 DOF models. The results for the pump (left) and the turbine speeds (right)— are shown for the linearized and the exact model. The zoomed plots for one second time range at 10 Hz and at 1 Hz are also presented. The transfer functions of the linearized model, $G = \Delta\omega_p / \Delta T_e$ and $G = \Delta\omega_t / \Delta T_e$, are shown at the bottom.

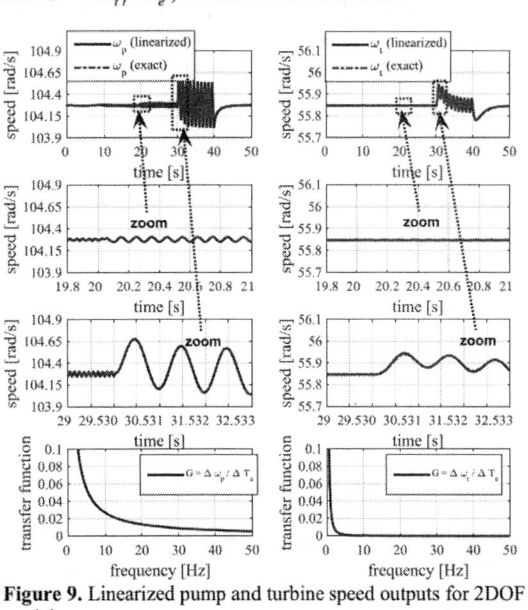

Figure 9. Linearized pump and turbine speed outputs for 2DOF model.

In both 2 DOF and 3 DOF models, the steady state values of the angular speeds are for the pump 104.3 rad/s and for the turbine 55.85 rad/s respectively. They match with the equilibrium point which is chosen for the simulations.

The results for the turbine and the pump speeds indicate that the higher frequency vibrations are damped by the torque converter, which can be seen also in the zoomed views in Figure 9 and Figure 10. Moreover, the amplitude of speed oscillations of the pump speed for the 3 DOF model is slightly higher than the ones for the 2 DOF model.

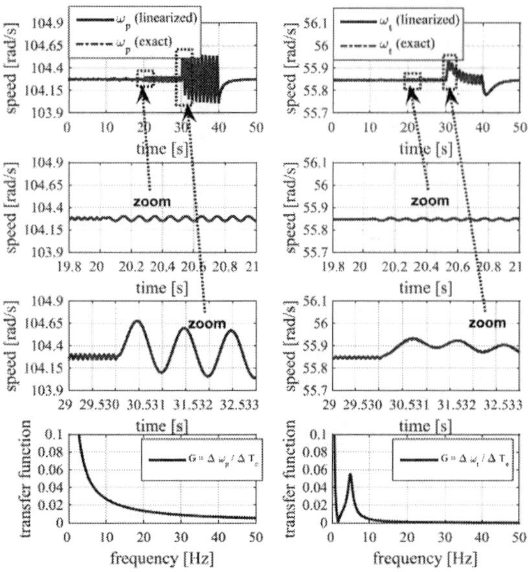

Figure 10. Linearized pump and turbine speed outputs for 3DOF model.

There is a significant difference between the transfer function plots of the 2 DOF and 3 DOF models at the natural frequency of the driveline.

The torsional mode resonance can be seen from the transfer function plot between the turbine speed and the engine torque for the 3DOF model in Figure 10. However, the transfer function looks different for the 2DOF model in Figure 9 which demonstrates only the fundamental mode. Since, the 2 DOF model has no flexibility, the effects of the natural frequency cannot be observed. Therefore, the 2 DOF model is inconvenient to consider excitations in the system around the natural frequency.

The exact and linearized solutions match perfectly for both the 2 DOF and the 3 DOF models.

6 Conclusions

The aim of the work was to analyse the damping characteristics of a torque converter in the low frequency domainby modelling a simple automotive driveline model. Since even the simplest torque converter model is described by a nonlinear relation, the linearization of the nonlinear differential equations is unavoidable in order to get a better understanding of which frequencies are mainly attenuated by a torque converter.

The driveline modelswith 2 DOF and 3 DOF are investigated. These systems are simulated by two methods; the linearization and the nonlinear dynamical modelling in *Simulink*. The system transfer functions are evaluated to observe the damping effect of torque converter on the small variations in the input of the engine torque.

Torsional mode resonance was observed in the transfer function plot between the turbine speed and the engine torque of the 3DOF model. However, the 2 DOF model was not sufficient to demonstrate the effects of the natural frequency.

In the frequency response analysis, it has been seen that the torque converter damps higher frequency vibrations in the low frequency domain effectively.

The linearized model used in the present studyprovided an accurate approximation of the system response around the chosen equilibrium point. It can be concluded that, by using the linearized system equations along with the reduced degrees of freedom models, the computational effort can be decreased significantly.

Acknowledgement

The research was funded by the K2-Comet Austrian Centre of Competence in Mechatronics ACCM (project number: C120405).

References

1. D. Hrovat, W.E. Tobler, J. Franklin Inst. **319**, 93–114 (1985)
2. M. Barglazan, U.P.B. Sci. Bull., Series D **72**, Iss.4 (2010)
3. P.D. Walker, N. Zhang, Mechanism and Machine Theory, **60**, 47-59 (2013)
4. S. Rosbi, T. Ryu, K. Matsuzaki, T. Nakae, A. Sueoka, Y. Takikawa, Y. Ooi, K. Kawazu, Int. Journal of Automotive Engineering **5**, 91-99 (2014)
5. H.A. Asl, N.L. Azad, J. McPhee, Int. J. Vehicle Systems Modelling and Testing, **9**, 36–55 (2014)
6. H.A. Asl, N.L. Azad, J. McPhee, ASME 2012 International Design Engineering Technical Conferences and Computers and Information in Engineering Conference (2012)
7. M.S. Castiglione, P.P. Pohl, R.A. Smithson, Technical Report, Southwest Research Institute IRD, **03**-9264 (2002)
8. K. Ogata, *Modern Control Engineering* (Prentice Hall, 2010)

DOI: 10.1051/matecconf/201710807002

Based on the traffic system tools magneto electric equipment design and application

Jian Sun, Lu Lyu, Yan Yang, Yan-ling Bai, Wen-li Jin, Yong Ren, and Lin-lin Zhang

Equipment manufacturing center, Institute of China Electronics technology group co., LTD. 28, Nanjng, China

Abstract. In transportation, the road is rough, the engine operation, the instability of the transmission system, and large vibration of the vibration system is to collect the vibration energy, and transformed into stable electricity, provide extra energy for transportation. Experiment device of the system is composed of three parts: simulate vibration test rig, vibration generating device, rectifier voltage regulator circuit. Vibration table is to simulate vehicle vibration in the road, cars, for example, through the check information that cars drive in ordinary road driving axle of some vibration data, thus to calculate all parameters of the vibration table. Vibration generating device is placed on the simulation vibration table, it can convert vibration energy into cars can use electricity. Rectifier voltage regulator circuit is generated by power generation device with a range of random variation of amplitude of electric energy through the rectifier voltage into a stable power output, for car use. The final design and making the vibration system, obtained by the method of experiment proved that the vibration of system feasible.

1 The background and significance

Vibration is ubiquitous in the transportation, it not only consume much energy, and bring the noise and vibration, reducing the comfort and stability. But the vibration is hard to avoid in the transportation, usually in order to reduce the loss of the vibration and the vibration will install some shock absorption on the mechanical design. In this paper, we study and design on the vibration of the vehicles produced in the process of driving the recovery and utilization system, through this system, the vibration in the vehicle running hard to eliminate forms of energy conversion, reasonably be can use energy. Compared with ordinary suspension mechanism, vibration system can be used to collect and convert vibration energy and can improve the efficiency of energy utilization if efficient use of the energy.

Because our country most of the pavement is class B (cement and asphalt) or C level road, therefore assume that car is in grade B road on GB7031 according to the road pavement power spectral density according to the rough degree is divided into 8, also stipulates that the scope of all levels of road roughness coefficient and the geometric average, the frequency of the grading road roughness index to 2, shown in the following table. Cars in the driving process, and different speed, different grades of road surface pavement roughness incentives are different. Road surface model is road roughness on the road grade and the mathematical description of the car speed variation. Table 2 as a vehicle for a domestic car technical parameters, using the simulation toolkit (Simulink), can be obtained under the system parameters and the road excitation Angle of sprung mass and the output of the sprung mass center displacement curve. Combining general speed is within the scope of the 36 ~ 108 km/h, the corresponding wheel input time between 0.1 ~ 75 hz frequency range, the inherent frequency of the suspension system in 1 ~ 2 hz, part of the suspension system inherent frequency in 10 ~ 16 hz. Shock absorber in the 10 ~ 16 hz frequency range when it sprung mass center displacement amplitude on - 40 db. After the suspension system is calculated the frame vibration amplitude is about 2 cm. So magnetoelectric vibration generator design work environment is: the automobile speed in the 36-108 km/h, on B road, vibration generated by the road and vehicle itself incentive the rear axle, produced by the vibration frequency range for 10 to 16 hz, vibration amplitude around 2 cm.

Table 1. The power spectral density coefficient of road surface roughness and the response of the body

road level	$G_q(n_0)\times10^{-6}m^{-3}$ $n_0=0.1m^{-1}$			$a_w/(m\cdot s^{-2})$								
				V=80km/h			V=100km/h			V=120km/h		
	1	2	3	1	2	3	1	2	3	1	2	3
A	8	16	32	0.183	0.256	0.366	0.203	0.287	0.406	0.218	0.309	0.436
B	32	64	128	0.366	0.518	0.732	0.406	0.574	0.812	0.436	0.617	0.873
C	128	256	512	0.732	1.035	1.464	0.812	1.148	1.623	0.873	1.234	1.744

DOI: 10.1051/matecconf/201710807002

Table 2. A domestic car vehicle technical parameters

parameter	The numerical	parameter	The numerical
M（kg）	975.37	$C_1(N\cdot s/m)$	2546.5
$m_1(kg)$	98	$C_1(N\cdot s/m)$	2480.6
$m_2(kg)$	98	$K_{t1}(N/m)$	604684.5
$K_1(N/m)$	45482.2	$K_{t2}(N/m)$	985865
$K_1(N/m$	52288.2	a（m）	1.1135
b(m)	1.5415	$J(Kg\cdot m^2)$	863.2

2 The design of the system vibration

Vehicle vibration system is shown in figure 1, includes the following parts: electromagnetic vibration generating device, rectifier voltage regulator circuit, simulation vibration table.

Figure 1. Vibration system.

3 Electromagnetic vibration generating device

Cars on the vibration of each part has its different directions, with similar principle to design the magnetoelectric sensor vibration collection and conversion of power generation equipment, through the different parts of the car, different directions, to install the device, to a certain direction vibration is converted into mechanical energy through the power of the device can use of electricity. Electromagnetic vibration generating device with the method of permanent magnet in place, it is composed of permanent magnet, spindle, coils, springs, linear bearing and the shell, etc. Permanent magnet is installed on the spindle, spindle on both ends of the spring end respectively fixed on the end cover shell, the other end, respectively, to resist the spindle, it plays the role balance reset, coil is fixed on the outside of the permanent magnet. Work in the process of permanent magnet reciprocating motion in the coil, an electrical current to cutting magnetic induction line. Because the car axle is directly from under such as uneven ground, engine running, the vibration incentives, such as transmission system is not smooth, so in the actual application can install multiple vibration generating device to the car's axle parts, so that it can be larger to collect vehicle vibration energy and convert it into electrical energy can be used. Vibration generating device of the three-dimensional figure as shown in figure 2, figure 3 for the real figure.

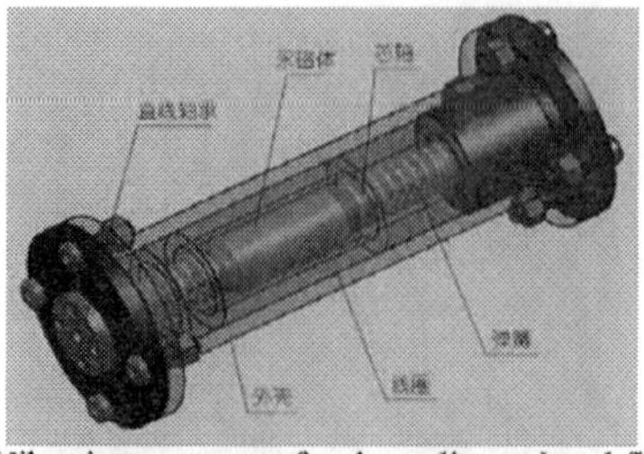

Figure 2. Vibration generator for three dimensional figure.

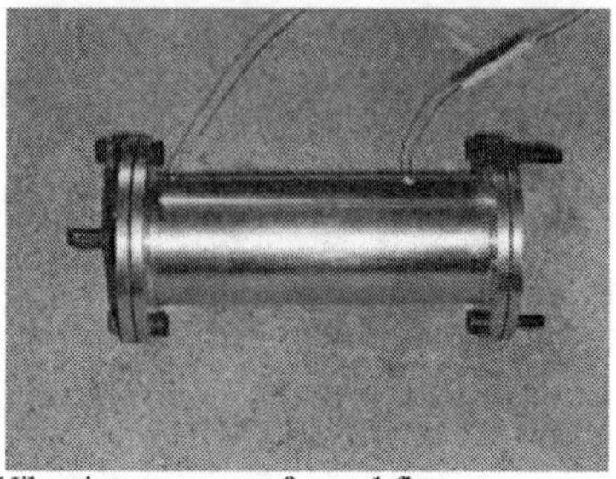

Figure 3. Vibration generator for real figure.

4 Voltage regulator rectifier device

Due to the different parts of the vibration amplitude is different, the resulting current size are different, the car can use voltage have special requirements, it is necessary to design a voltage regulator rectifier device, a voltage to integration of the magneto electric generator for producing, in order to meet specific application requirements. According to the design requirements of electromagnetic vibration generating device, you need to generator output alternating current (amplitude of 16 v) could satisfy the requirement of storage battery is converted into direct current (12 v).Thus we design a simple dc voltage system. Figure 4 is to convert alternating current (ac) to conform to the conditions of direct current (dc) diagram.

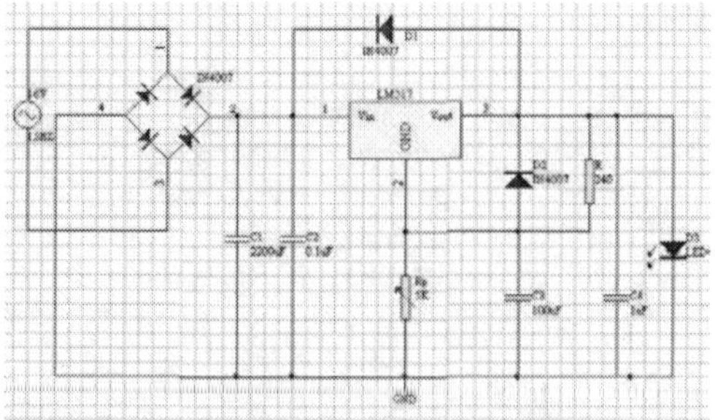

Figure 4. Circuit diagram.

5 simulation vibration table

Design the device in the product development phase, the purpose is approximately simulated vibration in the process of driving vehicle situation, it would not have to for real vehicle experiment, with the real vehicle experiment is very complex, and it is time consuming, high cost, bad for the project. As the automobile rear axle initial pavement displacement, encouragement is affected

DOI: 10.1051/matecconf/201710807002

by the surface roughness, and uneven road surface is a random process, with irregularity and self-similarity, showed obvious fractal characters, so the surface roughness have the fractal characteristics of chaos, so the simulation vibration table using chaotic vibration principle, the connecting rod slide block mechanism, the number of freedom degree is greater than the original dynamic mechanism of, by the mechanical principles known even move a deterministic input, output can still irregular vibration. Structure principle as shown in figure 5, Figure zhongyuan moving crank for uniform rotation driven by a motor. Uniform rotation, although a crank and connecting rod slider mechanism for no clearance, but the freedom degree is greater than the original number, Freedom degree of the organization $F=3n-2P_L-2P_h=3$, The number of components is n, Low number of vice is P_L, High number of vice is P_h. So, the design of the connecting rod slider vibrator, should be able to make the movement of the workbench has chaotic characteristics, to better simulate complex vibration when the car driving on the road.

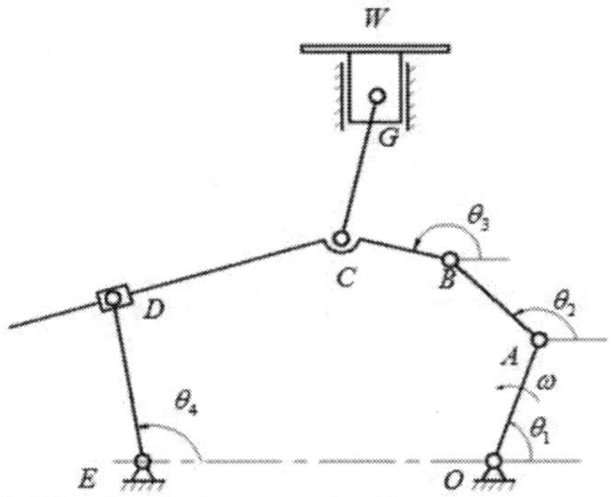

Figure 5. Vibration platform agencies diagram.

Frequency: $f=10\sim16Hz$

The amplitude: $A=0\sim20mm$

Under the above conditions by the calculation and simulation with given size parameters of the mechanism motion: OA=18mm, AB=3mm, BC=111.5mm, DE=3mm, EO=167mm, DC=180mm, CG=100mm.

Using Pro/E software to simulate the vibration test-bed for 3 d modeling, obtained by mechanism motion simulation vibration table of displacement, velocity and acceleration curve. As shown in figure 6-7.

By the simulation of displacement, velocity and acceleration curve, the average amplitude of vibration platform is $14.5mm$, the maximum amplitude of vibration platform is $20mm$, the minimum amplitude of vibration platform is $0.7mm$, the maximum speed of vibration platform is $771mm/s$, the maximum acceleration of vibration platform is $330709mm/s^2$.

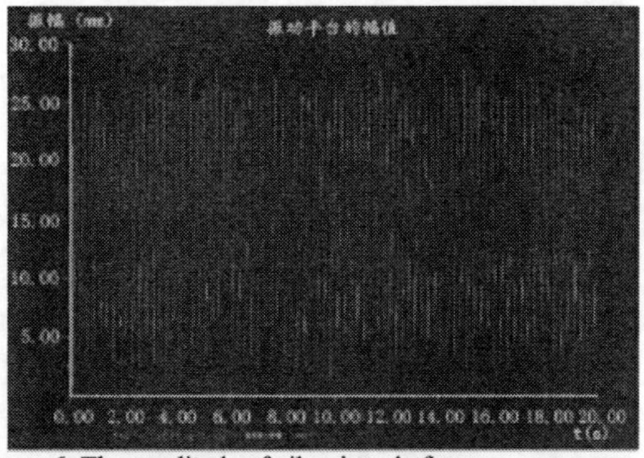

Figure 6. The amplitude of vibration platform.

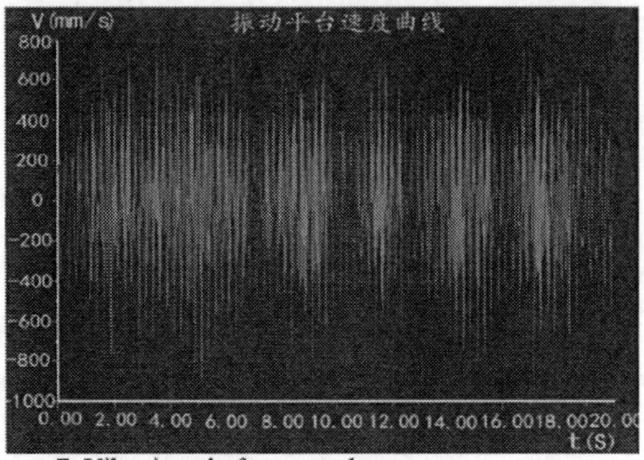

Figure 7. Vibration platform speed curve.

6 The experiment and analysis

Due to the different parts of the vibration amplitude is different, the resulting current size are different, the car can use voltage have special requirements, it is necessary to design a voltage regulator rectifier device, a voltage to integration of the magnetoelectric generator for producing, in order to meet specific application requirements. According to the design requirements of electromagnetic vibration generating device, you need to generator output alternating current (amplitude of 16 v) could satisfy the requirement of storage battery is converted into direct current (12 v).Thus we design a simple dc voltage system.

The determination of rectifier voltage part parameters: Rectification: the bridge type single-phase rectifier circuit. Known input maximum voltage = 16 v, maximum current input.IN4007 acuity 50 v st13, reverse breakdown voltage of the rated working current = 800 ma, reason of selecting IN4007 diode rectifier bridge. Filtering: because small output current in the circuit, capacitor filter circuit can be used. Capacitor filter on engineering application in general have formula or greater (3 ~ 5) / 2. We make = 5 x / 2 = 2.5 x = 2.5 / + = 2. 5 / (x) = 2.5 / (x) = 2083 uF, voltage = 1.2 = 13.6 V.So choose = 2200 uf, 50 v electrolytic capacitor. For load resistance, capacitor for ac voltage cycles, as ac voltage and frequency as the input voltage, input current. Voltage regulation: the adjustable voltage regulator integrated three-terminal voltage regulator LM317.LM317 output voltage can be adjusted through the potentiometer, approximation is expressed as: = 1.25 x (1 + /). = 12 v dc output voltage according to request, take = 240 Ω, computing available = 2064 Ω, take value for a 5 k Ω slide rheostat can meet. Is produced

DOI: 10.1051/matecconf/201710807002

to prevent self-excited oscillation, its values range between 0.1 ~ 1 u F, take in the design of the 0.1 u F; And, protection, which typically take 100 u F;Through the simple dc voltage system, we can get that could satisfy the requirement of automobile storage battery dc 12 v voltage, in order to realize to the battery charging process. In addition we will install a reserve battery, when the car battery charged to reserve battery recharged, make full use of so as to realize the electrical energy. For the diode).Through the simple dc voltage system, we can get that could satisfy the requirement of automobile storage battery dc 12 v voltage, in order to realize to the battery charging process. In addition we will install a reserve battery, when the car battery charged to reserve battery recharged, make full use of so as to realize the electrical energy.

7 Frequency domain characteristics analysis of generator

By the structure of the magnetoelectric vibration generator, the permanent magnet on receipt of the external random vibration frequency response can be used when the quality - spring - damper model, model diagram as shown in Figure 8.

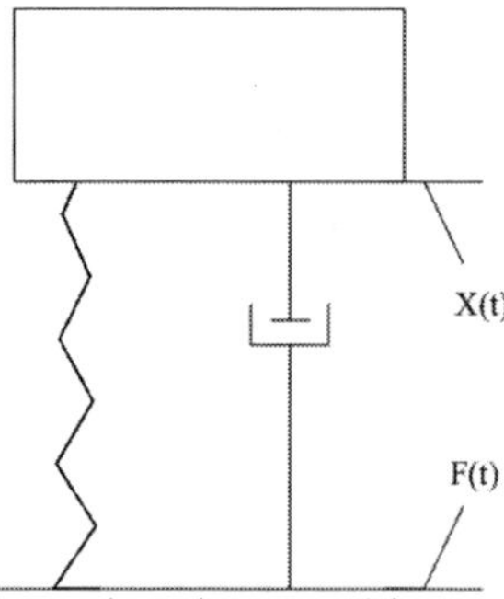

Figure 8. Quality - spring - damper model.

By Newton's laws of motion can be obtained under arbitrary excitation system of forced vibration differential equation for:

$$f(t) = m\frac{d^2x(t)}{dt^2} + c\frac{dx(t)}{dt} + kx(t) \qquad (1)$$

Type of axial displacement function for permanent magnets, sum of permanent magnet and the quality of the spindle, for the system damping coefficient, the elastic coefficient of spring. In the initial conditions to zero, to pull its transformation, the transfer function of system is:

$$G_1(s) = \frac{X(s)}{F(s)} = \frac{1}{ms^2 + cs + k} \qquad (2)$$

And by the principle of voltage available:

$$L\frac{dI(t)}{dt} + (R + R_c)I(t) = Bl\frac{dx(t)}{dt} \qquad (3)$$

Type of coil inductance, as induced current changing with the time function, load resistance to vibration generator is connected, for the coil resistance, as the remanence intensity, the length of the coil. To pull transformation type (3):

$$G_2(s) = \frac{I(s)}{X(s)} = \frac{Bls}{Ls + R + R_c} \qquad (4)$$

Because the voltage $V = I(t)R$, so permanent magnet displacement and the output voltage of the transfer function for:

$$\frac{V(s)}{X(s)} = \frac{BlRs}{Ls + R + R_c} \qquad (5)$$

Due to the induced current in coil will give a negative feedback system, for its transfer function

$$G_3(s) = \frac{Bl}{R} \qquad (6)$$

By type (2) and (4) (6) the system must transfer function is:

$$G(s) = \frac{V(s)}{F(s)} = \frac{G_1(s)G_2(s)}{1 + G_1(s)G_2(s)G_3(s)} \qquad (7)$$

Due to the small and can be ignored, after finishing the system overall transfer function is:

$$G(s) = \frac{V(s)}{F(s)} = \frac{BlRs}{[R(ms^2 + cs + k) + (Bl)^2 s]} \qquad (8)$$

According to the calculation formula of permanent magnet in the permanent magnet motor:

$$V_m = 51\frac{P_N \sigma_0 K_{ad} K_{dF}}{f K_\vartheta K_u C(BM)_{max}} * 10^9 \text{mm}^3$$

Automotive vibration is inevitable in the process of driving, and makes a considerable part of the energy loss, also has brought the noise and vibration at the same time, lower the car's comfort and stability. These can cause the vibration of the car, and are hard to use technical means to eliminate the vibration for the car run has brought many unfavorable factors such as inadequate performance play, economy becomes bad, through natural bad handling stability, peace, damaged parts, shorten service life and so on. So imagine a car that can collect the vibration of generating unit, the vibration of mechanical energy into electricity, as part of the car electrical system provides additional power, so as to achieve the aim of improve the efficiency of the automotive fuel use.

8 Conclusion

Automotive vibration is inevitable in the process of driving, and makes a considerable part of the energy loss, also has brought the noise and vibration at the same time, lower the car's comfort and stability. With the popularity of cars, the vibration energy losses have reached a certain proportion, in global energy increasingly scarce, and China is mainly under the background of construction of economical society, make full use of energy to become a problem we must solve. The vibration of the vehicle from several aspects such as the road is rough, the change of the speed and direction, wheels, the engine and transmission system is not smooth, the impact of the gear, etc. These can cause the vibration of the car, and are hard to use technical means to eliminate the vibration for the car run has brought many unfavorable factors such as inadequate performance play, economy becomes bad, through natural bad handling stability, peace, damaged parts, shorten service life and so on. So imagine a car that can collect the vibration of generating unit, the vibration of mechanical energy into electricity, as part of the car electrical system provides additional power, so as to achieve the aim of improve the efficiency of the automotive fuel use.

Vibration generating device could be common in the transportation of vibrations into usable energy, Using the built-in permanent magnet generating set to a larger extent realize energy conversion, also has a shaking table and adjustable rectifier voltage regulator circuits. Not only that it has a simple structure easy manufacturing, polluting the environment, changing volume, sustainable power and high efficiency etc., can be applied to a variety of mechanical vibration, the application prospect is broad.

References

1. TANG Koulin. Analysis of weather environment impact on electromagnetic situation [J]. Command Information System and Technology,2014,5(5) :20-24.(in Chinese)
2. CHEN ping. Global air traffic management(ATM)operational concept and its enlightenment[J].The command information system and technology, 2015, **6**(4) : 1-6. (in Chinese)
3. The command information system and technology, 2012, 3 (1) : 25 to 28
4. Cui Peng. The mechanism of debris flow initiation[Ph.D. Thesis][D]. Beijing : Beijing Forerstry University, 1990.(inChinese)
5. Cui P. Studies on condition and mechanism of debris flow initiationby means of experiment[J]. Chinese Science Bulletin, 1992, **37**(9), 759–763.
6. Mark E R, Richard G L H, Richard M I. Debris-flow initiationexperiments using diverse hydrologic triggers[A]. In : Chen C L ed.Debris-flow Hazards Mitigation Mechanics, Prediction, andAssessment[C]. New York : [s. n.], 1997. 1–11.
7. The command information system and technology, 2011, **02** (6) : 39-41
8. J. Bisplinghoff, J. Cormier, and S. Duma, Development and validation of eye injury and facial fracture criteria for the focus headform, Paper presented at the 26th Army Science Conference, Orlando, FL, 1–4 December, 2008.
9. C.K. Thorbole and H.M. Lankarani, Performance evaluation of computational HIC component tester for aerospace application, Paper No. 2008-01-2229, SAE International, Warrendale, PA, August 2008.
10. Chandler, R.F., "Human Injury Criteria Relative to Civil Aircraft Seat and Restraint System," SAE Paper 851847, society of Automotive Engineers, Warrendale,PA,1985
11. The command information system and technology, 2011, **02** (6) : 39-41
12. Lissner, H.R., Lebow, M., and Evans, F.G., "Experimental Studies on the Relation between Acceleration and Intracranial Pressure Changes in Man," Surgery, Gynecology, and Obstetrics, V. 111, 1960, pp 329-338.
13. Yin Zhen. Electrodynamics (third edition). Beijing: science press, 2010.2.
14. Sun Guangfei, strong Wen Jiang. Magnetic functional materials. Beijing: chemical industry press, 2007.2.
15. into the big, mechanical design manual (fifth edition). Chemical industry press. 2007.
16. Zhao Jihai road roughness measurement, analysis and application of Beijing university of science and technology press.

Research on Fairing design and CFD Analysis of Submarine Pipeline Inspection ARV

Xiaojian Jin[1], Shunqi Yang[2], Hai Chen[1]

[1]*Beijing University Of Posts And Telecommunications, Beijing, China*
[2]*Kunming Shipborne Equipment Research & Test Center, Kunming, China*

Abstract. Along with the fast development of the ocean exploitation, the cost-effective requirement of autonomous & remotely operated vehicle (ARV), which can perform more complicated missions such as the oil exploitation and the inspection of the submarine pipeline is more urgent. The submarine pipeline inspection ARV can help us better understand, protect and efficiently utilize them for human welfare. Fairing design of a new detection ARV are introduced in this paper. In order to select an appropriate thruster that will achieve the required speed of the ARV, the ANSYS-CFX tools are used to predicted the drag force. The CFD results reveal the distribution of velocity and pressure values of the ARV. In order to verify the CFD modeling process, a towed body was developed and analyzed, compared against the corresponding physical test data.

1 Introduction

Covering about two-thirds of the earth, the ocean has a great eff ect on our lives on land. With the substantial consume of energy sources and increase of population, ocean research and exploitation gradually become the new development space of people surviving. Meanwhile, along with the development of offshore oil and gas in our country is more and more active, especially the deep oil and gas development, the offshore platform under service is increasing year by year. The submarine pipeline is also increasing year by year as an important part of the offshore platform in development of sea oil. The environment of the submarine pipeline is strong tidal zone gulf of high risk, urgent tide, tidal range; the transmission medium is a high-sulphur crude oil. In such a complex and harsh service environment, safe and stable operation of the submarine pipeline is the most important issue to consider. Any failure of these submarine pipelines will not only affect productivity negatively but also cause tremendous environmental hazards. According to the present international and domestic requirement of the regulations for the underwater pipeline, routine or annual inspection to the pipelines should be taken to ensuring the safely or longer period of using of the pipeline.

In recent years, unmanned underwater vehicles (UUVs) have demonstrated their capabilities in many important applications in fields such as oceanographic research, offshore oil and gas industry, and military operations. Thus, as the necessary tool for submarine pipeline detection, the underwater robot submarine pipeline detection system is playing an increasingly important value. Usually, submarine pipeline inspections are performed by remote operated vehicle (ROV) driven by human operators on water surface. However, this task is often challenging, especially in conditions of poor visibility or in presence of strong currents; and ROV's working need support vessel support, low efficiency, high cost. Along with the fast development of the ocean exploitation, the cost-effective requirement of autonomous & remotely operated vehicle (ARV), which can perform more complicated missions such as the oil exploitation and the inspection of the submarine pipeline, is more urgent [1-3]. ARV has received much attention since they can be an effective tool to explore the ocean and efficiently utilize the ocean resources. ARV been used for various applications including military operations, environmental systems, scientific missions, ocean mining, and gas and oil industry. With the demand of ocean engineering and offshore petroleum exploitation, ARV which represents the direction of underwater vehicle has gained rapid development. As a sea power multiplier, ARV can not only be used in various important military applications, but is an important tool for peaceful exploitation of the ocean. ARV can carry the temperature, salinity, sonar detection instrument for collecting marine environment data and topography data, in order to provide information to the exploration of seabed resources, as well as in offshore drilling, submarine pipeline laying and maintenance of submarine pipeline. The ARV for Submarine Pipeline Inspection mission should exhibit compact size, which allows for economical overnight shipping, less manned deployment, and operation capability, together with eliminated need for large dedicated transporting vessels and costly special handling equipment. So ARV has been widely applied as an important component of ocean high-tech. The submarine pipeline inspection ARV cruises in the fixed date, which can independently find and track pipeline on the seabed. It not only can observe the state of pipeline and the circumstance near the pipeline, but also can warn the coming fault. Thus, the most effective way to ensure safe and efficient operation for the long-distance submarine pipeline is to perform periodic inspects for them by ARV. [4]

The submarine pipeline inspection ARV for oil industry has been developed and its design evaluated by CFD an analysis such as a resistance test. The development of commercial codes for CFD analysis now make it possible to predict drag and propulsion performance of an underwater robot without using a physical model test basin [5]. There are many efforts at predicting drag forces of underwater vehicles by CFD analysis and validating CFD simulation methods in design processes [6-7]. CFD simulations are also compared with in-service data for the self-propelled performance of an Autonomous Underwater Vehicle [8]. The present paper focuses on design issues with fairing of the ARV. A CFD analysis was first used to conduct the resistance test necessary to predict the total drag force for selecting an appropriate thruster that will achieve the required speed of the ARV. In order to verify the CFD modelling process, a model of a towed body was developed and analysed and compared against the corresponding physical test data.

The rest of this paper is organized as follows. Section 2 provides the mechanical design for submarine pipeline inspection ARV with 300 meters depth. The fairing design for ARV in Section 3. The CFD Setup for Predicting the ARV drag force is presented in Section 4. Section 5 presents some CFD Analysis results. Some conclusions are given in Section6.

2 Mechanical design for submarine pipeline inspection ARV

Mechanical systems that have excellent environmental resistance are needed. An underwater robot's mechanical structure can be divided into three functional subsystems: frame, pressure vessel, and fairing. A fairing that covers all or parts of an underwater robot offers some advantages, such as an aesthetically pleasing exterior, a decreased drag force, and protection. A fairing is desirable for ARV, due to limited on-board power sources for required hydrodynamic performance. The ARV, shown in Figure 1, was designed as a flat shape with hydrodynamic contour to keep stable and reduce the drag force on the body in the horizontal direction and designed for a maximum operating depth of up to 300 m, is composed of an titanium alloy frame that supports two cylinder-shaped hulls of 0.3 m in diameter and engineering plastic fairing skin in the outside cover. This design offers a good hydrodynamic performance and a large space for housing the equipment while maintaining a compact size that allows us to operate the vehicle from conventional ships.

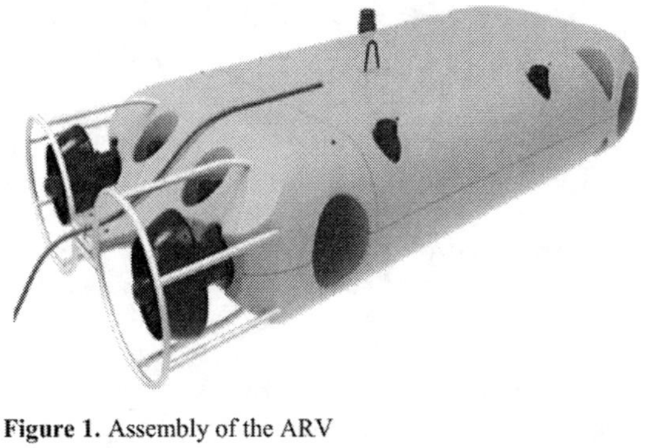

Figure 1. Assembly of the ARV

The ARV measures 2.8m in length, 1.3m width and 0.42m in height. The design speed for the initial design of the ARV is 2knots and the maximum speed is 6knot.

The ARV frame is made of TiAl6v4 titanium alloy and is composed of one I-shaped pillar screwed to three U-shaped profiles that serve as backbone to each one of the three cylinder-shaped hulls. The ARV pressure vessel is made of hard anodized 6082-T6 aluminium alloy and contains two cylindrical pressure housings which contain mainly the control electronics and battery cluster. The ARV flotation modules in between skin and pressure vessel, made of an epoxy composite foam with a density of 270 kg/m^3 , are placed on the top part of the vehicle. Their principal mission is to make the vehicle almost neutrally buoyant.

3 Fairing design

The frame, pressure vessel and flotation modules that compose the ARV vehicle are covered with a thermoformed ABS plastic skin whose streamlined shape is based on the Myring hull profile equations [9]. The ARV fairing skin provides protection to the sensible equipment and reduces the drag of the vehicle. The flow curve can be formulated as follows:

$$y = (x - f) \cdot \sqrt{L_1 \cdot X - X^2} \cdot D / (2 \cdot k) \tag{1}$$

Where, D is height of ARV. And the flow calculation plot is given in Figure 2.

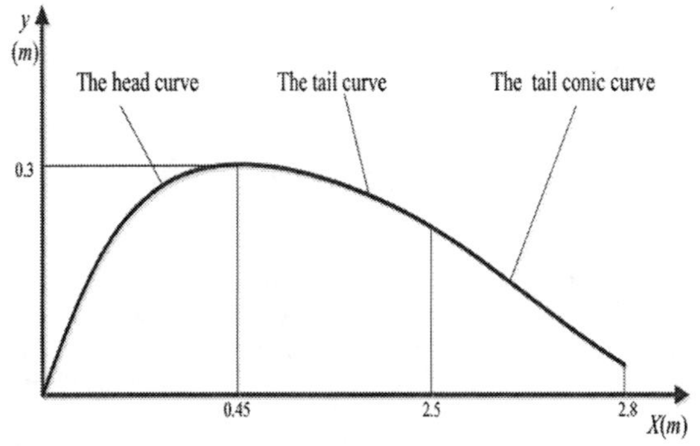

Figure2. Flow curve of the ARV

4 CFD Setup for Predicting the ARV drag force

The fluid flow around the ARV has been modeled using the commercial CFD analysis code ANSYS-CFX 14.0. For these calculations, the fluid's motion is modeled using the incompressible, isothermal Reynolds-averaged-Navier-Stokes (RANS) equations in order to determine the Cartesian flow field and pressure of the water around the ARV body. The equations consist of a general solution of the ensemble-averaged, steady-state, three dimensional Navier - Stokes equations, where the k-ε turbulence model (k is turbulence kinetic energy and ε is viscous dissipation rate) has been used to close the system of equations.

The continuity equation is given

$$\frac{\partial \overline{u_i}}{\partial x_i} = 0 \tag{2}$$

The c momentum equation is given

$$\rho \frac{\partial \overline{u_i}}{\partial t} + \rho \overline{u_j} \frac{\partial \overline{u_i}}{\partial x_j} = \rho \overline{F_i} - \frac{\partial \overline{p}}{\partial x_i} + \frac{\partial}{\partial x_j}\left(\mu \frac{\partial \overline{u_i}}{\partial x_j} - \rho \overline{u_i' u'_j}\right) \tag{3}$$

The intensity of turbulence k equation is given

$$\frac{\partial}{\partial t}(\rho k) + \frac{\partial}{\partial x_i}(\rho k u_i) = \frac{\partial}{\partial x_j}\left[\left(\mu + \frac{\mu_t}{\sigma_k}\right)\frac{\partial k}{\partial x_j}\right] + G_k - \rho\varepsilon \tag{4}$$

The turbulent dissipation ε equation is given

$$\frac{\partial}{\partial t}(\rho \varepsilon) + \frac{\partial}{\partial x_i}(\rho \varepsilon u_i) = \frac{\partial}{\partial x_j}\left[\left(\mu + \frac{\mu_t}{\sigma_\varepsilon}\right)\frac{\partial \varepsilon}{\partial x_j}\right] + G_{1\varepsilon}\frac{\varepsilon}{k}G_k - G_{2\varepsilon}\rho\frac{\varepsilon^2}{k} - R_\varepsilon$$

(5)

where,

$$\mu_t = \rho C_\mu \frac{k^2}{\varepsilon} \tag{6}$$

$$G_k = -\rho \overline{u_i' u'_j} \frac{\partial \overline{u_i}}{\partial x_j} \tag{7}$$

$$R_\varepsilon = \frac{C_\mu \rho \eta^3 (1 - \eta/\eta_0)}{1 + \beta \eta^3} \frac{\varepsilon^2}{k} \tag{8}$$

$$\eta = Sk/\varepsilon \tag{9}$$

and $\eta_0 = 4.48$, $\beta = 0.013$, $C_{1\varepsilon} = 1.45$, $C_{2\varepsilon} = 0.0845$, $\sigma_k = 1.0$, $\sigma_\varepsilon = 1.3$.

The plane mesh of the domain is shown in Figure 3. The length of the domain is 7 times longer than the ARV body and width and height are 6 times longer. The length of the tank was made long enough to allow the wake from the ARV body to be observed and the width and height of the tank are also long enough to reduce the wall effect. The meshing operation was performed using ICEM CFD as shown in Figure 4. Layered (5 layers) meshes were used for generating the boundary layer elements around the ARV body while unstructured (Tetrahedral) meshes were employed for the region far from the body as they are not suitable to resolve the boundary layers adjacent to a solid body [10]. The specifications for the pre-processing stage in CFX as shown in Figure 5-10, including the mesh generation, are summarized in Table 1.

Table 1 Principal conditions employed in the CFD analysis

Water tank size	**19.6m×7.2×2.5m**
Turbulence model	k-ε (Scalable wall function)
No. of the inflation layer	5
Total no. of elements (nodes)	320210 (63144)
No. of Tetrahedral elements	320210

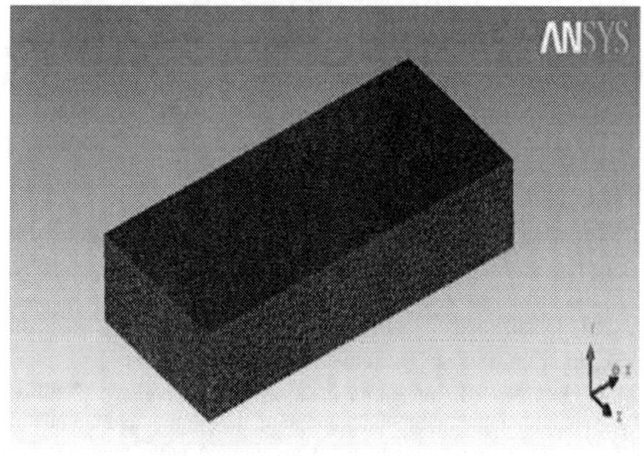

Figure 3. Review of the meshing results for ARV

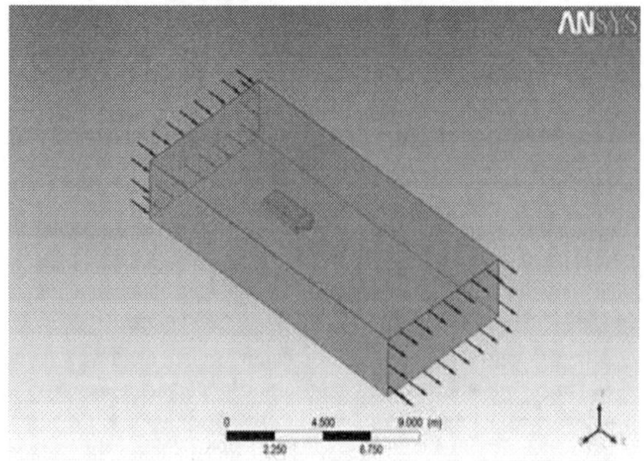

Figure 4. Pre-processing results for ARV

5 CFD Analysis results

A mesh convergence study was performed at the design speed (3knots), and the results are displayed in Table 2. It was observed that once the number of cells for the case is reached the size of the reference, the variation in the total drag force (RT) was small considering the significant changes made to the simulation. The reference mesh size for the case was therefore considered to be of sufficient

DOI: 10.1051/matecconf/201710807003

accuracy. The total drag force (RT) of the ARV body is obtained directly from the CFD analysis and shown in Table 3. Figure 5-10 shows velocity and pressure distribution at the lowest speed (1 knot), design speed (3knots) and highest speed (5knots) in this simulation test. In order to verify the CFD modeling method, a towed body was simulated and compared against the corresponding physical test data. It is found obviously that the calculation error between the simulation and the experimental results is less than 20%. Thus the model is validated.

Table 2 Result of convergence study with ARV CFD model at the design speed

Case	Number of cells Total	Drag Force(3kn)
Reference	320210	114 [N]
Reduced mesh size-1	91852	232 [N]
Reduced mesh size-2	215585	129 [N]
Increased mesh size-1	1125787	88 [N]

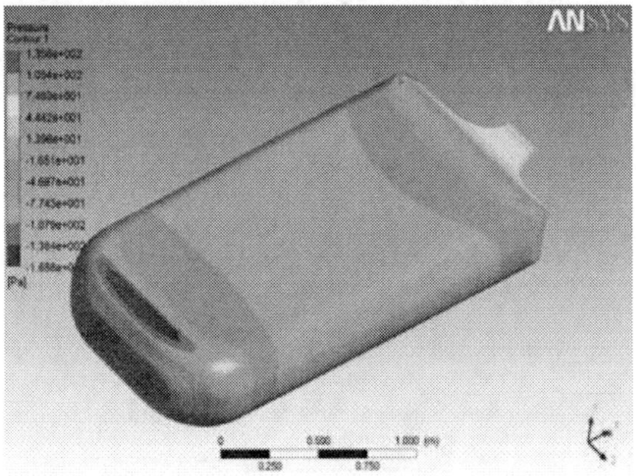

Figure 5. Velocity distribution at 1kn

Figure 6. Pressure distribution at 1kn

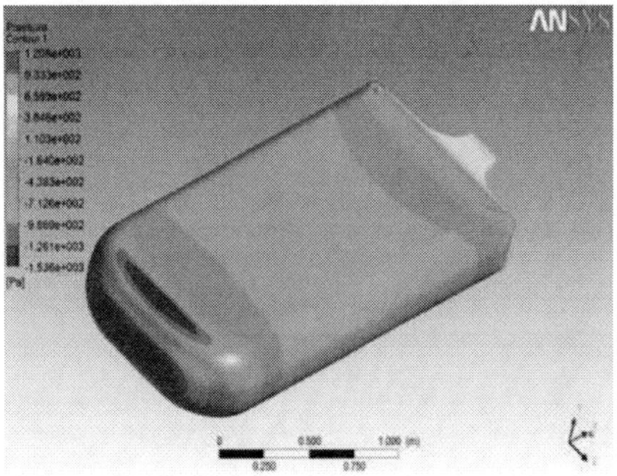

Figure 7. Velocity distribution at 3kn

Figure 8. Pressure distribution at 3kn

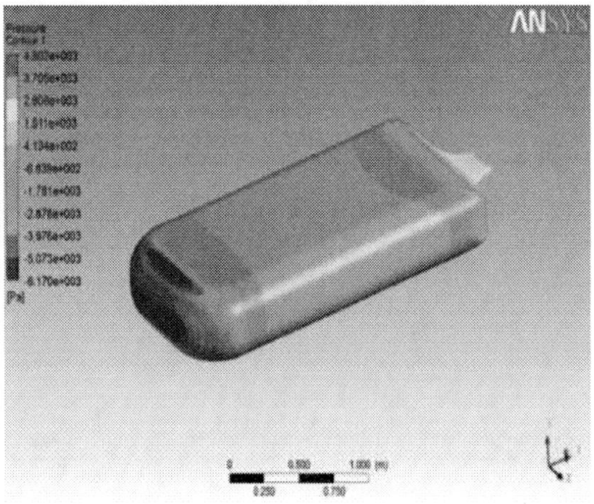

Figure 9. Velocity distribution at 6kn

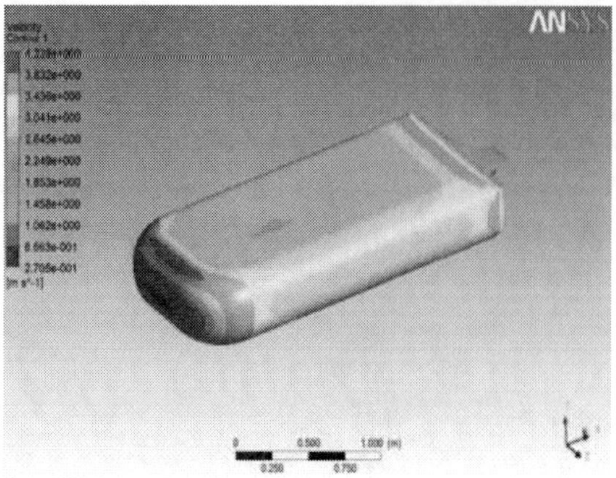

Figure 10. Pressure distribution at 6kn

Table 3 Result of drag force with ARV CFD model at different speeds

Speed	Drag Force	Speed	Drag Force
6knots	430 [N]	3knots	114 [N]
5knots	303 [N]	2knots	53 [N]
4knots	197 [N]	1knots	15[N]

6 Conclusion

A CFD procedure has been conducted on the proposed ARV design. The drag estimation was conducted first using CFD analysis with a mesh convergence study performed to validate the CFD analysis. The hydrodynamic forces on the ARV body were obtained from the CFD motion analysis. The CFD modeling approach was then compared with towed body which was compared against the experimental data. Based on the validation studies, the CFD results have been shown to be reliable and the CFD analysis techniques employed are shown to be suitable for initial concept analysis of underwater fairing.

References

1. A. Budiyono, Advances in unmanned underwater vehicle technologies. Modeling, control and guidance perspectives. Indian Journal of Marine Sciences, 38(3), pp.282-295 (2009).
2. J. Yuh, G. Marani, D.R. Blidberg. Applications of marine robotic vehicles, Intell. Serv. Robotics 4(4), pp.221–231 (2011).
3. S. Q. Yang, H. L. Yang and M. W. Cao. Reliability-based Design for Pressure Shell of Submarine Pipeline Inspection AUV with 1,500 Meters Depth, Proceedings of the MSMD 2015, Changsha, 12-13 December, pp320-326 (2015).
4. A. Budiyono. Advances in unmanned underwater vehicle technologies: Modeling, control and guidance perspectives. Indian Journal of Marine Sciences, 38(3), pp.282-295 (2009).
5. T.H. Joung, K. Sammut, F. He. Shape optimization of an autonomous underwater vehicle with a ducted propeller using computational fluid dynamics analysis. International Journal of Naval Architecture & Ocean Engineering, 4, pp.44-56 (2012).
6. X. Yu, Y. Su. Hydrodynamic performance calculation on mini-automatic underwater vehicle. Information and Automation (ICIA), 2010 IEEE International Conference, Harbin, 20-23 June, pp.1319-1324 (2010).
7. J.G. Bellingham, Y. Zhang, J.E. Kerwin. Efficient propulsion for the Tethys long-range autonomous underwater vehicle. Autonomous Underwater Vehicles (AUV), 2010 IEEE/OES, pp.1-7 (2010).
8. A.B. Phillips, R.T. Stephen and F. Maaten. Comparisons of CFD simulations and in-service data for the self propelled performance of an autonomous underwater vehicle. 27th Symposium on Naval Hydrodynamics, Seoul, Korea, 05-10 October 2008, pp.15 (2008).
9. D. F. Myring. A theoretical study of body drag in subcritical axisymmetric flow, Aeronaut. Q., vol. 27, no. 3, pp. 186–194 (1976).
10. Y. Nishi, M. Kashiwagi, and W. Koterayama,. Resistance and Propulsion Performance of an Underwater Vehicle Estimated by a CFD Method and Experiment, ISOPE '07, Lisbon, Spain, 1-6 July, pp.2045-2052 (2007).

Design and Modeling of a Novel Torque Vectoring Differential System

Yu-Fan Chen[1], Heng-Chuan Hsu [1], Cheng-Ping Yang[2] and Tyng Liu[1]

[1]*Department of Mechanical Engineering, National Taiwan University, 10617 Taipei, Taiwan*
[2]*Mechanical and Systems Research Laboratories, Industrial Technology Research Institute, Hsinchu 310, Taiwan*

Abstract. In this paper, a new concept torque vectoring differential (TVD) system is presented. It is shown that the structure and the mechanism of the system, the operating methods, and the parameters design by a simulation program, i.e. SimulationX. First of all, the structure of the new TVD system is introduced, as well as the relevant mechanic equations. Second, we attempt to verify the feasibility and accuracy of SimulationX through establishing a simple mechanical model by MATLAB, so that the further modeling and simulation results of the new TVD system will be credible. Then, the simulation results at the setting conditions are presented. Finally, the sensitivity of the design parameters is analyzed, including adjusting the braking torque and the dimensions of the gear sets in the differential. According to these results, the characteristics of the new TVD system can be derived in order to develop the whole system with vehicle dynamic model in the next stage.

1 Introduction

Of all vehicle dynamics, such as steering and braking system, are in pursuit of driving performance and safety. In vehicle stability system, torque vectoring is a new technology that has advanced most rapidly in recent years. In order to prevent vehicle accidents, vehicle dynamics control are widely developed and used since the 1980s [1]. For examples, direct yaw moment control (DYC) [2], electronic stability program (ESP) [3, 4], and traction control system (TCS) are control strategies for improving vehicle stability. In general, these control strategies adjust vehicle dynamics through braking system. However, there may be an enormous power loss due to brakes. Therefore, torque vectoring differential (TVD) [5] is proposed to solve the power loss issue and provided to adjust torque distribution. A torque vectoring differential is a system based on a conventional differential gear, and it is usually connected with one or more of different ratio speeds and clutches (or brakes) that can transfer input torque efficiently and direct the power flow separately into wheels while a vehicle is in acceleration or cornering. Torque vectoring differential is now a popular system and several different types of torque vectoring differential are invented by automobile companies such as Mitsubishi [6], Audi [7], and Honda [8]. However, in most current torque vectoring differentials, issues of structural complexity, space requirement, development cost, etc. are still high. Hence, with a simple structure, limited space, and low cost, a novel design of torque vectoring differential based on a conventional differential gear is proposed in this research.

2 Design and modeling

In this section, a new differential gear possesses torque vectoring differential function is proposed. Its mechanism and dynamics are introduced and analyzed, and then the mathematical form is derived. Before modeling the new TVD system, models of a conventional differential gear are established by SimulationX and MATLAB, and simulated results are also compared for verification.

2.1 New differential gear design

2.1.1 Function of torque vectoring differential

The new TVD system provided in this paper integrates the gear sets, actuator units, and required control methods. The main structure is shown in Fig. 1. It is a new concept differential with additional torque vectoring gears. This structure includes the outer and inner differential gears which are fixed with center pin; two side gears are connected with each output shaft; two torque vectoring gears which are concentric with each output shaft, and each of them stretches out of the differential case connecting with the brake.

In order to transfer torque from one side to the other side of the wheel, based on the tire model and the slip ratio model, the rotational speed of the wheel should be adjusted actively. To this new TVD system, it is easy to adjust the rotational speed of the output shafts by operating the brakes. For example, if the right-side output shaft needs to be accelerated, we can operate the left brake generating a braking torque into the left-side gear.

Then, the rotational speed of the left-side output shaft will decrease. On the contrary, the rotational speed of the right-side output shaft will increase. As a result, the drive torque will transfer from left-side output shaft to right-side output shaft.

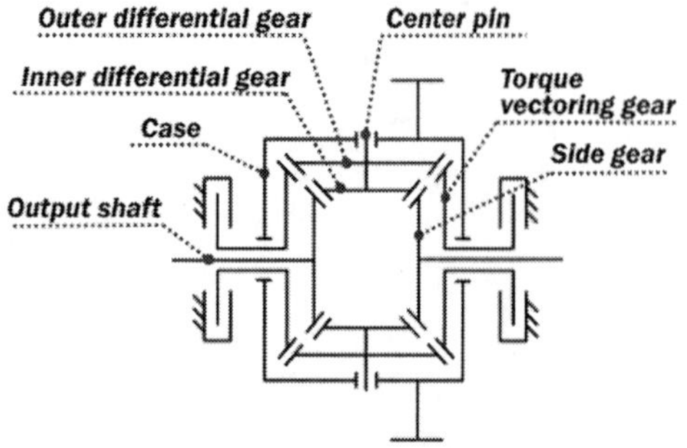

Figure 1. Schematic diagram of new TVD system.

2.1.2 Dynamic force analysis

According to the diagram in Fig. 2, equation of motion of each component can be derived as follows. Eq. (1) represents the motion of the case, Eq. (2) and (3) represent the motion of the wheels, Eq. (4) to (6) represent the motion of the gears, and Eq. (7) to (10) represent the differential of tangential velocity of the gear contacts. Therefore, a 10 x 10 matrix of the new TVD system is formed.

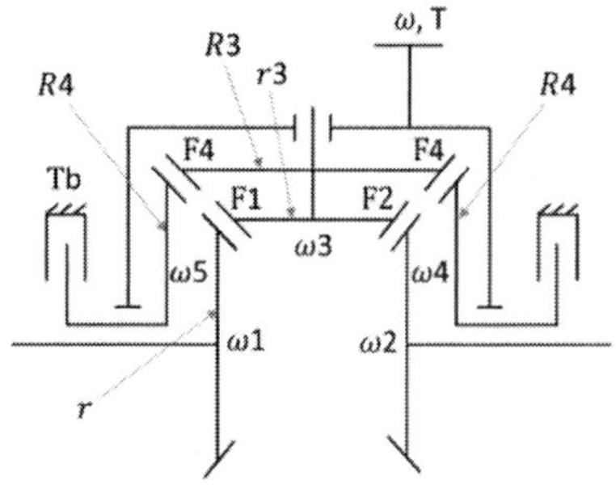

Figure 2. Free body diagram of new TVD system.

$$T + F_1 r - F_2 r - F_3 R_3 - F_4 R_4 = I\dot{\omega} \quad (1)$$

$$-F_1 r - T_1 = I_1 \dot{\omega}_1 \quad (2)$$

$$T + F_1 r - F_2 r - F_3 R_3 - F_4 R_4 = I\dot{\omega} \quad (3)$$

$$-F_1 r_3 - F_2 r_3 - F_3 R_3 - F_4 R_3 = I_3 \dot{\omega}_3 \quad (4)$$

$$F_3 R_4 = I_4 \dot{\omega}_4 \quad (5)$$

$$F_3 R_4 - T_b = I_4 \dot{\omega}_5 \quad (6)$$

$$\dot{\omega}_4 R_4 = \dot{\omega} R_4 + \dot{\omega}_3 R_3 \quad (7)$$

$$\dot{\omega}_5 R_4 = \dot{\omega} R_4 - \dot{\omega}_3 R_3 \quad (8)$$

$$\dot{\omega}_1 r = \dot{\omega} r - \dot{\omega}_3 r_3 \quad (9)$$

$$\dot{\omega}_2 r = \dot{\omega} r + \dot{\omega}_3 r_3 \quad (10)$$

$$\begin{bmatrix} I & 0 & 0 & 0 & 0 & 0 & -r & r & R_4 & R_4 \\ 0 & I_1 & 0 & 0 & 0 & 0 & r & 0 & 0 & 0 \\ 0 & 0 & I_2 & 0 & 0 & 0 & 0 & -r & 0 & 0 \\ 0 & 0 & 0 & I_3 & 0 & 0 & r_3 & r_3 & R_3 & -R_3 \\ 0 & 0 & 0 & 0 & I_4 & 0 & 0 & 0 & -R_4 & 0 \\ 0 & 0 & 0 & 0 & 0 & I_4 & 0 & 0 & 0 & -R_4 \\ R_4 & 0 & 0 & R_3 & -R_4 & 0 & 0 & 0 & 0 & 0 \\ R_4 & 0 & 0 & -R_3 & 0 & -R_4 & 0 & 0 & 0 & 0 \\ r & -r & 0 & -r_3 & 0 & 0 & 0 & 0 & 0 & 0 \\ r & 0 & -r & r_3 & 0 & 0 & 0 & 0 & 0 & 0 \end{bmatrix} \begin{bmatrix} \dot{\omega} \\ \dot{\omega}_1 \\ \dot{\omega}_2 \\ \dot{\omega}_3 \\ \dot{\omega}_4 \\ \dot{\omega}_5 \\ F_1 \\ F_2 \\ F_3 \\ F_4 \end{bmatrix} = \begin{bmatrix} T \\ -T_1 \\ -T_2 \\ 0 \\ 0 \\ -T_b \\ 0 \\ 0 \\ 0 \\ 0 \end{bmatrix} \quad (11)$$

2.2 SimulationX-based multi-physical model

2.2.1 Introduction to SimulationX

SimulationX is a multi-physics tool that can establish a three-dimensional model for dynamic simulation through physical modeling, signal modeling, and equation-based modeling. One of the advantages of SimulationX is to build a model by intuition, which means it is not necessary to derive equations for programming. For example, by using and connecting built-in modules from Table 1, then a diagram and three-dimensional view of a conventional differential gear model are established (Fig. 3 and Fig. 4). After an external force or torque is defined and applied, the kinematic and kinetic results of each component (module) can be calculated and measured by simulation. Fig. 5(a) shows an input torque applied on the case, and rotational speed and angular acceleration of the output wheels are shown in Fig. 5(b) and Fig. 5(c). The curves of right and left shafts are identical because the loading torque on each side is the same, which means there is no speed difference.

Table 1. Modules in SimulationX software.

Symbol	Description
cuboid1	This element type models a cuboid
cylinder1	This element type models a hollow cylinder
revoluteJoint1	This element type models a revolute joint with rotation around a selectable axis
gearExt1	This element type models an external teeth gearwheel
interfaceRotary1	This element type models an interface element with two mechanical connectors connectable to rotational elements
bodyTorque1	This element type models a body torque
curve1	This signal block permits the generation of signals from sampled data

DOI: 10.1051/matecconf/201710807004

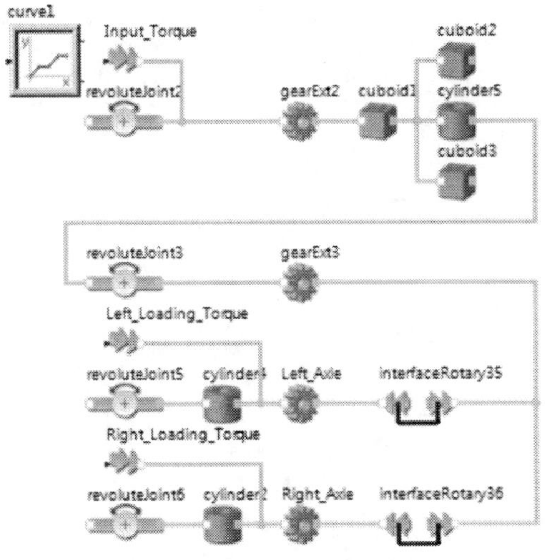

Figure 3. Model of a conventional differential gear.

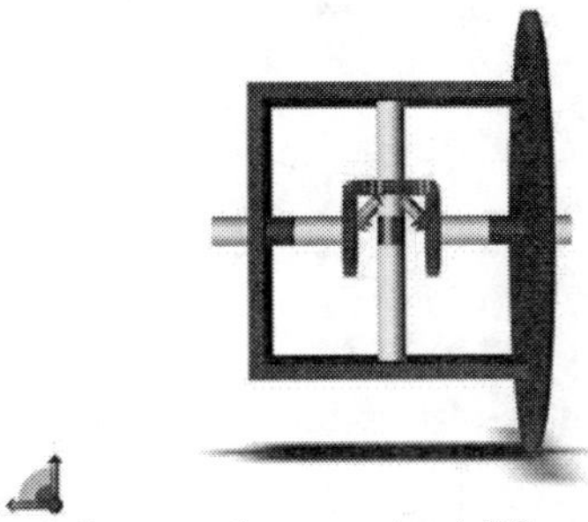

Figure 4. Structure of a conventional differential gear.

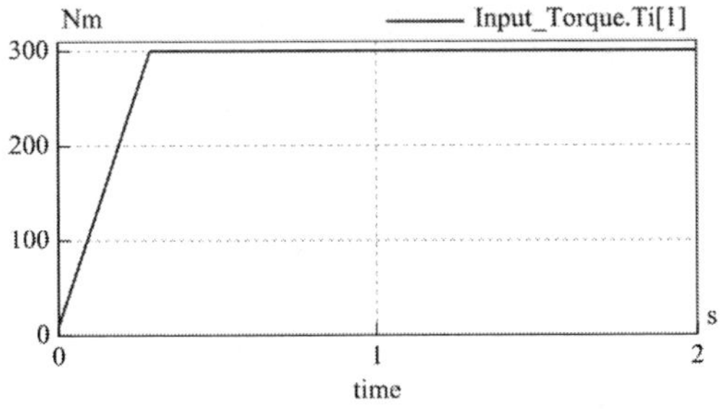

(a) Input torque applied on the ~~gear~~ case.

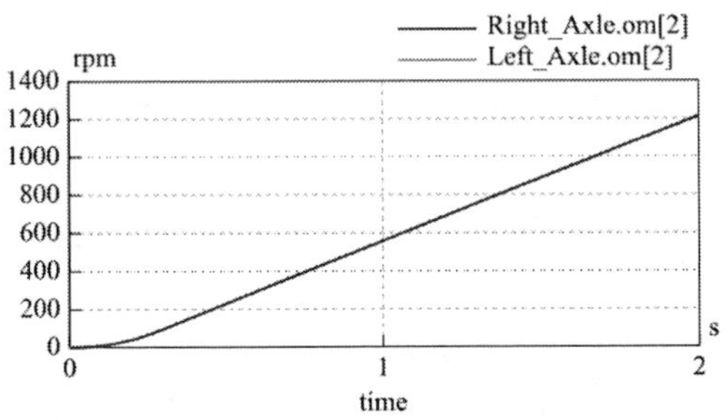

(b) Rotational speed of the two wheels.

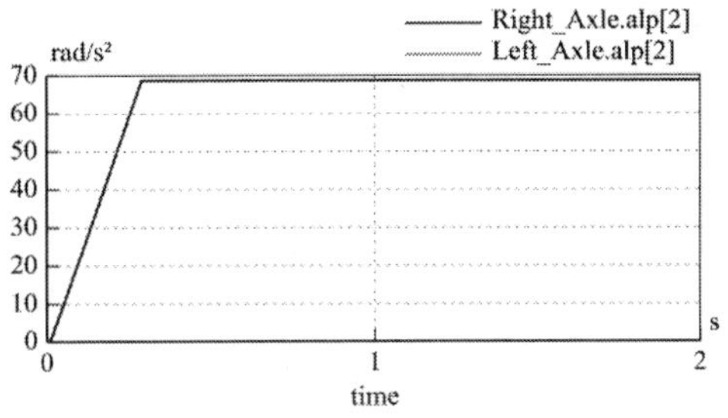

(c) Angular acceleration of the two wheels.

Figure 5. SimulationX simulation results of the conventional differential gear.

2.2.2 Verification by MATLAB programming

In this part, the previous SimulationX simulation results of the conventional differential gear model will be verified by MATLAB programming. Figure 6 shows a conventional differential gear connected with one shaft and two wheels. An input torque is applied on the shaft, and the two wheels are considered as outputs. Therefore, a mathematical matrix form can be derived as follows.

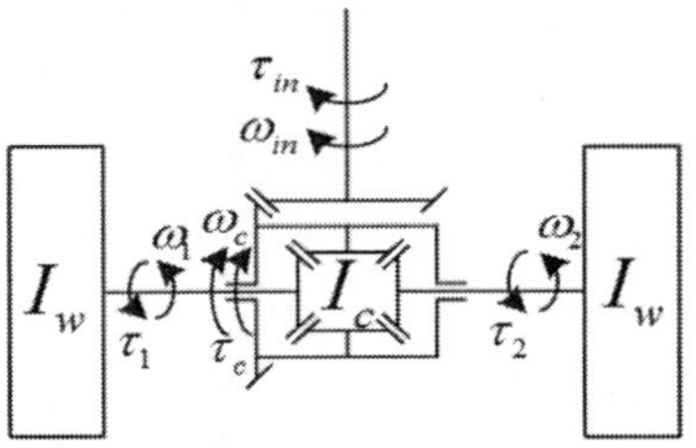

Figure 6. Free body diagram of a conventional differential gear.

$$\begin{bmatrix} I_W & 0 & 0 & 1 \\ 0 & I_W & 0 & 1 \\ 0 & 0 & I_c & -2 \\ 1 & 1 & -2 & 0 \end{bmatrix} \begin{bmatrix} \dot{\omega}_2 \\ \dot{\omega}_1 \\ \dot{\omega}_c \\ F \end{bmatrix} = \begin{bmatrix} 1 & 0 & 0 \\ 0 & 1 & 0 \\ 0 & 0 & 1 \\ 0 & 0 & 0 \end{bmatrix} \begin{bmatrix} -\tau_2 \\ \tau_1 \\ -\tau_c \\ 0 \end{bmatrix}$$

In MATLAB programming, all of the parameter values such as the inertia of the carrier and the powertrain I_c, the inertia of each wheel I_w, and the input torque τ_c are the same in the SimulationX model. Comparing these two simulation results, the curves illustrated by MATLAB in Fig. 7 nearly matches the curves illustrated by SimulationX in Fig. 5. Thus, it can be verified that SimulationX is a useful tool as MATLAB. Moreover, SimulationX enables to establish a multi-physical model to check design parameters such as size, position, and shape in three-dimensional space. It is more visual and sensible to perceive directly through human eyes. Therefore, analysis of the new TVD system is based on SimulationX software, and the simulated results will be discussed in next section.

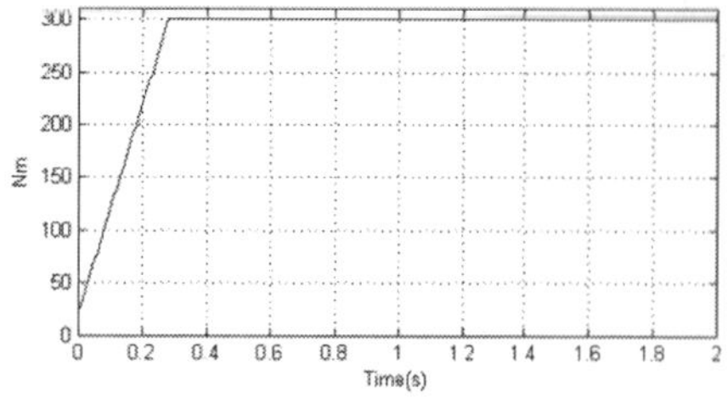

(a) Input torque applied on gear case.

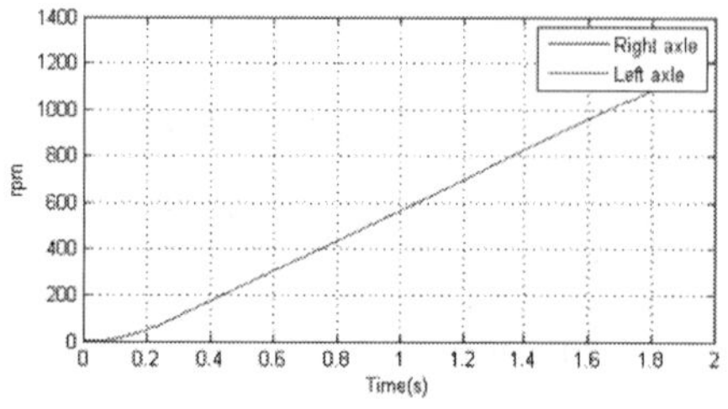

(b) Rotational speed of the two wheels.

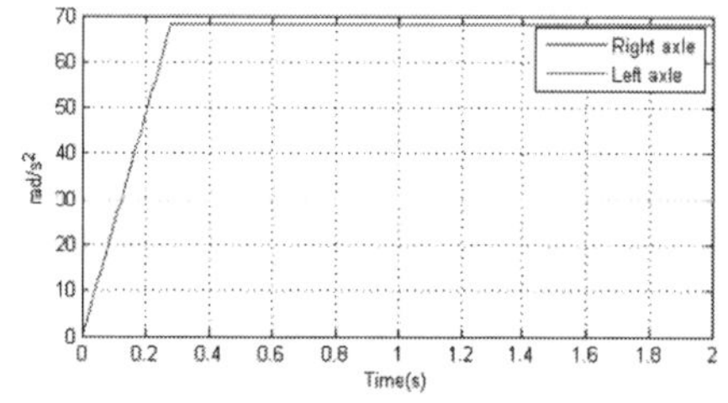

(c) Angular acceleration of the two wheels.

Figure 7. MATLAB simulation results of the conventional differential gear.

3 Simulation results

In this section, SimulationX is used to establish a three-dimensional structural model of the new TVD system, and the braking torque is applied to analyze the characteristics of the torque vectoring function. Then, various combinations of the parameter values are selected to observe the variation in the speed of right and left wheels. Therefore, the torque vectoring performance can be determined.

3.1 Simulation setup

3.1.1 Three-dimensional structural model

According to the schematic diagram of the new TVD system (Fig. 1), a prototype model of a diagram and three-dimensional view by using SimulationX is established as shown in Fig. 8 and Fig. 9. The new TVD system is composed of a conventional differential gear sets as two side gears and an inner differential gear (r and r3). Besides, there are additional outer differential gear and torque vectoring gears (R3 and R4) in the case. The output shafts fixed the side gears are represented as wheels (I1 and I2), whereas the shafts fixed the torque vectoring gears are able to be connected to brakes or clutches. All of them are rotating with respect to the same rotational axis but these shafts have a relative angular displacement that can be rotated in different rotational speed. Table 2 shows detailed values of parameters in the new TVD system simulation.

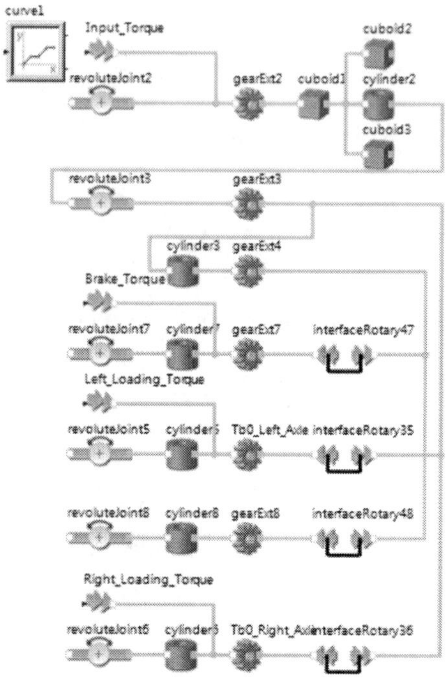

Figure 8. Model of new TVD system.

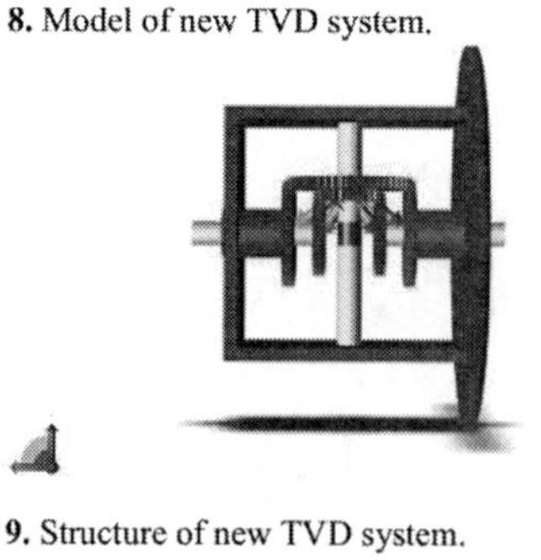

Figure 9. Structure of new TVD system.

Table 2. Parameter value settings in the simulation.

Parameters	**Value**
Inertia of the carrier and powertrain I (kg·m^2)	2
Inertia of each wheel I_1, I_2 (kg·m^2)	0.1
Inertia of the outer differential gear I_3 (kg·m^2)	0.01
Inertia of each torque vectoring gear I_4 (kg·m^2)	0.01
Radius of each side gear r (m)	0.032
Radius of the inner differential gear r_3 (m)	0.025
Radius of the outer differential gear R_3 (m)	0.05

DOI: 10.1051/matecconf/201710807004

Radius of each torque vectoring gear R_4 (m)	0.041
Input torque T (N·m)	300
Loading torque of each wheel T_1, T_2 (N·m)	10
Braking torque T_b (N·m)	20

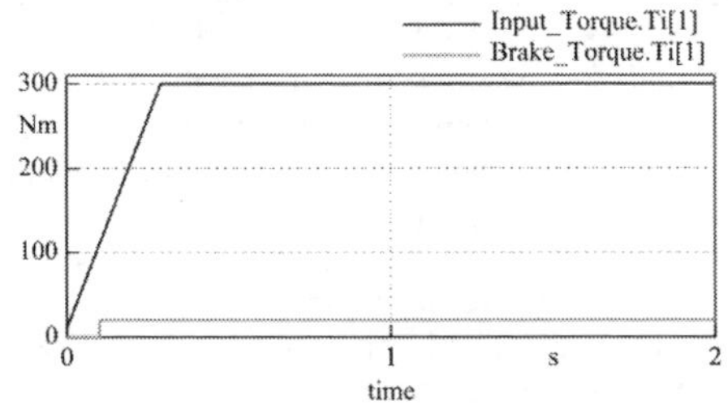

(a) Input torque and braking torque.

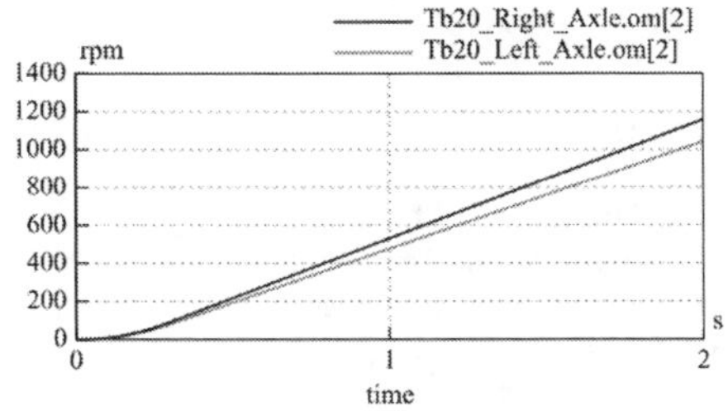

(b) Rotational speed of the two wheels.

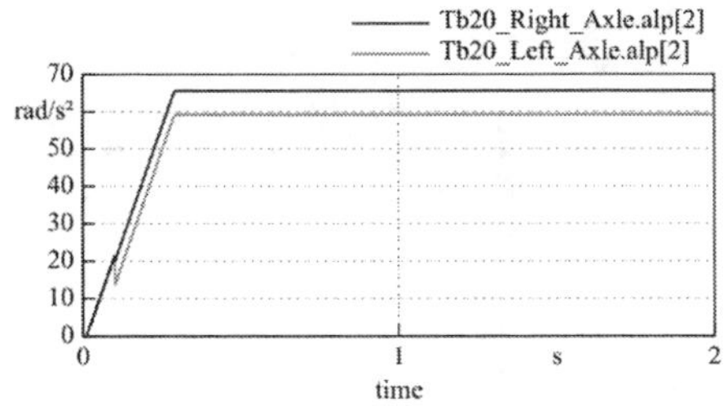

(c) Angular acceleration of the two wheels.

Figure 10. Characteristics of new TVD system.

3.1.2 Characteristics of New TVD system

In the dynamic simulation, an input torque is applied on the case, vehicle loading is represented as torque loadings applied on wheels, and a braking torque is applied on left side of the outer gear shaft. In Fig. 10(a), the input torque is applied from 0 N·m to 300 N·m in 0.3s, whereas the braking torque is applied by 20 N·m after 0.1s. The curves in Fig. 10(b) and Fig. 10(c) show there are rotational speed and angular acceleration differences between right and left wheels after the braking torque is applied. According to the simulation results, the new TVD system has the ability to transfer the torque power from one side to the other side, which means the new TVD system possesses torque vectoring characteristics.

3.2 Sensitivity analysis

3.2.1 Changing parameter T_b, R_3, and R_4

In order to understand the torque vectoring performance, different design parameters are changed. Therefore, the rotational speed and angular acceleration differences between the wheels are meaningful to determine the most important parameter to affect the torque vectoring performance. In this simulation, the model and initial parameter values are the same as the previous subsection. The parameters analyzed here include the braking torque T_b, the radius of the outer differential gear R_3, and the radius of the torque vectoring gear R_4.

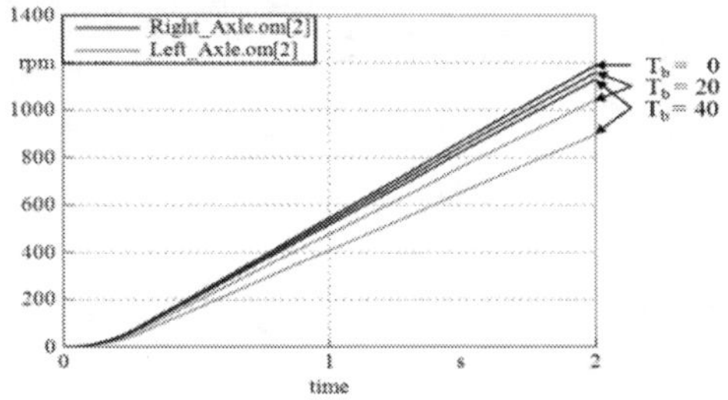

(a) Rotational speed of the two wheels.

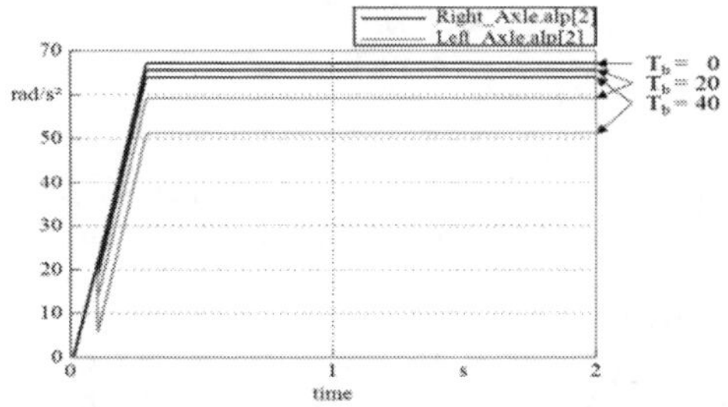

(b) Angular acceleration of the two wheels.

Figure 11. Change Parameter T_b.

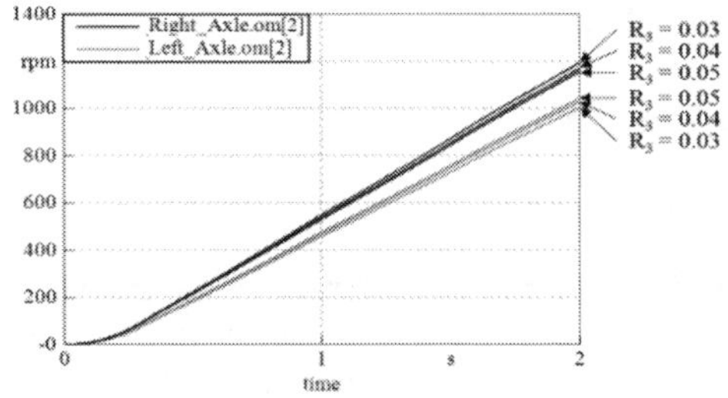

(a) Rotational speed of the two wheels.

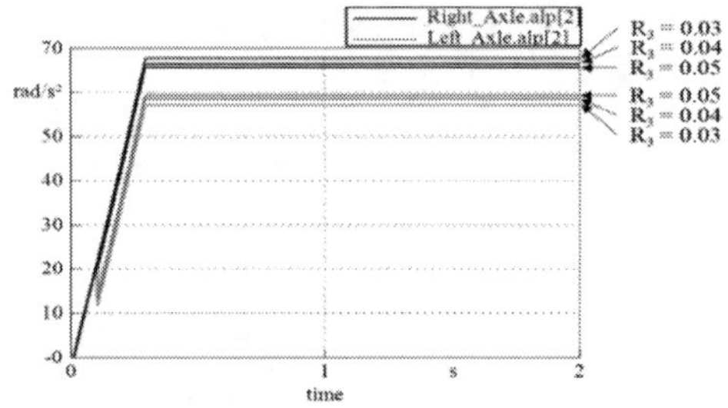

(b) Angular acceleration of the two wheels.

Figure 12. Change Parameter R_3.

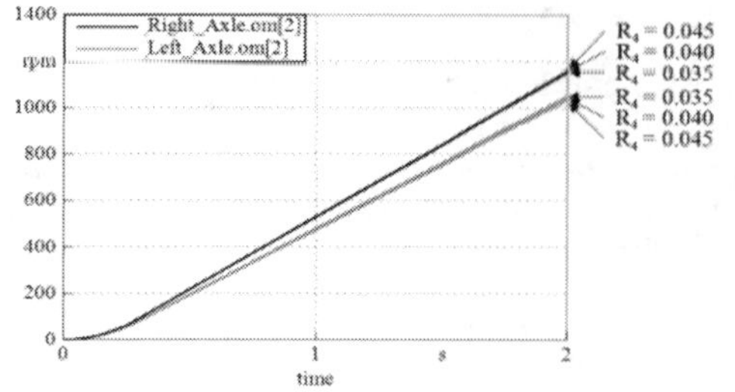

(a) Rotational speed of the two wheels.

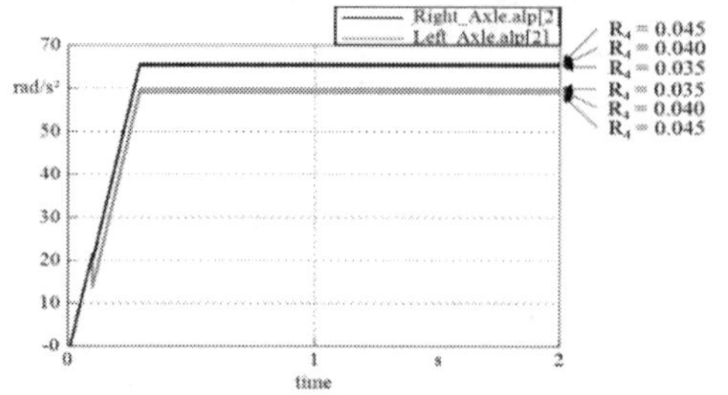

(b) Angular acceleration of the two wheels.

Figure 13. Change Parameter R_4.

3.2.2 Discussions

In changing the braking torque, T_b is set to be 0, 20, and 40. Figure 11 shows there is no speed difference between wheels while it is without braking torque, whereas the rotational speed of the left wheel is decreased immediately after the braking torque intervenes. The larger braking torque, the more speed dropped on the left wheel. In changing the radius of the outer differential gear, R_3 is set to be 0.03, 0.04 and 0.05, and the braking torque T_b is reset to 20. Figure 12 also shows dropped speed of the left wheel, however, there is no much difference while changing the parameter R_3. In addition, the similar simulated results can be observed while changing the parameter R_4 (Fig. 13).

3.2.3 Future work

The future work is to establish a whole vehicle model equipped with the new TVD system which is able to see the effect of the torque vectoring in vehicle dynamics simulation. In addition, important factors such as backlash and viscous friction will be taken into consideration.

4 Conclusions

This paper presents a new TVD system. A mathematical matrix form is derived, and a three-dimensional structural model is established, and a dynamic performance is illustrated based on the SimulationX multi-physics model. The simulation results demonstrate the torque vectoring capability and of the new TVD system that can transfer the torque power from one side to the other side while a braking torque is applied. In characteristics of sensitivity analysis, changing the braking torque is more efficient to affect the torque vectoring performance than changing the radius of the outer differential gears. As a result, a control strategy by engaging the braking torque is more effective way to distribute torque power on wheels, whereas the size of the new TVD system can be smaller due to its less sensitivity to dimension.

References

1. Y. Shibahata, K. Shimada, and T. Tomari, Vehicle System Dynamics, Vol. **22**, pp.465–481. (1993)
2. Geng, C., Mostefai, L., Denai, M. and Hori, Y., IEEE Transactions on Industrial Electronics, Vol. **56**, pp. 1411–1419. (2009)
3. Yim, S., Park, Y. and Yi, K., International Journal of Automotive Technology, Vol. **11**, pp. 147–153. (2010)
4. Furukawa, Y., Yuhara, N., Sano, S., Takeda, H. and Matsushita, Y., Vehicle System Dynamics, Vol. **18**, pp. 151–186. (1989)
5. J. Park and W. Kroppe, SAE Technical Paper 2004-01-2053. (2004)
6. T. Miura, Y. Ushiroda, K. Sawase, N. Takahahi, and K. Hayashikawa, Mitsubishi Motors technical review, pp. 23-26. (2008)
7. N. Mohd Zainal Abidin, S. Imamori and A. Alexander, SAE Technical Paper 2015-01-1098. (2015)
8. L. Kakalis, A. Zorzutti, F. Cheli and G. Travaglio, SAE Int. J. Passeng. Cars - Mech. Syst. **1**(1):514-525. (2009)

DOI: 10.1051/matecconf/201710807005

Development of Strength Test System for Structure under Large Uniform Surface Load

Yu Liu

China Ship Development and Design Centre, Hubei Wuhan 430064, China

Abstract. Large uniform load on plane widely exists in the structural design of ship equipment. In order to verify the reasonability of equipment structural design, a test system of structural strength is needed to make structural test. To this end, a strength test system is developed. In this test system, a concentrated load, which is equal to actual uniform surface load that acts on the equipment, is imposed on the loading part by Hydraulic Loading Method. Then, the concentrated load is turned into several concentrated loads uniformly distributed in loaded area of the equipment structure. Stress measurement system is applied to make a tracking measurement of the series of points in the equipment structure. The test data is professionally recorded with a real-time analysis. In order to verify the feasibility of the test system, a complete test platform for a ship equipment structure is designed, furtherly a contrastive analysis has been undertaken between the outcome of the test system and the one of finite element simulation, and it turns out that the application of the test system is feasible.

1 Introduction

Large uniform load on plane widely exists in the structural design of ship equipment, as which is well-known that it very complicated. In order to verify the reasonability of equipment structural design, making a test on the structural strength of a new equipment model is usually necessary. R. S. Dow (1996) puts large scale FRP structure into test to confirm the reasonability of marine structure design. Some researchers have developed a new inflatable/rigidizable hexapod structure testbed for shape and vibration control of large structures in space (O. Adetona et al. 2002). H. Jiao (2015) gives structural model test study of lightweight girder of bridge crane. W. Choi, H. B. Park, C. D. Kong (2014) have studied advanced propeller blade for next generation turboprop aircraft, discussed its static structural design and test.

In structural strength test, loading methods such as Gravity Loading Method (GLM), Hydraulic Loading Method (HLM), Mechanical Loading Method (MLM) and Pneumatic Loading Method (PLM) are usually used. Z. Li (2004), X. Cheng (1990) and S. J. Harrison (2004) explore all of the above-mentioned loading methods respectively. Z. Fang and J. Yu (2003) makes a further discussion of HLM method, and put forward determination of the parameters in the servo-hydraulic structure testing system. Airbag design is one of the key technology of PLM, design method of uniform loading system using Airbags in structure strength test is brought out in S. Chen, Z. Yang and B. Li (2012).

The GLM method can be easily achieved. Materials for load can be obtained from local sources, but loading weight on the structure surface is time-consuming. In HLM method, comparatively, high load can be easily obtained. Loading is not only high efficient but also safe, so that it deserves an extensive use. The loading system of MLM is the simplest, however, it has a flaw that the materials for high load can be hardly to be obtained. Loading in PLM method which can be loaded vertically on the sample's surface, the advantage of this loading method is stable, and the disadvantage is that the loaded surface is invisible. According to the above comprehensive analysis, it is supposed that HLM method is more suitable for the structural strength test of ship equipment with large uniform load on plane.

2 The HLM method

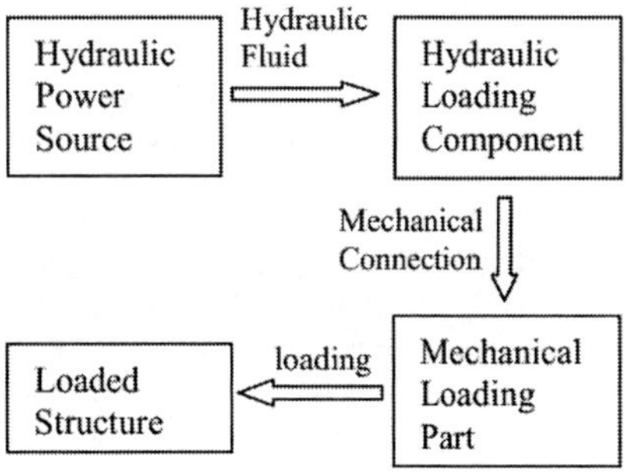

Figure 1. Loading sketch of HLM method

DOI: 10.1051/matecconf/201710807005

The HLM method is widely used in structure test. Figure 1 gives loading sketch of HLM method. In this method, there must be a Hydraulic Power Source. The Hydraulic Power Source presses hydraulic fluid into hydraulic loading component. Then, according to the mechanical loading part which connect to the hydraulic loading component, the structure is loaded.

In HLM method, the load is always concentrated load. When applied to structure test, the Mechanical Loading Part may need adaptive modification.

3 Test System

3.1 Equivalents from actual uniform load

Some concentrated loads take the place of the uniform load, which load on plane, as expressed in formula. (1). The alternative rule is shown as follows:

(1) the sum of all the concentrated loads is equal to the uniform load;

(2) the point of composition of the concentrated loads coincide with the point of composition of the uniform load;

(3) every concentrated load must be equal to each other, and point of the concentrated loads must be uniform distributed.

$$\begin{cases} P = S \cdot p \\ \sum_{i=1}^{N} P_i = P \\ \sum_{i=1}^{N} P_i L_i = P L \\ P_1 \approx P_2 \approx P_3 \approx \cdots P_N \end{cases} \quad (i = 1,2,3 \cdots N) \qquad (1)$$

Where P is concentrated load, S is plane area, p is uniform load, Pi is the concentrated load which act on point i, L is distance between acting point of P and the centre of rotation, Li is distance between acting point of Pi and the centre of rotation.

3.2 Theory of loading

As shown in Figure 2, concentrated load is acted on Rigid Loaded Plane. The concentrated load is equal to the uniform load which act on Tested Structure Plane. The acting point of the concentrated load and the uniform load coincide. There are elastomers between the Rigid Loaded Plane and the Tested Structure Plane. The elastomers are uniformly distributed throughout the Tested Structure Plane. In testing, the elastomer cannot be crushed, and the material of elastomer must be chosen to meet the testing requirement. When being tested, the concentrated load acts on the elastomers uniformly by the Rigid Loaded Plane. Because that every elastomer has identical size and performance, the load transmitted by the elastomer itself is unanimous. In this way, concentrated loads which act on the Tested Structure Plane are got, and they are uniformly distributed.

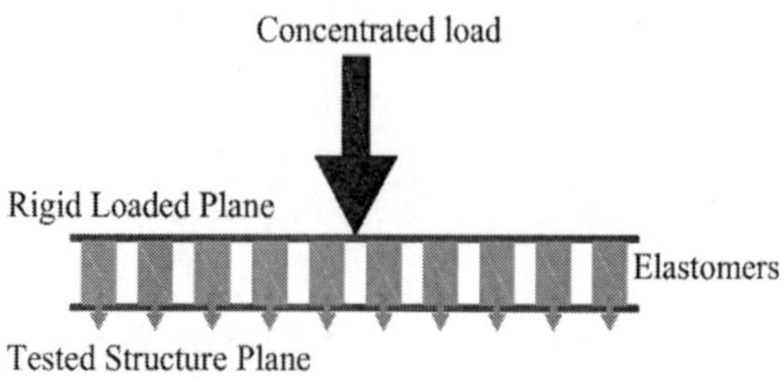

Figure 2. Sketch of loading

3.3 Composition of test system

The test system is comprised of hydraulic loading device, loading part, tested structure and test bed, and system of stress measurement as shown in Figure 3. The hydraulic loading device consists of a hydraulic power source and some hydraulic loading components. Concentrated load is supplied by this device, and tested structure would be under normal loading. Maximum load supplied by this device must be able to fulfil test requirement. The loading part has a plane structure at its end. There are many elastomers on this plane structure, and the elastomers are uniformly distributed. The plane structure of loading part and the tested structure plane have the same shape. In testing, tested structure must be fastened stably to test bed.

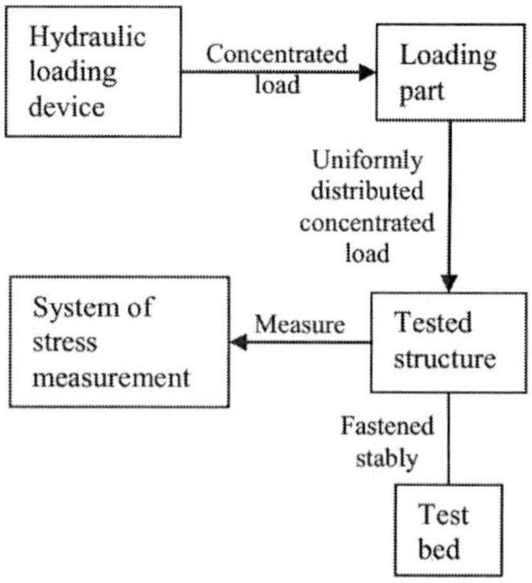

Figure 3. Constitution of the test system

3.4 Record and process of test data

There is a special system of stress measurement, which is used to record the test data and process the test data constantly.

4 Feasibility of the Test System

Plane structure as shown in Figure 4 is common in ship equipment. What we will discuss is that the application of the preceding test system to validate the feasibility of design of the plane structure.

4.1 Test platform

Figure 4. One plane loaded structure

Figure 5. Plane structure of the loading part and the elastomers

Figure 6. Loading condition of the plane loaded structure

A test platform is designed for the plane loaded structure as shown in Figure 5. In this test platform, hydraulic cylinder is the hydraulic loading component, and it is in charge of hydraulic loading. Structure of the loading part is specially designed for greater stiffness. Plane structure of the loading part and the loaded structure plane have the same shape. Also, there are many elastomers on this plane structure of the loading part, and the elastomers are uniformly distributed, as shown in Figure 5. An instrument of static strain gauge is used to measure the strains of the plane loaded structure while loaded, and then the strains are converted to stresses. When testing, the loading condition of the plane loaded structure can be seen in Figure 6. In Figure 7, the special system of stress measurement is shown.

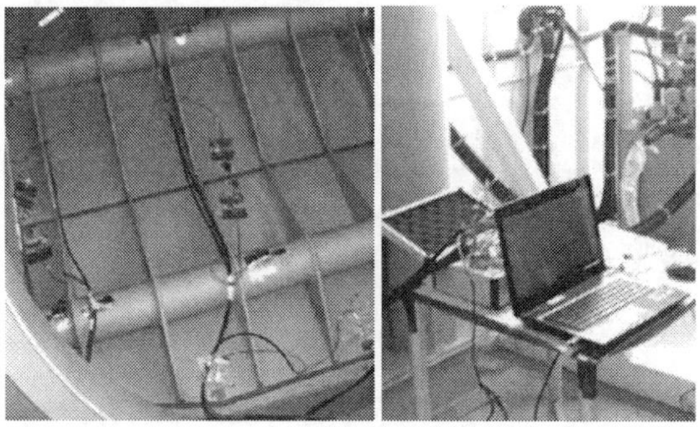

Figure 7. Special system of stress measurement

4.2 Analysis of test result

The plane loaded structure is fastened stably while testing. The 40%, 60%, 80% and 100% designing load tests are respectively completed, and every test is repeatedly carried out. In order to validate the reliability of the test result, finite element method is used to simulate the structural static stress of the plane loaded structure. Comparison of the result between the test and the simulation is conducted. Results of comparison of five points among them are shown in Figure 8- Figure 12.

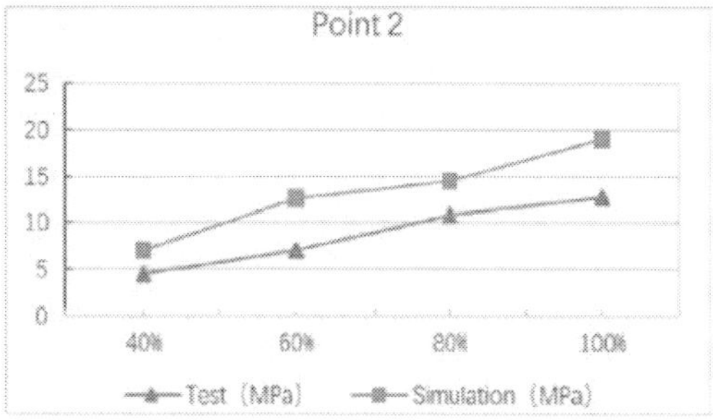

Figure 8. Data comparison of experiment and simulation at point 1

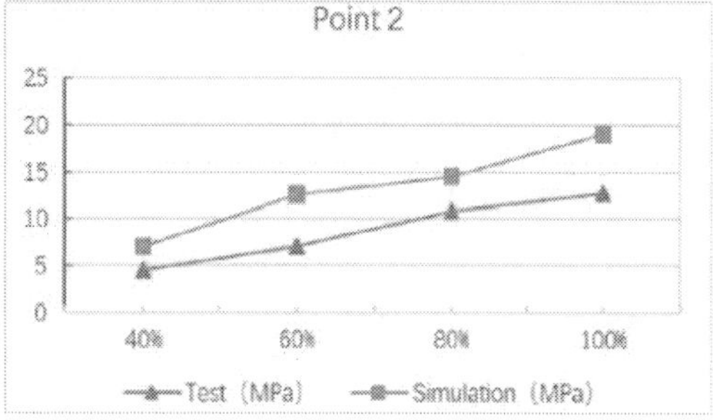

Figure 9. Data comparison of experiment and simulation at point 2

DOI: 10.1051/matecconf/201710807005

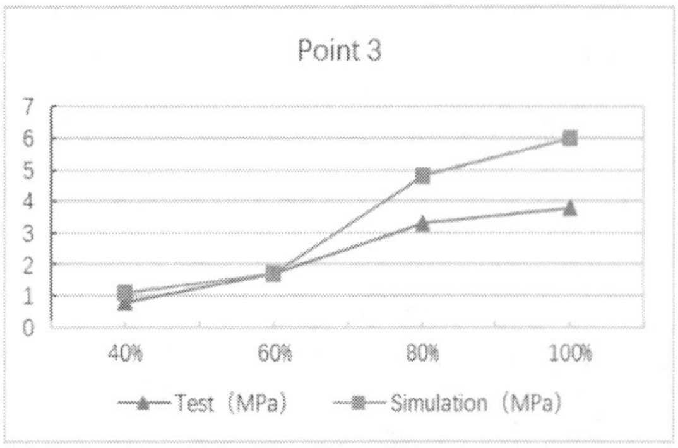

Figure 10. Data comparison of experiment and simulation at point 3

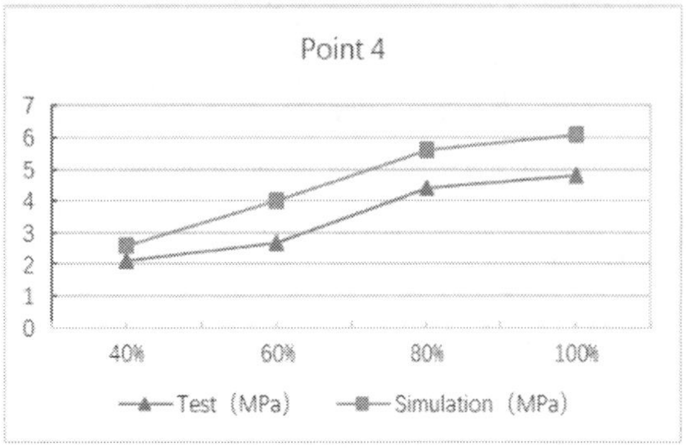

Figure 11. Data comparison of experiment and simulation at point 4

Whether the test or the simulation, the stress of the points increases with the increase of loading force from the figures, and this is accord with the actual situation. Comparison shows that the test results fit the finite element simulation very well, and maximum stress point of test and simulation coincide in Figure 12. Thus, we can say that this test platform can be used to accurately measure stress of the plane loaded structure which is common in ship device.

Because that the test platform can be used to accurately measure stress of the plane loaded structure which is common in ship device, it is shown that the strength test system for structure under large uniform surface load is feasible.

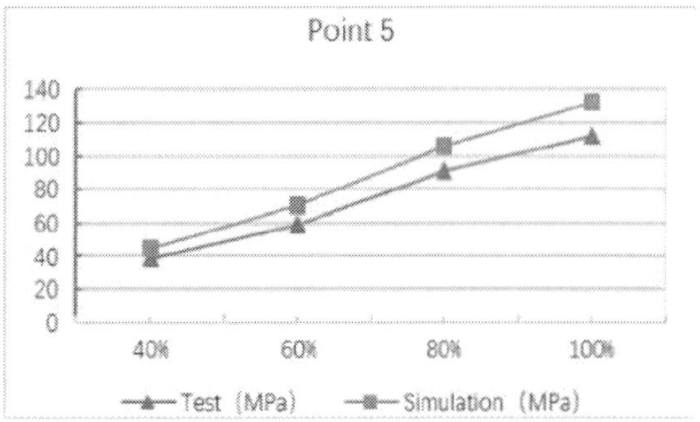

Figure 12. Data comparison of experiment and simulation at point 5

5 Conclusion

Hydraulic Loading Method is used to load on Rigid Loaded Plane. What it is loading is concentrated load, which is equal to the uniform load. Then, the concentrated load is turned into many concentrated loads which are uniformly distributed by loading part. Stresses of many points of device are measured through system of stress measurement, and the measured data is record and processed.

In order to validate feasibility of the test system, a test platform is designed for one plane loaded structure which common in ship device. Then data from test platform is compared with that of simulation. Results of comparison show that the strength testing system for structure under large uniform surface load is feasible.

References

1. R. S. Dow. "Large scale FRP structure tests", International Conference on Lightweight Materials in Naval Architecture: ICLMNA 1996. LONDAN: The Royal Institution of Naval Architects;1996.
2. O. Adetona, L. H. Keel, L. G. Horta, et al. "Description of New Inflatable/Rigidizable Hexapod Structure Testbed for Shape and Vibration Control", 43rd AIAA/ASME/ASCE/AHS/ASC Structures, Structural Dynamics, and Materials Conference Denver, 2002. American Institute of Aeronautics and Astronautics, p. 1451-1461.
3. H. Jiao. "Structural Model Test of Lightweight Girder of Bridge Crane", Journal of Mechanical Engineering, 2015, 51(23): 357-368.
4. W. Choi, H. B. Park, C. D. Kong. "The Study of Advanced Propeller Blade for Next Generation Turboprop Aircraft -Part II. Static Structural Design and Test", J. of The Korean Society for Aeronautical and Space Sciences, 2014, 42(4): 336-343.
5. Z. Li. "Theory and Technique of Engineering Structure Experiments", Tianjin: Tianjin University Press, 2004. (in Chinese).
6. X. Cheng. "Experimental Technique and Data Processing of Ship Structure Experiment", Shanghai: Profile of Shanghai Jiao Tong University Press, 1990. (in Chinese).
7. S. J. Harrison. "Military aircraft structural testing in Australia - supporting the ADF", Australia: University of South Australia, 2004.
8. Z. Fang, J. Yu. "The Ascertain of the Parameters in the Servo-Hydraulic Structure Testing System", Chinese Hydraulics & Pneumatics, 2003, 6: 12-14. (in Chinese).
9. S. Chen, Z. Yang, B. Li. "Design Method of Uniform Loading System Using Airbags in Structure Test", Engineering Mechanics, 2012, 29(6): 146 - 150. (in Chinese).

MATEC Web of Conferences **108**, 08001 (2017) DOI: 10.1051/matecconf/201710808001
ICMAA 2017

Calculation Methodology of Power Plant Universal Cycle Using Low-grade Heat of Refrigerant

S.V. Zaika , D.A. Uglanov and S.S. Dostovalova

Samara University, Heat engineering and heat engines Department, Samara, Russia

Abstract. The problems of high energy loss exists in many industries and in many technological processes. The problem of not effective use of the energy potential of the refrigerants is the main problem, because the refrigerant loses its potential and the potential energy at the regasification upon receipt of the liquefied natural gas. Solution of this problem is using low-grade heat of the refrigerant. This investigation describes the creation of a design procedure of the installation of the cycle that uses low-grade heat of the refrigerant. The cycle that uses low-grade heat of the refrigerant has been described and calculation method has been developed.

1 Introduction

Currently, the use of low-temperature heat of cryogenic liquids in the industry is spreading rapidly. This is due to increasing volumes of liquefied natural gas, the general trend of energy saving and energy potential of cryogen in form of energy which has been expended on liquefaction [1]. In a typical regasification of cryogen using environmental heat or special sources heat, the working fluid acquires parameters of environment and loses low-temperature potential. At the pressure equality, the working fluid wastes the potential energy of pressure gas [2]. Under real conditions, the conversion of thermal energy into other forms of energy takes place in the presence of the real environment with temperature T_e.

The environment can be used as a heat exchanger or a heat body. Figure 1 shows the Carnot cycle (A23BA and 1AB41). Cycles $T_1^+ > T_{o.c} = T_2^+$ are high-potential cycles. Cycles $T_1^- > T_{o.c} = T_2^-$ are low-potential cycles. Generally, the Carnot cycle has been used for an evaluation of thermal systems as a standard. Consider their effectiveness relative to the amount of heat input without regard to environmental effects [3-4].

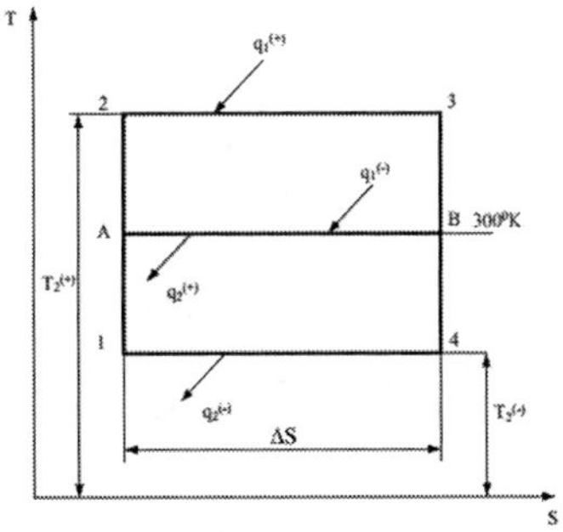

Figure 1. High-grade and low-grade Carnot cycles

The results indicate that Carnot cycle just serves as one form of the best cycles and only if the polytropic exponents (n) of the expansion and compression processes equal to each other, the efficiency of that cycle reaches the highest value, just that of Carnot cycle [5].

The purpose of this research is investigation of polytropic thermomechanical cycle working by low-grade heat of cryogen. The formula of the efficiency of thermal energy converters contain coefficient $\eta_O = \eta_{t_{n.o}} / \eta_{t_K}$. For its evaluation we consider the theoretical cycle efficiency dependence of the most important cycle parameters [6].

For this purpose it is sufficient to investigate the generalized polytropic cycle (Figure 2), which is limited by the following assumptions: 1) processes are reversible; 2) the specific heat, and hence the polytropic indices are constant values; 3) working body - ideal gas; 4) heat supply is carried out during II, removal - during IV; 5) heat transfer in the process I is absent, although the process is considered as a polytropic (a consequence of the IPOA specificity, in which energy components are supplied either liquid, or in a precompressed state); 6) Process III is considered to be adiabatic [7].

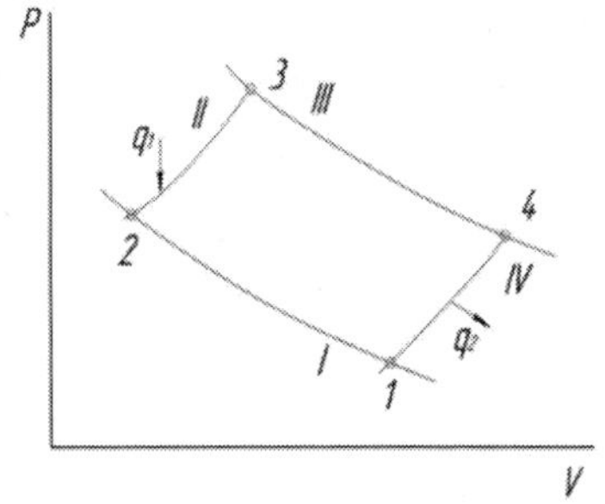

Figure 2. Polytropic cycle illustrated on a P-V diagram

The equation $\eta_{t_{n.o}}$ of polytropic cycle should include the maximum-to-minimum cycle temperature ratio T_3/T_1 cycle and the pressure at point 4 – p_4. For simplicity it can be taken $T_1 = 300$ K. When movement occurs in an environment with constant state parameters (on the surface of Earth, in space) backpressure $p_{o.c} \cong p_1 \leq p_4$, although ideally it should be equal $p_{o.c} = p_4$ [8]. This case will be considered.

2 Research of polytropic cycle

If the working fluid parameters were set at point 1 - p_1, V_1, T_1, polytropic index n_i, compression ε_{ci} or degrees of expansion ε_{pi}, or pressure ratio π ($\pi = 1/\varepsilon_p^n = \varepsilon_c^n$, the efficiency will be equal to:

$$\eta_{t_p} = 1 - \frac{c_{n_{IV}}\left(\pi_{IV}^{\frac{n_{IV}-1}{n_{IV}}} - 1\right)}{c_{n_{IV}}\varepsilon_{c_I}^{n-1}\left(\pi_{IV}^{\frac{n_{IV}-1}{n_{IV}}} - 1\right)}$$

Let's introduce the ratio of minimum and maximum temperature cycle T_1/T_3, and make the conversions. Because the

Because the

$$c_n = c_V\frac{n-k}{n-1},$$

and

$$\frac{T_3}{T_1} = \pi_I^{\frac{n_I-1}{n_I}} \cdot \pi_{II}^{\frac{n_{II}-1}{n_{II}}}$$

so

$$\eta_{t_n} = 1 - \frac{T_1}{T_3}\frac{\frac{n_{IV}-k}{n_{IV}-1}\left(\pi_{IV}^{\frac{n_{IV}-1}{n_{IV}}} - 1\right)}{\frac{n_{II}-k}{n_{II}-1}\left(\pi_{II}^{\frac{n_{II}-1}{n_{II}}} - 1\right)\cdot \pi_{II}^{\frac{n_{II}-1}{n_{II}}}} \quad (1)$$

or

$$\eta_{t_n} = 1 - \frac{T_1}{T_3}\frac{\frac{n_{IV}-k}{n_{IV}-1}\left(\varepsilon_{c_{IV}}^{n_{IV}-1} - 1\right)}{\frac{n_{II}-k}{n_{II}-1}\left(1 - \varepsilon_{p_{II}}^{n_{II}-1}\right)} \quad (2)$$

Formula (1) is convenient to use when $\varepsilon_{p_{II}} = \varepsilon_{p_{VI}} = 1$ and $n_{II} = n_{IV} = \infty$, at isochoric process II-III and IV-I, the expression (2) - when $\pi_{II} = \pi_{IV} = 1$ and $n_{II} = n_{IV} = 0$. All intermediate cases, they are equal and the choice should be made depending on the researcher's interests to influence π or $\varepsilon_c = 1/\varepsilon_p$. Substituting in (1) and (2) the values of the polytropic indexes and characteristics corresponding to the Carnot cycle, converts them to the Carnot efficiency $\eta_{t_n} = \eta_{t_k} = 1 - \frac{T_1}{T_3}$.

Let's investigate the expression (1). The cases has practical interest when $\pi_{II} > 1$, $0 < n_{II} < 1$; $1 < n_{II} \leq \infty$ and $\pi_{IV} > 1$, $0 < n_{IV} < 1$; $1 < n_{IV} \leq \infty$.

From the above $n_{IV} = f(\pi_I, \pi_{II}, \pi_{IV}, n_I, n_{II}, k)$. This relation could be easily represented in the form

$$n_{IV} = \frac{\lg\frac{p_4}{p_1}}{\lg\frac{\vartheta_1}{\vartheta_4}} = \frac{\lg \pi_{IV}}{\lg \pi_I^{(\frac{1}{n_I}-\frac{1}{k})} \cdot \pi_{II}^{(\frac{1}{n_{II}}-\frac{1}{k})} \cdot \pi^{\frac{1}{k}}} \quad (3)$$

It allows to identify and, most importantly, establish relationship with π_{IV} for each case. Communication between n_{IV} and T_1/T_3 can be obtained in the following form:

$$n_{IV} = \frac{\lg \pi_{IV}}{\lg\frac{T_1}{T_3}(\pi_I\pi_{II})^{1-\frac{1}{k}}\pi_{IV}^{\frac{1}{k}}} \quad (4)$$

For analysis of (1) values n_{IV} and π_{IV} can be specified, taking into account, that the ratios of (3) and (4) are not broken (since other parameters and characteristics can change accordingly). However, it's more interesting to obtaine the results, taking into account the impact of the dependencies (3), (4).

Considering, that under the real conditions $(\pi_I, \pi_{II}) \gg \pi_{IV}$, it is permissible to restrict the study in cases, where $(\pi_I, \pi_{II}) > \pi_{IV}$. The maximum take equal $\pi_{IV} = 1000$ (the case is unreal). We introduce into the formula (I) ratio:

$$C_c = \frac{1-\frac{1}{n_{II}}}{1-\frac{1}{k}} \cdot \frac{1-\frac{1}{n_{II}}}{1-\pi_{II}^{(\frac{1}{n_{II}}-1)}} = \frac{1-\frac{1}{n_{II}}}{1-\frac{1}{k}} \cdot \frac{1}{1-\varepsilon_{p_{II}}^{1-n_{II}}},$$

After that, the equation takes the form:

$$\eta_{t_n} = 1 - \frac{T_1}{T_3}\left[\frac{1-\frac{k}{n_{IV}}}{1-\frac{1}{n_{IV}}}(\pi_{IV}^{\frac{n_{IV}-1}{n_{IV}}} - 1)C_c\right] \quad (5)$$

Investigating n_{IV} change and choose C_c near $n_{II} = \infty$ and $n_{II} = 0$ due to the fact that these values are typical for gas and for vapor-liquid cycles. Change n_{IV} sign occurs in the transition of its value through $n_{IV} = \pm\infty$, is suitable, as is known, isochoric process. These values n_{IV} shifted depending on the temperature (and pressure π), which is easy to determine if the condition that $\lg\frac{T_1}{T_3}(\pi_I\pi_{II})^{1-\frac{1}{k}}\pi_{IV}^{\frac{1}{k}} = 0$.

So we can tacen this eqn:

$$T_3 = T_1(\pi_I\pi_{II})^{1-\frac{1}{k}}\pi_{IV}^{\frac{1}{k}}$$

From the equation it follows that

$$T_1 = \frac{T_3}{(\pi_I\pi_{II})^{1-\frac{1}{k}}\pi_{IV}^{\frac{1}{k}}} = \frac{T_3(\pi_I\pi_{II})}{\pi_{IV}^{\frac{1}{k}}}$$

Using the above relations, we can investigate $\eta_{t_{n.o}}$ according to the equation (5). It's enough to take n_{IV} in the range corresponding to π_{IV} and T_3 . Values are selected according to C_c that can be defined for known $\pi_I\pi_{II}$ and T_1/T_3 ($\frac{T_3}{T_1} = \pi_I^{\frac{n_I-1}{n_I}} \cdot \pi_{II}^{\frac{n_{II}-1}{n_{II}}}$).

$$\pi_{II} = \frac{1}{\varepsilon_{pII}^{n_{II}}} = \left[\frac{\pi_I\pi_{II}}{(\frac{T_1}{T_3})^{\frac{1}{\frac{1}{n_{II}}-1}}}\right]^{\frac{\frac{1}{n_{II}}-1}{(\frac{1}{n_{II}}-1)+(\frac{1}{n_I}-1)}}$$

$C_c \to \infty$ at $n_{II} \to k$ and $C_c \to 0$ for $n_{II} \to \infty$ and $n_{II} \to 0$. Increasing $\pi_{II} \gg 3 \div 10$, especially at high $n_{II} \gg k$, doesn't have significant impact on C_c . Conversely, at low $\pi_{II} \ll 3 \div 10$ absolute value C_c sharply increases. Since the process II-III can be adiabatic (as the heat is being supplied by condition) and can not be $n_{II} = k$. Therefore, it makes no sense to consider the case when $C_c = \infty$. Under the real conditions $n_{II} \neq \infty$ and $n_{II} \neq 0$ is sufficient to investigate the change η_{t_n} at $C_c = -50 \div +50$. If we consider that usually $n_{II} > k$, then the actual values can be limited by the quantities $-5 \div +5$.

In order to determine the range of variation C_c, we should determine the values C_{ck}, in which $\eta_{t_n} = \eta_{t_к}$, т.е.

$$C_{ck} = \left[\frac{1-\frac{k}{n_{IV}}(\pi_{IV}^{1-\frac{1}{n_{IV}}}-1)}{1-\frac{1}{n_{IV}}}\right]^{-1}.$$

3 Change polytropic cycle efficiency depending on conditions

Using these relations, curves were built (Figure 3 – 6), which express the relationship (the V), as an example for k = 1,6. If necessary, these charts can be constructed for any value of argument $\eta_{t_{n.o}}$. Two cases are considered: $n_{II} = 0{,}5$ and $n_{IV} = 0{,}55$ (Figure 3-4); $n_{II} = 1{,}2$ and $n_{IV} = 0{,}85$ (Figure 5-6).

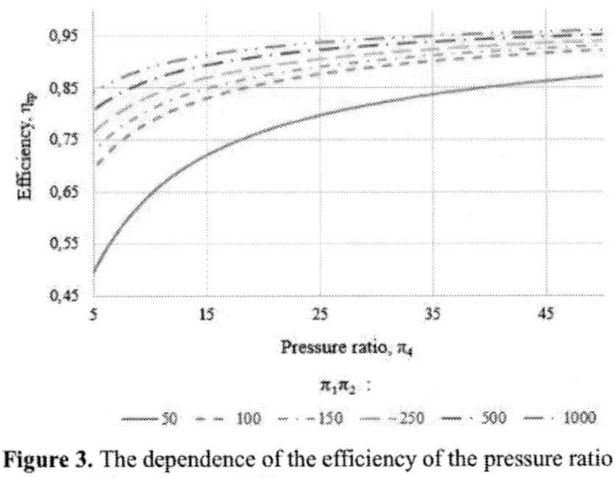

Figure 3. The dependence of the efficiency of the pressure ratio at $n_{II} = 0{,}5$ and $n_{IV} = 0{,}55$

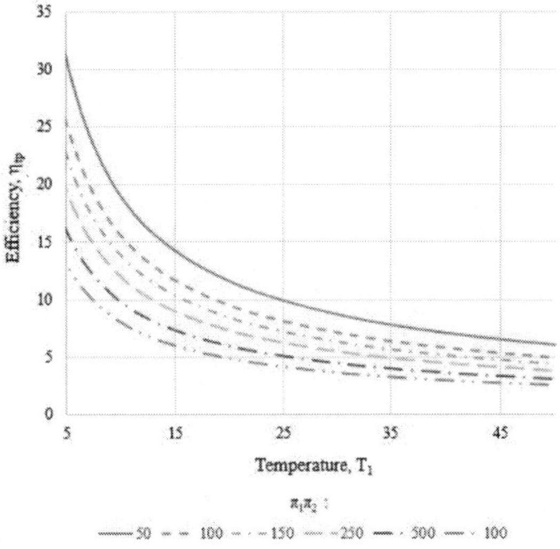

Figure 4. The dependence of the efficiency of the temperature at $n_{II} = 0{,}5$ and $n_{IV} = 0{,}55$

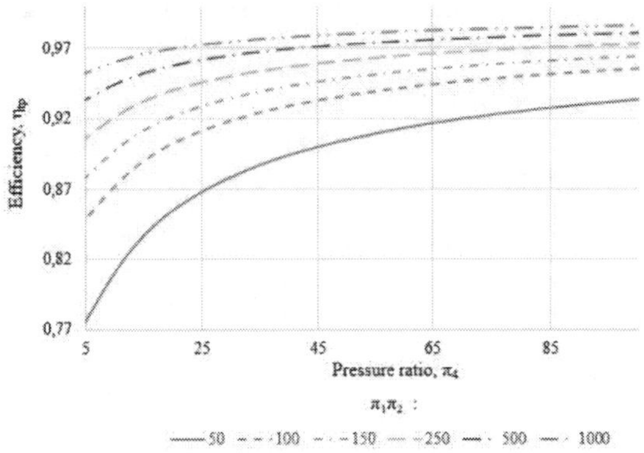

Figure 5. The dependence of the efficiency of the pressure ratio at $n_{II} = 1{,}2$ and $n_{IV} = 0{,}85$

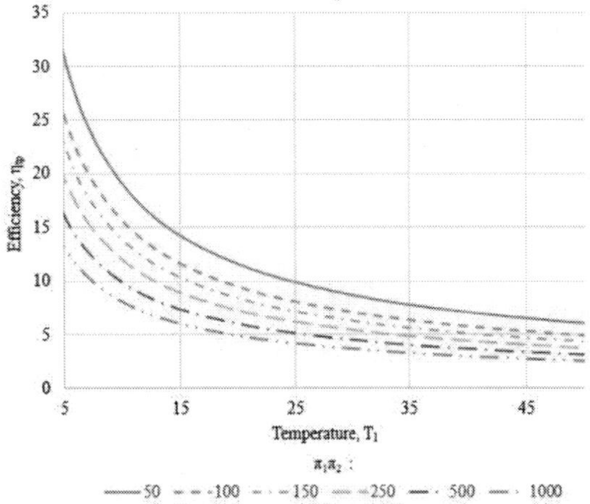

Figure 6. The dependence of the efficiency of the temperature at $n_{II} = 1{,}2$ and $n_{IV} = 0{,}85$

The above graph shows: 1) complex and different dependence of the efficiency η_{t_n} from π_{IV}, allowing in each case to find the optimal values k, π_I, π_{II} and T_1/T_3

for π_{IV} data; 2) significant differences of this dependence from the dependence $\eta_{t_k} = f\ (T_1, T_2, \Delta T)$. They show that the increase of economic efficiency is not enough to strive to increase the values of the polytropic exponent, since under certain conditions it can lead to the opposite result.

Analysis graphs below allows to choose the most advantageous values η_{t_n} and η_{t_k} for each particular case, and hence the value of the coefficient of completeness of the reversible conversion of energy η_o, as well as match PT (k, R). The relative efficiency of actual cycles depending on the working bodies compared to the Carnot cycle efficiency can be estimated by the number of work.

4 Conclusions

A method for calculating the universal thermodynamic cycle power plant using low-potential energy of cryogen was developed.

Main characteristics of dependence and, as well as their changes depending on the conditions were found. The following main conclusions can be distinguished on the basis of the study:

1. The calculations for two cases: $n_{II} = 0{,}5$ and $n_{IV} = 0{,}55$; $n_{II} = 1{,}2$ and $n_{IV} = 0{,}85$ were made. The results confirm the higher π_{IV}, the higher η_{t_n}. It can be concluded that increase T_1 leads to decrease π_{IV}, and hence to decrease η_{t_n}.

2. The higher n_{II} and n_{IV}, the higher the efficiency of the universal cycle power plant that uses low-grade heat of cryogenic.

Studies have been conducted polytropic cycle. Change polytropic cycle efficiency depending on conditions was analysed and described.

This research gives a new insight into cycles using low-grade heat of refrigerant, which is of great significance in to solve thermodynamic optimization problems. So this topic is really worth deep future investigations.

References

1. Hung T. Ahlborn B. Waste heat recovery of organic Rankine cycle using dry fluids// Energy Convers. Manage., 42 (2001), pp. 539–553
2. Walraven D., Laenen B., D'haeseleer W. Comparison of thermodynamic cycles for power production from low-temperature geothermal heat sources // Energy Convers. Manage., 66 (2013), pp. 220–233
3. Hansaem Park, Min Soo Kim. Performance analysis of sequential Carnot cycles withfinite heatsources and heat sinks and its application in organic Rankine cycles // Energy (V.99). 2016, pp. 1- 9
4. Dovgyallo A.I., Nekrasova S.O., Sarmin D.V., Uglanov D.A. Imitations tests of throttle cooling system with cryogenic refueling tank and its comparative characteristics // Applied Physics. 2013. Issue 4, 2013, Pages. P. 54-59
5. Rui Shang, Yin Zhang, Wenxing Shi, Xin Wang, Yinping Zhang. Fresh Look and Understanding on Carnot Cycle // Energy Procedia. 2014. Pages 2898-2901
6. Biryuk V.V., Gorshkalev A.A., Kayukov S.S., Uglanov D.A. Gas-Dynamic Analysis of Processes in a Small-Sizes Two-Stroke Combustion Engine // The Open Mechanical Engineering Journal. 2014. 8. P. 441-444
7. Biryuk V.V., Tsapkova A. B., Shimanov A.A. Numerical modeling of cyclone machine for cleaning gas generator gases, Key Engineering Materials. 2016. Volume 685. P. 153-157
8. Nekrasova, S.O., Uglanov, D.A.,Shimanov, A.A. Comparative analysis of calculation models of pulse tube thermoacoustic refrigerator, 4th International Conference for Young Scientists High Technology: Research and Applications, HTRA 2015, 21 - 24 April 2015, Publisher: Trans Tech Publications Ltd, 2016 г., P. 3-7

MATEC Web of Conferences **108**, 08002 (2017) DOI: 10.1051/matecconf/201710808002
ICMAA 2017

Optimal Power Flow-Based Calculation of Transmission Constraint Cost

Soonhyun Hwang and Balho H. Kim

Power Economics, Electronic and Electrical Engineering, Hongik University, South Korea

Abstract. As the construction of High Capacity transmission lines becomes a social issue, construction conditions of power system are being worsened, resulting in the decrease of power supply capacity and increase of supply costs. Because of this, feasibility for construction of High Capacity transmission lines needs to be explained logically, and the most effective method is to present economic effects such as the change of transmission constraint costs and power production costs. This paper intends to propose an Optimal Power Flow(OPF)-based method for calculation of constraint costs due to the construction of new transmission lines and to use it in order to perform case study for High Capacity transmission lines which will be constructed in Republic of Korea.

1 Introduction

Currently, Korean electric power industry faces the introduction and operation of competitive generation and greater importance of private generation companies, but as the construction of High Capacity transmission lines becomes a social issue, power system construction conditions are being worsened, making restrictions on power generation businesses of the transmission company as well as generation companies. It results in the decrease of power supply capacity, and unstable demand supply and increased supply costs in the long term. Because of this, feasibility for construction of High Capacity transmission lines needs to be explained logically to the public, and the most effective method is to present economic effects such as the change of transmission constraint costs and power production costs.[1]

Optimal Power Flow(OPF) is an advanced technique for performing Economic Dispatch and Power Flow calculations at the same time in order to determine output levels of each generators in the most economical way. It is effective for analyzing the economic effect of a system plan, as it can determine optimal(economical) output by considering transmission constraints and handle constraint cost calculation logically.[2]

This paper intends to propose an OPF-based method for calculation of constraint costs due to the construction of new transmission lines and to use it in order to perform case study for High Capacity transmission lines which will be constructed in Republic of Korea.

2 Method of calculating OPF-based constraint cost

Power supply costs with and without new transmission lines were compared to calculate constraint costs due to the construction of new transmission lines. Determining the Generator Maintenance Scheduling and the operation status(On/Off) of each unit and time may be an optimal method to calculate power supply costs for each case, but this paper ignores the Generator Maintenance Scheduling by considering practical limitations and replaces Unit Commitment with a plan to determine the operation status of units.

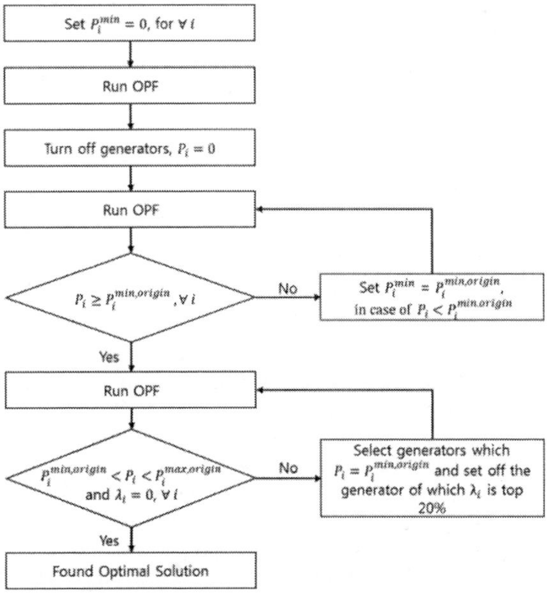

Figure 1. An algorithm for calculating the constraint cost of new transmission lines using OPF.

Figure 1 shows an algorithm for calculating the constraint cost of new transmission lines which uses OPF established in this study.

Where,
i : Generator number
P_i : Power produced by generator i
P_i^{min} : Minimum generation constraint of generator i
$P_i^{min,origin}$: Original minimum generation constraint of generator i
$P_i^{max,origin}$: Original maximum generation constraint of generator i
λ_i : Lagrange multiplier of generator i

The details of the algorithm process are as follows.

1) Minimum generation constraints of all units are set to zero and OPF is performed
- It is to determine the run(on) or stop(off) of units first and units of which generation is zero after OPF are targeted.
2) Units of which output is zero is stopped(off) on the basis of the result in step 1)
3) Units of which generation is below it's original minimum generation constraint are identified
- Units of which generation is below it's original minimum generation constraint(before set to zero) are selected among units which are running(On) in step 2). If all units generate power above it's original generation constraint of each, progress is made to step 5).
4) Minimum generation constraints are restored
- For units selected in step 3), minimum generation constraints are restored to the previous values(before set to zero). OPF is performed again and return is made to step 3) to check if there are units of which generation is below minimum generation.
5) Units restricted by minimum generation constraint are identified
- If all units are not restricted by minimum generation constraint, the current status is the optimal solution which can satisfy demand with the lowest cost. If there are units restricted by minimum generation constraint, however, these units can be stopped(off) to further reduce total generation costs. Lagrange multiplier(λ) is used as a criterion for determining which unit is to be stopped(off).[3]
6) Units to be stopped (off) are selected
- Units restricted by minimum generation constraint in step 5) are sorted in the order of λ, and among them, units of which the absolute value of λ belongs to top 20% are stopped. Then, return is made to step 5) and the step is performed repetitively. It is ideal that only one unit with the biggest λ is stopped and return is made to step 5), but this method takes very long time. On the contrary, if too many units are stopped to reduce processing time, total costs are made above the optimal solution. This paper proposes top 20% as a baseline, because repetitions proved empirically that it is most appropriate.

3 Case study

In this section, the procedure introduced in the above section is used to calculate transmission constraint costs for High Capacity transmission lines which will be constructed in Republic of Korea. In this study, transmission constraint cost is calculated as the difference in total generation costs between non-constrained supply and constrained supply which considers thermal capacity or stability constraints. This study uses PSS/E OPF as the OPF tool which solves a nonlinear problem consisting of an objective function and a linear combination of equality constraints to form a scalar Lagrangian function.[4] Other basic premises are as follows.

3.1 Basic premises

- Analysis is performed for the actual system in Republic of Korea and the High Capacity transmission lines are two lines with 7,290 MW capacity and 222 km length which will be completed in December 2019.
- Analysis period is total three years (2020 to 2022) and completion delay is set in the unit of six months to make total six scenarios.
- Load levels are 100%, 90%, 80%, 70% and 60% (total five levels), and averages of the recent three years are used for the number of days in a year for each load level.
- New and abolished facilities are based on the 6th Basic Plan for Long-Term Electricity Supply and Demand.[5]

New units which will be operated in the future are set discretionally as there is no correct information on the cost function. It is likely that new units will have highest priority of receiving Dispatching Instruction among those of the same type, as they clearly have better generation efficiency and lower generation costs than the currently operated units. Because of this, the above criterion has been applied to set the cost function of new units and the method is as follows.

- Facilities with the lowest generation cost are selected among generation source-specific units.
- The cost function coefficient of the selected units is decreased as much as 2% and then applied.

When selecting units with the lowest generation cost, cost functions of units are drawn in graphs and the graph in the lowest position is selected as a standard unit. As the graph of the selected cost function, however, does not always have the lowest value in all sections when compared with other facilities, the function coefficient is adjusted to generate the lowest value in all sections. At the time, 2% of decrease has been determined empirically to be optimum. If the decrease is smaller than 2%, Dispatching Instruction instructions have not been received preferentially sometimes, and if it is larger than 2%, total generation costs have been lowered too much. 2% set in this study, however, is not absolute and the cost function of a new generation unit has to be input more correctly in order to increase the reliability of the result value.

Table 1 shows the cost function of each generator type. Nuc, Ther, Com and CHP indicates Nuclear, Thermal, Combined power and Combined Heat and power.

Table 2 shows detailed scenarios for each load level and period. Timely completion(T/C) indicates the existence of new transmission lines in the corresponding period and delayed completion(D/C) indicates the non-operation of certain units in the corresponding period due to the delayed completion of new lines. Each scenario represents total generation cost for one hour in the corresponding load level.

Table 1. Cost function of new generator units.

Type		Cost Function
Nuc	Standard	$0.000105P^2 + 2.043505P + 267.815413$
	New	$0.000103P^2 + 2.002635P + 262.459105$
Ther	Standard	$0.000098P^2 + 1.856711P + 169.666484$
	New	$0.000096P^2 + 1.819577P + 166.273154$
Com	Standard	$0.000021P^2 + 1.620993P + 23.917894$
	New	$0.0000206P^2 + 1.58857P + 23.4395361$
CHP	Standard	$0.000382P^2 + 1.498898P + 62.204427$
	New	$0.000374P^2 + 1.46892P + 60.9603385$

Table 2. Detailed scenarios for each load level and period.

Year	2020'				2021'				2022'			
Load	First half year		Second half year		First half year		Second half year		First half year		Second half year	
90%	T/C	D/C	T/C	D/C	T/C	D/C	T/C	D/C	T/C	D/C	T/C	D/C
80%	T/C	D/C	T/C	D/C	T/C	D/C	T/C	D/C	T/C	D/C	T/C	D/C
70%	T/C	D/C	T/C	D/C	T/C	D/C	T/C	D/C	T/C	D/C	T/C	D/C
60%	T/C	D/C	T/C	D/C	T/C	D/C	T/C	D/C	T/C	D/C	T/C	D/C
50%	T/C	D/C	T/C	D/C	T/C	D/C	T/C	D/C	T/C	D/C	T/C	D/C

3.2 Results of case study

Table 3. Hourly generation costs for each scenario. (Unit: thousand KRW)

Year		2020	
Load	Case	First half year	Second half year
100%	T/C	3,602,298	3,801,023
	D/C	3,971,447	4,252,038
90%	T/C	2,647,932	2,813,253
	D/C	3,017,746	3,275,540
80%	T/C	1,782,651	1,847,402
	D/C	1,904,375	2,126,644
70%	T/C	1,454,792	1,513,545
	D/C	1,476,688	1,532,881
60%	T/C	1,137,197	1,185,121
	D/C	1,151,797	1,205,175
Year		2021	
Load	Case	First half year	Second half year
100%	T/C	3,624,048	3,731,809
	D/C	4,067,804	4,401,112
90%	T/C	2,658,845	2,736,267
	D/C	3,104,049	3,400,434
80%	T/C	1,797,932	1,813,831
	D/C	1,952,717	2,234,828
70%	T/C	1,464,208	1,476,342
	D/C	1,485,835	1,502,350
60%	T/C	1,141,364	1,149,327
	D/C	1,155,435	1,159,132
Year		2022	
100%	T/C	3,731,809	3,560,931
	D/C	4,401,112	4,371,199
90%	T/C	2,736,267	2,559,920
	D/C	3,400,434	3,375,452
80%	T/C	1,813,831	1,763,744
	D/C	2,234,828	2,206,481
70%	T/C	1,476,342	1,431,619
	D/C	1,502,350	1,460,355
60%	T/C	1,149,327	1,096,088
	D/C	1,159,132	1,113,034

Table 3 shows the PSS/E simulation results of for each scenario. Each value is divided into the first and second half for the total power generation cost for one hour. The analysis of detailed cost changes by each period is as follows.

3.1.1 First half of 2020/Second half of 2020

In the first half of 2020/second half of 2020, it is confirmed that the total cost of all the load units is increased when the new generator of 1000MW is invested and the load of specific level is increased. Compared with the first half of the year for 100% load, the total power generation of the coal generators was increased by 1000MW and the combined generators was increased by 1180MW since newly introduced coal generator achieved additional power generation in second half of the year during the timely completion in order to cover the increased load(There is no change in power generation for nuclear power and other generators). The total cost shows the increase of about 200 million won. In the case of delayed completion, some new equipments are not operated, and so there is more increase of cost than timely completion since combined power generator should cover all the load increase. In the case of 80% load, the difference between the total cost of the timely completion and delayed completion is large. It's because that in case of the timely completion, the increased load can be covered only by the coal generator whereas in case of delayed completion, combined power generator should cover the increased load due to non-operation of low cost power generator. In the case of 70% and 60% load, the increase in total cost is smaller than that of other load because the increase in power generation belongs only to coal-fired generators in both timely completion and delayed completion.

3.1.2 Second half of 2020/First half of 2021

In the second half of 2020 and the first half of 2021, The1460MW will be put into operation as new equipment. We can confirm that the total power generation cost is reduced in all load units as nuclear power generator of the lowest cost was newly introduced. Actually in case of 100%, the total power generation cost is reduced by 177 million won as the power generation of combined generator was decreased by 1460MW while the power generation by the nuclear power generator was increased by 1460MW. In case of 80% load, there is a large difference in the change of total power generation cost between timely completion and delayed completion in the first half/second half of 2020. This is because there is larger change of cost as the power generation of the combined generator decreases in delayed completion whereas there is small change of cost as the power generation of coal generator decreases as much as the increased power generation of nuclear power generator. In case of 70% and 60% load, the change of total cost is small in both timely completion and delayed completion due to the decrease of power generation of coal generator.

3.1.3 First half/Second half of 2021

In the first half/second half of 2021, 2460MW of new generators will be introduced and the abolition of some generators will be reflected, and the load will increase by 2040MW and the total power generation cost increased at all loads. In case of 100% load, the nuclear power plant is newly introduced but the total power generation cost has increased due to the abolition of some generators and the increase by 870MW in combined facility capability due to load increase. In case of 80% load, the power generation of the combined generator increased in delay completion. However, in case of the completion of the timed completion, there is no change in the power generation of the combined generator and the power generation increased just in nuclear and coal generators, which results in significant increase in change of total cost.

3.1.4 Second half of 2021/First half of 2022

In the case of the second half of 2021/the first half of 2022, there is no increase or decrease in the capacity of the entire power generation facilities, and also the load values are the same, which leads to the same result value.

3.1.5 First half of 2022/Second half of 2022

In the first half of 2022/the second half of 2022, there was no change in the total load and only 1460MW of new generators were introduced, which shows that the total generation cost is reduced in all load units. In case of 100% load, there was increase by 1460MW in nuclear power generation and decrease by 1460MW in combined power generation, which results in significant decrease in total power generation cost in timely completion. However, you can be see that the decrease in the total generation cost is small in the case of delayed completion, which means that for the delayed completion in the second half of 2022, one new nuclear power plant should be shut off and two coal generators should be shut off and therefore the power generation of nuclear power generators increases by 1460MW and the power generation of coal generators decreases by 1460MW when comparing with the second half of 2022. Total costs were reduced by about 30 million won in terms of cost. In case of 90% load, there is a difference in the total cost change between the timely completion and delayed completion due to the same reason as the 100% load, and the total load can be covered only by the nuclear power generator and the coal thermal power generator, which results in small change in total cost.

Table 4. Hours for each load level during one year

Load	First half year	Second half year	Sum
100%	132	63	195
90%	855	681	1536
80%	1768	1481	3249
70%	1229	1344	2573
60%	360	847	1207
Sum	4344	4416	**8760**

Table 5. Accumulated generation costs for total three years. (Unit: billion KRW)

	Timely completion	6 month delay	12 month delay	18 month delay
Total cost	47,600,512	48,212,805	49,012,567	49,757,097
Increased cost	0	612,293	1,412,055	2,156,585
Rate of Increased cost	0.00%	1.29%	2.97%	4.53%
	24 month delay	30 month delay	36 month delay	
Total cost	50,918,317	52,354,344	53,669,436	
Increased cost	3,317,805	4,753,832	6,068,924	
Rate of Increased cost	6.97%	9.99%	12.75%	

Table 5 shows the calculation of accumulated generation costs for total three years by multiplying hourly generation cost for each scenario(which is shown in table 3) and hours for which each load level is applied during one year(8,760 hours)(which is shown in table 4). The values of table 4 is the average value of past three year load data of Korea. Timely completion indicates total cost for three years if transmission lines are completed until 2020, and after that time, total costs are presented for delays in the unit of six months. Cost differences compared with timely completion are represented as increased costs if line completion is delayed. If the construction of new transmission lines are delayed one year, about 1,410 billion won of cost increase is made. It can be found that about 3,300 billion won and about 6,000 billion won of costs have increased for two and three years of delay, respectively.

4 Conclusion

This paper has proposed an OPF-based process for calculating constraint costs due to the construction of new transmission lines and used PSS/E OPF in order to perform case study for High Capacity transmission lines which will be constructed in Republic of Korea.

The OPF-based constraint cost calculation process which has been proposed by this paper may have theoretical limitations as it focuses on practical approach by replacing Unit Commitment with an algorithm to determine the operation status(On/Off) of units. It is judged, however, that this process can be a reasonable tool for future planning which considers uncertainty. It is also expected that accuracy can be improved by further study for linking with the Generator Maintenance Scheduling and Unit Commitment algorithms.

References

1. I.J Song, C.W Park, D.U Kang, M.J Bang, B.H Yun, Constrained Cost of the Transmission Line (1), KIEE.S.C, 154 (2006)
2. Y.C Kim, J.K Yoo, *Blackout and Power System Operation*, 515-516 (2015)
3. Dimitri P. Bertsekas, *Constrained Optimization and Lagrange Multiplier Methods*, MIT
4. SIEMENS, *PSS/E^{TM} 30.2 Users Manual*, 2005
5. ROK MOTIE, *6th Basic Plan for Long-Term Electricity Supply and Demand*, 2013

A Study on the Optimal Fuel Mix and ESS Capacity considering Environmental Policy

Seunghwan Lee, Soonhyun Hwang and Balho H. Kim

Power Economics, Electrical, Information and Control Engineering, Hongik University, South Korea

Abstract. Korea has been enforcing the environmental policies(ETS(Emission Trading Scheme), RPS(Renewable Portfolio Standard)) to reduce the greenhouse gas, and has been promoting the expansion of the ESS(Energy Storage System) supply through the K-ESS 2020 strategy. Accordingly, the role of the ESS in the power system has been attracting the recent interest and the effects are being verified. However, there is no standard for the optimal input capacity. It is rational to increase the ESS capacity as much as possible in order to optimize the proper function actualized by the ESS. However, since it is difficult to greatly increase the ESS capacity input as of this point due to high investment cost of the ESS, it is necessary to set up a standard for calculating the most economic input capacity. Accordingly, in this paper, in the process for establishing the generation plan, the environmental policies are considered into calculating the optimal fuel mix as well as the ESS optimal capacity per year.

1 Introduction

In Korea, approximately 87% of the greenhouse gas is emitted from the energy field, and approximately 42% of that is emitted from the power generation field[1]. Accordingly, in the power generation field, various greenhouse gas reduction policies(RPS, ETS and etc.) are enforced to reduce the greenhouse gas emission[2]. In addition, through the K-ESS 2020 strategy, an investment of approximately 6.4 trillion won in the ESS-related technology development and unit until 2020 is being promoted[3].

The ESS allows the generated power to be stored in the respective connection systems such as power plant, substation and transmission line, so that the stored power can be selectively/efficiently used to maximize the energy efficiency and stabilize the power system operatioin. According, to cope with the future power load increase and renewable energy expansion, the world has been promoting a strategy to develop and supply the ESS technology. As part of that strategy, diverse empirical projects are being performed through applying the ESS to substation, renewable energy complex, large-scale generator, house and building[3].

The main role of the ESS can be summarized into three processes: a peak load reduction after connecting the system; an ancillary service for the frequency regulation; and stabilization of the renewable energy output. Through such processes, the ESS contributes to enhancing the power quality[4]. However, as of this point, the price competitiveness of the ESS is insufficient in comparison to other power generation sources, and, therefore, in reality, there are difficulties in supplying the ESS. Accordingly, it is necessary to consider the strengths of the ESS into calculating the optimal capacity suitable for the installation purpose. However, as of this point, the economic feasibility based on the current supply of the ESS has not been analyzed, and the methodology to be used for calculating the optimal capacity does not exist.

Accordingly, in this paper, through the generation planning model to which the environmental policy is applied, the optimal fuel mix and the ESS optimal capacity per year are calculated.

2 Main subject

2.1 ESS capacity calculation methodology

Instead of directly generating the energy, the ESS is operated through storing and selectively using the energy generated from other power generation sources. Accordingly, it is difficult to view the ESS as other power generation sources such as nuclear, coal. In order to calculate the ESS capacity through the generation planning model, the ESS converted into the generation to which its characteristics are applied must be applied to the model. In the generation planning model, the non-dispatch such as wind power unit and photovoltaic unit contribute through modifying the shape of the load curve. Accordingly, in the generation planning model used in this paper, the ESS is treated and applied as the non-dispatch renewable generation, so that the load increases during charging and decreases during discharing.

DOI: 10.1051/matecconf/201710808003

2.2 Model for generation planning

2.2.1 Outline of model for generation planning

The power plant construction plan is an important part of the power industry investment plan. The power plant construction requires a considerable investment cost. Since the constructed power plant is operated for at least 30 years, it is influential on the operational cost of the overall power system throughout its lifespan. In addition, the generators for construction vary in terms of economic/technical characteristics. Accordingly, in the process for planning a generation unit, the generators specified above must be adequately combined to minimize the operation cost of the future power system.

The computer model used in this paper is the generation unit plan, and the purpose of this model is to consider the environmental policies(ETS and RPS) in Korea into deducting a plan for constructing the optimized power plant per year, and the functions/roles are as shown in Figire 1[5].

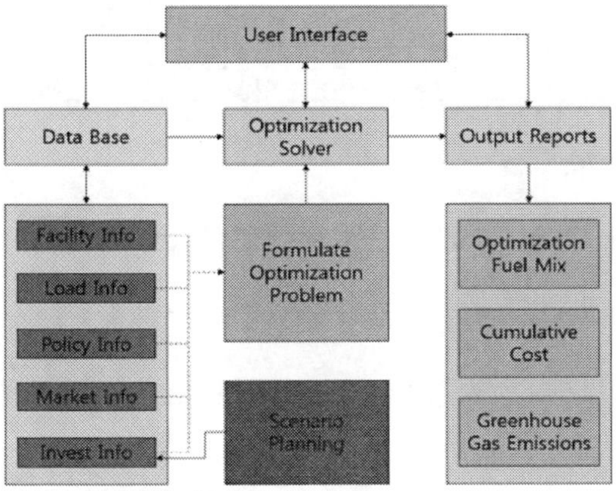

Figure 1. Generaion Planning Model.

2.2.2 Formulation

$$\mathrm{Min} \sum_y \left(C_y^{Gen} + C_y^{Env} + C_y^{Const}\right) \quad (1)$$

$$\mathrm{S.t} \sum_i X_{i,y,t} = D_{y,t} \quad (2)$$

$$X_{i,y,t} \leq I_{Cap,i} \quad (3)$$

$$\left(\sum_F \sum_t X_{F,y-1,t}\right) \times RPS_y^{Rate} \leq \sum_R \left[\sum_t \left(X_{R,y,t}\right) \times REC_R^{weight}\right] + REC_R^{Penalty} \quad (4)$$

$$ET_y^{Emission} = ET_y^{Trade} + ET_y^{Cap} \quad (5)$$

Where,

y : Year
t : Time
i : Generation Unit
C_y^{Gen} : Generation Cost in period y
C_y^{Env} : Environmental Policies Execution Cost in period y
C_y^{Const} : Construction Cost in period y
$X_{i,y,t}$: Generation for t, in period y
$D_{y,t}$: Demand in period t (MWh)
$I_{Cap,i}$: Capacity for Unit i (MW)
$X_{F,y,t}$: Generation Produced by Coal Generator
$X_{R,y,t}$: Generation Produced by Renewable Unit
RPS_y^{Rate} : RPS Obligation Rate
REC_R^{Weight} : REC Weight
$REC_R^{Penalty}$: Amount of Penalty in period y
$ET_y^{Emission}$: Emission in period y
ET_y^{Trade} : Emission Trading in period y
ET_y^{Cap} : Emission Criterion in period y

Formula (1) is the function for minimizing the overall cost, and it consists of generation cost, environmental policy execution cost and new unit construction cost. The generation cost, as shown in Formula (6), is formulated into a sum of fuel/maintenance costs per each unit/year/hour.

$$C_y^{Gen} = \sum_i \sum_y \sum_t \left(X_{i,y,t} \times C_i\right) + \sum_y \left(I_{Cap,i} \times M_i\right) \quad (6)$$

where,

C_i : Fuel cost of Conventional generator for i (won/MW)
M_i : Operating & Maintenance Cost for i (won/MW)

The environmental policy execution cost consists of RPS-nonfulfillment penalty cost and greenhouse gas emission trading cost, which are formulated into Formulas (7) and (8) respectively. The REC(Renewable Energy Certificate) penalty cost is 150% of the annual average REC trading cost.

$$\sum_y \left(REC_y^{Penalty} \times REC_y^{Penalty\_P}\right) \quad (7)$$

$$\sum_y \left(ET_y^{Trade} \times ET_y^P\right) \quad (8)$$

where,

$REC_y^{Penalty\_P}$: REC Penalty Cost in period y
ET_y^P : Emission Price in period y

The new unit construction cost, as shown in Formula (9), consists of new unit construction cost and salvage value difference. The unit construction cost and salvage value per unit are formulated into Formulas (10) and (11) respectively.

$$C_y^{Const} = C_y^C - C_y^S \quad (9)$$

$$C_y^C = \sum_i K_i \times Add_{i,t} \quad (10)$$

$$C_y^S = \sum_i K_i \times Add_{i,t} \times \left(\frac{P_{life\_i} - Y_T + y - y_s}{P_{life\_i}}\right) \quad (11)$$

where,

C_y^C : Construction Cost (won)
C_y^S : Salvage Value (won)
K_i : Unit Construction Cost for i (won/MW)
$Add_{i,t}$: New Unit Construction Cost in period y
P_{life_i} : Life Time for i
Y_T : Total Planning Period
y_s : Start year

Formula (2) signifies the restriction in supply and demand and Formula (3) signifies the restriction in generation per unit type. Formula (4) is the contraint condition for calculating the compulsory RPS-based fulfillment volume. The compulsory RPS-based fulfillment volume is calculated through multiplying the gross thermal power generation from the immediately preceding year by the compulsory execution volume from the involved year. The calculated value must be smaller than the sum of generation per renewable energy unit to which REC-issued weighted value is applied and RPS nonfulfillment penalty.

Formula (5) is the restriction in greenhouse gas emission per year. The gross greenhouse gas emission per year is the sum of emission trading volume and greenhouse gas emission permit from the involved year. The annual gross greenhouse gas emission, as shown in Formula (12), is calculated through multiplying the annual gross generation per generation unit by the greenhouse gas emission coefficient.

$$ET_y^{Emission} = \sum_i \sum_t (X_{i,y,t} \times Coef_i) \qquad (12)$$

where,

$Coef_i$: CO2 Emission Coefficient (tCO2/MWh)

3 Case study

3.1 Peak reduction

The ESS-based peak-load reduction stores the energy during light-load hours and discharges the energy during peak-load hours, and thereby standardizes the load and allows the system to be operated efficiently. Namely, through reducing the peak-load, the operating hours of the high-costing generator is shortened, the operational cost of the system is reduced, the power supply is stabilized, and the reliability is enhanced. At this point, since the operational effects vary depending on how the standard charge/discharge hours are set, it is necessary to determine an adequate charge/discharge pattern[4]. Accordingly, in this paper, a scaling is executed through using the load performance data from 2011, and the ESS charge/discharge pattern is constructed.

3.2 Scenario premise

In this paper, the input data from the 7th Basic Plan for Long-Term Electricity Supply and Demand[6] is used to conduct a case study. The key premises are as follows.

- All generation unit and load are connected to one node.
- The plan period ranges from 2015 to 2029, and the discount rate is 6.5%/year.
- The load data per 8760H time zone will be used.

Table 1. ESS Information[7].

Fuel Cost	0
Maintenance Cost (10^3won/MW)	70,708
Capacity(MW)	50
Greenhouse Gas Emission Coef	0
Construction Cost (10^3won /MW)	3,535,400

Table 2. Case Study Scenario.

Scenario	Environmental Policy	ESS	Note
1	X	X	-
2	O	X	-
3	O	O	-
4	O	O	ESS construction cost reduced 90%

3.3 Case study result

Table 3. Case Study Scenario Result (10^6won).

Scenario	Add ESS	Construction Cost	Generation Cost
1	-	4,856,614	305,736,458
2	-	10,266,737	304,423,288
3	-	10,266,737	304,423,288
4	600MW	10,440,705	304,213,269
Scenario	**Emission Trading Cost**	**RPS Penalty Cost**	**Total Cost**
1	-	-	310,593,072
2	19,846,540	22,607,071	357,143,635
3	19,846,540	22,607,071	357,143,635
4	19,873,131	22,611,565	357,138,670

3.3.1 Scenario 1

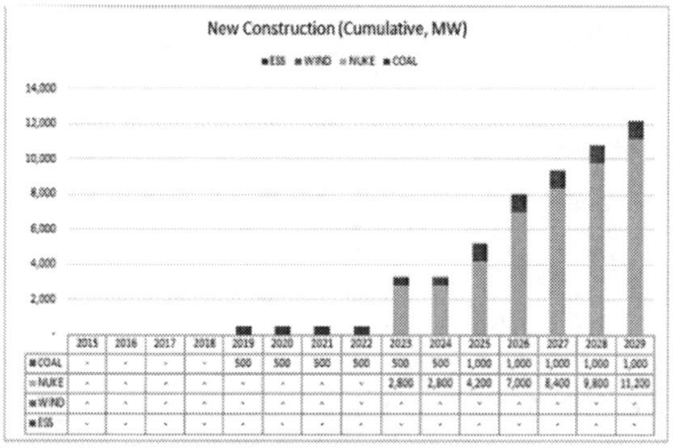

Figure 2. Scenario 1. New Construction Unit.

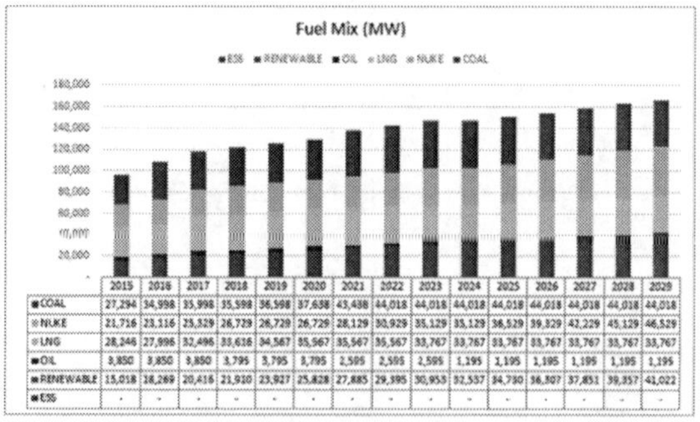

Figure 3. Scenario 1 Fuel Mix.

Scenario 1 is a scenario to which the environment policy and ESS are not applied, and the additionally constructed unit is mainly a neclear generator. In addition, since no environmental policy is executed, a tendency of contructing additional reneawable energy unit is not

DOI: 10.1051/matecconf/201710808003

displayed. As of this point, considering the fact that it is necessary to construct additional generators although the demand can be sufficiently satisfied through the confirmed unit and intention-reflected unit included in the 7th Basic Plan for Long-Term Electricity Supply and Demand, it is possible to deduct a conclusion that the overall operation cost can be reduced through constructing additional generators.

3.3.2 Scenario 2

Scenario 2, as the environmental policies (RPS and ETS) are included as the constraint conditions, shows a unit construction different from that of Scenario 1. Since the environmental policy is required to be executed, the renewable unit is additionally constructed, and no coal thermal power generator is additionally constructed to satisfy the target emission. However, even the results from the above scenario is not enough to completely satisfy the environmental policy, the environmental policy execution cost is additionally included in the overall operational cost, and, therefore, a rapid cost increase in comparison to Scenario 1 is caused. This signifies that not constructing additional generators, but imposing penalties may decrease the overall cost within the plan period.

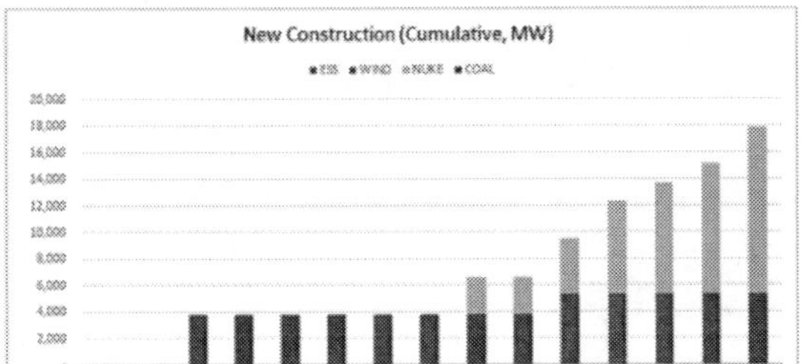

	2015	2016	2017	2018	2019	2020	2021	2022	2023	2024	2025	2026	2027	2028	2029
COAL	-	-	-	-	-	-	-	-	-	-	-	-	-	-	-
NUKE	-	-	-	-	-	-	-	-	2,800	2,800	4,200	7,000	8,400	9,800	12,600
WIND	-	-	3,800	3,800	3,800	3,800	3,800	3,800	3,800	3,800	5,300	5,300	5,300	5,300	5,300
ESS	-	-	-	-	-	-	-	-	-	-	-	-	-	-	-

Figure 4. Scenario 2 New Construction Unit.

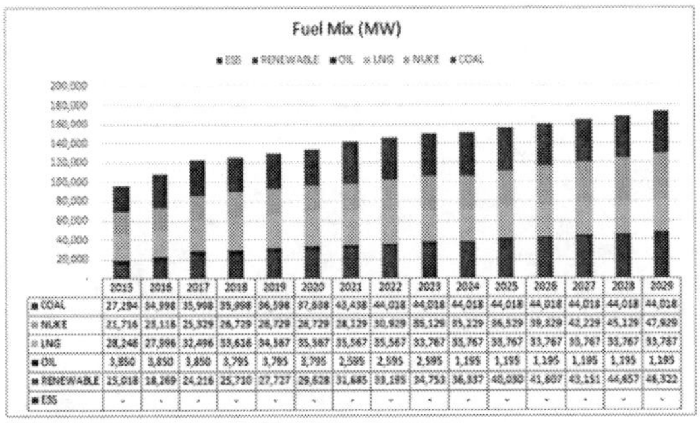

	2015	2016	2017	2018	2019	2020	2021	2022	2023	2024	2025	2026	2027	2028	2029
COAL	27,294	34,998	35,998	35,998	36,598	37,638	43,438	44,018	44,018	44,018	44,018	44,018	44,018	44,018	44,018
NUKE	21,716	23,116	25,329	26,729	26,729	26,729	28,129	30,929	35,129	35,129	36,529	39,329	42,229	45,129	47,929
LNG	28,246	27,996	32,496	33,616	34,567	35,567	35,567	35,567	33,767	33,767	33,767	33,767	33,767	33,767	33,767
OIL	3,850	3,850	3,850	3,795	3,795	3,795	2,595	2,595	2,595	1,195	1,195	1,195	1,195	1,195	1,195
RENEWABLE	15,018	18,269	24,216	25,710	27,727	29,628	31,685	33,195	34,753	36,337	40,030	41,607	43,151	44,657	46,322
ESS	-	-	-	-	-	-	-	-	-	-	-	-	-	-	-

Figure 5. Scenario 2 Fuel Mix.

3.3.3 Scenario 3

Scenario 3 is a case to which the ESS is applied, and the results are identical to those of Scenario 2. The reason for this is because the current cost of the ESS displays no economic feasibility. Namely, it is determined that applying the ESS for peak-load reduction to the system will only increase the overall operational cost.

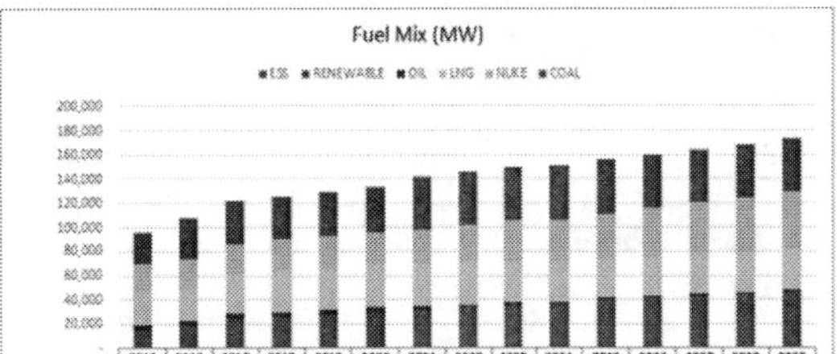

	2015	2016	2017	2018	2019	2020	2021	2022	2023	2024	2025	2026	2027	2028	2029
COAL	27,294	34,998	35,998	35,998	36,598	37,638	43,438	44,018	44,018	44,018	44,018	44,018	44,018	44,018	44,018
NUKE	21,716	23,116	25,329	26,729	26,729	26,729	28,129	30,929	35,129	35,129	36,529	39,329	42,229	45,129	47,929
LNG	28,246	27,996	32,496	33,616	34,567	35,567	35,567	35,567	33,767	33,767	33,767	33,767	33,767	33,767	33,767
OIL	3,850	3,850	3,850	3,795	3,795	3,795	2,595	2,595	2,595	1,195	1,195	1,195	1,195	1,195	1,195
RENEWABLE	15,018	18,269	24,216	25,710	27,727	29,628	31,685	33,195	34,753	36,337	40,030	41,607	43,151	44,657	46,322
ESS	-	-	-	-	-	-	-	-	-	-	-	-	-	-	-

Figure 6. Scenario 3 New Construction Unit.

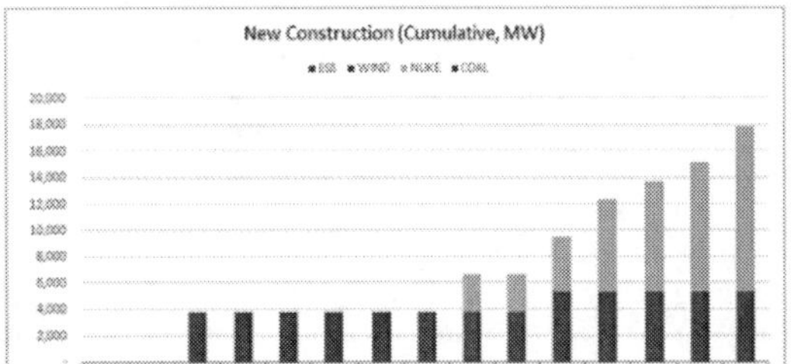

	2015	2016	2017	2018	2019	2020	2021	2022	2023	2024	2025	2026	2027	2028	2029
COAL	-	-	-	-	-	-	-	-	-	-	-	-	-	-	-
NUKE	-	-	-	-	-	-	-	-	2,800	2,800	4,200	7,000	8,400	9,800	12,600
WIND	-	-	3,800	3,800	3,800	3,800	3,800	3,800	3,800	3,800	5,300	5,300	5,300	5,300	5,300
ESS	-	-	-	-	-	-	-	-	-	-	-	-	-	-	-

Figure 7. Scenario 3 Fuel Mix.

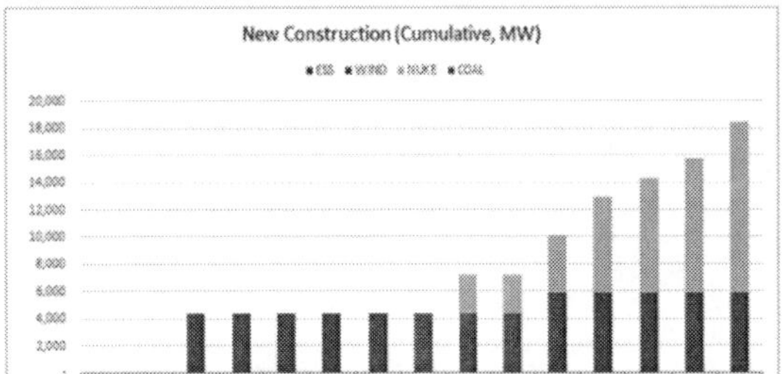

	2015	2016	2017	2018	2019	2020	2021	2022	2023	2024	2025	2026	2027	2028	2029
COAL	-	-	-	-	-	-	-	-	-	-	-	-	-	-	-
NUKE	-	-	-	-	-	-	-	-	2,800	2,800	4,200	7,000	8,400	9,800	12,600
WIND	-	-	3,800	3,800	3,800	3,800	3,800	3,800	3,800	3,800	5,300	5,300	5,300	5,300	5,300
ESS	-	-	600	600	600	600	600	600	600	600	600	600	600	600	600

Figure 8. Scenario 4 New Construction Unit.

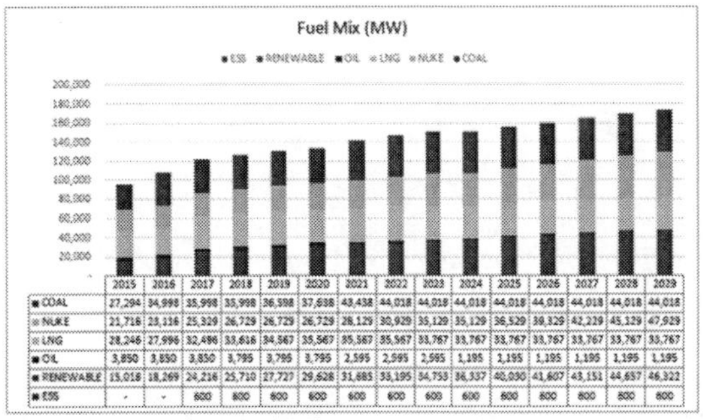

	2015	2016	2017	2018	2019	2020	2021	2022	2023	2024	2025	2026	2027	2028	2029
COAL	27,294	34,998	35,998	35,998	36,598	37,638	43,438	44,018	44,018	44,018	44,018	44,018	44,018	44,018	44,018
NUKE	21,716	23,116	25,329	26,729	26,729	26,729	28,129	30,929	35,129	35,129	36,529	39,329	42,229	45,129	47,929
LNG	28,246	27,996	32,496	33,616	34,567	35,567	35,567	35,567	33,767	33,767	33,767	33,767	33,767	33,767	33,767
OIL	3,850	3,850	3,850	3,795	3,795	3,795	2,595	2,595	2,595	1,195	1,195	1,195	1,195	1,195	1,195
RENEWABLE	15,018	18,269	24,216	25,710	27,727	29,628	31,685	33,195	34,753	36,337	40,030	41,607	43,151	44,657	46,322
ESS	-	-	600	600	600	600	600	600	600	600	600	600	600	600	600

Figure 9. Scenario 4 Fuel Mix.

3.3.4 Scenario 4

Scenario 4 was processed through decreasing the current cost by 90%. As a result, 600MW is applied to the system. It is determined that the cost for operating the ESS is more economical than the cost for constructing the generaion unit due to the technological enhancement of the ESS.

4 Conclusion

In this paper, through the generation planning model to which the environment policy is applied, the optimal fuel mix and ESS optimal capacity per year are calculated. Namely, the ESS capacity is calculated to optimize the overall cost for operating the system. When it is determined that there is no conomic feasibility, the cost variation is deducted to secure the economic feasibility of the ESS.

Based on the result of the scenarios processed in this paper, for the ESS reducing the peak-load within the system, it is confirmed that it is difficult to secure the economic feasibility due to the current installation cost. For the ESS to contribute to the operational cost reduction through reducing the peak-load, the current cost must be decreased by 90%. For the photovoltaic unit, considering that the unit cost has been decreased by approximately 80% for the past 5 years, it is determined that the economic feasiblity of the ESS will be secured in the future.

In this paper, of the roles of the ESS, only the peak-load reduction is considered. In the future, the frequency regulation as well as the renewable energy output smoothing must be considered into calculating the optimal capacity.

References

1. Greenhouse Gas Inventory & Research Center, 2015 *National Greenhouse Gas Inventory Report*, 5p, (2015.12)
2. Ministry of Environment, *Post-2020 Greenhouse Gas Reduction Goal Establishment & Promotion Plan*, (2015)
3. Ministry of Knowledge Economy, *Energy Storage Technology Development & Industrilization Strategy(K-ESS2020)*, (2011.5)
4. Korea Institute of Science Technology Evaluation & Planning, *Project for Empirically Researching ESS to Take Measures on Power Peak*, (2015)
5. S. H. Hwang, Development of a Generation Scheduling Simulator based on Multi-agent System, (2013)
6. Ministry of Trade, Industry & Energy, 7th *Basic Plan for Long-Term Electricity Supply and Demand*, (2015)
7. Ministry of Trade, Industry & Energy, *New & Analysis on Production Cost of Renewable Energy Facility-connected ESS(LiB)*, (2014)

DOI: 10.1051/matecconf/201710808004

Status Checking System of Home Appliances using machine learning

Chi-Yurl Yoon[1], Shin-Gak Kang[2]

[1]*University of Science and Technology, Information and Communication Network Technology, Daejeon, Republic of Korea*
[2]*ETRI, Infrastructure Standard Research Section, Protocol Engineering Center, Deajeon, Republic of Korea*

Abstract. This paper describes status checking system of home appliances based on machine learning, which can be applied to existing household appliances without networking function. Designed status checking system consists of sensor modules, a wireless communication module, cloud server, android application and a machine learning algorithm. The developed system applied to washing machine analyses and judges the four-kinds of appliance's status such as staying, washing, rinsing and spin-drying. The measurements of sensor and transmission of sensing data are operated on an Arduino board and the data are transmitted to cloud server in real time. The collected data are parsed by an Android application and injected into the machine learning algorithm for learning the status of the appliances. The machine learning algorithm compares the stored learning data with collected real-time data from the appliances. Our results are expected to contribute as a base technology to design an automatic control system based on machine learning technology for household appliances in real-time.

1 Introduction

Nowadays, most of electronics equipment and device can communicate with user or other things and these communications and its services among things are referred to Internet of Things (IoT). The user and things can get information about the equipment and control the equipment through the various communications technologies. Although IoT technology is raipidly advancing, but the application of IoT technologies to the traditional home appliances is a little bit slow in progress. Therefore, many of the existing home electronics appliances are not still connected to internet. In our study, we selected a washing machine as an experimental subject to develop a status checking system based on machine learning. Especially, washing machines in launderette are shared with other people, people who want to do the wash need to know when the washing will be ended and which washing machine is available for use to save their time and considerate of others. For these reasons, we developed the status checking system which consists of hardware board, data server and android application. The Arduino system equipped with wireless internet module, current sensor and vibration sensor. It collects data from the washing machines and transmits to server and recognizes the status of the washing machine. And the android application accesses the server to fetch the data and injects into a machine learning algorithm which will determine real time status of the washing machine. The machine learning algorithm returns a prediction result after comparing with the learned data and real time data.

This paper describes the system diagram, data flow and experimental result of the developed system. We analysed the results and the machine learning logic to improve the accuracy of the prediction of the washing machine status.

2 Related Work

2.1 Parse.com web server

The system introduced in this paper operates based on data communication with smartphone, data server and Arduino board. The server is Parse.com web server and the Parse.com provides proper APIs for android and Arduino [1], [2]. The web server is based on text data sheet format, and data requesting and saving actions are convenient by using the APIs. In this system, because the main data which are transmitted from Arduino to server are float format such as vibration data and current data, the Parse.com web server is proper to implement this system.

2.2 WEKA machine learning program

The Waikato Environment for Knowledge Analysis (WEKA) program is machine learning software developed by university of Waikato, New Zealand. Developers can choose a proper machine learning algorithm and modify the algorithms for their purposes as open source software. Because the software is developed using java code, it is compatible for android system by making the software library form.

DOI: 10.1051/matecconf/201710808004

2.3 RBF network algorithm

Radial Basis Function (RBF) network is one of the artificial neural network logic. The main feature of this algorithm is effect between input sources. In the Figure 1-(a), all of the input sources effect to all of the radial basis functions with each weights of input sources and the radial basis functions effect to the one output with linear weights. Because of the feature the RBF network has advantage not only in nonlinear system also linear system to derive a specific output by allocating independence weights to each input sources [3], [4], [5]. The Figure 1-(b) is expressing the probability distribution of two functions with equal value of weights in RBF network. In the system which is introducing in this paper, because the washing machine has four states of output such as staying, washing, rinsing and spin-drying, the functions should be four with equal weights and probabilities.

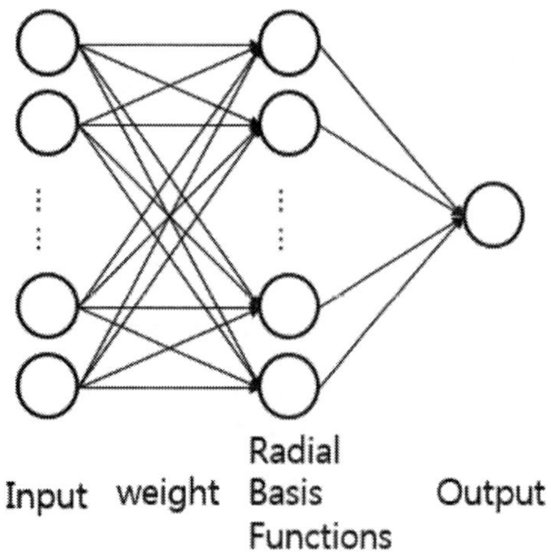

(a) The architecture of RBF network algorithm

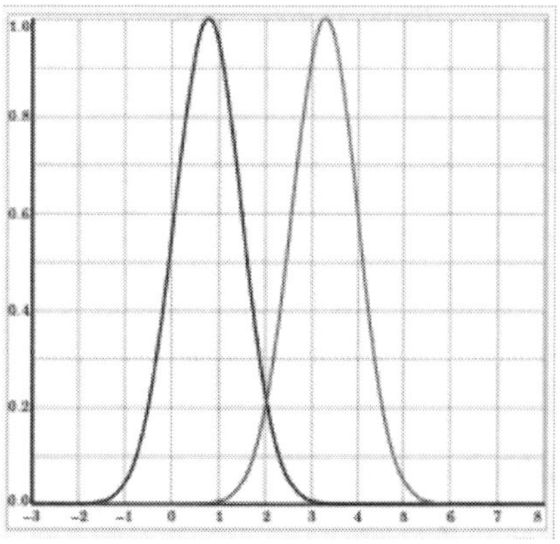

(b) The probability distribution of two RBFs

Figure 1. The principle of RBF network.

3 System Design

Figure 2 is a simple expression about the structure and data flow of this system. The sensor data are transmitted to server from washing machine using Arduino through wireless internet and the data are stored in the server. The stored data are transmitted to smartphone application and injected to a machine learning system.

In details, the process 1 is data flow from Arduino to web server. The first of the process in the Arduino is generating data using a current sensor and a vibration sensor. The current sensor is connected directly with washing machine and the vibration data is measured indirectly by detaching on the washing machine. To low the error range of the gap between analog data and digital data, the Arduino transmits the sensor data after calculating the average of the sensor values in a second. The data which transmitted from Arduino are accumulated in the server divided with learning data and real time data. The divided data are accessed by smartphone in the process 2. When user requests current status of washing machine, the application accesses the server and request learning data and real time data. And then, in the process 3, the data are injected machine learning to determine the current status of the washing machine. At last, the machine learning returns result to the application and the application displays the result.

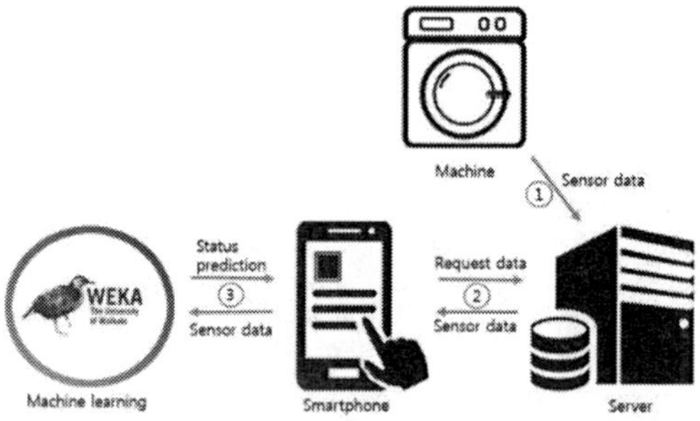

Figure 2. The structure of system and data flow

3.1 Arduino

We should create sensor that can extract data about four states of machine and we should create transmission system that can send the sensor data. First, it needs basic three buttons. One of them is stop button to avoid extracting wrong data from unpredicted situations that generate error. We use yellow LED to distinguish whether the stop button is on or off. If the yellow LED was on, the system is stopped. Another button is separating with testing data and learning data. It is also expressed by red LED and green LED. So, if green LED is on, the Arduino is extracting testing data. And red LED means extracting testing data. When the button is pushed, the state of extracting is changed. The other button is to distinguish learning data class. Because there are four states, the learning data are extracted separately according to each state. It is expressed by two green LED and the LED make binary number. We assume the LED on is 1 and off is 0. The each binary number means that 00 is staying, 01 is washing, 10 is rinsing and 11 is spin

DOI: 10.1051/matecconf/201710808004

drying. The Figure 3 is a simplified circuit diagram of the arduino board.

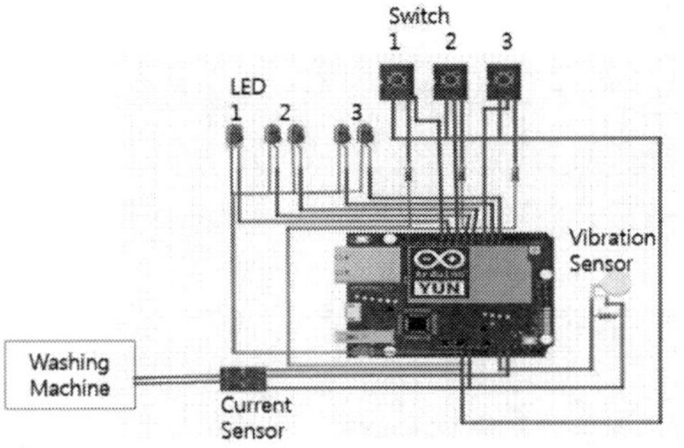

Figure 3. The Arduino system circuit

3.2 Server

In the server, there are two kinds of data storage as learning and real time. The learning data storage has four columns which are expressing dates, vibration values, current values and status. And, the real time data storage has dates, vibration values and current values.

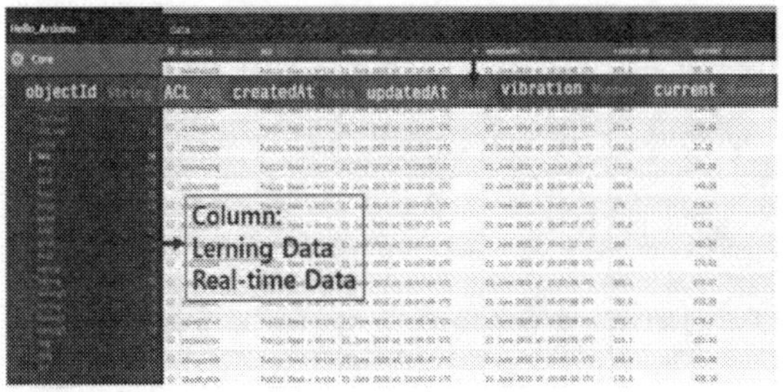

(a) Full scree of web server page

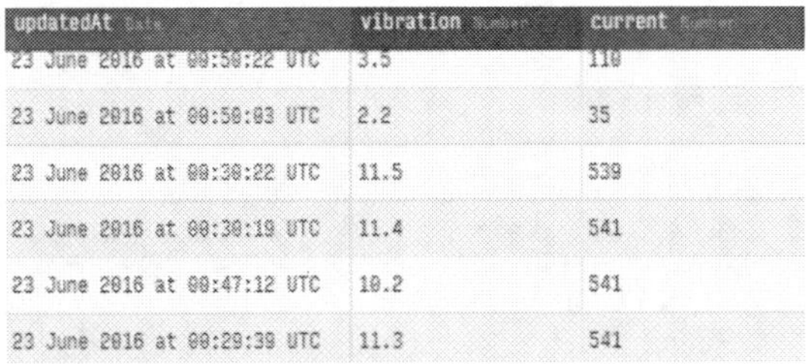

updatedAt	vibration	current
23 June 2016 at 09:58:22 UTC	3.5	110
23 June 2016 at 09:58:03 UTC	2.2	35
23 June 2016 at 09:30:22 UTC	11.5	539
23 June 2016 at 09:30:19 UTC	11.4	541
23 June 2016 at 09:47:12 UTC	10.2	541
23 June 2016 at 09:29:39 UTC	11.3	541

(b) Details in real time data

Figure 4. Parse.com web server

3.3 Android and machine learning

The android has a button for requesting the status. In the (a) of Figure 5, this capture of display is showing the status of rinsing. In the android application, there is determining logic with a WEKA machine learning algorithm library [6]. We choose a machine learning algorithm as the RBF network model which has some advantages in non-linear system, because we cannot assure that the sensor data show the linear trend at the moment of status change.

The (b) of Figure 5 is expressing the modified RBF network algorithm. This algorithm is simplified by allocating equal weight to each statuses and equal probability to 1. Because the washing machine has four statuses, the graph should be expressed with four functions. In the graph, y-axis means the probability of each status and x-axis means RBFs. When the RBF network learned from learning data, each statuses are defined 0, 1, 2 and 3 and determine the status by the probability of each RBFs. The x-axis value divide with 0.75 gaps between each functions and the boundaries on 0.75, 1.5 and 2.25 are standards of determining status in this algorithm. After input data passing this algorithm, the machine learning return the value of functions which is determined as the highest probability. And then, the machine learning find which the status' range is containing the value.

(a) Android application display

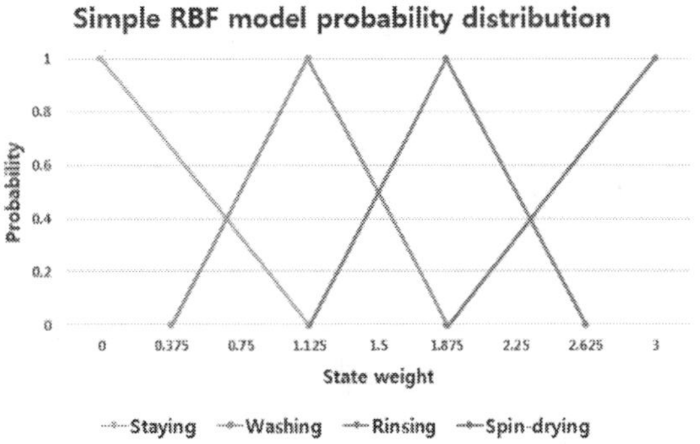

(b) Weight modified RBF network algorithm in application

Figure 5. Android system

4 Result

The experiment using this system is implemented two step as learning and real time for status determination. The Figure 6 is learning data which accumulated in server. The vibration data show the gathering aspect in each range of each statuses like (b). In the current data, the values of the staying and spin-drying statuses are

 DOI: 10.1051/matecconf/201710808004

stable, on the other hands, the washing and rinsing statuses show wide variations. The reason of the variations is, when washing and rising, the drum in the washing machine is turned direction periodically. When the drum is turned direction, the washing machine needs so much energy that makes need more current. The unstable graph aspect may cause misjudgement of machine learning between staying and rinsing in the moment at four seconds and six seconds. So, we should add the status judgement logic which counts the number of judgements in a moment and return the most number of judgements. To test the modified logic, we checked the android log in a moment which shows in Figure 7. The moment is changing status from staying to rinsing and the logs contain total nine judgements. As a result, because the nine judgements have three staying status and six rinsing status, the machine learning determines that the status is rinsing.

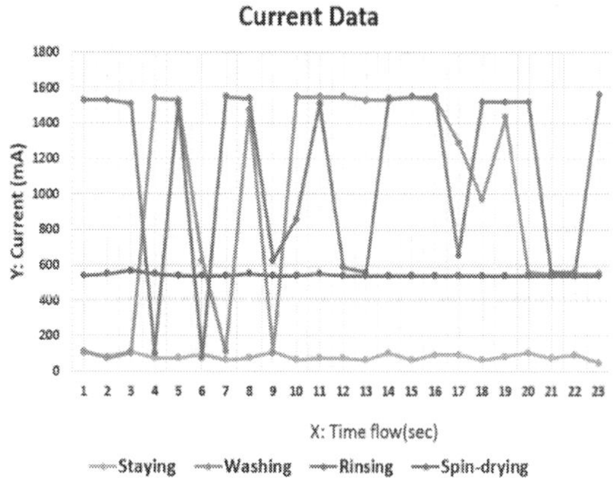

(a) Current sensor values

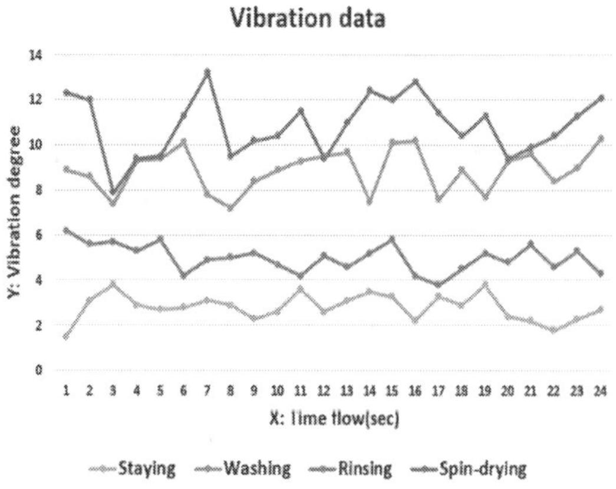

(b) Vibration sensor values

Figure 6. Learning sensor data

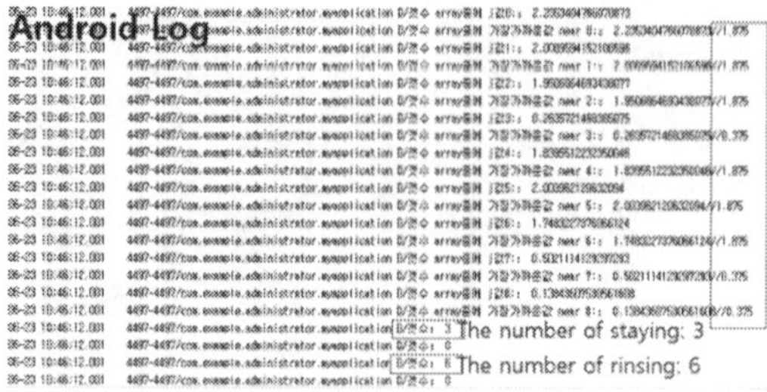

Figure 7. Modified judgement logic from android log

5 Conclusion

Nowadays, although many people could not experience the usability of the IoT technologies because of the high prices of the new home electronics appliances equipped with IoT technologies, the system introduced in this paper provides a method that the new IoT technologies can be applied to existing old-fashioned home appliances. These technologies can make home appliances automatic and it also can be used to other areas such as agriculture, effective management system of industrial field, if proper sensors are used and control software is developed for each application environments. In the transition phase of change to automation life, much research work on control technologies which can be applied to existing old-fashioned devices is needed to develop high quality of automation systems. Finally, our trials achieved in this paper may contribute to change our life to future IoT-based connected life.

Acknowledgement

This research was funded by Ministry of Trade, Industry and Energy. [10065742, Standardization of vertical and horizontal smart-factory integration]

References

1. https://parse.com/docs/android/api/
2. https://parse.com/docs/arduino/api/
3. Introduction to Radial Basis Function Networks, by Mark J L Orr, April (1996)
4. Radial Basis Function Neural Network for Approximation and Estimation of Nonlinear Stochastic Dynamic Systems, Sunil Elanayar
5. V.T. and Yung C. Sh, IEEE TRANSACTIONS ON NEURAL NETWORKS, VOL. 5, NO. 4 (1994)
6. Multivariable Functional Interpolation and Adaptive Networks. Complex Systems, Broomhead, D.S. and Lowe, vol.2, pp.321-355 (1988)
7. http://weka.sourceforge.net/doc.stable/weka/classifiers/functions/RBFNetwork.html

Effects of Waste Plastic Oil Blends on a Multi Cylinder Spark Ignition Engine

Kareddula Vijaya Kumar[1], Ravi Kumar Puli [2]

[1]*Research Scholar, MED, National Institute of Technology, Warangal, TS-506004, India*
[2]*Professor, MED, National Institute of Technology, Warangal, TS-506004, India*

Abstract. Existing fossil fuels are utilizing at their critical rate, leads to depletion of their reserves in a dramatic way. Generating alternative energy sources in a pragmatic way are necessitated, which demands the researchers to utilize the inherent energy of carbon based products as an energy source to the automobile sector. As a part of it, my research is focused on transforming and using the waste plastics as an alternative fuel in multi cylinder spark ignition engine. This paper aims to present the experimental investigations of performance and emission characteristics in an existing Maruti 800 petrol engine running with the blends of 5%, 10%, 15% and 20% of waste Plastic Pyrolysis Oil (PPO) with gasoline. From the results, it is noticed that hydrocarbon emissions are substantially reduced and oxides of nitrogen emissions are increased and petrol engine can operate with PPO blends up to 20% without any engine modifications.

1 Introduction

The rapid industrialization and rushing of automobile sector has led to a steep rise in the demand of petroleum based fuels. As a result, the reserves of the fossil fuels are rapidly exhausting. Scientist estimated that, known petroleum reserves to be depleted in less than 50 years with the present rate of consumption. It forewarns the research community regarding the shortage of energy by the shortage of finite fossil fuel reserves. Besides designing the efficient engines, it needs to search for the substitute of the current using fuels to fulfil future demands. Disposal of waste plastics is causing serious problem to the environment and mankind. Therefore, present work aims to conserving the energy from the waste plastic into needful form of fuel in internal combustion engines. Pyrolysis is the process of thermally degrading long chain polymer molecules into smaller, less complex molecules through heat and pressure. The process requires intense heat with shorter duration in the absence of oxygen. The three major products that are produced during pyrolysis are oil, gas and char which are valuable for industries especially production and refineries. Pyrolysis was chosen by many researchers since the process able to produce high amount of liquid oil up to 80 wt% at moderate temperature around 450-500^0 C and also reviewed the recycling of solid wastes and suggested chemical recycling to control the environmental issues [1-3]. [4-8] Experimental investigation was carried out on a compression ignition engine with waste plastic oil and tyre oil blends with and without exhaust gas recirculation. Their experimental report clearly indicates that the engines which are run with plastic oil and tyre oil releases large amount of NO_x emission, which can minimise by using exhaust gas recirculation. KareddulaVijaya Kumar et al. [9] experimental studies with distilled plastic pyrolysis oil as an alternate fuel for spark ignition engine. Their results show that 50 % distilled plastic pyrolysis oil was gave enhancement in performance compared with petrol and other blends even at full load conditions. The main objective of this research work is defined as an extension of earlier research, most of experimental investigations carried out with plastic pyrolysis oil operated on compression ignition engines and very minimum work was carried out on spark ignition engines. Analyse the engine performance and emission characteristics of three cylinder Maruti 800 petrol engine fuelled with waste plastic oil blends with gasoline at different volume fractions.

1.1 Pyrolysis process of waste plastics

Waste plastic is non-biodegradable nature and it contains toxic chemicals which consumption is harm full to cattle. According to the Central Pollution Control Board (CPCB), India produces 5.6 million tons of plastic waste yearly and solitary 60% of this waste is recycled presently. Even though there are eco-friendly substitutes for some plastics, they are not as versatile or durable as plastics. So it's becoming inevitable to use them in daily life, but need to reduce its utilization rate. Even though increase the reuse and recycling of plastic waste produced are crucial. Opportunely, there is thought-provoking idea

DOI: 10.1051/matecconf/201710808005

to convert carbon chain plastic waste into petroleum grade fuel as shown in Fig.1. Pyrolysis is the chemical decomposition of condensed organic materials by heating in a reactor, largely in the absence of oxygen. The Pyrolysis mainly uses a straw, branches, sawdust and other agricultural and forestry waste as raw material and, through high temperature and pressure, forms the raw materials into a variety of products. Depending on the pyrolysis method, residence time, the temperature to which the waste plastics are exposed, and the characteristics of the waste plastics, varying proportions of gas, liquid and solid products can be produced.

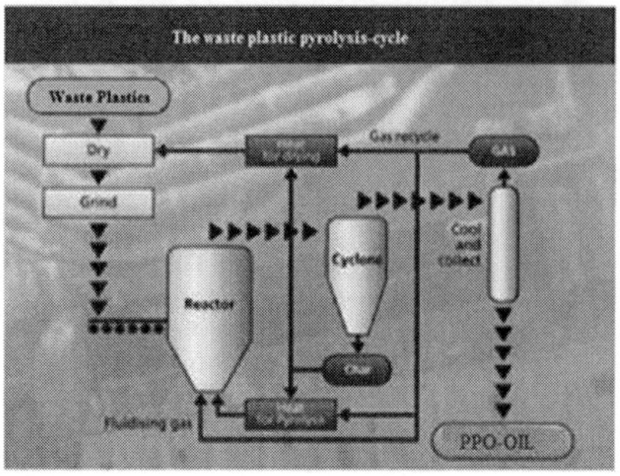

Figure 1. Pyrolysis process of waste plastics

1.2 Blending machine apparatus

The emulsifier is a blending apparatus as shown in Fig. 2, used for homogenize the waste plastic pyrolysis oil-gasoline blends at desired proportions. The homogeniser apparatus was supplied by Ormeroo Engineers Limited, England. So, the prepared plastic pyrolysis oil blends for investigating the nature and performance of emissions by a spark ignition engine.

Figure 2. Blending Machine Apparatus

1.3 Preparation of plastic pyrolysis oil blends

The basic properties of PPO-PF were measured and compared to conventional petroleum fuel. The Carbon content is higher for PPO than petrol fuel. In the present work 5 PPO, 10 PPO, 15 PPO and 20 PPO of PPO was blended with petrol fuel on a volume basis and observed for 15 days to check for any separation. No such separation was noticed. PPO blended with PF is indicated as 15 PPO. For example, 15% PPO blended with 85% PF is denoted as 15 PPO. The estimated PPO flash and fire point are 40 and 44 respectively. The basic properties of plastic oil compared with petrol are in Table 1.

Table 1. Basic properties of fuels

S. No.	Characteristics	Petrol	PPO
1	Specific Gravity	0.741	0.83
2	Kinematic Viscosity mm^2/sec	0.5	2.64
3	Calorific Value (KJ/kg)	43449.7	42807.5
4	Density (kg/m^3)	740.82	830

2 Experimental setup

Experiments were conducted on a three cylinder Maruti 800 petrol engine with a fixed compression ratio at a constant speed of 1500 rpm under different loads. The experimental setup is as shown in Fig. 3, it consists of three cylinder, 4-stroke, water cooled engine. A water brake dynamometer has been used in the present study to load the engine. The main principle of operation of the dynamometer is water acts as a cooling and loading medium. The running rotor causes the water to whirl in the chamber. The braking energy thus absorbed converts to heat, which is dissipated with circulating water. At a given constant speed of the dynamometer its reaction torque is a function of water value in the whirl chamber, which in turn is controlled by a sluice gate. It has a provision of digital load indicator for the display of load while loading the engine. A U-tube manometer is used to note the difference between the pressure head to calculate the air flow rate.

Figure 3. Schematic of Maruti 800 Petrol Engine Test Setup

2.1 AVL five gas analyser

The purpose of the AVL gas 444 is to measure the relative volumes of certain gases constituents in the five

 DOI: 10.1051/matecconf/201710808005

exhaust gas of motor vehicles has shown in Fig. 4. These gases are carbon monoxide, carbon dioxide, hydrocarbons, and oxygen nitric oxide. The air fuel ratio is calculated from the Carbon Monoxide (CO), Carbon Dioxide (CO_2), Hydrocarbons (HC) and Oxygen (O_2), constituents and displayed. The Liquid crystal display shows the constituents of the gas once the gas is allowed to flow in. It is a standard output device for the AVL gas 444.

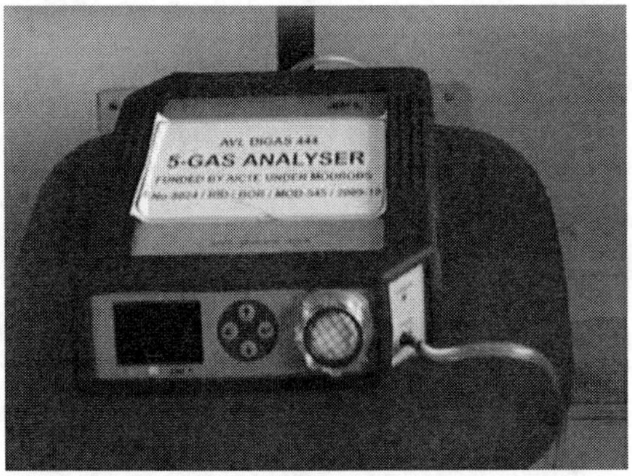

Figure 4. AVL Five Gas Analyser

3 Results and discussions

The influencing parameters such as Brake Specific Fuel Consumption (BSFC), Brake Thermal Efficiency (BTE) were presented on the performance of the petrol engine. Fig. 5 elucidates the influence of the brake power on the brake specific fuel consumption. The fuel consumption rate is increased uniformly as a function of blending proportions. At starting loading condition (0 to 0.5 kW), the variation of the BSFC is drastically increased after that a sudden decrease in BSFC along with the variation of brake power for petrol and other plastic oil blends. 25.53% BSFC augment was noticed at 20 PPO blend when compared with baseline petrol. Fig. 6 illustrates the brake thermal efficiency as a function of brake power. The BTE is linearly varied up to 2 kW brake power, after that it varies parabolically along with the brake power, and this variation is depending on blend proportions. At the full load conditions, the BTE was decreasing with increment in blend, proportions due to the lower calorific value of the plastic pyrolysis oil. Maximum decrement of 19.24% at 20 PPO compared to gasoline at full load conditions.

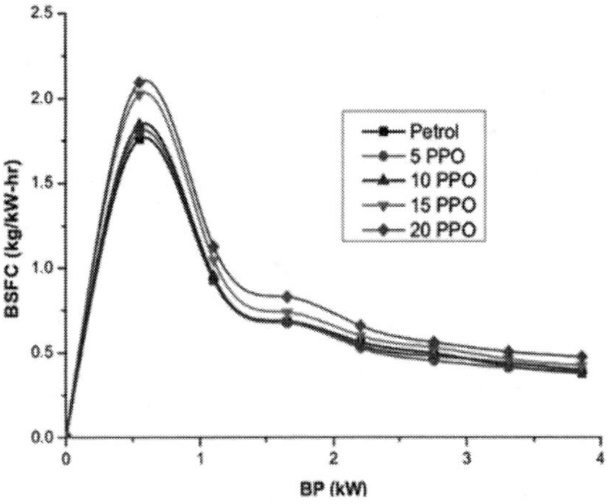

Figure 5. Variation of BSFC with Brake Power

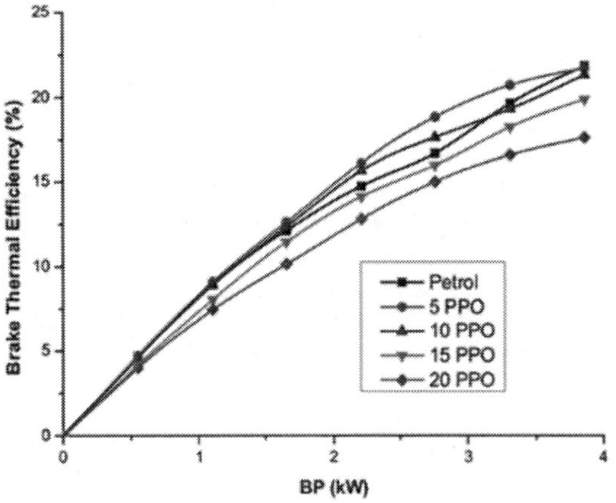

Figure 6. Variation of BTE with Brake Power

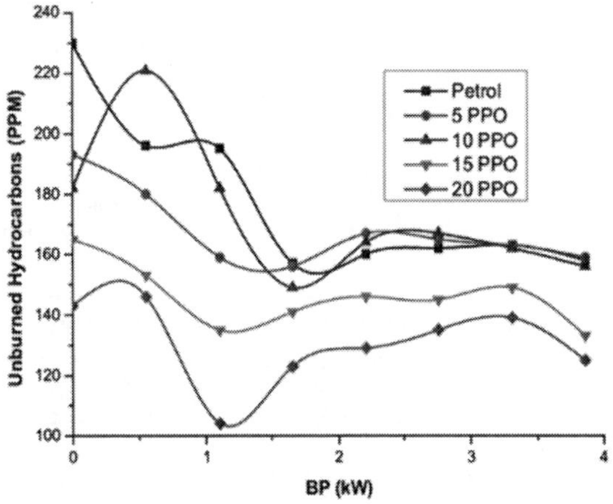

Figure 7. Variation of HC with Brake Power

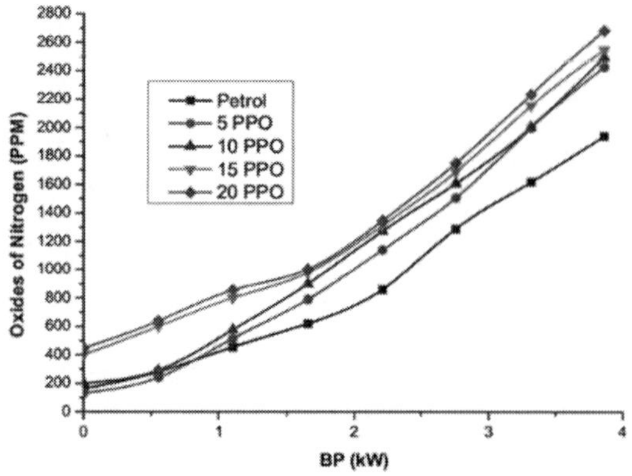

Figure 8. Variation of NO_x with Brake Power

Major constituents of engine emissions are CO, O_2, CO_2, unburned HC, NO_x and particulate matter. The effect of HC emissions using different gasoline blends is as shown in Fig. 7. The root cause for the formation of the HC emissions is unburned mixture of PPO blended gasoline. Some of the air fuel mixture pockets were not participating in the combustion hence, it gives rise to unburned HC emissions due to misfires and HC deposits. When compared to pure gasoline, there is a significant decrement in the unburned HC emissions for the all PPO blends. Fig. 8 it is noticed that, NO_x emissions are increasing with increase in the peak cylinder temperature by an increase in load. NO_x emissions are also varying with air fuel ratio, as the blend proportion increases air fuel ratio also increased and hence more amount of nitrogen is admitted into the engine cylinder in terms of more air intake. This admitted nitrogen will not participate in combustion process, but react with oxygen and generates oxides of nitrogen. These nitrogen oxide emissions are uniformly increasing with the blend proportion.

4 Conclusions

The engine performance and emission characteristics of PPO with gasoline blends are experimentally measured and reported. PPO blends are tested on three cylinder Maruti 800 petrol engine test rig and emissions are measured with AVL 5 gas analyser. To conduct the experiments, the plastic pyrolysis oil is blended with petrol using blending apparatus. The thermo physical chemical properties of all blends are closer to the gasoline properties. The experimental outcomes clears that, the brake specific fuel consumption of the engine fuelled with 20 PPO is 25.52 % higher than that of the engine fuelled with sole gasoline at full load. 19.24% lesser brake thermal efficiency is obtained for 20PPO compared to gasoline. The unburned hydrocarbon emissions are marginally reduced to 25.56% for 20PPO. The NO_x emissions are 29.13% more in the 20 PPO compared to that of the engine fuelled with 100% gasoline as fuel. In the view of performance, experiments can run beyond 20% plastic pyrolysis oil without altering the engine designs. But, as the PPO increased beyond 20% is responsible for higher NO_x emissions, which cross the EPA norms and augmentation in viscosity causes for higher surface tension and reduction in flowability of fuel in orifice.

References

1. BeheraPritinika, Murugan S, Combustion, performance and emission parameters of used transformer oil and its diesel blends in a DI diesel engine, Fuel 104, (2013) pp. 147-154.
2. Mehdi Sadat-Shojai, Gholam Reza Bakhshandeh, Recycling of PVC wastes, Polymer Degradation and Stability 96, Elsevier, (2011), pp. 404-415.
3. Murugan, S., Ramaswamy, M.C., Nagarajan, G., Production of Tyre Pyrolysis Oil from Waste Automobile Tyres, In the Proceedings of National conference on Advances in Mechanical Engineering, (2006), 899-906.
4. M Paul Daniel, KareddulaVijaya Kumar, B Durga Prasad, Performance and Emission Characteristics of Diesel Engine Operated on Tyre Pyrolysis Oil with exhaust gas recirculation, International Journal of Ambient Energy (IJAE), Taylor and Francis, August (2015), DOI: 10.1080/01430750.2015.1023837.
5. KareddulaVijaya Kumar, Ravi Kumar Puli, Monika Dixit, D Ravi Chandra, Performance and Emission Evaluation of Tyre Oil blended with Diesel Fuel in Compression Ignition Engine, International Conference on Environment and Energy (ICEE-2014), December (2014), JNTUH Kukatpally, Hyderabad, India.
6. M Paul Daniel, KareddulaVijaya Kumar, B Durga Prasad, Ravi Kumar Puli, Performance and Emission Characteristics of Diesel Engine Operated on Plastic Pyrolysis Oil with exhaust gas recirculation, IJAE, Taylor and Francis, September (2015), DOI: 10.1080/01430750.2015.1086677.
7. KareddulaVijaya Kumar, Ravi Kumar Puli, A VeereshBabu, J. A. RangaBabu, Performance and Emission Evaluation of Plastic Oil blended with Diesel Fuel in Compression Ignition Engine, International Conference on Environment and Energy (ICEE-2014), December (2014), JNTUH Kukatpally, Hyderabad, India.
8. K. Vijaya Kumar, Dr. Ravi Kumar Puli, D. Ravichandra, K. Abhishek, Investigation on a CI Engine Fuelled with blends of Waste Tyre oil, Third International Conference on Advances in Mechanical, Aeronautical and Production Techniques (MAPT), April (2015), Malaysia.
9. KareddulaVijaya Kumar, Ravi Kumar Puli, A SwarnaKumari, P Shailesh, "Performance and Emission Studies of a SI Engine using Distilled Plastic Pyrolysis Oil-Petrol Blends", MATEC Web of Conferences 45, 03002 (2016), DOI: 10.1051/matecconf/20164503002.

MATEC Web of Conferences **108**, 08006 (2017) DOI: 10.1051/matecconf/201710808006
ICMAA 2017

Control of 12-Cylinder Camless Engine with Neural Networks

"Moh'd Sami" Ashhab

Department of Mechanical and Industrial Engineering, American University of Ras Al Khaimah, UAE
Department of Mechanical Engineering, Hashemite University, Zarqa, Jordan

Abstract. The 12-cyliner camless engine breathing process is modeled with artificial neural networks (ANN's). The inputs to the net are the intake valve lift (IVL) and intake valve closing timing (IVC) whereas the output of the net is the cylinder air charge (CAC). The ANN is trained with data collected from an engine simulation model which is based on thermodynamics principles and calibrated against real engine data. A method for adapting single-output feed-forward neural networks is proposed and applied to the camless engine ANN model. As a consequence the overall 12-cyliner camless engine feedback controller is upgraded and the necessary changes are implemented in order to contain the adaptive neural network with the objective of tracking the cylinder air charge (driver's torque demand) while minimizing the pumping losses (increasing engine efficiency). All the needed measurements are extracted only from the two conventional and inexpensive sensors, namely, the mass air flow through the throttle body (MAF) and the intake manifold absolute pressure (MAP) sensors. The feedback controller's capability is demonstrated through computer simulation.

1 Introduction

One of the major challenges in the camless IC engine design is controlling the cylinder air charge (CAC) rapidly and accurately based on conventional measurements. A feedback control system that is responsible for this task has been developed in [1]. It consists of a feedforward controller, a CAC estimator and an on-line parameter estimator which is used to adapt the feedforward controller parameters using the CAC estimate. The performance of the controller requires accurate estimation of the CAC. It is very difficult to measure the cylinder air charge directly since there are no sensors that can be placed inside or close to the cylinders due to the high pressure and temperature of the cylinder. Thus, the CAC must be estimated based on conventional engine measurements.

It has been shown in [1] that the individual cylinder air charge measurements can be extracted from two conventional and inexpensive sensors, namely, the mass air flow through the throttle body (MAF) and the intake manifold absolute pressure (MAP) sensors. The estimated cylinder air charge is used therein to update the feedforward controller parameters via the on-line parameter estimator in order to account for modeling errors and engine parts degradation. The adaptation technique is based on shifting and scaling the inputs to the engine. These inputs are the intake valve lift (IVL) and intake valve closing timing (IVC) and they are set by electrohydraulic actuators. One disadvantage of the feedforward controller is that its operation is based on graphical techniques. In addition, the camless engine controller design becomes more challenging for a larger number of cylinders such as a 12-cylinder engine.

In this research, the 12-cyliner camless engine breathing process is modeled with artificial neural networks (ANN's). The inputs to the net are the IVL and IVC whereas the output of the net is the CAC. The ANN is trained with engine input output data at a constant speed of 1500 rpm. The ANN model forecasts the CAC in the future at this speed. It has been shown in [2] that the feedforward controller developed in [1] can be modeled by inverting the camless engine ANN model. The ANN based feedforward controller is more concise and easier to adapt. A method for adapting single-output feed-forward neural networks is proposed in this research and applied to the camless engine ANN model. As a consequence the overall 12-cyliner camless engine feedback controller is upgraded and the necessary changes are implemented in order to contain the adaptive neural network with the objective of tracking the cylinder air charge (driver's torque demand) while minimizing the pumping losses. The feedback controller's capability is demonstrated through computer simulation.

The camless engine model was initiated in [3] as a result of collaboration between academia and industry. Furthermore, the basic control system of such an engine was designed in [1]. Improvements of the initial control system design that enhanced the performance of the system through computer simulation were attempted later. In [2], graphical methods for the feedforward controller were replaced by inverting a neural network model for the engine. Adapting the neural network achieved good long term performance of the camless engine that

DOI: 10.1051/matecconf/201710808006

accommodates the changing environment and aging [4]. Analysis of the updated camless engine controller can be found in [5,6]. The 4-cylinder camless engine with the fully updated controller was presented in [7]. Similarly, the 8-cylinder engine with its updated controller was studied in [8]. Modeling and control of the 12-cylinder engine will be conducted in this research which represents the basis for this engine operation.

The camless engine performance depends on the design of electrohydraulic actuators for regulating the valve motion. Researchers have attempted different designs for the electrohydraulic actuators to achieve the best engine performance. Application of a robust motion control technique to an electrohydraulic camless engine valve actuator was presented in [9]. The research of [10] deals with a signal-based robust adaptive approximation technique for a proportional derivative (PD) regulator which is applied to an electromagnetic valve actuator control for camless internal combustion engines. Exhaust valve motion was analyzed towards optimum performance in [11].

Artificial neural networks (ANN's) have received a lot of attention in recent years due to their attractive capabilities in forecasting, modeling of complex nonlinear systems and control. Applications of neural networks include many engineering fields. For example, ANN's have been used for forecasting load [12] and glass/epoxy failure strength [13].

2 Intake process analysis

The camless engine is an internal combustion engine that operates without a camshaft. The intake valves are controlled by electrohydraulic actuators that activate the valves motion. In the case of the camless engine the valve motion is not tied to the engine rotation but it is set by an engine controller instead. Consequently the choice of the intake valve motion is very flexible which allows efficient engine performance characterized by satisfying the desired cylinder air charge (driver torque demand) and minimizing pumping losses (fuel economy) at all speeds. In order to design the appropriate controller that achieves these objectives, the engine breathing (intake) process needs to be analyzed and studied.

The thermodynamic equations that describe the intake process of internal combustion engines including the 12-cyliner camless engine are based on the principles of the conservation of mass and the ideal gas law [3]. More specifically, the following relations hold for the engine cylinders

$$\frac{dp_m}{dt} = \frac{RT}{V_m}\left(\dot{m}_{\varphi} - \sum_{i=1}^{n} \dot{m}_i\right) \tag{1}$$

$$\frac{dp_i}{dt} = \frac{1}{V_i}\left(RT\,\dot{m}_i - \dot{V}_i\, p_i\right), \tag{2}$$

where n is the number of cylinders, subscript m refers to manifold, subscript i refers to cylinder i (i=1,2 … n), p is the pressure, t is the time, V is the volume, T is the temperature, R is the gas constant, and $\dot{m}$ is the mass flow rate. The subscript ϕ refers to the manifold throttle. In the camless engine the manifold throttle is widely open all the time and not connected to the gas pedal since the air flow into the cylinders is controlled by the individual cylinder intake valves. On the other hand, the throttle opening changes in the conventional engine according to the driver demand due to the fact that this is the only means of controlling the air flow into the cylinders. The intake valve motion is fixed with the engine rotation in the conventional engine.

A simulation model in Matlab software was built for the engine based on Equations (1) and (2) and implementing the implicit relations of the terms within the equations [3]. The various implicit relations include parameters that needs to be selected by comparison with real experimental engine data available from the dynamometer test at an automotive company. From the engine equations the pressure versus crank angle (or time) are plotted and compared with real engine data under the same conditions. During this process the parameters of the engine mathematical model are set such that there is a good match between the simulation results and real experimental engine data. An example of such a comparison is given in Figure 1.

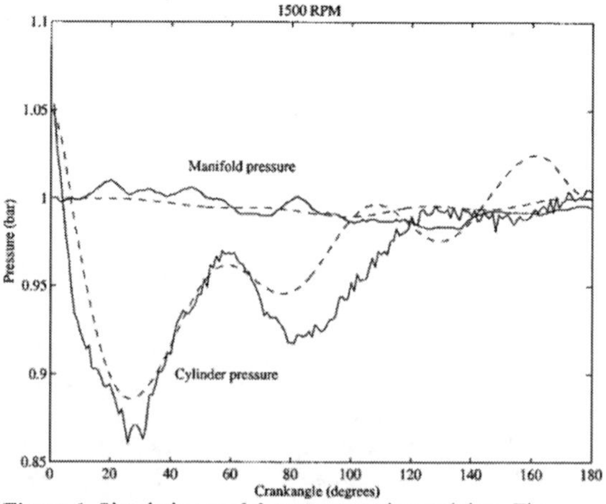

Figure 1. Simulation model versus experimental data. The continuous line represents the experimental data whereas the dashed line represents the model [1].

3 Neural net model

The breathing process of the 12-cylinder camless engine was modeled in [14] with artificial neural networks (ANN's) and brief results are given here. The neural net structure to be used for modeling the camless engine breathing process is shown in Figure 2. The inputs are the intake valve lift (IVL) and intake valve closing timing (IVC) whereas the outputs are the important variables for control design, namely, the cylinder air charge (CAC) and pumping loss (PL). CAC needs to be tracked based

DOI: 10.1051/matecconf/201710808006

on the driver's torque demand while minimizing the pumping loss. IVL and IVC feed forward through a hidden layer to the CAC and PL.

An artificial neural net mathematical model that represents the structure shown in Figure 2 is written as

$$Y = f(U) = W_o * \tanh(W_i * U + B_i) + B_o \qquad (3)$$

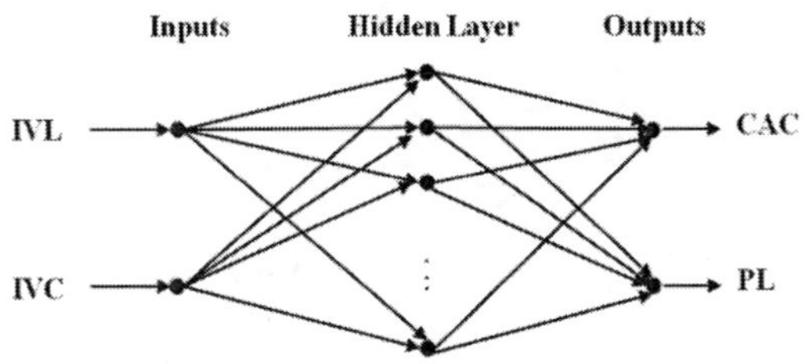

Figure 2. Camless engine neural net structure.

where, Y is a column vector which contains the 2 outputs of the process, U is a column vector that contains the 2 inputs of the process, W_o is a matrix of size 2xn that contains the weights of the neural net model from the hidden layer to the outputs with n being the number of neurons in the hidden layer, W_i is a matrix of size nx2 that contains the weights of the neural net model from the inputs to the hidden layer, B_i (not shown in Figure 2) is a column vector of size n that contains the biases from the inputs to the hidden layer and B_o (not shown in Figure 2) is a column vector of size 2 which contains the biases from the hidden layer to the outputs.

The cylinder air charge from the ANN is the first element of Y and thus Equation (3) can be used to write

$$\mathrm{CAC}(t+1) = \beta(U(t))\,\alpha) \qquad (4)$$

where,

$$\beta(U(t)) = \begin{bmatrix} \tanh(W_i * U(t) + B_i) \\ 1 \end{bmatrix}^{\mathrm{T}} \qquad (5)$$

$$\alpha = \begin{bmatrix} W_{o1}^{\mathrm{T}} \\ B_{o1} \end{bmatrix} \qquad (6)$$

with W_{o1} being the first row of W_o and B_{o1} the first element of B_o. It is clear that the estimated (neural network) future value of CAC is linear in the parameter α that is a column composed of the output weight vector and bias that correspond to CAC.

4 Feedback adaptive controller

One of the major challenges in the camless IC engine design is controlling the cylinder air charge (CAC) rapidly and accurately based on conventional measurements while minimizing the pumping loss (PL). The feedback control system that is responsible for this task is shown in Figure 3. It consists of a feedforward controller, a CAC estimator and an on-line parameter estimator which is used to adapt the feedforward controller parameters using the CAC estimate. It has been shown in [1] that the individual cylinder air charge measurements can be extracted from two conventional and inexpensive sensors, namely, the mass air flow through the throttle body (MAF) and the intake manifold absolute pressure (MAP) sensors. The estimated cylinder air charge is used therein to update the feedforward controller parameters via the on-line parameter estimator in order to account for modeling errors and engine parts degradation. The adaptation technique is based on shifting and scaling the inputs to the engine. These inputs are the intake valve lift (IVL) and intake valve closing timing (IVC) and they are set by electrohydraulic actuators. Two disadvantages of this feedforward controller are: 1) its operation is based on graphical techniques and 2) large amount of engine input-output data at various speeds needs to be stored in controller's memory.

The ANN is used to predict the cylinder air charge in the future. Due to possible gradual degradation of the camless engine, the short-term predictions will be more accurate than the long-term ones. The neural net of the camless engine is adapted in [4] using the Kaczmarz's algorithm in order to better accommodate the changing environment and improve the net's prediction accuracy over the long run. This adaptation scheme leads to less computational complexity and quick convergence. The real-time parameter estimation Kaczmarz's algorithm when applied to the ANN model described by Equation (4) gives the following recursive adaptation scheme for α

$$\alpha(t+1) = \alpha(t) + \frac{\beta(U(t-1))^{\mathrm{T}}}{\beta(U(t-1))\beta(U(t-1))^{\mathrm{T}}}\,e \qquad (7)$$

where the error, e, is the difference between the real and ANN estimated cylinder air charge at the time instant t.

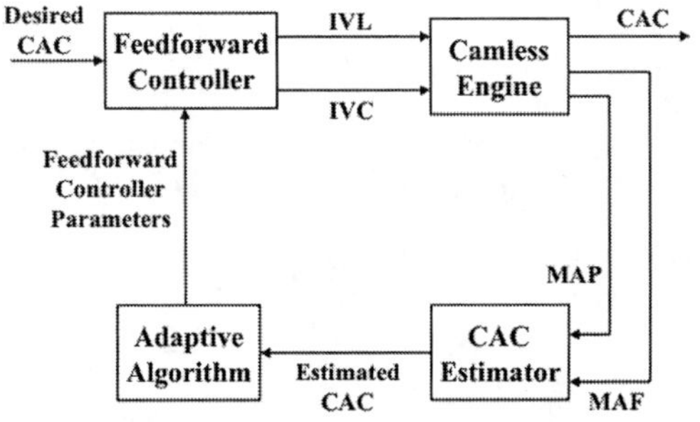

Figure 3. Camless engine adaptive feedback controller.

Note that the on-line adaptive algorithm described in Equation (7) assumes that the output cylinder air charge measurement at the current time t is available which is true in the camless engine. The estimated output of the ANN model is compared with the measured cylinder air charge of the engine. If there is a difference between the two values the parameter estimation Kaczmarz's algorithm updates α (according to Equation (7)) to be used in the subsequent predictions. The parameter estimator or adaptive algorithm compensates for engine parts degradation and modeling errors, and provides more accurate long-term predictions.

5 Simulation Results

The capability of the developed ANN based adaptive 12-cylinder camless IC engine controller is tested with the aid of computer simulation. Similar tests were performed for the 8-cylinder engine in [8]. However, the 12-cylinder engine case is more complicated due to the fact that there is a higher probability of overlapping cylinders during engine operation. We test three different cases. In the first case the controller tracks changing desired cylinder air charge. In the second case the controller is tested as the speed changes from 1500 RPM to 1700 RPM. Finally, in the third case deposits are assumed to be present on the intake valves.

In the first case of 1500 RPM engine speed, the controller achieves cylinder air charge tracking as it changes from 0.3 g to 0.4 g at 0.3 seconds, in addition to balancing by eliminating modeling errors as shown in Figure 4. Note that the 12 cylinders are plotted with different styles. The desired cylinder air charge is the solid line. The operation of the controller in this case depends almost completely on the feedforward controller that is the inverse of the camless engine ANN model. The feedforward controller selects the intake valve lift (IVL) and closing timing (IVC) that will assure that the cylinder air charge is equal to its desired value for the nominal situation. We have here a nominal engine since the speed is the same speed for which the ANN has been trained and there is no parts degradation or noise introduced in the system. A moderate adaptation effort is required to compensate for modeling errors. We conclude that the ANN based feedforward controller achieves its objective of satisfying the desired cylinder air charge.

The camless engine ANN model was trained using data at a speed of 1500 RPM. The feedforward controller performance will not satisfy the cylinder air charge demand at other speeds. The adaptive controller compensates for speed variations as shown in Figure 5. The speed changes from 1500 RPM to 1700 RPM at t = 0.24 seconds and the feedforward controller selects initially IVL and IVC from the nominal 1500 RPM data. These values fail to satisfy the desired cylinder air charge of 0.3 g and provide instead a lower cylinder air charge. The predicted cylinder air charge of the ANN model (approximately equals 0.3 g) is compared with the measured cylinder air charge of the engine which is estimated via the cylinder air charge estimator. The parameter estimation Kaczmarz's algorithm uses the error between the two values to update the ANN parameters for the current speed of 1700 RPM to be used in the subsequent predictions and feedforward controller. The adaptive controller achieves the desired cylinder air charge in one engine cycle.

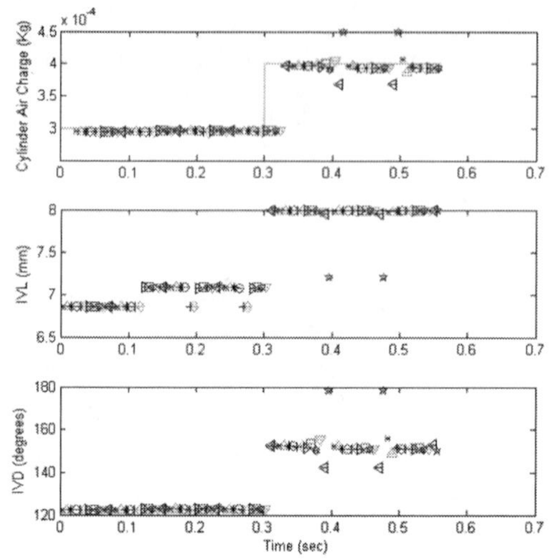

Figure 4. Cylinder air charge tracking at a fixed speed of 1500 RPM.

The designed adaptive feedback controller can compensate for engine parts degradation and thus keeps consistent and good engine performance over the long time horizon. We assume that intake valves allow less air into the cylinders due to deposits that accumulate with time. It is assumed that the effective area of the valve (through which air passes from the manifold to the cylinder) is reduced by 5% for all of the 12 cylinders due to the deposits. The simulation results of this case are shown in Figure 6. Note that the adaptive feedback controller overcomes the valve degradation and keeps the engine performance at the same high level. The feedforward controller initially does not satisfy the cylinder air charge. The deposits cause a CAC reduction. The adaptive algorithm uses the error in CAC to compensate for the degradation and the controller achieves CAC tracking in one engine cycle.

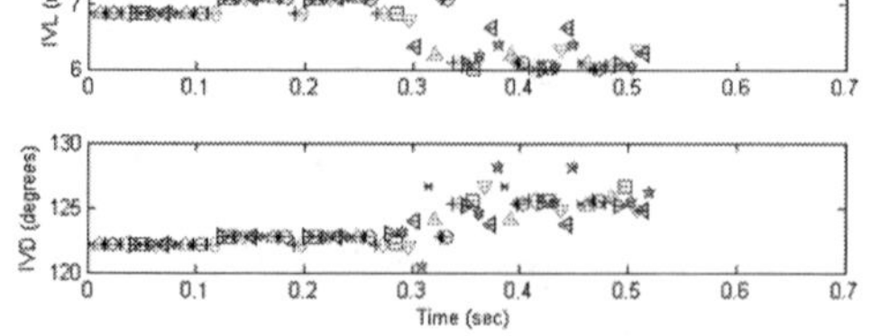

Figure 5. Cylinder air charge tracking with speed variation.

DOI: 10.1051/matecconf/201710808006

6 Conclusions

An adaptive feedback artificial neural net (ANN) based camless 12-cylinder engine controller with the objective of achieving the desired cylinder air charge (CAC) or driver's torque demand and minimizing the pumping loss (PL) was developed. The controller consists of a feedforward controller, cylinder air charge estimator, and on-line ANN parameter estimator. The feedforward controller is the inverse of the ANN camless 12-cylinder engine model which was developed at a speed of 1500 RPM. The feedforward controller provides the intake valve lift (IVL) and intake valve closing timing (IVC) that satisfy the desired CAC and minimize PL. The parameters of the feedforward controller are updated recursively using the error between the cylinder air charge measurement from the cylinder air charge estimator and its predicted value from the ANN. The adaptation scheme improves the ANN prediction accuracy when the engine parts degrade, speed changes and in the presence of modeling errors. The camless engine controller achieves its objective of CAC tracking and minimizing PL.

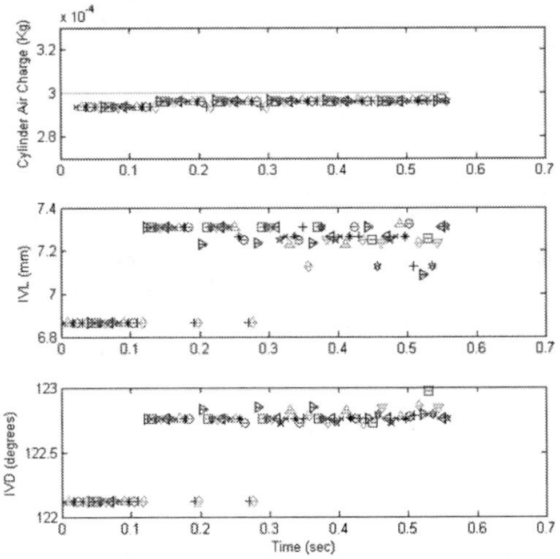

Figure 6. Cylinder air charge tracking with deposits on valves.

References

1. M. Ashhab, A. Stefanopoulou, J. Cook, M. Levin, Journal of Dynamic Systems, Measurement, and Control **122,** 131 (2000)
2. M. S. Ashhab, *Fourth IASTED International Conference on Modeling, Simulation and Optimization*, Kauai, USA (2004)
3. M. Ashhab, A. Stefanopoulou, J. Cook, M. Levin, Journal of Dynamic Systems, Measurement, and Control **122,** 122 (2000)
4. M. S. Ashhab, *Fifth IASTED International Conference on Modeling, Simulation and Optimization*, Oranjestad, Aruba (2005)
5. M. S. Ashhab, *International Conference on Dynamics, Instrumentation and Control*, Queretaro, Mexico (2006)
6. M .S. Ashhab, Energy Conversion and Management **49,** 365 (2008)
7. M. S. Ashhab, *IASTED International Conference on Applied Simulation and Modeling*, Palma de Mallorca, Spain (2007)
8. M. S. Ashhab, *Second International Conference on Kansei Engineering & Affective Systems*, Nagaoka, Japan (2008)
9. Y. Yoon, Z. Sun, IEEE Transactions on Industrial Electronics **63,** 5724 (2016)
10. P. Mercorelli, Asian Journal of Control **18,** 1299 (2016)
11. X. Y. Fan, L. Liu, S. Q. Chang, J. T. Xu, J. G. Dai, International Journal of Automotive Technology **17,** 361 (2016)
12. J. Massana, C. Pous, L. Burgas, J. Melendez, J. Colomer, Energy and Buildings **92,** 322 (2015)
13. C. Suresh Kumar, V. Arumugam, R. Sengottuvelusamy, S. Srinivasan, H. N. Dhakal, Applied Acoustics **115,** 32 (2017)
14. M. S. Ashhab, International Journal of Automation, Mechatronics & Robotics **3,** 99 (2016)

MATEC Web of Conferences **108**, 08007 (2017) DOI: 10.1051/matecconf/201710808007
ICMAA 2017

Power Optimization Configuration for Piezoelectric Cantilever Arrays

Bong Yu Jing[1] and Kok Swee Leong[1]

[1]*Faculty of Electronic and Computer Engineering, Universiti Teknikal Malaysia Melaka, Melaka, Malaysia*

Abstract. This paper investigates the changes in the output of the piezoelectric cantilever arrays when connected in different configurations. In this research matching load resistance determined and optimum output was measured by connecting the piezoelectric cantilever arrays to resistance ranging from 10 Ω to 1 MΩ while excited by constant vibration source at frequency of 300 Hz and acceleration of 1-g level. The result shows that matching load resistance for one single piezoelectric cantilever is 13 KΩ. When two, three and four cantilevers are connected in series, the matching load resistance is 26 kΩ, 39 kΩ and 52 kΩ respectively. While in parallel connection, matching load resistance reduced to 6.5 kΩ, 4.5 kΩ and 3.5 kΩ for two, three, and four connected cantilevers respectively. In series configuration, the voltage output produced is much higher as compared to the piezoelectric cantilever arrays that are connected in parallel connection. The voltage output of the piezoelectric cantilever increased from 3.41V to 6.09V when it is connected in series configuration with same polarity. Whereas in term of power output, piezoelectric cantilever arrays in parallel configuration produce higher power output as compared to piezoelectric cantilever arrays in series connection. The maximum power increased from 272μW to 521μW when two cantilevers are connected in parallel configuration with same polarity.

1 Introduction

Current researches and applications areas for piezoelectric materials are vast. The research stretches from the properties of piezoelectric materials, manufacturing, up to measurement techniques and practical applications; while in application wise, it is extended from sensor, actuator, transducer and now micro-power generator [1-3]. Recent researches had shown increasing interest in looking into methods to improve the performance of piezoelectric micro-power generator. Examples of the recent researches are the vibration-based MEMS piezoelectric energy harvester and power conditioning circuit which uses a MEMS piezoelectric power generator array for vibration energy harvesting [4]; wide-band energy harvesting using piezoelectric multi-cantilever which demonstrate the potential of improving the output of the piezoelectric when connected in alternating polarities [5]; and piezoelectric, solar and thermal energy harvesting for hybrid low-power generator systems with thin-film batteries which developed a hybrid power generator and storage system that has better functionality and robustness [6]. It is expected in the near future, piezoelectric material could be the replacement of battery to create a sustainable system.

This paper investigates the changes in the output of the piezoelectric cantilever arrays when connected in different configurations. In this research, piezoelectric cantilevers are connected in four configurations which are series with same polarity, series with alternating polarities, parallel with same polarity, and parallel with alternating polarities to observe the maximum power produced from each configuration. Matching load resistances were also identified to make sure optimum power was measured.

2 Experimental Set-up

A pre-mounted and wired standard quick-mount bending piezoelectric generator (Q220-A4-303YB) from Piezo Systems Inc [7], is used in this research as the test subject. The dimension of the piezoelectric cantilever is as shown in Fig. 1. The illustration of the experiment set-up is as shown in Fig. 2. Electrodynamics shaker was used to mimic the ambient vibration source in order to excite the cantilever, whereas, oscilloscope was used to measure the output of the piezoelectric cantilever.

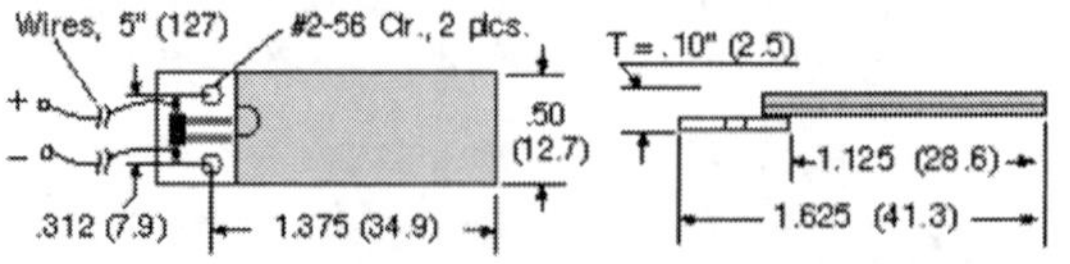

(a) Top View (b) Side View

Figure 1. Dimension of Piezoelectric Cantilever (Q220-A4-303YB) [7].

The matching load resistance of the cantilever can be identified by connecting the piezoelectric cantilever, named CL for reference convenient, to resistance from the resistance decade box ranging from 10 Ω to 1 MΩ while excited by vibration source fixed at the resonant frequency of the cantilever (300Hz) and constant magnitude 1-g (9.81 m/s2). The voltage output obtained from the piezoelectric cantilever when connected to various resistance values is observed and recorded in Fig. 4. The graph in term of voltage output is then converted to power output with the intention to identify the matching load resistance of the cantilever. The calculated result in term of power is recorded in Fig. 5.

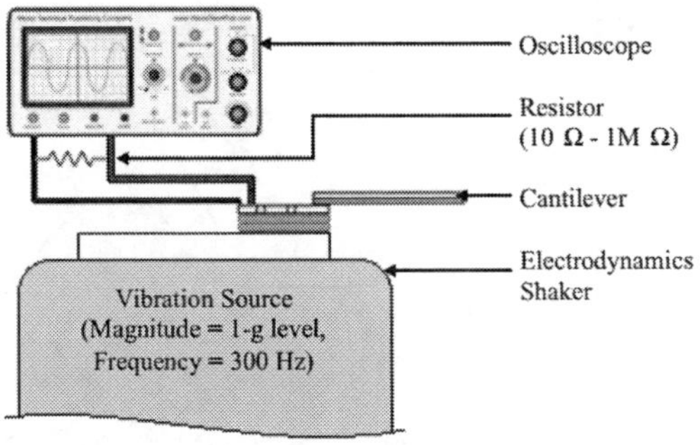

Figure 2. Illustration of the Experiment Set-up.

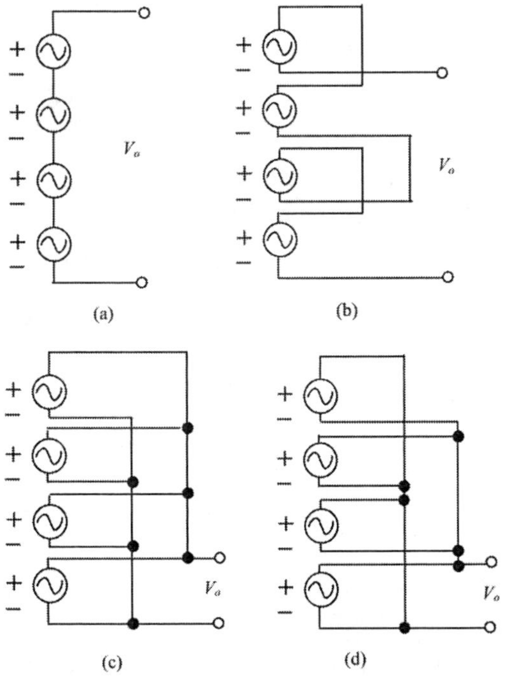

Figure 3. Illustrations for the type of Connections: (a) series, (b) series-alternating, (c) parallel and (d) parallel-alternating. [5].

In order to determine the maximum power of the piezoelectric cantilever arrays, the experiment is then continued by replacing the single piezoelectric cantilever with an array of cantilevers connected in four different configurations as shown in Fig. 3. The connections included series with same polarity, series with alternating polarities, parallel with same polarity, and parallel with alternating polarities. Piezoelectric cantilever is added one after another in order to examine the effect of the number of the connected cantilevers towards the total load resistance of the array. Three same sized cantilevers are used, named CL1, CL2 and CL3 for reference convenient. The output responses of the cantilever arrays in term of voltage and power are recorded and discussed in Section III.

3 Experimental Result

The output response in term of voltage and power for single piezoelectric cantilever CL is as shown in Fig.4 and Fig. 5 respectively. The graph in Fig. 4 shows that the output voltage produced increase in respect of the increasing of load resistance value. The output voltage increase slowly during low load resistance, then increased rapidly after 5 kΩ, then start to saturate at high load resistance. The maximum output voltage for CL is 3.41V. After converting the graph into power value, peak is formed as shown in Fig. 5. The result shows that peak formed at matching load resistance of 13 kΩ for CL with maximum of 272.25μW. This power value is the maximum power that Cantilever CL could achieve for vibration of 1-g level at frequency of 300Hz.

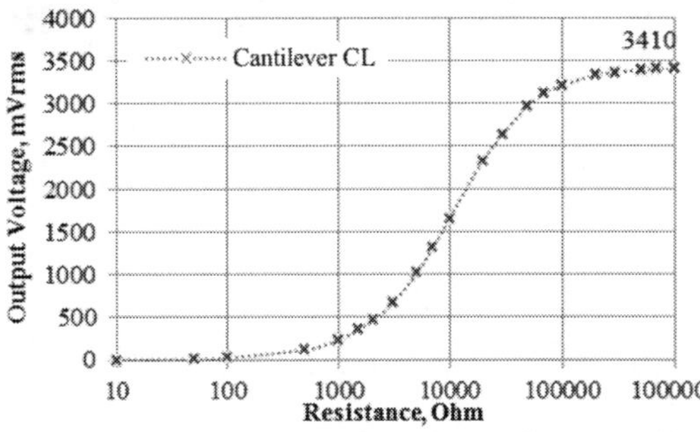

Figure 4. Output Responses (Voltage) for Cantilever CL with Varies Load Resistance.

The experiment is then continued by replacing the single piezoelectric cantilever with piezoelectric cantilever arrays connected in series with same polarity, series with alternating polarities, parallel with same polarity, and parallel with alternating polarities configurations.

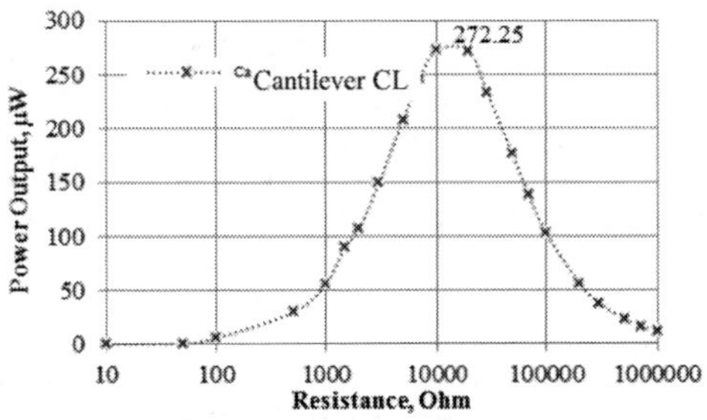

Figure 5. Output Responses (Power) for Cantilever CL with Varies Load Resistance.

3.1 Series Configuration with Same Polarity

The series with same polarity connected cantilevers are observed under various resistances in this subsection. The result in term of voltage output is shown in Fig. 6 while for the result in term of power value is shown in Fig. 7.

Similar to result in Fig. 4, the output increases slowly when low load resistance is connected to the cantilevers. It started to increase rapidly after 3 kΩ and started to saturate when connected to high load resistance. Notice that there is significant increase in maximum voltage output when cantilevers are connected as compared to the maximum voltage output produced by CL alone. Two series with same polarity connected cantilevers, CL and CL1 produced maximum voltage of 5.49V. While three series with same polarity connected cantilevers, CL, CL1, and CL2 produced maximum voltage of 5.73V. As four series with same polarity connected cantilevers, CL, CL1, CL2, and CL3 produced maximum voltage of 6.09V. The reason of the increasing of voltage output is due to the source are connected in series connection. When the sources are connected in series, the output voltage boosted up since the total voltage output would be same as adding all voltage output produced by each source.

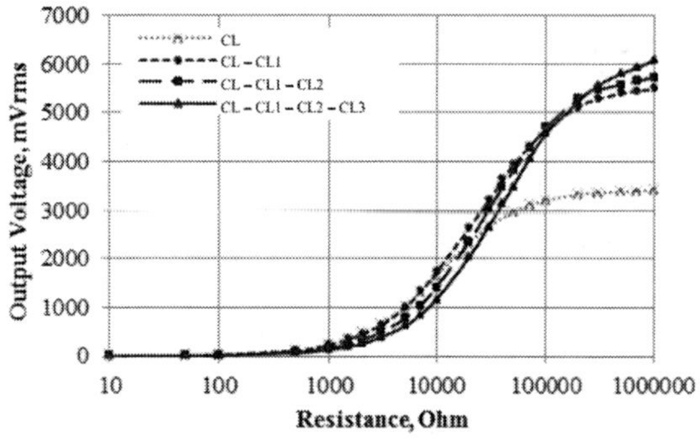

Figure 6. Output Responses (Voltage) toward Varies of Load Resistance for Cantilevers Connected in Series with Same Polarity Connection.

However, in series connection, the current output is expected to be lower and this would definitely affect the power output of the connected cantilevers. Based on the result obtained in Fig. 7, where the value is converted into power output, a peak is shown in each graph. The power generated is optimum at 346μW for two connected cantilevers during 26 kΩ load resistance, reduced to 302μW for three connected cantilevers during 39 kΩ load resistance, and further reduces to 245μW for four connected cantilevers during 52 kΩ load resistance.

The decreasing of maximum power is due to the fact that current value drops when connected in series because of the coupling effect from the piezoelectric cantilevers. Due to the fact that piezoelectric is conserver, the charge produced by the piezoelectric in direct piezoelectric effect will be dissipate through another conservative piezoelectric cantilever as there is no isolation between the cantilevers when connected together in series connection. Therefore adding more cantilevers in series with same polarity connection is not helpful in term of power output since it reduce the power produced. However the maximum power produced by two and three connected cantilevers are still higher than maximum power for single CL of 272μW during 10 kΩ load resistance, because of its high boost in voltage output when connected in series. The matching load resistance for a single cantilever is 13 kΩ. When it is connected in series connection with same polarity, matching load resistance for two connected cantilevers is 26 kΩ, for three connected cantilevers it is 39 kΩ, and for four connected cantilevers is 52 kΩ. The cantilevers produce maximum power when connected to respective load resistance value.

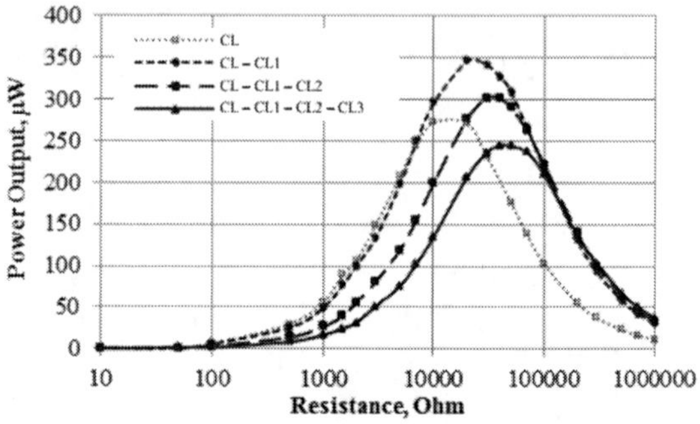

Figure 7. Output Responses (Power) toward Varies of Load Resistance for Cantilevers Connected in Series with Same Polarity Connection.

3.2 Series Configuration with Alternating Polarities

The series with alternating polarities connected cantilevers are observed under various resistances in this subsection. The result in terms of voltage output is shown in Fig. 8, while the result after converting to power value in shows in Fig. 9.

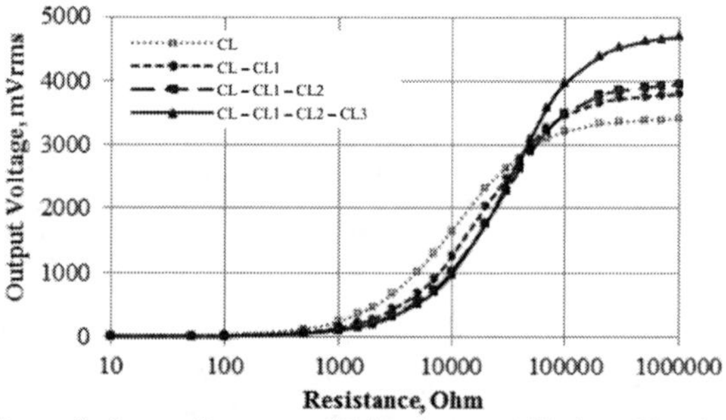

Figure 8. Output Responses (Voltage) toward Varies of Load Resistance for Cantilevers Connected in Series with Alternating Polarities Connection.

The result for voltage output is also similar to the result in Fig. 4, where the output increases slowly when low load resistance is connected to the cantilevers. It started to increase rapidly after 3 kΩ and starts to saturate when connected to a high load resistance. The maximum voltage outputs obtained from the multiple connected cantilevers are still higher than maximum voltage output produced from single CL, which is 3.79V for two connected cantilevers, 3.95V for three connected cantilevers, and 4.69V for four connected cantilevers. However, the voltage value is not as high as compared to the voltage result obtained from series with same polarities connection. This is due to the fact that piezoelectric cantilever generate output in alternating current (AC) form. Therefore, when connected in alternating polarities, the positive value of the output waveform is reduced by the negative value of the output waveform from the other cantilever.

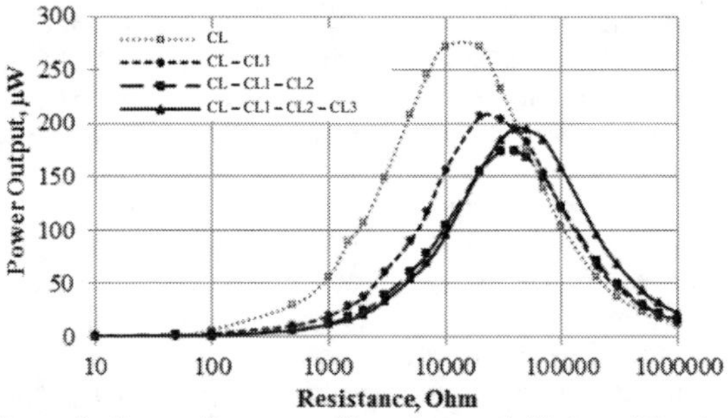

Figure 9. Output Responses (Power) toward Varies of Load Resistance for Cantilevers Connected in Series with Alternating Polarities Connection.

Same goes to its power output value, it is much lower compare to power produced by series with same polarities connection, in which two connected cantilevers produced maximum power of 206µW during 26 kΩ load resistance, three connected cantilevers produced maximum power of 173µW during 39 kΩ load resistance, and four connected cantilevers produced maximum power of 195µW during 52 kΩ load resistance. Notice that the matching resistances for piezoelectric cantilevers connected in series with alternating polarities connection are same as in series with same polarity connection.

3.3 Parallel Configuration with Same Polarity

As for parallel with same polarity connected cantilevers the result in terms of voltage output is shown in Fig. 10 while the result after converting it to power value in shown in Fig. 11.

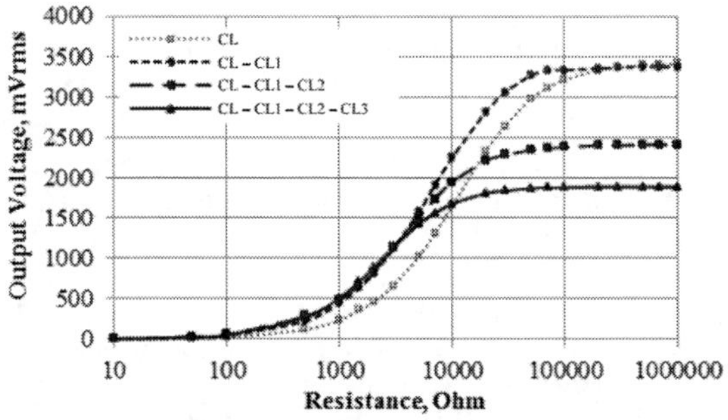

Figure 10. Output Responses (Voltage) toward Varies of Load Resistance for Cantilevers Connected in Parallel with Same Polarity Connection.

The voltage output also increased slowly at low load resistance and started to increase rapidly after 1 kΩ and starts to saturate when connected to high load resistance. The maximum output voltage achieved by two connected cantilevers is 3.37V, for three connected cantilevers is 2.41V, as for four connected cantilevers is 1.89V. When three other cantilevers are added to the connection, the total voltage output dropped because the voltage provided by cantilever CL1, CL2 and CL3 alone is lower than CL. Hence when average out, the total voltage output is decreased.

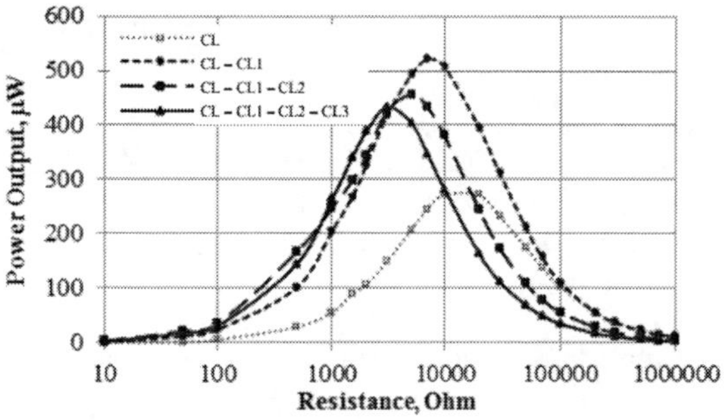

Figure 11. Output Responses (Power) toward Varies of Load Resistance for Cantilevers Connected in Parallel with Same Polarity Connection.

However, this is not the case for power output. The total power of the circuit was boosted up when power source are connected in parallel because current of the circuit increased when connected in parallel. This is shown in Fig. 11, where maximum power increased from 272µW to 521µW when two cantilevers are connected. However, the power drops to 456µW when three cantilevers are connected and to 433µW when four cantilevers are connected. This is due to the fact that total voltage output drops when connected in parallel. The increment in the current value is not high enough to cope with the dropping of voltage; hence the total power produced is lower. But still, the power outputs are higher than the output power produced by CL alone because of the current increased in parallel connection. The matching load resistances for two, three, and four connected cantilevers are 6.5 kΩ, 4.5 kΩ and 3.5 kΩ respectively since they reach maximum power when connected to those load resistance value.

3.4 Parallel Configuration with Alternating Polarities

The result in term of voltage output for parallel with alternating polarities connected cantilevers is shown in Fig. 12 while the result after converting to power value in shown in Fig. 13.

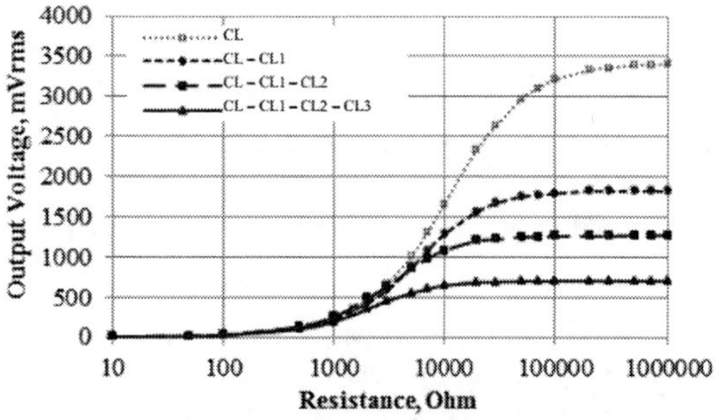

Figure 12. Output Responses (Voltage) toward Varies of Load Resistance for Cantilevers Connected in Parallel with Alternating Polarities Connection.

Again the result shows output increased slowly when low load resistance is connected to the cantilevers and started to increase rapidly after 1 kΩ and starts to saturate when connected to high load resistance. Notice that the maximum voltage output produced in this case is much lower compared to the voltage output produced by CL alone. The maximum voltage output produced by two connected cantilevers is 1.82V, three connected cantilevers are 1.27V, and for four connected cantilevers is 0.71V. This is expected since voltage would drop for parallel branches. Furthermore, the cantilevers are connected in alternating polarities which means its positive value would be cancelled out by the negative value of the other connected cantilevers.

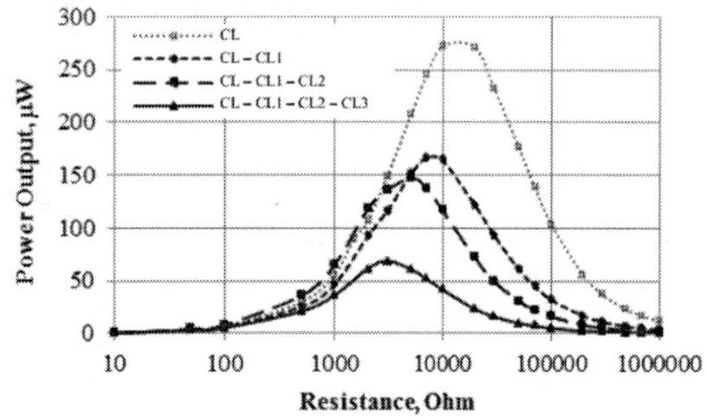

Figure 13. Output Responses (Power) toward Varies of Load Resistance for Cantilevers Connected in Parallel with Alternating Polarities Connection

Notice that the voltage value obtained in Fig. 12 for alternating polarities are much lower than the value obtained in Fig. 10 for same polarity. Even the converted power values shown in Fig. 13 are lower compared to power produced by cantilevers connected in parallel with same polarity connection. Where two connected cantilevers produced 167µW, three connected cantilevers produced 147µW, and four connected cantilever produced 69µW. This is simply because the increased current value cannot cope with the decreased voltage value of the overall connected cantilevers.

In this case the matching load resistance for two connected cantilevers is still 6.5 kΩ, for three connected cantilevers is 4.5 kΩ, and for four connected cantilevers is 3.5 kΩ.

4 Conclusion

The focus of this research is to investigate the changes in the output of the piezoelectric cantilever arrays when connected in different configuration. In term of voltage, all result shows that output voltage increase slowly during low load resistance, then increased rapidly after it reached certain resistance value, then starts to saturate at high load resistance. When converted to power values, the result shows that each array has one peak formed at respective matching load resistance, producing maximum power output. In series connections, the matching load resistance for two, three, and four connected cantilevers is 26 kΩ, 39 kΩ and 52 kΩ respectively. While in parallel connection, matching load resistance reduced to 6.5 kΩ, 4.5 kΩ and 3.5 kΩ for two, three, and four connected cantilevers respectively. These matching load resistance values are useful in order to maximize the output of the designed piezoelectric application. The resistance for the piezoelectric cantilever arrays posed the same behaviour as when 13 kΩ resistors are connected in series or parallel connection. The result also shows that when connected in series connection, the voltage output produced is much higher as compared to the piezoelectric cantilever arrays that are connected in parallel connection. The voltage output of the piezoelectric cantilever increased from 3.41V to 6.09V when it is connected in series configuration with same polarity. Whereas in term of power output, when connected to matching load resistance, piezoelectric cantilever arrays in parallel connection produce higher power output as compared to piezoelectric cantilever arrays in series connection. The maximum power increased from 272µW to 521µW when two cantilevers are connected in parallel configuration with same polarity.

Acknowledgments

The authors would like to thank the Ministry of Higher Education of Malaysia to sponsor this research project under the research grant no. FRGS/2/2014/SG02/FKEKK/02 as well as the facility support by the Advanced Sensors and Embedded Control System (ASECs) Research Group, Centre for Telecommunication Research and Innovation (CeTRI), UTeM.

References

1. Zhao, J., and You, Z., 2014. A Shoe-Embedded Piezoelectric Energy Harvester for Wearable Sensors. Sensor 2014, 14, pp.12497-12510.
2. Ansari, M.H., and Karami, M.A., 2015. Heartbeat Energy Harvesting Using the Fan-Folded Piezoelectric beam Geometry, In: Proceedings of ASME 2015International Design and Engineering Technical Conferences & Computers and Information in Engineering Conference (IDETC/CIE 2015), Boston, Massachusetts, USA, 2-5 Aug 2015. ASME Publisher.
3. Aidin, D., and Jeremie, V., 2014. Flexible piezoelectric energy harvesting from jaw movements. Smart Materials and Structures, 23.
4. Yu, H., Zhou, J., Deng, L., and Wen, Z., 2014. A Vibration-Based MEMS Piezoelectric Energy Harvester and Power Conditioning Circuit. Sensors 2014, 14, pp.3323-3341.
5. Bong, Y.J., Kok, S.L. and Thong, L.W., 2014. Experimental Verification of Wide-Band Energy Harvesting using Piezoelectric Multi-cantilever with Resonant Frequency Variation. In: Malaysian Technical Universities Conference on Engineering and Technology (MUCET) 2014, Melaka, Malaysia, 10-11 November 2014.

6. Gambier, P., Anton, S.R., Kong, N., Erturk, A., and Inman, D.J., 2012. Piezoelectric, solar and thermal energy harvesting for hybrid low-power generator systems with thin-film batteries. Measurement Science and Technology, 23.
7. Piezo Systems, Inc., 2011. Standard Quick-Mount Piezoelectric Bending Sensors (Generators). Available online: <http://www.piezo.com/prodbg7qm.html> [Accessed on 25 October 2016].

Study on the Influence of Blade Roughnesson Axial Flow Compressor Stage Performance

Xudong Shi[1], Hui Liu[2], Feiqi Long[1], Jie Tang[3] and Liwen Wang[4]

[1]*Electronic Information and Automation College, Civil Aviation University of China, China*
[2]*Sino-European Institute of Aviation Engineering, Civil Aviation University of China, China*
[3]*Mechanical Engineering Department, Civil Aviation University of China, China*
[4]*Aviation Ground Special Equipments Research Base, Civil Aviation University of China, China*

Abstract. A typically actual inlet stage NASA Stage 36 is chosen to study the influence of surface roughness on axial compressor performance. Firstly, a geometry model is created by blade design software BladeGen using transferred coordinates data of blade profile and flow path. Secondly, validation of simulation model is conducted by comparing computational data and field experiment data. Lastly, SST k-ω turbulence model is chosen to study the influence of blade surface roughness on performance parameters under different work points. It shows that adding roughness will significantly reduce axial compressor stage's adiabatic efficiency and total pressure ratio and cause stage characteristic map shift toward left. It should not neglect the influence of surface roughness of stator near stall region under 100% design speed; Mach number shows a big difference after adding surface roughness, and it can be considered as a sensibility parameter of roughness.

1 Introduction

The performance deterioration of compressor is directly connected to aircraft safety and efficiency, it has become more and more important to study the factors causing compressor performance deterioration and to explore the laws of performance deterioration. Blade damages like fouling, missing material or coating and tip curl change blade profile parameters and surface characteristics, the latter mainly acts as the increase of surface roughness, among which fouling is the main factor.

The roughness correction of turbulence flow dates back to the study of pipe flow conducted by Nikuradse [1] in 1933. From the view of boundary layer development, he discovered that it can be divided into three regimes to study the influence of roughness on pipe flow according to Reynold number value, pipe pressure loss coefficient depends on Reynold number and roughness height under low Reynold number condition, whereas friction resistance of pipe wall depends only on sand-grain diameter under high Reynold number condition. Schilichting [2] used Nikuradse's experiment data to correlate different rough surface types by which he came up with the concept of 'equivalent sand-grain roughness', which provided a method to describe randomly rough surface by an 'equivalent' uniform layer of actual sand-grain diameter, and his this methodology has gained a worldwide acceptance.

There is no doubt that roughness has a big influence on compressor performance as it changes the flow condition in compressor. Bammer [3] experimentally obtained the changes about the pressure distribution around blade profile, the boundary-layer development along profile contour, and the velocity distribution in the boundary layer due to increasing roughness on NACA 65-series compressor blade sections of different geometrical sizes. Syverud et al. [4-5] presented the deterioration of overall engine performance caused by spraying atomized droplets of saltwater into the axial compressor of engine GE J85-13, based on those test data, they also studied the influence of roughness on compressor performance from the perspective of friction loss, blockage and airflow separation, it shows that the most significant effect of salt deterioration is the reduction in flow coefficient. In order to study the influence of adding blade thickness and roughness on performance, Suder et al. [6] applied rough coatings to the pressure and suction surface of rotor blades, the results shows that roughness has the most significant influence on performance near the design point. Based on k- turbulence model, Mirko Morini et al. [7] chose NASA Stage 37 to study fouling influence on performance deterioration and flow blockage, according to their research, reductions in the flow coefficient and in the pressure coefficient were found to be the same order of magnitude of experimental results. Also based on k- turbulence model, Shi Hui et al. [8-9] chose NASA Stage 35 to study the influence of adding thickness and roughness caused by fouling on compressor stage performance, and it shows that the influence of roughness applied on rotor is nearly the same as that on whole stage. With the same method, Rotor 37 is chosen to study the

 DOI: 10.1051/matecconf/201710808008

influence of fouling on compressor performance, an emphasis was putted on studying the relationship between rotor performance and shock wave, shock wave/boundary interaction.

There are already many valuable results in terms of theory, experiment and numerical simulation about how surface roughness affects flow condition. In this paper, the effect of surface roughness on blades of NASA Stage 36 is investigated and SST k-ωturbulence model is chosen for three-dimensional numerical simulation. The feasibility of numerical model is conducted by comparing numerical results and experiment data. Different levels of surface roughness are considered to study fouling effects on compressor stage performance under different work points, and variations of input power and Mach number are also discussed.

2 Numerical simulations

2.1 Geometrical model

In order to study the influence of blade surface roughness on compressor stage's aerodynamic performance, a typically actual compressor stage NASA Stage 36 is employed. The profile, flow path, design point parameters and related experiment data are found in literature [10]. NASA Stage 36 is composed of 48 rotors and 62 stators, the aspect ratios for rotor and stator are respectively 1.63 and 1.78, design rotation speed is 17188.7 rpm with mass flow of 20.188 kg/s, design tip speed is 455.233m/s, stage total pressure ratio is 1.82 and adiabatic efficiency is 82.2%.

The geometrical modelling is briefly described as follows: data files of blade profile contours of different radius are obtained by coordinate transformation. Geometry of flow path is firstly obtained by fitting of discrete points using SolidWorks, then flow path including hub and shroud data file can be obtained by importing the geometry into blade generation software named BladeGen. Finally, we can obtain rotor or stator geometry by importing those three above data files. Fig. 1 shows the geometrical model of Stage 36.

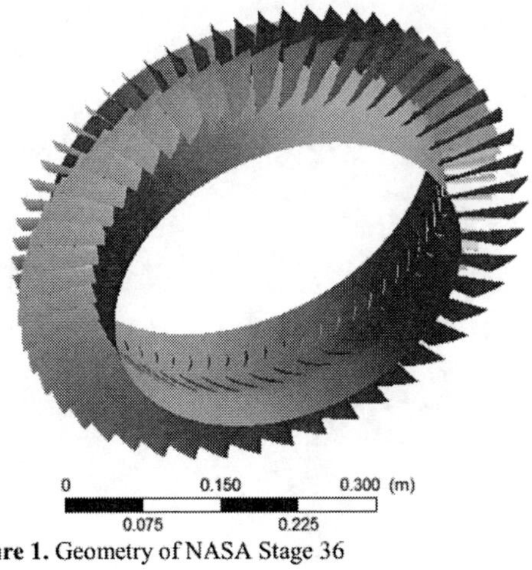

Figure 1. Geometry of NASA Stage 36

2.2 Turbulence and wall modelling

Only one passage is computed and the pitch ratio is then 1.29. The meshes were generated in Turbo Grid 14.0, using 'J-Grid' structured mesh for rotor and 'H/J/C/L-Grid' structured mesh for stator both with O-grid near blade wall, tip clearance is not taken into consideration. The total number of mesh elements is approximately 1000,000, the first mesh height normal to wall is fixed at 20μm, and numerical meshes are shown in Fig. 2. The reason for choosing the height of first grid will be discussed later. Final numerical calculations were done using commercial CFD software ANSYS CFX 14.0 whose calculations is based on Finite Volume Method, a second-order high-resolution advection is used in calculations.

Figure 2. Computational meshes of Stage 36

Surface roughness leads to an increase in turbulence production near the wall, which in turn results in an increase of wall shear stress and breaks up the viscous sublayer of turbulent flow, and it means a downward shift for logarithmic velocity profile. As a result, roughness effects on turbulent flow are realized by changing the logarithmic velocity profile, for ANSYS CFX14.0 [11], it has:

$$u^+ = \frac{1}{\kappa} ln(y^+) + 5.2 - \Delta B \quad (1)$$

$$\Delta B = \frac{1}{\kappa} \ln(1 + 0.3h_s^+) \quad (2)$$

$$h_s^+ = \frac{h_s u_\tau}{\nu} \quad (3)$$

where u^+ is the velocity near the wall, ΔB is the distance of downward shift of the logarithmic velocity profile, h_s is the sand-grain diameter, u_τ is the friction velocity, ν is the dynamic viscosity coefficient, h_s^+ is a dimensionless sand-grain roughness. Depending on the value of h_s^+, it can divide into three roughness regimes which are hydraulically smooth, transitionally rough and fully rough regimes, which comes to a corresponding relation with the studies of Nikuradse [1] and Schilichting [2].

According to open literatures, k- turbulence model is the most frequently employed model to study the influence of surface roughness on compressor performance, what's more, first mesh height is generally 50μm and maximum roughness value is about 100μm [7-9]. This paper determines to choose SST k-ωturbulence model, which to some degree covers the shortage of model choice. This model is only suitable for small values of h_s^+ with automatic rough wall treatment, when a

DOI: 10.1051/matecconf/201710808008

roughness height has been specified, then the automatic rough wall treatment is activated. Considering that the first mesh height should be the same order or bigger than the dimensionless sand-grain roughness h_s^+, the first mesh height is chosen as 20μm.

2.3 Boundary condition

Inlet flow direction is along the rotational axis and inlet total pressure and total temperature are fixed at 101315Pa and 288.15K respectively. An average static pressure is imposed at outlet, and different work point from near-choke flow region to near-stall flow regime can be obtained by gradually increasing pressure.

3 Results and discussion

3.1 Model validation

In Fig. 3, the numerical and experimental performance parameters at 100% and 70% design speed are presented. Comparing with the results at 100% design speed, those at 70% design speed are better consistent with experimental data, whereas all their mass flows at near-choke region are bigger than experimental value. There is a good enough agreement of total pressure ratio and adiabatic efficiency between numerical results and experimental results. The reasons for errors include no tip clearance, limited original data of geometry, one-passage calculation, existence of experimental errors and so on. Comparing with NASA Stage 35 and 37, blade number of this stage brings a bigger difficulty in modelling and calculation. Although there were some deviations from experimental results, the numerical results have well reproduced the variation characteristics of compressor stage performance under experiment condition. Therefore, it's feasible to use this model to study the influence of roughness on compressor's aerodynamic performance.

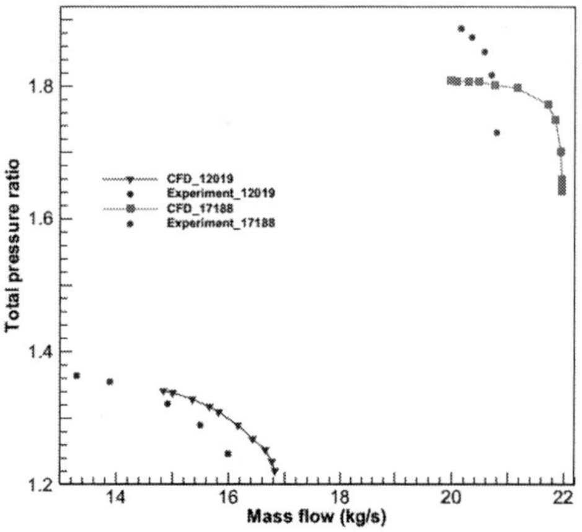

(3.a) Total pressure ratio

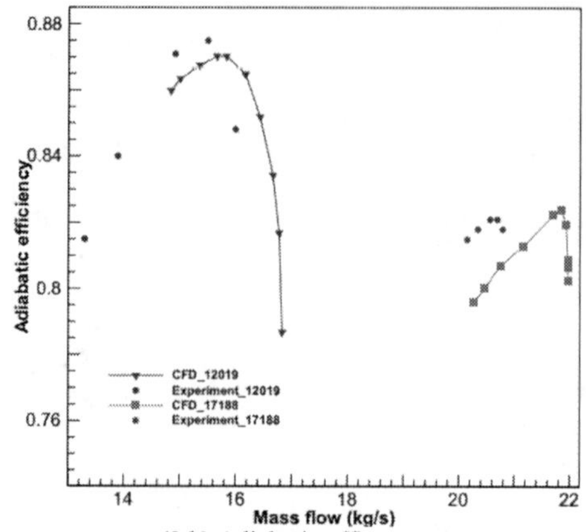

(3.b) Adiabatic efficency

Figure 3. Comparison of characteristic maps between computational and experimental results

3.2 Stage performance parameters

In Fig.4, the influences of surface roughness on stage total pressure ratio and adiabatic efficiency at 100% and 70% design speed are presented. It can be concluded that the stage performance maps move toward left after adding surface roughness, mass flows in near-choke region and in near-stall region have a remarkable decrease. Total pressure ratio and adiabatic efficiency also decrease, and their decreased magnitudes increase with roughness level and with rotation speed of stage's rotor. At 70% design speed, the variation of stage performance map has a similar tendency with ones obtained by using k- turbulence model, as the new blade-rough performance map has a similarity with the blade-clean performance map. At 100% design speed, it seems SST k-ωturbulence model has a stricter treatment for rough wall. Compared with clean blade condition at 100% design speed and with a same mass flow, there are five percent decrease both for total pressure ratio and adiabatic efficiency when the roughness value is 40μm.

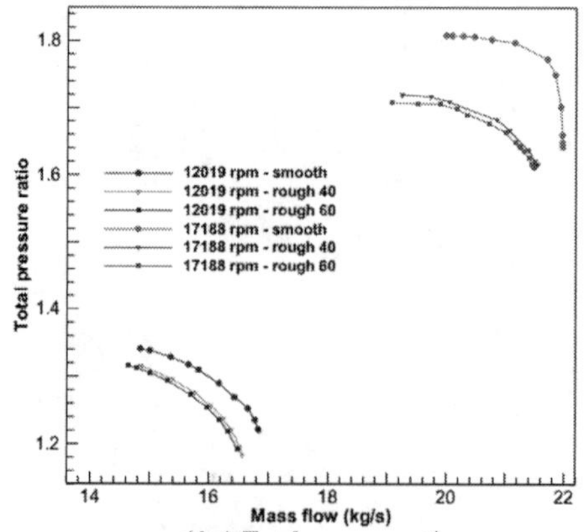

(4.a) Total pressure ratio

 DOI: 10.1051/matecconf/201710808008

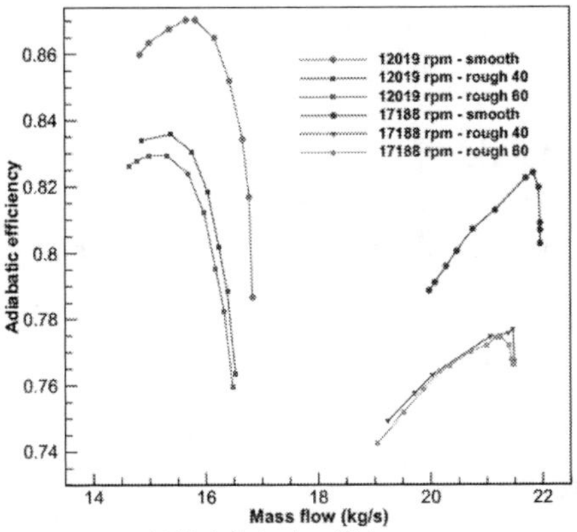

(4.b) Adiabatic efficiency

Figure 4. Effects of surface roughness on stage characteristic maps under different working points

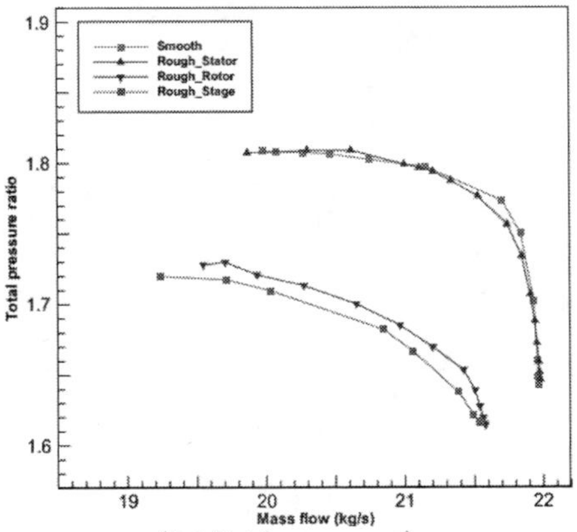

(5.a) Total pressure ratio

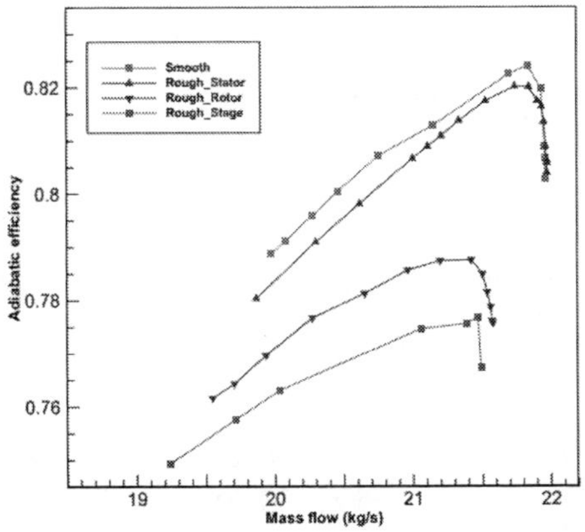

(5.b) Adiabatic efficiency

Figure 5. Sensibilities of rotor and stator to surface roughness at 100% design speed

In the above case, a same roughness is applied to both rotor and stator. In order to separately consider the influence of roughness on rotor or stator, Fig.5 shows the variation of stage performance at 100% design speed when a roughness 40μm is only applied to one of them. It can be concluded: an increase of roughness on stator has nearly no influence on total pressure, mass flow and efficiency in near-choke region, but it has a remarkable effect on adiabatic efficiency in far-choke region; the surface roughness applied to rotor occupies the main impact on stage performance deterioration.

The input powers for smooth blade and blade of roughness 40μm at 100% design speed are given in Fig.6. For a same pressure ratio, it demands more than 100 kW more input power for the compressor stage of rough blades; in the near-stall region, roughness causes a decrease of input power of approximately 150 kW. Total pressure ratio gradually increases as mass flow decreases, compared with smooth blade, rough blade aggravates flow separation and therefore impels the compressor to enter near-stall region in advance, which is the reason for an extra input power for a same pressure ratio and an decrease of maximum input power.

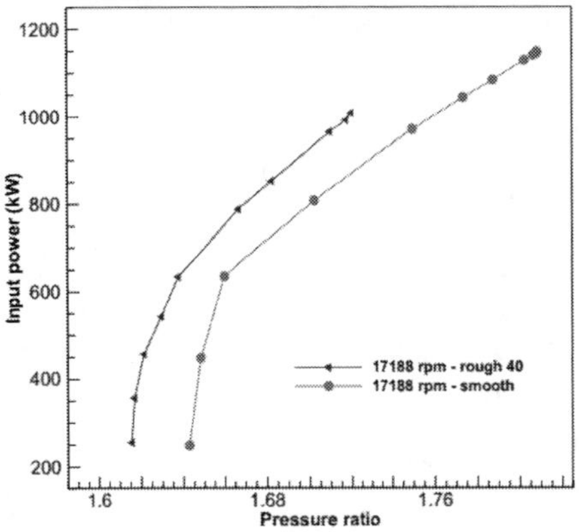

Figure 6. Variation of stage input power along with pressure ratio

3.3 Mach number

In near-stall region, the circumferentially averaged relative Mach number at leading edge of rotor along blade span is presented in Fig.7. In this paper, a near-stall point is taken as the work point where it can no longer endure 0.01atm static pressure rise on outlet. The circumferentially averaged relative Mach number decreases as blade roughness increase, a significant decrease of relative Mach number happens near 20% rotor tip.

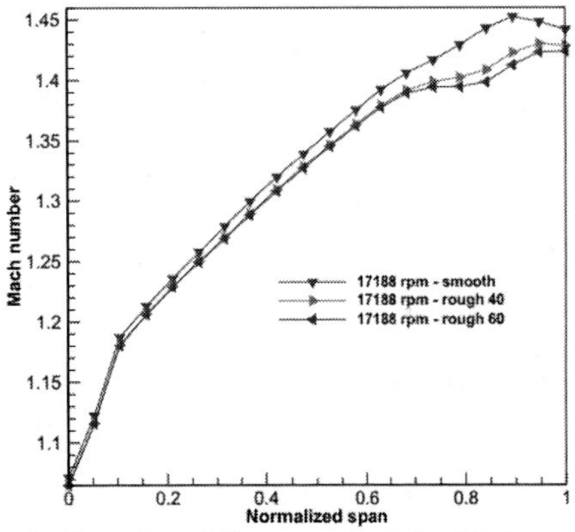

Figure 7. Circumferentially averaged relative Mach number at leading edge

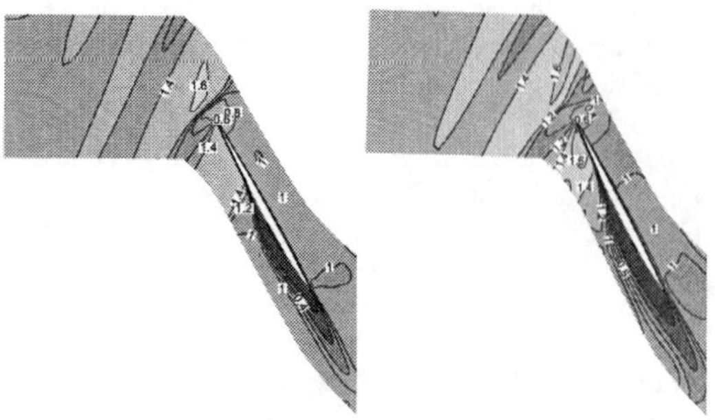

(8.a) Smooth (8.c) 60μm

Figure 8. Contour of Mach number at 80% span

Fig. 8 shows the distribution of Mach number at 80% span of rotor for near-stall point at 100% design speed. It can be included that roughness level has a big impact on the distribution of Mach number. Difference mainly manifests as an apparent increase of Mach number near leading edge on suction surface. What's more, the shock wave in the rear of suction surface moves both toward the front and far from blade wall, which will expand wake range of suction surface, create more choked region and further leads to an earlier stall occurrence. The variation of Mach number on pressure surface side is relatively not apparent in all three cases. Combined all analysis above about Mach number, it seems that Mach number is sensitive to surface roughness.

5 Conclusions

A geometrical modelling of NASA Stage36 based on original blade profile and flow path data is realized and model validation is also conducted. SST k-ω turbulence model is employed to study the influence of adding roughness caused by fouling on compressor performance. According to above results, it can conclude that:

Using original blade profile and flow path data, software BladeGen can realize a quick modelling of compressor, which greatly reduces time cost with a high enough accuracy;

Increase of blade roughness will remarkably decrease compressor's total pressure ratio and adiabatic efficiency, with stage characteristic map moving toward left; adding roughness on rotor has a bigger impact than on stator, whereas it cannot neglect the reduction of efficiency caused by roughness on stator under small mass flow situation.

With the same total pressure ratio, an increase of blade surface roughness leads to great increase of input power, and the maximum available input power significantly decreases, which is related to an earlier flow separation caused by surface roughness.

At near-stall point, there is a distinct decrease of the circumferentially averaged relative Mach number of leading edge near 20% rotor tip when surface roughness adds, meanwhile, Mach number near leading edge of rotor's suction surface increases, and shock wave in the rear position of suction surface moves both forward and far from blade wall.

Acknowledgement

This work was sponsored by science and technology project of Civil Aviation (20150218), Fundamental Research Funds for the Central Universities (3122016D017) and Tianjin Science and Technology Support Key Project (Grant no. #15ZCZDGX00350).

References

1. Nikuradse J. Laws of flow in rough pipes[C]//VDI Forschungsheft. (1933)
2. Schlichting H. Experimentelle untersuchungen zum rauhigkeitsproblem [J].Archive of Applied Mechanics, **7**(1): 1-34 (1936)
3. Bammert K, Milsch R. Boundary layers on rough compressor blades[C]//ASME 1972 International Gas Turbine and Fluids Engineering Conference and Products Show. American Society of Mechanical Engineers, (1972), V001T01A047-V001T01A047
4. Syverud E, Brekke O, Bakken L E. Axial compressor deterioration caused bysaltwater ingestion[C]//ASME Turbo Expo 2005: Power for Land, Sea, and Air. American Society of Mechanical Engineers, 327-337 (2005)
5. Syverud E, Bakken L E. The impact of surface roughness on axial compressor performance deterioration[C]//ASME Turbo Expo 2006: Power for Land, Sea, and Air. American Society of Mechanical Engineers, 491-501 (2006)
6. Suder K L, Chima R V, Strazisar A J, et al. The effect of adding roughness and thickness to a transonic axial compressor rotor[C]//ASME 1994 International Gas Turbine and Aeroengine Congress and Exposition. American Society of Mechanical Engineers, (1994), V001T01A113-V001T01A113
7. Morini M, Pinelli M, Spina P R, et al. Computational fluid dynamics simulation of fouling on axial compressor stages[J]. Journal of Engineering for Gas Turbines and Power, **132**(7): 072401 (2010)
8. CHEN S, ZHANG C, SHI H, et al. Numerical Study on the Impact of Fouling on Axial Compressor Stages [J]. Journal of Propulsion Technology, **3**: 006 (2012)
9. Shi H, Chen SW, Zhang C, et al. Numerical simulation of fouling deposition in compressor rotor[J]. Journal of Aerospace Power, **27**(5): 1061-1067 (2012)
10. Reid L, Moore R D. Design and overall performance of four highly loaded, high speed inlet stages for an advanced high-pressure-ratio core compressor[J]. (1978)
11. ANSYS CFX. Version 14.0[J]. Ansys Inc, (2011)

DOI: 10.1051/matecconf/201710808009

All-Coefficient Adaptive Control of Dual-Motor Driving Servo System

Haibo Zhao[1,2], Chengguang Wang[3]

[1]Engineering Technology Research Center of Optoelectronic Appliance, Tongling University, Tongling Anhui, 244061, China
[2]Department of Electrical Engineering, Tongling University, Tongling Anhui, 244061, China
[3]Sichuan Institute of Aerospace System Engineering, Chengdu Sichuan, 610100, China

Abstract. Backlash nonlinearity and friction nonlinearity exist in dual-motor driving servo system, which reducing system response speed, steady accuracy and anti-interference ability. In order to diminish the adverse effects of backlash and friction nonlinearity to system, we proposed a new all-coefficient adaptive control method. Firstly, we introduced the dynamic model of backlash and friction nonlinearity respectively. Then on this basis, we established the characteristic model when backlash and friction nonlinearity coexist. We used recursive least square method for parameter estimation. Finally we designed the all-coefficient adaptive controller. On the basis of simplex all-coefficient adaptive controller, we designed a feedforward all-coefficient adaptive controller. The simulations of feedforward all-coefficient adaptive control and simplex all-coefficient adaptive control were compared. The results show that the former has quicker response speed, higher steady accuracy, stronger anti-interference performance and better robustness, which validating the efficacy of the proposed control strategy.

1 Introduction

There are many nonlinearities which influencing the dynamic performance and steady-state accuracy of dual-motor driving servo system, such as backlash nonlinearity, friction nonlinearity, deadzone nonlinearity and so on in which the most obvious influence on system are backlash and friction nonlinearity. Thus, backlash and friction nonlinearity have always been caused extensive concern of academic and engineering at home and abroad. In terms of modeling of dual-motor driving servo system, many scholars at home and abroad have studied it. In [1], the dynamic model of dual-motor driving servo system containing backlash nonlinearity was established and the dynamic structure diagram of system was given. In [2], the simulation model of AC driving system based on dual-motor mutual feedback was established and the model was based on adaptive full order flux observer and torque sensor. In [3], a new compensation technique for friction that utilizes a backstepping control structure and an adaptive estimation of the friction force based on an observer for the class of dual-motor driving system was presented. In [4], the control model of machine tool feeding system driven by dual-motor containing backlash nonlinearity was established, using speed difference feedback to eliminate the synchronous error of dual-motor. In [5, 6], the mechanical model and experimental simulation block diagram of cement kiln driving system were given. But the system model established in the above literature are all dynamic models, and also the backlash and friction nonlinearity exist alone, without considering the system model when backlash and friction nonlinearity coexist.

In the compensate control strategies, many scholars at home and abroad have also studied them. In [7], the parameter identification of nonlinear dynamic systems with input saturation and output backlash using three-block cascade models was presented. Examples of three-block cascade system identification illustrated the feasibility of proposed method. In [8], aiming at unknown dynamic friction parameters and unknown load characteristic, the adaptive friction compensation was designed for the servo system. Simulation and experiment results were shown to effectively inhibit the disadvantageous influence of friction. In [9], an adaptive controller was designed for a class of periodically time-varying systems with input backlash, when the periodical uncertainties can be parameterized. The sign function was replaced by the hyperbolic function to ensure the differentiability for the controller and to avoid the chattering. In [10], an iterative learning control method was presented for a class of time-varying nonlinear systems with input backlash. In [11], an adaptive fuzzy controller was designed in the case of unknown system parameters and load torque. In [12], a friction torque compensation controller without distal-end feedback was designed to ensure that the output of the system can effectively track the expected torque trajectory, and the effectiveness of the scheme was validated by experiments. But most of scholars at home and abroad have studied the control strategies when backlash and friction nonlinearity exist alone, few scholars have studied the compensate strategies when backlash and friction nonlinearity coexist.

Taking a wide view of the development of backlash and friction nonlinearity in recent years, the main control algorithms are robust adaptive control, repetitive control, fuzzy control, sliding mode control, neural network control and so on. To some extent, the above proposed control methods have achieved good control effect, but still can't meet many problems in practical application, such as the dynamic tracking performance and control precision are not high, system parameter changes, etc. The all-coefficient adaptive control method has features of simple algorithms, strong robustness and so on, and has been successfully applied in actual industrial and aerospace control system, algorithm itself has a high maturity, but few scholars use it to control dual-motor driving servo system, so it is necessary to study it. Aiming at backlash and friction nonlinearity coexist, a new system model, namely characteristic model was established. We use all-coefficient adaptive control algorithm to control system and obtain the ideal control effect.

2 Characteristic Model

The so-called characteristic model is the modeling combining the dynamic characteristics of object, environmental characteristics, control performance requirements, not only depending on the accurate dynamic analysis of object. The features of characteristic model are as follows [13].

(1)Under the same control input, the object characteristic model is equivalent to the output of the actual object(namely in the dynamic process, system can keep within the allowable output error). In a steady state, the output are equal.

(2)The form and order of the characteristic model are mainly determined by the control performance requirements, in addition to the characteristics of the object.

(3)The form of characteristic model is simpler than the original object's dynamic equation, easy to design controller and easy to implement in engineering.

(4) The characteristic model is different from the reduced-order model of higher order system, which is to reduce the information of the higher order model into several characteristic parameters. In general, the characteristic model is described by the slowly time varying difference equation.

The characteristic model can be directly modeled by the object physical mechanism, and also can be derived from the original dynamic equation and the control performance requirements. In this paper, we adopt the latter method, firstly we establish the system dynamics model when backlash and friction nonlinearity coexist, then we design the system characteristic model on this basis.

2.1 System Dynamic Model

Fig. 1 is the schematic diagram of dual-motor driving servo system. O_0 is driven-subsystem, O_1 and O_2 are driving-subsystems.

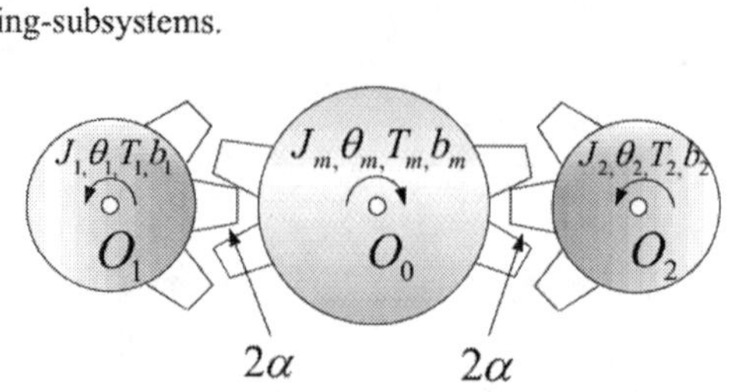

Figure 1. The schematic diagram of dual-motor driving servo system.

where J_1, J_2 are the rotational inertias of driving wheel. J_m is the rotational inertia of driven wheel. θ_1,θ_2 are the rotation angles of driving wheel shaft. θ_m is the rotation angle of driven wheel shaft. T_1,T_2 are the input torques of driving wheel. T_m is the load torque of driven wheel. b_1,b_2 are the viscous friction coefficients of driving wheel. b_m is the viscous friction coefficient of driven wheel. 2α is the backlash.

Assumption 1 In the control process, driven-subsystem O_1 and O_2 contact with O_0 alternatively. It was guaranteed by exerting some bias torques to driven-subsystem from output of two driving-subsystems [14]. The bias torques was equal and opposite. Because of the influence of backlash nonlinearity [15], $\tau_i(t)(i=1,2)$ can be expressed as follows.

$$\tau_i(t) = k\,f(z_i(t)) + cf(\dot{z}_i(t)) \tag{1}$$

where, k and c are stiffness coefficient and damping coefficient, driving wheel and driven wheel, respectively. $f(z_i(t))$ is the deadzone function and can be expressed as follows.

$$f(z_i(t)) = \begin{cases} z_i(t)+\alpha & z<-\alpha \\ 0 & -\alpha \le z \le \alpha \\ z_i(t)-\alpha & z>\alpha \end{cases} \tag{2}$$

where, $z_i(t)=\theta_i(t)-\theta_m(t)$ (suppose the transmission ratio is 1).

In dual-motor driving servo system, in addition to the influence of backlash nonlinearity, there is also the influence of friction nonlinearity. Let $\tau_n(t)(n=3,4)$ is the torque generated by friction in system (mainly on the driving wheel side). We use Stribeck friction model to model τ_n.

$$\tau_n(t) = [F_c + (F_s - F_c)e^{-(\dot{\theta}_i(t)/\dot{\theta}_s)^\delta}]\mathrm{sgn}[\dot{\theta}_i(t)] + b\dot{\theta}_i(t) \tag{3}$$

where, F_c, F_s, b, $\dot{\theta}_i(t)$, $\dot{\theta}_s$ are Coulomb friction, maximum static friction force, viscous friction coefficient, angular velocity of driving wheel and Stribeck velocity, respectively. $\dot{\theta}_s$ and δ are empirical constants. Usually $\dot{\theta}_s=0.05$, $\delta=2$.

The typical dynamic equation of dual-motor driving servo system is as follows.

$$\begin{cases} U_i(t) = R_i I_i(t) + L_i \dfrac{dI_i(t)}{dt} + C_{ie}\dot{\theta}_i(t) \\ J_i\ddot{\theta}_i(t) + b_i\dot{\theta}_i(t) = T_i - \tau_i - \tau_n \\ J_m\ddot{\theta}_m(t) + b_m\dot{\theta}_d(t) = \tau_1 + \tau_2 \\ T_i = k_{di} I_i(t) \\ \tau_i(t) = k f(z_i(t)) + cf(\dot{z}_i(t)) \\ f(z_i(t)) = \begin{cases} z_i(t) + \alpha & z < -\alpha \\ 0 & -\alpha \le z \le \alpha \\ z_i(t) - \alpha & z > \alpha \end{cases} \\ \tau_n(t) = [F_c + (F_s - F_c)e^{-(\dot{\theta}_i(t)/\dot{\theta}_s)^\delta}]\operatorname{sgn}[\dot{\theta}_i(t)] + b\dot{\theta}_i(t) \end{cases} \tag{4}$$

where, $U_i(t)$ is the input electromotive force of driving motor. R_i and I_i are the armature circuit resistance and armature current of driving motor, respectively. L_i is the armature circuit inductance of driving motor. C_{ie} is the back EMF coefficient of driving motor. k_{di} is the torque coefficient of driving motor.

2.2 System Characteristic Model

Suppose a nonlinear system is as follows:

$$\dot{x}(t) = f(x, \dot{x}, \cdots, x^{(n)}, u, \dot{u}, \cdots, u^{(m)}) \tag{5}$$

let

$$x = x_1,\ \dot{x} = x_2,\ \ldots,\ x^{(n)} = x_{n+1}$$
$$u = u_1,\ \dot{u} = u_2,\ \ldots,\ u^{(m)} = u_{m+1}$$

then Eqs.(5) can be expressed as follows:

$$\dot{x}_1(t) = f(x_1, \cdots, x_{n+1}, u_1, \cdots, u_{m+1}) \tag{6}$$

The following assumptions are made for the nonlinear system:

Assumption 2 The system is single input and single output.

Assumption 3 The order of control variable $u(t)$ is 1.

Assumption 4 If all the independent variables of $f(\cdot)$ are zero, then $f(\cdot) = 0$.

Assumption 5 All the independent variables in $f(\cdot)$ are continuous and differentiable, and all the partial derivatives are bounded.

Assumption 6 $|f(x(t+\Delta t), u(t+\Delta t))| - |f(x(t), u(t))| < M\Delta t$, where M is a positive constant, Δt is sampling period.

Assumption 7 All the variables x_i and u_i are bounded.

Theorem 1 If the controlled object can be described by nonlinear system Eqs.(5), and meet the above assumptions 2-5, under the condition of satisfying certain sampling period Δt, its characterisric model can be described by a second order time varying difference equation:

$$x(k+1) = f_1(k)x(k) + f_2(k)x(k-1) + g_0(k)u(k) + g_1(k)u(k-1) \tag{7}$$

When the controlled object is stable, and meet assumptions 5-6, then:

(1)In the dynamic process, the output of the characteristic model is equivalent to the actual output under the condition of the same input u (proper selection of sampling period Δt can guarantee the output error within the allowable range). The output are equal in steady state.

(2)The coefficients of $f_i(k)$ and $g_i(k)$ in eqs.(7) are slowly time varying.

(3) Coefficient range can be determined in advance.

By the theorem 1, we can see that only when all the independent variables of the system are continuous diferentiable, the characteristic model can be designed. But in the eqs.(4), the deadzone function and the sign function are not differentiable, so they must be smoothed. We use a smooth, continuous and differentiable function in place of deadzone function[16]:

$$f(z_i(t)) = z_i - 4\alpha\left(\frac{2}{1+e^{-rz_i}} - 1\right) \tag{8}$$

so,

$$\tau_i(t) = k\left(z_i - 4\alpha\left(\frac{2}{1+e^{-rz_i}} - 1\right)\right) + c\dot{z}_i\left(1 - 8r\alpha\frac{e^{-rz_i}}{(1+e^{-rz_i})^2}\right) \tag{9}$$

We use a smooth friction model to replace the Stribeck friction model[17], namely the introduction of driven wheel speed amplification coefficient, and adopt the arc tangent function of the driven wheel speed replace the symbol function sgn in Stribeck friction model. Smooth friction model is shown as follows.

$$\tau_n(t) = [F_c + (F_s - F_c)e^{-(\dot{\theta}_i(t)/\dot{\theta}_s)^\delta}]\frac{2}{\pi}\arctan(k_b\theta_i(t)) + b\dot{\theta}_i(t) \tag{10}$$

where $k_b > 0$ is the speed amplification coefficient of driven wheel.

After the smooth treatment of backlash and friction nonlinearity respectively, the dynamic equation of dual-motor driving servo system with backlash and friction nonlinearity coexist is as follows.

$$\begin{cases} U_i(t) = R_i I_i(t) + L_i \dfrac{dI_i(t)}{dt} + C_{ie}\dot{\theta}_i(t) \\ J_i\ddot{\theta}_i(t) + b_i\dot{\theta}_i(t) = k_{di} I_i(t) - \tau_i - \tau_n \\ J_m\ddot{\theta}_m(t) + b_m\dot{\theta}_d(t) = \tau_1 + \tau_2 \\ \tau_i(t) = k\left(z_i - 4\alpha\left(\dfrac{2}{1+e^{-rz_i}} - 1\right)\right) + c\dot{z}_i\left(1 - 8r\alpha\dfrac{e^{-rz_i}}{(1+e^{-rz_i})^2}\right) \\ \tau_n(t) = [F_c + (F_s - F_c)e^{-(\dot{\theta}_i(t)/\dot{\theta}_s)^\delta}]\dfrac{2}{\pi}\arctan(k_b\theta_i(t)) + b\dot{\theta}_i(t) \end{cases} \tag{11}$$

let $x_1 = \theta_i(t)$, $x_2 = \dot{\theta}_i(t)$, $x_3 = \theta_m(t)$, $x_4 = \dot{\theta}_m(t)$, $x_5 = I_i(t)$, then the state equation of system is as follows.

$$\begin{cases} \dot{x}_1 = x_2 \\ \dot{x}_2 = \dfrac{1}{J_i}(k_{di}x_1 - \tau_i - \tau_n - b_i x_2) \\ \dot{x}_3 = x_4 \\ \dot{x}_4 = \dfrac{1}{J_m}(\tau_i - b_i x_4) \\ \dot{x}_5 = \dfrac{1}{L_i}[U_i(t) - C_{ie}x_2 - Rx_5] \\ y = x_3 \end{cases} \tag{12}$$

From theorem 1, we can get the characteristic model of system is as follows.

$$\theta_m(k+1) = f_1(k)\theta_m(k) + f_2(k)\theta_m(k-1) + g_0(k)u(k) + g_1(k)u(k-1)$$

$$= \phi^T(k-1)\theta(k) \quad (13)$$

where,

$$\phi(k-1) = [\theta_m(k) \ \ \theta_m(k-1) \ \ u(k) \ \ u(k-1)]^T,$$

$$\theta(k) = [f_1(k) \ \ f_2(k) \ \ g_0(k) \ \ g_1(k)]^T$$

The recursive least square method is used to estimate the parameters $f_1(k)$, $f_2(k)$, $g_0(k)$, $g_1(k)$ (forgetting factor $\lambda = 0.97$).

$$K(k) = \frac{P(k-1)\phi(k-1)}{\lambda + \phi^T(k-1)P(k-1)\phi(k-1)}$$

$$\hat{\theta}(k) = \hat{\theta}(k-1) + K(k)[y(k) - \phi^T(k-1)\hat{\theta}(k-1)]$$

$$P(k) = \frac{1}{\lambda}[I - K(k)\phi^T(k)]P(k-1)$$

3 Controller Design

The main control object of all-coefficient adaptive control is the linear time-invariant system with unknown parameters or some nonlinear systems. Its main feature is to establish a mathematical model whose sum of all coefficients is equal to 1, and to design the adaptive controller on the basis of determining parameters range. Because this algorithm has the characteristics of debugging less parameters, simple algorithm and strong robustness, it has been successfully applied in engineering practice. In this paper, aiming at the characteristic model of dual-motor driving servo system, we combined the golden section adaptive control and the logical integral control. The system control block diagram is as follows.

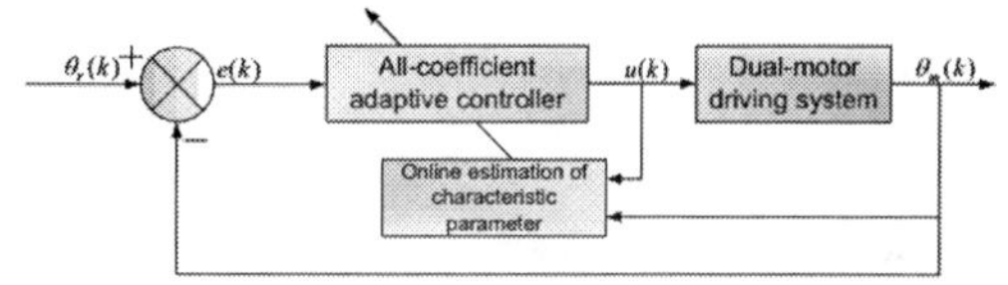

Figure 2. The block diagram of all-coefficient adaptive control.

where, $u(k)$ is the output control variable of all-coefficient adaptive controller, its expressions is as follows.

$$u(k) = u_1(k) + u_i(k) \quad (14)$$

where, $u_1(k)$ is the golden section control variable, $u_i(k)$ is the logical integration control variable.

$$u_1(k) = \frac{-1}{\hat{g}_0(k)+\eta}[l_1\hat{f}_1(k)e(k) + l_2\hat{f}_2(k)e(k-1) + \hat{g}_1(k)u_1(k-1)] \quad (15)$$

where, $\hat{g}_0(k)$ and $\hat{g}_1(k)$ are estimated values of $g_0(k)$ and $g_1(k)$, respectively. $e(k) = \theta_m(k) - \theta_r(k)$.

Let $k_i^+ = k_i^-$, $c = 1$, then the logical integration control variable is as follows [18].

$$u_i(k) = u_i(k-1) - k_i(\theta_m(k) - \theta_r(k)) \quad (16)$$

and when reference input $\theta_r(k) - \theta_r(k-1) = 0$, $u_i(k) = 0$.

If the feedforward control is added in Fig. 2, then system block diagram based on the feedforward all-coefficient adaptive control is as follows.

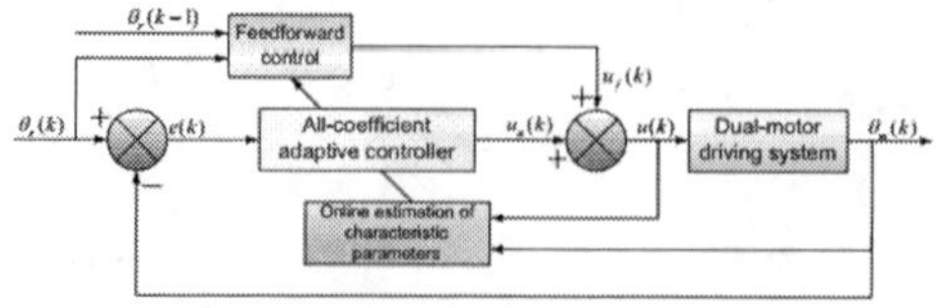

Figure 3. The block diagram of feedforward+all-coefficient adaptive control.

where, $u_f(k) = k_f(\theta_r(k) - \theta_r(k-1))$ is the feedforward control variable, k_f is the feedforward control coefficient, $u_a(k)$ is the output control variable of all-coefficient adaptive controller, the output control variable of feedforward all-coefficient adaptive controller is as follows.

$$u(k) = u_a(k) + u_f(k) = u_1(k) + u_i(k) + u_f(k) \quad (17)$$

4 Simulation Experiment

The physical parameters of system are shown in Table 1.

Table 1. The system physical parameters.

J_1, J_2	$0.185\text{kg}\cdot\text{m}^2$
J_m	$0.028\text{kg}\cdot\text{m}^2$
b_1, b_2	$1.2\text{N}\cdot\text{m}\cdot\text{s/rad}$
b_m	$1.3\text{N}\cdot\text{m}\cdot\text{s/rad}$
k	$[560+2\sin(\pi t)]\text{N}\cdot\text{m/rad}$
c	$[0.15+0.01\sin(\pi t)]\text{N}\cdot\text{m/rad}$
α	0.5rad
w	$50\text{N}\cdot\text{m}$

During the simulation, we compare the output response based on all-coefficient adaptive control (ACAC), feedforward all-coefficient adaptive control (FACAC), respectively. The parameters of all-coefficient adaptive controller $\eta = 1.8\times10^{-4}$, $k_i = 85$. The parameter of feedforward controller is $k_f = 115$.

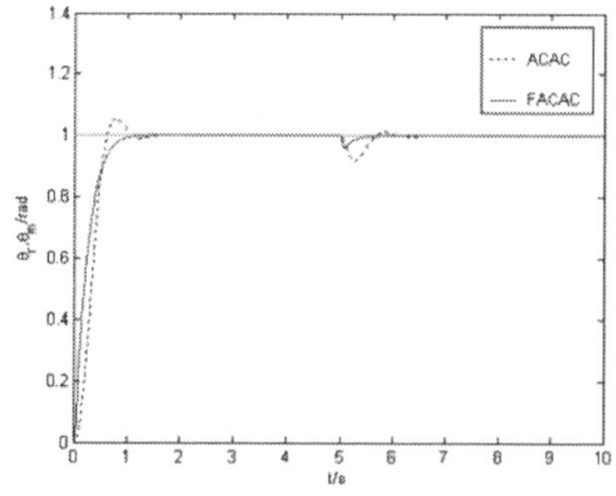

Figure 4. System tracking unit step response.

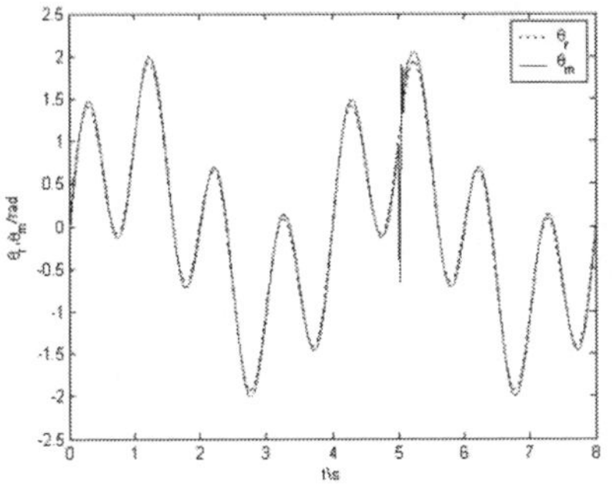

Figure 5. System tracking input response(ACAC).

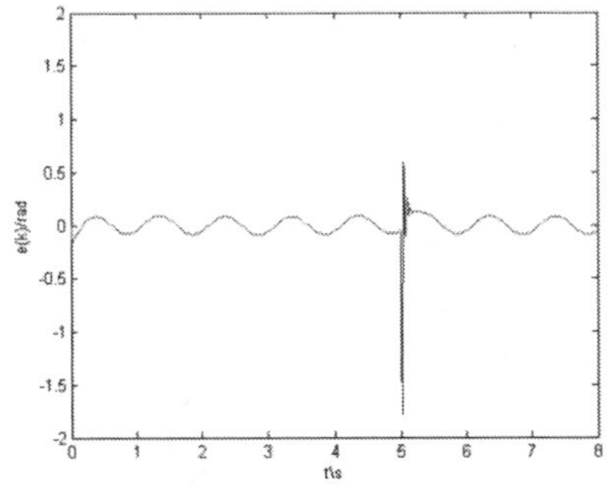

Figure 6. Tracking error curve(ACAC).

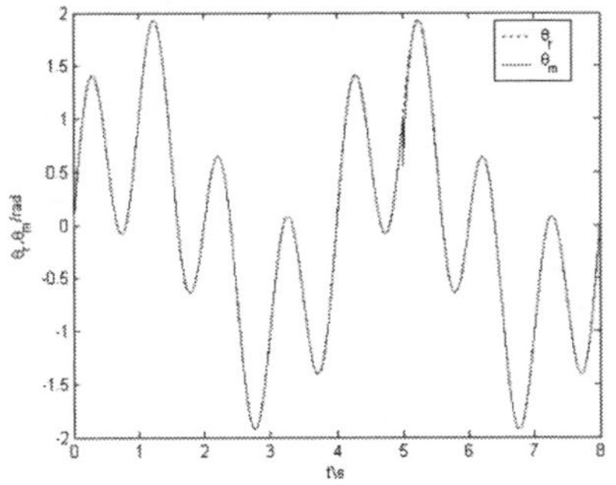

Figure 7. System tracking input response(FACAC).

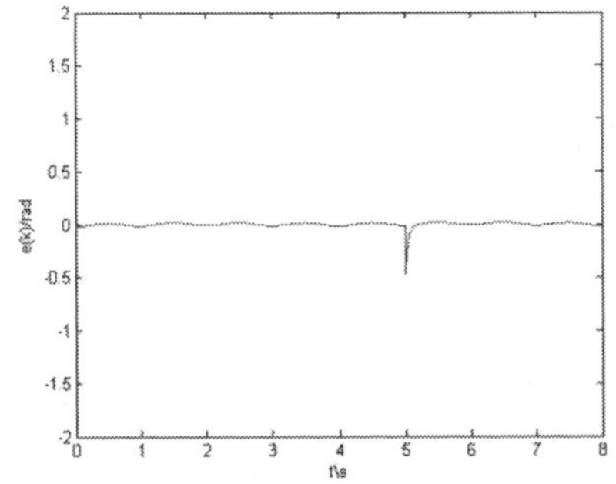

Figure 8. Tracking error curve (FACAC).

Fig. 4 is the output response of tracking unit step signal, we impose 5Nm disturbance on system at 5s; Fig. 5 and Fig. 7 are the output response of tracking input $\theta_r(t)=\sin(2\pi t)+\sin(0.5\pi t)$, their control strategies are ACAC and FACAC, respectively, and we impose 8Nm disturbance on system at 5s; Fig. 6 and Fig. 8 are tracking error curves using ACAC and FACAC, respectively. From Fig. 4 to Fig. 8, we can see that system has faster response speed, smaller tracking error, stronger anti-jamming capability and stronger robustness when FACAC method is used.

5. Conclusion

Aiming at the dual-motor driving servo system when backlash and friction nonlinearity coexist, the system characteristic model was established, the controllers based on all-coefficient adaptive control and feedforward all-coefficient adaptive control were designed, respectively. The simulation results show the superiority of the feedforward all-coefficient adaptive control method, which provides a reference for the further research of dual-motor driving servo system.

Acknowledgment

This work was supported by National Natural Science Foundation of China(No.61074023), Nature Science Foundation in Anhui Province of China(No.1508085MF130), Natural Science Research Key Project of Universities in Anhui Province of China (No.KJ2015A297), the Engineering Technology Research Center of Optoelectronic Appliance in Anhui Province of China, Sichuan Institute of Aerospace System Engineering in Sichuan Province of China.

The authors confirm that this article content has no conflicts of interest.

References

1. L. F. Sun, Research on motion characteristics and anti-backlash control of dual-motor system, Northeastern University, June 2012.
2. H. Z. Zhu, W. X. Song, Y. Dong, Modeling and simulation of coaxial connected dual motor drive system, Motor Contr. appl. 40(2) (2013) 23–27.
3. Y. Zhang, X. M. Ren, Adaptive backstepping control of dual-motor driving servo systems with friction, IHMSC 2014, 1, 214-217.
4. H. P. Ren, B. He, Anti-backlash control of machine tool feed system driven by dual-motors, Electr. Mach. Contr. 18(3) (2014) 60-66
5. I. X. Bogiatzidis, E. D. Mitronikas, A. N. Safaces, Investigation of backlash phenomena appearing in a twin AC cement kiln drive, Electrical Machines (ICEM), 2010, 1-6.
6. I. X. Bagiatzidis, A. N. Safacas, Vibration analysis and backlash identification of a twin AC drive for a cement kiln, Power Electronics, Machines and Drives (PEMD2010), 2010, 1-6.
7. Vrs, Jozef, Identification of nonlinear dynamic systems with input saturation and output backlash using three-block cascade models, J. Franklin Inst. 351(12) (2014) 5455-5466.
8. Y .F. Wu, D. W. Ma, J. Y. Yao, G. G. L, Nonlinear adaptive dynamic friction compensation for

mechatronic servo systems, J. Mech. Eng. 49(6) (2014) 89-94.
9. S. Zhu, M. X. Sun, X. X. He, Adaptive control of a class of periodically time-varying nonlinearsystems with input backlash" Contr. Theor. Appl. 29(4) (2012) 535-538.
10. S. Zhu, M. X. Sun, X. X. He, Iterative learning control of a class of nonlinear systems with input backlash, Acta Automat. Sin. 37(8) (2011) 1014-1017.
11. R. H. Du, Y. F. Wu, W. Chen, Q. W. Chen, Adaptive fuzzy control for the servo system with LuGre friction" Control and Decision, 28(8) (2013) 1253-1256.
12. Q. S. Wu, X. S. Wang, F. P. Du, L. Chen, Torque transmission characteristics and friction compensation for a tendon-sheath actuation system in pull-pull configuration" J. Mech. Eng. 51(5) (2015) 22-29.
13. H. X. Wu, J. Hu, Y. C. Xie, Intelligent adaptive control based on characteristic model, China Science and Technology Publishing House, 2009.
14. H. B. Zhao, Study on control and simulation of dual-motor driving servo system, Compu. Simul. 29(3) (2012) 262-265.
15. G. TAO, X. MA, Y. LING, Optimal and nonlinear control of systems with sandwiched backlash, Automatica, 37(2) (2001) 165-176.
16. R. MEZOUKI, J. A. DAVILA, L. FRIDMAN, Backlash phenomenon observation and identification in electromechanical systems, Control Eng. Pract. 15(4) (2007) 447-457.
17. Y. Y. Li, Design and experimental research of servo system based on characteristic model, Nanjing University of Science and Technology, March, 2014.
18. F. Q. Huang, Study of characteristic model base on multi-motor driving servo system, Nanjing University of Science and Technology, March, 2013.

MATEC Web of Conferences **108**, 08010 (2017) DOI: 10.1051/matecconf/201710808010
ICMAA 2017

Wind Turbine design and fabrication to power street lights

Mohammad Khan , Mohammed Alavi, Nithin Mohan, AbidAzeez, Abdul Shanif and Bilal Javed

College of Engineering, Abu Dhabi University, Abu Dhabi, UAE

Abstract. The objective of this work was to design and build a wind turbine which can be used to power small street lights. Considering the typical wind speeds in Abu Dhabi, UAE and ease of construction, the design of the wind turbine was chosen to be Sea Hawk design from vertical axis wind turbine category. A three phase AC generator was used for its availability over the DC motors within the region. A 12V battery was used for storage and a charge controller was used for controlling the charge flow into the battery and for controlling the turbine rotation when the battery is fully charged. The blades used in the turbine were made of foam board according to the NACA 0018 airfoil shape with a chord length of 15cm. The connecting shaft was made of stainless steel. Structural analysis and CFD analysis were performed along with other calculations. Testing was executed to calculate the voltage output from the turbine at different wind speeds. The maximum voltage the turbine produced at 6.4 m/s wind speed was 2.4Vand the rotational speed of the turbine was 60.3 rpm.

1 Introduction

Wind turbines typically are divided into two types based on their axis of rotation with respect to the wind flow direction; Horizontal Axis Wind Turbine (HAWT) and Vertical Axis Wind Turbine (VAWT). HAWT has blades that are similar to propellers that spin on horizontal axis. The position of main rotor shaft and electrical generators in horizontal axis turbine is at the top of a tower provided that this must be pointed in the direction of wind. Typically, the gearbox combined with the high-speed shaft increases the rotation of about 30-60 rpm to almost 1200-1500rpm. This is the typical rotation speed required for generators used in HAWT [1]. The offshore wind turbines have the advantage in case of noise pollution as it is far from shore without affecting the people and are less affected by turbulence whereas the onshore turbines create noise pollution and are affected more by turbulence [2]. In a VAWT, the main rotor shaft is fixed vertically. Vertical axis wind turbines can be subdivided into two kinds: Darrieus and Savonius [1]. Typical Darrieus wind turbine (different types shown in Fig. 1) contains 2 or 3 blades [3].

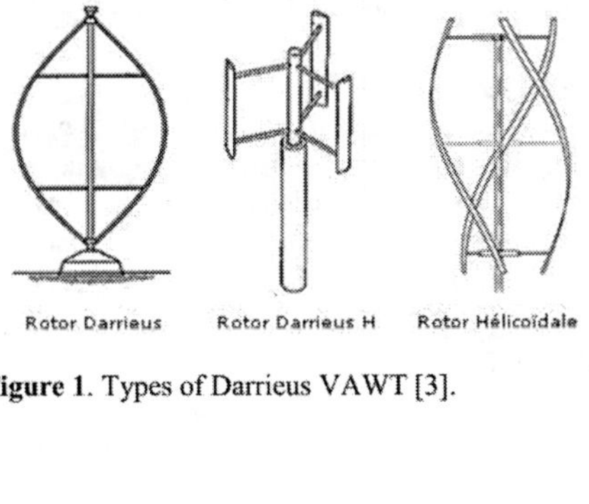

Figure 1. Types of Darrieus VAWT [3].

Savonius drag-based VAWT works similar to the paddle propelling in a boat. The drag is the pressure or thrust developed when wind passes through the blade. The blade is *S* shaped cross section; this is to implement the drag technique and also to provide certain lift [4].UAE is currently in the process of determining the viability in wind farm projects within the country. UAE Ministry of Energy is carrying out various research projects for further understanding the variability of wind from year to year [6].

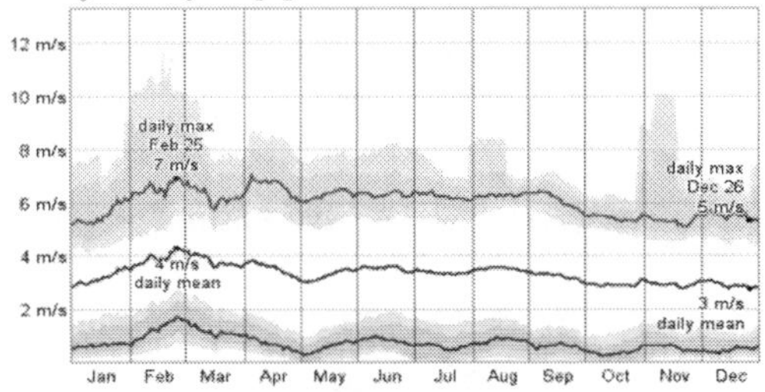

Figure2. Red is average daily minimum, Green is maximum, and black is average wind speed with percentile bands [7].

Figure 2 shows the wind speed data over a year in Abu Dhabi. Daily maximum wind speed ranges from lowest 5m/s to highest 7m/s. The lowest daily wind speed can go to a maximum of 2m/s. In an average curve, the maximum wind speed is found to be 5m/s and minimum is 3m/s. Hence the design constraint was selected to be a wind speed of around 5m/s [7]. The average wind speed in Abu Dhabi is less than 5m/s and VAWT can be utilized at such wind speeds. The start-up wind speed generally is 2m/s for the same. Sea hawk design as shown

 DOI: 10.1051/matecconf/201710808010

in Fig. 3 satisfies the regional conditions and is appropriate for Abu Dhabi weather conditions [5].

Figure3. Sea Hawk turbine [5].

The objective of the present work is to design and construct a wind turbine to power small street lights of typically 30 W lamps. The wind turbine is to be designed for wind conditions encountered in Abu Dhabi UAE. The turbine should run effectively with an average wind speed of 5-6 m/s. A battery will be used to store energy while the wind turbine is running during the day. The same can power the street light after sunset. In addition, the whole turbine should be modular and easily manufactured with inexpensive material requirements. Considering the use for street lighting, the typical wind velocities available in Abu Dhabi, and the ease of manufacturing Sea Hawk VAWT was selected as a good candidate to serve the purpose.

2 Design of the Wind Turbine

Swept area of the wind turbine is the area of the wind turbine which captures the wind. The power available from the wind turbine increases with an increase in the swept area. Another parameter that greatly affects the power available from the wind turbine is the wind speed [8]. Kinetic energy can be calculated as:

$KE = 0.5mV_\infty{}^2$; where V_∞ is the free stream velocity and m is the mass of air trans versing such that $m = \rho AV_\infty$ in which A = swept area and ρ = density of air

Therefore,

$$KE = 0.5\rho AV_\infty{}^3 \qquad (1)$$

Including coefficient of performance in equation (1), the power output can be determined as:

$$P_w = \frac{1}{2}C_p\rho AV_\infty^3 \qquad (2)$$

Power density is the terminology used to define the available power per unit area swept by the blade/airfoil [9]. Power density can be calculated as:

$$p = \frac{1}{2}C_p\rho V_\infty{}^3 \qquad (3)$$

Symmetrical shape is easier to manufacture than curved shapes. NACA 0018, 0012, 0015 are all symmetric in shape. NACA four series is recommended since symmetrical airfoils work better for higher Reynolds number; particularly NACA 0012, 0015, 0018. These are more appropriate for smaller angle of attack [8].

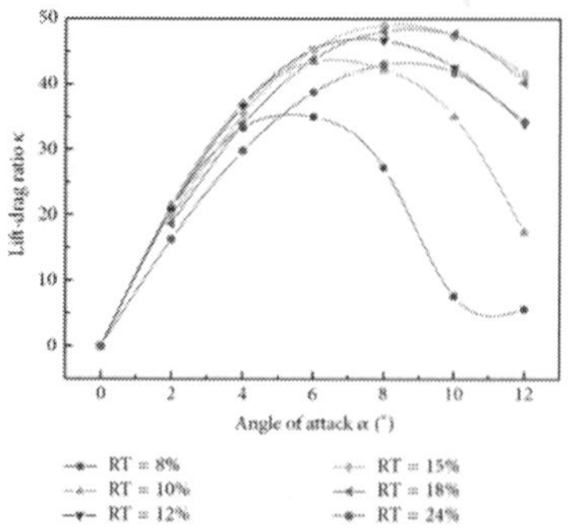

Figure 4. Lift-drag ratio of symmetric airfoils at various relative thickness [9].

Figure 4represents comparison between six different NACA airfoil thicknesses which include 8, 10, 12, 15, 18, and 24 percentages. The relative thickness 'RT' represents thickness / chord ratio (t/c ratio) and is the maximum thickness of the airfoil expressed as a percentage of the chord.Considering symmetric airfoil shapes, the airfoils with 15% thickness has the highest lift drag ratio. This means 15% thickness can provide the maximum lift and least drag. The second highest lift drag ratio is for 18% thickness NACA airfoil. It is evident in the graph that when angle of attack increases after a specific value, the drag element increases and the ratio decreases. When thickness increases beyond a limit, the drag also increases accordingly. Thus airfoil with thickness 15% or 18% is preferred. It should also be considered that 18% is thicker than 15% and is more robust than 15%. Another important finding is that thicker airfoil increases the startup torque. It was found that 19% thickness showed good start-up and traverse characteristics. Theoretically it is proven that 18% has better performance at low Reynolds number and highest power coefficient was achieved at 18% [9]. Studies [8] prove that a larger leading edge radius is preferred to be less sensitive to roughness. Thick blunt trailing edge is preferred for increasing lift [9]. Choosing an appropriate number of blades for the turbine is a balance between blade stiffness and aerodynamic efficiency. Increasing the blade number for a specific radius and thickness leads to sacrificing blades stiffness for a very slight increase in aerodynamic efficiency. Single bladed turbines suffer from a parasitic drag. Two bladed rotors are more sensitive towards mass imbalance vibrations and causes higher operating cost and maintenance compared to three blades. Minimum three blades are required to decrease vibrations due to mass imbalance issues [9].In the sea hawk type of vertical axis wind turbines, the blades are not cambered and are symmetric. The diameter of the turbine is 0.80m and 1.2m tall. All the blades positions are distributed evenly. Blades can be made from light and inexpensive materials. Sea Hawk design was chosen as the prototype design. This decision was made after comparing the score in efficiency, cost, ease of manufacturing, and ease of maintenance. The blades in this type of turbine can be designed in straight, symmetric, and anti-symmetric airfoil shapes.

Initial optimized parameters have been chosen to set the initial design of the turbine. Table 1 summarizes the initial dimensions of the turbine which were obtained based on best dimensions of the airfoil.

Table 1. Starting parameters.

Symbol	Parameter	Value
V∞	Wind speed	5m/s
TSR	Tip speed ratio	0.625
R	Rotor Radius	0.385m
C	Blade Chord	0.07m
L	Blade length	1.22m
N	Blade numbers	6, 8, 15, 16

Once the initial dimensions and TSR are specified the next step is to find the solidity which is a dimensionless parameter which will help to find the geometry of the VAWT. The blade solidity 'σ' can be calculated as:

$$\sigma = \frac{B\,C}{R} \quad (4)$$

where 'B' is the number of blades; 'R' is the radius of the rotor; 'C' is the chord length

An increase in solidity increases the production of the higher torque at the low tip speed ratios. Table 2 shows properties of different materials considered for the blade. The wind turbine structure is the main source of the stresses as well as the change of input forces [11].

Table 2. Material properties.

Material	Stiffness (GPa)	Tensile Strength (MPa)	Density (g/cm³)	Elongation at break (%)	Max temp (°C)
Steel	30	190	7.5	15	550
Aluminum	10	90	2.7	12	400
Glass-E	73	3500	2.54	3	350
Carbon	350	4000	1.75	1.8	500
Aramid	120	3600	1.45	11	250
Foam T90	0.5	1.5	0.065	0.2	70

The shaft of the turbine (1.05m long) is made up of stainless steel for its advantages such as durability and resistance to corrosion and rust. At some point the flow of charge from the generator to the battery should be regulated. According to the generator rated power output of 30W and voltage of 12V FW03-12A waterproof charge controller was chosen. The specifications of the controller are mentioned in Table 3.

Table 3. Controller specifications.

Specifications	
Rated power	300W
Battery Voltage	12V
Brake voltage	13.5V
Level of protection	IP67

3 Design Parameters

The main design objective of the turbine is to produce an output power of 30W or above to power LED street lights. The design constraints of the turbine are the turbine diameter, airfoil type, chord length, number of blades, and pitch angle. The input parameters of the design are the rotor radius, blade length, blade chord length, and number of blades. The input parameters of the wind are the free stream wind speed, and air density. The parameters such as swept area, power available from the wind, angular velocity, rated blade velocity, tip speed ratio, lift force, drag force, tangential force, normal force, torque and expected power output is calculated.

Using the equation for Swept Area,

$$\text{Swept Area (A)} = H \times D \quad (5)$$

$$A = 0.832\text{m}^2$$

Using equation (2),

$$\text{Power available from wind } (P_w) = 63.7\text{W}$$

From the literature review and based on the wind data, the cut in speed (V_d) was assumed to be $3.13\frac{m}{s}$

To find the Angular velocity(ω),

$$\omega = \frac{V_d}{R} = 7.825\frac{rad}{sec} \quad (6)$$

Using the equation for Blade velocity,

$$\text{Blade velocity } (V_b) = R \times \omega = 3.13\frac{\text{m}}{\text{s}} \quad (7)$$

Using equation, the magnitude of relative velocity is

$$|\vec{W}| = V_\infty\sqrt{\left(\frac{\omega R}{V_\infty} + \cos\theta\right)^2 + (\sin\theta)^2} \quad (8)$$

Finding the relative velocity, lift force, drag force, tangential force, torque, expected power output and coefficient of performance when the blade position is at 30 degrees and the pitch angle is at 10 degrees. Fig. 5 shows the velocity vector when the blade is at positioned at 30 degrees.

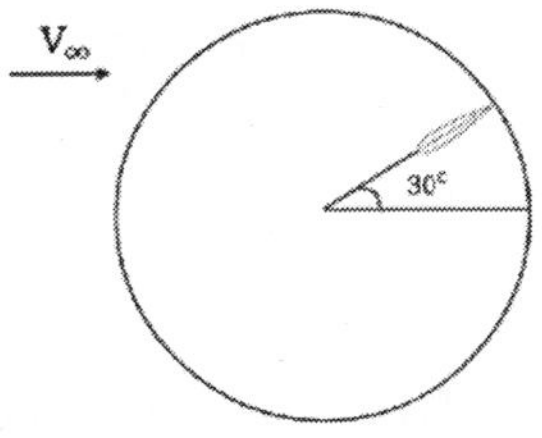

Figure 5. Velocity vector when blade is positioned at 30 Degree

When the blade is in position which is the azimuth angle (θ) is at an angle of 30°, $|\vec{W}| = 6.29\frac{\text{m}}{\text{s}}$

The angle of relative velocity can be found using equation,

$$\varphi = \tan^{-1}\left(\frac{\sin\theta}{\frac{\omega R}{V_d} + \cos\theta}\right) - \frac{\omega x_0 C}{W} - \frac{\omega C}{4W} \quad (9)$$

Where x_0 is the blade attachment at the middle of chord length, thus $x_0 = 0$. Therefore, $\varphi = 12.36^\circ$

Using the equation for tip speed ratio,

$$\lambda = \frac{R\omega}{V_\infty} = 0.624 \quad (10)$$

Angle of attack and pitching angle is related to each other and this can be mathematically modeled. By fixing the azimuth angle as 30° and varying the pitch angle -10° to 60° Fig. 6 was generated.

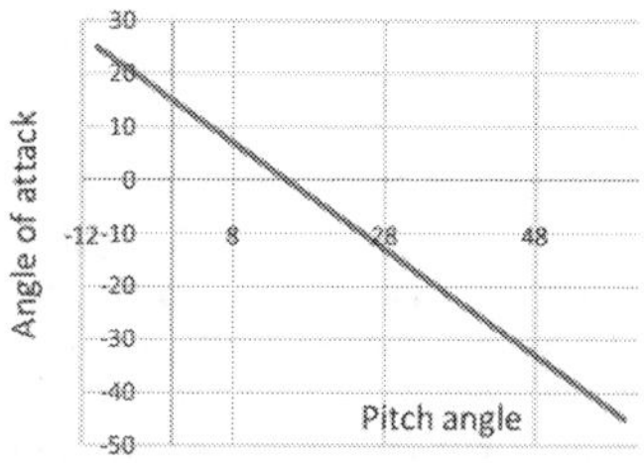

Figure 6. Angle of attack vs Pitch angle.

For the calculation, the pitch angle was assumed to be 10°. From the graph plotted in Fig. 6, the angle attack was found to be 5.025°.

RPM of the turbine

$$N = 7.825 \times \frac{60}{2\pi} = 74.72 \text{ RPM} \quad (11)$$

Using equation for Reynolds number,

$$Re = \frac{\rho WC}{\mu} = 58284.79 \quad (12)$$

The values for C_L and C_D are found to be 0.7352 and 0.03446 from the table for NACA 0018. Using the angle of attack (α) which is equal to 5.025°.

Using the equation for lift force,

$$\text{Lift Force } (F_L) = \frac{1}{2} \times C_L \times \rho \times W^2 \times A_P = 2.779\text{N} \quad (13)$$

Using the equation for Drag Force,

$$\text{Drag Force } (F_D) = \frac{1}{2} \times C_D \times \rho \times W^2 \times A_P = 0.13N \quad (14)$$

Using the equation for Tangential Force,

$$F_T = F_L \sin\varphi - F_D \cos\varphi = 0.64 \text{ N} \quad (15)$$

Chord length for the turbine blade was determined after comparing and analyzing existing VAWT turbine designs. It was found that most of the turbine blade chord length was somewhere between $\frac{1}{15}^{th}$ to $\frac{1}{10}^{th}$ of the turbine diameter. But based on the preliminary design and sketches performed it was found that the gap between the turbine blades was huge so the chord length was increased to 15cm from 7cm.According to the chosen NACA0018 blade, the thickness should be 18% of the chord length. The thickness of the blade was computed using equation (16) to be,

$$T = C \times 0.18 = 15 \times 0.18 = 3\text{cm} \quad (16)$$

The effect of coefficient of lift as well as coefficient of drag will affect the overall performance of the turbine. C_L and C_D are functions of *Re* and the particular airfoil section. The values of coefficient of lift and drag for the prototype blade with 15cm chord length was taken from the table for NACA 0018. Using the equation for the coefficient of performance,

$$C_p = \frac{\text{Expected power output}}{\text{Power available from the wind}} = \frac{P}{P_w} \quad (17)$$

$$C_p = \frac{C_T \times C \times W^2 \times B \times \lambda}{D \times V_\infty^2}$$

$$C_T = C_L \sin\alpha - C_D \cos\alpha = 0.03$$

$$C_p = \frac{0.03 \times 0.15 \times 6.29^2 \times 0.624}{0.8 \times 5^2} \times B$$

The graph shown in Fig. 7 was generated by fixing the wind speed, diameter of the turbine, tip speed ratio, and the only variation used was the blade number. This was performed in order to choose the best number of blades for the turbine. However, 16 blades were rejected as it had almost same value of C_p as 15 blades and also to reduce the added weight on the turbine.

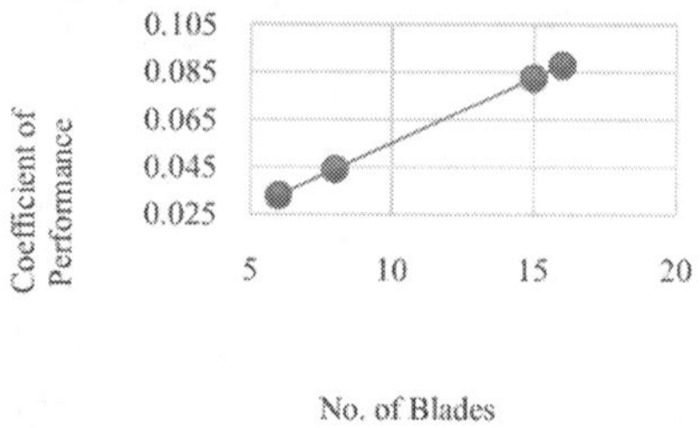

Figure 7. Graph of Coefficient of performance vs No. of blades.

The torque of the turbine was found using the equation for torque

$$Q = B \times F_T \times R = 15 \times 0.64 \times 0.4 = 3.84\text{Nm} \quad (18)$$

Using the equation for the power output,

$$P = \text{Torque} \times \text{Angular Velocity} = Q \times \omega = 30\text{W} \quad (19)$$

Using the equation for power,

$$C_p = \frac{\text{Expected power output}}{\text{Power available from the wind}} = \frac{P}{P_w} \quad (20)$$

$$C_p = 0.474$$

The above calculations for torque, expected power output, and power coefficient were calculated by assuming the blade position at 30 degrees from the reference and the pitch angle at 10 degrees.

4 Structural Analysis of the Turbine

Structural analysis of the turbine is one of the important parameters that need to be considered in terms of design. The factors considered in the structural analysis are maximum bending moment, normal and shear loads on the blades, allowable stresses, and moment of inertia on the blades.

The normal stress is calculated using equation

$$\sigma_N = \frac{\text{Normal force}}{\text{Area perpendicular to force}} \quad (21)$$

Normal force (F_N) is calculated using equation

$$F_N = \frac{1}{2} \times \rho \times C \times L \times W^2 \times C_N \quad (22)$$

Where C_N the coefficient of normal force, C is the chord length, ρ is the density of air, and W is the relative velocity of the blade.

The coefficient of normal force is calculated to be

$$C_N = C_L \sin\alpha + C_D \cos\alpha = 0.09844 \quad (23)$$

Therefore, the normal force from (22) is 0.372N

The perpendicular area is found by assuming the blade as cuboid. Thus, the perpendicular area is found using equation

$$A_d = \text{thickness of blade} \times \text{length of blade} \quad (24)$$

$$A_d = 0.0312m^2$$

Therefore, from (21),$\sigma_N = 11.923$Pa

Shear stress occurs due to tangential force. The tangential force was found to be, $F_T = 0.64$ N which is same as the shear force F_s . The shear stress is calculated using equation (25) as

$$\tau = \frac{F_s}{\text{Blade area}} = 4.1\ Pa \quad (25)$$

The bending moment is proportional to the centripetal acceleration. The centripetal acceleration is given by equation (26)

$$a_n = \frac{v_b{}^2}{R} = 24.5\text{m/s}^2 \quad (26)$$

Where, a_n is the centripetal acceleration, v_b is the blade velocity and R is the radius of the turbine.

$$\text{Bending moment} = F_N \times \text{Distance} = 0.558\text{Nm} \quad (27)$$

5 Structural Design of the Rotor

The material selection was done comparing stiffness, tensile strength, density, elongation at break, temperature up to which it can hold, cost and ease of manufacturing.

Table 4. Material Comparison [13].

Material	Stiff-ness (GPa)	Tensile Strength (MPa)	Density (g/cm^3)	Elongation at break (%)	Max temp (°C)
Steel	30	190	7.5	15	550
Al	10	90	2.7	12	400
Glass-E	73	3500	2.54	3	350
Carbon	350	4000	1.75	1.8	500
Aramid	120	3600	1.45	11	250
FoamT90	0.5	1.5	0.24	0.2	70

Foam board was selected as the material for the blades. Figure 8shows the schematic of NACA 0018 with 15cm chord length.

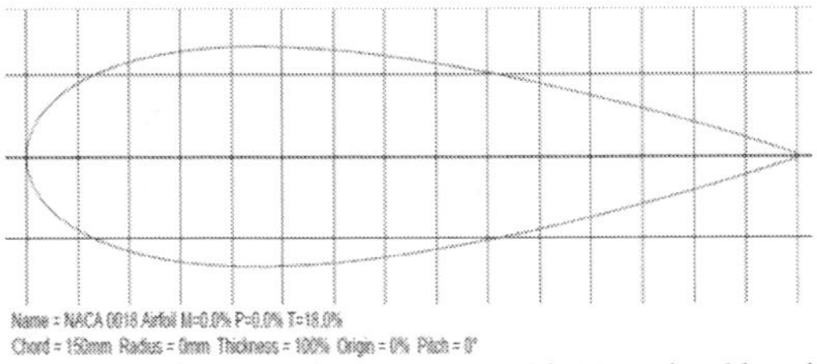

Figure 8. Schematic of NACA 0018 with 15cm chord length

The mass of the manufactured blade was found to be 0.169 kg.Acceleration analysis is done to determine theoretically that the turbine can achieve the desired RPM [14]. The inertial moments by approximating the blade shape as a cuboid; using Steiner theorem, is given using equation

$$I_{\text{blade}} = \frac{1}{12} m_{\text{blade}} \left[C^2 + (0.21C)^2\right] + m_{\text{blade}} R^2 \quad (28)$$

$$I_{\text{blade}} = 0.273\ kgm^2$$

Where m_{blade} is the mass of the blade, C blade chord is, R is rotor radius. To get an approximate value of total momentum, inertial momentum is multiplied by the number of blades. The angular acceleration (rad/s) can be calculated using equation

$$A_a = \frac{Q}{I} \quad (29)$$

Where Q is the torque and I is the rotor inertial moment.

Relative velocity, torque, tangential, normal, lift, drag force, and expected power output were calculated for one blade positioned at $\theta = 30°$ and $\delta = 10°$. For VAWT the torque varies according to position of the blade and angle of attack, unlike HAWT were torque remains constant. Relative velocity, torque, tangential force, normal force, lift force, drag force, and expected power out depends on angle of attack. To achieve a constant torque and power output angle of attack has to be fixed.

$$\begin{aligned}\text{Total moment of inertia} &= \text{moment of inertia of one blade} \\ &\times \text{number of blades} = 0.405kgm^2 \end{aligned} \quad (30)$$

Therefore, the angular acceleration is given by:

$$A_a = \frac{Q}{I} = \frac{3.84\ \text{Nm}}{0.405\ \text{kgm}^2} \times \frac{\text{kg}\ {}^{\text{m}}/_{\text{s}^2} \times \text{m}}{\text{kg m}^2} = 9.5\frac{\text{rad}}{\text{s}^2} \quad (31)$$

Load analysis for the blade dimensions can be divided into centrifugal and aerodynamic loads.

Centrifugal force is calculated by:

$$F_C = mr\omega^2 = 4.1128N \quad (32)$$

The aerodynamic load is normal force (F_N) and using equation (22) $F_N$was found to be 0.372N.

The resultant force for one blade in rotation can be calculated. The assumption made for this is that both centrifugal and normal force are in the same direction.

$$F_R = F_C + F_N = 4 \quad (33)$$

$$\text{Allowable stress} = \frac{\text{Yield Strength}}{\text{Tensile strength} \times \text{FS}} \quad (34)$$

Assuming factor of safety as 3, the allowable strength of the Styrofoam board was found to be 0.5MPa.

For 74.72 rpm, the angular velocity on the shaft can be calculated by:

$$\omega = \frac{2\pi N}{60} = \frac{2 \times \pi \times 74.72}{60} = 7.8\ rad/s \quad (35)$$

Torque on the shaft is based on the generator output power.

$$\text{Torque} = \frac{\text{Power output from the generator}}{\omega} = 3.8\text{Nm} \quad (36)$$

6 Prototype and its testing

A foundation is required to hold all the blades in place in a robust and straight manner. Two sheets 80cm diameter and 2cm thick was used on top and bottom made from Styrofoam sheets to satisfy these criteria. This adds strength to the shape and reduces the slanting of the blade while operating.The prototype has 15 blades with thickness of 3cm, height of 1m, and the material used was Styrofoam. The 3 phase AC PMA generator can produce 30watt. It is 20cm in diameter. The steel shaft was 1 meter long and with a diameter 1.2cm. The shaft goes through the Styrofoam sheets and covers almost all of the turbine height.The generator is fixed to the flange which is then fixed to the base. Flange acts as a mediator to the generator and base. The diameter of the flange is 9cm and height is 25cm.Turbine base helps to avoid turbine from tipping due to the force from the wind. The base was made from marine plywood. The dimensions of base were 60cm long, 54cm wide and 14cm. The turbine is bolted to base to ensure stability.Table 5 shows the specification of the generator.

Table 5. Generator Specification

Generator type	3 phase AC PMG
Rated voltage	DC 12V
Rated speed	300 rpm
Power output	30 W @10 m/s
Weight	4.5 kg

Start-up wind speed	2 m/s
Torque	9.74 Nm

Figure 9shows a 3D isometric view of the turbine respectively. A 15 blade model shown below was developed using Solid works. This model was later used to perform structural analysis.

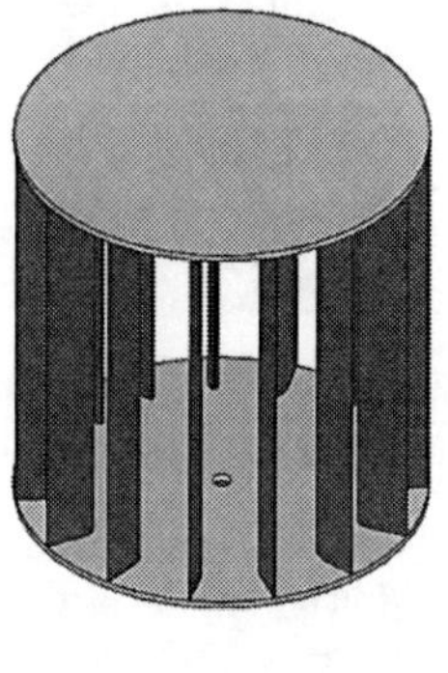

Figure 9. Engineering 3D isometric drawing of the turbine.

Wind turbine testing was done to ensure that it performs well at different wind speeds. Subjecting to various wind speeds the turbine voltage outputs were recorded and analyzed.Bending test was carried out in the lab using the method of three-point bending experiment. The strength and deformation capability of the blade was found out using this. Before performing the test, the blade was weighed and was found to be 0.169 kilogram The stress loaded on the blade was determined using the equation (37)

$$\sigma_b = \frac{3F_{Load\ ed}S}{2CT^2} \quad (37)$$

where σ_b stands for the stress on the blade F_{Loaded} is force loaded on blade, S means span, b is the width of the blade and T is the thickness of the blade.The blade broke at a load of 89.76N. Span was taken to 50 cm and 'C' the chord length was 15cm and thickness of the blade was 3cm. Therefore, the stress load came out to be 49.8 Pa. The turbine was subjected to different wind speeds by the help of wind tunnel facility. Fig.10compares the different wind speeds and voltages as output while the turbine rotates.

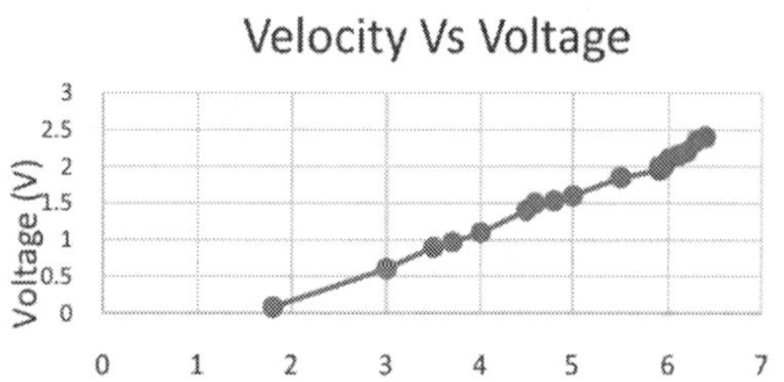

Figure10. Wind speed vs Voltage

As expected, when wind speed increases the voltage value increases. With the current facilities in wind tunnel the maximum voltage achieved was 2.4 volts with a wind speed of 6.4 m/s. The least voltage achieved was during 1.8 m/s wind with a voltage value of 0.08 volts. RPM of the turbine was calculated using tachometer counter and was found to be 60.3.

7 Conclusion

In conclusion, the project's initial intention was to design, create, and simulate a wind turbine which can be used to power LED street lights. The necessary approaches starting from literature review to find all the necessary background details required, collecting the wind data, selecting the appropriate turbine, blade type, determining the components required, calculating the power density, structural analysis, CFD analysis, building the prototype, till the testing of the turbine were performed. From the literature review it was found that VAWT works better due to the facts that the efficiency is better, works with lower wind speeds, cheaper, smaller, cheaper, and easier to manufacture. Operational parameters such as cut-in, cut-out speed, rated wind speed, rated power output, power coefficient, rpm and pitching of blade were analyzed for performing calculations. From wind data it was established that the average wind speed is 5m/s and relative humidity typically varies from the lowest value of 16% to the highest 85%. From detailed design, it was found that the power density increases as wind velocity increases. All the required design parameters such as the swept area, expected power output, turbine diameter, airfoil type, chord length, number of blades, angular velocity, rated blade velocity, tip speed ratio, lift force, drag force, tangential force, normal force, torque and expected power output were calculated by using the input parameters of the design such as the rotor radius, pitch angle, blade length, blade chord length and number of blades. In the structural analysis of the turbine normal stress, normal force, coefficient of normal force, perpendicular area, shear stress, and bending moment were determined. The maximum voltage the turbine achieved at 6.4 m/s wind speed was 2.4 volts and the rotational speed of the turbine was 60.3 RPM. Future development would be to build a mechanism to prevent the wind turbine from rotating at high wind speeds that may damage the blades.

References

1. P. S. Acosta, S. Tilvaldyev, L. G. G. Anaya, G. E. G. Sánchez and D. A. C. a. J. F. Garay, "ADVANCE HIGH EFFICIENT AERODYNAMIC BLADES FOR VERTICAL AXIS WIND TURBINE MODULAR AGGREGATE," International Journal of Arts & Sciences, 2012.
2. "Different Types and Parts of a Horizontal Axis Wind Turbines," 17 March 2011. [Online]. Available: https://dolcera.com/wiki/index.php?title=Different_Types_and_Parts_of_a_Horizontal_Axis_Wind_Turbines.
3. M. A. Rosen, J. O. Ajedegba and G. F. Naterer, "Predicting Vertical Axis Wind Turbine Behaviour: Effects of Blade Configuration on Flow Distribution and Power Output," Nova Science Publishers, Inc., Ontario, 2010.
4. M. Ragheb, "Vertical Axis Wind Turbines," University of Illinois at Urbana, Illinois, 2015.

5. Pac Wind, Sea Hawk Description, California: PacWind INC, 2007.
6. L. S. David A. Rivkin, Wind Energy, Burlington: JONES & BARTLETT LEARNING, 2013.
7. "UAE looks to tap vast potential of wind power," 14 June 2014. [Online]. Available: http://www.thenational.ae/business/industry-insights/energy/uae-looks-to-tap-vast-potential-of-wind-power.
8. L. Deisadze, D. Digeser, C. Dunn and D. Shoikat, "Vertical Axis Wind Turbine Evaluation and Design," WORCESTER POLYTECHNIC INSTITUTE, Worcester, 2013.
9. Y.-b. Liang, L.-x. Zhang, E.-x. Li, X.-h. Liu and Y. Yang, "Design Considerations of Rotor Configuration for Straight-Bladed Vertical Axis Wind Turbines," Advances in Mechanical Engineering, 2014.
10. "Abu-Dhabi-United-Arab-Emirates," 2016. [Online]. Available: https://weatherspark.com/averages/32854/Abu-Dhabi-United-Arab-Emirates. [Accessed 28 April 2016].
11. M. Islam, A. Fartaj and R. Carriveau, "Analysis of the Design Parameters related to a Fixed-pitch Straight-Bladed Vertical Axis Wind Turbine," WIND ENGINEERING, 2008.
12. Resource Supply, LLC, "IP67, What Does That Mean?" Tigard, 2008.
13. Y. Zhong-Jia, G. Yi-Zhuo*, L. Min, L. Yan-Xia, L. Jie and a. Z. Zuo-Guang, "Design and Analysis of Small-Scale Lift-Type Vertical-Axis Wind," Key Laboratory of Aerospace Advanced Materials and Performance, Beijing, 2014.
14. J. Castillo, "SMALL-SCALE VERTICAL AXIS WIND TURBINE DESIGN," Tampere University of Applied Sciences, Tampere, 2011.

Performance and Adaptive Surge-Preventing Acceleration Prediction of a Turboshaft Engine under Inlet Flow Distortion

Dalu Cao , Hailong Tang and Min Chen

Beihang University, Xueyuan Road No. 37, Haidian District, Beijing, China

Abstract. The intention of this paper is to research the inlet flow distortion influence on overall performance of turboshaft engine and put forward a method called Distortion Factor Item (DFI) to improve the fuel supply plan for surge-preventing acceleration when turboshaft engine suddenly encounters inlet flow distortion. Based on the parallel compressor theory, steady-state and transition-state numerical simulation model of turboshaft engine with sub-compressor model were established for researching the influence of inlet flow distortion on turboshaft engine. This paper made a detailed analysis on the compressor operation from the aspects of performance and stability, and then analyzed the overall performance and dynamic response of the whole engine under inlet flow distortion. Improved fuel supply plan with DFI method was applied to control the acceleration process adaptively when encountering different inlet flow distortion. Several simulation examples about extreme natural environments were calculated to testify DFI method's environmental applicability. The result shows that the inlet flow distortion reduces the air inflow and decreases the surge margin of compressor, and increase the engine exhaust loss. Encountering inlet flow distortion has many adverse influences such as sudden rotor acceleration, turbine inlet temperature rise and power output reduction. By using improved fuel supply plan with DFI, turboshaft engine above-idle acceleration can avoid surge effectively under inlet flow distortion with environmental applicability.

1 Introduction

Helicopters often encounter inlet flow distortion in flight. Inlet flow distortion can be caused by many sources such as crosswind-induced inlet separation, armament firing, and shock-boundary layer interaction in a supersonic inlet. Whatever the source, the performance and stability of the aeroengine generally reduced. Even the blade rupture fault occurred when the compressor surged caused by inlet flow distortion [1].

Previously, scholars have researched inlet flow distortion by using computational fluid dynamics (CFD) methods or experiments that concentrated on flow field analysis and distortion transfer in compressor. Chen [2] found the inlet flow distortion deteriorated the pressure ratio, flow range, and efficiency of the compressor. Yao [3-4] explained how the total temperature distortion phasing was generated and phase changes as the distortion transfers in the fan model. Sitaram [5] found that the circumferential distortion had more effect than the radial distortion on the mass averaged flow parameters. However, little research has been devoted to the influence of inlet flow distortion on the overall performance from the perspective of the whole aeroengine.

High Stability Engine Control (HISTEC) program [6] researched by NASA was to design, develop, and flight demonstrate an advanced, high-stability, integrated engine control system that uses measurement-based, real-time estimation of distortion to enhance engine stability. HISTEC installed on the F-15 Advanced Control Technology for Integrated Vehicles aircraft at NASA Dryden and the HISTEC technologies were successfully flight demonstrated. It indicated that inlet flow distortion parameters can be accurately estimated real-time in the future. The research of this paper is based on the assumption that the distorted inflow thermal parameters can be estimated.

Figure 1 shows that transition-state working line is closer to the surge boundary during acceleration process. It indicates that compressor surge margin is lower during the acceleration so that compressor surge is more likely to occur.

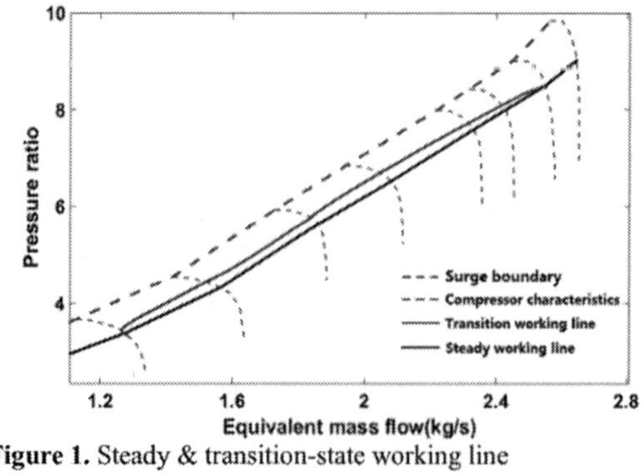

Figure 1. Steady & transition-state working line

DOI: 10.1051/matecconf/201710808011

Inlet flow distortion can be classified in various ways such as from the aspect of thermal parameters or spatial distribution. Due to the high radial gradient of pressure and temperature in compressor blade path, the flow mix more easily in radial direction so that the radial distortion is weakened obviously. Figure 2 shows the inlet flow distortion model that was researched in this paper. It is the simplest form of circumferential distortion, including both steady-state and dynamic-state.

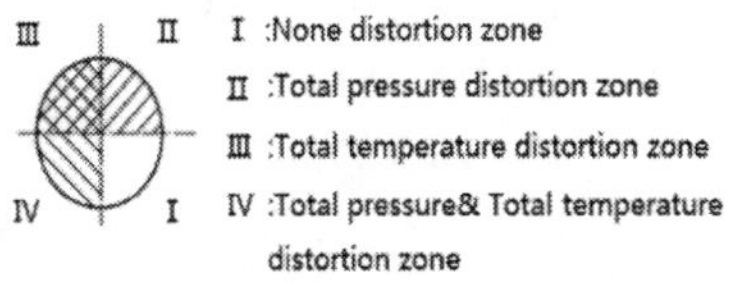

Figure 2. Division of circumferential inlet flow distortion

This paper is mainly divided into three major sections:

1) Section one introduced the modeling of turboshaft engine and sub-compressor.
2) Section two analyzed the influences of inlet flow distortion on the overall performance from the perspective of the whole turboshaft engine..
3) Section three presented the DFI method to improve fuel supply plan for surge-preventing acceleration and testified its environmental applicability.

2 Model development

Figure 3 shows the structure diagram of turboshaft engine. Based on a real turboshaft engine in service, the model was developed in Matlab/Simulink platform with the following characteristics [7]:

1) Physics based (e.g., thermodynamic, inertia, etc.)
2) Modular/component based construction.
3) Incorporating engine dimensions and component maps for sizing.

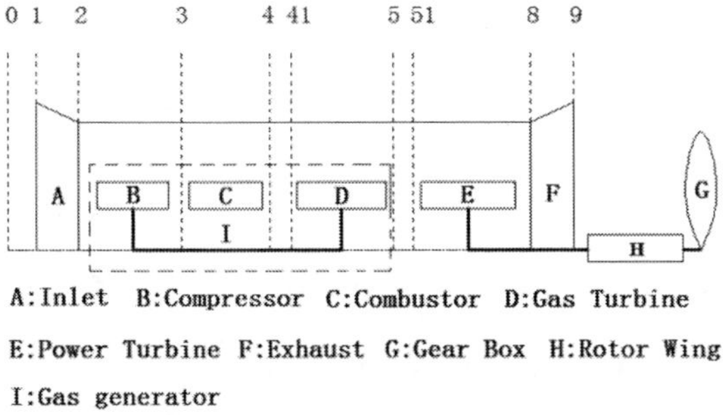

Figure 3. Simple layout of the turboshaft engine

2.1 Steady-state modeling

Figure 4 shows the Steady-state simulink model. Steady-state operation of the engine was described by Steady-state working equations, which were based on flow continuity of each components and rotor power balance. Physical speed of the gas generator (N) and power turbine (Np) were the control parameters in the steady-state control strategy.

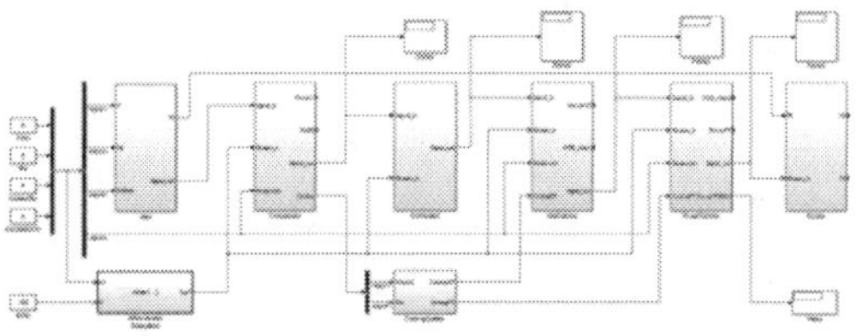

Figure 4. Steady-state simulink model

Four balance equations were used to described the steady-state operation.

1) Gas turbine flow continuity equation.

$$W_3 + W_{GT,Cooling} + W_f - W_5 = 0 \tag{1}$$

where W is mass flow; subscript *GT,Cooling* is Gas turbine cooling air from compressor.

2) Power turbine flow continuity equation.

$$W_5 + W_{PT,Cooling} - W_6 = 0 \tag{2}$$

where subscript *PT, Cooling* is power turbine cooling air from compressor.

3) Nozzle outlet static pressure balance equation.

$$Ps_9 - Ps_{amb} = 0 \tag{3}$$

where Ps is static pressure; subscript *amb* is

4) Gas generator power balance equation.

$$P_{GT}\eta_{GT} - P_{Compr} - P_{EXT} = 0 \tag{4}$$

where P is power; η is mechanical efficiency; subscript *GT* is gas turbine; subscript *Compr* is Compressor; subscript *EXT* is power extraction.

While in simulation calculation, the balance equations aren't strictly valid because the residuals of each balance equation always exist.

$$z_1 = W_3 + W_{GT,Cooling} + W_f - W_5 \tag{5}$$

$$z_2 = W_5 + W_{PT,Cooling} - W_6 \tag{6}$$

$$z_3 = P_{s9} - P_{amb} \tag{7}$$

$$z_4 = P_{GT}\eta_{GT} - P_{Compr} - P_{EXT} \tag{8}$$

If the residuals are sufficiently small, the balance equations could be valid approximately. To ensure the equations have the sole solution, four unknowns are needed, as shown in Table 1.

Table 1. Unknowns for balance equations

Unknow	Corrresponding variable	Meaning
x_1	Z_C	Compr Auxiliary line value
x_2	Z_{HPT}	GT conversion power
x_3	Z_{LPT}	PT conversion power
x_4	Wf_R	Relative fuel flow

The residual vector $Z=[z1, z2, z3, z4]^T$ and the unknows vector $X=[x_1, x_2, x_3, x_4]^T$ were defined.

DOI: 10.1051/matecconf/201710808011

Obviously, Z=F(X) and is calculated by thermodynamic calculation along the engine path. The balance equations solution algorithm is as follows: Assuming that k iterations have been made so that now $X^{(k)}=[x_1^{(k)}, x_2^{(k)}, x_3^{(k)}, x_4^{(k)}]^T$ and $Z^{(k)}=[z1^{(k)}, z2^{(k)}, z3^{(k)}, z4^{(k)}]^T$. The Newton-Raphson method is used to make Z's norm approach to zero. In the next iteration, X is updated by A (Jacobi matrix) and Z.

$$X^{(k+1)} = X^{(k)} - A^{-1}Z^{(k)} \tag{9}$$

$$A = \begin{bmatrix} \dfrac{\partial z_1}{\partial x_1} & \cdots & \dfrac{\partial z_1}{\partial x_4} \\ \vdots & \ddots & \vdots \\ \dfrac{\partial z_4}{\partial x_1} & \cdots & \dfrac{\partial z_4}{\partial x_4} \end{bmatrix} \tag{10}$$

If Z's norm meet the given accuracy requirement, the iteration would be ended and output X to accomplish the final thermodynamic calculation and output turboshaft engine performance calculation result. Figure 5 shows the algorithm flow chart of the steady-state calculation.

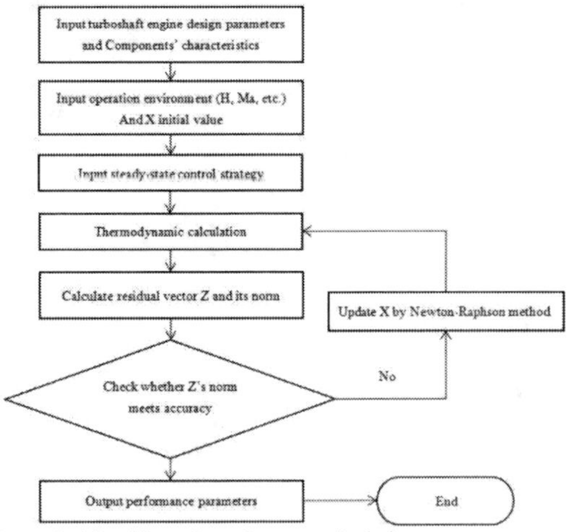

Figure 5. Algorithm of steady-state calculation

2.2 Transition-state modeling

Figure 6 shows the transition-state simulink model. In this model, the balance equation (1), (2) and (3) were still valid, while the balance equation (4) was not. Relative physical speed of power turbine (Np) and fuel supply (Wf) were the control parameters in the transition-state control strategy.

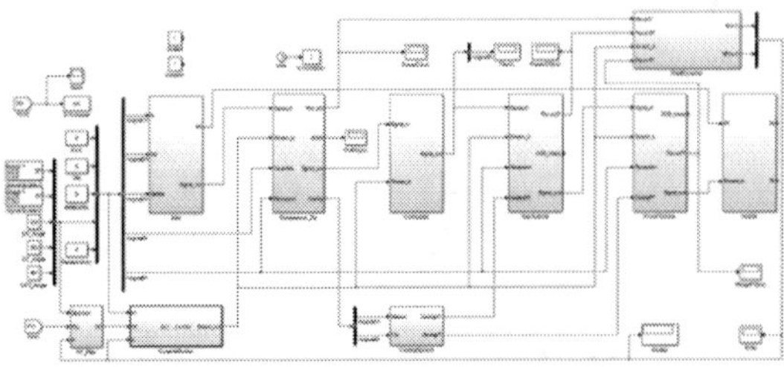

Figure 6. Transition-state simulink model

Unlike steady-state gas generator power balance equation, the inertia and dynamic characteristics of rotor was taken into consideration to describe the engine transition-state operation. The rotor motion was described by differential equations that solved by Euler method in fixed step size calculation.

$$P_{GT}\eta_{GT} - P_{Compr} - P_{EXT} - \left(\frac{\pi}{30}\right)^2 JN\frac{dN}{dt} = 0 \tag{11}$$

Figure 7 shows the algorithm flow chart of the transition-state calculation.

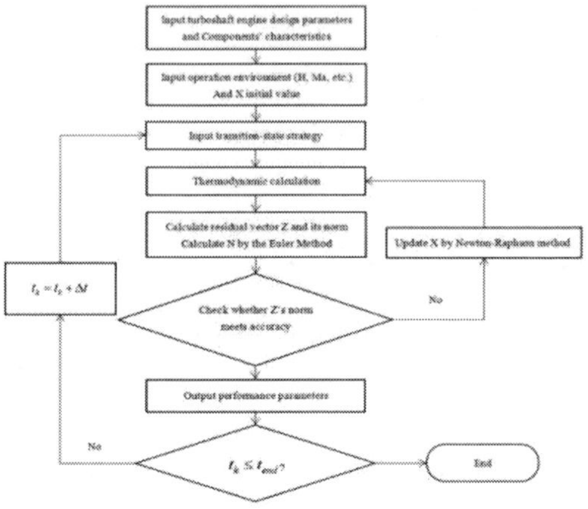

Figure 7. Algorithm of transition-state calculation

2.3 Inlet flow distortion parameters definition

1) Distortion level coefficient

DP and DT represent the level of total pressure and total temperature distortion respectively [8] whose definition are shown in the formula (12) and formula (13).

$$DP = \frac{P^*_{distorton} - P^*_{none\_distorton}}{P^*_{none\_distorton}} \times 100\% \tag{12}$$

$$DT = \frac{T^*_{distorton} - T^*_{none\_distorton}}{T^*_{none\_distorton}} \times 100\% \tag{13}$$

2) Distortion sector angle

θ_{DT}, θ_{DP} and θ_{DPT} represent the circumferential distribution of each distortion zone.

2.4 Sub-compressor modeling

The parallel compressor theory [8] can accurately explain the essence of inlet flow distortion and analyze its influence on the performance and stability of aeroengine. Its validity and accuracy have already been proved by experiments [9].

Figure 8 shows the algorithm flow chart of sub-compressor and its essential characteristics are shown as follow:

1) Internal flow in the sub-compressor is uniform.
2) No mass, momentum and energy exchange between sub-compressors.

DOI: 10.1051/matecconf/201710808011

3) The flow static pressure of each sub-compressor outlet is equal.
4) If any sub-compressor occurs surge, the compressor occurs surge.

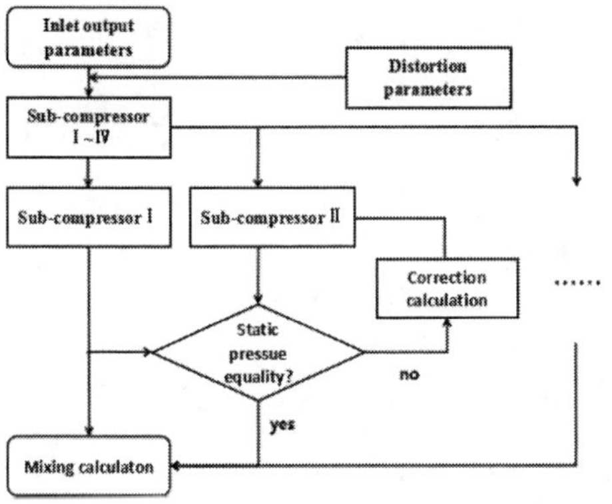

Figure 8. Algorithm of sub-compressor

2.5 Stability index

Conversion speed surge margin (SM) of compressor was chosen to be stability index in this paper. According to the fourth essential characteristic of the parallel compressor theory, the minimum SM value of each sub-compressor was considered as the SM of the whole compressor. SM is defined below:

$$SM = \frac{\pi_{C,s}/W_{cor,s} - \pi_{C,o}/W_{cor,o}}{\pi_{C,o}/W_{cor,o}} \quad (14)$$

where π is pressure ratio; W is mass flow; C is compressor; subscript s is surge boundary; subscript o is operating point; subscript cor means conversion value under international standard atmosphere (ISA).

3 Improved fuel supply plan with DFI

According to the acceleration performance requirements of this turboshaft engine, the fuel supply plan should be designed for satisfying those requirements, as shown in Table 2:

Table 2. Acceleration performance requirements

Parameters	Requirement
Surge margin (SM)	≮5%
Gas turbine inlet temperature (Tt41)	≯1430K
Relative rotational speed acceleration of gas generator (dN)	≯6%/s^2

3.1 Original fuel supply plan design

Fuel flow ratio (Wf/Pt3) is a widely used parameter in aeroengine control system [10]. Power extraction method [11] was used to design original fuel supply plan, as shown in Figure 9.

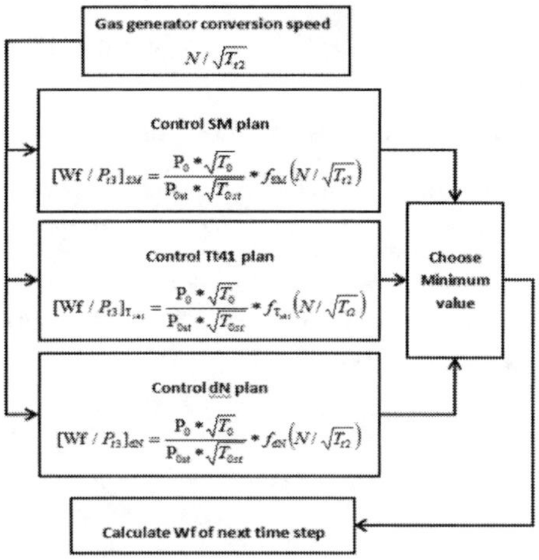

Figure 9. Original fuel supply plan

3.2 Distortion factor Item method

Distortion level coefficients were introduced into the control SM plan but without distortion Sector angles, Seen from formula (4). The SM of each sub-compressor was not affected by distortion sector angles because distortion sector angles had no impact on inlet flow thermal parameters.

$$[Wf/Pt3]_{\mathrm{SM}} = \frac{P_0 * \sqrt{T_0}}{P_{0st} * \sqrt{T_{0st}}} * f_{\mathrm{SM}}\left(N/\sqrt{T_{t2}}, 1+DP, 1+DT\right) \quad (15)$$

where subscript st is standard atmosphere.

The formula (5) was deduced by first order Taylor expansion and its high order items were neglected. The first item in bracket of the formula right side is original fuel supply plan. The second item and the third item are distortion factor items. According to the input signal of distortion level coefficients, the control SM plan can be adjusted adaptively by distortion factor items.

$$[Wf/Pt3]_{SM} = \frac{P_0 * \sqrt{T_0}}{P_{0st} * \sqrt{T_{0st}}} * \begin{bmatrix} f_{\mathrm{SM}}\left(N/\sqrt{T_{t2}}, 1, 1\right) \\ + \frac{\partial f_{\mathrm{SM}}}{\partial DP} * \mathrm{D}P \\ + \frac{\partial f_{\mathrm{SM}}}{\partial DT} * DT \end{bmatrix} \quad (16)$$

4 Results

4.1 Performance

Compared with the engine performance in the same operating condition without inlet flow distortion, a steady-state example was simulated to indicate the influence of the inlet flow distortion on overall performance.

Known conditions are shown as followed:
Environment: ISA, H=0km, Ma=0
Initial state: N=100%, Np=100%

Distortion level coefficient: DP=-6%, DT=4%
Distortion Sector angle: θ_{DP}=90°, θ_{DT}=90°, θ_{DPT}=90°

4.1.1 Compressor operation analysis

Figure 10 shows the positions of compressor's equivalent operating point (green circle) and the operating point (black circle) of the same operating condition under uniform inflow. The purple operating point of the sub-compressor which encountered combination distortion was very close to the surge boundary.

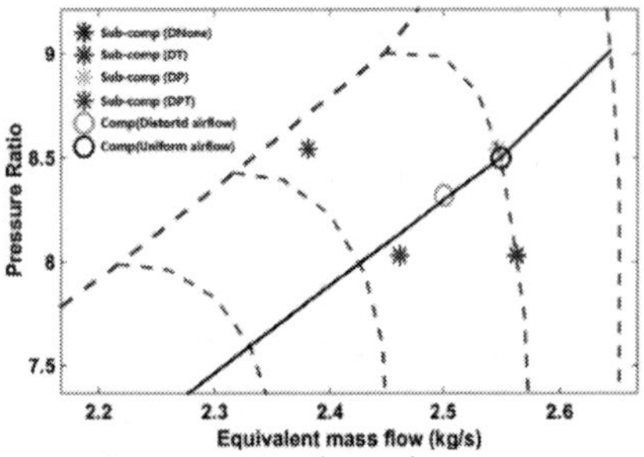

Figure 10. Positions of operating points

Inlet flow distortion led to drop of boosting capability and flow capacity. The SM had a great loss compared with uniform inflow (Table 3).

Table 3. Comparison of compressor performance

Parameters	Uniform	Distortion	change(%)
ΔSM(%)	100	16.7	-84.3
Pressure ratio	8.5	8.277	-2.62
Air inflow(kg/s)	2.537	2.379	-6.23

4.1.2 Overall performance

Compared with uniform inflow, total pressure dropped and total temperature rose in each section (Table 4). The overall performance was worse with the decrease of power output and the increase of specific fuel consumption (SFC) (Table 5). Drop of gas total pressure suggest that the gas expansion capacity decreased so that less gas thermal energy converted into mechanical energy in turbine. The increase of exhaust outflow static temperature suggest that the exhaust loss increased.

Table 4. Main section parameters

Section	Parameters	Uniform	Distortion	Change (%)
3	Pt (kpa)	857.0	809.7	-5.52
	Tt (K)	593.9	602.0	+1.36
4	Pt (kpa)	831.2	785.4	-5.51
	Tt (K)	1427.1	1447.3	+1.42
5	Pt (kpa)	289.1	273.3	-5.47
	Tt (K)	1142.5	1160.5	+1.58
6	Pt (kpa)	106.4	105.8	-0.56
	Tt (K)	920.7	947.2	+2.88
9	Pt (kpa)	103.5	103.3	-0.19
	Ts (K)	915.7	942.6	+2.94

Table 5. Overall performance parameters

Parameters	Uniform	Distortion	Change (%)
T_{t41} (K)	1405	1425	+1.42
π_{GT}	2.875	2.873	-0.07
η_{GT}	0.852	0.849	-0.35
T_{t51} (K)	1137	1155	+1.58
π_{PT}	2.672	2.531	-5.28
η_{PT}	0.893	0.890	-0.34
P_{PT} (kw)	643.0	581.5	-9.57
SFC (kg/(h*kw))	0.3266	0.3450	+5.63

4.2 Dynamic response

A transition-state example was simulated to indicate the dynamic respond of the engine supposed that it encountered inlet flow distortion in flight.

Known conditions are shown as followed:
Environment: ISA, H=0.5km, Ma=0.3
Initial state: N=100%, Np=100%
Control strategy: kept Wf and Np constant
Distortion level coefficient: seen from Figure 11
Distortion Sector angle: θ_{DP}=90°, θ_{DT}=90°, θ_{DPT}=90°

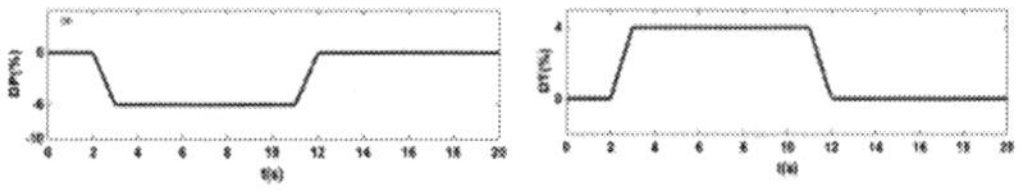

(a) DP (b) DT

Figure 11. Distortion level coefficient input signal

Figure 12 shows the dynamic respond of gas generator. The reduction of gas turbine power was smaller than that of compressor power load caused by gas generator acceleration. When the inlet flow distortion disappeared gas generator decelerated to the initial state.

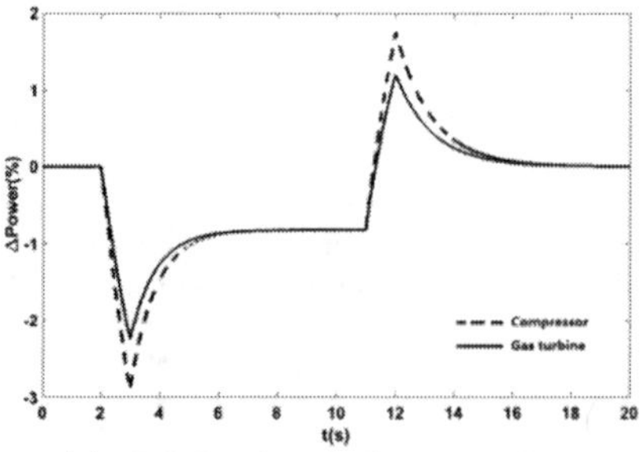

(a) Relative change of power and load

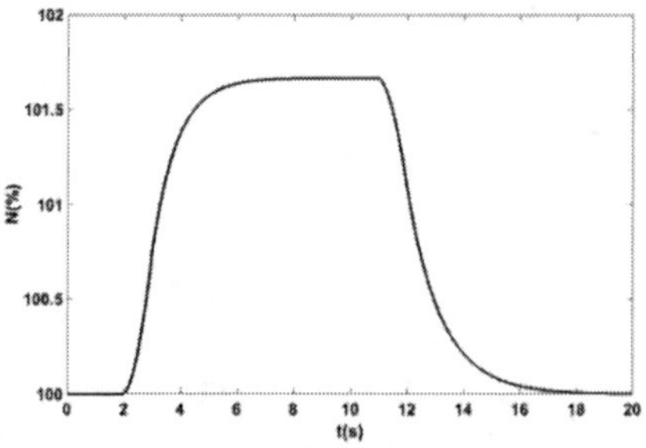

(b) N

Figure 12. Dynamic respond of gas generator

DOI: 10.1051/matecconf/201710808011

Figure 13 shows the dynamic respond of power output. The degree of gas mass flow decrease was much more than that of unit power increase so that power output dropped significantly.

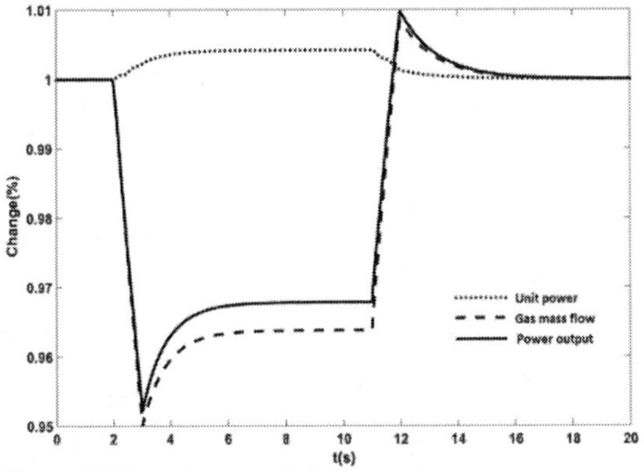

Figure 13. Dynamic respond of power output

4.3 Above-idle acceleration with DFI method

Several transition-state examples were simulated and shown below to demonstrate that DFI method has adaptability for surge-preventing acceleration from idle under different inlet flow distortion and has the advantage of environmental applicability. Wf was controlled by fuel supply plan with DFI.

4.3.1 Adaptive surge-preventing

Known conditions are shown as followed:
Environment: ISA, H=0, Ma=0
Initial state: N=75%, Np=100%

1) Inlet flow distortion Ⅰ: DP=-2%, DT=1%
2) Inlet flow distortion Ⅱ: DP=-3%, DT=2%
3) Inlet flow distortion Ⅲ: DP=-6%, DT=4%

(The waveform of the signals were the same as Figure 10)
Distortion Sector angle: θ_{DP}=90°, θ_{DT}=90°, θ_{DPT}=90°

Figure 14 shows the acceleration processes under three different inlet flow distortion.

1) Under inlet flow distortion I, Figure 14 (c) shows the engine surged with original fuel supply plan.
2) Under inlet flow distortion I, Wf increased more slowly for the SM requirement. The engine accelerated to expected operation condition directly but less slowly.
3) Under inlet flow distortion II, Wf stopped increasing temporarily and acceleration stopped. The engine accelerated to expected operation condition after the distortion disappeared.
4) Under inlet flow distortion III, Wf reduced and the engine decelerated. The engine accelerated to expected operation condition after the distortion disappeared.

Improved fuel supply plan with DFI could adaptively adjust fuel supply under different inlet flow distortion. The engine could maintain the expected surge margin and accelerated to the expected operating condition.

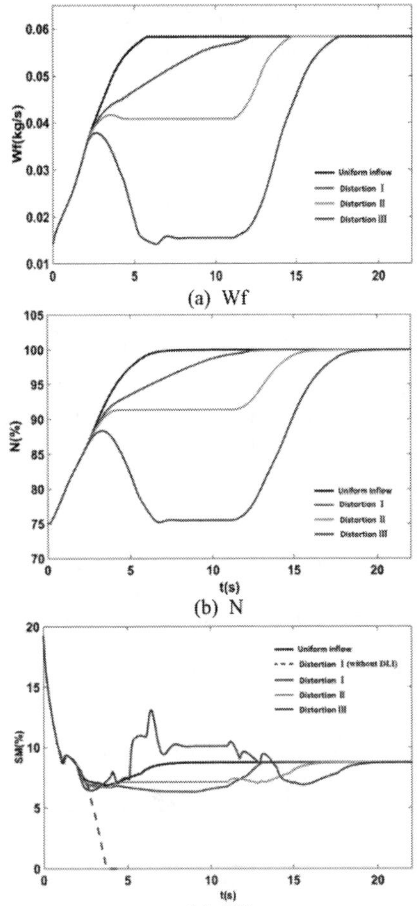

(a) Wf
(b) N
(c) SM

Figure 14. Acceleration under different inlet flow distortion

4.3.2 Environmental applicability

1) Plateau environment

Known conditions are shown as followed:
Environment: ISA,H=3km, Ma=0
Initial state: N=75%, Np=100%
Distortion level coefficient: DP=-6%, DT=4%
(The waveform of the signal was the same as Figure 10)
Distortion Sector angle: θ_{DP}=90°, θ_{DT}=90°, θ_{DPT}=90°

Figure 15 shows the acceleration process in Plateau environment.

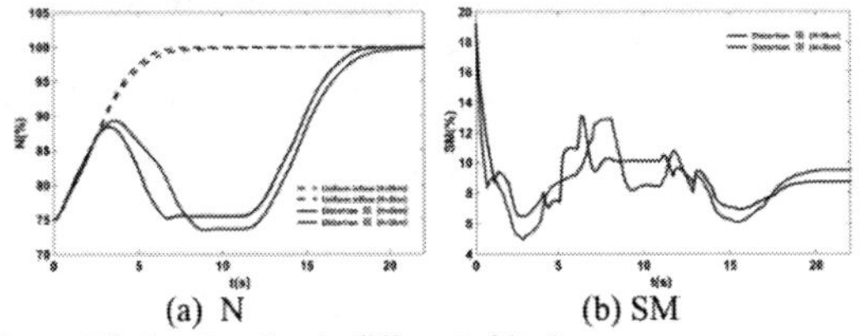

(a) N (b) SM

Figure 15. Acceleration in different altitude

2) High/ Low temperature environment

Known conditions are following below:
Environment: H=0km, Ma=0, T_{amb}=-30℃&50℃
Initial state: N=75%, Np=100%
Distortion level coefficient: DP=-6%, DT=4%

(The waveform of the signal was the same as Figure 10)
Distortion Sector angle: $\theta_{DP}=90^o$, $\theta_{DT}=90^o$, $\theta_{DPT}=90^o$

Figure 16 shows the acceleration process in high/ Low temperature environment.

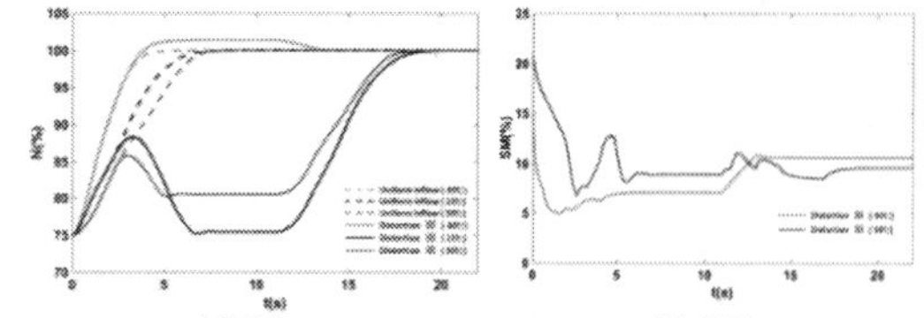

(a) N (b) SM

Figure 16. Acceleration in environment temperature

In extreme environment, the engine could still maintain the expected surge margin and accelerated to the expected operating conditions under inlet flow distortion.

5 Conclusion

The parallel compressor theory was applied to analyze inlet flow distortion influence on performance and dynamic respond of turboshaft engine. Acceleration was simulated with improved fuel supply plan under the condition of inlet flow distortion. The following conclusions were drawn from the analysis of the numerical simulation results:

1. Under condition of keeping rotor physical speed constant, inlet flow distortion weaken the flow capacity and boosting capacity of turbocharging system. Drop of total pressure weaken the energy conversion in turbine and leads to increase of exhaust loss. Surge margin has a great loss under inlet flow distortion so that the engine is more vulnerable to surge.
2. When fuel system supply constant fuel to the aeroengine without any intervention in flight, sudden encountering inlet flow distortion tips rotor power balance and makes rotor to accelerate. A rise in the gas total temperature of turbine inlet may lead to excessive thermal stress in turbine blades. The power output drops because the degree of working substance reduction is greater than that of unit power increase.
3. The Distortion Factor Item method presented here has adaptability for adjusting fuel supply when the engine encountering different inlet flow distortion. The Distortion Factor Item method is efficient for surge-preventing acceleration from idle under the circumstance of real-time distortion estimation, and its environmental applicability has been testified. This method could be beneficial to provide a solution to aeroengine control schedule when encountering flow distortion.

Acknowledgment

This research is funded by National Nature Science Foundation of China (NSFC) under Grants 51206005, and Project MIIT. The author is thankful for the support from Collaborative Innovation Center of Advanced Aero-Engine.

References

1. Mazzawy R S. Multiple Segment Parallel Compressor Model for Circumferential Flow Distortion[J]. Journal of Engineering for Gas Turbines & Power, **99**(2):288-296 (1977)
2. Chen J, Zheng X, Wang C, et al. Numerical Investigation of Influence of Radial Inlet Distortion on Compressor Performance in 2-Stage Turbocharging System[C]// ASME Turbo Expo 2014: Turbine Technical Conference and Exposition. American Society of Mechanical Engineers, (2014)
3. Yao J, Gorrell S E, Wadia A R. High-Fidelity Numerical Analysis of Per-Rev-Type Inlet Distortion Transfer in Multistage Fans: Part I—Simulations With Selected Blade Rows[J]. Journal of Turbomachinery, **132**(4):457-467 (2010)
4. Yao J, Gorrell S E, Wadia A R. High-Fidelity Numerical Analysis of Per-Rev-Type Inlet Distortion Transfer in Multistage Fans: Part II—Entire Component Simulation and Investigation[J]. Journal of Turbomachinery, **132**(4):457-467 (2010)
5. Sitaram N, Govardhan M, Murali K V. Effects of Inlet Pressure Distortion on the Performance and Flow Field of a Centrifugal Compressor at Off-Design Conditions[C]// ASME 2012 Gas Turbine India Conference. 35-50 (2012)
6. Orme J, Delaat J, Southwick R, et al. Development and testing of a High Stability Engine Control (HISTEC) system[J]. (2013)
7. Mink G, Behbahani A. The AFRL ICF Generic Gas Turbine Engine Model[C]// AIAA/ASME/SAE/ASEE Joint Propulsion Conference & Exhibit. (2013)
8. Cousins W T, Davis M W, Cousins W T, et al. Evaluating Complex Inlet Distortion With a Parallel Compressor Model: Part 1—Concepts, Theory, Extensions, and Limitations[C]// ASME 2011 Turbo Expo: Turbine Technical Conference and Exposition. 55-67 (2011)
9. Brennan B. Effect of a sinusoidal three-lobe total pressure distortion on a transonic compressor : distortion propagation, compressor performance, and blade passage events [D]. Massachusetts: Massachusetts Institute of Technology, Department of Aeronautics and Astronautics. 145-164 (2002)
10. Link C. Jaw, Jack D. Mattingly. Aircraft engine controls design, system, analysis, and health monitoring[M]. Aviation Industry Press, (2011)
11. CHEN Y C, XU S Y.Power extraction method of designing starting control strategy for turbofan engine[J]. Journal of Propulsion Technology, **30**(5):604-609 (2009)

Geometric Optimization of Turbocharger Compressor and Its Influence on Engine Performance

Fangming Zhang and Roland Baar

Technical University of Berlin, Institute for Internal Combustion Engines, 10587 Berlin, Germany

Abstract. This paper consists of two parts: aerodynamic and mechanical multi-objective optimization for centrifugal compressor impeller through combining the three dimensional fluid dynamic simulation module CFX 16.1, the static structure in the ANSYS Workbench and the optimization software optiSLang; and a comparison and analysis of the effects of the optimized compressor on the engine performance by the one dimensional simulation tool GT-Power. In the process of optimization, the compressor design point is regarded as the optimizing point, while impeller blades and hub line were parameterized through the Bezier curve. Pressure ratio, isentropic efficiency, quality and maximum deformation and maximum internal stress of the impeller were defined as the output conditions. MOP module was then adopted in optiSLang for the parameters sensitivity analysis and mapping relationship modeling between the impeller parameters and the objective functions. The genetic algorithm is applied to find out and validate the optimal design. Through 1D simulation tool GT-Power, the influence of the optimized compressor on rotational speed of the turbocharger, backpressure and pumping loss under different engine operating conditions is analyzed and compared.

1 Introduction

Under the increasingly strict environmental pollution and emissions law, interest in the development of turbocharged engines has grown rapidly in recent year due to their low fuel consumption and high power characteristics. At the same time the research on the turbocharger is also changing with each passing day. Attempts to overcome the hysteresis phenomenon, problems of high knock rating, and high loss rate have become a great challenge to the development of new turbocharged engines. The design of the compressor impeller has a direct influence on key factors including the maximum working range, exhaust gas utilization, and service life, as well as exerting an important influence on the power and economy of the engine. With the rapid development of computational fluid dynamics software and the rapid expansion of optimization algorithms in engineering, the development cycle of the turbocharger has been greatly shortened, and the costs were reduced accordingly.

Multi-objective optimization problems are widespread in the theoretical research and engineering application and widely popularized, their computation complexity will also be increased with the increase of dimensions, the quantity of sub-objective shows exponential growth with the rise of relevance. Maintaining the low dimension and controlling the number of objectives will greatly reduce complexity and computation burden. Correlation and relative importance of each sub-objective will be embodied through the weight coefficient. Besides, the overall multi-objective optimization problem will also be analyzed and solved through the resulting comprehensive evaluation function. A series of optimal solution set existed in multi-dimensional space according to different design objective, arbitrary units of this solution set will make each sub-objective value to achieve the overall optimum, this solution set is called as Pareto solution set. In order to solve multi-objective optimization problems, a series of algorithms have been carried out. Geller and Schemmann [1] conducted simulation analysis of three dimensional flow field and statics calculation on centrifugal compressor impeller on ANSYS Workbench platform, and combined with the genetic algorithm of optiSLang optimization software to conduct optimization design and analysis, the results showed that under guaranteed equal efficiency, maximum stress of impeller was reduced by 40%. Wei-guo Zhao et al. [2] combined the advantages of artificial neural network with the genetic algorithm, with the simulation results of CFD three-dimensional fluid as the optimal initial population, they carried out the optimization design of centrifugal pump impeller and overall simulation with redesigned volute, the efficiency at the design operating point improved by 1.59%, non-operating points are increased by 6%. But only through the three-dimensional aerodynamic optimization and ignored security such as blade mechanical strength limit, deformation, it will reduce the admissibility of some optimization results in the practical application. To this, Hildebrandt et al. [3] respectively conducted three dimensional flow field simulation, combined with three-dimensional fluid dynamics and simulation optimization of statics on centrifugal compressor, and verified through experimentation, the results showed that the optimal value difference was large, only the compressor performance obtained by the aerodynamic optimization of limiters obviously larger than the results obtained by two kinds of restriction of fluid optimal and security, and the working range was also greatly reduced. Dynaldo company [4] compared the Kriging approximation algorithm, support vector regression, artificial neural network and the MOP approximation (Metamodel of Optimal Prognosis) and evaluated their mapping effect of optimized variables and objective function, the results show that, the mapping effect established by MOP was not changed along with the increase of variable, while the mapping effect of other three kinds were less than 60% when the number of optimization variables more than 10. Therefore, the MOP algorithm will be applied in this paper. The main objective of this paper is to establish, using ANSYS Workbench a three-dimensional fluid and static model for optimizing centrifugal-flow compressor, combined with MOP and genetic algorithm module in optiSLang, optimizing the geometric size of impeller. The influences of an optimized compressor on engine performance were finally analyzed and summarized by the 1D-simulation tool GT-Power.

2 Optimization methods

2.1 Optimization object

This paper will carry on the optimization to the compressor impeller with splitter blades. Due to the geometric limit of compressor casing, the diameter and the outlet height of the impeller are fixed, and the shroud curve on impeller meridian plane is maintained constant in the process of optimizing. Optimizing parameters are the angle, thickness on the meridional curve of blade and the shape of the hub control curve. The compressor impeller used in the experiment is shown in Figure 1.

Figure 1. Impeller in the experiment

2.2 Optimization objective

The optimization of the compressor impeller is mainly carried out through two aspects of judging the operation safety and the work efficiency. Blade angle on leading edge and trailing edge, impeller clearance, blade slant Angle, and blade shape trajectory and so on will directly determine the direction of motion and path of flow field into the cavity of the impeller, including the growth and decline of secondary flow, the size and distribution of the entropy production, the pressure gradient distribution on the blade surface, etc. The increase in blade thickness can increase the impeller's mechanical strength to a certain extent. But due to the increase of quality, fluid blocking rate causes the growth of rotational inertia and the decrease of effective inlet area. The contradiction between the optimization objectives, the nonlinear response characteristics have become the main objectives need to be solved in multi-objective problem, while the judgment of the final optimization results will be embodied with weights w_i according to the requirements.

$$E_{max} = \Sigma w_i \cdot (T_i - T_{io})/T_i \quad (1)$$

This paper will outline the optimization of the design point condition of compressor (speed: 140 kr/min; flow: 0.08 kg/s). Optimization objective are detailed as the following objectives:

- Maximization of isentropic efficiency of whole turbocharging system
- Reduce the maximum stress lower than the allowable stress of impeller
- Reduce the maximum deformation and at least maintain original deformation
- Reduce the quality and at least maintain the original quality
- Increase and at least maintain the current maximum blocking flow
- Maintain or improve the supercharge ratio near surge area

2.3 Optimization process

The optimization process can be divided into three steps.

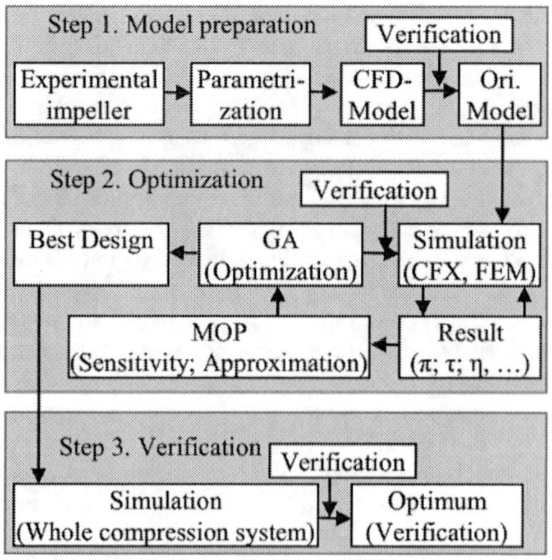

Figure 2. Optimization process

2.3.1 Step 1: Model preparation

For impeller optimization, following two premises must be met: first, the parametric modeling of impeller, by controlling parameters to establish different geometric features of the impeller model. Second, establish the same compressor system as the impeller in experiment, compared the experimental results through CFD and FEM simulation and ensure the correctness of the model. As a precondition for optimization calculation, blade parametric model will directly affect the optimal scale and complexity; also directly affect the accuracy of the optimization. In this paper, it will conduct recognition processing of structure angle and thickness on the original blade through the ANSYS Designmodeler 16.1 CAD-Import function, using Beta Angle and the thickness curve to describe blade type, all are adopted the 4-order Bezier curve fitting. Control points P_i of Bezier curve are uniformly distributed within the scope of blade design, each control point can be moved along the vertical or horizontal axis to determine the new blade model. The formal description of Angle and the thickness curve is given by:

$$P(x) = \Sigma P_i \cdot R_{i,x}(x) \quad x \in (0,1) \tag{2}$$

where $R_{i,x}(x)$ are Bernstein basis polynomials of degree i. The total of optimizing parameters is 47, of which the main blades and splitter blades including angle and thickness are 20 parameters of each.

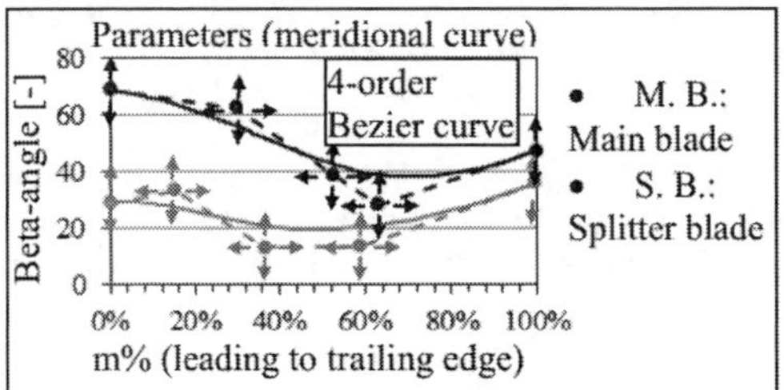

Parameters (hub curve)

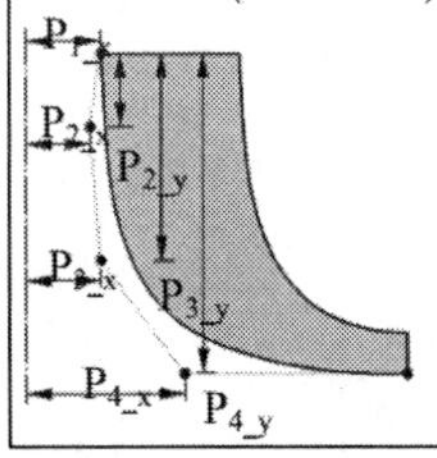

Number of Parameters

Input Parameters	n
M. B. (angle)	5*2
M. B. (thickness)	5*2
S. B (angle)	5*2
S. B (thickness)	5*2
Hub curve	7
total	47

Figure 3. Parameters on meridional and hub curve

For hub curve, the parametric model is established with the aid of SolidWorks, ANSYS is imported in for identification. The maximum characteristic of Bezier curve is that it can use less control point to describe and generate complex and smooth curve. Its convex hull property can make it always fall in the convex hull of control points, so the basic line type is unable to have fierce volatility [5], and operation is simple, these characteristics are beneficial to the parametric design of space curve and narrow space dimension.

2.3.2 Step 2: Optimization

optiSLang provided rich optimization schemes for solving the problem of multi-objective optimal value [6], mainly are three categories: gradient method, natural inspiration method (including genetic algorithm, particle swarm optimization (PSO) algorithm, etc.) and adaptive method. Genetic algorithm (GA) is suitable for solving the optimization problem of geometric parameters of the impeller with its high efficiency, global superiority, multi-parameter optimization, etc.

In the genetic algorithm, a conduct fitness analysis on new individual of each generation after restructuring is required, and variation, if 3-D flow field and mechanics analysis were carried on the individuals one by one in each generation, it is bound to increase operation difficulty and simulation time. By using MOP module in optiSLang, it will not only precisely establish the mapping relationship of optimization parameter and objective function, conduct fitness analysis on new individual by establishing approximation model, and nonlinear sensitivity analysis on optimization of parameters, so as to eliminate the interference of non-sensitive parameters on the approximation model, and improve the quality of optimization. While the initial population individuals shall be conducted 3-D flow field and statics calculation one by one, physical results such as the isentropic efficiency, compression ratio, and maximum stress are provided to the MOP as raw data. optiSLang software provides the CoP (Coefficient of Prognosis) parameters to judge the quality of the established approximation model, when the value is close to 100%, it indicates the mapping relationship is better. For nonlinear output, of course, the convergence results of the input will reduce the CoP value, which may cause the results obtained by approximation model deviate from the actual values, thus influence optimization results. optiPlug is respectively connected to ANSYS and optiSLang, convenient for the exchange, edit of input, output data under these two conditions. In addition, this paper adopts the optiSLang integrated Latin Hypercube method in optiSLang for random sampling of the overall design space. Compared with the Monte Carlo random sampling method, this method has advantages of more uniform coverage of the individual in the whole n dimensional space, number of samples are 10 times less than Monte Carlo method, etc. [7].

2.3.3 Step 3. Best design validation

Ultimately, verification of the optimal blade design will be performed. Based on the optimized model, the working performance of the blade at each operation point is calculated and analyzed.

2.4 3D simulation model

In order to improve the precision of the model, and shorten the simulation time, first of all, it conducts 3-D modeling on the whole turbocharging system, this model is determined as original model through steps such as non-dependence analysis of gridding and contrast experimental data. Secondly, a three-dimensional single passage optimization model is established based on the original model, with ANSYS Workbench as work platform for steady simulation and statics calculation of integral impeller. Finally, optimization is carried on through optimizing software optiSLang, and the optimal impeller model is substituted into the whole model, and then its simulation results are compared with original model. In Figure 4, the left picture is the 3-D whole compression model, and right picture is single passage impeller model.

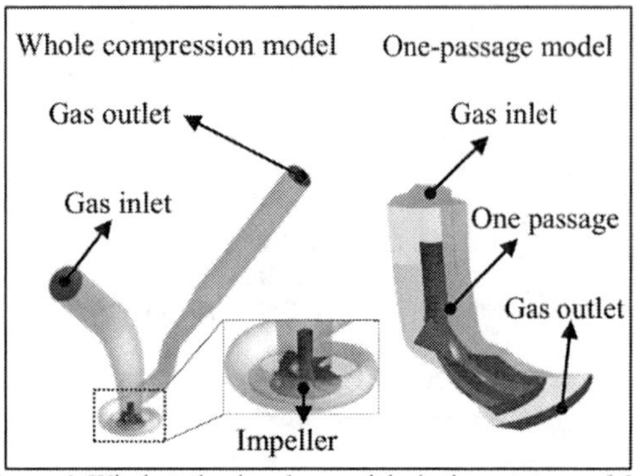

Figure 4. Whole turbocharging model; single passage model

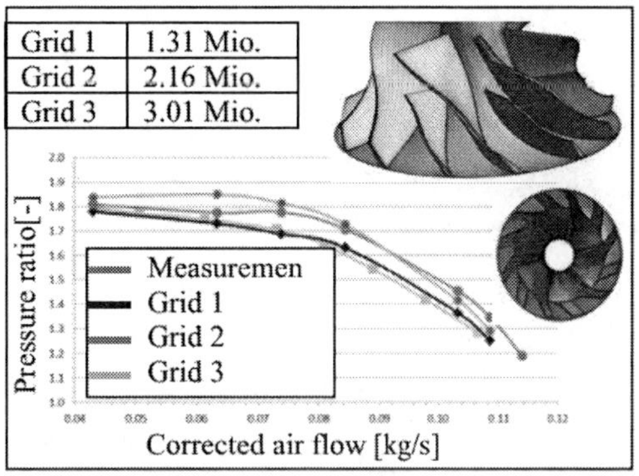

Figure 5. Compressor map under different grids

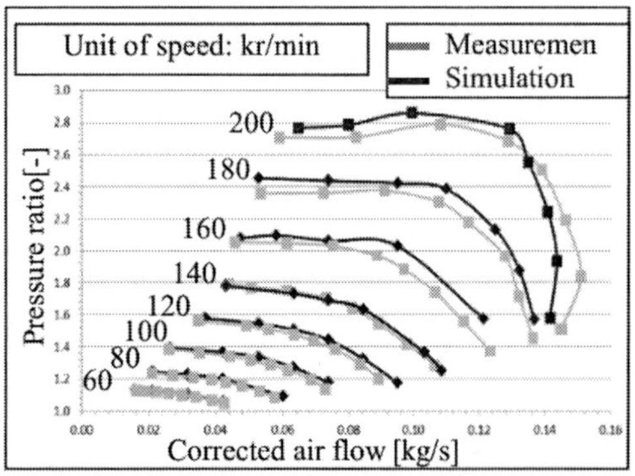

Figure 6. Comparison simulation results with measurement

Impeller passage grid is automatically generated by Turbogrid module. Turbulence model is chosen as the SST model, a more accurate description of near wall features and calculation of far field flow condition are advantageous to the accurate calculation of aerodynamic loss. Based on the finite volume method to solve the 3-D Reynolds average Navier - Stokes equation, second order center difference method is adopted to conduct space discrete, the viscous shear mode of Total Energy is added for precise calculation of the flow loss. The frozen rotor method is used to realize the combination of sensory transfer and localized flow field. The system boundary is set as adiabatic with no slip. The blade wall surface in flow field is dispersed by O-grid, inlet, outlet and the clearance of volute and flange are meshed by using H grid. It can be known from research on the impeller non-dependence grid that, as shown in figure 5, the simulation values in grid 2 and 3 were larger from experimental data, simulation results obtained through grid 1 were much better with the regional Y-plus values at less than 3, thus conforming to the requirements of the SST turbulence model. The working condition curve under each speed obtained by the first grid from the simulation was close to the experimental value, as shown in figure 6, and will be adopted.

3 Analysis of optimization results

3.1 Sensitivity analysis

The whole process of impeller geometry optimization adopted eight core CPU simulation, simulation time were 15 days, the original species are selected as 200, with a total of 20 simulation failures (<15%), after 90 generations of genetic variation, mutation rate is 20%, finally the optimization results obtained.

Figure 7 shows the MOP model of output parameters. Apart from to the pressure ratio, CoP value of other output parameters are higher than or equal to 90%, this indicates that the mapping relationship of optimization parameter and objective function is better. It can be known from impeller maximum stress, maximum deformation MOP model that the output values are mainly affected by the eight parameters, of which parameter FD1207 had the greatest influence, namely the geometric angle on top of main blade leading edge. The effect on isentropic efficiency is more than 60%. The influence of optimization parameters on the pressure ratio and quality is relatively uniform. The impeller quality is mainly affected by six control parameters of hub curves. Pressure increase caused by rotation of the impeller is mainly affected by the angle on the bottom of blade trailing edge.

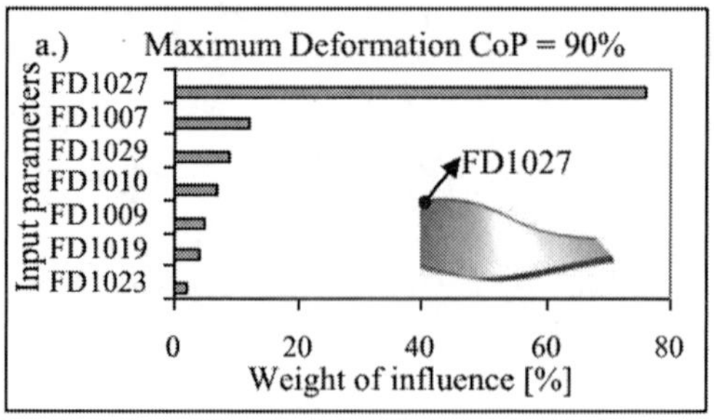

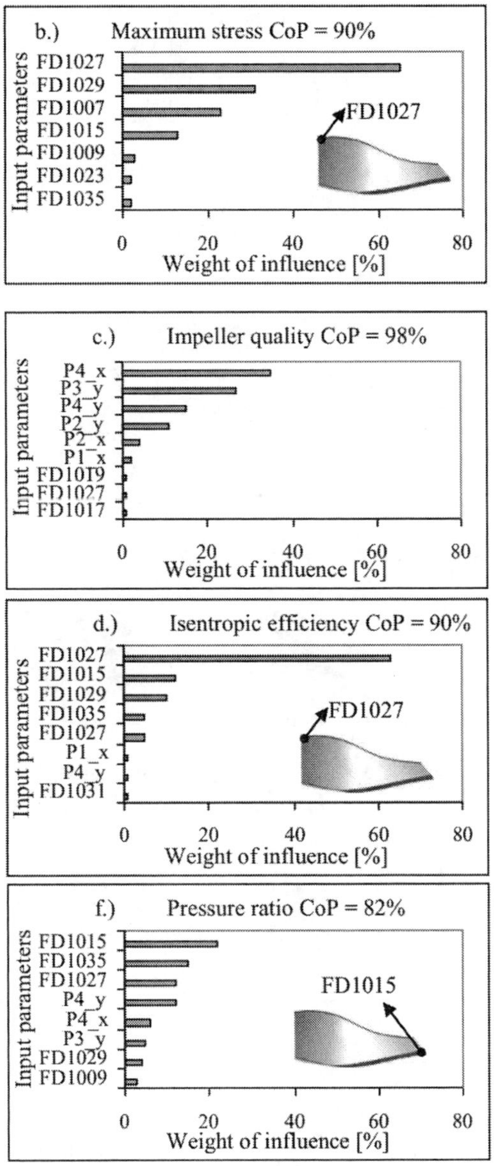

Figure 7. Result of sensitivity analyses

3.2 Optimization analysis and verification

In the process of global optimization, MOP model has created a total of 9000 design points, as shown on the bottom half of figure 8, pressure ratio is converged by generations, and close to the maximum.

Red marked points are ultimate Pareto solution sets. As a multi-objective optimization problem, the optimal solution for pressure ratio will be subject to other optimization objectives. Therefore, Pareto solution is not the maximum value of compression ratio. Similarly, the isentropic efficiency is also converged by generations and close to the maximum value.

As shown in Fig. 9, X-coordinate is the isentropic efficiency, Y-coordinate is the pressure ratio, Pareto solution set has a total of 1047 design points, focused on the right position of figure, and is superior to the obtained isentropic efficiency of original model and pressure ratio benchmark. Considering other optimization objectives, the efficiency of some optimal solutions is slightly lower than the highest value.

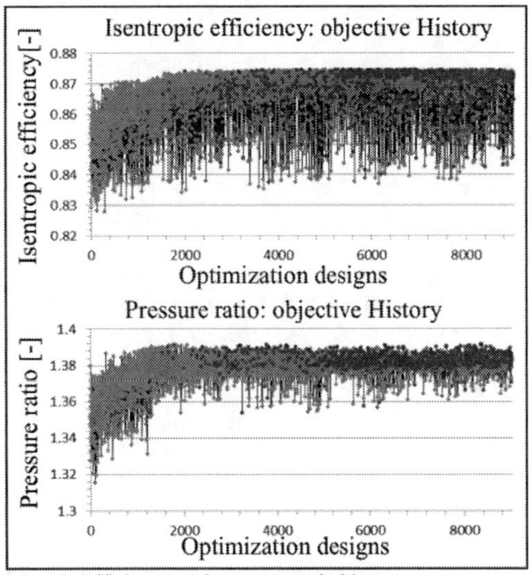

Figure 8. Efficiency and pressure ratio history

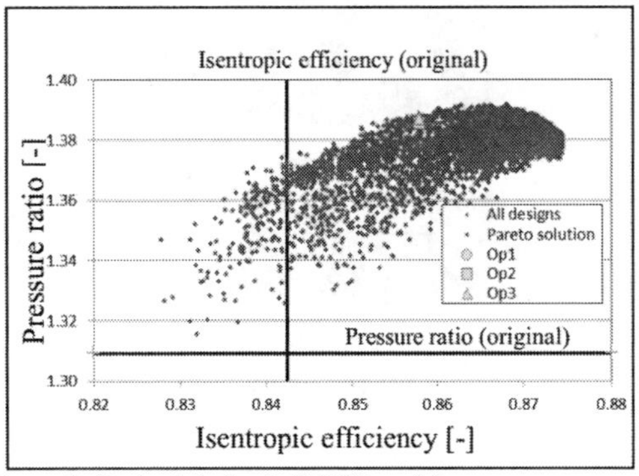

Figure 9. Optimal design in 'MOP' model

In the Pareto solution set, the optimal value will be found out by solving the equation 1 with the weight coefficient of all objectives. In view of the impeller job security and efficiency, the isentropic efficiency and the maximum stress are the most important optimization objectives.

Table 1. Comparison of three optimized compressors

	Op1	Op2	Op3
Pressure ratio	6.11%	5.88%	5.65%
Isentropic efficiency	1.31%	1.66%	1.66%
Impeller quality	-7.42%	-7.42%	-5.65%
Maximal miser stress	-1.68%	-3.74%	-5.52%
Maximal deformation	-3.87%	-0.26%	-0.26%

DOI: 10.1051/matecconf/201710808012

By the method of given weights on the objective functions, rating on the 1047 design points has been conducted. As a result, the rating of Op1, Op2 and Op3 design point is the highest. Therefore these three design point will be regarded as the final design points.

Table 1 gives the comparison of the three kinds of optimization results with the original model. A negative value indicates decrease. Research shows that the maximum stress, deformation and quality are decreased, the maximum loss is around 7%, and pressure ratio and isentropic efficiency also increase accordingly. In order to verify the accuracy of approximation model, 3-D single flow passage simulation and mechanical calculation on Op1, Op2 and Op3 has been conducted again. Figure 10 shows the comparison of verification results with approximate model results, the results are conform to expectation. Under the design point conditions, the maximum difference of pressure ratio for Op1, Op2 and Op 3 obtained by the approximation model and the results of verification simulation is about 0.003. Besides, the isentropic efficiency has a decrease of about 0.005, and the verification result is close to the approximation result, so the accuracy of the approximation model is higher.

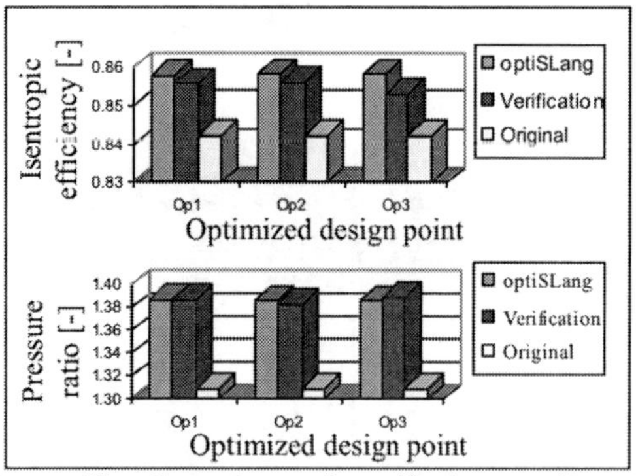

Figure 10. Comparison of verification results with approximation model result

3.3 Results comparison

Finally there is a whole compression system simulation with different mass flow in the speed of design point and comparing the simulation result of optimal impeller design and original model. As seen in Fig. 11, under the design speed, the isentropic efficiency of optimal impeller at design point has slight increase 2.24% and the increase of pressure ratio is 4.34%; for mass flow less than 0.083 kg/s (mass flow at design point) the optimized compressor maintains fast the same efficiency compared to the original model; when the optimization model has a large mass flow (greater than 0.083 kg/s), the increases of efficiency and pressure ratio are greater; the working range of the optimized compressor under design speed is obviously increased and the maximum bypassed flow into the impeller (choke flow) is 4.85% increase compared with original model. Known from the comparison of the maximum stress and the maximum deformation of the impeller (shown as Fig. 12), the maximum stress on the blade after optimization is reduced by 1.33% and the maximum deformation is reduced by 2.52%. In general, the optimization result conforms to the prediction. Due to the consideration of security, the growth of aerodynamic optimization is relatively small.

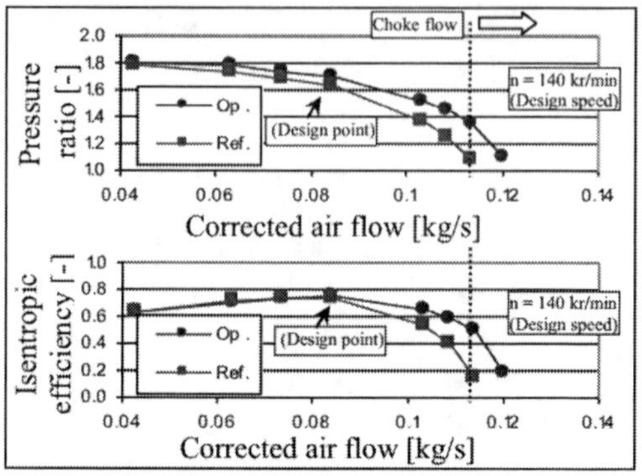

Figure 11. Comparison of efficiency and pressure ratio between optimized and original compressor

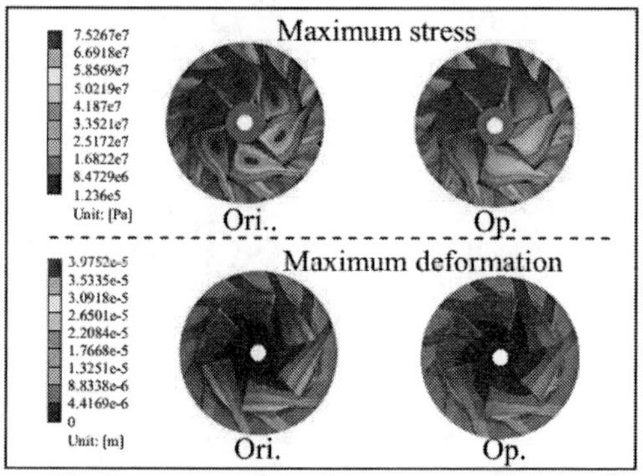

Figure 12. Comparison of maximum stress and deformation between optimized and original compressor

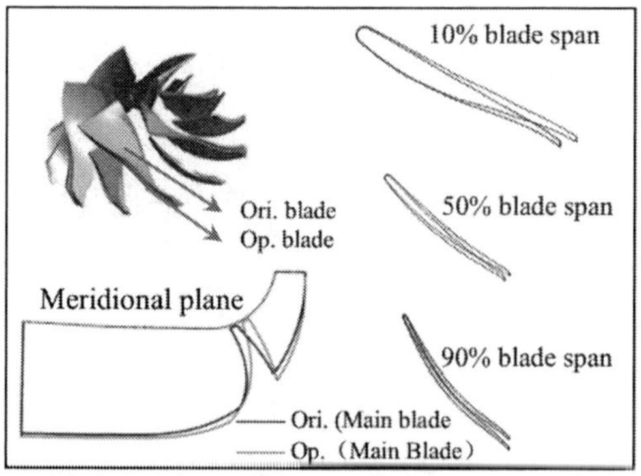

Figure 13. Comparison of blade profile between optimized and original compressor

The comparison of blade profile between original and optimized model is shown in Fig. 13: under design speed, the blade angle on the leading edge of optimized full blade is reduced, in order to meet the flow inlet angle and reduce the incident loss; after optimization the bottom of the trailing edge protrudes slightly backwards compared to the tip of trailing edge and the tile angle becomes small, which is results in increased pressure as well as isentropic efficiency (This conclusion is supported by sensitivity analysis). The middle of the optimized full blade is slightly inclined forward, so it can avoid low momentum fluids entering into the blade clearance as a result of centrifugal force to affect the increase of pressure. The excessive oblique of the trailing edge can not only affect the mechanical strength and it also leads to a decrease of the effective working area and an increase of force component for fluid in the passage, which is perpendicular to the blade; the tile angle of the optimized splitter blade is greater compared to the original model.

4 1D analysis of influence of optimized compressor on motor performance

4.1 Comparison of compressor operating map

Finally, the characteristics under each working conditions of the whole compression system are calculated. Figure 14 is the comparison of isentropic efficiency between original compressor and three optimized compressors. It can be seen that when the compressor produces pressure ration less than 1.8, the isentropic efficiency of all the three optimized compressors' are significantly higher than the that of the original compressor and the highest isentropic efficiency respectively is increased by 0.34%, 0.29% and 0.2%. However, when the operating point is located above the blue line, the original compressor shows better compression effect.

4.2 1D simulation model und comparison of results

To analyze the influence of the optimized compressor on engine performance and quantify the improvement resulting from the optimization of compressor, this article will conduct experimental measurement of the engine, with which the original turbocharger is fitted as well as a sequent one dimensional simulation.

Figure 15 modularly demonstrates the GT model of turbocharged engine. The simulated object is a four-stroke 2L-diesel engine with four cylinders. In this model, the experimental data replaces combustion model and the pressure curve caused by inter combustion in cylinder, which are determined by engine speed and torque. Injection timing and its duration, the trajectories of the intake valve and exhaust valve are modeled by experimental data. Boost Pressure, speed of the compressor, high- and low-pressure exhaust gas recirculation rates and exhaust gas flow are adjusted by control system, to make the adjustment value reach the target. This paper shows steady simulation of the engine under full load and transient simulation under acceleration and low load, and took the original engine's performance as a benchmark, making the degree of optimization of back pressure in the engine, compressor speed and pumping loss under different conditions as the research object. Figure 16 shows the influence of optimized compressor on its rotational speed under full load condition. The vertical coordinate illustrates the ratio of speed reduction after optimization to the original model. It can be clearly seen that except for the fourth and fifth operating points, the speed of Op1 is lower than that of original compressor.

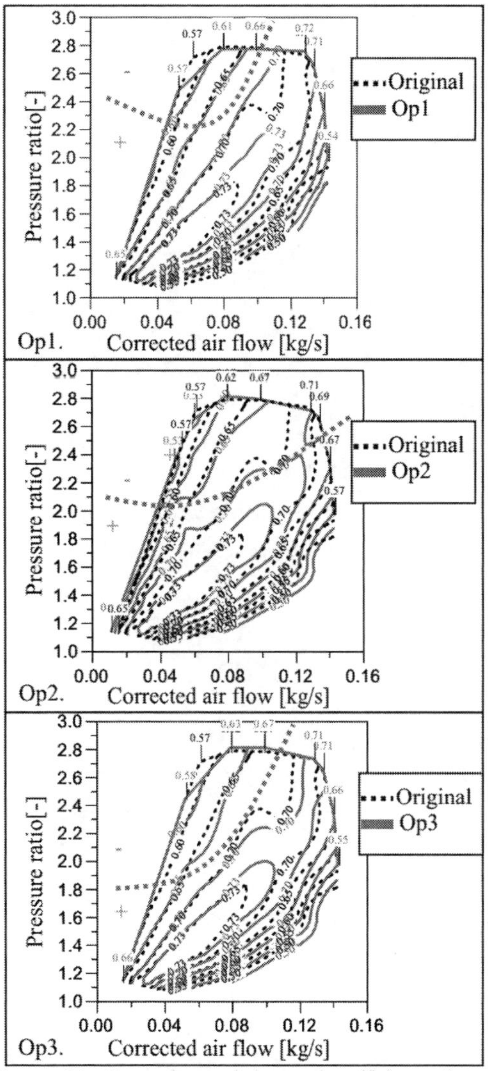

Figure 14. Comparison of isentropic efficiency between original compressor and three optimized compressors.

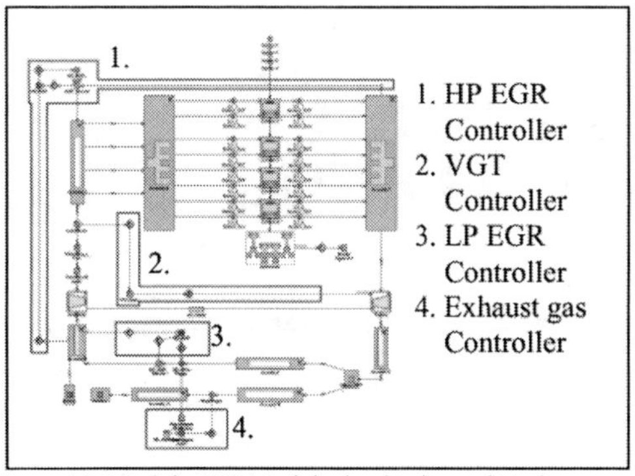

Figure 15. Modular GT-Model of turbocharged engine

DOI: 10.1051/matecconf/201710808012

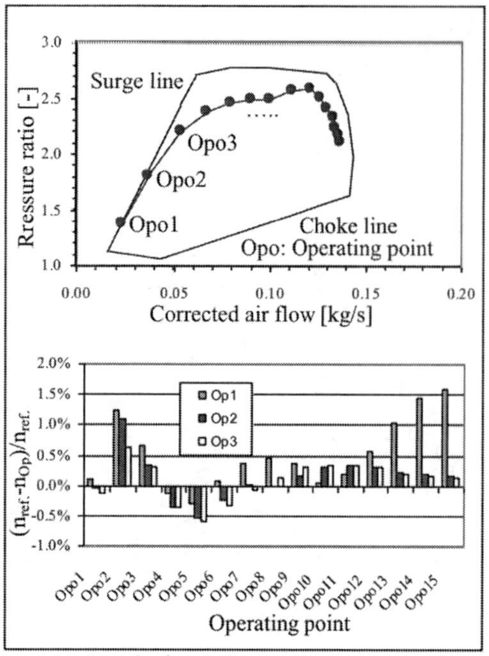

Figure 16. Speed of optimized compressor under full load condition

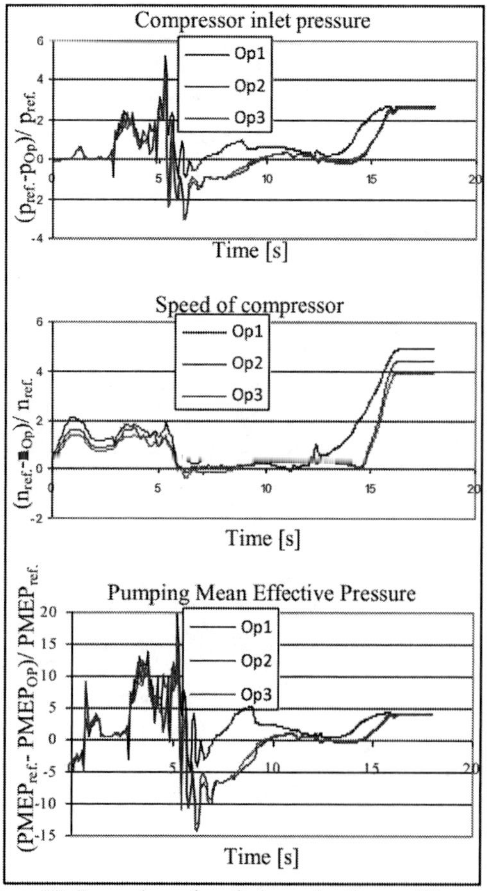

Figure 17. Comparison of results in acceleration condition

The increase of isentropic efficiency at these operation points also results in a reduction in the back pressure as well as in the load on the turbine blade. Figure 17 and 18 show the effect of three optimized compressors on engine performance in terms of acceleration and reaction under low load conditions. It can be seen that Op1 has a higher degree of optimization in the back pressure, compressor speed and the pumping loss.

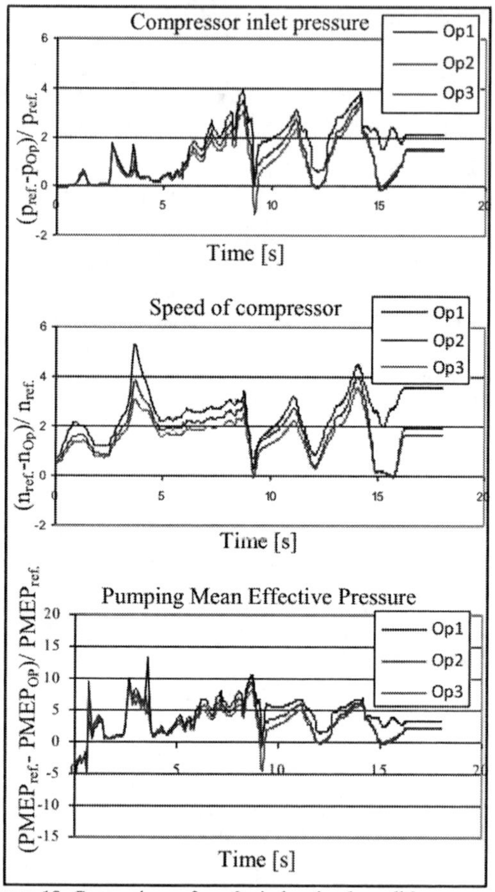

Figure 18. Comparison of results in low load condition

5 Conclusions

The multi-objective geometry optimization on compressor impeller was carried out through the coupling of CFD-simulation with Optimization method. The optimization process is as follows: conduct parametric modeling on impeller blade shape and hub curve by Bezier curve successful first, and verify the accuracy of three-dimensional CFD model, then using each module function in optimization software optiSLang. Using the Latin Hypercube method in optiSLang to establish the simulation sample set permits the advantages of less sample numbers and a more representative sampling. Through the sensitivity analysis of the results in aerodynamic and mechanics, the mapping relationship between the optimized parameters and the optimization objectives are established and the approximate model is obtained. Using the genetic algorithm in optiSLang to optimize and verify the model, finally the optimization model is compared with the original model, and an analysis and summary are carried out. The whole optimization system based on ANSYS and optiSLang coupling has features of simple operation, high efficiency, etc.

The aerodynamic performance and security of the optimized model were greatly improved. The optimization degree of isentropic efficiency and compression ratio was affected by different operating conditions. The maximum stress and maximum deformation of the impeller were also under control. A lower quality impeller will undoubtedly reduce its rotational inertia, and contribute to reducing the hysteresis phenomenon. Under the same condition of operation, the required speed of the optimized compressor is decrease. The pumping loss is reduced and the engine efficiency is increased. The optimized model has an overall smaller isentropic efficiency under high pressure ratio conditions.

References

1. M. Geller, C. Schemmann. Geometrieoptimierung eines Radialkompressorlaufrades mit Hilfe automatisierter Optimierungsmethode. 8. Weimarer Optimierungs- und Stochastiktage. 6 (2011)
2. W. Zhao, J. Sheng, J. Yang, J. Song. Optimization design and experiment of centrifugal pump based on CFD. Transactions of the Chinese Society of Agricultural Engineering (Transactions of the CSAE). 6 (2015)
3. Th. Hildebrandt, I. Gresser, M. Sievert. Kennfeldverbreiterung eines Radialverdichters fuer Abgasturbolader durch multidisziplinaere CFD-FEM-Optimierung mit FINETM/Turbo. 16.Aufladetechnische Konferenz. (2011)
4. Dynardo GmbH. Methods for multi-disciplinary optimization and robustness analysis. Weimar. February 21 (2016)
5. B. Liu, Q. Yang, Y. Cao, P. Zhang. Optimization Design of a Centrifugal Compressor with Splitters. Journar of propulsion technology. **35**. 7 (2014)
6. J. Will. OptiSLang inside ANSYS Workbench. Confidence by Design. Houston. June 20th (2012)
7. Dynamic Software and Engineering GmbH. Sensitivty Analysis. Optimization and Robust Design with optiSLang and ANSYS Workbench 12.Tutorial

MATEC Web of Conferences **108**, 09001 (2017) DOI: 10.1051/matecconf/201710809001
ICMAA 2017

A Study on FIDO Authentication System for Reinforcing the Security of Electronic Medical Records

Sujin Kim , Jinwoo Jung and Jungduk Kim

Security Convergence Department, Chung-Ang University, Seoul, Korea

Abstract. The target of compulsory certification in Information Security Management System has extended to medical institutions. This caused us to recognize the importance of information security in modern hospital information system that has changed from the medical record management that was recorded and managed largely in paper chart in the past to the Electronic Medical Record that medical personnel enter patient information into a computer directly for building a database. As medical institutions manage sensitive information like personal information basically, personal medical data infringement accident, if occurred can become a big social issue. Currently, the medical information in medical institutions are stored in electronic medical records and to access, user authentication is required by means of accredited certificate as security measure. Accredited certification has technical problems such as certificate storage method and security level of password and managerial problems such as certificate copy/leak/share. In this respect, this study proposes and presents how to build the FIDO-based authentication system that applies UAF or U2F authentication mechanism depending on the authority and work scope of medical personnel and medical support assistant like staffs, officers, licensed practical nurse and so on, within large medical institutions that use medical information system. The aim is to solve problems in accredited certificate authentication method in the existing medical institutions with the FIDO-based authentication system proposed in this study.

1 Introduction

Medical information management in medical institutions has changed from the medical record management that was recorded and managed largely in paper chart in the past to the Electronic Medical Record that medical personnel enter patient information into a computer directly for building a database. This caused people to keep raising the necessity of applying information security management method that was thought to be important in the existing companies and/or governmental agencies to medical institutions [1].

Information Security Management System authentication system is a system that screens information security management system that is established, operated, and managed to protect major information assets of companies and agencies from different kinds of threat [2]. Since medical institutions deal with sensitive information, especially health status, medical record, and other personal information, personal medical information infringement accident due to hacking, exposure to insider, malicious data leak and so on, can become a big social issue [3].

For this reason, the government revised the 'Act on Promotion of Information and Communications Network Utilization' in June to expand the target of compulsory certification in Information Security Management System authentication system to medical and educational fields. Thus a total of 43 large hospitals were included into the target of compulsory certification in Information Security Management System.

In line with this, medical institutions also tend to recognize the importance of information security. Currently, the medical information in medical institutions are managed through electronic medical records and as means for security authentication to access this, medical personnel use accredited certificate method. Medical institutions are using accredited certificate to acquire the legal force of digital signature and the reliability of records. Indeed, it has technical problems such as authentication certificate storage method and program installation and managerial problems such as certificate copy/leak/share [4].

This study provided the FIDO-based authentication system that applied UAF or U2F authentication mechanism depending on the authority and work scope of medical personnel and medical support assistant within large medical institutions, in other words, secondary and tertiary institutions that used the medical information system and proposed how to build such a system. This study aims to solve the problems of accredited certificate authentication method in the existing medical institutions by using the certification system proposed in this study.

2 Literature

2.1 Concept Range Setting of Electronic Medical Record and Medical Treatment Information

Electronic Medical Record (EMR) is a form that medical records recorded by the medium of paper are improved equally through computerization without any transformation of business transaction method, scope of information, and information content. In other words, it indicates a system that medical information such as business data, treatment, and surgery centering on patient treatment are input and stored into a computer [5].
The definition of medical information does not exist clearly in the current laws and regulations. In addition, most medical information do not have clear definitions and regulations although they recognize medical information as sensitive one. Instead, 'personal medical information' is used to indicate medical practice related information that medical personnel and medical treatment institutions use for the purpose of national health care and promotion in general and also understood as encompassing not only the data collected across the entire process of medical practice but also the information studied based on this [6].

This study defined the concepts of medical information that were not clearly defined as 'sensitive medical information' and 'universal medical information' by using the comprehensive concept of 'personal medical information' to identify and improve the problems in security of electronic medical records.

'Sensitive medical information' is defined as indicating the information that such medical personnel as doctor and nurse who treat patients among many health care providers for patients have access and use.

'Universal medical information' is defined as indicating the information that nurse assistants and individual responsible for medical support assistants who are in charge of office work and administration within medical institutions and do not engage in medical treatment have access and use for medical support assistant and business transaction purposes only, in other words, relatively less sensitive medical information rather than 'sensitive medical information.'

2.2 Authentication Certificate-based Certificate System Using Electronic Medical Records

In accessing the electronic medical records, one's identity is checked by verifying digital signature generated by using medical personnel's private key. If digital signature is done in accessing the electronic medical records, the system determines whether to approve or not by verifying the validity, status, and digital signature within the time limit of the accredited certificate [7].

However, authentication certificate-based authentication system has caused many inconveniences for medical institution users due to many other security limitations. Moreover, with the increasing number of various security related issues, there is a need to introduce a new certification system. The problems of the authentication certificate identified through previous related studies are analyzed in the table below.

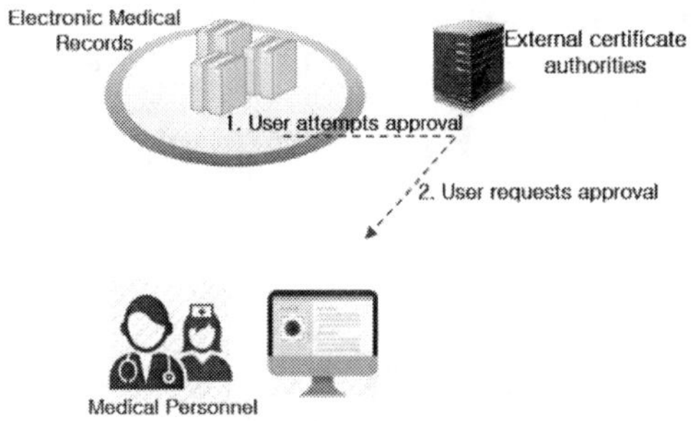

Figure 1. Authentication system based on PKI in medical institutions

Table 1. The problem of PKI based authentication system

Field	Item	Problems
Technical	Authentication certificate storage method	Authentication certificate is stored in non-standardized location (NPKI) where security is not secured[8]
	Certificate status check through external network	The validity of authentication certificate must be checked for every digital signature, but constraints still exist because EMR is a form of personal area network[9]
Managerial	Certificate duplication and leakage	As it is a form of file., anyone can copy and leak certificate private-key easily in the form of Cut & Paste even if it is not a security token[8]
	Password security level	Password that is made by using simple information or personal information that is easy to remember[9]
	Accredited certificate sharing	Vulnerable in managerial security as it is possible to share and use certificate without limitation in fact[10]

2.3 FIDO(Fast IDentity Online) Certification

FIDO is an abbreviation of Fast IDentity Online and indicates a technical standard of FIDO Alliance established in 2012 to support the safety, universality, and simplicity of user identity certification in online environment. FIDO Alliance enacted two standards: UAF, a certification mechanism by using individual user's unique biometric information and U2F, a certification mechanism to add secondary certification factor like security token including USB dongle that stores private key secondarily after getting an approval for the existing ID/password and announced on Dec., 2014.

The first certification mechanism is UAF(Universal Authentication Factor) and authenticates users by linking the authentication method provided in user's device with online service. UAF authentication procedure can be largely divided into twos: registration procedure that

 DOI: 10.1051/matecconf/201710809001

authentication token and public key are registered into FIDO server by executing biometric authentication via individual user's device and using generated key pairs and authentication procedure that the challenge value and authentication token transmitted when user requests login to server are transmitted to server after signature by using the private key stored in user's device and the values are verified by using the private key within server[11].

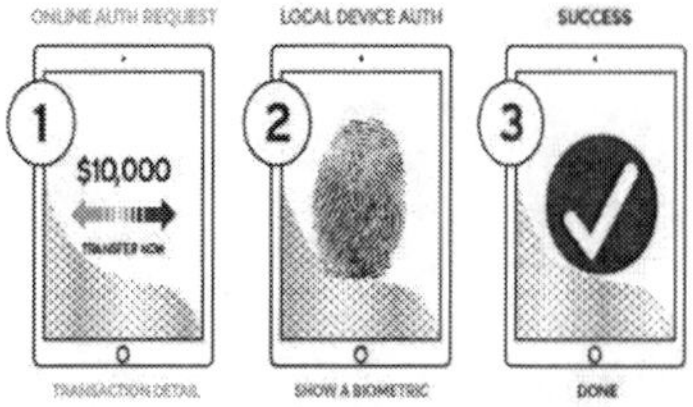

Figure 2. FIDO UAF Authentication Process.

The second is U2F(Universal 2nd Factor) and adds token (e.g. USB dongle) that applies encryption-based security technology in the online service that uses the existing ID/password as two-factor authentication factor. It identifies user by using the security module like security token and protects authentication value safely from various malignant codes by directly communicating the result values certified through PIN with web browser[12][13].

In addition, what is different from the existing authentication method is that authentication procedure cannot be performed if U2F dongle is not owned when other's password was seized. The currently released U2F dongle includes biometric authentication function and so if biometric information-based user authentication is not performed through the sensor fixated to dongle in case of connecting USB, dongle is not activated.

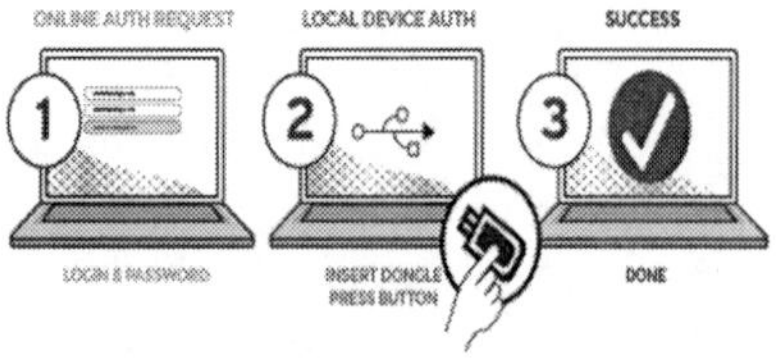

Figure 3. FIDO U2F Authentication Process.

The overall registration and authentication procedure of FIDO uses a public key-based encryption method similarly to the existing authentication certification system. However, the clear advantage of FIDO can be free from technical and managerial problems because it separates user verification from authentication process and uses biometric information when verifying user, but this biometric information is stored in user's device only without being stored and/or disperse-stored to server within service institutions like the existing Fast Identity Online. It is less likely to be abused because biometric authentication procedure must be preceded prior to FIDO server certification even if device was lost[14].

3 FIDO-based Authentication system

This study proposed a FIDO-based authentication system to improve the problems in the existing authentication system targeting large medical institutions.

3.1. FIDO-based Authentication System Electronic Medical Records System Building Plans in Large Medical Treatment-related Organization

In medical institutions, the FIDO-based authentication system building measure may depend on the size of medical institution, but this study selected large health organizations.

According to the Article 9 of the Medical Care Assistance Act, large medical institutions are considered as secondary if they hold over 30 nursing care beds and possess legal treatment department requirements and as tertiary if they hold over 500 nursing care beds and medical professionals for all treatment departments.
Such large medical institutions are recommended to build a FIDO authentication system separately and the specific measures are shown in the figure below.

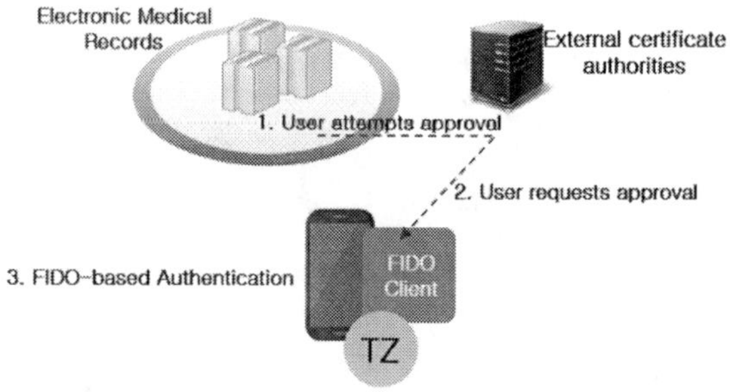

Figure 4. Proposed FIDO Authentication system.

The proposed FIDO authentication system applies public-key infrastructure (PKI), core technology of information security, performs digital signature within TZ(Trust Zone) safe against memory hacking to secure the reliability of user certification within various medical service environment, and supports strong security and non-repudiation against the existing authentication certificate environment.

It can also solve the problem of the present authentication certificate system, i.e. 'certificate check through external network' by locating FIDO authentication server within the personal area network. Given various environments and scales of domestic medical market, there are limitations in developing a PKI certification environment for medical purposes, and so building user certification system that observes the FIDO standards can be considered as reasonable measure[14].

3.2 FIDO-based Authentication System

Place the figure as close as possible after the point where it is first referenced in the text. If there is a large number

 DOI: 10.1051/matecconf/201710809001

of figures and tables it might be necessary to place some before their text citation. If a figure or table is too large to fit into one column, it can be centred across both columns at the top or the bottom of the page.

2.1.1 U2F-type Authentication System

U2F-type Authentication System targets medical personnel who need to access sensitive medical information like a medical treatment information and performs two-factor authentication by using the U2F token to store security key in the manner that authentication device is not included in private device. So it is expected to have an effect to reinforce the managerial security.

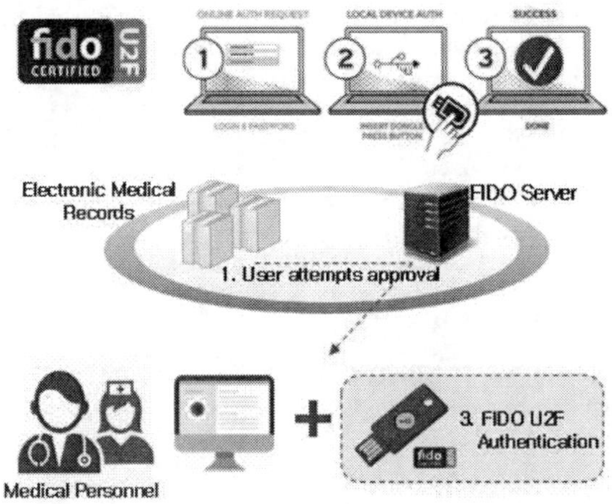

Figure 5. Proposed authentication system by FIDO U2F

If a medical personnel attempts approval to access sensitive medical information (treatment information), FIDO server requests user authentication to the medical personnel. Medical personnel perform FIDO authentication by using USB dongle that uses U2F after performing the existing ID/PW certification. This U2F dongle is attached with biometric certification sensor and the sensor is activated only by medical personnel who are approved or through biometric certification. This certification improves security by performing secondary certification with U2F dongle and at the same time clarifies accountability through biometric certification by using dongle.

2.1.2 UAF-type Authentication System

UAF-typed Authentication System targets medical institution employees including employees in the department of medical record and nursing assistant who need to access universal medical information(health information). The system uses the UAF-based biometric certification method that authentication device is included in private device and utilizes the mobile device into private device by considering rapid certification procedure and convenience, universality, and safety in access. The present FIDO 1.0 version is limited to mobile devices, but from the FIDO 2.0 version, FIDO UAF technology can be implemented in most devices like PC and tablet that are based on authentication through web.

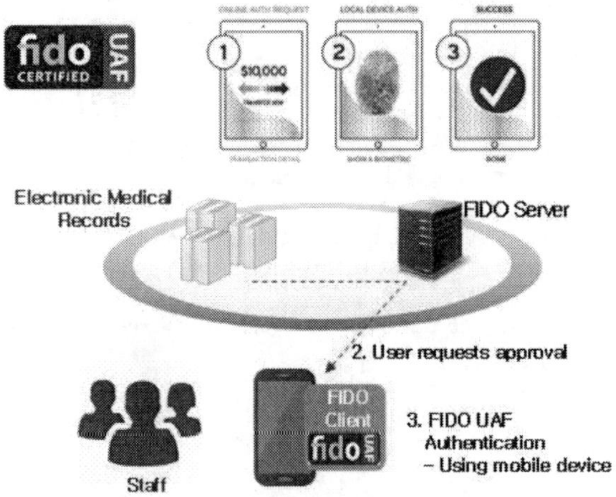

Figure 6. Proposed authentication system by FIDO UAF

If staff – employees in department of medical record and nursing assistant, and others – attempts approval for accessing the universal medical information (health information) rather than medical personnel within hospitals, FIDO server requests user approval to the staff. The staff performs biometric certification through personal device mounted with applications and FIDO clients and obtains certification from FIDO server if user identity is identified. As certification is proceeded by using the FIDO UAF-based private mobile devices, it provides immediacy, convenience, and universality. In addition, as certification is proceeded by using a personal device that is based on biometric certification, accountability can also be clarified.

4 Conclusion

The importance of medical data privacy protection has become an issue not only in laws and regulations but also in social dimensions. Currently, the medical information in medical institutions are computerized and stored in electronic medical records and accordingly for obtaining security and reliability, they use authentication certificate-based authentication system. However, the present authentication certificate-based authentication system has limitations in checking the status of authentication certificate by external network certificate authorities due to storage private key in places where security is not ensured, vulnerability in password security level in authentication certificate, and electronic medical records established in private networks. It also has managerial problems in that authentication certificate is easily copied and leaked because it is the form of a file and the non-repudiation is not valid actually as it shares the accredited certificate.

To solve these problems, this study proposed methods to introduce FIDO, the next-generation authentication system. FIDO-based authentication system uses public key-based authentication like the existing authentication certificate-based authentication system, but user verification and certification are separately proceeded and the biometric information used in generating private key for verification is not stored separately in server and rather stored within device only, which may increase the private key security. Also, differentiating the authority and role to access medical information system by using the FIDO-based authentication system and applying different UAF and U2F technologies can support strong security and non-repudiation and clarify responsibilities against the existing authentication certificate system. Furthermore, implementing simpler and stronger alternate authentication technologies in combination with biometric certification can provide convenience in user service access.

Nevertheless, the proposed FIDO-based authentication system is the one that proposed FIDO 1.0 technical specification and FIDO1.0 has limitations in that certification can be proceeded only by mobile devices if UAF technology is applied. This can be improved by the FIDO 2.0 to be enacted at the end of this year.

So, further studies need to verify case studies and empirical ones that addressed the actual application of the FIDO-based authentication system in the proposed medical fields.

References

1. J. Lee, H. Jo, S. Park, Y. Gang, KIISC Review **23**, 34 (2013)
2. J. Kim, K. Lee, KIISC Review, **18**, 1 (2008)
3. Y. Jung, Y. Lee, Health and welfare policy forum, **205**, 70 (2013)
4. C. Jung, *University of Soongsil, Seoul*, (2012)
5. Y. Jeon, The korean journal of health service management, **7**, 224 (2013)
6. G. Jung, Journal of korea information law, **6**, 3 (2002)
7. M. Fritscher, *Fifth Americas Conference on Information Systems*, 432 (1999)
8. G. Kim, People and ideas, **181**, 163 (2013)
9. Y. Lee, The KIPS transactionsty C, **18**, 379 (2011)
10. A. Salaiwarakul, M. Ryan, *IWSEC*, 231 (2008)
11. S. Kim, The Journal of The Korean Institute of Communication Sciences, **33**, 59 (2016)
12. FIDO Alliance, Universal 2nd Factor (U2F) Overview, 2015
13. FIDO Alliance, FIDO 1.0 Final Specifications have arrived, 2014
14. K. Lee, ICT standard & certification TTA journal, **165**, 12 (2016)

The Influence of Gate Scaling to Electrical Characteristics on n-MOS FinFET

Nuttapong Patchrasardtra , Weera Pengchan

Department of Electronics Engineering, Faculty of Engineering, King Mongkut's Institute of Technology Ladkrabang

Abstract. This paper investigates effects from gate scaling in Tri-gate FinFET structure by simulation method, to avoid problems and improve a structure to be good prototype. The experiments used GTS framework for simulation. Start from 20 nm device, then scaling to 22 nm 28 nm and 32 nm. Therefrom Minimos-NT function has used for biasing to giving two electrical characteristics as the drain current saturation and the threshold voltage. From these consequences can offer the subthreshold swing and the drain-induced barrier lowering by calculation. The results found that threshold voltage inversely proportional to saturated drain current, the subthreshold swing and the drain-induced barrier lowering. The short channel effect has affected to 20 nm model by highest DIBL. Therefore should be adjust the gate length and the oxide thickness properly to improve this effect.

1 Introduction

The introduced of Tri-gate transistor technology by Intel corporation in 2nd quarterly of 2011. Intel brought the new technology to giving good performance for them devices. 3-D transistor as known as FinFET. This is a choice to make new computer architecture. In 2011 M. Zakir Hossain and his researcher team shown electrical characteristics of SOI tri-gate FinFET. The result of this paper explain about basic characteristics such as the current and voltage characteristics (I-V), the threshold voltage, the electron mobility and field mobility [1]. In 2012 they proposed about the drain-induced barrier lowering (DIBL) and short channel effect (SCE). Influence of DIBL and SCE is coming together. DIBL make a drain to gate electron barrier lower. In a same way I-V and threshold voltage will strange from theory, and downward a subthreshold characteristics. It make device cannot change to cut-off stage at low voltage [2]. This concept make author interested in effects from scaling a device to solve problems for improved a structure of FinFET prototype and fabrication in the future.

2 Theory

2.1 Saturation Drain Current ($I_{ds,Sat}$)

While the drain current has no longer affected by the drain voltage, and where the FinFET acts more like a current source, has called the current saturation. This current value can be solve by [3]:

$$I_{ds} = \mu C_{ox} \frac{W_g}{L_g} \left(\frac{v_{gs} - v_{th}}{2m} \right)^2 \quad (1)$$

W_g, gate width (nm) L_g, gate length (nm) C_{ox}, Gate Oxide capacitance (F/nm)

2.2 Threshold Voltage (V_{th})

The minimum gate-to-source voltage differential that was needs to create a conducting path between the source and drain. This is the most significant characteristic that provide the value of power that used to switch the FinFET from off stage to on stage and show the switching performance of devices. The threshold voltage can giving from [4]:

$$V_{th} = \Phi + \frac{kT}{q} \ln \left(\frac{2C_{ox}kT}{q^{W} n_i t_{si}} \right) + \frac{h^2 \pi^2}{2mW_{si}^2} \quad (2)$$

n_i, Electron and Hole Concentration without Doping (cm^{-3}) t_{si}, Silicon Thickness (nm) h, Channel Height (nm) w_{si}, Silicon width (nm) η, Carrier Concentration (cm^{-3}).

2.3 Subthreshold Swing (SS)

The subthreshold swing is logarithm of drain current relationship of drain current and gate voltage that

DOI: 10.1051/matecconf/201710809002

determine the voltage per decade. The solution to find a subthreshold swing, can calculate with equation (3) [5] by methods in Figure 1.

$$SS = V_{gs2} - V_{gs1} \quad (3)$$

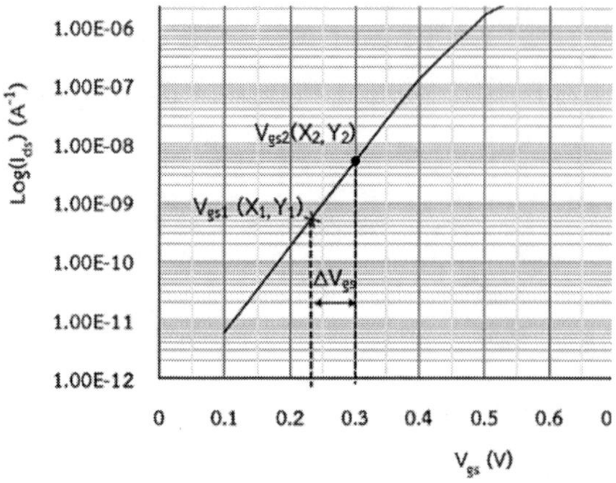

Figure 1. Relationship between the drain current and the gate-source voltage

2.4 Drain-Induced barrier lowering (DIBL)

The reduction of threshold voltage at higher drain voltages by the lowering of potential barrier on drain that affect to gate and Solve by [6]

$$DIBL = \frac{V_{th} - V_{th}^{Low}}{V_{dd} - V_{dd}^{Low}} \quad (4)$$

3 Experimental

This paper separated to two experiments part consist of the FinFET structure and the relationship between the electrical characteristics with the gate length. The first part, create a device structure on GTS framework [7] from 20 nm gate length and then increase L_g to 22 nm, 28 nm, and 32 nm 10 nm silicon layer width and 25 nm silicon thickness with 1 nm gate oxide thickness on the top of 70 nm p-type substrate with buried oxide (BOX). In second part, Minimos-NT has use for biasing to obtain the electrical characteristics by determined the drain voltage at 1 volt and sweep the gate voltage from 0 to 1 volt (0.1 volt per steps) for harvest the threshold voltage. Collect the I-V characteristics by fixed the drain voltage at 1 volt and vary the gate voltage from 0 to 1 volt (0.1 volt per steps) and calculate the saturation regions of drain current. After collected V_{th} and I-V, calculate the subtheshold slop from $\log(I_{ds})/V_{gs}$ and DIBL from V_{th} finally compare with gate length to giving the results of gate scaling effect.

4 Result and Discussion

4.1 FinFET Structure

From the first experiment part, created a FinFET structure that consist of the substrate, BOX, the silicon layer, channel, gate oxide and gate contact shown in Figure 2. (a) Side view of device with the gate length. (b) Cross section shown a fin width under oxide and contact, all of those are on buried oxide and the substrate. In this structure, determined the dopant concentration on channel at 1.5×10^{15} cm^{-3} and 3×10^{20} cm^{-3} on drain-source with poly-silicon gate contact therefore this part will provide four structures with difference of the gate length.

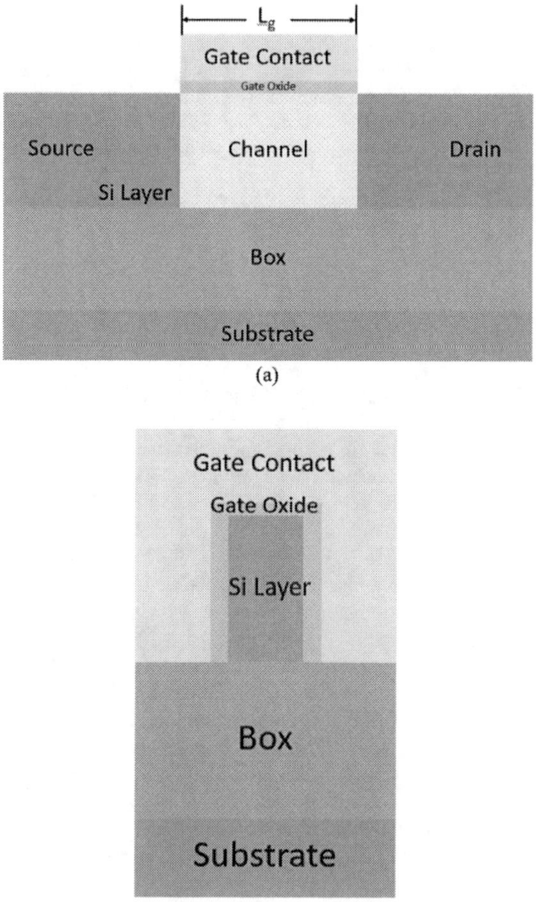

(b)

Figure 2. FinFET Structure (a) Side view (b) Cross Section

4.2 Characteristics with gate length relationship

The electrical characteristics of devices is the important point to determine a performance. From second part of experiment obtain the drain current saturation, the threshold voltage for calculate the subthreshold swing and the drain induced barrier lowering, to compare-between characteristics and enlargement of gate length.

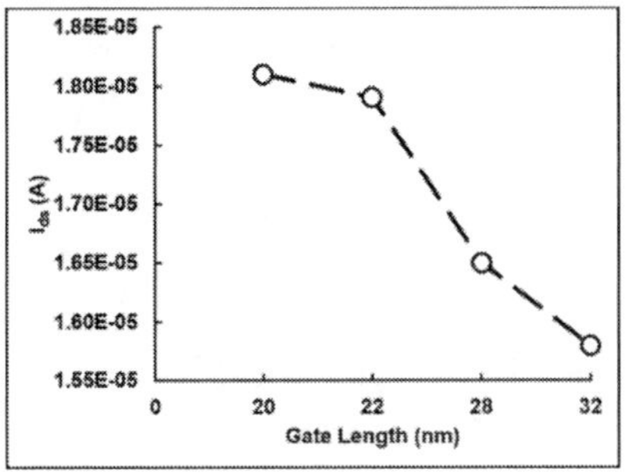

Figure 3. Drain current saturation with gate length

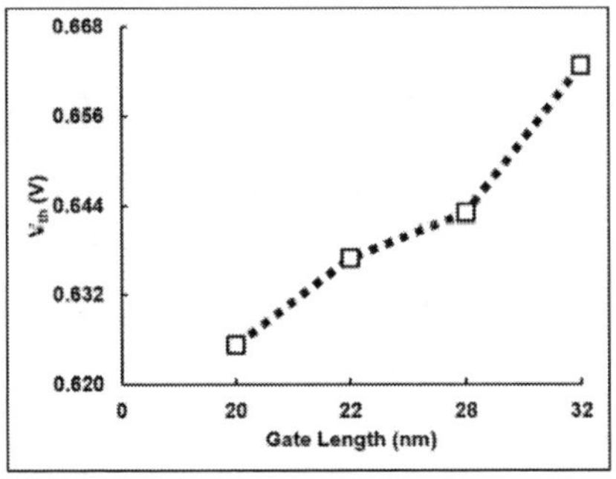

Figure 4. Threshold voltage with gate length

An enlargement of gate length determined a saturation current in Figure 3. The deportment of I_{ds} was lowering with increase of L_g. Due to adding of electron concentration in channel that affect to them mobility. This effect bring I_{ds} downward about 0.1 μA per increasing of L_g. It was giving I_{ds} on 20 nm ate length at 18.10 μA, 22 nm at 17.90 μA, 28 nm at 16.50 μA and 32 nm at 15.80 μA. Therefore $I_{ds\ (Sat)}$ behaviour on device with 20 nm L_g is better. In Figure 4, the threshold voltage has increased with L_g but after biased I_{ds} / V_{gs} giving amiss result of 20 nm. It was coming with a little SCE because in this experiment vary only L_g. The threshold voltage of 20 nm gate length about 0.625 volt, 22nm at 0.636 volt, 28 nm at 0.643 and 32 nm gate length has 0.662 volt thus SCE affect to this characteristics directly while gate length was increased. Normally, V_{th} will increasing with L_g since the gate length has increased and require more energy to turn device stage to turn-on.

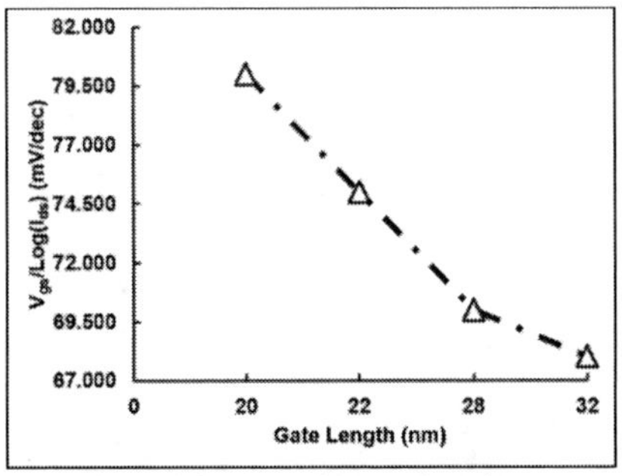

Figure 5. Subthreshold Swing with gate length

In Figure 5 shown the subthreshold swing has decreased by the gate length increasing, as more drain current passing in shorter L_g. From logarithm square root of drain current and gate-source voltage provide the SS on 20 nm L_g = 80 mV/dec, 22nm = 75 mV/dec, 28 mV = 70 and 32 nm gate length at 68 mV/dec.

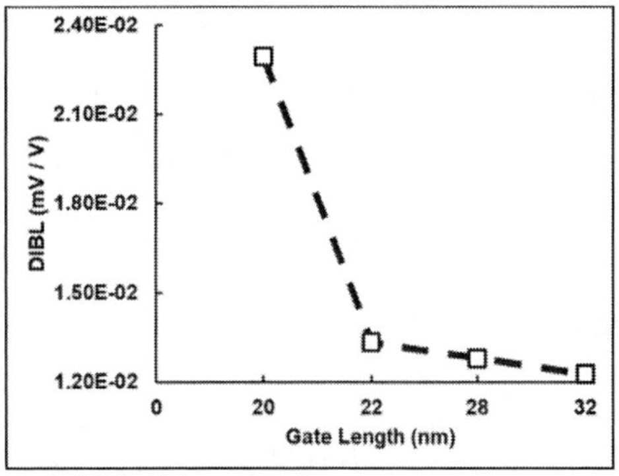

Figure 6. DIBL with gate length

The subthreshold swing is important attribute to obtain DIBL with gate length that shown in Figure 6. Influence of DIBL was directly to SS from lowering of barrier between channels and drain junction that affect to threshold voltage. DIBL on 20 nm gate length was about 23 mV/V, 22 nm at 13.30 mV/V, 28nm at 12.80 mV/V and 32 nm L_g approximate about 12.30 mV/V. The DIBL giving to known a problem from SCE. If it was high, this device will be out of characteristics.

Table 1. Electrical characteristics of FinFET

Gate Length (nm)	$I_{ds(sat)}$ (μA)	V_{th}(V) (V_{dd} = 1 V)	SS (mV/dec)	DIBL (mV/V)
20	18.10	0.62	80.00	23.00
22	17.90	0.63	75.00	13.30
28	16.50	0.64	70.00	12.80
32	15.80	0.66	68.00	12.30

From Table 1, Increasing of gate length will bring the threshold voltage up but the drain current, the subthreshold swing and DIBL are inversely proportional. The highest DIBL of 20 nm giving a least V_{th} that affected by short channel effect not from decreasing of gate length.

5 Summary

Investigate into FinFET structure simulation is best way to understand operating of device. Many of problems has found from this experiment. In this article provides experiments to bring up effects from designed device structure, which are two effects as the short channel effect and the drain-induced barrier lowering. From the result found relationship between the drain current saturation, the threshold voltage, the subthreshold swing and the drain-induced barrier lowering. Increasing of gate length directly with V_{th} but reverse with $I_{ds,sat}$, SS and

DIBL. The short channel effect has found in 20nm model therefore an adjustment of gate length will reach electrical characteristics suitability.

Acknowledgement

The authors would like to thank Electronic Research Center for many discussions and appreciate for software support from Global TCAD Solution.

References

1. M. Zakir Hossain and Md. Alamgir Hossain, Electrical Characteristics of Trigate Finfet Global, *Journals of researches in Engineering Electrical and Electronics Engineering* **11**, 8-10 (2012)
2. M. Zakir Hossain and Quazi D. M. Khosru Threshold Voltage Roll-Off Due to Channel Length Reduction for a Nanoscale n-channel FinFET, *IJETCAS* **13-125,** 152-156 (2012)
3. Viranjay M. Srivastava and Setu P. Singh, Analysis and Design of Tri-Gate MOSFET with High-k Dielectrics Gate I.J. *Intelligent Systems and Applications* **5**, 16-22 (2012)
4. Nour El Islam and Boukortt Baghdad and Hadri Salvatore Patanè, Effects of High-k Dielectric Materials on Electrical Characteristics of DG n-FinFETs, *International Journal of Computer Applications*, **139-10,** 28-32 (2016)
5. Isabelle Ferain Cynthia A. Colinge and Jean-Pierre Colinge, Multigate transistors as the future of classical metal–oxide–semiconductor field-effect transistors, *Nature* **479**, 310-316 (2011)
6. Yannis Tsividis, *Operational Modeling of the MOS Transistor*, 268 (1999)
7. Nuttapong Patcharasardtra and Weera Pengchan, Electrical Characteristics of Different Gate Geometry of FinFET, *Key Engineering Materials* **705**, 174-178 (2016)

Full 3D Simulation of the New Closed Shell-Electrode Detector

Mingfu Feng[1,2] , Zheng Li [1,2], Manwen Liu[1,2], Lipeng Tang[1,2]

[1] *Engineering laboratory of Special Functional thin film Materials, School (Faculty) of Materials and Engineering, Xiangtan University, Xiangtan 411105, China*

[2] *Center for semiconductor Particle and photon Imaging Detector Development and Fabrication, Xiangtan University, Xiangtan 411105, China*

Abstract. A new structure of 3D detectors has been proposed. In order to separate it from the non-etch-through 3D-Trench electrode detectors, we call it as the Closed Shell-Electrode Detector (CSED, Chinese Patent #ZL201620361767.1). The detector concept of the CSED will be described in detail here. Full 3D simulations of the performance behavior of the CSED will be carried out and presented. These simulations include detector potential, electric field, and electron (or hole) concentration profiles, as well as detector leakage current, capacitance, and charge collection properties. Comprehensive comparisons between the CSED and the non-etch-through 3D-Trench electrode detectors will be made. The novel CSED has much better electric field profiles near the backside and are much better isolated from neighboring cells than that in non-etch-through 3D-Trench electrode detectors.

1. Introduction

Detectors based on semiconductor materials have been used in various fields such as deep space imaging, medical imaging (e.g., positron emission tomography), tracking in high-energy physics, radiation-trace detection for the purpose of national security, etc. In the LHC (Large Hadron Collider) and HL-LHC (High Luminosity-LHC) at CERN, for example, the total radiation fluence can reach up to $1x10^{16} n_{eq}/cm^2$[1]. In such high radiation environment, high radiation will cause displacement damage in Si, which in turn will increase the defect concentration in the detector. Effects of such defects will change detector electrical properties such as increase of detector leakage current and effective space charge, and reduction in detector charge collection efficiency. The interest for 3D column electrode silicon detectors is continuously growing because of their many advantages as compared to conventional planar detectors due to the separation of detector electrode spacing from the detector thickness. For these detectors, if the column spacing is made much smaller than the detector thickness, they can be fully depleted at much lower bias voltages and exhibit higher radiation tolerance, as compared to planar detectors. However, the disadvantage of such detectors is the distribution of non-uniform electric field and the existence of vast low-field region[2, 3].

To improve electric field distribution and charge collection, a new type of 3D detector: 3D-trench electrode detector has been proposed at BNL in 2009[4], as shown Figure.1[1, 2, 5, 6]. The 3D-Trench Electrode detector not only can be radiation tolerant, but also it has more uniform electric field distribution as compared to the 3D column electrode detector. However, this 3D-Trench Electrode structure also has a deficiency: the trench electrode can not be fully etched through (or else the detector cell enclosed by the trench electrode will fall off the wafer). Although a support wafer or SOI technology may be used to overcome this deficiency[5], it will significantly increase the detector fabrication cost and complexity. The part of non-etching-through region (typically 10% of the detector thickness) gives rise to the low electric field region in the detector and provide an electronic connection between neighboring cells in the detector, thus compromising the detector charge collection efficiency and cell isolation. In order to overcome this deficiency in the non-etch-through 3D-Trench Electrode detector, a novel structure has been proposed in Xiangtan University (XTU) in 2016. To separate it from the non-etch-through 3D-Trench Electrode detectors, we call it as the Closed Shell-Electrode Detector(CSED) (Chinese Patent #ZL201620361767.1, Pub.Date: Aug. 31, 2016).
This article is mainly focused on the description of the

CSED structure concept and 3D performance simulations, including 3D distributions of electric potential, field and hole (electron) concentration; detector electrical properties such as detector leakage current (*I*) and capacitance (*C*) as a function of bias voltages (*V*) (*IV*, *CV*); and detector full depletion voltage.

2. Device description and simulation

For comparison purposes, a single cell of a non-etch-through square 3D-Trench Electrode detector is shown Figure.1. The non-etch-through portion is clearly shown in the bottom of the cell, which typically occupies 6% - 15% of the cell volume.

A single cell of a novel closed-shell electrode silicon detector with square shape is illustrated in Figure. 2. As shown Figure. 2a, this structure is composed of two different structures, square and octagonal, interleaved to each other. In our case shown here, the square trench is etched from the top side (from top down), but not etch through with a depth of $d_{top}<d$, where d is the thickness of the detector (typically $d \leq 500$ μm). The octagonal trench is etched from the bottom side (from bottom up), again not etch through (with a depth of $d_{bott}=\delta d$, $0<\delta<1$) but has an overlap in the thickness direction with the square trench. The depth of overlap part is defined as εd, where $0 \leq \varepsilon << 1$. As shown in Figure. 2b, the top and bottom trenches are disconnected in four corners. It is clear that we have the following relationship $d_{top}=(1-\delta+\varepsilon)d$ between various depths. The central column electrode can be etched all the way through the detector thickness, either from top down or bottom up depending on the specific processing need. The top-etched and bottom-etched trenches are doped the same way (in our case phosphorous or n^+-type), and the central column is doped in opposed way (in our case boron or p^+-type). In our example for simulations, Si bulk is chosen as p-type with a doping concentration of 1×10^{12} cm^{-3}, the edge length of the trench electrode is set at 100 μm, the width of all trenches are set at 10 μm, the width of the square central column is set at 10 μm, d =300 μm, δ =0.1, and ε=0.067.

Due to the arrangement of the top and bottom interleaved trenches, their enclosed cell bulk will not fall off the wafer, even though the overall trench electrode is effective "etch-through" with top and bottom trenches are connected in large areas (they are not connected in four corners) as shown in Figure. 2b. It is clear that, the larger the cell size, the bigger the connected portion of the top and bottom trenches. Neighboring cells are connected to each other and to the detector main body through the unconnected portion, and therefore will not fall off the wafer. The overlapping of the top and bottom trenches in the thickness direction ensures the electronic connection between the neighboring cells being minimum to achieving effective cell isolation.

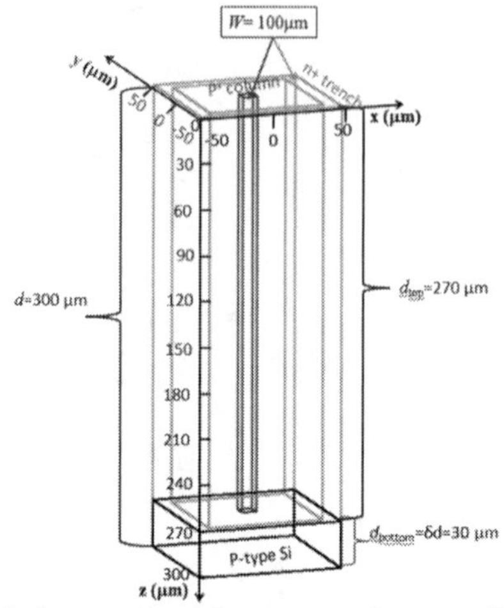

Figure 1. An example of the structure of a square 3D-Trench Electrode Silicon detector

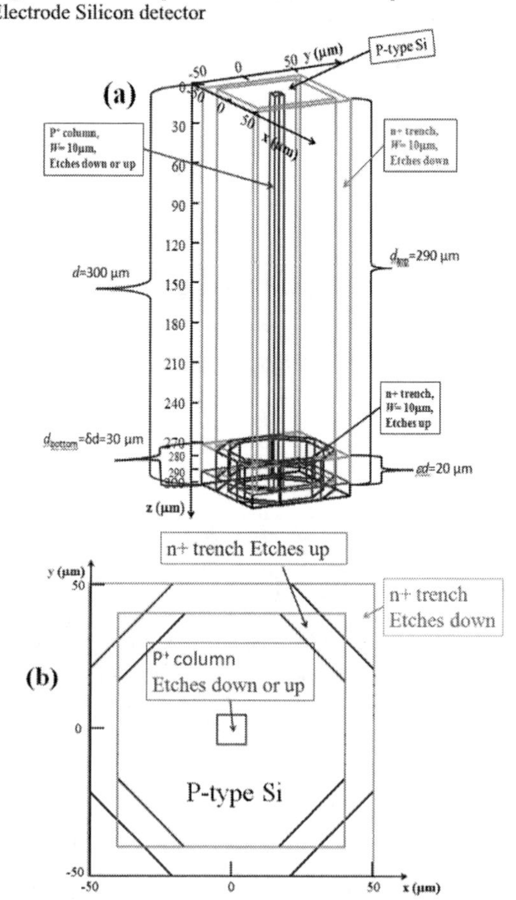

Figure 2. An example of the structure of a square shape novel CSED structure: a) the 3D view of the entire single cell, and b) the cross section view (x-y plane, at z = 270 μm)of the CSED in the portion of the interleaved trenches.

Full 3D simulations on a Closed-Shell Electrode Silicon detector shown in Figure.2 have been performed using the DEVICE 3D package of Silvaco's ATLAS. The n^+ outer trenches are doped with phosphorus of $10^{19}/cm^3$, and the p^+ central column is doped with boron of $10^{19}/cm^3$.

The choice of p-type bulk is for the radiation hardness consideration since under the high radiation influence, n-type silicon will space charge sign invert to p-type like[6]. Here, the p-n junction is formed between the bulk and the outer ring trench (ORJ, or outer-ring-junction). Compared to the case where the junction is formed at the center column electrode (CJ, center-junction), detectors with ORJ have more uniform electric filed profile and much smaller maximum electric field and full depletion voltage [1, 6]. From ref. [1], in fact one can write down the electric field profiles beyond full depletion for the perfect cylindrical geometry as the following:

$$\left.\begin{aligned} E(r) &= \frac{1}{2}\frac{eN_{eff}}{\varepsilon\varepsilon_0} r\left[1-\frac{r_c^2}{r^2}\right] - \frac{V-V_{fd}}{r\ln R/r_c} \quad (r_c \le r \le R, ORJ) \quad a \\ E(r) &= \frac{1}{2}\frac{eN_{eff}}{\varepsilon\varepsilon_0} r\left[\frac{R^2}{r^2}-1\right] + \frac{V-V_{fd}}{r\ln R/r_c} \quad (r_c \le r \le R, CJ) \quad b \end{aligned}\right\} \qquad (1)$$

Where Eq. (1)a is for the ORJ case and Eq. (1)b is for the CJ case, V_{fd} is the full deplete voltage, N_{eff} is the effect doping concentration, R is inner radius of outer trench (in this case, it is the radius of the fully depleted region), r_c is the radius of central column electrode, V_{bi} is the built-in potential, ε_0 is the vacuum dielectric constant, and ε is the relative dielectric constant for silicon. We can use Eq. (1) to qualitatively estimate the electric filed profiles in our non-perfect square and octagonal cases before full 3D simulations. For example, at full depletion voltage $V = V_{fd}$, the maximum electric field is at $r = R$ for the ORJ case at $E(r) = \frac{1}{2}\frac{eN_{eff}}{\varepsilon\varepsilon_0} r\left[1-\frac{r_c^2}{r^2}\right]$, which is much smaller than that for the CJ case at $E(r) = \frac{1}{2}\frac{eN_{eff}}{\varepsilon\varepsilon_0} r\left[\frac{R^2}{r^2}-1\right]$, if $R >> r_c$. In fact, for our case where $R = 50$ μm, $r_c = 5$ μm, it is about 9 times smaller!

After full 3D simulations, in order to clearly illustrated resulting distributions inside the detector cell, we have to plot 2D cut planes along the different sections the structure shown in Figure. 1 and Figure. 2a. For comparison purposes, we plot in Figure. 3 a 2D cut plane of the electric profile of a 3D-Trench Electrode detector shown in Figure. 1. The bias voltage here is set as 1V. The cut plane is in the y-z plane (x=0 μm). Shown in Figure. 4 is the same for a Closed-Shell Electrode Silicon Detector shown in Figure. 2a. Comparing simulation results in Figure. 3 and Figure. 4 one can clearly see that: 1) the electric field profile is more uniform in the bottom portion (from z=270 μm to 300 μm) in the CSED than that in the non-etch-through 3D-Trench Electrode detector; 2) the electric field is much higher in the bottom portion of the CSED as well; and 3) the cell is isolated from other cells where the trenches are connected in the CSED, while cells are connected by the bottom portion in the non-etch-through 3D-Trench Electrode detector. Therefore the novel CSED are much better than the non-etch-through 3D-Trench Electrode detector in these regards. Apparently CSED fabrication requires double-sided process, which is nowadays quite standard and routinely done in 3D detector fabrications [7, 8], Figure. 5 shows simulated electric field profiles in 2D cuts along the x-y plane at two different z positions the CSED shown in Figure. 2: z=100 μm (Figure. 5a) and z=270 μm (Figure. 5b). As shown Figure. 5a, high electric fields are near the outer square trenches (since the junction is near the trench electrode) with slight non-uniformity near the corners. In fact this electric filed profile should be no different from that in the CSED at the same z position in the non etch-through 3D-Trench Electrode detector since they have the same geometry at this z position. However, as shown in Figure. 5b, at z=270 μm, where the top square trench and the bottom octagonal trench are overlapped, the electric filed profile is quite different from that shown in Figure. 5a. First, the electric field profile in the bottom portion is more uniform since the octagonal shape is closer to the circle than the square shape; and 2) there exist four triangular regions near the square trench corners that are effectively cut off from the detector sensitive region, and can be considered as the dead region. We will discuss the percentage of this dead region as a function of the detector geometry in more details in the next section.

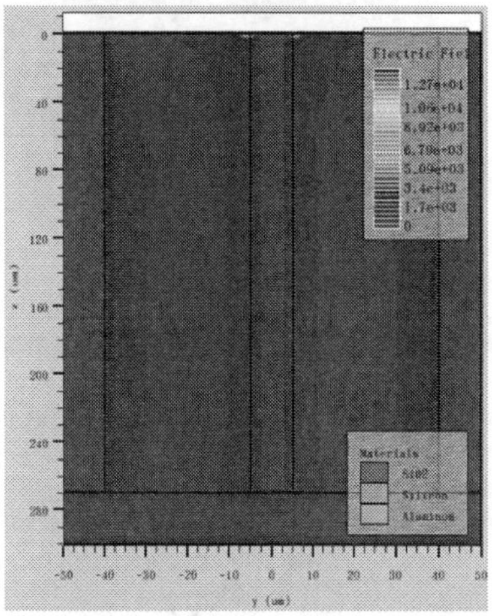

Figure 3. Simulated electric field profile along a 2D cut plane on the y-z plane at x = 0 for a non-etch-through 3D-Trench Electrode detector (see Figure. 1).

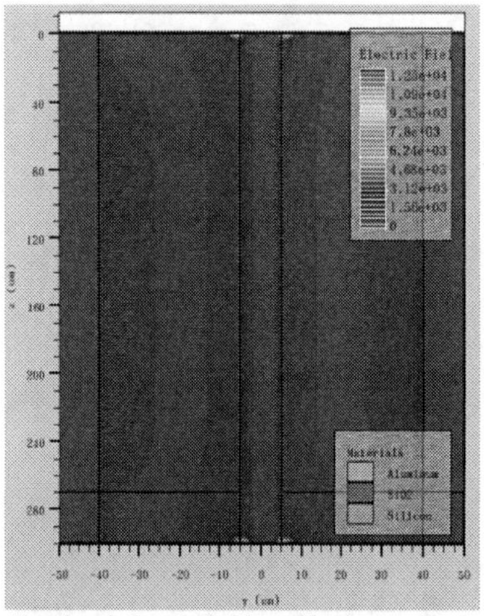

Figure 4. Simulated electric field profile along a 2D cut plane on the y-z plane at x = 0 for a CSED (see Figure. 2a).

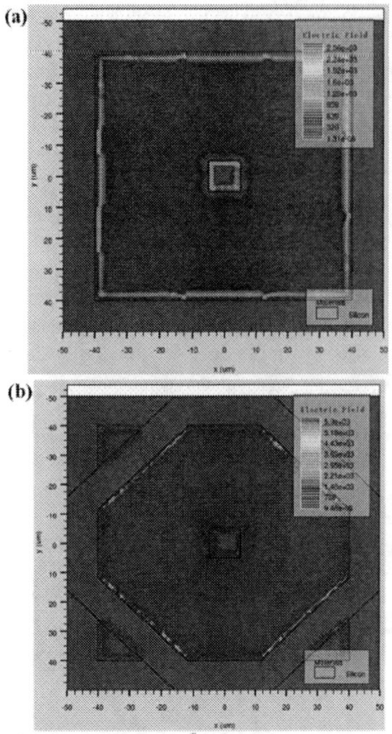

Figure 5. Simulated electric field profile for a CSED (see Figure. 2a): a) along a 2D cut plane on the x-y plane at z =100μm; and b) along a 2D cut plane on the x-y plane at z =270μm.

Figure. 6 shows the simulated detector leakage current as a function of the bias voltage (*I*-*V* curve) for the same CSED shown in Figure. 2. We can see from the *I*-*V* curve that the detector is fully depleted near 1 volt, which is very close to the value calculated using equations in ref. [1] for the cylindrical geometry. Also, the breakdown voltage for this particular CSED is about 174 volt, giving a large detector work range (1 volt<V<174 volt).

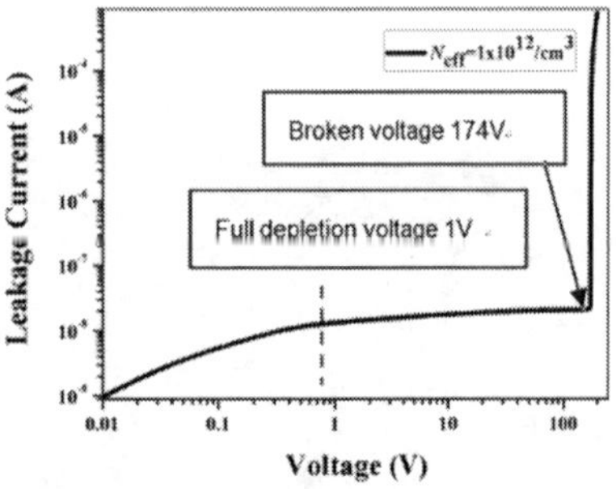

Figure 6. Simulated *I*-*V* curve the CSED under study

3. Full depletion voltage

As we know, the full depletion voltage of a planar detector is related to the detector thickness. For all types of 3D detectors, however, the detector full deplete voltage depends only on the electrode spacing [9, 10]. For the 3D-Trench Electrode detector, the detector full depletion voltage has been calculated for cylindrical geometry for the ORJ and CJ cases [1]:

$$\left.\begin{aligned} V_{fd} &= \frac{eN_{eff}}{2\varepsilon\varepsilon_0}\left[\frac{1}{2}\left(R^2-r_c^2\right)-r_c^2\ln\frac{R}{r_c}\right]-V_{bi}(ORJ) \quad a \\ V_{fd} &= \frac{eN_{eff}}{2\varepsilon\varepsilon_0}\left[R^2\ln\frac{R}{r_c}-\frac{1}{2}\left(R^2-r_c^2\right)\right]-V_{bi}(CJ) \quad b \end{aligned}\right\} \qquad (2)$$

Where Eq. (2)a is for the ORJ case and Eq. (2)b is for the CJ case.

In this paper, the detector full depletion voltage of the novel CSED has been extracted from the 3D simulation of electric field profiles. Figure. 7 shows under different bias voltages the electric field distribution with radius along different 1D cut lines in Figure. 5. Shown in Figure. 7a are 1D electric field profiles along radius from (0,0) to (50,0) in the CSED shown in Figure. 5, and shown in Figure. 7b are those along radius from (0,0) to (35,35) for the same CSED. It is clear that there exist electric field in the entire region between the trench electrode and central column electrode at a bias voltage of 1 volt in both cased, indicating the detector is fully depleted at V = 1 volt. This is in agreement with our previous estimations from calculation using Eq. (2)a and *I*-*V* curve in Figure. 6.

We notice here that volumes between the top and bottom trenches where they are not connected are in fact cut off from the detector sensitive volume, and therefore are dead space. One can easily calculate the percentage portion of this dead space in a single cell for different cell geometries. It is apparent that it will decrease as the cell size increases. It also depends on the depths of the top and bottom trenches and their overlap (ε). Details of dead volume study will be reported elsewhere in a different report. For our particular case shown in Figure. 5, it is only about 0.65%, which is nearly negligible.

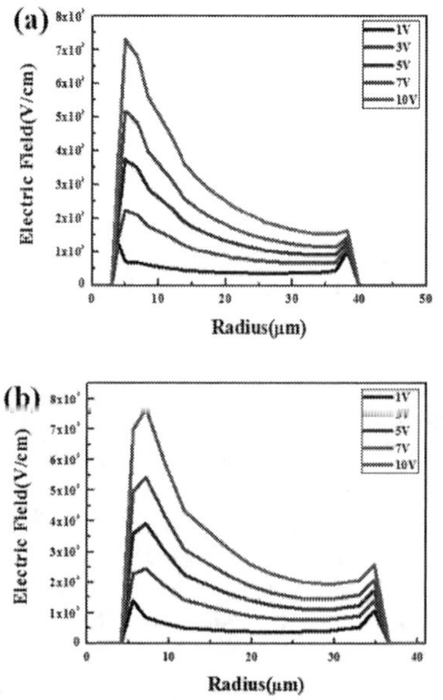

Figure 7. Electric field distribution under different bias voltages, a) along cutline from (0,0) to (50,0) and b) along cutline from (0,0) to (35,35) in Figure. 5.

4. Potential characteristics

The potential profile for this particular CSED structure is shown in Figure. 5 are also simulated using a the DEVICE

DOI: 10.1051/matecconf/201710809003

3D package of Silvaco's ATLAS simulation tool. Simulated electric potential profile is shown in Figure. 8 in a 3D way, which is along a 2D plane at z=100μm. The bias voltage is 1volt. It is clear that, the potential distribution is uniform without any saddle points as in the case of 3D column electrode detector [11].

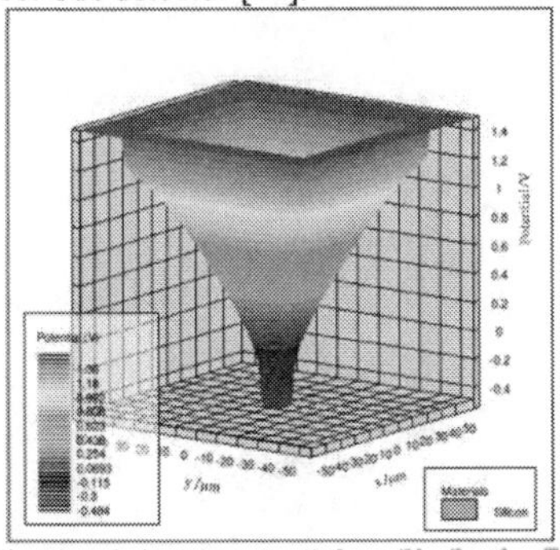

Figure 8. Simulated electric potential profile for the CSED suited here. The 2D cut plane is at z=100μm.

5. Capacitance- Voltage Characteristics

Capacitance is an important parameter for a detector, since the detector noise goes up with the detector capacitance. The capacitance of a fully depleted square CSED can be approximated theoretically by using the capacitance of a cylindrical capacitor[5, 12, 13]:

$$C_{cyl} = 2\pi\varepsilon_r\varepsilon_0 \frac{d}{\ln(R/r_c)} \qquad (3)$$

where C_{cyl} is the capacitance of the CSED.
Figure. 9 is the simulated *C-V* curve for the CSED shown in Figure. 5.

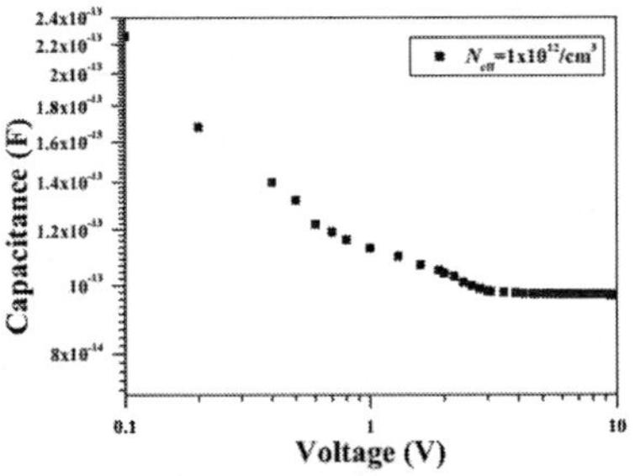

Figure9 . C-V characteristics without radiation.

As shown Figure.9, it is clear that the detector capacitance decreases with the bias voltage and becomes flat (reaching the geometry capacitance) around 2-3 volts, which is larger than that for a flat *I-V* curve and has been routinely observed before for planar detectors. The geometry capacitance is about 100 fF, which is close to calculated capacitance using Eq. 4.

6. Conclusion

A novel Closed-Shell-Electrode-Detector (CSED) has been proposed and described in details. Full 3D simulations have been conducted using a 3D-TCAD tool on the CSED, including electric potential and field profiles, detector leakage current, and detector capacitance. Compared to the non-etch-through 3D-Trench electrode detectors, the electric filed profile in the bottom portion (the non-etched region in the non-etch-through 3D-Trench electrode detectors and the bottom trench region in the CSED) is more uniform and much larger in the CSED, and cells are better separated in the CSED. The dead space is very small (only 0.65% in our case here in this study), and can be made even smaller by increasing the cell size. It requires a double-sided fabrication processing, which is quite standard in 3D detector fabrications.

References

1 Z. Li, Nucl. Instr. and. Meth. A 658, 90-97 (2011).
2 T. Grönlund, et al., Nucl. Instr. and. Meth. A 586, 180-189(2008).
3 Z. Li, T. Grönlund, Nucl. Instr. and. Meth. A 612, 509-515(2010).
4 Z. Li, New detectors with novel electrode configurations for applications in sLHC and photon sciences, presented at the Fifteenth Workshop of CERN RD50, CERN, Geneva, Switzerland, 16–18 November, 2009.
5 H. Ding, et al., Nucl. Instr. and. Meth. A 796, 29-33 (2015).
6 A. Montalbano, et al., Nucl. Instr. and. Meth. A 765, 23-28 (2014).
7 A. Zoboli, et al., IEEE TNS 55-5, p. 2775 (2008).
8 G. Pellegrini, et al., Nucl. Instr. and. Meth. A 592, 38-43 (2008).
9 J. Kalliopuska, S. Eränen, R. Orava, Nucl. Instr. and. Meth. A 568, 27-33 (2006).
10 S.I. Parkera, C.J. Kenneya, J. Segalb, Nucl. Instr. and. Meth. A 395, 328-343 (1997).
11 C. Piemonte, et al., Nucl. Instr. and. Meth. A 541, 441-448 (2005).
12 J. Chen, et al., Nucl. Instr. and. Meth. A 796, 34-37 (2015).
13 D.K. Schroder, T.T. Braggins, H.M. Hobgood, Journal of Applied Physics, 49, 5256 (1978)

 DOI: 10.1051/matecconf/201710809004

The Power Supply Module Design for Diesel Electronic Control Unit's Lift Time Test System

Jinguan Yin[1], Tiexiong Su[1], Xiaoyan Shen[2], Li Jia[2], Tao Liu[2] and Hui Lv[2]

[1] *College of Mechatronic Engineering, North University of China, Taiyuan, Shanxi 030051, China*
[2] *China North Engine Research Institute, Tianjin, 300400, China*

Abstract. In this article we focus on the voltage signals supply to power-on-off testing and some specific fault cases for diesel engine electronic control unit life time test system. According to the actual engineering needs we design six different voltage signals supply modules and use TESTBASE bench to supply power to diesel electronic control unit directly. These voltage signals supply modules are all used to supply varied power-on-off environment so we can observe the status of electronic control unit visualized. The accuracy and reliability of voltage power supply module are verified by simulation and experiment. The result shows it provides a good environment for diesel electronic control unit life time test.

1 Introduction

Almost all the diesel engine use electronic control nowadays. Facing the fault condition such as lightning, static shock, load dump et.al, the diesel engine electronic control unit (ECU) power supply circuit should protect circuit itself is not damaged by the excessive voltage surge, and guarantee the circuit running right at the same time. In order to ensure the reliability of ECU power supply system, on the one hand we designed related circuit to protect the power supply system; On the other hand we should carry on the relevant test to evaluate of the reliability and stability. Life time test system (LTT) is a real-time simulation system for the ECU product. And nowadays we use hardware-in-loop (HIL) to realize the LTT [1, 2], which can has test under different conditions especially the fault and limiting condition [3]. The voltage mutation that causes the single chip of supplementary equipment in and out of the external interrupt frequently, causes stack overflow, the network into a dormant state and couldn't be awakened, easy crash failures in the process of ECU power-on-off. The ECU system should be tested to ensure the reliability and stability of the ECU before it is out of factory. The test process is test system analysis, which is test the influence of wave source to test system using statistics method, giving the final report about the system whether pass muster or not. In engineering, using the repeatability and reproducibility of test system to analyse features usually. And when the ECU system is fault, we should have a fault recurrence to find what fault has happened on the ECU. For the power system, the fault is produced by voltage surge and electromagnetic interference. In order to test the state of ECU in the process of power-on-off and fault recurrence, we design power supply module contains six different voltage signals models satisfy LTT requirements.

2 Test environment

As shown in Fig. 1, Simulink model, VeriStand software as a PC interface, TestBase bench and real ECUs constitute the LTT environment.

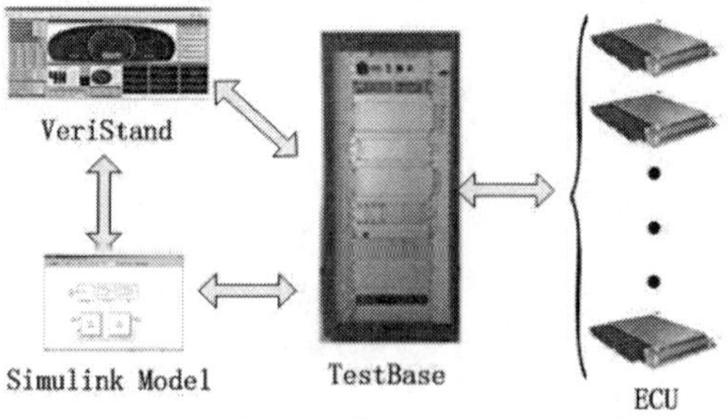

Figure 1. The environment of life time test system

A series of test and analysis hardware equipment comprise the TestBase bench as shown in Table 1. TestBase bench connects to VeriStand software using ethernet network.

TestBase bench use EA programmable power supply module which is used to simulate the ECUs power supply system. The voltage and current can be manual control, also can be controlled through real-time computer output port [4]. It can satisfy the ECU power test demand, such as voltage up and down surge, voltage over high and low limits, wave voltage etc. operating mode, especially the vehicle power is in short supply and voltage surge to power supply system when engine start. Fig. 2 (a) is a picture of EA programmable power supply module, Fig. 2 (b) is a picture of TestBase bench power supply management unit, and its main performance index as shown is Table 2.

Table 1. TestBase hardware component

Hardware	Application
Real-time component	Processor, I/O board
Signal management unit	Control board signal
Load simulation unit	Simulate real load
Fault injection unit	Electrical fault injection
Vehicle power simulator	Simulate vehicle power supply system
Power supply management unit	Manage bench power supply system

Table 2. Performance index of EA power

Version	PS8080-60
Voltage range	0-60V
Current range	0-80A

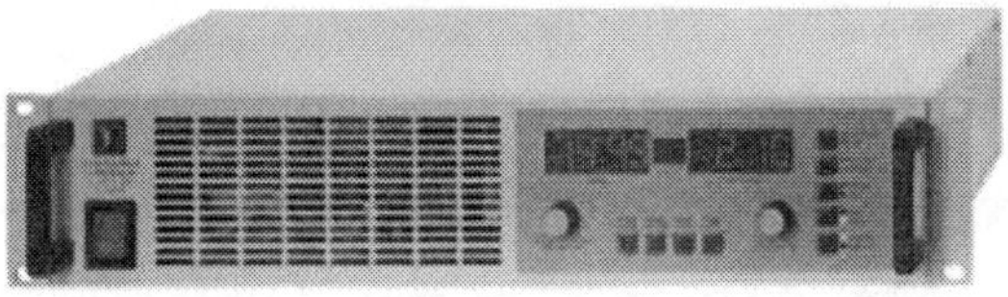

(a)

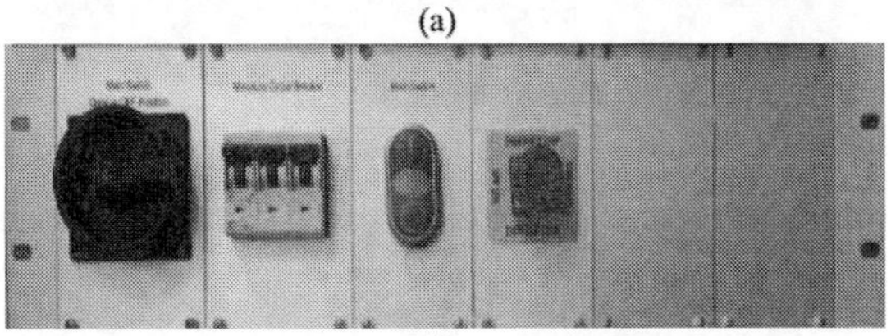

(b)

Figure 2. EA programmable power supply module and power supply management unit

The Simulink model includes the ECU control model, engine model, the I/O model and power control model. The engine model simulates the diesel engine, transmitting and receiving the signals from analogue input/output board and digital input/output board through I/O model. In the power control model we give our voltage signals arithmetic. In Simulink model, we can design arithmetic expediently and discretionarily.

VeriStand [5] software as a PC interface in the real-time simulation. We designed the power supply module in the Simulink software firstly, and download the arithmetic to TestBase bench. Then we control and observe test process on the VeriStand project. In the VeriStand project, the parameters of the voltage signals arithmetic can be changed easily. It can test more than a dozen ECUs at the same time.

3 Power control module design

We designed six modes to supply power for ECU power-on-off in LTT and fault recurrence, it can be arbitrary switched in the testing process. These power control modules are useful in the diesel ECU test, including the hardware and software. The six control mode are DC mode, high-low mode, fixed slope mode, random slope mode, cycle mode and normal mode. These different kinds of power control mode will be described in detail. The overview of the power control module in Simulink can be found in Fig. 3.

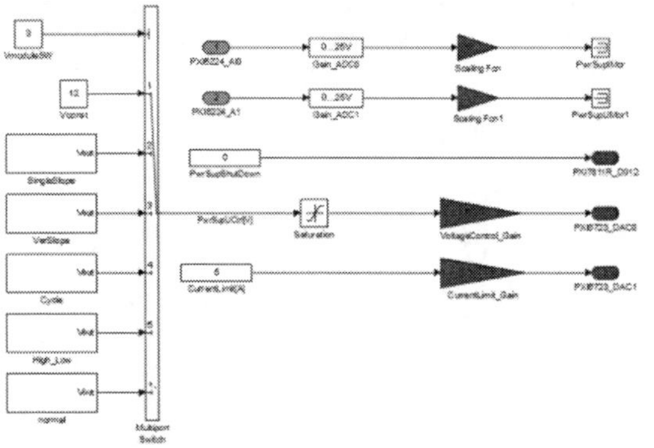

Figure 3. The overview of the power control module

3.1 DC mode

The most basic mode is DC power supply, which is giving a constant voltage to ECU. The main function of this mode is to simulate the operating voltage of the ECU in normal work.

$$V_k = C \quad (1)$$

As shown in Fig. 4, we appoint C is 12V, in the DC mode the experimental voltage output has an error with the ideal voltage output, but the error is negligible. The voltage can meet the demand of normal working. In practice, the electric wire between TestBase and ECUs should be as short as possible, since the electric wire has a voltage drop effect.

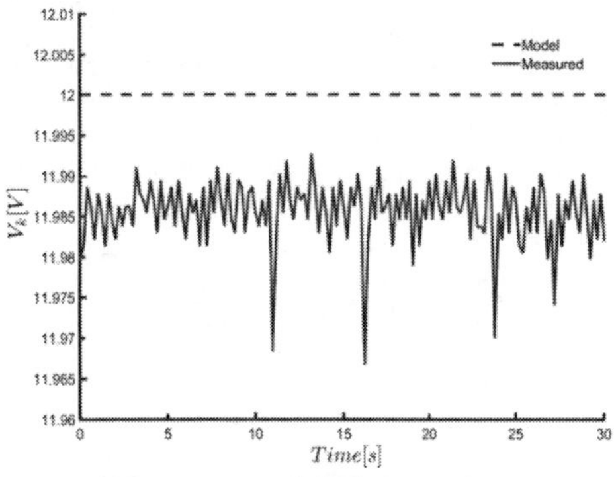

Figure 4. Voltage output to ECU in DC mode

3.2 High-Low mode

For simulating the ECU sudden power-on-off and some other special circumstances, we use high and low level voltage signals to ECU.

$$V_k = \begin{cases} V_{High}, t = t_{High} \\ V_{Low}, t = t_{Low} \end{cases} \quad (2)$$
$$t_{period} = t_{High} + t_{Low}$$

V_k is the current voltage output, V_{High} is the high level voltage and it's duration is t_{High}, V_{Low} is the low level voltage and it's duration is t_{Low}. All the parameters of voltage and time are variable in the process of test. The simulation and experiment result is shown in Fig. 5. This mode is used for testing how the ECU power sudden up and down influences the ECU hardware and software. If the software isn't designed well, it may causes stack overflow, the network into a dormant state and couldn't be awakened.

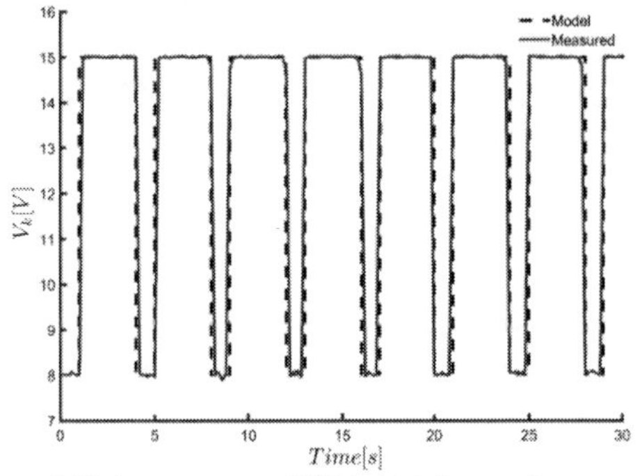

Figure 5. Voltage output to ECU in high-low mode

In this case, we appoint V_{High} is 15V, V_{Low} is 8V, $t_{High}$is 3s and $t_{Low}$is 1s. As shown in Fig.3 when the high-low level changed, the voltage is not changed immediately in experiment. But the voltage is changed very fast, so it can reach the design demand.

3.3 Fixed slope mode

In this mode, we set the minimum voltage and the maximum voltage. The TESTBASE's output voltage reciprocate between the maximum and minimum voltage, and have a fixed increase/decrease slope. Using a state machine controls the output voltage, when monitoring the last moment output voltage is less than the maximum voltage, then output voltage will be the sum of the last moment voltage and the rising slope until the output voltage over the maximum voltage. When the output voltage reached the maximum voltage, the output voltage will be the sum of the voltage and the down slope until the output voltage less than the minimum voltage. The next cycle will occur after a waiting time.

$$V_k = \begin{cases} V_{k-1} + UpSlope_{fixed}, V_{k-1} \leq V_{min} \\ V_{k-1} + DownSlope_{fixed}, V_{k-1} \geq V_{max} \end{cases} \quad (3)$$

where V_k is the current voltage output, V_{k-1} is the last moment voltage output, $UpSlope_{fixed}$ is the fixed increase slope, $DownSlope_{fixed}$ is the fixed decrease slope. Similarly we can appoint the values of these parameters on PC interface in the test. The simulation and experiment results as shown in Fig. 6.

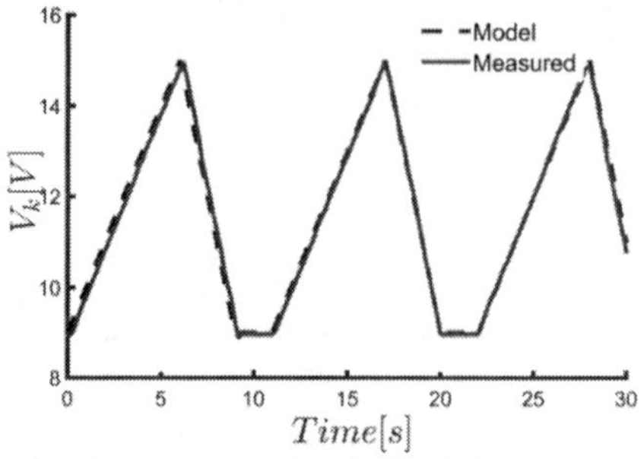

Figure 6. Voltage output to ECU in fixed slope mode

In this case we appoint V_{High} is 15V, V_{Low} is 9V, $UpSlope_{fixed}$ is increased 2V per minutes, $DownSlope_{fixed}$ is decreased 2V per minutes and the waiting time is 2s. Observing the status of ECU when the voltage up and down slowly. This mode is used for ECU power-on-off sluggish and the fault recurrence.

3.4 Random slope mode

Unlike the fixed slope mode, the random slope mode has a random increase/decrease slope. The step of model is 0.001s, so we give a random value as the slope in every step. Voltage is changed faster and the waiting time is shorter than other modes in random slope mode, this is in order to test the performance of the relay and the performance of ECU hardware and software. The formula of random slope mode is similar with fixed slope mode formula as:

$$V_k = \begin{cases} V_{k-1} + UpSlope_{random}, V_{k-1} \leq V_{min} \\ V_{k-1} + DownSlope_{random}, V_{k-1} \geq V_{max} \end{cases} \quad (4)$$

where V_k is the current voltage output, V_{k-1} is the last moment voltage output, $UpSlope_{random}$ is the random increase slope, $DownSlope_{random}$ is the random decrease slope. The simulation and experiment results as shown in Fig. 7.

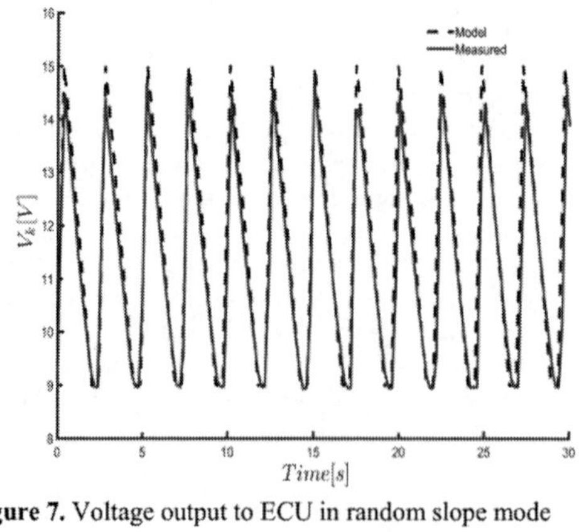

Figure 7. Voltage output to ECU in random slope mode

In the test we appoint V_{High} is 15V, V_{Low} is 9V, but in Fig. 7 we can see it cannot reach the highest voltage in experiment. This is because in most case the random number is too large results the calculative voltage is beyond the upper limit. The state machine determines the result is greater than the highest voltage and transfers the state to decrease.

3.5 Cycle mode

In some special occasions, we supply a cycle voltage wave to ECU. We also set the minimum voltage and the maximum voltage in a cycle, and numerable fixed slope mode have different fixed increase/decrease slopes compose one cycle voltage wave. Using this mode to test the reliability of ECU and to observe whether the ECU will fail in some special environment. And all the tests have good repeatability and automation in the HIL test[6,7].

$$V_{k,i} = \begin{cases} V_{k-1,i} + UpSlope_i, V_{k-1} \le V_{min} \\ V_{k-1,i} + DownSlope_i, V_{k-1} \ge V_{max} \\ (i = 2,3 \ldots) \end{cases} \quad (5)$$

In practice, we find that in one voltage wave cycle set up three or four different slopes generally can meet the requirement of testing work. The simulation and experiment results are shown in Fig. 8.

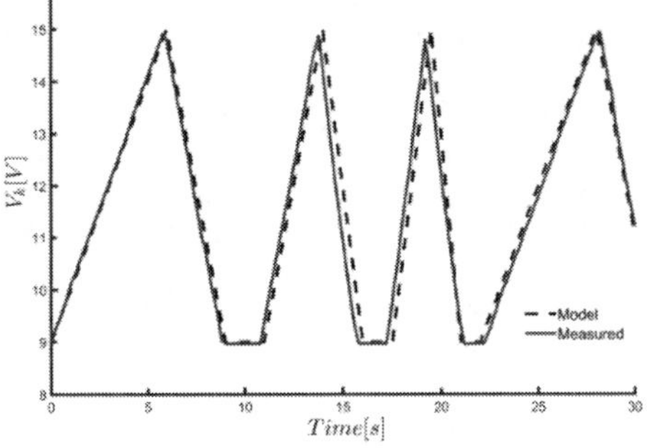

Figure 8. Voltage output to ECU in Cycle mode

In this article we set three different slopes, and V_{High} is 15V, V_{Low} is 9V. The first fixed slope mode's increase/decrease slope is 1/2, the second is 2/3 and the third is 3/4. The waiting time is 1.5s.

3.6 Normal start mode

In diesel engine normal start process, there has a surge voltage since the starter working or increasing load over the range of working voltage suddenly. The single chip of supplementary equipment would be in and out of the external interrupt frequently, causes stack overflow, the network into a dormant state and couldn't be awakened, easy crash failures in the process of ECU power-on-off since the voltage mutation. These faults lead to ECU product quality decline. So the test about ECU power in normal start status should be carried though.

In normal start mode, the voltage will decrease to 6V and back to normal voltage in a short time. Using this mode to test the reliability of ECU circuit and to observe whether the ECU will fail in the start condition. The simulation and experiment results are shown in Fig. 9. For the purpose of simulate real diesel engine start process, we give a white noise to the voltage since the voltage from TestBase bench and simulation result is tight agreement.

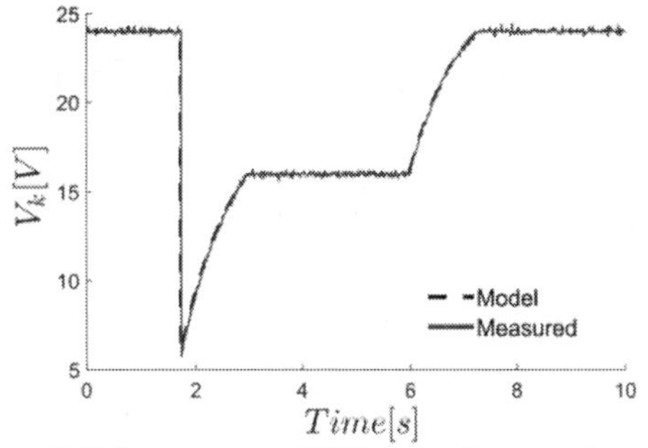

Figure 9. Voltage output to ECU in normal start mode

4 Conclusion

According to the engineering needs we designed six kinds of power supply mode for ECU life time test and fault recurrence. DC mode supplies constant voltage for ECU normal work. High-Low mode make ECU power-on-off conventional, since the high and low level voltage transform instantaneously and the high level is hold on for a period of time. Fixed slope mode and Cycle mode have a andante slope, we designed these for make ECU power-on-off sluggish and the fault recurrence. Random slope mode is used for testing the performance of the relay and ECU hardware and software, and this mode has a fast up/down slope. The normal start model simulates the real engine start process, it can test ECUs using surge voltage. The modes and parameters in each mode can be changed manually on PC interface, or they can be written in automated test cases. The above six kinds of modes can meet the requirements of the life test for power supply basically, and we can add more mode if we have other requirement.

References

1. K. Lamberg, M. Beine, M. Eschmann, etc. SAE Paper, 01-1593 (2004)
2. C. Washington, J. Dolman. *IEEE Autotestcon,* 1-6 (2010)
3. W. Bi, JM. Wang. *3rd International Conference on Advanced Computer Theory and Engineering*, 177-181 (2010)
4. CP. Young, WL. Juang, MJ. Devaney. IEEE T INSTRUM MEAS, **49**, (3), 579-584 (2000)
5. Nation Instruments White Paper, https://decibel.ni.com/content/docs/DOC-5684
6. A. Palladino, G. Fiengo, F. Giovagnini. *IEEE European Control Conferenc*, (2009)
7. Z. Liu, N.Gu. *Proceedings of 2005 The Fifth International Conference on Computer and Information Technology*, 1-3 (2005)

Design of the patent evasion and rescue manipulator based on TRIZ

Hong-liang Liu , Li-yan Chen

Changchun University of Technology, Jilin Province， China

Abstract. In this paper, the theoretical system and technical contradiction of TRIZ theory are analyzed. Combined with the patent law and patent law implementation rules, to determine the avoidance strategy. And through the goal of patent retrieval and determine the technical barriers, technical problems and solutions, according to the working mechanism and task, complete to the rescue robot manipulator mechanical structure design. The test results show that the stability and reliability of the manipulator are obviously improved.

1 Introduction

Due to the background of the global natural disasters, rescue work, showing a diversity and complexity, the development of rescue equipment proposed new requirements, the robot manipulator, which is able to detect the rescue rescue equipment instead of the natural person, the technology in nearly two years at a fast development in high speed. Study of manipulator through the biomechanics, mechanics, mechanics, electronics, material science, many areas of computer science and robotics, which belongs to the interdisciplinary field. In recent years, with the publication of relevant patents and papers, the robot has become a hot research topic in the field of international robotics.

Due to the recent emergence of robotic products, as well as a substantial increase in the amount of patent applications, in a number of areas formed a large area of technical barriers, restricting the development of the robot technology. Therefore, in the process of product design, it is needed to avoid the related patent reasonably, to find out the new technical problems and to solve the problem. But the TRIZ theory system is perfect, has the important instruction function to the risk aversion. TRIZ theory is applied to the design and improvement of robot, which can effectively improve the speed of updating and the speed of the development of the technology.

In this paper, we first analyze the whole theoretical system and technical contradiction of TRIZ theory. Patent circumvention combined with TRIZ theory, based on TRIZ patent to circumvent the robot product innovation design process, and through the goal of patent retrieval, determine the technical barriers, technical problems and solutions, to complete the design of manipulator is presented[1,2].

2 Overview of TRIZ

2.1 Overview of TRIZ theory

The core idea of TRIZ theory is mainly embodied in 3 aspects. First of all, both a simple product or a complex technology system, its core technology is to follow the objective law of development and evolution. Secondly, a variety of technical problems, contradictions and conflicts continue to be solved is to promote the driving force of this process of evolution. Then the ideal condition of the development of the technology system is to achieve as much as possible with the least resources.

TRIZ theory at the beginning of the problem solving, must first put aside all the objective conditions, through an idealized to define member end ideal solution, to define the ideal solution where the direction and position, assurance in problem solving process along the forward and the ultimate ideal solution is obtained, thus avoiding the defects of traditional methods of innovation design lack of goals, which improves the efficiency of design innovation. If the method of creative problem solving is likened to a bridge to success, then the ultimate ideal solution is the bridge pier. Ideal final result has four characteristics: to maintain the advantages of the original system, and eliminate the deficiencies of the original system; (3) did not make the system become more complex, 4 no province introduces new defects[3,4].

2.2 Contradiction and its solution

According to the TRIZ theory, the core of the invention is to solve the contradiction, and the design of the contradiction is not a creative design.Technical contradiction is the contradiction between the two parameters, improve the system of a certain parameter, leading to the deterioration of the other parameters. As shown in the following Figure 1.

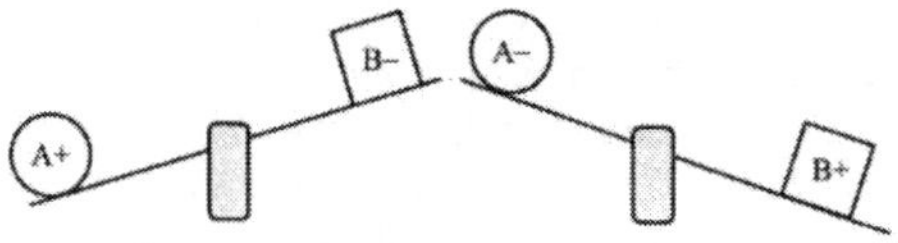

Figure 1. Technical contradiction

In order to solve the contradiction between technology and TRIZ defining the 39 technical parameters, these parameters can be broadly divided into three categories, namely: physical and geometrical parameters (including weight, length, etc.), positive parameter technology (accuracy and reliability, etc.) and negative to the parameters (energy consumption),

When a technical contradiction is identified for the specific problem, the contradiction is described in a specific term in the technical field of the problem. Then, the description of the contradiction is translated into general terms, which can be used to select the general engineering parameters from these general terms. Once a certain or several invention creation principle is selected, it is necessary to transform the invention creation principle according to the specific problem and produce a specific solution. The following Figure 2. for the whole process of the problem solving[1-3].

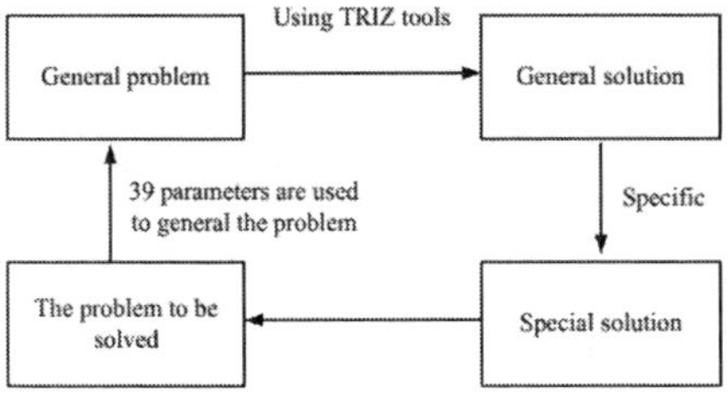

Figure 2. Problem Solution

3. Patent infringement and patent evasion

3.1 Basic knowledge of patent

Patent, also known as patent, refers to the state in accordance with the patent law granted to the applicant in a certain period of time and a certain region of its invention to create exclusive, use and dispose of the rights. At the same time, a patent document refers to the document containing the invention to create a detailed content and technical scope, but also can be used as a reference for the technical personnel to solve technical problems in this field.

A patent application to the grant of the patent right, in addition to meet with the object of the patent law in Article 2 of the protection and Patent Law Article 5 stipulated in does not violate the law, and are not in Article 25 of the patent law does not grant patent invention of some special creativity preconditions, the most important is to have the novel, creative and practical. That solves the different technical contradiction, or uses the obvious difference in the past technical plan, has obtained the unexpected technical effect.

3.2 Patent evasion

Due to the concept of patent evasion design is derived from the patent infringement judgment procedure in the legal science, the existing patent evasion strategy is also based on the infringement judgment. At present, avoidance strategy is mainly used to remove the component design method, substitution method and the method based on the old estoppel. Here are some of the common design strategies to avoid a brief introduction. Delete method, is refers to the already effective patent on the subtraction of one or more of the necessary technical features, the transfer of its function to other system components; delete some components or auxiliary function. The legal basis of the deletion method is based on the comprehensive coverage principle and the doctrine of equivalents in the infringement judgment. It should be noted that, if all the necessary technical features in the patent right requirements of the existing patent, adding new technical features, it will fall into the scope of protection of the patent right. When using the method of deleting, it often produces new problems. At this time, we need to set up a new problem model, and solve the problem in the course of solving the problem. Alternative law refers to the use of different means (technology, methods, principles), so that the system has the same function, to achieve the same effect. The legal basis of alternative sources is also based on the principle of comprehensive coverage and the doctrine of equivalents. The use of alternative methods, it is necessary to pay attention to the need to take a different technology, methods or principles, otherwise it will be regarded as equivalent, thus falling into the scope of patent infringement. Estoppel method, namely the use of patent rights in the process to apply for a patent or follow-up procedures in the process of give up the rights of the original requirements. The legal basis of estoppel is law infringement estoppel principle based on. By this way, the need to apply for or download to avoid the patent review files, search for evidence of estoppel. If the patent has to avoid the content, then can be used directly and does not constitute infringement, otherwise, it must be turned to other strategies to avoid.

4 Application of patent evasion based on TRIZ in the design of rescue robot system

4.1 Research background

It is one of the common problems in the field of rescue at present that the huge volume of rescue equipment and limited operating range is one of the common problems. The research and development of micro rescue robot, which has the advantages of flexible operation and wide working range, is of great significance to improve the efficiency of rescue work and reduce the loss.

Micro rescue manipulator is an important part of the rescue robot. Rescue manipulator (big arm, small arm, wrist, end clamping mechanism) by imitating the function and structure of human hand, and combined with the measurement system, and ultimately the end of the clamping mechanism to achieve accurate target location. The design and development of the rescue manipulator in the stiffness and strength to meet the requirements of the rescue work on the premise, improve the control accuracy and stability, and to achieve miniaturization and lightweight.

4.2 Functional analysis

Through the analysis of the function of the five main components of the rescue manipulator and the interaction, it is found that there may be some harmful effects or surplus function, so as to find the direction to avoid the improvement. Found the transmission device structure is too complex, affecting the accuracy of positioning, through the analysis and for end effector position positioning error may exist, and will increase the rate of vibration and fault can thus be identified it as the role of excess.

Provisions on the implementation of the patent law, the second paragraph of the twentieth paragraph provides the necessary technical features of the independent claims records to solve technical problems. To determine the role of analysis of these effects can be identified excess non essential technical features, the effect generally only in the review process or invalid program, facing the scope of protection had to shrink when, as modifying a source based on, at the same time in the product upgrading a process can also be replaced by elements or elements omitted the were removed or replaced, and the excess function replacement and also derived from new inventions create.

In the process of determining the scope of protection, the establishment of a mechanical hand function model, as shown below.

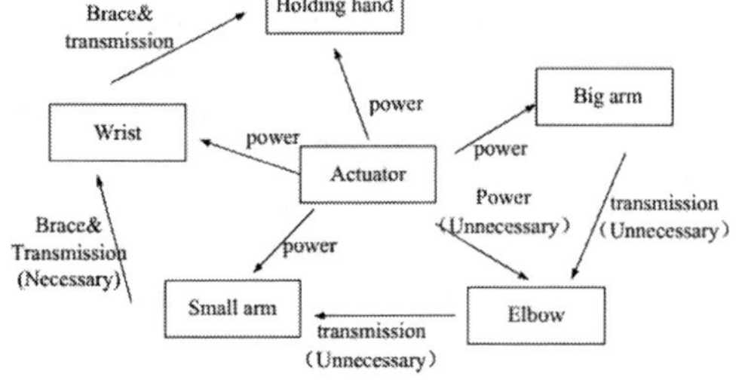

Figure 3. Mechanical hand function model

4.3 Retrieval analysis

Patent retrieval is the most commonly used method of retrieval: keyword search, IPC search and the use of it in order to get a comprehensive and accurate patent information. To the main classification number "(B25J1/00) and its classification, the following group and group of integrated retrieval. Finally, 412 patent, the patent for invention or utility model 372, according to the key words and the same or similar technical characteristics were screened to find the closest prior art, ultimately determine the" 101780673A "is the closest prior art. The patent deadline is still in the effective state.

The target document is a total of 3 rights requirements. The rotary joint is composed of a rotary joint, a connecting plate and a base, wherein the five rotary joint structures are similar, and are composed of a reducer, a flange, a motor, a box body and an encoder. Unique body design and ingenious motor, a retarding mechanism coordinating scheme of the rotary joint has light weight, simple and compact structure, convenient wiring, high strength, modular degree higher characteristic, so as to solve the general service robot arm because of weight problems and bring about the action is not flexible, inertial motion problem[3,4].

4.4 Design in the process of patent evasion

In order to avoid the patent of technical barriers, you need to increase to be solved technical parameters of the contradictory ways, selected a technology negative to the technical parameters and (31 objects produce harmful factors) positive parameters (29 manufacturing precision and 35 adaptability and the use of) to solve. In this application, increasing the pre tightening device for the manipulator to suppress negative to parameters in the vibration and improve manufacturing precision, and by increasing the end manipulator system increases a clip to function, so as to solve the existing technology in the three technical contradiction is executed.

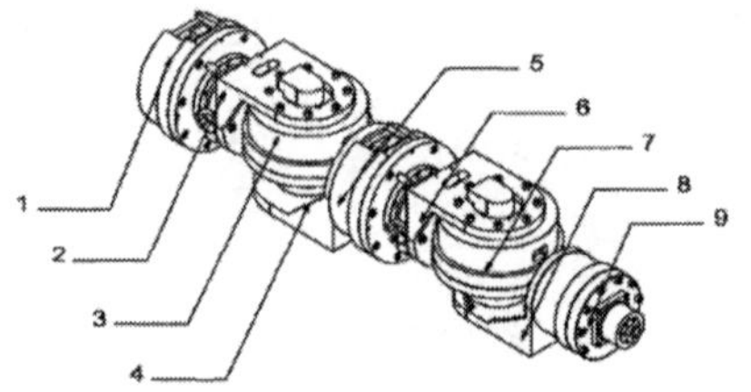

a Target patent

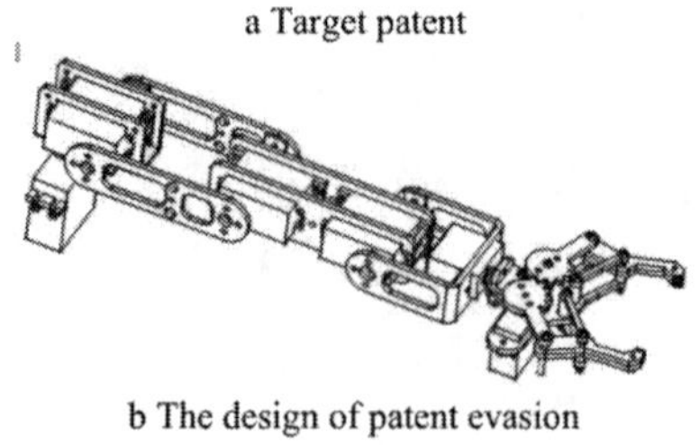

b The design of patent evasion

Figure 4. Design comparison

The patent after to avoid the robot by a rotary pair and swing side basal joint, with a swing side of the elbow joint, with turning pairs and the swinging side of the wrist joint, the connection between the base joint and elbow joint arm, connecting between the elbow and wrist arm and wrist joint is connected with the clamping hands. Each joint rotation pair and swing pair, and the connecting rod drives the holding hand independent movement respectively by a stepper motor drive. The robot can move to a place where a person can't safely arrive at the time of danger. Through the mechanical arm and the chassis with the action, and through the increase of the number of gaskets with the connection, so that the device is more compact, reduce assembly clearance and vibration. As shown in the following Figure 4[4,5].

Summary

Robot technology in the past two years, the rapid development of the relevant technical documents open number multiple areas of increased year by year, in addition to the rescue, services, entertainment, rehabilitation, robot products and technology has also been widely used. Correct use of patent information technology, can reduce the technology development cycle and development costs, and based on the patent of TRIZ to avoid, can effective technical protection and patent layout, at the same time, through the technical contradictions continue to analyze and solve, can also promote the update and improvement of robot and robot technology, greatly improved the reliability and stability of the robot.

Acknowledgement

This work was financially supported by the Jilin province education department Twelfth Five-Year Scientific research project (No.[2013]153).

References

1. QIAN Wei-miao,LI Gui-ping,Research on Product Innovation Design Based on QFD,TRIZ and Patent Knowledge Mining. [J]. Light Industry Machinery, (2011,29)
2. Wang Rui, The Research and Application of TRIZ Based Patent Around Product Innovative Design, A Dissertation Submitted to Guangdong University of Technology for the Degree of Master [D], (2015)
3. Vinodh.S,Kamala.V,Jayakrishna.K.Integration of ECQFD,TRIZ,and AHP for innovative and sustainable product development [J]. APPLIED MATHEMATICAL MODELLING, (2014), 38 (11-12) : 2758-2770.
4. Jiang Ping, etc,Method about Patent Design Around Based on Function Trimming [J],JOURNAL OF MECHANICAL ENGINEERING, (2012,11)
5. Guidelines for patent examination, State Intellectual Property Office of P.R.C, (2010)

3D Simulation and Modeling of Ultra-fast 3D Silicon Detectors

Manwen Liu [1,2], Zheng Li [1,2], Mingfu Feng [1,2]

[1]*School of Materials Science and Engineering, Xiangtan University, Xiangtan, 411105, China*
[2]*Center for Semiconductor Particle and photon Imaging Detector Development and Fabrication, Xiangtan University, Xiangtan, 411105, China*

Abstract. 3D detectors with very small electrode spacing can provide ultra-fast detection due to their extremely small charge collection time. Since the detector full depletion voltage and charge collection time are independent to the detector thickness, ultra-fast 3D detectors can be made relatively thick (or not too thin, ~200 μm) to ensure a large signal. The results of the 3D simulations and modeling of 3D silicon detectors with very small electrode spacing and relatively large thickness will be shown in this paper. The column spacing LP is in the range of 5 μm to 10 μm. At a bias voltage of only a few volts, the electric field in the detector can be large enough to ensure the carrier saturation drift velocity in most volume of the detector, and the detector charge collection time there can be as short as 10's of ps. In this paper, we will analysis the simulated electrical characteristics of this detector structure through systematic 3D simulations using the Silvaco's TCAD tool. Profiles of detector electric potential and electric field will be presented. We will investigate the region of low electric field (the "slow region") in the detector. We will also exam whether the detector reach the breakdown condition at operation voltages suggested in this work.
Keywords. 3D simulation, Ultra-fast, 3D silicon detector, Electric field

1 Introduction

In the middle 1990's, S.I. Parker and his colleagues proposed a structure of solid-state radiation detectors using a 3D array of electrodes that penetrate into the detector's bulk [1]. In recent years, researchers of 3D detectors have mainly focused their efforts on improving detector radiation hardness by proposing various 3D structures as tracking detectors for the high luminosity upgrade of the Large Hadron Collider (HL-LHC) [2-8]. Examples of them are 1) In 2008, a non-fully penetrating, dual-column structure was proposed [6], with an electrode spacing of 50μm; and 2) a double-side, dual column structure [7]. In 2010, it was pointed out that in future development for detectors irradiated up to 1×10^{16} n/cm^2, to fully depleted a dual-column 3D detector with a reasonable bias, the columns' spacing should be further reduced to 30μm [8].

In this work, we further reduced the column spacing to obtain ultra-fast detectors with reasonable signal size. An all column electrode 3D silicon detector with very small electrode spacing and a thickness of 200μm is shown here in Figure. 1. For simplicity, we chose the etch-through columns in the simulations, which will not affect the conclusion obtained in this work about the property of ultra-fast charge collections of this type of detectors. In practice, the etch-through detectors can be realized by using a support wafer or SOI. The column spacing is chosen in the range of 5μm to 10μm. As we will show, at a bias voltage of only a few volts, this structure's charge collection time can be as fast as 10's of ps. We undertook full 3D simulations of dual-column 3D detectors with the Silvaco DEVICE3D package within ATLAS [9].

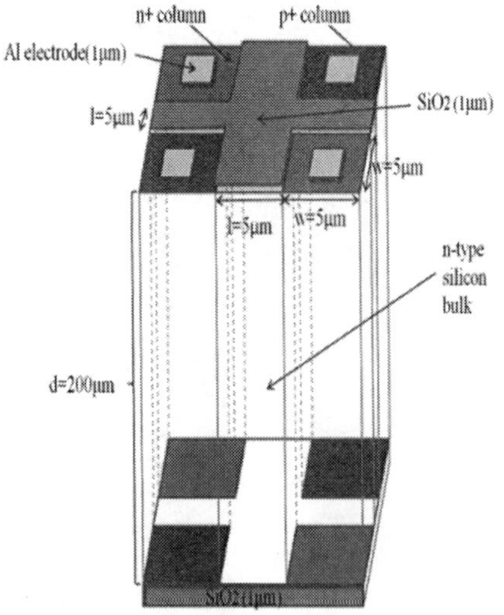

Figure 1. 3D device structure in square shape.

2 The device description and 3D simulation

Figure. 2 is the cell of the detector simulated in this work. This structure consists of two n-type- and two p-type-doped columns on n-type substrate with an initial phosphorous doping concentration of 1×10^{12} cm^{-3}. The width of all columns is 5 μm, which can be realized with today's new etching technology on wafers with relatively thin thickness (≦200 μm). The distance between the electrodes is 5 μm or 10 μm. The doping concentrations of the columns are 1×10^{19} cm^{-3}. (boron for p^+ columns and phosphorous for n^+ columns, respectively). Depths of the n^+ column electrode and the p^+ column electrode are all 200 μm, the same as the detector thickness. With improved processing technology at Fondazione Bruno Kessler (FBK), Trento, Italy, columns with 5 μm width can be etched up to 200 μm using their Deep Reactive Ion Etching system with aspect ratio up to 30:1 [10]. For a detector with column spacing of 10 μm, collection electrodes can be read out using a bump bonding technology of 10 μm bumps. Although technically bump bonding technology of 10 μm bumps may not be used for a detector with column spacing of 5 μm, with future improvements of bump bonding technology, it may also be realized. We therefore went ahead to simulate this extreme situation for future applications. Anode and cathode electrodes are formed by covering 1 μm aluminum on n^+ and p^+ columns on the top side of the detector. Other parts of the top side (area between electrodes) and the entire area of the bottom side of the detector are covered by 1 micron thick SiO_2 layers. Figure. 2 is the top view (although it looks up) of square column of columns spacing in 10 μm and 5 μm simulated using Silvaco's 3D TCAD tool. The aim of the numerical simulations is to investigate the minimum bias voltages needed for carriers, generated by incident particles/photons, to reach the saturation drift velocity in the most part of the detector volume. Detector depletion voltages will also be determined.

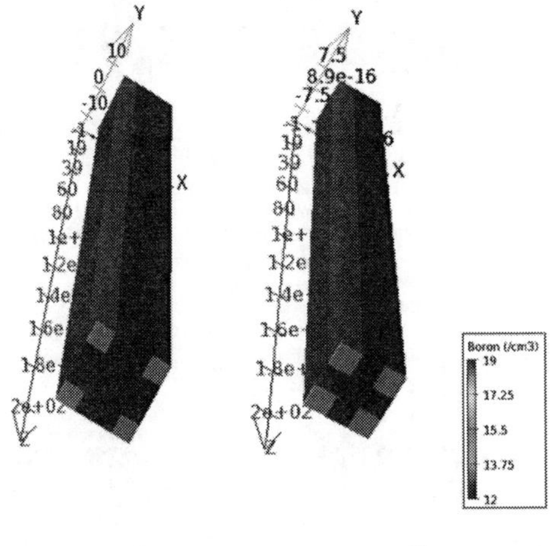

a) b)

Figure 2. Detector structure simulated in this work with a)10 μm and b)5 μm electrode spacing.

3 Detector electrical characteristics

Before the full 3D simulation, let us first estimate the minimum bias voltage needed to reach the carrier saturation drift velocity in the detector. Carrier (electrons here) drift velocity can be written as:

$$v_{dr} = \frac{\mu_e E}{1 + \frac{\mu_e E}{v_s}} \tag{1}$$

Where v_{dr} is the drift velocity of electrons induced by the incident particle, v_s **错误！未找到引用源。** is the saturation velocity for electrons, and E is the electric field in the detector.

To reach saturation velocity, $v_{dr} = v_s = 1 \times 10^7$ cm/s, we must have $E >> \frac{v_s}{\mu_e}$ **错误！未找到引用源。**, from $\mu_e = 1450$ $cm^2/V/s$ **错误！未找到引用源。**, we can get $E >> 10^4$ V/cm. So we chose the minimum electric field to reach the carrier saturation drift velocity as 10^4 V/cm. When the carrier saturation drift velocity is reached, the charge collection time is then $t_c = L_p / v_s$.

The relationship between the average electric field E and the bias voltage V can be written as: $E = \frac{V}{L_p}$ **错误！未找到引用源。**, then we have:

$$\frac{V_{\min}}{L_p} = \frac{v_s}{\mu_e} \tag{2}$$

where $V_{\min}$ is the minimum voltage needed for carriers to reach the carrier saturation drift velocity:

$$V_{\min} = \frac{v_s}{\mu_e} \times L_p \tag{3}$$

错误！未找到引用源。

For example, for L_P = 5 μm, $V_{\min}$ is about 3.5 volts, and the charge collection time is about 50 ps. Similarly, for L_P = 10 μm, $V_{\min}$ is about 7 volts. The charge collection time is about 100 ps. While for L_P = 20 μm, $V_{\min}$ is 14 volts, and the charge collection time is about 200 ps, etc. In the full 3D simulations of this work, we carried out simulations around these estimated minimum voltages, and use the electric field value of 10^4 V/cm to determine if most of the detector volume reaches the carrier saturation drift velocity condition. On the other hand, our full 3D simulations of the detector can verify what we calculated above. Profiles of simulated detector electric potentials, electric fields will be shown below. Detector full depletion voltages are obtained from simulated electric potential and electron concentration profiles.

To illustrate the detailed electric potential and electric field profiles in our 3D Si detectors, we made 2D cut planes along the detector thickness (Z-direction, at z=103 μm), as depicted in Figure. 3 to Figure. 8. Figure. 3 and Figure. 4 are the electric potential profiles of two structures with different electrode spacing. We put negative voltages on cathode electrodes, while zero bias on anode electrodes. Bias voltages are -5 and -8 volts for

Figure. 3 and Figure. 4, respectively. As one can see in Figure. 3, the potential near p^+ electrodes is about -5 V, and that near n^+ electrodes is 0 V. The potential at the center of the detector cell is -2.5 V. These potential distributions clearly define a saddle shape potential, with zero electric field at the center regardless of what value the bias voltage is. Similar conclusions can be easily obtained from Figure. 4, where the potential is -8 V near p^+ electrodes and that near n^+ electrodes is 0 V.

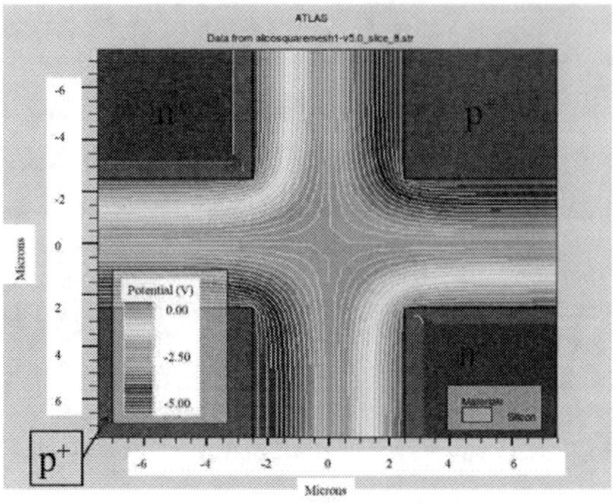

Figure 3. Electric potential profiles of detector with a column spacing of 5 μm.The bias voltage is -5 volts.

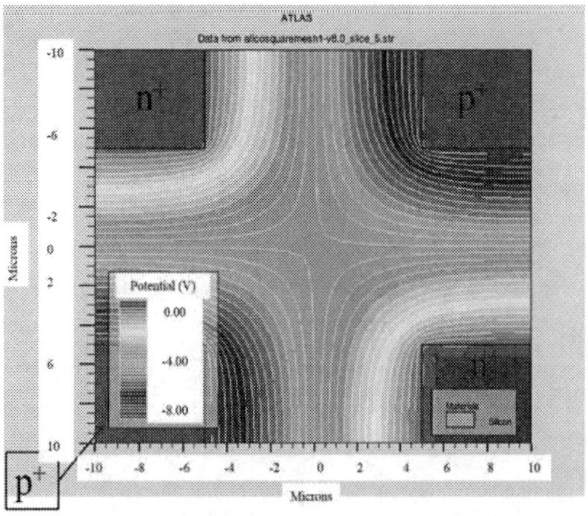

Figure 4. Electric potential profile of a detector with a column spacing of 10 μm. The bias voltage is -8 volts.

From simulations of the electric field profiles, as shown from Figure. 5 to Figure. 8, we can observe that in the majority regions of detectors, the electric field already reaches to 10^4 V/cm when the absolute value of bias voltage V is larger than 4 and 8 volts for L_P = 5 μm and L_P = 10 μm, respectively. For example, as shown in Figure. 5, where L_P is 5 μm and bias voltage V is -4 volts with the maximum electric field (or the upscale of electric field display for better clarity) shown as $1.5x10^4$ V/cm, the electric field in regions between p^+ and n^+ electrodes is larger than 10^4 V/cm. As we calculated above, we can therefore get a minimum carrier drift time of about 50 ps in these regions of the detector at a bias voltage of $|V| \geq$ 4 V for L_P = 5 μm. As shown in Figure. 6, when the bias voltage goes to -6 volts, the electric field reaches $2x10^4$ V/cm in regions between p^+ and n^+ electrodes of the detector, which will further guarantee the carrier saturation drift velocity condition. Shown in Figure. 7-8 are simulated electric field profiles at two different bias voltages for a detector with electrode spacing of 10 μm (L_P = 10 μm). Similarly, as shown in Figure.7 and Figure.8, the electric field reaches 10^4 V/cm in regions between p^+ and n^+ electrodes of the detector when the absolute value of bias voltage is larger than 8 volts, which ensures the carrier saturation drift velocity condition, and therefore a fast charge collection time of about 100 ps.

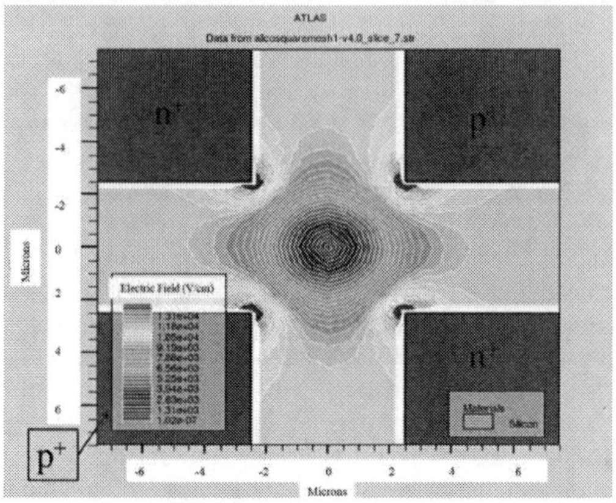

Figure 5. Electric field profile of a detector with a column spacing of 5 μm. The bias voltage is -4volts. Maximum electric field shown is $1.5x10^4$ V/cm.

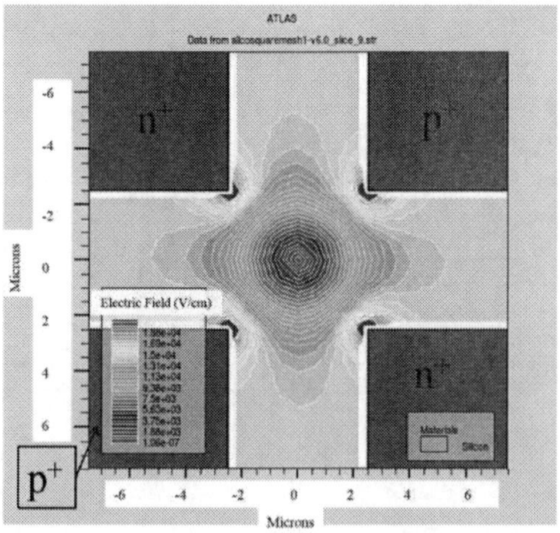

Figure 6. Electric field profile of a detector with a column spacing of 5 μm. The bias voltage is -6volts. Maximum electric field shown is $2x10^4$ V/cm.

However, as we can see from Figure. 5-8, the electric field in the central region of the detector is very low even at high bias voltages. This is due to the symmetry structure of this type of 3D detector (all column electrode detectors). Usually, a slow region is defined as the region near the cell center where there is an electric potential saddle point with zero electric field. Due to the low field, the carrier drift velocity will be low as well

If we use an electric field value of $\leq 5x10^3$ V/cm as

the low electric field limit that defines the border with the ultra-fast charge collection region, then we can estimate the percentage of the region with low electric field (the "slow region") as the following:

for L_P = 5μm, the proportion of the "slow region" is:9.07% $((1.9 \times 1.9 \times 3.1416)/(5 \times 5 + 4 \times 5 \times 5))$, and for L_P = 10μm, the proportion of the "slow region" is:9.42% $((3 \times 3 \times 3.1416)/(10 \times 10 + 4 \times 10 \times 5))$.

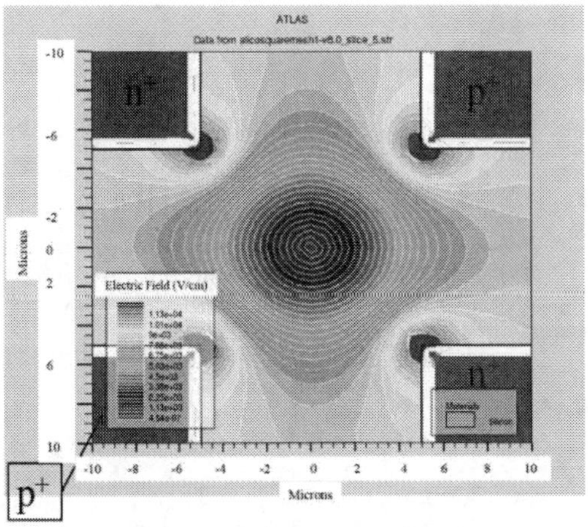

Figure 7. Electric field profile of a detector with a column spacing of 10 μm. The bias voltage is -8 volts. Maximum electric field shown is 1.2×10^4 V/cm.

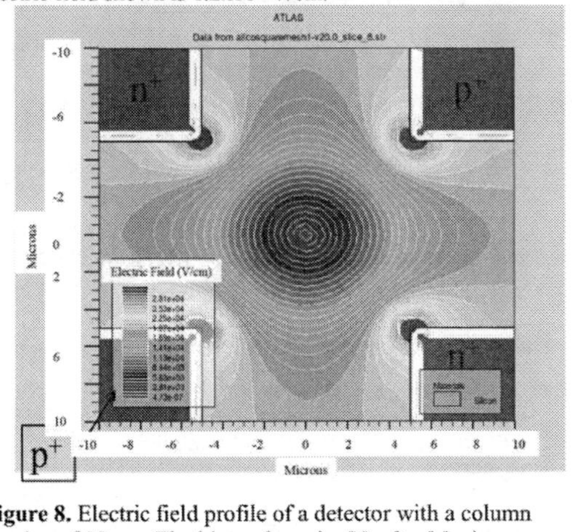

Figure 8. Electric field profile of a detector with a column spacing of 10 μm. The bias voltage is -20volts. Maximum electric field shown is 2×10^4 V/cm.

In both cases, they are below 10%. Meanwhile, since the electrode spacing is extremely small, in the order of the size of electrode columns, the relative dead region (non-sensitive region) will be large. In fact, for L_P = 5 μm, the dead region is 25% (in a unit cell, the dead column area is $4 \times 1/2 \times 25$ μm^2 =25 μm^2, and the sensitive area is $(4 \times 1/2 + 1) \times 25$ μm^2=75 μm^2), and for L_P = 10 μm, it is 12.5%. Therefore there is a trade-off between the ultra-fast and large relative dead-space.

The full depletion voltage can be estimated as:

$$V_{fd} = \frac{qN_D L_p^{\ 2}}{2\varepsilon_0 \varepsilon} \qquad (4)$$

Here N_D is the detector bulk doping concentration. The doping of the silicon bulk is 1×10^{12} cm^{-3}. When the column spacing is 5 μm or 10 μm, the depletion voltage is very small.

4 Breakdown Consideration

Even though the absolute values of applied bias voltages are small, (≤ 10 volts), due to the nature of small device with very small electrode spacing, there could be local high field regions where intrinsic breakdown field for Si (~300 KV/cm) may be reached. Such high fields can exist especially near the corners of p+ columns, just under oxide layers. We cut the profile just under the oxide layer (Z-direction, at z=1 μm) as shown in Figure.9 (L_P = 5 μm, the bias voltage is -8 volts) and Figure.10 (L_P = 10 μm, the bias voltage is -9 volts). From the cutline between two p+ columns, presented in Figure.11 and Figure.12, we can read that the largest value of the electric field is about 9×10^4 V/cm and 6.7×10^4 V/cm respectively, which are all smaller than 3×10^5 V/cm, the intrinsic breakdown filed in Silicon. This means that the detector does not reach breakdown even when the saturation velocity is reached at biases voltage conditions as suggested in this work.

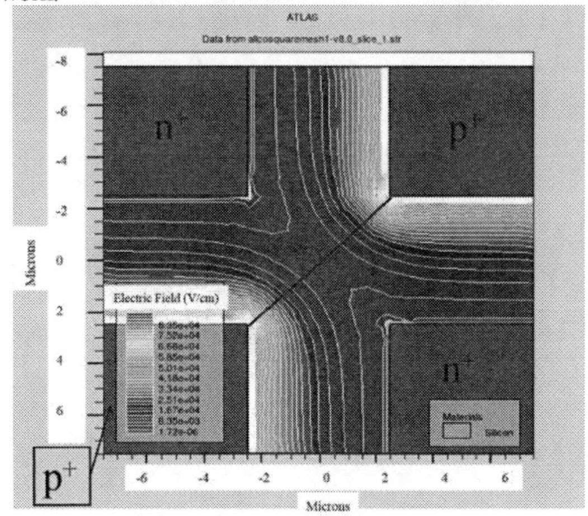

Figure 9. Electric field profile of a detector with a column spacing of 5 μm.The bias voltage is -8 volts.

5 Summary

From calculations and full 3D simulations, we conclude that the ultra-fast charge collection time (≤ 100 ps) can be obtained at a bias voltage of only a few volts (≤10 volts) for small 3D detectors with electrode spacing less than or equal to 10 μm. As we simulated in this study, we can get a minimum carrier drift time of about 50 ps in most part of the detector when L_P is 5 μm, and a minimum carrier drift time of about 100 ps in most part of the detector when L_P is 10μm. The full depletion voltages for these types of detectors are less than 1 volt.

However, because of the symmetry structure of the all column electrode type 3D detectors, a low electric field region exists in the center of the detector regardless of the detector bias voltage. This is the shortcoming of this type of 3D detector. We are working on a new structure to solve this problem [2-5]. Also, we investigated the possible breakdown situation in this type of detectors, and found that at the minimum voltages to reach carrier drift velocity saturation, the highest electric field in the detector is much below the value of the Si intrinsic breakdown field.

We note here that those type of 3D detectors are intrinsically rad-hard due to their very small electrode spacing, and we will study this further.

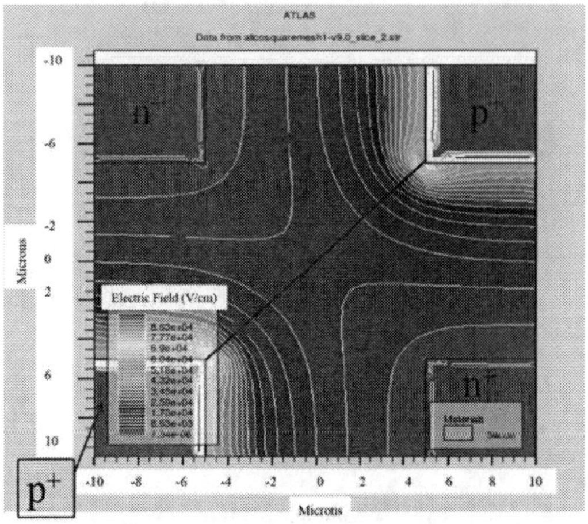

Figure 10. Electric field profile of a detector with a column spacing of 10 μm.The bias voltage is -9 volts.

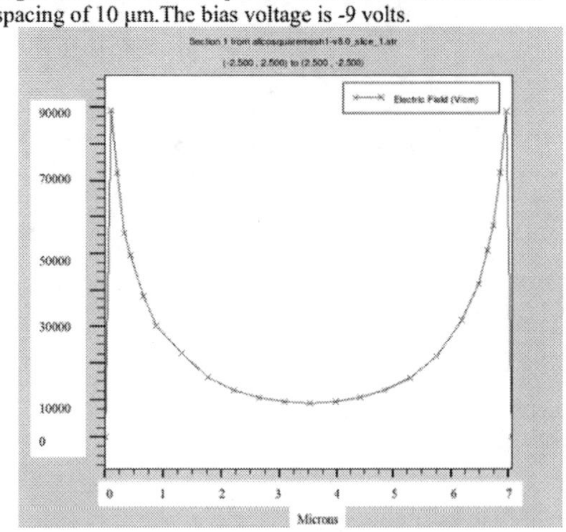

Figure 11. Electric field cutline of a detector with a column spacing of 5 μm.The bias voltage is -8 volts.

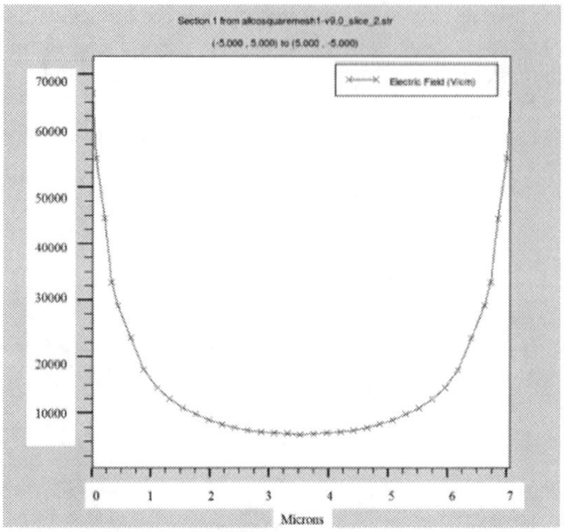

Figure 12. Electric field cutline of a detector with a column spacing of 10 μm.The bias voltage is -9 volts.

References

1. S. Parker, C. Kenney, and J. Segal, 3D—A proposed new architecture for solid-state radiation detectors, Nuclear Instruments and Methods in Physics Research A 395 (1997) 328–343.
2. Z. Li, New BNL 3D-Trench electrode Si detectors for radiation hard detectors for sLHC and for X-ray applications, Nuclear Instruments and Methods in Physics Research A 658 (2011) 90–97.
3. A. Montalbano, D. Bassignana, Z. Li, et al., A systematic study of BNL's 3D-Trench Electrode detectors, Nuclear Instruments and Methods in Physics Research A 765 (2014) 23–28.
4. J. Chen, H. Ding, Z. Li, et al., 3D simulations of device performance for 3D-Trench electrode detector, Nuclear Instruments and Methods in Physics Research A 796 (2015) 34–37.
5. H. Ding, J. Chen, Z. Li, et al., Modeling and simulation of charge collection properties for 3D-trench electrode detector, Nuclear Instruments and Methods in Physics Research A 796 (2015) 29–33.
6. T. Grönlund, Z. Li, G. Carini, and M. Li, Full 3D Simulations of BNL One-Sided Silicon 3D Detectors and Comparisons with Other Types of 3D Detectors, Nuclear Instruments and Methods in Physics Research A 586 (2008) 180-189.
7. A. Zoboli, M. Boscardin, L. Bosisio, et al., Functional characterization of 3D-DDTC detectors fabricated at FBK-irst, IEEE Transactions on Nuclear
8. Z. Li, T. Gronlund, 3D simulations of irradiated one-sided dual-column 3D silicon detector, Nuclear Instruments and Methods in Physics Research A 612 (2010) 509-515.
9. Atlas User's Manual—device simulation software, Silvaco International.
10. G. F. Dalla Betta, dFondazione Bruno Kessler (FBK), Trento, Italy, private conversions (2014).

Characteristics of Timbers Dried Using Kiln Drying and Radio Frequency-Vacuum Drying Systems

Zairul Amin Rabidin[1], Gan Kee Seng[1] and Mohd Jamil Abdul Wahab[2]

[1]*Wood Drying Laboratory, Forest Research Institute Malaysia, 52109 Kepong, Selangor Darul Ehsan, Malaysia*
[2]*Timber Engineering Laboratory, Forest Research Institute Malaysia, 52109 Kepong, Selangor Darul Ehsan, Malaysia*

Abstract. Heavy hardwoods are difficult-to-dry timbers as they are prone to checking and internal stresses when dried using a conventional kiln drying system. These timbers are usually dried naturally to reach 15% to 19% moisture content with an acceptable defects. Besides long drying time, timbers at these moisture contents are not suitable for indoor applications since they will further dry and causing, for example, jointing and lamination failures. Drying to a lower moisture content could only be achieved in artificial drying kilns such as conventional kiln, dehumidification kiln, solar kiln, radio frequency-vacuum, etc. The objective of this study was to evaluate the characteristics of 30 mm and 50 mm thick kekatong (*Cynometra* spp.) timber dried using kiln drying (KD) and radio frequency-vacuum drying (RFV) system. The investigation involved drying time, moisture content (MC) variations between and within boards, drying defects, shrinkage, and drying stress. Drying defects include checks (surface, end, and internal checks) and warping (bowing, cuping, spring, and twisting). The results showed that RFV drying time was reduced to 50% compared to the KD. RFV dried boards demonstrated a more uniform MC between and within boards. Shrinkage in width and thickness, as well as tangential/radial and volumetric shrinkages were substantially less in RFV boards. The amount of cupping, bowing and spring were very low and negligible in all drying runs. There was no twisting observed in all drying methods. The number of stress-free RFV board was higher than KD. With proper procedure, the RFV technology could be used for drying heavy hardwoods which are difficult to dry in conventional kilns due to excessive drying times and degradation.

1 Introduction

Freshly sawn timbers have to be dried before they can be efficiently processed and used. Timber which has not been dried will gradually dry during service. Therefore, the primary reason for drying of wood is to reduce dimensional changes. Excessive dimensional changes will cause distortion and checking in wood, causing both appearance and structural issues in furniture and other wooden products. To minimize extreme changes in MC, and subsequently restrict the movement of wood in service, wood should be dried to its equilibrium moisture content. Timbers for furniture and other interior woodwork should be dried to a targeted MC of 15 to 19% (depending on the timber species), which can be achieved by air drying. Timbers in air-conditioned environment should be dried artificially in a kiln to 12% MC. Although air drying is the cheapest method of drying timber, it is a long process and the MC could not be reduced to 12%.

The common method of drying timber to 12% MC is using the conventional kiln drying system (KD). Generally, light and medium hardwoods are easy to dry with an acceptable degrade [1]. However, drying heavy hardwoods, especially of the thicker boards, is difficult since the timber is susceptible to excessive degrade and take much longer drying time [2].

Previous studies indicated that thick and refractory timbers can be successfully dried using radio frequency-vacuum (RFV) drying system which may be difficult to dry when using KD [2, 3]. The advantages of employing RF heating are rapid and fairly uniform heat transfer often to solidly stacked timbers, high drying rates, and less drying defects such as case-hardening and oxidative discoloration of the wood [4, 5, 6, 7). [3] also reported that 30 mm mixed oak species can be dried in RFV kiln from green to 8% MC in 65 hours compared to 42 days for kiln drying. According to [8], drying sugi boxed-heart timber in RF after kiln drying to 20% MC improved the drying rate and reduced surface checks.

The objective of this study is to assess the drying characteristics of heavy hardwood timbers of two different thicknesses (30 and 55 mm thick) dried using conventional kiln drying and radio frequency-vacuum drying systems.

2 Materials and methods

2.1 Preparation of samples

DOI: 10.1051/matecconf/201710810001

Two kekatong logs were sawn into 30 mm x 120 mm and 55 mm x 120 mm according to cutting pattern as shown in Figure 1. Flat sawn boards, or tangential cut, was used because this cutting is easier or cheaper to produce and is susceptible to deformation during drying. Each board then was cut into 990 mm and 1,700 mm long which was the length required for KD and RFV, respectively. Two 25 mm thick strips were cut between samples, 100 mm from both ends, for estimating MC for each sample at time of sawing. Samples measuring 20 mm x 20 mm x 20 mm were obtained from each logs for determination of moisture content and density variations in cross section. Initial weight of each sample was weighed and its oven-dry weight was estimated in accordance with [1]. Samples for KD were end-coated with bituminous paint in order to minimize effect of end drying.

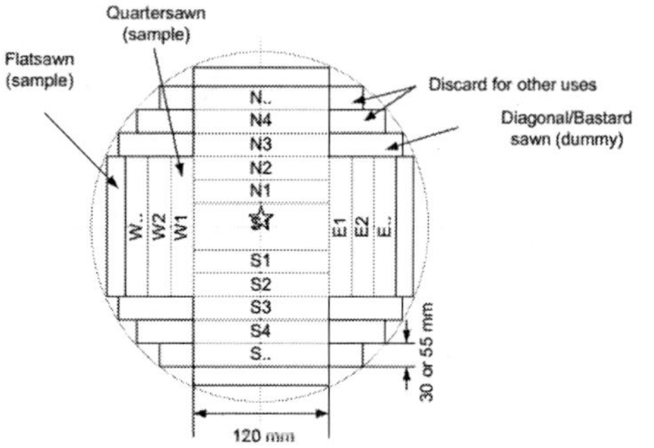

Figure 1. Sawing pattern of a log (E=East, N=North, W=West, S=South).

2.2 Kiln drying

Drying was conducted in a 0.21 m^3 capacity experimental-scale conventional kiln drying. Drying condition was controlled by setting the dry-bulb temperature and the wet-bulb temperature. Humidity was provided by steam and controlled using an automatic damper during drying. Heat was supplied by passing steam through heating coils. Air flow through the stack was provided by the fan located at one side of the kiln with air velocity of 1.5 m/s. The samples were stacked on 25 mm by 25 mm stickers. The stacks were restrained by placing iron block on top of the stack to minimize warping. Drying condition was set according to Table 1. Drying was conducted in different batch for each thickness. High steaming was provided during drying to reduce surface checking, 24 hours in the initial stage of drying and 4 hours for every 2 days.

Table 1. Kiln drying schedule.

MC (%)	DBT (°C)	WBT (°C)	RH (%)	EMC (%)
Starting	40.5	40.5		
	40.5	40		
	40.5	39.5		
Green	40.5	39	90	19.8
60	40.5	38	85	16.9
60	43.5	40	80	14.9
40	43.5	39	75	13.3
35	46	40.5	70	12.2

DBT = dry bulb temperature WBT = dry bulb temperature
RH = relative humidity EMC = equilibrium MC

2.3 Radio frequency-vacuum drying

Drying was carried out in an experimental-scale RFV dryer of 0.54 m^3 capacity. RF generator operates at a fixed frequency of 6.7 Mhz with maximum output 6kW at a maximum electrode voltage of 7.3 kV. The samples were stacked in the drying chamber without stickers between three 1.5 mm thick aluminium plates. Drying was conducted under continuous mode of heating with input current to the RF generator was set to 0.4 A and voltage of HFV tube set at 5.5 kV. Heating temperature and pressure were set to 40°C and 50 torr, lowering the boiling point of water to approximately 38°C. During drying process, sample boards were weighed every two days until the MC of between 10 to 14 % is reached.

2.4 Drying quality

The following drying characteristics were investigated in this study:
- Drying time.
- Moisture content variation within and between boards.
- Shrinkage of boards in width and thickness, determined by measuring its dimension before and after drying.
- Cupping, bowing, and spring, measured on each sample before and after drying by measuring deflection from a straight line between two ends.
- Twisting, quantified by holding three corners of the lumber against a flat surface and the rise in the fourth corner was measured.
- The residual drying stress (casehardening), evaluated by cutting 30 mm thick sections into three equal sized prongs and the central portion were removed as described by [9]. After 24 hours, the severity of casehardening, given by degree of turning in, was evaluated and classified into four classes, namely not casehardened, slightly casehardened, casehardened and reverse casehardened [9].
- Surface checks and end checks were assessed by estimating the percentage of checked area.

3 Results and discussion

The physical properties and drying characteristics of 30 mm and 55 thick kekatong boards dried using KD and RFV system are summarized in Table 2 and Table 3, respectively.

Table 2. Density and moisture content of kekatong.

	Green density (kg/m³)	Oven-dry density (kg/m³)	Basic density (kg/m³)	MC (%)
Mean	1165	970	821	42
Min	863	822	708	21
Max	1425	1189	895	59
SD	117	111	38	11

Table 3. Summary of drying characteristics of kekatong.

Characteristics		Kiln		RFV	
Thickness (mm)		30	55	30	55
Drying time (day)		14.2	27.3	6.7	16.1
Initial MC (%)		34.3 (5.4)	39.0 (9.6)	35.0 (3.4)	39.5 (6.7)
Final MC (%)		10.2 (1.4)	16.2 (2.39)	13.3 (0.9)	13.9 (2.0)
MC variation within boards	Inner part MC (%)	9.5 (1.3)	14.8 (1.4)	13.2 (0.9)	11.4 (2.1)
	Outer part MC (%)	10.5 (2.8)	18.8 (3.7)	13.2 (1.1)	14.4 (2.3)
Shrinkage	Tangential (%)	6.4 (0.9)	6.9 (0.8)	4.7 (0.8)	4.6 (1.3)
	Radial shrinkage (%)	3.3 (0.9)	3.5 (0.8)	2.8 (0.9)	3.0 (0.7)
	T/R (%)	2.1 (0.7)	2.2 (0.4)	1.5 (0.6)	1.6 (0.5)
	Volumetric shrinkage (%)	9.5 (1.0)	9.5 (1.1)	7.4 (1.4)	7.4 (1.5)
Surface checks	No. of boards with checks/Total boards	21/60	20/20	25/32	15/16
	Percentage of boards with checks[a]				
	≤ 5%	88%	55%	84%	100%
	6% to ≤ 10%	0%	5%	13%	0%
	11% to ≤ 15%	2%	20%	0%	0%
	>15%	10%	20%	3%	0%
End checks	No. of boards with checks/ Total boards	11/60	20/20	31/32	13/16
	Percentage of boards with checks[a]				
	≤ 5%	100%	55%	80%	100%
	6% to ≤ 10%	0%	35%	19%	0%
	11% to ≤ 15%	0%	7%	1%	0%
	>15%	0%	3%	%	0%
Internal checks (honeycomb)	Boards with checks (%)	0%	0%	0%	0%
Collapse	Boards with checks (%)	0%	0%	0%	0%
Cupping	No. of boards	0/60	0/20	0/32	2/16
	Deflection/Width (%)	0	0	0	0.1
Bowing	No. of boards	13/60	3/20	15/32	5/16
	Deflection/Length (%)	0.1	0.1	0.1	0.1
Spring	No. of boards	10/60	3/20	16/32	8/16
	Deflection/Length (%)	0.1	0.1	0.1	0.1
Drying stress (%)	Absent	74	35	100	90
	Slightly	7	65	0	10
	Casehardened	19	0	0	0
	Reversed casehardened	0	0	0	0

[a] Percentage of surface area with checks

3.1 Moisture content and density

The basic density of kekatong was 821 ± 38 kg/m^3 (Table 2). The average air dry density of kekatong at 15% MC is between 880 and 1,155 kg/m^3 [10]. The mean green MC was 42 ± 11%. The results obtained in this study indicated that kekatong had a low green MC but dense, which could be difficult to dry.

3.2 Drying time

Figure 2 shows the drying curve of kekatong. The 30 mm boards achieved average moisture content of 10.2% after drying in a kiln for 14.2 days while 55 mm boards achieved average MC of 16.2% after 27.3 days. RFV drying time for 30 mm boards was 6.7 days while 55 mm boards were dried to 13.9% MC in 16.1 days. Drying time of kekatong in RFV was about half of the kiln drying time.

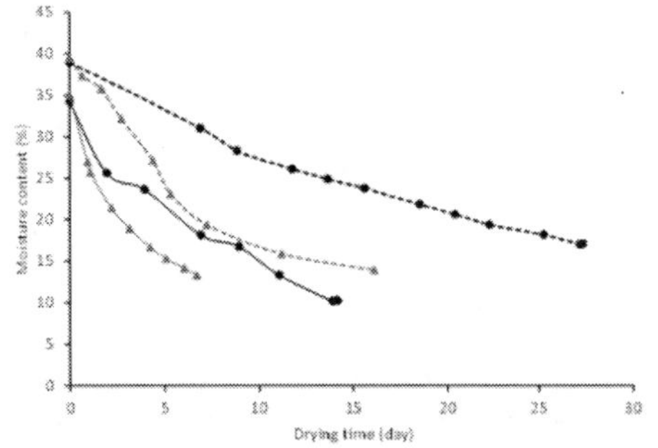

Figure 2. Drying curve of 30 mm and 55 mm thick kekatong.

3.3 Moisture content variation between samples

Final MC for RFV ranged from 11.2 to 13.9 for 30 mm and 9.5 to 15.2%for 55 mm boards, respectively while for KD, MC ranged from 10.1 to 16.2% (30 mm boards) and 13.3 to 18.7 (55 mm boards). It was observed that large variation in MC was observed for kiln dried samples and the variation of MC however was reduced in the RFV runs. The percentage boards between 10 to 14% MC, below 10% MC and above 14% is given in Table 4. RFV had more boards between 10% and 14% MC than KD. The variation of MC in RFV drying was reduced by placing the wettest lumbers at the centre of the stack, near to the positive electrode plate before and during drying. In the case of kiln drying, more even MC could be achieved by conditioning treatment at the final stage of drying.

Table 4. Percentage of boards at difference MC range.

Final MC range	KD30	KD55	RF30	RF55
< 10%	61	0	0	9
10% to 14%	34	15	86	66
14%	5	85	14	25

3.4 Moisture content variation within samples

The MC difference between the outer and inner parts for 30 mm boards were comparatively low compared to 55 mm boards for both drying methods. RFV dried samples had relatively lower average inner-outer MC difference than KD with the variation below than 2%. MC difference of up to 7% were observed in some 55 mm boards dried using KD. It was also found that a number of 30 mm boards dried using RFV had a drier inner MC compared to the outer part. This could be due to the fact that RF wave selectively heats the wetter areas in a lumber, which most of the time are in the inner.

3.5 Shrinkage

The shrinkage values for 30 and 55 thick boards dried using the same methods were essentially the same. Shrinkage in width (tangential) and thickness (radial) were substantially less for RFV. The shrinkage values for 30 mm thick boards dried using RFV were reduced to 27% tangentially and 15% radially compared to KD while for 55 mm thick boards, the reduction were about 33% and 14%. Both reductions in tangential shrinkage and in radial shrinkage resulted in approximately 30% less in T/R ratio and 22% less in volumetric shrinkage in the RFV dried boards. The reduction of T/R ratio for RFV dried timber could be primarily contributed by reduction of tangential shrinkage compared to radial shrinkage. [6] reported that, the low T/R ratio of RFV dried wood might be desirable to help prevent warping and checking during drying and improvement in dimensional stability might be expected.

3.6 Surface checks and end checks

RFV had substantially less surface and end checks than KD. Thinner boards also had lower checks than thicker boards. In KD30, RF30 and RF55, none of the surface checks were deep enough (over 4 mm deep) to cause manufacturing problems. It was observed that some surface checks developed in the initial stage of RFV drying stage were closed towards the end of drying (Figure 3). These surface checks can be easily removed by planing the boards 3 mm from the surface. However, deep checks were found on about 5% of the boards in KD55. In all drying run, the end checks were less than 50 mm long which were not observed after cross cutting 100 mm from both ends. The lower checks in RF was contributed by the lower drying stress developed due to differential MC in outer and inner surface. This may also be contributed by reduction in T+R and T/R values in RFV drying compared to kiln drying method. There were no internal checks (honeycomb) and collapse in both drying methods.

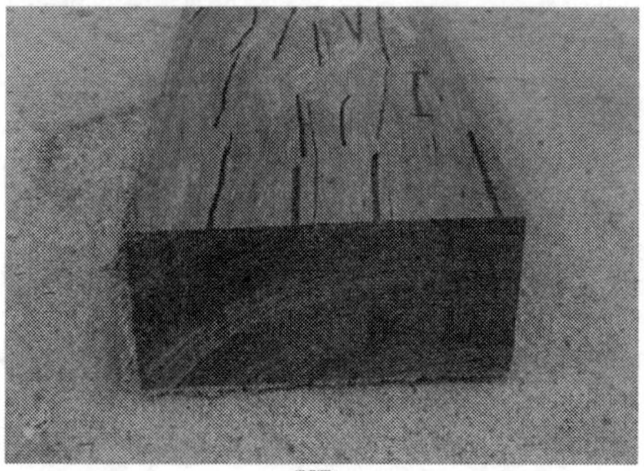

KD.

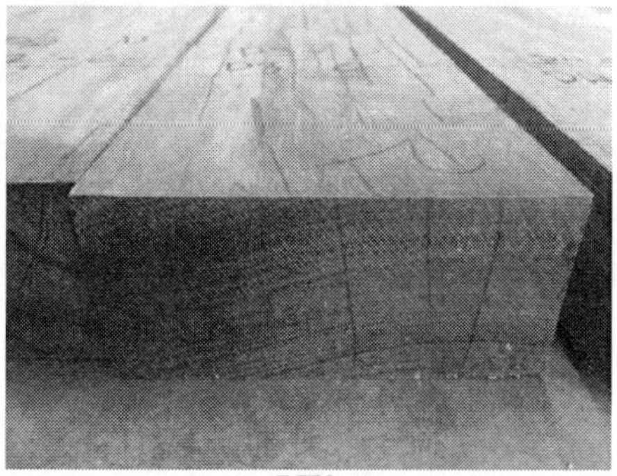

RFV.

Figure 3. Surface checks and end checks in 55 mm thick sample

3.7 Cupping, bowing, spring and twisting

The amount of cupping, bowing and spring were very low and negligible in all drying run. There was no twisting detected in all drying methods. In this study, distortion due to the differential shrinkage in radial and tangential grain orientations was restrained mechanically by hydraulic pressing during RFV drying and placing a concrete block on top of the timber stack during KD.

3.8 Drying stress

The percentage of boards with drying stress were lower in RFV compared with KD (Table 3). This was contributed by a lower moisture gradient across the timber thickness in RFV. Excessive residual drying stresses causes machining problems especially during ripping.

4 Conclusions

Drying kekatong timber using RFV drying system significantly reduced the drying time by 50% of that using conventional kiln-drying system. The quality of timber in terms of the uniformity of MC, shrinkage, surface- and end-checks, warping (cupping, bowing, spring), and drying stress was better than KD. This study has demonstrated that the RFV drying system may be used not only where quality improvement is possible, but also where kiln drying methods become very slow or inefficient toward the end of drying process. Further investigation should be conducted on improving procedure of drying other refractory and thicker heavy hardwoods including possibility to combine RFV drying with kiln drying to take advantages and limitations of these two drying technologies.

The present study was conducted on heavy hardwood species which is difficult to dry. The results however can be used as a guide for the timber manufacturer to explore the possible use of RFV for drying other timber of different thicknesses. Since the RFV drying system totally uses electricity for generating high frequency wave as well as operating the chiller, vacuum pump and hydraulic press, it is possible to incorporate RFV with KD in the initial or later stages of drying. Optimizing the drying parameters and improving the permeability of heavy hardwoods are recommended for future study.

Acknowledgement

The research was financially supported by the Ministry of Science, Technology and Innovation Malaysia (Science Fund Project No. 06-03-10-SF0214). Acknowledgement is due to Tanjong, Jamal, and Ascem for their technical assistance.

References

1. G.S. Grewal. Kiln drying characteristic of some Malaysian timbers. Timber Trade Leaflet No. 42. Ministry of Primary Industries, Malaysia (1998)
2. S. Avramidis, Liu F. Drying characteristics of thick lumber in a laboratory radio-frequency/vacuum dryer. Drying Technology **12**(8): 1963-1981 (1994)
3. R.A. Harris, M.A. Taras. Comparison of moisture content distribution, stress distribution, and shrinkage of red oak lumber dried by a radio-frequency/vacuum drying process and a conventional kiln. For. Prod. J. **34**(1): 44–54 (1984)
4. R.A. Harris, A.W.C. Lee. Properties of white pine lumber dried by radiofrequency/ vacuum process and conventional kiln process. Wood and Fib. Sci. **17**(4): 549–552 (1985)
5. H.S. Jung, C.D. Eom, B.J. So. Comparison of vacuum drying characteristics of radiata pine timber using different heating methods. Drying Tech. **22**(5): 1005–1022 (2004)
6. N. Lee, H. Jung. Comparison of shrinkage, checking, and absorbed energy in impact bending of Korean ash squares dried by a radio-frequency/vacuum process and a conventional kiln. For. Prod. J. **50**(2): 69–72 (2000)
7. R.A. Harris. Dimensional stability of red oak and eastern white pine dried by radiofrequency/vacuum and conventional drying process. For. Prod. J. **38**(2): 25–26 (1988)
8. J. Piao, N. Fujimoto, Y. Yamamoto, S. Nagata. Development of hybrid kiln drying system with radio frequency heating for the sugi heart timber. Journal of the Faculty of Agriculture, Kyushu University **52**(1): 117-121 (2007)
9. W.T. Simpson. Dry Kiln Operator's Manual Handbook AH-188, USDA Forest Service, Forest Products Laboratory: Madison, WI. Pp 11–150 (1991)
10. P.K.B. Menon. Structure and identification of Malayan woods. Revised by S. Ani, S.C. Lim. Malayan Forest Records No. 25. Forest Research Institute Malaysia: Kepong. P. 55 (1993)

Improved State Feedback H$_\infty$ Control for Flexible Air-Breathing Hypersonic Vehicles on LMI Approach

Xue Zhang , Xiaogeng Liang

Luoyang Opto-Electro Technology Development Center, China

Abstract. Focusing on a nonlinear longitudinal dynamical model for Air-breathing Hypersonic Flight Vehicles (AHFV), a linearized model on a nominal trim condition is proposed. To stabilize the flight of an AHFV in the presence of external disturbances and actuator uncertainties, a state feedback H_∞ control is designed. With bounds on the uncertainties, a feedback stabilization problem is converted to an optimal control problem and the cost function is minimized by solving a set of linear matrix inequalities. Since uncertainties in the design of AHFV are inevitable, to make a comparison, a general H_∞ robust controller is constructed by only considering the disturbances firstly. Then the results are extended by incorporating the actual existing uncertainties as well as the external disturbances in the AHFV system. Numerical simulation shows that the controller, which takes both disturbances and uncertainties into account, can effectively stabilize the AHFV system.

1 Introduction

Considering the high demand for extremely fast global strike capability and development of defense systems, the research on hypersonic vehicles has been of a great importance for many researchers. Since the 1970s, the hypersonic speed has generally been defined as at least the value of Mach 5 (five times the speed of sound). High velocity may cause a hypersonic vehicle to be highly sensitive to the change of flight conditions [1], such as the Mach number, and the angle of attack. A comparison among subsonic, supersonic and hypersonic speeds shows that the drag and lift forces become non-linear functions of the angle of attack at hypersonic speeds [1]. The loss of linearity of some parameters may lead the system to be less robust and unstable. An Air-breathing Hypersonic Flight Vehicle (AHFV), such as X-43A, usually is equipped with an air breathing engine which can scoop oxygen from the air. Because of the highly integrated design between airframe and engine, the dynamic characteristics of hypersonic vehicles change drastically with the operating conditions and mass distribution. Meanwhile, the inaccuracy of a mathematical model or control law may also add some uncertainties. In addition, the disturbances from outside atmosphere and density variation at a high Mach number also add uncertainties in the performance of air-breathing propulsion systems [2]. Therefore, a design of a robust and performance oriented control system is the key issue in the design of an Air-breathing Hypersonic Flight Vehicle (AHFV).

Many strategies related to efficient controllers for the hypersonic vehicles have been presented, including adaptive control [3] [4], sliding mode control [5] [6], and H_∞ control [7] [8]. Considering uncertainties due to both aerodynamic and center of gravity, adaptive controllers are suitable for command tracking [9]. However, for an unknown input matrix, the results of the adaptive control have only local stability. The sliding mode control can preferably solve the model uncertain problem [5], and it is insensitive to the out-side interference [6]. But more control force is needed in this method. Due to high parameter perturbation of controller, the air-breathing hypersonic flight model is extremely complex. Therefore, although the single H_∞ controller or μ −synthesis methods may achieve the desired performance [10] [11] [12] [13] [14], it is not easy for these controllers to guarantee the stability and desired performance.

In this paper, the problem of stability is solved by designing an improved state feedback H_∞ control law for a rigid body AHFV. Therefore, the stabilization design problem of a nonlinear system is converted into a problem of optimal control. Since uncertainties are inevitable, to make a comparison, a general controller is constructed by only considering the disturbances firstly. Then the results are extended by incorporating the actual existing uncertainties as well as the external disturbances. Ideally, the control model is MATLAB-based and the simulation results show that the control law is effective and can facilitate the development of the control system.

This paper is organized as follows: Section II presents a linearized AHFV dynamical model. In Section III, by considering only disturbances, a general H_∞ robust controller is presented. Moreover, this control law is further developed by taking both disturbances and uncertainties into account. The simulation results are shown in Section IV, which demonstrates the effectiveness of the proposed control law. Section V draws conclusions.

Notation: The notation used in this paper is fairly standard. The superscript 'T' indicates matrix transposition; $\mathbb{R}$ denotes the n-dimensional Euclidean space; sym(P) denotes $P + P^T$); The asterisk (*) is used to represent a term that is introduced by symmetry. For a real symmetric matrix P , $P > 0$ means P is positive definite and $P < 0$ means P is negative definite.

2 Longitudinal dynamical model of hypersonic flight

In this paper, the longitudinal model of an AHFV is adopted from [15]. The aerodynamic coefficients are the typical cruising flight coefficients with Mach number 15. Flight velocity, altitude, path angle and the pitch rate are respectively 15060 ft/s, 110000 ft, 0 deg, 0 deg/s governed by the following equations:

$$\dot{V} = (T\cos\alpha - D)/m - \mu\sin\theta/r^2$$

$$\dot{\theta} = (L + T\sin\alpha)/(mV) - [(\mu - V^2 r)\cos\theta]/(Vr^2)$$

$$\dot{h} = V\sin\theta$$

$$\dot{\alpha} = q - \dot{\theta}$$

$$\dot{q} = M_{yy}/I_{yy}$$

Where V(ft/s) , θ(deg) , h(ft) , α(deg) , q(deg/s) , m(slug) , $\mu(\mathrm{ft}^3/\mathrm{s}^2)$, and $I_{yy}(\mathrm{slug}\cdot\mathrm{ft}^2)$ denote velocity, flight-path angle, altitude, angle of attack, pitch rate, mass, gravitational constant, and moment of inertia, respectively. The equations for lift L(1bf) , drag D(1bf) , thrust T(1bf) , pitching moment M_{yy}(1bf·ft) , and the radial distance from earth's center r(ft) are:

$$L = \frac{1}{2}\rho V^2 S C_L$$

$$D = \frac{1}{2}\rho V S^2 C_D$$

$$T = \frac{1}{2}\rho V S^2 C_T$$

$$r = h + R_E$$

$$M_{yy} = \frac{1}{2}\rho V^2 S\bar{c}[C_M(\alpha) + C_M(\delta_e) + C_M(q)]$$

with the following lift coefficient C_L , drag coefficient C_D , thrust coefficient C_L ,and pitching moment coefficients C_α, C_{δ_e}, C_q:

$$C_L = \alpha(0.493 + 1.91/M)$$

$$C_D = 0.0082(171\alpha^2 + 1.15\alpha + 1) \times (0.0012M^2 - 0.054M + 1)$$

$$C_T = \begin{cases} C_{*T}(1+0.15)\beta & \text{if}\beta < 1 \\ C_{*T}(1+0.15\beta) & \text{if}\beta \ge 1 \end{cases}$$

$$C_M(\alpha) = 10^{-4}(0.06 - e^{-M/3}) \times (-6565\alpha^2 + 6875\alpha + 1)$$

$$C_M(\delta_e) = 0.0292(\delta_e - \alpha)$$

$$C_M(q) = \frac{q\bar{c}}{2V}(-0.025M + 1.37) \times (-6.83\alpha^2 + 0.303\alpha - 0.23)$$

where

$$C_{*T} = 0.0105[1 - 164(\alpha - \alpha_0)^2](1 + 17/M)$$

M is the Mach number,

$$M = V/a \quad (a = 8.99\times10^{-9}h^2 - 9.16\times10^{-4}h + 996)$$

and ρ denotes the air density $\rho = 2.38\times10^{-3-(h/24000)}$.

The longitudinal model of the hypersonic vehicle can be transformed into a general MIMO nonlinear system form [16]:

$$\dot{X} = f(X) + \sum_{k=1}^{2} g_k(X)U_k = F(X,U)$$

where the state and control vectors are $X = [V, \theta, h, \alpha, q]^{\mathrm{T}}$ and $U = [\beta, \delta_e]^{\mathrm{T}}$.

In order to simplify the design of controller for the longitudinal model, the model is linearized which is represented by:

$$\begin{cases} \dot{x}(t) = Ax(t) + B_1 u(t) \\ z(t) = Cx(t) \end{cases}$$

$$A = \frac{\partial F(X,U)}{\partial X}\Big|_{X=X_0,U=U_0} = \begin{bmatrix} a_{11} & a_{12} & a_{13} & a_{14} & 0 \\ a_{21} & a_{22} & a_{23} & a_{24} & 0 \\ a_{31} & a_{32} & 0 & 0 & 0 \\ -a_{21} & -a_{22} & -a_{23} & -a_{24} & 1 \\ a_{51} & 0 & a_{53} & a_{54} & a_{55} \end{bmatrix}$$

$$B_1 = \frac{\partial F(X,U)}{\partial U}\Big|_{X=X_0,U=U_0} = \begin{bmatrix} b_{11} & 0 \\ b_{21} & 0 \\ 0 & 0 \\ -b_{21} & 0 \\ 0 & b_{52} \end{bmatrix}.$$

and

$$a_{11} = \frac{\rho S}{2m}\left[(2VC_T + V^2\frac{\partial C_T}{\partial V})\cos\alpha - 2VC_D - V^2\frac{\partial C_D}{\partial V}\right],$$

$$a_{12} = \frac{-\mu\cos\theta}{r^2},$$

$$a_{13} = -\frac{\rho V^2 S}{48000m}(C_T\cos\alpha - C_D) + \frac{2\mu\sin\theta}{r^3},$$

$$a_{14} = \frac{\rho V^2 S}{2m}(1 + \frac{17}{M})(1+0.15)\beta \times 0.0105\{-328(\alpha - \alpha_0) \times\cos\alpha - [1 - 164(\alpha-\alpha_0)^2]\sin\alpha\} - \frac{\rho V^2 S}{2m} \times 0.0082(0.0012M^2 - 0.054M + 1)(342\alpha + 1.15),$$

 DOI: 10.1051/matecconf/201710810002

$$a_{21} = \frac{-1}{mV^2}(L + T\sin\alpha) + \frac{1}{mV}(\rho VSC_L + \rho VSC_T \sin\alpha) + \frac{\mu\cos\theta}{V^2 r^2} + \frac{\cos\theta}{r},$$

$$a_{22} = \frac{\mu\sin\theta}{Vr^2} - \frac{V\sin\theta}{r},$$

$$a_{23} = -\frac{\rho VS}{48000m}(C_L + C_T\sin\alpha) + \frac{2\mu\cos\theta}{Vr^3} - \frac{V\cos\theta}{r^2},$$

$$a_{24} = \frac{1}{2mV}\left[\rho V^2 S(0.493 + \frac{1.19}{M}) + 3.9606\rho V^2 S \times (1 + \frac{17}{M})\beta(\alpha - \alpha_0) + 2T\cos\alpha\right],$$

$$a_{31} = \sin\theta, \quad a_{32} = V\cos\theta, \quad a_{51} = \frac{1}{I_{yy}} M_{yy_v},$$

$$a_{53} = \frac{1}{I_{yy}} M_{yy_h}, \quad a_{54} = \frac{1}{I_{yy}} M_{yy_\alpha}, \quad a_{55} = \frac{1}{I_{yy}} M_{yy_q},$$

$$M_{yy_v} = \rho VS\bar{c} \times [C_M(\alpha) + C_M(\delta_e) + C_M(q)] + \frac{1}{2}\rho V^2 S\bar{c}\left(\frac{\partial C_M(\alpha)}{\partial V} + \frac{\partial C_M(\delta_e)}{\partial V} + \frac{\partial C_M(q)}{\partial V}\right),$$

$$M_{yy_h} = \frac{1}{2}\frac{\partial\rho}{\partial h} V^2 S\bar{c}[C_M(\alpha) + C_M(\delta_e) + C_M(q)] + \frac{1}{2}\rho V^2 S\bar{c}\left(\frac{\partial C_M(\alpha)}{\partial M}\frac{\partial M}{\partial h} + \frac{\partial C_M(\delta_e)}{\partial M}\frac{\partial M}{\partial h} + \frac{\partial C_M(q)}{\partial M}\frac{\partial M}{\partial h}\right),$$

$$M_{yy_\alpha} = \frac{1}{2}\rho V^2 S\bar{c}[\frac{\partial C_M(\alpha)}{\partial\alpha} + \frac{\partial C_M(\delta_e)}{\partial\alpha} + \frac{\partial C_M(q)}{\partial\alpha}]$$

$$M_{yy_q} = \frac{1}{4}\rho VS\bar{c}^2(-0.025M + 1.37) \times (-6.83\alpha^2 + 0.303\alpha - 0.23),$$

$$b_{11} = \frac{1}{m}T_\beta\cos\alpha, \quad b_{21} = \frac{1}{mV}T_\beta\sin\alpha, \quad b_{52} = \frac{1}{I_{yy}}M_{yy\delta},$$

$$T_\beta = 0.0060375\rho V^2 S[1 - 164(\alpha - \alpha_0)^2](1 + \frac{17}{M}),$$

$$M_{yy\delta} = 0.0146\rho V^2 S\bar{c}.$$

3 An improved state feedback H_∞ controller design

In this work, the control objective is to robustly stabilize the flight of AHFV in the presence of disturbances and uncertainties. To transform a feedback stabilizability problem into an optimal control problem, an improved state feedback H_∞ control is proposed by considering the Linear-Quadratic Regulator (LQR) control, which involves finding a controller to minimize the value of following cost function:

$$J_L = \int_0^\infty [u^T(t)Ru(t) + x^T(t)Qx(t)]dt,$$

where $R \in \mathbb{R}^{2\times2}$ and $Q \in \mathbb{Q}^{5\times5}$ are given positive-definite symmetric matrices.

The conventional LQR control does not consider disturbances which may significantly compromise the stability of system. In order to increase the robustness, the external disturbances $\omega(t)$ should be taken into account. For a guaranteed cost controller $u(t) = Kx(t)$, a system with disturbances can be expressed as follows:

$$\begin{cases} \dot{x}(t) = (A + B_1K)x(t) + B_2\omega(t), \\ z(t) = C_1x(t), \end{cases} \quad (1)$$

where B_2 is a constant matrix and $z(t)$ is the controlled output. The objective is to find a desired controller $u(t) = Kx(t)$ to make the system (1) asymptotically stable and minimize the cost function. Hence, the analysis of stability performance is required.

Theorem 1 Consider the AHFV dynamics (1) with given parameters. The system is asymptotically stable with a given H_∞ performance if there exists a positive definite sym-metric matrix $P = P^{\mathrm{T}} > 0$ such that the following condition holds:

$$\begin{bmatrix} \mathrm{sym}(P(A+B_1K)) & PB_2 & \begin{bmatrix} Q^{\frac{1}{2}} & 0 \end{bmatrix} + K^{\mathrm{T}}\begin{bmatrix} 0 & R^{\frac{1}{2}} \end{bmatrix} \\ * & -\gamma^2 I & 0 \\ * & -\gamma^2 I & -I \end{bmatrix} < 0 \quad (2)$$

Proof. Define a Lyapunov function as

$$V(t) = x^{\mathrm{T}}(t)Px(t), \quad (3)$$

where P is a positive-definite symmetric matrix. The H_∞ cost function is as follows:

$$J_H = \int_0^\infty [z^{\mathrm{T}}(t)z(t) - \gamma^2\omega^{\mathrm{T}}(t)\omega(t) + \dot{V}]dt,$$

where $\dot{V}$ is the derivative of $V(t)$.

In order to connect the H_∞ performance index with LQR control, a new vector is defined:

$$z(t) = \begin{bmatrix} Q^{\frac{1}{2}} & 0 \end{bmatrix}^{\mathrm{T}} x(t) + \begin{bmatrix} 0 & R^{\frac{1}{2}} \end{bmatrix}^{\mathrm{T}} u(t).$$

It can be proved that

$$z^{\mathrm{T}}(t)z(t) = u^{\mathrm{T}}(t)Ru(t) + x^{\mathrm{T}}(t)Qx(t)$$

Thus, by considering the disturbances under the zero initial condition, the improved LQR cost function with a prescribed H_∞ performance can be rewritten as

$$J = \int_0^\infty [u^{\mathrm{T}}(t)Ru(t) + x^{\mathrm{T}}(t)Qx(t) - \gamma^2\omega^{\mathrm{T}}(t)\omega(t) + \dot{V}]dt.$$

The integrand of cost function is denoted as J_*

$$J_* = z^{\mathrm{T}}(t)z(t) - \gamma^2\omega^{\mathrm{T}}(t)\omega(t) + \dot{V},$$

$$= \begin{bmatrix} x(t) \\ \omega(t) \end{bmatrix}^{\mathrm{T}} \begin{bmatrix} \mathrm{sym}(P(A+B_1K)) + C_1^{\mathrm{T}}C_1 & PB_2 \\ B_2^T P & -\gamma^2 I \end{bmatrix} \begin{bmatrix} x(t) \\ \omega(t) \end{bmatrix},$$

where

$$C_1 = \begin{bmatrix} Q^{\frac{1}{2}} & 0 \end{bmatrix}^{\mathrm{T}} + K^{\mathrm{T}}\begin{bmatrix} 0 & R^{\frac{1}{2}} \end{bmatrix}^{\mathrm{T}}.$$

According to the Schur complement, the conditional in equality can be replaced by (23):

$$\begin{bmatrix} \mathrm{sym}(P(A+B_1K)) & PB_2 \\ B_2^TP & -\gamma^2 I \end{bmatrix} - \begin{bmatrix} C_1^{\mathrm{T}} \\ 0 \end{bmatrix} [-I][C_1 \quad 0] < 0.$$

Then

$$\begin{bmatrix} \mathrm{sym}(P(A+B_1K)) + C_1^TC_1 & PB_2 \\ B_2^{\mathrm{T}}P & -\gamma^2 I \end{bmatrix} < 0$$

Therefore:

$$J_* < 0,$$

and

$$\begin{aligned} &\mathrm{sym}(P(A+B_1K)) + C_1^TC_1 < 0, \\ &\mathrm{sym}(P(A+B_1K)) < 0. \end{aligned} \tag{4}$$

The quadratic Lyapunov function candidate is shown as (3) and taking the time derivative of V (t) will result:

$$\dot{V}(t) = x^{\mathrm{T}}(t)\mathrm{sym}(P(A+B_1K))x(t).$$

Thus $\dot{V} < 0$, which means the system is stable.

On the other hand, since $J_* < 0$, the following inequality holds:

$$\int_0^{\infty} \dot{V} dt + \int_0^{\infty} (z^{\mathrm{T}}(t)z(t) - \gamma^2 \omega^{\mathrm{T}}(t)\omega(t))dt < 0,$$

Assuming zero condition, integration from 0 to ∞ yields:

$$\| z(t) \|_2^2 < \gamma^2 \| \omega(t) \|_2^2 .$$

Therefore, the H_∞ performance is guaranteed. The proof is completed.

Now, the minimum optimal problem is converted into a problem of solving a prescribed H_∞ control, which will significantly enhance the robustness of system. In this control problem, transfer function from disturbances $\omega(t)$ to output $z(t)$ would satisfy the H_∞ norm bound restriction. At the same time, by substituting the control input $u(t)$ into (1), the control system is turned into:

$$\begin{cases} \dot{x}(t) = (A+B_1K)x(t) + B_2\omega(t), \\ z(t) = \begin{bmatrix} Q^{\frac{1}{2}} & 0 \end{bmatrix}^{\mathrm{T}} x(t) + K^{\mathrm{T}} \begin{bmatrix} 0 & R^{\frac{1}{2}} \end{bmatrix}^{\mathrm{T}} x(t). \end{cases}$$

Furthermore, in order to obtain desired dynamics of the closed-loop systems, usually some pole placement constraints are considered. In this paper a disc pole placement region is chosen [17].

Lemma 1 (Disc region) Let $\mho(\eta, r)$ denotes any disc region centered in η with radius r in a complex plane ($\eta, r \in \mathbb{R}$ and $r > 0$). Then all the eigenvalues of A in (1) lie in the region $\mho(\eta, r)$ if and only if there exists a matrix $P > 0$ satisfying

$$\begin{bmatrix} -P & P(\bar{A} - \eta I) \\ * & -r^2 P \end{bmatrix} < 0. \tag{6}$$

where $\bar{A} = A + B_1K$.

Now, a multi-objective controller can be obtained by setting $P = X^{-1}$.

Theorem 2 Consider the system in (1). If the scalars r and η are known and there exist matrices $X > 0$ and Y with appropriate dimensions and also the following LMIs are satisfied

$$\begin{bmatrix} AX + XA^{\mathrm{T}} + B_1Y + Y^{\mathrm{T}}B_1^{\mathrm{T}} & B_2 & X\begin{bmatrix} Q^{\frac{1}{2}} & 0 \end{bmatrix} + Y^{\mathrm{T}}\begin{bmatrix} 0 & R^{\frac{1}{2}} \end{bmatrix} \\ * & -\gamma^2 I & 0 \\ * & * & -I \end{bmatrix} < 0 \tag{7}$$

and

$$\begin{bmatrix} -X & AX + B_1Y - \eta X \\ * & -r^2 X \end{bmatrix} < 0, \tag{8}$$

Then there is a proper controller K which can make system described by (1) asymptotically stable and the poles lie in a disc region. The nominal controller gain matrix is given by:

$$K = YX^{-1}. \tag{9}$$

Proof. In order to get the state feedback controller K, congruence transformation is performed on (4) by P^{-1}

$$P^{-1}[P(A+B_1K) + (A+B_1K)^{\mathrm{T}}P^{\mathrm{T}}]P^{-1} < 0.$$

Accordingly,

$$(A+B_1K)P^{-1} + P^{-1}(A+B_1K)^{\mathrm{T}} < 0.$$

Let $X = P^{-1}$ ($X = X^{\mathrm{T}} > 0$),therefore:

$$AX + B_1KX + XA^{\mathrm{T}} + XK^{\mathrm{T}}B_1^{\mathrm{T}} < 0.$$

To calculate the inequality, it is denoted that $Y = KX$, thus

$$AX + B_1Y + XA^{\mathrm{T}} + Y^{\mathrm{T}}B_1^{\mathrm{T}} < 0,$$

and the controller is:

$$K = YX^{-1}.$$

When $K = YX^{-1}$, performing the congruence transformation to (7) by diag $\{P, I, I\}$, condition (2) is obtained. On the other hand, performing congruence transformation to (8) by diag$\{P, P\}$, one can get (6). The proof is finished.

Remark 1 In the problem formulation above, only external disturbances are considered. However, the actuator uncertainties are inevitable in the practical control system. Therefore, it is quite necessary to incorporate both external disturbances and actuator uncertainties factors to the system model.

A controller is designed to make AHFV asymptotically stable with a given H_∞ attenuation level γ considering external disturbances as well as actuator uncertainties. Assuming the actuator uncertainties can be denoted by ΔA and actual system matrix $\bar{A} = A + \Delta A$, the AHFV dynamical system can be written as

$$\begin{cases} \dot{x}(t) = (A+\Delta A)x(t) + B_1u(t) + B_2\omega(t), \\ z(t) = C_1x(t). \end{cases} \tag{10}$$

The uncertain matrix ΔA is practically assumed to be norm bounded, which means

$$\Delta A = LN(t)E, \tag{11}$$

where L and E are known constant matrices of certain appropriate dimensions. $N(t)$ is an unknown time-varying variable which satisfies $N^T(t)N(t) \le I$.

In order to deal with the uncertainties, the following lemma is used [18]:

Lemma 2 Let $\Theta = \Theta^T$, $\bar{L}$ and $\bar{E}$ be real matrices with compatible dimensions and $N(t)$ be time-varying variable which satisfies (11). Then the following condition

$$\Theta + \bar{L}N(t)\bar{E} + \bar{E}^{\mathrm{T}}N^{\mathrm{T}}(t)\bar{L}^{\mathrm{T}} < 0, \qquad (12)$$

holds if and only if there exists a positive scalar $\varepsilon > 0$ such that

$$\begin{bmatrix} \Theta & \varepsilon\bar{L} & \bar{E}^{\mathrm{T}} \\ * & -\varepsilon I & 0 \\ * & * & -\varepsilon I \end{bmatrix} < 0. \qquad (13)$$

The following theorem is derived to eliminate the influence from the actuator uncertainties.

Theorem 3 Consider the AHFV dynamics in (10) with given parameters. The system is asymptotically stable with a given H_∞ performance γ if there exists a positive-definite symmetric matrix $P = P^{\mathrm{T}}$ such that the following inequality holds

$$\begin{bmatrix} \mathrm{sym}(P(A+B_1K)) & PB_2 & \begin{bmatrix} Q^{\frac{1}{2}} & 0 \end{bmatrix} + K^{\mathrm{T}}\begin{bmatrix} 0 & R^{\frac{1}{2}} \end{bmatrix} & PL & \varepsilon E^{\mathrm{T}} \\ * & -\gamma^2 I & 0 & 0 & 0 \\ * & * & -I & 0 & 0 \\ * & * & * & -\varepsilon I & 0 \\ * & * & * & * & -\varepsilon I \end{bmatrix} < 0 \qquad (14)$$

Proof. According to Lemma 2, the system described by (10) is asymptotically stable with a given H_∞ performance γ if the condition in (13) holds. Therefore, the condition (14) can be changed into

$$\begin{bmatrix} \mathrm{sym}(P(A+B_1K)) & PB_2 & \begin{bmatrix} Q^{\frac{1}{2}} & 0 \end{bmatrix} + K^{\mathrm{T}}\begin{bmatrix} 0 & R^{\frac{1}{2}} \end{bmatrix} \\ * & -\gamma^2 I & 0 \\ * & * & -I \end{bmatrix} + \begin{bmatrix} PL \\ 0 \\ 0 \end{bmatrix} N(t)\begin{bmatrix} E & 0 & 0 \end{bmatrix} + \begin{bmatrix} E^{\mathrm{T}} \\ 0 \\ 0 \end{bmatrix} N^{\mathrm{T}}(t)\begin{bmatrix} L^{\mathrm{T}}P^{\mathrm{T}} & 0 & 0 \end{bmatrix} < 0. \qquad (15)$$

By means of Lemma 2, the matrix inequality (14) can be obtained. This completes the proof.

Theorem 4 (Disc region) Let $\mho(\eta, r)$ denotes any disc region centered in η with radius r in a complex plane ($\eta, r \in \mathbb{R}$ and $r > 0$). Then all the eigenvalues of $A + \Delta A$ in (10) lie in the region $\mho(\eta, r)$ if and only if there exists a matrix $P > 0$ satisfying

$$\begin{bmatrix} -P & P(\bar{A} - \eta I) & \varepsilon PL & 0 \\ * & -r^2 P & 0 & E^{\mathrm{T}} \\ * & * & -\varepsilon I & 0 \\ * & * & * & -\varepsilon I \end{bmatrix} < 0. \qquad (16)$$

Proof. When taking the uncertain matrix ΔA into account, Lemma 1 can be changed into

$$\begin{bmatrix} -P & P(\bar{A} + \Delta A - \eta I) \\ * & -\gamma^2 P \end{bmatrix} < 0, \qquad (17)$$

since $\Delta A = LN(t)E$, (17) can be rewritten as

$$\begin{bmatrix} -P & P(\bar{A} - \eta I) \\ * & -r^2 P \end{bmatrix} + \begin{bmatrix} PL \\ 0 \end{bmatrix} N(t)\begin{bmatrix} 0 & E \end{bmatrix} + \begin{bmatrix} 0 \\ E^T \end{bmatrix} N^{\mathrm{T}}(t)\begin{bmatrix} L^{\mathrm{T}}P^{\mathrm{T}} & 0 \end{bmatrix} < 0. \qquad (18)$$

By means of Lemma 2, the matrix inequality (16) can be obtained. This completes the proof.

Under an uncertain precondition, the H_∞ based LQR control design with pole placement can be presented in the following theorem.

Theorem 5 Consider the system in (10). If the scalars r and η are known and there exist matrices $X > 0$ and Y with appropriate dimensions, the following LMIs are satisfied

$$\begin{bmatrix} AX + XA^{\mathrm{T}} + B_1Y + Y^{\mathrm{T}}B_1^{\mathrm{T}} & B_2 & X\begin{bmatrix} Q^{\frac{1}{2}} & 0 \end{bmatrix} + Y^{\mathrm{T}}\begin{bmatrix} 0 & R^{\frac{1}{2}} \end{bmatrix} & \varepsilon L & XE^{\mathrm{T}} \\ * & -\gamma^2 I & 0 & 0 & 0 \\ * & * & -I & 0 & 0 \\ * & * & * & I & 0 \\ * & * & * & * & -\varepsilon I \end{bmatrix} < 0, \qquad (19)$$

and

$$\begin{bmatrix} -X & AX + B_1Y - \eta X & \varepsilon L & 0 \\ * & -r^2 X & 0 & XE^{\mathrm{T}} \\ * & * & -\varepsilon I & 0 \\ * & * & * & -\varepsilon I \end{bmatrix} < 0. \qquad (20)$$

Then there is a proper controller K can make system (10) asymptotically stable with poles lying in a disc region. The nominal controller gain matrix is given by:

$$K = YX^{-1}. \qquad (21)$$

Proof. Applying congruence transformation to (14) by diag $\{P^{-1}, I, I, I, I\}$, and set $X = P^{-1}, Y = KP^{-1}$, will result:

$$\begin{bmatrix} AX + XA^{\mathrm{T}} + B_1Y + Y^{\mathrm{T}}B_1^{\mathrm{T}} & B_2 & X\begin{bmatrix} Q^{\frac{1}{2}} & 0 \end{bmatrix} + Y^{\mathrm{T}}\begin{bmatrix} 0 & R^{\frac{1}{2}} \end{bmatrix} & L & X\varepsilon E^{\mathrm{T}} \\ * & -\gamma^2 I & 0 & 0 & 0 \\ * & * & -I & 0 & 0 \\ * & * & * & -\varepsilon I & 0 \\ * & * & * & * & -\varepsilon I \end{bmatrix} < 0, \quad (22)$$

Then, left-multiply (22) by diag $\{I, I, I, I, \varepsilon^{-1}\}$ and right-multiply (22) by the transpose of diag $\{I, I, I, I, \varepsilon^{-1}\}$. Next, replace ε^{-1} by ε_1 and multiply the resulting LMI by diag $\{I, I, I, \varepsilon_1, I\}$ on the left and its transpose on the right to obtain (19). Additionally, by application of congruence transformation on (20) using diag $\{P, P, I, I\}$, the condition (16) can be obtained.

Theorem 5 indicates that even in the presence of unavoidable actuator uncertainties, the robustness of the real system can be guaranteed by computing the inequalities.

4 Simulation results

To illustrate the performance of improved state feedback H_∞ controller, some numerical simulations are done using MATLAB. The parameters used in this study are adopted from [15] as listed in Table 1.

Table 1. Trim Condition States Value.

Symbol	Value	Symbol	Value
h	110000ft	θ	0rad
S	3603ft^2	β	0.183
$\bar{c}$	80ft	δ	-0.0066
m	9375slug	V	15060ft/s
I_{yy}	7×10^6slug·ft^2	α	0.0315rad
R_E	20903500ft	M	15
μ	1.39×10^{16} ft$^3/s^2$	q	0rad/s
α_0	0.0315rad		

Under a straight-and-level flight trim condition, the values for state vectors and input control are:

$$x_0 = [15060 \quad 0 \quad 110000 \quad 0.0315 \quad 0]^{\mathrm{T}},$$

$$u_0 = [0.183 \quad -0.0066]^{\mathrm{T}}.$$

The values of A, B_1 are shown as follows:

$$A = \begin{bmatrix} -3.4700\times10^{-5} & -31.4788 & -7.2040\times10^{-6} & -47.8373 & 0 \\ 2.7135\times10^{-7} & 0 & -5.3151\times10^{-8} & 0.0406 & 0 \\ 0 & 15060 & 0 & 0 & 0 \\ -2.7135\times10^{-7} & 0 & 5.3151\times10^{-8} & -0.0406 & 1 \\ 5.5375\times10^{-6} & 0 & -1.4010\times10^{-7} & 0.5923 & -0.0682 \end{bmatrix},$$

$$B_1 = \begin{bmatrix} 27.2963 & 0 \\ 5.7113\times10^{-5} & 0 \\ 0 & 0 \\ -5.7113\times10^{-5} & 0 \\ 0 & 3.3168 \end{bmatrix}.$$

When there are only disturbances, the gain matrices K_1 can be computed by (9). The external disturbances are modeled by Dryden turbulence model, and the turbulence spectrum is defined by the weight [13]:

$$\omega(s) = \sqrt{\frac{2V_0\sigma_u^2}{L_u}}\frac{1}{s + V_0 / L_u},$$

$$\sigma_u = 5.8ft/s, L_u = 65574ft/s.$$

The matrix of external disturbances is $B_2 = [0 \quad 0.5 \quad 0 \quad 0.5 \quad 0]^{\mathrm{T}}$. Within a desired disc region, assuming $Q = 0.0001I$, $R = 0.01I$, controller $u(t) = Kx(t)$ is designed to make the system asymptotically stable. It is assumed that the disc region is $\mho(-8,10)$ and $Q = 0.0001I$, $R = 0.01I$. The gain matrices K_1 is

$$K_1 = \begin{bmatrix} -0.5636 & -32.7615 & -0.0089 & 1.6366 & -0.0023 \\ 0.0454 & -2.1398\times10^4 & -4.7814 & -132.1430 & -9.7263 \end{bmatrix} \quad (23)$$

Minimum disturbance attenuation level is $\gamma_1 = 3.4652$.

The uncertainties inherently exist in the system and will inevitably degrade the performance of AHFV. Hence, it is important to consider these uncertainties which may significantly affect the stability of system.

By using the approach developed in this paper, the gain matrix K_2 can be computed by (21).

$$K_2 = \begin{bmatrix} -0.6245 & 1.8893\times10^3 & -0.1323 & 16.5366 & 0.5268 \\ 0.8663 & -2.0088\times10^4 & -3.1857 & -136.7458 & -1.1669 \end{bmatrix}, \quad (24)$$

The minimum disturbance attenuation level is $\gamma_2 = 5.4144$.

Figures 1 and 2 show the simulation results when uncertainties exist. As shown, the controller developed in [19] cannot stabilize the system while the controller developed in this paper is able to attenuate the uncertainties and can quickly enter the stable state when uncertainties happen.

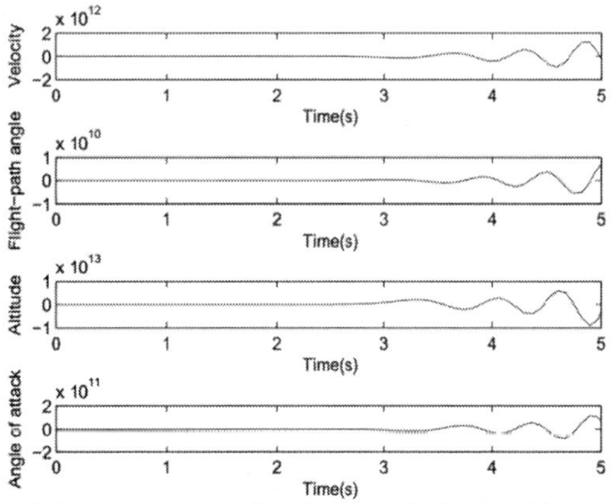

Figure 1. Output responses using the approach developed in [19].

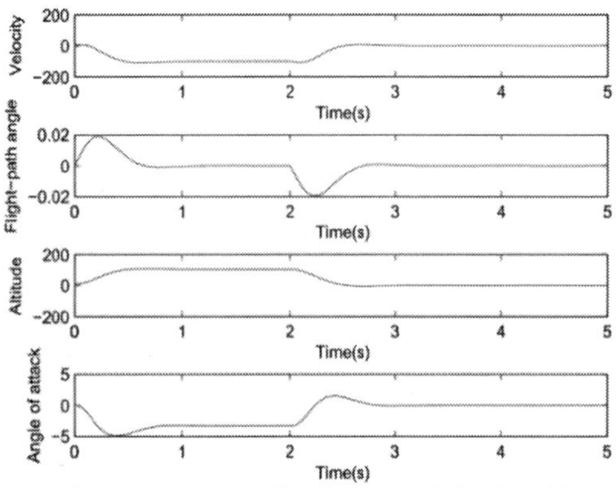

Figure 2. Output responses using the approach developed in this paper.

Remark 2 This paper is motivated by [19]. Gao et al presenting an LQR controller but the actual existing disturbances $\omega(t)$ and uncertainties ΔA are not considered. In this paper, firstly, when there are only disturbances, an improved state feedback H_∞ controller is developed by considering the H_∞ and LQR indexes, which can increase the robustness of system. Secondly, since the actual existing disturbances is inevitable, it is critical to take the actual existing disturbances into account in terms of both theoretical merits and practical application perspective. When there are both disturbances and uncertainties, the output responses in the control mode of [19] cannot converge to an equilibrium state, however, using the control law in this paper, the output responses can rapidly converge to equilibrium states, which means the stability performance of AHFV system is significantly improved.

5 Conclusions

In this paper, a technique of designing improved state feedback H_∞ controller has been developed. Indexes for H_∞ performance, quadratic performance and pole placement are concerned together. By converting the mixed cost function, the multi-objective controller can be deduced by LMIs. Moreover, in terms of a more complete consideration on both external disturbances and actuator uncertainties, the simulation results in [19] and in this paper are presented to make a clear comparison. The results show that, the controller developed in this paper, which takes the actual existing actuator uncertainties into account, can stabilize the flight of AHFV successfully. Performances of flexible air-breathing hypersonic vehicles demonstrate the capability of the algorithm and the feasibility of the controller.

References

1. C. C. Coleman and F. A. Faruqi, On stability and control of hypersonic vehicles, Australian Government Department of Defence, PO Box 1500 Edinburgh South Australia 5111 Australia, Technical Report,(2009).
2. I. M. Gregory, R. S. Chowdhry, J. D. McMinn, and J. D. Shaughnessy, Hypersonic vehicle model and control law development using H_∞ and synthesis, National Aeronautics and Space Administration, Hampton, Virginia, Technical Memorandum, (1992)
3. T. E. Gibson, L. G. Crespo, and A. M. Annaswamy, Adaptive control of hypersonic vehicles in the presence of modeling uncertainties, in *American Control Conference*, pp. 3178–3183(2009)
4. H. Xu, M. D. Mirmirani, and P. A. Ioannou, Adaptive sliding mode control design for a hypersonic flight vehicle, *Journal of Guidance*, Control, and Dynamics, vol. 27, no. 5, pp. 829–838(2004)
5. J. Liu and F. Sun, Research and development on theory and algorithms of sliding mode control, *Control Theory & Applications*, vol. 24, no. 3, pp. 407–418(2007)
6. H. Li, W. Sun, and Z. Li, Index approach law based sliding control for a hypersonic aircraft, in *Systems and Control in Aerospace and Astronautics*, ISSCAA 2008. 2nd International Symposium on, pp. 1–5(2008)
7. H. Buschek and A. J. Calise., Fixed order robust

control design for hypersonic vehicles, in *AIAA Guidance, Navigation and Control Conference*, pp. 1094–1103(1994)
8. P. Lohsoonthorn, E. Jonckheere, and S. Dalzell, Eigenstructure vs constrained H_∞ design for hypersonic winged cone, *Journal of Guidance, Control, and Dynamics*, vol. 24, no. 4, pp. 648–658(2001)
9. H. Zhang, Y. Shi, and A. Saadat Mehr, Robust static output feedback control and remote PID design for networked motor systems, *IEEE Transactions on Industrial Electronics*, vol. 58, no. 12, pp. 5396–5405(2011)
10. ——, Robust H_∞ PID control for multivariable networked control systems with disturbance/noise attenuation, *International Journal of Robust and Nonlinear Control*, vol. 22, no. 2, pp. 183–204, 2012.
11. H. Zhang, Y. Shi, A. Saadat Mehr, and H. Huang, Robust energy-to-peak FIR equalization for time-varying communication channels with intermittent observations, *Signal Processing*, vol. 91, no. 7, pp. 1651–1658, 2011.
12. H. Zhang, Y. Shi, and A. Saadat Mehr, On H_∞ filtering for discrete-time takagi-sugeno fuzzy systems, *IEEE Transactions on Fuzzy Systems*, vol. 20, no. 2, pp. 396–401(2012)
13. C. Dong, Y. Hou, Y. Zhang, and Q. Wang, Model reference adaptive switching control of a linearized hypersonic flight vehicle model with actuator saturation, *Proceedings of the Institution of Mechanical Engineers*, Part I: Journal of Systems and Control Engineering, vol. 224, pp. 289–303(2010)
14. H. Zhang, Y. Shi, and A. Saadat Mehr, Robust weighted H1 filtering for networked systems with intermitted measurements of multiple sensors, *International Journal of Adaptive Control and Signal Processing*, vol. 25, no. 4, pp. 313–330(2011)
15. Q. Wang and R. F. Stengel, Robust nonlinear control hypersonic aircraft, *Journal of Guidance, Control, and Dynamics*, vol. 23, no. 4, pp. 577–585(2000)
16. Z. Gao, B. Jiang, P. Shi, J. Liu, and Y. Xu, Passive fault-tolerant control design for near-space hypersonic vehicle dynamical system, *Circuits, Systems and Signal Processing*, vol. 31, no. 4, pp. 565–581(2012)
17. H. Li, Y. Si, L. Wu, X. Hu, and H. Gao, Guaranteed cost control with poles assignment for a flexible air-breathing hypersonic vehicle, *International Journal of Systems Science*, vol. 42, no. 5, pp. 863–876(2011)
18. H. Zhang, Y. Shi, and A. S. Mehr, Robust non-fragile dynamic vibration absorbers with uncertain factors, *Journal of Sound and Vibration*, vol. 330, no. 4, pp. 559–566(2010)
19. H. Gao, Y. Si, H. Li, X. Hu, and C. Wang, Modeling and control of an air-breathing hypersonic vehicle,in *Proceedings of the 7th Asian Control Conference*, pp. 1114–1119 (2009)

DOI: 10.1051/matecconf/201710810003

Air cooling analysis and design of impacted flexible actuator's servo controller based on PLECS

Mingjun Zhu, Liqun Xia, Yixue Hu, Wei Zhang, Duohong Ji and Xin Song

Flight Automatic Control Research Institute, 710129Dianzi Yi Road Xi'an ShaanxiChina

Abstract. At present, SUAV (small unmanned aerial vehicle) is developing with small shape, high dynamic performance and distributed structure. Meanwhile the power by wire, which can replace traditional hydraulic system to lighten weight and improve reliability, is important developing direction of actuator technology in future. When a plane which applies power by wire is doing high-g routine, the power devices' loss is giant because of the high power. Thus it is necessary to design cooling structure to carry out heat from controller inner. To avoid over design and analyze more exactly, this paper introduces a method based on MATLAB and PLECS union Simulink to calculate controller's loss. By power devices' thermal resistance and working temperature referred from products manual, it is possible to define cooler's size. Then we can achieve wind pressure and flow characteristic curve, and select the cooling fan. Compared with theory calculation, this method can achieve exact controller loss in different working conditions and take effects which influenced by temperature in consideration, by what we can avoid cooling structure's over design.

1 Introduction

At present, SUAV (small unmanned aerial vehicle)is developing with small shape, high dynamic performance and distributed structure. While aerospace servo actuator technology is approaching to power by wire, high temperature and pressure resistant actuator or combination with advanced material. Future actuator will have such features: high power -weight ratio, high reliability and long server life. In this background, distributed impacted flexible actuator is come up, whose drive is not hydraulic but electric power and controller has obvious optimization in volume, power-weight ratio, reliability and serve life. However, controller's cooler design is far more difficult.

In controller, IGBT power module is major energy loss source, if its cooling design has limitation, the power components' juction temperature will rise up immediately and out of allowance range, which decreases of performance and reliability, and even leads to components' burning. Thus, the cooling design of impacted flexible actuator's controller is extremely important.

The traditional method to calculate the size of heat sink is as followed, formula is used to calculate the conduction loss and switching loss on the most harsh working conditions, then calculate the total loss, the size of heat sink is determined according to this value. this paper presents a method of co-simulation for loss calculation of power devices, and then the maximum resistance is used to determine the size of heat sink. The loss of power devices is also obtained in different working conditions, the value calculated by this method is more close to the real value, it can also avoid the over design of heat sink.

2 Loss analysis of controller

The diagram of impacted flexible actuator discussed by this paper is shown in Figure 1. It consists of high speed wet motor, hydraulic pump, check valves, stress sensors, relief valves, SOVs (solenoid operation valve), anti-cavitation valves, MSV (mode selected valve) and actuator ram. The controller is used to realize the motor power supply and transform voltage from DC to AC, to achieve the closed-loop control of motor current loop and speed loop, and to realize over-voltage, over-current, over-temperature protection functions. Controller power supply voltage is 270V and 28V, rated current is 60A. Because of the design requirement of the actuator power to weight ratio, cooling design is the key technology in this project.

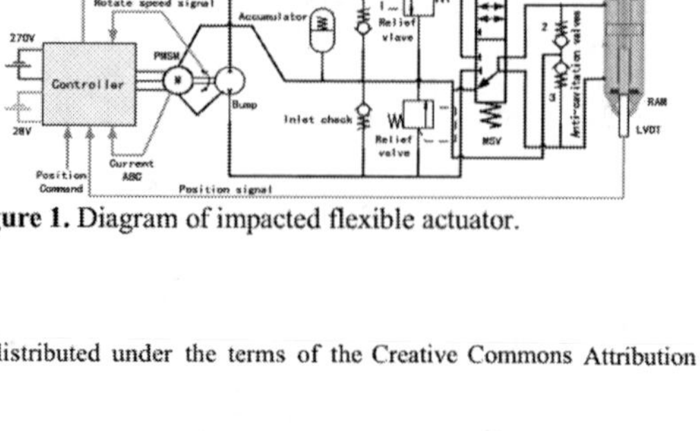

Figure 1. Diagram of impacted flexible actuator.

2.1 Switching loss

Switching loss includes turn-on loss and shut-off loss. Turn-on loss refers to the non-ideal switches in open state, voltage is not immediately drops to zero, but there is a decline in time, at the same time, current also not immediately rises to the load current, there is a rising time. During this period of time, the current and voltage of the switch tube have an overlapping area, which will produce the loss. The reason of producing turn-off loss is similar to the turn-on loss. .

$$P_{sw-Tr} = f_s(E_{on} + E_{off})\frac{\sqrt{2}}{\pi}(\frac{I_{out}}{I_{rated}})^{K_{swTr-I}}(\frac{V_{cc}}{V_{rated}})^{K_{swTr-V}} \quad (1)$$
$$[1 + K_{swTr-T}(125^\circ C - T_{j-Tr})]$$

$$P_{sw-D} = f_s E_{rr}\frac{\sqrt{2}}{\pi}(\frac{I_{out}}{I_{rated}})^{K_{swTr-I}}(\frac{V_{cc}}{V_{rated}})^{K_{swTr-V}}$$
$$[1 + K_{swD-T}(125^\circ C - T_{j-D})] \quad (2)$$

Among them:

P_{sw-Tr}	IGBT's switching loss;
f_s	Carrier frequency;
E_{on} / E_{off}	IGBT's single impulse turn-on and shut-off loss in rated condition;
I_{rated}	Reference current;
V_{cc}	Bus voltage;
V_{rated}	Reference voltage;
K_{swTr-I}	Coefficient which current amplitude influences P_{sw-Tr};
K_{swTr-V}	Coefficient which bus voltage influences P_{sw-Tr}
K_{swTr-T}	Coefficient which temperature influences P_{sw-Tr}
P_{sw-D}	Fly-wheel diode's switching loss;
E_{rr}	Fly-wheel diode's single impulse conduction and shut-off loss in rated condition;
K_{swD-I}	Coefficient which current amplitude influences P_{sw-D};
K_{swD-V}	Coefficient which bus voltage influences P_{sw-D};
K_{swD-T}	Coefficient which temperature influences P_{sw-D}

2.2 Conduction loss

When switch is in conducting state, there is conduction voltage at the two ends of the switch tube, because of this reason, the loss produced by this switch tube is called conduction loss. The loss is proportional to voltage, time and the square of the current. And the value is also related to temperature.

$$P_{cond-Tr} = I_{out}\sqrt{2}(\frac{1}{2\pi} + \frac{M\cos\psi}{8})[V_{ce-25} + K_{v-tr}(T_{j-Tr}$$
$$-25^\circ C)] + 2I_{out}^2(\frac{1}{8} + \frac{M\cos\psi}{3\pi})[R_{ce-25} \quad (3)$$
$$+ K_{r-Tr}(T_{j-Tr} - 25^\circ C)]$$

$$P_{cond-D} = I_{out}\sqrt{2}(\frac{1}{2\pi} - \frac{M\cos\psi}{8})[V_{F-25} + K_{v-D}(T_{j-D}$$
$$-25^\circ C)] + 2I_{out}^2(\frac{1}{8} - \frac{M\cos\psi}{3\pi})[R_{F-25}$$
$$+ K_{r-D}(T_{j-D} - 25^\circ C)] \quad (4)$$

Among them:

$P_{cond-Tr}$	IGBT's conduction loss;
I_{out}	Actual output of current virtual value;
M	PWM's modulation degree(phase voltage peak value divides half bus voltage);
ψ	Phase degree between actual current vector and actual voltage vector;
V_{ce-25}	IGBT's rated conduction voltage in 25 degree centigrade;
T_{j-Tr}	IGBT's actual juction temperature;
R_{ce-25}	IGBT's rated conduction resistance in 25 degree centigrade;
K_{r-Tr}	Coefficient which temperature influences IGBT's conduction resistance;
K_{V-Tr}	Coefficient which temperature influences IGBT's conduction voltage;
P_{cond-D}	Fly-wheel diode's conduction loss;
V_{F-25}	Fly-wheel diode's rated conduction voltage in 25;
K_{V-D}	Coefficient which temperature influences conduction voltage;
T_{j-D}	Fly-wheel diode's actual juction temperature;
R_{F-25}	Fly-wheel diode's rated conduction resistance in 25 degree centigrade;
K_{r-D}	Coefficient which temperature influences Fly-wheel diode's conduction resistance.

3 Inverter loss analysis based on PLECS

In PLECS emulation, the values of power model's switching loss andstate loss are modified, the simulation model can be used calculate the loss of power module at every moment in the working state of the device. At the same time, the external dissipation structure of power model can be designed, and the influence of temperature on the loss of power model is considered. The steps of simulation analysis are as follows. Firstly, according to the working condition of the controller, including cooling loop model, simulation is designed, and then according to the data sheet of power model, all kinds of loss parameters can be fond and set up. Finally simulation analysis is done to get the simulation results.

3.1 Building emulation model

The tested power module is applied to a large power motor controller, and the power module is used as the inverter to complete the DC to AC conversion function. So the simulation model needs to include the motor control algorithm model, three phase inverter model and motor model, as shown in Figure 4. The inverter model is built by the module in PLECS, which is because the power module model of PLECS contains thermal analysis function, and the model of the motor is used in PLECS.

The control strategy model is completed by the mathematical model of MATLAB, it needs to realize the closed-loop control of motor current and speed loop, which is used to drive the motor to always work in the maximum power state.

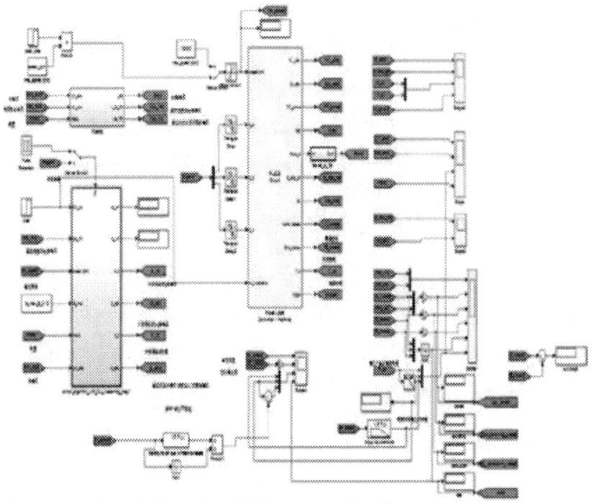

Figure 2. Emulation model for controller loss.

PLECS in the calculation of power device heat dissipation performance is more prominent, and it can be combined with MATLAB simulation, in this paper a single bridge of the power device is built in Figure 5. The driving signal is obtained by MATLAB simulation module, the dead time is generated through the simulation module. The purple area is heat loss calculation region. Several modules in the lower right corner is the standard module for calculating power device conduction loss and switching loss.

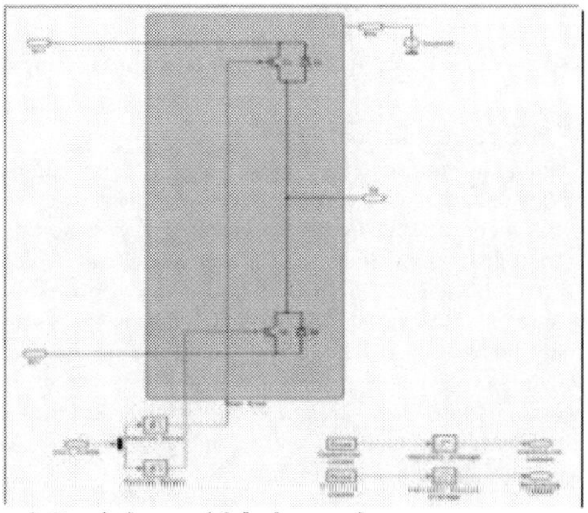

Figure 3. Emulation model for inverter loss.

3.2 Power components loss parameters configuration

In loss of power device, conduction and switching losses accounte for most of the total loss and other loss is negligible. temperature and current change lead to the change of this two kinds of losses. With data sheet, switching loss and conduction loss curves can be obtained as shown in Figure 4.

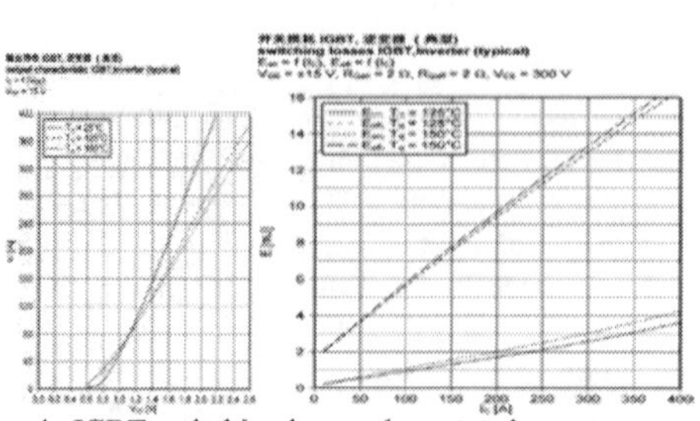

Figure 4. IGBT switching loss and on-state loss curve.

According to the curves shown in Figure 4, using the traditional method of tracing points,the values ofIGBT's and parallel diode's switching loss and condition loss can be obtained at different working condition points. Loss curves at different temperature are shown in Figure 5 and Figure 6.

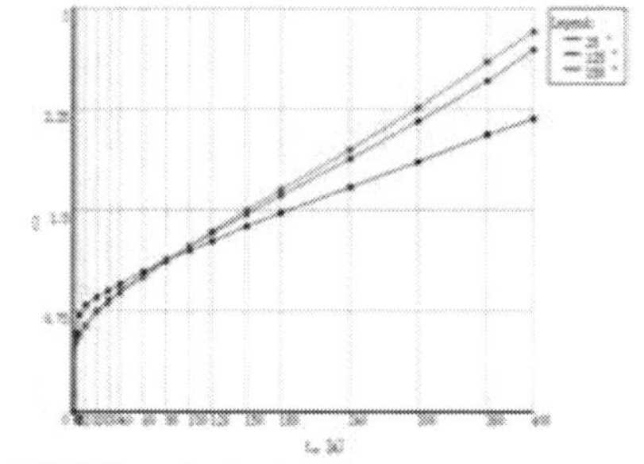

Figure 5. IGBT confuction loss curve.

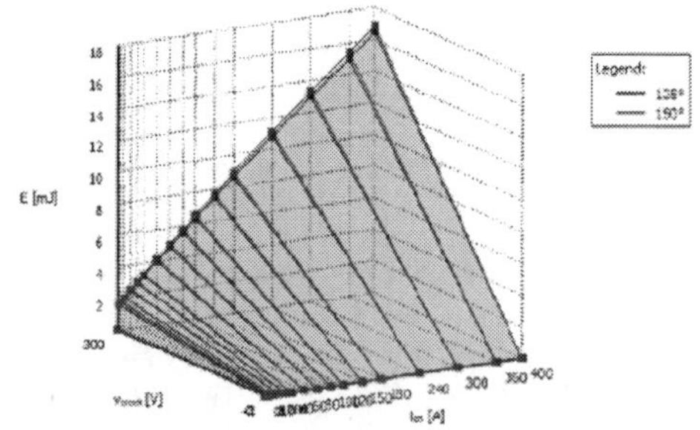

Figure 6. IGBT shut-off loss curve.

3.3 Emulation and analysis results

Through the control strategy, it can ensure that the motor always works in the maximum load state, the motor speed is from 0 rpm up to maximum speed. When motor reaches the maximum speed, reduce the current amplitude to reduce torque. Look up the loss of power device in different operating conditions, and then determine the maximum loss value, which is used to calculate the subsequent heat dissipation structure parameter.

Through the simulation analysis, the results are obtained as shown in Figure 9, considering the theoretical analysis, the conduction loss is proportional to motor current, having nothing to do with the switching frequency, and switching loss is not only proportional to the current amplitude but also associated with the switching frequency. The motor works at the maximum power point when motor speed runs from 0 rpm to maximum speed. After speed reach the maximum value, motor's electromagnetic torque will decline and current amplitude will decrease, so the total loss of power module will also decline. The loss of 3 up bridge arms is 207W, so the total loss is 414W. When the motor's actual rotate speed achieves given rotate speed, switching loss and conduction loss decrease because of inverter current being small. The emulation results are shown in Figure 7.

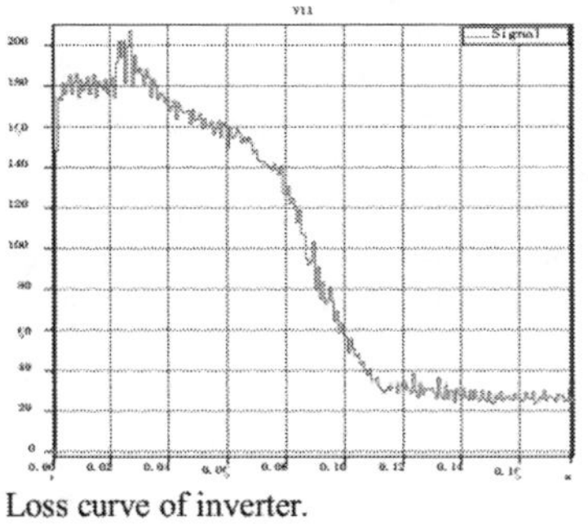

Figure 7. Loss curve of inverter.

4. Design of inverter's heat sink

In the actual use of heat sink, there are many factors affecting the thermal resistance. In the choice of heat sink's size, the max needed thermal resistance is used to determine the heat sink's structure, and also other factors are considered as follow, like stricter, cost, wind pressure, cooling effect and so on. The thinner fin the better cooling is, but while its effect is better, it is more difficult and cost more. So far, there is no accurate mathematical formula to calculate the thermal resistance of the heat sink, the size of this heat sink used in this project is shown in Figure 8.

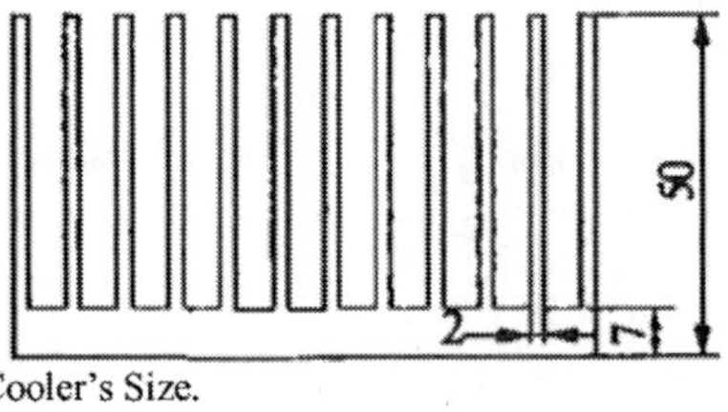

Figure 8. Cooler's Size.

We can calculate the maximum heat sink required thermal resistance to determine the size of heat sink. When the thermal resistance of actual heat sink is less than the required maximum thermal resistance, even if the controller works in the most harsh environment, the controller's temperature will not exceed allowable working temperature of power device. The maximum heating amount obtained by simulation is 414W. Through look up the data sheet, we find the follow information, IGBT's highest acceptable working temperature is Centigrade 150 degrees, thermal resistance which from IGBT's tablet core to shell is: RIthCH=0.085K/W. Which from IGBT's shell to cooler is: RIthJC=0.25K/W. Which from diode's tablet core to shell is: RDthCH=0.15K/W, Which from diode's shell to cooler is:RDthJC=0.45K/W.

According to thermal resistance values above, the thermal resistance from power module to cooler's surface is:

$$R_I = \frac{1}{6}(R_{IthJC} + R_{IthCH})//(R_{DthJC} + R_{DthCH}) = 0.0358K/W \quad (5)$$

Suppose:

1. air temperature in controller box is worst to Centigrade 70 degrees
2. TIjuction≥0.9TImax=135℃.

So we can get:

$$R_H \leq 0.12K/W \quad (6)$$

According to conclusion above, when thermal resistance of cooler selected is smaller than 0.12K/W, the design can meet cooling requirement. The cooler's size is 180mm × 100mm × 50mm, which has 25 cooling air ducts, and every fin's length is 43mm, the distance between two fin is 5mm.

4.1 Fan's selection and working point's check

Traditional cooling technology, including air cooling, water cooling or oil cooling, has different application occasion. When used in aircraft servo actuator system, forced air cooling is the best controller cooling style based on weight limit. This paper selects 4 fans produced by SANYO, whose type is 9CRD0412P5J03. The fans work in constant power, which means when wind pressure produced by fan increases, the wind flow rate will decrease. Thus the fans' working point should be made sure for cooling efficiency.

4.2 Wind resistance characteristic of cooler

Suppose the fans and the cooler selected can bring total heat out of controller box, then based on heat balance equation we can get:

$$L = \frac{Q}{\rho C_p(t_0 - t_i)} = \frac{414}{1.13*1.005*(70-50)} = 1.09m^3/\min \quad (7)$$

L	Cold air flow;
Q	Loss of controller;
ρ	Air density;
C_p	Air specific heat;
t_i	In port temperature of air;
t_0	Out port temperature of air.

To ensure reliability of cooling, there is additional 20% cooling margin needed, thus:

$$L_{\min} \geq 1.2 * L = 1.31m^3/\min \quad (8)$$

So the flow every fan is required to produced is not less than 0.328m3/min.

When flowing, the air duct, whose length is y, has wind resistance of:

$$R = \frac{\lambda}{4R_s}\frac{v^2}{2}\rho y = \frac{0.022}{4\times 2.24\times 10^{-3}}\frac{v^2}{2}\times 1.13\times 0.1 = 0.139v^2 \quad (9)$$

R	Resistance of cooler's air duct;
λ	Coefficient of frictional resistance;
ρ	Air density;
v	Mean velocity of air in air duct;
R_s	The equivalent hydraulic radius of air duct;
y	Length of air duct;

The cooler's total width is 180mm, which have 25 air ducts. The relationship between cooler flow and mean velocity of air in air duct is that:

$$v = \frac{L}{60\times 25ab} = \frac{L}{60\times 25\times 5\times 43} = 3.1L \quad (10)$$

Combine equation (10) into equation (9), we can get:

$$R = 0.139v^2 = 0.139\times(3.1L)^2 = 1.336L^2 \quad (11)$$

When fans make air flows in, wind pressure must achieve to meet cooling requirement.

$$R_{\min} \geq 1.336{L_{\min}}^2 = 2.29Pa \quad (12)$$

By referring the fan's working characteristic curve as shown in Figure 9, it is obvious that when drive PWM's (fan's) duty cycle is 50%, the air flow is able to get, while fan's wind pressure is 45Pa, which meets the requirement.

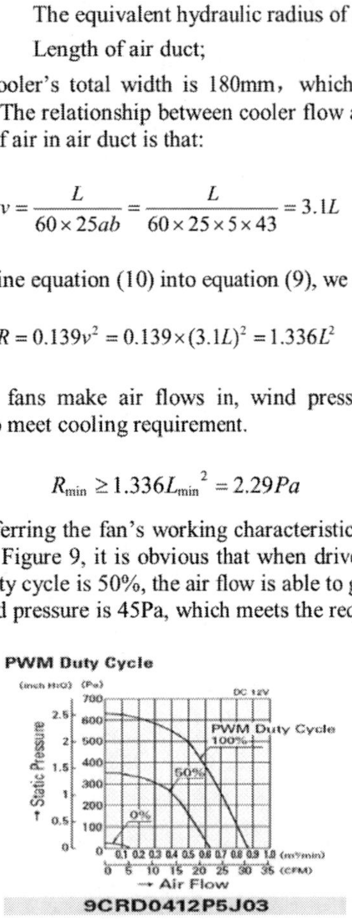

Figure 9. Working characteristic curve of fan.

According to method put forward by this paper, the cooling structure is as following:

Figure 10. Picture of cooler structure.

5. Conclusion

This paper introduces a compact smart actuator structure, in order to improve the controller's power to weight ratio, we need ensure the performance of heat sink, and reduce the size of heat sink as possible. Then it introduces the formula of power device loss to obtain the power device loss factors. Finally, it presents a method of calculating the total loss of power device based on PLECS and MATLAB, according to this method, switching loss, conduction loss and total loss curves are obtained, they are used to determine the size of heat sink, and select the fan. At the last checking the cooling system's effect need to be done to check whether heat sink meets the thermal design requirement. Using this method described in this paper, the loss curves in different working conditions and considering the influence of temperature are obtained.

References

1. Chang Y J, Wang C C. A generalized heat transfercorrelation for louver fingeometry [J]. InternationalJournal of Heat and Mass Transfer, 1997,40 (3):533-544.
2. J. Y. Wang, W. R. Zhong, W, J, Wu, "Research for Improving High Power Anti-explosion Inverter 's Cooling Performance ," Xi'an: Journal of Xi'an polytechnic university , 2006, vol. 1, pp. 42-45.
3. Masud Behnia, David Copeland, Denpong Soodphakdee. A Comparison of Heat Sink Geometries for Laminar Forced Convection:Numerical Simulation of Periodically Developed Flow [C] 1998 Inter Society Conference on Thermal Phenomena, 310 - 315.
4. Dvinsky A, BarCohen A, Strelets.Thermal fluidAnalysis ofStaggered and Inline Pin Fin sinks[C] 2000 Inter Society Conference on Thermal Phenomena, 157- 164.
5. Wieting A R. Empirical correlations for heat transfer and flow friction characteristics of rectangularoffsetfin heat exchangers[J]. Heat Transfer,1997(3): 488-490.
6. Manglik R M, Bergles AE. Heat transfer and pressure drop correlations for the rectangular offset stripfin compact heat exchangers[J]. Exp Thermal andFluid Sci, 1995,10: 171-180.
7. Ren N, Gu B. Experimental study and numericalsimulation of heat and mass transfer on plain fin inwet conditions[J] . Journal of Chemical Industry andEngineering , 2007, 58(7) : 1 626 1 631.
8. PHILIPS A M, JANKOVIC M, BAILEY K E, Development issues of an ISG PM machine and control system [C] 7th international Conference on Power Electronics, Oct. 22-26,2007,Daegu, Korea. 2007:922-929.
9. Flannery J, Cheasty P, Meinhardt M, Present practice of power packaging for DC/DC converters[J] Integrated Power Packaging, 2000(7):3-7.
10. Darbandi M, Naderi A. Multiblock hybrid gridfinite volume method to solve flow in

irregulargeometries[J]. Computer Methods in AppliedMechanics and Engineering, 2006,196(1/3):321-336.

11. DE Backer J W, Vos W G, Germonpre P,Flow analyses in the lower airways:patientspecificmodel and boundary conditions[J]. Medical Engineeringand Physics, 2008,30(7): 872-879.
12. Jafaria, Tynjala T , Mousavis M, Simulation of heat transfer in aferrofluid usingcomputational fluid dynamics technique[J] .Inter national Journal of H eat and Fluid Flow, 2008,29(4) : 1197 1202.
13. Tsuji T, Kajitani T,Nishino T. Heat transferenhancement in a turbulent natural convection boundarylayer along a vertical flat plate[J]. International Journalof Heat and Fluid Flow, 2007, 28(6): 1472-1483.
14. Thoben M , Mainka K, Bayerer R, Fromvehicle drive cycle to reliability testing of powermodules for hybrid vehicle inverter [C] PCIM Europe2008: International Exhibition & Conference for PowerElectronics Intelligent Motion Power Quality, May 27-29,2008,Nuremberg,GermanyIEEE,2008:651-656.

Design and Implementation of A-SMGCS Based on ICE

Jing Li[1], Xuguang Zhao[1], Min Wu[1], Ergai Yu[2], Li Jiang[1] and Pengfeng Sheng[1]

[1]*The Second Research Institute of Civil Aviation Administration of China (CAAC), Chengdu, China*
[2]*Chengdu Civil Aviation Information Technology Co., Ltd, Chengdu, China*

Abstract. This paper aims at the problems of safety and efficiency of airport, designs and implements an Advanced Surface Movement Guidance and Control System (A-SMGCS) terminal, discusses the goals and key technologies about the system, including communication module design based on Internet Communications Engine, cache design of aircraft targets, and efficiency improvement of a large number of targets on the aerodrome. Finally, we give the test and conclusions of the system at a international airport in China.

1 Introduction

With the increasing demand for air transport all over the world, pressure is growing for airport systems to safely handle greater capacity in all weather conditions, improve traffic distribution and management, maximize use of existing infrastructure and so on [1]. The number of accidents and incidents during airport surface movements, including runway incursions, is increasing. Contributing factors include the increasing number of operations that take place in low visibility conditions, the progressive increase in traffic, the complexity of aerodrome layouts, and the proliferation of capacity-enhancing techniques and procedures. Therefore, advanced capabilities are needed to ensure spacing when visual means are not adequate and to maintain aerodrome capacity in all weather conditions. In addition, as visibility is gradually reduced, the ability of controllers and pilots to carry out visual surveillance becomes increasingly impaired [2].

A-SMGCS utilizes multiple surveillance sources such as ADS-B (Automatic Dependent Surveillance - Broadcast), MLAT (Multilateration), PSR (Primary Surveillance Radar), SSR (Secondary Surveillance Radar), SMR (Surface Movement Radar), and other fusion data, and provide routing, guidance and surveillance for the control of aircraft and vehicles in order to maintain the declared surface movement rate under all weather conditions within the aerodrome visibility operational level (AVOL), especially in low visibility, while maintaining the required level of safety [3].

2 Goals for the system

A-SMGCS is a kind of software that makes use of all kinds of data like sensors data, fusion data, alarms data, flight plan, meteorological data, to surveillance and control the traffic on the aerodrome [4]. The main functions of our A-SMGCS consist of:

1) Real-time presentation of surveillance sensor data and fusion data, such as ADS-B, MLAT, SMR, PSR/SSR;

2) Real-time display runway incursion, taxiway alerting, alerting for loss of data and so on;

3) Maintenance and check of flight plan, meteorological data and stand information;

4) Recording and playback of surveillance information.

In addition to the main functions provided by A-SMGCS described above, functions of data acquisition, fusion and communication with other systems must be considered. Concretely speaking, when monitoring the real-time traffic state on the aerodrome, our A-SMGCS terminal mainly obtains single-sensor data, such as ADS-B, SMR, MLAT, SMR, PSR/SSR from data pretreatment system, and obtains fusion data from fusion system [5]; when processing conflict alert information, the terminal obtains corresponding alert information from alert system, after the alert information are processed, the alert information could be feedback to alert system; flight plan and meteorological data of the terminal come from flight plan system and meteorological system.

3 Internet communications engine overview

The Internet Communications Engine (ICE), is an open source RPC framework developed by ZeroC. It provides SDKs for C++, C#, Java, JavaScript, Objective-C, PHP, Python, and Ruby, and can run on various operating systems, including Linux, Windows, OS X, and Android. ICE implements a proprietary communications protocol, called the ICE protocol that can run over TCP, TLS, UDP, and WebSocket. As its name indicates, ICE can be suitable for an application that communicate over the

 DOI: 10.1051/matecconf/201710810004

Internet, and includes functionality for traversing firewalls [6].

The ICE architecture provides a number of benefits to application developers: object-oriented semantics, support for synchronous and asynchronous messaging, support for multiple interfaces, machine independence, language independence, implementation independence, operating system independence, threading support, transport independence, location and server transparency, built-in persistence, source code availability and so on. Overall, ICE provides a state-of-the art development and deployment environment for distributed computing that is more complete than any other platform we are aware of.

4 Key technologies

4.1 Communication module design

Figure 1 illustrates the communication module of the terminal system, in which, the system use ICE for communication.

Figure 1 gives a schematic overview of the communication module. The data transmission of the terminal system mainly consists of fusion data, associated data, single sensor data, alarm information, flight plan, meteorological information, gate information and other message transmission. Among them, fusion data and artificial association data module primarily communicate with fusion system; other single sensor data module chiefly communicates with data preprocessing system; the alarm information module mostly communicates with the alarm system; the flight plan module mainly communicates with the FDP system; meteorological information module is mainly to communicate with the meteorological system.

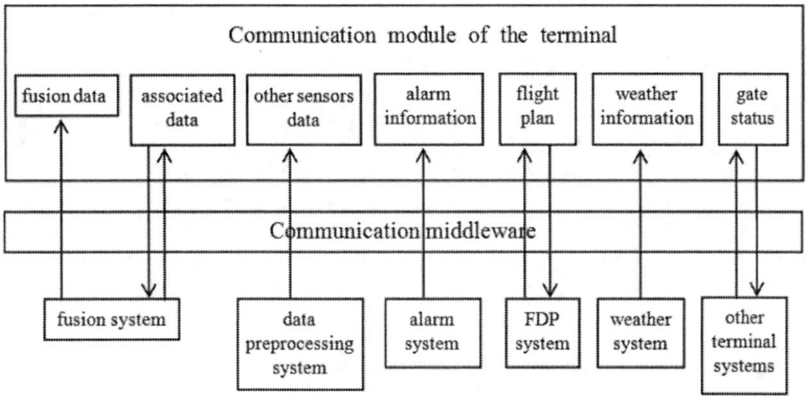

Figure 1. Communication module of the terminal system

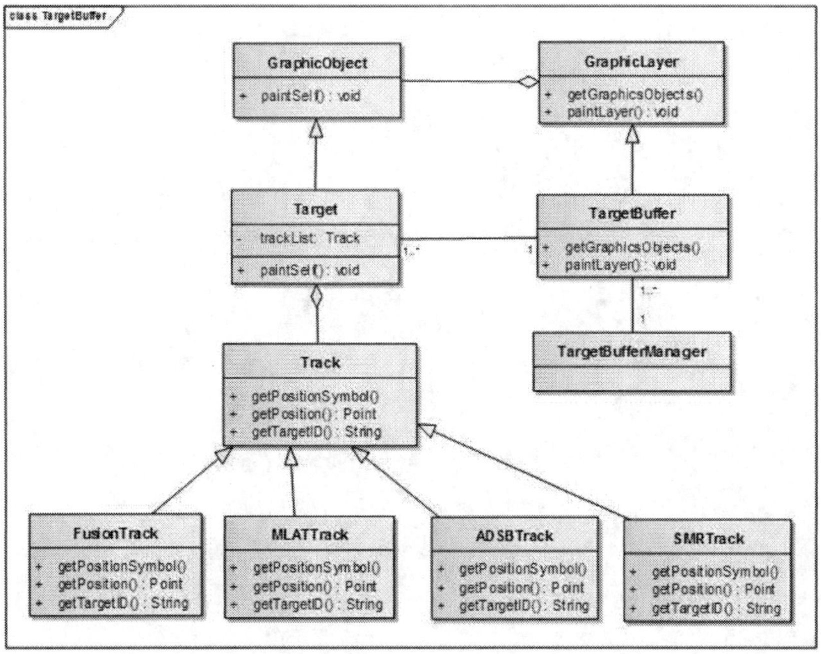

Figure 2. Structure design of aircraft targets cache

In the process of research and development for middleware, the system has designed two kinds of maintenance for messages: shared memory mode and message queue mode, which are used in the field of civil aviation data transmission due to large amounts of data and high real time.

4.2 Cache design of aircraft targets

Receipt of track data in our terminal is divided into two categories: data form single sensor and fusion data. The terminal firstly decodes data through decoder interface after accepting the two types of data; secondly, the system stores the track data in the cache of the aircraft target. Meanwhile, our system renders the aircraft target according to its cache. Therefore, it is extremely important to design targets cache that is easy to extend and stable, and Figure 2 shows the design of targets cache for airplanes.

4.3 Efficiency improvement of a large number of targets

When the number of aircraft targets on the aerodrome exceeds 400 at the same time, the efficiency of system will reduce, we analyze the reason for efficiency reduction, and find that because the work to render each tag for each aircraft is time consuming. Therefore, we provide solution with respect to:

① Increase efficiency of rendering for single tag: we render target by lightweight component, and the tag inherits Class Lightweight Component, which applies double buffer in implementation. Therefore, we could raise the efficiency and stability of drawing.

② Drawing range: we only draw the target on the visible region or nearby region, which not only cuts down the workload of massive drawing of needless, but also raises the drawing efficiency for the entire system.

On these basis, a pressure test about 600 airplane targets running at the same time on the aerodrome is made, the display and related system functions operation are still smooth.

5 Site proof

A-SMGCS terminal system has been tested at a domestic international airport for more than one year. Practical running and testing proves the design's feasibility and reliability. The performance is stable, and enjoys a high evaluation and accreditation from the airport users. The screen-capture of our system is as follows (Figure 3).

After a comprehensive internal test, the system test results are excellent, such as: the system capacity is greater than 300 aircraft and 300 vehicles, the location report accuracy of targets is less than 7.5 meters, the average processing time of aircraft and vehicle identification data is less than 3 seconds, the average processing time of aircraft and vehicle location data is less than 1 second, and the response time of human machine interface is less than 1 second. All the performance indexes meet the technical requirements of CAAC.

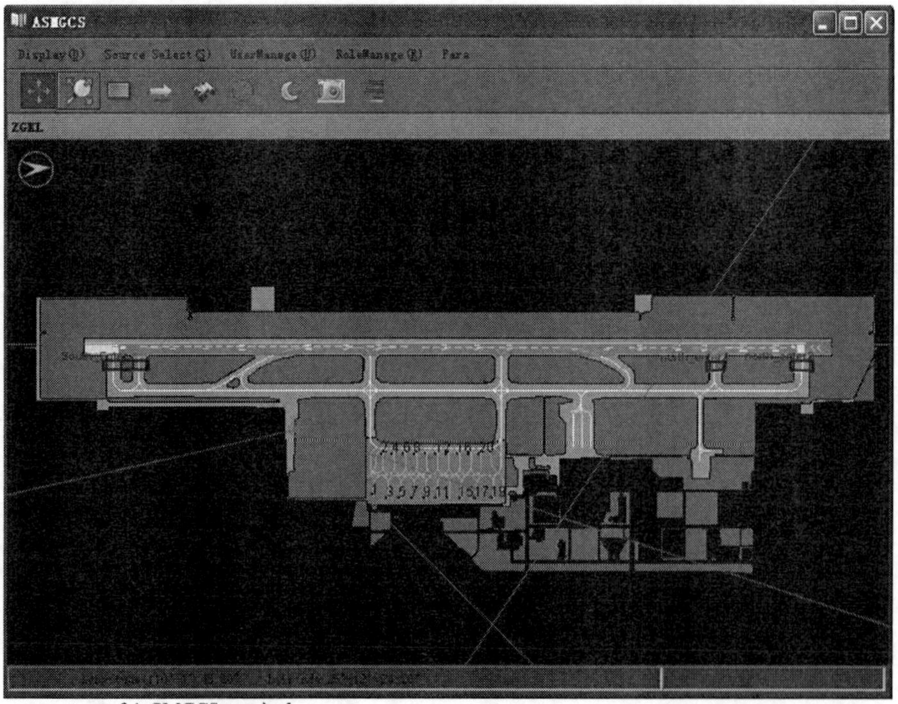

Figure 3. Screen-capture of A-SMGCS terminal

6 Conclusions

ICE is a comprehensive RPC framework with support for C++, C#, Java, JavaScript, Python, and more. The ICE core provides a sophisticated client-server platform for distributed application development. Communications between client and server can be fully secured with strong encryption over SSL, so applications can use unsecured public networks to communicate securely.

Due to the growth in density of traffic flow and capacity of airport, problems of runway incursion, conflicts on the aerodrome, aviation safety and flight delay become more and more prominent. To solve these problems, we design and implement A-SMGCS terminal system, which shows wider application prospect after test at a domestic airport for more than one year. Our system is a modular system consisting of different functionalities to support the safe, orderly and expeditious movement of aircraft and vehicles on aerodromes under all circumstances with respect to traffic density and complexity of aerodrome layout, taking into account the demanded capacity under various visibility conditions.

Acknowledgement

This paper is supported by the National Science and Technology Support Program of China (No. 2011BAH24B06), the National Science and Technology Support Program of China (No. 2011BAH24B05), and the Technology and Innovation Fund Program of CAAC (MHRD20130206).

References

1. Qingchuan Feng. Aerodrome Control,Southwest Jiaotong University Press. (2004)
2. Xuan Wu. Probing into integration technology of tower control positions, Air Traffic Management. (2003)
3. International Civil Aviation Organization. ICAO: Doc9830 AN/452-2004 Advanced Surface Movement Guidance and Control Systems(A-SMGCS) Manual. (2004)
4. The United States Department of Transportation. Challenges Facing the Implementation of FAA's Automatic Dependent Surveillance-Broadcast Program. (2007)
5. Teotino D, Ruggieri M, Nuzzo A. European Airport Movement Management: The Italian Challenge. IEEE Aerospace and Electronic Systems Magazine. (2008)
6. http://zeroc.com/products/ice#everywhere. (2015)

Graphical user interface for input output characterization of single variable and multivariable highly nonlinear systems

M. D. Shahrukh Adnan Khan[1,2], Sharsad Kara Kuni[2], G. K. M. Sadikul Amin[3], Mukul Agarwal[4]

[1]*Assistant Professor, University of Asia Pacific, Dhaka, Bangladesh*
[2]*University of Nottingham, JalanBroga, Selangor, Malaysia*
[3]*Director, Epic Airways Limited, Dhaka, Bangladesh*
[4]*Managing Director, MukTar Enterprises Pte Ltd, 320 Serangoon Road, 4-46 Serangoon Plaza, Singapore*

Abstract. This paper presents a Graphical User Interface (GUI) software utility for the input/output characterization of single variable and multivariable nonlinear systems by obtaining the sinusoidal input describing function (SIDF) of the plant. The software utility is developed on MATLAB R2011a environment. The developed GUI holds no restriction on the nonlinearity type, arrangement and system order; provided that output(s) of the system is obtainable either though simulation or experiments. An insight to the GUI and its features are presented in this paper and example problems from both single variable and multivariable cases are demonstrated. The formulation of input/output behavior of the system is discussed and the nucleus of the MATLAB command underlying the user interface has been outlined. Some of the industries that would benefit from this software utility includes but not limited to aerospace, defense technology, robotics and automotive.

1 Introduction

Mathematical model is a prerequisite in most of the cases for a controller to begin a design. In real life systems, mathematical models are quite sophisticated and most of the times it could be represented only by using computer codes [1]. With the fact that all practical real life systems contain nonlinearities, describing function approach has attracted considerable interest when conventional controller design techniques were not able to satiate the designers. Though this paper is not intended to discuss on the limitations and advantages of describing function method, it is desirable to point out a few of them. The primary motivation of the use of describing functions is to extend the very powerful transfer function approach of linear systems to nonlinear systems [2]. One of the most important characteristics of nonlinear system is the dependency on the amplitude of excitation. This dependency on amplitude of nonlinear plant can be well represented via the sinusoidal-input describing function model of the plant [3]. In general, most of the models can be presented in the following form;

$$\dot{x}=f(x,u,t) \tag{1}$$

$$Y=g(x,u,t) \tag{2}$$

where x is the state vector of dimension n, u is the input vector of dimension m, y is the output vector of dimension l, and t represents the time.

Since numerous literatures have been produced discussing the describing function model concept, only some of the key points underlying this approach will be reviewed here. A describing function can be defined as the ratio of the fundamental component of the output of a nonlinear process to the sinusoidal input signal [2, 4]. With the modern formulation of SIDF modelling (Taylor (1983), Taylor and Strobel (1985), Nassirharand (1987)), based on evaluating 'Fourier Integrals' by integrating the equations of motion, most of the earlier restrictions such as the requirement of sinusoidal input to each nonlinearity are lifted. The only assumption left with the describing function model is that it forms a better representation of the nonlinear system. However, an input/output formulation of following general framework is obtained using this approach [3];

$$x=Xsin\omega t \tag{3}$$

$$y\approx a\ Re[G(j\omega,a)ej\omega t] \tag{4}$$

2 Application of nonlinear control system

To put in layman's words, the input signals are more often towards minimum and maximum values than in the middle [1]. Thus, need for control design techniques, where robustness as the primary design criterion is imposed. It has been well established that sinusoidal input describing function models provides an excellent

 DOI: 10.1051/matecconf/201710810005

platform for robust control system design due to the fact that no other conventional design techniques exist so far; that may attribute to its ability to capture the amplitude sensitive characteristics of nonlinear plant [1] [3-6]. In this work, particular interest will be on formulation of input/output behaviour of such industrial areas as aerospace, defence and robotics where the amplitude distribution of excitation signals are similar to sinusoidal case. A 3 seat, piston engine based, fixed wing British light aircraft was designed by Mr. John Edgily [7]. It first flew on 14th December 1979, achieving certification in the year of 1985, on February 8. With a visibility of a Helicopter and stability of a fixed wing, the aircraft namely OPTICA is happened to be much safer and cost-effective [8]. With a stall speed of 108km/hr, service ceiling of 4275m and range of 1056km, the carries focuses dual control capability for pilot training. The proprietor of the architectural design of carrier Optica, MukTar Enterprises, proposed to establish a nonlinear control system for their aircraft [9]. Aircraft flight control design is conventionally created on linear control principle. However, to reach tactical benefits, modern aircrafts struggle towards performing exercises outside the region in which the dynamics of flight are nonlinear. Hence, the necessity for nonlinear tools arises.

Figure 1 and 2 shows the proposed set-up spot for establishing a GUI based nonlinear controlling system in carrier Optica.

Figure 1. 3 Seat, Piton Engine Fixed Wing OPTICA.

Figure 2. Interior of OPTICA.

3 Graphical user interface

A Graphical User Interface (GUI) has been developed in the MATLAB environment which would enable the user to derive the input/output characterization of highly nonlinear single variable and multivariable nonlinear systems through generating Sinusoidal input describing function models of the plant. The GUI shown in Fig. 3 is designed for MATLAB R2011a platform. The main utility of the GUI are 1) generate the SIDF model from the state-variable models 2) output the pseudo frequency bode plots 3) automate time-data plots. The User Interface has no restriction on nonlinearity type provided system output(s) for input(s) must be obtainable by either simulation or experiments.

Apart from the user friendly environment it provides, the developed GUI also allows user to save the SIDF developed each time for different desired set of operational conditions and can obtain a combined pseudo bode plots which may be used for comparing the system behaviour under different excitation frequencies and amplitudes . The other features enable the user to save the plots and output data which may be retrieved for later use. The software utility may be better demonstrated by an example from both single-variable and multi-variable case.

4 Single-Input Single-Output Case

Before demonstrating an example problem, it is worthwhile to discuss some of the important input/output characterization techniques of single variable system followed by an insight to the user interface environment. The formulation of input/output characterization technique for single variable is based on Nassirharand's work which was originally based on Taylor's Fourier's analysis approach to develop the sinusoidal input describing function models [3]. The plant is excited by the sinusoidal input of the following form.

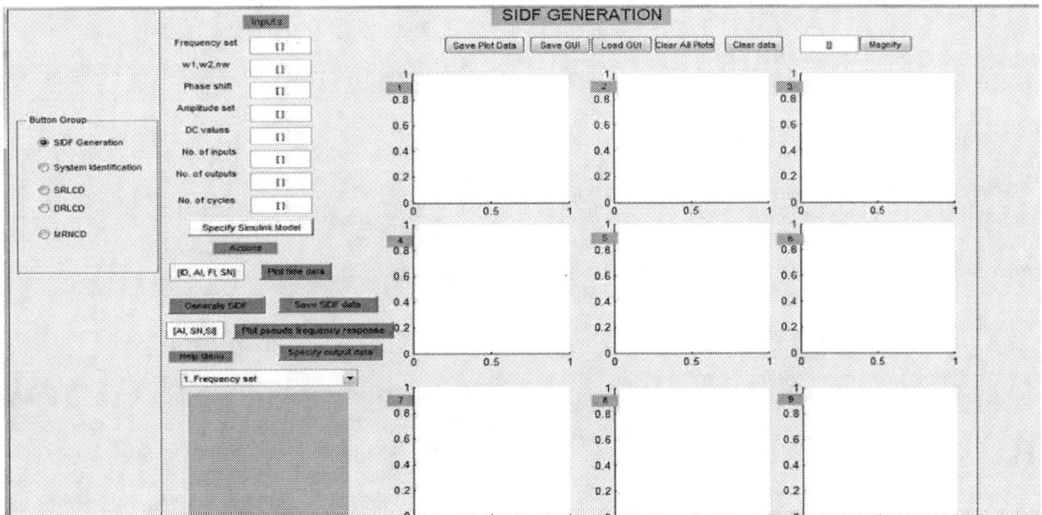

Figure 3. Developed GUI for Nonlinear Control System.

$$u(t)=u_0+a\cos(\omega t) \quad (5)$$

Here, u is the input to the system, u0 is the DC value of the input signal, a is the amplitude of the excitation and ω the frequency of the excitation signal. The approach adapted herein, compute the output y (t) as a function of time by numerically integrating the equations of motion. Fourier integrals of the following form are evaluated once y (t) approaches steady state.

$$I_{m,k}=\int_{(k-1)t}^{kt} y(t)e^{-jm\omega t}dt \quad (6)$$

Here, period index=1,2,3.....,harmonic index=0,1,2...., and period T is given by 2π/ω. The DC component of the response is $I_{0,k}$ and the pseudo transfer function is given as follows.

$$G_{m,k}(jm\omega,u_0,a)=\omega I_{m,k}/a\,\pi.\ [3] \quad (7)$$

The GUI requires the user to input the following parameters and file(s) in order to obtain the corresponding sinusoidal input describing function models of the nonlinear plant.

(1) Either desired frequency array in rad/sec or range of the excitation frequency of interest.
(2) An array of desired excitation amplitude set.
(3) For single variable case, phase shift values must not be entered.
(4) An array of DC values of the input sinusoids. The size of this vector must be equal to the size of input amplitudes.
(5) The number of inputs and outputs should be entered as 1.
(6) Desired Fourier integration cycles for each frequency range in the ascending order.
(7) Simulink file corresponding to the system for which SIDF model to be generated.

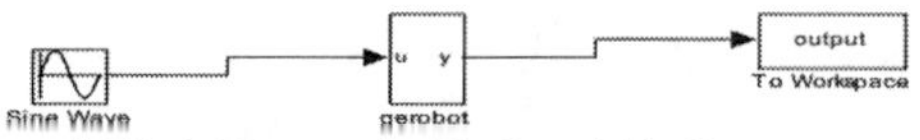

Figure 4. Block Diagram for Single variable Case.

In many cases where mathematical models are not available, the user may want to generate sinusoidal input describing model where system output(s) is obtained from experimental results. The GUI presented also can be well used with such conditions where Simulink file is not required. In such cases, the user must provide the output data in a prescribed format where first column is time, the second column is the first output, and the third column is second output (in case of multivariable) and so on. The nucleus of the underlying software of the GUI is a MATLAB function that generates the real and imaginary parts of the describing function models as a function of excitation frequencies and amplitudes. Consider an example problem of the sort encountered in robotics. Figure 4 represents the user input Simulink file for the nonlinear single variable plant. We may obtain the SIDF models of the single-input single-output system around the following operating conditions for this example.(1) excitation frequencies of [1 2 3];(2) excitation amplitudes of [3 4 5];(3) DC values of [0 0 0];(4) NCYCLE inputs array covering the frequency range as [1 2 3 3 5]. Once the desired values are entered, the user may generate the SIDF model using the GUI. The software utility will output the sinusoidal input describing function as gr and gi (see Appendix 1). The software will also generate the pseudo frequency bode plots which are shown in Fig.6 as well as time-data plots shown in Fig. 5. Time-data plots can be analyzed in row wise to see that the first plot corresponds to all amplitudes and all corresponding frequencies and go through maximum NCYCLE value (that is 5 for this example problem). The second graph for time-data plot represents the first frequency and it cover through its Fourier integral cycle range of 3. Similar analysis can be done for plot 3 and 4 which constitute for second and third frequencies plotting through their NCYCLE values respectively.

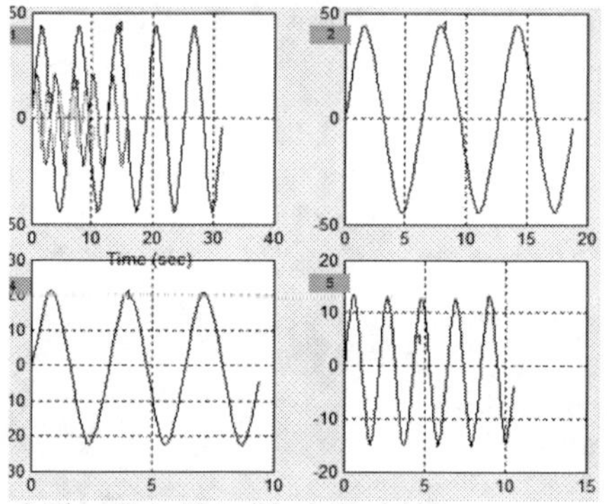
Figure 5. Time-Data Plots (Single Variable).

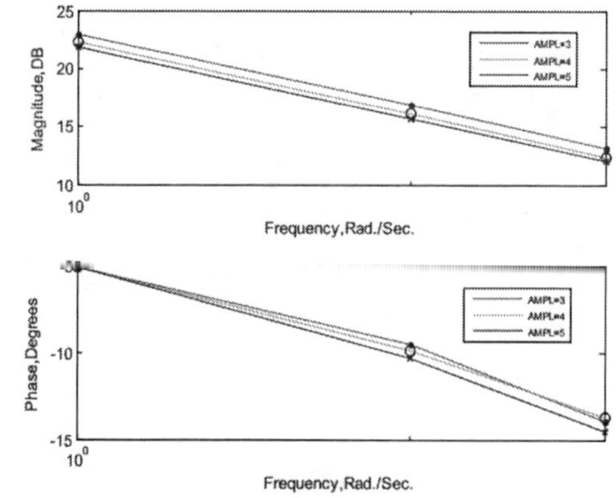

Figure 6. PSEUDO-BODE Plots (Single Variable).

5 Multi-input multi-output case

Let us discuss some of the main input/output characterization techniques involved in the case of multi-input multi-output system as well as GUI environment for multivariable plant before coming to the example problem. For multivariable case, one cannot be able to determine the effects of each input to the corresponding

DOI: 10.1051/matecconf/201710810005

outputs; hence the principle of superposition is not valid. Furthermore, it has to be noted that the separate effects of each input on each output cannot be determined if any of the input frequencies are same. Keeping these points in mind, the input/output behavior of a multivariable system can be formulated by assuming the input to the system is of following general form [1].

$$up(t)=u0,p+apcos(\omega pt+\theta p); \quad p=1,2....m \tag{8}$$

Here, u is the input to the system, u0 is the DC value of the input signal, a and ω are the excitation amplitude and frequency respectively, θ is the phase shift and p is the input channel index. Output as a function of time, $y_q(t)$ is obtained by numerically integrating the equations of motion and thus the Fourier integral matrix, $I_{q,p}^{h,k}$ are evaluated as below.

$$I_{q,p}^{h,k} = \int_{(k-1)t}^{kt} y_q(t)exp(-jh(\omega_p t+\theta_p)dt \tag{9}$$

Here h is the harmonic index, k the period index, q the output channel index and T is the overall period. The pseudo transfer function $(G_{q,p}^{1,k})$ is given as follows.

$$G_{q,p}^{1,k}(j\omega_q;u_0,a,\theta)=(2/a_pT)\,I_{q,p}^{1,k} \tag{10}$$

The convergence criteria for Fourier Integrals are as follows;

$$\left|\frac{M_{q,p}^{k}-M_{q,p}^{k-1}}{M_{q,p}^{k}}\right|<\varepsilon_M \tag{11}$$

$$\left|\phi_{q,p}^{k}-\phi_{q,p}^{k-1}\right|=\varepsilon_\varphi \tag{12}$$

Here, $M_{q,p}^{k}$ is the log magnitude and $\phi_{q,p}^{k}$ is the phase of $G_{q,p}^{k}$ [1]. In order to obtain the sinusoidal input describing function models and thus the corresponding pseudo bode plots and time data plots, the user needs to input all the required parameters as discussed above in the case of single-input single-output system. Those are not repeated here except some of the points peculiar to the multivariable case.1) the number of rows of frequency array input must equals the number of inputs. 2) the amplitude array column size must be same as the number of inputs. 3) Unlike single variable case, user must enter an array containing phase shift values and the size of this matrix has to be identical to the size of frequency input matrix. 4) the number of inputs corresponding to the multivariable plant and the desired number of outputs need to be provided by the user .5) Fourier integral cycle input must cover all frequencies of first channel (or first row of frequency matrix). The rest of the requirements would remain unchanged. The key MATLAB command takes the excitation amplitude array and excitation frequency array as inputs and its main outputs are the real and the imaginary parts of the describing function models (pseudo frequency response). The other outputs are the value of gain error convergence parameter and the value of phase error convergence parameter.

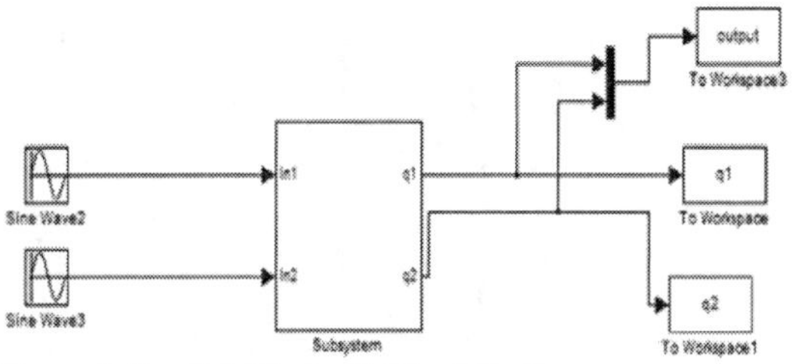

Figure 7. Simulink File for Multivariable Case.

6 Simulink file for multivariable case

Consider the nonlinear multi-input multi-output plant with two inputs as shown in Fig. 7 which represents the user-input Simulink file. For the example purpose, the pseudo frequency response data(SIDF function) of the system can be obtained for the following input values: 1) excitation array frequency values of [0.1 0.2 0.3;0.4 0.5 0.6]; 2)array of phase shift values is given by [pi/2 pi/2 pi/2; pi/2 pi/2 pi/2]; 3)excitation amplitude set is given by [7 8]; 4) DC value array of [0 0]; 5)number of inputs and outputs are entered as 2; 6)Input for No. of cycles are [0.1 0.2 2 0.3 3]. At this stage, the user may execute the software utility for generating the describing function. The pseudo-bode plots for the given example generated using the GUI is shown in Fig. 8 and time-data plots can be seen from Fig. 9. Similar analysis as in the case of single-input single-output can be done here for time-data figures to see how different frequencies for its corresponding Fourier integral cycles are plotted for time-data analysis.

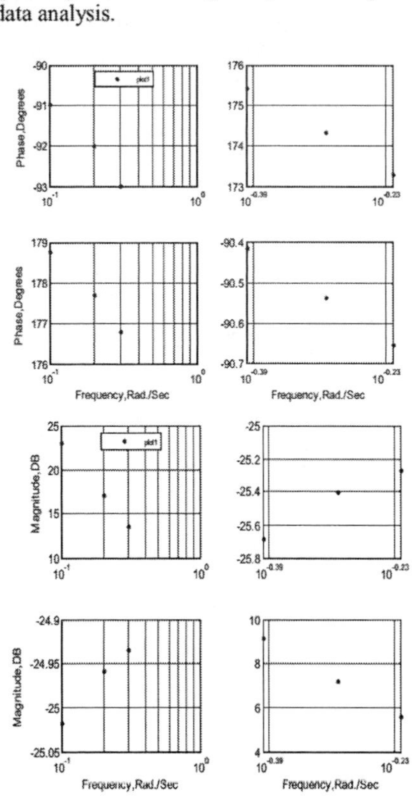

Figure 8. PSEUDO-BODE Plots for Multivariable.

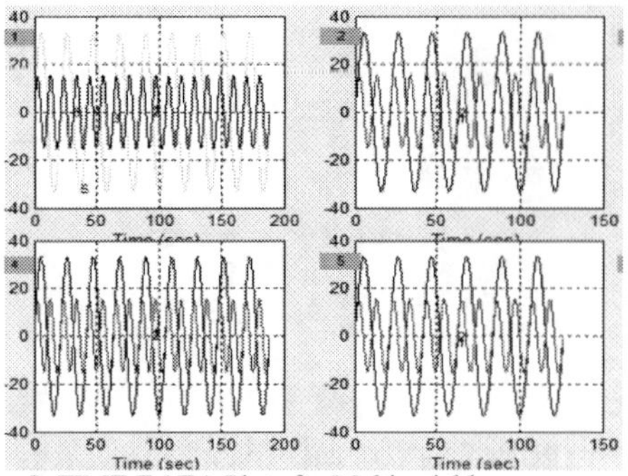
Figure 9. TIME-DATA Plots for Multivariable.

7 Conclusions

The proposed objective of the study was to develop a Graphic user interface utility for developing sinusoidal input describing function for both single variable and multivariable nonlinear systems. The aim was achieved. The formulation of input/output characterization is based on Fourier analysis where equations of motion are integrated to obtain the Fourier integrals. This procedure is repeated for different frequencies and amplitudes to obtain the sinusoidal input describing function model of the plant. Once the SIDF models are generated, system characteristics can be improved using conventional control design theories [3]. As a future work, the development of a user interface utility for the design of single range controller design for both single-input single-output case and multivariable case using describing function model can be proposed. With higher degree non-linear equations, the future calibration with GUI could be a realistic control system to work in the aircrafts. Implementing this proposed GUI interface for nonlinear controlling system in Optica, we can possibly find some credible features of characteristics in controlling modality of Optica flying Nonlinear equations of motion in aircraft; for example computation of exact flight paths with motions detection, optimization of flight paths along with nonlinear, time-varying large amplitude motions and significant change in mass in the aircraft could be a future scope in this newly developed interface considering higher degree non-linear equations.

References

1. Nassirharand, A. Karimi, H. 2002, Input/output characterization of highly nonlinear multivariable systems, AdvEngng software 33(2002).
2. Shinners Stanley, Describing function concept, Chap.5 section 5.7 of advanced modern control system theory and design, pp. 334-36, 1998.
3. Nassirharand, Input/output characterisation of highly nonlinear systems, .AdvEngng software, 6(3), pp. 129-33, 1987.
4. Nagrath, I. J. Gopal, The describing function method: Basic concepts. Chap.15 section 15.7 of control Systems Engineering, Fifth Edition, pp. 742-44, 2010.
5. Nassirharand, Design of dual-range linear controllers for nonlinear systems. ASME Trans DynSyst, Measure, Control, 1991.
6. Taylor JH, Strobel KL, Applications of a nonlinear controller design approach based on quasilinear system models, Proceedings of American Control Conference. San Diego, pp. 817, 1984.
7. Ola H¨arkegard and S. Torkel Glad, A backstepping design for flight path angle control. In Proceedings of the 39th Conference on Decision and Control, pages 3570–3575, Sydney, Australia, December 2000.
8. Ola H¨arkegard and S. Torkel Glad, Flight control design using backstepping. Technical Report LiTH-ISY-R-2323, Department of Electrical Engineering, Link¨opingsuniversitet, Sweden, December 2000.
9. WWW.flyoptica.com

Transformer fault diagnosis based on chemical reaction optimization algorithm and relevance vector machine

Wei Luo[1], Minfang Peng[1], Xun Wan[2], Shengquan Wang[3] and Xiupu Liao[1]

[1]*College of Electrical and Information Engineering, Hunan University, Changsha 410082, Hunan Province, China*
[2]*State Grid Hunan Electric Power Corporation Research Institute, Changsha 410007, Hunan Province, China*
[3]*State Grid Jiangsu Electric Power Corporation Research Institute, Nanjing 211103, Jiangsu Province, China*

Abstract. Power transformer is one of the most important equipment in power system. In order to predict the potential fault of power transformer and identify the fault types correctly, we proposed a transformer fault intelligent diagnosis model based on chemical reaction optimization (CRO) algorithm and relevance vector machine(RVM). RVM is a powerful machine learning method, which can solve nonlinear, high-dimensional classification problems with a limited number of samples. CRO algorithm has well global optimization and simple calculation, so it is suitable to solve parameter optimization problems. In this paper, firstly, a multi-layer RVM classification model was built by binary tree recognition strategy. Secondly, CRO algorithm was adopted to optimize the kernel function parameters which could enhance the performance of RVM classifiers. Compared with IEC three-ratio method and the RVM model, the CRO-RVM model not only overcomes the coding defect problem of IEC three-ratio method, but also has higher classification accuracy than the RVM model. Finally, the new method was applied to analyze a transformer fault case, Its predicted result accord well with the real situation. The research provides a practical method for transformer fault intelligent diagnosis and prediction.

1 Introduction

The power transformer is the core equipment in power system, so it is very important to analyze its operating parameters and make a correct judgment on the internal latent faults. The research on intelligent fault diagnosis of transformer is the key of ensuring power system safety [1].

The operating environment of power transformers are complicated, which has subjected by heat and electricity. Transformer insulation oil will be decomposed to corresponding gases by the process of heat and electricity. The gas types and their volume fraction can distinguish the fault types effectively. Now, dissolved gas analysis (DGA) is a convenient way to analyze and diagnose the potential fault of transformer [2, 3]. The classical IEC three-ratio method is widely used because of its simple principle. However, its coding mode is too absolute, which causes missed diagnosis. Therefore, how to improve diagnosis accuracy becomes a research focus in the field of transformer fault diagnosis [4].

In recent years, with the rapid development of artificial intelligence, many scholars are committed to the research of the transformer fault diagnosis based on artificial intelligence algorithm combined with DGA. For example, artificial neural network(ANN) was widely used in Transformer fault diagnosis [5]. However, it requires a lot of training samples and its convergence rate is too slow. Also, support vector machine(SVM) has advantages in dealing with nonlinear and small sample data classification problems[6, 7]. Thus it has made some achievements in transformer fault diagnosis. But its deficiency is that the number of support vectors will increase sharply with the sample size expanding, and the kernel function parameters are restricted by Mercer condition [8].

Relevance Vector Machine (RVM) is a new supervised learning method, which is proposed by Michael E. Tippping [9]. RVM is a sparse learning algorithm based on Bayesian framework. Compared to SVM, the number of relevance vectors is greatly reduced, and the kernel function is no longer constrained by the Mercer condition. At present, RVM has been applied to signal processing and power load forecasting, but it is rarely used in transformer fault diagnosis. This paper presents a multi-layer classifier model based on relevance vector machine, then combined with chemical reaction optimization algorithm to optimize the parameter of kernel functions in each layer. Finally, this paper collected 300 sets of measured data, then we use IEC three ratio method, the standard RVM model and the CRO-RVM model to test these data sets. The results show that the diagnostic model presented in this paper has obvious advantages over other models in diagnostic accuracy. And through some diagnosis cases, we found it is suitable for practical application.

2. Introduction of relevance vector machine multi-classification model

2.1 Relevance vector machine (RVM)

Hyper parameter is introduced into RVM machine learning algorithm to ensure its sparsity. It can assign zero-mean Gaussian prior distribution to the weight vector. The maximizing marginal likelihood function method is applied to estimate hyper parameter, and the automatic regulation mechanism is adopted to avoid the problem of the difficulty in determining the regularization coefficient [10, 11].

Let $\mathbf{t}=[t_1,t_2,\cdots t_N]^T$ be objective function and $\{x_i\}_{i=1}^N$ be input vector, then the RVM classification model can be expressed as the following equation (1):

$$y(\mathbf{x},\mathbf{w})=\sum_{i=1}^{N} w_i K(\mathbf{x},\mathbf{x}_i)+w_0 \tag{1}$$

where $\mathbf{w}$ is weight vector, $\mathbf{w}=[w_0,w_1,\cdots,w_N]^T$; and $K(\mathbf{x},\mathbf{x}_i)$ is kernel function .

For a two classification problem, the target value is only 0 or 1, so we can use formula (2) to characterize the likelihood function of the whole data sets:

$$P(\mathbf{t}\mid\mathbf{w})=\prod_{i=1}^{N}\sigma[y(\mathbf{x}_i,\mathbf{w})]^{t_i}\{1-\sigma[y(\mathbf{x}_i,\mathbf{w})]\}^{1-t_i} \tag{2}$$

Where $\sigma(\bullet)$ is a sigmoid function.

Sparse Bayesian method is used to give zero-mean Gaussian prior distribution for weight vector **w**, which can avoid over-fitting and keep the model sparse:

$$p(\mathbf{w}\mid\boldsymbol{\alpha})=\prod_{i=0}^{N}N(w_i\mid 0,\alpha_i^{-1}) \tag{3}$$

where α is a N+1 dimensional hyper parameter vector, $\boldsymbol{\alpha}=(\alpha_0,\ \alpha_1,\cdots,\ \alpha_N)^T$; and $N(\bullet)$ represents the normal distribution function.

For a new input vector **x**, the probability prediction expression of objective value *t* is given by the following equation (4) :

$$p(\mathbf{t}_*\mid\mathbf{t})=\int p(\mathbf{t}_*\mid\mathbf{w},\boldsymbol{\alpha})p(\mathbf{w}\mid\mathbf{t},\boldsymbol{\alpha})p(\alpha\mid t)dwd\alpha \tag{4}$$

Since the weighted posterior probability $p(\mathbf{w}\mid\mathbf{t},\boldsymbol{\alpha})$ and the marginal likelihood function $p(\boldsymbol{\alpha}\mid\mathbf{t})$ in formula (4) can not be solved by integration, the approximate solution proposed by Mackay needs to be used. The specific steps of this method are as follows.

1)Firstly, assign a value to the current hyper parameter $\boldsymbol{\alpha}$, then the "most likely" weight $\mathbf{w}_{MP}$ is estimated by using the maximum weight posterior probability function $p(\mathbf{w}\mid\mathbf{t},\boldsymbol{\alpha})$. since $p(\mathbf{w}\mid\mathbf{t},\boldsymbol{\alpha})\propto p(\mathbf{t}\mid\mathbf{w})p(\mathbf{w}\mid\boldsymbol{\alpha})$, the following equation (5) can be used to estimated the weight value of $\mathbf{w}_{MP}$.

$$\log[p(\mathbf{t}\mid\mathbf{w})p(\mathbf{w}\mid\boldsymbol{\alpha})=\sum_{i=1}^{N}[t_i\log Y_i+(1-t_i)\log(1-Y_i)]-\frac{1}{2}\mathbf{w}^T\mathbf{A}\mathbf{w} \tag{5}$$

where, $\mathbf{A}=\mathrm{diag}\,(\alpha_0,\ \alpha_1,\cdots,\alpha_N)$, $Y_i=\sigma[y(\mathbf{x}_i,\mathbf{w})]$. And $\mathbf{w}_{MP}$ is solved by the second-order Newton method, the gradient vector of **w** can be represented as equation (6). Similarly, the Hessian matrix and in formula (5) is given by equation (7) and (8):

$$\mathbf{g}=\nabla_w\log[P(\mathbf{t}\mid\mathbf{w})p(\mathbf{w}\mid\boldsymbol{\alpha})]=\boldsymbol{\Phi}^T(\mathbf{t}-\mathbf{Y})-\mathbf{A}\mathbf{w} \tag{6}$$

$$\mathbf{H}=\nabla_w\nabla_w\log[P(\mathbf{t}\mid\mathbf{w})p(\mathbf{w}\mid\boldsymbol{\alpha})=-\boldsymbol{\Phi}^T\mathbf{B}\boldsymbol{\Phi}-\mathbf{A} \tag{7}$$

$$\begin{cases}\Delta\mathbf{w}=-\mathbf{H}^{-1}\mathbf{g}\\ \mathbf{w}_{MP_{new}}=\mathbf{w}_{MP}+\Delta\mathbf{w}\end{cases} \tag{8}$$

where, $\mathbf{Y}=[Y_1,Y_2,\cdots,Y_N]^T$; and $\mathbf{B}=diag(\beta_1,\beta_2,\cdots,\beta_N)$, $\beta_i=[Y_i(1-Y_i)]$; $\boldsymbol{\Phi}$ is the design matrix, with $\boldsymbol{\Phi}=[\phi(x_1),\phi(x_2),\cdots,\phi(x_N)]^T$.

2) The posterior probability of Gaussian approximate weight is given by $p(\mathbf{w}\mid t,\alpha)\approx N(\mathbf{w}_{MP},\Sigma)$, where $\mathbf{w}_{MP}$ is posterior probability mean value of weight, and $\Sigma=(-\mathbf{H}\mid\mathbf{w}_{MP})^{-1}$ is a covariance matrix.

3) Using formula (9) to update the hyper parameter α:

$$\begin{cases}\alpha_i^{new}=\dfrac{\gamma_i}{w_{MPi}^2}\\ \gamma_i=1-\alpha_i\Sigma_{i,i}\end{cases} \tag{9}$$

Where the $\Sigma_{i,i}$ is the ith diagonal element, and w_{MPi} is the ith element for weight vector $\mathbf{w}_{MP}$.

When the hyper parameter α^{new} is updated, we need re-estimate the posterior mean of weight and calculate the covariance matrix. Repeat steps 1) to 3) until the convergence condition is satisfied. After the update completes, the value of most α_i will be close to infinity, the corresponding value of w_i will be 0. As for the remaining α_i will take a limited value, and the collection of the corresponding $\mathbf{x}_i$ is called Relevance Vector (RV).

2.2 RVM multi-classification model

Single relevance vector machine is mainly used for binary classification problem. In this paper, the transformer has been divided into five states. So we built four RVM classifiers based on binary tree architecture to identify the five types:normal state (NS), middle and low thermal overheating (LT), high thermal overheating (HT), high-energy discharge (HD) and low-energy discharge (LD). The RVM1 classifier is trained to distinguish between normal state and fault state. When the input sample is in normal state, output "0"; otherwise, output "1". The RVM2 classifier is used to separate the overheat fault from the discharge fault. When input a overheating

DOI: 10.1051/matecconf/201710810006

sample to RVM2, output "0", otherwise, output "1". The RVM3 classifier is used to separate middle and low thermal overheating from high thermal overheating. When input a middle and low thermal overheating sample to RVM3, output "0"; otherwise, output "1". The RVM4 classifier is used to separate low-energy discharge from high-energy discharge. When input a low-energy discharge sample to RVM4, output "0"; otherwise,output "1". The flowchart of multi-layer classification model based on RVM is shown in Figure 1.

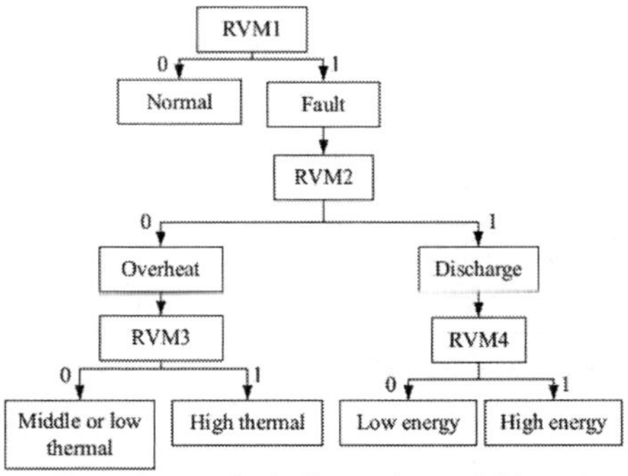

Figure 1. Transformer fault diagnosis model based on RVM.

3 Transformer fault diagnosis based on CRO-RVM model

3.1 Fault feature vector extraction

In this paper, we chose 5 diagnostic gases as input vector of RVM classifier. These gases include hydrogen (H_2), methane (CH_4), ethane (C_2H_6), ethylene (C_2H_4) and acetylene (C_2H_2). In order to reduce the calculation error caused by the difference of the DGA data, the training set data and testing-set data should be standardized, it is shown as the equation (10):

$$x_{ij}^{*} = \frac{x_{ij}}{x_j} \tag{1}$$

Where, x_{ij} is the volume fraction of the jth gas in the ith sample, and $j \in \{1,2,\cdots,5\}$. x_j is alarm value of the jth gas refer to *GB/T 7252-2001 Guide to the Analysis and the Diagnosis of Gases Dissolved in Transformer Oil,* and x_{ij}^{*} is called gas deterioration degree.

3.2 Chemical reaction optimization algorithm (CRO)

CRO is a meta-heuristic algorithm which has been developed in recent years. It has been put forward by Albert Y.S.Lam and Victor O.K.Li from the University of Hong Kong [12]. It simulates molecular motion and energy conversion processes in chemical reactions. Compared with genetic algorithm and particle swarm optimization algorithm [13, 14], it is characterized by not only the dynamic change of group size, but also the rich information interactive modes among individuals.

There are two basic arithmetic units in CRO algorithm, one is called molecule (ω) and the other is named container wall (buffer). Molecules have two kinds of energy are called kinetic energy (KE) and potential energy (PE). Buffer represents the reaction environment, PE is the measure index in CRO algorithm. When CRO algorithm is applied to solve a real problem, PE represents the objective function. KE is the quantitative value to judge whether the molecular reaction can be carried out. There are four basic reaction operators in CRO: single-molecule collision, single-molecule decomposition, inter-molecular collision and molecular synthesis[15, 16].

The chemical reaction algorithm first used the strength strategy formed by invalid collision to search the adjacent region. When a lower energy state cannot be found in the neighborhood, it will use the diversification strategy formed by inter-molecular collision and molecular synthesis to jump to a relatively remote areas continue to search. In the search process, the energy will be redistributed among molecules in various ways.

The choice of the initial value of the algorithm has a great influence on the results. In order to make the algorithm have better convergence speed, search precision and optimization ability, the initial molecular population is generated in a random way. The related parameters include molecular population size (Popsize), central energy buffer (Buffer), molecular initial kinetic energy (initialKE), molecular kinetic energy loss rate (KelossRate), etc. The specific steps of parameter setting are as follows:

1) Form a random molecule in a random manner within the range of molecular properties;

2) Judge whether the random molecule to meet the constraints or not, if satisfied, then enter step 3), otherwise, return to step 1);

3) Calculate the molecular potential energy, set the initial kinetic energy of the molecule, and save the molecule into the reaction container. At the same time, verify whether the initial molecular population in the container to achieve the initial set size or not, if reached, then finish the initialization process, otherwise, return to step 1).

In this paper, the main parameters of CRO algorithm are assigned as follows: Popsize=10; initialKE=1500; Kelossrate=0.25 and Buffer=0.

3.3 The diagnosis steps of transformer fault based on CRO-RVM model

Transformer fault diagnosis based on CRO-RVM model used CRO algorithm to optimize the kernel function parameters of the RVM classifiers. The flowchart is shown in Figure 2, the specific steps are described as follows:

1) The sample data are selected and normalized to obtain the feature vector. We divided the sample data into training set and test set according to the ratio of 2: 1.

2) Use the binary tree method shown in Figure 1 to establish a multi-level transformer fault diagnosis model based on RVM, and a set of kernel function parameters are initialized , then inputting the training set to train the model.

3) Initialize the chemical reaction optimization algorithm, select the mean square error of the RVM's actual output and expected output as the objective function.

4) Calculate the initial potential energy and initial kinetic energy of the molecule. Optimize the kernel function parameters by iterative calculation.

5) When the PE of molecule is smallest, parameters will obtain the global optimal solution. Assign these kernel function parameters to corresponding RVM classifiers.

6) Input the test data sets into the CRO-RVM model, and test its diagnostic accuracy.

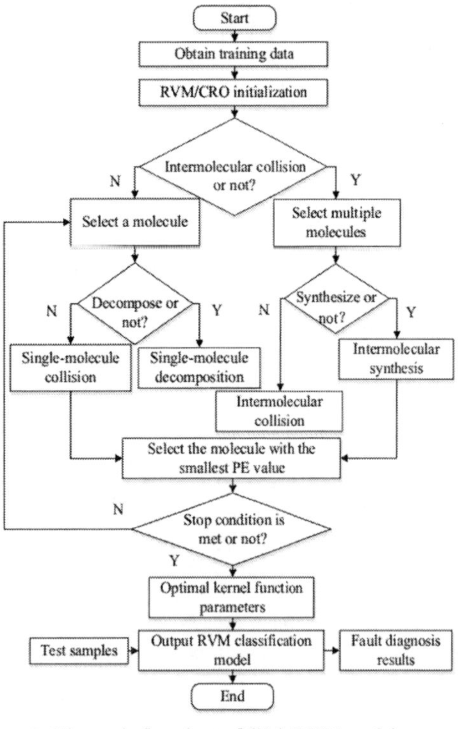

Figure 2. Diagnosis flowchart of CRO-RVM model.

4 Comparison of different diagnostic models and example analysis

4.1 Comparison of different diagnostic models

In this paper, we collected 300 sets of measured DGA data from the 220kV, 330kV and 500kV transformers. The 15 sets of measured data samples shown in Table 1 are part of them. We divided all of them into training set and testing set according to the ratio of 2: 1. We used IEC three-ratio method, RVM model and CRO-RVM model to train and diagnose these data samples. The kernel functions of RVM model and CRO-RVM model both are RBF, and the kernel function parameters of the RVM model all are 0.5. The four classifiers' parameters of the CRO-RVM model are optimized by CRO algorithm, they are [8.625, 0.264, 0.516, 1.246]. Figure 3 shows the mean square error performance of the CRO-RVM model is changed with the iteration number. The results of the three diagnostic methods are shown in Table 2 as follows.

Table 1. DGA data samples of power transformer

Transformer state	Diagnosis gas content (μL/L)				
	H_2	CH_4	C_2H_6	C_2H_4	C_2H_2
NS	13.6	5.3	8.2	29	2.1
NS	32	11.38	8.2	30.6	3.9
NS	5.8	2.7	1.8	0.8	0
LT	13.1	44.7	91.5	213.3	4.2
LT	70.2	246.6	171.5	371.6	0
LT	93	58	43	37	0
HT	39.4	40.6	17.2	51.4	4.3
HT	8.4	28.7	13	105	2.1
HT	170	320	53	520	3.2
HD	263	628	16.8	46	855.4
HD	354	387	65.8	101.2	189
HD	158	71	1.3	146	685
LD	97.8	15.9	2.7	8.1	24.3
LD	66	8.3	8.2	9.2	8.2
LD	59	10.4	4	10	12.7

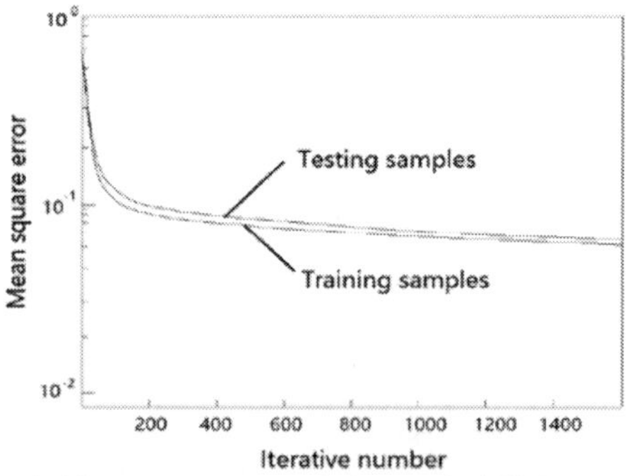

Figure 3. Mean square error variation trend of CRO-RVM model.

Table 2. Comparison of fault diagnosis results.

state	Sample Size	IEC/%	RVM/%	CRO-RVM/%
NS	10	60	100	100
LT	16	100	81.3	93.8
HT	20	80	80	90
HD	40	37.5	72.5	87.5
LD	14	42.9	57.1	100
Total	100	64.08*	78.18	94.26

(The symbol * indicates the presence of unrecognized samples.)

We can get the following conclusions from the diagnostic results based on different models:

1) The IEC three-ratio method has the problem of missing codes. For example, [354, 387, 65.8, 101.2, 189] is a group of high-energy discharge data. According to the coding rules shown in table 3, it is obviously out of the coding range. So the IEC three-ratio method can't diagnose it. The RVM model and the CRO-RVM model

DOI: 10.1051/matecconf/201710810006

solve this problem perfectly, they can identify this kind of data sample correctly so as to decrease the misdiagnosis rate.

Table 3. Coding rules of the IEC three-ratio method

Fault types	C_2H_2/C_2H_4	CH_4/H_2	C_2H_2/C_2H_6
LT	<0.1	>1	1-4
HT	<0.2	>1	>4
HD	0.6-2.5	0.1-1	>2
LD	>1	0.1-0.5	>1
unrecognized sample	1.53	1.09	2.87

2) The kernel function parameters in RVM model are fixed, so it is difficult to ensure the classifiers in different levels can achieve the best classification results at the same time. So the diagnostic accuracy of the RVM model is only about 78%.

3) After using CRO algorithm to optimize the kernel function parameters of the RVM classification model, the diagnostic accuracy has been obviously improved, reached more than 94%. The correct recognition rate meets the requirement of practical application.

4.2 Transformer fault diagnosis case

The No.1 main transformer in a 330kV substation was broken down in June 25, 2014. With the transformer heavy gas operation, its three side switch tripped. The main transformer was made by Xi'an XD Transformer Co., Ltd, its model number is OSFPSZ9-M-360000/330. It has been put into operation since 2006. Oil chromatographic test data is shown in Table 4.

Table 4. DGA test data

Test time	Gas content(μL/L)					Remark
	H_2	CH_4	C_2H_6	C_2H_4	C_2H_2	
2014-5	8.0	2.6	0.6	1.0	0	Pre-fault
2014-6	2142	234	4.5	57.4	86.9	After-fault

The CRO-RVM model proposed in this paper is applied to analyze these DGA test data. The output diagnosis result is HD, that means the fault type of No.1 transformer is high-energy discharge. After disassembling inspection, we found there were discharge marks at the bottom of transformer bushing and the end of current-carrying rod. And there was insulation breakdown in B phase winding. So the real fault type of this transformer is high-energy discharge. It perfectly confirms the prediction of the CRO-RVM model. The disassembling picture of No.1 transformer is shown in Figure 4.

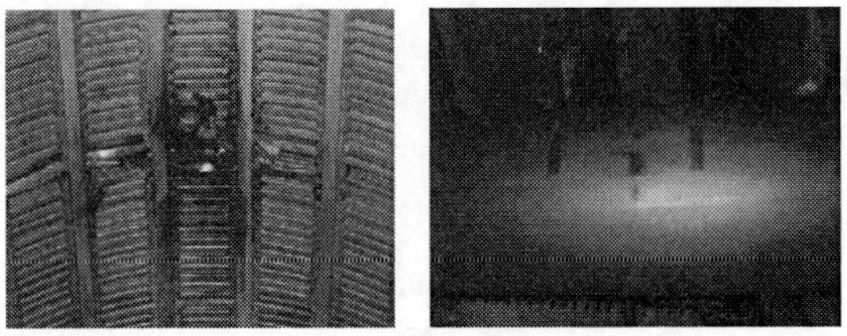

Figure 4. Disassembling picture of transformer 1.

The analysis of above case indicates that the transformer fault diagnosis method we proposed not only has good theoretical feasibility, but also has splendid ability of practical application. The new method can be applied to identify multiple fault types of large power transformer.

5 Conclusions

Relevance Vector Machine (RVM) is one of the emerging algorithms in machine learning field. The use of RVM in transformer fault diagnosis has been paid more and more attention. Based on the standard RVM classification model, this paper proposed a CRO-RVM model for transformer fault diagnosis. Using the CRO algorithm to optimize RVM kernel function parameters and improve the classification correct rate. Through the comparative analysis of the diagnosis results by different methods, we found that the CRO-RVM model can fully consider the individual differences of different transformers, and makes a relatively accurate diagnosis. Compared with the IEC three-ratio method, it doesn't has the problem of missing codes. And its diagnostic accuracy is higher than the standard RVM model. And a fault diagnosis case shows that the new method we presented is suitable for practical application. Its predicted results accord well with the real situation. The research indicates that it is a more useful tool for transformer fault diagnosis. In the future, we will try to combine DGA data with the routine electrical test data as fault features to further improve the diagnostic performance of the CRO-RVM model. In addition, we will study how to apply it to diagnose the fault of other oil-filled electrical equipment.

References

1. C.X. Guo, Y. Wang, M. Wang, "Time-varying Outage Model for Transformers Representing Internal Latent fault," Proc. Chin. Soc. Electrical. Eng. **33**, 1 (2013)
2. F. Jakob, J.J. Dukarm, "Thermodynamic Estimation of Transformer Fault Severity," IEEE Trans. Power. Del. **30**, 4, 1941-1948 (2015)
3. S. Bhowmick, S. Nandi, "Online Detection of an Interturn Winding Fault in Single-Phase Distribution Transformers Using a Terminal Measurement-Based Modeling Technique," IEEE Trans. Power. Del, **30**, 2, 1-9 (2015)
4. Z.X. Liu, B. Song, E.W. Li, et al. "Study of "code absence" in the IEC three-ratio method of dissolved gas analysis," IEEE Electron. Insulat. Maga. **31**, 6, 6-12 (2015)
5. P.V.B. Reddy, C.R.V. Kumar, K.H. Reddy, "Modeling of Surface Roughness in Wire Electrical Discharge Machining Using Artificial Neural Networks," Int. J. Mech. Eng. &Rob. Res. **2**, 1, 57-64 (2013)
6. K. Bacha, S. Souahlia, M. Gossa, "Power transformer fault diagnosis based on dissolved gas

analysis by support vector machine,". Electron. Power. Sys. Res. **83**, 1, 73-79 (2012)
7. A.M. Shah, B.R. Bhalja, "Discrimination Between Internal Faults and Other Disturbances in Transformer Using the Support Vector Machine-Based Protection Scheme," IEEE Trans. Power. Del. **28**, 3, 1508-1515 (2013)
8. S. Hwang,M.K. Jeong, B.J. Yum, "Robust Relevance Vector Machine With Variational Inference for Improving Virtual Metrology Accuracy," IEEE Trans. Semiconduct. Manu. **27**, 1, 83-94 (2014)
9. M.E. Tipping, "Sparse bayesian learning and the relevance vector machine," J. Mach. Learn. **1**, 3, 211-244 (2001)
10. J. Yin, Y. Zhu, Y.U. Guoqin. "Relevance vector machine and its application in transformer fault diagnosis," Power. Sys. Techno. **32**, 12, 2239-2247 (2012)
11. T. Wang,H. Xu,J. Han, et al. "Cascaded H-Bridge Multilevel Inverter System Fault Diagnosis Using a PCA and Multi-class Relevance Vector Machine Approach," IEEE Trans. Power. Electron. **30**, 12, (2015)
12. A.Y.S Lam, V.O.K Li, "Chemical-reaction-inspired metaheuristic for optimization," IEEE Trans. Evol. Comput. **14**, 3, 381-399 (2010)
13. P. Sreeraj, T. Kannan, S. Maji, "Simulation and Parameter Optimization of GMAW Process Using Neural Networks and Particle Swarm Optimization Algorithm," Int. J. Mech. Eng. &Rob. Res. **2**, 1, 130-146 (2013)
14. M. Adinarayana, G. Prasanthi, G. Krishnaiah, "Optimization for Surface Roughness, MRR, Power Consumption in Turning of En24 Alloy Steel Using Genetic Algorithm," Int. J. Mech. Eng. &Rob. Res. **3**, 1, 20-26 (2014)
15. H.L. Ao, J. Cheng, J. Zheng, et al. "Roller Bearing Fault Diagnosis Method Based on Chemical Reaction Optimization and Support Vector Machine," J. Comput. Civil. Eng. **29**, 5, (2014)
16. H. Duan, L. Gan, "Orthogonal Multiobjective Chemical Reaction Optimization Approach for the Brushless DC Motor Design," IEEE T. Magn. **51**, 1, 1-7 (2015)

Responses' Prediction Standard Error Analysis in Pareto Solutions

Nuno Costa[1,3], João Lourenço[1,2]

[1]Instituto Politécnico de Setúbal – Escola Superior de Tecnologia de Setúbal, Setúbal, Portugal; e-mail: nuno.costa@estsetubal.ips.pt,
[2]INESC-ID, Rua Alves Redol 9, 1000-029 Lisboa, Portugal; e-mail: joao.lourenco@estsetubal.ips.pt
[3]UNIDEMI, Faculdade de Ciências e Tecnologia-Universidade Nova de Lisboa, 2829-516 Caparica, Portugal

Abstract - A Pareto front provides to the decision-maker a set of optimal solutions from where a unique solution should be selected to solve a multiresponse problem. However, the responses' prediction standard error for each one of the optimal solutions is different, which impacts on the reproducibility of Pareto solutions. This means that the decision-maker has to take into account the responses' prediction standard error in the solution selection process. For this purpose, a metric is presented. A bi-objective problem is used as example and results displayed graphically.

1. Introduction

Industrial problems are multidimensional so the optimization of multiresponse problems has been an active research field. The simultaneous optimization of multiresponses is an often used practice to generate solutions for problems with conflicting responses, which are usual in real-life problems. A desired condition for any candidate solution for these type of problems is that the solution is non-dominated. This means that no better values for one or more responses can be achieved in another solution without degrading the value of, at least, another response.

A representative set of optimal or non-dominated solutions constitutes a Pareto front, and provide the most favorable alternative choices for solving a multiresponse optimization problem. However, assuming that the Pareto front was appropriately generated, the decision-maker faces another issue: to choose an optimal solution from the generated set. This is not a trivial task, because the reproducibility of optimal solutions cannot be ignored by the decision-maker. In fact, there is no guarantee that product or process performs as expected when is run at a chosen input variable location due to the existence of a natural process variability and uncertainty in the estimated responses. This means that the responses' prediction standard error in each optimal solution has to be evaluated and considered in the decision process, which has been ignored so far. Therefore, the objective of this manuscript is to introduce a metric to assess the prediction standard error of each optimal solution to help the decision-maker in making a more informed decision in the solution selection process.

2. Pareto Optimality-Dominance Relation

A multiresponse optimization (MRO) problem can be defined as follows:

$$\underset{\mathrm{x}}{\text{Minimize}}\, f(\mathbf{r}, \mathrm{M}) \tag{1}$$

subject to

$$\mathbf{g}(\mathbf{x}, \mathbf{b}) \leq 0$$

$$\mathbf{h}(\mathbf{x}, \mathbf{b}) = 0$$

$$\mathrm{x}_l \leq \mathbf{x} \leq \mathrm{x}_u$$

where f is a scalar valued function, M represents the set of user-specified or preference parameters (weights, priorities, shape factors, ...), and $\mathbf{r}$ is the vector of responses ($r = 1, ..., p$). Parameters do not controlled by the decision-maker are denoted by $\mathbf{b}$. Design or input variables, denoted by vector $\mathbf{x}$, with lower bound x_l and upper bound x_u. Constraints are represented by $\mathbf{g}(\mathbf{x}, \mathbf{b})$ and $\mathbf{h}(\mathbf{x}, \mathbf{b})$.

In contrast to single response optimization problems, where the optimal solution is defined easily, for MRO problems a solution is more a concept than a definition [1]. In practice, a compromise solution has to be selected, because the utopia point (the variables setting that yields a solution with all responses at their target value) for formulation (1) cannot be achieved.

A predominant concept in defining a non-dominated solution is that of Pareto optimality. For a minimization problem like that formulated in Eq. (1), a solution (a vector of responses; r_1) dominates another one (r_2), is Pareto optimal, if both the following conditions are true:

a) the value of any response in r_1 is no worse (is lower or equal) than that of r_2;

b) the value of at least one response in r_1 is strictly better (is lower) than those of r_2.

Pareto frontiers can be displayed graphically only for problems with two or three responses, and consensus on what qualities a representation of the non-dominated set should possess do not exist. According to [2], quality measures that have been proposed and may be useful to this end are the following: 1- measures of cardinality (which refer to the number of points in a representation); 2- coverage (which refer to the regions of the outcome set that are represented); 3- spacing (which refer to the distance between points in the representation). Hybrid measures which overlap the above three categories have been also introduced in the literature. An example is the hypervolume measure, a well-established indicator of a front's quality [3-4].

A representative set of non-dominated solutions for MRO problems is helpful as it provides a broad overview of alternative solutions. However, in practice, only one solution is selected. This leads to a critical question: which optimal solution must be chosen?

3. Reproducibility of Non-dominated Solutions

A discrete representation of a Pareto frontier (a collection of solutions distributed along the Pareto front that provide a finite and manageable alternative solutions to the decision-maker) is, in general, sufficient for selecting a solution for MRO problems. Majority of the literature has focused on how to find the Pareto frontier with its most promising choices without providing more insights on how to proceed from those choices to a final decision [5]. However, it is important to be aware that some Pareto solutions may lead to operation conditions more hazardous, more costly or more difficult to implement and control than others. In fact, the responses' prediction standard error is not the same for all the Pareto solutions and, if it is ignored, can potentially lead to suboptimal choices and selected solution yield, in practice, unexpected results. To minimize the gap between the theoretical and practical results, the decision-maker must assess the prediction standard error. For this purpose, one can use Eq. (2),

$$se = \sqrt{MSE(1 + \mathrm{x}_0^T(X^TX)^{-1}\mathrm{x}_0)} \quad (2)$$

which is the second term of a two-sided prediction interval for the r-th predicted response at optimal location x_0. Note that prediction intervals give a range for the y-value of the next observation given specific x-values, and is defined as

$$\mathrm{x}^T\hat{\beta} \pm t_{\alpha,n-z}\sqrt{MSE(1 + \mathrm{x}^T(X'X)^{-1}\mathrm{x})} \quad (3)$$

where the mean square error $MSE = SS_E/(n-z)$, SS_E is the error sum of squares, n is the number of experimental runs in the model matrix X, z is the number of terms in the response's model, and α is the significance level.

4. Case Study

The objective of this example is to determine the settings for reaction time (x_1), reaction temperature (x_2), and amount of catalyst (x_3) to maximize the conversion (y_1) of a polymer and achieve a target value for the thermal activity (y_2). A central composite design with four center points was run and the (mean) models fitted to responses are

$$\hat{\mu}_1 = 81.0943 + 1.0290\,x_1 + 4.0426\,x_2 + 6.2060\,x_3 - 1.8377\,x_1^2 + 2.9455\,x_2^2 - 5.2036\,x_3^2 + 2.1250\,x_1x_2 + 11.3750\,x_1x_3 - 3.8750\,x_2x_3$$

$$\hat{\mu}_2 = 59.8505 + 3.5855\,x_1 + 0.2547\,x_2 + 2.2312\,x_3 + 0.8360\,x_1^2 + 0.0742\,x_2^2 + 0.0565\,x_3^2 - 0.3875\,x_1x_2 - 0.0375\,x_1x_3 + 0.3125\,x_2x_3$$

The range values for y_1 and y_2 are [80, 100] and [55, 60], respectively. Assuming that y_1 is a Larger-The-Better response type - the estimated response value is expected to be equal or larger than an upper bound, the target value is set equal to 100; y_2 is a Nominal-The-Best response type - the estimated response value is expected to be equal to a target value), the target value is set equal to 57.5. The constraints for the input variables are $-1.682 \leq x_i \leq 1.682$ $(i = 1, 2, 3)$.

Figure 1 displays the Pareto front generated with the compromise programming-based criterion proposed in Ref. [6], varying the shape factor $1 \leq \omega_i \leq 4$ in increments of 0.01 and a Sequential Quadratic Programming algorithm. This example deals with the optimization of two responses so it was possible to display the Pareto frontier graphically, and one can see that responses are in conflict: when $\hat{\mu}_1$ increases, $\hat{\mu}_2$ deviates from target and vice-versa.

Figures 2 and 3 show that pick up a solution based on technical, economic or decision-maker preferences may not produce the expected results in practice, because responses' prediction standard errors are not homogeneous. Prediction standard error ranges from 6.5 % to 9.5 % for the polymer conversion and from 2.4 °C to 3.6 °C in thermal activity. These values range is not as small as desired and may impact on process or product performance. The normalized responses' prediction standard errors (relative magnitude of the prediction error) for each solution $[(s.e.(\hat{\mu}_r))/\hat{\mu}_r]$ is also presented in Figures 2-3, and it is interesting to note that the values trend in both graphs are similar and analogous to those of $s.e.(\hat{\mu}_r)$. This means that the same decision would be taken from both pair of graphs, $s.e.(\hat{\mu}_1)$ versus $s.e.(\hat{\mu}_2)$ or $s.e.(\hat{\mu}_1)/\hat{\mu}_1$ versus $s.e.(\hat{\mu}_2)/\hat{\mu}_2$.

 DOI: 10.1051/matecconf/201710810007

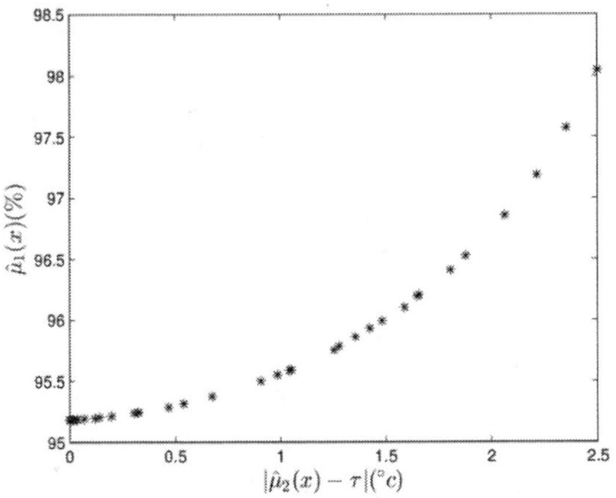

Figure 1. Pareto front

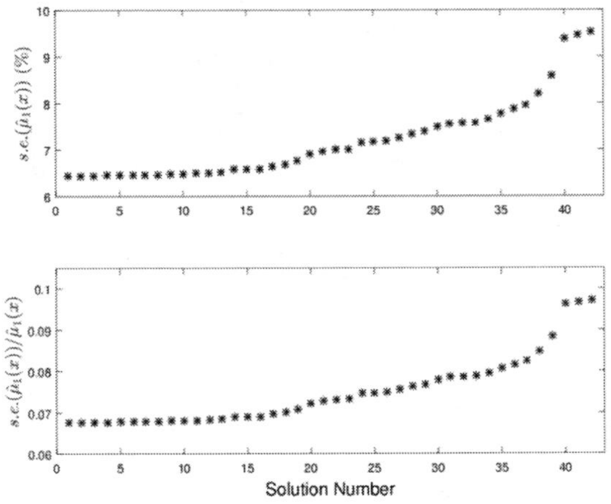

Figure 2. Standard error: Polymer conversion

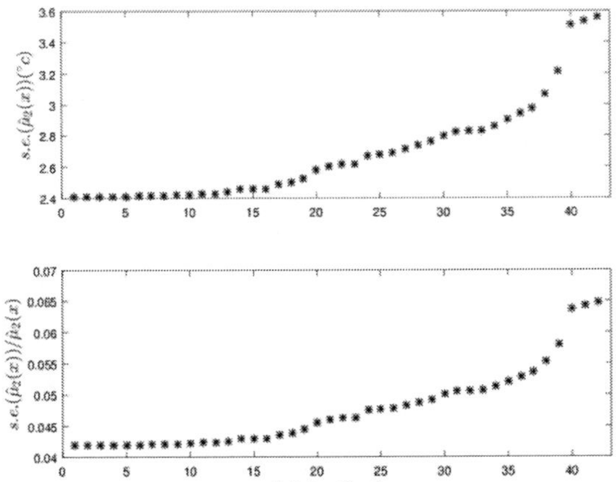

Figure 3. Standard error: Thermal activity

5. Results Discussion

A compromise solution to solve a MRO problem must be selected among those of the Pareto front. However, those solutions may not yield the expected practical result, because responses' prediction standard error of a selected solution at optimal variable settings can be (excessively) high. The magnitude of responses' prediction standard error can be displayed graphically, such as shown in Figures 2 and 3, and may help decision-maker in making a more informed optimal solution selection.

Optimal solution selection is more complicated and Pareto front visualization impossible for more than four responses so alternative tools such as tabular lists are necessary to highlight more favorable solutions in an effective manner. Nevertheless, the decision-maker needs to balance the conflict between responses' priority and responses' prediction standard error. For instance, if priority is setting $\hat{\mu}_2$ on target value ($\hat{\mu}_2 = \tau$; see Figure 1), one can see from Figure 3 that lower prediction standard error values for thermal activity $[s.e.(\hat{\mu}_2)]$ are found for the first solutions, where $\hat{\mu}_2 = \tau$, as well as for solutions where $\hat{\mu}_2$ is slightly deviated from its target ($\hat{\mu}_2 \neq \tau$), namely the solutions number 1 up to 12. Moreover, it is interesting to note that $s.e.(\hat{\mu}_1)$ values are also low for these solutions. If priority is setting $\hat{\mu}_1$ on target value, Figure 2 shows that $s.e.(\hat{\mu}_1)$ value is undesirably higher for solution number 42 (the solution with the highest $\hat{\mu}_1$ value) than for other solutions, for instance, the thirty-fifth solution, though its $\hat{\mu}_1$ value are slightly worst (lower than 96.5%). Note that $s.e.(\hat{\mu}_2)$ value for solution number 42 is also undesirably higher. This confirms that no conflict exists between $s.e.(\hat{\mu}_1)$ and $s.e.(\hat{\mu}_2)$ values if decision-maker assigns a higher priority either to $\hat{\mu}_1$ or $\hat{\mu}_2$. Nevertheless, this case study shows that only by chance a better solution would be selected from the simple analysis of a list or a graph with a representative set of Pareto (optimal) solutions. Some solutions should be avoided because their prediction standard error is too high. This confirms that a compromise is necessary to select a solution for MRO problems and information about solutions reproducibility ($s.e.(\hat{\mu}_i)$ or $s.e.(\hat{\mu}_i)/\hat{\mu}_i$ values) cannot be ignored. In fact, responses are usually in conflict and *s.e.* values for some solutions in either one or both responses is undesirably high and likely lead to less favorable results in practice. Confirmatory runs can help in the decision-making process, and decision-maker must be aware that solutions whose response values are slightly deviated from target and $s.e.(\hat{\mu}_i)$ or $s.e.(\hat{\mu}_i)/\hat{\mu}_i$ values are lower can be effective alternatives for solving MRO problems.

6. Conclusions and Future work

This work investigates the Pareto front solutions in terms of responses' prediction standard error, and shows that responses' prediction standard error is different for each solution. From a theoretical point of view, it is desirable to select Pareto solutions with the lowest responses' prediction standard error. However, this may not be possible due to the conflict between the responses' priority and the responses' prediction standard error. In fact, the prediction standard error may be higher for desired response values. Thus, in practice, a compromise between technical and economic considerations will dictate the final decision about the most favorable solution.

To help the decision-maker in making more informed decisions, future work shall investigate the impact of models coefficient uncertainty in Pareto fronts. Optimal solutions built on the worst-case estimated responses must be generated and other metric(s) developed to evaluate optimal solutions.

Acknowledge

The authors thanks for IPS/ESTSetubal and NOVA ID-FCT/UNIDEMI for the support received.

References

1. Marler, R., Arora, J. Survey of multi-objective optimization methods for engineering; Struct Multid Opt, 26(1), 369-395, 2004.
2. Faulkenberg, S., Wiecek, M. On the quality of discrete representations in multiple objective programming; Opt Engng, 11(3), 423-440, 2010.
3. Cao, Y., Smucker, B., Robinson, T. On using the hypervolume indicator to compare Pareto fronts: Applications to multi-criteria optimal experimental design; J Stat Plan Inf, 160(8), 60-74, 2015.
4. Lu, L., Anderson-Cook, C., Adapting the hypervolume quality indicator to quantify trade-offs and search efficiency for multiple criteria decision making using Pareto fronts; Qual Reliab Engng Int, 29(8), 1117-1133, 2012.
5. Chapman, J., Lu, L., Anderson-Cook, C. Impact of response variability on Pareto front optimization; Statistical Analysis and Data Mining: The ASA Data Science J, 8(5-6), 314-328, 2015.
6. Costa, N., Pereira, Z. Multiple response optimization: a global criterion-based method; J Chemom, 24(6), 333-342, 2010.

A Function to Data Matrix (FDM) Approach for Mission Variables Consideration

Mauricio Fernández-Montoya, Ricardo Mejía-Gutierrez and Gilberto Osorio-Gómez

Design Engineering Research Group (GRID), Universidad EAFIT. Cra 49 N. 7 Sur 50, Medellin, Colombia

Abstract. Precise control of technical system variables in industrial processes and manned and unmanned missions is critical as their results should be achieved plenty. This takes place in Control Rooms (CR) where Operators make decisions based on the large amount of received data. This work proposes a method for weighting variables of a technical system based on their impact on the mission objectives. The relationships between variables and mission functions are defined using a Function to Data Matrix (FDM) in order to allow Operators to determine their criticality. The proposed method was applied for the mission control of the Racing Solar Vehicle *Primavera1* which participated in the World Solar Challenge (WSC) 2013 in Australia.

1 Introduction

In a Control Room (CR) or Mission Control Center (MCC) Operators control a Technical system trying to reduce unwanted variabilities related to its dynamics and manage its operational phases in order to get expected results. In steel processing facilities, nuclear power plants, orbital and planetary missions or top end racing vehicles, there are several components working together as subsystems, interacting each other and performing a sequence of steps. Proper operation of those components depends on accurate measuring, monitoring and deep knowledge about the process and large amounts of data are processed to make decisions about mission development. Since objectives of the mission define the nature of implemented technical systems, Operators should be aware of data that have direct impact with mission operations, especially if they can not only monitor but influence them by changing system parameters. As the systems become more complex, the number of underlying information flowing between subsystems or devices increases, masking any behaviour out of which Operators must be aware of. Analysing a mission with a proposed Function to Data Matrix (FDM), data can be linked with their role and relevance to mission functions based on objectives to be achieved. Operators can use FDM to decide if any system variable impacts in one or more functions of the mission and its relevance based on Variable Relevance Indicator (VRI) and Function Relevance Indicator (FRI).

2 Previous studies and models

Several studies have focused on developing mission analysis methods to improve mission execution by studying which information should be presented to operators and how to show it. Most developed work is on space missions like Space Mission Analysis and Design (SMAD), implemented by NASA [1] but it can be adopted in every process requiring a CR as well. Non-Parametric methods like Expected Value of Displayed Information (EVDI) and Expected Value of Revealed Information (EVRI) were used. Besides, the amount of data actually used by Operators during decision making can be measured using probabilistic methods. Since levels of expertise are different between Operators, the results of the method may vary and several iterations may be performed [2]. Ecological Interface Design (EID) defines semantic fields as a Work Domain Analysis (WDA) in a constrained base style, and it is especially useful for non-expert users with no in-deep knowledge about the technical system [3][4]. Canadian *BlueSky* Solar Car Team implemented an EID Approach for their racing Vehicles [5]. The proposed approach is focused on Functional Structure as a tool for better understanding of system internals linked to mission basic functions that must be covered to get satisfactory results and objectives accomplishment.

3 Approach based on Function to Data Matrix (FDM) applied to Mission Control Rooms

The proposed approach, presented in Fig. 1, considers four steps. It starts analysing the internal of the technical system using documentation such as datasheets, mechanical and electrical plans, and software. Then, general function carriers are defined by Reverse Engineering methods. Also, a mission analysis is required, starting from a statement and being complemented by additional documentation as Product Design Specifications (PDS), regulations or manuals. Finally, the FDM is created.

3.1 Step 1. Mission Essential Analysis.

The *Mission Description* is denoted by context, required and available resources, tasks to be performed and timelines. The purpose of the mission task execution and its expected results can be defined as Mission Objectives or *Mission Goals*. Although they are set out qualitatively, they can be measured in a quantitative way defining performance indicators or *Success Criteria Indicators*, showing the degree of compliance or failure of the tasks.

3.2 Step 2. Goal to Function Analysis.

Certain functions must be completed in a specific order and in its entirety. They are related to operational capabilities, devices and tasks to be performed, which are designed for mission execution by the use of a Goal to Function Tree (GFT) [6]. The highest level function to be covered (Mission Goal) can be broken down into lower level functions. While going down the tree, functions become simpler and give detailed indications on how higher level functions should be performed. If functions cannot be broken, they become *Basic Functions* [7] and they appear at the end of each branch.

3.3 Step 3. Technical System Analysis: Functional Diagram (FD).

A simple way to analyse technical systems is through the division of its parts and definition of interactions between them as Functional structures manage it [8]. Generally used as product design tool, could be used to analyse existing systems by the use of Function Structure Transformation (through definition of basic classes and functions) [9] as Reverse Engineering process does [10] and starting from a physical block diagram of the process obtained by observation [11]. In a FD the interactions (Matter, Energy and Information) between each subsystem emerges. From the point of view of the Operators, certain information cannot be observed or does not exist. Only emerging information flows are visible to them, and interact with the environment. By themselves, underlying information flows can be used as performance indicators for a particular subsystem adding a function carrier block called *Sensor*. A sensor does not transform neither incoming energy nor matter, but quantifies crossing energy or matter through an information flow, which can be conducted outside the FD, thus, making it explicit to Operators. Theoretically, a *Sensor* does not disturb Energy or matter flows, but, some aspects of adding a *Sensor* must be considered:

Feasibility. Knowledge, tools and skills should be available to install a sensor. If not, consider the possibility to generate or create them within time, technical and budget constraints.

Need. As success criteria, is the information necessary for the fulfilment of the mission objectives and to obtain performance indicators of them?

Degradation. Would installed sensors affect the technical system performance in terms of power consumption and signal distortion? If they would, they affect mission objectives accomplishment.

In the context of a Technical System, Information Flows lead *Raw Data* and a post-processing of these data, expressed as variables, is required. Then, relationships between variables are created as *Mission Information* and shown to Operators using a Graphic User Interface (GUI) or Human Machine Interface (HMI). *Mission Variables* can be classified by four attributes.

***α* or *Variable Type*.** It takes a value XX where the first digit is defined by 0:Continous, 1:Nominal, 2:Dichotomous, and second digit is defined by 0: Observable, 1: Controllable. **β or *Magnitude*.** It describes variable units. If it does not apply it takes a value of 0. **γ or Interval**, is defined by Eq. 1

$$\begin{aligned} A &= \{x \in \mathbb{R} / a \leq x < c\} \wedge \\ B &= \{x \in \mathbb{R} / b < x \leq d\}, \\ \gamma = A \cap B &= \{x \in \mathbb{R} / b < x < c\} \end{aligned} \tag{1}$$

Any value in the interval $\gamma = (b, c)$ is considered within mission operational range being b Minimum Mission Value and c Maximum Mission Value. Values $(A \cup B) - \gamma$ are out of operational range and they are useful to Operators as failure indicators as a indicates Minimum Absolute Values an d Maximum Absolute Values of the variable. **δ denotes *Sampling Frequency*** and it is expressed in Hertz *(Hz)*. With asynchronous sampling variables it is marked as 0.

3.4 Step 4. Function to Data Matrix (FDM).

Basic functions (rows) and Mission Variables (columns) are linked in FDM by Variable Relevance Indicator (VRI), denoted by P_{fj}, and Function Relevance Indicator (FRI), denoted by P_{vj}. VRI ranks certain variable in terms of its relevance along and within mission functions, and it is presented in Eq. 2 FRI shows to Operators which function requires especial attention by the amount of data related to it, and it is presented in Eq. 3

$$P_{fj} = \frac{1}{9} \sum_{i}^{n} k f_j V_j \tag{2}$$

$$P_{vj} = \frac{1}{9} \sum_{j}^{m} k f_j V_j \tag{3}$$

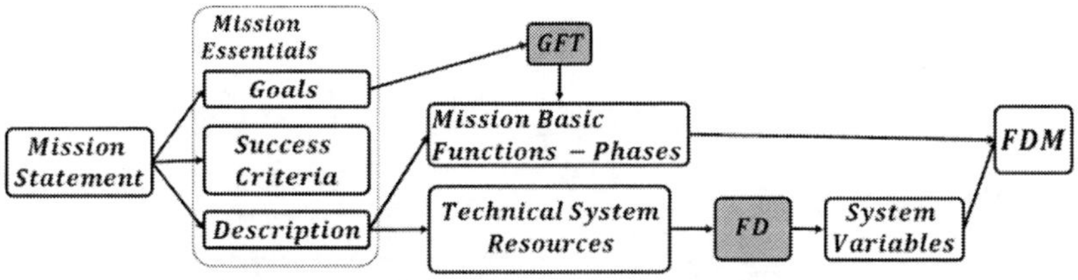

Figure 1. Function to Data Matrix construction from a mission statement analysis

It is required that Operators define the degree of relationship between both, using a relationship descriptor kfjvj with four categorical values: kfjvj=0 or Null, kfjvj=1 or Weak, kfjvj=3 or Medium and kfjvj=9 or Strong. Both VRI and FRI are normalized.

4 Case study: Racing solar electric vehicle *Primavera1*

A Solar Electric Vehicle (SEV) harvests irradiated energy from the sun using a Photovoltaic (PV) array. High Electrical and Aerodynamic efficiencies must be accomplished due to race conditions. Then, constant monitoring of vehicle variables and strategy should be done aboard an escort vehicle, forcing Operators to receive and analyse large amounts of information in a limited time. *Primavera1,* shown in Fig.2, is the first Colombian solar car competed during World Solar Challenge (WSC) 2013 in a 2505km self-powered travel.

Figure 2. Primavera 1, First Colombian solar electric vehicle, which competed in the World Solar Challenge 2013.

4.1 Mission statement for WSC.

The WSC mission statement is defined as *Departing from city of Darwin and arriving to the city of Adelaide in less than 50 hours in a SEV* [12]. Then, Primary Objective is to *Transport a Driver from Darwin to Adelaide in a Solar Electric Vehicle* taking into account all restrictions and conditions stated in the WSC Event Regulations Document (as complementary Mission Description). Success Criteria is measured by time it takes to cross the finish line, with a reference maximum time limit of 50 hours.

4.2 Goal to Function Tree (GFT).

In the simplified GFT for the mission, presented in Fig.3, highlighted boxes show Basic Functions. Due to nature of the race and Technical specifications of the Vehicle, most of the functions are focused on collect, store and manage available energy resources (Driving Strategy).

4.3 Primavera 1 Functional Diagram (FD).

For this Solar Vehicle a simplified Structural Analysis is presented in Fig. 4, Vehicle uses a communication bus based on *CANBusV2.0*, and both Control commands and system data are sent to MCC trough a RS232-Based Radio Frequency Module. Data packages between devices are represented by information flows in Fig.4, not the actual Bus layout. An additional current integrator and voltage sensor, g_1 was added between the battery pack leads and the ESC Module, due to mismatch lectures from battery current integrator.

4.4 Function to Data Matrix (FDM).

A summary of Proposed FDM is presented in Table 1. *For Primavera1*, 279 variables were identified. kf_jv_j values were obtained by Expert Engineers from Photovoltaics, Electronics and Computational Mechanics areas. Solar Array to Solar ECU, Battery-Pack to Solar ECU and Solar ECU to Electronic Speed Controller were selected as they are representative in the vehicle. Also, information flow, from g_1 sensor, is included in the analysis.

5 Conclusions

FDM can be useful to Operators as a tool for easy understanding of the technical system in context of mission goals and requirements. All relevant information is shown in a simple way and can be used as a Product Design Specification for GUI designers and Operators during Mission Execution. FDM can be hard to implement by hand in high complex systems so computer aided FDM tool approach will be implemented for future work. Several interpretations can be made while constructing GFT and FD. So, more than one iteration of the method is recommended. Both VRI and FRI values depends on the amount of analysed variables and basic functions. It shows no absolute values, but relative to each other, changing as the analysed amount of variables changes. There is no differentiation between α values shown in VRI for each type of variable, so an additional ponderation factor is required. This is especially evident with Alarms and Flags Variables.

Operational stages analysis will be also implemented in the future.

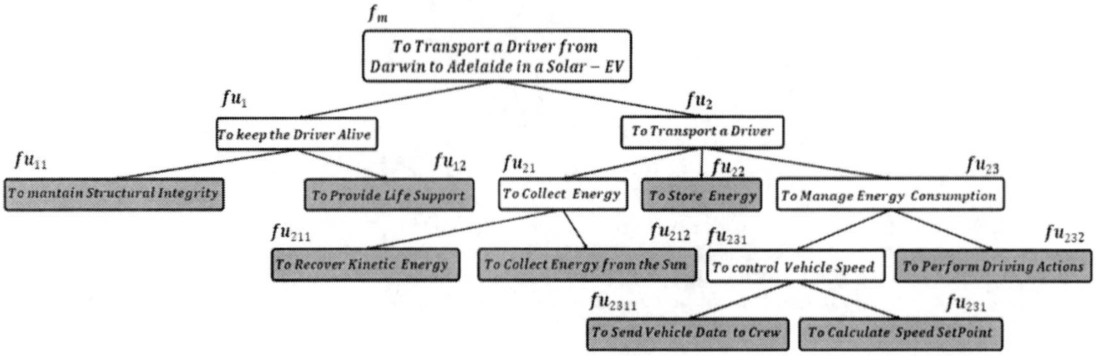

Figure 3. GFT for a WSC Class Vehicle, based on Mission Statement Analysis.

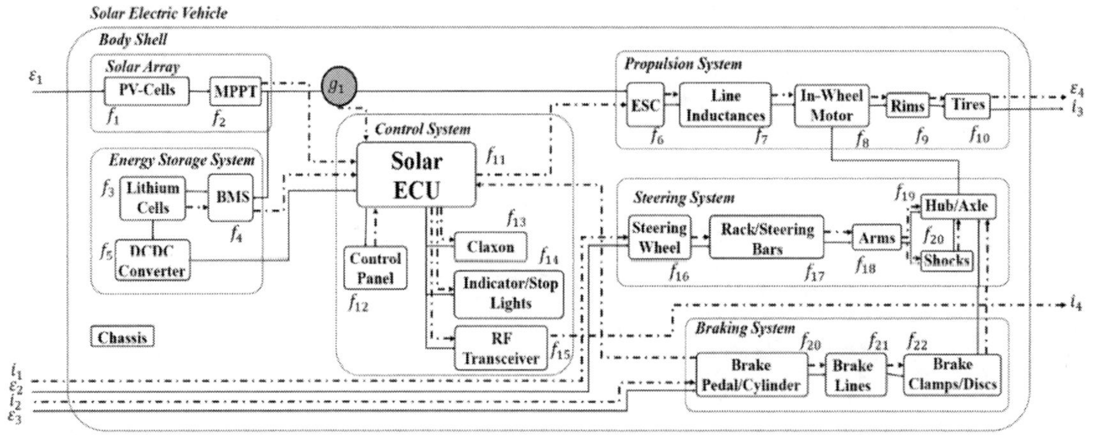

Figure 4. Simplified FD of *Primavera 1* based on the analysis of its inner systems.

Table 1. Summary of FDM for Primavera 1 Selected Information flows f_2~ f_{11}, f_4~f_{11}, f_{11}~f_6 and g_1~f_{11} across basic functions are obtained from GFT presented in Fig.2

Flow	Variable Name	Attributes				Basic Functions								V.Rel.
		α	β	γ	δ	f_{u11}	f_{u12}	f_{u22}	f_{u211}	f_{u212}	f_{u232}	f_{u2311}	f_{u2312}	P_{fj}
$f_2 \sim f_{11}$	State of Charge	00	$[Ah]$	(0,40.8)	1	0	0	9	9	3	3	0	9	3.6
	Total pack capacity	00	$[Ah]]$	(0,40.8)	10	0	0	9	3	0	0	0	9	2.3
	Min. Cell Voltage	00	$[V/1000]$	(2.5,4.2)	10	0	0	9	3	3	3	0	3	2.3
	Max. Cell Temperature	00	$[°C/10]$	(20,55)	1	0	0	9	0	0	0	0	1	1.1
	Battery Voltage	00	$[V/1000]$	(90,147)	10	0	0	9	9	9	3	0	9	4.3
	Battery Current	01	$[A/1000]$	(0,30)	10	0	0	9	9	9	3	0	9	4.3
	Status Flags	10	0	*NA*	1	0	0	9	3	0	3	0	3	2
$f_{11} \sim f_6$	Motor Current	01	$[A]$	(0,30)	4	0	0	0	0	0	0	0	9	1
	Motor Velocity	01	$[(\pi/30)Rad/s]$	(850,1100)	4	0	0	0	1	0	3	0	9	1.4
	Limit Flags	10	0	*NA*	5	0	0	0	9	0	9	0	0	2
	Vehicle Velocity	01	$[m/s]$	(16.6, 33.3)	5	0	0	0	3	0	9	0	9	2.3
	Odometer	00	m	*NA*	1	0	0	0	0	9	9	0	9	3
$f_2 \sim f_{11}$	Input Voltage	00	$[V]$	(36-141.4)	0	0	0	0	0	9	0	0	9	2
	MPPT Input Current	00	$[A]$	(0,4)	0	0	0	0	3	9	0	0	9	2.3
	MPPT Output Voltage	00	$[V]$	(90,147)	0	0	0	1	0	9	0	0	0	1.1
	MPPT Temperature	00	$[°C]$	(30,70)	0	0	0	0	0	3	0	0	0	0.3
$g_1 \sim f_{11}$	Battery Voltage	00	$[V/1000]$	(90,147)	0	0	0	9	9	3	3	0	9	3.6
	Integrator Current	01	$[A]$	(0,30)	0	0	0	9	9	3	3	0	9	3.6
	Capacity	00	$[Ah/1000]$	(0,40.8)	0	0	0	9	1	3	3	0	9	2.7
	Temperature	00	$[°C]$	(30,70)	0	0	0	3	0	0	0	0	0	0.3
	⋮									⋮				
	Function Relevance P_{vj}					0	0	35.8	16.5	13	8.8	0	19.5	

References

1. W.J. Larson, J.R.Wertz, Tech. rep., Microcosm, Inc.,Torrance, CA (US) (1992)
2. E. Horvitz, M. Barry, Display of information for time-critical decision making, in Proceedings of the Eleventh conference on Uncertainty in artificial intelligence (Morgan Kaufmann Publishers Inc.,1995), pp. 296–305
3. J. Rasmussen, K.J. Vicente, International Journal of Man-Machine Studies 31, 517 (1989)
4. C.M. Burns, J. Hajdukiewicz, Ecological interface design (CRC Press, 2004)
5. A. Hilliard, G.A. Jamieson, Ergonomics in Design: The Quarterly of Human Factors Applications 16, 6 (2008)
6. J.D. Patterson, S.B. Johnson (2013)
7. N. Viola, S. Corpino, M. Fioriti, F. Stesina, Functional Analysis in Systems Engineering: methodology and applications (InTech, 2012)
8. G. Pahl, W. Beitz, Engineering design: a systematic approach (Springer Science & Business Media, 2013)
9. C. Kirschman, G.M. Fadel, Journal of mechanical design 120, 475 (1998)
10. K.L. Wood (1997)
11. M.A. Viscio, N. Viola, R. Fusaro, V. Basso, Acta Astronautica 114, 79 (2015)
12. W.S. Challenge, World solar challenge race overview, http://www.worldsolarchallenge.org/page/viewbyid=76

Designing a Quenching System for Grinding Steel Balls - Case Study

Loice Keresia Gudukeya , Charles Mbohwa

University of Johannesburg, Faculty of Engineering and the Built Environment, Auckland Park 2006, P.O Box 524, Johannesburg, South Africa

Abstract. The availability of grinding balls of superior quality than those produced at a Harare company in Zimbabwe, resulted in loss of business. In order to retain business, a new ball that has a similar chemical composition to that of imported grinding balls was developed. Proper thermo-mechanical treatment has to be done after casting, in order for the ball to be of satisfactory quality. In this research, heat treatment tests were done on the developed ball to determine a suitable quenching medium for the heat treatment cycle. Using the results of the heat treatment tests obtained, a suitable heat treatment cycle which utilises optimum quenching medium and has effective agitation for minimal defects was designed.

1 Introduction

A Harare company's core business is the manufacturing of grinding steel balls used in mining operations. Cast steel balls are used to grind run of mine coarse materials in a ball mill. The balls vary in size from 15mm to 120mm in in diameter and are manufactured from scrap metal. However, the company now faces stiff competition from imports of the same product that is coming in with better quality. There is a need to counter this competition by producing steel balls that are of at least the same quality or better. The chemical composition of a locally developed steel ball is already in place. What is required is the heat treatment cycle which include quenching and tempering process ideal for the developed ball to give the desired results.

2 Literature review

2.1 Quenching

Quenching is the process of cooling a metal at a rapid rate from a predetermined high temperature. This process is important as it is key in addressing the main challenge of manufacturing steel balls of high quality. When hardening metallic materials, cooling is an important and decisive element in determining the properties of the part such as it hardness. [3]

2.2 The quenching process

When heat treatment fluids are used to quench metals, cooling occurs in three distinct stages: film boiling, nucleate boiling and convective heat transfer.[1]

Film Boiling - Film boiling, also known as the "vapor blanket" stage, occurs upon initial immersion. Contact between the hot metal surface and quenchant creates a layer of vapor (known as the Leidenfrost phenomenon) due to the supply of heat being greater than that which is carried off. The stability of the vapor layer, and thus the ability of the oil to harden steel, is dependent on the metal's surface irregularities, oxides present, surface-wetting additives (which accelerate wetting and destabilize the layer), and the quench oil's molecular composition (including the presence of more volatile oil degradation by-products). Cooling in this stage is a function of conduction through the vapor envelope and is relatively slow since the vapor blanket acts as an insulator.

Nucleate Boiling - As the part cools, the vapor blanket collapses at points and nucleate boiling (violent boiling of the quenchant) results. Heat transfer is fastest during this stage, with heat transfer coefficients sometimes over two orders of magnitude higher than during film boiling, largely due to the heat of vaporization.

Convective Heat Transfer - When the part has cooled below the boiling point of the quenchant, slow cooling occurs by convection and conduction (also called the "liquid" stage). Cooling rate during this stage is slow, and is exponentially dependent on the oil's viscosity, which varies with the degree of oil decomposition. Heat-transfer rates increase with lower viscosities and decrease with increasing viscosity.

The ideal quenchant is one that exhibits little or no vapour stage, a rapid nucleated boiling stage, and a slow rate during convective cooling. [4]

2.2.1 Quench cracks

Cracking may happen during quenching or sometimes after quenching. The causes of cracking are [2]:

- Excessive amount of non-metallic inclusions in the steel
- Quenching from a high temperature – high austenitizing temperatures lead to grain coarsening of austenite resulting in fine grained, highly dislocated martensitic structure which is more likely to crack.
- Improper selection of quenching medium – By using a quenching fluid which gives higher cooling rate than the one necessary may lead to cracking.
- Uneven quenching in the quenching bath
- Improper entry of the component into the quenching media causing eccentric and non-uniform cooling

Therefore the selection of quenching medium, the design of the quench tank as well as agitation system has to be done carefully in order to prevent or minimise quench cracks.

2.3 Heat treatment test

2.3.1 Materials

The materials used in the experiments are: the steel balls obtained from the company (its chemical composition is shown in Table 1); quenching oil sourced from Jacob Bethel Corporation.

Table 1: Chemical Composition of the ball by percentage

C	Si	Mn	P	S	Cr	Ni	Cu	Fe
2.4 2	0.8 5	0.39 7	0.02 6	0.02 8	18.8 7	0.26 5	0.13 3	76. 9

Equipment

The equipment used in the experimental procedure include:

- Heat treatment furnace
- Hardness test machine
- Tongs
- Grinding media
- Oil Quench Tank

2.3.2 Procedure

1. Two samples were cut from the steel grinding ball. Both samples were heated in a heat treatment furnace to 10000C for 4 hours.
2. One sample was quenched in an oil bath of Castrol HT 5 heat transfer fluid and the other was quenched in plain water.
3. The hardness of the samples were then measured using the Rockwell hardness test. A preliminary test force is first applied to a sample using a diamond indenter. This load represents the zero or reference position that breaks through the surface to reduce the effects of surface finish. After the preload, an additional load, call the major load, is applied to reach the total required test load. This force is held for a predetermined amount of time to allow for elastic recovery. This major load is then released and the final position is measured against the position derived from the preload, the indentation depth variance between the preload value and major load value. This distance is converted to a hardness number.
4. The samples were returned into the furnace for tempering at 2500C and held at this temperature for one and half hours. Tempering is done to relieve the metal of any residual stresses that might be set up during the quenching process, thereby reducing the likelihood of quench cracks developing in the metal. The tempering process also has the effect of softening the surface of the metal.
5. The samples were taken out of the furnace and left to cool in open air for an hour.

2.3.3 Design procedure

Data collection
Before the design of the system, data pertaining to the design parameters was gathered. This data was used in the sizing and selection of various equipment of the system. This data included capacity of the system, desired daily output, temperature parameters for the quench oil and properties of the quench oil.

Evaluation of cooling methods
After determining the quenchant to be used and collecting data pertaining to design parameters, three cooling methods of the quenchant were evaluated. Quench bath temperature control is of great importance in that it is the final parameter determining the mechanical properties of the grinding balls. The methods were then evaluated using a selection matrix table, where predetermined parameters are given different weights.

System design
All the data collected as well as the chosen method of cooling were then used in the system design. In the design, the sizing of the equipment such as the heat exchanger and pumps was done.

3 Results and discussion

After the heating and soaking, the samples were quenched in water and oil separately. This was done to compare the efficiency of oil against water. The hardness tests were taken using the Rockwell hardness test. The results of the hardness values taken were recorded in Table 2.

Table 2: Hardness test results

	Water quenched sample/HRC	Oil quenched sample/HRC
First batch	64	62
Second batch	65	63
Third batch	64	63
Average hardness	64.3	62.7

The results of these tests were then used to determine the quench fluid for the system. The resulting hardness values obtained after the Rockwell hardness test was done on both samples and are presented in Table 2. The water quenched sample shown in Figure 1 gave a higher hardness value possibly linked to the fact that water has a lower viscosity than heat transfer of oil and also that the initial temperature of water was lower than that of the oil.

Figure 1: Water quenched sample showing quench cracks

The water quenched sample also had quench cracks which may be attributed to the high heating temperature which may be unsuitable for quenching in water. During quenching, water has a tendency to cause vapour entrapment on the surface of the metal causing uneven hardness and therefore unfavourable stress distribution. This uneven stress distribution may be responsible for the cause of cracks which occur as residual stress in the metal. This is subsequently relieved by tempering in order to get a more even stress distribution throughout the metal.

Oil is more viscous than water, thereby producing a sample with improved toughness and ductility. Oil has a slower cooling rate than water, which means that there is greater depth of hardness and less residual stress in the ball. Slower cooling rate also means that there is a more even stress distribution in the sample, which explains why the sample quenched in oil did not develop cracks. Although hardness value obtained after quenching in oil is lower than that obtained after quenching in water, it also meets the target hardness of at least 60HRC. Hence Castrol HT 5 Heat Transfer oil was found to be the most suitable quenching medium and will be adopted for use in the quenching system.

4 Quench system design

From the experiments conducted, the quenching system will consist of the following components:

- Main quench tank (Size)
- Quench tank components such as draft tubes, marine impeller agitators, heaters and oil fume hood.
- Oil storage tank
- Heat exchanger
- Pumps which will circulate the quenchant

This design covers the Sizing of the Quench tank, Rotation of balls, Sizing of the agitators, Design of gearbox for agitators, Draft tube design, Quench Tank Layout, Sizing of the heat exchanger, Sizing of the pump to heat exchanger, Sizing of the pump from storage tank to quench tank, sizing of Heaters and Filtration Equipment. Figures2, 3 and 4 show the top view of the quench tank, the side view of the quench tank and the system layout respectively.

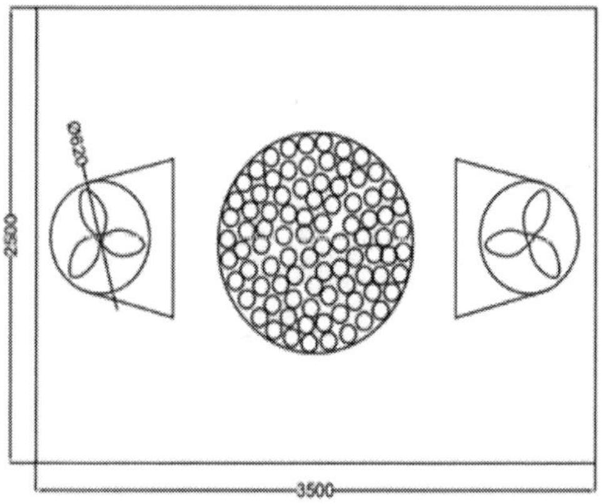

Figure 2: Top view of the quench tank

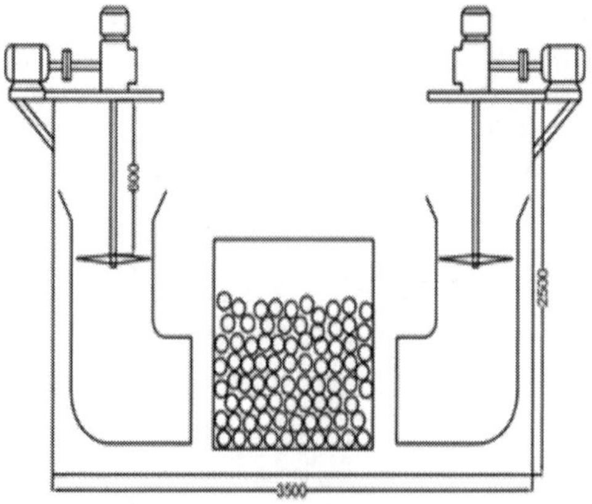

Figure 3: Side view of the quench tank

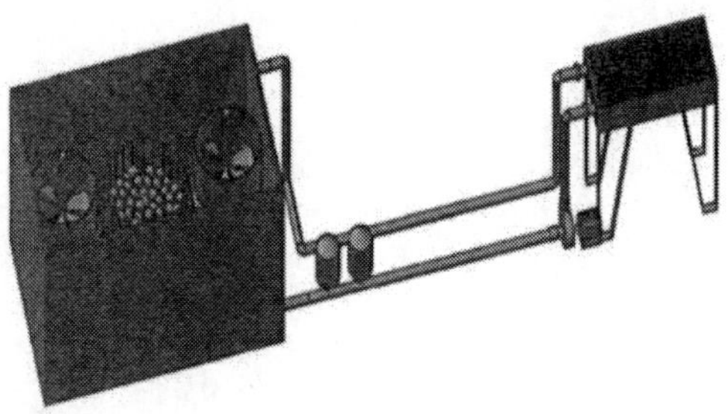

Figure 4: System layout

5 Conclusion and recommendations

The uniform quenching achieved results in a minimum number of rejected parts. The mathematical calculations related to the sizing of the equipment as well as design of the fluid flow dynamics in the quench tank show that the grinding balls quenched using this system will be of comparable quality with that of imported balls. An accurate cost benefit analysis of the project will need to be done versus the expected growth in demand for the new ball. Purchasing of the oil will be part of running costs therefore ways of maintaining the oil need to be established in order to increase its running life. These may include regularly checking for sludge content and moisture content as these contaminants may affect the quench uniformity if present in the oil in large quantities. Accelerants may be added to the oil in order to prolong its life in use.

References

1. CE Bates, G. T. N. C., 1993. Handbook of Quenchants and Quenching Technology. Ohio, USA: ASM International.
2. Kumar, S. S., 1999. Material and Murtallurgical Science. 1st ed. s.l.:Anuradha Agencies.
3. Segerbag S, 2001. Industrial cooling – some problems and their solutions. IVF industrial research and development corporation Gothenburg Sweden.
4. A. A. Yekinni, S. K. Bello, L. O. Mudashiru, L. A. Animashaun. 2015. Modification and Fabrication of Quenching Machine: Influence of Agitation and Temperature Gradient on Hardness of Al 6050 Alloy. International Journal of Materials Engineering.

Sensible Performance Analysis of Multi-Pass Cross Flow Heat Exchangers

Karthik Silaipillayarputhur[1], Tawfiq Al-Mughanam[2], Abdulelah I. Al-Niniya[2]

[1]*PO Box 380, King Faisal University, Dept. of Mechanical Engineering, Al-Ahsa, Kingdom of Saudi Arabia*
[2]*PO Box 380, King Faisal University, College of Engineering,, Al-Ahsa, Kingdom of Saudi Arabia*

Abstract. In this paper, a steady state sensible performance analysis of a multi-pass cross flow exchanger exhibiting various flow circuiting is considered. Counter cross flow, parallel cross flow and pure cross flow (where the flow circuiting is neither in parallel nor in counter flow) are considered in this paper. A previously developed matrix approach is used to study the heat exchanger performance at each individual pass. The equations required for modeling a cross flow heat exchanger for each flow arrangement are presented. Thereafter, a baseline heat exchanger geometry was selected and performance of the heat exchanger for each flow circuiting was described. As expected, the best thermal performance was seen in a counter cross flow heat exchanger and the performance of pure cross flow was intermediate between that of a parallel and a counter cross flow heat exchanger.

1 Introduction

This paper presents results from a steady state sensible performance investigation of a multi-pass cross flow exchanger possessing various flow circuiting. The analysis is based on the matrix approach previously reported in Silaipillayarputhur and Idem [1]. Silaipillayarputhur and Idem [2] considered practical validation of the matrix performance model. In that study the governing equations required to model a multi-pass counter cross flow heat exchanger with continuous wavy fins were presented. The heat exchanger selected for validation was a chilled water coil used at a chemical facility in Chattanooga, TN, USA. The predictions obtained from the performance model were compared with actual data from the chilled water coil and the theoretical performance data from the manufacturer. Based on the comparisons, it was concluded that the matrix heat exchanger performance model predicted the performance of a counter cross flow heat exchanger with at least 95% accuracy.

In process industries, several heat exchanger flow configurations occur, due to existing piping connections and space constraints. However, the thermal advantage of using counter flow heat exchangers is well known. In the present paper a baseline, finned six-pass cross flow heat exchanger presented in [2] is considered for further analysis. A steady state thermal performance comparison is conducted between overall counter flow, overall parallel flow, and pure cross flow. In each instance, the matrix approach is employed to study the heat exchanger performance for each individual pass. Thereafter, a parametric study is performed on the same six-pass cross flow heat exchanger subjected to counter, parallel, and pure cross flow circuiting by varying the NTU and capacity rate ratios. For each case, the effectiveness of the heat exchanger is plotted against the significant dimensionless parameters.

Numerous other papers have been reported in the literature pertaining to steady state heat exchanger performance modeling, and only the most relevant papers are reported herein. Pignotti and Shah [3] considered the effectiveness and NTU relationships for heat exchangers with complex flow arrangements. Heat exchanger terminal temperatures, surface area, and fluid flow rates were presented in dimensionless form in terms of heat exchanger effectiveness, number of transfer units, and heat capacity rate ratios. Explicit effectiveness-NTU relationships were obtained for a total of 18 complex heat exchanger flow arrangements, and the results were summarized in tabular format. Domingos [4] presented general method for calculating the total effectiveness and intermediate temperatures of assemblies of heat exchangers. The assemblies could consist of associations of any types of heat exchanger. The method utilized a transformation that related the inlet and outlet temperatures of the fluid streams, and thus permitted the derivation of closed form expressions. Shah and Pignotti [5] examined complicated heat exchanger flow arrangements and related them to simple forms for which either a solution existed, or an approximate solution could be obtained. Chen and Hsieh [6] developed a simple and systematic procedure to determine the effectiveness and exit fluid temperatures of complex

DOI: 10.1051/matecconf/201710811002

assemblies of identical heat exchangers. Three complex assemblies were chosen to illustrate the procedure. The assembly with non-identical heat exchangers was also studied to examine the general applicability of the present procedure. Baclic [7] proposed a simplified formula for cross flow heat exchanger effectiveness. The formula related effectiveness NTU and heat capacity rate ratio for cross flow heat exchangers when both the fluids are unmixed.

2 Nomenclature

A – Heat transfer surface area
C – Heat capacity rate of a fluid
c – Specific heat at constant pressure
$\dot{m}$ - Mass flow rate
n – Number of passes
NTU – Overall number of transfer units
r – Capacity rate ratio
T_A - Temperature of the external (hot) fluid
T_{Ai} - Inlet temperature of the external (hot) fluid
T_{Ao} - Outlet temperature of the external (hot) fluid
T_w - Temperature of the tube-side (cold) fluid
T_{wi} - Inlet temperature of the tube-side (cold) fluid
T_{wo} - Outlet temperature of the tube-side (cold) fluid
U_o - Overall heat transfer coefficient
Greek Letters
ε - Effectiveness, matrix approach
Subscripts
A – External fluid (hot fluid)
i – running index (1 through n)
i - Inside
j – running index (1 through n)
min – Minimum
max – Maximum
o - Overall
r – Row
sum - Sum
w – Tube-side fluid (cold fluid)
Superscripts
" – Quantity expressed on per pass basis

3 Steady state performance model

A baseline heat exchanger presented in [2] is considered for further analysis. Table 1 presents the geometry of the baseline heat exchanger, as well as other fundamental operating characteristics. The baseline heat exchanger is a six-pass cross flow heat exchanger, and it is employed for a quenching process in a chemical plant in Chattanooga, TN. The baseline heat exchanger has air in the gas-side of the heat exchanger and has chilled water in the tube-side of the heat exchanger. In this paper, the thermal performance of the baseline heat exchanger is compared between overall counter and parallel flow, and pure cross flow. Only sensible heat transfer between the fluids is considered, and there is no phase change. The number of transfer units is assumed to be uniformly distributed among the heat exchanger passes. Although fouling is a common occurrence in practice, the effects of fouling are not considered in the current study.

For the overall heat exchanger, the external fluid is assumed to be the minimum capacity rate fluid, since this is a commonly encountered situation in process industries. For a parallel and counter cross flow heat exchanger, the overall capacity rate ratio is equal to the capacity rate ratio per pass. This is because each pass encounters the same full mass flow rate of the external fluid and the tube side fluid. However, for a pure cross flow heat exchanger, the tube-side fluid is assumed to be evenly split among the tube passes. Considering the baseline heat exchanger with six passes, for a pure cross flow configuration on a per pass basis the tube-side fluid encounters one-sixth of the overall mass flow rate. Thus for a pure cross flow configuration, on a per pass basis the tube-side fluid can be the minimum capacity fluid.

Heat exchanger overall performance

For the overall heat exchanger, the external fluid (air), designated through subscript "A" is the minimum capacity rate fluid and while the tube-side fluid (chilled water) designated through subscript "w" is the maximum capacity rate fluid. The capacity rate ratio for the overall heat exchanger is given by [8,9]

$$r = \frac{C_{min}}{C_{max}} = \frac{(\dot{m} \cdot c)_A}{(\dot{m} \cdot c)_w} \quad (1)$$

Likewise, the overall NTU for the heat exchanger is expressed as [8,9]

$$NTU = \frac{U_o \cdot A_o}{C_{min}} = \frac{U_o \cdot A_o}{(\dot{m} \cdot c)_A} \quad (2)$$

The effectiveness for the overall heat exchanger, assuming both fluids to be unmixed, may be calculated as [8,9]

$$\varepsilon = 1 - \exp\left[\frac{(NTU)^{0.22}}{r}\left\{\exp\left[-r(NTU)^{0.78}\right] - 1\right\}\right] \quad (3)$$

The calculated effectiveness of the heat exchanger is related to rate of heat transfer by

$$\varepsilon = \frac{q}{q_{max}} \quad (4)$$

The heat transfer is determined by means of the following energy balance expressions [8,9]

$$q = \dot{m}_A c_A (T_{Ai} - T_{Ao}) \quad (5)$$

and:

$$q = \dot{m}_w c_w (T_{wo} - T_{wi}) \quad (6)$$

The inlet temperatures of both fluids are known quantities. In that case the maximum heat transfer is given by [8,9]

$$q_{max} = C_{min}(T_{Ai} - T_{wi}) = \dot{m}_A c_A (T_{Ai} - T_{wi}) \quad (7)$$

Employing Equations 4 and 7, the rate of heat transfer between the two fluids in the heat exchanger can be determined, and from that calculation the discharge temperatures of both the fluids can readily be evaluated.

Intermediate thermal performance for parallel and counter cross flow heat exchangers using the matrix approach

The baseline heat exchanger as described in Table 1 is a six-pass cross flow heat exchanger. Figures 1 and 2 depict the flow circuiting of a six-pass parallel and counter cross flow heat exchangers.

Table 1. Baseline heat exchanger operating characteristics

Baseline heat exchanger	
Item	**Description**
Number of passes	6
Fin material	Aluminium
Fin thickness	0.00024 m
Fin density	315 fins/m
Tube material	Copper
Tube OD	0.016 m
Tube thickness	0.0008 m
Number of rows per pass	1
Number of tubes	46
Tube length	1.65 m
Longitudinal pitch	0.038 m
Transverse pitch	0.033 m
Mass flow rate of water	9396 kg/hr
Mass flow rate of air	17863 kg/hr
Inlet air temperature (T_{Ai})	296.3 K
Inlet water temperature (T_{wi})	284.1 K

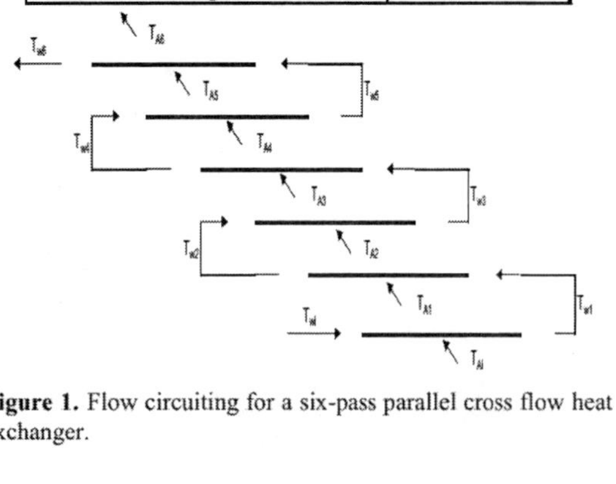

Figure 1. Flow circuiting for a six-pass parallel cross flow heat exchanger.

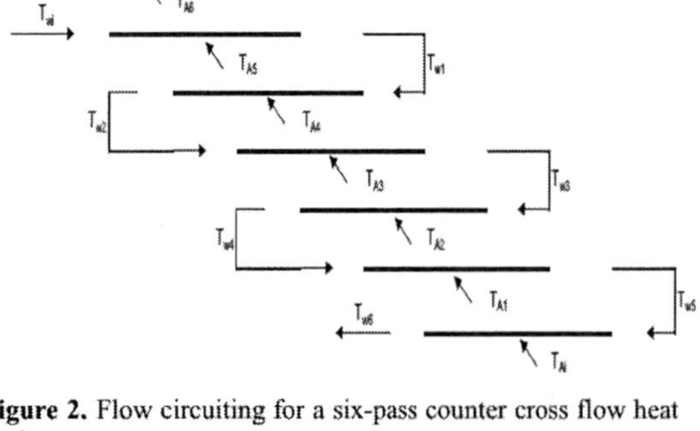

Figure 2. Flow circuiting for a six-pass counter cross flow heat exchanger.

Each pass of the cross flow heat exchanger encounters the full mass flow rate of the external fluid and tube-side fluid. Hence, the capacity rate ratio per pass may be given as

$$r'' = r = \frac{(\dot{m}\cdot c)_A}{(\dot{m}\cdot c)_w} \tag{8}$$

The surface area of the heat exchanger is assumed to be uniformly distributed among the heat exchanger passes. Therefore, the NTU per pass may be expressed as

$$NTU'' = \frac{U_o \cdot A_o / n}{C_{min}} = \frac{U_o \cdot A_o}{n(\dot{m}\cdot c)_A} \tag{9}$$

Thus by Equation 3 the heat exchanger effectiveness per pass is

$$\varepsilon'' = 1 - \exp\left[\frac{(NTU'')^{0.22}}{r''}\left\{\exp\left[-r''(NTU'')^{0.78}\right]-1\right\}\right] \tag{10}$$

The matrix approach [1] uses the concept of an energy balance and effectiveness applied to each pass of a cross flow heat exchanger. Thus if a cross flow heat exchanger has "n" number of passes, there will be "2n" simultaneous equations to be solved. For every pass there are two unknowns, namely the discharge temperature of the external fluid and the discharge temperature of the tube-side fluid. By solving the system of simultaneous linear equations the unknown discharge temperatures exiting each pass can be readily determined. For a multi-pass parallel cross flow heat exchanger, the following generalized equations are employed to determine the intermediate and final temperatures of both the fluids in the heat exchanger. If "m" corresponds to the number of passes, "n" corresponds to twice the number of passes, and j corresponds to anywhere between "1" and "n", then referring to [1]:

For j = 1

$$r''T_{Aj,j+1} + T_{Bj,j+1} = r''T_{Ai} + T_{wi} \tag{11}$$

For j = 2;m-1

$$r''T_{Aj-1,j} - r''T_{Aj,j+1} + T_{wj-1,j} - T_{wj,j+1} = 0 \tag{12}$$

For j = m

$$r''T_{Aj-1,j} - r''T_{Ao} + T_{wj-1,j} - T_{wo} = 0 \tag{13}$$

For j = m+1

$$T_{A12} = (1-\varepsilon''_{r1})T_{Ai} + \varepsilon''_{r1}T_{wi} \tag{14}$$

For j = m+2;n-1

$$(1-\varepsilon''_{rj-m})T_{Aj-(m+1),j-m} - T_{Aj-m,j-m+1} + \varepsilon''_{rj-m}T_{wj-(m+1),j-m} = 0 \tag{15}$$

For j = n

$$(1-\varepsilon''_{rj-m})T_{Aj-(m+1),j-m} - T_{Ao} + \varepsilon''_{rj-m}T_{wj-(m+1),j-m} = 0 \tag{16}$$

Likewise, for a multi-pass counter cross flow heat exchanger, the following generalized equations can be utilized [1] to determine the intermediate and final temperatures of both the fluids in the heat exchanger:

For j = 1

$$T_{Aj,j+1} - \varepsilon''_{rj}T_{wj} = (1-\varepsilon''_{rj})T_{Ai} \tag{17}$$

For j = 2;m-1

$$(\varepsilon''_{rj}-1)T_{Aj-1,j} + T_{Aj,j+1} - \varepsilon''_{rj}T_{wj,j+1} = 0 \tag{18}$$

For j = m

$$(\varepsilon''_{rj}-1)T_{Aj-1,j} + T_{Ao} = \varepsilon''_{rj}T_{wi} \tag{19}$$

For j = m+1;n-2

$$-r''T_{Aj-m,j-m+1} + r''T_{Aj-m+1,j-m+2} + T_{wj-m,j-m+1} - T_{wj-m+1,j-m+2} = 0 \quad (20)$$

For j = n-1

$$-r''T_{Aj-m,j-m+1} + r''T_{Aj-m+1,j-m+2} + T_{wj-m,j-m+1} = T_{wi} \quad (21)$$

For j = n

$$r''T_{A12} + T_{w12} + T_{wo} = r''T_{Ai} \quad (22)$$

Intermediate thermal performance for pure cross flow heat exchangers using the matrix approach

Figure 3. depicts the flow circuiting of a six-pass pure cross flow heat exchanger. Each pass of the cross flow heat exchanger encounters the full mass flow rate of the external fluid and one sixth of the overall mass flow rate of the tube-side fluid.

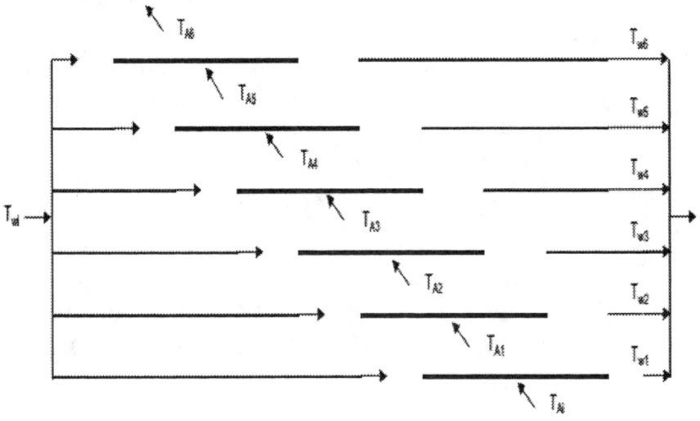

Figure 3. Flow circuiting for a six-pass pure cross flow heat exchanger.

In his analysis, it is reasonable to assume that on a per pass basis, the tube-side fluid is the minimum capacity rate fluid. Thus, the capacity rate ratio per pass is given by

$$r'' = \frac{(\dot{m}\cdot c)_w}{n(\dot{m}\cdot c)_A} = \frac{1}{n}\cdot\frac{1}{r} \quad (23)$$

Likewise, the NTU per pass can be expressed as

$$NTU'' = \frac{U_o \cdot A_o / n}{C_{min}} = \frac{U_o \cdot A_o / n}{(\dot{m}\cdot c)_w / n} \quad (24)$$

Or:

$$NTU'' = \frac{U_o \cdot A_o}{(\dot{m}\cdot c)_w} = NTU \cdot r \quad (25)$$

Assuming both fluids to be unmixed in a given pass the heat exchanger effectiveness per pass may be calculated using Equation 10.

For a multi-pass pure cross flow heat exchanger, the following generalized equations [1] are applied to determine the intermediate and final temperatures of both the fluids in the heat exchanger:

For j = 1

$$T_{Aj,j+1} + r''T_{wj} = T_{Ai} + r''T_{wi} \quad (26)$$

For j = 2;m

$$-T_{Aj-1,j} + T_{Aj,j+1} + r''T_{wj} = r''T_{wi} \quad (27)$$

For j = m+1

$$T_{w1} = (1-\varepsilon''_{r1})T_{wi} + \varepsilon''_{r1}T_{Ai} \quad (28)$$

For j = m+2;n

$$-\varepsilon''_{rj-m}T_{Aj-(m+1),j-m} + T_{wj-m} = T_{wi} - \varepsilon''_{rj-m}T_{wi} \quad (29)$$

4 Baseline heat exchanger performance

The baseline six-pass cross flow heat exchanger performance was studied for overall counter and parallel flow and pure cross flow circuiting configurations. Therein, the heat exchanger overall performance was described through the heat exchanger effectiveness. The matrix approach was employed to study the intermediate thermal performance of the heat exchanger. A Matlab code was developed to solve the system of simultaneous linear equations and to determine the intermediate and the overall thermal performance of the cross flow heat exchanger. In every instance, the external fluid was assumed to be the minimum capacity rate fluid. The external fluid and the internal fluid were considered unmixed in the analysis.

Table 2. Performance of baseline cross flow heat exchanger

Baseline cross flow heat exchanger					
6 pass Counter Flow		**6 pass Parallel Flow**		**6 pass Pure Cross Flow**	
Temp Label	**Temp (K)**	**Temp Label**	**Temp (K)**	**Temp Label**	**Temp (K)**
T_{Ai}	296.3	T_{Ai}	296.3	T_{Ai}	296.3
T_{A1}	293.4	T_{A1}	292.3	T_{A1}	293.5
T_{A2}	291.1	T_{A2}	290.2	T_{A2}	291.3
T_{A3}	289.3	T_{A3}	289.1	T_{A3}	289.6
T_{A4}	287.9	T_{A4}	288.5	T_{A4}	288.3
T_{A5}	286.8	T_{A5}	288.2	T_{A5}	287.4
T_{A6}	285.9	T_{A6}	288.1	T_{A6}	286.6
T_{wi}	284.1	T_{wi}	284.1	T_{wi}	284.1
T_{w1}	284.5	T_{w1}	286.0	T_{w1}	286.2
T_{w2}	285.0	T_{w2}	286.9	T_{w2}	286.8
T_{w3}	285.7	T_{w3}	287.4	T_{w3}	287.6
T_{w4}	286.5	T_{w4}	287.7	T_{w4}	288.7
T_{w5}	287.5	T_{w5}	287.8	T_{w5}	290.1
T_{w6}	288.9	T_{w6}	287.9	T_{w6}	291.9
$\varepsilon_{overall}$	**0.86**	$\varepsilon_{overall}$	**0.67**	$\varepsilon_{overall}$	**0.80**

A comparison of steady state performance was conducted for cases where the baseline heat exchanger was operated either with overall parallel or counter flow circuiting, or otherwise for pure cross flow circuiting. The specific input conditions are outlined in Table 1. A complete accounting of the equations used to characterize the heat exchanger geometry, and to evaluate such dimensionless parameters as the particular capacity rate ratio and NTU, are provided in [2], and are not included herein. The results of the present study are presented in Table 2 in terms of the overall and intermediate steady state sensible performance of the baseline heat exchanger. As expected the counter cross flow heat exchanger yielded the best heat exchanger performance, i.e., the maximum overall effectiveness. Likewise, the performance of the pure cross flow heat exchanger was intermediate between that for overall counter and parallel cross circuiting.

5 Summary

Steady state sensible performance of a cross flow heat exchanger was studied in this paper. A baseline heat exchanger presented in [2] was analyzed subject to various flow circuiting considerations, i.e., overall parallel and counter flow, as well as pure cross flow. The baseline heat exchanger's performance was determined for each flow circuiting, assuming operating conditions typical of design conditions.

In every instance, the heat exchanger considered in this study had six passes. However, the number of passes considered is arbitrary (subject to mandatory performance requirements), and the analysis can be readily extended to any number of passes by employing the matrix approach. The matrix approach uses fundamental analytical relationships to study the intermediate and the overall performance of a cross flow heat exchanger. The matrix approach is very simple to use, and by solving a set of generalized equations, the intermediate and the overall performance of the heat exchanger can be readily determined. Matrix analysis uses physically significant parameters such as NTU and capacity rate ratio to evaluate overall heat exchanger effectiveness, and thus provides clear information to the engineers during the initial design or selection of the cross flow heat exchanger. With the availability of intermediate temperatures, the matrix approach can be utilized to optimize size, material, weight, and initial cost of the cross flow heat exchanger. Furthermore, the matrix method can assist with thermal stress analysis of heat exchanger, since it provides detailed information pertaining to local temperature variations in the heat exchanger.

References

1. K. Silaipillayarputhur and S. Idem, 2013, A general matrix approach to model steady state performance of cross flow heat exchangers, Heat Transfer Engineering, Vol. **34**, No. 4, pp. 338-348.
2. K. Silaipillayarputhur and S. Idem, 2013, Practical Validation of a Matrix Approach Steady State Heat Exchanger Performance Model, Journal of Applied Global Research, ISSN: 1940-1841, Vol. **6**, No. 17, pp. 1-22
3. A. Pignotti and R. K. Shah, 1992. Effectiveness-number of transfer units relationships for heat exchanger complex flow arrangements, Int. J. Heat Mass Transfer, Vol. **35**, No. 5, pp. 1275-1291.
4. J. D. Domingos, 1969, Analysis of Complex Assemblies of Heat Exchangers, Int. J. Heat Mass Transfer, Vol. **12**, pp. 537-548.
5. R. K. Shah and A. Pignotti, 1993, Thermal Analysis of Complex Crossflow Exchangers in Terms of Standard Configurations, J. Heat Transfer, Vol. **115**, pp. 353-359.
6. J. D. Chen and S. S. Hsieh, 1990, General procedure for effectiveness of complex assemblies of heat exchangers, Int. J. Heat Mass Transfer, Vol. **33**, No. 8, pp. 1667-1674.
7. B. S. Baclic, 1978, A Simplified Formula for Crossflow Heat Exchanger Effectiveness, Trans. ASME, Vol. **100**, pp. 746- 747.
8. W. M. Kays and A. L. London, 1984, Compact Heat Exchangers, 3rd Ed., McGraw-Hill, New York.
9. F. P. Incropera, D. P. Dewitt, T. L. Bergman, and A. S. Lavine, 2006, Fundamentals of Heat and Mass Transfer, 4th Edition, John Wiley & Sons, Inc., New York.

Numerical Modelling of Evaporation Process in Cryogenic Systems

A.A. Shimanov , D.A. Uglanov, D.V. Sarmin and A. B. Tsapkova

Samara National Research University, Heat engineering and heat engines Department, Samara, Russia

Abstract. This article deals with a problem of evaporation numerical modeling in cryogenic systems of constant volume. Particularly, a method of nitrogen evaporation modeling in cryogenic fueling tank was considered. User-defined functions of nitrogen thermodynamic properties in temperature range of 70...400 K and pressure range of 0.1...30 MPa. The result of this research is calculation algorithm which allows to model processes of phase change in cryogenic systems with constant volume. A comparison obtained between results of numerical modeling with experimental and analytical data is presented. Deviation of results from experimental data is 5-10%.

1 Introduction

A wide range of optical-electronic devices and systems, which require a cryogenic temperatures for its normal operation, is used onboard of aircrafts. Refrigerations systems vary in mass, consumed power, dimensions, durability cold capacity and temperature.

One of the type of onboard refrigeration systems for keeping temperature in aircraft is throttle refrig system. This refrigeration system is based on Joule-Thomson effect. Such refrigeration systems can provide refrigeration on temperature level 20-100 K with cold working fluid flow rate 0.1-1.5 kg per hour (depending on working fluid type). High pressure gas is required for throttle refrigeration system operation. Launder containing moist consuming fluid and filter are installed in high pressure line for removal of undesired примесей and moisture. Throttle refrigeration systems can be divided on systems with open and closed cycle. Throttle system with closed cycle includes compressor, evaporator and condenser which significantly increases mass of system and decreases its durability. Another disadvantage is high consumed power.

In open throttle systems gas from high-pressure tank passes from the throttle and cools an object of refrigeration. These systems utilize cryogenic working fluid in both critical and supercritical conditions as well s liquid working fluid in some special cases. Main advantages of these systems are simplicity, durability, reasonable economic efficiency and small energy consuming. Disadvantages are significant mass of high-pressure tank (including mass of working fluid) and change of the throttling temperature due to decrease of working fluid pressure in tank. In systems with open cycle (tank systems), mass of the fueled gas and working fluid parameters determine their exploitation properties and realistic characteristics. It is a reason why creation of new type of tank able to operate as a part of onboard refrigeration system with improved characteristics (improvement must be achieved by advantages of tank) is important task [1].

2 Problem

Cryogenic fueling tank (CFT), which is developed in Samara National Research University and can be used as element of onboard throttle refrigeration system, has a new functional properties such as increased lifetime and utilizing of both liquid and gaseous working fluids.

Tank design scheme is presented on figure 1. Cryogenic volume (Dewar) is located inside the high-pressure tank. Cryogenic working fluid is filled inside the Dewar. Gas chamber is created between Dewar and tank inner wall which serves as a heat insulation. Relation of the Dewar and inner volume dimensions is selected in such manner that if cryogenic working fluid is fully gasified i.e. gas will embrace entire tank volume, it will corresponds to a nominal pressure and nominal fueling mass of gaseous product.

Optimal point to start CFT is a mass of fueled working fluid. Considering mass of working fluid in liquid condition, volume of the CFT is calculated and then Dewar is selected. Considering final pressure in all volume of CFT, volume of CFT itself, containing both Dewar and screen insulation, is selected. Dependently on final parameters of temperature and pressure of single-phase working fluid in all volume of CFT both heat insulation and number of screens is selected.

To increase a lifetime of throttle system of onboard energy plant or decrease of mass and dimensions of existing system, it is suggested to use given tank during partial gasification of cryogenic product. It is necessary to know accurate duration and intensity of evaporation for this purpose. Complexity of the task is caused by the fact that considered thermodynamic system "gas chamber-liquid" is not in equilibrium. Gas chamber as subsystem is open (i.e. with changing mass of working fluid) and heat transfer is executed by heat conductivity through multilayer gas wall with counter gas flow rate varying along the radius. All of these processes are not stationary.

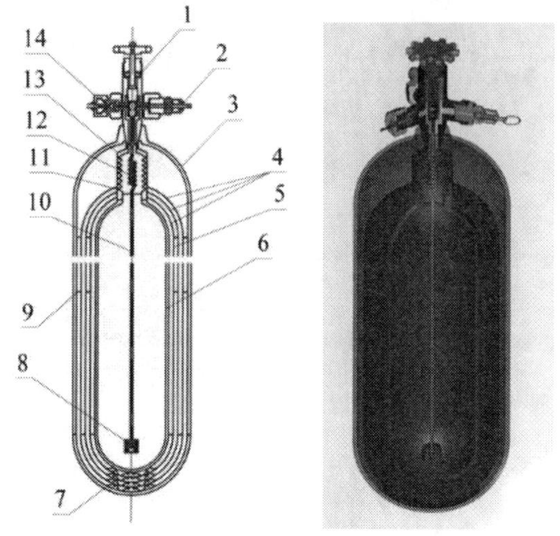

Figure 1. Reservoir with cryogenic filling:
1 – shutoff valve; 2 – safety valve; 3 – reservoir; 4 – insulation; 5 – insulation of internal reservoir; 6 – internal reservoir; 7 – bucking spacer plates; 8 – filter; 9 – radial spacer plates; 10 – barrel; 11 – filling pipe; 12 – coil; 13 – bellows; 14 – flow nipple

A process of cold energy distribution along the tank design elements starts after filling of inner volume by cryogenic fluid. Obviously, in case where there is no outside heat fluxes finite temperature condition of tank is determined by relation of total enthalpy of tank warm wall and cold energy of cryogenic working fluid. If amount of heat determined by enthalpy of wall is enough for evaporation of fueled mass of cryogenic working fluid, the task is converged to a solution of energy equations but process dynamic and duration of full evaporation are determined from the heat transfer equations. Conditions on boundaries of gas volume on every time step corresponds to a first-type boundary conditions. It is necessary to consider changes of gas and liquid heat and thermodynamic properties on every time step of the calculation.

In case of heat fluxes to external surface of the tank on every time step of the calculation, the heat transfer task is solved with third-type boundary condition from external side and first-type boundary conditions from the side of the liquid. In both cases, the processes which occur inside the tank are next.

Liquid starts to evaporate and evaporated part of cryogenic working fluid is distributed along the layers of gas chamber. Pressure in initial moment of time can even decrease which will promote further liquid evaporation. However, after sometime pressure may increase and evaporation will stop. Features of the initial processes and further cryogenic working fluid behavior will be caused by relation of liquid and gas phases volumes. Thus, if specific volume of two-phase mixture in isochoric process is higher than critical one, liquid will continuosly evaporate until full transfer to a condition of dry saturated vapor.

Provided discussion is based on analysis of so called classic two-phase systems were temperature and pressure of saturated vapor and liquid are equal. But in described case where temperature in the gas chamber significantly differs from the temperature of the liquid, the only condition which can be called equilibrium is condition of vapor and liquid in inner volume of tank.

Thus, Clapeyron-Clausius equation is not applicable for such thermodynamic system (entire tank) and analytical solution will not be correct.

Algorithm of this problem solution was suggested in works [1,2].

3 Numerical modeling

To increase accuracy and velocity of different cryogenic processes calculations it was suggested to use numerical modeling. Numerical modeling of the heat processes in cryogenic fueling tank was carried out by ANSYS Fluent software. 2D model was used to increase the calculation speed. CFT 2-D model was created by geometric parameters of tank which was used in the experiment [2] (figure 2) in ANSYS Design Modeler. For the numerical simulation a geometric model of CFT has been simplified. Then model was meshed by ANSYS Meshing module. Figure 2b shows a section of the CFT with applied mesh.

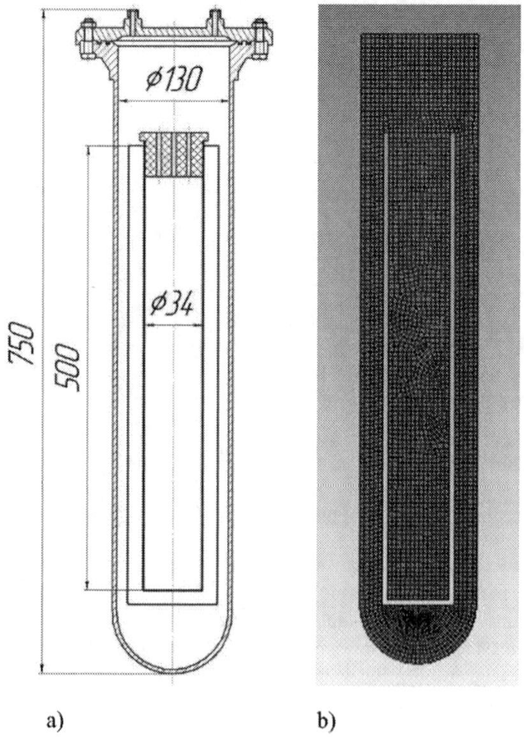

Figure 2. Model of the CFT with applied finite-element mesh.

Nitrogen was used as a working fluid with phase change in this task. Model Volume of Fluid (VOF) [3, 4, 5] was used for modeling of two-phase condition. The methods and results of evaporation modeling with constant pressure are presented in [6, 7, 8, 9]. The main difficulty of the CFT modeling is the fact the nitrogen evaporation takes place in constant volume [10, 11]. Pressure inside the tank is increased from fueling pressure 0.1 MPa up to 1.2 MPa and temperature increases from 77 K to environment temperature during the process of nitrogen evaporation. Thermodynamic properties and nitrogen evaporation temperature during the gasification will significantly change. Option "floating operating pressure" was used for calculation pressure rise from the integral mass balance, separately from the solution of the pressure correction equation. This option can only be operated by using the ideal gas model.

User-Defined Functions (UDF) were written to describe thermodynamic properties which depend on temperature and pressure (specific heat, heat conductivity, viscosity, density) of liquid and gaseous phase. These functions allows to determine and calculate complicated process of heat-and-mass transfer between nitrogen phases in temperature range of 70 K to 400 K and pressure range from 0.1 to 30 MPa.

Function, which describes evaporation temperature inside the tank dependency on pressure inside the tank, was derivated:

$$T_{sat} = 11{,}546 \cdot ln(P) - 53{,}844 \quad (1)$$

Temperature of phase change become constant when pressure inside the tank reaches the value of critical pressure.

According to experimental data [1], Dewar chamber is filled by liquid nitrogen in initial moment of time with mass of 0.25 kg, temperature 77 K and pressure 0.1 MPa. This amount of nitrogen corresponds to one-third filling of tank. Temperature of external wall is set equal to 293 K. Temperature of inner wall is set variable corresponding to calculation results [1,2]. In initial moment of time temperature of the Dewar inner wall is set 2 K higher than evaporation temperature for given pressure. Time Step Size in initial moment of time is set less than 0.001 second due to complexity of heat-and-mass change process and solution stability. After evaporation stabilization time step size can be increased up to 0.01 sec.

4 Results

Figure 3 shows intermediate results of gasification process calculation. Nitrogen evaporation and transformation to a vapor is clearly seen as well as phase division line. Volume boiling takes place in CFT Dewar where vapor phase occurs directly in the volume of liquid as separate vapor bubbles. Thus, it can be stated that mechanism of phase change in the constant volume evaporation is bubble boiling.

At the initial time gasification process occurs in stages (Figure 4). When the liquid reaches the boiling point, the liquid mass is evaporated, and the CFT pressure increases. Boiling point increases and the boiling process is stopped. The process is repeated until the critical boiling point (126.25 K). As can be seen from the Figure 4 the intensity of the boiling with the course of time is reduced. This is due to the decrease of the temperature gradient between the CFT wall and the nitrogen in CFT. On reaching a critical temperature of the boiling process is stabilized.

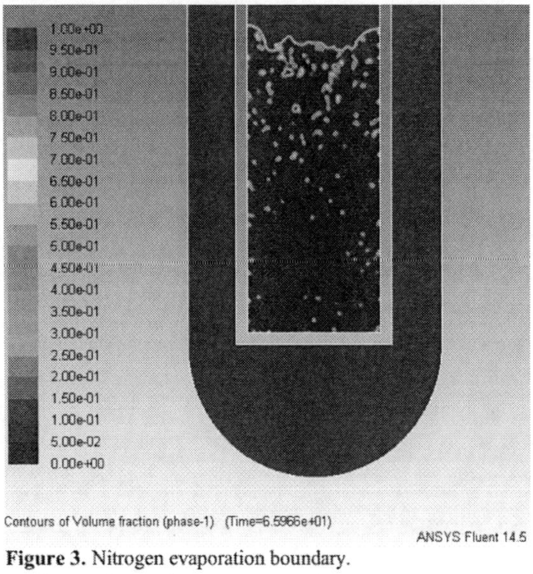

Figure 3. Nitrogen evaporation boundary.

A comparison of the calculation and experimental results dependencies [1,2] with the results of numerical modelling is presented on figure 5. As it can be seen from the plot, pressure rises quicker on the initial stage of the gasification (up to 0.9 MPa for 30 minutes). It is connected with intensive nitrogen transfer from liquid state to a gaseous one. Then pressure of the gas mass inside the tank increases due to increase of the evaporated mass of the gas. This causes boiling point "higher" and slowdown gasification process. 1440 hours of real time was spent on modeling of 3300 second of gasification process.

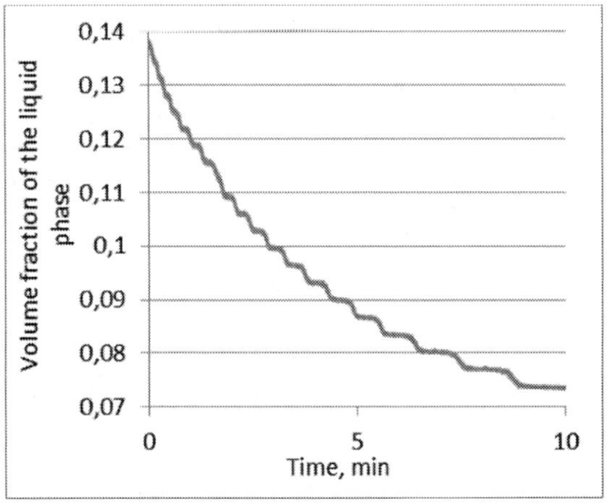

Figure 4. Volume fraction of the liquid phase depending on time.

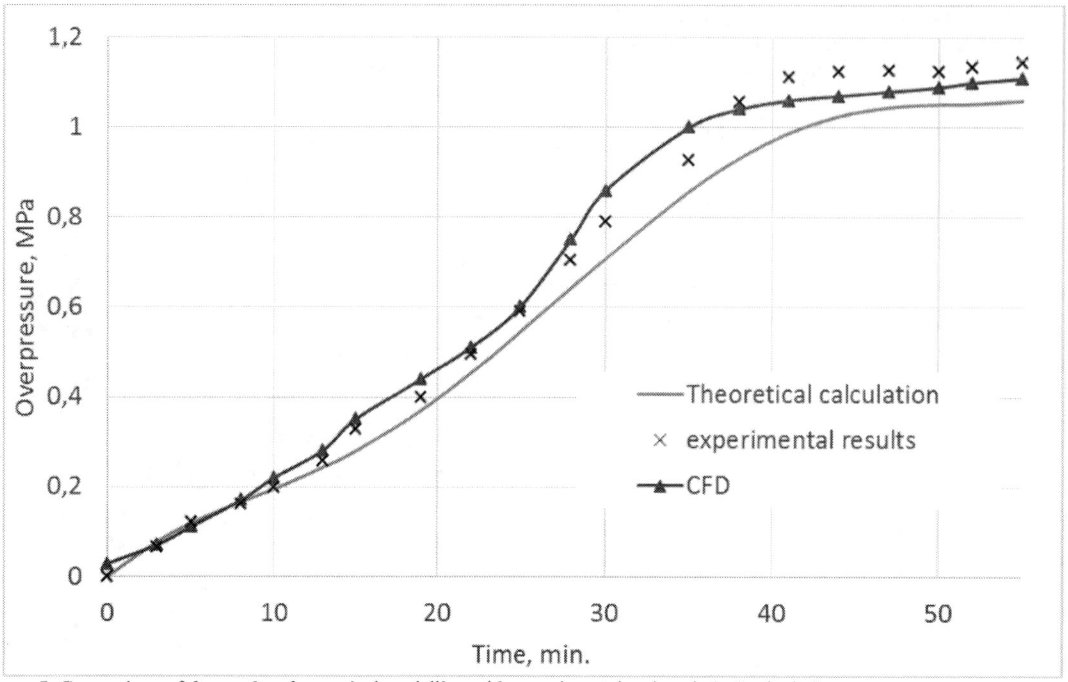

Figure 5. Comparison of the results of numerical modelling with experimental and analytical calculations.

5 Conclusions

The result of this research is calculation algorithm which allows to model processes of phase change in cryogenic systems with constant volume.

Algorithm is based on integration of the equation which describe thermodynamic properties of nitrogen depending on condition parameters in ANSYS Fluent software.

Method is verified by comparison of numerical calculation results with the results of experimental and analytical research carried out earlier. Deviation of results from experimental data is 5-10% which allows to make a conclusion that given method is suitable for modelling of heat processes in various cryogenic systems. Better results can be obtained by increase of thermodynamic properties number which are described in user-defined function.

This work was supported by the Ministry of Education and Science of the Russian Federation in the framework of the implementation of the Program ''Research and development on priority directions of scientific-technological complex of Russia for 2014–2020.

References

1. A.I. Dovgyallo, S.O. Nekrasova, D.V. Sarmin, D.A. Uglanov, Appl. Phys., **4**, 54-59 (2013)
2. E.V. Blagin, A.I. Dovgyallo, S.O. Nekrasova, D.V. Sarmin, D.A. Uglanov, Applied Thermal Engineering, **101**, 537-544 (2016)
3. Y. Liu, T. Olewski, L.N. Véchot, Journal of Loss Prevention in the Process Industries, **35**, 125-134 (2015)
4. A. Hens, G.Biswas, S. De., Physics of Fluids, **26**, 012105 (2014)
5. X. Li, R. Wang, R. Huang, Y. Shi, Applied Thermal Engineering, **26**, 2425-2432 (2006)
6. M.H. Yuan, Y.H. Yang, T.S. Li, Z.H. Hu, International Journal of Heat and Mass Transfer, **51**, 1646-1657 (2008)
7. ANSYS Fluent Theory Guide, Ansys Inc.(2014)
8. J.H. Lee, Y.J. Kim, S. Hwang, Ocean Engineering, 2015, **106**, 77–86 (2015)
9. M. Ahammad, Y. Liu, T. Olewski, L.N. Véchot, M.S. Mannan, Industrial and Engineering Chemistry Research, **55**, 7548-7557 (2016)
10. Z. Xu, Y. Zhang, B. Li, J. Huang, International Journal of Heat and Mass Transfer, **101**, 614-619 (2016)
11. B. Fadhl, L.C. Wrobel, H. Jouhara, Applied Thermal Engineering, **60**, 122-131 (2013)

DOI: 10.1051/matecconf/201710811004

Development of Universal Exergy Analysis to the Selection of Cryogenic Cooling System

S. V. Zaika , D. A. Uglanov, S. S. Dostovalova

Samara University, Heat engineering and heat engines Department, Samara, Russia

Abstract. Efficiency is one of the most important parameters that apply to the cooling system. This paper describes a development of universal methodology to evaluate the effectiveness of cryogenic cooling systems. The problem of the lack of a universal method of comparison of cryogenic cooling systems was solved. This was achieved by creating a universal method of exergy analysis. In addition, as an example were examined three of the cooling system, which applies the universal exergy analysis. Basic universal parameters that characterize cooling system level of sophistication were defined, as well as main calculated. The result of this study is a new universal method of exergy analysis, which can be used to select a cryogenic cooling system.

1 Introduction

The most important requirement for many energy projects is providing a cooling of IR-receivers, elements of optical systems and other devices to be installed on power plant to cryogenic temperature. When choosing an cryogenic cooling system one should take into account not only the energy waste, but also waste that arise due to the cooling system organizational management, such as: acquisition and transportation of the working fluid waste, the heat flows waste, waste due to equipment imperfection [1].

Evaluation of any thermodynamic system power resources, as well as energy transformation, should take into account the influence of environmental parameters. Therefore, using the concept of energy as a common measure of the matter motion in considered system is insufficient when solving engineering problems. The technology development makes us consider the fact that not any kind of energy and not under any conditions can be entirely suitable for practical use. Technical value of energy depends on its kind and parameters, as well as on the environment parameters.

From the standpoint of technical feasibility any kind of energy value is determined not only by the energy amount, but also by the extent to which it can be used in set conditions, i.e. transformed into other kinds of energy. It is important to study transformed energy waste when performing thermodynamic analysis of technical processes.

Mechanical or electrical energy, fully converted into another form of energy, can be taken as the usage measure of any kind of energy. Therefore, if a certain amount of energy can be transformed into electrical or mechanical energy in the set conditions, that fully guarantees the possibility of complete transformation of this amount of energy in any other kind of organized energy. Conditions of such transformation are determined by the second law of thermodynamics.

2 The basic equation of exergy

In the light of the foregoing, it is necessary to introduce a common measure for the energy resources capable of transformation into other forms of the organized energy when constant parameters are interacting with the environment. It is clear that the ultimate energy transformation can only occur when all the processes within the system, as well as its interaction with the environment are reversible.

Forenamed measure of transformed energy resources of the system is referred to as an exergy of the system. In the mentioned state it is measured by the amount of mechanical, or other fully transformed energy, that can be obtained from this system due to its reversible transition from this state to the state of equilibrium with the environment.

It directly follows from the first and second laws of thermodynamics that in each state system's exergy, as well as energy, has a certain fixed value.

Exergy of the system, existing in the environment of constant parameters, remains unchanged only in case of reverse processes, occuring within it, as well as in case of interacting with the environment. If any of these interactions are irreversible, exergy of the system decreases accordingly.

This basic property of exergy allows us to use it as the measure of the process' reversibility. The difference between total amount of the exergy input of the system (E_{in}) and amount of the exergy output (E_{out}), determines

DOI: 10.1051/matecconf/201710811004

the total value of waste due to the system irreversibility by eqn (1):

$$\sum D = \sum E_{in} - \sum E_{out} > 0 \qquad (1)$$

The exergy balance equation can be written expressing the useful work for a general process as depending on the useful work for the corresponding reversible process and the exergy destruction term. However, this can be made only if the terms (other than the exergy destruction term) on the exergy balance equation for the general process lead to the same result as the terms on the exergy balance equation for the corresponding reversible process [2].

The exergy loss is similar to the entropy, which increase in a closed system also reflects waste due to system irreversibility. However, the practical advantage of the exergy concept is that it gives a decrease in the value of the transformed energy waste right away and allows you to compare waste value to the available amount of transformed energy (i.e. to find both absolute and relative waste values). The ratio of the exergy output of the system (E_{out}) for the evaluation of exergy input (E_{in}) is represented by the efficiency - the exergy efficiency, which characterizes the process degree of approximation to the ideal is determined by the eqn (2):

$$\eta = \frac{\sum E_{out}}{\sum E_{in}} = \frac{\sum E_{out} - \sum D}{\sum E_{in}} = 1 - \frac{\sum D}{\sum E_{in}} \qquad (2)$$

In ideal process $\eta=1$, in real process $\eta<1$.

Waste (D) can be divided into two groups in terms of their distribution:

- internal, associated with the irreversibility of the processes, occurring within the system. They are usually denoted by the letter i (D_i);
- external, related to the terms of system interaction with the environment and other sources and receivers of energy. They are denoted by the letter e (D_e).

In heat transformers and refrigeration units, in particular, waste due to the throttling, hydraulic resistance, friction in machines, heat and mass transfer at finite temperature and concentration head, etc. may serve as an example of internal waste. External waste include those associated with temperature difference between the working fluid and the coolant, as well as waste through the heat insulation. Waste of the products, leaving the plant, which exergy is not consumed also belong to this group, e.g. heated cooling water, waste nitrogen in oxygen plants, and others.

Internal and external waste separation can be easily produced by the same formula. If it contains assumed exergy values, taken according to the parameters of the plant, the obtained value D_i will correspond only to the internal waste. If the equation contains exergy values either given to or received from the external sources and receivers of energy, then D includes both external and internal waste. Consistently applying the equation to the individual elements of the plant, it is easy to determine the distribution of the plant internal waste.

The energy balance of this system is determined by the eqn (3):

$$(L+E+Q+I)_{in} = (L+E+Q+I)_{out} \qquad (3)$$

The energy balance of the system is determined by the equation (1), which shows that in real processes part of the energy is converted into inoperative form.

By the exergy the energy efficiency is meant, i.e. the amount of work_operation that can be obtained with the set energy as a result of reversible interaction with the environment.

In case of mechanical and electrical kinds of energy, exergy and energy values are equivalent, as in an ideal process these kinds of energy can be transformed into work operation entirely. For other energy kinds, such as the heat flux, exergy value is determined by the amount of energy that can be transformed into work operation in the perfect tune [3].

The first law of thermodynamics imposes no restrictions on the transition of one kind of energy, it is only important that the amount of total remain unchanged; the first law of thermodynamics imposes certain restrictions on the energy transformation.

For the exergy balance of the system it is necessary to use the method of determining the exergy of each kind of energy.

3 Exergy analysis of major cryogenic cooling systems

It is necessary to consider the most common system before proceeding with the development of calculation methods.

Exergy analysis gives the presentation of a system relative to its best performance. In addition, the exergy destructed can react with its surrounding and harm environment processes [4].

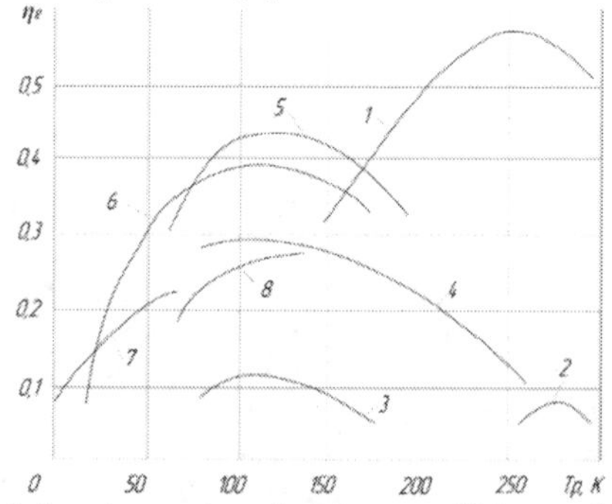

Figure 1. Low-temperature plants' exergy efficiency relation to cryostatting temperature.

Figure 1 shows the exergy efficiency relation on cryostatting temperature for some cycles and patterns of cryogenic plants. The graph shows that for each cooling method there is a relatively narrow temperature range, within which a high performance of the system remains. Maximum efficiency in the temperature range of 70-150 K belongs to the gas cryogenic machines, running on the Stirling cycle (curves 5,6), minimum - to the throttle system, in which nitrogen is used as a refrigerant (curve 3). The use of thermoelectric coolers (curve 2) in microcryogenic systems is limited by the temperature conditions. Throttle cryogenic systems are most easily amenable to miniaturization, but have the lowest efficiency in comparison with other types of closed systems due to the imperfections of the throttling process and small isothermal efficiency. However, the effectiveness of throttle microcryogenic systems can be significantly increased by the coolants' thermal properties improvement [5].

In case of miniaturization, Stirling cycle systems retain the highest efficiency as compared with other cycles and cooling methods.

1 - steam compressor units; 2 - thermoelectric coolers; 3 - throttle system with nitrogen as a refrigerant; 4 - regenerative systems with turbomachines; 5,6 - horizontal forging machine working on Stirling cycle; 7 - expanders; 8 - throttle system using gas mixtures.

3.1 Thermoelectric cooling system

Thermoelectric cooling system (electronic refrigerator) is optimal device for receiver of radiative energy. Electronic refrigerators have several advantages compared with other cooling systems, if the receiver is to be cooled to 180 K. They have low weight and dimensions, low energy consumption, high operational reliability.

Electronic refrigerators have elements which consist of two series-connected semiconductor branches with electronic conduction (n) and hole conduction (p). All thermocouples are connected in series or in parallel to form a thermopile (Figure 2) [6].

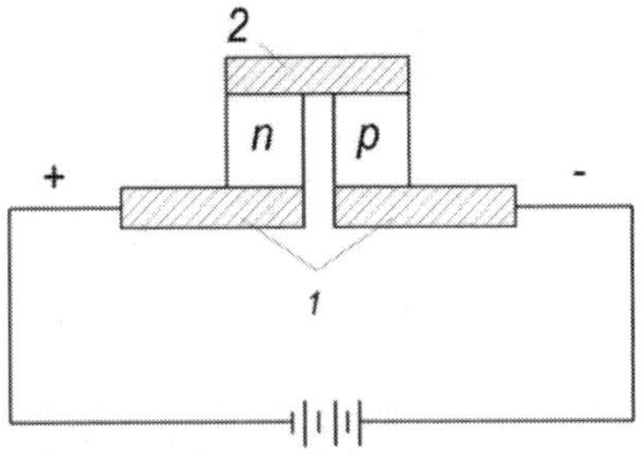

Figure 2. Elementary thermopile.

We need to use multi-stage thermopile (thermoelectric module) to provide the required temperature conditions (Figure 3). Selection standard thermoelectric modules depend on at least three parameters. This is temperature of hot and cold side of the module as well as the heat load that should be removed from the cold side.

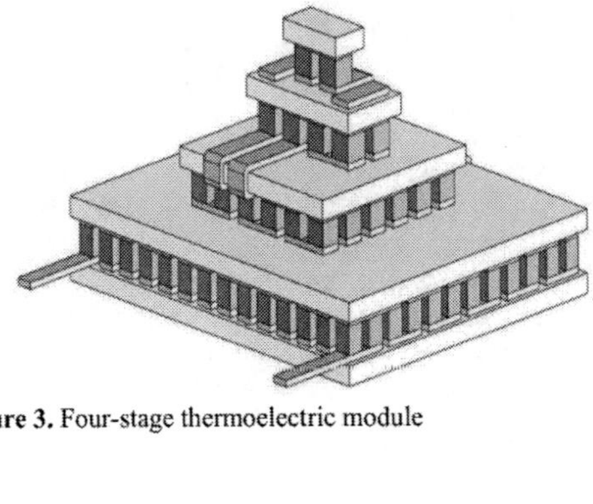

Figure 3. Four-stage thermoelectric module

3.2 Throttle cooling system

Outer contour of this type cooling system includes the following devices: throttle, tank for storing high-pressure gas, filter-adsorber (Figure 4).

According to calculations, using nitrogen as a refrigerant for cooling in the temperature range 200 - 250 K and obtain cooling capacity of about 350 W, it is necessary to provide the gas flow rate of about 15 kg / h .

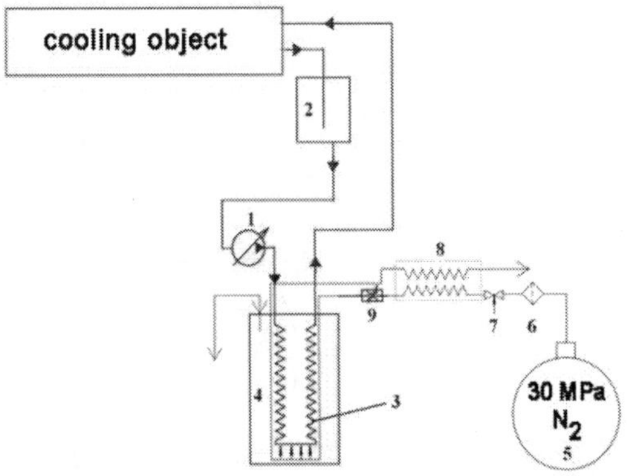

Figure 4. Throttle cooling system.
1 - pump; 2 - a broad tank; 3 - heat exchanger; 4 - an external coolant tank; 5 - a cylinder for storing high-pressure gas; 6 - filter-adsorber; 7- flow control valve; 8 - exchanger of throttle system; 9 - throttle

Practical implementation of such a system makes it possible to cool not only the laser, but infrared receivers that are part of the complex laser.

3.3 Refrigerating machines working on Stirling cycle

Refrigerating machines working on Stirling cycle most widely used to provide the required temperature conditions. They have a high efficiency and very compact.

In a closed cycle Stirling working fluid is involved in a cyclic process consisting of four stages (two isochor and two isotherms).

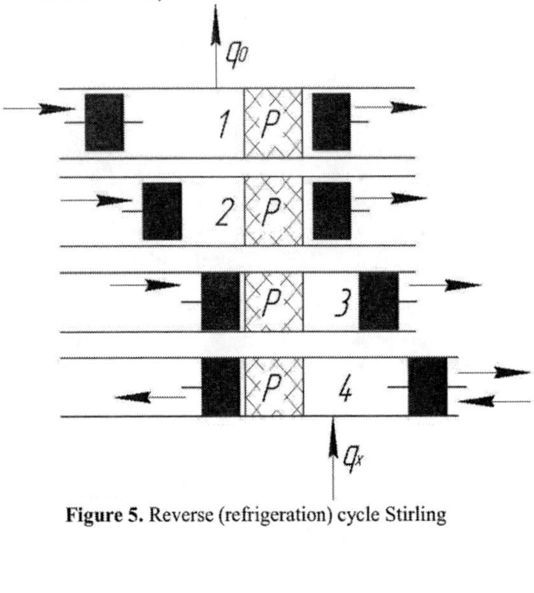

Figure 5. Reverse (refrigeration) cycle Stirling

Heat is removed from the cycle into the environment occur at a temperature T_0 in the compression. Heat is removed from the cooling body is passed into the cycle at the temperature T during expansion. The regeneration of the heat in cycle is carried out in the process V=const (Figure 5).

The work of compression (1-2) is determined by eqn:

$$l_{сж} = l_{1-2} = RT_0 \ln\frac{\upsilon_2 - b}{\upsilon_1 - b} - a\left(\frac{1}{\upsilon_1} - \frac{1}{\upsilon_2}\right) \quad (4)$$

The work of expansion (3-4) is determined by the eqn:

$$l_p = l_{3-4} = RT \ln\frac{\upsilon_1 - b}{\upsilon_2 - b} - a\left(\frac{1}{\upsilon_2} - \frac{1}{\upsilon_1}\right) \quad (5)$$

The amount of heat taken away to the environment is determined by the eqn:

$$q_{oc} = RT_0 \ln\left[\frac{(\upsilon_2 - b)}{(\upsilon_1 - b)}\right] \quad (6)$$

The amount of heat reported to the working fluid in the expansion process is determined by the eqn:

$$q = RT \ln\left[\frac{(\upsilon_1 - b)}{(\upsilon_2 - b)}\right] \quad (7)$$

Refrigeration Stirling cycle efficiency is determined by the eqn:

$$\varepsilon_{Str} = \frac{q}{l_\Sigma} = \frac{T}{(T_0 - T)} = \varepsilon_C \quad (8)$$

The change in internal energy in the compression (1-2) is determined by the eqn:

$$\Delta u_{1-2} = \Delta u_{T_0} = u_1 - u_2 = a\left(\frac{1}{\upsilon_2} - \frac{1}{\upsilon_1}\right) \quad (9)$$

The cooling capacity of the ideal machine Stirling with the real body work of expansion is determined by the eqn:

$$q = l_p - \Delta u_{T_0} \quad (10)$$

The fundamental way to implement the reverse Stirling cycle in a piston machine illustrated in Figure 5. Two pistons move in cylinder intermittently utlom phase shift β. In the working space between the piston positioned regenerator P which divides the working chamber into two parts - warm and cold. Sometimes they are conventionally called cavities of compression and expansion. In the warm cavity maintained at a constant temperature T_0 by removing heat q_{OS} refrigerated. The cold temperature of the cavity X is constant (due to the heat supply through the cooler q).

The above graph shows: 1) complex and different dependence of the efficiency η_{t_n} from π_{IV}, allowing in each case to find the optimal values k, π_I, π_{II} and T_1/T_3 for π_{IV} data; 2) significant differences of this dependence from the dependence $\eta_{t_k} = f\ (T_1, T_2, \Delta T)$.

They show that the increase of economic efficiency is not enough to strive to increase the values of the polytropic exponent, since under certain conditions it can lead to the opposite result.

Analysis graphs below allows to choose the most advantageous values η_{t_n} and η_{t_k} for each particular case, and hence the value of the coefficient of completeness of the reversible conversion of energy η_o, as well as match *PT* (*k*, *R*). The relative efficiency of actual cycles depending on the working bodies compared to the Carnot cycle efficiency can be estimated by the number of work.

3.4 Development of a methodology of exergy analysis

During the research, it was found that the exergy efficiency of the cryogenic cooling system is as follows:

$$\eta_e = \frac{Q_0 \cdot \tau_e}{L} \quad (11)$$

where Q_0 – cooling capacity of the system; τ_e – exergy temperature function; L – cycle operation.

In this case, taking into account the costs for the preparation of the working fluid, exergy efficiency is determined by eqn:

$$\eta_{ep} = \frac{Q_0 \cdot \tau_e}{\sum E + L} \quad (12)$$

where $\sum E$ – total energy input.

The total exergy efficiency, taking into account the costs for the preparation of the working fluid, as well as the costs for cooling system transportation on the power plant is determined by eqn:

$$\eta_{et} = \frac{Q_0 \cdot \tau_e}{\sum E + L + e_m} \quad (13)$$

where e_m – exergy mass ratio.

$$e = k_m \cdot m_{cs} \quad (14)$$

where k_m - mass ratio, m_{cs} – mass cooling system.

$$k_m = \frac{m_f \cdot H}{m_{pp} \cdot V} \quad (15)$$

where m_f – fuel mass; m_{pp} – power plant mass; H – combustion value; V – power plant work time.

The mass ratio k_m and the exergy mass ratio e_m are the new values that need to be taken into account in the calculations.

4 Conclusions

The result of this study is a new universal method of exergy analysis, which can be used to select a cryogenic cooling system. Applying the obtained methodology allows us to perform comprehensive analysis of the cryogenic cooling systems effectiveness. This, in turn, allows us to choose the most effective cooling system and to measure the performance of power plants in the industry, aviation, mechanical engineering. In addition, this technique allows to evaluate the effectiveness of the special measuring equipment.

References

1. Dovgyallo A.I., Nekrasova S.O., Sarmin D.V., Uglanov D.A. Imitations tests of throttle cooling system with cryogenic refueling tank and its comparative characteristics // Applied Physics. 2013. Issue 4, 2013, Pages. P. 54-59
2. Samad Jafarmadar, Peyman Nemati. Exergy analysis of diesel/biodiesel combustion in a homogenous charge compression ignition (HCCI) engine using three-dimensional model // Renewable Energy (V.9). 2016. Pages 514–523
3. Biryuk V.V., Tsapkova A. B., Shimanov A.A. Numerical modeling of cyclone machine for cleaning gas generator gases, Key Engineering Materials. 2016. Volume 685. Pages 153-157
4. V.A.F. Costa. On the exergy balance equation and the exergy destruction // Energy (V.116). Pages 824-835
5. Nekrasova, S.O., Uglanov, D.A.,Shimanov, A.A. Comparative analysis of calculation models of pulse tube thermoacoustic refrigerator, 4th International Conference for Young Scientists High Technology: Research and Applications, HTRA 2015, 21 - 24 April 2015, Publisher: Trans Tech Publications Ltd, 2016 г., P. 3-7
6. Biryuk V.V., Gorshkalev A.A., Kayukov S.S., Uglanov D.A. Gas-Dynamic Analysis of Processes in a Small-Sizes Two-Stroke Combustion Engine // The Open Mechanical Engineering Journal. 2014. 8. P. 441-444

Characterisation of the dynamic characteristics of a complex jointed structure

M. S. A. Sulaiman, M. A. Yunus, M. N. Abdul Rani, A. Mohd Saman, Z. Mohamed and M.A.S. Aziz Shah

Center of Excellence of Dynamics and Control, Faculty of Mechanical Engineering, Universiti Teknologi MARA (UiTM), 40450 Shah Alam, Malayisa

Abstract. Every structure that assembled with the joints has a significant effect on the dynamic behaviour of the structure. Therefore, the overall dynamic behaviour of the jointed structure highly relies on these joints. Finite element method (FEM) is commonly used to model an element connector for the jointed structure due to its because every jointed structure has its owns joint properties. The aims of this research are to identify the potential element connectors that are available in commercial finite element analysis package in order to represent the joints for spot welds and adhesive joints. The reliability of potential connector elements are quantified by comparing the predicted result calculated from commercial finite element analysis package with the experimental result that obtained from the LMS Scadas. The model of element connectors have been constructed based on three types of the case studies. The comparison reveals that the combination of Type 2 (ACM2) and Type 3 (CWELD) have shown a better capability to represent the joints of the complex jointed structure.

1 Introduction

In automotive engineering, the complex jointed structure, such as car hood are assembled by an arrangement of various sizes of metal sheets and they are joined by a thousand number spot welds and adhesives. Spot welds are also recognised as resistance spot welding (RSW) is a preferable choice for joining method the car hood due to its simple fabrication process. Nowadays, the modern adhesive have been used in automotive industry to reduce the number of spot welds or combined these two joints for assembling the car hood. Understandably, imperfections in assembled and jointed the structure can cause variations in dynamic characteristics, for examples, mode shapes and natural frequencies of the car hood. Besides, the reliability and the dynamic characteristics of the jointed structure greatly relies on these joints [1], [2]. For some significant time, research has demonstrated that the huge discrepancies between finite element analysis and experimental results because of the invalid assumption on modelling the joints in the finite element [3].

The main challenges to predict the dynamic behaviours is a due to the joined structure. To model a connector element for the jointed structure is very complex because every jointed structure has its own joint properties. For instance, to model the spot welds, Donders et al. had used CWELD to model spot welds for jointed the structure [4]. Besides, Heiserer et al. have proposed to use Area Contact Model 2 (ACM2) for the connector element to represent the spot welds [5]. Many researchers use this type of model to represent their model of spot welds. On the other hand, to represent the adhesive joints, Tahmasebi was proposed their model for adhesive joints to NASA Goddard Space Flight Centre [6]. Their model is in 1D element and a combination between rigid element and a spring element. However, the result of the dynamic behaviour of the model still has occurred an error.

Therefore, a continuous study about potential element connectors to represents the joints (spot weld and adhesive joints) for connector element in finite element need to be investigated. It is because the overall dynamic behaviour of the assembled structure depends on the model of the connector element [7], [8].

2 Finite element

Finite element analysis (FEA) or Finite element method (FEM) is a method to discretize differential operators. It is a technique used for analysis of structure by the modelling of structure. Furthermore, it is applied by engineers to finish the issues that identified with an exceptionally complex structure by utilising mathematical model and numerical method. Generally, the finite element analysis are useful to be applied in production design as to guarantee the design is a reliable model according design specifications before the manufacturing process conducted.

The complex jointed structure was constructed using finite element analysis based on a model of the hood of the car. It was developed using Altair HyperMesh software and OptiStruct was used to analyse the natural frequencies and mode shapes. The complex jointed structure was discretized into a shell element type which represents 1mm for their thickness. The frequencies of interest are set between 1Hz to 200Hz and only first five modes were identified.

In this study, three types of the element connectors in FEM will be used to model the spot welds and two types for adhesive joints. Each type of element connectors will be named as Type 1, Type 2, Type 3 and Type 4 connector. Table 2 and Table 3 shows the detail of the properties and FE model for the joints.

Table 1. Details of the FE model of the complex jointed structure without the joints (Car Hood)

Type of Elements	No of Elements	Properties	Value
QUAD4	28287	Young's Modulus	210 GPa
TRIA3	1700	Poisson's Ratio	0.33
		Density	$7900 kg/m^3$

Table 2. Details of the properties for the joints.

Properties	Adhesive	Spot Welds
Young's Modulus	5.96 MPa	210 GPa
Poisson's Ratio	0.49	0.3
Density	$1000 kg/m^3$	$7900 kg/m^3$

Table 3. Details of the FE model for the joints.

Type of Elements	No of Elements
HEXA solid element	93
WELD (1D)	24

Type 1 connector
Type 1 element connector as shown in Table 4(a) is the rigid body element type of joint. In this approach, since only rigid body element (RBE2) was used, there are no mechanical properties and mass apply to the joint. Basically, it consists of one independent node and one or more dependent nodes. This element connector was used as the initial model for spot welds and adhesive joints and to compare with the measurement.

Type 2 connector
Type 2 element connector as shown in Table 4(b). The model of element connector is for adhesive joint, and only one model has been designed. It was designed as an Area Contact Model 2(ACM2). This type of element connector is the combination of HEXA solid combined with RBE2. The patch will be jointed with the rigid element to the component. The idea of using this type of connector is the patch of solid element is represented the adhesive so that it can input and update their properties.

Type 3 connector
Type 3 connector represented for spot welds is used CWELD as type 3 connector as shown in Table 4(c); the model has already existed at HyperWorks. It is a 1D element and to construct the model is based on node to node. This element connects between two components at each node that represents the joint. The size of spot welds diameter and their properties can be updated. The detail or the properties as shown in Table 2.

Type 4 connector
Table 4(d) shows the 1D element that represents for spot welds. Same as Type 3, this type of connector has used the CWELD element, but it is modelled based on the node to the surface. The name of this model is GA-GB ELPAT. This type of connector can also input their properties and size of spot welds diameter.

Table 4. (a) Type 1 connector; (b) Type 2 connector; (c) Type 3 connector; (d) Type 4 connector.

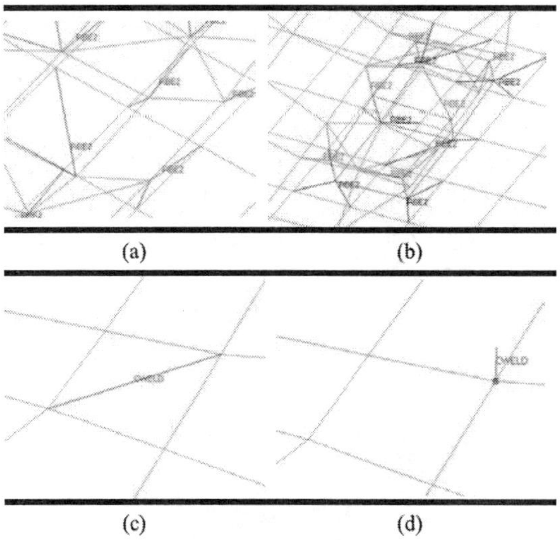

It has three case studies to represent the joints for adhesive joints and spot welds. The model of element connectors has been constructed based on three types of the case study, firstly is case study 1, secondary is case study 2 and third is case study 3;

- Case study 1 – Type 1
- Case study 2 – Type 2 and Type 3 connectors
- Case study 3 – Type 2 and Type 4 connectors

3 Experimental works

Experimental modal analysis (EMA) is a technique or a process to determine the dynamic behaviour of the structure, which is mode shapes, natural frequencies and damping ratios from vibration test [9]. It is also mentioned to as model testing or modal analysis. Basically, this test measures modal properties such as natural frequencies, mode shape and damping. Another explanation behind vibration testing is to check a predicted model proposed for the test framework. The result from finite element model will be verified and compared to experimental data to get a better outcome. If the result from experimental data, such as, mode shapes and natural frequencies agree with those predicted by the finite element model, the model is confirmed and can be used in design and response prediction with some

DOI: 10.1051/matecconf/201710812001

confidence. Niw Chang Chee has used this method to verify their model is reliable, and their model can use for another experiment [10].

Prior to experimental setup, all of this process must be consider to get the acceptable result. It is the hanging orientation of the structure, the number of measuring points or nodes of the structure, the excitation method and the number of the accelerometers. Meanwhile, the roving accelerometer method and shaker testing were used to excite the structure.In this paper, the experimental work of the complex jointed structure was using shaker testing method and roving accelerometer. The structure was tested on free-free boundary conditions as shown in Figure 1 where the damping will be neglected because the structure is lightly damp. Four nylon string with springs were used to simulate the free-free boundary conditions. The frequencies of interest of the structure were between 1Hz to 200Hz. As a guideline, the results of finite element analysis was utilised to decide the frequencies of interest and the excitation point. Besides, the mode shapes from finite element will be used as an initial idea how the mode shapes behave. The response signal for this experiment was interpreted by LMS SCADAS analysis.

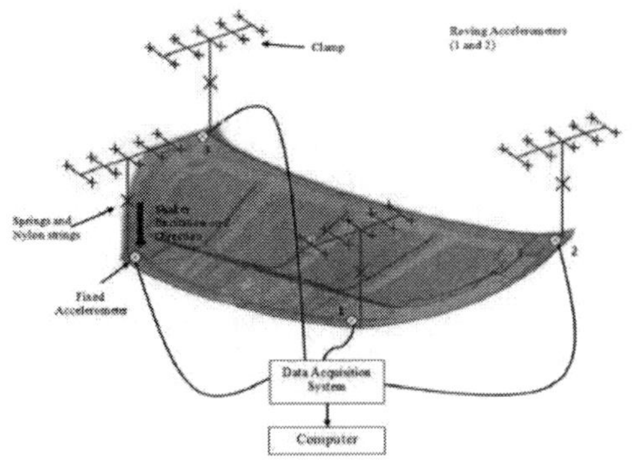

Figure 1. Schematic diagram of the complex jointed structure set up.

4 Results and discussion

In this complex jointed structure, the results are shown based on three types of the case study. Table 5 and Table 6 had shown the comparison of the results between test data, case study 1, case study 2 and case study 3. The comparison between test data and case study 1, clearly shown that the percentage of error is very high other than case study 2 and 3 because the joint is defines a rigid body. Therefore, the complex jointed structure are very sensitive with the design of joint which it highly depend on their stiffness.

Table 5. The comparison of the natural frequencies between test data and case study 1.

I Mode No.	II Test Data (Hz)	III Case Study 1 (Hz)	IV Error (%)	V MAC Value
1	27.163	41.406	52.435	0.98
2	50.642	71.414	41.017	0.61
3	60.164	83.902	39.455	0.68
4	80.445	120.666	49.998	0.94
5	106.536	161.318	42.277	0.35
Total Error (%)			273.535	

Table 6. The comparison of the natural frequencies between test data and case study 2.

I Mode No.	II Test Data (Hz)	III Case Study 2 (Hz)	IV Error (%)	V MAC Value
1	27.163	29.480	8.530	0.97
2	50.642	51.230	1.108	0.73
3	60.164	64.277	6.836	0.79
4	80.445	82.930	3.089	0.84
5	106.536	101.972	4.284	0.81
Total Error (%)			23.847	

Table 7. The comparison of the natural frequencies between test data and case study 3.

I Mode No.	II Test Data (Hz)	III Case Study 3 (Hz)	IV Error (%)	V MAC Value
1	27.163	29.629	9.079	0.97
2	50.642	51.265	1.230	0.73
3	60.164	64.350	6.958	0.79
4	80.445	83.231	3.463	0.85
5	106.536	102.299	3.977	0.82
Total Error (%)			24.707	

Table 8. The comparison of the mode shapes between measured and predicted for case study 1.

Mode Shapes	Case Study 1	
	EMA	**FE**
1	 27.163Hz	 41.406Hz
2	 50.642Hz	 71.414Hz
3	 60.164Hz	 83.902Hz
4	 80.445Hz	 120.666Hz

5	 106.536Hz	 158.048Hz

Table 9. The comparison of the mode shapes between measured and predicted for case study 2.

Mode Shapes	Case Study 2	
	EMA	FE
1	 27.163Hz	 29.480Hz
2	 50.642Hz	 51.203Hz
3	 60.164Hz	 64.277Hz
4	 80.445Hz	 82.930Hz
5	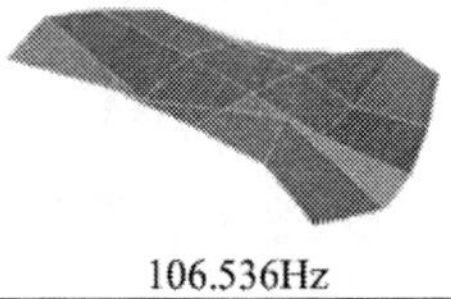 106.536Hz	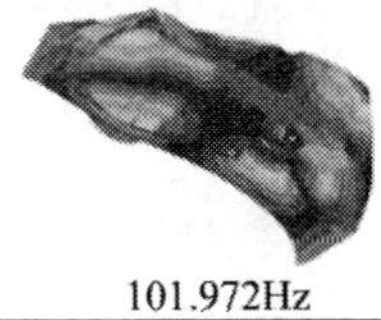 101.972Hz

Table 10. The comparison of the mode shapes between measured and predicted for case study 3.

Mode Shapes	Case Study 3	
	EMA	FE
1	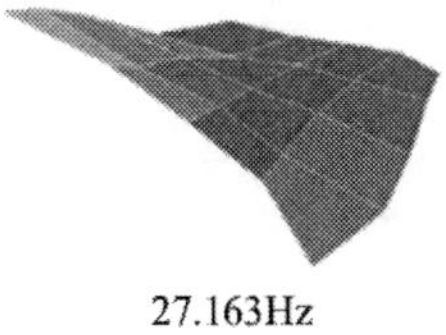 27.163Hz	 29.629Hz
2	 50.642Hz	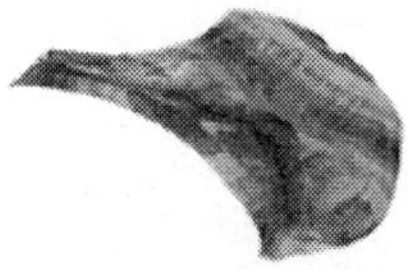 51.265Hz
3	 60.164Hz	 64.350Hz
4	 80.445Hz	 83.231Hz
5	 106.536Hz	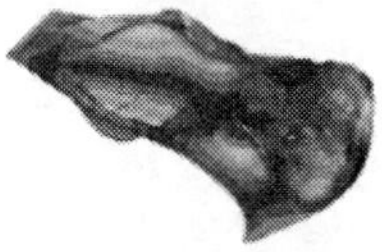 102.299Hz

However, based on Table 6 and Table 7, shown that the natural frequencies of case study 2 have lower natural frequencies than the case study 3 which is 35.550 percent and 36.559 percent respectively with the MAC value above 0.8. Besides, that its shown the highest error which in mode six. Meanwhile the mode two shown the least of 1.108 percent with a total of MAC value 0.73. Next, for case study 3, the highest and least error is same with case study 1 which is 11.853 percent and 1.230 percent. It happen because the overall dynamic behaviour of the assembled structure depends on the model of the connector element and the reliability a composition of the structure is greatly dependent on the joints [1]. By observing the Table 5, Table 6 and Table 7, it is agreed that the predicted natural frequencies can be vary by manipulating the element connectors. Therefore, the data for a finite element in case study 2 shows more reliable than the case study 1 and 3. This indicates a good selection of element connectors can produce a reliable results. The result will be validated and confirm the accuracy of the finite element analysis with the experimental result by comparing the mode shapes [9]. Therefore, Table 8, Table 9 and Table 10 shows the mode shapes based on case study 1, case study 2 and case study 3 that were did from test data and finite element results.

5 Conclusion

The first five frequencies of a complex jointed structure of car hood were investigated experimentally and numerically. Besides, the potential element connectors that are available in the commercial finite element analysis to represent the joints for spot welds and adhesive joints were identified. The reliability of potential element are quantified by comparing the predicted result with experimentally based on three case studies. The comparison reveals that the case study 2, the combination of Type 2 (ACM2) and Type 3 (CWELD) have a better capability to represent the joints for spot welds and adhesive joints of the complex jointed structure. However, the correlation between the numerical and the experimental data can be improved by applying the finite element reconciliation method to the finite element model of the complex jointed structure which will be applied to the next stage. Moreover, the continuous study in modelling the jointed structure was recommended.

Acknowledgement

The authors gratefully acknowledge the Kementerian Pendidikan Malaysia (KPM) and Research Management Centre (RMC) UiTM for the Fundamental Research Grant Scheme (FRGS) with the file number 600-RMI/FRGS 5/3 (78/2014) for the sponsorship of this research. The authors also would like to express appreciation to Mohamad Azwan Abd Khair from the Centre of Structural Dynamics and Vibration Engineering (SDAV) UiTM for their technical assistance and laboratory facilities.

References

1. M. A. Yunus, M. N. A. Rani, H. Ouyang, H. Deng, and S. James, "Identification of Damaged Spot Welds in a Complicated Joined Structure," J. Phys. Conf. Ser., vol. **305**, p. 12057, (2011)
2. R. A. Ibrahim and C. L. Pettit, "Uncertainties and Dynamic Problems of Bolted Joints and Other Fasteners," J. Sound Vib., vol. **279**, no. 3–5, pp. 857–936, (2005)
3. M. A. Yunus, "Finite Element Modelling and Updating of Structure of Sheet Metal with Bolted and Welded Joints," (2011)
4. Y. Xia and H. Hao, "Statistical Damage Identification of Structures with Frequency Changes," J. Sound Vib., vol. **263**, no. 4, pp. 853–870, (2003)
5. D. D. Heiserer, M. Chargin, and D. J. Siela, "High Performance , Process Oriented , Weld Spot Approach," 1st MSC Worldw. Automot. User Conf., pp. 1–14, (1999)
6. F. Tahmasebi, "Finite Element Modeling of an Adhesive in a Bonded Joint," (1999)
7. G. Lasker, J. G. Maloney, M. T. Shelton, and D. A. Underhill, "Structural Dynamic Properties of Tactical Missile Joints-Phase 3," Ntis, no. May, p. 186, (1974)
8. D. J. Ewins, J. M. Silva, and G. Maleci, "Vibration analysis of a helicopter with an externally-attached structure," Shock Vib. Inf. Cent., vol. **50**, pp. 155–170, (1980)
9. K. A. Ramsey, S. M. Systems, and S. Jose, "Experimental Modal Analysis , Structural Modifications and FEM Analysis on a Desktop Computer mathematical dynamic models of structures so that they could," Sound Vib., no. February, pp. 1–10, (1983)
10. N. C. Chee, A. Bakar, and R. Holloway, "Finite element modeling of arc welded joints," J. Mek., vol. **23**, no. June 2007, pp. 15–30, (2007)
11. Y. Zhao, C. Yang, L. Cai, W. Shi, and Z. Liu, "Surface Contact Stress-based Nonlinear Virtual Material Method for Dynamic Analysis of Bolted Joint of Machine Tool," Precis. Eng., vol. **43**, pp. 230–240, (2016)
12. A. A. S. M. Darwish, Sh. M. Elseufy and Advanced, "Finite Element Analysis of an Automobile Engine Finite Element Analysis of an Automobile Engine Hood," no. MARCH 2013, pp. 1–8, (2016)
13. L. Majkut, "Acoustical diagnostics of cracks in beam like structures," Arch. Acoust., vol. **31**, no. 1, pp. 17–28, (2006)
14. N. Rieger, "Relationship between finite element analysis and modal analysis.," Sound Vib., pp. 16–31, (1986)
15. Stephen A. Dyer, *Survey of Instrumentation on Measurement.* Wiley, (2004)
16. W. Kuntjoro, *An Introduction to the Finite Element Method.* McGraw-Hill Education (Asia, 2005)
17. N. S. Gokhale, *Practical Finite Element Analysis.* (2008)
18. D. Hutton, *Fundamental of Finite Element Analysis.* Mc Graw Hill, (2004)
19. He, J. and Z.F. Fu, *Modal analysis.* (Butterworth-Heinemann, (2001)

Molecular Clustering and Velocity Increase in Converging-Diverging Nozzle in MD Simulation

Jeoungsu Na[1], Jaehawn Lee[2], Changil Hong[2], Suhee Kim[2]

[1]*R&D Department of NaJen, Co. LTD, Korea*
[2] *Dept. of Computer Engineering, Hoseo University, Asan Korea*

Abstract. A Molecular dynamics simulation in a converging-diverging nozzle was performed to study molecular collisions and their influence to average flow velocity according to a variety of vacuum levels. The static pressures and the dynamic pressure exerted by the molecule collision on the selected walls were compared to figure out the intensity variances of the directional flows. With pressure differences constant between the entrance and the exit of the nozzle, The numerical experiment was performed for molecular velocities and directional flows. The result shows that the velocities are increased at the nozzle exit as the vacuum level gets higher in that area because of less molecular collisions

1 Introduction

Hydroelectric generator converts potential energy of high positioned water into electricity using gravitational force with over 90% of total efficiency. It is high value compared to other electric power generation such as 38% efficiency of thermal power plants. However, for stable operation of the power plants and to make sufficient electricity, thermal or nuclear power plants are used widely even though the low efficiency. They use high pressured steam jet to rotating turbine by the Rankine cycle, which has condensing stage for recirculation of liquid water. In the condensing state, however, we have been ignored the discharging heat energy without doubt. Especially, for these power plants, the energy of discharging cooling water is over 50%, that is the energy lease to sea or atmosphere affecting harmful damage to environment.

We focus the unused energy of discharging cooling water for transforming useful energy with Cavitation phenomenon of water. Cavitation is an inevitable phenomenon that occurs often in hydrodynamic machine or draft tube of hydroelectric generator[1]. In hydro machine, the abrupt event causes severe damage to surface of hydro-blades operating below saturated vapor pressure. In the short moment, internal energy of combined water molecules makes free movement of molecules for creating bubbles and they implode to one point on a surface if the collision cannot support pressure by surrounding liquid[2]. Since the implosion is in fact rapid concentration of molecules to one point, it causes intense shockwave. At the concentration process, if we placed a nano-sized turbine generator in the passing of moving molecules, it could make electricity.

Converging-Diverging nozzle is mainly used in compressible gas jet industry. However, if liquid water passes through the nozzle, some of the liquid boils partially so that we can see mixed two phase flows of water. The void by the nozzle boiling is also a cavitation that is little different from undesirable cavitation. The induced cavitation makes the volume expanded in the fixed nozzle space so that it let average velocity of flow increased. Since fluid is simply moving group of molecules colliding each other, each molecule has its own colliding velocity. Therefore, we have been studyed each molecular velocity to achieve maximum kinetic energy by collision based on conservation of momentum.

Since molecules in the liquid state are binding with neighboring molecules tightly, we cannot approach to the velocity of each molecule. However, it is possible that the water molecules generate cavitation bubbles so that we can approach to each velocity of molecules, when the liquid pass through the nozzle and go into vacuum. Then, we might think of getting momentum by colliding with rotating turbine blades.

In this paper, ideal molecular dynamics simulation for nozzle is introduced for possibilities of getting more energy over high potential energy. Nozzle would be part for separating of liquids and making straight directional flow of molecules. Back pressures in between the nozzle exit area and in front of turbine blades interferes main flow. Since low back pressure means high level of vacuum, moving molecule from nozzle throat may go longer without collisions of random molecules.

Therefore, we simulate nozzle first with molecular dynamics. The goal of this study is investigating whether or not the velocity of molecules at the nozzle exit increases as the vacuum level gets higher.

2 Configurations of MD Simulation and Implementation

Molecular dynamics simulation is performed to study velocity changes according to degrees of vacuum level in the molecular world.

We design a 2-dimensional nano-sized shape of nozzle and implement it. We generate moving molecules with initial velocities in the nozzle. We simulate molecular movements with simple numerical methods such as the Verlet integrations and the Lennard-Jones potentials. Instead of water molecules, argon molecules are used in this simulation. Since argon molecule is symmetric and monatomic, it is desirable for a starting stage. To grasp general tendency of pressure and velocity changes, 2-dimensional simulation is used with a personal computer.

2.1 Nozzle geometry and measurement positioning

Nozzle geometry for our molecular dynamics simulation is converging-diverging type as shown in Fig. 1. The converging region is marked as A and the straight moving region after diverging is marked as B. To measure static pressures, we select the arc L and the line M for the front part and the rear part of the nozzle throat, respectively. The vertical line N is selected to measure dynamic pressure and average exit velocity.

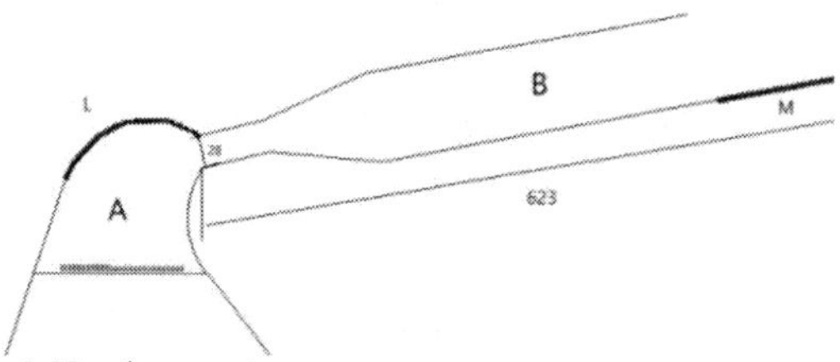

Figure 1. Nozzle geometry

We define base unit as the same size of diameter for one molecule. The length from the throat to exit is 623 base units. The throat diameter is 28 base units and the diverging angle of expansion part is 12°. For a argon molecule, 623 base units is 212 nanometers in physical dimension.

2.2 Defining pressures for regions A and B

Static pressure is normal force per unit area that we usually use it with pressure gage in fluid machinery. To find out pressure drop when the fluid is passing throat, we define two dimensional pressure which is sum of collision force with chosen line divided by length of the chosen line. Because this simulation is 2D, it is computed as follows:

$$P_{2d} \cong \frac{\Sigma m_i a_i}{length\ of\ chosen\ curve\ or\ line} \qquad (1)$$

The arc L is a semi-circle before the throat. The length of the arc L is equal to the length of the selected line M.

Dynamic pressure is the kinetic energy of fluid particles per unit volume. To compute the dynamic pressure for the vertical line N, we get the exit velocity of each particle passing through the vertical line N and calculate the dynamic pressure. Let q_{2d} be pressure for the vertical line N. It is expressed as follow:.

$$q_{2d} = \frac{1}{2}\Sigma \rho {v_i}^2 \qquad (2)$$

2.3 Placing molecules for pressure difference configuration

The most important target of this study is investigating whether or not the velocity of molecules at the nozzle exit increases as the vacuum level gets higher. Since the vacuum level is closely related to the number of molecules in the nozzle, we set the number of molecules generated in region A to be different from the number of molecules in region B to have different vacuum levels. That is, we assign different number of molecules to regions A and B to keep 4000 molecules difference for experimental sets. Molecules in region A are created with random velocities at the bottom of the nozzle and they collide with each other and the closed walls. Molecules in region B are generated with random velocities at end of the nozzle, to exert some level of pressure on region between the nozzle exit and the rotating blades. These random molecules generated in region B behave as obstructer to the moving fluids to the nozzle exit.

2.4 Collisions of molecules with sloped nozzle walls

Here, the colliding nozzle walls in Fig. 1 are sloped. So, the normal force won't be along the conventional Y axis. It makes sense to resolve the velocity along the normal and tangential directions of point of contact. While the velocity along the tangential component would remain the same, the velocity along the normal would reverse in direction. Collisions of molecules with sloped nozzle walls are implemented to calculate new velocities[3].

2.5 The Lennard-Jones potential and Verlet integration

The Lennard-Jones Potential is a mathematically simple model that approximates the interaction between a pair of molecules[4]. The Lennard-Jones Potential formula is expressed as follows:

$$V_{LJ} = 4\varepsilon\left[\left(\frac{\sigma}{r}\right)^{12} - \left(\frac{\sigma}{r}\right)^{6}\right] \qquad (3)$$

where ε is the depth of the potential well, σ is the finite distance at which the inter-molecule potential is zero, r is

DOI: 10.1051/matecconf/201710812002

the distance between the molecules. The attractive r^{-6} term in Eq. (3) represents the long-range behavior of the van der Waals force between molecules. The attractive r^{-12} term represents repulsive particles in this simulation system. force. This is implemented to calculate the potential between molecules. Verlet integration is a numerical method used to integrate Newton's equations of motion[5,6]. It is used to calculate trajectories of molecules

2.6 Implementing the cell list method

As algorithm that scales with the number of molecules is the cell list method[7,8,9,10]. The idea of the cell list is illustrated in Fig.2. Region of nozzle is divided into cells with a size slightly larger than the cutoff radius r_c. Each particle in a given cell interacts with only those particles in the same or neighboring cells. Since the allocation of a particle to a cell is an operation that scales with the number of molecules and the total number of cells that needs to be considered for the calculation of the interaction is independent of the system size, the cell list method scales as with the number of molecules. The cell list method is implemented to save CPU times.

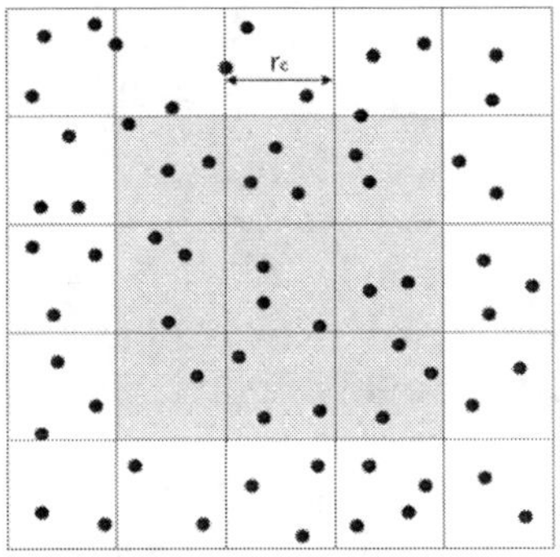

Figure 2. The Cell List (in 2D, 9 neighboring cells)

3 Result of the simulation

3.1. Running the simulation

The nozzle experiment is shown in Figure3. The number of molecules in regions A and B is 8000 and 4000, respectively and the random velocity generated is around 10m/s ~ 15m/s. As starting the simulation, molecules in region A become equilibrium state after 1 nanosecond and start to pass the nozzle throat. Generated molecules moving in region B for static pressure at the exit play well as obstacles. After 1.5 nanosecond, we can see the moving flow by moving the group in nozzle but collide soon with the random molecules. Static pressure in the line M is lculated by the (1). In low pressure area B, we observe that very active clustering occurs and see clear clusters as shown in Fig. 3, after 2 nanosecond. They are moving to the exit area as colliding with neighboring particles. Hardware specification for running the simulation is shown in Table 1.

Table 1. Hardware specification

	System Information
CPU	intel i-7 extream 5960X (haswell-E)
RAM	DDR4 PC4-22400 CL15 (8Gx2) x2 = 32GB
GPU	NVIDIA GEFORCE GTX1080 super JETSTREAM D5X 8GB

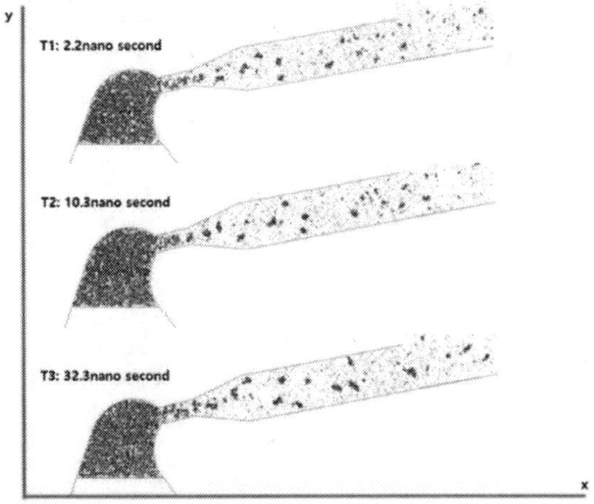

Figure 3. Running the simulation

3.2 Case study according to vacuum levels

The most important target of this simulation is analyzing static and dynamic pressure changes and their general trend in case study of different pressure configuration in regions A and B. The experimental sets are shown in Table 2 for the configuration. To see static pressures of these regions, we directly compare P_{2d} in Figure4. The graph shows that the pressure in the arc L is higher than the pressure in the line M. the pressure difference of the arc L and the line M is more than 10 time in the nozzle.

Figure 5 shows static pressure comparison at L part, the semi-circle before throat of nozzle. The graphs show general trend of pressure in converging part of the nozzle. The more number of molecules, the more number of collisions are detected in the simulation.

Figure 6 shows that static pressure comparison at M part, the line part after throat of nozzle. The graphs show general trend of pressure in diverging part of the nozzle. The more number of molecules, the more number of collisions are detected in the simulation. Different from Figure 5, however, trend of graph according to the number of random molecules for back pressure setting shows interesting results.

Figure 7 shows that dynamic pressure comparison at N part, the vertical line part near the exit of the nozzle. The graphs show general trend of dynamic pressures that can convert useful power. The more number of molecules, the more number of collisions are detected in the beginning stage. However, according to the number of random molecules for back pressure setting, very different results are acquired. When random molecules of back pressure collide with directional molecules, the dynamic pressure reduced. Even though this simulation setting can be ideal, it is evident that the vacuum level affect to the velocity of exit. The numerical setting is that for each case, they have the same number of molecules that is they have same difference of pressure

Table 2 Number of molecules for regions A and B

Case	8000 by 4000	7000 by 3000	6000 by 2000	5000 by 1000	4000 by 0
Symbol	——	......	— —	—— · ·	▬ ··

Captions should be typed in 9-point Times. They should be centred above the tables and flush left beneath the figures.

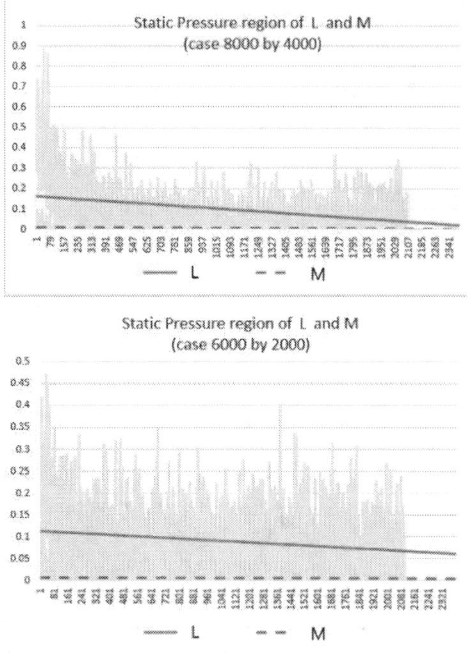

Figure 4. Static Pressures of regions L and M

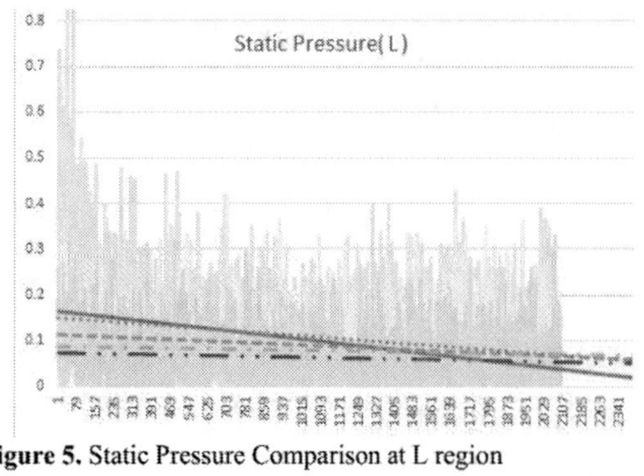

Figure 5. Static Pressure Comparison at L region

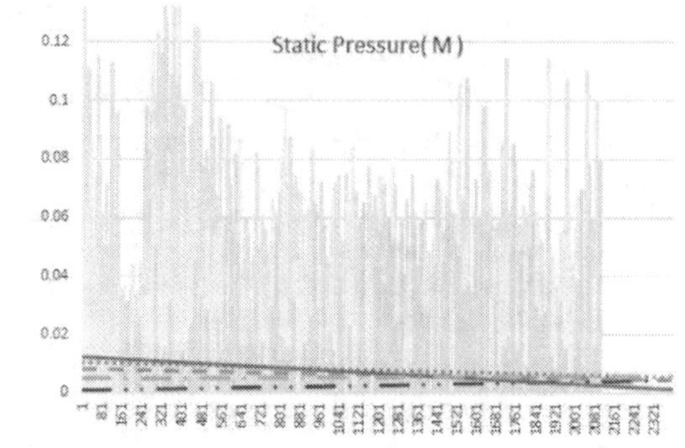

Figure 6. Static Pressure Comparison at M region

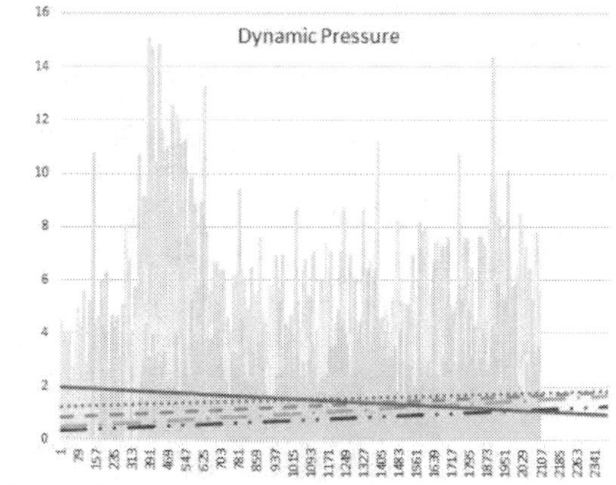

Figure 7. Dynamic Pressure Comparison near N region

4 Review of the clustering and increasing mean velocity

4.1 Clustering in the nozzle

When high-pressured liquid expands to vacuum region, it is difficult to think for fluid dynamics researchers that the molecules are gathered each other and become state like liquid. However, in the experiments in high-degree of vacuum space, we can observe a new phenomenon of clustering. The clustering is made by the Lennard-Jones potential, which is weak attraction between near molecules. If the molecules are expanded to vacuum but governed by specially designed shape of wall, they close to each other in same directional flow, so that they gather and are weakly bonded. Fig.3 shows well developed clusters. Nuclear physics researchers use the clustering phenomenon to major relative velocity of ion-beam[11]. In view of conservation of momentum, it is expected that if the clustered molecules collide with fixed object, then they bound with scattered gas. However, if they collide to moving object with relatively long time, they might lose the momentum and reduce their velocities. Fig.3 and Fig.7 show the degree of clustering according to vacuum level. Degree of clustering related to moving mass increase without loss of velocity by collision in equilibrium space.

4.2 Increase of average velocity of the flow

The average velocity increase appeared in Fig.8, accordin g to vacuum level. In real physics, pressure in near exit of nozzle means disturbing the flow by the random moving. Therefore the collision less moving group gives possibilit y of converting power by the rotating turbine and electric generator.

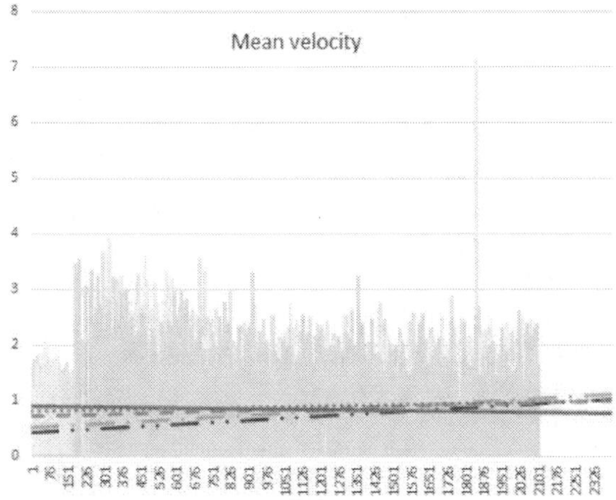

Figure 8. Mean Velocity Comparison near N region

5 Conclusion

We simulated the pressures and velocities of moving molecule in a nozzle to approach each molecule's velocity. Different from general molecular study of high-pressure jet, we performed case study of nozzle eruption according to degree of vacuum level. The result shows that the velocities are increased at the nozzle exit as the vacuum level gets higher in that area because less molecular collisions.

6 Future works

To use this study in engineering scale, enlarging of nozzle size at least a micro level, and modifying the shape of nozzle for better performance are essential. So, we started developing for the code in graphic user interface environment. In diverging part of the nozzle, we are going to investigate relationship between diverging shape and speed of the flow to out-normal direction. According to the increasing nozzle size, a huge number of molecular collisions are inevitable problem so that we plan to acclerate the calculation code with GPU before using supercomputer.

References

1. S. Khurana, Navtej and Hardeep Singh, "Effect of Cavitation on Hydraulic Turbines-A Review," INPRESSCO(2012).
2. M. Parlange, M. Farhat, O.Minster, D.Obreschkow, J.R. Thome, "The effect of pressure gradient on the collapse of cavitation bubbles in normal and reduced gravity," 22-23, 2013
3. D. John Cutnell and W. Kenneth Johnson, "Essential of Physics," John Wiley & Sons, Inc., 2006.
4. Lennard-Jones, "On the Determination of Molecular Fields", *Proc. R. Soc. Lond. A*, **106** (738): 463–477, 1924
5. Verlet, Loup, "Computer "Experiments" on Classical Fluids. I. Thermodynamical Properties of Lennard−Jones Molecules". *Physical Review*. **159**: 98–103, 1967.
6. V. Daniel Schroeder, "Interactive molecular dynamics,"Am. J. Phys., Vol. 83, No. 3, 210-218, March 2015.
7. D. Frenkel and B. Smit, Understanding Molecular Simulation(Academic, New York, 2002).
8. B. R. Holstein,"The van der Waals interaction," Am. J. Phys. 69(4), 441–449 (2001).
9. K. A. Milton, "Resource Letter VWCPF-1: van der Waals and Casimir-Polder forces," Am. J. Phys. 79(7), 697–711 (2011).
10. G. C. Maitland et al., Intermolecular Forces: Their Origin and Determination (Clarendon Press, Oxford, 1981). This monograph makes an exhaustive assessment of the limitations of the Lennard-Jones 6-12 potential.
11. A. Giannuzzi; F.A. Stevens (2004). *Introduction to Focused Ion Beams: Instrumentation, Theory, Techniques and Practice. Springer Press*

Interactions of double slanted cracks under mode I loading

Al Emran Ismail, Muhd Hafeez Zainulabidin, Fazimah Mat Noor, Hanani Abd Wahab, Maznan Ismon, Zaleha Mohamad, Mohd Amran Madlan

Crashworthiness and Collision Research Group (COLORED), Mechanical Failure Prevention and Reliability (MPROVE), Faculty of Mechanical and Manufacturing Engineering, Universiti Tun Hussein Onn Malaysia, Batu Pahat, 86400 Johor, Malaysia

Abstract. This paper presents numerical analysis of double slanted central cracks in square plates under mode I tension stress. ANSYS finite element program is used to model and solve the problem of slanted cracks where the stress intensity factor is determined according to J-integral approach. Three relative crack length, a/W = 0.1, 0.2 and 0.3 is used while relative spacing between the cracks, s/2a are ranged between 0.1 to 0.6. In order to study the roles of slanted cracks on the stress intensity factor, several slanted angles are used such as 0°, 15°, 30°, 45°, 60° and 75°. According to the numerical analysis, all parameters above played an important role in determining the stress intensity factors. As expected, longer crack produced higher stress intensity factor. For the cracks with a/W $\leq$ 0.2, insignificant crack interaction is observed however when crack reached a/W=0.3, the interaction is gradually increased. Increasing the slanted angles slightly increased the interaction factor however the effect of relative crack spacing became insignificant.

1 Introduction

The formations of cracks are strongly depend on the type of materials, loading, environmental aggression and design condition of mechanical components. In general, crack grown perpendicularly to the axis of loading however under circumstances, crack also propagates obliquely especially as a result of hydrogen induced cold crack after the metal welding. In this case, most of cracks appeared not in a single crack but in the form of multiple cracks.

Tremendous amount of works can be found in analysing the normal cracks in open literature [1-3]. For examples Matsumto et al. [1] evaluated the stress intensity factor of normal crack along two different materials using an interaction energy release rate and boundary element method. In order to indicate the reliability of their method, some numerical models are used and compared. The present method capable to produce more accurate results with a coarse mesh than the method based on the displacement extrapolation which is based on the boundary element method.

Hammond and Fawaz [2] reviewed stress intensity factors of various size single edge-cracked tension specimens. Finite element method is used to calculate the stress intensity factors for wide ranges of crack and sample geometries. Comparison is also performed between the existing and their results generally satisfactory correlation and large differences about 12.7% is observed for short cracks in small plate aspect ratio.

Ismail et al. [3] investigated stress intensity factors of double edge cracks in large groove plate under mode I tension. They used finite element method to determine the stress intensity factor via domain integral method. It is found numerically that the SIFs are strongly affected by the relative crack depth and the groove geometries. It is also found that the large groove is capable of reducing the SIFs in comparison with the circular notched due to lower stress concentration factors.

Several works discussed on the multiple normal cracks can also be found in [4-8]. For examples Yan and Miao [4] studied the interaction of multiple cracks in a rectangular plate using boundary element method. They reported that the boundary element method is simple yet accurate for determining the stress intensity factors of multiple crack problems. While, Yang and Soh [5] developed finite element method using complex potential and the conformal mapping technique to study the multiple crack problems. It is found that the method developed useful for modelling the crack interactions between many cracks and defects.

Shu et al. [6] studied the problems of three cracks on both edges of finite width sheet under mode I loading. They modelled the multiple cracks using finite element method and due to symmetrical effect only half of model is developed. They found that when there are several cracks existed, the flexibility of the plate increased and the stress intensity factors at the crack tips decreased.

Ismail [7] studied the stress intensity factors of three parallel edge cracks under bending moments. Cracks are modelled using finite element method and different relative crack length and spacing between cracks are used. It is found that due to the presence of multiple crack edge cracks, the stress distribution is relaxed and therefore, the stress intensity factors for all cracks decreased and when the distance between the cracks is increased, interaction is seemed to be diminished and it is can be neglected especially for swallow crack. The behaviour of offset crack under mixed mode loading can also be found in [8].

In term of slanted cracks, several works are found such as [9-11]. Kuang and Chen [9] used a displacement extrapolation method to investigate the mixed mode crack problems under mode I loading. There are two important parameters are used such as crack length and size of element at the crack tip. However, the crack is only slanted at limited angle then it is hard to understand the role of angles on the stress intensity factors.

Albinmousa et al. [10] provided solution of stress intensity factors for mixed mode I-II single edge notched tension specimen. Wide range of crack geometries and inclined angles are used and stress intensity factors are determined based on such parameters. Then, curve fitting is performed to simplify the stress intensity factors for prediction purposes. They found that their model can be used to predict the stress intensity factors very well. Another study on the slanted crack can be found in [11-13]. However, this work focused on the three dimensional cracks.

However, there are lack number of publications available on the multiple slanted cracks. Therefore, this paper presents the numerical analysis on the double slanted cracks embedded in a plain strain plate under mode I tension pressure. Double slanted cracks are positioned in the central of 100×100mm plate where the relative crack length, a/W = 0.1, 0.2 and 0.3. The cracks are also oriented using different angles such as 0°, 15°, 30°, 45°, 60° and 75°. The spacing distance between two slanted cracks are varied in the range of 0.1 to 0.7. The stress intensity factors calculated from single and multiple cracks are analysed with the relations of parameters mentioned previously.

2 Finite element modelling

ANSYS finite element program is used to model and calculate the stress intensity factors (SIF) at the crack tips. In this work, SIFs are based on the J-integral calculated around the several integral contour around the crack tip. Figure 1 shows the schematic diagram of the double slanted cracks in square plate under mode I pressure with the assumption that the plate is fulfilled the plain strain condition. The dimension of plate is height, H = 100mm while width, W = 100mm. On the other hand, crack length, 2a is varied by normalizing with W to produce a/W = 0.1, 0.2 and 0.3. The distance between slanted cracks, 2a/S are also varied between 0.1 and 0.6. The orientations of cracks are also varied in order to study the effect of such angles on the crack behaviour. The values of angles used are 0, 15°, 30°, 45°, 60° and 75°. These angles are then normalized against the angle of 90°.

Two types of cracks are used such as normal and slanted cracks as in Figures 2 and 3, respectively. For normal cracks, quarter model is used to take an advantage of symmetry while full model approach is implemented to model the slanted cracks. SIFs obtained numerically are then normalized according to Equation (1).

$$F = \frac{K}{\sigma\sqrt{\pi a}} \quad (1)$$

where, F is a normalized SIF or geometrical correction factor, K is a stress intensity factor, σ is tension stress and a is a crack length. Before the model is further used, it is a crucial task to compare the current with the existing models. Figure 4 indicates the validation of normal cracks when compared with the results obtained from Murakami [12]. For this kind of cracks, the present model is well validated and therefore it can be used for further analysis. In order to analyse the crack interaction, another factor is considered as in Equation (2).

$$\psi = \frac{K_{\text{two-crack}}}{K_{\text{single-crack}}} \quad (2)$$

where $K_{two\text{-}cracks}$ is stress intensity factor obtained from interacting cracks and $K_{single\text{-}crack}$ is stress intensity factor of single crack. According to Figure 1, there are four crack tips. Based on the preliminary work, the stress intensity factors at tip A is equal with tip D and on the other hand the stress intensity factor at tip B is similar with tip C. Then, for the sake of simplification on the stress intensity factors at crack tips A and C are presented and analysed.

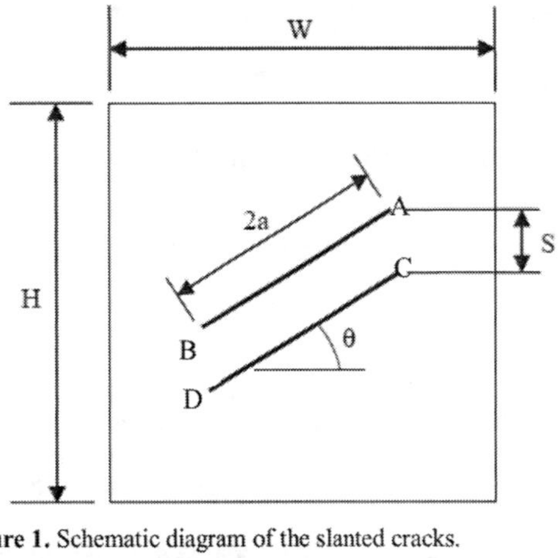

Figure 1. Schematic diagram of the slanted cracks.

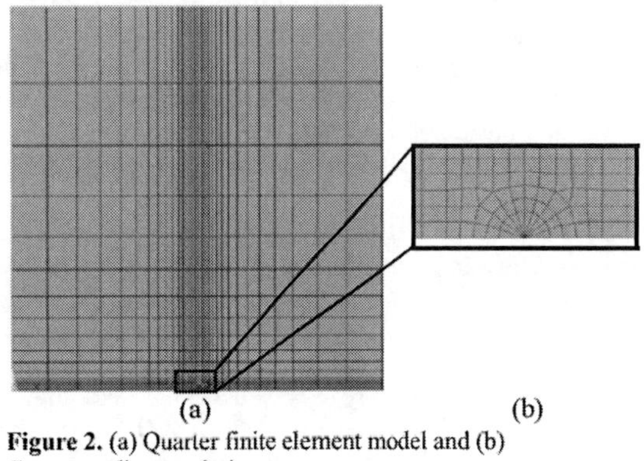

Figure 2. (a) Quarter finite element model and (b) Corresponding crack tip.

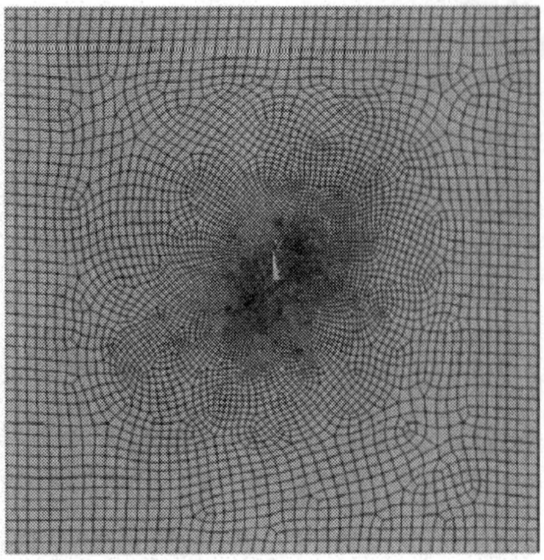

Figure 3. Double slanted crack.

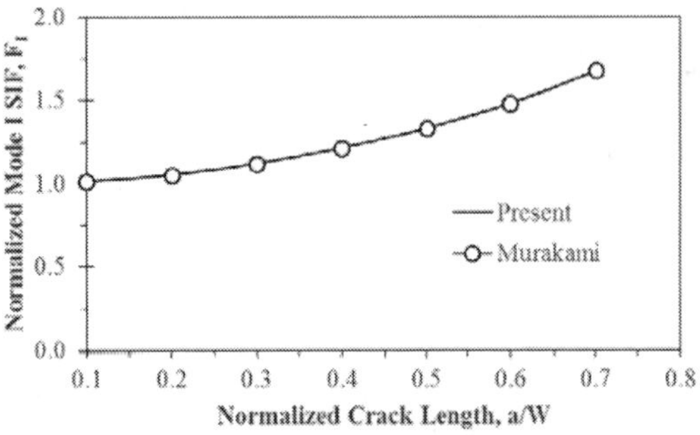

Figure 4. Validation of normal cracks.

3 Results and discussion

3.1 The effect of crack spacing on the interaction factors

Figure 5 shows the effect of the relative crack spacing on the interaction factors of different relative crack length when different slanted angles are used for a crack tip A. Figure 5(a) reveals the interaction factors of normal cracks which is the principal stress is perpendicular to the cracks. It is indicated that the interaction are almost uniform when crack length is increased since the stress around the crack is uniformly distributed. However when slanted angles are introduced, the effect of crack length is obvious as in Figure 5(b)-5(d) where for the crack length less than 0.2, the factor of interaction are almost similar indicating the shielding effect. Whereas for crack length greater than 0.3, the interaction factor increased gradually however it is still lower than the case of normal cracks. This is due to the fact that when the cracks are inclined, the capability of mode I loading to open the cracks are insignificant since the formation of mode I stress intensity factors are strongly related with such angles. Higher slanted angles higher shear stress are induced resulting lack of capability to open the crack faces. Figure 6 indicates the behaviour of interaction factors at the crack tip B. The role of such interactions are considerably similar however it is lower than the interaction for the case of crack tip A.

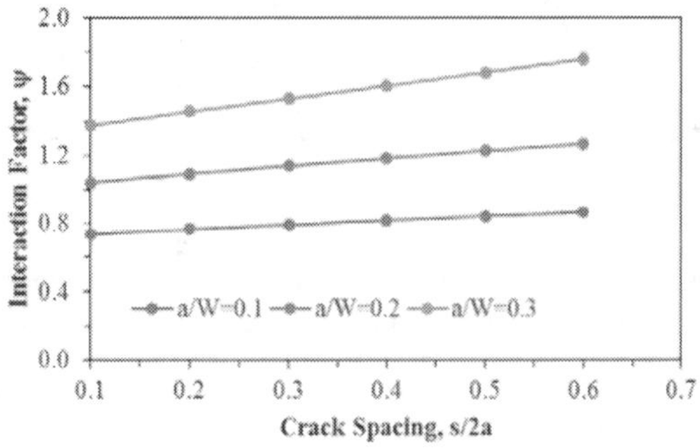

(a)

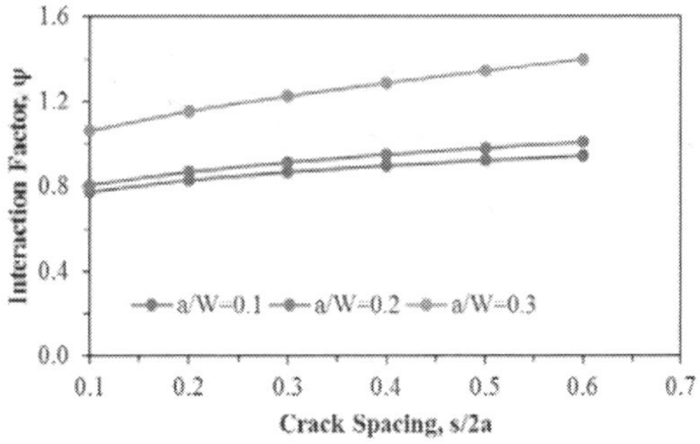

(b)

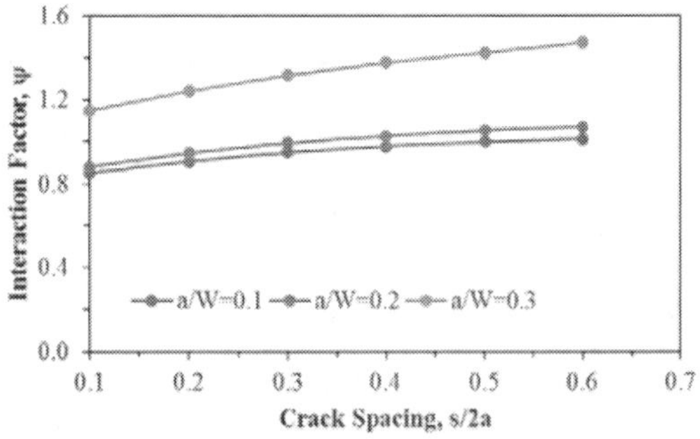

(c)

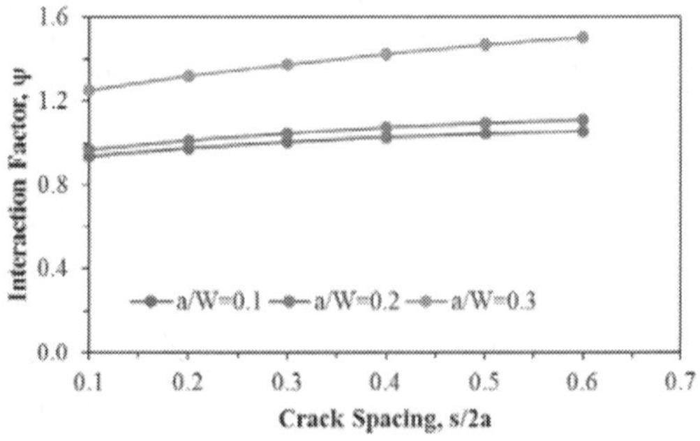

(d)

Figure 5. Effect of crack spacing on the interaction factor of (a) 0°, (b) 15°, (c) 45° and (d) 60° slanted angle cracks for crack tip A.

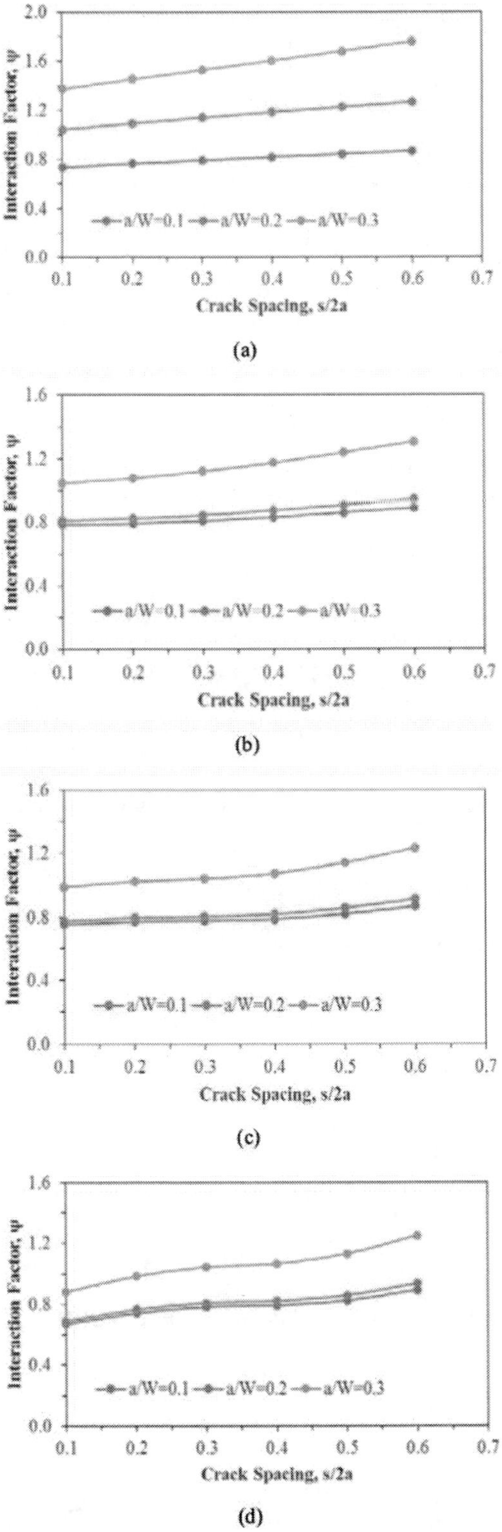

Figure 6. Effect of crack spacing on the interaction factor of (a) 0°, (b) 15°, (c) 45° and (d) 60° slanted angle cracks for crack tip C.

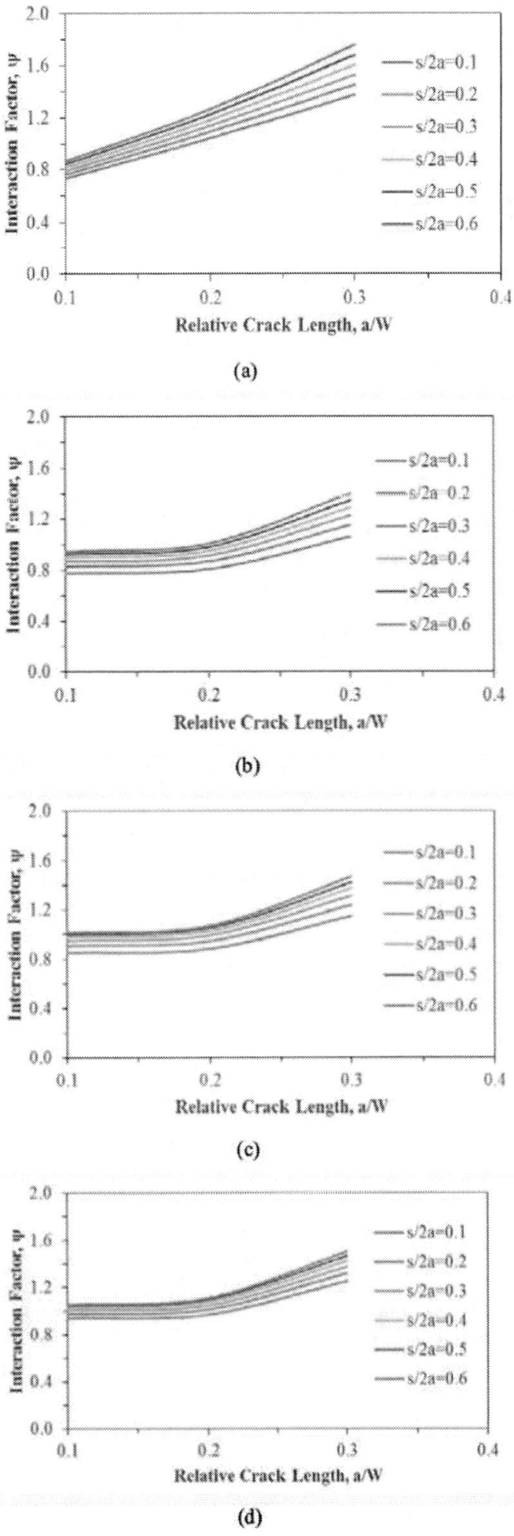

Figure 7. Effect of relative crack length on the interaction factor of (a) 0°, (b) 15°, (c) 45° and (d) 60° slanted angle cracks for crack tip A.

3.2 The effect of relative crack length on the factor of crack interactions

Figures 7 shows the roles of relative crack length on the interaction factors when different slanted angles are used for crack tips A and C, respectively. In general for all cases of crack tips, higher the distance between the cracks produced higher interaction factors. It is indicated that if the cracks are closed together shielding effect is produced while if the cracks are positioned away from each other, higher interaction factors are produced. It is indicated that the amplification effect is induced. In addition for crack length of 0.1, the interaction factors are insignificantly varies however when relative crack length increased,

Figures 7(a) and 7(b) show the interaction factors of normal cracks. It is revealed that such interaction increased as relative crack length increased. Every increment of relative crack distance, increased the interaction factors indicating the existence of amplification effect. It is occurred when the cracks are away from each other and their stress intensity factors became similar to those of single cracks. It is also indicated that when crack length increased, the interaction factors are also increased. On the other hand, the interaction factors diverged slightly as relative length is increased. This is due to the fact that when longer crack length is used, the flexibility of the ligament between the cracks is lowered and therefore giving an advantage for the cracks to widely open.

The effect of slanted angles on the interaction factors can be observed in Figure 7(b)-7(d). Additionally, the distribution of interaction factors are strongly related to the positions of crack tips. For the case of crack tip A if the slanted angle is increased, the interaction factor is also increased slightly. This is due to the increment of shear force when the cracks are tend to the right angle. Insignificant effect on the interaction factors are observed when different relative crack distances are used. It is believed that at the crack tip A, the flexibility around the tip is unchanged leading to the insignificant variations of interaction factors.

Whereas at the crack tip C, the behaviour of interaction factor is significantly different where increasing the slanted angles have dispersed the distribution of such factors. This is probably due to the fact that the crack tip C is located below a crack tip A. therefore varying the relative crack distances varied the flexibility of the ligaments between them and then affecting the stress intensity factors.

4 Conclusion

In general, the interaction factors of normal cracks under mode I loading can be easily predicted since the stress intensity factors are the function of relative crack length. However for the case of slanted cracks, the crack interactions strongly related with the relative distance among them. If the cracks are positioned close to each other shielding effect took place and if it is far away amplification effect is induced. On the other hand, position of crack tips are also played an important role since each tip produced considerably different characteristics.

Acknowledgement

Author acknowledges Universiti Tun Hussein Onn Malaysia for financially supporting this work.

References

1. T. Matsumto, M. Tanaka, R. Obara. Engineering Fracture Mechanics **65**, 683-702 (2000)
2. M.J Hammond, S.A Fawas. Engineering Fracture Mechanics **153**, 25-34 (2016).
3. A.E Ismail, MA Azlan, AL Mohd Tobi, MH Ahmad. IOP Conf. Series: Materials Science and Engineering **152** 012049 (2016).
4. X. Yan, C. Miao. Applied Mathematical Modelling **36**, 5727-5740 (2012).
5. C.H Yang, A.K Soh. Theoretical and Applied Fracture Mechanics **38**, 81-101, (2002).
6. H.M Shu, J. Petit, Z.D Jiang, G. Bezine. Engineering Fracture Mechanics **45**, 407-414 (1993).
7. A.E Ismail. IOP Conf. Series: Materials Science and Engineering **50**, 012020 (2013).
8. A.E Ismail, A.K Ariffin, S. Abdullah, M.J Ghazali. International Journal of Automotive Technology, **12**, 225-232 (2011).
9. L.S Chen, J.H Kuang. International Journal of Fracture **57**, 51-58 (1992).
10. J. Albinmousa, N Merah, S.M.A Khan. Engineering Fracture Mechanics **78**, 3300-3307 (2011).
11. A.E Ismail, A.L Mohd Tobi, N.H Mohd Nor. AIP Conference Proceedings **1660**, 070027 (2015).
12. Y. Murakami, Stress Intensity Factors Handbook Volume 1, Pergamon Press (1987).
13. A.E Ismail, A.K Ariffin, S. Abdullah, M.J Ghazali. HKIE Transaction **18**, 13-18 (2011).

A Relationship between Tyre Pressure and Rolling Resistance Force under Different Vehicle Speed

Apiwat Suyabodha

Department of Automotive Engineering, Rangsit University, Lak-hok, Pathumthani, Thailand

Abstract. Tyres are a final output torque of the vehicle before all the forces are distributed to the road surface as a tractive force. When the tyre pressure changed, it had a direct effect on rolling resistance force of the vehicle and reduced the tractive force. Moreover, it had an effect on a rate of fuel consumption and driving comfort. This research had tested the effect of tyre pressure of 25 psi and 45 psi to determine the total resistance force between the vehicle speed from 40 km/hr to 100 km/hr and compare with the reference tyre pressure of 35 psi by using the "coast-down technique". The results from calculation showed the total resistance force when the vehicle speed was zero from tyre pressure of 25 psi increased by 48.52% from reference value and the total resistance force from tyre pressure of 45 psi decreased by 13.46% from the reference tyre pressure. The different resistance force from tyre pressure of 35 psi with the other 2 values of tyre pressure in the research were called the different rolling resistance force because the vehicle changed only tyre pressure which had no effect on aerodynamics drag. The relationship of the different rolling resistance force along the test speed from tyre pressure of 25 psi and 45 psi decreased while the vehicle speed increased. The relationship can be fitted with a linear equation. The slope from tyre pressure of 25 psi was 0.2358 which was steeper than the slope of 0.0828 from 45 psi tyre pressure. These were the effect of tyre characteristic that made from rubber. When the vehicle speed was low, the frequency of tyre deformation was also low and resulted in high energy loss from hysteresis phenomena. Therefore high rolling resistance force was generated. Considering the effect of tyre under high vehicle speed, the frequency of tyre deformation increased and made the molecule of rubber vibrate more than low vehicle speed. These effects made the tyre stiffer than low frequency and resulted in decreasing the rolling resistance force. In additional, high frequency that caused by high tyre pressure had effected on driving comfort in a real driving condition.

1 Introduction

A vehicle uses an internal combustion engine as a power unit to produce a torque for pushing the vehicle body moving forward. The torque from engine is generated by pressure from burning fuel and air in the combustion chamber during power stroke [1,2]. Then, the pressure pushes the piston move down to the bottom dead centre. Next, the piston moves up to the top dead centre for release the exhaust gas. The process continues to draw the fuel and air in the combustion chamber and the mixture of air and fuel are compressed in the compression stroke for the next power stroke. The cycle of combustion make the piston move up and move down in the combustion chamber. The movement of piston is converted to rotation by crankshaft and generates output torque (T_{engine}) as can be expressed by (1)

$$T_{engine} = mep \cdot V_d / (6.28 \cdot n_R) \quad (1)$$

where

mep = mean effective pressure (Pa)

V_d = engine displacement (m^3)

n_R = number of crank revolution for each power stroke (for 4-stroke engine is equal to 2)

Torque from the engine is transmitted to a gearbox for developing a sufficient torque for moving and accelerating the vehicle from standstill to the design speed [3]. In gearbox, there are different size of gears that are meshed in series for varying torque and the speed of vehicle as explain in (2) and (3).

$$T_{wheel} = T_{engine} \cdot i_{total} \quad (2)$$

$$n_{wheel} = n_{engine} / i_{total} \quad (3)$$

where

T_{wheel} = output torque to the wheel (N.m)

n_{engine} = engine speed (rpm)

n_{wheel} = wheel speed (rpm)

i_{total} = total gear ratio

The output torque for moving the vehicle forward on the flat road surface needs to meet a minimum resistance force as showed in Fig 1. which is a combination between an aerodynamics drag (Rd) and a rolling resistance force (R_r) following (4) and (5) respectively [4];

$$R_d = (1/2) \cdot \rho \cdot C_d \cdot A \cdot v^2 \quad (4)$$

$$R_r = K_r \cdot W \quad (5)$$

where;

ρ = density of air (kg/m^3)

C_d = coefficient of aerodynamics drag

A = frontal area of vehicle (m^2)

v = vehicle speed (m/s)

K_r = coefficient of rolling resistance

W = weight of vehicle (N)

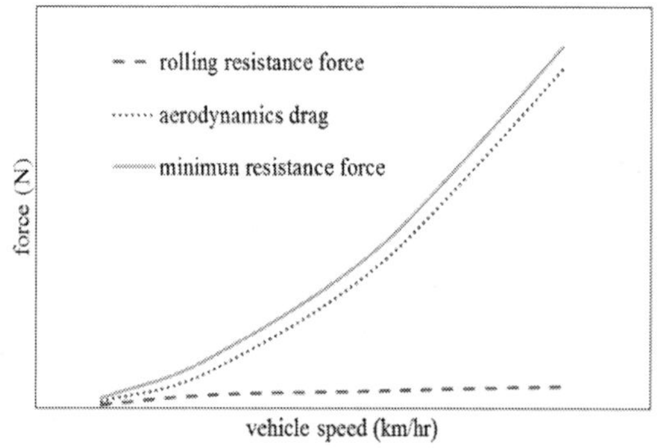

Figure 1. A minimum resistance force for driving a vehicle at constant speed which is a combination between rolling resistance force and aerodynamics drag.

Considering the rolling resistance force, it is a force that acts on the opposite of the direction of tyre motion. The rolling resistance force is not constant and mainly depends on tyre speed, tyre pressure and properties of tyre as it made from rubber. Fig 2. showed the mechanism of tyre which is rolling on the flat road surface, the front edge of tyre contact area has high stress compare to the tailing edge. These results of the total reaction force (F_z) that locates in the front edge of the tyre contact area and generated counter moment in the opposite direction of the tyre motion [5,6].

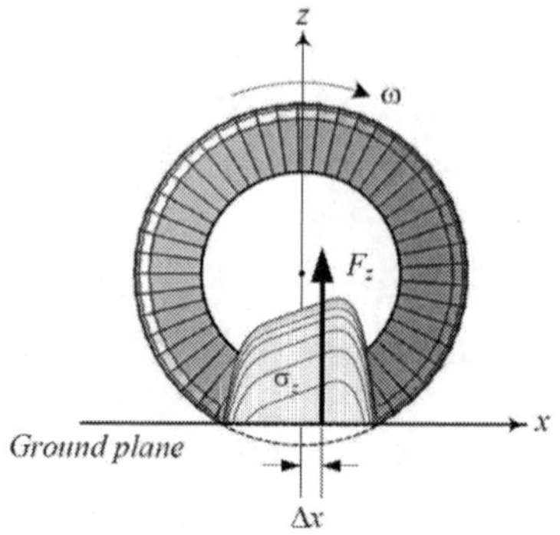

Figure 2. Side view of normal stress distribution of tyre [5]

In 1999, the research proved that the coefficiency of rolling resistance reduced while the tyre pressure increased. The relationship of both parameters was fitted with an exponential function [7]. Another research focused on measuring the friction torque of vehicle tyre and claimed the friction torque which referred to the rolling resistance force reduced while the vehicle speed increased under constant tyre pressure [8-10].

In additional, the rolling resistance force is defined as an energy loss from tyre deformation. When the tyres rotate on the road surface under loading, the contact area as showed in Fig 2. is compressed and cause tyre deformation. Under unloading condition, the tyre is uncompressed and returns to original position. Considering the energy as showed in Fig 3., the deformation force for tyre under loading condition is greater than the deformation force for unloading condition. The area in the loop is the hysteresis loss [11].

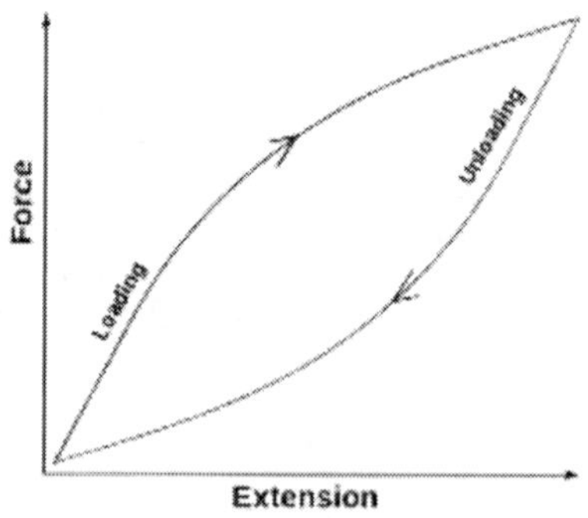

Figure 3. Hysteresis Loop of tyre [11]

Low tyre pressure has hysteresis loss more than high tyre pressure because the tyre has high deformation. The relationship of hysteresis loss proved that the loss gradually reduced when the tyre pressure increased [12]. Moreover, low tyre pressure had low tyre stiffness because the molecule of rubber vibrates more than high tyre pressure [13]. The stiffness increased as a linear

DOI: 10.1051/matecconf/201710812004

function with the tyre pressure [12]. In the other word, low hysteresis loss means low rolling resistance force which results in increasing the vehicle power but the tyre is very stiff so it has an effect on driving comfort.

2 Aim of research

The research is aimed to study the effect of tyre pressure on the rolling resistance force by comparing with the tyre pressure that giving by the automotive manufacturing as a reference value. The coast-down technique had been used to determine the resistance force of the vehicle. The research also narrowed down to the result and discussion on the force of different tyre pressure under the same vehicle speed which is difference from the previous research.

3 Coast - down technique

The test in this research was carried out by using "coast-down technique" [14]. The basic principle of this technique is to accelerate the vehicle on the flat road to a predetermined speed and let the vehicle decelerate in neutral gear down to the predetermined speed and record the time during decreasing speed. The total resistance force (Ftotal) is calculated from Newton's second law by using the vehicle weight and the different speed and time according (6)

$$F_{total} = m \cdot \Delta v / \Delta t \qquad (6)$$

where;

m = weight of vehicle (kg)

v = vehicle speed (m/s)

t = time (s)

This technique is used for estimation the total resistance force which is the aerodynamics drag and the rolling resistance force. This research applied the technique for estimation the rolling resistance force. As the vehicle was the same condition, the tyre pressures were the only factor that was changed. Therefore the different resistance force from tyre pressure were claimed to be the changing of rolling resistance forces by this technique.

4 Methodology

The research used the pick-up truck vehicle that had front engine with 2500 c.c. included turbocharged system. The transmission was a real wheel drive with a manual gearbox. The engine had 4 cylinders with water cooled system. The total weight of vehicle was 1560 kilograms. The size of tyre was 215/70 R15.

The test condition used the flat road surface which made from asphalt. The test was carried out when the speed of atmosphere wind was zero in order to avoid the extra force from aerodynamics drag. The test procedure was;

1) Set up the tyre pressure to 25 psi for all 4 tyres. Start the engine and run the vehicle until the speed achieved 110 km/hr and then disengaged the gearbox from engine by put the gear to neutral position and let the vehicle slow down by itself. Start to record the time when the speed of vehicle was 100 km/hr and recorded the time every changing speed of 5 km/hr until the vehicle speed slow down to 40 km/hr.

2) Repeat the test by changing the tyre pressure to 35 psi according to the vehicle manufacturing and 45 psi respectively.

4 Results and discussions

According to the coast-down test, the vehicle speed decreased very fast with the tyre pressure of 25 psi but under the tyre pressure of 45 psi, the vehicle speed decreased slightly slower than 35 psi trye pressure. The relationships between time and vehicle speeds were plotted and were fitted with exponential equations as shows in Fig 4.

The time for decreasing the vehicle speed from 100 km/hr to 40 km/hr with the tyre pressure of 25 psi, 35 psi and 45 psi were 58.90 seconds, 78.69 seconds and 85.57 seconds respectively. For more analysis, the results in Fig 4 had been converted to the total resistance force and fitted with the exponential equations as can be seen from Fig 5.

The exponential equations showed the value of resistance forces which were the combination between aerodynamics drag and rolling resistance force at different vehicle speed. When the vehicle speed was zero, the line of relationship crossed the total resistance force axis which represented only the rolling resistance force of the tyre. With zero vehicle speed, the value of rolling resistance force from 35 psi tyre pressure was 86.489 N. Under the tyre pressure of 25 psi, the exponential showed the results of the rolling resistance force increased by 48.52% while the rolling resistance force from tyre pressure of 45 psi decreased by 13.46%. Considering the different resistance force by using the 35 psi as a reference value of tyre pressure from the vehicle speed of 40 km/hr to 100 km/hr in Fig 6. The different values represented the only changing in rolling resistance force for each vehicle speed because changing tyre pressure had no effect on aerodynamics drag.

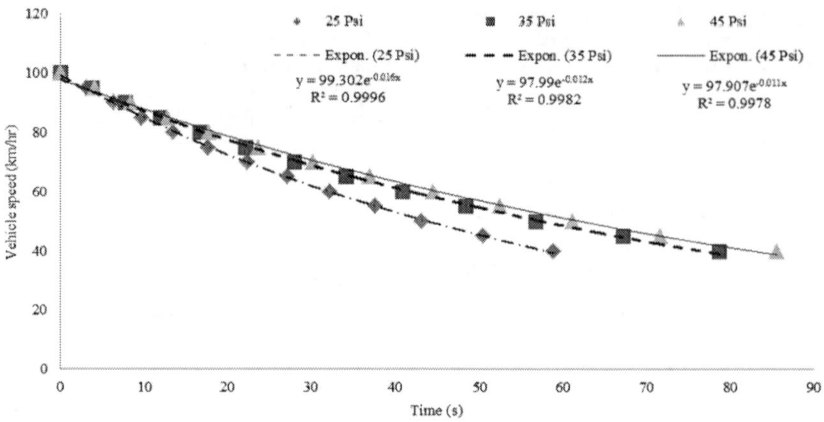

Figure 4. A relationship between time and vehicle speed under different tyre pressure.

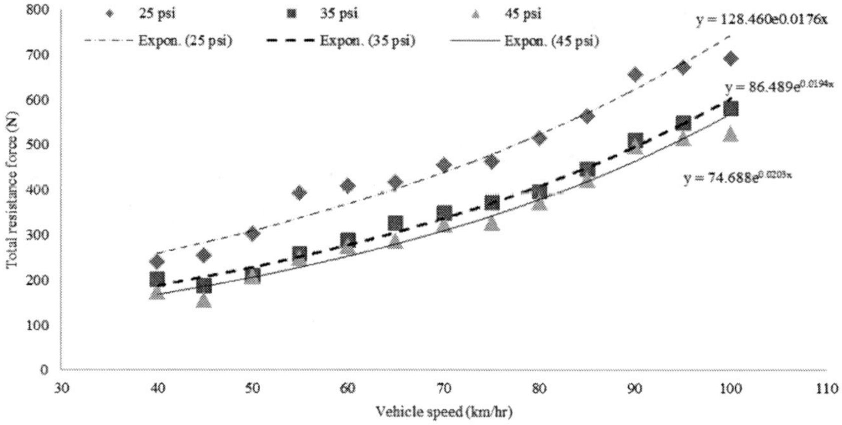

Figure 5. A relationship between vehicle speed and total resistance force under different tyre pressure.

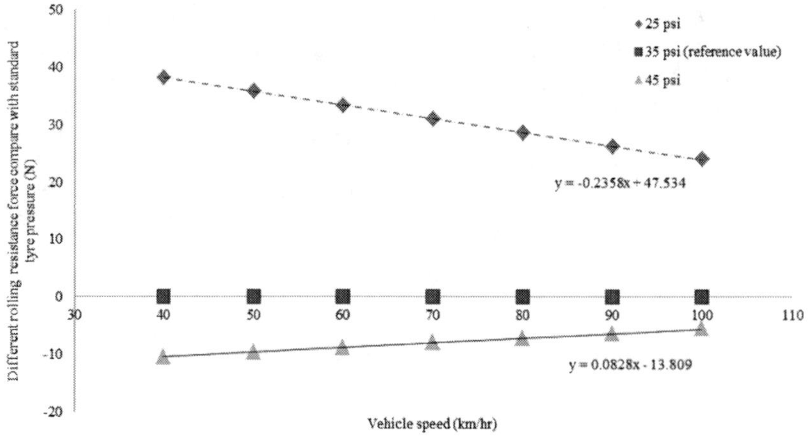

Figure 6. The different resistance force compare to the resistance force from reference tyre pressure of 35 psi

The different resistance force of the trye pressure of 25 psi and 45 psi in Fig 6. were fitted with linear equations while the tyre pressure of 35 psi was used as a reference value. Under 25 psi of tyre pressure, the rolling resistance force decreasing with the slop of 0.2358 while the slope of 45 psi tyre pressure was only 0.0828. This phenomenon was the effect of tyre properties and the rate of tyre deformation. Under low vehicle speed, the rate of tyre deformation was high which increased hysteresis loss and resulted in high rolling resistance force. When the vehicle speed increase, the molecules of tyre moving faster than low speed which made the tyre stiffer than low vehicle speed. In additional, high tyre stiffness results in low hysteresis loss. These physical phenomenon had high effect on 25 psi of tyre pressure but under trye pressure of 45 psi, the tyre was already stiffer than the other pressure. The results also proved that the stiffness of the tyre increased with the vehicle speed.

5 Conclusion

This research was focused on the effect of tyre pressure and rolling resistance force. The coast-down technique was used to determine the resistance force of vehicle under tyre pressure of 25 psi and 45 psi by comparing with the reference tyre pressure of 35 psi which was set from the vehicle manufacturing. The results showed the rolling resistance force increased 48.52% when the pressure of tyre was 25 psi. Under tyre pressure of 45 psi, the rolling resistance force reduced 13.46% from the reference value. These results proved the low tyre pressure made the tyre deformed more than the high tyre pressure during rotating on the flat road surface and caused high hysteresis loss. However, the value of rolling resistance force reduced when the vehicle speed increased because the tyre contacted the road surface in a short period of time. Therefore, the frequency of tyre that contacted on the road surface was high and caused the molecule of the rubber tyre moved faster than low vehicle speed. This phenomenon made the tyre stiffer than low vehicle speed and made the value of rolling resistance force reduced. Considering the hysteresis phenomena, the deformation of tyre under high vehicle speed was less than low vehicle speed. Therofore, the energy loss from hysteresis phenomena was low and allowed more output torque to be transmitted to the road.

6 Future works

According to this research, the relationship of rolling resistance force and tyre pressure were explained. For the future test on this research, the tyre pressure of 25 psi, 35 [15] psi and 45 psi are used for conducting the test on acceleration rate and fuel consumption rate. The results are used to explain an effect of different tyre pressure under real driving condition which would be benefit for daily use.

Acknowledgment

I would like to acknowledge Rangsit University for supported some fund for conducting this research. Also, I would like to thanks Damrong Limsompom for helping me to conduct the test and collected the data.

References

[1] J.B. Heywood, "*Internal Combustion Engine Fundamentals*", International Edition, McGraw-Hill Book Company (1988)

[2] R.Stone, "*Introduction to Internal Combustion Engines*", 3rd, Society of Automotive Engineers, Inc (1999)

[3] H. Heisler, "*Vehicle and Engine Technology*", 2nd , Butterworth-Heinemann (1999)

[4] Robert Fischer R., Küçükay F., Gunter Jürgens G., Najork R. and Pollak B., "*The Automotive Transmission Book*", Springer International Publishing Switzerland (2015)

[5] R.N. Jazar, "*Vehicle Dynamics : Theory and Appication*", Springer Science+Business Media, LLC (2008)

[6] J. Reimpell, H. Stoll and J.W. Betzler, "*The Automotive Chassis : Engineering Principles*", Ed 2nd,Butterworth-Heinemann (2001)

[7] F. Grappe, R. Candau, B. Barbier, M.D. Hoffman, A. Belli and J.D. Rouillon, "*Influence of tyre pressure and vertical load on coefficient of rolling resistance and simulated cycling performance*", Ergonomics, **Vol 10**, page 1361-1371 (1999)

[8] Pedro de F.V. Carvalheira, "*An Experimental Method to Measure The Tire Rolling Resistance Coefficient of An Ultra-Efficient Vehicle*", 15th International Conference on Experimental Mechanics, 22-27 July (2012)

[9] V.V. Mozharovskii, S.V. Shilko, S.B. Anfinogenov and A. V. Khotko, "*Determination of resistance to rolling of tires in dependence on operating conditions. part 1. method of multifactorial experiment*", Journal of Friction and Wear, Vol 28, page 154-161 (2007)

[10] J.C. Páscoa, F.P. Brójo, F.C. Santos and P.O. Fael, "*An innovative experimental on-road testing method and its demonstration on a prototype vehicle*", Journal of Mechanical Science and Technology, **Vol 26**, page1663 – 1670 (2012)

[11] D. Mukherjee, "*Effect of Pavement Conditions on Rolling Resistance*", American Journal of Engineering Research (AJER), Vol 3, Issue 7, page 141-148 (2014)

[12] P.S. Pillai, "*Inflation pressure effect on whole tyre hysteresis ratio and radial spring constant*", Indian Journal of Engineering & Materials Sciences, **Vol 13**, page 110-116 (2006)

[13] P. Haney, "*The Racing & High-Performance Tire*", TV Motorsports abd SAE (2003)

[14] P. Norrby, "*Prediction of coast-down test results : A statistical study of environmental influences*", Master of Science Thesis, Chalmers University of Technology (2012)

The Effect of Position and Different Size of Radial Hole on Performance of Cannulated Pedicle Screw

F. A. Zakaria[1], R. Daud[1], H. Mas Ayu[1], S. H. Tomadi[1], M. S. Salwani[1], Mohammed Rafiq Abdul Kadir[2]

[1]*Faculty of Mechanical, Universiti Malaysia Pahang, 26600 Pekan, Pahang, Malaysia*
[2]*Faculty of BioSciences & Medical Engineering, Universiti Teknologi Malaysia, 81310 Skudai, Johor, Malaysia*

Abstract. Cannulated Pedicle Screw (CPS) is the instrumentation that has been introduced in Biomedical field in order to stabilize the spine structure and also to enhance the pullout strength in osteoporotic patient. The cement is injected through the CPS which is flow from the screw head to the radial hole and then distributed around the CPS to give higher performance of pullout strength. However, the fracture still reported to happened around the radial hole. This paper is aim to focus on the effect of the position and different diameter of radial hole on the CPS. Three designs were constructed using SolidWork software where two of it are the current models whereas the third was the proposed design which the radial holes were position differently (skewed to each other) and its size is smaller than the current model, 1.0 mm and 1.5mm respectively. The software used for Finite Element Analysis (FEA) is Ansys workbench 16.0. The Finite Element Models were verified with FEA result of the previous research. The FEA results of torsional stress for current and proposed design were then compared to obtain the maximum torque that both designs are able to resist before fracture. It can be concluded that the performance of CPS can be improved by the proposed design which is 2 radial hole skewed to each other and smaller radial hole diameter (1.0mm) than the current cannulated pedicle screw (1.5mm).

1 Introduction

The normal bone mineral densities is between -1 and above, low bone density also called as Osteopenia is in range -1 to -2.5 whereas -2.5 and below than that is in the category of Osteoporosis. The decreasing value of the bone mineral density decreased the holding strength of the bone. Thus, the normal pedicle screw is not suitable for the osteoporosis bones due to low holding strength. Cannulated pedicle screw (CPS) with cement augmented has been introduced as a solution for osteoporosis bones to increase the pullout strength [1].

Pullout strength of a pedicle screw is significantly correlated with the screw design whereas the more interface between screw and bone tissues can provide more pullout strength. To increase the interface, radial holes could be an option which allows bone growth through the holes. Mckoy et al. compared the pullout strengths of CPS with radial holes and normal pedicle screw on osteoporotic human vertebrae [2]. They found that radial holes had increased the amount of the cement exuded from the CPS, thus it caused pullout strength of CPS is 2.78 times higher than standard pedicle screw. Chen et al. found that the position of the radial hole primarily limited to the distal one third of the shaft to avoid the leakage of the cement into the spinal canal [3]. They proved that the performance of CPS is depends on the position of the radial holes but not on the number of the radial holes. They also found that the failure rate of CPS can be reduced by reducing the number of the radial hole. Although the effect of parameters of the radial holes (diameter and position of radial holes) on failure rate of CPS have been proof experimentally and numerically by several researchers [4-10], the optimum position and diameter of radial holes still becoming one of the gap in this research field. Thus, the aim of this paper is to propose the design optimization of cannulated pedicle screw which focuses more on optimization of diameter and position of radial holes.

2 Methodology

2.1 Construction of Cannulated Pedicle Screw (CPS)

The solid models of CPS were constructed and assemble by using Solidwork Software 2015. The geometry and the dimensions of the screw were followed the standard size of the cannulated pedicle screw. Three different 3D CPS were created where the first and second one is following the current design of CPS with different size and position of radial hole while another one is the proposed design. By referring to the research entitled 'Biomechanical Performance of Various Cement-Augmented Cannulated Pedicle Screw Design for Osteoporotic Bones' by Toga Tolunay et. al [1] and the 'Effect of different radial hole designs on pullout and structural strength of Cannulate

Pedicle Screw' by Hsin-Chang Chen et al. [3], two current model of Cannulated Pedicle Screw was selected with higher strength compared to other and was constructed using Solidworks software:

i) CPS with 4 radial hole positioned 90° to each other, descending in a counterclockwise direction according to the thread (figure 1).

ii) CPS with 2 radial hole row unilaterally (figure 2).

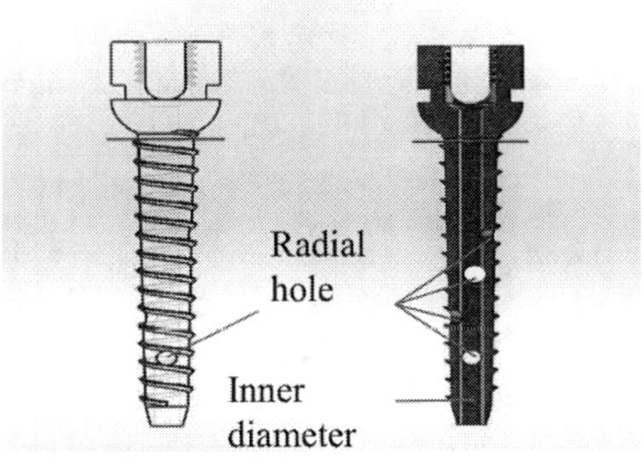

Figure 1. Front and sectional view of current model of CPS with 4 radial hole positioned to each other, descending in a counter clockwise direction according to the thread

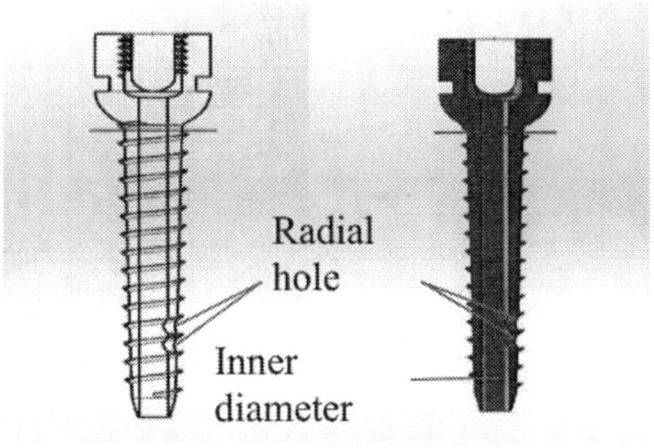

Figure 2. Front and sectional view of CPS with 2 radial hole row unilaterally.

The proposed design of the CPS is designed with different sizes and positions of the radial holes. By comparing the performance of CPS with different sizes and positions of radial holes, we obtained the optimum diameter is 1.0 mm and the optimum position of radial holes is skewed to each other (Figure 3).

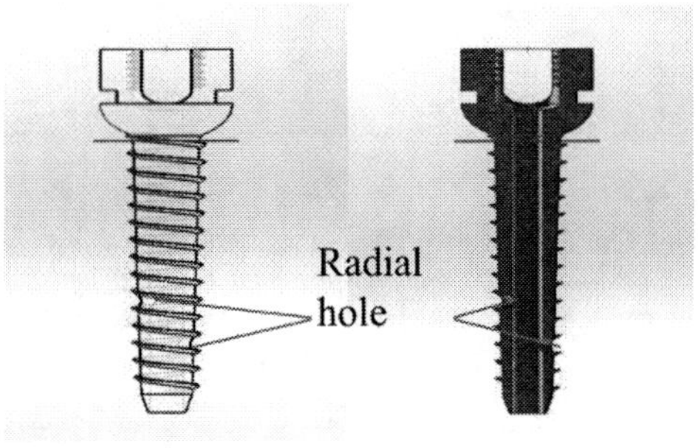

Figure 3. The proposed design with 2 radial hole positioned skewed to each other

2.2 Finite Element Analysis (FEA)

The Finite Element Analysis (FEA) is conducted by FEA Engineering Software. Ansys Workbench 16. The material used is Titanium Alloy. In order to perform finite element analysis of each model, the model of the screw need to be saved in parasolid x_l. file in solidworks and then imported as the modeler in Ansys software. The geometry firstly need to be mesh before it is proceed into analysis. The finer the meshing size the precise the value of the Equivalent Stress (Von Mises Stress) also the longer the duration of the analysis to finished. The fine meshing size is shown in the Figure 4 below.

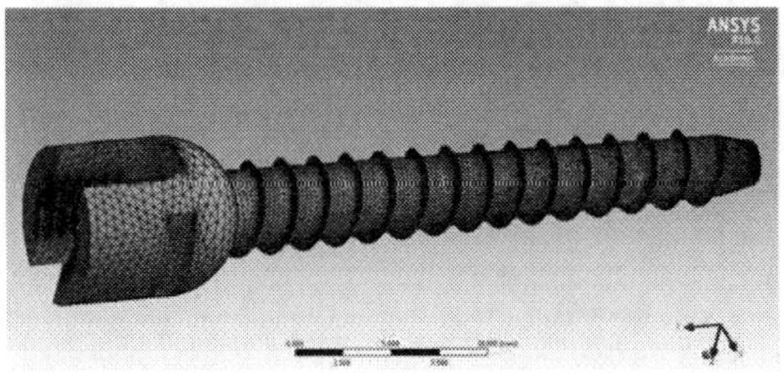

Figure 4. Meshing of the Cannulated Pedicle Screw using Ansys Software.

The boundary condition of the screw applied on the meshed model in order to results in stress analysis act on the screw. For the environment condition, the temperature is set 37 °C following the normal body temperature. To perform static structural of torsion test, as shown in Figure 5 below, the screw tip was completely fixed in all direction and an initial torque of 1.29 Nm was applied to the body of the screw and was repeated in increment of 0.3 Nm until the Von Mises stress approximately similar or approaching the yield stress. The yield stress was set at 790 MPa.

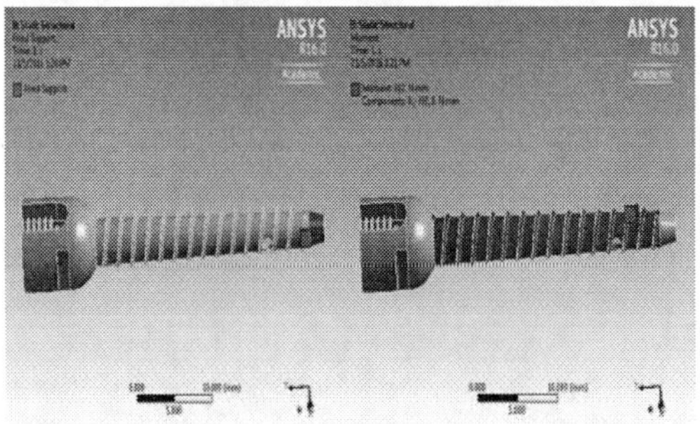

Figure 5. Fixed point applied on the screw tip and torque applied on the screw body.

Both proposed and current 3D results of analysis are compared to see either the proposed or the current design have the better performance. The comparison of both current and proposed design is based on the maximum stress that each of the cannulated screw can bear with the different torque applied without exceeding the yield stress of 790MPa.

3 Results and Discussion

3.1. Radial hole positioned 90° to each other, descending counter clockwise according to the thread

The results of equivalent stress that approaching yield stress obtained for this design is 770.47 MPa with maximum allowable torque of 3.39 Nm. The stress focused around the radial hole which it means if the stress applied more than that can cause fracture around the radial hole (Figure 6).

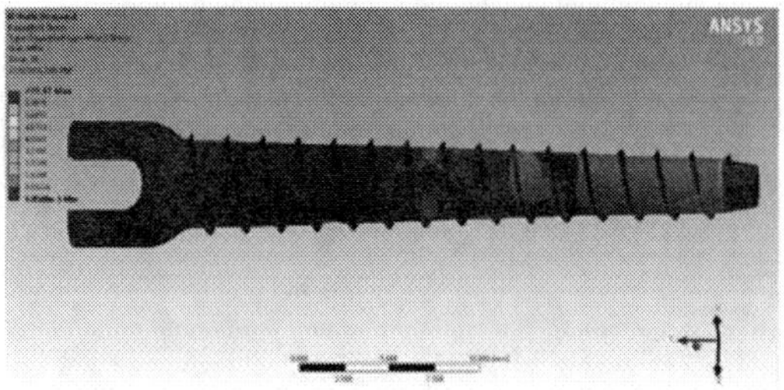

Figure 6. Equivalent Stress (Von Mises Stress) of CPS with 4 radial holes

3.2 Radial hole row unilaterally

From the FEA result (Figure 7)of this design shown that the highest allowable torque that the screw can bear is 3.2 Nm before it is approaching the yield strength of the CPS, 790Mpa (maximum Von Mises stress 778.51 MPa).

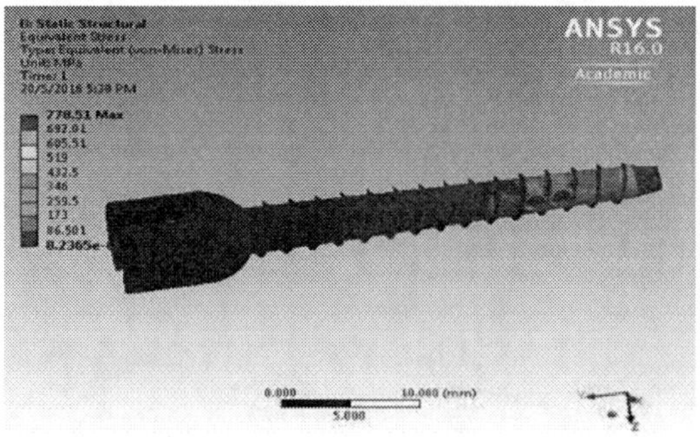

Figure 7. Equivalent Stress (Von Mises Stress) of CPS with 2 radial holes

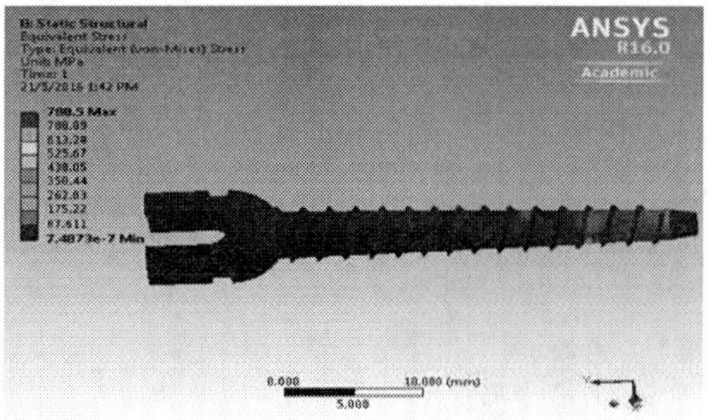

Figure 8. Equivalent Stress (Von Mises Stress) of CPS with 2 radial holes skewed to each other

3.3 Proposed Design of CPS

From the analysis, the results were obtained from the structural analysis for this design shown that the maximum stress with the highest torque of 3.92 Nm applied is 788.5 MPa (Figure 8). The highest allowable torque that the screw can bear is 3.92 Nm, and higher than that value, the deformation of the screw might be happen.

3.4 Comparison Between the Current Designs and the Proposed Design

In order to find the better design of CPS, each of the CPS is compared between each other in aspect of their maximum torque that can be applied. The detailed of the results obtained from the analysis is recorded as in Table 1 below.

Table 1. Moment and Maximum stress between the current design and the proposed design

Current Design/Previous Research by Chen at el.				Proposed Design	
4 Radial hole positioned 90° to each other		2 Radial hole row unilaterally		2 Radial hole positioned skewed to each other	
Moment (Nm)	Maximum Stress (MPa)	Moment (Nm)	Maximum Stress (MPa)	Moment (Nm)	Maximum Stress (MPa)
1.29	293.19	1.1	501.79	2	402.28
1.59	361.37	1.4	553.6	2.3	462.63
1.89	429.55	1.7	605.51	2.6	522.98
2.19	497.74	2	657.41	2.9	583.33
2.49	565.92	2.3	761.21	3.2	643.67
2.79	634.1	2.6	769.86	3.5	704.01
3.09	684.1	2.9	775.05	3.8	764.36
3.39	770.47	3.2	778.51	3.9	784.47
				3.92	788.5

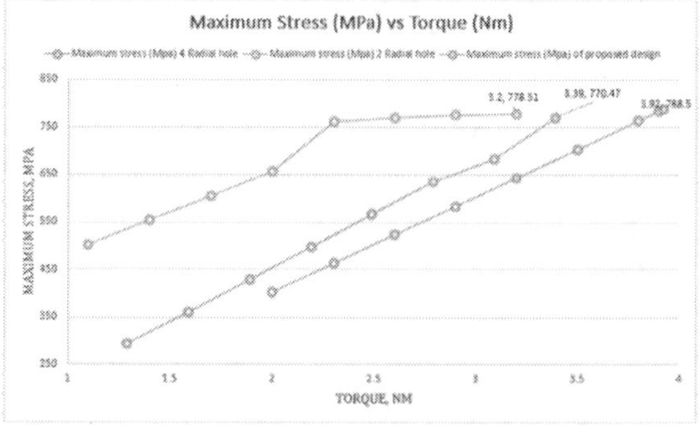

Figure 9. Comparison on the performance of CPS between the current designs and the proposed design.

From the results obtained, the recorded value of different moments that were applied on each design was then interpreted into the graph form as shown in the Figure 9. The red line on the graph represents current CPS designed with two radial hole row unilaterally while for the CPS with four radial hole positioned 90° to each other descending counter clockwise according to the thread represented by blue line. Another one is the proposed design with two radial holes (1 mm) positioned skewed to each other represent by green line.

By referring to the graph, 2 radial holes row unilaterally shows the reading of the maximum stress (Von Mises Stress) of 778.51 MPa with the highest applied moment of 3.2 Nm. Whereas the maximum stress (Von Mises Stress) reading for the 4 radial hole Cannulated Pedicle Screw is 770.47 MPa with applied highest moment of 3.39 Nm. This is means, between this both design, CPS with 4 radial hole is better than 2 radial hole since the screw can bear the higher torque 3.39 Nm and 3.2 Nm respectively. However, among the three designs, the proposed design shows an improvement with the highest torque (3.92 Nm) that can be applied before it fractures. It can be concluded that based on FEA results, the position and size of radial holes effect the performance of CPS where as the number of radial holes is not the factor that effected the performance of CPS same as reported by Chen at el. The research gap regarding on optimum of position and size of radial holes is closed since the proposed design with optimum position (2 radial holes skewed each other) and size of radial hole (1.00mm diameter) shown better performance of CPS than previous design by Chen at el.[3].

References

1. Tolunay, T., Arslan, K., Yaman, O., Dalbayrak, S., & Demir, T. Biomechanical Performance of Various Cement-Augmented Cannulated Pedicle Screw Designs for Osteoporotic Bones. Spine Deformity, 3(3), 205-210 (2015)
2. B.E McKoy and Y. H An. An injectable cementing screw for fixation in osteoporotic bone. Journal of Biomedical Materials Research A, 53(3), 216-220 (2000)
3. Chen, H. C., Lai, Y. S., Chen, W. C., Chen, J. W., Chang, C. M., Chen, Y. L., Cheng, C. K.. Effect of different radial hole designs on pullout and structural strength of cannulated pedicle screws. Med Eng Phys, 37(8), 746-751 (2015)
4. Abshire BB, McLain RF, Valdevit A, Kambic HE. Characteristics of pullout failure in conical and cylindrical pedicle screws after full insertion and back out. Spine J 1:408–414 (2001)
5. Bianco RJ, Arnoux PJ, Mac-Thiong JM, Wagnac E, Aubin CE. Biomechanical analysis of pedicle screw pullout strength. Comput Methods Biomech Biomed Eng 16(Suppl 1):246–248 (2013)
6. Chao CK, Hsu CC, Wang JL, Lin J Increasing bending strength and pullout strength in conical pedicle screws: biomechanical tests and finite element analyses. J Spinal Disord Tech 21(2):130–138 (2008)
7. Chao KH, Lai YS, Chen WC, Chang CM, McClean CJ, Fan CY, et al. Biomechanical analysis of different types of pedicle screw augmentation: cadaveric and synthetic bone sample study of instrumented vertebral specimens. Med Eng Phys.35:1506-12. (2013)
8. Chatzistergos PE, Magnissalis EA, Kourkolis SK A. parametric study of cylindrical pedicle screw design implications on the pullout performance using an experimentally validated finite element model. Med Eng Phys 32:145–154 (2010)
9. Paré, P. E., Chappuis, J. L., Rampersaud, R., Agarwala, A. O., Perra, J. H., Erkan, S., & Wu, C. Biomechanical evaluation of a novel fenestrated pedicle screw augmented with bone cement in osteoporotic spines. Spine, 36(18), E1210-E1214 (2011).
10. Patel, R. D., Graziano, G. P., Vanderhave, K. L., Patel, A. A., & Gerling, M. C. Facet. Violation With the Placement of Percutaneous Pedicle Screws. Spine, 36(26), E1749-E1752 (2011)

Effect of using different types of methods for the derivation of elastic modulus of bone - A critical survey

Zartasha Mustansar[1, 2], Arslan Shaukat[1] and Lee Margetts [2, 3, 4]

[1]*National University of Sciences and Technology (NUST), H-12, Islamabad, Pakistan.*
[2] *School of Earth, Atmospheric and Environmental Science, The University of Manchester, Williamson Building, Oxford Road, Manchester, M13 9PL, UK.*
[3] *Research Computing, The University of Manchester, Devonshire House, Oxford Road, Manchester, M13 9PL, UK.*
[4] *Oxford eResearch Centre, University of Oxford, 7 Keble Road, Oxford, UK.*

Abstract. The purpose of this paper is to critically review and address the cause of variation in the values of elastic modulus (E) of bone. Properties used for materials (especially) bones is one of the crucial factors in modeling and simulation experiments with sensitive procedures such as Finite Elements. The values of elastic modulus acquired from secondary databases, literature and online repositories differ in many ways. This is due to an underlying fact that different methods are used for one definite set of species and vice versa. According to a human understanding, this might bar the perceptive of use of elastic modulus in the research when (especially) insightful procedures are in question for example Finite Element Analysis (FEA) of materials with sensitive procedures. The means and the methods, through which the values are obtained, do not follow a standard methodology. Thus, this area needs attention. Despite the advances in technology and state-of-the-art methods in the field of biomechanics and biomedical engineering there is a need to formulate a standard method which can produce_reliable set of E-values. Through this paper, we seek to raise an important scientific question as to what systemic methodology should be adapted to achieve a value of elastic modulus that is not only reliable, but reasonable as well. The paper also, focus briefly on the sources of error responsible for this variation and the significance of elastic modulus values in important numerical methods such as Finite Elements (FE).

1 Introduction and Background

Young's modulus or elastic modulus is one of the fundamental inputs in the constitutive analysis of bones using Finite Element Method (FEM). It is therefore important (at minimum) to get the linear part of the FE analysis correct. Elastic modulus can be obtained experimentally from a material that is subjected to a load, and the strain and stress are determined concurrently [1].

Bone, consist of combination of ingredients, details of which can be obtained from [2]. The shape of bone is determined by the arrangement of the HA crystals. The arrangement of the collagen fibrils and HA crystals enhances bone strength [3]. Due to a higher elastic modulus of HA crystals, bone is stiffer to bear the external fractures [4]. Bone has two distinct components. Cortical bone which constitutes the outer thick layer providing strength to the inner porous structure. Whereas the shape of trabeculae bone is related to porosity. Trabecular orientation is commonly studied nowadays mainly as a composite with anisotropic and open porous cellular solid from an engineering point of view [5-7].

Bones possesses specific material properties. They, fundamentally help determining an overall mechanical behaviour. The Young's modulus and the Poisson ratio are widely considered. The Young's modulus is already described above and the Poisson ratio (ν) is the negative ratio of transverse strain to the longitudinal strain [1]. When a specimen is loaded in compression, it will shorten in the direction of the load and will elongate in width. In other case if the specimen is loaded in tension it enlarges in the direction of the load but become narrower in width [8]. The ratio of change of these quantities describes how Poisson ratio is related to bones and usually ranges from 0.1 – 0.33. A vast variation can be seen in the results of the values of Young's modulus (from 1GPa – 20GPa) in different analytical bone models (Please see section methods). One of our previous papers [9] discusses the effect of different loading tests in the derivation of modulus of elasticity.

This paper particular address one of the factors i.e.types of loading tests used for the derivation of elastic modulus (E) value. Another objective of this paper is to report an evaluation of the degree of refinement of biomechanical experimental data. Despite the advances in technology and state-of-the-art methods in the field of computational mechanics and biomedical engineering in general, there is a need to devise a standard method which can validate predicted results from the numerical methods.

2 Methodology

This section describes the rationale behind the survey in order to identify published literature for the determination of bone's Young's modulus. Authors points out how the selection of different testing methods for the derivation of material properties of bone may result in huge variation. We use a search method with a typical inclusion and exclusion criteria. Search methods consist of the relevant articles retrieved from published literature. Any information with incomplete citation on the web is excluded. The discussion includes only peer- reviewed journals, conference papers, conference proceedings, books and edited book sections, high impact letters and internationally accepted reports to address various data sets that report a typical value for elastic modulus. In this article, we will only focus on one factor i.e. relationship of testing method on elasticity (Elasticity–Type of method).

3 Results and Observations

Authors investigated, methods reported in literature to find out what should be the reasonable way of choosing one method over another while handling procedures used for the derivation of Young's modulus in bones. This section is divided in two parts, first the methods in literature used for derivation of elastic modulus are reported, then the authors discuss their own recommendations briefly.

3.1 Methods used in Literature

In this subsection, author describes various methods used in the literature for derivation of elastic modulus. For each method, a review of how the method works (principle of method); some limitations where required and for what purpose was it used, is described. In the end, we discuss the values of Young's modulus typically obtained from these methods (Figure 1) and then a profound discussion on why the variation exists in the literature library.

3.1.1 Compression

A compression test determines behaviour of materials under crushing loads. The specimen is compressed and deformation at various loads is recorded. It is expressed in Eq. (1) as:

$$\sigma_{comp} = \frac{F}{A_{cort}}, \qquad (1)$$

where σ_{comp} is compressive stress, F is the applied force and A_{cort} the cross-sectional cortical area [10]. Compressive stresses and strains are calculated and plotted as stress-strain diagram which is used to determine elastic limit ;yield point, yield strength and for some materials compressive strength.

Compression has been used in studies involving bones, by [8, 11-14] and [15]; biomedical research Boyd et al., [16]; [6, 17-19] in natural sciences [20]. In some studies it is used for the study of local deformation [16],[21].

3.1.2 Tension

A tensile test, also known as tension test, is probably the most fundamental type of mechanical testing in materials. The test is performed by pulling the material to determine how the material will react to forces which are applied in tension test. Pulling material will give the strength of the material as well as strain (i.e. how much the material went elongation). The tension test usually is plotted as a graph showing the relationship of stress (applied force) to the strain (change in length/observed elongation) – the gradient the curve determines the value of the Young's modulus. It is a widely used method for testing bones [19, 22-24].

3.1.3 Bending

This test is basically used to test the ductility, strength, stiffness and modulus of elasticity. Based on the type of testing (3-point and 2-point) the direction of forces can be located in order to predict stress-strain response of the material. The gradient of the stress-strain curve obtained, gives the value of the Young's modulus of the material. Mostly the specimens with long cylindrical shapes are tested in bending because, it is simple to set up and predict reasonably accurate response.

3.1.4 Buckling

Buckling is a mathematical instability which leads to a failure mode in a specific material. This phenomenon usually occurs in thin bones and they tend to break the structure into two. It is a kind of failure mode that causes incapability to the function of the bone. Theoretically, it is caused by a bifurcation (in two). Detailed example can be seen in [25].

3.1.5 CT scanning

Computed tomography is a non-destructive evaluation technique that allows the internal structure of an object to be imaged by reconstructing the spatial distribution of the local linear X-ray absorption coefficients of the materials/phases contained within [26]. In one of the studies [27] a method of theoretical derivation was proposed. The method aimed to determine a relationship between voxel-specific tissue density and micro CT scan data (i.e., microCT derived tissue mineral density (TMD), mgHA/cm3) and bone constituent properties. 'Voxel based tissue density derivation' is used from the scan data to derive an equivalent hydroxyapatite density using a manufacturer-specific calibration curve. This equivalent mineral density is then employed to determine elastic modulus of the individual mineral content. With an increase in the tissue density Young's modulus increases, which was one way of deriving value for the modulus of elasticity. For details of the method reader is referred to [27].

3.1.6 Ultrasonic

Natural resonance frequencies can be employed for a single small sample of bone to measure the elastic modulus of material. Such method can be seen in [28]. Typically this method is based on the measurement of wave propagation time in specimen [29] and then deriving Young's modulus. For detail on methodology reader is directed to [37].

This method outputs a 'dynamic Young's modulus E [MPa]' which is calculated by using Eq. (2):

$$E = v^2 \rho. \tag{2}$$

3.1.7 Multi-scale modelling

The multi-scale modelling approach consists of successive homogenization steps. Homogenization is the method of treating different scales/levels within the same specimen at individual phases. The method is described in [30]. In this study the trabecular bone was modelled as a nano-composite with hierarchical structure. This model was then used to predict its elastic properties at different structural scales. Their analysis made used of bottom-up multi-scale approach such that the starting phase is initiated with nanoscale (mineralized collagen fibril) and therefore onwards moving up to the sub-microscale (single lamella), microscale (single trabeculae) and mesoscale (trabecular bone) levels. Other examples can be seen in [4, 30-32].

3.1.8 Finite Element Method

By using CT based data, it is possible to convert a digital model into a numerical model. For example using Simpleware suite Ltd,. The numerical model (FE- based model) is used to determine the Young's modulus by testing the specimen under compression, tension and/or bending by using a computer solver (e.g. Abaqus). The constitutive models therefore obtained from this method are used to calculate elastic modulus via stress-strain curves.

3.1.9 Cantilever bending with finite element analysis

This method uses cantilever beam type bone specimen. The actual material properties are obtained by using the initial input estimates of the bone tissue material properties which were later employed in the finite element model. Difference between the calculated finite element displacements and the experimentally observed displacements help in the determination of actual Young's modulus.

3.1.10 Back-calculating from finite element models

By using the deformation values i.e. the change in length to the original length of the samples, finite element models can be used to determine the elastic properties 'Young's modulus' of the materials. Mente and Lewis [33] used cantilever bending with finite element analysis for elastic properties determination and Williams and Lewis [34] used back-calculation from FE models.

3.1.11 Micro hardness

This method uses the static indentations made with loads not exceeding than 1 kg. In bone mechanics this method is seen in [35]. However, it *re*quires a metallographic finish for the surface to be tested. Smaller loads tend to use higher finish.

3.1.12 Using stochastic operators

Wille et al., [36] used the stochastic operations to pool data cited in the literature based on the relationship between Young's modulus and bone density. All the experiments were pooled over time regarding the E-d (Young's modulus-density) relationship a stochastic relationship between them was developed.

3.1.13 Nanoindentation

Nanoindentation is used to measure the intrinsic elastic properties of individual micro structural bone components (here trabecular bone and cortical bone) in various directions. This method uses a load which is applied to an indenter of known properties. This leads to a formation of residual impression in the test material. The dimensions of the residual impression and the *contact area* are then measured directly to determine the hardness of the specimen.This method has the ability to give a spatial resolution of around 1 μm [37] to determine bone's mechanical properties and to achieve bone's nearly accurate young's modulus in several directions because of the highly anisotropic [6, 38-44] nature of trabecular bone.Nanoindentation also has the capability of producing large indent matrices in a very short period of time with stability especially for mapping the mechanical properties of complex surfaces [45]. Moreover it uses advance depth sensing indentation method, to produce a load displacement curve which is used to measure the stiffness and elastic properties of the sample using various approaches [46].

3.1.14 Digital volume correlation (DVC)

Digital volume correlation (DVC) is a measurement technique that quantifies strains throughout the interior of a specimen unlike traditional strain gauge methods where only few points on the surface of the materials are considered. Through DVC it is possible to generate stress-strain curve, using a software suite like Daviz. The gradient of stress-strain curve gives value of the Young's modulus.

3.2 Recommendations

In comparison to all the methods listed above, authors reserves their own opinion on how a large sets of different methods could affect the values of the elastic modulus obtained.

Through careful investigation, at first all the values of elastic modulus of bones are fetch from the literature against different types of species and methods used are populated in the form of the graph/illustration as shown in Figure 1. Figure 1 reviews the author's own survey on trend/behavior of huge variation in the values of Young's modulus from the literature of bones. It is clear from the chart that the variation in the method of determination yields a large variation in the spread of elastic modulus values. Eventually, all the biomechanical analyses which require elastic modulus as one of their important parameters can be affected. This means, a careful consideration is required in choosing a specific value for the elastic modulus in any type of analysis where elastic modulus is one of the inputs. To be more precise, a systematic procedure is inherent to introduce so that this variation can be doomed to a typical consistency level as required originally.

The bone's elasticity is important because, it gives the bone ability to dissipate energy during deformation phenomena's. Methods like Finite Elements, which requires the value of elastic modulus as one of the major inputs predicts results of simulation on the basis of E-value that is used as an input. If one value goes wrong the whole simulation will result into an inaccurate and ambiguous situation. Apparently, the criticality of this basic factor is under estimated and it might result into the situations which do not relate to any real physical phenomena's at all.

3.3 Limitations

In this study, detailed studies on review of the papers that uses Finite Element Method as a method of derivation is not reported.

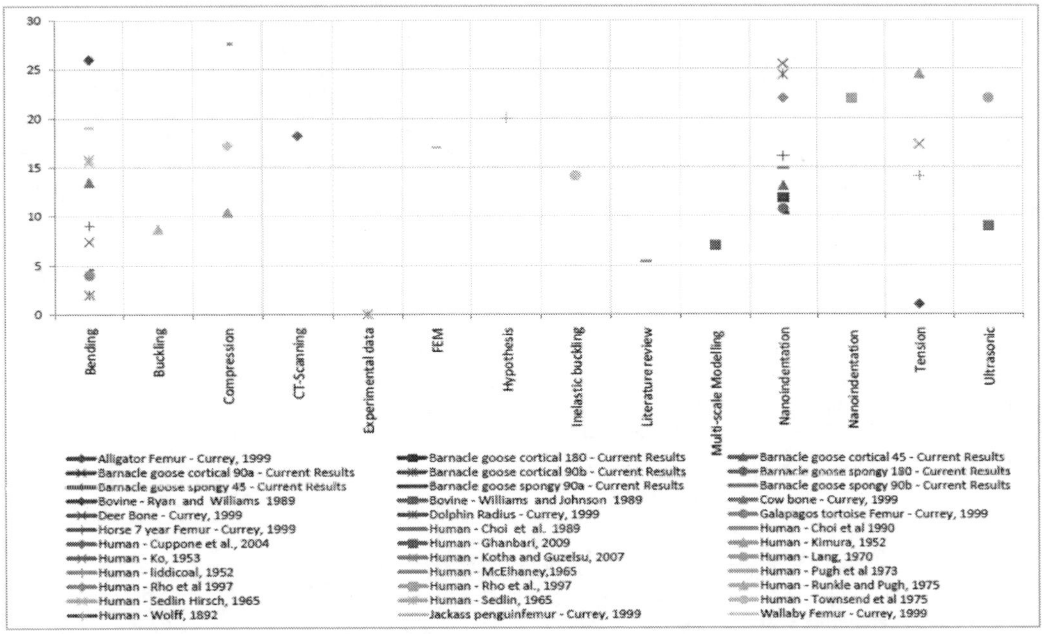

Figure 1. Chart showing spread of material properties compared to results obtained from Barnacle goose (Branta leucopsis) specimen.

This might have helped to validate our argument on how the variation in the values of elastic modulus, might affect the simulation results. This is something we are hoping to include in our next coming papers.

Secondly, the study takes into account only the affect of loading methods, used for the derivation of elastic modulus. Such type of tests were briefly discussed in one of our previous papers as well [9] but only for the non homogenous materials and the methods were also not discussed in much detail. Here we report almost every single method found from literature to address variation in the elastic modulus.

Additionally, apart from the types of loading tests, there are several other factors, for example, type of mechanical environment, bone density, bone age, specie type, physical conditions in which bones are prepared for the mechanical testing and treated bones, which adds to the variation among the E-values. It might be good to include those studies as well in future.

4 Discussions and Conclusions

Due to the complex geometry of bone it is not possible to measure the exact Young's modulus. The way bone samples are prepared and the methods used for testing, equally contribute to the variation of the elastic modulus values achieved for bones. A correlation between methods used to determine Young's modulus with respect to the factors affecting E value is a 'must' required procedure.

Different testing methods tend to yield different values of Young's modulus. For example, a bone in compression will yield a different value of Young's modulus compared to the bone tested in tension under the same physical conditions. Therefore, it is important to hypothesize a 'generalized method.

By using similar methods and techniques to find an answer for material property can be helpful in maintaining a consistency and reliability among these values. Authors also conclude that there is a need to hypothesize a 'generalized method' which can give reasonable results under similar physical conditions (for example, specific laboratory standard should be practiced at least for one type of bone specimen.

As said in limitations, there are number of factors which can affect the method of determining value of elastic modulus that may in turn affect the original elastic properties of bone when it was alive. For example a healthy will function much better than a damaged bone whereas a damaged bone is likely to cause a functional disability. So if the condition of specimen changes, its functionality also changes.

The factors effecting value of E can be biological, mechanical or experimental. These parameters, if considered in detail in future will reduce the risk of false generation of data for the elastic modulus of bone. Increasing water loss from bones also make them lose elasticity due to breaking bonds within the elastic trabeculae. It should also be noted that bigger animals tend to have stronger bones due to thicker trabeculae [47]. This means size of the bone might also be one limiting factor as well to generate a real values for Young's modulus.

In this paper we have presented the scheme of the method to be used in future for the selection of Young's modulus in numerical methods (specially). This is quite a challenging task though crucially very basic. Therefore there is a need to validate this by using finite element method which is a reliable method to address such problems virtually by replicating the reality in a virtual environment.

Therefore we propose a new methodology, by using E-values from ten different species initially. All specie bones will be prepared under different physical conditions and mechanical testing methods. Additionally, all bones will be of different age and bone density. Under such conditions, a finite element analysis will be conducted under one set of boundary conditions and i.e. force and the base conditions remains the same throughout the ten species. The analysis will entail the difference in the deformation scale factor and the amount of stress due to different values of elastic modulus (E). To our opinion, this will be a very useful step to address any kind of ambiguity in the future of biomechanics.

Acknowledgments

Authors would like to thank The Microsoft-Dorothy Hodgkin Postgraduate award (FA01546) for their financial support.

References

1. A. Sharir, *et al.*, "Whole bone mechanics and mechanical testing," *The Veterinary Journal*, vol. **177**, pp. 8-17, (2008).
2. J. D. Currey, *Bones: Structure and Mechanics.*: Oxford, Princeton University Press,(2000).
3. J. D. Currey, "Collagen and the Mechanical Properties of Bone and Calcified Cartilage," in *Collagen structure and mechanics*, ed: Springer US, (2000), pp. 397-420.
4. J. Ghanbari and R. Naghdabadi, "Nonlinear hierarchical multiscale modeling of cortical bone considering its nanoscale microstructure," *Journal of Biomechanics*, vol. **42**, pp. 1560–1565, (2009).
5. P. J. Kolston, "Finite element modeling: a new tool for the biologist.," *Philosophical Transactions of the Royal Society (Series A: Mathematical, Physical and Engineering Sciences)*, vol. **358**, pp. 611-631, (2000).
6. T. M. Keaveny, *et al.*, "Biomechanics of trabecular bone," *Annu. Rev. Biomed. Eng.*,vol. **3**, pp. 307–33, (2001).
7. S. Sobieszczyk, *et al.*, "FEM Approach to estimate the behaviour of Bio-Composite metal-surface coating systems.," *Advances in Materials Sciences*, vol. **8**, pp. 166-172, (2008).
8. A. Benoit, *et al.*, "3D analysis from micro-MRI during in situ compression on cancellousbone," *Journal of Biomechanics*, vol. **42**, pp. 2381–2386, (2009).
9. R. Khan and Z. Mustansar, "Reliability of using elastic modulus for non- homogeneous materials.," in *6th International Conference on Bioscience, Biochemistry and Bioinformatics (in press)*, Pattaya, Thailand, (2016).
10. Gere J and G. B., "Mechanics of materials.," *Stamford, CT: Cengage Learning.*,(2012).
11. R. Müller, *et al.*, "Micro-compression: a novel technique for the nondestructive assessment of local bone failure," *Technology and Health Care*, vol. **6**, pp. 433-444, (1998).
12. S. Nagarajaa, *et al.*, "Trabecular bone microdamage and microstructural stresses under uniaxial compression," *Journal of biomechanics*,vol. **38**, pp. 707–716, (2005).
13. P. J. Thunrner., *et al.*, "High-Speed Photography of Human Trabecular Bone during Compression," in *In: Material Research Society*vol. **874**, ed, (2005).
14. R. Das and W. Cleary, Paul. , "Uniaxial Compression Test And Stress Wave Propagation Modelling Using Sph," CSIRO Mathematical and Information Sciences, Clayton, Victoria 3168, Australia15-16 December 2006 (2006).
15. R. Rodríguez-Martín, *et al.*, "Use of electronic speckle pattern interferometry in the detection of fatigue failure in high strength steels," *Engineering Failure Analysis*, vol. **17**, pp. 226–235, (2010).
16. S. Boyd, *et al.*, "Measurement of cancellous bone strain during mechanical tests using a new extensometer device," *Medical Engineering & Physics*, vol. **23**, pp. 411–416, (2001).
17. E. Perilli, *et al.*, "Structural parameters and mechanical strength of cancellous bone in the femoral head in osteoarthritis do not depend on age," *Official journal of the international bone and mineral society*, vol. **41**, pp. 760-768, (2007).

18. G. Y. H. Lee and C. T. Lim, "Biomechanics approaches to studying human diseases," *Trends in Biotechnology,*vol. **25**, pp. 111-118, (2007).
19. S. P. Kotha and N. Guzelsu, "Tensile behavior of cortical bone: dependence of organic matrix material properties on bone mineral content.," *Journal of Biomechanics,* vol. **40**, pp. 36-45, (2007).
20. D. D. Deligianni and C. A. Apostolopoulos, "Multilevel finite element modeling for the prediction of local cellular deformation in bone.," *Biomechanics and Modeling in Mechanobiology,* vol. **7**, pp. 151–159, (2008).
21. E. Verhulp, *et al.*, "Indirect determination of trabecular bone effective tissue properties using micro-finite element simulations "*Journal of biomechanics,* vol. **41**, pp. 1479-1485, (2008).
22. D. R. Carter, *et al.*, "Tensile fracture of cancellous bone," *Acta Orthop. Scand,* vol. **51**, pp. 733-741, (1980).
23. S. J. Kaplan, *et al.*, "Tensile strength of bovine trabecular bone," *Journal of biomechanics.,*vol.**18**, pp. 723-727, (1985).
24. C. B. Ryan and J. L. Williams, "Tensile testing of rodlike trabeculae excised from bovine femoral bone. ," *Journal of Biomechanics,* vol. **22**, pp. 351-355, (1989).
25. X.-P. Zheng, *et al.*, "Determining the elastic modulus of thin films using a buckling-based method: computational study," *Journal of physics D: Appplied physics.,* vol. **42**, (2009).
26. J. C. Elliott and S. D. Dover, "X-ray Microtomography," *Journal of Microscopy-Oxford,* vol. **126**, (1982).
27. W. D. Wagner, *et al.*, "Deriving tissue density and elastic modulus from microCT bone scans," *BONE,* vol. **49**, pp. 931–938, (2011).
28. R. B. Ashman and J. Y. Rho, "Elastic modulus of trabecular bone material. ," *Journal of Biomechanics,* vol. **21**, (1988).
29. H. Nìmcová, *et al.*, "*Comparison of methods for dynamic young's modulus determination in gypsum materials,*" presented at the The 4th International conference: Modelling Of Mechanical And Mechatronic Systems 2011, The 4th International conference, (2011).
30. E. Hamed, *et al.*, "Multi-scale modelling of elastic moduli of trabecular bone," *Journal of royal society interface,* vol. (in press), (2012).
31. D. Porter, "Pragmatic multiscale modelling of bone as a natural hybrid nanocomposite," *Materials Science and Engineering,* vol. **365**, pp. 38-45, (2004).
32. V. Sansalone, *et al.*, "Multiscale modelling of mechanical properties of bone: study at the fibrillar scale," *Comptes Rendus Mecanique* vol. **335**, pp. 436–442, (2007).
33. P. L. Mente and J. L. Lewis, "Experimental method for the measurement of the elastic modulus of trabecular bone tissue.," Journal *of Orthopaedic Research,* vol. **7**, (1989).
34. J. L. Williams and J. L. Lewis, "Properties and an anisotropic model of cancellous bone from the proximal tibial epiphysis," *Journal of Biomechanics and Engineering,* vol. **104**, pp. 50-56, (1982).
35. R. Hodgskinson, *et al.*, "Hardness, an indicator of the mechanical competence of cancellous bone.," *Journal of Orthopaedic Research,* vol. **7**, pp. 754-758, (1989).
36. H. Wille, *et al.*, "Prediction of the mechanical response of the femur with uncertain elastic properties," *Journal of Biomechanics,* vol. **45**, pp. 1140–1148,(2012).
37. J. Y. Rho, *et al.*, "Young's modulus and hardness of trabecular and cortical bone in various directions determined by nanoindentation.," in *In: Transactions of the 43rd Annual Meetings of the Orthopaedic Research Society.*, (1997), p. 891.
38. Yang G, *et al.*, "The anisotropic hooke's law for cancellous bone and wood," *Journal of Elasticity,* vol. **53**, pp. 125-46, (1999).
39. J. Seto, *et al.*, "Tough Lessons From Bone: Extreme Mechanical Anisotropy at the Mesoscale," *Advance Function Materials,* vol. **18**, pp. 1905-1911, (2008).
40. S. Dendorfer, *et al.*, "Anisotropy of the fatigue behaviour of cancellous bone, Journal of Biomechanics," *Journal of Biomechanics,* vol. **41**, (2008).
41. P. Fratzl, "Collagen and the Mechanical Properties of Bone and Calcified Cartilage," in *Collagen structure and mechanics*, J. D. Currey, Ed., ed: Springer US, (2008), pp. 397-420.
42. J. Y. Rho, *et al.*, "Mechanical properties and the hierarchical structure of bone," *Medical Engineering and Physics,* vol. 20, pp. 92–102, 1998.
43. J. Zhang, *et al.*, "Mechanical property determination of bone through nano- and micro-indentation testing and finite element simulation.," *Journal of Biomechanics,* vol. **41**, pp. 267–275, (2008).
44. P. K. Zysset, *et al.*, "Elastic modulus and hardness of cortical and trabecular bone lamellae measured by nanoindentation in the human femur," *Journal of Biomechanics,* vol. **32**, pp. 1005–12, (1999).
45. X. N. Randall, *et al.*, "Nanoindentation analysis as a 2-dimensional tool for mapping mechanical properties of complex surfaces. ," Journal of Materials Research, vol. **24**, (2009).
46. C. W. Oliver and G. M. Pharr, "An improved technique for determining hardness and elastic modulus using load and displacement sensing indentation experiments," Journal of Materials Research, vol. **7**, (1992).
47. M. Doube, *et al.*, "Trabecular bone scales allometrically in mammals and birds," *Proceedings of Royal society, (*2011).

MATEC Web of Conferences **108**, 13003 (2017) DOI: 10.1051/matecconf/201710813003
ICMAA 2017

WheelWalker: a foot-wheel driving exoskeleton for the alternate walk of paraplegic patients

Qingchuan Ma , Linhong Ji and Rencheng Wang

State Key Laboratory of Tribology, Department of Mechanical Engineering, Tsinghua University, Beijing 100084, China

Abstract. Upright walk is the fundamental need of paraplegic patients. In this study, we proposed a novel lower-limb exoskeleton for assisting the patients to walk alternately and further improve their engagement in the rehabilitation training. The exoskeleton was driven by a hub motor which was mounted under the foot, and a pair of custom-made crutch with wireless controller was used to control the walking condition of exoskeleton and maintain balance. This paper mainly introduced the general designing concept and detailed specifications of each modules of the exoskeleton. A pilot using experiment was performed on two healthy subjects to qualitatively analyse the capability of this exoskeleton when adopting different gaits. The results showed that the subjects with different body characteristics could complete continued walk by wearing this exoskeleton either in a three-point gait mode or four-point gait mode.

1 Introduction

Paraplegia affects the lives of millions of people and most of them have to rely on the conventional assistive device like wheelchair or reciprocating gait orthosis (RGO) to achieve the basic lives[1, 2]. However, long-term use of the wheelchair could induce several physical and psychological problems due to maintaining of the sitting posture[3]. And the physical energy consumption of RGO is too remarkable often leading to a relative high abandon rate[4]. Portable exoskeletons have become the promising intelligent rehabilitation devices to assist patients walking upright without causing problems that wheelchair and RGO have. Several successful prototypes were developed including ReWalk (Israel)[5], Ekso (USA)[6], WPAL (Japan)[7]. Whereas, the complicated controlling system, short battery time and the price still restrict current portable exoskeleton from the widely application in clinic[8-10].

From the perspective of mechanical structure, most of the current portable exoskeletons belong to joint-driving, some researchers believe that such mechatronic design have fundamental shortcoming and need to be revolutionary updated to satisfy the future requirement[11, 12]. Li et al also suggested that researchers in this area should downgrade the automation level and design the exoskeleton as a supplementary equipment rather than a robot-like machine[13]. These researches indicate that the development of a simple, reliable and affordable portable exoskeleton is a more practical way for the patient-oriented clinical device.

Wheeled mechanism has better power efficiency and easier control experience than joint driving as it utilized in wheeled robot[14]. Therefore, based on the analysis of current rehabilitation study, we developed a foot-wheel driving exoskeleton for paraplegic patents which uses simplified mechatronic system to realize the basic needs of lower-limb disabled individuals for upright walking.

2 System overall designing concept

The main goal of this exoskeleton is using the simplest mechatronic part to realize the basic upright walking, furthermore, downgrading the automation level of exoskeleton and using the least sensor technologies to empower the user fully control of their walking procedure and improve their rehabilitation engagement. This system mainly consists of two parts as shown in Fig. 1, the foot-wheel driving exoskeleton and the wireless controlling crutch. The powered exoskeleton could immobilize and support user's body and drive the lower limb moving forward by the wheeled foot. The custom-made crutch could maintain the balance when walking and control the moving status of the exoskeleton by a wireless controller which was embed in the handlebar of the crutch.

The limitations for target users including: The user should have no spastic cerebral palsy to avoid the unpredictable falling; the user should have an abled upper limb to use the crutch and the paraplegia level of subjects should better than T10; the user's height and weight are less than 185cm and 100kg, respectively.

Several basic functions should be satisfied according to the project planning:

(a) The exoskeleton should assist user to complete the independent walk in a standing posture on a level or inclined road at a maximum speed of 2.1km/h;

(b) The user could fully control the walking status in real time including speed, step length, stop and start;
(c) The exoskeleton's wheel should automatically maintain a secure stop when there is no signal received from the crutch and the maximum step length also should be limited to protect the patient from the potential electronic system failures or misoperations;
(d) The length of the mechanical frame should be adjustable according to the specific body characteristic of the user.

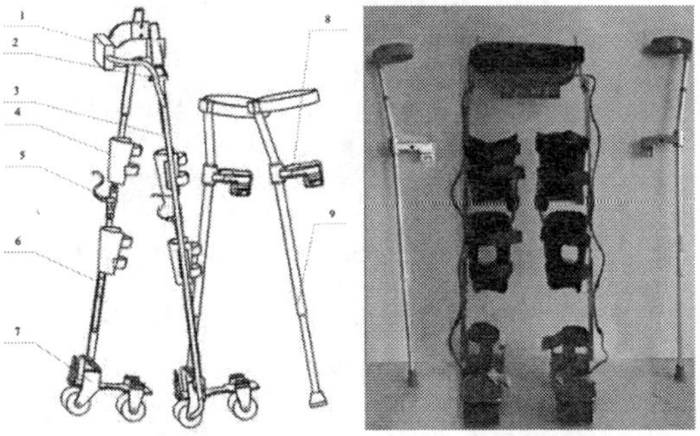

(a) (b)

Figure 1. The 3D designing model (a) and front appearance of the exoskeleton system (b).

3 Technological details

3.1 Mechanical specifications

The hip and ankle of this exoskeleton have one rotatory degree of freedom in sagittal plane respectively, while the knee joint can be manually locked or unlocked in standing or sitting posture. The degree of freedom is a crucial issue in the design of rehabilitation device as it can directly affect the performance of an exoskeleton. Here we developed this full Hip-Knee-Ankle-Foot (HKAF) exoskeleton which only enable the rotation of joint in sagittal plane while mechanically restrict movement in coronal plane.

Typically, human has a 15 ° of maximum ankle dorsiflexion angle and 10-20° of plantar flexion angle at the normal walking gait. Therefore we limited the maximum forward and backward rotation of this exoskeleton's ankle at 15° separately for safety reason. For a better adaption in users with different height, the length of the exoskeleton's shank could be adjustable in a range of 70 mm.

Different materials were used at different part of this exoskeleton to balance the weight and the strength. The exoskeleton's thigh and shank part were made of high-strength steel bars to withstand the supporting load of user. The pelvis, hip and ankle were all made by aluminium alloy. While the knee joint was made by steel as it has to provide a secure lock of the lower-limb. The knee fixator was made of PE (polyethylene) to achieve a light and high-stiffness supporting of leg, while sponge cloth was installed at the inner surface of this fixator enabling adequate comfortable using experience.

3.2 Foot-wheel driving unit

The frame of the exoskeleton's foot consisted of six aluminium plates and had two Velcro bands to securely fix the foot of user (Fig.2). The ankle joint of the exoskeleton connected the foot with two thrust bearings and a custom-made connecting bolt. We made a restricting groove at ankle bar and used a restricting bolt at the rear plate of the foot to limit the total rotation angle of the ankle in a range of 30° . Eight screwed holes with a 10mm interval were designed at the ankle joint bar to provide 70mm adjustable length for the different height of users.

Each foot-wheel driving unit uses two wheels, the front one is unpowered assistant wheel and the rear one is a custom-made powered hub motor. The hub motor is specially designed and manufactured for this study, which adopted an external BLDC motor and two-level planetary reducer to achieve a high-torque and small-diameter performance. As shown in Fig. 2(b), the hub motor was covered by 8mm thick anti-slip rubber with a final rotation speed of 150 r/min, which is 2.1 km/h in linear moving, approaching the maximum speed of ReWalk (2.6 km/h). The high reducing rate of this hub motor could enable a remarkable stopping torque functioning as the brake. Cooperated with the custom-made motor controller, in our test this hub motor could realize a reliable stop.

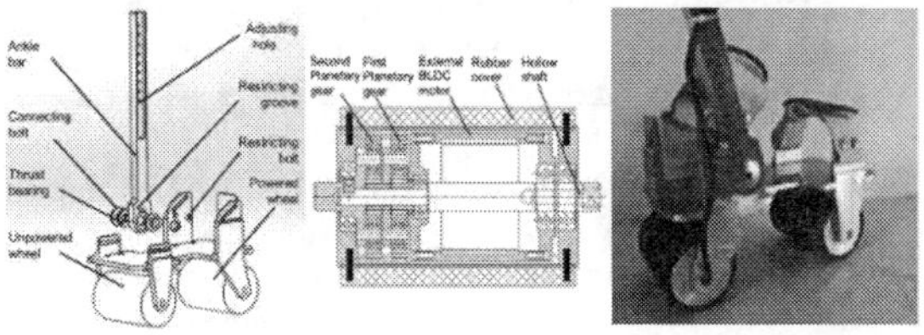

(a) (b) (c)

Figure 2. The detailed 3D designing model (a), section view of the hub motor (b), and the overall appearance (c) of the foot-wheel driving unit. In 3D model, the fixing bands were removed for a better illustration of the mechanical details.

3.3 Wireless controlling crutch

A pair of custom-made arm crutch were used in this system to provide balance maintaining and control of exoskeleton. The handlebar of this crutch was made by aluminium alloy which mainly consisted of three parts, the upper shell, the lower shell and the button (Fig. 3). The upper shell could provide the weight support for the balance maintaining and a wireless controller was embedded in it. The lower shell covered the wireless controller and had a higher platform at the end of handlebar to protect the button from the misoperation of the user. The button had a slope with 30° of inclination which could transfer the rotatory speed control of joystick into a 20mm linear control. The button could slide in the guiding slot of lower shell and would be recoiled by a

DOI: 10.1051/matecconf/201710813003

spring. The handlebar was mounted at the upper body of crutch and the length of this crutch could be adjusted according to the height of the user.

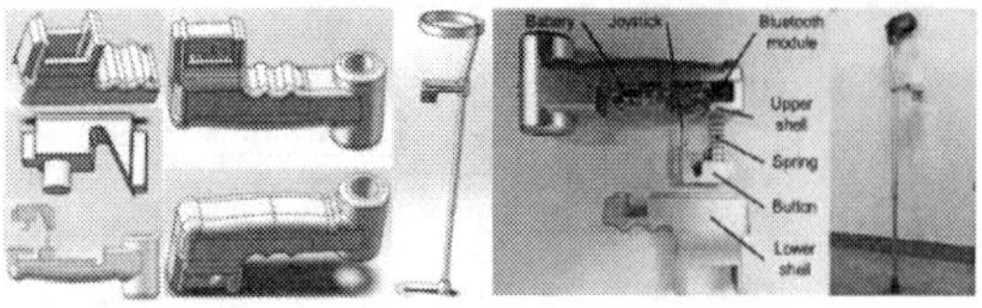

(a) (b)

Figure 3. The detailed 3D designing model (a) and explosive view (b) of the wireless controlling crutch.

3.4 Electronic system

Fig. 4 showed the electronic modules we utilized in this prototype. The wireless controller is a STM32F103 based Bluetooth transmitter which is powered by a rechargeable 4.2V*630mA Li-ion battery. The joystick is the crucial part of the controller which could control the moving status of the exoskeleton. The rotational angles of joystick directly reflect to the speed of motor and the brake will be enabled when the joystick maintain the 0 degree. The wireless receiver has a same Bluetooth module with the transmitter, which could output the PWM signal to the motor controller for speed control. The motor controller is a custom-made driver for a 24V Hull-sensor based BLDC motor. When the PWM signal was detected, this controller will actuate the motor, otherwise it will induce an electronic brake. A 24V*5000mAH rechargeable Li-ion battery pack was chosen as energy source to power the exoskeleton. All electronic parts were packaged in a 220 * 122 * 32 mm aluminium box which was mounted at the pelvis of the exoskeleton. The wire of motor was fixed along the mechanical frame and connected the motor controller at the lateral side of electronic box with the aviation plugs.

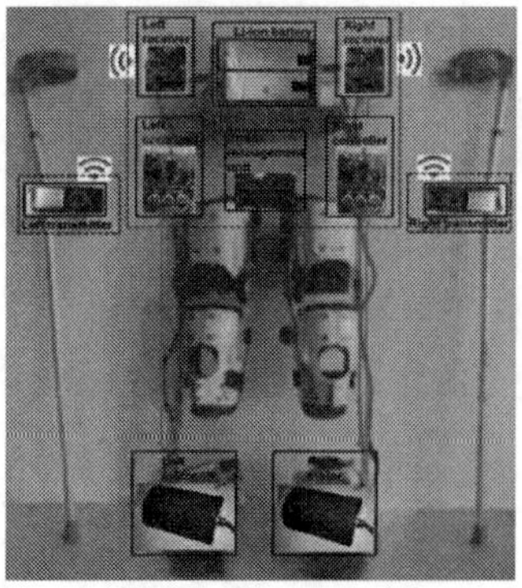

Figure 4. Electronic system block diagram (the circuit parts were enlarged). The red dashed box indicates the electronic parts in the main electronic box, blue dashed box indicates the electronic parts in the crutch, green linking arrow indicates the wire connection among parts.

4 Pilot tests on healthy subjects

To qualitatively analyse the capability of this exoskeleton, we conducted the indoor using tests on healthy subjects. The preparing procedure should be finished before functionally operating this exoskeleton. First, the user should wear the equipment at sitting posture, unlocking the knee joint of the exoskeleton and immobilizing the lower limbs with the fasten band. Secondly, switch on the exoskeleton and crutch, then stand up and lock the knee joint. Thirdly, press the button of crutch by the index finger to control the leg moving forward.

The main testing objective for this exoskeleton is its usability when adopting the four-point gait and three-point gait, which frequently appeared in the common crutch users. As shown in the Fig. 5, the motion sequence of the four-point gait is that the left crutch move forward first and the right leg step later, then the right crutch move forward and the left leg step, then repeat this sequence. In three-point gait, the two crutches move forward simultaneously and step left leg later, then crutches move forward again and step the right leg, then repeat this sequence. Although the foot will keep touch the ground in the whole moving procedure in this exoskeleton, while like the normal crutch user the leg of the exoskeleton would also step alternatively to walking forward. Therefore, theoretically either four-point gait or three-point gait could be adopt in this exoskeleton.

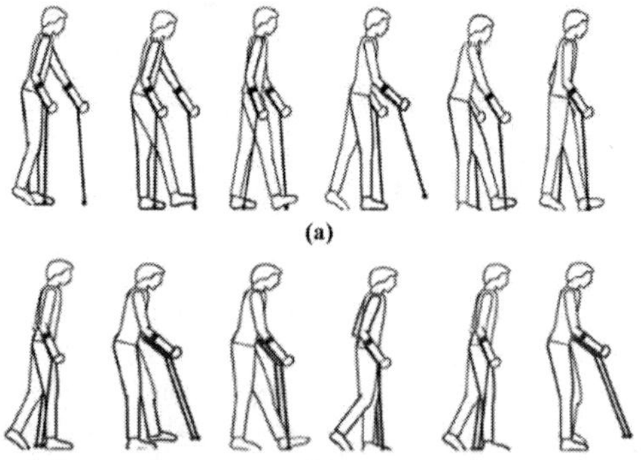

(a)

(b)

Figure 5. The typical operational procedure of the crutch when adopting four-point gait (a) and three-point gait (b)[15].

As shown in Fig. 6a, a healthy male subject with the height and weight 180cm, 70kg was recruited to operate the exoskeleton in four-point gait mode. The user moved the right crutch first and then pressed the button to control the left leg moving forward, and then repeated this procedure in right leg; A healthy subject with the height and weight 168cm, 60kg tested the three-point gait mode (Fig. 6b). Similarly with the common crutch user, the subject moved two crutches forward and then pressed one of the controlling button to move one leg, then repeated this procedure to move another leg.

(a)

DOI: 10.1051/matecconf/201710813003

(b)

Figure 6. Snapshots of using experiments on two healthy subjects with four-point gait (a) and three-point gait (b).

It should be noted that we tested two different gait modes by different subjects mainly for the consideration of testing diversity, while the effectiveness of different operation modes will not be affected by the different physical characteristics of user. Our test showed that both the user could individually operating the exoskeleton in four-point gait mode or three-point gait mode without the help from others. An additional fact we found in our tests is that although the capability of length adjusting for this exoskeleton's shank is 70 cm, whereas the wider scope of application than the range of 70 mm is still acceptable. This most because the exoskeleton do not need to consider about the lower limb alignment as joint-driving exoskeleton since the knee joint is locked when walking. In our practicing test, both the subjects with the height of 180 cm and 168 cm could operate this exoskeleton freely without uncomfortable using experience.

4 Conclusions

Upright walk is the most desirable needs of paraplegic patients. In this study we proposed a wheeled foot exoskeleton which uses a custom-made hub motor as driving unit and the crutch based wireless controller for controlling, enabling the patient to walk in a standing posture. The general designing concept and detailed specifications of each module were presented in this paper. Pilot experiments on two healthy subjects were performed and the results showed that the user could successfully complete the individual upright walking. In future study, we plan to perform a motion analysis experiment in a 3D gait analysis system to obtain more objective and evaluable data of this exoskeleton.

References

1. S. Hussain, Q. X. Sheng, P. K. Jamwal, J. Parsons, Med. Eng. Phys., **34**, 1448–1453 (2012)
2. R. E. Cowan, B. J. Fregly, M. L. Boninger, L. Chan, M. M. Rodgers, D. J. Reinkensmeyer, J. Neuroeng. Rehabil, **9**, 971-981 (2012)
3. A. Kittel, M. A. Di, H. Stewart, Disabil. Rehabil. **24**, 106-114 (2009)
4. V. Castellano, D. Coratella, and F. Felici, J.Electromyogr. Kines. **9**, 149-157(1999)
5. S. Hussain, S. Q. Xie, P. K. Jamwal, J. Parsons , Med. Eng. Phys., **34**. 1448-1453 (2012)
6. K. A. Strausser, T. A. Swift, A. B. Zoss, H. Kazerooni, B. C. Bennett, K. A. Strausser, *ASME 2011 Dynamic Systems and Control Conference and Bath/ASME Symposium on Fluid Power and Motion Control*, 419-425 (2011)
7. S. Tanabe, S. Hirano, E. Saitoh, NeuroRehabilitation, **33**, 99-106(2013)
8. K. H. Low, *Defense Science Research Conference and Expo*, 1 - 10 (2011)
9. R. Bogue, Industrial Robot, **42**, 5-10(2015)
10. JL. Contreras-Vidal, A. B. N, J. Brantley, JG Cruz-Garza, Y. He, J. Neural Eng., **13**, 3(2016).
11. D. Sasaki, T. Noritsugu, M. Takaiwa, *IEEE/RSJ International Conference on Intelligent Robots and system*, 4440-4445 (2013)
12. J. L. Contreras-Vidal and R. G. Grossman, *IEEE*, 579-1582013(2013)
13. Z. Li, H. Xie, W. Li, Z. Yao, Chin. J. Mech. Eng-En, **27**, 437-447(2014)
14. R. P. M. Chan, K. A. Stol, C. R. Annu. Rev. Control, **37**, 89-103(2013)
15. Z. Yang, *Gait planning and human-machine coordination study of walk assisting exoskeleton for paraplegics* (Master Thesis, Mechanical Engineering, Tsinghua University, Beijing, 2014).

DOI: 10.1051/matecconf/201710813004

Modelling and Simulation of Vessel Surgery based on Mass-spring

Da Ren, Yiming Chen, Bojun Lin, Fanjin Zeng, Jiaming Huang and Jinfeng Wang

Department of mathematics and informatics, South China Agricultural University, Guangzhou, 510642, China

Abstract: As the technology developing, precision and accuracy required in medical surgery can be realized through virtual-reality technology in computer aided systems, so that it can satisfy with medical experiments and teaching. In surgery simulation on soft tissues, Mass-spring takes the important roles on simulating the surface transformation of tissues. In this article, we established an intelligent simulation platform for surgery of vein in which includes the transformation based on Mass-spring. This platform can provide good human-computer interface and control some simple motions. It is convenient for medical teaching to instruct the operation scene.

1 Introduction

Medical simulation mainly produce virtual projects by using 3D reconstruction, new Photoelectric display, human-computer interaction, computer graphical technology and medical data visualization. The surgery simulation and visualization can provide an actual and immersed virtual environment for doctors by fusing multiple technologies. It can imitate the operation scenes from visual, audition and tactile angles. Surgery simulation provides strong support on the stages of preoperative planning, Intraoperative assistance and postoperative rehabilitation, in which Geometry modeling and physical modeling are fundamental technologies [1, 2]

In the early development, transformation of soft tissues is the key technology in virtual surgery simulation. Transformation modeling includes two kinds of methods: Finite element and Mass-spring model. Finite elements method has good stability and high accuracy. But it is hard to implement and compute. Comparably, Mass-spring is easier and has little computation. Mass-spring has been applied into many fields, such as transformation of soft tissues, cloth deformation and water surface [3]. Mollemans proposed Mass-spring model based on tetrahedron [4], in which each vertex corresponds to each particle and each edge corresponds to each spring. This model was close to elastic structure. In this article, our group proposed an intelligent platform for simulating the vein surgery based on Mass-spring. In traditional method, the force coefficient is set manually. In our model, we tried to tune the parameters using optimal algorithms.

The article is structured as follows. The model based on Mass-spring will be presented in section 2. Then, a virtual surgery simulation platform will be introduced in the next section. In final section, analysis and conclusions will be drawn.

2 Physical Model of Vein based on Mass-spring

2.1 Mass-spring model

Mass-spring can be treated as a piece of cloth like a new container of many masses. Each point has own quality without size. It is just the point mathematically and all points are distributed evenly. One spring connects two points and one point can connect multiple springs. These springs are linear and follow to Hooke's law. The topology structure is shown as Figure 1.

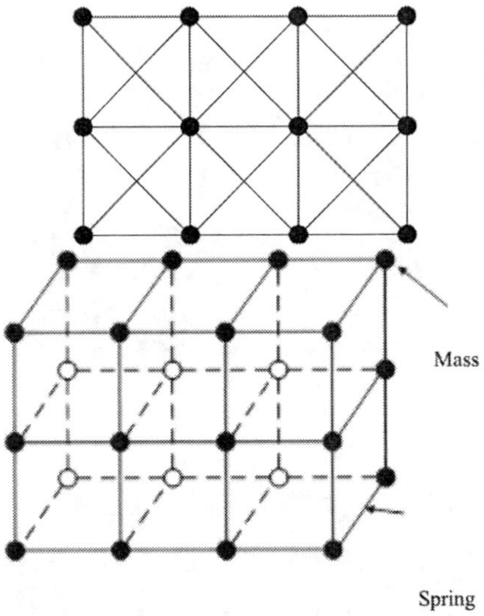

Figure 1. Topology structure of Mass-spring

Mass-spring includes twist spring, structure spring and tension spring as shown in figure 2. Twist spring means to the spring between diagonal vertices which can prevent tissues twisting abnormally. Structure spring can fix the whole tissue by connecting adjacent nodes. Tension spring connect every two points in order to help folding smoothly. Displacement of points is affected by many factors [5, 6]. One part is outer force which maybe single gravity, wind power, force from users or resultant of forces. The other means the elastic forces coming from the internal of Mass-spring. The affection to mass-point is the joint forces of external and internal ones. Accelerated velocity of some mass-points can be computed [7, 8].

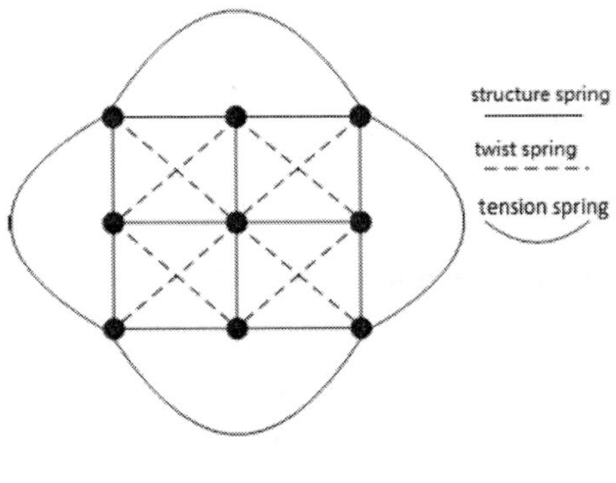

Figure 2. Fundamental structure of Mass-spring

In this article, we consider vessel as soft tissue surface model with uniform mass distribution, which can be partitioned equably into grids. We subdivided the grids further for improving accuracy and adopted grid topology structure as Figure 1 for enhancing real time performance in our model.

2.2 Setting parameters

Given that elastic modulus E, the corresponding spring constant k can be induced by $k = E\frac{A}{H}$, in which A is the cross-sectional area and H is the height. So we can obtain $k = E\frac{V_T}{L^2}$, in which V_T equals to the third of total volume of tetrahedrons which contain some points; L is the distance between the vertices of tetrahedron and this point. Based on these inferences, all parameters can be placed into the mechanical equation of Mass-spring. Let mass-point be i, each force on i would be decomposed respectively along x-axis, y-axis and z-axis. We take x-axis for example to explain as follows.

$$F_{ij} = \begin{cases} \frac{x_j - x_i}{\|x_j - x_i\|}\left(k_i\left(\|x_j - x_i\|\right) - L_{ij}\right) \Delta x_i < \Delta x_{ic} \\ \frac{x_j - x_i}{\|x_j - x_i\|}\left(k_1\|\Delta x_c + k_2\left(\|x_j - x_i\| - L_{ij} - \|\Delta x_c\|\right)\|\right) \Delta x_i \geq \Delta x_{ic} \end{cases} \quad (1)$$

in which x_i, x_j correspond to vectors of vertices on spring, L_{ij} is the static distance between two points.

Supposed that there are N tetrahedrons related to one point i, the volume of these tetrahedrons is V_i, the mass and volume of vessel is M and V, so the mass of this point can be computed as:

$$M_i = M \times \frac{V_i}{V \times N \times 3} \quad (2)$$

Then, we can attribute the mass of vessel into each point uniformly, which means the mass of each point is confirmed.

2.3 Force analysis

For any point in Mass-spring model, its dynamic characteristics can be represented as $m_i \ddot{x}_i + \beta_i \dot{x}_i + \sum_j^n F_{ij} + F_{ex} = 0$ according to Newton principle, in which m_i is the mass of point i, x_i is the coordinate vector, β is the damped coefficient of spring

DOI: 10.1051/matecconf/201710813004

and $\sum_{j}^{n} F_{ij}$ is the sum of forces affected by other points. When one point was stressed, the deformation depends on the external force and acting forces from other points.

3 Virtual surgery simulation platform in VTK

We established an intelligent platform based on VTK (Visualization Toolkit) using Visual C++. The vessel model was induced into system and each point can be extracted and incorporated in to Mass-spring model. After constructing the model, when the medical instruments like needle collide to the vessel, we need produce force on the touching point and make it move in a certain domain. Under Mass-spring action, the movement of touching point presents a satisfied visualization with high precision and accuracy. The flowchart of building platform is shown as Figure 3.

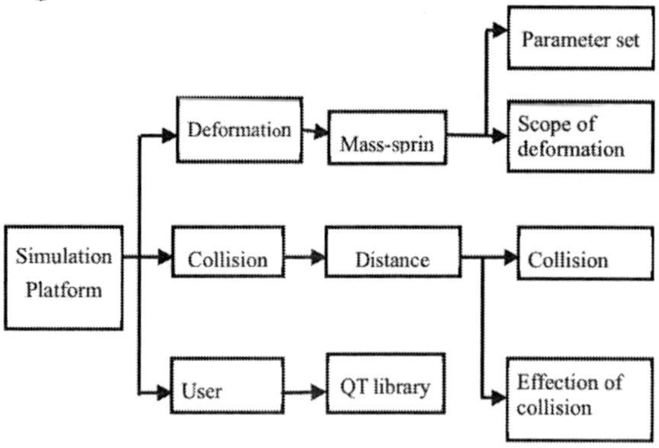

Figure 3. The flowchart of intelligent platform

We adopted classical programming structure for Mass-spring model in constructing model process. The location of point was described by coordinates and springs were established between points. When point was stressed to move, one triggered function can correct the location of the point according to the Mass-spring theory. For describing elastic behavior better, we added the body spring into system, which can keep tissue out of excessive deformation and ensure the model return to the original status. In our platform, we take needle for example to show the simulation process. The effect of this system is realistic enough and acceptable for operators. Several screenshots were selected to show as figure 4(a) and (b).

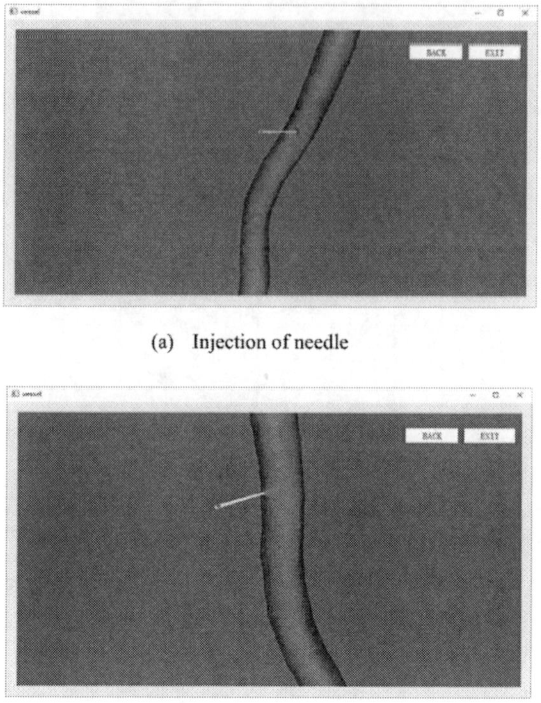

(a) Injection of needle

(b) Extraction of needle

Figure 4. Screenshots of intelligent platform

4 Analysis and conclusion

Mass-spring model have been developed greatly. The degree of simulation is improving, so the application fields are very wide. We focus on the surgery simulation based on Mass-spring and have realized an intelligent platform for vessel operation. There is still large space worth to explore and research.

In future work, we hope to solve three problems as follows.

1) Tuning parameters using Genetic Algorithm or Neural Network. Traditionally, the spring constants were set manually. It cannot satisfy the requirements for high precision and accuracy. So it is meaningful adopting machine learning algorithms to confirm those parameters;

2) 3D reconstruction for vessel model. As we all know, 3D model can provide intuitive knowledge for operators. However, the contrastographic picture of vessel is usually 2D graph [9]. How to reconstruct 3D vessel model based on contrastographic pictures from different angles? It will be an important problem.

3) Improving our intelligent platform by incorporating many different instruments simulation. Existing simulation

software was limited one or few surgery instruments [10, 11]. But different instruments have different interaction performance when colliding with tissues. We need consider different situation to construct different models. Intelligent adapting to different medical tools will be our research target.

Acknowledgment

This research is supported by the National Natural Science Foundation of China (No.: 61202295), and the Technology Planning Project of Guangdong Province (No.: 2015A020209150), and the Technology Planning Project of Guangdong Province (No.: 2015A030401081) and College Students' Science and Technology Innovation Project of Guangdong Province (No.: 201510564283).

References

1. T. W. Sederberg. Free-form deformation of solid geometric models [J], Acm Siggraph Computer Graphics, 20(4):151-160 (2010)
2. S. F. Gibson.3D chainmail: a fast algorithm for deforming volumetric objects [J], Symposium on Interactive 3d Graphics, 149-154, 195(1997)
3. Louchet J., Provet X., Crochemore D.. Evolutionaly identification of cloth animation models[C]// computer Animation and Simulation '95 Maastricht: 44-54(1995)
4. W. Mollemans, F. Schutyser, J. V. Cleynenbreugel, P. Suetens. Tetrahedral Mass-spring Model for Fast Soft Tissue Deformation[J], International Conference on Surgery Simulation & Soft Tissue Modeling, 2673:145-154(2002)
5. Coutinho B. B., Oliveira A. A. F, Atencio Y. P., et al. Rainscene animation through particle systems and surface flow simulation by SPH [C] // 23th SIBGRAPIGramado, Brazil, 255-262(2010)
6. Thürey N., Wojtan C., Gross M., et al. A multiscale approach to mesh-based surface tension flows [J]. ACM Transactions on Graphics, 29(4): 48-56(2010)
7. L.P. Nedel, D. Thalmann. Real time muscle deformation using mass-spring systems [J], Computer Graphics International(1998)
8. Iker A., Basil F. Filtering of high modal frequencies for stable real-time explicit integration of deformable objects using the Finite Element Method [J]. Progress in Biophysics and Molecular Biology, 103: 225-235 (2010)
9. Chris Guy D. F. An introduction to the principles of medical imaging [M]. London: Imperial college Press, 95-98(2005)
10. Sun S. H., Milan S. Lung segmentation refinement based on optimal surface finding utilizing a hybrid desktop virtual reality user interface [J]. Computerized Graphics, 37:15-27(2013)
11. Lam C. K., Sundraj K. Virtual simulation of eyeball and extra ocular muscle reaction during cataract surgery [J]. Procedia Engineering, 41:150-155(2012)

The simulation for study of von Mises stress sensitiveness to the change of coronal conformity in knee prosthesis

Usman[1], Shyh Chour Huang[2]

[1,2]*National Kaohsiung University of Applied Sciences, Mechanical Engineering Department, 80778 Kaohsiung, Taiwan, R.O.C.*

Abstract. Two-dimensional finite element contact modelling is often conducted to simulate the contact mechanism in knee prosthesis. This is based on the assumption that the contact stresses are more sensitive to change in the coronal plane than in the sagittal plane. Therefore, the simulation is then can be simply conducted only in the coronal plane with neglecting the sagittal dimension. Thus, the model is only a two-dimensional model. Based on this issue, the investigation of the response of contact stress (von Mises) to the coronal and sagittal conformity variation was conducted. For the need of investigation, ten three-dimensional finite element models of knee prosthesis have been created. From the result, the gradient of von Mises stress when the conformity varied in the coronal plane was -12.015, and when the conformity was varied in the sagittal plane, the gradient was -9.774. This meant that the von Mises stress had high sensitiveness to the change of the coronal conformity variation of knee prosthesis than to the change of the sagittal conformity. Therefore, the contact modelling of knee prosthesis can be simply carried out in the coronal plane.

1 Introduction

Contact stresses in knee prosthesis are more sensitive to change in coronal plane [1] than in sagittal plane. Therefore, the investigation of contact stresses in knee prosthesis is said simply can be carried out only in the coronal plane, without involving the sagittal plane. Thus, the contact stresses in knee prosthesis can simply be attained by only two-dimensional simulation which only simulated in the coronal plane. Therefore, the purpose of this research is to examine the response of von Mises stress, as one of the contact stresses, to coronal and sagittal conformity variation. Ten finite element models of knee prosthesis had been created for the investigation.

1.1 Knee joint

Among the joints in the human body, the knee joint is the largest and the most complicated joint [2]. The contact category in a knee joint is non-conforming. The non-conforming articular counterface leads to a point contact, which is a knee joint load is applied to a small area. Therefore, the high contact pressure occurs in the knee joint.

There are four bones make up the knee joint. Two of them are the thigh bone and shin bone, which are so called femoral and tibial bone, respectively. The other two are the fibula and kneecap bone (patella). The fibula is positioned along the lateral side of the shin bone [2]. And the patella as its knee joint shielding function, it is located in front of the knee joint [3]. The knee joint bones aforementioned are connected each other by mean of the ligaments. And for the need of joint movement, they are also linked by the tendons to the leg muscles [2].

There are three ligaments in the knee joint; the anterior cruciate ligament, the posterior cruciate ligament, the medial collateral ligament, and the lateral collateral ligament. The anterior and posterior cruciate ligaments avoid the thigh bone to dislocate posteriorly and anteriorly, respectively. And the medial and lateral collateral ligaments prevent the thigh bone to slide laterally and medially, respectively [2]. In some knee prostheses design, both the anterior and posterior cruciate ligaments are removed. Instead, to constraint the knee prosthesis from sliding anteroposteriorly and mediolaterally, the stabilizer is created. Besides, the cam shape of knee prosthesis also behaves as anterior and posterior cruciate ligaments. Meanwhile, in some other knee prostheses design, the cruciate ligaments are retained to keep natural stability. For a cushion and shock absorber in the knee joint, the medial and lateral menisci exist. In total knee arthroplasty, the menisci are removed and the cushion and shock absorber task is substituted by the insert.

1.2 Indication for total knee arthroplasty

Severe damage of the knee joint due to the osteoarthritis, the rheumatoid arthritis or some other knee joint degenerative diseases, indicates the total knee arthroplasty. The main objective of the total knee arthroplasty is to bring back the knee joint loss function due to severe knee arthritis that inflicts a serious pain and loss of stability of the knee joint. Otherwise, they can reduce the quality of life, or indeed, to some extent, they can disable the patient. The total knee arthroplasty is usually performed to older patients, but the procedure may also be implemented to young people considering that young people also have the possibility of suffering from the degenerative knee joint diseases [4].

Three among the many knee joint degenerative diseases are the osteoarthritis, the rheumatoid arthritis, and the post-traumatic arthritis. The osteoarthritis is induced by continued wear and damage of knee joint. The rheumatoid arthritis is occurred due to the inflammation of the membrane encompassing the knee joint. Both osteoarthritis and rheumatoid arthritis, at a certain level, tear the knee join cartilage. And thenceforth, the patients undergo a serious pain. Post-traumatic arthritis is the arthritis generated by harsh knee joint trauma. The bone fracture and ligament damages can lacerate the cartilage [4].

From these explanations of knee joint diseases, the serious pains in the knee joint appear after cartilage deterioration. This is because they are cartilages (distal femoral and proximal tibial cartilage) that rub each other during human locomotion. Therefore, the cartilage damage is a reason for the total knee arthroplasty.

1.3 Knee prosthesis

There are two popular designs of the knee prosthesis; the posterior-stabilized and the cruciate-retaining prostheses design. In the posterior-stabilized prostheses design, both anterior and posterior cruciate ligament are removed. The cruciate ligament functions are, then, substituted by the stabilizer and cam mechanism created in the knee prosthesis. Meanwhile, the cruciate-retaining prostheses design maintains the cruciate ligament on the existing position [4].

There are three main parts in knee prosthesis: condyle, insert, and tray (Figure 1). Condyle and tray are made of metal, and insert is made of polyethylene. Condyle is mounted on distal femur and tray is on proximal tibia. Meanwhile, the insert serves as a cushion between condyle and tray. When the condyle is brought into contact with the insert, the contact area will occur around the contact region. There are two kinds of contact area in knee prosthesis contact mechanism: initial contact area before deflection when there is not load is applied to the knee prosthesis; and the contact area after deflection when load is applied. The initial contact area is a result of conformity. Meanwhile, the contact area after deflection is a result of insert deformation.

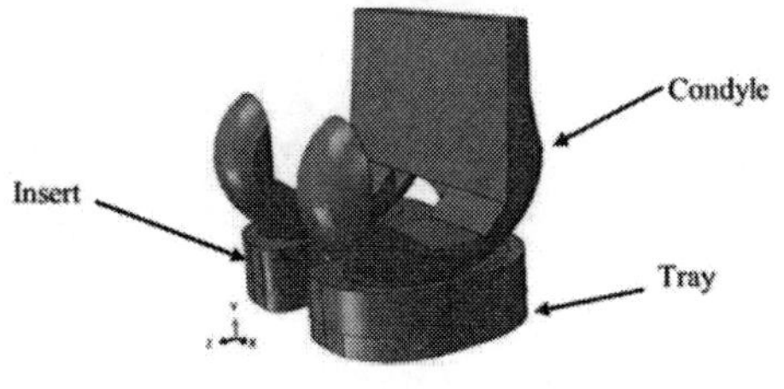

Figure 1. Parts of knee prosthesis

2 Materials and methods

The preprocession, simulation, and postprocession are three stages of finite element analysis in Abaqus software. The preprocession is the stage when the models are defined. There are seven steps have to be determined in this stage: creating the parts, defining the material property, parts assembly, deciding the step size, making the interaction between parts, applying loads, and discretizing the parts (meshing). In the simulation stage, the numerical problems of the model are solved. The time needed for the simulation depends on the complexity of models. The results of simulation are, then, evaluated in the postprocession stage [5].

Solidworks software was used to create the models and subsequent analysis was carried out in Abaqus finite element analysis software by importing the models into it. The coronal radius of condyle was 23 mm and the sagittal radius was 37 mm [6]. The condyle dimensions were not varied. Instead, to obtain the difference value of contact articular conformity, the radius of insert for both in coronal and sagittal plane was varied.

Due to the focus of our design was a symmetric knee prosthesis, that is the lateral and medial section of knee prosthesis are equal in size, we simplified the knee prosthesis model by only taking the medial section of knee prosthesis. We chose the medial section in the analysis due to the fact that the medial section supports 60% of total load applied to knee joint [7].

We used the knee joint absolute maximum load of four times bodyweight which occurred on the activity of descending stair [8]. And, the bodyweight of 700 N was chosen. Thus, the normal load yielded 1680 N.

When the conformity in sagittal plane was varied, the coronal conformity was set to be maximum value at 1 (full conforming). And conversely, when the conformity in the coronal plane was varied, the sagittal conformity was set to be maximal. The conformity was varied from 0.2 until 1 with interval of 0.2, namely 0.2, 0.4, 0.6, 0.8, and 1.

The condyle was assumed as a rigid body due to its much higher elastic modulus compared with UHMWPE as the material of the insert (E=550 MPa), and the Poisson's ratio of UMWPE of 0.44 was chosen [9]. The nonlinear behavior of true stress and the corresponding true strain of UHMWPE model are given in Figure 2. The friction coefficient of 0.04 was chosen [10].

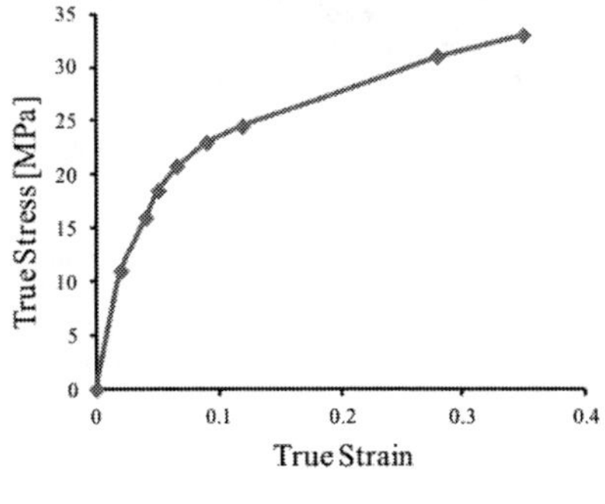

Figure 2. The nonlinear behavior of stress-strain of UHMWPE [10].

The condyle models, as a rigid body, were meshed with four-noded bilinear rigid quadrilateral element with global edge size of 0.75mm. Meanwhile, the inserts, as a deformable body, were mashed with eight-noded linear brick element with global edge size of 1mm. One of the finite element models for both condyle and insert models are shown in Figure 3.

3 Results and discussions

Figure 4 shows the conformity versus the von Mises stress for the coronal and sagittal conformity variation. The figure depicts the von Mises stress versus conformity equipped with the linear regression line. By the regression line, we know that the sensitiveness of von Mises stress to the variation of the coronal conformity was greater (with gradient of -12.015) than that occurred when conformity was varied in the sagittal plane, where the gradient (-9.774) was smaller than that found when the coronal conformity was varied.

The results assured us that the contact simulation in knee prosthesis can be decisively done by using simple two-dimensional simulation in the coronal plane. This can reduce not only the complexity of creating the biomechanical model -which is very difficult to create with the existing software because the shape is always irregular, but also can reduce the computation time and cost.

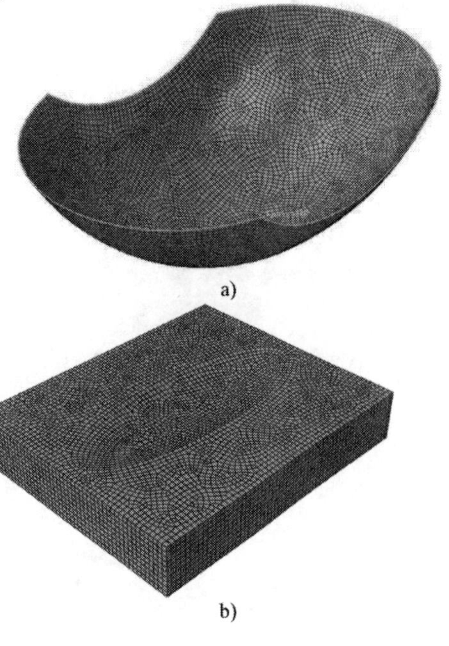

Figure 3. a) Condyle model with bilinear rigid quadrilateral element; (b) Insert model with eight-noded linear brick element.

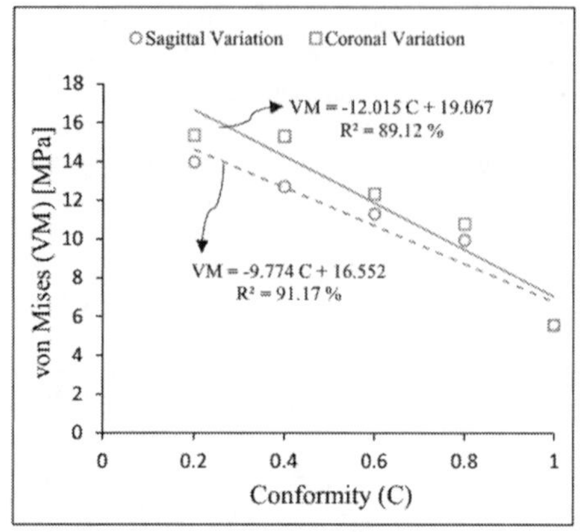

Figure 4. The von Mises stress versus the coronal and sagittal conformity variation with linear regression line.

Figure 5 shows the von Mises stress in insert model at lowest conformity level. The most important phenomenon that has to be noted here was the highest von Mises stress took place at the edge of insert dish. Meanwhile, at the center of contact, the highest von Mises stress took place below the insert surface. This meant that the sharp edge in the insert edge dish design has to be avoided. These phenomena existed in both coronal and sagittal conformity variation simulation.

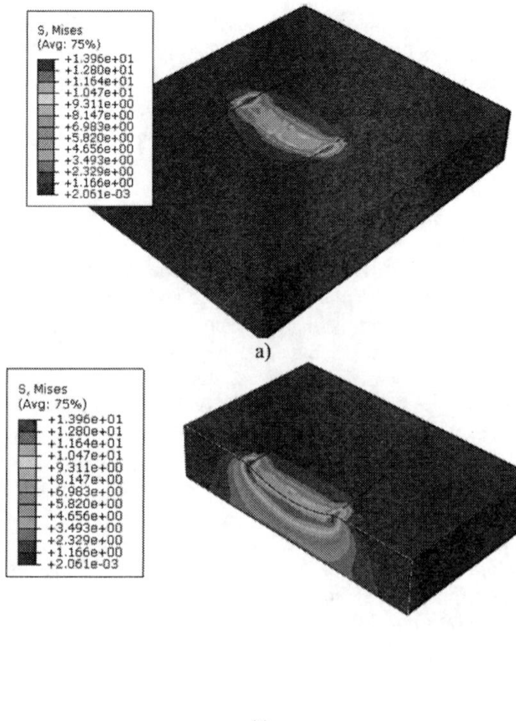

DOI: 10.1051/matecconf/201710813005

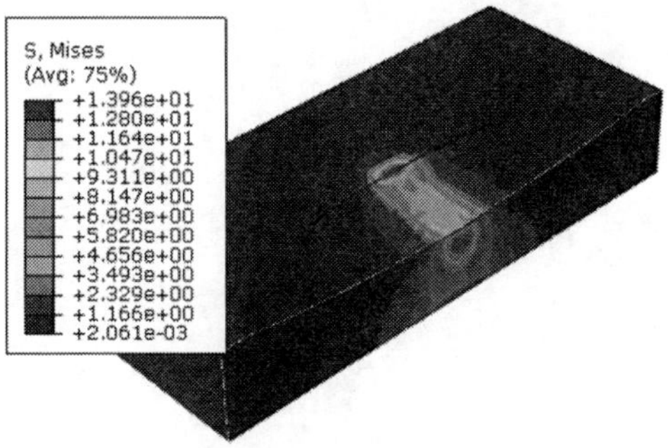

c)

Figure 5. a) One of the insert models with the von Mises stress appears at the surface; b) The coronal cut view of the insert model. It shows that the highest von Mises stress occurred below the surface along the coronal plane; c) The sagittal cut view of the insert model.

4 Conclusions

The investigation for proving the sensitiveness of the contact stress to change in the coronal plane has been conducted. We have revealed that the coronal variation of the conformity gave a more considerable increasing in von Mises stress, as one of contact stresses, than that given by the sagittal conformity variation.

Therefore, the study of the contact stresses in knee prosthesis can be only simulated in the coronal plane. So that, we can save time and computation cost because the simulation is conducted in only two-dimensional analysis.

Acknowledgement

The authors acknowledge and thank the Ministry of Science and Technology of the Republic of China for their partial financial support of this study with the Contract No. MOST 105-2221-E-151-016.

References

1. J. J. Rawlinson, D. L. Bartel, J Biomech **35**, 27-34 (2002).
2. Information on http://www.webmd.com.
3. Information on http://orthoinfo.aaos.org.
4. S. Affatato, *The history of total knee arthroplasty (TKA)* (Elsevier Ltd., 2015).
5. Simulia, *Getting Started with Abaqus: Interactive Edition* (Dassault Systèmes, 2014).
6. Y. F. Huang, S. C. Huang, *Optimal Shape Wear Analysis of Total Knee Replacement* (Department of Mechanical Engineering, National Kaohsiung University of Applied Sciences, Master Thesis, 2015).
7. E. A. Morra, A. S. Greenwald, J Bone Joint Surg. **85**, 111 -114 (2003).
8. I. Kutzner, B. Heinlein, F. Graichen, A. Bender, A. Rohlmann, A. Halder, A. Beier, G. Bergmann, J Biomech **43**, 2164-2173 (2010).
9. R. Willing, I. Y. Kim, J Biomech **42**, 2520–2527 (2009).
10. M. Bahraminasab, B.B. Sahari, M. R. Hassan, M. Arumugam, M. Shamsborhan, Trends Biomater Artif Organs **25**, 95-100 (2011).

Effect of Average Conformity on Contact Stresses in Total Knee Replacement: Finite Element Analysis

Usman[1] and Shyh Chour Huang[2]

[1,2]*National Kaohsiung University of Applied Sciences, Mechanical Engineering Department, 80778 Kaohsiung, Taiwan, R.O.C.*

Abstract. The excessive contact stresses have been proved by many investigations for their role in the tibial insert failure. One parameter in the total knee replacement (TKR) design that influences excessive contact stresses is the average conformity. Thus, the aim of this study was to investigate the effect of the average conformity on the contact stresses in the TKR tibial insert. Ten three dimensional-finite element models with different conformity level were created. The conformity values of 0, 0.2, 0.4, 0.6, 0.7, 0.8, 0.9, 0.95, 0.99, and 1 were applied to the sagittal and frontal plane of the TKR models. The results showed that the contact pressure and the von Mises stress were decreased if the average conformity were increased, and vice versa. And, the average conformity has a larger effect on the contact pressure than on the von Mises stress. Also, from these simulations, we were informed that the developing of the contact pressure and the von Mises stress in the TKR tibial insert was greatly influenced by the average conformity. And, the von Mises stress and the contact pressure had the same trend in response to the average conformity.

1 Introduction

The previous reports show that there is a positive correlation between wear intensity and contact stress level. The wear intensity is greater in low conformity than in high conformity total knee replacement (TKR) design [1]. The conformity is defined as the difference of radius between two articular profiles involved in a contact [2]. The conformity is attained by dividing the femoral component radius by tibial insert profile radius, to be between 0 and 1. Where, 0 is the minimum conformity and 1 is the maximum conformity [3].

Although some other investigations reveal that the decreasing conformity can lead to lower wear, the low conformity may produce the other failure mode of TKR due to the excessive contact pressure [3]. Therefore, the conformity is still important to study in TKR design. Thus, the aim of this study is to investigate the effect of TKR conformity on contact stresses. In this study, the conformity value was represented by the average conformity of frontal and sagittal conformity.

Ten three dimensional-contact models with different conformity value, both in frontal and sagittal plane, were created in Solidworks and then simulated statically in Abaqus software. Due to the purpose of this simulation was to study the conformity effect phenomenon in TKR, the contact models of TKR were simplified by only using two parts: a curved indenter as the model for femoral component and a plane with dish-out-path for the tibial insert. The contact pressure and von Mises stress were then extracted from output database for analysis.

2 Finite element analysis

2.1 Models design

Ten three dimensional-finite element models with different conformity level were created. The sagittal (RFS) and frontal (RFF) profile radius of femoral component were kept constant at 37 and 23 mm [4], respectively (Figure. 1a). Instead, the sagittal (RTS) and frontal (RTF) profile radius of tibial insert dish (Figure. 1b) were varied to create the conformity level between femoral component and tibial insert dish. The conformity values of 0, 0.2, 0.4, 0.6, 0.7, 0.8, 0.9, 0.95, 0.99 [5], and 1, for both in sagittal and frontal plane of TKR, were chosen. The conformity values were then randomized by using Microsoft Excel to give all frontal and sagittal conformities the equal chance to pair [6], as we can see in Table 1 (column two and three).

By using the conformity values, the variation of profile radius of tibial insert dish, in sagittal and frontal plane, can be determined by dividing the femoral profile radius (RFS or RFF) with the corresponding conformity value. For example, model 1, RTF was obtained by dividing RFF (23 mm) by the conformity in frontal plane (0.7), it equals to 32.9 mm. And, RTS was obtained by dividing RFS (37 mm) by sagittal plane conformity (0), it

DOI: 10.1051/matecconf/201710813006

equals to infinity. The same way was used for the rest of models.

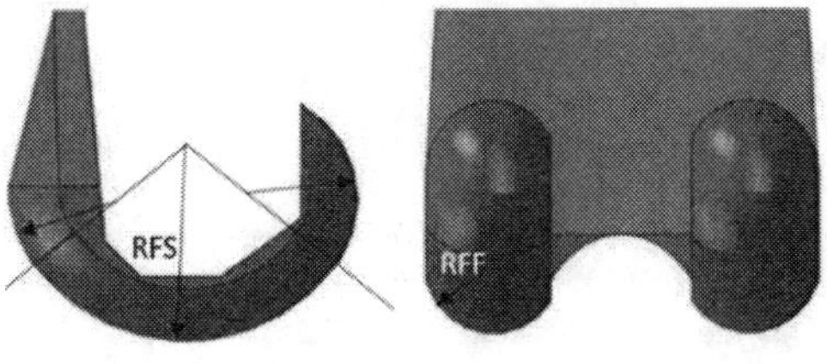

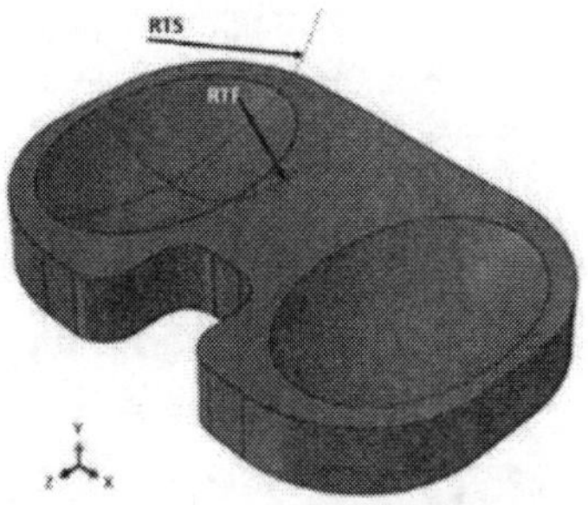

Figure 1. a) RFS and RFF are the profile radius of femoral component in sagittal and frontal plane, respectively; b) RTS and RTF are the profile radius of tibial insert dish in sagittal and frontal plane, respectively.

Table 1. The conformity and measurement of tibial insert dish both in sagittal and frontal plane.

Model No.	Conformity plane		RTS [mm]	RTF [mm]	Average conformity
	Frontal	Sagittal			
1	0.7	0	~	32.9	0.7
2	0.9	0.7	52.9	25.6	0.8
3	0.95	0.8	46.3	24.2	0.875
4	0.6	1	37.0	38.3	0.8
5	1	0.4	92.5	23.0	0.7
6	0.99	0.9	41.1	23.2	0.945
7	0.2	0.2	185.0	115.0	0.2
8	0	0.6	61.7	~	0.6
9	0.4	0.95	38.9	57.5	0.675
10	0.8	0.99	37.4	28.8	0.895

The models were simplified by only using the medial part of TKR (Figure. 2) due to symmetric TKR design. This is because of the medial part supports larger portion of knee joint load than lateral part at 60-40% ratio as also used by Halloran et. al. [7]. The femoral component was represented by curved shell with sagittal and frontal profile radius of 37 and 23 mm, respectively. And the tibial insert was represented by a plane containing a dish with sagittal and frontal profile radius as tabulated in Table 1 (column four and five), respectively. This way of simplifying model followed the work of Wang et. al [8]. The minimum thickness of tibial insert of 8 mm was maintained to follow Bartel et al.'s recommendation for the minimum thickness of tibial insert [9].

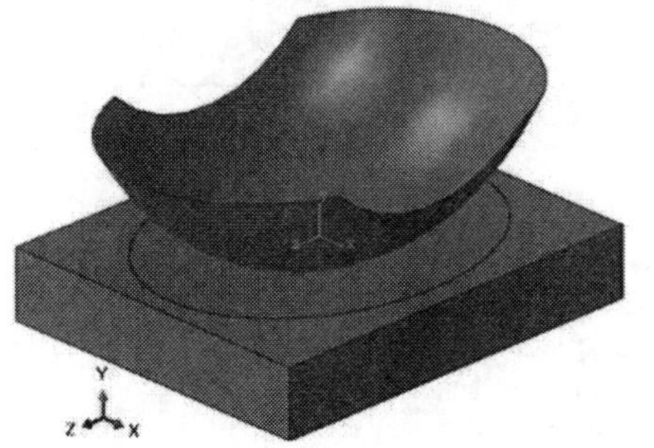

Figure 2. The simplified of TKR contact model for finite element analysis.

2.2 Material properties

The tibial insert was modeled as a deformable body with UHMWPE as its material (E=550 MPa [7], ν = 0.44 [3]). Meanwhile, a rigid body assumption was applied to the femoral component because the CoCr (the material of femoral component) modulus of elasticity is much higher than UHMWPE. The nonlinear true stress versus true strain and coefficient of friction of UHMWPE were used as in the previous work of Halloran et. al. [7].

2.3 Loads and boundary conditions

The vertical knee joint load during descending stair of 400% bodyweight [10] was applied at the femoral component reference point. The bodyweight of 700 N was chosen. Due to the simulation was only in medial part, 60% of load portion was used. Thus, the total vertical load equaled 1680 N (400% x 60% x 700). The femoral component was only allowed to translate in vertical direction, while the tibial insert was fixed in all directions.

2.4. Finite element Models.

The finite element analysis was conducted in ABAQUS software. The Solidworks was used for creating the models and then inputted into ABAQUS. A discrete rigid element (R3D4: 4-node 3-D bilinear rigid quadrilateral) with 0.75 mm of global mesh size was used for femoral component. And the tibial insert was discretized by using 8-node linear brick element (C3D8) with 1 mm of global mesh size. The finite element is shown in Figure. 3.

Figure 3. The finite element model.

2.5 Results

The Figure 4 shows the contact pressure (left) and von Mises stress (right) for all models. From the figure, we can see that the contact stresses have the same trend with the von Mises stresses. As also can be seen in the figure, the von Mises stress will increase if the contact pressure increases, and vice versa. The smallest contact pressure and von Mises stress of 11.44 and 9.62 MPa, respectively, were occurred in the model 6 with the average conformity of 0.945. By this result, we got the evidence that the average conformity has a great influence in developing contact pressure and von Mises stress in the TKR tibial insert. The largest contact pressure and von Mises stress were occurred in model 7. This model has the lowest average conformity of 0.2. Therefore, the contact pressure and von Mises stress increased to 37.48 and 20.46 MPa, respectively.

The average conformity was a new parameter we introduced in TKR design. Compared with the only conformity which merely represents the conformity in one plane (frontal or sagittal) of TKR, the average conformity is more important to study because it represents the conformity in both frontal and sagittal plane at the same time.

Through this study, we propose the average conformity as a parameter that should be considered in TKR design in the future. This is because, to the authors' knowledge, the average conformity has never been used in TKR design.

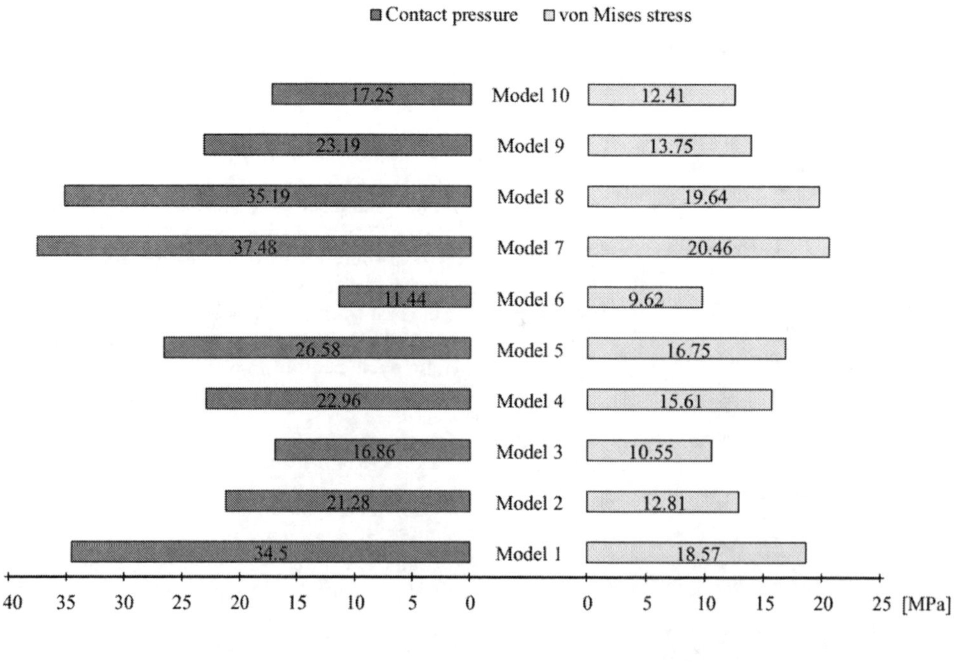

Figure 4. Contact pressure and von Mises stress in all model.

By using the linear regression (Figure. 5) we can see that the contact pressure (CP) and the von Mises Stress (VMS) decreased by the increasing of average conformity (AC). It also showed that 93.9% of contact pressure in tibial insert was explained by the average conformity, while the von Mises stress (VMS) was at 85.75%. From the gradient of the regression line, we know that the contact pressure was more sensitive to the average conformity than the von Mises stress. For the contact pressure, the gradient was – 31.152, while - 12.952 was for the von Mises stress.

In the figure, we found that the data are only concentrated around the tip of regression line. This is because the conformity values used in this simulation were taken from the previous literature where the conformity is selected from the rational value of the conformity in TKR. So, the data did not spread along the regression line. However, with the limited conformity data presented in the literature, it is enough to be used to know the average conformity effect on the contact stresses as shown in Figure 5.

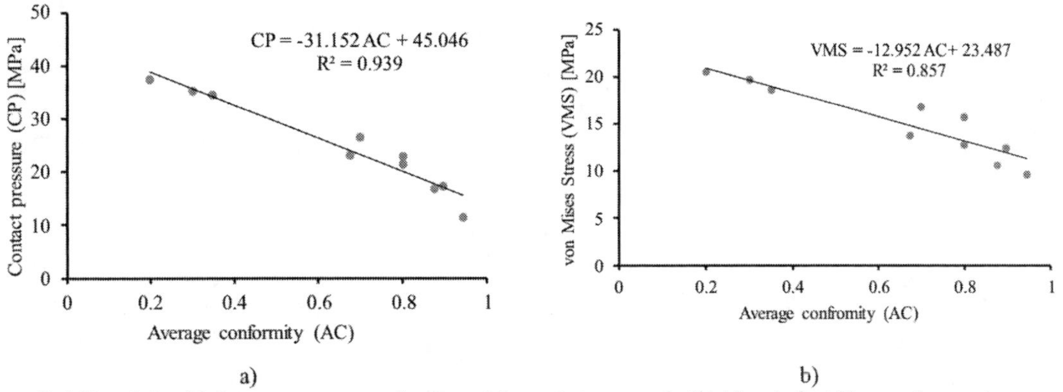

Figure. 5. a) The relationship between average conformity and the contact pressure in tibial insert of total knee replacement. Meanwhile b) is the relation of average conformity with von Mises stress.

3 Conclusions

We have conducted the finite element simulation of ten models with different value of average conformity. From the result, we found that the contact pressure and the von Mises stress were decreased by the increasing of the average conformity, and vice versa. Compared with the von Mises stress, the contact pressure was more sensitive to the average conformity. Along with these simulations, we also got the clarification that the developing of contact pressure and von Mises stress in the TKR tibial insert was greatly influenced by the average conformity. And, the von Mises stress and the contact pressure had the same trend in response to the average conformity. Based on its effect on developing contact stresses in the TKR and its representation of sagittal and frontal conformity, the average conformity should be considered as design parameter in TKR design in the future. To the authors' knowledge, the average conformity has never been used in TKR design.

Acknowledgment

We would like to show our gratitude to the Minister of Sciences and Technology of the Republic of China for their partial financing this study under Contract Number MOST 105-2221-E-151-016.

References

1. J. P. Collier, M. B. Mayor, J. L. McNamara, V. A. Surprenant, R. E. Jensen, Clin. Orthop. Relat. Res. 273, 232-242(1991).
2. M. M. Ardestani, M. Moazen, Z. Jin, Knee 22, 217–224 (2015).
3. R. Willing, I. Y. Kim, Struct. Multidiscipl. Optim. 38, 105-414 (2009).
4. H. C. Jhao, J. Y. Wang, *Construction of Custom-made Artificial Knee Joint by Means of Contact Information* (Master Thesis in Institute of Electrical and Mechanical Engineering, Cheng Shiu University, Taiwan, 2013).
5. M.S. Kuster, S. Horz, E. Spalinger, G.W. Stachowiak, A. Gächter, Clin. Orthop. Relat. Res. 375, 302-312 (2000).
6. D. S. Moore, G. P. McCabe, B. A. Craig, *Introduction to the Practice of Statistic, sixth edition* (W. H. Freeman and Company, United States of America, 2009).
7. J. P. Halloran, A. J. Petrella, P. J. Rullkoetter, J Biomech. 38, 323-331 (2005).
8. C. Y. Wang, H. C. Jhao, C. F. Huang, Life Sci J. 10, 254-258 (2013).
9. D. L. Bartel, V. L. Bicknell, T. M. Wright, J Bone Joint Surg Am. 68, 1041 -1051 (1986).
10. I. Kutzner, B. Heinlein, F. Graichen, A. Bender, A. Rohlmann, A. Halder, A. Beier, G. Bergmann, J Biomech. 43, 2164-2173 (2010).

Development and validation of a C0-C7 cervical spine finite element model

Wei Wei[2], Yin Liu[1], Xianping Du[3], Na Li [1]

[1]*Department of Radiology, The Third Xiangya Hospital, Central South UniverstiyChangsha, 410012,China City*
[2]*Laboratoire de Biomécanique Appliquée, MRT24 IFSTTAR-Aix-Marseille Université, Bd. P. Dramard, Faculté de Medecinesecteur-Nord, Laboratory, Marseille, 13916, France*
[3]*Department of Mechanical Engineering, Embry-Riddle Aeronautical University, Daytona Beach, FL 32114, USA*

Abstract. A finite element model (FEM) of human C0-C7 cervical spine has been developed as a baseline to study the biomechanical factors in spinal surgical intervention. To accurately simulate the anisotropic properties, the intervertebral disc was divided into nucleus, annulus matrix and annulus fiber, and it was simulated with viscoelastic and fabric material models. The nonlinear force-displacement curves of ligament experiments were implemented to simulate the elastic zone and neutral zone at low loads. The model was validated with experimental data on range of motion (ROM) for normal, non-degenerated cervical spines tested in flexion and extension, lateral bending, and axial rotation at loads of 0.33, 0.5, 1.0, 1.5, and 2.0 Nm, and intra-discal pressure (IDP) was validated against experiment data at loads of 2Nm and 5Nm. For lateral bending and axial rotation, the model was well within in vitro experimental standard deviation corridors for the whole load range. For extension and flexion, however, the error index of C3-C4, C4-C5, C5-C6 flexion were 0.268, 0.03, 0.124 and the C6-C7 extension was 0.046 respectively. The IDP result was in well agreement with the experiments. These results indicated the C0-C7 cervical spine model was biofidelic for static simulation of these motions.

1 Introduction

With advancement of clinical and experimental studies on the pathogenesis of cervical spine diseases and injuries, biomechanical factors have been found to play an essential role as a series of parameters were found to change in injury and degeneration[1]. Comparing to the limited availability of specimen and poor repeatability of experiment in clinical and laboratory study, mathematical modeling, especially finite element modeling, is an important complement method in spine biomechanics study as it could quantify both internal and external response of cervical spine under various loading scenarios, which may hopefully improve our understanding of the mechanism of cervical spine diseases and injury, as well as prevention or treatments.

As the spinal degeneration and surgical treatment are in a quasi-static condition and under this condition, accurate geometry and reliable soft tissue response are of primary concern in investigating the biomechanical responses of cervical spine. In the last two decades, finite element modeling for static cervical spine diseases has made great progress. On the basis of segmental models[2][3] in the early time, occipito-atlanto-axial joint was included to simulate whole head and neck, as structure integrity of whole cervical spine is highly important for accurate prediction made by FEM. Minor soft tissues structures, such as cartilage endplate and facet joint cartilages, which were also not included.

Despite the improvement mentioned above, validation of FEMs response against experimental data is essential for model bio-fidelity as it defines the conditions under which the model can be applied and the confidence of the prediction results is guaranteed. For model validation, the computer model should produce similar results as that of experiment under identical boundary and load conditions. The level of validation is assessed by the degree to which computer model response matches experimental response and the quality of experimental data. However, not all finite element models (FEMs) of the human cervical spine have been subjected to rigorous validation and therefore have limited application to the study of normal and abnormal cervical conditions. In this study, a FE model of C0-C7 human cervical spine was developed and extensively validated with segmental ROM of flexion, extension, lateral bending and axial rotation, as well as inter-vertebral disc pressure at various momentum loading condition. Hopefully, this FE model may be used for patient-specific adjustment in spinal degeneration and surgical intervention studies.

2 Method

DOI: 10.1051/matecconf/201710813007

The geometry was defined by CT and MRI image information of a 24-year-old healthy male with average height and weight. After three dimensional osseous structure of cervical spine was reconstructed with MIMICS (Version 12, Materialise Inc., Leuven, Belgium), the elements were meshed with HYPERMESH (Version 10, Altair, Troy, MI). Soft tissues including intervertebral disc, ligaments and articular cartilage were then incorporated into the model. The entire model consisted of 86,286 nodes, 106,311 elements, including 64,926 solid elements and 40,564 shell/membrane elements, 821 beam elements. The length, width, and height of most of the elements ranged from 0.75 to 1.25mm, as shown in Figure 1.

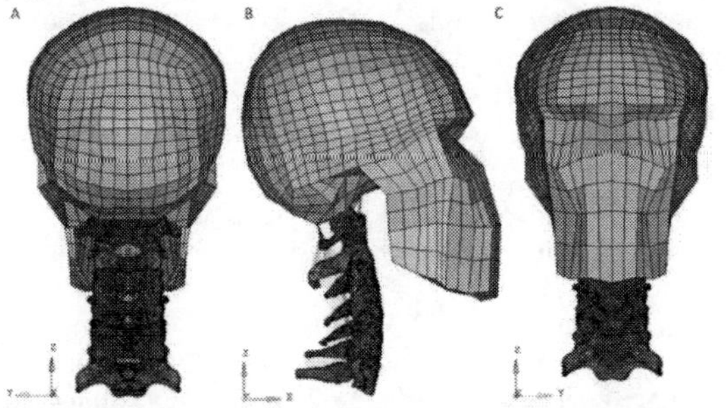

Figure 1. The head-neck model in Cartesian coordinate system

The material type and properties of this model were listed in Table 1. adjust the margins to those shown in Table 1.

Table 1. Material Properties of Cervical Spine Finite Element Modeling[4].

Tissue names	Element type	Material type	Material Parameters
Cortical bone	Shell	power-law plasticity	ρ=1.61e-9 t/mm^3, E=16700 MPa, μ=0.3, K=354.8 MPa, N=0.2772
Cancellous bone	hexahedron	power-law plasticity	ρ=8.77e-10 t/mm^3, E=291 MPa, μ=0.3 K=5.7 MPa, N=0.2741
Endplate	Shell	power-law plasticity	ρ=1.61e-9 t/mm^3, E=5600 MPa, μ=0.3, K=153.2 MPa, N=0.2772
Matrix of annulus fibrosus	hexahedron	Hill Foam	m=3, n=2, C1=2.1857 MPa, b1=1, C2=-2.36 MPa, b2=2, C3=0.891 MPa, b3=3
Annulus fibrosus fibres	Shell	Fabric	Strain-Stress Curve
Nucleus	hexahedron	General viscoelastic	N=4, K=1.72 GPa, G_1=0.5930 kPa, β_1=0.001477 1/s, G_2=0.6763 kPa, β_2=0.061524 1/s, G_3=0.9516 kPa, β_3=1.017893 1/s, G_4=2.0384 kPa, β_4=13.20041 1/s,
Cartilage endplate	hexahedron	Isotropic elastic	ρ=1.36e-9 t/mm^3, E=25 MPa, μ=0.4
Facet articular cartilages	Hexahedron	Isotropic elastic	ρ=1.36e-9 t/mm^3, E=10.4 MPa, μ=0.4
Ligaments	Beam	Non-linear	Displacement-Force Curve

For each motion segment, the model is validated in flexion, extension, lateral bending and axial rotation under quasi-static tensile loading with moments of 0.33, 0.5, 1.0, 1.5, and 2.0 Nm. The flexion and extension responses of this model were compared with previous cadaveric studies reported by Nightingale et al.[5] and Wheeldon et al.[6]. In addition, the inter-vertebral disc pressures(IDP) were compared to 2Nm flexion/extension quasi-static experiment in C2-C3, C6-C7 segments reported by Dmitriev[7] and the 5Nm axial rotation and lateral bending quasi-static experiment in C4-C5, C5-C6 segments reported by Kretzer[8].

To evaluate the accuracy of the model, the evaluation index on ROM response for all loadings was calculated in gross correlation index (GCI)[9] equation(1).

$$F(E)=\frac{1}{6}\sum_{n=1}^{m}\left[\left(\frac{F_i^m-F_i^0}{F_i^m}\right)^2\right] \quad (1)$$

The goodness of fit (or correlation) between the simulated results and test data was evaluated using a GCI developed by Deb et al.[9]. Here F_i^m presents the mean value of experiment data, while F_i^0 was the predicted simulation value. This algorithm was utilized to assess the accuracy of FE models simulations in previous literatures [10][11].

3 Result

The ROM of each segment from C0 to C7 in this model were compared to the studies from Nightingale et al.[5] and Wheeldon et al.[6] of flexion/extension, as shown in Figure 2. According to the gross correlation index (GCI) equation, the error index of C3-C4, C4-C5, C5-C6 flexion were 0.268, 0.03, 0.124 and the C6-C7 extension was 0.046 respectively. The extension of C0-C2, C2-C3,

C3-C4, C4-C5, and the flexion of C0-C2, C2-C3, C6-C7 agreed well with the experimental data.

As cadaveric research on the IDP was quite scarce, IDP of the single segment models was validated with corridors obtained from a pure moment study of Dmitriev 2005[7] on 2Nm flexion/extension, and the study of Kretzer 2012[8] on 5Nm axial rotation and lateral bending. The model also matched very well with results obtained in these literatures, as shown in Figure 3.

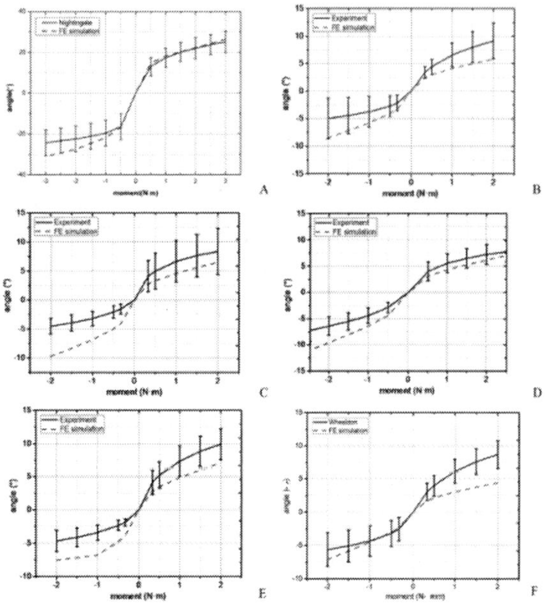

Figure 2. Flexion-extension response of each segments (A. C0-C2, B. C2-C3, C. C3-C4, D. C4-C5, E. C5-C6, F. C6-C7) for inter-segment models under a range of quasi-static flexion and extension moments. For sub-axial segment validation, the simulation results were compared to experiment of Nightingale et al, 2007(A. C0-C2) and Wheeldon et al, 2004(B. C2-C3, C. C3-C4, D. C4-C5, E. C5-C6, F. C6-C7).

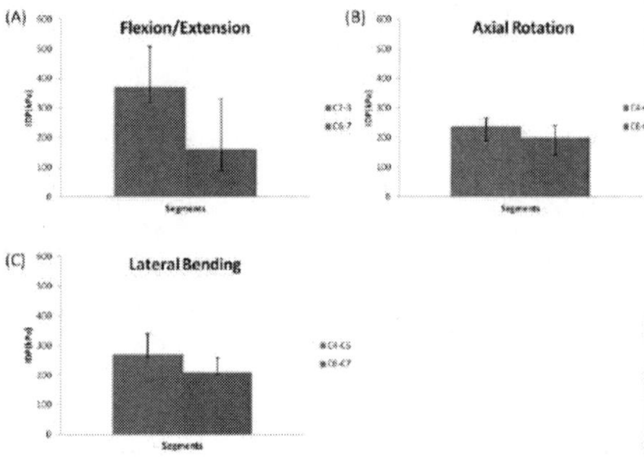

Figure 3. IDP response for inter-segment model in different motion compare to literatures (A. Flexion in moment 2Nm. B. Axial rotation in moment 5Nm. C. Lateral bending in moment 5Nm).

4 Discussion

Several quasi-static C0-C7 cervical FE models were developed and validated in last few years. Del Palomer et al.[12] developed a C0-C7 cervical spine model for quasi-static simulations, but only flexion-extension moment-rotation curves were validated with previous cadaveric experimental literatures; similarly, Panzer et al.[4] developed a C0-C7 cervical spine model for frontal impact injury, while only the flexion and extension quasi-static experiments reported by Nightingale et al.[5] and Wheeldon et al.[6], 2004 were utilized in validation. Toosizadeh et al.[13] also developed a C0-C7 cervical spine with muscle for the study of neck pain, and only single torque responses in 1Nm loading were validated with all the motions. To our knowledge this is the only C0-C7 FE model that was accurately validated against quasi-static ROM and IDP experimental data of normal specimens.

In this study, an FE cervical model is developed as a baseline to investigate the tissue-level biomechanical change in static simulation of various motions including flexion/extension and IDP. The C0-C7 FE model of human cervical spine without any adjustments was exercised in 24 loading cases without numerical difficulties to ensure its precision in multiple cervical motion simulations. Based on the model developed in this study, patient-specific FE models could be developed to further investigate the biomechanical response induced by structural change during surgical intervention and/or degeneration. As a matter of fact, the baseline model is being used for patient-specific modeling of patients after surgical intervention and hopefully more accurate biomechanical data might be obtained to investigate the cervical spine response after surgery.

Acknowledgement

The authors gratefully acknowledge the support of the National Education Department Key Laboratory through the research projects cqu-moekl-br-2011-3 and 2012QNZT161.

References

1. Katsuura A, Hukuda S, Saruhashi Y, Mori K. Kyphotic malalignment after anterior cervical fusion is one of the factors promoting the degenerative process in adjacent intervertebral levels. European Spine Journal. 2001;10:320-324.
2. Yoganandan N, Kumaresan S, Pintar FA. Biomechanics of the cervical spine part 2. Cervical spine soft tissue responses and biomechanical modeling. Clinical Biomechanics. 2001;16:1-27.
3. Ha SK. Finite element modeling of multi-level cervical spinal segments (c3–c6) and biomechanical analysis of an elastomer-type prosthetic disc. Medical engineering & physics. 2006;28:534-541.

4. Panzer MB, Fice JB, Cronin DS. Cervical spine response in frontal crash. Medical engineering & physics. 2011;33:1147-1159.
5. Nightingale RW, Winkelstein BA, Knaub KE, Richardson WJ, Luck JF, Myers BS. Comparative strengths and structural properties of the upper and lower cervical spine in flexion and extension. Journal of biomechanics. 2002;35:725-732.
6. Wheeldon JA, Pintar FA, Knowles S, Yoganandan N. Experimental flexion/extension data corridors for validation of finite element models of the young, normal cervical spine. Journal of biomechanics. 2006;39:375-380.
7. Dmitriev AE, Cunningham BW, Hu N, Sell G, Vigna F, McAfee PC. Adjacent level intradiscal pressure and segmental kinematics following a cervical total disc arthroplasty: An in vitro human cadaveric model. Spine. 2005;30:1165-1172.
8. Kretzer RM, Hsu W, Hu N, Umekoji H, Jallo GI, McAfee PC, Tortolani PJ, Cunningham BW. Adjacent-level range of motion and intradiscal pressure after posterior cervical decompression and fixation: An in vitro human cadaveric model. Spine. 2012;37:E778-E785.
9. Deb A, Haorongbam B, Diwakar J, Chou C. Efficient approximate methods for predicting behaviors of steel hat sections under axial impact loading. 2010.
10. Jiang B, Zhu F, Jin X, Cao L, Yang KH. Computational modeling of the crushing behavior of skydex® material using homogenized material laws. Composite Structures. 2013;106:306-316.
11. Zhu F, Chou CC, Yang KH, Chen X, Wagner D, Bilkhu S. Application of am60b magnesium alloy material model to structural component crush analysis. International journal of vehicle safety. 2012;6:178-190.
12. Del Palomar AP, Calvo B, Doblaré M. An accurate finite element model of the cervical spine under quasi-static loading. Journal of biomechanics. 2008;41:523-531.
13. Toosizadeh N, Haghpanahi M. Generating a finite element model of the cervical spine: Estimating muscle forces and internal loads. Scientia Iranica. 2011;18:1237-1245.

MATEC Web of Conferences **108**, 13008 (2017) DOI: 10.1051/matecconf/201710813008
ICMAA 2017

Microfluidic Blood Typing Chip with Voltammetry Measurement

Yaw-Jen Chang , Xin-Miao Zhou, Hsiu-Rong Yen and Chih-Ling Yeh

Department of Mechanical Engineering, Chung Yuan Christian University, Chung Li District, Taoyuan City, Taiwan

Abstract. Blood typing is a critical test in many medical procedures, especially for blood transfusion. Incompatible red blood cells (RBCs) in transfusion might result in intravascular hemolysis and even cause recipients death. The test result of blood typing can be determined by observing the agglutination of RBCs. In this study, a microfluidic blood typing chip with interdigitated electrodes is presented. ZnO nanowires were synthesized on the interdigitated electrodes so that voltammetry measurement was employed to determine the test results of blood typing. The experimental results show that the discrimination between agglutination and non-agglutination was obvious and successful determinations of RBC agglutinations in blood typing tests were achieved by means of voltammetry measurement.

1 Introduction

Blood is a body fluid in humans for delivering nutrients and oxygen to the cells and, simultaneously, transporting metabolic waste products away. In case of heavy bleeding, blood transfusion is a medical treatment to replace the lost blood components. Before the transfusion, blood typing is a critical test to ensure the serological compatibility of a donor and a recipient [1]. Incompatible red blood cells (RBCs) in transfusion, i.e. mistyping of the ABO blood group, might result in intravascular hemolysis and even cause recipient's death [2]. Blood typing is based on the antigen-antibody interaction. If antigens on the surface of RBCs correspond to antibodies in the serum, agglutination of RBCs occurs. Therefore, the test result of blood typing can be determined by observing the agglutination of RBCs with the naked eyes.

Currently, many traditional blood typing methods are available [3,4]. Fully automated blood typing systems are used in blood banks for routine tests. However, manual operation using 96-well microtiter plate is still the main approach in hospital transfusion services or laboratories. Besides, many researchers proposed different approaches for blood typing. Kline et al. [1] utilized the microfluidic droplet technique to encapsulate RBCs, antibody, and buffer in droplets to perform multiple agglutination assays. Makulska et al. [5] presented a method to determine the agglutination by measuring the speed of flow of microdroplets, containing RBCs and antibodies, in the microchannel.

Microfluidic systems have many advantages, including smaller consumption volume of reagents, shorter reaction time, and parallel operation in a single chip. Recently, microfluidic techniques attract a lot of interest and find important applications in chemistry and biology. Microfluidic chips for blood typing allow the operation sequence to be conducted automatically through the manipulation of microfluids. Nevertheless, a robust determination of the extent of RBC agglutination, instead of visual judgment, remains to be solved.

Over the past few years, interdigitated electrode (IDE) is a widely used structure as sensing component. Alternatively, nanowires (NWs) are also thought of as ideal elements for sensing devices and have been widely employed in a wide variety of fields, such as gas sensing [6,7], chemical sensing [8], and biomolecule sensing [9,10]. Among the nanowire materials, zinc oxide (ZnO) is a chemically and thermally stable n-type semiconductor material. Besides, this semiconductor is environmentally friendly and has several favorable properties, including high electron mobility, high melting point, good transparency, and large piezoelectric response. Many formation methods of ZnO nanowires have been developed [11]. Hydrothermal synthesis is a simple and popular method [12].

In this study, a microfluidic blood typing chip was designed for measuring RBC agglutination with voltammetry technique. Moreover, the voltammetry measurement was carried out using an interdigitated electrode on which ZnO nanowires were synthesized by hydrothermal method.

2 Materials and methods

2.1 Chip design

Blood typing is based on the antigen-antibody interaction. Agglutination is then observed to determine the serological compatibility of bloods. In this study, biomaterials, including A cell, B cell, anti-A, and anti-B, were purchased from Formosa Biomedical Technology Corp. (Taiwan) for carrying out the bioassays.

This microfludic blood typing chip, as shown in Figure 1, is comprised of five injection inlets for loading the necessary reagents (blood cell and antibody) and one detection chamber where the voltammetry measurement is performed. The injected reagents traverse the main channel to the detection chamber. On the way, the reagents pass through the special shaped mixing channels allowing both reagents to be mixed thoroughly and rapidly. Inside the detection chamber, there exists an interdigitated electrode array acting as sensor for voltammetry measurement of blood agglutination. The IDE array was characterized by 4 interdigitated fingers on each side of device with finger width of 250 μm, length of 3.9 mm, and gap of 250 μm. On the surface of IDE array, ZnO nanowires were synthesized to promote the sensitivity of measurement. This nanowires-synthesized IDE array is denoted as NW-IDE. The voltammetry measurement was performed using 4145B Semiconductor Parameter Analyzer.

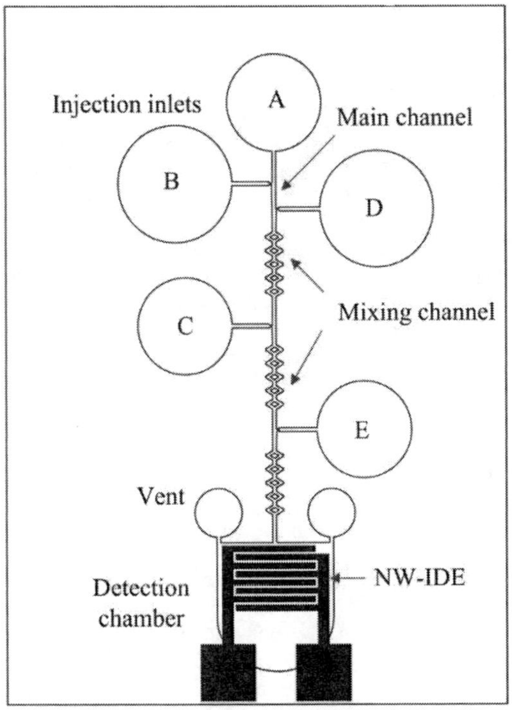

Figure 1. Design of microfluidic blood typing chip.

The width of the main channel is 200 μm. All the microfluidic components such as channel and chamber have depth of 200 μm. The reagents were driven using a syringe pump. Figure 2 shows the experimental setup.

2.2 Chip fabrication

Figure 2. Experimental setup.

This proposed chip includes a microchannel layer allowing microfluids to flow and a substrate with NW-IDE. This chip was produced by bonding the microchannel layer on the substrate.

2.2.1 Fabrication of microchannel layer

To fabricate the microchannel layer, the technique of standard SU-8 molding was used, as shown in Figure 3. A SU-8 mold with the opposite pattern of all microfluidic components was prepared first and, thus, the microfluidic layer was formed by casting polydimethylsiloxane (PDMS) in the mold. Then, the PDMS microchannel layer was left to cure at 95°C in an oven for 1 h. Finally, it was carefully peeled from the SU-8 mold. This layer is denoted as PDMS1.

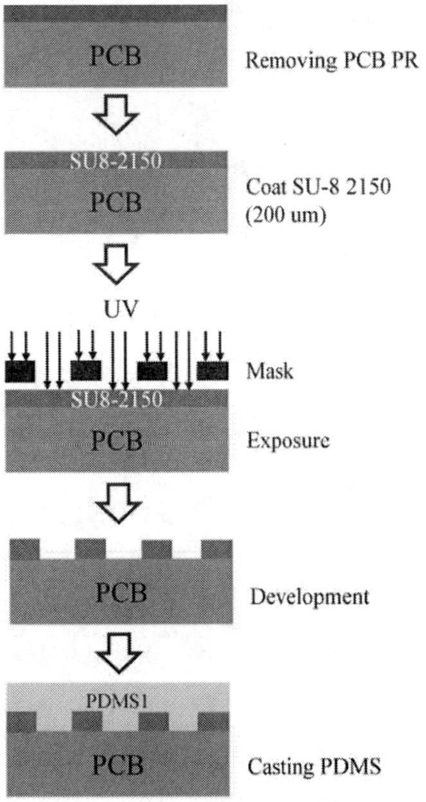

Figure 3. Fabrication process of microchannel layer.

2.2.2 Fabrication of NW-IDE

Figure 4 shows the fabrication steps of NW-IDE. Printed circuit board (PCB) has a conducting layer typically made of thin copper foil. Hence, PCB was used as the substrate and its Cu film was etched to form the interdigitated electrodes. The designed pattern of IDE array was first transferred onto the photoresist layer of PCB by photolithography process. Then, etching the copper-conducting layer of PCB through the conventional PCB treatment generated the IDE structure.

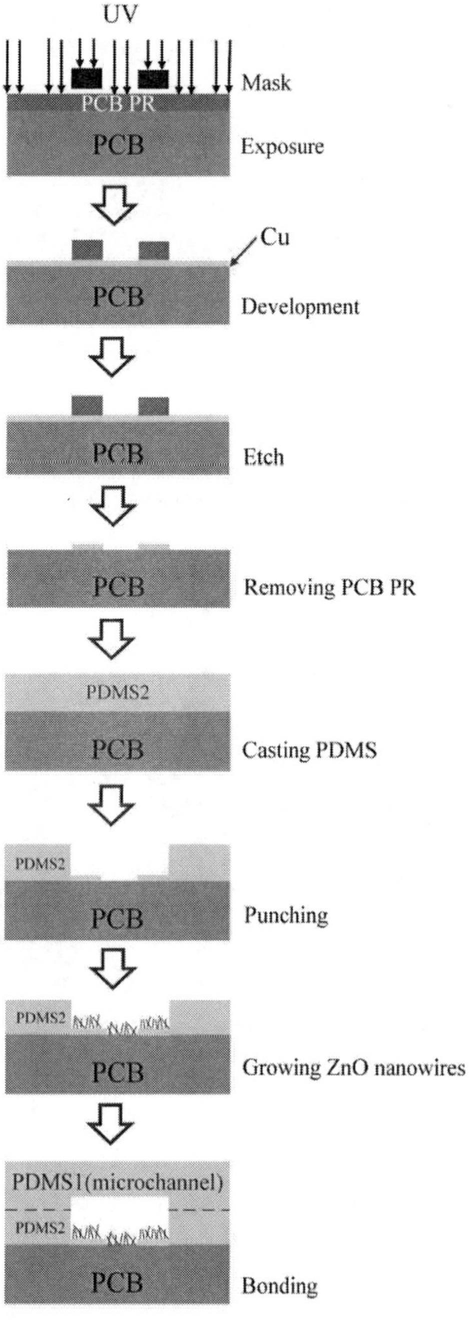

Figure 4. Fabrication process of NW-IDE.

The next step is the crystal growth of ZnO nanowires on the entire area of IDE array. However, ZnO must be constrained to grow only at the region of detection chamber. Otherwise, nanowires would influence the bonding between the microchannel layer and the substrate. Thus, prior to the growth of ZnO nanowires, a thin layer of PDMS was spun on the PCB substrate. This layer is denoted as PDMS2. After PDMS2 was cured in an oven, we opened a round hole in the PDMS2 using a hole punch, which is a kind of craft tool, to make the IDE array exposed for the growth of ZnO nanowires.

The conventional hydrothermal method was adopted. First, ZnO seed solution was prepared by mixing 0.328 g of zinc acetate dehydrate (J.T.Baker®, purchased from Avantor Performance Materials, USA) with 150 mL of methanol, and stirring and heating. After the temperature of mixture raised to 60°C, 0.12 g of NaOH and 100 mL of methanol were added slowly. Heating was stopped when the solution became white muddy mixture. However, stirring was continued until the solution cooled down to the room temperature. Thereupon ZnO seed solution was spread on the surface of IDE array, followed by fifteen-minute dehydration baking in a 130°C oven. After the chip surface was dried, careful check on whether the electrode surface is entirely covered by ZnO seed film was made using microscope.

ZnO growing solution was formulated with 0.46 g of $Zn(NO_3)_2$ (analytical reagent grade, 98%, Acros Organics, USA) and 0.2 g of $C_6H_{12}N_4$ (Showa Chemical Industry Co., Ltd., Japan) dissolved in 80 mL of DI water. Then, the ZnO seeded substrate was fixed facing down and placed in the growing solution. The lid of the vessel was sealed to prevent evaporation. Reaction was carried out at 95°C for 4 h. An additional 4-h reaction was conducted after the solution was changed. Finally, the substrate was taken out from the vessel and rinsed with DI water, followed by drying with nitrogen.

Finally, the microchannel layer (PDMS1) was bonded on the PDMS layer of substrate (PDMS2) using oxygen plasma treatment.

2.3 Bioassay

Four sets of blood typing bioassays were conducted: (1) A cell and anti-A; (2) A cell and anti-B; (3) B cell and anti-A; (4) B cell and anti-B. For each experimental design, three chips were prepared for the purpose of repetitious experiments. The voltammetry measurement was then performed.

In this study, only two injection inlets were used, i.e., inlet A and inlet D. The other three inlets can be utilized for the introduction of other reagents. For example, the manual polybrene (MP) technique for blood typing, which is widely adopted in Asia, requires other reagents such as low ionic medium (LIM), polybrene, and resuspension solution. The LIM can promote the reaction between RBC antigens and antibodies if they are associated. The function of polybrene is to cause non-specific RBC aggregation. However, the resuspension solution is for reversing the non-specific polybrene-induced aggregation. These reagents must be added in a

DOI: 10.1051/matecconf/201710813008

specific order. The design of this proposed chip is ready for conducting the MP test.

3 Results and discussion

If type A RBCs expose to anti-A, agglutination must occur. Similar result must be observed for the experiment of type B RBCs and anti-B. However, for non-agglutination situations, such as type A exposing to anti-B, most of the reactant solution should not display evident agglutinated blood cells.

3.1. Photo image

Photo image is a common evidence of blood typing result. Figures 5-1 to 5-4 show the photo images of our bioassays. Part (a) in each figure is the image of entire chip ; while part (b) and (c) are the detailed images of detection chamber. In addition, front light source was adopted for taking photo of part (b). On the contrary, back light source was adopted for part (c).

As shown in Figures 5-1 and 5-4, red spots on NW-IDE can be observed. They are the RBC agglutinations. However, it is difficult to determine the blood typing result with naked eyes. Since small amount of blood (3.6 μL) was used in each bioassay, agglutination became unobvious.

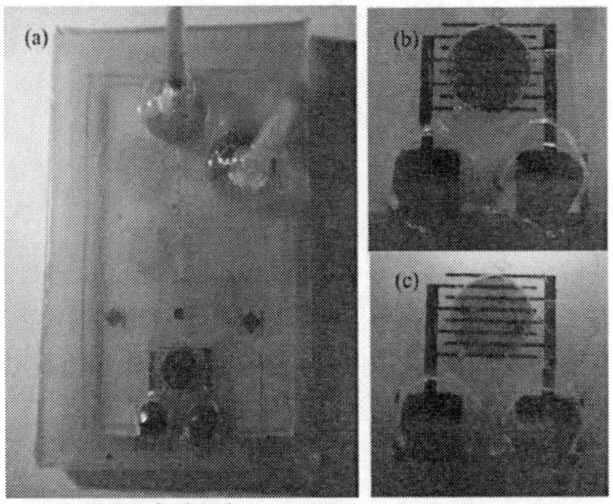

Figure 5-1. Agglutination (A cell and anti-A).

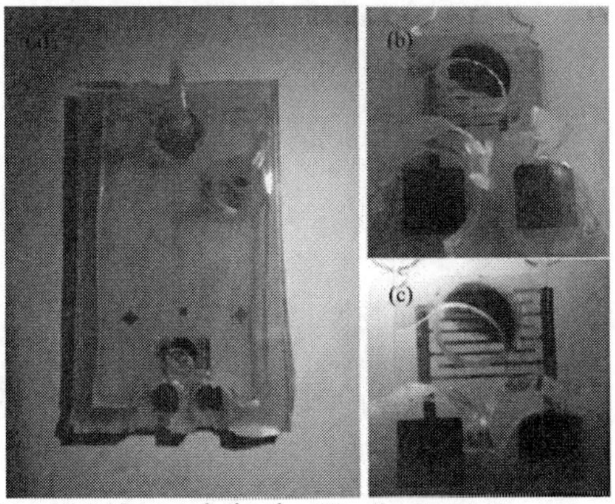

Figure 5-2. Non-agglutination (B cell and anti-A).

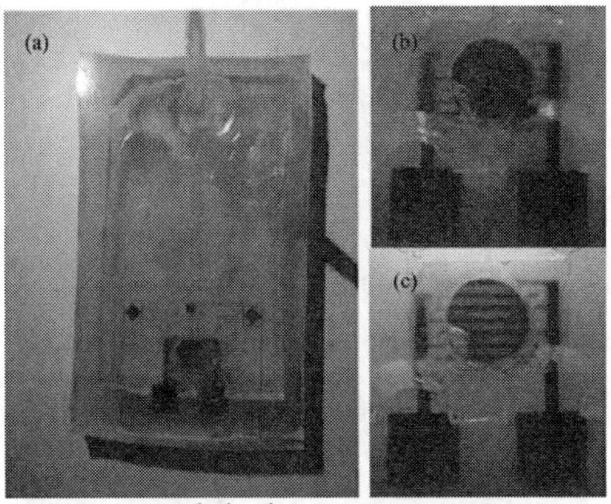

Figure 5-3. Non-agglutination (A cell and anti-B).

Figure 5-4. Agglutination (B cell and anti-B).

Moreover, the image in Figure 5-4(b) is similar to those in Figures 5-2(b) and 5-3(b). Under the front light source, small RBC agglutination is not evident. Hence, false conclusion might be determined. Although the photo image is a fundamental method to indicate the blood type, a precise method deserves to be explored.

3.2 Voltammetry

Voltammetry is an electroanalytical method used in analytical chemistry. In voltammetry, information about an analyte is obtained by measuring the current as the potential is varied. Based on this principle, RBC agglutination was measured using voltammetry technique.

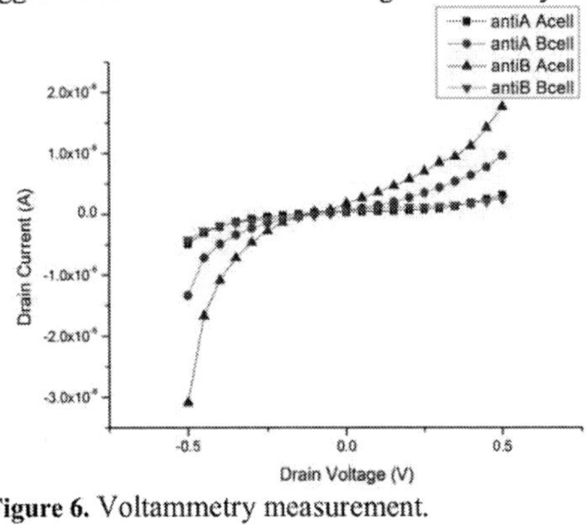

Figure 6. Voltammetry measurement.

As shown in Figure 6, the curve of RBC agglutination (A cell and anti-A) coincides with that of agglutination of B cell and anti-B. However, the discrimination between agglutination and non-agglutination is obvious. Although small amount of blood was used in bioassay, accurate result can be obtained.

3.3 Experimental summary

To observe a small quantity of RBC agglutination, photo image is not a perfect method. However, voltammetry can provide a precise measurement so that agglutination and non-agglutination can be discriminated.

4 Conclusions

This proposed microfluidic blood typing chip has NW-IDE for conducting voltammetry measurement of RBC agglutination. Usually, determination of agglutination relies on photo images. False determination might occur if minute amount of blood is used. However, the discrimination between agglutination and non-agglutination was obvious when the voltammetry measurement was employed. This approach provides more accurate result.

References

1. T.R. Kline, M.K. Runyon, M. Pothiawala, R.F. Ismagilov, *Analytical chemistry*, **80** (2008)
2. P.P. Goodell, L. Uhl, M. Mohammed, A.A. Powers, *Am. J. Clin. Pathol.*, **134** (2010)
3. W. Malomgré, B. Neumeister, *Anal. Bioanal. Chem.*, **393** (2009)
4. S.G. Sandler, A. Langeberg, N. Avery, P.D. Mintz, *Transfusion*, **40** (2000)
5. S. Makulska, S. Jakiela, P. Garstecki, *Lab Chip*, **13** (2013)
6. M. Cuscunà, A. Convertino, E. Zampetti, A. Macagnano, A. Pecora, G. Fortunato, L. Felisari, G. Nicotra, C. Spinella, F. Martelli, *Appl. Phys. Lett.*, **101** (2012)
7. C. V. Tuan, M.A. Tuan, N.V. Hieu, T. Trung, *Curr. Appl. Phys.*, **12** (2012)
8. V.A. Antohe, A. Radu, S. Mátéfi-Tempfli, L. Piraux, *IEEE Trans. Nanotechnol.*, **10** (2011)
9. A.A. Ansari, A. Kaushik, P.R. Solanki, B.D. Malhotra, *Bioelectrochemistry*, **77** (2010)
10. H.-K. Chang, F.N. Ishikawa, R. Zhang, R. Datar, R.J. Cote, M.E. Thompson, C. Zhou, *ACS Nano*, **5** (2011)
11. Y.W. Heo, D.P. Norton, L.C. Tien, Y. Kwon, B.S. Kang, F. Ren, S.J. Pearton, J.R. LaRoche, *Mater. Sci. Eng. R-Rep.*, **47** (2004)
12. J. Song, S. Lim, *J. Phys. Chem. C*, **111** (2007)

DOI: 10.1051/matecconf/201710813009

Effect of the Periodontal Ligament of the Bilateral Support Teeth on the Stress Analysis of Dental Implant

Yanhua Xie, Xi Lu , Jianrun Zhang and Beibei Sun

School of Mechanical Engineering, the University of Southeast, Nanjing, China

Abstract. The aim of this work is to analyze the function of natural teeth's periodontal ligament and the effects of periodontal ligament on implants by the finite element method (FEM), when static functional loads occur. The finite element analysis models are established, which consist of fragment of mandible, natural teeth, periodontal ligament, and implant. Finite element software Abaqus is used to analyze the stress transfer in models with or without periodontal ligament, and mandibular first molar is implant. The implant obtained the maximum stress value of 87.71MPa when periodontal ligament of natural tooth is absent, but the maximum value reduces to 38.43 MPa with the action of periodontal ligament. It illustrates that periodontal ligament has significant effects on stress transfer. When the finite element model of single natural tooth or dentition with implant is generated, periodontal ligament should be taken into account.

1 Introduction

Periodontal ligament can buffer chewing force and reduce the stress that spread to jaw bone, which has been extensively recognized [1]. The property of periodontal ligament has been researched continuously, and the values of elasticity modulus and poisson's ratio are various [2]. The linear elastic [3], bilinear elastic, nonlinear hyperelastic [4] or viscoelastic [5, 6] constitutive models have been extensively adopted to simulate the behaviour of the periodontal ligament. There are very few in vivo measurements have been performed. However, the specimens of the periodontal ligament were obtained from human dead bodies or animals [7]. There would be the inevitable errors and uncertainties of the values of periodontal ligament. Since the intensive and extensive investigation of the periodontal ligament is limited [2] and in vivo experiment is hardly carried out. It's difficult for us to obtain the accurate values.

There have been many studies about the effects of the positions, numbers [8] or geometrical parameters [9] on the dental implant, while the effects of the support of natural teeth on the side of denture were neglected [10-12] during the analysis of stress and deformation of implants. And in a lot of finite element models that stress of dental implant is analysed periodontal ligament is excluded.

In this paper, the specialized models with or without periodontal ligament are created and denture is supported by natural teeth, which are applied to evaluate and compare the values of the stress that transferred to the jaw bones under the same load.

2 Materials and methods

2.1 Plan design

For there is a variability in teeth and mandible bone morphology, the geometric model should be necessary to refer to the actual mandible shape and teeth morphology. The geometry of the model is produced with CT image slices of an adult human mandible obtained with a CT scanner. Scanned images are imported into the image segmenting and calculating software MIMICS (Materialise's interactive medical image control system). Based on the different values of Hounsfield of muscle, mandibular bone and teeth are segmented from the whole geometry image, and teeth are separated from alveolar bone in the same way. By performing calculate 3D order, the geometry models are reconstructed, while triangulated files are exported in STL format. Then, the files are inputted in the reverse engineering software Geomagic Studio 10. According to the measurement values of thickness of periodontal ligament from the faultage image, 0.2mm is selected as the average thickness of the periodontal ligament.

When the dental crown is supported by the natural teeth, the stress distribution and transmission of the implant is analyzed.

In model 1, the two natural teeth that on the side of implant have periodontal ligament and support the crown of prosthesis. The finite element model consists of natural teeth with periodontal ligament, dental implant (mandibular first molars) and fragment of mandible as seen in Fig. 1.

In model 2, the two natural teeth have no periodontal ligament and the finite element model includes natural teeth without periodontal ligament, dental implant (mandibular first molars) and fragment of mandible, which is shown in Fig. 2.

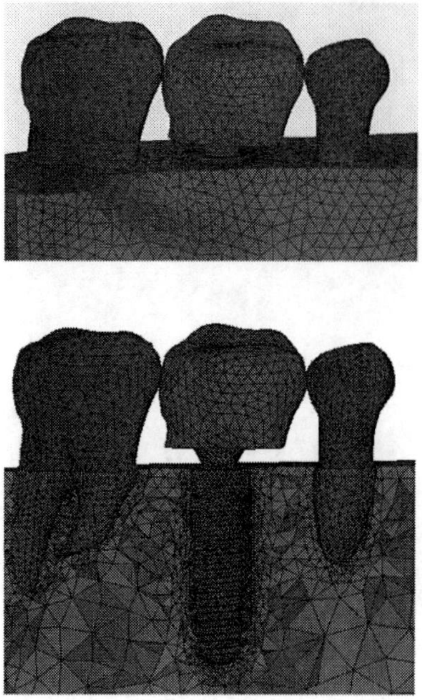

Figure 1. The finite element model with periodontal ligament.

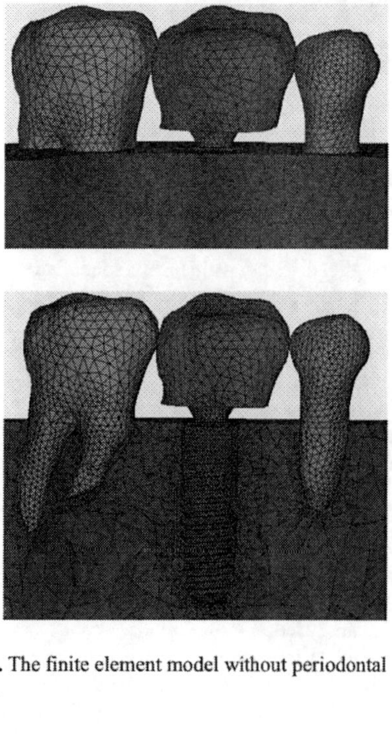

Figure 2. The finite element model without periodontal ligament.

2.2 Material properties

The materials of periodontal ligament, the second premolar (natural teeth), the second molars (natural teeth), titanium implants, mandible bone and metal-ceramic were considered continuous, homogeneous, isotropic linear elastic materials. Material properties are present in Table 1 [13-15].

Table 1. Material properties.

Material	Modulus of elasticity (MPa)	Poisson's ratio	Density (g/cm^3)
natural teeth	18600	0.31	3.1
periodontal ligament	40	0.45	1.05
mandible bone	13700	0.3	1.67
titanium implants	110000	0.35	4.5
metal-ceramic	84000	0.28	2.4

2.3 Boundary conditions

In model 1, this section involves a total of nine contact pairs are generated in second premolar and internal surface of periodontal ligament, outside surface of periodontal ligament of the second premolar and the surface on mandible, the outside surface on second molar and internal surface of periodontal ligament, outside surface of periodontal ligament of the second molar and the surface on mandible, the surface on mandible and the surface on implants, the surface on implant and the surface on abutment, the surface on abutment and the inner surface on crown of molar, mesial surface of crown of molar and distal surface of second premolar, distal surface of crown of molar and mesial surface of second molar. Additionally, stresses of five surfaces that showed in Fig. 3 are analysed.

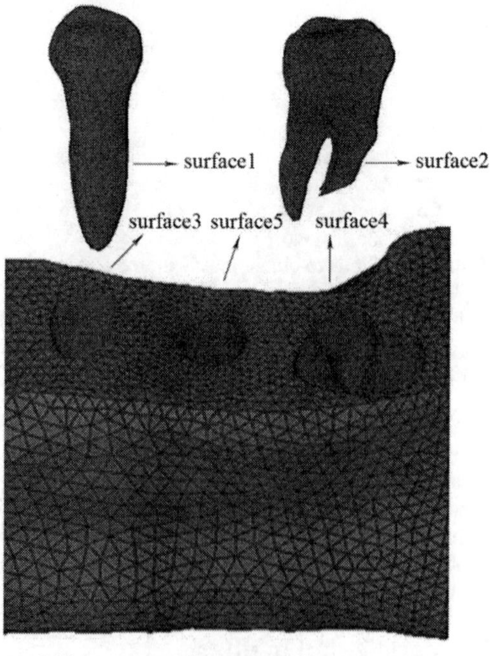

Figure 3. Five contact surfaces of model.

DOI: 10.1051/matecconf/201710813009

Since the periodontal ligament is excluded in model 2, there are only seven contact pairs that between the outside surface of the second premolar and the surface on mandible, outside surface of the second molar and the surface on mandible, the surface on mandible and the surface on implants, the surface on implant and the surface on abutment, the surface on abutment and the inner surface on crown of molar, mesial surface of crown of molar and distal surface of second premolar, distal surface of crown of molar and mesial surface of second molar.

The contact surface of mandible and implant is assumed as Osseo integration and treated as tie constraint. The rest contact pairs keep close contact without gaps, and friction coefficients of 0.3 is considered for these contact surfaces.

The model 1 and model 2 are constrained at subjacent outer surface of mandible in all directions. The upper of the outer surface of is not exerted on any constraints.

Contact points are used to replace occlusal contact surface [16] and to simulate the surface to suffer from the load that perpendicular to the occlusal surface and located on the buccal cusp and tongue cusp of the teeth. Two contact points are selected on the second premolar. Four contact points are selected on the second molar. And four contact points are selected on crown on the top of the implant.

The value of applied force is 250N, which is regarded as the average force loading on each tooth [17, 18]. Boundary conditions and loads are shown in Fig. 4.

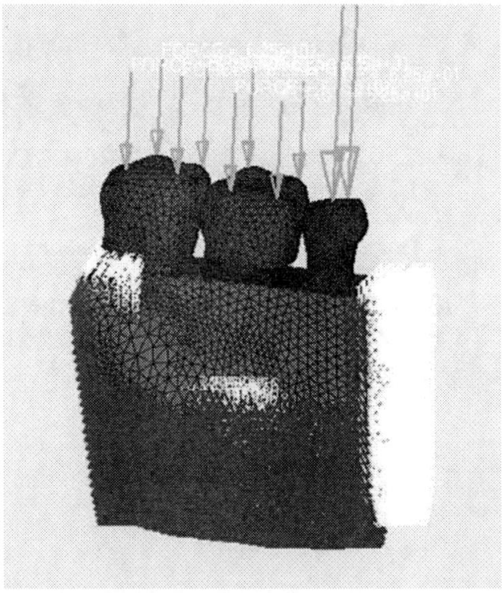

Figure 4. Forces and constraints.

3 Results and discussion

The maximum values of stress on the five surfaces and implant in model 1 are figured out and presented in Table 2. Additionally, the stress distributions on the five surfaces and implant in model 1 are shown in Fig. 5-7.

It can be seen that under the effect of the periodontal ligament the magnitude of stress on the surface on mandible is smaller than those on the root of the teeth.

On the position of second premolar, the maximum magnitudes of stresses on surface 1 and surface 3 are 9.45MPa and 2.27MPa respectively. In comparing the surface 1 and surface 3, the maximum value of stress on the surface 1 is obviously higher than that in the surface 3. And the value of stress on surface 3 declined 75. 98 percent. On the position of second molar, the maximum magnitudes of stresses on surface 2 and surface 4 are 40.88MPa and 6.90MPa. The maximum value of stress on surface 2 is many times greater than that on surface 4. It's apparent that the values declined steeply almost 83.12 percent. Additionally, the maximum value of stress on surface 5 and implant are 13.28MPa and 38.43MPa. The maximum value of stress on surface 5 is apparently decreased compare to that on the implant. And the values declined almost 83.12 percent.

As it shown in Fig 8, and the magnitude of stress on implant in the model 1 is 38.43MPa, while that in the model 2 is 87.71MPa. It's obvious that the value of the stress on implant in the model 1 is much less than that in the model 2. The model without the periodontal ligament would induce higher value of stress.

The present research is performed to analyse the function of natural teeth's periodontal ligament and the effects of periodontal ligament on implants. Two representative models are reproduced. Model 1 is used to simulating the biomechanical behaviour of implant under the condition that the nature teeth without periodontal ligament, while in the model 2 both the support teeth have the function of periodontal ligament. The analyses results adequately illustrate that periodontal ligament can buffer chewing force and reduce the stress [1].

Table 2. The values of stresses on the five surfaces

Upper surface	Maximum stress(MPa)	Lower surface	Maximum stress(MPa)	Proportion of stress decline
surface1	9.45	surface3	2.27	75.98%
surface2	40.88	surface4	6.90	83.12%
implant	38.43	surface5	13.28	65.44%

Inaccurate results may result in the cases that the denture without the support of natural teeth on the side [10-12] during the analysis of stress and deformation of implants. Since the periodontal ligament is absent, the analysis results may much different.

The finite element model that included natural teeth with periodontal ligament is crucial in the biomechanical research of implant.

The magnitudes of the stresses on surfaces that located in alveolar bone are almost three times more than that on the root of teeth. And the maximum value of stress on implant in model 1 is up to 38.43MPa. However, in comparing the maximum magnitude of stress on implant in the model 2 and that in the model 1, the value of the stress on implant in the model 2 is much higher.

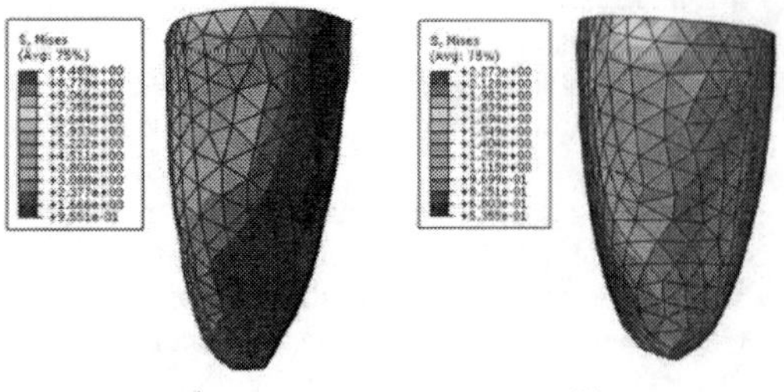

(a) (b)

Figure 5. Stress distribution of the surfaces. (a) surface 1, (b) surface 3.

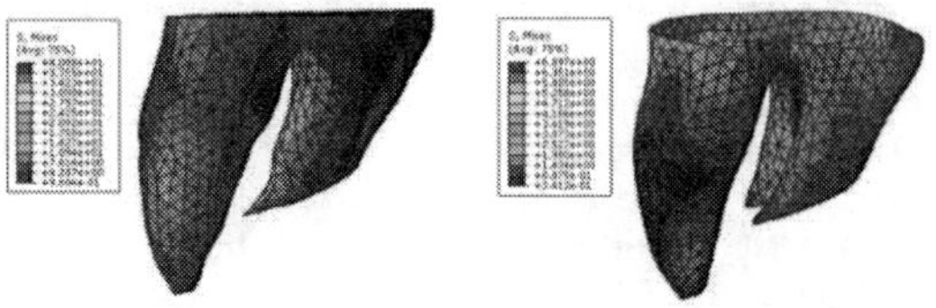

(a) (b)

Figure 6. Stress distribution of the surfaces. (a) surface 2, (b) surface 4.

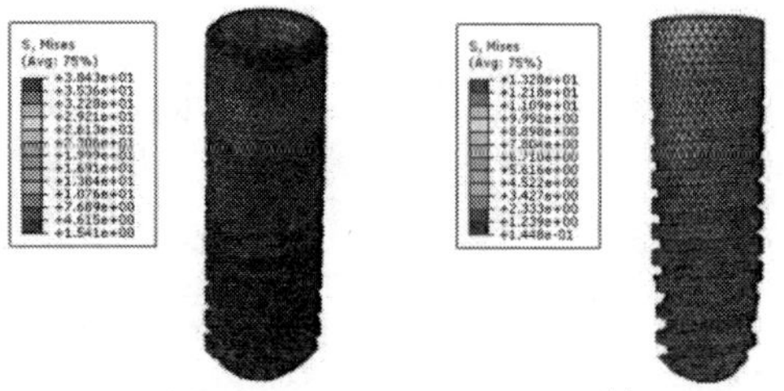

(a) (b)

Figure 7. Stress distribution of the surfaces. (a) implant, (b) surface 5.

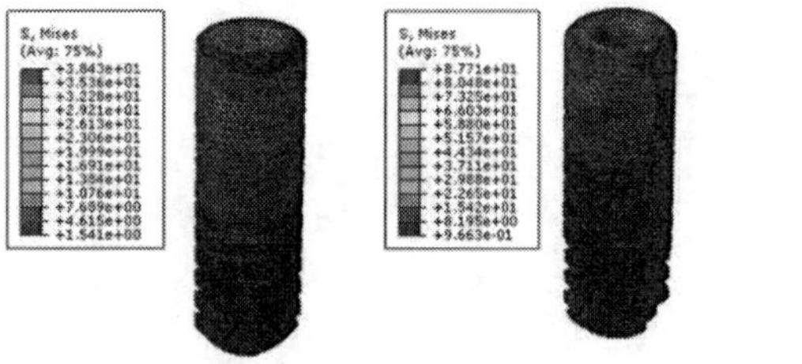

(a) (b)

Figure 8. Stress distribution of the implants. (a) model 1, (b) model 2.

4 Conclusions

It clearly demonstrates that periodontal ligament has significant effect on stress transfer. The values of the stress of implant that with both side support of natural teeth will significantly increase from 38.43MPa to 87.71MPa, when the periodontal ligament of natural teeth is not taken into consideration. It explicitly illustrates that the periodontal ligament of natural teeth that supporting the denture has significant influence on the stress analysis of the implant and should be take into consideration during the modelling.

References

1. M. Tunan, E. Sunbuloglu, E. Bozdag. J. B **47**, 2883-2890 (2014)
2. Xia Z, Jiang F, Chen J. A. J. O. D. O **4**, 486-491 (2013)
3. I. A. V. P. Poiate, A. B. Vasconcellos, M. Mori. C. M. P. B **2**, 292-299 (2011)
4. Qian L, Todo M, Morita Y. J. D. M **10**, 1285-1292 (2009)
5. S. R. Toms, G. J. Dakin, J. E. Lemons. J. B **10**, 1411-1415 (2002)
6. A. N. Natali, P. G. Pavan, C. Venturato, C. M. P. B **2**, 193-198 (2011)
7. N. Yoshida, Y. Koga, C. L. Peng. M. E. P **8**, 567-572 (2001)
8. T. Topkaya, M. Y. Solmaz. J. B **10**, 2102-2109 (2015)
9. M. B. Toniollo, A. P. Macedo, R. C. S. Rodrigues. J. P. R **3**, 206-212 (2013)
10. A. N. Natali, E. L. Carniel, P. G. Pavan. D. M **4**, 561-569 (2008)
11. Lin D, Li Q, Li W. J. M. B. B. M, **5**, 410-432 (2009)
12. L. Baggi, I. Cappelloni, F. Maceri. S. M. P. T **8**, 971-987 (2008)
13. P.M. Tulimar, B. Estevam. J. Med. Biol.Eng.Comp **48**, 1107-1113 (2010)
14. T. Topkaya, M.Y. Solmaz. J. B **48**, 2102-2109 (2015)
15. J.S. Ree, P.H. Jacobsen. J. B **18**, 995-999 (1997)
16. Ding, X, et al. J. M. D **10**, 168-175 (2010)
17. J. Middleton, M.L. Jones, J. Hiekman, etal. *10th comference of the European Society of Biomechanic*, (1996)
18. G.G. Robert, M.P. John. *Dental restoration materials science* (2006)

Influence of epoxy resin as encapsulation material of silicon photovoltaic cells on maximum current

David Acevedo-Gómez , Alejandro Velásquez-López, Gilberto Osorio-Gómez and Ricardo Mejía-Gutiérrez

Design Engineering Research Group (GRID), Universidad EAFIT. Cra 49 N. 7 Sur 50, Medellin, Colombia

Abstract. This work presents an analysis about how the performance of silicon photovoltaic cells is influenced by the use of epoxy resin as encapsulation material with flat roughness. The effect of encapsulation on current at maximum power of mono-crystalline cell was tested indoor in a solar simulator bench at 1000 w/m² and AM1.5G. The results show that implementation of flat roughness layer onto cell surface reduces the maximum current inducing on average 2.7% less power with respect to a cell before any encapsulation. The losses of power and, in consequence, the less production of energy are explained by resin light absorption, reflection and partially neutralization of non-reflective coating.

1 Introduction

Even though, since mid 80's, humans are seeking solutions for climatic and energetic problems, levels of Green House Gases (GHG), such as CO and CO2 caused in big proportion by the fossil fuels used to provide global energy, have grown inordinately in the last 40 years. Unquestionably, harnessing renewable energy sources is playing an important role. In fact, for closing the gap between needing energy and actual generation and reducing GHG, solar energy has the potential to provide clean energy because of its availability and versatility that allowed decentralized and small scale generation. With this aim in view, the evolution in both materials and fabrication methods for PhotoVoltaic (PV) cells has been motivated in order to increase the rate of generated energy per kiloWatt-peak (kWp), and to reduce its cost per kiloWatt-hour (kWh), as a result of efficiency improvement. Nevertheless, worldwide PV cells production keeps mainly dominated by Silicon (Si), due to abundance of material and its relation between cost and performance [1]. Anyway, for commercial purposes of crystalline modules, Si PV cell needs to be encapsulated, bringing protection against biological agents and causing low impact in optical properties such as reflection and absorption. The most commonly materials used in the industry are Ethylene Vinyl Acetate (EVA) and silicone with non-reflective glass [2]. Some authors have performed experiments showing that the impact of these and other materials combined with different non-reflective glasses have influence onto current at Maximum Power Point response [3, 4, 5]. However, for other products design whose application and its life cycle require different characteristics in topics such as optics and manufacturing, it is necessary to use different kind of materials, and these projects involve solar energy coming with encapsulation developments with semi-transparent rigid epoxy resin. This study analyzes the behavior of current when this material is used as encapsulation. On the next section a review of the state of the art is presented, followed by a description of experimental and methodology approach. Then, the results are presented in graphical way, followed by corresponding discussion. Finally, some conclusions and further research are presented.

2 Background

Lab efficiency measured and reported in several reviews takes into consideration solar cells before they are integrated in products. In consequence, to integrate cells into products, an encapsulation process has to be accomplished. Such process consists in put a coating of transparent material on the silicon face of solar cell in order to protect it against weather, biological agents, mechanical failures, among others. Therefore, selected material must be durable (25 years at least) and it should allow light going through it. Tao & Du [4] investigated through reflection test how optical properties are influenced by presence of encapsulation, and they found that EVA mixed with different class glasses reduces up to 6% the effective light. As a result of use of encapsulation material, the positive effects generated by non-reflective coating is partially neutralized when PV cells are processed for commercial purposes. Peter et al. [5] published a detailed study about how performance is reduced through losses for reflection and absorption of encapsulation materials, where 9% of current losses are causing by absorption and reflection of EVA and glass, it

 DOI: 10.1051/matecconf/201710814001

means 1.1 mA/cm². The losses are partially neutralized implementing textured glasses. Schneider et al. [3] evaluated electric performance with different textured glasses and results showed that glass can reduce influence of encapsulation by 5%. Furthermore, a manufacturing company of PV cells evaluates material in order to improve their products. The BP solar company made a comparison [2] with outdoor PV arrays during one year where performance of silicone showed advantage of 1% with respect to EVA. Moreover, Berlin University, trough simulation, noticed that 1.3% losses in Q-cells modules are given due to EVA absorption.

3 Experimental proposal

In literature is evident that encapsulation material causes losses in power and current output, but analyses are reduced to commercial purposes. Nevertheless, when product life cycle is short, e.g. competition for one week, other kind of encapsulation materials become visible. As a result of new products development, Sumiglas R rigid epoxy resin was used as encapsulation material in modules for a competition solar vehicle. This research tests the hypothesis that usage of epoxy resin reduces the output power and studies the effect on current at Maximum Power Point (MPP) (IMP P) through the usage of non conventional material for solar cell encapsulation. Fig. 1 presents how the implementation of any coating onto a solar cell surface (see Fig. 1(b)) reduces the probability (also named efficiency) of converting power inlet in power outlet with respect to a non encapsulated one (see Fig. 1(a)).

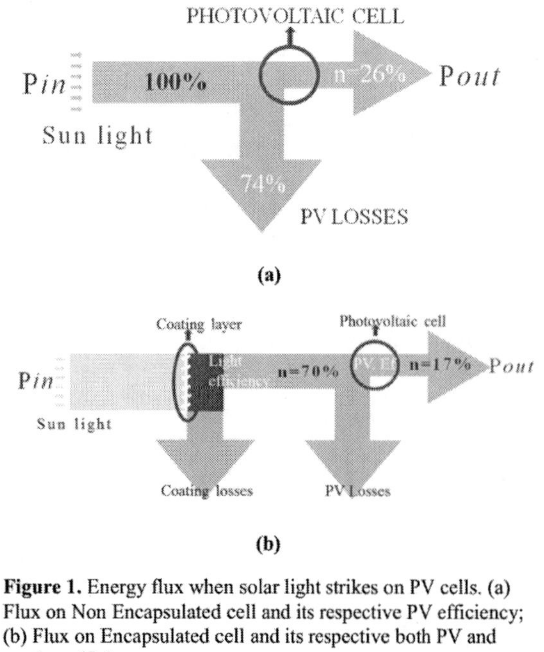

Figure 1. Energy flux when solar light strikes on PV cells. (a) Flux on Non Encapsulated cell and its respective PV efficiency; (b) Flux on Encapsulated cell and its respective both PV and coating efficiency.

3.1 Sample preparation

The cell is encapsulated by molding method using vacuum bag, which is a process that consist in expel the air present between mold surface and the bag [6], employing epoxy resin and glass as mold for finishing. The module sandwich is completed by two layers of fiber glass as back-sheet. The electrical connection was performed with tin ribbon on 5 x 5 inches mono-crystalline cell. The piece stays six hours inside the bag and after that, it is post-cured for two days. There were prepared 17 samples in total: nine samples of Group NE (non encapsulated cells), eight samples of group E (encapsulated cells with 1mm thickness of rigid epoxy resin and flat roughness). General description is presented in Table 1.

3.2 Methodological approach

Measurements of Maximum Power Point (Pmpp) and their respective current-voltage (I-V) curve for each cell, were performed using a high-power LED-based solar simulator test bench [7], where the mean radiation is 1000 w/m2 . Cells were tested one by one under same conditions such as encapsulation material, thickness, radiation and time, among others. Power is proportional to voltage and current, but due to dependence of voltage with increments of temperature, which is not a controlled variable, it was decided to take current at maximum power (Imp) from I-V curve, because of relation between radiation and current is proportional to radiation and power. The data set was tested for normal distribution assumptions and a respective statistic analysis was carried out.

4 Results

Fig. 2 presents a distribution with left tail which does not present any violation to assumptions either homocedasticity or normal distribution (Levene and shapiro test p-value>0.05).

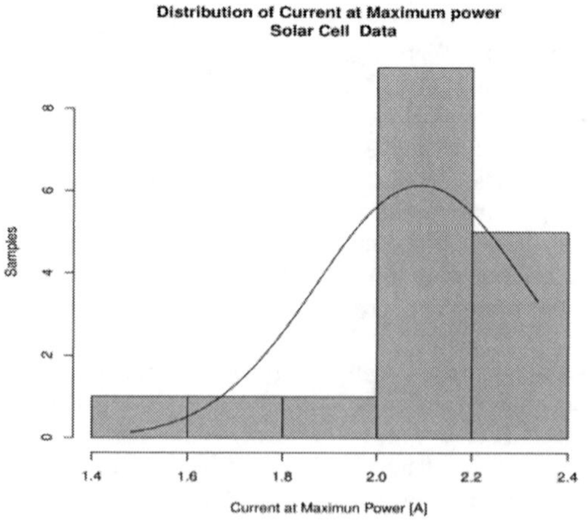

Figure 2. Data set: distribution of Current at maximum power.

DOI: 10.1051/matecconf/201710814001

Besides, residuals analysis in linear model are fitted to curve in the Q-Q Plot (see fig. 3). In consequence, current at maximum power is selected as data set for ANOVA. Fig. 4 presents data for finished surfaces and it is possible to infer that the usage of epoxy resin works as a shield which leads to reduce the mean power and current that can be converted. Table 2 proves initial hypothesis with p-value<0.05 and it is accepted that encapsulation with resin, in the same way that EVA and silicone, reduces the probability to convert light in energy.

Table 1. Description of PV cell samples.

	Category	PV cell Surface [20μm]	Description	Mean Maximum Power Measured [mW]
NE	Non-Encapsulated (Control)	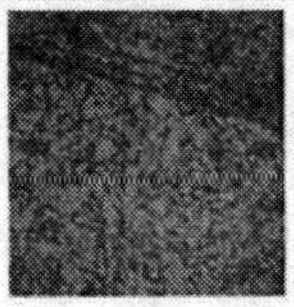	Non encapsulated mono-crystalline cell and non backsheet. Image at 20μm	670
FR	Flat Roughness		Semi-transparent rigid epoxy resin flat roughness encapsulated mono-crystalline cell with fiber glass backsheet. Image at 20μm	645

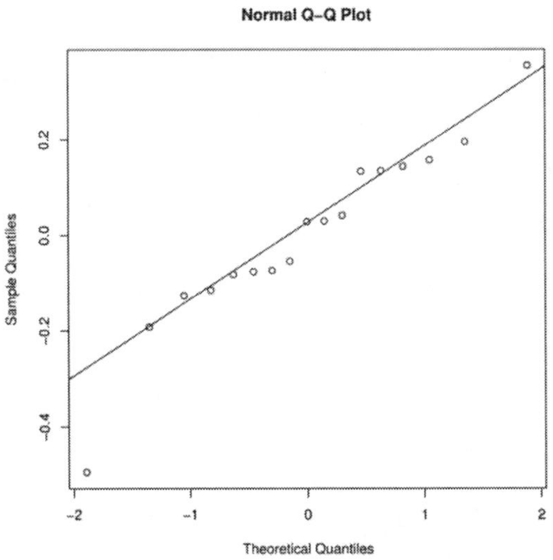

Figure 3: Data set: distribution of Current data residuals and fit curve.

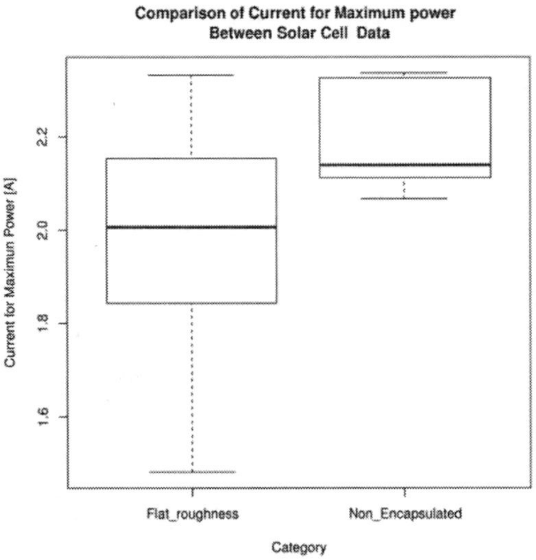

Figure 4: Data set: observed data by category

Table 2. Analysis of Variances.

	Degree freedom	Sum Square	Mean Squarere	F Value	Pr(> F)
Treatment	1	0.1974	0.19738	5.095	0.0393*
Residuals	15	0.5811	0.03874		

5 Conclusions

The usage of resin epoxy as encapsulation impacts the power output of solar cells, generating a decrement of 2.7% in Encapsulated cell with respect to a Non Encapsulated one. The phenomenon is explained since encapsulation works as a shield which interferes the light way until the PV cell surface. Further research should include an approach to implement Textured encapsulation in order to increment the power output, as glasses according to literature. In addition, resin epoxy losses more energy than EVA and silicone but it works better in short life cycle. For further work, we will implement other methods to collect information about solar cells and perform experiments with natural light sources in order to validate the data set. Besides, we keep researching about materials which allow short life cycle, good efficiency and low cost, and texture for surface finishes. It would be interesting to test bigger modules in order to know losses in wire and other components.

References

1. B. Parida, S. Iniyan, and R. Goic, "A review of solar photovoltaic technologies," Renewable and sustainable energy reviews, vol. 15, no. 3, pp. 1625–1636, 2011.
2. B. Ketola, C. Shirk, P. Griffith, D. W. Cunningham, and B. Solar, "Encapsulation performance comparison of pv modules in outdoor arrays,".
3. A. Schneider, R. Harney, and S. Koch, "Impact of cell texturing quality on cell to module losses," Energy Procedia, vol. 77, pp. 414–419, 2015.
4. W. Tao and Y. Du, "The optical properties of solar cells before and after encapsulation," Solar Energy, vol. 122, pp. 718–726, 2015.
5. I. M. Peters, Y. S. Khoo, and T. M. Walsh, "Detailed current loss analysis for a pv module made with textured multicrystalline silicon wafer solar cells," IEEE Journal of Photovoltaics, vol. 4, no. 2, pp. 585–593, 2014.
6. T. E. Miller and N. Y. C. Institute, Introduction to composites: reference handbook. Composites Inst., 1998.
7. V. Toro-Betancur, A. Velasquez-López, D. Velásquez, and D. Acevedo-Gómez, "Optical design ′ of a high-power led-based solar simulator," in SPIE Photonics Europe, pp. 98890Q–98890Q, International Society for Optics and Photonics, 2016.

MATEC Web of Conferences **108**, 14002 (2017) DOI: 10.1051/matecconf/201710814002
ICMAA 2017

Improving energy performance power station of ship with integrated electric propulsion

Andrey Dar'enkov[1], Ivan Samoyavchev[1], Oleg Khvatov[2], Valeriy Sugakov[2]

[1]*Nizhny Novgorod State Technical University n.a. R.E. Alekseev, Institute of Electric Power Engineering, 603950 Nizhny Novgorod, Russia*
[2]*Volga state university of water transport, Electromechanical Faculty, 603950 Nizhny Novgorod, Russia*

Abstract. The article presents the method of calculating the power plant fuel consumption on the basis of diesel-generating installation (DGI) with variable speed for ships with integrated electric propulsion. Application DGI with variable speed allows saving fuel and therefore reducing harmful emissions into the atmosphere. In the paper calculated the fuel efficiency of unified power station on the basis of a variable speed DGI power of 1000 kW.

1 Introduction

On water transport are increasingly using integrated electric propulsion. This propulsion motors (PM) can be supplied together with other consumers from a unified power station (UPS). Application UPS simplifies maintenance and improves the reliability of the energy system of autonomous object by reducing the total number of system elements. The primary engine UPS widely used internal combustion engines (ICE).

As a rule, UPS are based on diesel DGI of constant rotation speed rotation. The electric power station gain in performance is possible due to variable speed rotation of DGI application. Fuel saving in this case is performed by setting for each load power value the optimal ICE shaft rotation frequency corresponding to the minimal fuel-consumption rate [1, 2].

The UPS - DGI with variable speed rotation is a relatively new technical trend in small power. Research this systems carried out several foreign companies such as Fubag, Honda, Hyundai, Kypor, ABB, Wartsila etc. In Russia research carried out – JSC "R&D Center for power engineering" (Moscow), JSC "ZVEZDA" (St. Petersburg), JSC "Sigma" (Kovrov), VSUWT and NNSTU (N.Novgorod).

The use of UPS - DGI can reduce the absolute and relative Fuel consumption and ensure optimal thermal operation of DGI, improve service life and reduce the amount of harmful emissions into the atmosphere.

2 Fuel saving calculation of UPS with variable speed rotation

Determination the relative and absolute indexes of fuel consumption of UPS with a variable frequency of rotation DGI has been carried out as shown by UPS ship with electric propulsion [4]. The calculation takes into account the energy loss of each element of the power structure of the UPS (fig. 1) when the load power changing over the entire range. On fig. 1 energy diagram, where G - absolute fuel consumption rate, P_m – power of DGI, P_{SG} - power of SG, P_{FC} – power of frequency converter, P_T – power of transformer, ΔP_m, ΔP_{SG}, ΔP_{FC}, ΔP_T – power loss.

It should be noted that the change load power in UPS causes a change speed shaft of DGI [3]. When calculating UPS efficiency it is taken into account those electrical equipment elements, comprised in UPS with a variable frequency of rotation DGI (synchronous generator (SG), PM, transformer (T), frequency converter (FC)) operate at variable frequency and voltage range. The ICE fuel consumption rate is determined according to the multivariate ICE characteristic [1].

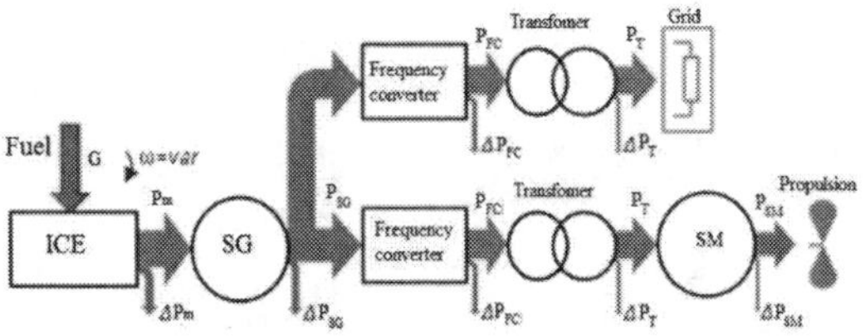

Figure 1. Energy diagram of power circuit UPS with variable speed DGI.

3 UPS elements efficiency calculation

To calculate the efficiency of the UPS are used depending on the efficiency of the voltage frequency for each power element of UPS.

3.1. Efficiency of synchronous machine

The synchronous machine efficiency (SM) is determined through the following equations

 DOI: 10.1051/matecconf/201710814002

$$\eta_{SM} = 1 - \frac{p_\Sigma}{P_1},$$
$$p_\Sigma = p_i + p_b + p_c + p_{add} + p_{mech},$$
$$p_{add} = 0{,}005 P_1 \left(\frac{I_1}{I_{1n}} \right),$$
$$p_i = p_{1,0/50} (f/50)^\beta K (B_{s1}^2 m_{s1} + B_{s2}^2 m_{s2}), \quad (1)$$
$$p_b = U_b I_f,$$
$$p_c = m_1 I_1^2 r_1 m_t + m_2 I_2^2 r_2 m_t,$$
$$p_{mech} = 8(n_1/1000)^2 (D_1/100)^3.$$

In (1) are used η_{SM} – synchronous machine efficiency P_1 – power consumption, p_Σ – total losses in a machine, p_i – iron losses, p_b – brush friction losses, p_c – copper losses, p_{add} – additional losses, p_{mech} – mechanical losses, $p_{1,0/50}$ – specific magnetic losses, f – stator voltage frequency, B_{s1}, B_{s2} – induction corresponding to stator finger and lamination, K – correcting coefficient, β — coefficient, considering the steel losses dependency on the magnetization ratio; m_{s1} — stator finger mass; m_{s2} — stator lamination mass, r_1 – effective resistance of stator coil phase, m_t – ratio considering the coil heat.

Fig 2 shows the efficiency dependencies on stator primary voltage frequency for a synchronous machine MCK1250-1500 (P_N = 1000 kW, n = 1500 min^{-1}). The dependencies correspond to load power P_1 = 0,25P_N; 0,5P_N; 0,75P_N; P_N.

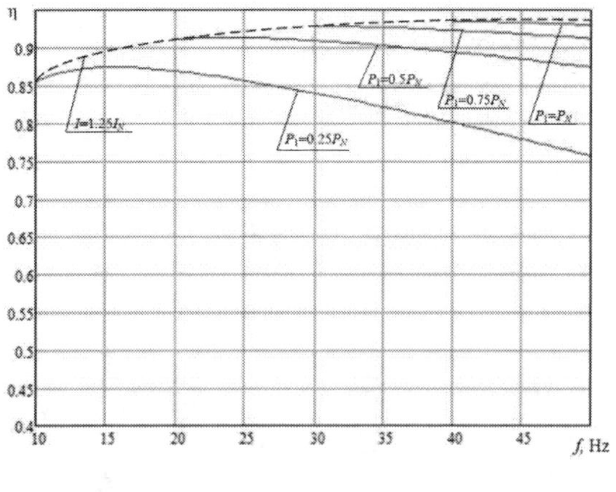

Figure 2. Dependency of MCK1250-1500 SM efficiency on frequency of power voltage at variable power P_1.

According to [1] synchronous machines of MCK series the operate duration at stator current I=1,25I_N is not to exceed 0,5 h. Fig.2 and the following ones the dotted line limits the area of currents control $I \le 1{,}25 I_N$.

3.2. Efficiency of transformer

The transformer efficiency is determined by mainly the following equations

$$\eta_T = 1 - \frac{p_\Sigma}{P_1},$$
$$p_\Sigma = p_{w1} + p_{iT} + p_{w2},$$
$$p_{iT} = v'(f/50)^\beta B^\alpha G_c, \quad (2)$$
$$p_{w1} = m I_1^2 r_1 m_t,$$
$$p_{w2} = m I_2^2 r_2 m_t.$$

In (2) the following notations are used η_T – transformer efficiency; p_Σ – total losses, p_{w1}, p_{w2} - electrical losses ones in winding, p_{iT} – steel losses, υ'- specific steel losses [1, 5], G_C –core steel mass, m – the transformer phase number, m_t – coil heat ratio; I_1, I_2, r_1, r_2 – respectively currents and resistance values of primary and secondary transformer windings.

Fig.3 shows dependencies of the efficiency of voltage frequency of transformer TC3M 1000-75.OM5.

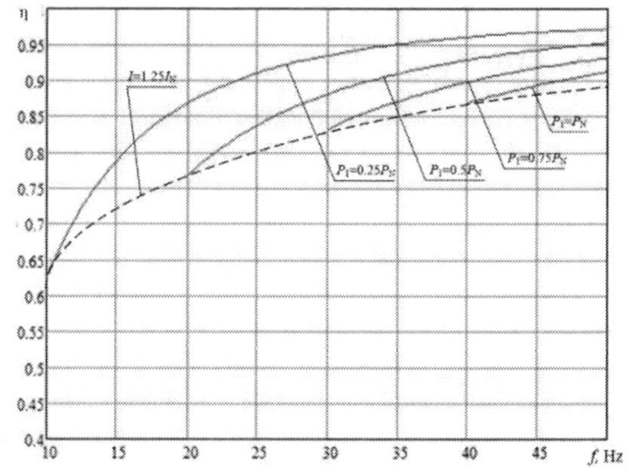

Figure 3. Dependency of transformer efficiency TC3M 1000-75.OM5 of supply voltage frequency.

3.3. Efficiency of frequency converter

Frequency converter (Fig. 4) efficiency as shown are calculated according to the following equations

$$\eta_{FC} = \eta_{rec} \eta_{pwc} \eta_{f1} \eta_{inv} \eta_{f2},$$
$$\eta_{rec} = \frac{P_d}{P_d + \Delta P_{rec}}, \Delta P_{rec} = \Delta P_{sem} + \Delta P_{sub},$$
$$\Delta P_{sem} = (U_{T0} I_{Fav} + r_T k_\phi^2 I_{Fav}^2) \cdot 6, \quad (3)$$
$$\Delta P_L = R_L I_D^2, \eta_{f1} = \frac{P_d}{P_d + \Delta P_L},$$
$$\Delta P_{IGBT} = P_D + P_{SW} + P_{DF}, P_D = I_{cav} U_{CE}(I_c),$$
$$P_{SW} = E_{ts} f, P_{DF} = I_{FM} U_{FM}.$$

In (3) are used η_{FC} – efficiency of frequency converter, η_{rec} – efficiency of rectifier, η_{pwc} – efficiency of pulse-width converter, $\eta_{f1,2}$ – efficiency of filter, η_{inv} – efficiency inverter, P_d – active power at the rectifier outlets, ΔP_{rec} – rectifier losses, P_{sem} - semi-conducting equipment devices losses, ΔP_{sub} – cooling system, system protection and control system losses, Δ. P_{FAV} – diode losses resulting from direct current flow I_{FAV}, k_ϕ – current form ratio, r_T – differential resistance, U_{T0} – threshold voltage, ΔP_L –filter inductance losses, ΔP_{IGBT} – IGBT losses, P_D - dynamic in IGBT transistor losses, P_{SW} - static in IGBT transistor losses, P_{DF} - built-in IGBT diode losses, I_{CAV} – the average impulse current conductivity collector value, $U_{CE(Ic)}$ – saturation voltage at the given collector current , E_{ts} – total energy losses at the collector losses current together with block drive circuit , I_{FM} – diode average current; U_{FM} – diode voltage at current I_{FM}.

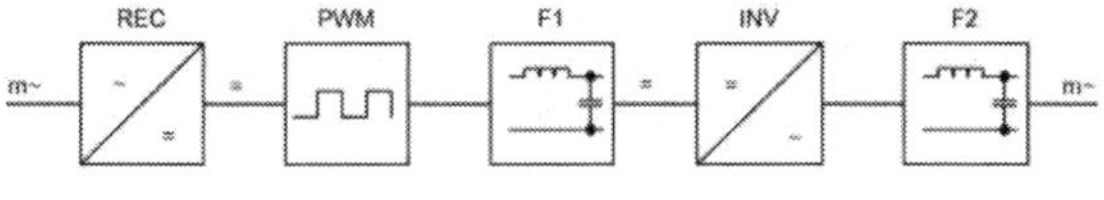

Figure 4. Flowchart of UPS frequency converter: rec – out-of-control rectifier, PWM – pulse width converter, F1, F2 – filters, Inv – stand-alone inverter.

Fig. 5 shows the dependencies of the efficiency of voltage frequency for the frequency converter having the nominal power 1000 kW.

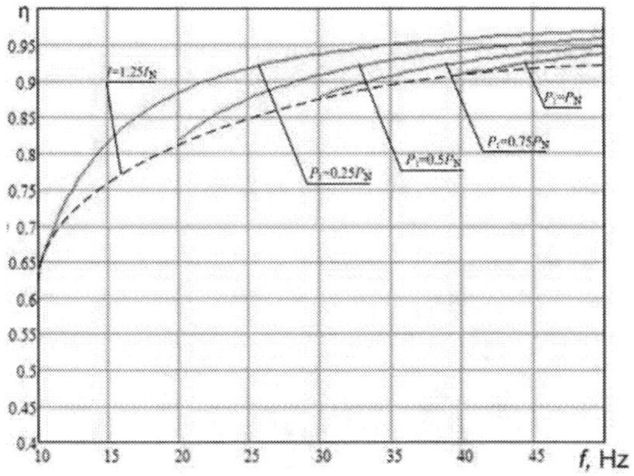

Figure 5. The dependence of the efficiency of the frequency converter power 1000 kW from the utility power frequency.

4 UPS fuel economy estimate at load power «propulsion engine»

Installed power of the SM is the sum propulsion power and the onboard ship power (fig. 1).

Fuel consumption of ICE is calculated in a range of screw-propeller (SP) speeds from 75 to 300 min^{-1} in increments of 25 min^{-1}.

To determine the ICE fuel consumption is used multivariable characteristic [6, 7]. ICE with a nominal speed of 400 min^{-1} and a power 1100 kW is selected to perform the calculations [1].

The shat moment on the SP is determined according to the equation

$$M = K\rho n^2 D_p^5 \quad (4)$$

where K–torque coefficient SP, ρ – medium density, n – frequency of rotation SP, D_p – diameter SP.

The supply voltage frequency SM is determined through the expression

$$f = f_0 \frac{n_{SP}}{n_{SP\_0}} \quad (5)$$

where f_0 – nominal frequency SM, n_{SP} – current frequency SP, n_{SP_0} – nominal frequency of DGI.

Determine the efficiency of the FC, T, SM for each specified speed of SP.

The power SM is determined through the following equation

$$P_{SM} = P_{SP}/(\eta_{SM} \cdot \eta_{FC} \cdot \eta_T). \quad (6)$$

There have been obtained the graphs of specific fuel consumption rate (G_e) on the SP rotation (n) for the two UPS modes with the controlled and uncontrolled DGI (Fig. 6-9). The obtained figures show UPS with the controlled rotation frequency of DGI as compared to UPS without frequency control of DGI.

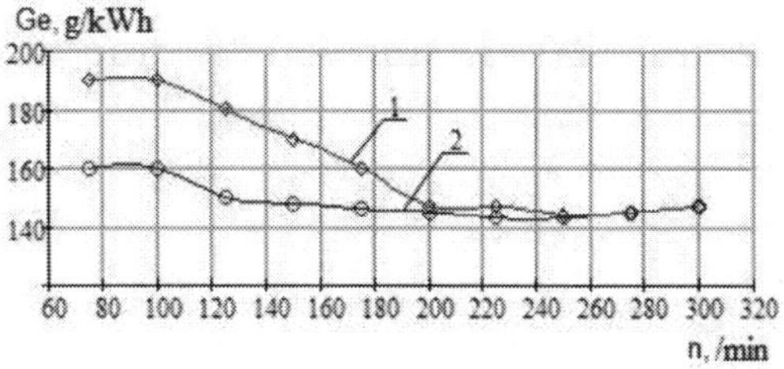

Figure 6. ICE fuel consumption rate of UPS power 1000 kW: 1 –without regulation of speed DGI, 2 – with frequency regulation of speed DGI.

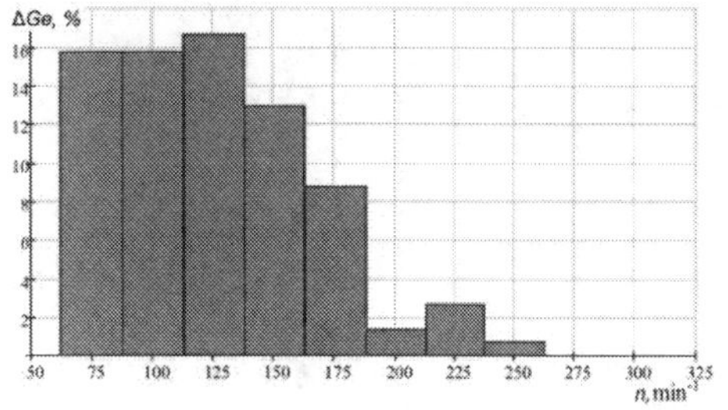

Figure 7. ICE fuel consumption rate UPS powered by 1000 kW.

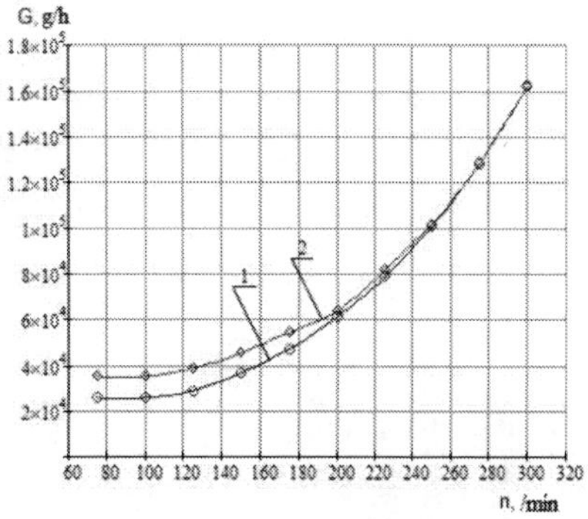

Figure 8. ICE absolute fuel consumption of UPS powered by 1000 kW: 1 – with the regulation of frequency DGI; 2 – without the frequency regulation of DGI.

DOI: 10.1051/matecconf/201710814002

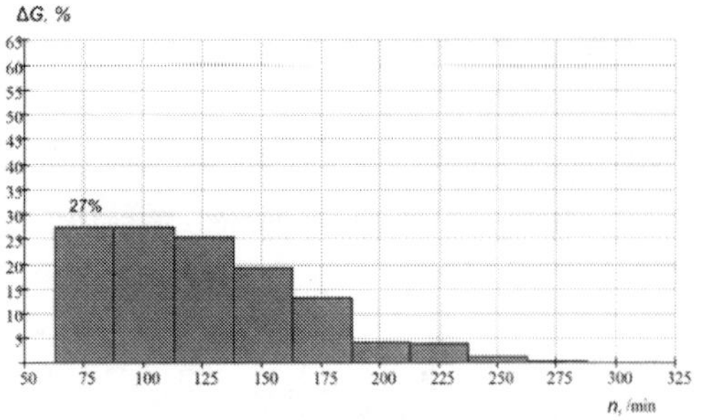

Figure 9. Absolute fuel consumption of ICE in UPS of 1000 kW.

In absolute fuel consumption UPS with regulation frequency of DGI economical system without regulation by 25-27%.

5 UPS calculation of fuel economy considering at the SM control

The above characteristics of specific and absolute fuel consumption rate have been obtained without taking into account SM excitation control. To enhance the economy of UPS, it is necessary to control the SM with respect to load power as related to the expression [1, 8, 9]

$$i_{ex} = \sqrt{\frac{1 + p_1^2 x_d^2 + 2p_1 x_d \sqrt{1-\cos^2\varphi}}{1 + x_d^2 + 2x_d\sqrt{1-\cos^2\varphi}}} \tag{7}$$

where $i_{ex}=I_{ex}/I_{ex\_nom}$, $p_1=P_1/S_{NOM}$.

As shown in graphs (Fig.10, 11), at the excitation control (Fig.10, curves 1,3), the fuel consumption rate increases by the range of light and average loads.

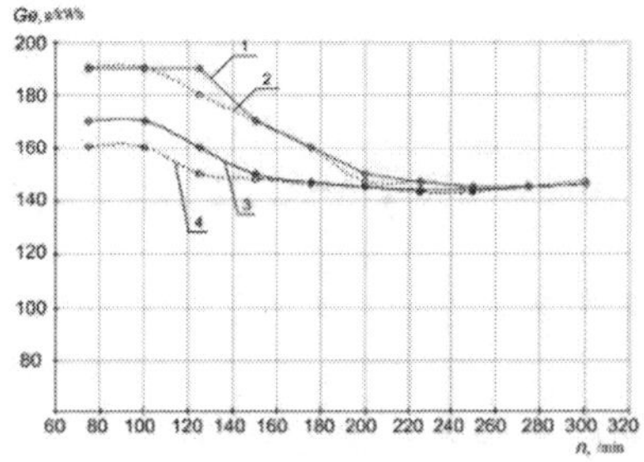

Figure 10. ICE average consumption rate of UPS power 1000 kW: 1,2 –without the speed regulation of DGI; 3,4 - with speed regulation of DGI; 1,3 – with regulation current excitation of SM; 2,4 – without regulation current excitation of SM.

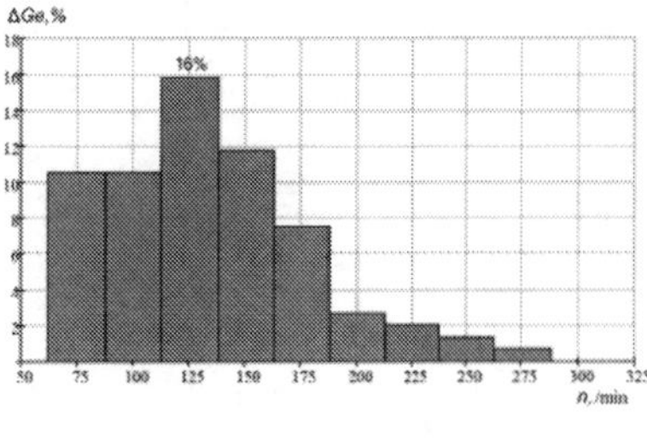

Figure 11. The ICE fuel consumption rate economy in UPS power 1000 kW with SM excitation control.

According to a multivariate characteristic, within the range of ICE light loads is displaces into the range of higher fuel consumption rate. The fuel calculation data are shown in fig.12, 13.

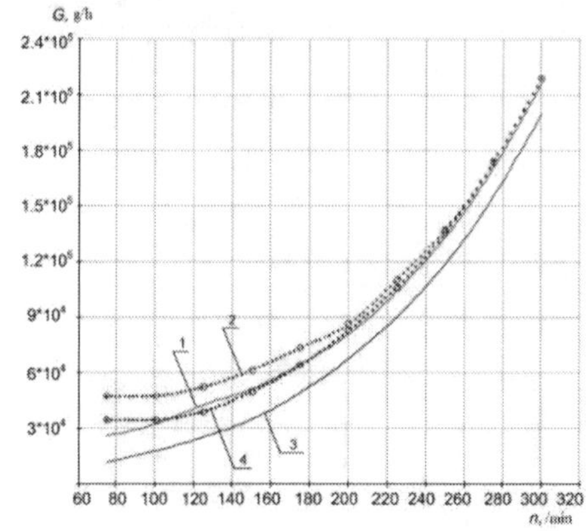

Figure 12. Absolute fuel rate consumption of ICE in UPS: 1,2 – without regulation speed of ICE, 3,4 – with speed regulation of ICE, 1,3 – with regulation current excitation of SM, 2,4 – without with regulation current excitation of SM.

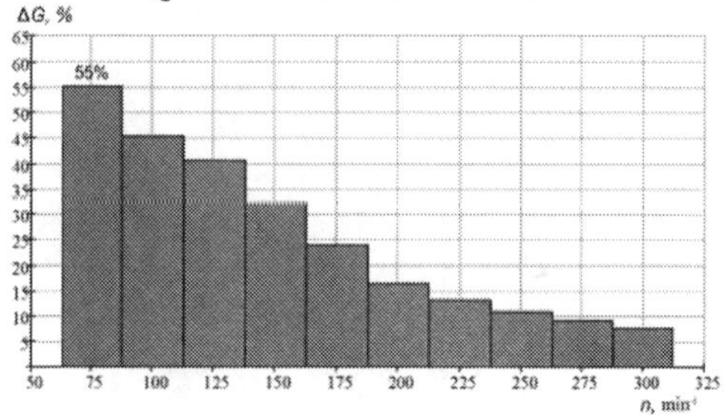

Figure 13. . Absolute fuel consumption of ICE in UPS power 1000 kW with regulation current excitation of SM.

6 Conclusion

Application UPS on the variable speed of ICE allows to efficiently economizing fuel consumption rate and reduces air emissions. Providing optimal mode of ICE in terms of fuel efficiency requires a new approach to the fuel supply control and can be implemented on modern management tools, including intelligent neural network-type control systems.

In the paper calculated the fuel efficiency of unified power station on the basis of a variable speed DGI power of 1000 kW. On the basis of the calculation results obtained by the comparative characteristics of the specific and absolute fuel consumption for the two modes of operation: a speed control of rotation DGI, depending on the load and a constant frequency of rotation of the DGI. In UPS of 1000 kW fuel savings can be up to 31% in absolute consumption.

Acknowledgment

The research was performed with the support of the State Research Program (project № 8.2668.2014/K) on department «Electrical equipment, electric drive and automatics» of Nizhny Novgorod State Technical University n.a. R.E. Alekseev.

References

1. O.S. Khvatov, A.B. Daryenkov, I.S. Samoyavchev. Jekspluatacija morskogo transporta. *The fuel profitability of unified electric power station of ship based on a explosion engine by alternating frequency rotation of shaf*, **71**, 47 (2013)
2. O.S. Khvatov, A.B. Daryenkov, I.S. Polyakov. Vestnik IGEU. *Neural Network Algorithm of Control System of Fuel Supply of Frequency Rotation Diesel Generator Set*, **3**, 50 (2013)
3. O.S. Khvatov, A.B. Daryenkov. Promishlennaya energetika. *A stand-alone diesel electric power station with a neural network economy mode cof neural network economy mode setting device*, **12**, 26 (2013)
4. O.S. Khvatov, A.B. Daryenkov. Elechtotechnika. *Power plant based on a variable-speed diesel generator*, **3**, 28 (2014)
5. O.S. Khvatov, A.B. Dar'enkov, Russian Electrical Engineering. *Power plant based on a variable-speed diesel generator*, **85**, 145 (2014)
6. A.B. Daryenkov, O.S. Khvatov, F.F. Yurlov, N.V. Usov. Vestnik of Astrakhan state technical university. Series: Marine engineering and technologies. *Definition of economic efficiency of diesel electric power stations with alternating frequency of rotation of the shaft*, **3**, 64 (2014)
7. A.B. Daryenkov, O.S. Khvatov. News of the Tula state university. Technical sciences. *Intelligent controls of highly effective the diesel-generator installation of variable frequency of rotation*, **3**, 126 (2010)
8. S.M. Dmitriev, V.V. Konratyev, O.S. Khvatov, A.B. Daryenkov, E.A. Chernov. Proceedings VII International Conference on Power Drives Systems. *Intellectual servo-drive of fuel supply of a variable rotation speed stand-alone electric power station*, 592 (2012)
9. O.S. Khvatov, A.B. Daryenkov, I.S. Samoyavchev. Proceedings VIII International Conference on Power Drives Systems. *High-efficiency electric propulsion system of a stand-alone object*, **2**, 396 (2014)

MATEC Web of Conferences **108**, 14003 (2017) DOI: 10.1051/matecconf/201710814003
ICMAA 2017

Experimental research of variable rotation speed ICE-based electric power station

Andrey Dar'enkov

Nizhny Novgorod State Technical University n.a. R.E. Alekseev, Institute of Electric Power Engineering, 603950 Nizhny Novgorod, Russia

Abstract. Developing variable rotation speed ICE-based stand-alone electric power stations which can supply distant regions and autonomous objects with electricity are of scientific interest due to the insufficient study. The relevance of developing such electric power stations is determined by their usage is to provide a significant fuel saving as well as increase ICE motor service life. The article describes the electric station of autonomous objects with improved fuel economy. The article describes multivariate characteristic. Multivariate characteristic shows the optimal frequency of rotation of the internal combustion engine. At this rotational speed there is the greatest fuel economy.

1 Introduction

A major part of the Russian Federation territory is not connected to the central energy power supply system. Such territories include Russian Far East, North, Siberia larger territories and other regions. In these territories are living 10 million people. The main electric power sources of these are electric power stations based on the system "internal combustion engine (ICE) – generator" (E-G) [1]. The existing diesel generators operate at a constant (nominal) rotational speed at different loads. Operation of ICE at a constant rotational speed at variable load is characterized by a non-optimal fuel consumption. This reduces efficiency E-G power station. The article describes the electric station of autonomous objects with improved fuel economy. Improved efficiency is achieved by setting the optimal ICE rotational speed for each value of load power. This optimal speed corresponds to the lowest specific fuel consumption.

2 E-G power station

E-G electric power stations, having high reliability, sufficient motor capacity and long period of service is indispensable as stand-alone of main and back-up energy power supply sources. However, the fuel necessary for E-G functioning is transported form distant regions by sea, automobile means and occasionally even by helicopters, which significantly increases its cost. Besides the fuel transportation to distant regions often depends on weather conditions, the season thus it becomes not ever possible. To provide the above regions with electric energy there are used over 50 000 of E-G having in total 17 million kW with electric energy generation of over 50 billion kWh a year. These power stations fuel consumption is about 6 million tons of oil equivalents [1].

E-G electric power stations are also widely used as back-up energy supply sources on the territory of Russia having a single energy system. E-G power stations are as a rule both the main and back-up energy sources on stand-alone objects, such as merchant and combat vessels, locomotives, automobile transportation etc.

Thus a considerable part of electricity is produced by electric power stations stand-alone E-G. One of the most important tasks of modern Russia is resource-saving, in particular efficient fuel resources utilization. However the overwhelming majority of the existing E-G function at the constant (nominal) shaft speed rotation frequency within the whole range of the load power change. In such E-G, the AC stator synchronous generator (SG) frequency stabilization is provided by the ICE shaft speed stabilization whereas SG stator voltage amplitude stabilization is carried out by the current change in the exciting coil [2]. At this stand-alone E-G function as a rule in the fractional modes their load being within the range of 30 up to 70% of the nominal one. It is common knowledge that ICE at a constant rotation rate but at an alternating load generally functions with non-optimum fuel consumption (with non-optimum efficiency coefficient) [2, 3]. To provide the optimal (from the point of view of the fuel consumption) ICE operating mode, it is necessary to change its rotation frequency depending on the load power [3, 4].

To choose the economical ICE operating mode, functioning at the changing load, it is practical to use its multivariate characteristic. Fig. 1 shows a typical multivariate ICE characteristic on which by the dash-dotted line is shown the effective pressure p_e and the effective power N_e dependency on the frequency of ICE shaft rotation n at the minimal fuel consumption rate g_e.

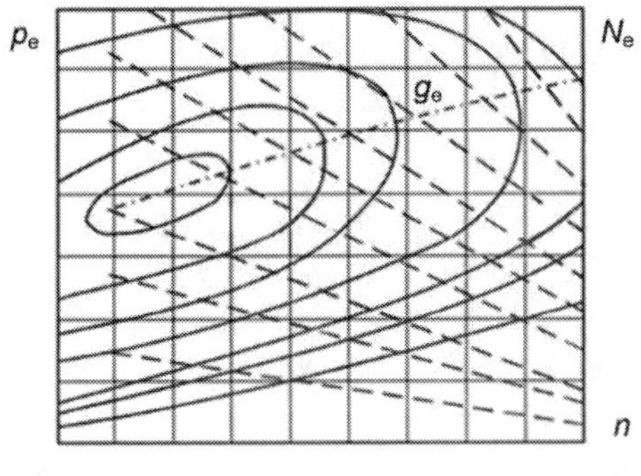

Figure 1. Multivariate characteristic of ICE.

As shown in Fig. 1, at the changing SG load , being rotated by ICE, the electric current is produced, its voltage and frequency being changed, as a rule, lower than nominal values. This requires an innovative approach to building the E-G system able to supply the voltage and frequency output SG frequency stabilization.

The designs of variable speed E-G systems frequency voltage are the prospective ones. The resources to develop these lie within the resource and energy saving, being currently the acute ones. In Russia the following problems are dealt with by S.Petersburg (JSC Zvezda), Moscow (Ltd. NTC Malaya Energetica), Kovrov of Vladimirskaya oblast' (JSC Signa), Tomsk (Tomsk Polytechnic University), in Nizhny Novgorod (Nizhny Novgorod State Technical University after R.E. Alexeev) etc. [4-12].

The existing E-G systems of the variable rotation speed are designed according to the scheme as shown in Fig.2. [5-7,11]. In such systems (see Fig. 2) the voltage frequency at the outputs 6 are stabilized at a set level through the frequency converter (FC) 5, comprising the rectifier 3 and voltage invertor 4. The stabilization of SG 2 stator voltage amplitude, connected to the ICE 1shaft, is carried out by the impact of control system 7 on the excitation coil rate SG 2, being such systems deficiency. From the point of view of such system fuel efficiency at the change of load power, the ICE shaft speed is to be calculated within a wide range, the minimal shaft speed being several times different from the nominal one. Thus SG voltage amplitude might change within the wide range, its minimal value being several times lower than the nominal one. At this the SG stator voltage stabilization method, based on the impact of the excitation coil current will not provide the fact of its being supported on its nominal value.

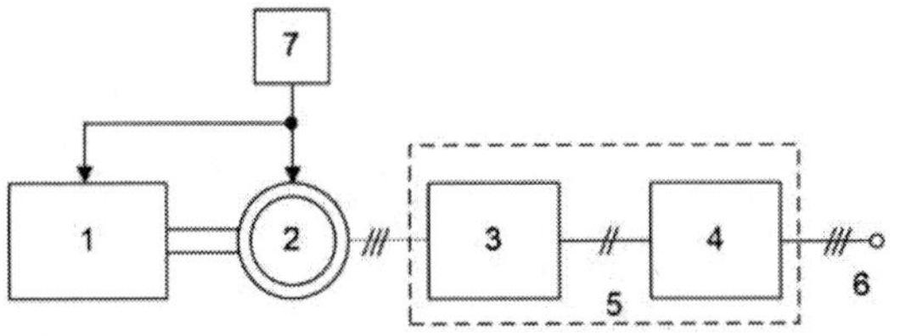

Figure 2. E-G flowchart of variable speed (1 – ICE; 2 – SG; 3 – rectifier; 4 – voltage invertor; 5 – FC; 6 – outputs; 7 – control system).

Of interest are also some foreign E-G producers manufacturing invertor power stations, such as SDMO, Honda, Fubag, Hyundai, Kypor etc. These provide lower fuel consumption due to the ICE switch function into the economy saving mode. Invertor power station flowchart is shown in Fig. 3 [9]. Its operation principle is the following: ICE rotates the multipolar generator, its rotor being a permanent magnet. AC three-phase voltage of stator is rectified and DC voltage is converted by an inverter into a sinusoidal voltage with a stable frequency and amplitude.

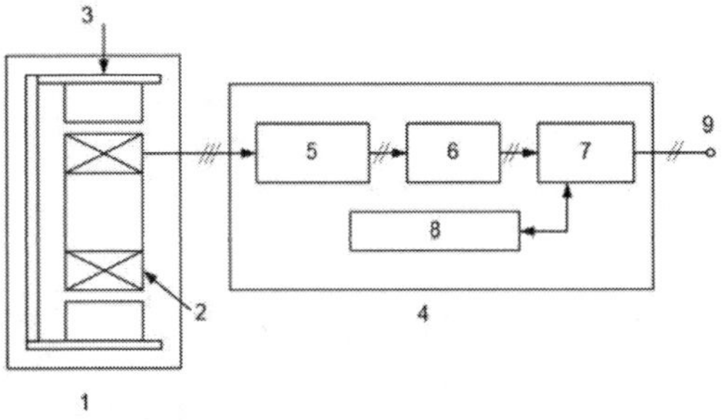

Figure 3. Invertor electric power station flowchart (1 – multipolar generator (in section); 2 – generator stator; 3 – generator rotor; 4 – DC; 5 – rectifier; 6 – filter; 7 – invertor; 8 – microprocessor control system; 9 – outputs).

In this case the economy mode means the ICE transfer to a lower rotating speed at also lowering the load beyond the set limit (about beyond 50% of the nominal value). An invertor power station is equipped by a special throttle regulating the ICE rotating frequency depending on the load power, which allows reducing the fuel consumption rate by 20 % (see Fig. 4). The parameters stability of the electric power station voltage output is provided by the invertor unit. Voltage amplitude fluctuations are not over 0.5% [9].

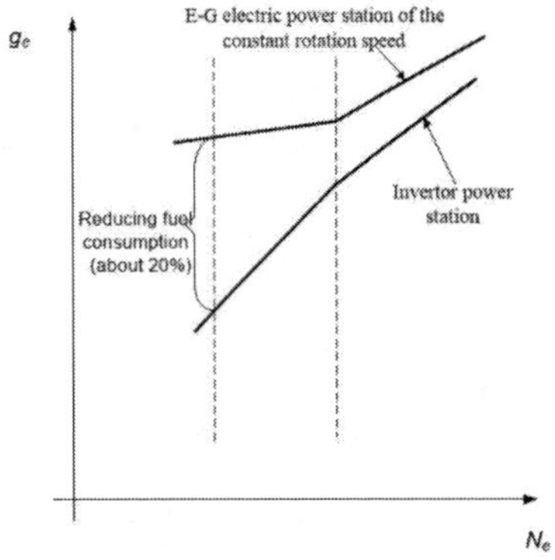

Figure 4. The fuel consumption dependence on the load power of the invertor electric power station and E-G electric power station of the constant rotation speed.

Develop E-G electric power stations of variable speed of rotation is possible ac-cording to the scheme given in Fig. 5 [13-15]. The mechanism functions as follows. The intellectual servomotor of fuel supply control shown by unit 7 receives a signal from the load power evaluator 15, connected to the outputs of voltage sensors 14 and current sensor 4, measuring the voltage and current at the FC 3 outlet respectively. Depending on the load power value, unit 8 setting the economy mode (SEM) forms a signal, proportionate to the optimal ICE 1 shaft speed. This signal comes to the electric drive 9 inlet of the control rack displacement, supporting the ICE 1 frequency rate at a level set by SEM (unit 8).

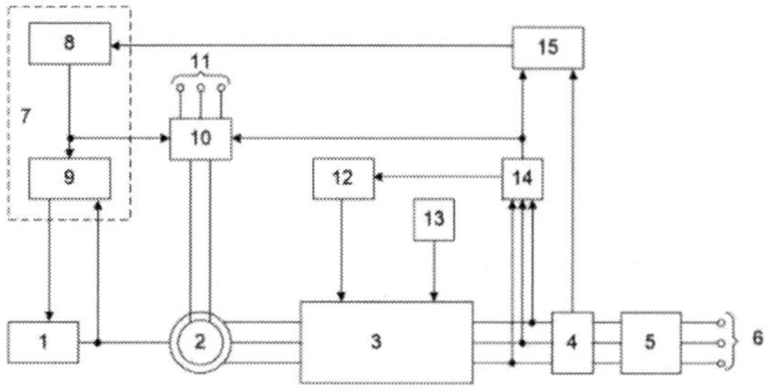

Figure 5. System flowchart of E-G variable rotation speed (1 – ICE; 2 – SG; 3 – FC; 4 – current sensor; 5 – step-up transformer; 6 – outputs; 7 – the intellectual servomotor of fuel supply control; 8 – SEM; 9 – electric drive of the control rack displacement; 10 –SG excitation unit; 11 – outlets of SG excitation unit connecting points; 12 – voltage stabilization unit; 13 – output voltage frequency setter; 14 – voltage sensor; 15 – load power evaluator).

The AC amplitude stabilization at the outputs 6 at the nominal SG 2 value level is performed by a FC 3 and the step-up transformer 5. SG 10 excitation unit of SG 2 forms current in excitation coil SG 2 at the signal from SEM 8 unit and voltage sensor 14. The output voltage frequency at outputs 6 with the change of ICE 1 shaft speed rotation frequency is guaranteed the same with the help of FC 3 at the level deter-mined by output voltage frequency setter 13.

The elementary variant of SEM algorithm might comprise the ICE multivariate characteristic. However multivariate characteristic for each ICE is individual (characteristics of two ICE of the same type and power differ), and ICE manufacturers usually don't provide such characteristics. It is impossible to calculate multivariate characteristic. It is possible to obtain only through the experimental procedure. Besides an ICE multivariate characteristic will vary depending on the outer (pressure, temperature, humidity) and inner (ICE amortization, make and fuel quality) of ICE functioning. So an SEM unit should at the absence of multivariate characteristic automatically calculate the optimal frequency value for the current load power value at a given time with the changing outer and inner ICE functioning conditions. Thus, SEM should be a self-taught system.

SEM flowchart is shown in Fig 6. SEM consists of four basic elements: main controller (MC), content addressable memory (CAM), teaching control (TC) and non-volatile memory (NVM), connected together by a common data bus, which provides a two-way traffic and further system expansion in case of controlling several E-G [16-20].

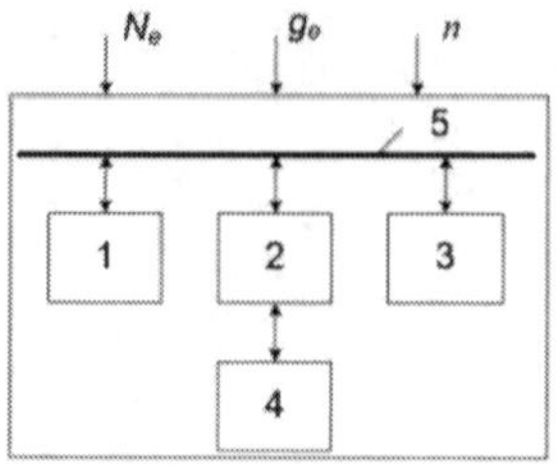

Figure 6. SEM flowchart (1 – MC; 2 – TC; 3 – NVM; 4 – data memory; 5 – data bus).

NVM of SEM is a neural network, which performs the task of table function approximation $n=f(N_e)$, supplemented by the interpolation (extrapolation) procedure. The backpropagation is used for neural network training [16-19].

An E-G experimental research of a variable speed of rotation of the ICE-based electric power station with nominal voltage of 60 kW has been carried out. The main goal of the research was to determine ICE optimal modes at its speed of rotation fluctuations from 1200 up to 4000 min^{-1} within the full range of possible load power.

3 ICE multivariate characteristic

As a result there have been obtained the sets of experimental motor characteristics comprising the following:

- ICE load-voltage characteristics set for seven speed modes: 1200, 1500, 1600, 2000, 2500, 3200, 3800 min-1;
- ICE outer speed characteristics;
- ICE idling characteristic.

The experimental results give the following:

- multivariate characteristics of fuel-consumption rate within "rotation frequency n – mean effective value p_e";
- multivariate characteristics of fuel combustion heat-availability ratio within "rotation frequency n – net torque M_e".

ICE multivariate characteristic (Fig.7), being a part of the researched electric power station, has been obtained through processing the above experimental characteristics set. The optimal ICE rotation frequency dependence on the load power from the point of view of fuel consumption is shown in Fig. 7 by the red dotted line.

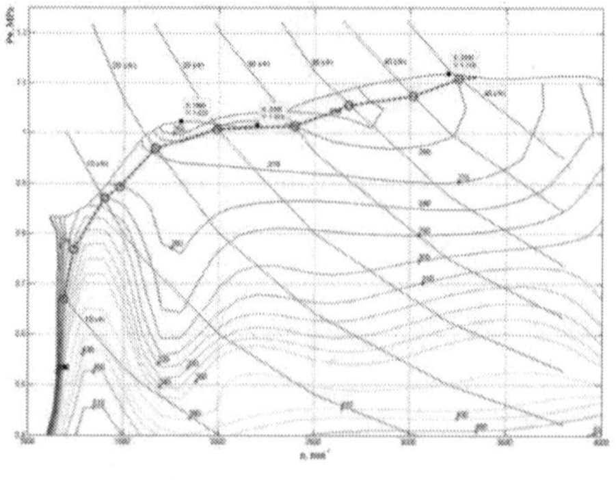

Figure 7. ICE multivariate characteristic with power of 60 kW.

According to the results of the research the following conclusions have been made:

1. Experimental areas of consumption rate g_e show the necessity of wide frequency fluctuations of the researched E-G electric power station within the range of 1500 - 3500 min^{-1} to optimize the fuel consumption while changing the electric power station capacity.

2. In the case of optimal choice of ICE rotation frequency variation it is possible to significantly improve the E-G electric power station efficiency.

As an example, for the researched electric power station Fig. 1 shows the following:

- dependence of fuel consumption rate g_e^{3000} on power at the ICE constant rotation frequency 3000 min^{-1};
- dependence of fuel consumption rate g_e^{VAR} on power N_e at the ICE variable rotation frequency.

The lines corresponding to 2-fold power control reduction are shown in table 1. Herewith, the fuel saving at controlling the ICE rotation frequency is 26% as compared to the experiment in which the ICE rotation frequency was leveled off at 3000 min^{-1}.

Research shows that the ICE rotation frequency control, being a part of E-G electric power station, at load power change allows reducing the fuel consumption rate to 20-30%. Consequently the E-G variable frequency system operation significantly economizes the use of expensive fuel resources. Simultaneous change of rotation frequency and load power provides the optimal ICE thermal conditions operation, and thus an increase of its motor service life.

N_e, kW	G_E^{3000}, G/KWH	g_e^{VAR}, g/kWh	g_e^{VAR}/g_e^{3000}	Fuel saving, %
45	260	260	1	0
40	258	258	1	0
35	264	250	0.95	5
30	288	260	0.90	10
25	315	262	0.83	17
20	360	268	0.74	26
15	460	280	0.61	39

4 Conclusion

Experimental research of a variable speed of rotation of the ICE-based electric power station with nominal voltage of 60 kW has been carried out in accordance with the developed and approved program and test pattern. The algorithm of the SEM variable frequency ICE-based electric power station operation has been developed on the basis of neural networks apparatus. The back propagation has been used for neural network training.

There have been obtained multivariate fuel consumption rate areas, as well as fuel economy estimate for the variable rotation frequency 60 kW ICE-based power stations.

Using variable frequency ICE-based electric power stations let's get the reduction of fuel rate consumption of up to 30%. The simultaneous change of rotation frequency and load power provides the optimal ICE thermal conditions operation, and thus increases its motor service life.

Acknowledgment

The research was performed with the support of the State Research Program (project № 8.2668.2014/K) on department «Electrical equipment, electric drive and automatics» of Nizhny Novgorod State Technical University n.a. R.E. Alekseev.

References

1. O.V. Fedorov, A.B. Dar'enkov. *Energy saving politics* (Knorus , Moscow, 2015)
2. V.A. Michaylov. *Automated electric power systems of vessels* (Sudostroenie, Leningrad , 1977)
3. A.S.Orlin, M.G. Kruglov. *Internal Combustion Engines: Theory of reciprocating and compound engines* (Mashinostroenie, Moscow, 1983)
4. O.S. Khvatov, A.B. Dar'enkov, I.M. Tarasov. Vestnik IGEU. *Diesel-generator Power Plant with Shaft Frequency Rotation*, **2**, 53 (2010)
5. O.A. Aleshkov. *Primary diesel motor fuel consumption economy enhancement as a part of multifunctional energotechnological complex by optimizing the speed mode*, dissertation work, Barnaul, (2009)
6. N.M. Parnikov. *Decentralized power supply energy efficiency complexes enhancement as exemplified in Saha Republic (Yakutia)*, dissertation work, Tomsk (2009)
7. M.A.Surkov. *Wind-diesel electrotechnical complexes energy performance optimization*, dissertation work, Tomsk (2011)
8. I.A.Tarpanov. *Stand-alone generator complexes of variable rotation frequency*, dissertation work, N. Novgorod (2012)
9. I.S. Polakov. *Diesel driven generator*, dissertation work, N. Novgorod (2013)
10. I.S. Samoyavchev. *Stand-alone objects electric power station on variable rotation frequency-based diesel driven generator*, dissertation work, N. Novgorod (2014)
11. A.E. Zagorsky, U.G. Shakaryan. *Independent fix frequency power supply source (variants)*. RF patent 34817, 10 Oct 2003
12. Honda Information, http://www.honda.co.ru/power-products/documents
13. A.B. Dar'enkov, O.S. Khvatov. *AC stand-alone electric power station*, RF patent 2412513, 10 Dec 2011
14. A.B. Dar'enkov, O.S. Khvatov. *AC stand-alone electric power station*, RF patent 137701, 05 Feb 2014
15. A.B. Daryenkov, O.S. Khvatov. News of the Tula state university. Technical sciences. *Intelligent controls of highly effective the diesel-generator installation of variable frequency of rotation*, **3**, 126 (2010)
16. O.S. Khvatov, A.B. Daryenkov, I.S. Polyakov. Vestnik IGEU. *Neural Network Algorithm of Control System of Fuel Supply of Frequency Rotation Diesel Generator Set*, **3**, 50 (2013)

17. O.S. Khvatov, A.B. Daryenkov. Promishlennaya energetika. *A stand-alone diesel electric power station with a neural network economy mode cof neural network economy mode setting device*, **12**, 26 (2013)
18. O.S. Khvatov, A.B. Dar'enkov, Russian Electrical Engineering. *Power plant based on a variable-speed diesel generator*, **85**, 145 (2014)
19. A.B. Dar'enkov, V.I. Erofeev. Proceeding of International conference on informatics, networking and intelligent computing. *Semiconducting invertor generators with minimal losses*, 227 (2015)

MATEC Web of Conferences **108**, 14004 (2017) DOI: 10.1051/matecconf/201710814004
ICMAA 2017

Design of High Efficient MPPT Solar Inverter

K. A. Sunitha , G Prem Kumar, Nidhi Priya and Jatin Verma

Department of Electronics and Instrumentation, SRM University, India

Abstract. This work aims to design a High Efficient Maximum Power Point Tracking (MPPT) Solar Inverter. A boost converter is designed in the system to boost the power from the photovoltaic panel. By this experimental setup a room consisting of 500 Watts load (eight fluorescent tubes) is completely controlled. It is aimed to decrease the maintenance cost. A microcontroller is introduced for tracking the P&O (Perturb and Observe) algorithm used for tracking the maximum power point. The duty cycle for the operation of the boost convertor is optimally adjusted by using MPPT controller. There is a MPPT charge controller to charge the battery as well as fed to inverter which runs the load. Both the P&O scheme with the fixed variation for the reference current and the intelligent MPPT algorithm were able to identify the global Maximum power point, however the performance of the MPPT algorithm was better.

1 Introduction

Energy plays vital role for development in all sectors. With depletion of fossil fuels used for power generation and increase in demand for power, the gap between supply and demand is becoming more. Renewable energy sources can only provide solution to it. Sun radiates 180 billion MW of energy over Earth. Just one hour of this energy could meet power needs of entire planet for a year. India receives 5000 Trillion KWh of energy from Sun per annum. This energy is clean, pollution free and inexhaustible.

Photovoltaic (PV) is a method of generating electrical power by converting solar radiation into direct current electricity using semiconductors that exhibits the photovoltaic effect.

PV panel is connected to the charging circuit. In a photovoltaic system the maximum power output changes with load, temperature and the solar intensity.

In this paper we have designed an efficient MPPT solar inverter. Purpose of investment in solar power project is to enter in development of green energy technology, which is the only ultimate source of energy for future generations. This system is easy to fit and install in the existing systems [1]. The higher performance of the proposed intelligent system has been verified in [2]. The solar tracking itself can be continuous or step tracking, depending on the application. Continuous tracking is needed for highly accurate systems [3]. The controller aims at maximizing the solar PV cell's efficiency by forcing sunlight to be incident perpendicularly [4].

The solar power system is also used in the solar powered vehicles [8]; it can also be used in residential applications [9].

2 The proposed PV system

Photovoltaic (PV) is a method of generating electrical power by converting solar radiation into direct current electricity using semiconductors that exhibits the photovoltaic effect. The intensity of sun's light is the reason for generation of electricity and not the heat. For example, compared to the month of January, April will provide less power as heat is more, that may affect the working of device.

An Arduino is programmed in such a way to rotate the stepper motor to the desired angles. The motor drives the panel to the maximum power output positions. The generated voltage goes to the voltage regulator, passing through the control unit it goes to the inverter. From the inverter we get the AC output power.

A 22 Volt, 120 Watts solar panel gives around 7.59 ampere current in bright light. The solar panel with the inverter can produce a maximum power continuously. The batteries used are connected in parallel. The battery will store the energy produced by the panel. This battery is connected to the Inverter. Battery used in our system is standard 12 V battery. Inverter is used to convert the input DC voltage from the battery into output AC voltage.

The MPPT algorithm acts as the interface between the PV array and the energy storing device. For maximum power point tracking (MPPT) perturb and observe method (P&O) is used.

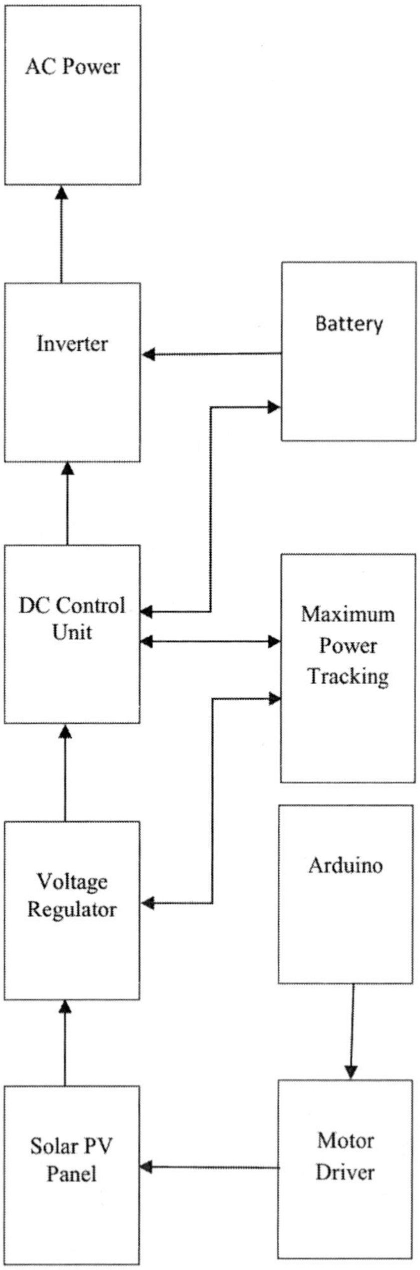

Figure 1. Block diagram of the solar inverter.

3 Flow chart

The flow chart shows the sequential movement of the whole process. The program checks the time and decides the position of the panel. Four time domains are taken into consideration i.e. 10 am, 12 noon, 2pm (14 hours in the 24-hour clock) and 17 hours or 5 pm. Position 1 is maintained till 10am, after 10 the position changes to 2 till 12 noon. After that the panel takes position 3 till 5pm. After this the panel comes back to position 1 and remains at that position till 10am. And the process continues the same way.

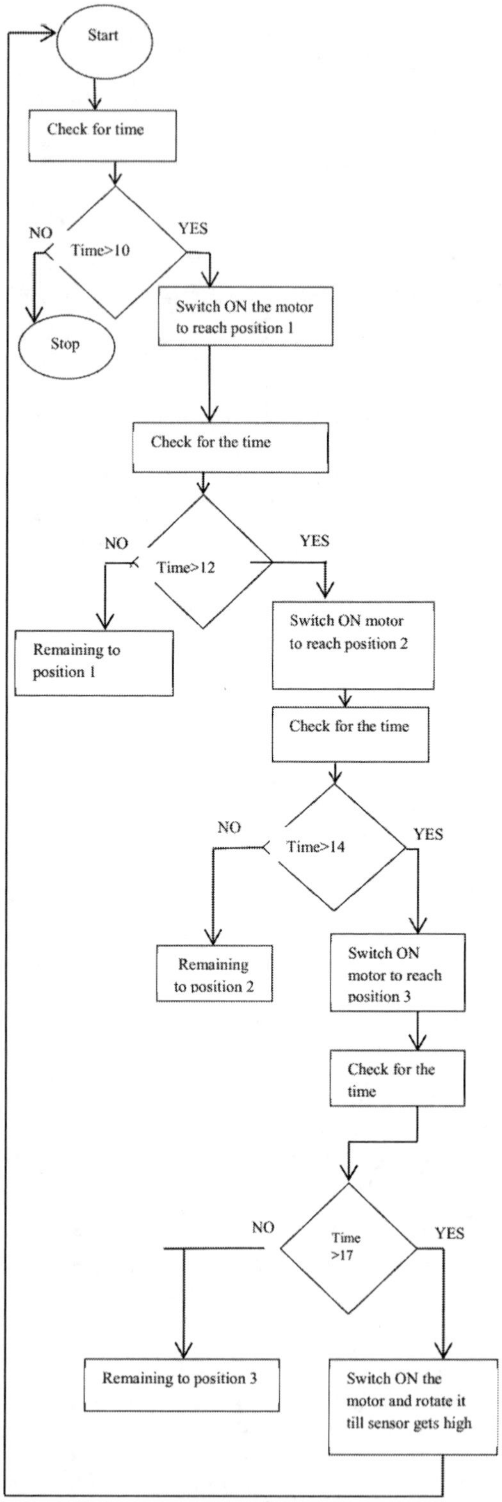

Figure 2. Persistent loop of the solar inverter.

4 Theory and modeling

When the light falls on the solar panel, it produces a dc current. As the light falling on the panel is not constant the dc voltage produced by the panel is also not a constant value. This goes to the voltage regulator which regulates and stabilizes the dc output. The regulated voltage then goes to the dc control unit. The controller consists of a chopper circuit. Depending on its duty cycle, it transfers the input power to the output power. It is also called the buck boost converter. The output then goes to the inverter circuit. In the inverter circuit the dc output is converted to AC output. This AC power can directly be used.

Apparently, the dc output thus generated can be stored in the battery. The inverter circuit was made using transformer, capacitor, dc Battery etc.

4.1 PV panels

Photovoltaic (PV) is a method of generating electrical power by converting solar radiation into direct current electricity using semiconductors that exhibits the photovoltaic effect. Photovoltaic power generation employs solar panels comprised of an array of cells containing a photovoltaic material. The PV generator is formed by the combination of many PV cells connected in series and parallel to provide the desired value of the output voltage and current. PV Panel is connected to the charging circuit.

PV Panel Specifications: -

1) General specifications:
   - Manufacturer: AMMVEE
   - Model: D2160Wp
   - Solar cell: Multi crystalline.
   - Solar cell shape: Rectangle
2) Electrical specifications:
   - Open circuit voltage: V_{oc} = 22.0volt.
   - Short circuit current: I_{sc} =6.0 ampere.
   - Maximum voltage: V_{mp} =36.7 volt.
   - Maximum current: Imp =7.59 ampere.
   - Maximum power at STC: P_{max}= 120 Wp.
   - Operating Temperature/Humidity = 47Â˘rC + 2.

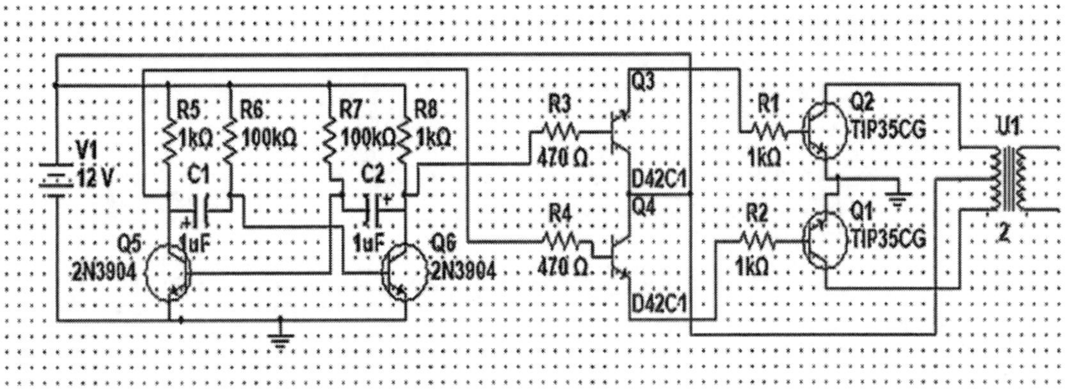

Figure 3. Inverter circuit.

4.2 Inverter

It is used to convert the DC input voltage from battery or PV panel into AC voltage. Inverter used here does not produce a pure sinusoidal output rather it generates square pulses which is connected to almost all the appliances now-a-days except some high frequency appliances. Power transistors are used to drive the transformer.

After adding protection circuits/components for over voltage protection, over current protection, over load protection this project will be a complete solution to the energy crisis problem.

There is only one variable resistance in this circuit diagram which is used to adjust frequency of 240V AC output voltage. The frequency meter is adjusted to frequency of 50HZ to 60HZ as per the requirement.

4.3 Battery

Here, Battery means the series of batteries that are connected in parallel. The battery will store the energy produced by Panel. This block shown above (Fig.01.) is connected to the Inverter. Battery used in our system is standard 12 V of 45 Ah.

Table 1. Power values at different time and positions.

VOLTAGE MEASURED VARYING THE LOAD	10AM				3PM				12 Noon	
	STATIONARY		TRACKING		STATIONARY		TRACKING			
(V)	I(Amp)	P (W)	I(Amp)	P(W)	I(Amp)	P(W)	I(Amp)	P(W)	I(Amp)	P(W)
20	00	00	00	00	00	00	00	00	00	00
18	1.43	25.74	1.48	26.64	1.44	25.92	1.52	27.36	2.20	39.6
16	2.96	47.36	3.26	52.16	2.47	39.52	3.1	57.6	3.67	58.72
14	4.50	63	4.51	63.14	2.89	40.46	4.7	65.8	4.79	67.07

12	5.44	**65.28**	5.75	**69**	3.15	37.8	5.42	65.04	5.05	60.6
10	5.82	58.2	6.42	64.2	3.43	38.9	5.85	58.5	5.38	53.8
08	6.01	48.08	6.78	54.24	3.89	31.12	6.30	50.4	6.06	48.48
06	6.14	36.84	6.80	40.74	4.22	25.32	6.56	39.36	6.46	38.46
04	6.20	24.8	6.85	27.2	4.74	18.96	6.78	27.12	6.89	27.56
02	6.27	12.54	6.87	13.7	4.66	8.64	6.79	12.81	7.15	14.3
00	6.31	00	6.91	00	4.55	00	6.72	00	7.59	00

*Maximum power has been tracked and bolded

5 Simulation results and analysis

The experiment was performed by keeping the panel at 3 different angles. Each time it was made facing the sun in a way that most of the sun rays fall perpendicularly on it. The 1st angle is 45 degrees facing east, the 2nd one is 180 degrees and the 3rd angle is 45 degrees facing the west. The voltage and current are measured in both stationary and tracking positions. By the data's received we plot a graph and obtain the maximum point tracking graph. By these graphs we see that the tracking system produces a higher output power than the stationary set up. In the table below 'I' denotes current value in amperes and 'P' denotes the output power produced in watts.

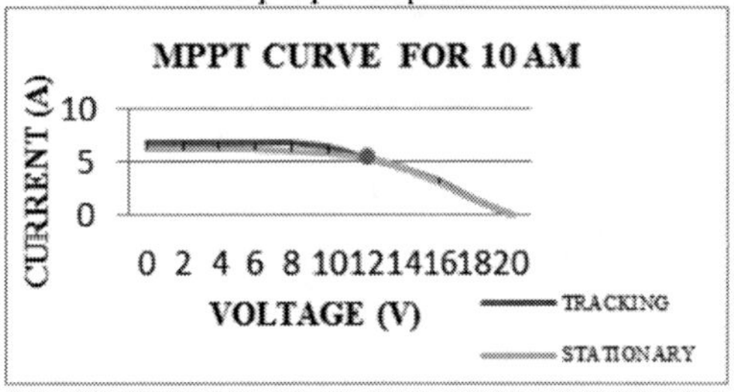

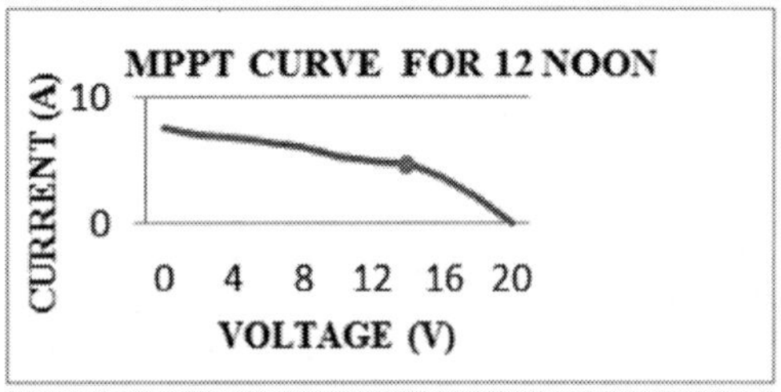

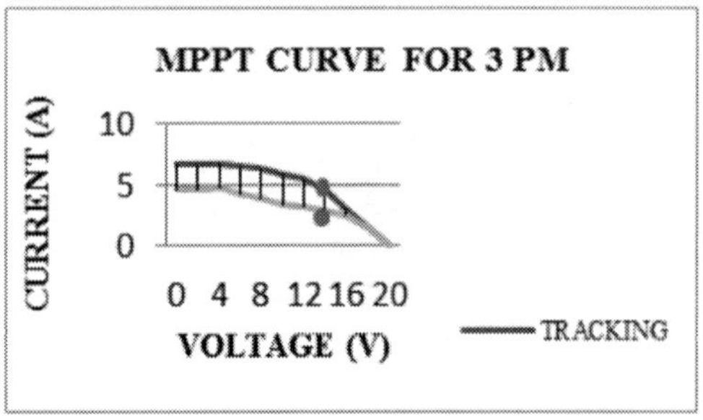

*Maximum power marked in each case
Figure 4. Graphs showing the V-I curve.

6 Conclusions

In this paper, we presented that a tracking system is more efficient than the existing stationary one. The output power is more in all the time domains in which the experiment was carried out. In the developed MPPT solar inverters the power consumed by the tracking system is found to be more, which can be reduced in the future works.

7 Future scope

The future work consists of making a solar tracking system based on Fuzzy Logic which could help it to improvise the efficiency and output power generated.

References

1. M. G. Simoes and N. N. Franceschetti, IEE Proc.-Electr. Power Appl., **146**, no. 5, (September 1999)
2. N. S. D'Souza, L. A. Lopes and X. Liu, Concordia University Department of Electrical and Computer Engineering, IEEE, (2005)
3. M. Alata, M..A. Al-Nimr, Y. Qaroush, Energy Conversion and Management, **46** (2005) 1229–1245
4. W. Batayneh, A. Owais, M. Nairoukh, Elsevier, Automation in Construction **29** (2013) 100–106
5. C. S. Chiu, IEEE transactions on Energy conservation, **25**, No. 4, (December 2010)
6. J. S. Choi, D. Y. Kim, K. T. Park, C. H. Choi and D. H. Chung, International Conference on Smart Manufacturing Application (April. 9-11,2008) in KINTEX, Gyeonggi-do, Korea
7. M. Taherbaneh, H. Ghaforifard, A. H. Rezaie and S. Karbasian, (2007) IEEE
8. B. K. Bose, P. M. Szczesny, R. L. Steigerwald, Proceedings of the IEEE Industry Applications Society annual meeting, (Sept. 1984), pp. 852-859
9. C. R. Sullivan, M. J. Powers, Proceedings of the IEEE-PESC Power Electronics Specialists conference, (1993), pp. 574-580
10. C. Y. Won, D. H. Kim, S. C. Kim, W. S. Kim, and H. S. Kim, Proceedings of the IEEE Power Electronics Specialists conference, (1994), pp. 396403

Copper-indium-gallium-diselenide nanoparticles synthesized by a solvothermal method for solar cell application

Chuan-Sheng Chiou and Huan-Chih Penga

Department of Mechanical Engineering, Yuan Ze University, No.135, Far-East Rd., Chung-Li, Taoyuan 32003, Taiwan

Abstract. Chalcopyrite copper-indium-gallium-diselenide (CIGS) nanoparticles are useful for photovoltaic applications. In this study, the synthesis of CIGS powder was examined, and the powder was successfully synthesized using a relatively simple and convenient elemental solvothermal route. From the reactions of elemental Cu, In, Se and Ga(NO3)3 powders in an autoclave with ethylenediamine as a solvent, spherical CIGS nanoparticles, with diameters ranging from 20-40 nm, were obtained using a temperature of 200°C for 36h. The structure, morphology, chemical composition and optical properties of the as-synthesized CIGS were characterized using X-ray diffraction, transmission electron microscopy, selected area electron diffraction, scanning electron microscopy, inductively coupled plasma-mass spectrometry. In this sample, the mole ratio of Cu:In:Ga:Se was equal to 0.89:0.71:0.29:2.01, and the optical band gap was found to be 1.18 eV. The solar cell obtained a power conversion efficiency of 5.62% under standard air mass 1.5 global illumination.

1. Introduction

The simplest scenario to stabilize CO2 emissions by the middle of this century is one in which photovoltaics and other renewables are used for electricity, hydrogen is used for transportation, and fossil fuels are used for residential and industrial heating [1]. Chalcopyrite semiconductors have recently attracted intense interest due to their appropriate band gaps and high absorption coefficients, which are ideal for solar cell applications. Recent studies have demonstrated efficiencies as high as 20.3% in thin-film copper-indium-gallium-diselenide (CIGS)-based solar cells [2]. Co-evaporation or sputter deposition is normally employed to fabricate the absorption layer of thin-film CIGS-based solar cells. A conventional vacuum-based process is thus required to obtain these CIGS films. However, this type of processing is complex, expensive and difficult to scale-up [3]. Many researchers have, thus, recently tried to fabricate various types of absorbers for thin-film photovoltaics using relatively simple procedures, which will make semiconductor preparation more cost-effective. The creation of diverse morphologies in different chalcogenide semiconductors has been accomplished via several methods, including solvothermal routes [4], spray pyrolysis [5], electrochemical deposition [6], sol–gel methods [7], chemical bath deposition [8] and the hot injection method[9]. The ink is made from low-cost commercially available precursors with relatively low toxicity, and the fast blading process can deposit uniform layers over large areas. Using CIGS as an initial demonstration, we made solar cells with an absorber layer that is flat, contaminant-free. As an important technique for wet chemistry, the solvothermal method has been employed for the preparation of different kinds of materials [10], due to the low temperature required, the simplicity and good yield.

In this study, CIGS nanoparticles were also synthesized using the solvothermal method at reaction temperatures in the range of 160-200°C for 36h with the intention of synthesizing a single phase CIGS powder. The structure, morphology, chemical composition and optical properties of the nanoparticles were studied, contributing to the exploration of new photovoltaic materials.

2. Experimental procedure

In this method, Cu, In, Ga (NO3)3 and Se powders were used as raw materials, and ethylenediamine (EN) was used as the solvent. Analytical grade reactants of Cu (0.904 g), In (1.270 g), Ga(NO3)3 (1.878 g) and Se (2.496 g) powders were loaded into a 100 ml Teflon lined autoclave, which was then filled with EN up to 50% of its total volume. The sealed autoclave was maintained at a reaction temperature in the range of 160-200°C for 36h.

After cleaning, the CIGS granulates were mixed with propanol and a cationic dispersant in a ball mill for grinding. After initial grinding with 0.1 mm zirconium for one hour, these zirconium beads were replaced with smaller ones (0.05 mm diameter) and ground for three more hours. The entire milling process was carried out with cooling by a water bath at 40°C. After 4 h of milling, the required viscosity of 3~5cP was obtained for the nano-scale CIGS ink.

Using a doctor blade, the CIGS nanoparticle-ink was then coated on 3.2 mm thick soda-lime glass substrates containing a 350 nm thick molybdenum (Mo) layer. After coating, solvents were removed by placing the sample in an oven at 100°C for 5 min, yielding the required dry precursor films. The Figure 1 shows the novel nontoxic selenium system used in this study this route shortens the reaction pathway for the fabrication of CIGS thin films.

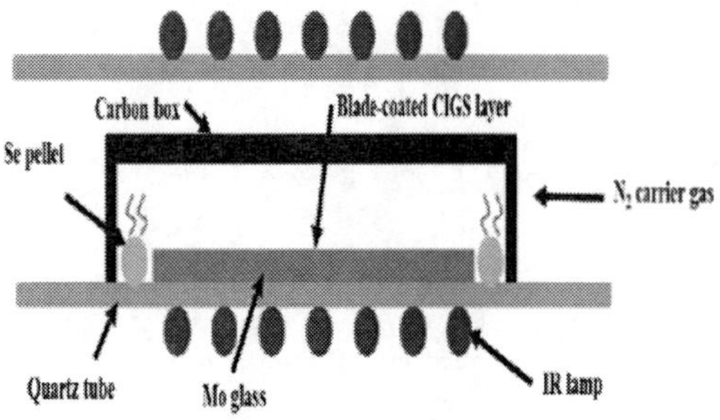

Figure 1. Schematic illustration of apparatus.

The Figure 2 show the precursor film selenization was then performed at 350°C for 5 min, followed by 540°C for 15 min under a Se vapor environment with an N2 gas flow using a furnace with a quartz tube.

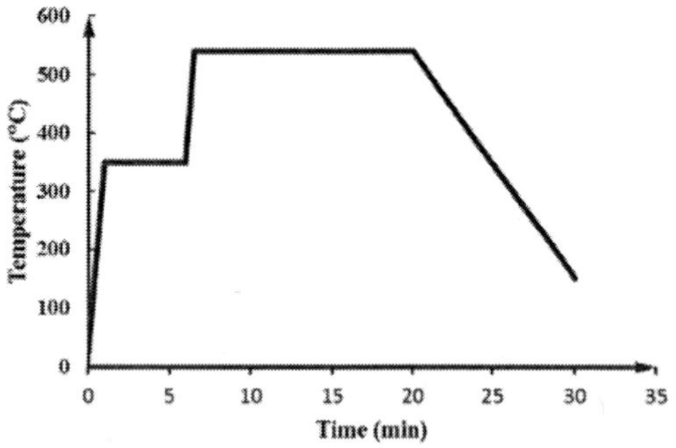

Figure 2. Temperature profile of selenization treatment.

Finally, the solar cell devices were fabricated according to the conventional configuration (Mo/CIGS/CdS/i-ZnO/n-ZnO:Al). The phase and the crystallographic structure of the products were identified using X-ray diffraction (PANalytical, X' pert pro). The particle size and morphology of the products were investigated using High-resolution transmission electron microscopy (JEOL, JEM-2100) and selected area electron diffraction. The morphologies of the samples were investigated using Scanning electron microscopy (HITACHI, S-4200). The chemical composition properties were analyzed using an Inductively coupled plasma-mass spectrometer (Perkin Elmer, SCIEX ELAN 5000)The solar cell was analyzed by measuring I-V under illumination (PV Measurement, Inc, PVM 415), with its I-V characteristics obtained by using simulated standard test conditions of AM1.5G and 1000 W/m2 at 25°C.

3. Results and discussion

3.1 Chemical composition

CIGS nanoparticles have previously been obtained using EN as a solvent [11]. As a strong polar solvent, EN plays an important role in the formation of CIGS nanoparticles in the synthesis process. A nucleophilic attack by EN can activate elemental selenium to form Se2-. In this electron transfer reaction, Cu2+ is also reduced to a Cu+ ion by EN. The Cu+ ion with EN forms [Cu(EN)2]+, which can effectively prevent the formation of binary copper chalcogenides. The reactions shown in Eqns. (1)-(5) may thus occur in the solvothermal process:

$$Se+x(EN) \rightarrow Se(EN)x \quad (1)$$

$$2Cu+Se(EN)x \rightarrow 2[Cu(EN)_2]^{+}+2Se^{2-} \quad (2)$$

$$2Ga(NO_3)_3+3Se^{2-} \rightarrow Ga_2Se_3+6(NO_3)^{-} \quad (3)$$

$$2In+3Se(EN)x \rightarrow In_2Se_3+x(EN) \quad (4)$$

$$20[Cu(EN)_2]^{+}+7In_2Se_3+3Ga_2Se_3+5Se^{2-} \rightarrow 20[(CuIn_{0.7}Ga_{0.3}Se_2)] \quad (5)$$

Stoichiometries compositions of the CIGS samples synthesized at 200℃ were analyzed by an inductively coupled plasma-mass spectrometer. Philip Jackson et al.[4] reported that the optimal ratios were approximately 0.3 and 0.80-0.92 for Ga/(In+Ga) and Cu/(In+Ga), respectively. A constant heating temperature of 200℃ for 36h was considered optimal for the samples since the resulting mole ratios were 0.29 for Ga/(In+Ga) and 0.90 for Cu/(In+Ga). The sample prepared at this temperature formed a CIGS film with a Cu:In:Ga:Se mole ratio of 0.89:0.71:0.29:2.01 compared to the 1:0.7:0.3:2 reported earlier[12].

3.2 Structural characterization

Figure 3 presents the XRD patterns of samples that were synthesized at various temperatures (160°C, 180°C and 200℃), with a holding time of 36h. XRD measurements were made by scanning the diffraction angles from 20° to 60° , using a grazing angle of incidence of 1° .

The XRD patterns show the variance in a crystal structure under different temperatures. For a temperature of 160°C, the crystalline structures of CIGS and Se show many diffraction peaks at (112), (220)/(204), (312)/(116) and (210)/(211). For temperatures below 180°C, the crystalline structures of CIGS, InSe and CuSe2 display many diffraction peaks at (112), (220)/(204), (312)/(116), (110) and (210)/(211).

DOI: 10.1051/matecconf/201710814005

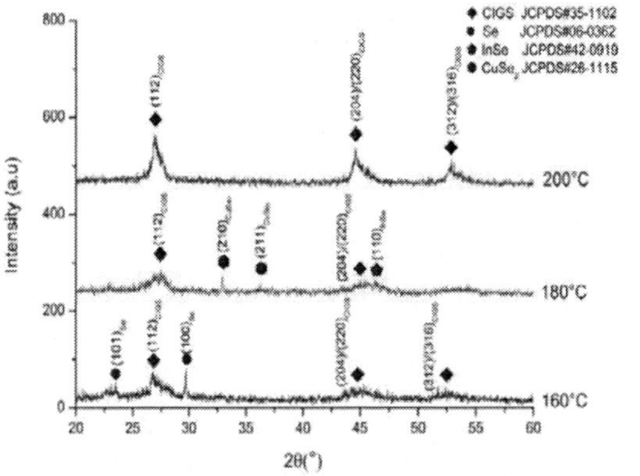

Figure 3. XRD patterns of samples that were synthesized with various temperatures of 160°C, 180°C and 200°C with holding time of 36h.

Furthermore, the XRD patterns show that a secondary phase exists at 160°C and 180°C. This decreases the capacity of EN to chelate with In3+ and Ga3+ at a low temperature. At a higher temperature of 200°C, the crystalline structures of CIGS shows many diffraction peaks at (112), (220)/(204) and (312)/(116). A secondary phase does not exist because the In2Se3 and Ga2Se3 dissolved it. There was a [Cu(EN)2]+ reaction under high temperature, and then, near-stoichiometries chalcopyrite structured CIGS nanoparticles were obtained.

According to the Joint Committee of Powder Diffraction Standard (JCPDS) card number 00-035-1102, the optimal angles for Cu(Ga0.3In0.7)Se2, with 2 θ of (112), (220)/(204) and (312)/(116) are 26.89° , 44.64° and 52.94° , respectively. As seen in Figure 3, this corresponds well to the 2 θ data from the sample synthesized at a temperature of 200℃, with angles of 26.89° , 44.64° and 52.94° , respectively.

The full width at half the maximum (FWHM) of the (112)-peak of the CIGS samples synthesized at various reduction temperatures. As is well known, the FWHM corresponds to the reduction of point defects and grain growth [13]. For the sample synthesized at a heating temperature of 200℃ for 36h, the peak of the preferred orientation (112) has a smaller FWHM (0.82°) and a larger grain size than those of the other samples. Analysis of the XRD patterns reveals that this sample is highly crystallized, and its Cu/(In+Ga) ratio is 0.90.

Figures 4 (a) and (b) show the TEM micrograph and SAED image of the sample synthesized at 200°C with a holding time of 36h. Figure 4 (a) shows a TEM micrograph of large agglomerates of CIGS nanoparticles. The CIGS nanoparticles are slightly dispersed, with some overlap. This confirms that spherical nanoparticles with diameters in the range of 20-40 nm are obtained from a reaction at 200℃ for 36h. Moreover, the SAED image in Figure 4 (b) matches the structure of the CIGS, as indicated by the diffraction spots corresponding to the (112), (220)/(204) and (312)/(116) planes.

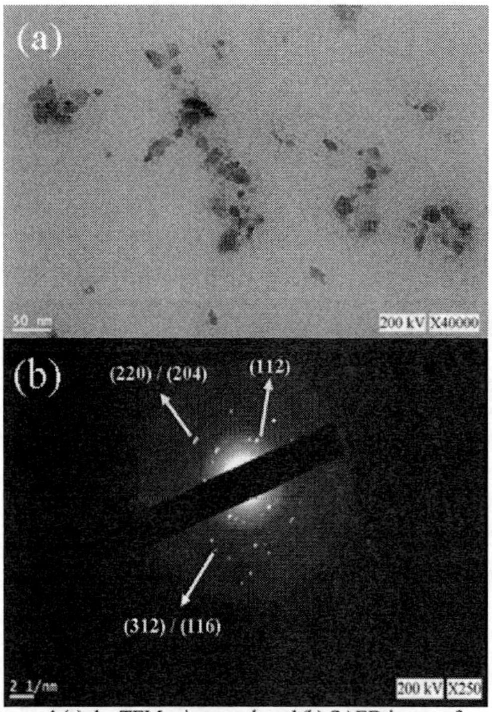

Figure 4 (a) the TEM micrograph and (b) SAED image of sample that was synthesized with a temperature of 200°C with holding time of 36h

3.3 Photovoltaic applications

The conventional CIGS solar cell has an Al:ZnO/ZnO/CdS/CIGS/Mo/glass structure, and in this study, a solar cell with an active area of 0.45 cm2 was fabricated. Figure 5 shows a cross-section of the CIGS film observed using SEM, revealing that the thickness of the CIGS layer is ~2.0 μm.

Figure 5 The cross-section of the CIGS film using SEM.

Figure 6 (a) show the cross-section of CIGS cell observed using SEM, and the Figure 6 (b) plots the illuminated I-V characteristics of such a device, fabricated on an absorber. This device has a conversion efficiency of 5.62% (open circuit voltage (Voc) = 0.48 V, current density (Jsc) = 27.32 mA/cm2, fill factor (FF) = 41.15%).

Although the Jsc of our device is very close to the 20.3% efficiency solar cell reported by ZSW, the Voc and FF are still inferior to those of the record cell. Based

DOI: 10.1051/matecconf/201710814005

on a model calculation, Shunt resistance and diode ideality factor are two key factors that affect the Voc value. Low shunt resistance resulting from pin-holes, cracks, or impurities in and near the junction leads to an increase of shunt current and a decrease of Voc. Therefore, the elimination of pin-holes, cracks, or impurities is essential for the fabrication of highly efficient solar cells. In addition, high Ga content near the CIGS/CdS interface and low interfacial roughness should further improve the Voc as well. As to the series resistance of device, it is 100 times larger than the series resistance of the high efficiency solar cells[5] which results in the softness in the knee of I－V curve and the degradation of FF and η values. Considering the high resistance resulting from reduce the thick layer of MoSe2[4], it is reasonable to assume that the cell efficiency could be improved by a thicker Mo layer.

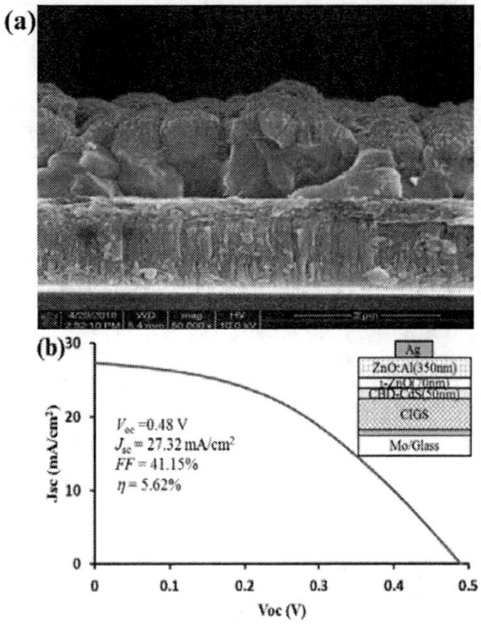

Figure 6 (a) the cross-section of the CIGS film using SEM, and (b) illuminated I-V characteristics of such a device fabricated on an absorber such a device, which was fabricated on an absorber

4 Conclusion

CIGS nanoparticles were synthesized through the solvothermal method at a reaction temperature of 200°C for 36h, with the intention of synthesizing a single phase CIGS powder. Near-stoichiometric chalcopyrite structured CIGS nanoparticles were obtained. The narrow FWHM of the XRD pattern and the clear crystal lattices of the CIGS nanoparticles revealed by HR-TEM indicate their high quality. This further confirmed that CIGS nanoparticles with diameters in the range of 20-40 nm were obtained. In this sample, the synthesized raw CIGS powder mole ratio of Cu:In:Ga:Se was equal to 0.89:0.71:0.29:2.01. Analysis of the performance of the obtained solar cell under a standard AM1.5 G illumination revealed a conversion efficiency of 5.62%. In summary, these results support the notion that CIGS nanoparticles synthesized through the solvothermal method may be useful as the absorption layer in CIGS solar cells.

References

1. T.M. Razykov, C.S. Ferekides, D. Morel, E. Stefanakos, H.S. Ullal, H.M. Upadhyaya, Solar photovoltaic electricity: Current status and future prospects, Solar Energy, **85** (2011) 1580-1608.
2. P. Jackson, D. Hariskos, E. Lotter, S. Paetel, R. Wuerz, R. Menner, W. Wischmann, M. Powalla, New world record efficiency for Cu(In,Ga)Se_2 thin-film solar cells beyond 20%, Progress in Photovoltaics: Research and Applications, **19** (2011) 894-897.
3. M. Kaelin, D. Rudmann, F. Kurdesau, H. Zogg, T. Meyer, A.N. Tiwari, Low-cost CIGS solar cells by paste coating and selenization, Thin Solid Films, **480–481** (2005) 486-490.
4. W.-L. Lu, Y.-S. Fu, B.-H. Tseng, Preparation and characterization of $CuInSe_2$ nano-particles, Journal of Physics and Chemistry of Solids, **69** (2008) 637-640.
5. J.M. Peza-Tapia, A. Morales-Acevedo, M. Ortega-López, Electrical characterization of Al, Ag and In contacts on $CuInS_2$ thin films deposited by spray pyrolysis, Solar Energy Materials and Solar Cells, **93** (2009) 544-548.
6. M. Benaicha, N. Benouattas, C. Benazzouz, L. Ouahab, Effect of bath temperature and annealing on the formation of $CuInSe_2$, Solar Energy Materials and Solar Cells, **93** (2009) 262-266.
7. P. Luo, R. Zuo, L. Chen, The preparation of $CuInSe_2$ films by combustion method and non-vacuum spin-coating process, Solar Energy Materials and Solar Cells, **94** (2010) 1146-1151.
8. A. Wangperawong, J.S. King, S.M. Herron, B.P. Tran, K. Pangan-Okimoto, S.F. Bent, Aqueous bath process for deposition of Cu_2ZnSnS_4 photovoltaic absorbers, Thin Solid Films, **519** (2011) 2488-2492.
9. A. Shavel, J. Arbiol, A. Cabot, Synthesis of Quaternary Chalcogenide Nanocrystals: Stannite $Cu_2Zn_xSn_ySe_{1+x+2y}$, Journal of the American Chemical Society, **132** (2010) 4514-4515.
10. Y.-L. Zhou, W.-H. Zhou, Y.-F. Du, M. Li, S.-X. Wu, Sphere-like kesterite Cu_2ZnSnS_4 nanoparticles synthesized by a facile solvothermal method, Materials Letters, **65** (2011) 1535-1537.
11. Y.G. Chun, K.H. Kim, K.H. Yoon, Synthesis of $CuInGaSe_2$ nanoparticles by solvothermal route, Thin Solid Films, **480–481** (2005) 46-49.
12. C. Suryanarayana, E. Ivanov, R. Noufi, M.A. Contreras, J.J. Moore, Phase selection in a mechanically alloyed Cu2013;In–Ga–Se powder mixture, Journal of Materials Research, **14** (1999) 377-383.
13. T. Schlenker, M.L. Valero, H.W. Schock, J.H. Werner, Grain growth studies of thin Cu(In, Ga)Se_2 films, Journal of Crystal Growth, **264** (2004) 178-183.

Application of solar photovoltaic generation in the world

Zheng Hongxia[1], Tian Yuqing[2] and Chen Deying[1]

[1]Training Center of State Grid Jiangxi Electric Power Company, Nanchang, 330000, China
[2]Education and Evaluation Center of Yunnan Power Grid Corporation, Kunming, 650000, China

Abstract. Recent years,the rapid development of solar photovoltaic has become a new hope to save the environment pollution and resource shortage in the electric power era.Countries have introduced relevant policies to support the development of photovoltaic power generation industry actively.Independent household solar photovoltaic system and Large-scale photovoltaic grid-connected system have been used successfully.However,development of countries' photovoltaic inductry is uneven due to synchronized technology and resources.This paper describes the status quo of PV industry in major PV application countries.

1 Introduction

Energy is the basic of social development.Since the industrial revolution,we have experienced steam age and electrical age.Every change is accompanied with a change in the pattern of energy and marks great progress of human civilization.In the last centuries, electricity is the basis of global social and economic development accompanied with the improvement of industrialization and energy application.During this period,mining non-renewable resources and consuming fossil energy continuously leads to resource tension,air pollution,greenhouse effect,climate change and so on.Primary energy consumption is mainly concentrated in Asia Pacific,European continent,North America.As of 2013,Coal production in the Asia-Pacific region accounts for 26.7% of global coal production,but its coal use accounts for 68.8% of global consumption;Europe and the Eurasian coal production accounted for 42.4% of global coal production, but its coal use accounts for only 11.6% of global consumption.The origin and consumption areas of coal is seriously uncoordinated as traditional power production materials,therefore,the exploition and transportation of coal may pollutes environmental.Indeed,the gap between fossil fuel reserves has caused disputes in the Middle East countries.Energy application model is changing which is aimed at reducing the impact of fossil energy on human beings.Clean energy substitution and electric energy substitution is the core of energy consumption patterns transformation.At present, wind power and photovoltaic power is the most widely used cleaning power.Wind power is limited by its geographical distribution of resources.Solar photovoltaic has such features as simple installation,small floor area,noise free,individual household or photovoltaic grid-connected,and that it has been popularized widely[1-3].Photovoltaic power generation will become the mainstream.

2 Distribution of global solar radiation

The distribution of solar energy resource is evidently regional which is affected by climate.Solar radiation in North Africa,Middle East,Southwestern United States,Mexico,Southern Europe,Australia, South Africa and Western China is strongest.The solar energy in the areas is best both duration and intensity of sunlight so these areas have a unique advantage in using solar power.Total solar radiation(year) in some areas was shown in Table 1.The solar radiation in North Africa and part of American is the most abundant.The distribution of solar radiation in these areas is not significant as Southern Europe,the Middle East,Australia and China.The following application of the current situation is based on Table 1 whose solar radiation is abundant.Photovoltaic power generation of countries in the areas is not all going well although those solar radiation is abundant.

3 Applicatin of solar photovoltaic generation

Photovoltaic power generation includes independent photovoltaic system and photovoltaic grid connected system according to whether the grid is connected to the

Table 1. Total solar radiation(year) in major countries

country Radition and area	**Total solar radiation(year)/ (MJ/m^2)**	**distribution area**
Algeria	9720	
Morocco	9360	
Tunisia Libya	>8280	
Spain	8100	
Italy	7200	
Greece	6840	
Portugal	7560	
Israel Jordan Saudi Arabia	8640	
United Arab Emirates, Iran	7920	
United States	9198-10512	First area(9.36% of US area)
	7884-9198	Second area(35.67% of US area)
	6570-7884	Third area(41.48% of US area)
China[4]	6680-8400	Western China
Australia	7621-8672	First area(54.18% of Australia's area)
	6570-7621	Second area(35.44% of Australia's area)

grid.Independent photovoltaic includes photovoltaic system with storage battery and photovoltaic system without storage battery.The cost of solar photovoltaic generation system is composed of solar panels,inverters, batteries, post-maintenance costs.The cost of solar photovoltaic generation is much higher than traditional power generation system which is counted base on the life of solar panel.Although photovoltaic grid connected technology has been put into application,it is still in the early stages of development and the technology is immature.Countries have introduced relevant supporting policies,however,the support gradually weakened in recent years.In addition,government support is not a permanent solution, because each photovoltaic system has its service life.

Factors that restricted photovoltaic power generation industry development are as follows:

-The main raw material for solar cell silicon crystal is scarce in prophase.

-The conversion efficiency of solar panel sold on market is low as the coversion efficiency of mainstream monocrystalline silicon cell and polycrystalline silicon cell 15% -21% and the coversion efficiency of thin-film solar battery is just 6% -13%,the differences of solar cells sold on market is shown in Table 2.

-The development of solar cell manufacturing technology is uneven, the production technology is not shared.

-Photovoltaic power generation has a typical periodic,but storage battery can not achieve a large number of energy storage.

Table 2. Comparison of photovoltaic cells

Performance Types of solar cell	**Monocrystalline silicon cell**	**Polycrystalline silicon cell**	**Thin film solar cell**
photoelectric conversion efficiency	15%-21%	15%-18%	6%-13%(The cell can still work with weak light.)
manufacturing cost	The production cost is high, and can not be widely applied at present	Mass production can be carried out due to its low production cost with simple manufacturing process and low power consumption.	The production process is completely different from the crystal silicon battery, the process is greatly simplified, the silicon material consumption is low, and the power consumption is low
Service life	It is packaged by toughened glass and waterproof resin.Its service life is about 15-25 years	It is packaged by EVA.Its service life is shorter than Monocrystalline silicon cell.	About 10 years.Its conversion efficiency decreases obviously with time.

Photovoltaic cells on market is less redundant because the longer the time,the lower the efficiency is.Polysilicon is the main raw material for photovoltaic panels.We can get it that polysilicon industry is glut base on the data during 2010-2015.It reflects the supply of raw materials for photovoltaic cells is sufficient, while the production technology is not advanced.

3.1 Application of photovoltaic power generation in USA

United States has introduced a new bill which is aimed at solving the problem that new energy enterprise is lacking motivation in developing new technologies.New energy plan and harsh environmental requirements for fossil fuels are presented.The plan proposed that goverment will offer capital and technology to photovoltaic power generation industry by subsidizing raw material procurement and production[5].Renewable Energy Power Generation Tax Credit Policy (PTC) has been extended to 2020 and Federal Commercial Energy Investment Tax Credit Policy (ITC) has been extended to 2022.New gross capacity of photovoltaic power generation in United States is 7.286GW in 2015.The number is 14.726GW in 2016.PV installed capacity accounted for 39% in the whole year in 2016.The accumulated photovoltaic power installed capacity of American is 25GW by 2015.

3.2 Application of photovoltaic power generation in Germany

Germany is undoubtedly the leader in the photovoltaic industry.Germany demands the most in Europe market demand for photovoltaic power generation products which is benefited from that German government is forward-looking in terms of new energy inductry and advanced industrial system and semiconductor technology.It has obvious advantage in photovoltaic inductry due to its key production technology and production technics.German government revises the German renewable energy law continuously to make it better serve the photovoltaic industry in according to domestic and international economic environment[6].The goverment has provided cridit to feed-in-tariffs.Germany has a lot of home photovoltaic power generation system, which gave birth to the development of household energy storage market.More and more industries are participating in this market.The accumulated photovoltaic power installed capacity of Germany is 39.40GW by 2015.

3.3 Application of photovoltaic power generation in Japan

Japan is a country that is less dense and lacks natural resources.Japan is one of the countries that started the photovoltaic industry earlier.Japan has re-planned its energy development structure after Fukushima nuclear meltdowns.Subsequently,it introduced a 10-year bill in regard to renewable energy subsidies[7].The bill aims to promote the development of photovoltaic power generation industry.However,solar companies appear bankruptcy peak in Japan due to government support policy tightening and poor management.So the future of Japan's PV development trend is still unknown

3.4 Application of photovoltaic power generation in China

China's photovoltaic manufacturing production plays original equipment manufacturer in global.Failed to grasp key production technology and production technics of photovoltaic industry contributed its passive position.In view of environmental problems such as resource shortage and severe haze,China has introduced a series of supporting policies of which the core is subsidizing photovoltaic power generation equipment installation and feed-in tariffs.The pre-subsidy policy only takes into account the installed capacity without taking into account the power generation.The electricity price subsidies is reducing gradually in order to reduce the dependence of the photovoltaic industry on the policy.Industrial development focus from the previous expansion of the scale of development to improve efficiency and prrofit[8-10].The accumulated photovoltaic power installed capacity of China is 43GW by 2015.

4 Development direction of photovoltaic power generation

The count shows that global power demand showed a sustained growth and areas have a shortage of electricity.Electricity is not enough to rely on traditional electricity generation,the shortage must be supplied by renewable energy power generation.Solar photovoltaic generation is bound to be the main power supply form in the future due to its innate advantages.The main factors that limit the large-scale application of photovoltaic power generation are:

-Long-term effective encouraging policies and unified planning is lacking.

-There is no uniform technical standards.

-The construction cost of photovoltaic power generation system is high.

-Conversion efficiency of photovoltaic power generation inverter is low.

-Post-maintenance costs are high and maintenance-related statutes are immature[11-12].

5 Conclusion

Solar radiation on earth is renewable and ample.Solar photovoltatic will certainly be a major player because of its simple installation,small floor area and noise free.Photovoltaic power generation has been applied in many countries in the past decades.Independent household photovoltaic power generation system can solve the problem of power supply in remote area effectively and large-scale photovoltaic grid-connected system can effectively alleviate the shortage of power supply and environmental pollution.However,PV industry still has a series of problems in terms of policy and technology, such as high production cost of power generation system,low conversion efficiency of photovoltaic inverter , grid fluctuation caused by grid connection,large amount of electric energy storage etc.At present,regardless of which country,its rapid development of photovoltaic power generation scale and policy support are closely tied.And then,what we have to do is reducing the installation cost,generation cost and maintenance cost of photovoltaic generation.It's no longer a dream that solar photovoltaic generation become the dominant as science and technology is developing.

References

1. Lv Bei,Qiu Hemei,Zhang Yu.Present Status and Development of Solar Photovoltaic Power Generation Industry [J].Huadian Technology,2010,**01**:73-76+82.
2. Wang Yimin.Concept and Prospects of Global Energy Internet [J].China Electric Power,2016,**03**:1-5+11.
3. Liu Zhenya.Study and Prospect of Global Energy Internet between country and between continental [J].Proceeding of the CSEE,2016,**19**:5103-5110+5391.
4. Zhu Weigang,Lin Yanmei,Zhou Lei.Application of Solar Photovoltaic Power Peneration in China [J].Modern electric power,2007,**05**:19-23.
5. Xiao Li.Study on The International Competitiveness of Chinese Photovoltaic Industry [D].Hainan University, 2016.
6. Kong Xiangxin.Study of Photovoltaic Industry Trade Conflict between China and Europe and United States [D].Shandong Normal University,2014.
7. Liu Qingrong,Gu Qunyin,Ruan Yingjun,Ren Jianxing,Long Youer,Gao Weijun .Policy and Example of Japanese Solar Photovoltaic Power Generation System [J].East China Electric Power,2009,**02**:279-283.
8. Wang Sicheng, Wu Dacheng. Review and Prospect of photovoltaic policy in China (Part I) [J]. Solar Energy, 2016,**06**: 19-24+39.
9. Wang Sicheng, Wu Dacheng, Review and Prospect of photovoltaic policy in China (Part II) [J]. Solar Energy, 2016, **07**: 5-11.
10. Zhang Pengchao. Research on incentive policy of PV industry in China [D]. North China Electric Power University, 2015.
11. Shao Riming,Mary KAYE,Chris DIDUCH, Zhang Liuchen. Basic function components of distributed photovoltaic inverters in smart grid [J]. Proceedings of the Chinese society of electrical engineering, 2013, **34**: 83-90+15.
12. Lu Meiling, Zhang Geng, Liu Song, Lv Lei. Optimal scheduling model for economic benefit of home photovoltaic power generation system based on energy Internet [J]. Power System and Automation, 2016, **S1**: 141-146.

An Economic Solution to Optimize Performance of Photovoltaic Modules under Partial Shading

Paoyuan Lin[1], Kok Chin Chai[1], Herchang Ay[1,2]

[1]*Department of Mold and Die Engineering, National Kaohsiung University of Applied Sciences, Taiwan*
[2]*St. John's University, Taiwan*

Abstract. This paper investigates the effects of partial shading pertaining to photovoltaic (PV) modules and subsequently a solution that aims to minimize the effects and optimize the performance of PV modules is proposed. The concept of the solution is to integrate boost and buck converters into an electronic control system. The proposed system is simple and economic solution as it integrates cheap and easy available electronic components such as microcontrollers, capacitors, transistors and inductors. To validate the effectiveness of proposed system, a measuring instrument, namely NI cFP-1808 is used to perform the electrical characteristic analyses or known as I-V and P-V curve analyses. Furthermore, an outdoor experiment was conducted to examine the practicality of the proposed system in tackling real-world conditions. Experimental results showed that the proposed system improves the performance of the PV modules by 82.9% compared to the controlled experiment. In short, the effectiveness and practicality of the proposed system is proven with experiment results.

1 Introduction

Photovoltaic (PV) system has received prominent attention over the last decade as it offers great advantages such as low maintenance cost, absence of moving or rotating parts and freedom from environmental pollution [1,2]. Though the PV system has been widely implemented, there are still numbers of ongoing limitations that need to be investigated and improved. As an example, partial shading on PV system is one of the ongoing issue [3,4]. Partial shading is defined as the PV arrays that are partially shadowed by the neighboring trees, buildings, towers and even passing clouds [3]. Under partial shaded conditions, the PV characteristics get more complex with multiple peaks. A research from Germany shows that the effect of partial shading can reduce the output energy up to 20% [4].

Literature reveals that numbers of researches have been conducted to tackle partial shading conditions. These researches can be categorized into maximum power-point tracking (MPPT) methods [2], modeling and simulation based approaches [3, 5, 6], DC-DC converter [7]. In [2], a metaheuristic optimization algorithm namely, firefly algorithm, is used to track the maximum power point under partial shading condition in a PV system. A comprehensive study on MPPT methods has been discussed in [8, 9]. In [3, 5], MATLAB-based modeling and simulation approach studies the effect of shading patterns on PV panels pertaining to different configurations. In [6], PSpice-based modeling and simulation approach investigates the effects of bypass diode configurations on PV modules. In [7], a DC-DC converter is introduced to boost the output energy in partial shading condition. While many solutions pertaining to partial shading are available, an economic and simple solution is relatively new.

The aim of this paper is to propose an electronic system that improves the performance of photovoltaic modules under partial shading. It is worth-mentioning that the proposed system is built with cheap and easily available components such as microcontrollers, capacitors, transistors and inductors. The system consists of two parts namely, boost converter and buck converter. In general, the boost and buck converters are used to increase and decrease the output voltage level pertaining to input voltage, respectively. The performance of the proposed system is validated using a measuring instrument, namely NI cFP-1808. NI cFP-1808 is designed to measure electric current, voltage and power. Besides that, it can also perform the electrical characteristic analyses or known as I-V and P-V curve analyses with simple configurations.

In this paper, the partial shading effect is investigated through different shading portions. To effectively model the shading portion, an opaque tape with good thermal insulation is used to shade the PV cell. The materials used to shade the PV cell should be carefully considered else it may collect heat that will indirectly affect the performance of PV cell. The effectiveness and practicality of the proposed system is further evaluated with an outdoor experiment.

The rest of the paper is organized as follows. Section 2 introduces the notations with respective descriptions and some important equations. Section 3 presents the proposed system. Section 4 presents the experiment setup pertaining to two different investigations. Section 5 discusses the experiment results from different investigations. Finally, Section 6 presents the conclusion and future works.

2 Preliminaries

The related notations and their descriptions are listed in Table 1 as follows. As an example, V_{oc} represents open circuit voltage.

Table 1. List of notations

Notations	Descriptions
V_{OC}	Open circuit voltage
I_{sc}	Short circuit current
V_{pm}	Maximum power voltage
I_{pm}	Maximum power current
FF	Fill factor
P_{max}	Maximum power
P_{in}	Input power
η	Efficiency of photovoltaic (PV) cell
ΔI_{sc}	Declined percentage for I_{sc}
ΔI_{pm}	Declined percentage for I_{pm}
ΔP_{max}	Declined percentage for P_{max}

Besides several important definitions and equations used in this paper are defined as follows.

Fill factor, FF is defined as the ratio of maximum obtainable power to the product of the open-circuit voltage and short-circuit current. FF can be calculated using Eq. (1).

$$FF = \frac{I_{mp} \times V_{mp}}{I_{SC} \times V_{OC}} \tag{1}$$

Maximum power, P_{max} is product of V_{OC},I_{SC} and FF as simplified in Eq. (2).

$$P_{max} = V_{OC} I_{SC} FF \tag{2}$$

The efficiency of a PV cell is an important parameter to compare the performance of PV cell. It is defined as the fraction of energy output from the PV cell to input energy from the sun which can be simplified into Eq. (3)

$$\eta = \frac{P_{max}}{P_{in}} \tag{3}$$

where the input power per meter square of PV cell is 1 kW/m^2.

3 The proposed system

The proposed system consists of two parts, i.e., buck converter and boost converter. Both converters are similar in design. The general concept of the converters is illustrated in the block diagram as shown Figure 1. It comprises three modules namely, voltage regulator, pulse width modulation (PWM) generator and inductor-based switching regulator (SR). The first module is used to maintain a constant voltage at 10V. The second module is known as PWM to generate 40 kHz frequency and the third module is used to step up/down of voltage source.

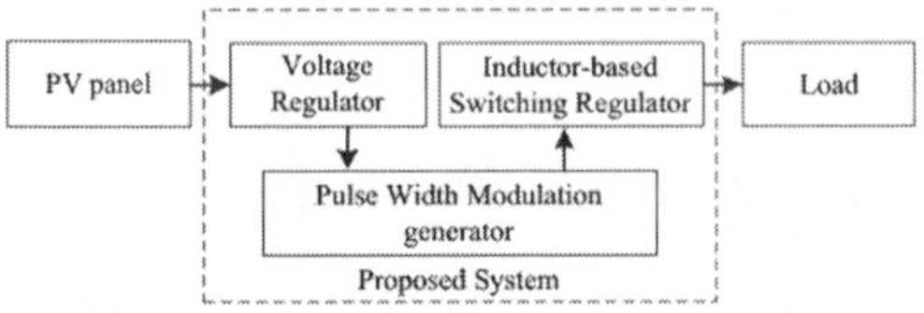

Figure 1. Block diagram of the proposed system

The three modules are integrated and designed in a single printed circuit board (PCB). An example of PCB layout for the proposed buck Converter is depicted in Figure 2.

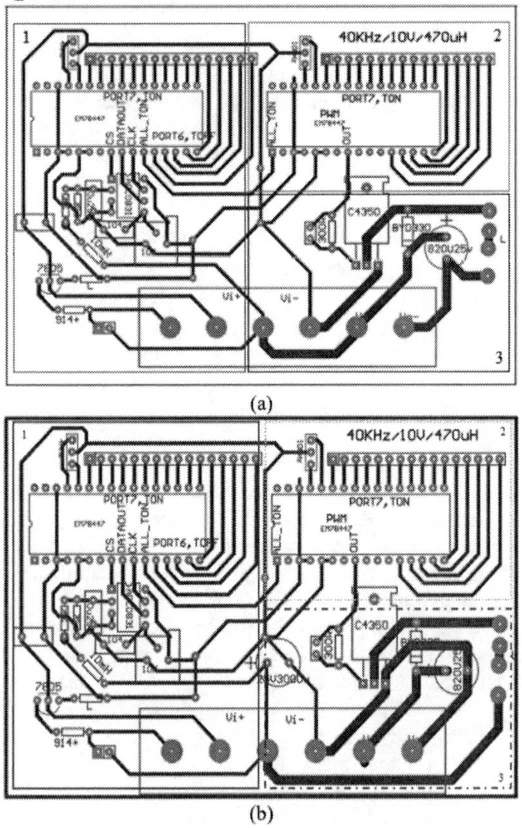

Figure 2. PCB layout diagram of the proposed (a) buck Converter (b) boost Converter

The buck converter consists of few electronic components such as two microcontrollers, EM78447, two crystal oscillators (i.e., 4MHz and 10MHz), Analog to Digital convertor (ADC), capacitors, transistors, inductors and etc. As in Figure 2, it is observed that the square outline labeled with "1" indicates the voltage regulator module, "2" indicates the PWM generator module and "3" indicates the inductor-based SR module. In general, the circuit connection for both converters is similar, except for the configuration in third module. The difference for boost and buck inductor-based SRs is indicated in Figure 3. Technically, the buck inductor-based SR reduces a DC voltage to a lower DC voltage and boost inductor-based SR provides an output voltage that is higher than the input. With these boost and buck inductor-based SR, voltages can be regulated consistently. It is worth-mentioning that the inductor-based SR is beneficial as it has minimal power loss and provides high power efficiency [10].

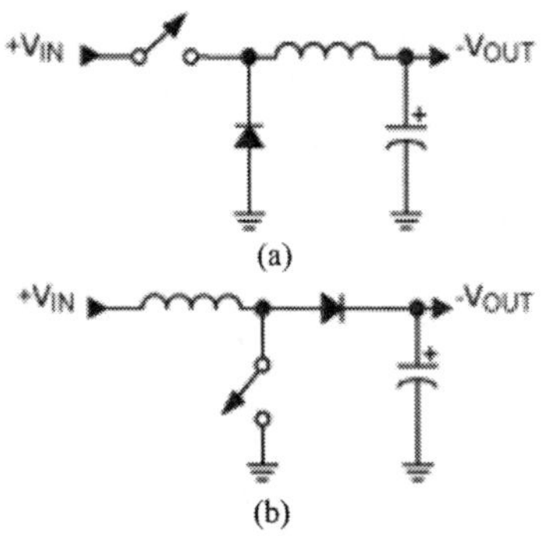

Figure 3. Inductor-based switching regulators (a) buck and (b) boost [11]

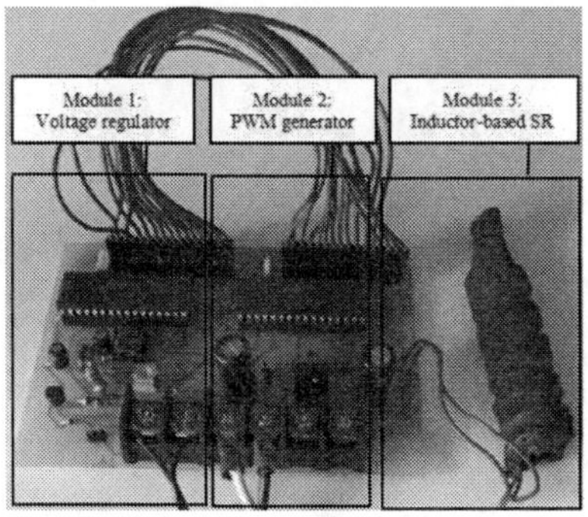

Figure 4. The proposed buck converter

4 Experiment setup

In this experiment, a PV cell is an electrical device with dimension 19.5×52 mm that converts the light energy into electricity energy. Therefore, the total power generated from a PV cell per meter square is 1.014W. A PV module is a packaged, connected assembly via the arrangement of 5×4 PV cells i.e., 5 cells in row and 4 cells in column as illustrated in Fig.5 (a). The total area of a PV module that occupy with PV cell is 20×19.5×52 mm or 20280 mm^2. Total power generated from a PV module is 20.28W. A PV array is defined as numbers of PV modules connected in serial/parallel. This research used a notation "SxPy XXX" to explain the connection of PV array where S and P indicate serial and parallel connections, respectively while x and y indicates the number of PV modules that connect in either serial or parallel connection and XXX indicates the shaded portion. For examples, "S1P1 6.25" represents a single PV module with 6.25% shaded portion and "S3P1 25" represents three PV modules connected in serial with 25% shaded portion.

It is worth-mentioning that our experiment focuses on two investigations, (1) investigate the effects of different shaded portions pertaining to a PV module and (2) investigate the practicality of the proposed method in improving the performance of PV module under partial shading effects. These two investigations are presented in detail as follow.

4.1 Investigate the effects of different shaded portions

A PV module pertaining to different shaded portions (i.e., 6.25%, 12.5%, 18.75% and 25%) as shown in Figure 5, is investigated under a controlled experiment. The performance of the PV module is examined by using a photovoltaic module tester (SPI-SUN SIMULATOR 350i). SPI-SUN SIMULATOR 350i detects the efficiency of PV module. The tester uses 1.3M Xenon lights to test PV module with dimension 2200mm × 1200mm. Technically, its requirements can meet with ASTM E927 Class A rating which is a standard specification for solar simulation used for photovoltaic testing. SPI-SUN SIMULATOR 350i measures the important parameters of PV module such as V_{oc}, I_{sc}, V_{pm}, I_{pm}, P_{max}, FF, η, ΔI_{sc}, ΔI_{pm}, ΔP_{max}. Also it provides important visualization i.e., I-V and P-V characteristic curves. The results and discussions from this experiment are further discussed in Section 5.1.

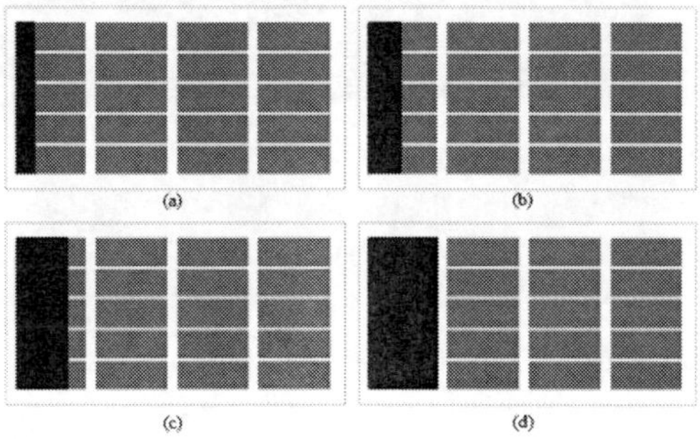

Figure 5. A PV module with different shaded portions (a) 6.25% (b) 12.5% (c) 18.75% (d) 25%

4.2 Investigate the proposed system pertaining to shaded portion

To validate the practicality of the proposed system, an outdoor experiment is essential. The experiment was conducted on a sunny day from 0540 until 1840, at rooftop of Mold and Die Department, National Kaohsiung University of Applied Science, Taiwan. The setup of the experiment is shown in Figure 6. It consists of two groups of experiment, namely experiment and control groups. Both groups use three PV modules connected in serial (i.e., S3P1, in short). One of the three PV modules is shaded with 18.75%, i.e., a total of 6.25% shaded portion for a PV array connected in S3P1. Besides that, both groups are controlled and applied that the same PV material and configurations, i.e., monocrystalline silicon, tilt angle 23.5° and azimuth angle 175° with 5° facing south east.

The results from the experiment and control groups are collected using data acquisition system (i.e., National Instruments, NI cFP-1808). NI cFP-1808 has Input and

Output (I/O) interface for Ethernet and RS232 networks. It can be easily read via the networked host PC. On the other hand, the solar irradiance of this experiment is measured using solar power meter (TEES-1333R) as depicted in left-corner of Figure 6.

Subsequently, the results from both groups are analyzed and compared. In this way, the practicality of the proposed system pertaining to shading effect of PV is validated. The results and discussions from the experiment are further discussed in Section 5.2.

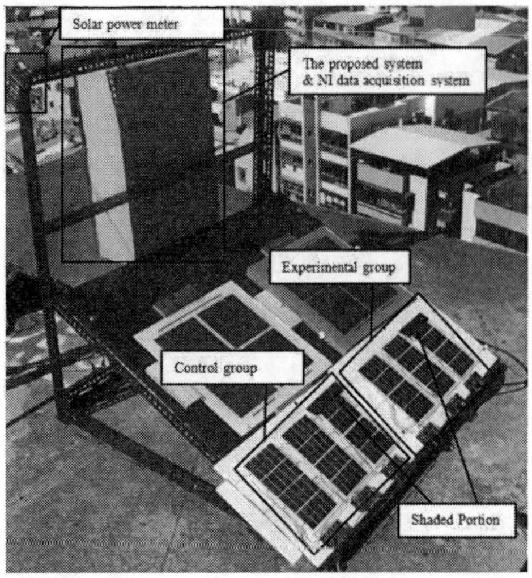

Figure 6. Setup for outdoor experiment

5 Results and discussions

The experimental results of the two investigations are presented in two sub-sections as follows.

5.1 Results pertaining to different shaded portions

Through SPI-SUN SIMULATOR 350i, PV without shaded portion obtains V_{oc} is 12.35V, I_{sc} is 350.8mA, V_{pm} is 10.25V, I_{pm} is 327.9mA, P_{max} is 3.362W, FF is 0.776 and η is 16.58%. While the rest of the results for 6.25%, 12.5%, 18.75%, 25% are tabulated in Table 2. It is obvious that the efficiency of PV cell decreases drastically with the increase of shaded portions as in row "η" it is observed that efficiency drop from 16.58% to 13.18%, 8.955%, 4.655% and 0% for shaded portions of 6.25%, 12.5%, 18.75%, 25%, respectively. Also, it is observed that the declined percentages pertaining to I_{pm} i.e., ΔI_{pm} are increasing with respective to increase in shaded portions.

I-V and P-V characteristic curves in Figure 7 further illustrate the effect of different shaded portions. Again, the effect is significant as the increasing of shaded portions will lead to reduce in both current (A) and power (W), correspondingly. When the shaded portion is up to 25%, it is observed that the power generated is reduced drastically to about 0W.

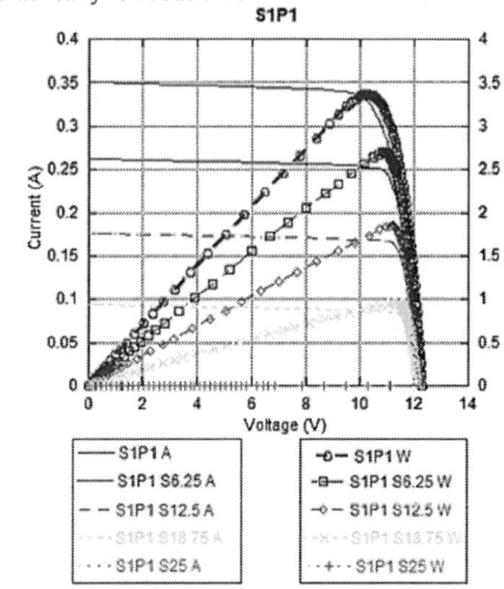

Figure 7. I-V and P-V characteristic curve of a PV module tested under different shading percentages

Further to this, other configurations such as S2P1 and S3P1 are investigated. The results are tabulated in Table 2. The results obtained from S2P1 and S3P1 are in line with the observations in S1P1. It is observed that the efficiency of PV cell reduces with the increase of shaded portions. It indicates that the increase of shaded portions significantly reduce the efficiency of PV module. Such deduction is also applicable to different configuration of PV module. As such, a research that improves the performance of the PV modules under partial shading is important.

Table 2. Experimental results of different PV configurations pertaining to different shading percentages

Parameters	S1P1 with different shading percentages					S2P1 with different shading percentages					S3P1 with different shading percentages				
	0%	6.25%	12.5%	18.75%	25%	0%	6.25%	12.5%	18.75%	25%	0%	6.25%	12.5%	18.75%	25%
V_{oc}	12.35	12.30	12.24	12.14	11.24	24.76	24.72	24.70	24.51	17.50	37.15	37.10	37.08	37.02	29.72
I_{sc}	350.8	262.5	176.7	94.40	1.100	350.7	351.6	351.0	350.8	350.9	350.9	350.8	351.0	350.9	350.7
V_{pm}	10.25	11.23	11.32	11.11	6.181	6.662	5.293	3.707	3.037	3.017	10.00	8.185	6.368	6.347	6.340
I_{pm}	327.9	238.1	160.4	85.00	0.600	20.52	20.80	20.51	9.393	9.331	30.76	31.61	19.69	19.60	19.81
P_{max}	3.362	2.673	1.816	0.944	0.004	324.6	254.5	180.7	323.4	323.4	325.1	259.0	323.5	324.0	323.3
FF	77.62	82.78	83.98	82.46	27.55	76.73	60.89	42.75	29.09	49.13	76.72	62.90	48.93	48.60	60.83
η	16.58	13.18	8.955	4.655	0	16.43	13.05	9.140	7.488	7.438	16.28	13.32	10.47	10.43	10.42
ΔI_{sc}	0	25.17	49.63	73.09	99.69	0	−0.257	−0.086	−0.029	−0.057	0	0.029	−0.029	0	0.057
ΔI_{pm}	0	32.13	51.08	74.08	99.82	0	21.60	44.33	−0.370	−0.37	0	20.33	0.492	0.338	0.554
ΔP_{max}	0	20.49	45.99	71.92	99.88	0	20.55	44.36	54.41	54.71	0	18.15	36.32	36.53	36.60

5.2 Results pertaining to outdoor experiment

The results of the outdoor experiment are summarized in Table 3. The maximum solar irradiance of the day is 960 Wh/m^2 at time 12:22:58. The maximum power collected through the experiment and control groups are 5.089W and 2.742W, respectively. The maximum power collected through the experiment group is higher than the control group. This is the first evidence that the proposed method installed in experiment group has improved the power efficiency of PV modules under partial shading.

Further to this, the cumulative total generated power from the experiment group is 30.0Wh compare to the control group 16.4Wh. It shows that the proposed system improved the total accumulated power by 13.6Wh, i.e., the overall power efficiency improves by 82.9%. Besides that, the performance of both groups is visualized in Figure 8. The cumulative total generated power of the experiment group is always higher than the control group throughout entire the experiment.

Table 3 Results of outdoor experiment

Details	Results
Duration of experiment	05:40:00~18:30:00
Maximum solar irradiance	960Wh/m^2 (12:22:58)
Maximum power collected through the experiment group	5.089Wh (12:22:58)
Maximum power collected through the control group	2.742Wh (12:22:58)
Cumulative total solar irradiance	4921Wh/ m^2
Cumulative total generated power from experiment group	30.0Wh
Cumulative total generated power from control group	16.4Wh
Percentage of power restored via the proposed system	**82.9%**

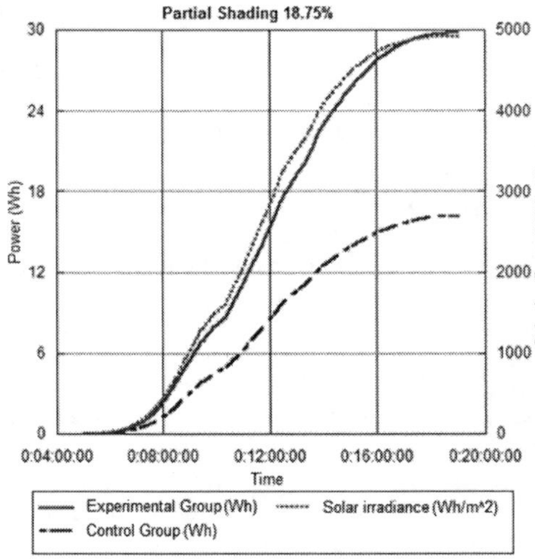

Figure 8. Cumulative generated power and solar irradiance

6 Conclusion and future works

In this research, the effects of different shading portions are investigated. The results show that the performance of PV cell is significantly affected. As such, a new solution that optimizes performance of photovoltaic modules under partial shading is proposed. Through the experiment, the proposed system has proven to improve the performance of the PV modules by 82.9%. It is worth-noting that the proposed solution is important as it is simple and economic.

In future, it is important to further investigate the practicality of the proposed system pertaining to PV array. It is interesting to investigate the durability of the proposed system in real implementation.

References

[1] M. Z. S. EL-Dein, M. Kazerani, and M. M. A. Salama, "Optimal photovoltaic array reconfiguration to reduce partial shading losses," IEEE Trans. Sustain. Energy 4, 1 (2013)

[2] K. Sundareswaran, S. Peddapati and S. Palani, "MPPT of PV systems under partial shaded conditions through a colony of flashing fireflies", IEEE Trans. on Energy Convers. 29, 2,(2014)

[3] H. Patel and V. Agarwal, "MATLAB-Based Modeling to Study the Effects of Partial Shading on PV Array Characteristics", IEEE Trans. on Energy Convers. 23, 1 (2008)

[4] U. Jahn and W. Nasse, "Operational performance of grid-connected PV systems on buildings in germany", Prog. Photovolt: Res. Appl. 12, 6 (2004)

[5] K. Ishaque and Z. Salam,. "A comprehensive MATLAB Simulink PV system simulator with partial shading capability based on two-diode model", Sol. Energy 85, 9 (2011)

[6] S. Silvestre, A. Boronat, A. Chouder, "Study of bypass diodes configuration on PV modules", Appl. Energy 86 (2009)

[7] S. Strache et al., "Photovoltaic output power improvement applying DC–DC converters on submodule level," in Proc. IEEE Int. Conf. Smart Grid Technol. Economics Policies (2012)

[8] A. Bidram, A. Davoudi and R. S. Balog, "Control and circuit techniques to mitigate partial shading effects in photovoltaic arrays". IEEE J. Photovolt. 2, 4 (2012)

[9] B. Subudhi and R. Pradhan, "A comparative study on maximum power point tracking techniques for photovoltaic power systems," IEEE Trans. Sustain. Energy, 4, 1 (2013)

[10] A. Wang and S. Naffziger, Adaptive Techniques for Dynamic Processor Optimization: Theory and Practice. Springer Science & Business Media (2008)

[11] "DC-DC Converter Tutorial - Tutorial – Maxim". *Maximintegrated.com*. (2016)

Model-based Pose Estimation for Texture-less Objects with Differential Evolution Algorithm

Linh Tao , Tinh Nguyen and Hiroshi Hasegawa

Graduate School of Engineering and Science,Shibaura Institute of Technology,Japan

Abstract. This paper proposes a novel object-tracking method to estimate three dimensions position of texture-less objects using one camera system and 3D model. The system uses efficient chamfer matching method to calculated distances between 2D edge templates of pose hypotheses with edges from the Canny edge query image. Differential Evolution algorithm uses those distances as inputs to ensure the close optimum results and find the most suitable position of objects. For initialization the exhaustive searching is employed. With the good initialization, a smaller searching space is set to guaranty the online tracking ability. The first results showed the potential of the method in solving object tracking and detection problem.

1 Introduction

In the last decade, object detection and recognition have gained significant improvement by using keypoint features [1]. Since geometric transformations and illumination changes has no effect on finding keypoints, they have been widely used for matching images from slightly different viewpoints [2]. Keypoint-based approaches work well in textured objects but texture-less objects. Textured objects have various keypoints, those have high potential appearing on both images. After finding keypoints, sample consensus such as RANSAC [3] calculates the most suitable transformation of the object from reference position to current position. The more matched keypoints, the more accurate the transformation is.

On texture-less objects lack of keypoint repeatability and stability on texture-less regions neither reduces the accuracy of sample consensus method nor leads to wrong results. Like keypoints, edges are also invariant to general geometric transformations and illumination changes [4]. Using edges are more suitable as a general approach even with texture-less objects.

In early computer vision research, to find the best alignment between two edge maps, a given priori set of edge templates compare their suitability to the current edge maps to draw the mot suitable transformation. The current proposed method of chamfer distance matching [5] enhances the cost functions enable for applying global searching algorithm into object tracking problem.

Harris [6] and various proposed edge-based tracking system such as [7] used edges and contours for visual tracking task. One drawback of using edges is that they are not distinctive enough to provide effective discrimination in complex background or occlusions, there have been efforts to enhance the previous one by unifying interest points or considering multiple but limited hypotheses on edge correspondences. For consideration of multiple hypotheses in a more general sense global searching algorithm should come into consideration.

We propose an approach of using Differential Evolution (DE) [8] as the global searching method to continuously search for the 3D position of object in camera coordinate.

2 Methodology

Initialization is the most important step, a subject of the paper's interest, in tracking algorithm. Following steps presents the implementation pipeline of the initialization:
- Canny edge image is employed to archive edge images from query images.
- Distance maps or chamfer matching maps is calculated from edge images.
- Differential Evolution is employed to calculate the best fit pose of the object which create an 2D edge images fitted into the chamfer matching maps for initialization.

After initialization, narrower searching boundary is used to get the accurate results at on-line speed. If the cost function goes large, initialization is required.

2.1 Chamfer matching maps from query images.

DOI: 10.1051/matecconf/201710815001

2.1.1 Canny edge detection

The white lines in Fig1 are output of Canny [9] edge detection method. Canny method includes five different steps:

1. Apply Gaussian filter to smooth the image in order to remove the noise.
2. Find the intensity gradients of the image
3. Apply non-maximum suppression to get rid of spurious response to edge detection
4. Apply double threshold to determine potential edges.
5. Track edge by hysteresis: Finalize the detection of edges by suppressing all the other edges that are weak and not connected to strong edges.

Figure 1. An edge image from Canny method

OpenCV [10] library documentation gave us above implementation [11] in detail.

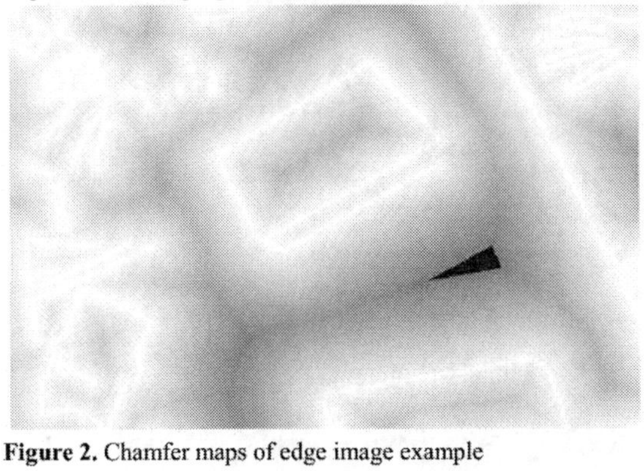

Figure 2. Chamfer maps of edge image example

2.1.2 Chamfer matching maps

In the comparison step, points get higher score when they come closer to the edge point. By indexing the value of pixel with its distance to the closest edge point, we get the distance map as shown in Fig 2.

2.2 Camera model and edges from CAD

From a camera with prior-known configuration and object CAD model, we are able to archive ideal visible edges of objects by using camera model matrix. This matrix convert a point with coordinate of (x, y, z) in real coordinate to a image point (u, v) as Equation 1.

$$\begin{bmatrix} u \\ v \\ 1 \end{bmatrix} = \begin{bmatrix} f_x & 0 & c_x \\ 0 & f_y & c_y \\ 0 & 0 & 1 \end{bmatrix} \begin{bmatrix} x/z \\ y/z \\ 1 \end{bmatrix} \quad (1)$$

To determine visibility of object edges we use edge features based method from [12]. Figure 3 shows an edge between adjacent faces $A =< v_0, v_1, v_2 >$ and $B =< v_0, v_1, v_3 >$ with unit face normal n_A and n_B calculated as in Equation 2,3.

$$n_A = \text{normalize}([v_1 - v_0] \times [v_2 - v_0]) \quad (2)$$

$$n_B = \text{normalize}([v_3 - v_0] \times [v_1 - v_0]) \quad (3)$$

To determine the visibility of edge $E =< v_0, v_1 >$, we uses additional vector v_e with direction from v_0 to the camera center. E is visible edge if cross manipulation value of v_e with n_A or v_e with n_B positive.

2.3 Initial pose searching

2.3.1 Cost function calculation

The cost function for global searching algorithm is a comparison result between edge maps from image and edge maps from CAD model. To gain a equivalent between ideal edge maps, a re-sampling step is applied, so the number of edge points in different ideal maps is set equally at N=200 points.

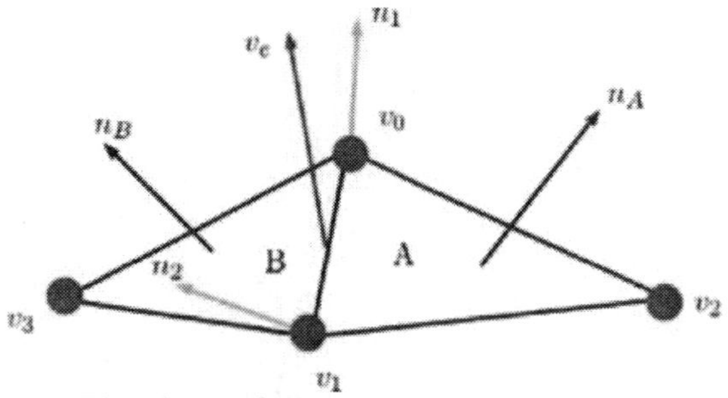

Figure 3. Edge of two surfaces

The error cost function is calculated as in Equation 2,

$$F(R,t) = f(n)\frac{1}{n^2}\sum_1^n (E_i - M_i)^2 \quad (4)$$

where $f(n)$ is function depended on number of inlier (n) as in Equation 3. E_i is value of real edge images at inlier i, M_i is value of CAD model edge images at inlier number i.

$$f(n) = 1 - \frac{n}{N} \quad (5)$$

2.3.2 Differential evolution

Differential evolution (DE), proposed by Storn and Price, is a very popular EA. Like other EAs, DE is a population-based stochastic search technique. It uses mutation, crossover and selection operators at each generation to move its population toward the global optimum minimum.

a) Initialization in DE

The initial population was generated uniformly at random in the range lower boundary (LB) and upper boundary (UB).

$$X_{i,j}^{G=0} = lb_j + rand_j(0,1)(ub_j - lb_j) \tag{6}$$

where $rand_j(0,1)$ a random number in [0,1].

b) Mutation operation

In this process, DE creates a mutant vector, $X_i^G = (X_{i,1}^G, X_{i,2}^G, \ldots, X_{i,n}^G)$. For each individual at each generation X_i^G is a target vector in the current population.

There are several variants of DE based on mutation schemes, which are: DE/rand/1, DE/best/1, DE/current to best/1, DE/rand/2, DE/best/2, DE/rand to best/1.

c) Crossover operation

After mutation process, DE performs a binomial crossover operation on X_i^G and V_i^G to generate a trial vector $U_i^G = (U_{i,1}^G, U_{i,2}^G, \ldots, U_{i,n}^G)$ for each particle i as shown as Equation 5.

$$U_i^G = \begin{cases} V_{i,j}^G & \text{if } rand_j \le CR \text{ or } j = j_{rand} \\ X_{i,j}^G & \text{otherwise} \end{cases} \tag{7}$$

where i=1, ..., NP. j=1,..., D is randomly chosen integer in $[1, D]$, $CR \in [0,1]$ is the crossover control parameter.

d) Selection

The selection operator is to select the better vector between the target vector $X_{i,j}^G$ and the trial vector $U_{i,j}^G$ to enter the next generation.

$$X_i^{G+1} = \begin{cases} U_i^G & \text{if } f(U_i^G) \le f(X_i^G) \\ X_i^G & \text{otherwise} \end{cases} \tag{8}$$

where i =1, ..., NP, X_i^{G+1} is target vector in the next generation.

3 Experiment and results

3.1 Experiment setup

To implement the algorithm, the system hardware included a iBuffalo BSW20KKM11BK camera. We used small box as a tracking object. All code is implemented on C++ code on a standard Desktop Computer powered with Intel Core i7-4790 CPU 3.6x8.

In the experiment, we used the box object with size of 145x95x40(mm) in white colour as the tracking object. The object "*.ply" extension is use the input model.

The searching boundary for the objects translation are in [-100, 100] and $[-\pi/2, \pi/2]$.for rotation angles of roll-pitch-raw.

Parameters for Differential Evolution searching algorithm are presented in Table 1.

Table 1. DE parameters

F0	Cr	Cross over	Max gen	Pop
0.8	Cr	DE/rand/1/bin	150	300

3.2 Results

Fig. 4 and Fig. 5 show results from Canny edge detection method.

Figure 4. A canny edge detection result

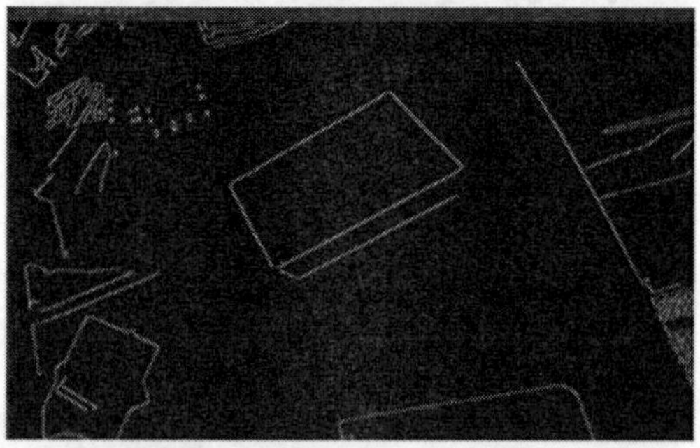

Figure 5. A canny edge detection result

Figure 5. A canny edge detection result

Fig. 6 and Fig. 7 show results of boundary search of the box with different positions from Fig 4 and Fig 5. The box boundary is in blue colour in the left image.

The result images showed that, the method were able to find the position of the object so the boundary could fit into the edge maps.

Depend on the objects, symmetric or non-symmetric, we need to set different searching boundaries for the searching algorithm (DE). Table 2 shows consuming time depend on number of population of DE.

Table 2. Time consuming on population size

Pop size	400	300	200	100	20	10
ms	2817	2190	1404	741	161	76

4 Discussion and conclusions

Object tracking has been always a challenging task in computer vision. Recently, evolution based global searching methods have proved its potential of tackling the tracking problem with ability of finding robust and accurate global optima solutions.

We proposed a novel approach of using differential evolution as a global searching method to find the best 3D position of objects. The experimental results showed promising results.

In the future work, we would like to improve cost function with the method to narrow searching area for more accuracy but smaller generation of searching. By doing so, we expect to reduce the runtime but remains the accuracy and robustness.

References

1. D. G. Lowe, "Distinctive image features from scale-invariant key- points," IJCV, vol. 60, no. 2, pp. 91–110, 2004.
2. K. Grauman and T. Darrell, "The pyramid match kernel: Discrimina- tive classification with sets of image features," in ICCV, vol. 2, 2005, pp. 1458–1465 Vol. 2.
3. M. A. Fischler, R. C. Bolles. Random Sample Consensus: A Paradigm for Model Fitting with Applications to Image Analysis and Automated Cartography. Comm. of the ACM, Vol 24, pp 381-395, 1981.
4. Luc Vosters, Caifeng Shan, Tommaso Gritti, Real-time robust background subtraction under rapidly changing illumination conditions, Image and Vision Computing, Volume 30, Issue 12, December 2012, Pages 1004-1015, ISSN 0262-8856, http://dx.doi.org/10.1016/j.imavis.2012.08.017.
5. H. Barrow, J. Tenenbaum, R. Bolles, and H. Wolf, "Parametric correspondence and chamfer matching: Two new techniques for image matching," in IJCAI, 1977, pp. 659–663.
6. C. Harris, "Tracking with Rigid Objects". MIT Press, 1992.
7. A. I. Comport, E. Marchand, and F. Chaumette, "Robust model-based tracking for robot vision," in IROS, vol. 1, 2004.
8. Storn, Rainer "Differential Evolution – A Simple and Efficient Heuristic for global Optimization over Continuous Spaces", Journal of Global Optimization, doi: 10.1023/A:1008202821328, 2007.
9. Canny method, http://docs.opencv.org/master/da/d22/tutorial_py_canny.html#gsc.tab=0
10. Bradski, G. "opencv_library", Dr. Dobb's Journal of Software Tools, 2000.
11. OpenCV Edge Detection Implementation. http://docs.opencv.org/2.4/doc/tutorials/imgproc/imgtrans/canny_detector/canny_detector.html
12. Morgan McGuire and John F. Hughes, "Hardware-Determined Feature Edges", Proceedings of the 3rd international symposium on Non-photorealistic animation and rendering, 2004, doi: http://doi.acm.org/10.1145/987657.987663

Automatic Puncture System Based on NIR Image and Ultrasonic Image

Zhendong Qiao, Yuhe Li , Zhenhuan Wu, Jiahua Kou

Department of Precision Instrument, Tsinghua University, Beijing, China

Abstract. In view of modern clinical treatment and care, venipuncture has a very important position, but in order to train a medical staff with high level skill of puncture need to spend a lot of time, human and material resources. For this reason, in this paper, an automatic system that can achieve venipuncture is proposed, which can replace the medical staff and achieve the puncture process. The automatic venipuncture system, combined with NIR imaging and ultrasound imaging technology, the overall plane information is obtained by NIR image, depth information and fine positioning of the plane information is obtained by ultrasound image. By way of image processing, the NIR image is enhanced and segmented, getting the spatial location of the vein. And the most suitable blood vessel segments were selected by piecewise straight fitting. Then the location ultrasonic probe should be placed is determined. Then the ultrasound image with the blood vessel in the vicinity of the longitudinal centreline can be obtained. All the information of blood vessel have been obtained, then the machine can be driven to achieve venipuncture.

1 Introduction

Venipuncture as one of the basic methods of clinical treatment, is of important significance in the present medical examination and treatment. It is a critical step in blood sample collection, infusion and interventional therapy. In order to ensure rapidity and accuracy, medical staff should go through long-term training to get acceptable level of venipuncture. To obtain high level of puncture technology, at least several years of clinical training is needed. And for patients with no obvious blood vessels (such as obese patients), it is difficult to puncture successfully. [1] However, an auto puncture system, can be quickly applied in the clinical fields once in mass production. Despite of saving manpower, it also has the advantage of avoiding reduce human error from the medical staff and other external interference.

Currently the clinical products related to puncture are mainly auxiliary equipment, which enhances the contrast, so that the medical staff can find the blood vessels easily. As infrared projection equipment and ultrasound-guided peripherally inserted central catheter (PICC).[2-9] In the laboratory research, has been able to carry out venous puncture automation equipment. In 2006, Western University successfully developed an automatic puncture device using ultrasound guidance similar to the PICC, however, different from the PICC, this system used machine instead of manpower to insert a needle.[10]In 2013, Rutgers University successfully developed a Near-infrared(NIR) and ultrasound-based automatic venipuncture system, which allows the puncture needle into the blood vessels automatically, [11]and in 2015 a comprehensive upgrade of the system has been realized.[12]

2 System design

Venipuncture in this paper is not only aimed to collect blood samples, but also aimed to administrate intravenous infusion. Therefore, after the needle has entered the blood vessel at an angle of about 45°, it is necessary to adjust the direction of the needle so as to coincide with the direction of the blood vessel and proceed a predetermined distance. This requires the needle can turn around the tip after it has entered the blood vessel in the mechanical design. Also, the blood vessel segment selected for puncture should be longer than the puncture needle, and it should be as straight as possible to avoid being pierced by the needle. As shown in Figure 1, the system obtains the plane information of vessel by NIR camera and then move the B-ultrasonic with the puncture unit fixed together onto the vessel obtained to get the depth information, completing the three-dimensional positioning of blood vessels. At this moment, the vessel is under the center of B-ultrasonic probe, the relative position of the needle to the vessel is shown by the dotted line in Figure 1. Then drive the needle downwards ΔH to make the tip reach the vessel center O', then rotate β around O', as previously stated to make the direction coincide with the direction of the vessel, then drive the needle forward ΔL completing the entire puncture process.

DOI: 10.1051/matecconf/201710815002

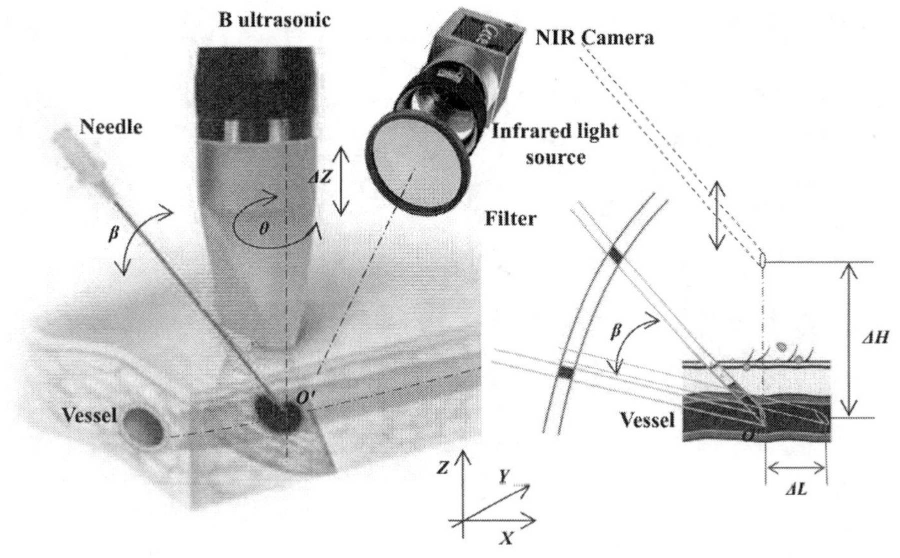

Figure 1. Puncture process demonstration

3. Unit design

3.1 NIR unit design

Near infrared spectroscopy is an imaging technique using the spectrum which wavelength range of 750nm~1500nm. Compared to other bands, NIR has a strong ability in penetrate human tissue, the distance can be up to 10mm.[13] And as shown in Figure 2, the absorption characteristic curve for the NIR of different material,[14] it can be seen that when the wavelength range is 760nm~850nm the absorption coefficient of reduced hemoglobin reaches a peak. Considering the spectral characteristics of NIR camera and other effects, eventually using a wavelength of 850nm.[15][16]

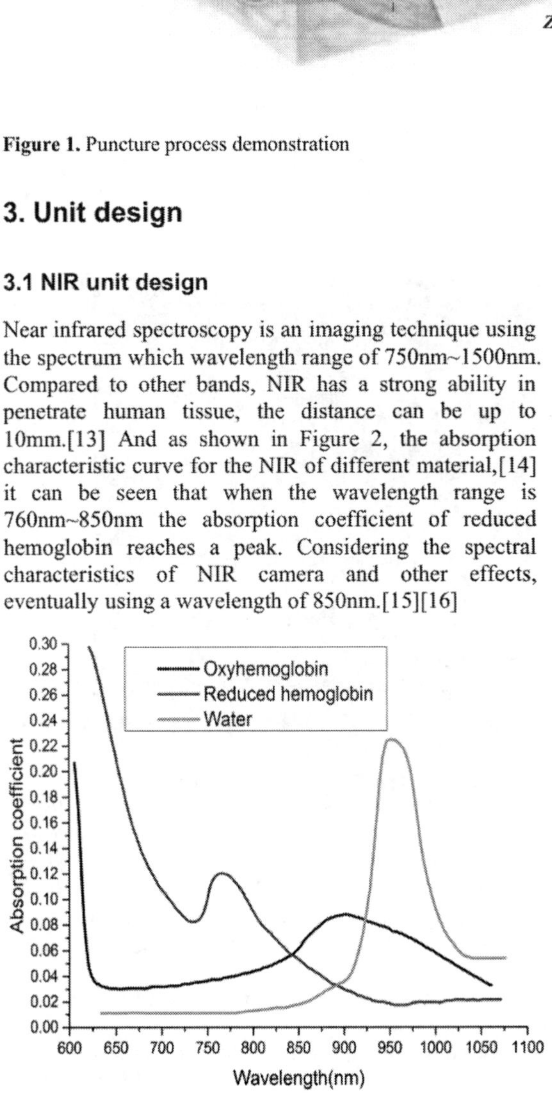

Figure 2. The absorption characteristic curve for the NIR of different material

The comparison between NIR image and visible light image is shown in Figure 3. It shows that in the NIR image the details of image under visible light are filtered out, but the vessel information is enhanced. However, if a lot of subcutaneous fat or other reasons resulting in poor image quality, it's more difficult to get the vessel information when compared with retina image or other image with high contrast. For this reason, Convex Region Gradient (CRG) algorithm is used to enhance the vessel, which is laboratory preparatory work.

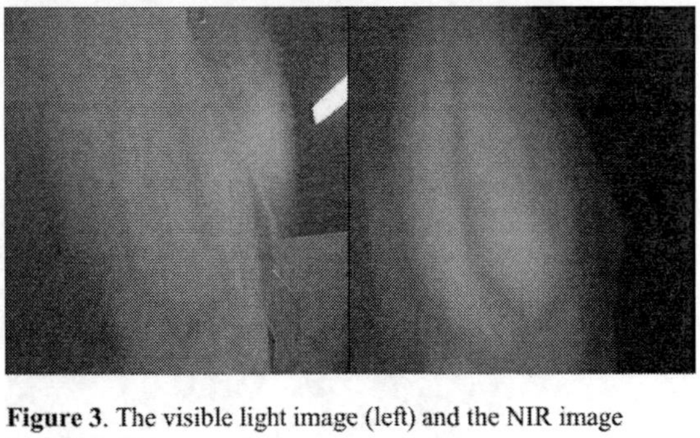

Figure 3. The visible light image (left) and the NIR image (right)

The CRG value is defined as the following equation

$$F(x)=\sum_{n=0}^{N_V/2}\left(f(x+n\Delta x)+f(x-n\Delta x)\right)-\sum_{n=N_V/2+1}^{N_V+1}\left(f(x+n\Delta x)+f(x-n\Delta x)\right) \quad (1)$$

Where Δx represents the width of single pixel, and N_V represents the count of pixel that a vessel model contained. Figure 4 shows the original image with the result from CRG algorithm and histogram equalization. It shows that the vessel information obtained by CRG algorithm is more obvious.

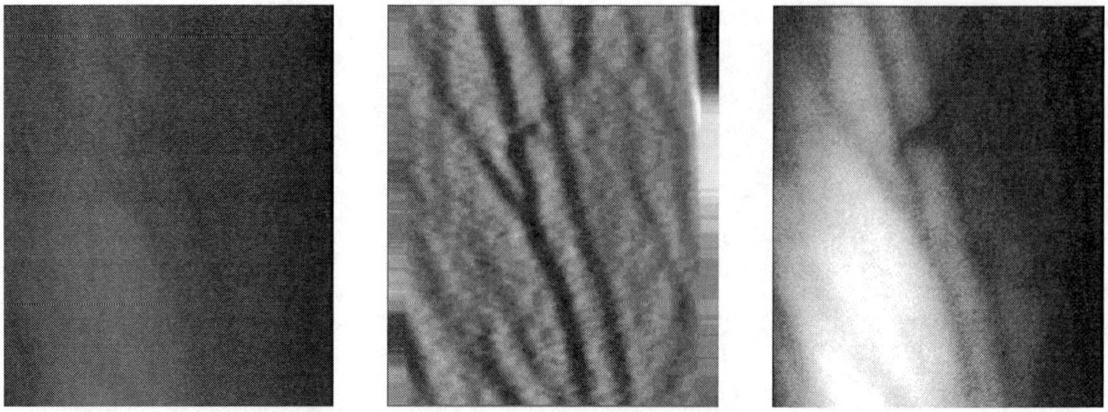

Figure 4. The origin image (left), the CRG model image (middle) and the histogram equalization image (right)

After obtained the centre line of vessel, it should be identified that which vessel segment is the most suitable one for puncture, it contains the following features:

1) Length: longer than the needle;
2) Straightness: needle can go through without touching vessel wall.

Based on the constraints above, a piecewise linear fit was made to the found centerline of the blood vessel (the segment length is greater than the length of the needle), and then the centerline segment with the largest correlation coefficient is selected as the destination location for puncture. As shown in Figure 5, the linear correlation coefficient is 0.996398, the maximum value in the same length line segments.

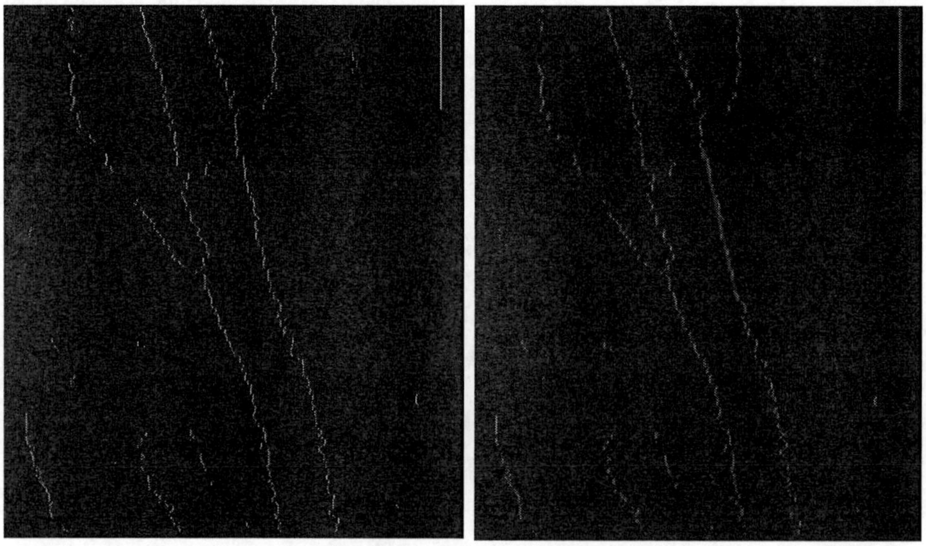

Figure 5. The centre line (left) and the target vessel segment (right)

3.2 Ultrasound unit design

Using an external image acquisition box, the ultrasonic signal acts as an external camera input, so the ultrasound image can be called directly for a real-time processing.

Unlike NIR, ultrasound imaging is a contact imaging process. Coupling agents are required to make the ultrasonic probe in contact with human tissue intact. So as to introduce ultrasonic waves into the human body to obtain an echo image of the human body tissue. When the ultrasonic probe is operated manually, medical staff can find the suitable location, according to their experience. So that it is possible to introduce the ultrasonic wave into the human body tissue without causing the extrusion deformation. Due to different measurement targets have different surface shapes, when using automation equipment, easily lead to poor ultrasound signal or squeeze the blood vessels that can not be identified. As shown in Figure 6.

Because plane position information has obtained by NIR unit, compared with feedback control using sensors, increasing the thickness of the coupling agent to make the unevenness of the skin surface compensated is more suitable and simple. However, considering the fluidity of the coupling agent, how to concentrate it between the ultrasound probe and the skin is a tricky problem. The method presented in this paper is that placing a soft rubber film on the ultrasonic probe, in order to perfect fit with the skin. Ultrasound can be good import without squeezing the skin. The operation process and the acquired image is shown in Figure 7.

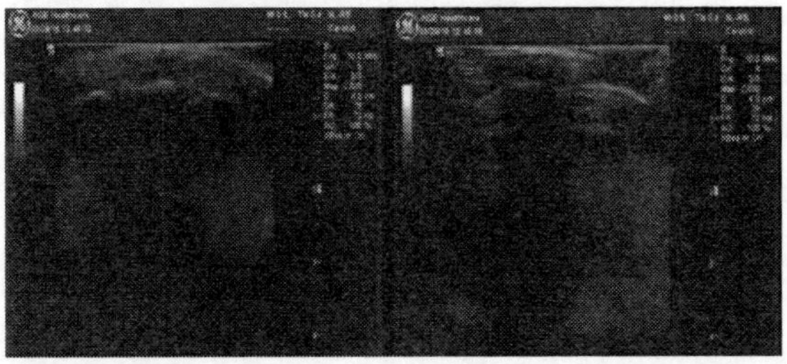

Figure 6. The condition that vessel has been squeezed(left) and poor ultrasound signal(right)

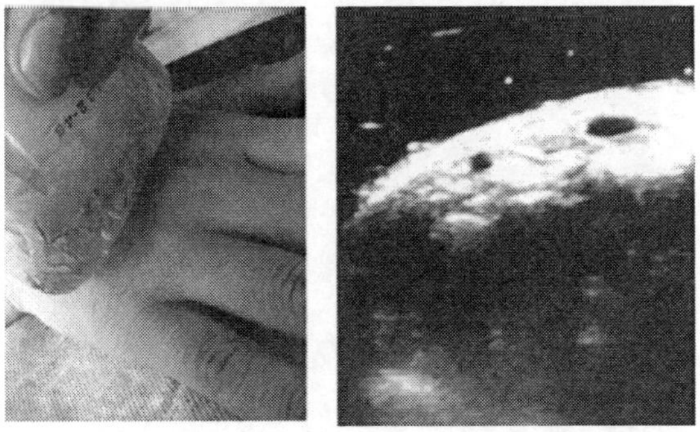

Figure 7. The operation process of the method with a soft rubber film (left) and the acquired image (right)

After obtaining the ultrasound image, the following work is extracting vessel information. Since the plane location has been got by NIR image, the ultrasound image can be processed relatively easy, as following:

1) Based on the position and width of the blood vessel obtained from the infrared image, combining with the error of aligning the ultrasound probe with the vessel, determining the transverse position range in which the blood vessels appear in the ultrasound image.
2) Find the boundary between the skin and the coupling agent, and the vessel position is below the boundary.
3) In the previously determined interval, through the threshold segmentation method to find the darkest point at the top.
4) Output vessel coordinates.

4. Conclusion

This paper designed a venipuncture system to perform vessel localization combining NIR imaging with ultrasound imaging, and proposed methods in image acquisition and processing. Compared with the existing research results, the trajectory of puncture needle is more in line with clinical practice. However, considering the conditions of laboratory, we were unable to carry out a complete validation test, some problems may not be exposed. In addition, during the selection of the final puncture of vessel segment, we only consider the straightness of the center line, the width, direction, and bifurcation were not considered. In the follow-up work, these problems will be further investigated in order to find a resolution.

References

1. Walsh G. Difficult peripheral venous access: recognizing and managing the patient at risk[J]. Journal of the Association for Vascular Access, 2008, 13(4): 198-203.
2. Ernst D J. Take the guesswork out of venipunctures[J]. MLO Med Lab Obs, 2009, 41(4): 18-20.
3. Zeman H D, Lovhoiden G, Deshmukh H. Design of a clinical vein contrast enhancing projector[C]//BiOS 2001 The International Symposium on Biomedical Optics. International Society for Optics and Photonics, 2001: 204-215.
4. Zeman H D, Lovhoiden G, Deshmukh H. Optimization of subcutaneous vein contrast enhancement[C]//BiOS 2000 The International Symposium on Biomedical Optics. International Society for Optics and Photonics, 2000: 50-57.
5. Zeman H D, Lovhoiden G. Enhancing the contrast of subcutaneous veins[C]//BiOS'99 International Biomedical Optics Symposium. International Society for Optics and Photonics, 1999: 219-230.
6. Zeman H D, Lovhoiden G, Deshmukh H. Design of a clinical vein contrast enhancing projector[C]//BiOS 2001 The International Symposium on Biomedical Optics. International Society for Optics and Photonics, 2001: 204-215.
7. Lovhoiden G, Deshmukh H, Zeman H D. Clinical evaluation of vein contrast enhancement[C]//International Symposium on Biomedical Optics. International Society for Optics and Photonics, 2002: 61-70.
8. Zeman H D. Contrast enhancing illuminator: U.S. Patent 5,969,754[P]. 1999-10-19.
9. Lovhoiden G, Deshmukh H, Vrancken C, et al. Commercialization of vein contrast enhancement[C]//Biomedical Optics 2003. International Society for Optics and Photonics, 2003: 189-200.
10. Waspe A C, H Jason C, Lacefield J C, et al. Design, calibration and evaluation of a robotic needle-positioning system for small animal imaging applications.[J]. Physics in Medicine & Biology, 2007, 52(7):1863-78.
11. Alvin Chen, Kevin Nikitczuk, Jason Nikitczuk, et al. Portable robot for autonomous venipuncture using 3D near infrared image guidance[J]. Technology, 2013, 1(1).
12. Balter M, Chen A, Maguire T. The System Design and Evaluation of a 7-DOF Image-Guided Venipuncture Robot[J]. IEEE Transactions on Robotics, 2015.
13. Kessoku S, Maruo K, Okawa S, et al. Influence of Blood Glucose Level on the Scattering Coefficient of

the Skin in Near-Infrared Spectroscopy[C]// Asme/jsme 2011 Thermal Engineering Joint Conference. 2011:T10007-T10007-7.
14. Bashkatov A N, Genina E A, Kochubey V I, et al. Optical properties of human skin, subcutaneous and mucous tissues in the wavelength range from 400 to 2000 nm[J]. Journal of Physics D: Applied Physics, 2005, 38(15): 2543.
15. Li C, Zhang Y, Wang M, et al. In vivo real-time visualization of tissue blood flow and angiogenesis using Ag 2 S quantum dots in the NIR-II window[J]. Biomaterials, 2014, 35(1): 393-400.
16. Shimawaki S, Sakai N. Change in blood vessel images of the human finger using near-infrared radiation while compressing the upper arm[C]//6th World Congress of Biomechanics (WCB 2010). August 1-6, 2010 Singapore. Springer Berlin Heidelberg, 2010: 1262-1265.

DOI: 10.1051/matecconf/201710815003

Multi-UAV joint target recognizing based on binocular vision theory

Zhang Yuan, Wang Jun

AVIC Fai, Avonic Department, 710089 Xi' an, China

Abstract. Target recognizing of unmanned aerial vehicle (UAV) based on image processing take the advantage of 2D information containing in the image for identifying the target. Compare to single UAV with electrical optical tracking system (EOTS), multi-UAV with EOTS is able to take a group of image focused on the suspected target from multiple view point. Benefit from matching each couple of image in this group, points set constituted by matched feature points implicates the depth of each point. Coordinate of target feature points could be computing from depth of feature points. This depth information makes up a cloud of points and reconstructed an exclusive 3D model to recognizing system. Considering the target recognizing do not require precise target model, the cloud of feature points was regrouped into n subsets and reconstructed to a semi-3D model. Casting these subsets in a Cartesian coordinate and applying these projections in convolutional neural networks (CNN) respectively, the integrated output of networks is the improved result of recognizing..

1 Introduction

A common method of target recognizing for UAV is to input a single image containing suspected target to the recognizing system and compare to template in the library. Affect by lighting, view point position of EOTS or noise etc., the recognizing accuracy is lower than ideal. Despite of these distractions, the suspected target also has the exclusive 3D model (Considering there are no identical targets in the image). Reconstructing the 3D model of suspect target by cloud of feature points provides extra information could be more accurate than 2D recognizing method. Based on the binocular vision theory, the 3D model also could be reconstructed by UAVs with single EOTS. Assuming the image from each EOTS was grouped into an image set. As long as the UAV group does share overlapping field of view, which require enough matched points (the number of points is agile and determined by complexity of suspected target) in the image group, the 3D model was able to be reconstructed. Figure 1 shows a scenario of 3 EOTS joint recognizing.

Although these matched feature points is able to reconstruct a coarse 3D model, the recognizing system is not in need. In the recognizing procedure, the semi-3D model is satisfied and lower resource consuming. The semi-3D model is a layer group which depth value on the light axis of EOTS is compute from matched points in couple image. This layer group is a degenerating of 3D model. Casting this layer group on Z plane and X plane of Cartesian coordinate, which the light axis of image sensor is X axis, these 2 projections contained all 3 dimension features. After identifying these 2 projections with trained convolutional neural networks (CNN), fusing outputs of CNN is indispensable and final procedure in the recognizing system. The system architecture is shown in Figure 2. The first step is data assembly. When recognizing, the center node UAV, called the image processing node assemble all sensor images (at least there are 2 images in the assembly set). The second step is the pre-processing of reconstructing. By matching feature points and computing the depth information, the space relationship of layer group was get. In this procedure, mismatching of feature points which leads to inaccurate semi-3D model should be avoiding. Matching points should be grouped into n (n is re-set and decided by complexity of targets) sets. Each set represents one feature zone which the depth of zone is represented by the center point depth itself. In the last step, all these n feature point sets reconstruct the semi-3D model of suspected target and recognize with 2 projections.

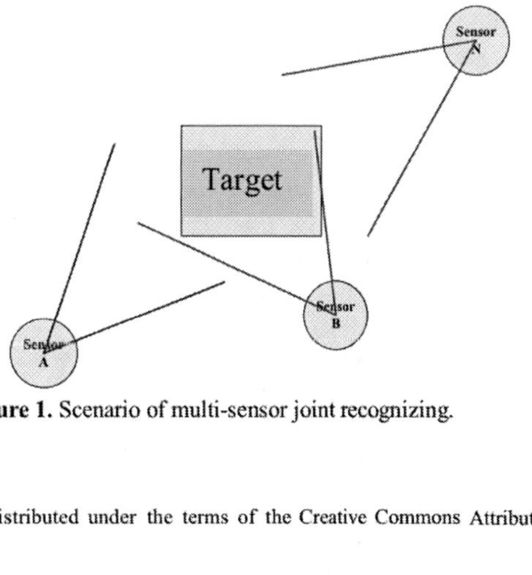

Figure 1. Scenario of multi-sensor joint recognizing.

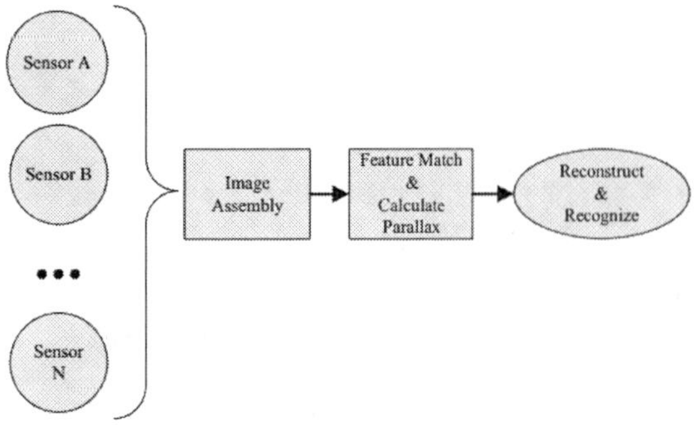

Figure 2. Recognizing System Architecture.

2 Target processing and recognizing

2.1 Matching feature points

In recognizing, the first important prerequisite is the field of view of each couple of EOTS is overlap and the suspected target should be in this field. Otherwise, there are no matched points in the field of view to identify the suspected target or even reconstruct the 3D model. In figure 1, both couples (A-B, B-N and A-N) and the UAV group (including A, B and N) could be joint recognizing. But the couple of A and N, which has a few overlap fields of view, could not reconstruct the semi-3D model for lacking or missing feature points.

In case of matching 2 images whose size is M*N, the image from UAV A was called image A and the image from UAV B was called image B. After extracting and matching feature points with SIFT, the mismatching points in result should be eliminate for badly reconstructing. Assuming U and V are column and row coordinates respectively. Based on the hypothesis that the layout relationship of matched feature points is almost coincident (electric-optic turret of UAV is able to keep the image plane from view rotating, shown in figure 3 a) and b)), set up a vector as follows: $f = [l, k]$ where:

$$l = (U_A + N - U_B)^2 + (V_A + M - V_B)^2 \quad (1)$$

$$k = \frac{U_A + N - U_B}{V_A + M - V_B} \quad (2)$$

For the layout relationship of feature points is stable, the vector f has a narrow variable range. Computing the density of sub-space constructed by each couple points with follows:

$$D_{k,l} = \frac{\Sigma i}{\pi r^2} \quad (3)$$

Where: The subscript k and l is the center point and edge point index respectively, this point couple construct a circle area as the selected space. πr^2 = the number of matched feature points in selected space, Σi = the area of selected space and the radius r is the distance between point k and l. Computing each space density and selecting the most density space as the room which the matched feature points in. The feature points scattered before and after eliminating were shown in figure 3 c) and d).

Figure 3. Feature points scatter: a) image A; b)image B; c) before eliminating; 4)after eliminating

2.2 Reconstructing semi-3D model and Casting

According the binocular vision theory, the depth information of feature points which is the foundation to reconstruct 3D model of suspected target is exclusive and certain. The abundant 3D information benefits the accurate rate of recognizing system but consuming amount of memory and computing resource. Therefore 3D model is not required and even harmful to system in the target recognizing task. For avoiding useless depth information overload the computer resource, the feature points will be divided into n subsets which each subset just has single depth value. K-means cluster algorithm which took the distance between each point into account is used to decide the subset. Figure 4 shows the clustering result for the image in sensor A. In this case, n was set to 4.

Figure 4. The clustering result in image from sensor A.

 DOI: 10.1051/matecconf/201710815003

After clustering, each subset is confirmed and its depth value is represented by the point nearby the geometry center point of subset. The depth information in EOTS A coordinate system computing follows（Origin of this coordinate is the center point of sensor A imaging plane, x axis is the light axis of sensor A, y and z axis are parallel to the imaging plane row and column.）:

$$\frac{y_{OA}}{U_{OA}} = \frac{x_{OA}}{f_{OA}} \tag{4}$$

$$\frac{z_{OA}}{V_{OA}} = \frac{x_{OA}}{f_{OA}} \tag{5}$$

$$tan\theta = \frac{x_{BA}y_{OA}-y_{BA}x_{OA}}{y_{BA}y_{OA} + x_{OA}x_{BA}} \tag{6}$$

Where: x_{OA}, y_{OA}, z_{OA}, x_{BA} and y_{BA} is the target and UAV B position in A coordinate, θ is the angle between line of sensor A-target to sensor B-target. The position of UAV B in geographic coordinate is come from navigation system.

For target position in sensor B coordinate, the position is computing as follows:

$$\frac{y_{OB}}{U_{OB}} = \frac{x_{OB}}{f_{OB}} \tag{7}$$

$$\frac{z_{OB}}{V_{OB}} = \frac{x_{OB}}{f_{OB}} \tag{8}$$

$$tan\theta = \frac{x_{OB}y_{AB}-y_{OB}x_{AB}}{y_{OB}y_{AB} + x_{OB}x_{AB}} \tag{9}$$

Where: x_{OB}, y_{OB}, z_{OB}, x_{AB} and y_{AB} is the target and UAV A position in B coordinate. The position of UAV A in geographic coordinate is come from navigation system.

Transforming the point in B to A coordinate with equation (10), the target point coordinate should be similar with its position coordinate in A coordinate.

$$\begin{bmatrix} x_A \\ y_A \end{bmatrix} = \begin{bmatrix} cos\gamma & sin\gamma \\ -sin\gamma & cos\gamma \end{bmatrix} \begin{bmatrix} x_B \\ y_B \end{bmatrix} + \begin{bmatrix} x_{BA} \\ y_{BA} \end{bmatrix} \tag{10}$$

Where: γ is the angle from light axis of EOTS B to EOTS A.

As the position of each UAV is given (the navigation system of each UAV broadcast their self-positions in the group), after transforming the geographic coordinate system to Cartesian coordinate system, depth of each feature point could be compute for combining the equation (4) to equation (10).

2.3 Reconstructing semi-3D model and casting

After reconstructing, analyzing the semi-3D model of suspected target with CNN require the model 2D features. Casting the semi-3D model on planes parallel to light axis and vertical to light axis, figure 5 shows the projection on z plane and projection on x plane. The projection on z plane, which represents the grey value accumulation of points with same z position, shows the geometry relationship of each layer and the scatter of feature points on z plane. The grey scale value of each pixel of the projection on the x plane is multiply the original grey scale value by the depth value after normalizing.

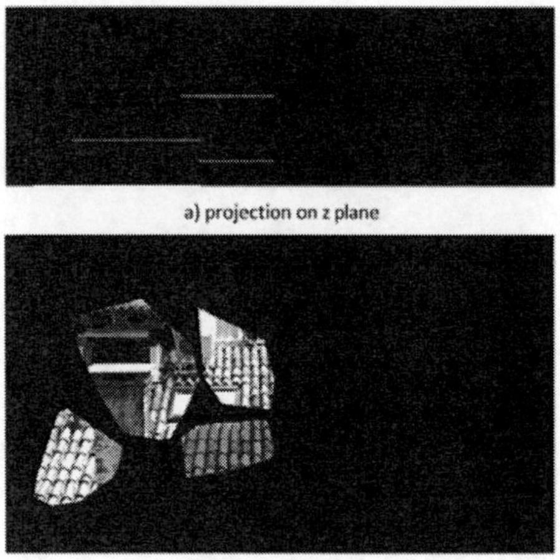

Figure 5. Projection of semi-3D model: a) projecting on z plane; b) projection on x plane.

Setting up CNN (in this case, LeNet-5 was apply to recognizing system) networks to analyzing these 2 projections and fusing each output together, the final outputs is the recognizing outputs of suspected target. To fuse the result, Establish the fuse principle follows:

$$minJ = (y - a_1y_1 - a_2y_2)^2 \tag{11}$$

Where: y is the standard output of target model in the library, a1 is the weight of CNN output for z plane projection, y1 is the output of CNN output for z plane projection, a2 is the weight of CNN output for x plane projection, y2 is the output of CNN output for x plane projection. a1 and a2 should be training with gradient descent method before recognizing suspected target.

3 Simulation and analyzing

Set three kinds of building as targets for testing this recognizing system: department, mall and bridge. Stp1, establishing the target library by these VR models. Stp2, arranging 2 or 3 virtual moving cameras from different view point. Stp3, processing these three models follows the procedure shown in figure 1and training CNN networks. Stp5, set up a test set included 3 models above and castle and tank. The recognizing result is shown in table I and table II. The recognizing rate is improving with the increasing of feature point number. The complicated surface structure (complicated surface texture, the shortage of valid depth feature point information, the onefold spatial target structure, etc.)

requires more feature points in overlap field for archiving similar recognizing rate which means the larger overlap field and smaller angle between target to sensors. Table II shows the increasing of recognizing rate with the number of feature points. After appending one EOTS to recognizing system, the overlap field is larger and the feature points in each 2 field of view could be reconstructing a model in larger scale. Feature points in the public field of 3 EOTS, after reconstructing by its depth information, could be compensate the depth value which provide a more precision semi-3D model. The recognizing rate difference is shown in Table 1 and Table 2.

Table 1. Recognizing results of 2 EOTS joint recognize.

	Result		
	department	mall	bridge
False alarm rate(%)	12	10	15
Recognizing rate(%)	67	70	66
Feature points in total	35	42	37

Table 2. Recognizing results of 3 EOTS joint recognize.

	Result		
	department	mall	bridge
False alarm rate(%)	12	10	15
Recognizing rate(%)	71	72	70
Feature points in total	52	60	49

4 Conclusions

This paper provides a method of recognizing suspected target with multi-image sensor. By reconstructing a semi-3D model of suspected target, the exclusive semi-3D geometry features was set up and represented by layer. The recognizing system not only takes layers constructed by feature points as input but also the projection of these layers which is represent the position relationship of these layers. Benefited with the abandoned geometry feature provided by semi-3D model, the multi-image sensor joint recognizing can improve the estimation. Though the joint recognizing, a more accurate and reliable result can be achieved.

References

1. 1.Hartley Zisserman, "Multiple View Geometry in Computer Vision" Phil. Cambridge University Press, New York, pp. 237 – 360, 2000.
2. 2.Ke Y, Su kthankar R, "PCA- SIFT: a more distinctive representation for local image descriptors", Proceedings of IEEE Conference on Computer Vision and Pattern, Washington D.C.506-513, 2004.
3. Hinton G E, Salakhutdinov R R, "Reducing the dimensionality of data with neural networks", Science, **313**(5786):504-507, 2006.
4. Rumelhart D E, Hinton G E, Williams R J, "Learning representations by back-propagating errors", Nature, 323:533-538,1986.
5. Krizhevsky A, "Learning multiple layers of features from tiny images", University of Toronto, Toronto, Canada, 2009.

 DOI: 10.1051/matecconf/201710815004

Research on group enterprise multimedia information publishing system

Dong Zhimin

Hengshui power supply branch, Hebei province, State grid, China

Abstract. A electric power group enterprises to actively explore the innovation of enterprise culture management, making full use of modern information and communication technologies, construction of trans-regional multimedia information publishing platform. Construction of a municipal pilot units in Group region, for example, through consolidation, Office LANs, corporate networks in electric power communication network, cable TV network, realized with pictures, video, PPT, FLASH animations, WORD documents, WEB pages, video conference streams, radio, television, and other media as the carrier's digital communications.

1 Introduction

A large-scale electric power group company for the process of integration of enterprise culture construction, make full use of information and communications data network construction, build internal trans-regional enterprises multimedia information publishing platform, linking the Group Headquarters and city and County grades[1]. Currently, group a city level pilot units regional construction has completed, in suburban and transforming centralized control station deployed a total of 23 multimedia network TV and terminal multimedia broadcasting machine， has achieved enterprise internal group, and layered, and points regional clock work life, and Humanities thought, and education training, various propaganda, has created knowledge shared, and learning exchange, and image show, and information interactive of platform, the platform support pictures, and video, and PPT, and FLASH animation, and WORD document, and WEB page, and video conference flow, and Radio, television and other media, has achieved notice, training, news, technical reports, poetry, on demand, brand planning, cultural salons, professional educational organization, production planning, field work, fact sheets and other information shown.

2 Multimedia information publishing system architecture

Multimedia information publishing system architecture is composed of hardware and software system architecture below introduces basic features of hardware infrastructure and software systems.

2.1 Hardware infrastructure

The system used B/S architecture, by terminal office computer, and room server, and information communications network, and broadcast controlled machine, and displayed equipment composition, information released by server concentrated management control, Terminal Office personnel through browser landing server management software, making various program, via Group internal network transmission to each displayed end broadcast controlled machine, broadcast controlled machine through VGA line, and audio line, and HDMI HD digital line released to each displayed equipment[2-3]. As shown in Figure. 1.

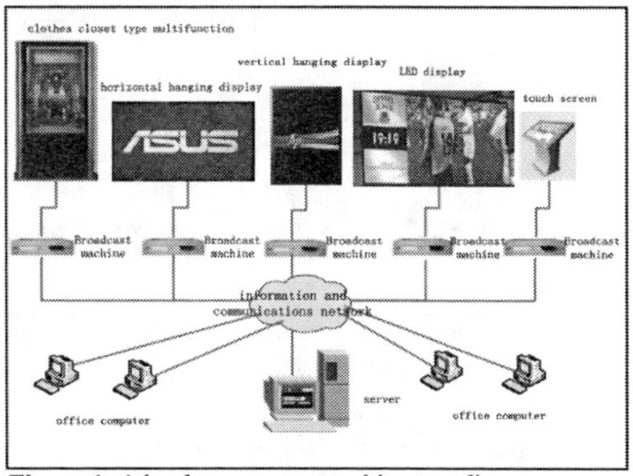

Figure 1. A hardware system architecture diagram.

Server using DELL2850,2G memory 300G-RAID1 hard drive, dual supply. Grew the business to increase in the future, will replace minicomputers.

Terminal published computer: Lenovo's widescreen desktop computers, monitor aspect ratio 16:9, core 2 Duo dual-core 3.2GHz,RAM2G/HD500G, adapter boards support the TCP/IPv4 protocol.

Broadcast machine for ultra-thin embedded computers and embedded display devices such as televisions or placed in the top of the ceiling at the back. Via VGA cable, audio cable, HDMI high definition digital cable connections, such as LCD or plasma TV, professional landscape, portrait displays, touch screens, LED screens, stitching large-screen display devices, through the VGA-AV module is connected directly to your computer host and monitor[4-5]. NIC board support the TCP/IPv4 protocol, parameter configuration: CPU 1.6GHz, memory, 1G, hard 80G.

Display device: 5mx0.5m plasma TV screen ;4mx4m stitching large screen; 32 inch LED thin LCD TV;55 inch LED thin LCD TV.

Information and communication networks: the group domain information in a city traffic network as in Figure 2, is divided into a core layer, convergence layer and access layer, core router with Huawei NE40E-X8, aggregation layer router Huawei NE20E-X8, access layer using Huawei S9303 switch line using 1G fibre connection, access network using the OSPF protocol, bringing together core network using the BGP protocol to transmit data. The VPN dedicated communication network through the establishment of information channels transmitting multimedia information publishing data to ensure transmission reliability and bandwidth and QOS quality.

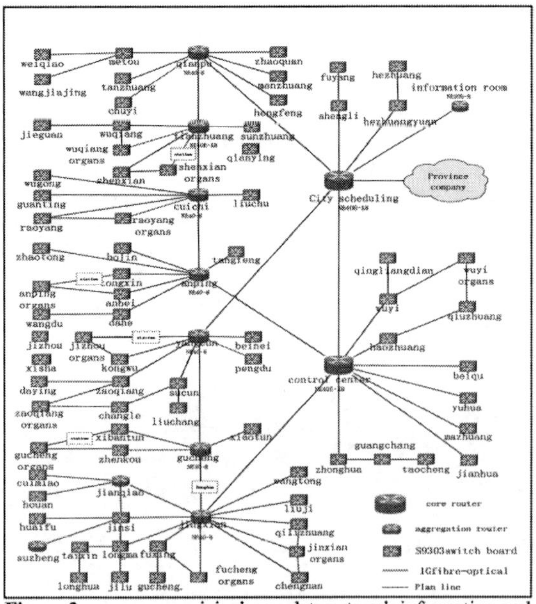

Figure 2. group a municipal area data network information and communication.

2.2 Software systems basic functions

Management end software: will was installation to multimedia information released system server, general user and the administrator are can through browser landing this software to upload pictures, and text, and PPT, and FLASH animation, and music, and video, various material file, making program scene, management program list, will latest of program released to play end; administrator also can group, and grading, and points domain set different user of permission, on various program for audit, and remote monitoring broadcast controlled machine state and log information, and set timing opened and closed[6].

Broadcast software: Studio Adone System broadcast software needs to be installed according to instructions downloaded from the management side of the server needed programming, called broadcast machine installed WORD, EXCEL, PPT format, FLASH Player plug-in, achieve a variety of dynamic transitions, page and scroll through results, and publish it via VGA or HDMI HD digital lines to the specified set.

3 System construction and maintenance

Group pilot municipal units of multimedia information issuing system construction in two phases, the first phase of construction of 9 points, the second phase construction of 14 points, a total of 23 points. As shown in Table 1.

Table 1. Construction of the multimedia information publishing system

locations	display	install style	count
building elevator hall	32inch LED TV	solid wall-mounted	2
scheduling elevator hall	32inch LED TV	solid wall-mounted	1
marketing office	55inch LED TV	solid wall-mounted	1
marketing office fees	55inch LED TV	solid wall-mounted	1
organs restaurant	55inch LED TV	hang ceiling	2
hezhuang restaurant	55inch LED TV	solid wall-mounted	1
repairing east building	55 inch LED TV	decorative wall embedding	1
line west building east	55 inch LED TV	decorative wall embedding	1
line west building west	55 inch LED TV	solid wall-mounted	1
operation second building	32 inch LED TV	solid wall-mounted	1
logistics service center	55 inch LED TV	decorative wall embedding	1
qianpu operations center	55 inch LED TV	hang ceiling	1
yangcun operations center	55 inch LED TV	solid wall-mounted	1
jingxian operations center	55 inch LED TV	solid wall-mounted	1
anping operations center	55 inch LED TV	hang ceiling	1
hengshui operationscenter	55 inch LED TV	hang ceiling	1
building management	55 inch LED TV	solid wall-mounted	1
electric company	55 inch LED TV	solid wall-mounted	1
industry company	32 inch LED TV	hang ceiling	1
office loudaping	plasma TV	solid wall-mounted	1
schedule loudaping	plasma TV	solid wall-mounted	1

After the multimedia information publishing system up and running, due to broadcast controlled machine long-term high load run in buildings ceiling, and

DOI: 10.1051/matecconf/201710815004

substation shipped dimension Center Hall, and restaurant storage between, and hollow decorative wall, and stairs sandwich suite, and corridor wall body wall, dust more, work environment bad, unmanned unattended of industrial production field, easy appeared built-in fan blocking turned, led to thermal system failure, makes system automatically shut machine or overheating burned CPU and motherboard[7].

4 Not ash radiator fan-free design principles

The result has applications for invention patents, provides a dense ash cooling systems, including heating1, thermal devices2, chassis3, cooling structure, thermal2 conductivity equipment end connected to heating1 and cooling structure, heating1 parts located inside the chassis3, heat sink structure outside the chassis3. Figure 3

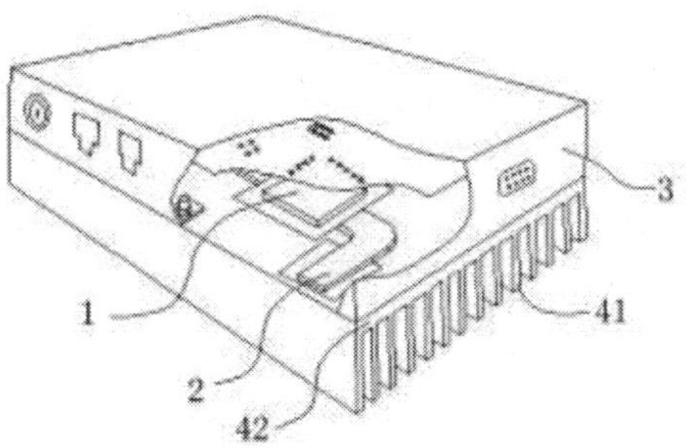

Figure 3. No ash cooling without fan schematic.

Thermal conductivity device can be copper, it can contain other convective type heat spreader. Copper strips, the best use of copper sheet with a higher thermal performance, and according to the device space to determine the dimensions bar. If used to control machines, bar dimensions length x width x thickness =70mmx50mmx12mm. Meanwhile, after taking into account costs and cover both dust effects, thermal solution optional length x width x thickness =220mmx150mmx30mm special heat conducting aluminum heat sink profiles. In addition, the parts joints to choose higher technical parameters of silicon or silicone grease.

Heating and cooling structure in individual cases is in contact, because the heat sink structure actually played a role in conduction of heat, can be seen as a collection of thermal structure and thermal solution.

Further, the foregoing not fouling in cooling system, heat sink structure for the Panel41. Is actually in the original equipment or multilateral set heat sink, or a direct replacement for the original equipment rear cover plates, achieve the purpose of cooling.

Further, the foregoing not fouling in cooling system, heat sink with fins42 on the Board41. Fin42 extended the cooling area, better cooling effect can be achieved. In order to achieve a better cooling effect, where you can change the size and arrangement of fins42, and even changes the fin42 array arranged acicular structure.

Further, the foregoing not fouling in cooling system, finned42 heating Panel41 centered on the axis of radial distribution. This form applies to fans in the central situation.

Further, the foregoing not fouling in cooling system, arranged parallel to the fins on the Panel. This form applies to fans in the central situation.

Further, the foregoing not fouling in cooling system, heat sink away from the side of the chassis with cooling fan. The cooling fan is external to the chassis, auxiliary heat sink structure for cooling. If the heat sink with fins on the Board, fan is located in the thermal center of the plate.

Further, the foregoing not fouling in cooling system, thermal structure and chassis side cover a whole set or fixed connections, or thermal structure and cover plate fixed on the side of the chassis structure, can replace the chassis side cover. For example, thermal structures and chassis cover the same shape, buckle and bolts position the same structure, you can replace the cover to the chassis

Meanwhile, heat sink structure outside the chassis, and heat sink structure can replace the cover on the side of the chassis, in order to minimize heat transfer path length, equipment boards front and thermal structure, site of CPU heat dissipation structure.

Further, the foregoing not fouling in cooling system, heat conduction and thermal solution structure is arranged between the semiconductor refrigeration piece. Semiconductor refrigeration piece actually played a role in enhance the effect of thermal conductivity, heat ends and the thermal conductivity of a semiconductor refrigeration device connected, radiating side connected to the thermal structure of semiconductor refrigeration, thermoelectric power on the device.

Further, the foregoing not ash cooling systems used for control or projector.

Provides of a not product gray of thermal system will thermal structure placed chassis zhiwai, chassis within no fan, chassis both inside and outside no too much of of air exchange, chassis within on not has dust gathered; and thermal structure Shang if has product gray also without on equipment for demolition, directly cleanup can, and convenient; while moved except chassis within of thermal fan, save has chassis within of space, conducive to equipment of miniaturization; and structure simple, can directly through on existing equipment modified achieved, applies range wide.

5 Physical processing and production of cooling system for broadcast machine

Renovation of the Studio machine to fanless cooling system in embedded computer systems (broadcasting) within a small space, effective and bridge CPU chip heat generated while working in the fanless cooling system under the conditions of export. Working under the temperature 75℃rise in temperature under conditions of balance. In accordance with the broadcasting machine lumen design of a thermal block, without damaging the structure of the broadcasting cases, using broadcast machine heat pipe fixtures, to thermal block fixing on the original duct position. Block's thermal dissipation space expanded, connected with an external heat sink, cooling area, in order to achieve the intended results.

Concrete step is to broadcast original dismantle the fan cooling system. Bridge pads and uniform piece of the silicon wafer coated with grease and the grease evenly to the surface of the CPU. Machined cooling block with 4 screws on the original location of the duct, screws tightened as much as possible to make blocks with a bridge with CPU surface in close contact. Aluminum processing for the fin structure, as an external heat sink, and it is fixed with 4 screws to the base pieces that have a good coating of silicone grease uniformly apically. Figure 4

Cooling block select red copper material with high heat dissipation. Dimensions length x width x thickness =70x50x12mm. Taking into account the cost and effect of dust-proof cover, fins use length x width x thickness =220x150x30mm good thermal conductivity of aluminum heat sink extrusions[8].

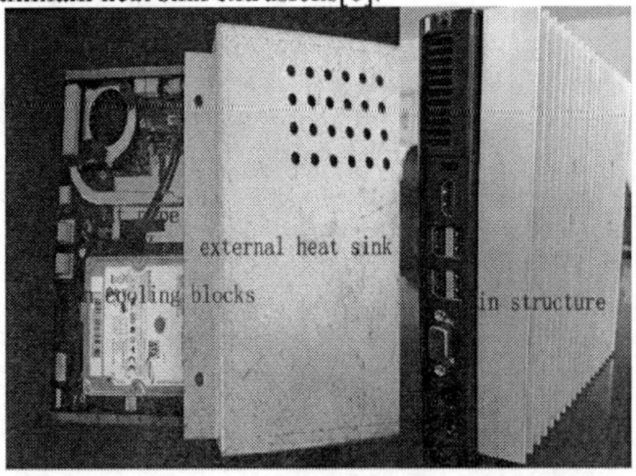

Figure 4. no fan for heat broadcast mechanism

The transformation is complete, Master Lu temperature monitoring software. Test conditions for fanless heat sink for broadcasting and broadcasting with fan cooling after the machine has been running for 30 days at the scene, together play a 8-minute video clip, then Master Lu measuring software testing[9].

Fan heat sink for broadcasting video playback after 5 minutes, CPU occupied 25%, CPU temperature 82℃, the hard disk temperature 43℃, motherboard temperature of 46℃ and CPU temperatures exceed 80℃, Master Lu started the alarm, alarm message "CPU temperature is too high, check that the heatsink or fan is normal!"

Fanless heat sink for broadcasting is complete, CPU occupied 23%, memory usage 39%,CPU temperature 45℃, the hard disk temperature 41℃, motherboard temperature of 46℃. Master Lu revealed a normal temperature, heat sink in good condition.

6 Summary

Based power Group pilot municipal units of multimedia information publishing system, for example, describes Group multimedia information publish system design architecture and building. According to production requirements of practice and for multimedia information issuing system of broadcasting and control technology the machine cooling system. Next group corporate communications and information network will be based on the future requirements of the transition from IPv4 to IPv6, and to optimize the multimedia information publishing system upgrades.

References

1. Kampichler, W., Goeschka, K.M.: Plain End-to-End Measurement for Local Area Network Voice Transmission Feasibility. In: Proceedings 9th International Symposium on Modeling, Analysis and Simulation of Computer and Telecommunication Systems, pp. 235–240 (2001).
2. Q. Zheng and K.G. Shin, On the ability of establishing real-time channels in point-to-point packet-switched networks, IEEE Trans. Commun. 42(2/3/4) (1994) 1096–1105.
3. Wang, Q., Abu-Rgheff, M.A.: Integrated Mobile IP and SIP Approach for Advanced Location Management. In: 4th International Conference 3G Mobile Communication Technologies, pp. 205–209 (2003).
4. Sklar, B.: Digital Communications Fundamentals and Applications. Prentice-Hall International Editions (1988)
5. Mete Feridun and Stelios Karagiannis (2009) Growth Effects of Information and Communication Technologies: Empirical Evidence from the Enlarged EU, Transformations in Business and Economics, 8(2), 86-99.
6. Loeb, S.: Delivering Interactive Multimedia Documents Over Networks, IEEE Commun. Magazine 30 (1992) 52–59
7. Little T.D.C. and Ghafoor A.: Multimedia Synchronization Protocols for Broadband Integrated Services, IEEE J. on Sel. Areas in Commun. 9 (1991) 1368–1382
8. Bertsekas, D. and Gallager, R.: Data Networks, Prentice Hall (1992) 510–513
9. Norwicki,B.: NFS: Network File System Protocol Specification, RFC 1094.

Comparison and Realization of Fast Acquisition Method for GPS Signal under High Dynamic

Chan Zeng [1], Weimin Li [1], Bo Bi [1] and Qiliang Chen [1]

[1] *Dept. Navigation Chip, Beijing Microelectronics Technology Institute, Beijing, China*

Abstract. In order to solve the problem of fast acquisition for GPS signal under high dynamic, the paper add window in traditional Partial Matched Filter (PMF) plus Fast Fourier Transform (FFT) for fast acquisition with the purpose of reducing the scalloping loss and improving the acquisition performance. And comparing the parallel code phase acquisition based on FFT with the new acquisition method based on the improved PMF+FFT. The high dynamic digital intermediate frequency signal is generated based on high dynamic trail. For the quicker acquisition speed, a method of fast blind search combining pseudo code correlation is used in acquisition. The results of simulation based on Matlab show that the new acquisition method can achieve the performance of 100g dynamics.

1 Introduction

The change of Doppler frequency is very large if the time of acquisition is long under high dynamic, so the traditional receiver have been unable to meet the demand of high-speed run-time of navigation and positioning. The receiver has a very large velocity or acceleration and jerk which leading to a sharply change of Doppler and difficult to capture the GPS signal because of the bad environment under high dynamic. The traditional capture algorithms [1] such as sliding correlating which is two-dimension searching for the Doppler frequency and code phase, the method has large calculations and long search time so that it can't adapt to the high dynamic environment, the signal is easy to lose lock even if acquired, not to mention stably tracking. The receivers studied are in missile, rocket and satellite, therefore achieving fast and stable capture is the primary task of the study of the high dynamic receiver.

In order to achieve the acquisition of high dynamic, the goal is to capture as much as possible in shortest time, incorporating the speed of acquisition [2] and the resource consumption of hardware, the paper improves the structure of the PMF combining FFT of the laboratory. The PMF can effectively reduce the subsequent points of FFT, and FFT algorithm can compensate the problem of correlation peak steeply attenuated caused by the increase of Doppler frequency shift. So the PMF combining FFT not only greatly reduces the consumption of the hardware resource, but also speeds up the capture.

Finally, the paper compares the common acquisition method of parallel code phase based on FFT [3] with the proposed the improved method based on PMF+FFT. The performance comparison show that the new fast acquisition method can make the rapid and accurate capture for GPS signal under high dynamic.

2 Digital intermediate frequency signal

2.1 Signal Model of GPS

There are orthogonal modulated with C/A code and P(Y) code on GPS L1 signal, it takes the form:

$$s_{L1}(t_j) = A\cdot C[t_j-\tau_1(t_j)]\cdot D[t_j-\tau_1(t_j)]\cdot \sin[2\pi f_{L1}t_j+\phi_1(t_j)]+ \frac{A}{\sqrt{2}}\cdot P[t_j-\tau_1(t_j)]\cdot W[t_j-\tau_1(t_j)]\cdot D[t_j-\tau_1(t_j)]\cdot \cos[2\pi f_{L1}t_j+\phi_1(t_j)]+n_1(t_j) \tag{1}$$

Where t_j is the time of sampling, A is the amplitude of L1 C/A carrier, C is C/A PRN code, D is the 50 Hz navigation data bit stream, τ_1 is the Pseudo Random Noise (PRN) code phase of L1, f_{L1} is the intermediate value of the L1 carrier frequency, ϕ_1 is the carrier phase of L1, P is the known P PRN code, W is the encrypted W PRN code, n_1 is zero-mean discrete time Gaussian white noise of L1.

2.2 Simulating the Doppler Frequency of Carrier and Code

The core problem of generating intermediate frequency is simulating the Doppler frequency shift of carrier and pseudo code under high dynamic. The Doppler frequency shift of satellite signal is caused by the relative motion between the connection of user and satellite receivers [4].

$$f_d = f_r - f = f \cdot v \cdot (\cos\beta) / c \qquad (2)$$

Where f_r is the reception frequency, f is the transmitting frequency of satellite, v is the velocity of the user receiver, β is the angle between the direction of the receiver and the incoming signal, c is the velocity of light, f_d is the Doppler frequency.

The Doppler Effect has significant impact on carrier and pseudo code, the changes of Doppler are corresponding with the dynamic model reception moment in high dynamic environment. Assuming that the receiver has the acceleration a and jerk a', the changes of Doppler can be expressed as following:

$$\begin{cases} f_d = f_{RF} \cdot v / c \\ f'_d = f_{RF} \cdot a / c \\ f''_d = f_{RF} \cdot a' / c \end{cases} \qquad (3)$$

Where f_{RF} is the RF frequency, f'_d is the change rate of the Doppler frequency, f''_d is the second order change rate of the Doppler frequency.

So the computation formula of f_{L1} in (1) is as following:

$$f_d(t) = f_{dinit} + 1/2 \cdot f'_d t + 1/6 \cdot f''_d t^2 \qquad (4)$$

$$f_{L1}(t) = f_{IF1} + f_d(t) \qquad (5)$$

Where f_{dinit} is the initial value of the Doppler frequency, f_{IF1} is the intermediate frequency of L1.

In this paper, it is only the delay Δd caused by the Doppler frequency shift to be considered, the delay is as follows:

$$\Delta d = \frac{v(t) \cdot t}{c} = \frac{t}{c} \cdot (v + at + 1/2a't^2) \qquad (6)$$

So the formula of the Doppler frequency shift for pseudo code is as follows.

$$f_{cd} = f_C \cdot v / c = \frac{f_d}{f_{RF}} \cdot f_C \qquad (7)$$

Where f_C is the nominal frequency of the pseudo code, f_{cd} is the code frequency deviation of pseudo code. For the C/A code of L1:

$$f_{cd} = f_d / 1540 \qquad (8)$$

It is obvious that the code frequency deviation of pseudo code should not be neglected under high dynamic.

The generation of pseudo code uses the code NCO to output the dynamic code rate. The code NCO is the phase accumulator composed by adder and register, which the working process is: first accumulation until overflow to zero, M is the phase control word, f_s is the driven clock frequency, the register's count value of NCO increase M in each clock cycle, the value overflow to zero when it is maximum, it is a cycle, so the frequency of the cycle signal is:

$$f_o = \frac{M \cdot f_s}{2^N} \qquad (9)$$

Where N is 32 to improve the frequency resolution, the formula (8) is the ode frequency deviation of pseudo code, so the frequency control word of PN code NCO:

$$M = \frac{2^N \cdot (f_C + f_{cd})}{f_s} \qquad (10)$$

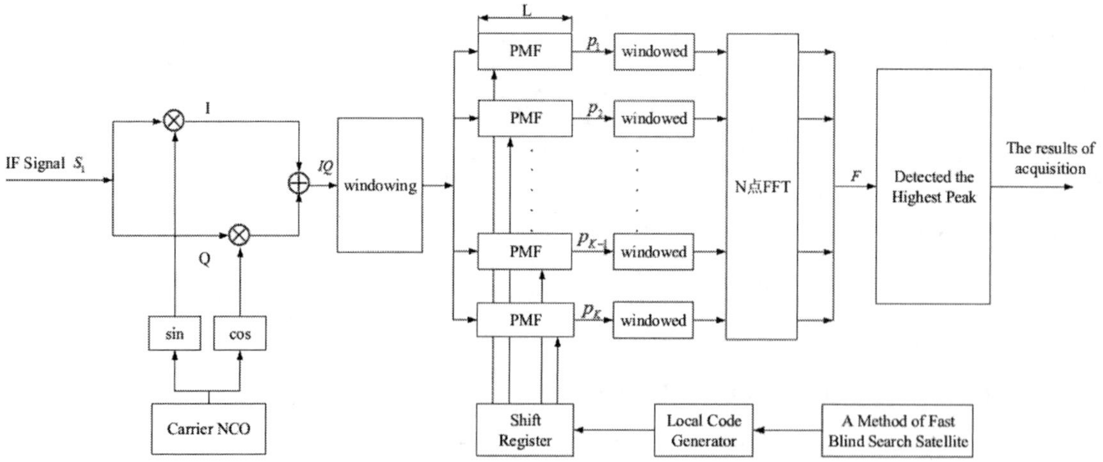

Figure 1. the structure of the improved PMF+FFT

3 Fast acquisition

3.1. The Improved PMF+FFT

The acquisition method of PMF+FFT has been adopted in the laboratory, the PMF means to divide the coherent integration time T which decided by the frequency resolution into K, it is done by the number of K PMF.As shown in Fig. 1 is the structure of the improved PMF+FFT, the intermediate frequency signal is demodulated by local carrier, and then by local pseudo-code, the range of Doppler frequency determine the N points of FFT. $M = T \cdot f_C$ is the total relevant calculations in the time of integration time T, so the length of each matched filter is $L = T \cdot f_C / K$.

The code phase shifter will move code phase values to continue the above operations until all the code phase done. The correlative value of all code phase and frequency bands will be obtained, then use the adaptive threshold method to decide the correlation peak. It is through the principle of opening window to continue to detect the second peak when detected the highest peak. The ratio of the highest and the second peak is compared with threshold, if the ratio is greater, this satellite is visible. But the local code's phase is close to the received, so the peak detection need to set a window near the highest peak, the highest peak is as the center.

Assuming that the local pseudo-code is synchronized with received pseudo-code, the output of the n-th PMF is as follows:

$$p(n) = \frac{1}{M} \sum_{k=(n-1)L+1}^{nL} e^{j(2\pi \frac{f_d}{f_c} k + \varphi)} = \frac{1}{M} \frac{\sin(w_d L)}{\sin(w_d)} \cdot e^{j[2w_d((n-1)L+1)+\varphi]} \quad (11)$$

Where $w_d = \pi \cdot f_d / f_C$. The result of the number of K PMF are operated by N points FFT. It is available that the output normalized amplitude response of the FFT.

$$|F(w_d,k)| = \left|\sum_{n=1}^{K} p(n) e^{-j\frac{2\pi}{N}kn}\right| = \left|\frac{\sin(w_d L)}{\sin(w_d)}\right| \cdot \left|\sum_{n=1}^{K} e^{j(2w_d L - 2\pi \frac{k}{N})n}\right| = \frac{1}{M}\left|\frac{\sin(w_d L)}{\sin(w_d)} \cdot \frac{\sin(w_d M - \pi \frac{K}{N} k)}{\sin(w_d L - \pi \frac{k}{N})}\right| \quad (12)$$

The formula $F(w_d,k)$ can be divided into two parts:

$$F_1(w_d) = \frac{\sin(w_d L)}{L \cdot \sin(w_d)} \quad (13)$$

$$F_2(w_d,k) = \frac{\sin(w_d M - \pi \frac{K}{N} k)}{K \cdot \sin(w_d L - \pi \frac{k}{N})} \quad (14)$$

Where $F_1(w_d)$ is the influence of the amplitude-frequency response to the whole by PMF; $F_2(w_d,k)$ shows the FFT arithmetic contribution to the overall amplitude-frequency response.

Fig. 2 shows the normalized amplitude frequency response of the traditional PMF+FFT [5], as known from the figure, the FFT has played an important role in extension on the frequency for the amplitude-frequency response of PMF. The range of the Doppler frequency is expanded twice.

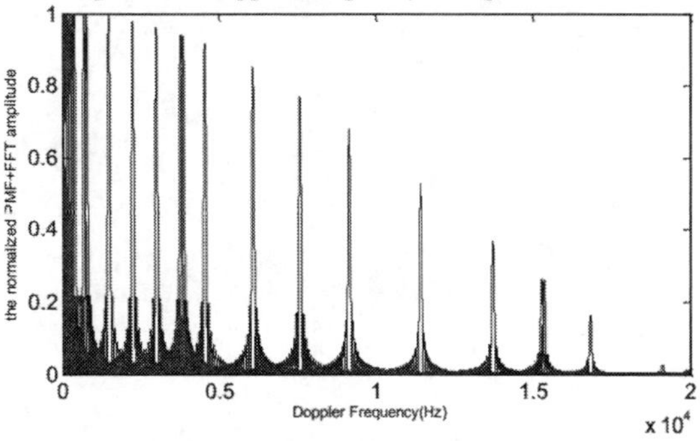

Figure 2. the normalized PMF+FFT amplitude frequency response

The shortcoming of PMF+FFT is the scallop loss [6] which makes the high false alarm rate and low performance of capture, so the improved is to reduce the scallop loss by using adding window. The window function can increase the main lobe width, thus to reduce the loss of scallops. For instance of hanning window:

$$w(n) = \frac{1}{2} - \frac{1}{2}\cos(\frac{2\pi n}{N}) \quad n = 0,1,2,\ldots N- \quad (15)$$

The amplitude frequency response function of hanning window can be expressed as follows:

$$W(w) = \frac{1}{2} W_R(w) + \frac{1}{4}[W_R(w - \frac{2\pi}{N}) + W_R(w + \frac{2\pi}{N})] \quad (16)$$

Where $W_R(w)$ is the amplitude characteristic of rectangular window which the length is N.

Adding window on PMF can improve the low-pass characteristic. The same as hanning window, the parameters of system are multiplied by two to avoid amplitude attenuation. So the normalized frequency response of the input of PMF after adding window is:

$$F_{1w}(k, f_d) = F_1(f_d) + \frac{1}{2}[F_1(f_d - \frac{1}{LT_c}) + F_1(f_d + \frac{1}{LT_c})] \quad (17)$$

For the module of FFT, the introduction of window function is to reduce the scallop loss, the normalized frequency response of the output of FFT after adding window is:

$$F_{2w}(k, f_d) = F_2(k, f_d) + \frac{1}{2}[F_2(k, f_d - \frac{1}{MT_c}) + F_2(k, f_d + \frac{1}{MT_c})] \quad (18)$$

The formula (18) shows that the time-windowing of signal can be realized by linear operation in frequency domain.

[a] Corresponding author: 13011083453@163.com

As shown in Fig. 3, it is the comparison the amplitude frequency response after adding window on PMF+FFT with no window. Considering the PMF and FFT after windowing, the simulation result shows that the improved algorithm not only reduce the scallop loss, but also improve the capture performance under high dynamic.

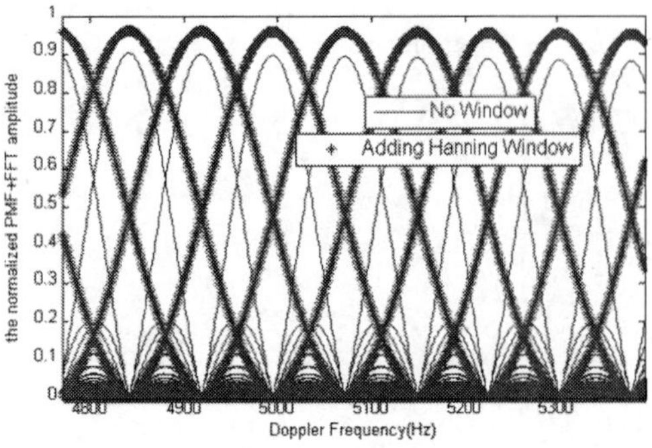

Figure 3. the normalized PMF+FFT amplitude frequency response after adding hanning window

3.2 The Method of Fast Blind Searching for Satellite

The receivers studied are in a state of blind search, there is no satellite ephemeris and almanac information to assist in search, the traditional search method is a waste of time that is searching 32 satellites in serial. Therefore, a method of fast blind search satellite combining pseudo code correlation is put forward. First, searching multiple satellites at the same time, if detected a correlation peak, then searching a single satellite from it, calculating the ratio between the highest peak and the second peak, and comparing this ratio with the threshold of acquisition to decide whether it is successful capture, until there are four satellites acquired. If searching four satellites at the same time, the search time will reduce 1/4 at least for 32 satellites.

$$C^{(com)}(n) = \sum_{i=1}^{k} C^{(i)}(n) \tag{19}$$

Where $C^{(i)}(n)$ is the pseudo code of satellite i, $C^{(com)}(n)$ is the combination pseudo code of k satellites.

The pseudo-code generator copy the combination code of multiple satellites, then are correlated with received signal, which is equivalent to search and detect for the k satellites. If there is without a visible, the k satellites can be considered non-existent. Then the receiver can be turned to another k satellites, if one of more correlation peak, it can be decided that one or more satellite from the k satellites are visible, thus capturing in the k satellites until four satellites acquired.

4 The parallel code phase acquisition

4.1 The Principle of Parallel Code Phase Acquisition

The essence of the parallel fast capture [7] method based on FFT is that the convolution algorithms of time domain is converted into multiply arithmetic in frequency domain by Fourier transformation. It is that the two sequences $x(n)$ and $y(n)$ correlation arithmetic in the time domain are equivalent to their Fourier $X(k)$ and $\overline{Y(k)}$ multiplied in frequency domain. The method can calculate the all correlation value of code phases in each frequency band, equivalent to parallel correlator to improve the efficiency of the capture.

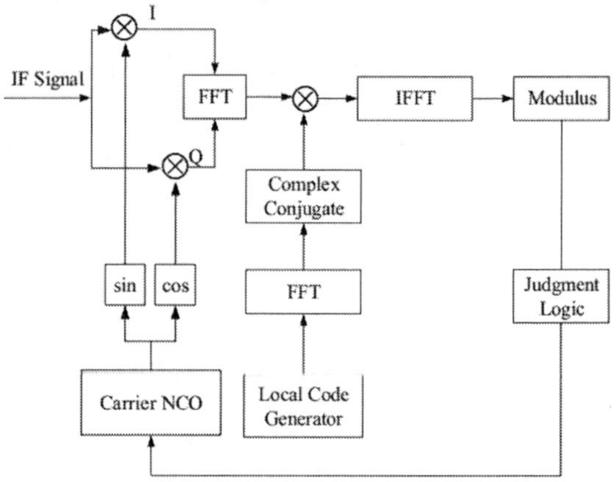

Figure 4. The parallel code phase search method based on FFT

As shown in Fig. 4, the digital intermediate frequency signal after mixing with local I and Q road carrier signal are combined into a plural form $I + jQ$ for FFT arithmetic. The transformation results are multiplied with the Fourier transform complex conjugate of local C/A pseudo-code, the product by IFFT operation can obtain the correlation result in time domain. Finally, detecting the peak of the correlation result after modulus, the peak compared with capture threshold method, if more than the capture threshold criterion captures the satellite signal. Then step the carrier frequency, continuing to the next band on the above process. The results by IFFT operation is all code phase correlation values on one frequency band after mixing with the local copied C/A pseudo code. In searching for the same satellite, its local copy C/A code of pseudo-code phase is constant;it will transform the code phase of the local copy C/A pseudo-code when searching different satellites.

4.2 The Comparison of Two Methods of Acquisition

Considering the acquisition in high dynamic environment, so the complexity time and space are used to evaluate the algorithm. The coherent integration time T_{coh}, the range of doppler frequency f_{band} is about -20 kHz~20 kHz,and the sampling clock is f_s.

(a) Parallel Code Phase Acquisition based on FFT

If the T_{coh}=1ms , the number of sampling points are $N_{sampl} = f_s / 1000$. A search for a frequency and need two times FFT of N points and one time IFFT operations, so the computation is:

$$N_{sampl} + 3N_{FFT} = \frac{f_s}{1000} + \frac{9N}{2}\log_2 N \qquad (20)$$

The total number of band are $N_{freqbin}$. If the search time of a band is the coherent integration time, the total time of searching all frequency bands is

$$T_{count} = N_{freqbin} \times T_{coh} \qquad (21)$$

(b)The improved PMF+FFT

Assume that the number of PMF and FFT points are same as N. The computation of searching for one code phase is:

$$N_{sampl2} + N_{FFT2} = \frac{T_{coh}}{N} \times f_s + \frac{3N}{2}\log_2 N \qquad (22)$$

For C/A the code chips are 1023, so if searched out all code phases the computation required is as following:

$$N_{count2} = 1023 \times (T_{coh} \times f_s + \frac{3N}{2}\log_2 N) \qquad (23)$$

As all known, the all code phase of PMF is operated by pipelining in the coherent integration time, the operation of FFT can get the correlation value under all code phase and all frequency bands, so the time of PMF+FFT to complete a searching field is the coherent integration time:

$$T_{count2} = T_{coh} \qquad (24)$$

Through the above analysis of the computational cost and the time required, the parallel code phase acquisition based on FFT has small hardware resource consumption, but the points of FFT is very large and the time of search is long due to the serial searching for frequency. To the contrary, the PMF in the improved PMF+FFT achieve the purpose of the down sampling to greatly reduce the points of FFT, and adding window expand the range of frequency, the pipelining of PMF also accelerate the speed of acquisition.

To sum up, the improved acquisition method of PMF+FFT is the better choice to realize fast acquisition under the high dynamic.

5 Experimental test results

Assume the dynamism of the GPS receiver is that the initial velocity is 400 m/s which corresponds to the initial Doppler frequency is 2100 Hz, the acceleration is 100g which corresponds to the rate of the Doppler frequency is 5146.372 Hz/s, g=9.8 m/s^2, the jerk is zero.

This paper simulates the improved algorithm PMF+FFT and the parallel code phase acquisition based FFT. Data about 10 ms was collected,theoretical value of the Doppler frequency is $2100 + 0.5 \cdot 5146.372 \cdot 0.01 = 2125.73Hz$.

Fig. 5 and Fig. 6 show the acquisition reslut of improved PMF+FFT which has the obvious peak, the code phase offset and doppler frequency shift can be obtained by threshold decision. The code phase offse is correct and the doppler frequency shift is about 2075 Hz, the error is within the frequency resolution about 76.85 Hz.

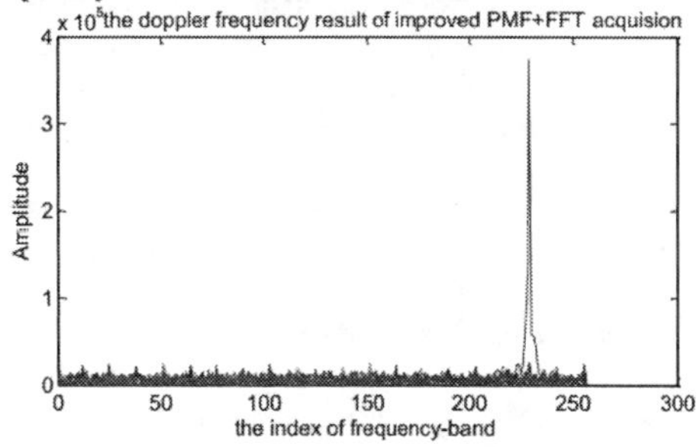

Figure 5. The Doppler frequency result of improved PMF+FFT acquisition

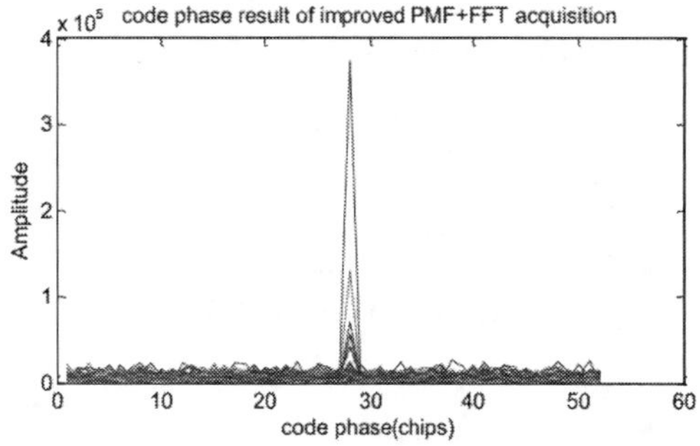

Figure 6. Figure of code phase result of improved PMF+FFT acquisition

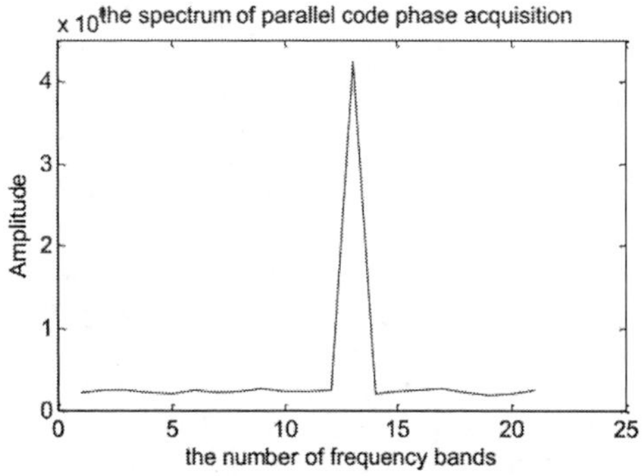

Figure 7. The Spectrum of parallel code phase acquisition

As shown in Fig. 7 ,it is the doppler spectrum figure after parallel code phase acquisition based on FFT. Set the the capture range of 20 KHz, and frequency step is 1000 Hz, eventually the doppler frequency is 2000 Hz, error about

[a] Corresponding author: 13011083453@163.com

126 Hz, it is visible that acquisition accuracy of the parallel code phase trapping method based on FFT is not high.

6 Summary and conclusions

For the problem that rapidly and accurately acquisition GPS signal under high dynamic, there is a new method of high dynamic acquisition which is the fast acquisition based on improved PMF+FFT. The system simulation is tested in Matlab simulation platform using the source signals generated by simulator of the high dynamic signal. The simulator of the high dynamic signal can set dynamic movement of arbitrary acceleration and jerk. The improved PMF+FFT show that the scallops loss after adding window decrease by half, so the method can short acquisition time and improve the performance. Finally, the simulation results show that the proposed method is more accurate than the parallel code phase acquisition based on FFT. And it lays a good foundation for subsequent high dynamic tracking. The future work can fine spectrum on the basis of the results of the PMF+FFT, it will further improve the precision of acquisition.

References

[1] Zhou San-wen, Huang Long, Lu Man-hong. Acquisition of High Dynamic Spread Spectrum Signals with FFT [J]. Journal of Spacecraft TT&C Technology, Vol24,P61-64 (2005).
[2] Hu Qiong. Research on the Acquisition Algorithm of High Dynamic and Weak GPS Signal [D].Shanghai: Shanghai Jiao Tong University (2013).
[3] Xie Gang. Principles of GPS and Receiver Design [M]. Beijing: Publishing House of Electronics Industry : P31 (2009).
[4] Bian C L,Ge H B,Chen R J.High-precision algorithm in PN code acquisition based on PMF-FFT[J]. Video engineering ,40(1):62-66 (2016).
[5] Nunes, F. D., Leitao J. M. N. A new fast code/frequency acquisition algorithm for GPS C/A signals[C]. Vehicular Technology Conference, 766-770 (2003).
[6] Zeng Qingxi,Tang Linlin,Wang Qing.Weak GPS L1 Signal Acquisition based on BPDC.[J].Transactions of Nanjing University of Aeronautics and Astronautics, Vol31,Jun, 3:331-337(2014).
[7] Li Yang, Xi XiaoLi. Combining FFT and Circular Convolution Method for High Dynamic GPS Signal Acquisition[J]. The Eighth International Conference on Electronic Measurement and Instruments, 159-162(2007).

 DOI: 10.1051/matecconf/201710815006

A numerical coupling model for a multibody system with multiple lubricated clearance joints

Zhao Bo [1], Shen Fei [2]

[1] School of Mechanical Engineering, Shanghai Jiao Tong University, 200240, Shanghai, China

[2] School of Mechanical and Aerospace Engineering, Nanyang Technological University, 639798, Singapore

Abstract. In machines and mechanisms, most of the clearance joints are designed to operate with lubricant fluid, which is an effective way of ensuring better performance of the mechanical systems. The differently located lubricated joints interact with each other, influencing the total performances of the mechanical system. In this study, a numerical coupling model for the dynamics and lubrication analyses of the multibody system with multiple lubricated clearance joints is established. This approach couples the dynamics model of the multibody system with the lubrication models of translational and revolute clearance joints. The dynamics model is established by the Lagrange's method, while The lubrication models are solved using the finite element method. The hydrodynamic forces built up by the lubricant fluid are evaluated from the motion states of the system and then input into dynamics model of the multibody system. Finally, the proposed approach is applied to a four-stroke gasoline engine with the lubricated piston skirt-liner subsystem and lubricated bearing at the big end of the connecting rod.

1 Introduction

Most of the joints in machines and mechanisms are designed to operate with lubricant fluid, which is an effective way of ensuring better performance of the mechanical systems [1-3]. The differently located joints with lubricated clearances depend on and work on each other, influencing the total performances of the system. Therefore, an improved model of the multibody system with the consideration of the interactions among the multiple joints is required to achieve a better understanding of the dynamics and tribology performances of the mechanism.

Internal combustion engines (ICEs) are important applications, and the lubricated joints in them play crucial roles in maintaining their stability of motion and reducing the friction power loss. The translational joints at the piston assembly and the revolute joints at the bearings interact each other and cooperate to affect the dynamics and tribological performances of engines.

Over the last several decades, considerable studies have been carried out on the lubrication of the piston skirt-liner system and the performances of the dynamically loaded bearing, respectively. To illustrate for the investigation on the piston skirt-liner system, Li et al. [4] modeled the dynamics of the piston by incorporating the hydrodynamic lubrication model into the dynamics equations of the piston. Zhu et al. [5, 6], Keribar and Dursunkaya [7], Wong et al. [8] and Liu et al. [9] established more perfect dynamics models of the piston based on Patir and Cheng's average Reynolds equation [10, 11].

By taking dynamically loaded bearings as the lubricated revolute clearance joints, many endeavors have also investigated the performances of bearings. Flores et al. [12] presented hydrodynamic lubrication models for lubricated revolute joints in constrained rigid planar multibody systems by considering the infinitely long and short journal bearing conditions. Daniel et al. [13] proposed a model to analyze the dynamics performance of a slider–crank mechanism with lubricated clearance joint. Machado et al. [14] studied the effect of the lubricated revolute joint parameters and hydrodynamic force models on the dynamic response of planar multibody systems. Zhao et al. [15, 16] established a dynamics model of the piston–connecting rod–crank system in a framework of the multibody mechanical system with lubricated joint.

The above investigations just focused on the mechanisms with single clearance joint. The investigation on the mechanisms with lubricated multiple joints, including the translational and revolution clearance joints, have been rarely reported.

The current study proposed a new numerical coupling method for the multibody mechanical system with lubricated translational and revolute clearance joints. This approach is established by coupling the dynamics model of the multibody system with the mixed lubrication models of multiple clearance joints. The Reynolds equation is solved based on the finite element method (FEM) due to its advantage of the flexibility in dealing with irregular domains [17]. In the end, the proposed approach is applied to a four-stroke gasoline engine with lubricated piston skirt-liner system and lubricated revolute clearance joint at the big end of the connecting rod.

2 The lubrication models of lubricated clearance joints in ices

In this section, the piston skirt-liner systems in ICEs can be considered as lubricated translational joints with clearance, while the bearings at the big end of the connecting rod can be considered as lubricated revolute clearance joints.

2.1 Hydrodynamic lubrication model of the piston skirt-liner system

In order to take the influences of the roughnesses of mating surfaces into account, the oil film pressure between the skirt and liner will be evaluated based on the average Reynolds equation [10, 11] as

$$\frac{\partial}{\partial x}\left(\phi_x \frac{h^3}{12\mu}\frac{\partial p}{\partial x}\right)+\frac{\partial}{\partial y}\left(\phi_y \frac{h^3}{12\mu}\frac{\partial p}{\partial y}\right)= -\frac{u_p}{2}\left[\phi_c\frac{\partial h}{\partial y}+\sigma_s\frac{\partial \phi_s}{\partial y}\right]+\phi_c\frac{\partial h}{\partial t}, \quad (1)$$

where μ is the dynamic viscosity of the oil, and u_p is the relative velocity between the piston and liner. The variables h and p are the oil film thickness and pressure. The local coordinate axis x and y are respectively along the circumferential direction and axial direction of the piston skirt as shown in Fig. 1. The pressure flow factors ϕ_x and ϕ_y, as well as the shear flow factor ϕ_s and contact factor ϕ_c, are defined in the terms of surface roughness parameters [10, 11, 18]. σ_s is the variance of the composite surface roughness.

The oil film thickness distribution between piston skirt and liner can be expressed as [19-21]

$$h(\alpha,y)=c_p+h_{\text{skirt}}(\alpha,y)+[e_0+(y-a_p)\gamma]\cos\alpha+d(\alpha,y,t), \quad (2)$$

where e_0 and γ denote the lateral displacement and tilt angle of the piston, respectively.The function $h_{\text{skirt}}(\alpha,y)$ is used to define the piston skirt profile (Fig. 1 (d)), and α is the angular coordinate of the piston skirt from the thrust side (Fig. 1 (c)).

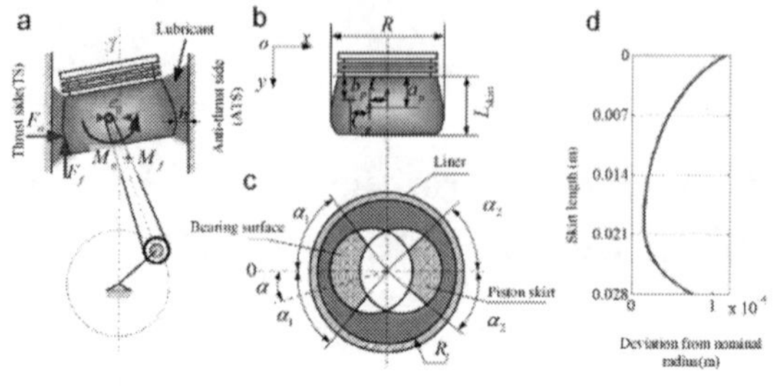

Figure 1. Piston skirt-liner system: (a) the variables of the system; (b) the geometry parameters of the piston; (c) skirt-liner; and (d) the piston skirt profile.

According to Eq. (2), the squeeze part of the Reynolds equation can be defined as

$$\frac{\partial h}{\partial t}=\left[\frac{de_0}{dt}+(y-a_p)\frac{d\gamma}{dt}\right]\cos\alpha, \quad (3)$$

The dimensionless process for Eq. (1) can be written as

$$\begin{aligned}&\bar{x}=\alpha=x/R,\quad \bar{y}=y/R,\quad \bar{h}=h/c,\\&\bar{\sigma}_s=\sigma_s/c,\quad \bar{u}=u/L_1\omega_c,\bar{a}_p=a_p/R,\\&\bar{e}_0=e_0/c\omega_c,\quad \bar{\gamma}=\gamma R/c\omega_c,\quad \bar{p}=pc^2/\mu\omega_c L_1 R\end{aligned}, \quad (4)$$

where R is the radius of the piston and L_1 is the length of the crank. In this study, the angular velocity of the crank is chosen as the reference angular velocity ω_c. The dimensionless Reynolds equation can be written as

$$\frac{\partial}{\partial\bar{x}}\left(\phi_x\bar{h}^3\frac{\partial\bar{p}}{\partial\bar{x}}\right)+\frac{\partial}{\partial\bar{y}}\left(\phi_y\bar{h}^3\frac{\partial\bar{p}}{\partial\bar{y}}\right)=-6\bar{u}\left[\phi_c\frac{\partial\bar{h}}{\partial\bar{y}}+\bar{\sigma}_s\frac{\partial\phi_s}{\partial\bar{y}}\right]+\frac{12R}{L_1}\phi_c\left[\frac{d\bar{e}_0}{dt}+(\bar{y}-\bar{a}_p)\frac{d\bar{\gamma}}{dt}\right]\cos\alpha \quad (5)$$

Considering that the distribution of oil film pressure is symmetrical about the xoy plane, so only the zone from $\alpha=0$ to $\alpha=\pi$ is solved. The boundary conditions of the piston skirt lubrication model can be expressed as [21, 22]

$$\begin{cases}p(x,y)=0, & y=0 \text{ or } L_{\text{skirt}}\\ p(x,y)=0, & \alpha_1\le\alpha\le\alpha_2\\ \partial p/\partial x=0, & \alpha=0 \text{ or } \pi\end{cases} \quad (6)$$

where L_{skirt} is the length of the piston skirt.

2.2 Hydrodynamic lubrication model of the dynamically loaded revolute lubricated joint

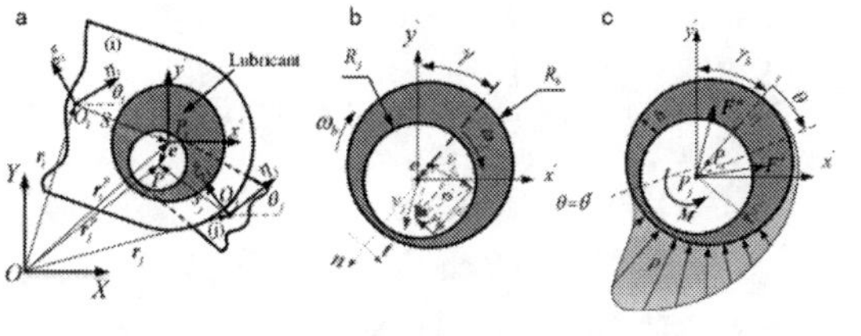

Figure 2. Lubricated dynamically loaded joint: (a) Genetic configuration the joint; (b) relative motion of the joint; and (c)oil film forces

Figure 2 (a) depicts a cross section of a lubricated revolute clearance joint. The space between the bearing and journal is filled with lubricant. Two bodies i and j are connected by a lubricated revolute joint where the clearance is filled with lubricate. The centers of mass are denoted by O_i and O_j for bodies i and j, respectively. The centers of bearing and journal are represented by P_i and P_j, respectively. Body-fixed coordinate systems $\eta o \xi$ are attached at their center of mass, while the global coordinate system of the multibody system is represented by XOY.

The eccentricity vector connecting the centers of journal and bearing can be calculated as

$$\boldsymbol{e} = \boldsymbol{r}_j^P - \boldsymbol{r}_i^P , \tag{7}$$

where $\boldsymbol{r}_i^P$ and $\boldsymbol{r}_j^P$ are respectively the position vector of P_i and P_j in the global coordinate frame XOY.
The magnitude of the eccentricity vector can be obtained by

$$e = \sqrt{\boldsymbol{e}^T \boldsymbol{e}} . \tag{8}$$

The unit eccentricity vector along the eccentricity direction can be expressed as

$$\boldsymbol{n} = \frac{\boldsymbol{e}}{e} . \tag{9}$$

The eccentricity ratio can be defined as

$$\varepsilon = \frac{e}{c_b} , \tag{10}$$

The time rate of eccentricity ratio variation $\dot{\varepsilon}$ can be expressed as

$$\dot{\varepsilon} = \frac{\dot{\boldsymbol{e}}\boldsymbol{n}}{c_b} \tag{11}$$

where $\dot{\boldsymbol{e}}$ is the velocity of P_j relative to P_i and can be obtained by differentiating Eq. (7) with respect to time. The average Reynolds equation is also adopted to evaluate the hydrodynamics oil forces at the joint and can be expressed as [15]

$$\frac{\partial}{\partial x}\left(\phi_x \frac{h^3}{12\mu}\frac{\partial p}{\partial x}\right) + \frac{\partial}{\partial y}\left(\phi_y \frac{h^3}{12\mu}\frac{\partial p}{\partial y}\right) = \frac{u_b}{2}\left[\phi_c \frac{\partial h}{\partial y} + \sigma_b \frac{\partial \phi_s}{\partial y}\right] + \phi_c \frac{\partial h}{\partial t} ,$$

(12)
where $u_b = \omega_r R_b$ is the relative velocity between journal and bearing, and ω_r is relative angular velocity. The parameter R_b is the radius of the bearing. The variance of the composite surface roughness σ_b is the standard deviation of the roughness height.
The oil film thickness distribution can be expressed as

$$h(\varphi, t) = c_b\left(1 + \varepsilon \cos\theta\right) , \tag{13}$$

where θ is angular coordinate in $\xi' P_j \eta'$ which is a local coordinate system used to evaluate hydrodynamic forces.

The derivative of oil film thickness with respect to time can be defined as

$$\frac{\partial h}{\partial t} = c_b\left(\cos\theta \dot{\varepsilon} + \varepsilon \sin\theta \dot{\gamma}_b\right) , \tag{14}$$

where, the parameter $\dot{\gamma}_b$ can be expressed as

$$\dot{\gamma}_b = \frac{\dot{\boldsymbol{e}}\boldsymbol{t}}{c_b} , \tag{15}$$

where $\boldsymbol{t}$ is the unit tangential vector defined by rotating the vector $\boldsymbol{n}$ by an angle of 90° in the counterclockwise direction as shown in Fig. 2 (b).
The dimensionless process for Eq. (12) can be written as

$$\bar{x} = \theta = \frac{x}{R_b}, \bar{h} = \frac{h}{c_b} = 1 + \varepsilon\cos\theta, \bar{y} = \frac{y}{b}, \quad \bar{\sigma}_b = \sigma_b / c_b, \bar{\omega} = \frac{\omega}{\omega_c}, \bar{p} = \frac{p c_b^{\,2}}{6 R_b^{\,2} \mu \omega_c} , \tag{16}$$

where b is the width of the bearing.

The dimensionless Reynolds equation for Eq. (12) can be written as

$$\frac{\partial}{\partial\theta}\left(\phi_x \bar{h}^3 \frac{\partial \bar{p}}{\partial\theta}\right) + \left(\frac{R_b}{b}\right)^2 \frac{\partial}{\partial\bar{y}}\left(\phi_y \bar{h}^3 \frac{\partial\bar{p}}{\partial\bar{y}}\right) = \bar{\omega}\left[-\phi_c \varepsilon \sin\theta + \bar{\sigma}_b \frac{\partial\phi_s}{\partial\theta}\right] + \frac{2\phi_c}{\omega_c}\left(\cos\theta\dot{\varepsilon} + \varepsilon\sin\theta\dot{\gamma}_b\right) . \tag{17}$$

It is assumed that the pressure gradient is equal to zero in the middle section due to the symmetry, and the pressure at both edges will decrease to zero. Thus, the boundary condition in the axial direction can be expressed as [15, 23]

$$\begin{cases} p=0 \quad , \quad y=\pm\dfrac{b}{2} \\ \dfrac{\partial p}{\partial y}=0 \;, \quad y=0 \end{cases} . \tag{18}$$

In the circumferential direction, the Reynolds boundary condition is adopted due to its accuracy and can be written as [15, 23]

$$\begin{cases} p=0 \quad\quad\quad , \quad \theta=0 \\ p=0, \dfrac{\partial p}{\partial \theta}=0 \;, \quad \theta=\theta^{*} \end{cases}, \tag{19}$$

where θ^{*} is a floating boundary [23].

2.3 Finite Element Method for solving the lubrication models

Equations (5) and (17) are nonhomogeneous partial differential equations of elliptical type. Such equations will be solved currently by way of using FEM due to its flexibility in solving the elliptic equation with irregular domains [17]. Eqs. (5) and (17) can be rewritten with the same form as

$$\frac{\partial}{\partial \bar{x}}\left(\xi\frac{\partial \bar{p}}{\partial \bar{x}}\right)+\alpha\frac{\partial}{\partial \bar{y}}\left(\zeta\frac{\partial \bar{p}}{\partial \bar{y}}\right)-\psi=0, \tag{20}$$

where for the piston skirt-liner system,

$$\begin{aligned} &\xi=\phi_x\bar{h}^3, \quad \zeta=\phi_y\bar{h}^3, \alpha=1 \\ &\psi=-6\bar{u}\left[\phi_c\frac{\partial \bar{h}}{\partial \bar{y}}+\bar{\sigma}\frac{\partial \phi_s}{\partial \bar{y}}\right]+ \\ &\frac{12R}{L_1}\phi_c\left[\frac{d\bar{e}_0}{dt}+(\bar{y}-\bar{a}_p)\frac{d\bar{\gamma}}{dt}\right]\cos\alpha \end{aligned} . \tag{21}$$

while for the lubricated revolute joint,

$$\begin{aligned} &\xi=\phi_x\bar{h}^3, \quad \zeta=\phi_y\bar{h}^3, \alpha=\left(\frac{R_b}{b}\right)^2 \\ &\psi=\bar{\omega}\left[-\phi_c\varepsilon\sin\theta+\bar{\sigma}\frac{\partial \phi_s}{\partial \theta}\right]+\frac{2\phi_c}{\omega_c}(\cos\theta\dot{\varepsilon}+\varepsilon\sin\theta\dot{\gamma}_b) \end{aligned} . \tag{22}$$

According to the Galerkin theory, the integration over their respective whole lubrication domain Ω shown in Fig. 3 can be built as

$$\iint_{\Omega}\boldsymbol{N}\left(\frac{\partial}{\partial \bar{x}}\left(\xi\frac{\partial \bar{p}}{\partial \bar{x}}\right)+\alpha\frac{\partial}{\partial \bar{y}}\left(\zeta\frac{\partial \bar{p}}{\partial \bar{y}}\right)-\psi\right)\mathrm{d}\bar{x}\mathrm{d}\bar{y}=0 , \tag{23}$$

where $\boldsymbol{N}$ is the shape function of discrete elements.

After the integration using the Green's theorem, Eq. (23) can be expressed as

$$\int_{\Gamma}\begin{matrix} \boldsymbol{N}\xi\dfrac{\partial \bar{p}}{\partial \bar{x}}\mathrm{d}\bar{y}\text{-}\alpha\displaystyle\int_{\Gamma}\boldsymbol{N}\zeta\dfrac{\partial \bar{p}}{\partial \bar{y}}\mathrm{d}\bar{x}- \\ \displaystyle\iint_{\Omega}\left(\xi\frac{\partial \boldsymbol{N}}{\partial \bar{x}}\frac{\partial \bar{p}}{\partial \bar{x}}+\alpha\zeta\frac{\partial \boldsymbol{N}}{\partial \bar{y}}\frac{\partial \bar{p}}{\partial \bar{y}}+\boldsymbol{N}\psi\right)\mathrm{d}\bar{x}\mathrm{d}\bar{y} \end{matrix}=0, \tag{24}$$

where Γ is the boundary of each integral domain. In order to solve the above equation, the domain is discretized into a number of triangular elements (Fig. 3). The shape function of the triangular element can be expressed as

$$\boldsymbol{N}=[N_1, N_2, N_3]^T , \tag{25}$$

where

$$\begin{aligned} &N_i=a_i+b_i\bar{x}+c_i\bar{y}, \quad i=1,2,3 \\ &a_i=\frac{1}{2\Delta}\left(\bar{x}_j\bar{y}_k-\bar{x}_k\bar{y}_j\right), \\ &b_i=\frac{1}{2\Delta}\left(\bar{y}_j-\bar{y}_k\right), \\ &c_i=\frac{1}{2\Delta}\left(\bar{x}_k-\bar{x}_j\right) \\ &\Delta=\frac{\left(\bar{x}_j-\bar{x}_i\right)\left(\bar{y}_k-\bar{y}_i\right)-\left(\bar{x}_k-\bar{x}_i\right)\left(\bar{y}_j-\bar{y}_i\right)}{2} \end{aligned} , \tag{26}$$

For the inner elements Ω_i^e of each domain, only the third item in Eq. (24) should be considered and can be expressed as

$$\begin{cases} -\displaystyle\iint_{\Omega_i^e}\left(\xi\frac{\partial \boldsymbol{N}}{\partial \bar{x}}\frac{\partial \bar{p}}{\partial \bar{x}}+\zeta\frac{\partial \boldsymbol{N}}{\partial \bar{y}}\frac{\partial \bar{p}}{\partial \bar{y}}\right)\mathrm{d}\bar{x}\mathrm{d}\bar{y}=K_i^e\cdot\bar{\boldsymbol{p}}_i^e \\ \displaystyle\iint_{\Omega_i^e}\boldsymbol{N}\psi\mathrm{d}\bar{x}\mathrm{d}\bar{y}=\boldsymbol{F}_i^e \end{cases}, \tag{27}$$

where K_i^e and $\bar{\boldsymbol{p}}_i^e=[\bar{p}_1^i, \bar{p}_2^i, \bar{p}_3^i]^T$ are element stiffness matrix and dimensionless nodal pressure vector of the element i, respectively. Force vector $\boldsymbol{F}_i^e$ can be calculated using numerical integration algorithm.

For the boundary elements with the second kind boundary conditions stated in Eqs. (6), (18) and (19), it can be easily proved that the integrations on the boundaries are equal to zero. While the first kind boundary conditions can be considered by setting the oil film pressure on the corresponding node to zero.

DOI: 10.1051/matecconf/201710815006

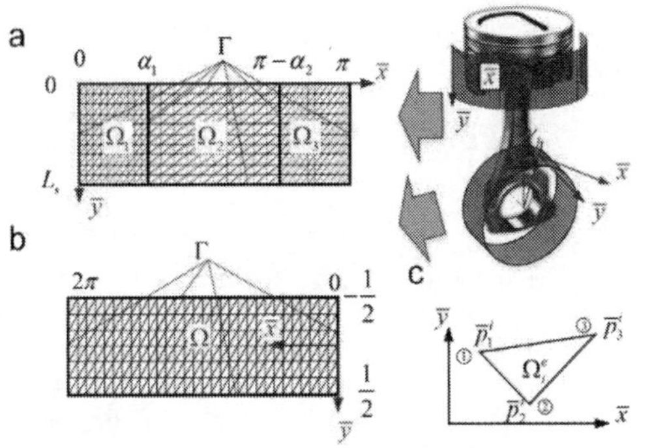

Figure 3. The meshing diagram: (a) the meshing for the skirt domain; (b) the meshing for the bearing domain; (c) the diagram of an element.

Based on the above statement and by assembling the corresponding items of the element in Eq. (27), Eqs. (5) and (17) can be transformed into:

$$K\bar{p} = \boldsymbol{F} \quad . \tag{28}$$

The hydrodynamic oil film pressure distribution on the both skirt and bearing can be obtained by solving this equation with the consideration of the boundary conditions.

2.4 Asperity contact Forces

When the oil film thickness ratio h/σ between two mating surfaces is below 4.0, the effect of solid-to-solid contact in mixed lubrication could be considered [24, 25]. The asperity contact pressure $p_c(x,y)$ is given as

$$p_c(h) = KE^{'}F_{5/2}\left(\frac{h}{\sigma}\right), \tag{29}$$

where $E^{'}$ is the composite elastic modulus. The dimensionless parameter K is assumed to be a constant with the value of 1.198×10^{-4} [25, 26]. The switch function $F_{5/2}(H)$ can be written as [24]

$$F_{5/2}(H) = \begin{cases} 4.486\times10^{-5}(4-H)^{6.804} & H<4 \\ 0 & H\geq 4 \end{cases} . \tag{30}$$

The friction forces between two mating surfaces result from the lubricant shear and asperity contacts in mixed lubrication. The contact friction can be simply calculated with a boundary friction coefficient. While the average shear stress can be expressed as [10, 11]

$$\tau = -\frac{\mu u}{h}\left(\Phi_f + \Phi_{fs}\right) - \Phi_{fp}\frac{h}{2}\frac{\partial p}{\partial y}, \tag{31}$$

where Φ_f, Φ_{fs} and Φ_{fp} are shear stress factors [10, 11].

3 Dynamic analysis of the piston-connecting rod-crank system

Figure 4 depicts a piston-connecting rod-crank multibody system with the lubricated revolute and translational joints at the piston and big end bearing of the connecting rod, respectively. All the bodies are assumed to be rigid, and other joints are ideal without considering the existence of the friction and clearance. Subsequently, the dynamics equations that govern the of system is first formulated.

The generalized Cartesian coordinates of the system can be written as

$$\begin{aligned} \boldsymbol{q} &= \left[\boldsymbol{q}_1^{\mathrm{T}} \quad \boldsymbol{q}_2^{\mathrm{T}} \quad \boldsymbol{q}_3^{\mathrm{T}}\right]^{\mathrm{T}}, \\ \boldsymbol{q}_i &= \left(x_i \quad y_i \quad \theta_i\right)^{\mathrm{T}}, (i=1,3) \end{aligned}, \tag{32}$$

where $(x_i, y_i)^{\mathrm{T}}$ denotes the position vector of the reference point of the body i, and θ_i is the angular coordinate in the global coordinate system XOY.

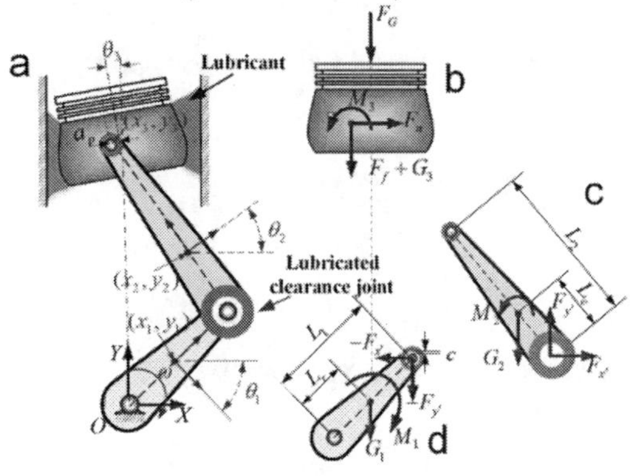

Figure 4. Crank-connecting rod-piston system: (a) the geometry and generalized coordinates of the system; (b) external forces of the piston; (c) external forces of the rod; (d) the external forces of the crank.

The ideal joints in the system provide kinematic constraints to the connected components, while the clearance joints provide the dynamics constraints instead of the kinematic constraints [27]. The kinematic constraints $\boldsymbol{\Phi}$ in this system due to the mechanical joints and specified motion trajectories can be written in a compact form as

$$\boldsymbol{\Phi}(\boldsymbol{q},t) = \begin{pmatrix} x_1 - L_r\sin\theta_1 \\ y_1 - L_r\cos\theta_1 \\ \theta_1 - \omega t \\ x_2 - x_3 - (y_3 - y_2)\tan\theta_2 \end{pmatrix} = 0, \tag{33}$$

where L_r determines the location of mass center on crank.

Differentiating Eq. (33) twice with respect to time yields the acceleration constraint equations:

$$\ddot{\Phi} = \left(\Phi_q \dot{q}\right)_q \dot{q} + \Phi_q \ddot{q} + 2\Phi_{qt}\dot{q} + \Phi_{tt} = 0, \quad (34)$$

where Φ_q is the Jacobian matrix of the constraint equations. $\dot{q}$ and $\ddot{q}$ are vectors of the system velocity and acceleration, respectively.

The dynamics equations for the rigid bodies in a constrained multibody system can be written as [27, 28]

$$\mathbf{M}\ddot{q} = Q - \Phi_q^T \lambda, \quad (35)$$

where the vector λ denotes the Lagrange multipliers. The term $-\Phi_q^T \lambda$ can be used to express the joint reaction forces [29]. The system mass matrix $\mathbf{M}$ is a diagonal matrix with diagonal elements as $[m_1, m_1, J_1, m_2, m_2, J_2, m_3, m_3, J_3]$. Here, m_i and J_i are the mass and inertia moment about the point O_i on the body i, respectively. The generalized external force vector Q can be expressed as

$$Q = \begin{bmatrix} -F_{x'}, -G_1 - F_{y'}, M_1, \\ F_{x'}, F_{y'} - G_2, M_2, F_n, F_f + F_c + G_3, M_3 \end{bmatrix}^{\mathrm{T}}, \quad (36)$$

where G_1, G_2 and G_3 are the gravitations of the crank, connecting rod and piston, respectively. The combustion gas force F_c acts on the top of the piston. The moments M_1 and M_2 are developed at the bearing and acts on the mass center of the connecting rod and crank, respectively. While M_3 is the total moment about the piston pin and equal to the sum of M_n and M_f (Fig. 1).

In multibody mechanical systems, a lubricated clearance joint does not produce any kinematic constraint. Instead, it acts as a force element producing time-dependent forces. Thus, the lubricated joint produces, the so-called, force constraints. The hydrodynamic forces built up by the lubricant fluid are evaluated from the knowledge of the system variables as stated in Section 2 and then introduced as external generalized forces into the dynamics equations of the mechanical system.

Combining Eqs. (34) and (35), the dynamics equations of a system subjected to constraints can be stated in the form [1, 27-31]:

$$\begin{bmatrix} \mathbf{M} & \Phi_q^T \\ \Phi_q & \mathbf{0} \end{bmatrix} \begin{bmatrix} \ddot{q} \\ \lambda \end{bmatrix} = \begin{bmatrix} Q \\ \gamma \end{bmatrix}, \quad (37)$$

where γ can be calculated from Eq. (34) and expressed as

$$\gamma = \Phi_q \ddot{q} = -\left(\Phi_q \dot{q}\right)_q \dot{q} - 2\Phi_{qt}\dot{q} - \Phi_{tt}. \quad (38)$$

According to Eq. (37), $\ddot{q}$ and λ can be solved with the proper initial conditions. However, the equation does not use the explicit position and velocity equations associated with the kinematic constraints, which implies that during simulation, the original constraint equations will be violated [1, 14, 30, 32]. To control the constraints violation during numerical integration, the Baumgarte stabilization technique is adopted to damp out the acceleration constraint violations by feeding back the violations of positions and velocity constraints [1, 14, 30, 32-34]. Eq. (37) can be modified as

$$\begin{bmatrix} \mathbf{M} & \Phi_q^T \\ \Phi_q & \mathbf{0} \end{bmatrix} \begin{bmatrix} \ddot{q} \\ \lambda \end{bmatrix} = \begin{bmatrix} Q \\ \gamma - 2\vartheta\left(\Phi_q \dot{q} + \Phi_t\right) - \varsigma^2 \Phi \end{bmatrix}, \quad (39)$$

where ϑ and ς are feedback parameters and should be adequately chosen [34, 35].

4 Results and discussion

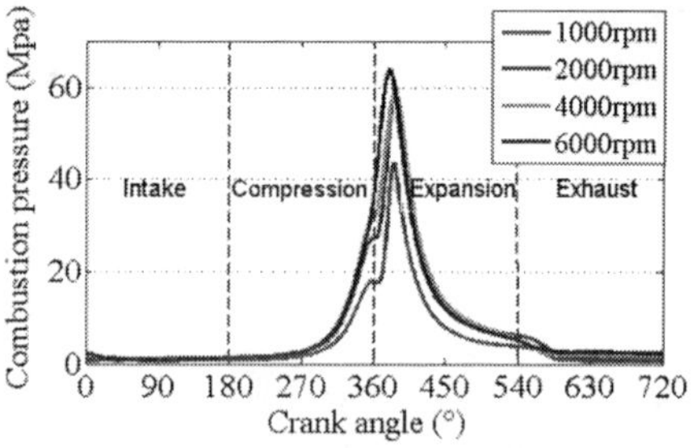

Figure 5. Cylinder pressure acting on the piston crown.

The proposed model is performed on a four-stroke gasoline engine in this section. Some necessary input data for the simulation are listed in Table 1. Fig. 5 shows the combustion gas pressures acting on the piston crown respectively at the speeds of 1000, 2000, 4000 and 6000 rpm, which were obtained from experiments. The simulation parameters for this system are shown in Table 2. Here,the parameter N_y^s demotes the number of the elements along the axial direction of the piston skirt, while N_{x1}^s, N_{x2}^s and N_{x3}^s are the number of elements along the circumferential direction in lubrication domains Ω_1, Ω_2 and Ω_3 (Fig. 3), respectively. N_x^b and N_y^b are respectively the number of elements along the circumferential and axial direction in lubrication domain

Ω of the bearing. In the following calculations, the solving efficiency for hydrodynamic lubrication with the above mesh density is improved by about 65% compared with the case with the doubled element numbers, but the results remain about the same. Therefore, it can be proved that such mesh density is convergent and can guarantee a good accuracy.

TABLE I. Basic input data of the four-stroke gasoline engine.

Parameter	**value**	**Parameter**	**value**
a_p	11.8mm	J_3	1.29E-4 kgm^2
b	22mm	R	36.475mm
b_p	-1.737mm	α_1	46°
c_b	0.1mm	α_2	46°
c_g	0.13mm	L_{skirt}	28mm
c_p	0.025mm	R_l	36.50 mm
C_p	0.7mm	R_b	15.00 mm
L_1	39.93mm	R_j	14.95 mm
L_2	129.5mm	μ	0.02Pas
L_r	20.00mm	μ_f	0.13
L_c	28.49mm	σ_{skirt}	0.615E-6m
m_2	0.345kg	σ_{liner}	0.815E-6m
m_3	0.264kg	$\sigma_{journal}$	0.308E-6m
J_2	9.67E-4kg m^2	$\sigma_{bearing}$	0.408E-6m

TABLE II. Simulation parameters for the system.

Baumgarte coefficients ϑ, ς	10000
Integration step size	1.0E-007s
$N_{x1}^s, N_{x2}^s, N_{x3}^s, N_y^s$	30,2,30,40
N_x^b, N_y^b	50,10
Convergence criterion	1%

The case with the constant crank velocity of 6000 rpm is first simulated. Several numerical results are presented and analyzed to demonstrate the computational implementation of the proposed methodology, as well as to reveal the dynamics and lubrication performances of the system. In order to reveal the interactions between the lubricated piston skirt-liner system and lubricated bearing at connecting rod big end (LBRBE), several other previously proposed models are also introduced: the first model is named as "Only-Skirt model" in which only the piston skirt-liner joint is taken into account and other joints are considered to be ideals without clearances and friction as proposed in study [21]. The second model just considers the LBRBE in the multibody system as stated in [15, 16] and is called as "Only-LBRBE model". The current coupling model considers both the lubricated piston skirt-liner system and LBRBE and is named as "Skirt-LBRBE model" for short.

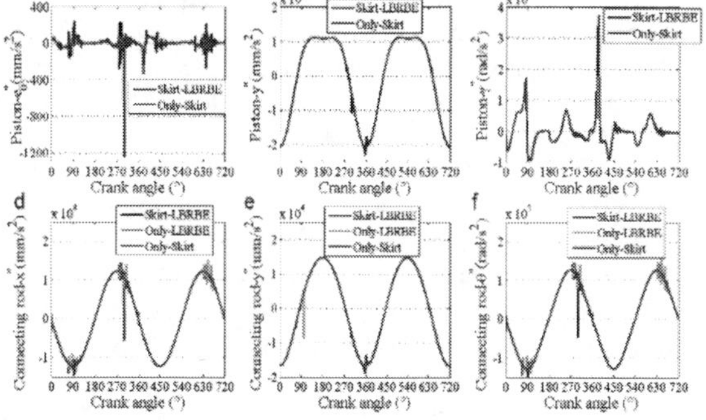

Figure 6. The kinematics results of the piston and connecting rod.

Figure 6 shows the kinematics results of the piston and connecting rod. All the results presented are relative to two full crank rotations after steady-state has been reached. It can be found that the translation accelerations of the piston (i.e., $\ddot{e}_0$ and $\ddot{y}$) obtained from the Skirt-LBRBE model have more obvious oscillations than those from the Only-skirt model, especially in the horizontal direction. The reason may be that the LBRBE has influences on the oil film force acting at the piston skirt, or the reaction forces between the piston and connecting rod.

In order to better explain the influence of the LBRBE, Figs. 7 (a)-(c) shows the hydrodynamic oil film forces acting on the piston skirt. It is obvious that the plots of oil film forces from both Skirt-LBRBE and Only-Skirt models, including total normal thrust force, friction force as well as their moments relative to the center of the piston pin, overlap very well. This means that the LBRBE almost has no influence on the lubrication performance of the piston skirt-liner system.

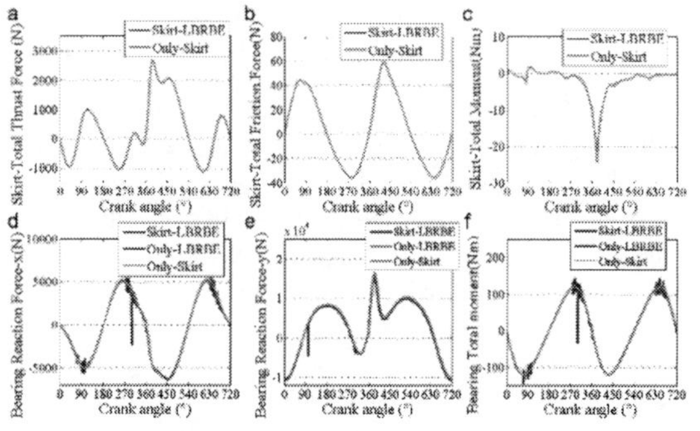

Figure 7. The oil film forces acting on the piston skirt and LBRBE.

Figures 7 (d)-(f) plot the reaction forces at the revolute joint of the big end of the connecting rod. The results are calculated based on three different models: the Only-Skirt model, the Only-LBRBE model, and the Skirt-LBRBE model. By way of comparison, it can be found that compared with the Only-Skirt model, the other models with the LBRBE will introduce the extra degrees of freedom for the connecting rod, which will result in the oscillation of the reaction forces at the revolute joint, and thus affect the accelerations of the connecting rod (Figs. 6 (d)-(f)), as well as the reaction forces between the rod and piston.

By comparing the Skirt-LBRBE model with the Only-LBRBE model, it can be found that the absence of the clearance at piston skirt-liner will decrease the accelerations oscillation of connecting rod and thus the reaction forces at bearing (Figs. 6 and 7). In addition, from the comparison between the Skirt-LBRBE model with the Only-Skirt model, it can be found that due to the fact that rod and piston are connected by the piston pin, although the LBRBE will result in the oscillation of the connecting rod acceleration and reaction forces at the joint of the big end of the connecting rod, no extra moment about the piston pin will be introduced into the piston. This can also be observed in Figs. 6 (c) and 7 (c).

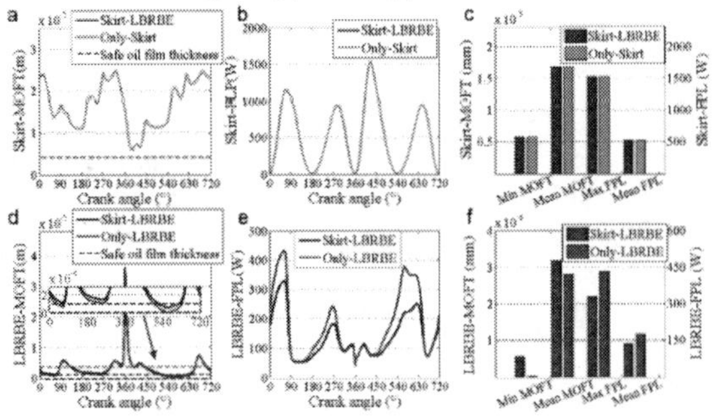

Figure 8. The comparisons of lubrication performances at the skirt and LBRBE.

Figure 8 presents the MOFTs and the friction power losses (FPL) of the piston skirt-liner system and the LBRBE. It is clear that the existence of the LBRBE almost has no effect on the lubrication of the piston skirt as observed in Fig. 7. However, for the lubrication performances of the LBRBE, there are obvious differences between the Skirt-LBRBE model and the Only-LBRBE model. Fig. 8 (d) shows that for most period of the time, MOFTs obtained from the both models agree well with each other. While at some special moments when the MOFT has very small values (i.e., at 62, 252 and 562 ° Crank Angel (CA)), the MOFTs derived from the Only-LBRBE model have the smaller values.

The above phenomenon can also be inferred from Fig. 9 which presents the center trajectory of journal relative to the center of the bushing. It can be found that except for the moments of 62, 252 and 562 °CA where there are some obvious differences, the trajectory from the both models almost overlap. In addition, the journal will quickly pass by the center of bearing at about 361°CA, which causes the MOFT a transient sharply increase as shown in Fig. 8 (d).

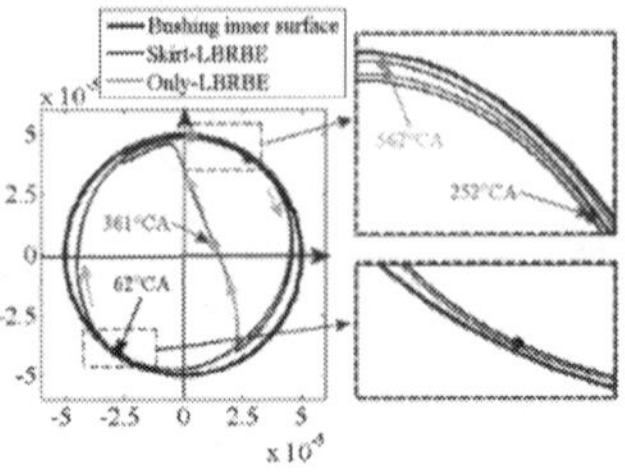

Figure 9. Journal center trajectory relative to the bearing center at the LBRBE.

Figure 8 (e) displays the comparisons of the FPL at the LBRBE. The obvious deviation of FPL from Skirt-LBRBE and the Only-LBRBE models mainly occurs at about 62 and 562 °CA where FPL have the highest values and MOFTs have the smallest values, respectively. Fig. 10 displays asperity contact pressures, as well as the oil film and asperity friction stresses at these two moments. It can be found that due to the smaller oil film thickness in the Only-LBRBE model, the higher oil film shear stress and more surface asperity contacts will be obtained, which will generate the larger friction force and FPL (Fig. 8 (e)).

From the comparison between Figs. 10 (c) and (f), it can be found that although the MOFT at 62 °CA is larger than that at 562 °CA, the friction from the oil film shear stress is smaller at the latter moment. The reason may be that the relative slip velocity between the surfaces of the journal and bearing at 562 °CA is smaller than that at 62 °CA , which limits the contribution of the lubricant shear stress to the friction as stated in Eq. (31). Furthermore, the lower relative slip velocity at 562 °CA will also result in the lower PLF (Fig. 8 (e)).

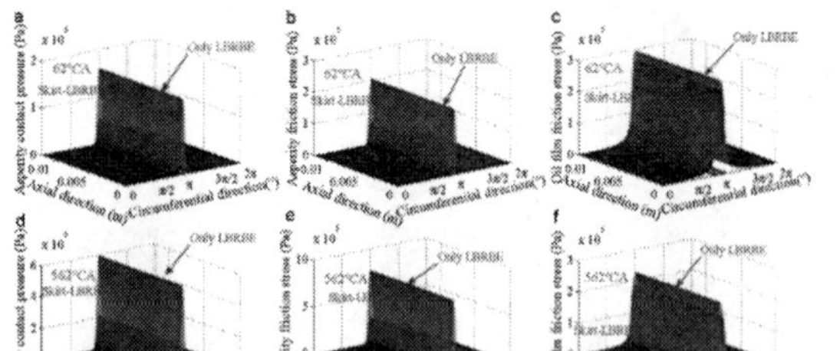

Figure 10. The normal and friction asperity contact pressure and the oil friction stress of LBRBE at 62 and 562 °CA.

From the quantitative comparisons presented in Figs. 8 (c) and (f), it can also be concluded that the influence of LBRBE on the lubrication performance of skirt-liner

DOI: 10.1051/matecconf/201710815006

system is very limited. However, the lubricated skirt-liner system is obviously conducive to improving the lubrication performances of the LBRBE (i.e., the MOFT is increased and the FPL is reduced). This conclusion is consistent with that shown in Fig. 7.

Subsequently, the dynamics and tribological performances for different crank speeds are investigated based on the Skirt-LBRBE coupling model. Fig. 11 presents the comparisons of the secondary motions of the piston for different speeds. It can be found that the amplitudes of piston secondary velocity (i.e., $\dot{e}_0$ and $\dot{\gamma}$) tend to increase with the speed, and both reach the maximum values at about 387 °CA. The reason may be that in the case with higher speed, a smaller period of motion is required to complete the secondary motion of the piston, and thus a larger secondary velocity of the piston will be produced. Moreover, slap noise has a direct bearing on the secondary movement velocity of the piston. This means that the increase of the speeds will cause higher slap noise.

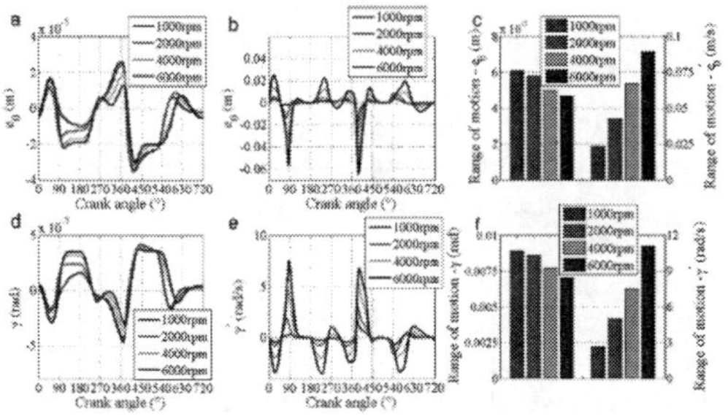

Figure 11. The comparison of kinematic results of the piston for different crank speeds.

With the increase of speed, the ranges of secondary motion of the piston (i.e., e_0 and γ) will also increase for the most of the time. While during the period from about 270 to 450 °CA, the amplitudes of secondary motion will decrease with the increase of the speed. In fact, e_0 and γ depend on the combined effects of both the piston secondary velocity and the thrust forces acting on the piston skirt. Moreover, the higher secondary velocity and smaller thrust force will result in the decrease of secondary motion.

Figure 12 (a) presents the thrust force normally acting on the piston skirt. By way of comparison, the same phenomenon as that of e_0 and γ can also be observed. This is due to the fact that the thrust forces are mainly used to balance the forces which act on the piston in the horizontal direction and are mainly induced by the combustion force and the inertias of the piston and connecting rod (Fig. 13 (a)). Compared with the forces induced by the combustion force, the forces induced by the inertias of the piston and the connecting rod are dominant and increase with the crank speed for most of the time.

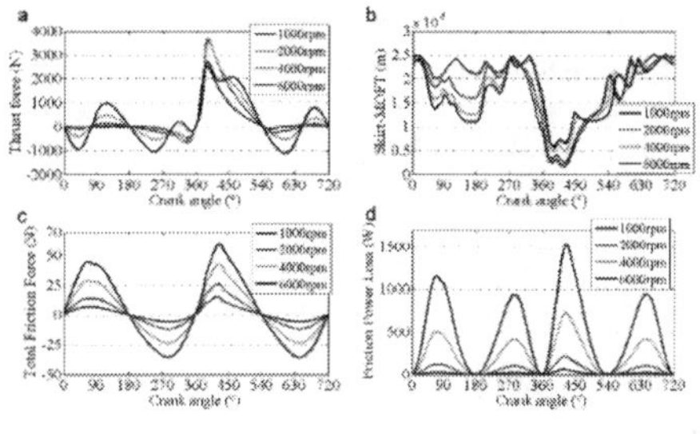

Figure 12. The comparison of lubrication performances of the skirt-liner for different crank speeds.

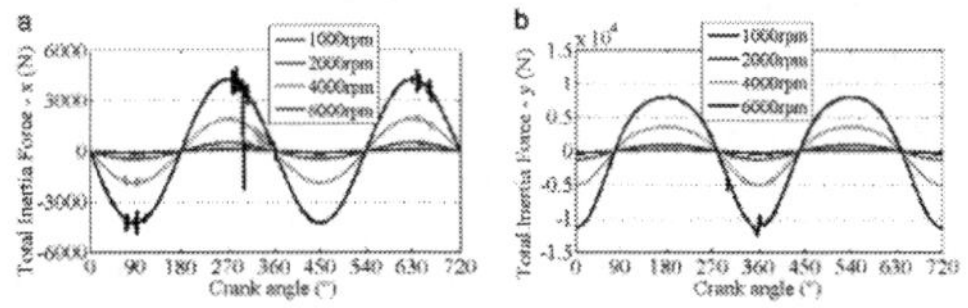

Figure 13. The total inertia forces of the piston and connecting rod in for different crank speeds.

However, during the period from about 270 to 450 °CA, the combustion forces have high values and the thrust forces are mainly used to balance the forces induced by them. Simultaneously, the directions of the forces induced by the inertias of the piston and connecting rod are horizontally opposite with the directions of the forces induced by combustion forces, which will weaken the influence of the combustion force. With the increase of the crank speeds, this weakening effect will increasingly obvious and result in the decrease of the thrust forces, especially as the speed increase from 2000 to 6000 rpm.

Figure 12 (b) lays out the comparison of the MOFT for different speeds. It can be found that the higher speed is accompanied with the smaller MOFT for most of the time. As stated in Eq. (31), the decrease of the oil film thickness and the increase of the relative velocity can increase the lubricant shear stress and thus leads to a rise in the friction force and friction power loss (Figs. 12 (c) and (d)). However, in the range from about 270 to 450 °CA, a reserve trend of the MOFT can be found, and the asperity contacts will occur in a short period in the cases with the speeds of 1000 and 2000 rpm. Although the direct contact among micro-asperities will dramatically increase the friction force and friction power loss, the higher average lubricant shear stress induced by the higher relative velocity as stated in Eq. (31) is still the main source of the friction. Therefore, the friction forces as well as the friction power loss still increase with speeds during this period.

Figure 14 sets out the comparison of 3D phase portraits of ε, $\dot{\varepsilon}$ and $\dot{\gamma}_b$ for different speeds. With the decrease of the speed, the values of ε, $\dot{\varepsilon}$ and $\dot{\gamma}_b$ tend to reduce, and the dynamic behavior of the LBRBE is more regular and stable. This also means that the higher running speed will result in the decline of the dynamics performance of the LBRBE.

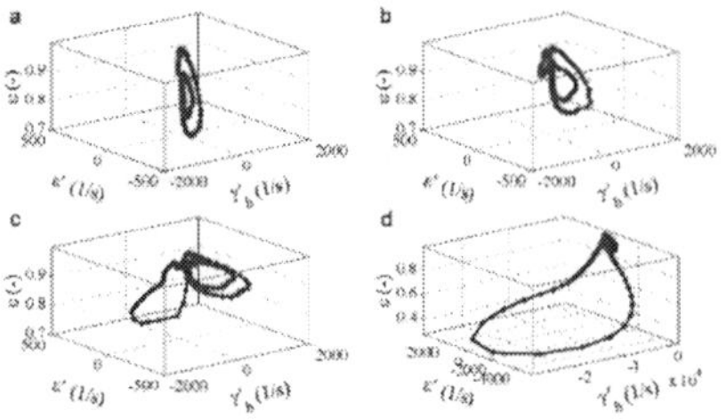

Figure 14. The comparison of kinematic results of the LBRBE for different crank speeds: (a)1000 rpm; (b) 2000 rpm; (c) 4000 rpm; (d) 6000 rpm.

Figures 15 (a) and (b) lay out the reaction forces at the LBRBE in horizontal and vertical directions. In the horizontal direction, both the thrust forces acting on the piston (Fig. 12 (a)) and the reaction forces at the LBRBE are also used to balance the horizontal inertia forces of the piston and connecting rod (Fig. 13 (a)). It can be found that the horizontal reaction forces increase with the speeds.

In the vertical direction, it is obvious that the plot of the reaction forces has the same tendency with that of combustion pressures when the system has a small speed. However, with the increase of the speed, the tendency of the reaction forces comes to resemble that of the inertial forces (Fig. 13 (b)), especially in the case with the speed of 6000rpm. Therefore, it can be concluded that the reaction forces at the LBRBE result from the joint effects of both combustion and inertial forces. In the case with the higher crank speed, the higher reaction forces at the LBRBE will be generated for most of the time (Figs. 15 (a) and (b)). Thus, the smaller MOFTs will be required so as to produce higher oil film pressure to separate the journal from the bushing. While during the period from about 270 to 450 °CA, the MOFT will depend on the joint effects of both combustion forces and inertial forces rather than decrease with the increase of the speeds.

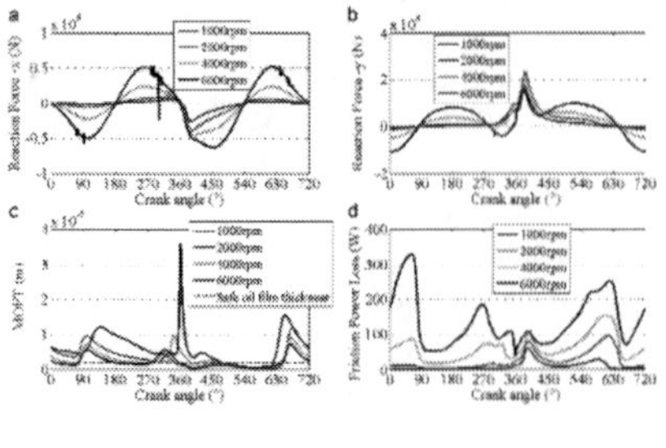

Fig. 15. The comparisons of lubrication performances of the LBRBE for different crank speeds.

Figure 15 (d) lays out the comparison of the friction power loss for different speeds. From the figure, it can be observed that the friction power loss increases with the speeds. The main reason may be that the higher speed will generate higher lubricant shear stress and thus higher friction force and friction power loss. Although in other moments, the cases with the lower speeds will have the thinner oil film thickness and thus more asperity contacts (Fig. 15 (c)), the lubricant shear stress induced by the relative velocity as stated in Eq. (31) is still the main source of the friction. Therefore, the friction forces as well as the friction power loss still increase with speeds during this period.

5 Conclusions

This study develops a new numerical model for the dynamics and lubrication analyses of a multibody system with multiple lubricated clearance joints. This method is established by coupling the dynamics model of the multibody system with the mixed lubrication models of translational and revolute clearance joints. The multibody dynamics model is established by Lagrange's method, while the lubricated clearance joints are built based on the average Reynolds equation and solved using the finite element method. The hydrodynamic forces built up by the lubricant fluid are evaluated from the motion state of the system and then input into the dynamics equations of the multibody systems.

The proposed approach is applied to a four-stroke gasoline engine with the lubricated piston-liner subsystem and bearing at the big end of the connecting rod. Both the dynamics and lubrication performances are also analyzed. From the results obtained, it can be found that there are obvious interactions between the piston skirt-liner system and the LBRBE. In the end, the performances of the system with different crank speeds are also analyzed. With the increase of the speeds, the dynamics and lubrication performances of the system will decrease.

Acknowledgment

The authors are grateful to the China Postdoctoral Science Foundation (2015M57036).

References

[1] Flores P, Ambrósio J, Claro JP. Dynamic analysis for planar multibody mechanical systems with lubricated joints. Multibody System Dynamics. 2004;12:47-74.

[2] Hamrock BJ, Schmid SR, Jacobson BO. Fundamentals of fluid film lubrication: CRC press; 2004.

[3] Flores P, Lankarani HM, Ambrósio J, Claro JCP. Modelling lubricated revolute joints in multibody mechanical systems. Proceedings of the Institution of Mechanical Engineers, Part K: Journal of Multi-body Dynamics. 2004;218:183-90.

[4] Li DF, Rohde SM, Ezzat HA. An Automotive Piston Lubrication Model. A S L E Transactions. 1983;26:151-60.
[5] Zhu D, Cheng HS, Arai T, Hamai K. A numerical analysis for piston skirts in mixed lubrication—Part I: Basic modeling. Journal of tribology. 1992;114:553-62.
[6] Zhu D, Hu Y-Z, Cheng HS, Arai T, Hamai K. A Numerical Analysis for Piston Skirts in Mixed Lubrication: Part II—Deformation Considerations. Journal of tribology. 1993;115:125-33.
[7] Keribar R, Dursunkaya Z. A comprehensive model of piston skirt lubrication. SAE Technical Paper; 1992.
[8] Wong VW, Tian T, Lang H, Ryan JP, Sekiya Y, Kobayashi Y, et al. A numerical model of piston secondary motion and piston slap in partially flooded elastohydrodynamic skirt lubrication. SAE Technical Paper; 1994.
[9] Liu K, Xie Y, Gui C. A comprehensive study of the friction and dynamic motion of the piston assembly. Proceedings of the Institution of Mechanical Engineers, Part J: Journal of Engineering Tribology. 1998;212:221-6.
[10] Patir N, Cheng H. An average flow model for determining effects of three-dimensional roughness on partial hydrodynamic lubrication. Journal of Tribology. 1978;100:12-7.
[11] Patir N, Cheng H. Application of average flow model to lubrication between rough sliding surfaces. Journal of Tribology. 1979;101:220-9.
[12] Flores P, Ambrósio J, Claro JCP, Lankarani H, Koshy C. Lubricated revolute joints in rigid multibody systems. Nonlinear Dynamics. 2009;56:277-95.
[13] Daniel GB, Cavalca KL. Analysis of the dynamics of a slider–crank mechanism with hydrodynamic lubrication in the connecting rod–slider joint clearance. Mechanism & Machine Theory. 2011;46:1434–52.
[14] Machado M, Costa J, Seabra E, Flores P. The effect of the lubricated revolute joint parameters and hydrodynamic force models on the dynamic response of planar multibody systems. Nonlinear Dynamics. 2012;69:635-54.
[15] Zhao B, Zhou K, Xie Y-B. A new numerical method for planar multibody system with mixed lubricated revolute joint. International Journal of Mechanical Sciences. 2016;113:105-19.
[16] Zhao B, Zhang Z-N, Fang C-c, Dai X-D, Xie Y-B. Modeling and analysis of planar multibody system with mixed lubricated revolute joint. Tribology International. 2016;98:229-41.
[17] Wang N, Chang C. An application of Newton's method to the lubrication analysis of air-lubricated bearings. Tribology transactions. 1999;42:419-24.
[18] Wu C, Zheng L. An Average Reynolds Equation for Partial Film Lubrication With a Contact Factor. Journal of Tribology. 1989;111:188-91.
[19] Meng X, Xie Y. A new numerical analysis for piston skirt–liner system lubrication considering the effects of connecting rod inertia. Tribology International. 2012;47:235-43.
[20] Meng X, Ning L, Xie Y, Wong VW. Effects of the connecting-rod-related design parameters on the piston dynamics and the skirt–liner lubrication. Proceedings of the Institution of Mechanical Engineers, Part D: Journal of Automobile Engineering. 2012:0954407012464025.
[21] Zhao B, Dai X-D, Zhang Z-N, Xie Y-B. A new numerical method for piston dynamics and lubrication analysis. Tribology International. 2016;94:395-408.
[22] He Z, Xie W, Zhang G, Hong Z, Zhang J. Piston dynamic characteristics analyses based on FEM method Part I: Effected by piston skirt parameters. Advances in Engineering Software. 2014;75:68-85.
[23] Pinkus O, Sternlicht B. Theory of hydrodynamic lubrication. New York: McGraw-Hill; 1961.
[24] Greenwood J, Tripp J. The contact of two nominally flat rough surfaces. Proceedings of the institution of mechanical engineers. 1970;185:625-33.
[25] Akalin O, Newaz GM. Piston ring-cylinder bore friction modeling in mixed lubrication regime: Part I—Analytical results. Journal of tribology. 2001;123:211-8.
[26] Hu Y, Cheng HS, Arai T, Kobayashi Y, Aoyama S. Numerical simulation of piston ring in mixed lubrication—a nonaxisymmetrical analysis. Journal of tribology. 1994;116:470-8.
[27] Shabana AA. Dynamics of multibody systems: Cambridge university press; 2013.
[28] Nikravesh PE. Computer-aided analysis of mechanical systems: Prentice-hall Englewood Cliffs; 1988.
[29] de Jalón JG, Bayo E. Kinematic and dynamic simulation of multibody systems. Mechanical Engineering Series, Springer, New York. 1994.
[30] Muvengei O, Kihiu J, Ikua B. Numerical study of parametric effects on the dynamic response of planar multi-body systems with differently located frictionless revolute clearance joints. Mechanism and Machine Theory. 2012;53:30-49.
[31] Ravn P. A Continuous Analysis Method for Planar Multibody Systems with Joint Clearance. Multibody System Dynamics. 1998;2:1-24.
[32] Muvengei O, Kihiu J, Ikua B. Dynamic analysis of planar rigid-body mechanical systems with two-clearance revolute joints. Nonlinear Dynamics. 2013;73:259-73.
[33] Baumgarte J. Stabilization of constraints and integrals of motion in dynamical systems. Computer methods in applied mechanics and engineering. 1972;1:1-16.
[34] Flores P, Machado M, Seabra E, da Silva MT. A parametric study on the Baumgarte stabilization method for forward dynamics of constrained multibody systems. Journal of computational and nonlinear dynamics. 2011;6:011019.
[35] Kim J, Chung I, Lee B. Determination of the feedback coefficients for the constraint violation

DOI: 10.1051/matecconf/201710815006

stabilization method. Proceedings of the Institution of Mechanical Engineers, Part C: Journal of Mechanical Engineering Science. 1990;204:233-42.

Analysis of displacements in beam structures and shells with middle developable surfaces

Marina Rynkovskaya

RUDN University, Engineering Department, 6 Miklukho-Maklaya st., Moscow, Russia, 117198

Abstract. Generally the displacements of the beams are calculated by the traditional methods of structural mechanics with the help of approximate formulas which provide good results for linear tasks. The author shows that the approximate formula is not suitable for analysis in the case of geometrical nonlinearity which occurs during the manufacturing process of bending developable shells. The comparison of results got by methods of structural mechanics and shell theory for the shells with developable middle surfaces is given.

1 Introduction

The displacements of the beams are traditionally calculated by approximate equations of the methods of the structural mechanics, while the displacements of flexible structures are mostly based on mathematical models with the systems of nonlinear differential equations [1-4]. Professor Krivoshapko S.N. [5] suggested an exact method of the determination of internal bending moments which occur during the process of parabolic bending of thin developable shells with the help of differential geometry methods, Hooke's law and Love's formulas for the elastic shell theory. In a paper [6], the shell stress-strain state in view of very large displacements and very slight normal and shear strains is studied.

In a paper [7], Elenitsky E.Ya. suggested the algorithm of static calculation of geometrically nonlinear compound thin structures with the help of linear differential equations of moment theory and considered the nonlinearity by assigning unknown initial angular displacement of each segment retaining the form of the generating line. Unknown values of algebraic equations resolving system are the arbitrary constants of the general solution and the initial angles of generating lines rotation, while linearization is realized by Newton-Raphson iterative method. It is shown in the paper on examples of the console beam and the hinged annulus that such approach provides the high precision of results.

2 Methods applied to the beam

2.1 Approximate method applied to the beam

It is well known that displacements in the beam in traditional linear theory of structural mechanics [8] can be found by the formula of the advanced mathematics

$$\frac{y''}{\left(1+y'^2\right)^{3/2}} = k = \frac{1}{R}, \quad (1)$$

where k is a curvature of the curve; R is a radius of the curve; y=y(x) is a deflection equation; y' is a tangent of the angle of inclination of the tangent to the elastic line of a beam axis.

In practice, these angels are very small and one can neglect their squares in comparison with the unit that allows getting the famous equation named the main differential equation of the curved axis of a beam:

$$y'' = \frac{M}{EI}, \quad (2)$$

(where E is a modulus of elasticity; I is a moment of inertia, M is an inner bending moment), which is approximate, because the equation of the axis curvature is replaced by an approximate one, and deformations caused by shear forces are also not considered. In most cases, this influence is insignificant and can be neglected. However, let us describe the equation without any approximations and analyse the accurate results.

2.2 Precise method applied to the beam

To simplify the calculations, we will consider the load to be consistent and consist only of the converging bending moment attached at the end of a console beam (Fig. 1). In this case, the Eq. 1 will be

$$\frac{EIy''}{\left(1+y'^2\right)^{3/2}} = M, \quad (3)$$

where M = const is an inner bending moment.

On integrating the Eq. 3, one can obtain the following equation of the deflection of an axis:

DOI: 10.1051/matecconf/201710816001

$$y = \frac{1}{M}\sqrt{(EI)^2 - (Mx)^2} - \frac{EI}{M}, \qquad (4)$$

which is an equation of a circle, while the approximate method (Eq. 2) gives us the deflection equation in the form of parabola:

$$y = \frac{Mx^2}{2EI}. \qquad (5)$$

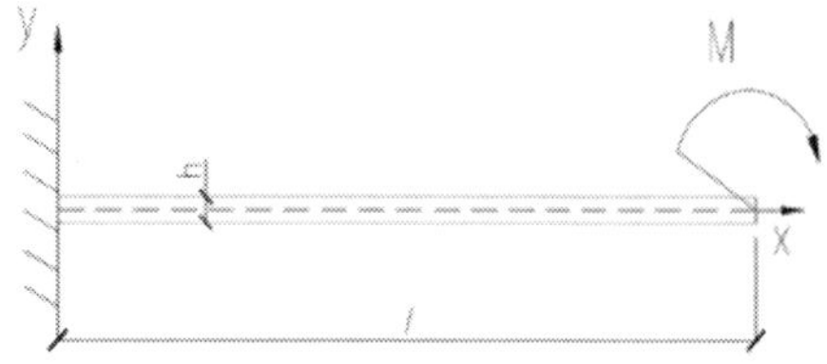

Figure 1. The analyzed console beam

Test. Let us analyze an example calculated by the means of two approaches (approximate and precise). The steel beam with the following geometrical and physical characteristics: section height is h = 0.01 m; section width is b = 1 m; the length is l = 6 m; Young's modulus is $E = 2.06 \cdot 10^5 MPa$, the yield point stress is $\sigma_y - 245MPa$; was calculated. The displacements are shown on Fig. 2 and in the Table 1, where y(x) was calculated by precise method (Eq. 4) and y1(x) was calculated by approximate method (Eq. 5).

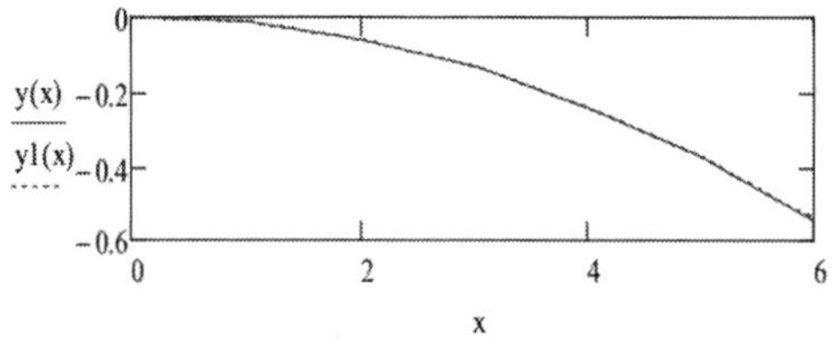

Figure 2. Displacements *y(x)* and *y1(x)* which are calculated by precise (Eq. 4) and approximate (Eq. 5) methods respectively.

Table 1. Displacements *y(x)* and *y1(x)* which are calculated by precise (Eq. 4) and approximate (Eq. 5) methods respectively and the difference δ in the results of displacements *y(x)* and *y1(x)*.

y(x) =	y1(x) =	δ(x) =
0.000000	0.000000	$3.385 \cdot 10^{-5}$
-0.015003	-0.015000	0.023
-0.060054	-0.060000	0.09
-0.135274	-0.135000	0.203
-0.240870	-0.240000	0.363
-0.377133	-0.375000	0.569
-0.544446	-0.540000	

The results of displacements on Fig. 2 are shown in meters. The displacements are more than 50 thicknesses of the beam, so the task is geometrically nonlinear and cannot be analyzed by linear theory.

The differences in results of displacements which are calculated by approximate and precise methods

$$\delta(x) = \frac{y(x) - y1(x)}{y1(x)} \cdot 100$$

are shown in Table 1. The biggest difference on the edge of the beam is 0.569 per cent that can be considered as a small difference, but in the case of bigger displacements such as 100 thicknesses the difference is considerable which will be shown further.

3 A precise method applied to the developable shells

According to [9], a developable surface can be flattened onto a plane without stretching, tearing or creasing, and tangent lines of the edge of regression (cuspidal edge) of this surface generate it. Since this type of surfaces is widely used in manufacturing [10-15] in many fields such as shipbuilding [16, 17], engineering [18] and architecture [19, 20] and products are mostly obtained by the means of bending, the problem of plastic deformations should be taken into account. However, most works about developable surfaces do not concern to this point [21-27].

In the case of manufacturing of developable surfaces by applying the bending moment My to the plane plate, one can be interested in obtaining the minimum radius Rmin of the bending until which the plate works in elastic stage. The plastic moment My, created by the normal stresses, occurs when maximum normal stresses mount to the yield point stresses ($\sigma_{s,\max} = \sigma_y$) is

$$M_y = \frac{\sigma_y h^2}{6}.$$

After rewriting the Eq. 4 into more traditional form for the circle

$$y = \sqrt{\frac{(EI)^2}{M^2} - x^2} - \frac{EI}{M},$$

one can obtain the evaluation of the radius of the deformation axis of the beam

$$R = \frac{EI}{M} = const. \qquad (6)$$

The Eq. 6 can be called a radius of a curve of the developable surfaces while bending.

On the other hand, it was shown by the author in [28] that the minimum radius of the developable surface while manufacturing can be also calculated by another approach with the help of the shell theory

$$R_{\min} = \frac{EI}{M_y} \frac{1}{\left(1 - \nu^2\right)}. \qquad (7)$$

According to the Eq. 6 and Eq. 7 the difference between standard methods of resistance of materials and the method of the shell theory leads to different results. Thus, in the case when Poison's ratio is $\nu = 0.3$, the minimum radius occurs to be almost 10 per cent higher than the radius which is calculated by standard methods of structural mechanics.

4 Practical applications

Let us take the steel rectangular plate with the following geometrical and physical characteristics: section height is h = 0.01 m; section width is b = 1 m; the length is l = 6 m; Poison's ratio is $\nu = 0.3$; Young's modulus is $E = 2.06 \cdot 10^5 MPa$, the yield point stress is $\sigma_y = 245 MPa$. Let it be required to produce a product in the shape of a half circle (C-shaped plate) with the radius R = 5 m (Figure 3) by bending.

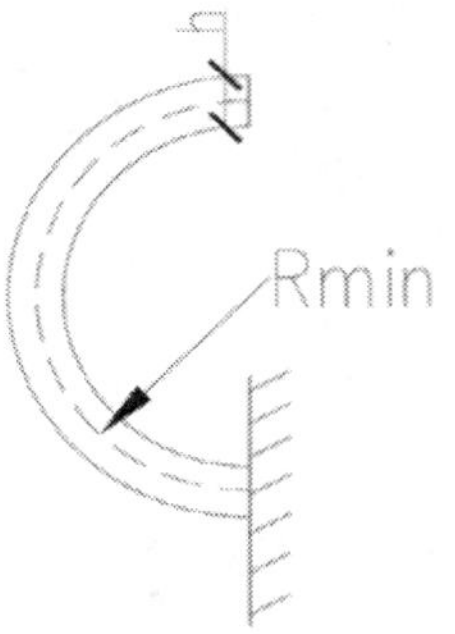

Figure 3. The required C-shaped product

We calculate the minimum value of the radius ($R_{\min}$) of bending at which the plastic deformations will not occur (the shaped plate will turn around into a plane plate again). In other words, the plate is bent without plastic deformations ($\sigma_{s,\max} \leq \sigma_y$).

The plastic bending moment My, created by the normal stresses, occurs when maximum normal stresses mount to the yield point stresses ($\sigma_{s,\max} = \sigma_y$), is

$$M_y = \frac{\sigma_y h^2}{6} = 4038 N \cdot m / m.$$

According to Eq. 6 and Eq. 7, the minimum value of the radius of bending without plastic deformations will respectively be Rmin = 4.082 m for the method of resistance of materials and Rmin = 4.485 m for the shell theory method. This means that required product cannot be manufactured by simple bending because after removing the clamps product will turn back into the plane plate.

5 Results and discussion

In the case when the biggest displacement y(x) is less or the same as the thickness h of the beam, the difference between displacements which are calculated by accurate and approximate methods is very small (less than 0.0002 per cent) which coincides with linear theory. In the case when the biggest displacement y(x) is less than 50 thicknesses, the difference is also small and shown on Fig. 3 (δ max = 0.569 per cent).

However, in the case of developable shells which are manufactured by bending the displacements are much bigger and the difference in results can be considerable. For example, in the case when the biggest displacement y(x) is more than 100 thicknesses the difference is 2.36 per cent, while in the case of displacements more than 200 thicknesses the mistake may achieve even 10 per cent.

Obviously, the suggested analysis is not absolutely accurate as far as it does not take into account the displacements along the coordinate x, but it generally shows that once displacements are very big, for example, in the case of manufacturing a product from the thin plate by bending, the traditional approximate equations cannot be used. In addition, the suggested equations can be improved by further research and the displacements along the line x can be also be included in analysis.

For future researchers, it can be interesting to conduct an experiment that will confirm the suggested theoretical results, the minimum value of the bending radius. There have not been found any information in this field of research, except [7] where some approaches for calculation of this experimental task are presented.

Generally, most engineers consider the task of bending of the thin plate to the task of bending of the middle surface, and do not take into account the normal strains, because as it is well known normal strains do not occur on the middle surface. In practice, since the margin of safety is usually very high for such products as steel tubes, such approach does not lead to any problems. However, it can be useful to consider plastic deformations when plate is being bent into the product.

6 Conclusion

The investigation presented shows some problems in usage of linear and geometrically nonlinear theories for beams and shell structures with developable middle surface. There are some results of analysis of the console beam which was calculated by both the approximate and precise methods of structural mechanics as well as by the shell theory. Since the plastic deformations can occur during the manufacturing process of bending, the determination of these deformations in the shells with developable middle surfaces is a considerable point of engineering task, and it seems to be rather reasonable to obtain the simple formula for such analysis.

It is shown in the paper that the difference between results of displacements which are calculated by standard methods of resistance of materials and shell theory achieves 10 per cent. Since geometrically nonlinear task usually occurs in manufacturing process, the analysis is worth to be conducted with precise method of structural mechanics and shell theory.

As a future research topic, the suggested calculations can be improved by taking into account the displacements along the line x, as well as the comparison with the results which can be obtained by numerical methods. The two presented tasks can also be used as test examples for numerical methods.

References

1. Ogibalov P.M., Koltunov M.A., *Shells and Plates*, Moscow Univ., 695pp. (1969).
2. Karmishin A.V., Lyaskovets V.A., Myachenkov V.I., Frolov A.N., *Statics and dynamics of thin-wall shell-type structures*, Moscow, Mashinoctroenie, 376 pp. (1975).
3. Valishvili N.V., *Methods of analyzing shells of revolution on a digital computer*, Mashinostroenie, 279pp. (1976).
4. Timoshenko S.P., Woinowsky-Krieger S. *Theory of Plates and Shells*, New York, McGrawHill, 595pp (1959).
5. Krivoshapko S.N., Parabolic bending of thin elastic plane or developable half-finished products into developable shell, Structural Mechanics of Engineering Constructions and Buildings, Vol. **12,** P. 12-19 (2003).
6. Krivoshapko S.N., Gbaguidi G.L.A., Developable shell products made by parabolic bending of thin metal slabs, Journal of the Ghana Institution of Engineers, Vol. **6&7**, № 1, p. 51-56 (2009).
7. Elenitsky E.Ya., Boundary value problem for the flexible axially loaded compound shells of revolution and beams systems, Vestnik of Samara State Techn. Univ. (Techn. Scien. Series), **4 (29)**, pp. 122-130, (2012).
8. Krivoshapko S.N., *Structural Mechanics: theory and practice*, 2 edit., Moscow, Urait, 392 pp. (2014).
9. Krivoshapko S.N., Ivanov V.N., *Encyclopedia of Analytical Surfaces*, Springer International Publishing Switzerland, 752 pp. (2015).
10. Haeberli P., Modeling and fabrication of objects represented as developable surfaces, US Patent 6493603, USA (2002).
11. Rynkovskaya M.I., Analytical Methods of Analysis of Right and Open Helicoids, *ASEE 2009 Conference: Engineering in the New Global Economy*, USA, (2009).
12. Liua Y.-J., Tangb K., Gongc W.-Y., Wuc T.-R., Industrial design using interpolatory discrete developable surfaces, Computer-Aided Design, Vol. **43**, Iss. 9, p. 1089–1098 (2011).
13. Krivoshapko S.N., *Geometry of Ruled Surfaces with Edge of Regression and Linear Theory of Analysis of Developable Shells*, Monograph, Izd-vo RUDN, 357pp. (2009).
14. Krivoshapko S.N., Static analysis of shells with developable middle surfaces, Appl. Mech. Rev., Vol. 51, №12, Part 1, pp. 731-746 (1998).
15. Redont P., Representation and deformation of developable surfaces, Computer Aided Geometric Design, Vol. **21 (1)**, p. 13-20 (1989).
16. Oetter R., Barry C.D., Duffty B., Welter J., Block Construction of Small Ships and Boats Through Use of Developable Panels, Journal of Ship Production, Vol. **18**, №2, p. 65-72 (2002).
17. Steel-Chalfant J., Analysis and design of developable surfaces for shipbuilding, Master's thesis (1997).
18. Gotman A.S., Use of Developable Surfaces for Designing Well-Stream Shapes, Oceanic Engineering International, Vol. **9**, NUMB 2, p. 81-96 (2005).
19. Koschitz D., Demaine E.D., Demaine M.L., Curved Crease Origami, Abstracts from Advances in Architectural Geometry (AAG 2008), p. 29–32 (2008).
20. Tevfik A., Ahmet K., Ergun A., *Developable sculptural forms of Ilman Koman*, Tarquin Publications, London (2006).
21. Solomon J., Vouga E., Wardetzky M., Grinspun E., Flexible Developable Surfaces. Computer Graphics Forum, Vol. **31**, Iss. 5, p. 1567-1576 (2015).
22. Yoon J.S., Ryu C., Lee J.H., Developable polynomial surface approximation to smooth surfaces for fabrication parameters of a large curved shell plate by Differential Evolution, Computer-Aided Design, V. **40**, Iss. 9, p. 905-915 (2008).
23. Xia M., Yan Q., Zuo D., Xie J., An investigation on multistage bending of blank sheet into cylindrical tube by experiment and numerical simulation, Int. J. of Adv. Manuf. Tech., Vol. **53**, Iss. 1-4, p. 145-155 (2011).
24. Kilian M., Flory S., Chen Z., Mitra N.J., Sheffer A., Pottman H., Curved Folding, ACM Transactions on Graphics (TOG), Vol. **27**, Iss. 3, Article № 75 (2008).
25. Rose K., Sheffer A., Wither J., Cani M.-P., Thibert B., Developable Surfaces from Arbitrary Sketched Boundaries, *SGP '07 Proceedings of the fifth Eurographics symposium on Geometry processing*. P. 163-172 (2007).
26. Dias M. A., Dudte L. H., Mahadevan L., Santangelo C. D., Geometric Mechanics of Curved Crease Origami, Physical Review Letters 109, 114301 (2010).
27. Jung A., Hahmann S., Rohmer D., Begault A., Boissieux L., Cani M.-P., Sketching Folds: Developable Surfaces from Non-Planar Silhouettes, ACM Transactions on Graphics (TOG), Vol. **34**, Iss. 5, Article No. 155 (2015).
28. M. Rynkovskaya. Plastic deformations in the shells with developable middle surfaces while bending, ICONTES 2016 (to be published).

DOI: 10.1051/matecconf/201710816002

Development of Constructions of Mobile Road Overpasses

Adil Kadyrov, Aleksandr Ganyukov and Kyrmyzy Balabekova

The Karaganda State Technical University, 100024 Karaganda, Kazakhstan

Abstract. The new type of engineering constructions that is mobile overpasses is developed. The offered mobile overpasses are applied for elimination the arisen traffic jams along city highways, when repairing city municipal networks, during emergency situations, etc. They allow to carry out continuous movement of motor transport through various obstacles: traffic jams, ditches and trenches formed because of underground repair work or emergency situations (earthquakes, floods, etc.). The issues of development theoretical bases of calculation and design of mobile overpasses are considered.

1 Introduction

In the conditions of heavy city car traffic there are jams and traffic blocks on roads because of various reasons: car accidents, rush hours, repair work on highways, city municipal networks. In many large cities there is a problem of the solution of jams during everyday traffic and repairing of city infrastructure. For big cities it is a great transport problem. As it is well-known jams and traffic blocks along city highways worsen the transport logistics of the city, ecology, cause economic damage to businessmen, city budget, etc.

For the solution of this problem the design of mobile overpasses is offered by us.

They represent collapsible mobile bridge overpasses which are installed over road strips where there was a traffic block of transport or on the section where there is underground repair of municipal networks under highways that promotes movement of transport on the emergency or repair section and, respectively, eliminates traffic jams and removes the necessity of bypass road of repair sections of roads [1]. (See Fig. 1 and Fig. 2)

Use of the overpasses offered by us considerably improves transport logistics in the city during traffic jams or repair work on municipal networks: formation of traffic jams is reduced, compelled bypass roads of repair sites are absent, inconveniences for car drivers and residents of city districts that can cause the compelled transport bypass are eliminated.

Figure 1. Mobile road overpass

Figure 2. Modular municipal overpass

Such mobile overpasses can be used not only during elimination of traffic jams and at underground repair work in the city conditions, but also during emergency situations caused by flood, earthquake, caused collapse of various infrastructure – roads, bridges, crossings, etc.

Distinctive feature of the offered bridge overpass is its mobility – conveyance on own chassis by means of the automobile trailer or by freight vehicles. Quick assembling and disassembling at the place of its installation due to use of the unified collapsible modules and modes of their fixing between each other and on the ground base. It provides immediate delivery to necessary sites with autojams, repair sites of municipal underground networks or sites with the damaged infrastructure caused by various emergencies [2].

The analysis of the condition of this subject in RK, CIS and abroad showed almost total absence of existence of similar mobile bridge constructions for elimination traffic jams, city repair needs and conditions of emergency.

Separate elements of similar subject are partially developed in military bridge construction – cross over trawls, automobile bridge layers, etc. The matter isn't developed in civil transport engineering and demands comprehensive study and formation of scientific bases of mobile overpasses design for various conditions of their operation. Thus, the mobile overpass has no analogs in the world.

2 Main results

2.1 Types of overpasses

We offer several versions of overpasses: road, municipal and overpasses for conditions of emergency situations.

The road mobile overpass is intended for elimination traffic jams along city highways. It consists of the horizontal module and two inclined modules equipped with own wheel chassis for transportation and sliding bearing parts. In case of necessity, modules are transported to the crossroad, assembled and fixed among themselves by special captures into a uniform construction, where sliding bearing parts lean on the road base (Figure 3).

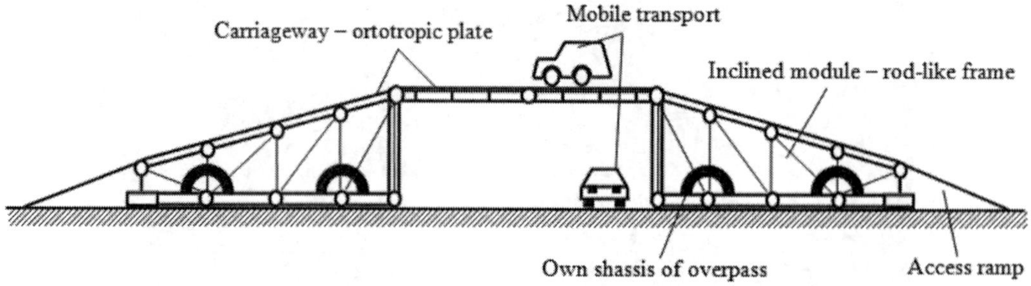

Figure 3. Mobile road overpass. Profile view (facade).

The road overpass has to meet dimensional conditions of a carriageway: journey height under them is not more than 4.5m, the width of a transport strip in one direction is not less than 3.5m (for conditions of the Republic of Kazakhstan). After elimination of a traffic block, the overpass is disassembled into several modules and towed to its base place. Operation of assembling and disassembling of the overpass is automated and it doesn't take much time due to its specially developed gears of fastening the overpass parts among themselves and a carriageway [1,3].

The municipal overpass is used during repair or replacement of city municipal networks: thermal, water, cable, etc. The city municipal networks are usually located under highways. During their repair there is a separation of repair ditches and trenches on the place of highways that causes their blanking for a long time and need to organize various long inconvenient bypass roads. It causes formation of traffic jams, worsens transport logistics of the city, increases accident rate level, etc.

In these conditions, instead of the compelled ways of bypass roads of municipal ditches and trenches we suggest to organize direct bridge crossings over them - modular municipal overpasses. Such overpass is installed through repair ditches of municipal networks and allows not to stop transport streams near repair sections of roads for all the time of underground repair (Figure 4). The use of the municipal overpass naturally improves transport condition in the city.

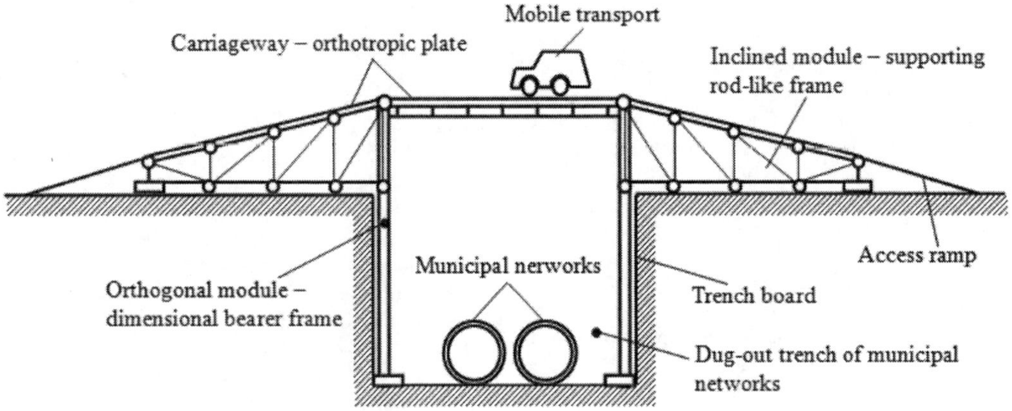

Figure 4. Modular municipal overpass. Side view (facade).

The municipal overpass assembled from the unified collapsible modules of two types: one orthogonal focused module and two inclined modules.

The orthogonal focused module represents a spatial steel frame which base is fastened at the bottom of a repair trench by means of special modes. The top part of the frame presents a carriageway in the form of orthotropic plate with the supporting reinforcement ribs .

The inclined module represents the steel rod-like frames bearing a carriageway on themselves in the form of ortotropic plate. The base of rod-like frames leans on the ground base, other part of rod-like frames leans on a bearer frame.

The linking of inclined modules with a bearer frame, allows to avoid loading on the edges of boards of repair trenches and reduces probability of their collapse while using the overpass. However the second basic devices of inclined modules are in power interaction with the natural ground base in this connection, there is a problem of additional research of deformed state of the intense ground massif and prevention of trench boards destruction [4].

Bridge moving is delivered to the installation site by separate modules and assembled in "one-way" or by hitch method in "two-way" construction.

Outline dimensions of "one-way" modules are the following: length of the orthogonal focused module – 8m; width – 3,5m; height (taking into account their installation at the bottom of trenches) – 5m.

The length of the inclined module – 4m; width – 3,5m; greater height – 2m; smaller height– 0,2m.

After use of the municipal overpass its disassembling into separate modules in places of "hitch" is made, and then they are delivered to the place of their storage.

2.1 Theoretical bases of overpass calculation

Our scientific group developed the scientific-theoretical bases of calculation and design of mobile and modular overpasses of various types including design decisions for creation of all its parts and nodes.

Space-planning and design solutions are developed for creation of the unified modules of the mobile bridge crossings providing the required bearing ability of its carriageway for the admission of the types and number of transport determined by norms of the automobile bridge building. In this case preliminary selection of forms and the sizes of lateral sections of the bearing constructions is also made to perform their further calculations.

The calculations of the orthotropic plates of the carriageway of bridge crossings by various numerical methods, such as the method of final elements, final differences and so forth are carried out.

The intense deformed condition of dimensional frames of the orthogonal focused modules of overpasses on the basis of calculations of power state is investigated by the analytical method of conveyances with formation of the general matrixes considering parametrical variability of geometrical and rigidity characteristics of bearing components of spatial frames [5].

It allows to carry out automatic calculation of similar frames for bridge crossings of various constructive solution and intention. Here the final selection of lateral sections of the bearing constructions of spatial frame on the basis of standard conditions of strength, rigidity and stability taking into account specific requirements is carried out for designs of automobile bridge building.

The work of the bearing flat statically indefinable rod-like frames of inclined modules of overpasses is investigated by analytical method of forces. Here, we studied the influence of rigidity characteristics of rod-like frame cores which changes are able to make an essential result on power and tension rod-like frame state in general.

Interactions of basic devices of the overpass with the natural ground base are investigated and the analysis of ground massif stress condition in the zone of trench boards is developed. On this basis practical bases of calculation for strength and stability of the corresponding volume of the ground massif which is under loading from pressure of basic devices of the overpass and the recommendation for prevention of their destruction in the process of the overpass use are developed.

Different types of bearing parts of overpasses, chassis constructions, gears of modules fastening are developed for which the corresponding patents and certificates of intellectual property are received [1].

For creation of the effective design solution of mobile overpasses, on the basis of the mathematical theory of optimization the methods of optimizing calculations for the purpose of minimization of metal consumption of designs are developed for reduction of their cost.

2.2 Constructive decisions

The bearing constructions of all types of overpasses are carried out from the strengthened brands of steel. The flooring of carriageway is carried out from sheet steel 20 mm thick. The cloth of carriageway is made of thin corrugated steel or rubber cloth for ensuring the required grip of tires of the motor transport.

Preliminary selection of lateral sections of bearing components of bridge crossing modules of all types is made on the basis of strength conditions, rigidity, stability with a sufficient stock taking into account possible overloads and dynamic effect of the movement on the carriageway of automobile stock. The size of bridge loadings is accepted under the terms of euronorms.

Geometrical and rigidity characteristics of the bearing constructions are generalized and provided to the leading parameters for the possibility of calculations of elements of demountable modules of different function and intention functions [5].

Resting of racks of orthogonal focused module of municipal overpass on the bottom of the dug-out ditches and trenches is made through fixed concrete piles racks.

3 Conclusion

The theoretical bases of calculations developed and described above, allow to design overpasses of various types on a practical basis, since the development of their constructive solution, selection of lateral sections of the bearing constructions under the terms of their strength, rigidity and stability taking into account various loadings: both static and dynamic, arising from the movement of the automobile stock on the carriageway. Using methods of optimum design allows to create economic and reliable designs of overpasses.

The theoretical and applied developments described in this work allow to receive reliable and economical decisions in design and constructing of practical mobile and modular bridge crossings.

References

1. K.G. Balabekova, A.S. Kadyrov, *in Proceedings of the International scientific-practical conference "Integration of science, education and production basis for the implementation of the Plan of Nation: selected papers,* **vol 2**. 241-243 (2015).
2. L. Tasci, N. Kuloglu. The Balt. J of R. and B. Eng., **6 (1),** 23–29 (2011).
3. J. Hu. *Mechanics of Sheet Metal Forming,* 211 (2010).
4. A.B. Darkov. *Structural mechanics*. 656 (2010).
5. S.K. Ahmadiev. *Teoreticheskie osnovy konstuirovanie mehanicheskoi inzhenerii*. 99 (2010)

Discrete element analysis of the excavation effect of cross-river tunnel on the surroundings

Dongyuan Liu[1, 2], Xiedong Zhang[1] and Zhihua Zhang[1]

[1]*School of Transportation, Wuhan University of Technology, Wuhan 430063, China*
[2]*CCCC First Highway Engineering Co.,LTD, Beijing 100024, China*

Abstract. Cross-river tunnel as one of the underground constructions is complicated during its construction. For stability of tunnel excavation, it is emergency to analyze the dynamic characteristics of tunnel deformation under high water pressure. Therefore, a cross-river tunnel model is proposed based on DEM in this paper. Stiffness of particles decreases during excavation process which is as one of excavation methods. Porosity ratio of original porosity over its value at different excavation time has been considered. Radial displacements of particles at different angle around the tunnel are detected during excavation process. It shows that large deformation occurs at the vault of the excavation zone which accompanies with large radial displacement. The upper half of the tunnel performs larger deformation than the lower half part which results in many cracks in the concrete lining, the high water pressure may play an crucial role in it.

1 Introduction

Substructures have been developed rapidly with the demand of transportation in recent decades. These substructures, such as basement parking, highway tunnels, metros and et al, play crucial roles in people's life. Designers and engineers have payed more attention to these constructions due to the high stress environment which may result in inconceivable disasters.

For the influence of tunnel excavation process on the surroundings, a large numbers of works have been conducted by scholars in the world. Yang[1] proposed an analytical model for the evolution of fractured zone in viscoelastic surrounding rock mass combining the sliding crack model and equivalent crack method, and the displacement after lining was set up is far more than the one at the moment of excavation just finished. Ding[2] summarized the safety risks and risk factors of the metro construction, interpreted the acquisition process of risk identification rules, and the production of rule-extensions and the storage of rule structures. Martino[3] conducted experimental analysis and showed that a zone of irreversibly damaged rock surrounding the excavation, which may include failed zones, inner and outer damage zones and a zone of excavation disturbance where the in situ stresses were influence by the excavation no damage was measurable.

For cross-river tunnel, the influence factors for tunnel safety, besides the above, the variation of water level should be taken into account. The tunnel excavation under high-water pressure has significant influence on the surrounding rock or soft clay, which may lead to collapse or piping. In view of past researches, however, little literature conducted by discrete element method exists on this subject.

In this paper, a large scale cross-river tunnel of two dimensional model is proposed based on discrete element method, which has large advantages in simulating large deformation and stress distribution in detail. Wuhan metro 2nd line is as the case study. Excavation process has analyzed under high water pressure. The deformation and displacement of particles in different angles of the tunnel have been detected during excavation. The purpose of this paper is trying to reveal the dynamic characteristics of excavation effect on the surroundings when water pressure is taken into account.

2 Discrete element method

In the DEM, the interaction of the particles is treated as a dynamic process with states of equilibrium developing whenever the internal forces balance. The contact forces and displacements of a stressed assembly of particles are found by tracing the movements of the individual particles. Movements result from the propagation through the particle system of disturbances caused by specified wall and particle motion and/or body forces. This is a dynamic process in which the speed of propagation depends on the physical properties of the discrete system.

The calculations performed in the DEM alternate between the application of Newton's second law to the particles and a force-displacement law at the contacts[4], shown in Figure 1. Newton's second law is used to determine the motion of each particle arising from the contact and body forces acting upon it, while the force-displacement law is used to update the contact forces arising from the relative motion at each contact. The presence of walls in PFC2D requires only that the force-displacement law account for ball-wall contacts. Newton's second law is not applied to walls, since the wall motion is specified by the user.

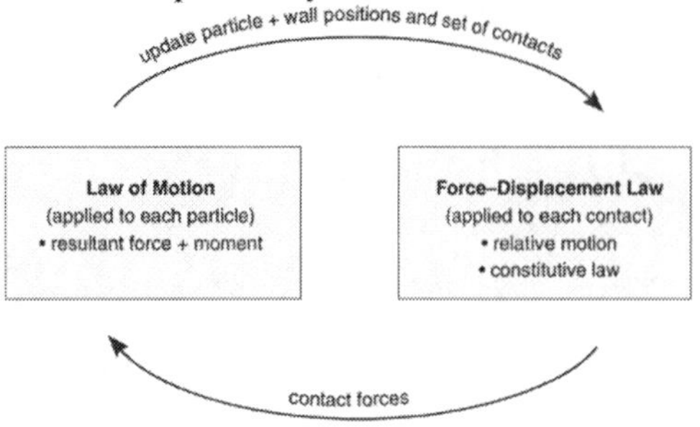

Figure 1. Calculation cycle in PFC.

The magnitude of the normal contact force is calculated by:

$$F^n = K^n U^n \tag{1}$$

where K^n is the normal stiffness at the contact.

The overlap, U^n, defined to be the relative contact displacement in the normal direction, is given by:

$$U^n = \begin{bmatrix} R_1 + R_2 - d(ball - ball) \\ R_1 - d(ball - wall) \end{bmatrix} \tag{2}$$

where R_1, R_2 are the radii of adjacent particles, respectively. d is the distance between the ball centers.

The shear component of the contact displacement increment occurring over a timestep of Δt is calculated by:

$$\Delta U^s = V^s \Delta t \tag{3}$$

and is used to calculate the shear elastic force increment as:

$$\Delta F^s = -k^s \Delta U^s \tag{4}$$

where k^s is the shear stiffness at the contact.

3 Cross-river tunnel model

Alternating the three-dimensional method, the sample establishment, parameters of each strata and micromechanics of 2D numerical model of cross-river tunnel can be easily controlled. With the assumption of isotropic subsurface soils with finite thickness along the Yangtze River, the numerical model can be conducted through the direction view of the tunneling of STM. In order to analyze the influence of excavation process on cross-river tunnel, a cross-section surface of cross-river tunnel in Wuhan metro 2nd line is chosen to be the case study, the detail of the cross-section is shown in Figure 2. Since hydraulic coefficient of concrete lining is about 10^{-9} m/s[5] and duration of maintenance for each time was relatively short, the effect of seepage on tunnel face stability is expected to be small before the concrete lining cracking. Thus, the effect of seepage is not taken into account in this study. As the silty clay being hard-solved problem in the world due to its low strength and strong plasticity[6], it is necessary to take silty clay into account in the model. Parameters of each stratum are shown in Table 1.

The excavation processes are as follows:

• Step 1. Deleting the particles in the range of the tunnel area.

• Step 2. Generating particles with smaller radii in the deleting area.

• Step 3. Decreasing particle stiffness gradually from the original to zero with thickness of concrete lining left.

Table 1. Macro parameters of each stratum in DEM modeling

Symb	Soil	γ	w	e	c	φ	Ec
B	Silty clay	16.7	49.5	1.40	14.0	10.5	2.17
C	Coarse sand(grav	20.1	16.1	0.44	---	42.0	35.4
D	Coarse sand	20.7	18.0	0.51	---	40.0	38.1
E	Mud rock	24.18	12.5	0.31	112.6	25.72	305.65

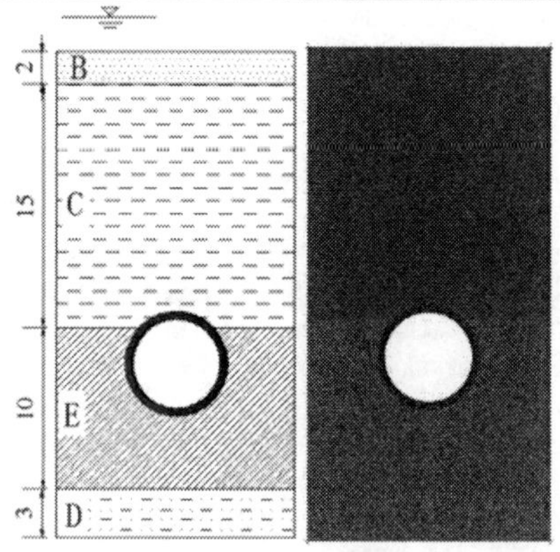

(a) (b)

Figure 2. Cross-river tunnel model.(a): strata information of tunnel;(b) DEM model.

4 Effect of excavation process on the surroundings

4.1 Porosity

Many studies have been conducted around EDZ in formation and long-term processes[7]and area of damage[8]. In present model, the construction of lining would be completed in 2400 cycles. The variation of porosity of the surroundings at the distance of within 4.0 m has been tracked, the results are shown in Table 2.

As shown in Table 2 that the porosity ratios of original porosity over its value of forthwith cycles are quite different after 2400 cycles and 22000 cycles were executed. Particles at the vault of the tunnel are

DOI: 10.1051/matecconf/201710816003

compacted each other which the porosity ratio decrease as shown in Table 2 of angle 90°. The adjacent area at the upper half of the tunnel is compacting as well based on the data of porosity ratio of angle 45° and 135°. The situation of deformation of the lower half of the tunnel in the surroundings is asymmetrical based on the data of angle 225° and 315°. It is extremely visible to reveal the characteristics of porosity evolution that blue and red area represent surrounding material more loose and dense, respectively.

Table 2. Variation of porosity ratio between data at 2400 cycles and 22000 cycles

Distance from out edge of lining		0.1 m	4.0 m
Porosity ratio (%) (2400cycles/22000 cycles)	0°	23.5/58.2	-4.4/-5.3
	45°	-12.1/27.2	-7.8/-42.8
	90°	7.6/-12	8.6/-3.6
	135°	-16.3/-24.6	5.1/-34.1
	180°	-15.4/-13.2	5.2/-16.1
	225°	-3.8/5.7	-7.9/-3.4
	270°	-7.4/4.6	-1.1/0
	315°	-16.4/-10.4	0.2/-0.7

As shown in Table 2 that the porosity ratios of original porosity over its value of forthwith cycles are quite different after 2400 cycles and 22000 cycles were executed. Particles at the vault of the tunnel are compacted each other which the porosity ratio decrease as shown in Table 2 of angle 90°. The adjacent area at the upper half of the tunnel is compacting as well based on the data of porosity ratio of angle 45° and 135°. The situation of deformation of the lower half of the tunnel in the surroundings is asymmetrical based on the data of angle 225° and 315°. It is extremely visible to reveal the characteristics of porosity evolution that blue and red area represent surrounding material more loose and dense, respectively.

4.2 Radial displacement

The dependence of radial displacement on time around tunnel at distance from the lining outer edge of 0.5 m is depicted in Figure 3. As shown in Figure 3, after excavation, radial displacement grows as the iteration cycles increasing. The variation of radial displacements of angle 45°, 90° and 135° are the largest in the surroundings. These points are located in the upper half circle around the tunnel, which is consistent with the characteristics of porosity and stress distribution. After 5000 cycles, radial displacement of the particle at the distance of outer edge of the concrete lining of 0.5 m at the other angles gradually tends to be constant. It should be noted that radial displacements of measurement points of 45°, 90° and 135° at 22000 cycles increase.

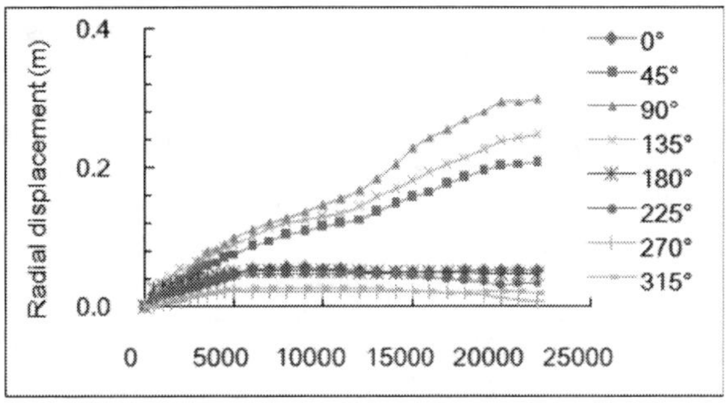

Figure 3. Radial displacement of the surroundings versus cycles at distance of 0.5 m.

4.3 Cracks tracking in the concrete lining

Although the excavation disturbance zone problem has been conducted in many literatures, the information is mainly focus on the surrounding rock mass. However, as far as the crack weakened rock mass sensitive to the rock structure is considered, (which is sensitive to the rock structure,) the growth process of cracks should be taken into account, especially cracks occur in the concrete lining. As we known, concrete lining has large cohesive bond and high strength which results in low permeability, however, the infiltration effect will be imposed when cracks are generated in it. In present model, crack property has been tracked during excavation process, the illustration of crack distribution is shown in Figure 4.

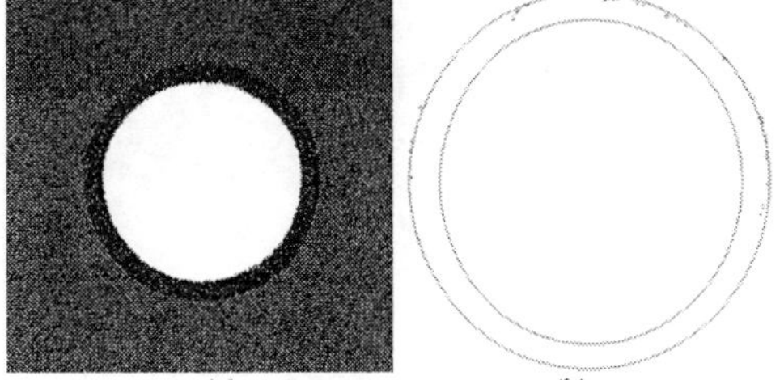

Figure 4. Cracks occur in the concrete lining after 22000 cycles.

As shown in Figure 4, cracks occur in the out edge of the concrete lining, it is remarkable to reveal the high stress of cross-river tunnel which may lead great strength of concrete lining to break. There are 55 cracks generated between particles when large compressive stress exceeds the bond strength as the process executing 22000 cycles. It is better understanding that cracks at the edge of the concrete lining have little influence on the stability of the tunnel, whereas the tunnel will be collapse or submerged when cracks running through the lining.

5 Conclusions

To investigate the influence of excavation process of cross-river tunnel on the surroundings, two dimensional discrete element model is conducted. As the excavation process executing with cycle increases, different variables of particles at different angles in the surroundings are detected. The following conclusions could be made:

(1) During tunnel excavation, the vault of the tunnel in the surroundings is compressive as the porosity of local area decreases. Based on the data of porosity ratio in the surroundings, the deformation of the tunnel is not symmetrical along the center of the tunnel, which will lead the stress distribution of concrete lining to perform asymmetrical deformation.

(2) Based on radial displacement of particles around the excavation zone, large collapse occurs at the vault of the tunnel which is consistent with the data of porosity variation. Therefore, for stability of tunnel excavation, advanced reinforcement should be needed when the shield tunneling machine excavates the low-strength layer.

(3) Large deformation results in many cracks between particles, however, it is relative safe when the cracks just occur at the outer edge of the concrete lining.

References

1. H.Q. Yang, D. Huang, X.M. Yang, X.P. Zhou. Tunn. Undergr. Sp. Tech. **35**: 78-88 (2013)
2. L.Y. Ding, H.L. Yu, H. Li, C. Zhou, X.G. Wu, M.H. Yu. Automat. Constr. **27**: 120–137 (2012)
3. J.B. Martino, N.A. Chandler. Int. J. Rock Mech. Min. **41**: 1413– 1426 (2004)
4. Itasca Consulting Group. Inc. PFC2D, Version 4.0. Minneapolis: *Itasca* (2008)
5. F.L. Min, W. Zhu, X.R. Han, X.C. Zhong. Geotech. Spec. Publ. **204** GSP: 210–215 (2010)
6. Z.H. Zhang, X.D. Zhang, H.S. Qiu, M. Daddow. Constr. Build. Mater. **125**: 564-573 (2016)
7. P. Blümling, F. Bernier, P. Lebon, C.D. Martin. Phys. Chem. Earth, Parts A/B/C. **32**:588-599 (2007)
8. G. Bäckblom. *Swedish Nuclear Fuel and Waste Management Co* (2008)

Indonesian Maritime Logistics Network Optimization Using Mixed Integer Programming

Arry Rahmawan, Komarudin and Natasha Angelina

Department of Industrial Engineering, University of Indonesia, Kampus Baru UI Depok 16424, Indonesia

Abstract. Some causes of the high logistic costs and significant disparity in Indonesia's commodity price is the imbalanced distribution of the flow of goods between west and east regional of Indonesia and the slow development infrastructure. This issue can be happened because maritime transport infrastructure in Indonesia have not served effectively yet,. Therefore, the provider of shipping services should be able to design the Indonesian maritime logistics network that best suits with the current maritime conditions and provide the greatest benefits. This study aims to obtain a model of decision-making at all levels (strategic, tactical, operational) to determine the type of vessel, the shipping routes, and cargo allocation that gives the maximum profit. Main purpose from this study is to develop decision-making model for maritime logistics network design at all levels (strategic, tactical, operational) by using Mixed Integer Programming which can solve the combination of fleet-design, ship-scheduling, and cargo routing problems simultaneously.

1 Introduction

Based on data from State of Logistics Indonesia (2015), the proportion of Indonesian logistics cost is equivalent to 24.66% of Gross Domestic Product (GDP) in 2011. Indonesia's Logistic Performance Index (LPI) is on the 5th rank in ASEAN and 53rd of 160 countries in the world which is conducted by the World Bank in 2014.

Indonesia's economic growth in the second quarter of 2015 is increased by 4.67%. From year to year, Indonesia's economic growth tends to be dominated by west regional only which furtherly caused an imbalance distribution of goods between east and west Indonesia. These make costs of logistics and maritime transportation in Indonesia significantly high and resulting in disparity of commodity price in various regions of Indonesia.

Shipping logistics through sea shipping can be divided into three types: liner shipping, tramp shipping, and industrial shipping (Lawrence, 1972). Liner shipping is cargo shipments that have fixed shipping schedule and follow the schedule as the freight shipments schedule. In contrast, tramp shipping does not have a schedule and send cargo shipping available to maximize profits, as well as taxi transportation services. Meanwhile, industrial shipping is managed by shipping cargo-owner who is looking for the lowest cost to send the cargo. In tramp shipping and industrial shipping, type of cargo can be sent in form of container and non container (bulk cargo). This paper will be focused and limited to liner shipping, which container is the type of cargo shipped.

The imbalanced flow of goods, increased logistic and transportation cost is a few of major challenges for stakeholders in the logistics sector, especially the shipping company to design maritime logistic network. In designing the maritime logistics network, particularly in the liner shipping, there are three issues that need to be resolved, namely: fleet design, ship scheduling, and cargo routing (Meijer, 2015). Figure 1 outline the key decisions that need to be made at different levels of the planning horizon. (Agarwal and Ergun, 2008).

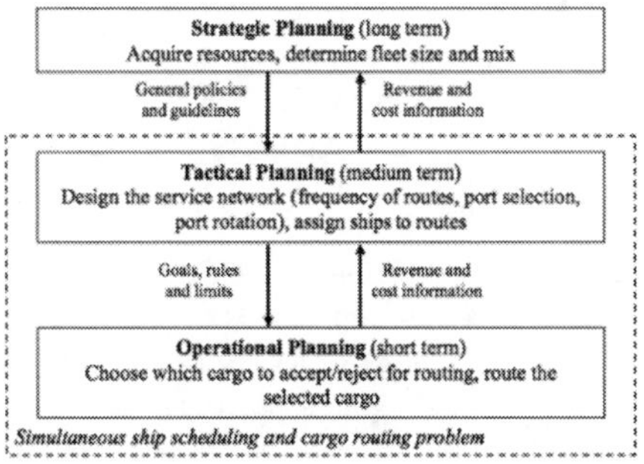

Figure 1. Planning Horizon in Maritime Logistics Network Design

Decision-making horizon is divided into three levels: strategic (3-5 years), tactical (4-12 months), and operational (1-4 weeks) level (Pesenti, 1995). At the strategic level, the decision to be taken is determining fleet design, where the optimal number and type of fleet that is appropriate with maritime conditions need to be determined, as well as in terms of the economy and efficiency. Furthermore, the tactical level, the network should be established shipping routes, especially the allocation of shipping routes and ship on each route. Then, at the operational level, the questions to be solved include the allocation of cargo on each shipping route and the number of available fleet. This level needs to decide the most efficient cargo allocation to avoid empty cargo when shipping operations are held.

2 Research method

Any problems are interconnected and linked with other problems. Thus, the decision for any problems that are on every level will affect decisions on other issues. Therefore, it is important for these three problems to be solved simultaneously, namely by taking decision simultaneously to all levels of strategic planning, tactical, and operational (Mulder and Dekker, 2013).

The combination of these three problems can be solved by using the Mixed-Integer Programming to determine the type of vessel, the ship scheduling, and cargo allocation. By using the basic principle of an investigational formulation from Mulder and Dekker (2013) which was then developed by Meijer (2015) by adding the calculation of fixed costs and the cost of gasoline consumption for each route on the objective function. Therefore, this study used a formula by Meijer (2015) as a basis to complete the combination of these three issues-design fleet, ship-scheduling, and cargo-routing problems.

This study uses 13 ports that have the capacity and facilities to cater for loading and unloading, and classified to the main class and class I port according to the IPC. The ports used in this study are: 1) Belawan ; 2) Tanjung Priok ; 3) Tanjung Perak ; 4) Banjarmasin ; 5) Makassar ; 6) Sorong ; 7) Panjang ; 8) Palembang ; 9) Tanjung Emas ; 10) Pontianak ; 11) Samarinda ; 12) Ambon ; 13) Bitung.

Main point that distinguishes this research with Meijer (2015) is the number of ports used for data input models. Port number used in Meijer (2015) is 6 Indonesian ports. Meanwhile, this study used the whole container terminal in Indonesia, which have 13 ports in total.

Sets:

$h \in H$	Set of ports
$t \in T \subseteq H$	Set of transhipment ports
$s \in S$	Set of ship routes
$j \in J$	Indicator set denoting whether ship passes both ports $h1 \in H$ and $h2 \in H$ on ship route $s \in S$, where *j* = *(h₁, h₂, s)*
$k \in K$	Indicator set denoting whether port $h2 \in H$ is directly visited after port $h1 \in H$ on ship route $s \in S$, where *k* = *(h₁, h₂, s)*

Parameters:

$r_{h1,h2,s}$	Revenue of transporting one TEU from $h1 \in H$ to port $h2 \in H$
c_t^t	Cost of transhipping one TEU in transhipment port $t \in T$
c_h^h	Cost of (un)loading one TEU in origin or destination port $h \in H$
$d_{h1,h2}$	Demand with origin port $h1 \in H$ and destination port $h2 \in H$
b_s	Capacity on ship route $s \in S$
$I^{path}_{h1,h2,h3,h4,s}$	0/1 parameter that takes the value 1 if a ship passes consecutive ports $h2 \in H$ and $h1 \in H$ when sailing from port $h1 \in H$ to port $h2 \in H$ on ship route $s \in S$
f_s	Fixed cost of using route $s \in S$
$dist_{h1,h2}$	Distance from origin port $h1 \in H$ to destination port $h2 \in H$
f_s^f	Fuel price of ship $s \in S$ per *nautical mile (nm)*

Variables:

$x_{h1,h2,s}$	Cargo flow on ship route $s \in S$ between consecutive ports $h1 \in H$ and $h2 \in H$
y_s	Integer variables that denotes the number of time the routes $s \in S$ is used
$x^{od}_{h1,h2,s}$	Direct cargo flow on ship route $s \in S$ between ports $h1 \in H$ and $h2 \in H$
$x^{ot}_{h1,t,h2,s}$	Transhipment flow on ship route $s \in S$ between port $h1 \in H$ and transshipment port $t \in T$ with destination port $h2 \in H$
$x^{td}_{t2,h2,s1,s2}$	Transhipment flow on ship route $s_2 \in S$ between transhipment port $t \in T$ and destination port $h2 \in H$ where the flow to transhipment port $t \in T$ was transported on ship route $s_1 \in S$
$x^{tt}_{t1,t2,h,s1,s2}$	Transhipment flow on ship route $s_2 \in S$ between transhipment port $t_1 \in T$ and transhipment port $t_2 \in T$ with destination port $h2 \in H$, where the flow to transhipment port $t_1 \in T$ was transported on route $s_1 \in S$

Objective Function:

$$\max \sum_{h1\in H}\sum_{h2\in H}\sum_{s\in S} r_{h1,h2,s}\left(x^{od}_{h1,h2,s}+\sum_{t\in T}x^{ot}_{h1,t,h2,s}\right)-\sum_{h1\in H}c^{h}_{h1}\left(\sum_{t\in T}\sum_{h2\in H}\sum_{s\in S}\left[x^{ot}_{h1,t,h2,s}+x^{td}_{h2,t,h1,s}\right]+\sum_{h2\in H}\sum_{s\in S}\left[x^{od}_{h1,h2,s}+x^{od}_{h2,h1,s}\right]\right)$$

$$-\sum_{t1\in T}c^{t}_{t1}\left(\sum_{t2\in T}\sum_{h2\in H}\sum_{s1\in S}\sum_{s2\in S}\left[x^{tt}_{t1,t2,h2,s1,s2}+\sum_{h2\in H}\sum_{s1\in S}\sum_{s2\in S}x^{td}_{t1,h2,s1,s2}\right]\right)-\sum_{s\in S}f_s y_s-\sum_{s\in S}\sum_{k\in K}dist_{h1,h2}\,y_s f^{f}_s$$

s.t :

$$\sum_{t\in T}\sum_{s\in S}x^{ot}_{h1,t,h2,s}+\sum_{s\in S}x^{od}_{h1,h2,s}\le d_{h1,h2,s} \qquad h1\in H,h2\in H \qquad (1)$$

$$x_{h1,h2,s}\le b_s y_s \qquad (h1,h2,s)\in K \qquad (2)$$

$$\sum_{h1\in H}x^{ot}_{h1,t,h2,s1}+\sum_{t2\in T}\sum_{s2\in S}x^{tt}_{t2,t1,h2,s2,s1,s}-\sum_{s2\in S}x^{td}_{t1,h2,s1,s2}-\sum_{t2\in T}\sum_{s2\in S}x^{tt}_{t1,t2,h2,s1,s2}=0 \qquad (t1,h2,s)\in J \qquad (3)$$

$$x_{h1,h2,s}-\sum_{h3\in H}\sum_{h4\in H}x^{tot}_{h3,h4,s}I^{path}_{h3,h4,h1,h2,s}=0 \qquad (h1,h2,s)\in K \qquad (4)$$

$$x^{tot}_{h1,h2,s1}-x^{od}_{h1,h2,s1}-\sum_{h3\in H}x^{ot}_{h1,t1,h2,s1}-\sum_{s2\in S}x^{td}_{t1,h2,s2,s1}-\sum_{h3\in H}\sum_{s2\in S}x^{tt}_{t1,t2,h2,s2,s1}=0 \qquad (5)$$

$$x_{h1,h2,s}\ge 0 \qquad (h1,h2,s)\in K \qquad (6)$$

$$x^{od}_{h1,h2,s}\ge 0 \qquad h1\in H,h2\in H,s\in S \qquad (7)$$

$$x^{tt}_{t1,t2,h,s1,s2}\ge 0 \qquad h\in H,s1\in S,(t1,t2,s)\in J \qquad (8)$$

$$x^{td}_{t,h,s1,s2}\ge 0 \qquad s1\in S,(t,h,s2)\in J \qquad (9)$$

$$x^{ot}_{h1,t,h2,s}\ge 0 \qquad h2\in H,(h1,t,s)\in J \qquad (10)$$

The objective function of this research is to maximize corporate profits, which functions above reflect all revenues minus total cost: fuel cost, transhipment cost, loading / unloading cost, and fixed costs. While the constraints number in this model is 10 (ten) functions, which has the explanations for each function as follows :

Table 1. Constraint Definitions

Constraint	Remarks
1	Ensure all cargo shipped on each route combinations between ports does not exceed demand level at the port.
2	Ensure all transshipped cargo on each route does not exceed the capacity of the ship sailing this route.
3	Ensure all cargo which have to be transshipped, will also be loaded on another route
4	Define the amount of flow between two consecutive ports
5	Define total flow between each two ports in the same cycle
6, 7, 8, 9, 10	Ensure cargo flow is nonnegative

Optimization modeling is using Netbeans IDE 8.0 software and Gurobi. When develop this model, the author makes the set, parameters, variables, objective function and constraint functions as described before. However, to get some parameters in the model, the author calculate some parameter which are different than those described above for input into the model as follows.

1. Calculate the optimum speed by using a formula Brower et al. (2014) for each of these combinations using all types of vessels.

$$F_s(v) = 600x\left(\frac{v}{v_s}\right)^3 x\, f_s\, x\, numberdays\, of\, sailing + 600\, x\, f_{idle} x\, number\, of\, days\, in\, port$$

$F_s(v)$ indicates gasoline consumption for each ship routes on the vessel type s with sailing speed v *(nm/hour)* and has a design speed of v_s^* . F_s is gasoline consumption (tons/day) when the ship is sailing and f_{idle} gasoline consumption (tons/day) when the vessel is not used, conducting loading and unloading, or wait for the next trip. The cost of gasoline consumption was $ 600 / ton. The formula above consists of two parts: the calculation of fuel consumption when cruising and calculation of fuel consumption while it is docked in the harbor.
2. Calculate the number of ships needed for each route based on the duration of the trip.
3. Calculate fuel costs per type of vessel for each route by using optimum speeds that have been obtained previously.
4. Calculate weekly fixed cost per type of vessel for each route.

   In addition, these models are created using several assumptions:

   a. Each port is assumed to have unlimited capacity, so it can serve all types of ships.
   b. All types of vessels can bypass each route.
   c. Time used by ships while docked to unload or load containers are assumed constant for 24 hours.
   d. Sailing speed assumed to be constant, there are no external factors that affect the speed of the ship.
   e. The level of demand at a port which can not be met by the company will be met by other ports are still in one of the region

3 Research result

There are two scenarios in running the optimization model: scenario 1 is an operational condition of the company in general, a condition in which the company decided to serve requests that are considered beneficial and scenario 2 is the condition where the company serves the entire demand arising on all cargo terminals in Indonesia. All model optimize the combination of these 390 round-trip originating from the 13 container ports in Indonesia.

3.1 Ship-allocation and routes used for Scenario 1

Optimization models use Netbeans IDE 8.0 software and Gurobi within 11 minutes 40 seconds. Table 2 shows the results of optimization regarding the ship type, routes used, and number of ships needed to meet the weekly demand from origin to destination port in Indonesia. In addition, the optimization model also recommends the speed of the ships based on the distance and time travel for each of these vessels and routes.

Table 2. Ship Types and Routes Used in Scenario 1

Ship Type	Routes Used	Number of Ships Needed	Optimum Speed [kn]
Feeder_450	3 – 11 – 3	1	10,69
Feeder_800	1 – 3 – 1	2	10,33
Feeder_800	3 – 9 – 3	1	10
Panamax_1750	1 – 2 – 1	2	17,73
Panamax_1750	2 – 3 – 2	1	12
Panamax_1750	2 – 5 – 2	1	13,43
Panamax_1750	3 – 4 – 3	1	12
Panamax_1750	3 – 6 – 3	2	12,61
Panamax_2400	2 – 4 – 2	1	12
Panamax_2400	3 – 5 – 3	1	12
Total	**10**	**13**	-

According to the table above, if the model is executed on the scenario 1, the shipping company requires 13 logistics ships consisting of 1 feeder_450, 3 feeder_800, 7 panamax_1750, and 2 panamax_2400. There are 10 shipping routes by serving 8 container ports, which are supposed to provide maximum benefit to the company. The route used can be seen in Table 2 by using the sequence of number corresponding to section 2 above. The ports that can be served by the shipping companies are Belawan, Tanjung Priok, Tanjung Perak, Banjarmasin, Makassar, Sorong, Tanjung Emas, and Samarinda.

The high operating costs and low demand will occur if the company serves some ports outside the served ports above, the remaining 5 ports are not served because they are considered resulting a not optimum condition. Even though Panjang, Palembang, Pontianak, Ambon, and Bitung port are not served by the shipping company, but the demand to the area considered to be supplied by the served ports and is geographically located around the remaining ports.

3.2. Ship-allocation and routes used for Scenario 2

Optimization models use Netbeans IDE 8.0 software and Gurobi within 1 minute 27 seconds. Scenario 2 is made with the intention to see how operational and financial condition of the company if required or seek to serve all demand that appear throughout the container port in Indonesia. If this condition is executed, it is expected to balance the distribution of the flow of goods in the country and eventually reduce domestic logistics costs.

Same as the table 2 on scenario 1, table 3 shows the results of the optimization regarding ship type, routes used, number of ship needed to meet the weekly demand from the origin to destination port in Indonesia, and recommendation on the speed of vessels used.

Table 3. Ship Types and Routes Used in Scenario 2

Ship Type	Routes Used	Number of Ship Needed	Optimum Speed [kn]
Feeder_450	2 – 7 – 2	1	10
Feeder_450	2 – 10 – 2	1	10
Feeder_450	6 – 13 – 6	1	10
Feeder_450	12 – 13 – 12	1	10
Feeder_800	1 – 8 – 1	1	11,76
Feeder_800	3 – 4 – 3	1	10
Feeder_800	3 – 8 – 3	1	12,15
Feeder_800	3 – 9 – 3	1	10
Feeder_800	3 – 11 – 3	1	10,69
Panamax_1200	3 – 4 – 3	1	12
Panamax_1750	1 – 2 – 1	2	17,73
Panamax_1750	2 – 3 – 2	1	12
Panamax_1750	3 – 6 – 3	2	12,61
Panamax_2400	2 – 4 – 2	1	12
Panamax_2400	3 – 5 – 3	1	12
Total	**15**	**17**	-

Based on Table 3, if the model is executed on the scenario 2, the shipping company requires 17 logistics ships consisting of 4 feeder_450, 5 feeder_800 ships, 1 panamax_1200, 5 panamax_1750, and 2 panamax_2400. There are 15 shipping routes, which are supposed to provide maximum benefit to the company by meeting the entire demand from all 13 container ports in Indonesia. This condition is indirectly forcing shipping companies to serve all of the container port, so that the number of ships required increases compared to scenario 1 in order to facilitate shipping operations across the service delivery logistics in Indonesia.

3.3. Financial and operational scenario comparison

Financial and operational are two main cosiderations for deciding liner shipping network design.

Figure 2 is a financial comparison of shipping companies in scenario 1 and scenario 2. Scenario 2 had greater revenues and total costs rather than scenario 1 based on Figure 2 because the condition of the companies that seek to serve the entire port container in Indonesia. However, the profit gain is 4.29% lower than scenario 1 because the shipping company also serve ports which have low demand rate and a very far distance, especially in east regional of Indonesia.

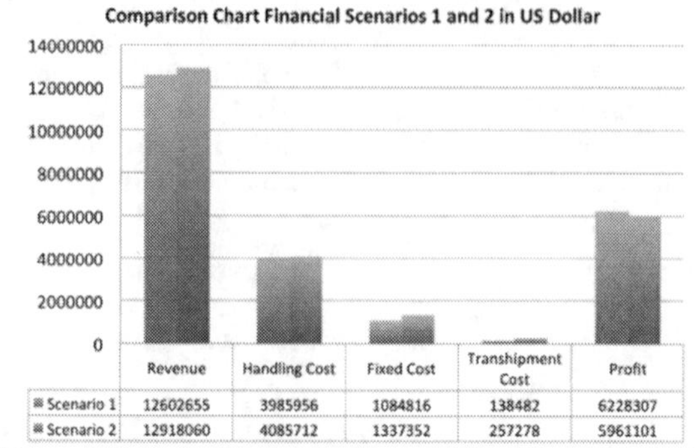

Figure 1. Optimization Result - Financial Condition Scenario 1 and 2

Figure 3 shows the condition for weekly operation scenario 1 where the shipping company is suggested to serve 10 shipping routes from 8 ports by using 13 logistics ships. This condition is the best condition for shipping company to maximize the profit.

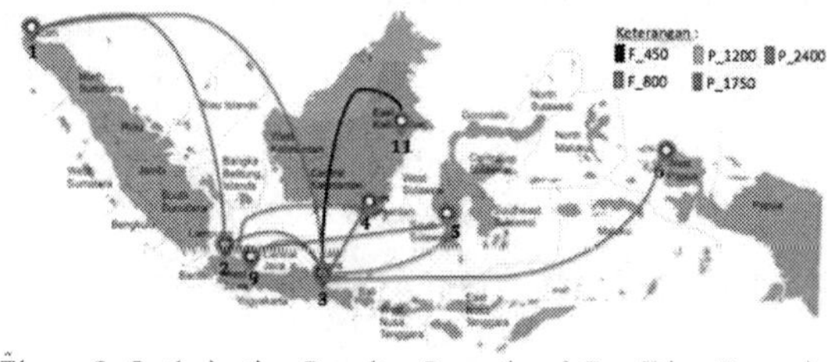

Figure 3. Optimization Result – Operational Condition Scenario 1

Figure 4 shows the operational condition of a shipping company serve the entire container port in Indonesia by using 17 logistics ship and serving 15 shippingg routes.

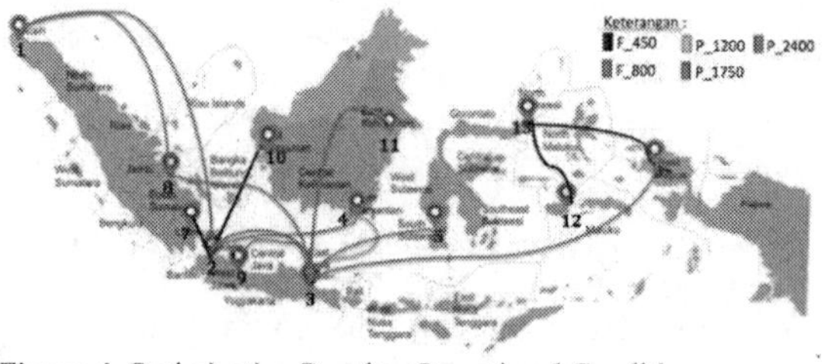

Figure 4. Optimization Result – Operational Condition Scenario 2

Scenario 2 is made with the intention to see how the company's operational and financial condition if companies is trying to serve all demand that appear throughout the container port in Indonesia. In addition, scenario 2 may occur if the government requires the company to serve all ports in west and east regions in Indonesia. It is expected to balance the distribution of goods in the country and eventually reduce domestic logistics costs.

However, after seeing the results issued by the optimization model, it can be seen that scenario 2 shows a greater amount of ship investment rather than scenario

1. This is why not many shipping companies wishing to serve the entire container port in Indonesia. Therefore, the authors provide recommendations to the government to review the regulation and to consider giving incentives to shipping companies that are willing to serve the container ports that are less profitable, especially those in the East of Indonesia.

4 Conclusion

Based on the results and analysis has been explained in previous chapters, mathematical model is the main output from this study. Its function as the decision-making model to design maritime logistics network design at all levels (strategic, tactical, operational) to determine the type of vessel, the ship scheduling and cargo allocation using mixed integer programming. The model optimization was used to compare the impact of route changes on liner shipping company profits.

Based on model output, the maximum benefits can be obtained if the shipping company serve 8 ports : Belawan, Tanjung Priok, Tanjung Perak, Banjarmasin, Makassar, Sorong, Tanjung Emas, and Samarinda with 15 shipping routes and 13 logistics ships consisting of 1 feeder_450, 3 feeder_800, 7 panamax_1750 and 2 panamax_2400. But, if company wants to serve the entire container port in Indonesia, the company is suggested to serve 15 shipping routes and make investments on 17 vessels consisting of 4 feeder_450, 5 feeder_800, 1 panamax_1200, 5 panamax_1750 and 2 panamax_2400.

References

1. Agarwal R, Ergun O. Ship Scheduling and Network Design for Cargo Routing Liner Shipping. Transportation Science. (2016); 00(0):1-3.
2. Badan Pengawas Keuangan dan Pembangunan. Ceta Biru Pengembangan Sistem Logistik Nasional. Jakarta: Badan Pengawas Keuangan dan Pembangunan; 2016.
3. Badan Pusat Statistik [Internet]. Bps.go.id. 2016 [cited 4 November 2016]. Available from: http://www.bps.go.id/linkTabelStatis/view/id/1277
4. Brouer B. A Matheuristic for The Liner Shipping Network Design Problem. Transportation Research Part E. 2014; 42-44.
5. Chiang J. Outlook of Global Container Port Market with Focus on Asia. Dreary Maritime Advisors; 2013 p. 1-28.
6. Chiang J. Market Briefing: Indonesia's Masterplan for Ports, Shipping, and Logistics. Drewry Maritime Advisors; 2015 p. 1-23.
7. Dekker R, Mulder J. Design and Analysis of Container Liner Shipping Networks. Presentation presented at; Erasmus School of Economics, Rotterdam, The Netherlands.
8. Indonesia Sebagai Poros Maritim Dunia [Internet]. Presidenri.go.id. 2016 [cited 5 March 2016]. Available from: http://presidenri.go.id/maritim/indonesia-sebagai-poros-maritim-dunia.html
9. Dekker R, Mulder J, Kalem H, Meijer J, Warden W, Van Rijn L. Liner Shipping Network Design: The Case of Intra Indonesia Connection. Presentation presented at; Erasmus School of Economics, Rotterdam, The Netherlands.
10. Guericke STierney K. Liner Shipping Cargo Allocation with Service Levels and Speed Optimization. Transportation Research Part E. 2015:40-48.
11. Potret Daya Saing Logistik Indonesia [Internet]. Kemenkeu.go.id. 2015 [cited 13 March 2016]. Available from: http://www.kemenkeu.go.id/Artikel/potret-daya-saing-logistik-indonesia
12. Kementerian Dalam Negeri. Cetak Biru Pengembangan Sistem Logistik Nasional. Kementerian Dalam Negeri; 2012.
13. Pusat Informasi Harga Pangan Strategis Nasional. (2016, February 15). *Beranda: Harga Rata-Rata Komoditas per Provinsi*. Retrieved February 15, 2016, from PIHPS Nasional: www.hargapangan.id
14. Kemme N. Container-Terminal Logistics. In: Kemme N, ed. by. Design and Operations of Automated Container Storage Systems. 1st ed. Springer; (2013). p. 9-18.
15. Meijer J. Creating a Liner Shipping Network Design [Post Graduate]. Erasmus University Rotterdam; 2015.
16. Mulder J, Dekker R. Methods for Strategic Liner Shipping Network Design. European Journal of Operational Research. (2013);:367-371.
17. Power T. Proposals for a New Domestic Container Shipping Network in Indonesia. Drewry Maritime Advisory; 2015 p. 1-25.
18. State of Logistics Indonesia 2015. Bandung: Centre of Logistics and Supply Chain Studies, Institut Teknologi Bandung, Asosiasi Logistic Indonesia, STC-Group, World Bank; 2016.
19. Van Rijn L. Service Network Design for Liner Shipping in Indonesia [Postgraduate]. Erasmus University; (2015).

Logistics Indicators Could Improve Logistics Performance of Hospitals

Andre M. R. Wajong

Faculty of Engineering, Industrial Engineering Department, Bina Nusantara University, Indonesia
Bina Nusantara University (Syahdan Campus), Jl. K.H. Syahdan No. 9, Palmerah, Jakarta Barat 11480, Indonesia

Abstract. Hospitals are institutions in which numerous transactions take place. A significant amount of those transactions happen in the realm of logistics. Logistifics is one of the concern of the hospital's managing entities. Hospitals must provide a variety of logistics within a limited amount of time. The implementation of hospital logistics indicators were expected to assist the management in improving performance in the logistics units of hospitals in the Province of Jakarta. It was expected that by attending to a number of variables, namely information technology support, logistics management, logistics business intelligence, and logistics performance, such improvement could be achieved. The result of this study provided a description of major indicators that could improve the logistics performance in hospital industry located within the Province of Jakarta. Index Terms: hospital logistics indicator, information technology support, logistics business intelligence, logistic management, logistics performance.

1 Introduction

Globalization has resulted to many alterations within the practice of free trade which, consequently, brought about positive impacts for the international community and Indonesia. Those alterations, combined with the level of competitiveness of a country's industry, constitute a factor which can influence a country's trade position. Indonesia, considering its current context and situation, has seen its industry spending a lot more in terms of production of goods and services in comparison to other countries. One of the factor that underlies such phenomenon is the slow development of logistics performance or capacity that is implemented within the country [1].

Logistics Performance Index was introduced by World Bank through one of its study [2]. The study on logistics performance was explored further, by focusing on time, cost, and reliability of logistics [3].

Asian Development Bank (ADB) conducted a research on a macro level which reviewed a country's condition. On the other hand, a country's logistics activities could also be observed from many industries, such as logistics activities in hospital industry. An example for such observation is documented in a study of several hospitals in ASEAN countries which was conducted by Hoeur and Kritchanchai (2015) [4].

Hospital logistics cover a variety of activities, such as those related to medicine logistics, medical devices, food and beverage, consumable materials, office stationeries, and linen.

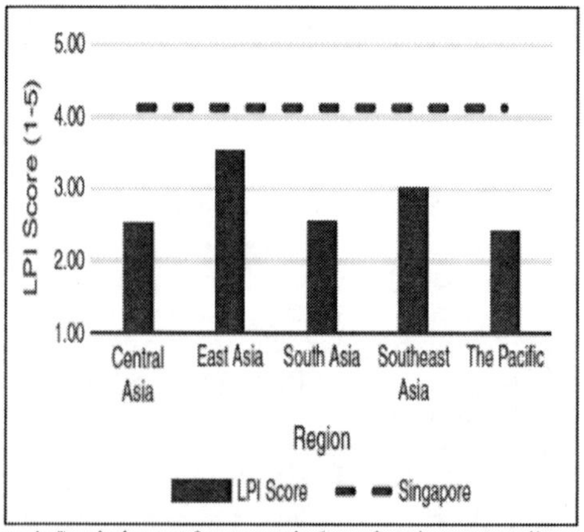

Figure 1. Logistics performance index of regions according to ADB [3]

Considering the fact that the majority of funding in a hospital is allocated to logistics, amounting to 46% of the total funding in a hospital [5], inefficient logistics activities would spike the cost charged to the patients, as well as making the hospital performance as a whole unsatisfactory. As a consequence, there might be a chance for a hospital to shut down due to the poor condition of its logistics financing.

Olaru (2014) mentioned that a factor which influences a hospital's logistics performance lies on its ability to implement a good business intelligence [6]. Meanwhile, Coyle, et al. (2013) [7] and Taylor (2007) [8] claimed that the implementation of a wholistic logistics management would enable the hospital to manage the total cost of logistics, therefore influencing the logistics performance of hospitals. On the other hand, logistics management and business intelligence require support from information technology in order to make the existing data processible and meaningful for the managing entities of the hospital [9].

2 Problem formulation

The root of the problem in hospital logistics in particular stems from its procurement activities. This is due to the fact that purchasing department needs to be able to determine materials and products required by the hospital, as well as pointing suppliers who are able to fill those deficits, aside from purchasing activities conducted in relatively routine manner.

By taking into account the connection between IT Support, Logistics Management, Logistics Business Intelligence, and Logistics Performance, the aforementioned variables would be able to be observed through a correct and accurate operationalization of those variables.

Figure 2 provides an explanation regarding literature study and theories which are used, as well as setting the foundation and framework in giving recommendations of solutions for existing problems.

3 Literature review

3.1 Information system

According to O'Brien and Marakas (2013), a system is a set of related components with clear distinctions which aim to reach a desired goal by gaining inputs and providing outputs in a process of transformation.

Processed data and data which are meaningful to humans are referred to as information [9]. According to O'Brien and Marakas (2013), information is data which has become a useful context for the end users of the system. According to Stair and Reynolds (2012), information is a collection of facts with added values to them [10].

According to several definitions regarding information system, it can be inferred that information system comprise of several components, such as software, hardware, human, communication network as well as coordinated data to conduct information gathering, processing, and storing in order to support decision-making process, control activities, conduct coordination, analyze problem, and conduct visualization inside an organization.

3.2 Hospital information system

Information system inside hospitals are specific developments of information system which are implemented within the hospital industry. This system will gather, process, store, and spread real-time information according to the data entered by every operator within the hospital's information system, specifically the hospital's supporting department which is responsible in managing patients' medical data. These medical data will then be used in the process of making coordinated decision regarding a patient's health. Currently, many hospitals are willing to move from paper-based information systems to electronic information systems in order to support patients' care [11].

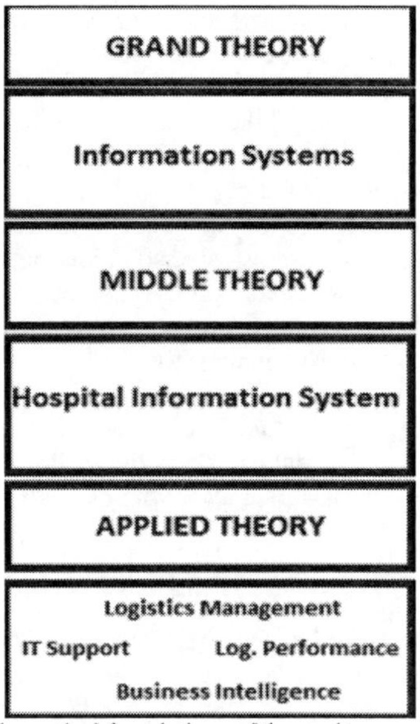

Figure 2. Theoretical foundations of the study

Hospital information system has a significant potential in reducing the cost of patient's care, as well as providing a satisfactory end result [12]. It is also predicted to provide similar positive outcomes for the logistics department of the hospital.

3.3 Logistics performance

Gil-Saura, et al. (2010) claimed that logistics performance is not merely about logistics operational activities, but that it serves as a main factor and strategic variable in fulfilling consumers' satisfaction [13]. The fact is that if a consumer's needs is not fulfilled, the result of that dissatisfaction will lead the consumers to search for other service providers which will be able to provide for their needs, due to the negative experience that the consumers felt with the initial service provider [14].

Logistics performance is a result of a good logistics activities aimed at optimizing quality, cost, and time.

3.4 Logistics management

It is a field of science concerning the process behind planning and conducting needs assesment for procurement, distribution, maintenance, and storage, as well as rescission of materials/instruments from the logbook. Therefore, logistics management will be able to explain the means to achieve the goal of procurement of materials/products when they are needed and can be used optimumly [15].

Logistics management is a logistics activity such as planning, executing, and monitoring of incoming and exitting items in order to create an organization that runs optimally.

3.5 Business intelligence

It is a means to gather, store, organize, reshape, and summarize data, as well as providing information, be it company's internal business activities, or its external activities, including the business activities of its competitors which are easy to access and analyze for many managerial activities [16].

Business Intelligence is an analytical instrument used to consolidate, analyze, store, and access data in order to assist the decision making process, akin to a software for query database and reporting, multidimensional data analysis tool and data mining [9].

Logistics business intelligence is an activity of collecting, storing, and analyzing logistics data in hospitals as well granting access to user for the sake of a better decision-making process within a certain period.

3.6 Information technology support

According to Raymond McLeod, IT Support is a collection of data which can be processed in order to provide a certain meaning as well as values [17]. Meanwhile, according to James O'Brien, it is a form of data existing within a context, which indicates that it has meaning and is useful for end user [18]. According to Laudon, it is a form of data which has been processed and the outcome is a meaningful and useful data for human lives [9].

Information Technology Support is every kind of technology implemented to process, form, shape, and spread information electronically.

4 Methodology

4.1 Framework

Inferring from the literature review conducted in this study, there several gaps that serves as grounds for furthere research. These gaps include variables of information technology support, logistics management, logistics business intelligence and logistics performance. The discussion within each reviewed article only touched one or two of the aforementioned variables. In this article, however, the discussion of all four variables will be advanced.

Logistics performance is a logistics activity aimed to optimize quality, cost, and time in order to ensure that patient's needs are fulfilled through the existence of logistics management which consist of logistics activities such as planning, executing, and monitoring the movement of incoming and exitting logistics supply in order to ensure a more optimum service for the patients. Due to the variety and large amount of logistics activities data, an implementation of logistics business intelligence is needed, namely an activity of collecting, storing, and analyzing logistics data as well as granting access to user in order to create a good decision-making process, which will certainly require an IT support which is every kind of technology implemented to process, form, shape, and spread information electronically.

4.2 Model Analysis

Data collected from 45 hospitals located in the Province of Jakarta by simple random sampling from the population of 81 hospitals, ranging from those of class A, B or C, were processed using SmartPLS 2.0. After running the application all the indicators were valid (loading factor > 0,5 and t-statistics >2,0.

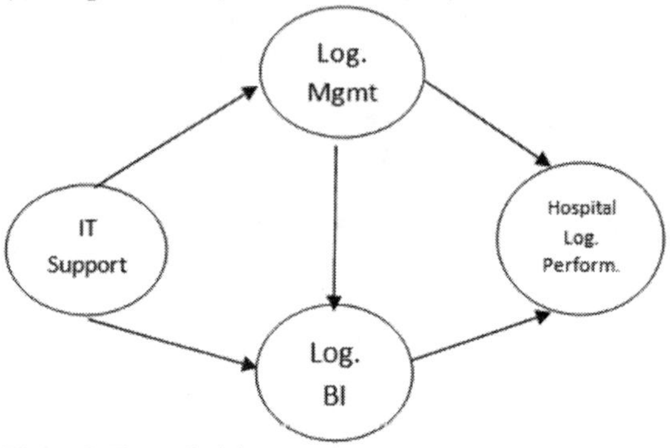

Figure 3. Theoretical framework

After checking the loading factor, the next process is the resampling process using bootstrapping with 1000 samples.

From the evaluation of convergent validity through the states of evaluation, validity indicator and reliability construct (variables) that indicates condition of available data from variable information technology support has the biggest impact, variable logistics management does not have much influence in this research, variable logistics intelligence with indicator adding service value has the biggest impact in the model and indicator regularly is the biggest impact for logistics performance variable (Table 1).

5. Conclusion

The study has identified three major indicators that could significantly affect the performance of logistics, including reliability tracking which will greatly assist the hospital in the process of tracking, in order to find an item / product that is needed by patients, to determine the existence. And indicator adding service value could be considered as an important factor in creating customer satisfaction (patients and families) in Jakarta. In additioin to that, the quality of information provided on a regular basis to solve the problem together with third parties.

By implementing those indicators (Table 1), it is expected that hospitals will be able to have references in calculating their logistics performance, because previously, the system which is used in hospitals have yet to have a standard of reference in terms of logistics. Yet by identifying and defining the dominant indicator that affects the dimensions and variables could increase the performance of logistic. As a consequence, hospitals will be able to monitor and control the logistics performance in order to reach an improved optimum level. On a macro level, Indonesian government will be able to have an higher LPI in contrast to other countries within the same region.

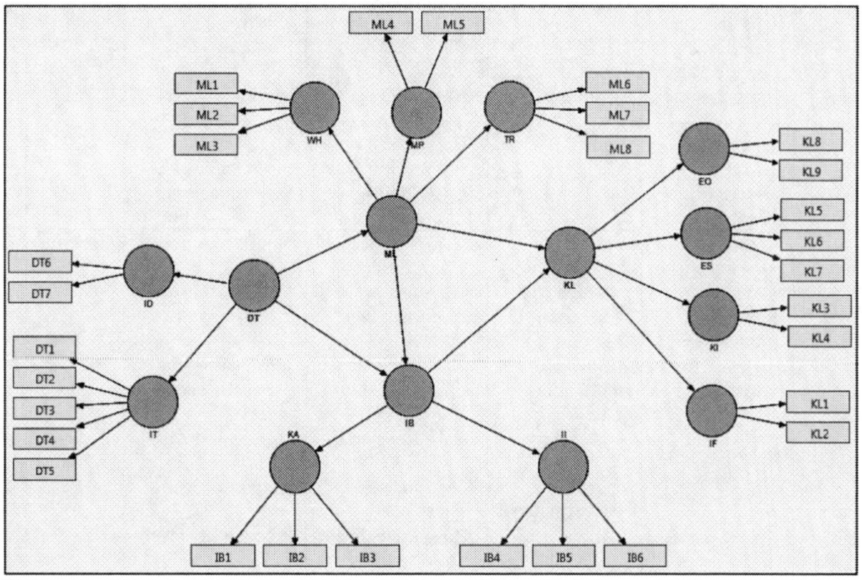

Figure 4. Path Diagram

Table 1. Variables operationalization

Variable	Concept	Dimension	Indicator	Measurement	Loading Factor	T-Statistics
Information technology support (DT)	Every kind of technology implemented to process, form, shape, and spread information electronically	-. Information and technology management (IT)	DT1 - Use of technology	Level of use of technology	0,599	4,612
			DT2 - Product identification	Identification level	0,647	6,523
			DT3 - Tracking accuracy	Accuracy level	0,860	22,992
			DT4 - Reliability tracking	Reliability level	0,871	21,798
			DT5 - Information availability	Information availability level	0,773	10,968
		-. Data integration (ID)	DT6 - Condition of available data	Data condition level	0,897	33,226
			DT7 - Consistency of data source	Data source level	0,851	15,221
Logistics management (ML)	logistics activities such as planning, executing, and monitoring the movement of incoming and exitting logistics supply in order to ensure a more optimum service for the patients	-. Storage (WH)	ML1 - Spatial utilization	Utilization level	0,513	2,209
			ML2 - Categoriztn	Categorization level	0,695	6,997
			ML3 - Acceptance note	Acceptance level	0,869	21,024
		-. Supply management (MP)	ML4 - Inventory visibility	Visibility level	0,738	3,963
			ML5 - Inventory availability	Availability level	0,771	3,673
		-Transportation and distribution (TR)	ML6 - Product condition	Product condition level	0,653	3,383
			ML7 - Delivery order	Order level	0,815	14,302
			ML8 - Received	Date level	0,583	2,896

			date			
Logistics business intelligence (IB)	An activity of collecting, storing, and analyzing logistics data as well as granting access to user in order to create a good decision-making processwithin a certain period	-. Analytical skills (KA)	IB1 - Interactive report	Interactive report level	0,735	4,647
			IB2 - Analytical application	Analytical application level	0,824	8,191
			IB3 - Dashboard	Dashboard level	0,773	9,836
		-. Information implementat ion (II)	IB4 - Building innovation	Innovation level	0,617	5,343
			IB5 - Altering strategies	Strategic alterations level	0,772	6,132
			IB6 - Adding service value	Service level	0,824	10,952
Logistics performance (KL)	A result of a good logistics activities aimed at optimizing quality, cost, and time.	-. Level of information which will be given (IF)	KL1 - Notification of information	Notification level	0,850	14,116
			KL2 - Increasing coordination	Coordination level	0,837	11,379
		-. Quality of information (KI)	KL3 - Complete information	Information completeness level	0,844	11,452
			KL4 - Regularly	Degree of order	0,886	31,460
		-. System effectivity (ES)	KL5 - Increasing response	Response level	0,849	21,003
			KL6 - Increasing productivity	Productivity level	0,817	12,579
			KL7 - Decrease time of delivery	Time delivery level	0,883	20,013
		-. Organizatinl effectivity (EO)	KL8 - Target market	Target market level	0,805	6,999
			KL9 - Innovation	Innovation level	0,888	19,001

References

1. Dedi Mulyadi, "Pengembangan sistem logistik yang efisien dan efektif dengan pendekatan supply chain management, (*Development of logistics system efficient and effective approach with supply chain management*)," Jurnal Riset Industri, **vol. 5**, no. 3, pp 275 – 282 (2011)
2. J.F. Arvis, Y. Duval, B. Shepherd, and C. Utoktham, "Trade cost in the developing world 1995-2010," World Bank Policy Research Working Paper No. 6309, World Bank. Washington, DC. (2013)
3. Ben Shepherd, and Shintaro Hamanaka, "Overcoming trade logistics challenges: Asia-Pacific experiences," Asia Pacific Journal of Marketing and Logistics (Emerald), **vol. 27**, Iss 3, pp. 444 – 466, Des (2015)
4. Soriya Hoeur, and Duangpun Kritchanchai, "Key performance indicator framework for measuring healthcare logistics in ASEAN," Toward Sustainable Operations of Supply Chain and Logistics Systems, p. 37 – 50 (2015)
5. Sylvain Landry, and Richard Philippe, "How logistics can service healthcare," Supply Chain Forum, an International Journal **vol. 5**, no 2 (2004)
6. C. Olaru, "Business intelligence in telecommunications industry," International Journal of Economic Practices and Theories, **vol. 4** no.1, pp 89-100 (2014)
.7. J.J Coyle, C.J. Langley, R.A. Novack, and B.J. Gibson, *Supply chain management: a logistics perspective*, 9th edition. South-Western / Cengage Learning, Mason, Ohio (2013)
8. GD. Taylor, "*Logistics engineering handbook*," CRC Press, Boca Raton, Florida (2007)
9. Kenneth C. Laudon, and Jane P. Laudon, "*Management information systems*," Pearson Prentice Hall, 13th edition (2014)
10. Ralph M. Stair, and George W. Reynolds, "*Fundamentals of information systems*," Course Technology - Cengage Learning, 6th edition (2012)
11. Hesamaddin K. Takhti, Azizah A. Rahman, Samireh Abedeni, and Sadigheh Abedeni, "Impact of hospital information systems on patient care: Nurses's perception," International article (2010)
12. Robert G. Fichman, Rajiv Kohli, and Ranjani Krishnan, "The role of information systems in healthcare: Current research and future trends," Information Systems Research, **vol. 22**, no 3, pp 419 – 428 (2011)
13. I. Gil-Saura, D. Servera-Frances, and M. Fuentes-Blasco, "Antecedents & consequences of logistics value: & empirical investigation in the Spanish market," Industrial Marketing Management, **vol. 39** No. 3, pp. 493-506 (2010)

14. O. Bouzaabia, A.C.R. Van Riel, and J. Semeijn, "Managing in-store logistics: a fresh perspective on retail service," Journal of Service Management, **Vol. 24** No. 2, pp. 112-129 (2013)
15. Gianpaolo Ghiani, Gilbert Laporte, and Roberto Musmanno, "*Introduction to logistics systems management*," John Wiley & Sons, Ltd, second edition (2013)
16. Edward David Moreno, Marina L. Gavrilova, and C. J. Tan, Kenneth, "Transactions on computational science XI," Special Issue on Security in Computing (Part 2) (2013)
17. Raymond McLeod, and George Schell, "*Management information systems*," Pearson Int'l Edition (2007)
18. James A. O'Brien, and George M. Marakas, "*Introduction to information systems*," McGraw-Hill/Irwin, 16th edition (2013)

AUTHOR INDEX

AUTHOR INDEX

AUTHOR INDEX

AUTHOR INDEX